42ND MECHANICAL WORKING AND STEEL PROCESSING CONFERENCE PROCEEDINGS

Volume XXXVIII
Toronto, Ontario, Canada
October 22-25, 2000

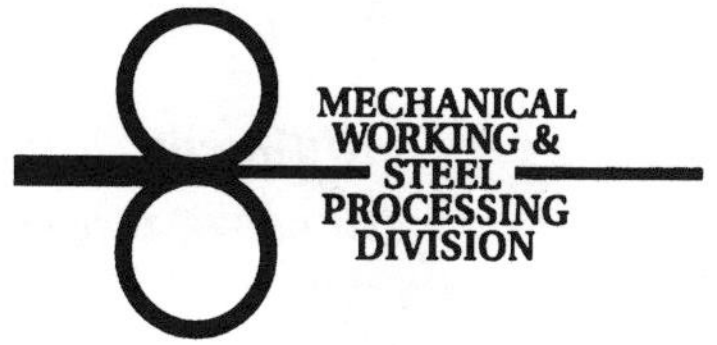

Sponsored by the
Mechanical Working and
Steel Processing Division
of the
Iron & Steel Society

David L. Kanagy
Publisher

Dennis J. Fuga
Manager, Book Publishing

Margaret A. Baker
Editorial Assistant

42nd Mechanical Working and Steel Processing Conference Proceedings, Vol. XXXVIII
These proceedings are also available in CD-ROM format.

ISBN: 1-886362-45-9
ISSN: 1075-878X

ISS Publications...Knowledge fulfillment with a commitment to the quality, integrity and timeliness of information transfer.

Iron & Steel Society
186 Thorn Hill Road
Warrendale, PA 15086-7528
Phone: (724) 776-1535, ext. 1
Fax: (724) 776-0430
E-mail: custserv@iss.org
Web site: www.iss.org

SUBCOMMITTEES

BAR, ROD AND SEMI-FINISHED PRODUCTS AND FORGINGS

R. B. Bertolo
G. R. Boal
E. B. Damm
J. S. Doolittle
R. P. Foley
M. J. Leap
D. K. Matlock
N. K. Mehra
R. Shivpuri
S. V. Subramanian
S. Yue

FLAT ROLLED PRODUCTS

D. Aichbhaumik
B. Allen
P. J. Belanger
D. Bhattacharya
M. Blankenau
A. J. Boucek
L. A. Burroughs
A. J. DeArdo
J. R. Fekete
I. Gupta
J. Hiam
R. Krause
K. W. McCallum
B. D. Nelson
T. M. Osman
D. E. Overby
K. Poole
R. Pradhan
G. Smith
J. G. Speer
G. A. Tither
L. Zheng

PRODUCT PHYSICAL METALLURGY

D. Aichbhaumik
R. I. Asfahani
D. Bhattacharya
R. L. Bodnar
A. J. Boucek
D. Boyd
J. W. Boyd
B. L. Bramfitt
E. R. Case
C. V. Darragh
A. J. DeArdo
E. Essadiqi
R. P. Foley
C. I. Garcia
M. J. Leap
M. J. Merwin
J. R. Paules
A. C. Perry
R. Pradhan
P. Purtscher
F. Rana
M. Schmidt
J. G. Speer
W. Sun
K. A. Taylor
S. W. Thompson
G. A. Tither
K. M. West
Z. Yao
S. Yue

ROLL TECHNOLOGY

T. P. Adams
F. J. Barchfeld
J. W. Boyd
J. Breyer
D. Carman
R. Cummins
W. Davies
M. J. Empel
P. W. Everman
R. M. Fleig
F. X. Goyanes
P. R. Heckman
S. A. Heim
W. Heiser
J. Herman
L. E. Hiller
E. J. Kerr
M. J. Kirwin
G. Lee
C. L. Lentz
K. W. Marsden
G. A. Ott
A. G. Payling
P. C. Perry
S. M. Purdy
J. J. Robb
J. M. Senne
T. Smith
J. R. Valentine
K. L. Zeik

FOREWORD

Welcome to the Iron & Steel Society 42nd Mechanical Working and Steel Processing Conference, being held October 22-25, 2000 at the Sheraton Centre Toronto Hotel in Toronto, Ontario, Canada.

This year the proceedings will be distributed in both book and electronic formats. While the CD-ROM format continues to grow in interest, many still prefer the traditional book format.

More than 100 papers have been selected for presentation during 17 conference sessions. Sessions have been organized by the standing committees of the division including: Product Physical Metallurgy; Flat Roll Products; Roll Technology; Bar Products and Forgings; Tubular Products; and Student Research; along with a session of General Ferrous Metallurgy. This year two sessions have been organized in conjunction with the ISS Process Technology Division on the topic of Hot Rolled Metallurgy — Thermomechanical Processing.

Highlights of the 42nd MWSP Conference include:

- Strengthening Mechanisms in Steel Products, a memorial seminar to Paul Repas who authored numerous associated works. The seminar will begin with a keynote address highlighting Paul's works and presenting some previously unpublished work.
- Topical sessions focusing on:
  - Surface quality of strip products and roll bite conditions associated with scale and strip defects.
  - Temperature, mechanical and processing control in strip rolling for the development of material properties.
- A continuing education course on Nondestructive Testing.
- Plant tours at Dofasco, Inc. of the Galvanizing Line, the Hot Strip Mill and Roll Shop.

Again this year, the society maintains a commitment to supporting the conference expenses of college students and young engineers. The budget for student travel has been increased to direct the interest of students toward the steel industry as a career. The division initiative funds have been used to support the attendance of three young engineers who are working within the steel industry.

I would like to thank the organizing committees, the authors and presenters, session chairs, and growing number of exhibitors for committing and contributing to the success of the conference. Many thanks also, to the staff of the Iron & Steel Society for their organization and professionalism.

Enjoy your experience in technical and information exchange, along with the fellowship with friends and industry peers.

Thomas P. Adams
Chairperson
2000 Mechanical Working
and Steel Processing Division
Iron & Steel Society

ISS President-Elect Designate E. M. O'Donnell presents the 1998 Jerry Silver Award to A. C. Perry (Bethlehem Steel Corp.)

Mechanical Working and Steel Processing Division Chair, C. V. Darragh (The Timken Co.) (center) presents the Michael Tenenbaum Award to Ph. Hartlet (left) and P. Cantinieaux (Cockerill Sambre SA)

Mechanical Working and Steel Processing Chair, C. V. Darragh (The Timken Co.) (left) Presents the Gilbert R. Speich Award to R. L. Bodnar (Bethlehem Steel Corp.)

Mechanical Working and Steel Processing Division Chair, C. V. Darragh (The Timken Co.) (center) presents Mechanical Working and Steel Processing Division Appreciation Certificates to M. J. Leap (The Timken Co.) (right) and J. M. Senne (Xtek Inc.)

Mechanical Working and Steel Processing Division Chair C .V. Darragh (The Timken Co.)(right) presents the Meritorious Award for Roll Technology to T. P. Adams (National Roll Co.)

Mechanical Working and Steel Processing Division Chair-Elect, T. P. Adams (National Roll Co.) (left) presents the Mechanical Working and Steel Processing Conference Meritorious Award for Bar Products and Forgings to R. Shivpuri (The Ohio State University)

R. Issa (Danieli Corp.) (left) accepts the Mechanical Working and Steel Processing Meritorious Award for Minimills (on behalf of E. Crisa, E. Donini and M. Rotti) from Mechanical Working and Steel Processing Division Chair-Elect, T. P. Adams (National Roll Co.)

Mechanical Working and Steel Processing Division Chair-Elect, T. P. Adams (National Roll Co.) (right) presents the Mechanical Working and Steel Processing Conference Meritorious Award for Product Physical Metallurgy to R. L. Bodnar (Bethlehem Steel Corp.)

Mechanical Working and Steel Processing Division Chair, C. V. Darragh (The Timken Co.) (right) presents the Past Chair Award to T. P. Adams (National Roll Co.)

CONTENTS

BAR PRODUCTS AND FORGINGS I

BAR PRODUCTS AND FORGINGS II

BAR PRODUCTS AND FORGINGS III

FLAT ROLL PRODUCTS I/ROLL TECHNOLOGY II (Joint)

FLAT ROLL PRODUCTS II

FLAT ROLL PRODUCTS III

GENERAL FERROUS METALLURGY

PRODUCT PHYSICAL METALLURGY I

PRODUCT PHYSICAL METALLURGY II

PRODUCT PHYSICAL METALLURGY III

PRODUCT PHYSICAL METALLURGY IV

ROLL TECHNOLOGY I

ROLL TECHNOLOGY III

STUDENT RESEARCH

THERMOMECHANICAL PROCESSING I (PTD)

THERMOMECHANICAL PROCESSING II (PTD)

TUBULAR PRODUCTS

BAR PRODUCTS AND FORGINGS I

INFLUENCE OF SILICON ON THE KINETICS OF METADYNAMIC RECRYSTALLIZATION IN MICROALLOYED HIGH CARBON STEELS

A. M. Elwazri[1], P. Wanjara[2] and S. Yue[1]
[1]Department of Mining and Metallurgical Engineering, McGill University 3610 University St. Montreal, QC, Canada, H3A 2B2.
E-mail: **steve@minmet.lan.mcgill.ca**
[2]IVACO Rolling Mills, 1040 HWY 17
Box 322, L'Orignal, Ontario, Canada, K0B 1K0

Key Words: Metadynamic Recrystallization, Silicon, Microalloyed high carbon steels

ABSTRACT

The occurrence of metadynamic recrystallization in vanadium microalloyed high carbon steels was investigated under isothermal conditions. Compression tests were performed using double hit schedules at temperatures between 900 °C to 1050 °C, strain rates of 0.01/s, 0.1/s and 1/s, and inter-pass times of 0.1 to 30 seconds. The results revealed that Si has a strong solute drag effect, on the the kinetics of metadynamic recrystallization. A kinetic model is proposed which takes the V and Si concentrations into account.

INTRODUCTION

Dynamic recrystallization occurs during straining. Once the critical strain for dynamic recrystallization is surpassed during deformation, the dynamically recrystallized nuclei continue to grow when the specimen is unloaded. This process is called metadynamic recrystallization and does not involve an incubation time, since the nuclei are already present upon termination of the deformation.

The kinetics of metadynamic recrystallization are rapid and sometimes completion occurs during quenching after deformation [1]. Nevertheless, the quantitative determination of the kinetics is necessary for accurate control of the microstructure. An Avrami-type equation is generally used to describe recrystallization processes involving nucleation and growth, and metadynamic recrystallization has also been successfully described using an Avrami equation, even though metadynamic recrystallization does not involve nucleation. [2-4]

Published solubility data [5] indicate that vanadium is the most soluble of the conventional microalloying elements, and quantities in excess of 0.15% can be dissolved at normal reheating temperatures (e.g. 1200 °C) regardless of carbon and nitrogen content.

In fact vanadium has a rather high solubility in austenite even at temperatures as low as 1050 °C. This is important for maximizing the precipitation strengthening effect, as the microalloying element must be completely dissolved at the reheating temperature and remain in solution until it precipitates in ferrite in the form of finely dispersed particles. However, vanadium in solid solution has little effect on the kinetics of recrystallization of the austenite of low carbon steels.

The purpose of the present investigation is to examine the effect of vanadium and silicon additions on the kinetics of metadynamic recrystallization in high carbon steels by using compression tests.

EXPERIMENTAL PROCEDURE

Materials- The steels used in the course of this work were prepared and hot rolled at the Materials Technology Laboratory, CANMET, (Ottawa, Ontario, Canada). The chemical compositions of the steels used in this study are given in **Table I**. In this work, V has been chosen as the key microalloying element rather than Nb, since it is more likely to precipitate in the ferrite, leading to more effective precipitation strengthening. In addition, Si additions are being investigated since it has been linked with decreasing pearlite interlamellar spacing, leading to possible improvements in drawability.

Table I Chemical composition of experimental steels

Steels	A	B	C
C	1.1	1.0	1.08
Si	0.23	0.99	0.78
Mn	0.63	0.77	0.75
Cr	0.04	0.066	0.049
Ni	0.038	0.039	0.039
Cu	0.036	0.037	0.037
V	0.17	0.078	0.26
P	0.010	0.0086	0.009
S	0.007	0.008	0.009
N	0.007	0.013	0.005
Austenite grain size D, μm	94	92	97

To determine the soaking temperature to fully dissolve the vanadium prior to deformation testing, the following equations were used [6]:

$$\log [V][C] = -\frac{9500}{T} + 6.72 \tag{1}$$

$$\log [V][N] = -\frac{8330}{T} + 3.40 \tag{2}$$

Table II gives the solution temperatures calculated for the three steels used in this study.

Table II Calculated solution temperatures of the steels

Steels	Temperature of solubility of VN (°C)	Temperature of solubility of VC (°C)
A	1044	1000
B	1029	940
C	1050	1033

Experimental equipment- Compression testing was carried out using an MTS (Materials Testing System) machine at the CSIRA (Canadian Steel Industry Research Association) Laboratory at McGill University, (Montreal, Quebec, Canada). The compression test specimens of 11.4 mm in height and an aspect ratio of 1.5 were machined from hot rolled plates with the longitudinal direction in the rolling direction.

Single hit compression tests- To determine the dynamic recrystallization characteristics, the specimens were heated at a constant rate of 1.5 °C/s to 1200 °C for 20 minutes to enable significant amounts of microalloying elements to dissolve. Then the temperature was decreased at a rate of 1 °C/s to the test temperature (900, 950, 1000, 1050 °C) and held for 5 minutes to homogenize the temperature within the specimen. The specimens were deformed isothermally at a strain rate of either 0.01, 0.1 or 1/s. This was followed by quenching, to examine the recrystallized structure. Figure (1) illustrates the deformation and thermal cycles applied.

Double hit compression tests- To study the progress of metadynamic recrystallization, double hit tests were performed. The specimens were again austenitized at 1200 °C for 20 min. A slow cooling rate of 1 °C/s was used from the solutionizing temperature to the testing temperatures (Figure (1)). The first deformation was interrupted at the peak strain of dynamic recrystallization. After unloading, the specimen was held at the test temperature for a time between 0.1 to 30 s to enable metadynamic recrystallization to progress. A second deformation was then applied to measure the amount of softening, and then the specimen was quenched in water. The deformation temperature was varied from 900 to 1050 °C in 50 °C increments and strain rates of either 0.01, 0.1 or 1/s were used.

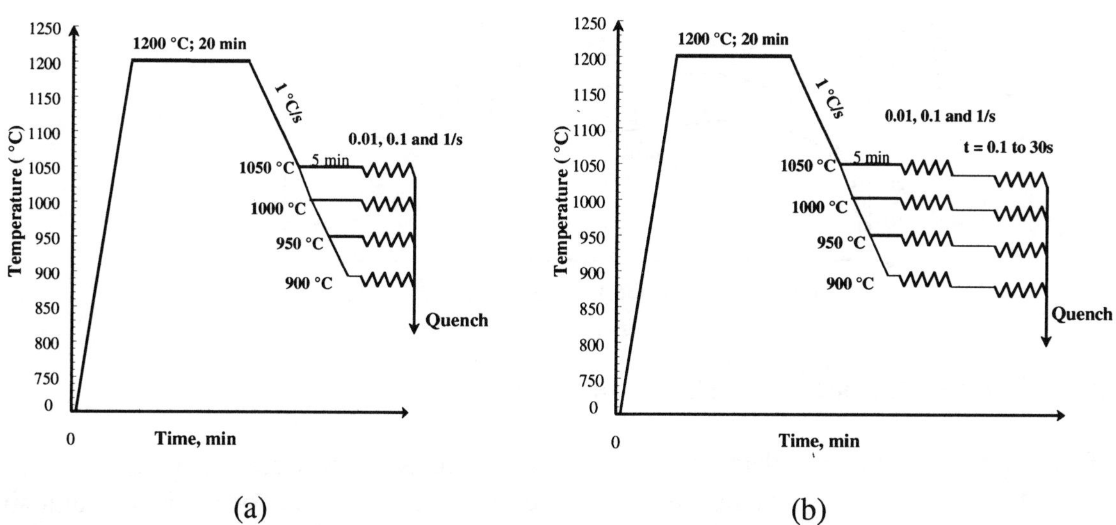

Figure (1) Schematic illustration of the TMP schedule for characterizing (a) dynamic recrystallization and (b) metadynamic recrystallization.

Measuring the softening- The interrupted deformation method is based on the principle that the yield stress at high temperatures is a sensitive measure of the structural changes. In this work, the 0.2% offset yield strength was used to determine the softening due to metadynmic recrystallization. The softening, X, is measured by:

$$X = \frac{[\sigma_m - \sigma_1]}{[\sigma_m - \sigma_2]} \times 100 \tag{3}$$

where: σ_m is the flow stress at the interruption
σ_1 is the offset stress (0.2%) at the first hit
σ_2 is the offset stress (0.2%) at the second hit

RESULTS

Ausenite grain size- In order to determine the austenite grain size before deformation, the three steels were held at the austenitization temperature (1200 °C) for 20 minutes before quenching. The austenite grain sizes were determined by the intercept method (ASTM E112). The grain sizes determined in this way were shown in the **Table I**. As can be seen, there is no major difference between these steels.

Single hit flow curves- The flow curves obtained for the three steels deformed to a strain of 0.7 were plotted for the deformation temperatures (900 °C-1050 °C) for various strain rates. All flow curves displayed a rapid initial increase to a stress maximum, characterized by a peak strain and peak stress, followed by a gradual fall to a constant stress level (steady state stress). Examples of the flow stress/strain curves are given in Figure (2) for steel A.

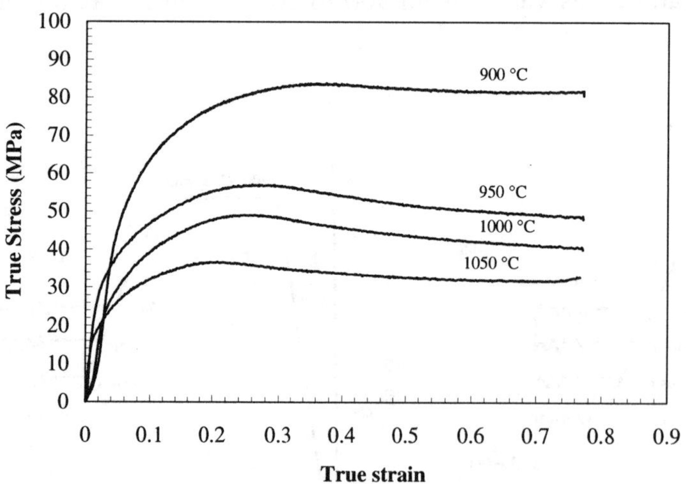

Figure (2) Stress-Strain for the steel A at a strain rate 0.01/s.

Double hit flow curves- Using the data from the single hit tests, tests were designed to initiate dynamic recrystallization in the first hit, so that metadynamic recrystallization would occur in the unloading (static) period. Interrupted compression stress-strain curves with increasing unloading times are given in Figure (3) for steel A. As expected, when the unloading time is short, little softening occurs. As a consequence of the lack of softening, the second hit curve displays little further work hardening. When the unloading time was increased, metadynamic recrystallization occurred and diminished the dislocation density somewhat, leading to increased work hardening on reloading. However, if the unloading time is large enough to allow the softening mechanisms to destroy the dislocation structure established during the first deformation, then the second hit curve work hardens in a similar way to the first hit curve. Note that the peak stress and strain are much smaller after full metadynamic recrystallization because of the grain refinement that has occurred.

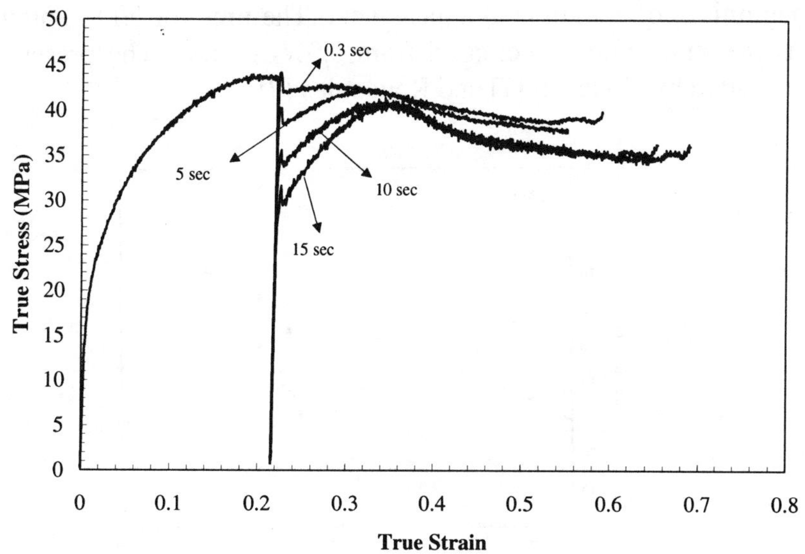

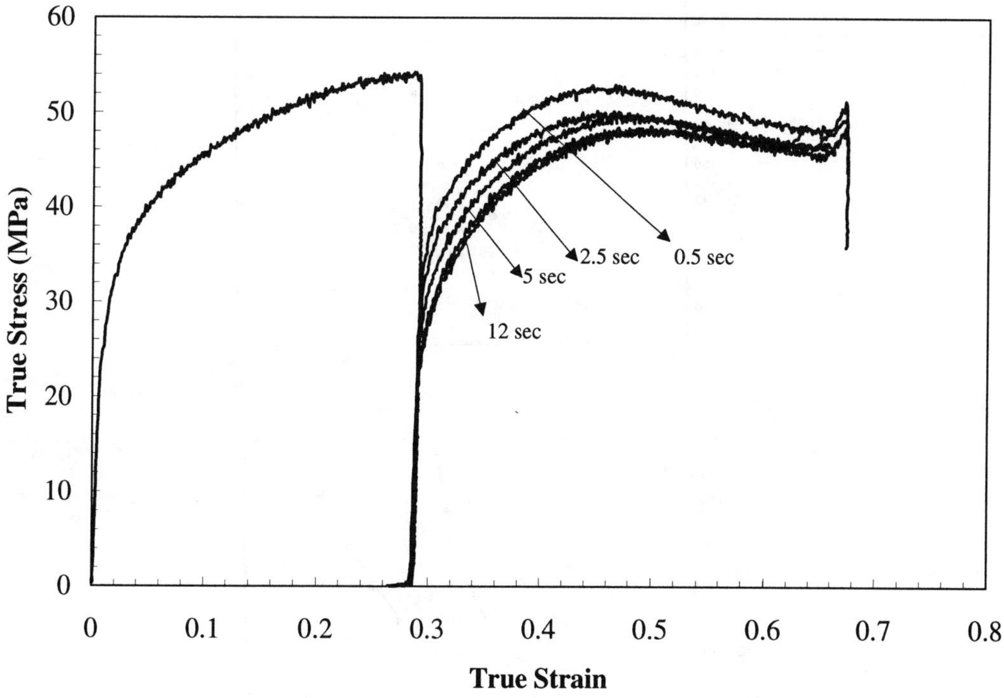

Figure (3) Stress-strain curves determined for steel A showing the softening taking place at 1050 °C for different unloading times (a) 0.01/s and (b) 0.1/s.

Effect of strain rate on softening- In general, increasing the strain rate increases the flow stress. Increasing the flow stress is associated with an increase in the dislocation density, a decrease in the subgrain size and an increase in the stored energy. The driving force for recrystallization, therefore, will increase and the time for 50 % recrystallization will thus decrease with strain rate.

The effect of strain rate was investigated over the range 0.01/s to 1/s at 1050 °C. The measured fractional softening is plotted as a function of the logarithm of the holding time as shown in Figure (4). For all the steels, these plots have the beginnings of a sigmoidal appearance. The time for 50 % softening decreases from about 5 s to less than 1 s as the strain rate is increased from 0.01/s to 1/s. The present observations are in good agreement with those reported by Hodgson [7] and Roucoules [2].

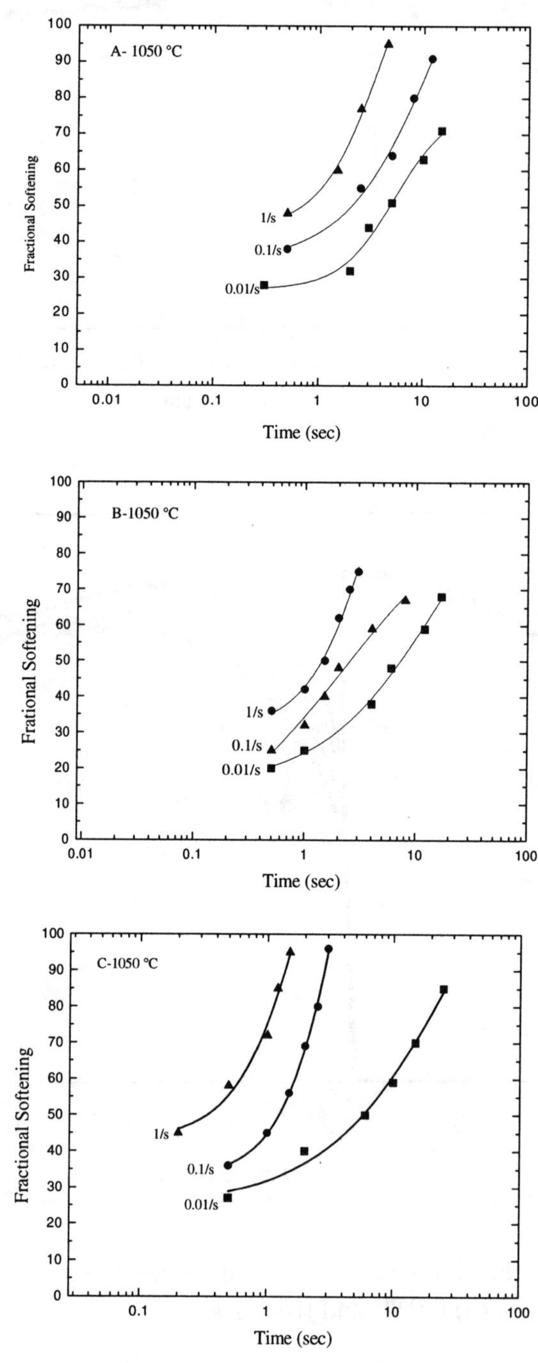

Figure (4) Effect of strain rate on softening in steels A, B & C.

Effect of the deformation temperature on softening- Figure (5) shows the effect of deformation temperature on softening at a constant strain rate. These softening curves are also sigmoidal in appearance. Both steels show that the rate of softening increases with increasing temperature.

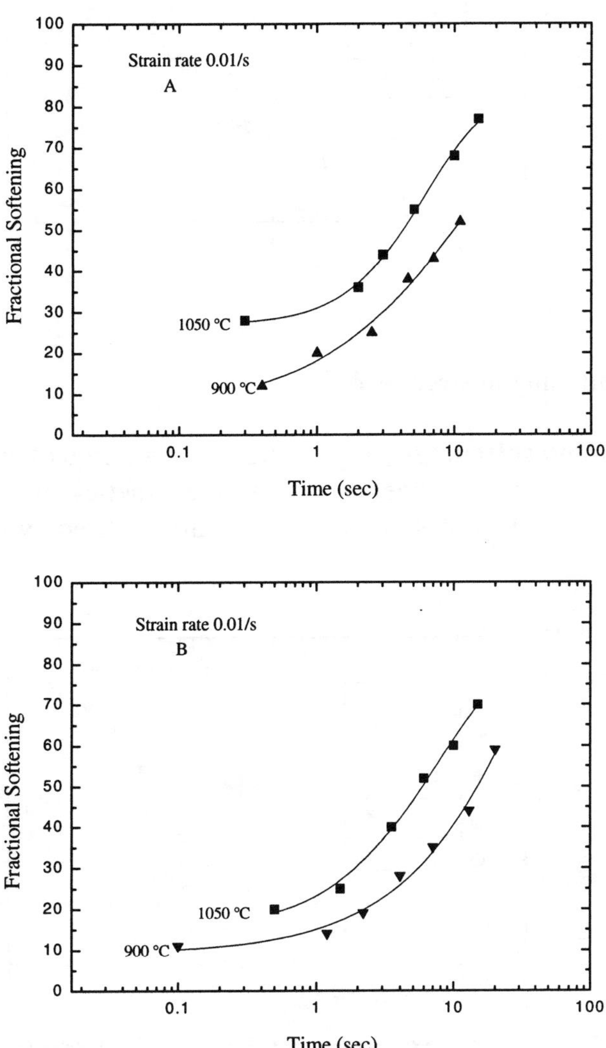

Figure (5) Effect of temperature on softening in steels A & B.

Effect of strain on softening- The softening curves shown in Figure (6) indicate that strain has little influence upon the fractional softening of metadynamic recystallization. This observation agrees with that of Hodgson [7], Roucoules [8] and Bai [9].

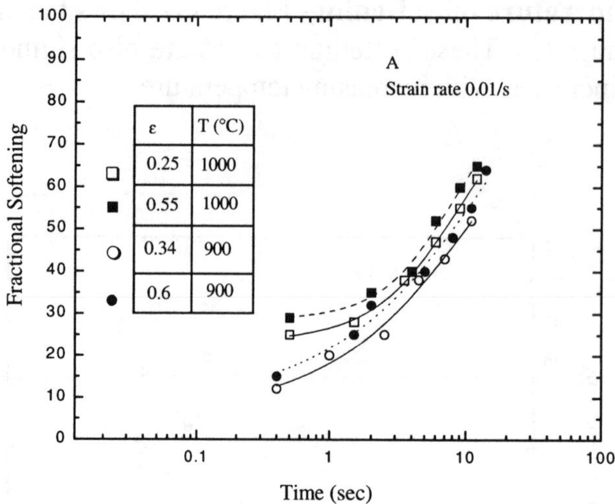

Figure (6) Effect of strain on softening in steels A & B.

Effect of chemical composition on softening-. The recrystallization rate of steel B is slower than that of steel A (Figure 7). The higher Si contents of steel B retards the kinetics of metadynamic recrystallization in microalloyed high carbon steels. This is probably due to solute drag caused by silicon.

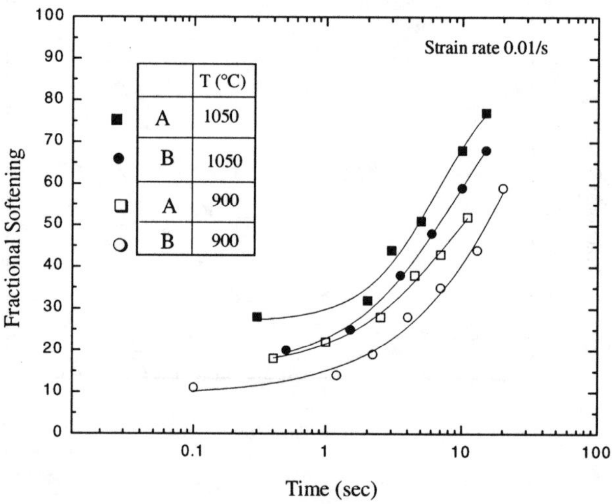

Figure (7) Effect of chemical composition on softening in steels A & B.

DISCUSSION

The kinetics of metadynamic recrystallization are usually described by an Avrami equation {4}, which incorporates an empirical time constant for 50% recrystallization, $t_{0.5}$.

$$X = 1 - \exp\left[-0.693\left[\frac{t}{t_{0.5}}\right]^n\right] \qquad (4)$$

As the data generated in this work were for a constant prior austenite grain size, a simplified expression of the following form was derived to describe the data:

$$t_{0.5} = A\,\dot{\varepsilon}^{p} \exp\left[\frac{Q_{app}}{RT}\right] \quad (5)$$

where

X	fraction recrystallized
t	times (s)
$\dot{\varepsilon}$	strain rate (s^{-1})
A, n and p	material dependent constants
$t_{0.5}$	hold time for 50% softening to take place
Q_{app}	apparent activation energy of recrystallization
T	absolute temperature
R	gas constant

The strain rate exponent can be obtained from the ln ($t_{0.5}$) – ln (strain rate) plots (Figure 8). Although steel C seems to have a somewhat different behaviour, in this paper, it was decided to generate an average value for the exponent p. This gives a value of –0.6, which is close to that of Roucoles et al. [9] for metadynamic recrystallization, and is somewhat lower than that observed by Hodgson et al. [10]. Note that metadynamic recrystallization exhibits a strain rate dependence about twice as strong as conventional static recrystallization [8,11,12].

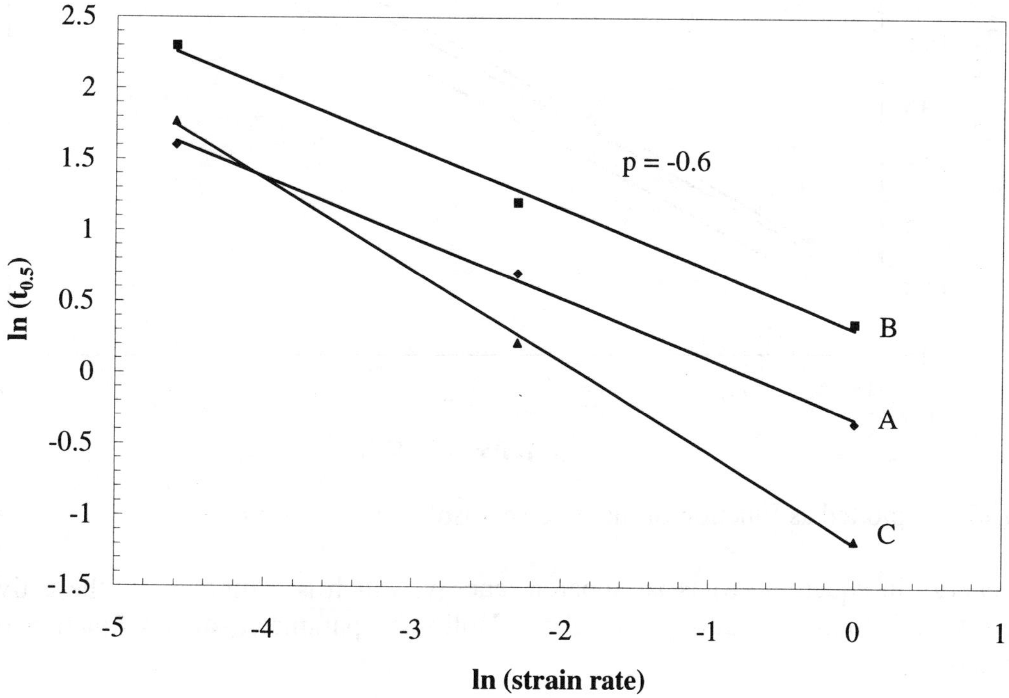

Figure (8) Effect of strain rate on the metadynamic $t_{0.5}$.

Activation energy of metadynamic recrystallization- The activation energies were determined using the following Arrhenius relationship:

$$\ln\left(t_{0.5}/Z^{-0.6}\right) = \ln(A) + \left(\frac{Q_{mdrx}}{R}\right)\left(\frac{1}{T}\right) \tag{6}$$

$$Z = \dot{\varepsilon}\ \exp\left(Q_{def}/RT\right) \tag{7}$$

where $\dot{\varepsilon}$ is the strain rate, T is the absolute temperature, R is the gas constant. Q_{def} is the activation energy derived from the steady state stress, and was found to be 350 kJ/mol for these microalloyed high carbon steels.

The parameter ln ($t_{0.5}/Z^{-0.6}$) is plotted as function of the inverse absolute temperature, as shown in Figure (9). The activation energy for metadynamic recystallization can be determined from the slopes and intercepts of these plots. A value of Q_{mdrx} = 380 kJ/mol was found for the three steels.

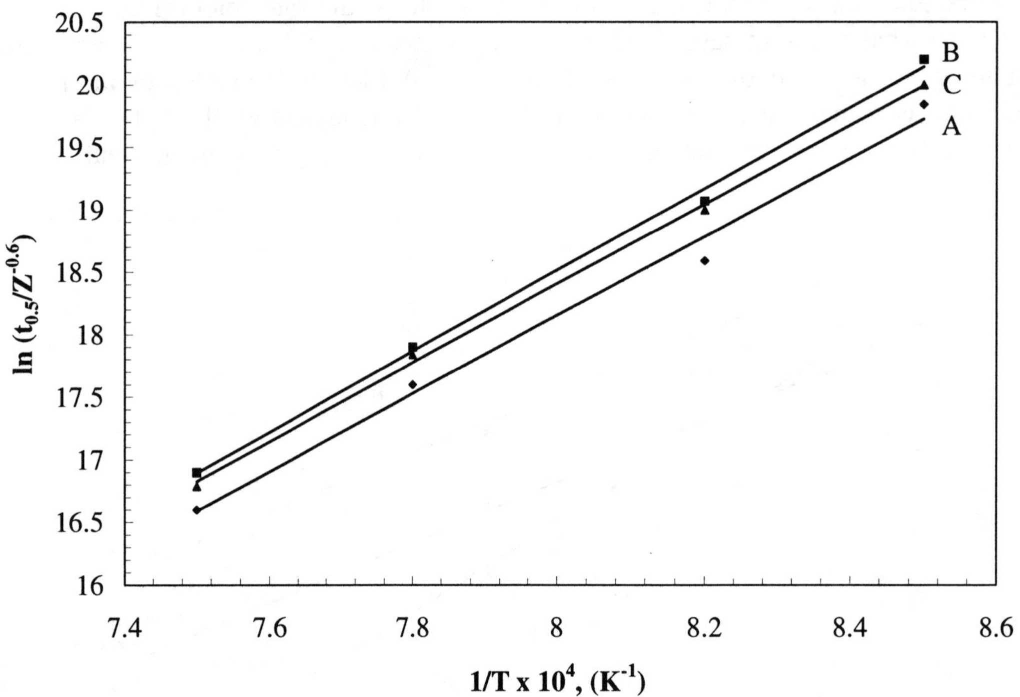

Figure (9) ln ($t_{0.5}/Z^{-0.6}$) plotted as function of the inverse absolute temperature.

The activation energy in equation {8} is an apparent energy, which is a function of the activation energy of metadynamic recrystallization and, through the Zener-Hollomon parameter, also a function of the activation energy of deformation [13].

$$Q_{app} = Q_{mdrx} - 0.6\ Q_{def} \tag{8}$$

Q_{app} is 170 kJ/mole, which is considerably lower than for conventional static recrystallization [8,11,12], which means that metadynamic recrystallization is not as strongly temperature dependent as static recrystallization.

In order to determine n, the softening results were plotted as ln(ln(1/X)) vs. ln(time) plots. The results are presented for different conditions of temperature and strain rate, as shown in Figure (10). The average values of n were found to be 0.95, 0.95 and 0.93 for steels A, B and C, respectively, which approach the range of values of 1-1.6 observed by other workers [8,14-16]. The dependence of A_{mdrx} on Si + V concentrations, is shown in Figure (11).

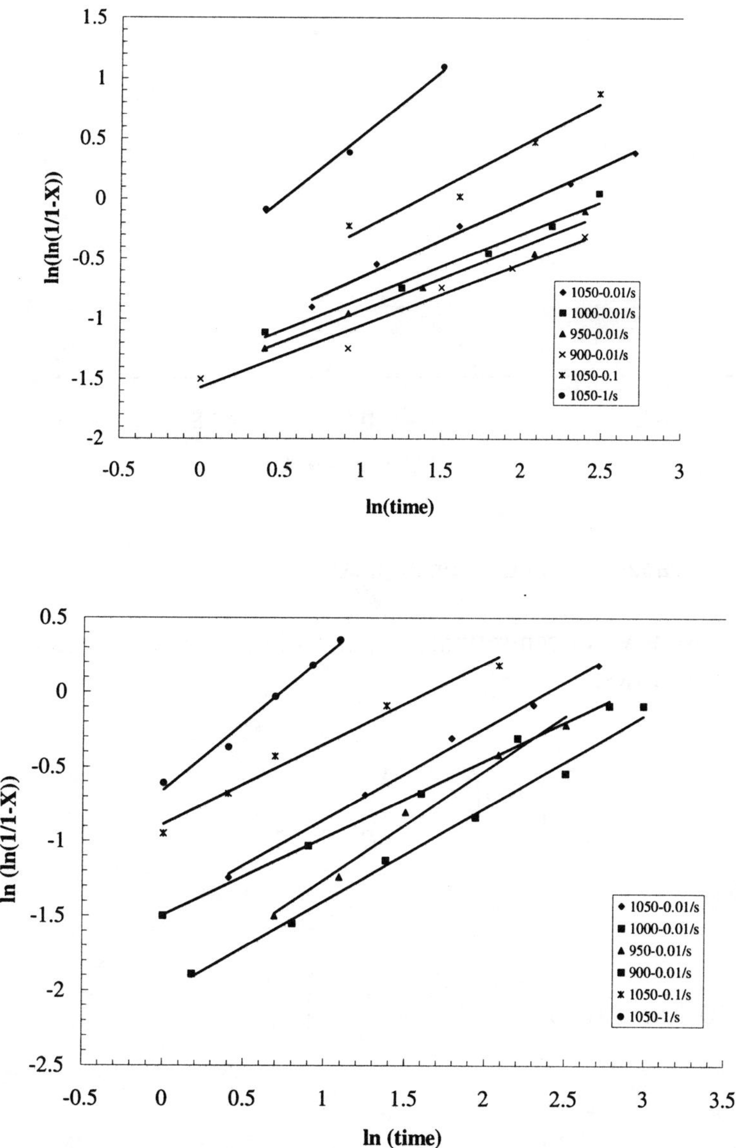

Figure (10) Dependence of ln(ln(1/1-X)) on ln(time) for steels A and B.

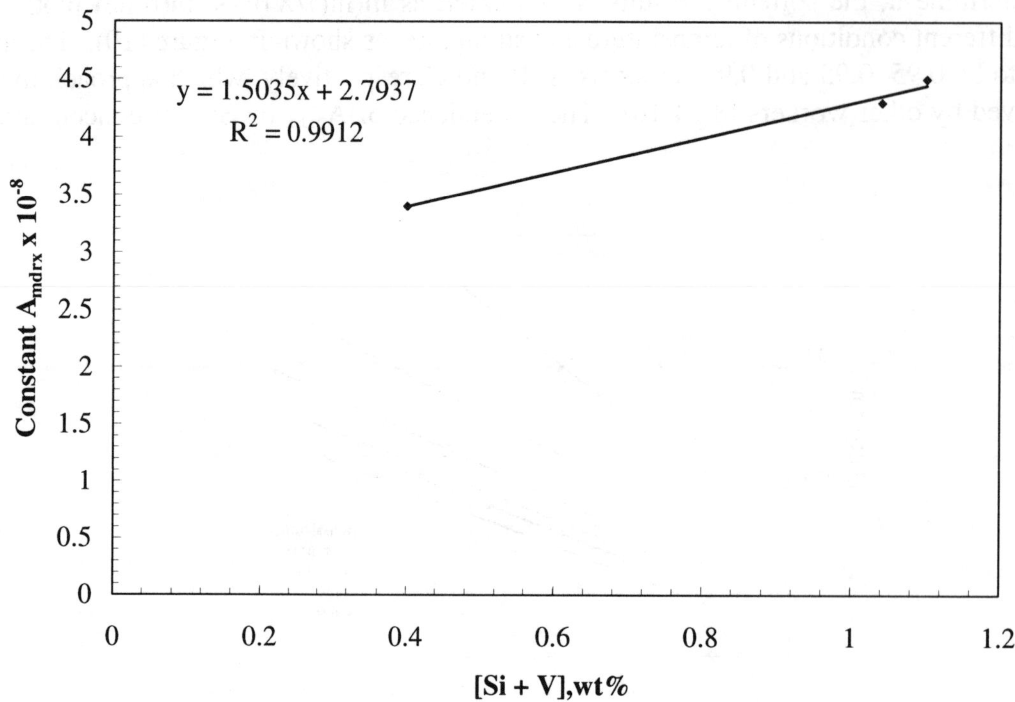

Figure (11) Effect of silicon and vanadium on the constant $A\varepsilon_p$.

Thus, the dependence of $t_{0.5}$ on Si + V concentrations, Zener-Hollomon parameter, strain rate and deformation temperature can be expressed as follows:

$$t_{0.5} = (1.5\,([Si]+[V]) + 2.8)\times 10^{-8}\ Z^{-0.6}\ \exp\left(\frac{380 \left({kJ}/{mol} \right)}{RT} \right) \tag{9}$$

$$t_{0.5} = (1.5\,([Si]+[V]) + 2.8)\times 10^{-8}\ \dot{\varepsilon}^{-0.6}\ \exp\left(\frac{170 \left({kJ}/{mol} \right)}{RT} \right) \tag{10}$$

A comparison between the predicted and measured values of $t_{0.5}$ is made in Figure (12). As can be seen, the developed relationship models the behaviour quite well.

It is generally accepted that the effect of the alloying and microalloying elements on retarding the onset of metadynamic recrystallization is related to the atomic size difference between γ-Fe and Mn, Si, V, Mo, Ti and Nb, which increases in the order listed. The effect of Mn could not be determined from these alloys, but it generally plays only a minor role in retarding recrystallization due to its similar atomic size and diffusion rate compared to iron.

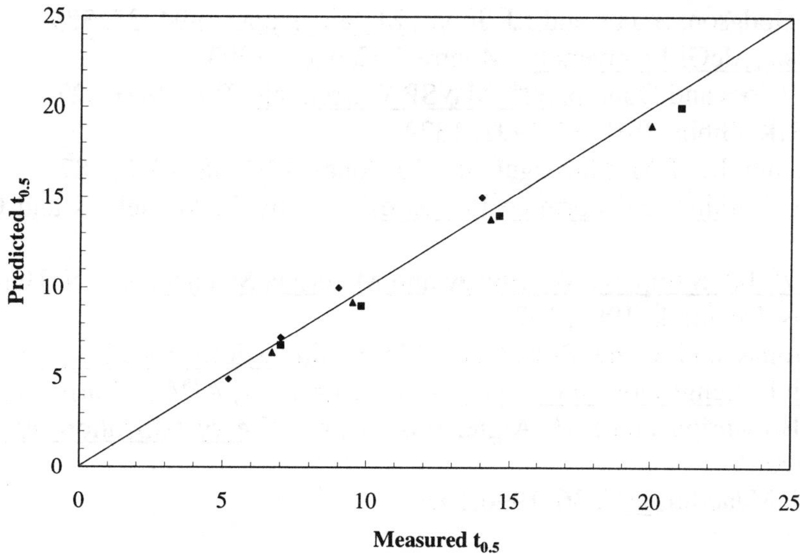

Figure (12) Comparison between predicted and measured $t_{0.5}$ values.

CONCLUSIONS

The effects of Si and V additions on the kinetics of metadynamic recrystallization were investigated during the hot working of steels. Three steels containing different levels of V and Si were tested using the double hit schedules.

1. The additions of Si and V decrease the rate of metadynamic softening.
2. Equations to predict metadynamic recrystallization kinetics for high carbon microalloyed steels have been generated.

ACKNOWLEDGMENTS

The authors would like to thank the Canadian Steel Industry Research Association (CSIRA) and the Natural Sciences and Engineering Research Council of Canada (NSERC) for their financial support. A. Elwazri thanks with gratitude the scholarship received from the University of Garyounis, Benghazi, Libya.

REFERENCES

1. I. Tamura, C. Ouchi, T. Tanaka and H. Sekine, "Thermomechanical Processing of High Strength Low Alloy Steels", London, Butterworths, 1988, 32
2. C. Roucoules, Ph.D. Thesis, McGill University, Montreal, 1992
3. I.P. Kemp, P.D. Hodgson, and R.E. Gloss, Proc. Conf. "Modelling of Metal Rolling Processes", London, September 21-23, 1993, 149
4. C. Roucoules, S. Yue and J.J. Jonas *ibid.* 165.
5. K.J. Irvine, F.B. Pickering and T. Gladman J. Iron and Steel Inst. 205 1967 161-182.
6. H. Ohatani and F. Nakasato American Society for Metals, Metals Park, Ohio, 1985, 169

7. P.D. Hodgson Ph.D. Thesis, University of Queensland, Australia, 1993
8. C. Roucoules, P.D. Hodgson, S.Yue and J.J. Jonas, Metall. Trans. 1994, 25, 389
9. D.Q. Bai Ph.D. Thesis, McGill University, Montreal, Canada, 1993
10. P.D. Hodgson, R.E. Gloss and Dunlop, 32nd MWSP Warrendale, PA, 1990, 527
11. P.D. Hodgson and R.K. Gibbs, ISIJ, 32, 1992, 1329
12. A. Kirihata, F. Siciliano, Jr., T.M. Maccagno and J.J. Jonas, ISIJ, 38, 1998, 187
13. C. M. Sellars, "Hot Working and Forming Processes", ed. by C. M. Sellars and G. J. Davies, Met. Soc., London, 1980, 3.
14. C.M. Sellars, Proc. 7th Int. Symp. on Metallurgy and Materials Science, eds. N. Hansen et al., Riso National Laboratory, Roskilds, Denmark, 1986, 167
15. P.D. Hodgson, J.J. Jonas, and S.Yue, Proc. Conf. "Processing Microstructure and properties of microalloyed and other modern high strength low alloy steel", Pittsburgh 1991, AIME, Warrendale, PA, 41
16. P.D. Hodgson, A. Brownrigg and S.H. Algie, Proc. Conf. "Recrystallization 90", ed. T. Chandra, TMS, Warrendale, PA, 1990, 41
17. S.F. Medina and J.E. Mancilla, ISIJ, 36, 1996,1070

A REAPPRAISAL OF DUCTILE FRACTURE IN HIGH-STRENGTH STEELS

M. J. Leap and J. C. Wingert
Materials Science Department
The Timken Company
Canton, Ohio 44706

Numerous investigations have focused on the degradation in toughness that results from the presence of large second-phase particles such as MnS and TiN in the microstructure of high-strength steels (1-4). The results of these investigations illustrate the beneficial effects of increasing the interparticle spacing of large second-phase particles on ductile fracture resistance. Consistent with a characteristic distance concept for strain-controlled fracture (5), the Rice-Johnson model (6) has been traditionally manipulated into a linear relationship between sharp-crack initiation toughness and the interparticle spacing of void-initiating particles:

$$\delta_{Ic} = C\chi_S (R_V / R_I)\Big|_{R_o,f_I}, \qquad [1]$$

where δ_{Ic} = critical crack-tip opening displacement, χ_S = interparticle spacing of primary void-initiating particles, R_V = void radius, R_I = inclusion radius, R_o = mean value of true inclusion radii, f_I = volume fraction of primary void-initiating particles, and C = constant of proportionality.

The linear material model specifies local microstructural ductility in terms of the limiting extent of primary microvoid growth, R_V/R_I (7). While this parameter represents *one* measure of local ductility, the interruption of primary microvoid growth is typically associated with the initiation and growth of secondary microvoids at much smaller second-phase particles in the microstructure. However, very little attention has been given to the specific effects of small particles on the degradation in the toughness of tempered martensite. Among investigations focusing on the secondary fracture process (8-10), Garrison (11) showed that the addition of molybdenum and vanadium degrades the toughness of 0.35% C steel tempered to the 1650 MPa yield strength level. Gore, Olson, and Cohen (12) discussed the potentially detrimental effects of small secondary particles on the toughness of powder-metal and VAR 4340 steel. Their results indicate that increases in austenitizing temperature promote the dissolution of $M_{23}C_6$, (Ti,Mo)C, and Ti(C,N) to varying degrees, although shear testing illustrated a dominant effect of grain size on the instability strain with increases in austenitizing temperature. Handerhan and Garrison (13) illustrated the same effect of austenitizing temperature on precipitate dissolution in as-quenched HP9-4-20 steel, but in this case changes in the dispersions of Ti(C,N) and V(C,N) produced increases in δ_{Ic} with increasing amounts of microstructural coarsening. Leap and Wingert (14,15) have also evaluated the potent effect of a variety of grain-refining precipitates on the toughness of tempered martensite, particularly with respect to the processing-induced increases in toughness that result from precipitate refinement. More recent work (16,17) has highlighted the interrelationships between the size and density of small secondary particles and the ductile fracture resistance of high-strength steels containing different types and contents of grain-refining precipitates and/or iron-based carbides retained through the hardening operation.

The linear interpretation of the Rice-Johnson model (6) provides a reasonable description of the ductile fracture process for steels in which relatively low matrix ductility limits the extent of primary microvoid growth (2,3,18-20). However, Garrison and coworkers (21,22) have shown that the linear representation for δ_{Ic} does not adequately describe the ductile fracture process in ultrahigh strength steels. The inability of the linear material model to functionally describe the dependence of δ_{Ic} on relevant microstructural parameters stems from several sources. First, the linear interpretation of the Rice-Johnson mechanics analysis is based on the tacit assumption that the volume fraction of void-initiating particles, f_I, is constant. Most investigations (2,3,18-22) are based on data sets in which the volume fraction of non-metallic inclusions is not constant, such that initiation toughness as defined by the linear model is related to only one of two stereological parameters that are required in defining the dispersion of void-initiating particles. Second, the linear model does not provide a functional mechanism to delineate the effects of crack tip deformation mode on the development of initiation toughness (23,24). Finally, the model does not consider interactions between primary microvoids, particularly with respect to the potential for sampling of the inclusion population in steels with small interparticle spacings.

The purpose of this investigation is to consider the combined effects of the primary and secondary fracture processes on the development of toughness in high-strength steels. The description of the ductile fracture process is based on an alternative interpretation of the Rice-Johnson mechanics analysis (6), and the applicability of the resulting material model is demonstrated with the results of several independent investigations (2,3,8,11,13,18-26). Secondary fracture is considered in terms of a microstructurally relevant strain that is dependent on the size and density of small second-phase particles that initiate secondary microvoids, and this measure of local strain is shown to be related to the microstructural fracture strain as defined by the extent of void growth at primary microvoids, $\ln(R_V/R_I)$. Finally, the effects of particle sampling during ductile fracture are assessed with respect to existing experimental data for primary particles that are "weakly bound" to the matrix (3,11,13,22,24,25) and secondary particles that exhibit substantial resistance to microvoid initiation (16,17).

THE PRIMARY FRACTURE PROCESS

The analysis of Rice and Johnson (6) defines the onset of flow localization between the crack tip and a void located below the crack tip as:

$$\delta_I F(\psi) - R_X = R_Y, \qquad [2]$$

where δ_I = crack-tip opening displacement, $F(\psi)$ = integrated result of a dimensionless function of the tangent angle, ψ, that indicates position on the fracture plane below the blunted crack tip, R_X = void radius in the fracture plane along the direction of crack extension, and R_Y = void radius in the direction of applied loading. Rice and Johnson utilized a modified form of Equation 2 to numerically evaluate the relationship between the dimensionless parameters δ_I/χ_V and χ_V/R_i:

$$\frac{\delta_I}{\chi_V}\frac{R_i}{2R_V}F(\psi) = R_i/\chi_V, \qquad [3]$$

where R_V = average void radius (i.e., $2R_V = R_X + R_Y$), R_i = initial void radius, and χ_V = distance from the undeformed crack tip to the first primary microvoid below the crack tip.

A linear material model can be derived from the mechanics analysis by substituting the interparticle spacing of void-initiating particles, χ_S, for χ_V, the particle size, R_I, for R_i, and the cube root of the volume fraction of void-initiating particles, $f_I^{1/3}$, for R_I/χ_S (27). The fracture criterion, Equation 3, then degenerates into Equation 1 when f_I is constant. An alternative form of the linear model can be expressed as:

$$\delta_I = \frac{C\chi_S(R_V/R_I)}{f_I^{1/3}}. \quad [4]$$

However, since R_i/χ_V is redundantly incorporated into the fracture criterion as a means of numerically evaluating the relationship between δ_I/χ_V and R_i/χ_V, the linear material models derived from Equation 3 do not reflect the implicit dependencies of δ_I and R_V on ψ and the stress-strain behavior of the matrix material for an initial geometry specified in terms of χ_V and R_i. McMeeking's (28) analysis of the Rice-Johnson model, on the other hand, predicts a linear dependence of δ_I/χ_V over a limited range of R_V/R_I for various types of blunting behavior, and this correlation provides an appropriate basis for the linear material model, Equation 1.

The Rice-Johnson numerical analysis, Figure 1, alternatively suggests that $\ln(\delta_I/\chi_V)$ at fracture is linearly related to $\ln(\chi_V/R_i)$ over a substantial range of χ_V/R_i for both the small geometry change and fully plastic solutions:

$$\ln\left[\frac{\delta_I}{\chi_V}\right] = \alpha + \beta\ln\left[\frac{\chi_V}{R_i}\right], \quad [5]$$

where α and β are fitting parameters. This expression degenerates directly into the fracture criterion, Equation 3, when $\alpha = \ln(2R_V/R_iF(\psi))$ and $\beta = -1$. Equation 5 also degenerates into a linear material model, Equation 1, for any constant value of β when $\alpha = \ln(R_V/R_IF(\psi))$ and f_I = constant (i.e., χ_S/R_I = constant). These algebraic manipulations are based on the partitioning of the fitting parameters into a form where the functional dependence of δ_I on ψ and the stress-strain behavior of the matrix material is entirely incorporated into α when f_I is constant. However, the fully plastic and small geometry change cases shown in Figure 1 exhibit β values of ~1.10 and ~1.23, respectively. These results suggest that β is dependent on the stress-strain response of the matrix regardless of whether the dependence of δ_I on ψ is only incorporated into α or partitioned in some manner between α and β. Accordingly, β may not be constrained to a single value for materials with different strain hardening rates, strain hardening capacities, and crack tip deformation modes, such that a more general material model can be developed without having to impose a condition of constant f_I.

Although the Rice-Johnson mechanics analysis does not directly incorporate a measure of matrix ductility, Equation 5 can be manipulated into a more general form based on the sole assumption that α contains a local strain term that represents the limiting extent of primary microvoid growth. This assumption, which is consistent with the results of McMeeking's (28) analysis, is justifiable in terms of the functional form of the fracture criterion that provides the basis for the numerical results in Figure 1.[1] Thus, if χ_V is replaced with χ_S, R_i replaced with R_o, and the appropriate void growth equation for $\ln(R_V/R_I)$ utilized as a measure of the average local strain around a primary microvoid (29), Equation 5 can be rewritten as:

---

[1]The incorporation of a matrix ductility parameter or microstructural fracture strain into the model is generally justifiable from the standpoint that primary microvoid growth is typically limited by the onset of a secondary fracture process in high-strength steels. Since the latter process is defined by the ductility of the matrix material, the description of ductile fracture will necessarily rely on some measure of microstructural ductility for the case where fracture occurs prior to the complete coalescence of primary microvoids.

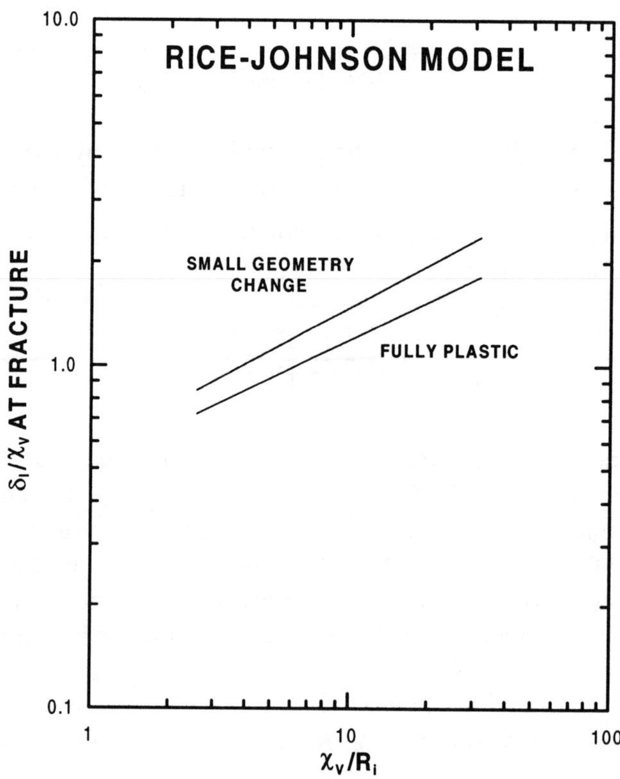

Figure 1. Calculated estimates of δ_I/χ_V at fracture are shown as a function of χ_V/R_i for the small geometry change and fully plastic cases, where δ_I = critical crack opening displacement, χ_V = distance from the crack tip to a microvoid located below the crack tip, and R_i = initial void size (i.e., inclusion size). Numerical results of Rice and Johnson (6).

$$\ln \delta_I = C + \ln\left[R_o^{-\beta}\chi_S^{1+\beta}\right] + 0.322 \int_{\varepsilon=0}^{\varepsilon=\varepsilon_P} e^{\left[\sqrt{3}\sigma/\tau_o\right]} d\varepsilon, \qquad [6]$$

where ε_P = plastic strain, σ = effective stress, τ_o = yield strength in shear, and C = constant. Although Equation 6 is applicable to the limits of primary microvoid growth that define crack extension in the Rice-Johnson model, a secondary fracture process typically interrupts primary microvoid growth well prior to coalescence in high-strength steels. In this connection, the widespread initiation of the secondary fracture process occurs when the applied strain-energy density exceeds the load bearing capacity of the intervoid material ahead of the crack tip, and δ_I approaches δ_{Ic} as the integral term in Equation 6 approaches a critical level of strain. If the limiting material response is defined in terms of a microstructural fracture strain (29,30), ε_C, Equation 6 can be rewritten as:

$$\ln \delta_{Ic} = C_1 + \ln\left[R_o^{-\beta}\chi_S^{1+\beta}\right] + \kappa\varepsilon_C, \qquad [7a]$$

or:

$$\ln \delta_{Ic} = C_2 + \ln[f_I^{-\beta/3}\chi_S] + \kappa\varepsilon_C, \qquad [7b]$$

where C_i and κ are constants. The applicability of the power law representation of the ductile fracture model, Equation 7, can be illustrated with the data from several independent investigations.

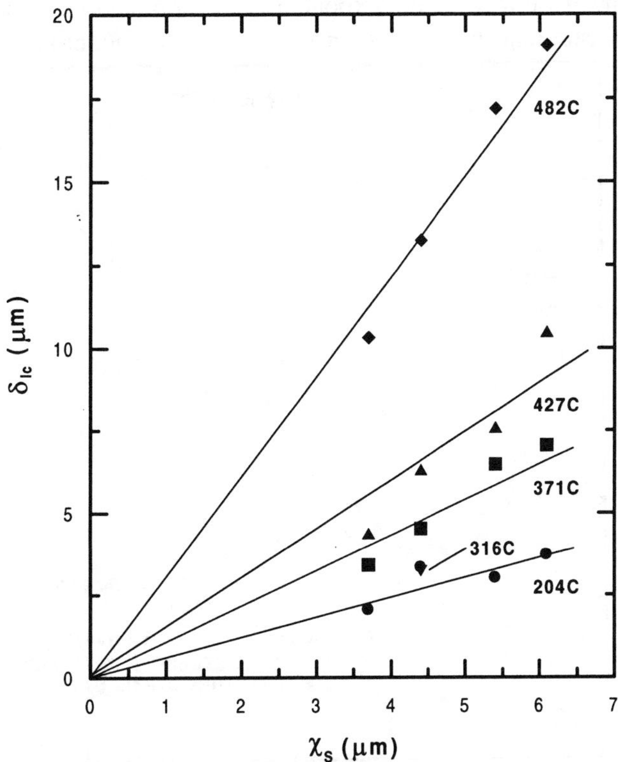

Figure 2. Estimates of δ_{Ic}, obtained from measurements of K_{Ic}, are shown as a function of inclusion interparticle spacing, χ_S, and tempering temperature for longitudinally-oriented specimens of 0.45% C-Cr-Ni-Mo steels containing 0.008-0.049% S. Data of Birkle, Wei, and Pellissier (18).

Birkle, Wei, and Pellissier (18) generated both microstructural and mechanical property data for a 0.45% C-Cr-Ni-Mo steel containing 0.008-0.049% S. Estimates of δ_{Ic}, calculated from measured values of K_{Ic}, are shown as a function of χ_S in Figure 2 for specimens tempered at temperatures in the 204-482°C range. These data suggest that the linear material model provides a good description of the ductile fracture process at each level of matrix ductility (i.e., tempering temperature), although the correlation between δ_{Ic} and χ_S violates the condition of constant f_I implicit in the derivation of Equation 1.

The data for each tempering temperature can be combined and analyzed as a single set if a suitable representation of the matrix ductility parameter is defined in terms of material strength. Based on the assumption that the matrix ductility is inversely related to material strength, ε_C is defined for the purpose of this analysis as:

$$\varepsilon_C \approx C_3 \sigma_o^{-\varsigma}, \qquad [8]$$

where σ_o = yield strength and C_3 and ς are constants. Neglecting the restriction regarding constant f_I, the linear interpretation of the model is evaluated from Equations 1 and 8:

$$\delta_{Ic} = C_4 \chi_S e^{(C_3 \sigma_o^{-\varsigma})}. \qquad [9]$$

Based on the data of Spitzig (19), specimens tempered at 204°C, 427°C, and 482°C were alternatively evaluated with a matrix ductility parameter defined as the exponential of the plane strain tensile ductility. The results of these analyses, Figure 3a, indicate that the linear interpretation of the model provides an adequate description of

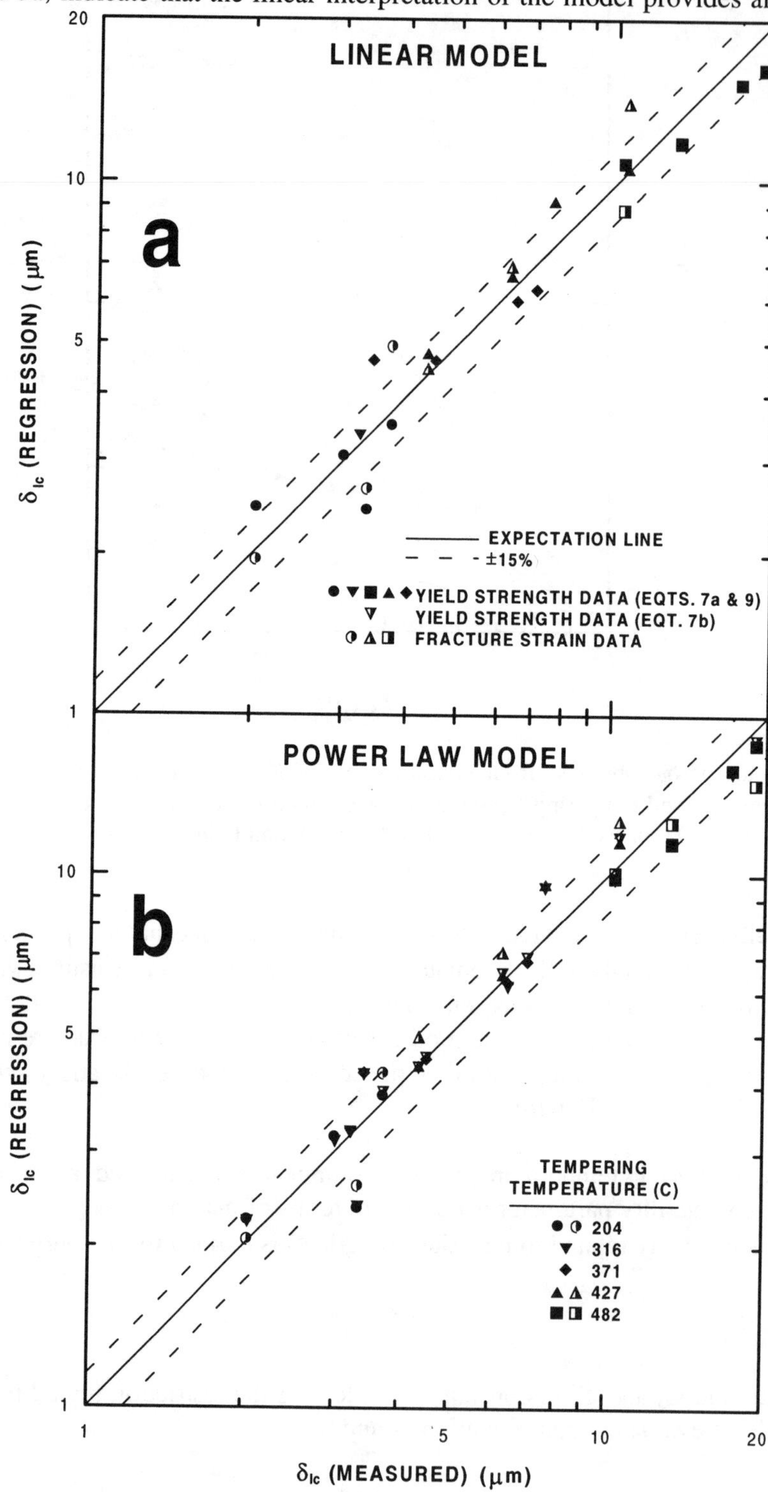

Figure 3. Calculated values of δ_{Ic} obtained by regression analysis are shown as a function of measured values of δ_{Ic} for longitudinally-oriented specimens of 0.45% C-Cr-Ni-Mo steels containing 0.008-0.049% S (18): (a) linear material model, Equation 9, and (b) power law material model, Equation 7. The matrix ductility parameter for specimens tempered in the 204-482°C range is defined in terms of material yield

strength, Equation 8, and plane strain tensile ductility based on the data of Birkle, Wei, and Pellissier (18) and Spitzig (19), respectively.

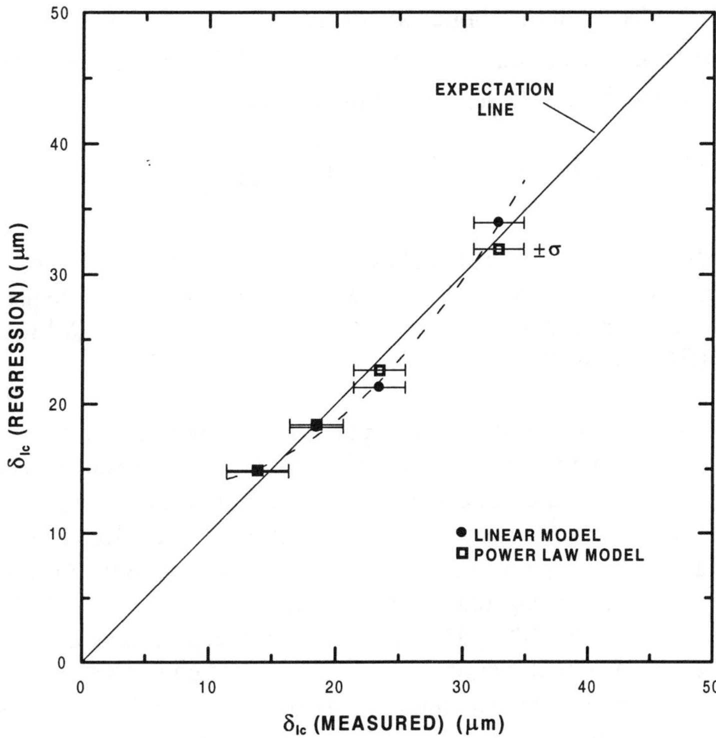

Figure 4. Calculated values of δ_{Ic}, obtained by regression analysis of the functional forms defined in Equations 1 and 7, are shown as a function of measured values of δ_{Ic} for longitudinally-oriented specimens of 0.2% C-12% Cr steels containing 0.004-0.035% S. Data of Raghupathy, Srinivasan, Krishnan, and Chandrasekharaiah (20).

the ductile fracture process in the 0.45% C-Cr-Ni-Mo steel regardless of the specific form of the matrix ductility parameter. The entire set of data was also regressed to the functional forms in Equations 7a and 7b using the same definitions for the matrix ductility parameter. Since inclusion size data were not reported in Reference 18, regression to the functional form of Equation 7a is based on the assumption that R_o is constant for the four steels. In contrast, Equation 7b was evaluated in terms of the experimental values of δ_{Ic} and χ_S and estimates of f_I based on steel composition (31). The results of these calculations, Figure 3b, suggest that the general functional form of Equation 7 provides an equally adequate description of the ductile fracture process in the 0.45% C-Cr-Ni-Mo steel.

Raghupathy et al. (20) evaluated the effects of MnS on the toughness of 0.2% C-12% Cr steels containing 0.004-0.035% S. Estimates of δ_{Ic}, calculated from measured values of K_{Ic}, are shown as a function of χ_S in Figure 4 for specimens austenitized at 1020°C, oil quenched, and tempered at 670°C. Since relevant measures of matrix ductility (i.e., R_V/R_I or plane strain tensile ductility) are not reported in Reference 20, the matrix ductility parameter is assumed constant in the analysis of these data. This assumption is reasonable from the standpoint that, with the exception of sulfur content, the steels exhibit similar compositions and were heat treated in an equivalent manner. The linear model, applied neglecting the restriction of constant f_I, adequately describes the process of ductile fracture in this steel, although the trend between δ_{Ic} and χ_S exhibits curvature (i.e., the steel with the largest MnS interparticle spacing deviates from the trend predicted by Equation 1). Measured values of δ_{Ic} also exhibit a good correlation with calculated values based on a fit of the MnS interparticle spacing data to the general functional form of Equation 7a. These results suggest that either interpretation of the Rice-Johnson model provides a good description of the ductile fracture process in the stainless steel.

The results of Garrison and coworkers (2,3,8,11,13,21-25) on a variety of alloy and ultrahigh strength steels are summarized in Reference 26. The tabulated data include average values of δ_{Ic}, χ_S, R_I, and R_V/R_I for each steel.[1] The linear interpretation of the Rice-Johnson model is evaluated from Equation 1, although the requirement of constant f_I is neglected in the application of the model to these data. Moreover, since the average value of R_V/R_I, $\overline{R_V/R_I}$, provides a measure of the critical condition for fracture in the intervoid material, the power law interpretation of the model is evaluated from Equation 7a as:

$$\ln \delta_{Ic} = C_5 + \ln\left[R_o^{-\beta}\chi_S^{1+\beta}\right] + C_6 \ln(\overline{R_V/R_I}), \qquad [10]$$

where C_i = constant. In order to account for differences in base composition and microstructure, the steels were evaluated in terms of sets comprising data for (i) lightly-tempered 0.35% C alloy steels; (ii) as-quenched HP9-4-10 and HP9-4-20 steels; (iii) quenched and aged 0.1% C steels (HP9-4-10 and HY180), and; (iv) quenched and aged 0.16-0.20% C steels (AF1410 and HP9-4-20).

The predicted values of δ_{Ic} for some of the steels scale in rough proportion to the measured δ_{Ic} values when evaluated with the linear model, Figure 5a. However, a substantial difference between the measured and calculated values of δ_{Ic} exists in many cases, and the scaling of the predicted δ_{Ic} values within some of the data sets (e.g., the lightly-tempered 0.35% C alloy steels) is inconsistent with the trend predicted by the model.[2] In contrast, values of δ_{Ic} calculated from Equation 10 scale in direct proportion to the measured values of δ_{Ic}, and the agreement between the measured and calculated values is extremely good for both the lightly-tempered alloy steels and as-quenched specimens of the HP9-4-10 and HP9-4-20 steels, Figure 5b.[3] The toughness data for the AF1410, which exhibits a particularly high degree of non-linearity when evaluated in terms of Equation 1, are more effectively fit by the power law model, although two of the aged (510°C) specimens exhibit large differences between measured and calculated values of δ_{Ic}. While the cause of the poor correlation is not explicitly known, one of the steels appears to exhibit anomalously high toughness relative to the other steel irrespective of which model is applied to the data.[4] If the data for this steel are omitted from the analysis, the power law model yields a substantially improved fit to the censored data set, whereas no improvement is realized in the fit of the data to the linear model. Considering the numerous combinations of steel composition and microstructure represented by these data, the results in Figure 5b suggest that the alternative interpretation of the Rice-Johnson model provides a realistic description of the ductile fracture process in high-strength steels.

---

[1]Data for smaller second-phase particles are limited to measurements of mean size and particle density for (i) (Ti,V)(C,N) and M_2C (and possibly AlN) in the as-quenched HP9-4-20 (Heat #1) (13); (ii) V(C,N) in the as-quenched HP9-4-20 (Heat #2) (24), and; (iii) titanium-rich carbonitrides in the HY180 steels (25). However, these data do not provide any quantitative information relevant to evaluating the secondary fracture process since extraction replicas were obtained from polished metallographic specimens. Vanadium-rich (V,Mo)(C,N) was also identified as the predominant precipitate phase in the Base + Ni + Si + MoV steel (11), although the source of the extraction replicas utilized in this investigation was not specified.

[2]These data were also evaluated with a linear function containing two fitting parameters:

$$\delta_{Ic} = C_7 + C_8\chi_S(R_V/R_I),$$

where C_i = constant. This analysis yielded roughly the same degree of correlation as Equation 1 for each of the data sets.

[3]The power law model fit to the data for the quenched and aged HY180 and HP9-4-10 steels (i.e., three data points) is exact since Equation 10 is defined by three fitting parameters.

[4]The two AF1410 steels designated Heat #2 (510°C) and Heat #3 (510°C) in Reference 26 exhibit δ_{Ic} values of 66 μm and 61 μm, respectively. Based on the scaling of microstructural parameters in the linear and power law models, the predicted ratio of initiation toughness for the two steels (0.37/0.45) is significantly less than the experimentally observed value of 1.08. Alternatively, if it is assumed that the toughness of the steels aged at 510°C scales in the same proportions as the toughness of the same steels aged at 425°C, the toughness of Heat #2 is predicted to be on the order of 54-56 μm after aging at 510°C. This range of values is in general accord with the trend line in Figure 3 of Reference 26.

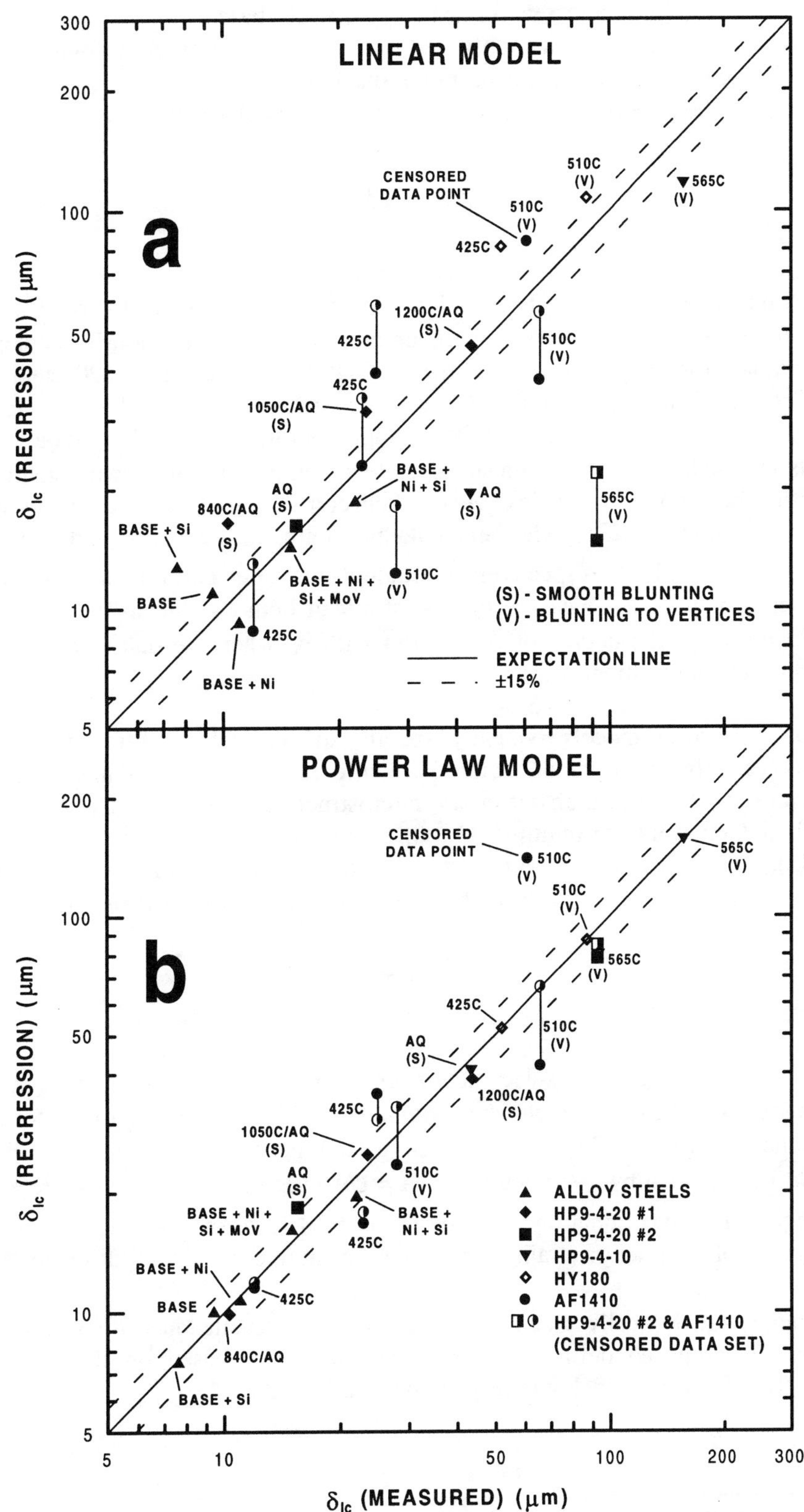

Figure 5. Calculated values of δ_{Ic} obtained by regression analysis are shown as a function of measured values of δ_{Ic} for longitudinally-oriented specimens of a variety of high-strength and ultrahigh strength steels (3,8,11,13,21,22,24-26): (a) linear material model, Equation 1, and (b) power law material model, Equation 10.

The partitioning of these data into sets for regression, while based on differences in steel composition and microstructure, also represents a categorization based on differences in crack tip blunting behavior. Handerhan and Garrison (24) utilized a modified form of Equation 4 to characterize the toughness of several ultrahigh strength steels in terms of the effects of matrix microstructure on crack tip blunting behavior:

$$\delta_{Ic} = f\left\{\frac{\chi_S \varepsilon_{PS}}{f_I^{1/3}}\right\}, \quad [11]$$

where ε_{PS} = plane strain tensile ductility. The functional form of the independent variables, which can be obtained by substituting ε_{PS} for R_V/R_I in Equation 4, was apparently derived from a non-linear dependence between the aforementioned variables for as-quenched specimens of HP9-4-10 and HP9-4-20 steel (refer to Figure 6 of Reference 26). Their analysis clearly illustrates a large difference in the level of toughness exhibited by specimens that blunt to vertices and specimens with smooth blunting behavior, Figure 6a.[1] However, these results are limited to steels in which δ_{Ic} exhibits a linear dependence on χ_S (i.e., steels with small values of χ_S), and the distinction in blunting behavior is much less clear when companion steels with larger inclusion spacings are included in the comparison (13,21,22). The same data are plotted in Figures 6b and 6c with the matrix ductility parameter defined as $e^{\varepsilon_{PS}}$ and R_V/R_I, respectively. Although the latter definitions for matrix ductility parameter are functionally consistent with the measure of true strain in both the linear and power law models, neither provides any improvement in the degree of correlation with δ_{Ic} when the data are partitioned with respect to blunting behavior or steel composition/microstructure.

Micromechanistic fracture models typically specify the critical condition for crack initiation as the attainment of a representative measure of material fracture resistance over a microstructural distance that is relevant to the fracture process (i.e., a characteristic microstructural distance). These parameters are defined as the fracture strength and a non-integer multiple of grain size for the case of propagation-limited brittle fracture in ferritic materials (32). The critical driving force for ductile fracture initiation is specified in terms of the generally accepted expression relating δ_{Ic} to the product of fracture strain, ε_F, and characteristic microstructural distance, ℓ_o (5):

$$\delta_{Ic} \propto \ell_o \varepsilon_F. \quad [12]$$

This relationship was derived from the displacement expression for a center crack in an infinite elastic body and Irwin's plastic zone size estimate (33), which when combined yield an expression of the form $\delta_I = 2\pi r_P \varepsilon_Y$, where r_P = plastic zone size and ε_Y = yield strain. Wells (34) suggested that the range of applicability for the small strain approximation could be extended by assuming that $r_P/a = \varepsilon/\varepsilon_Y$, such that $\delta_I \propto \varepsilon a$, where a = crack length. This latter expression degenerates into Equation 12 when the critical level of crack-tip opening displacement, δ_{Ic}, is defined as the attainment of the fracture strain over some relevant microstructural distance.

Although ε_F and ℓ_o are typically taken as the plane strain tensile ductility and interparticle spacing of inclusions, Garrison (7) proposed definitions for these parameters based on the partitioning of the linear interpretation of the Rice-Johnson model with respect to Equation 12:

[1]The data for AF1410 #1 (425°C), HP9-4-20 (200°C), and HY180 (425°C) are included in Figure 9 of Reference 26 and Figure 14 of Reference 24, but the blunting behavior is not specified in either Reference 26 or the source references for these steels (22,24,25). However, based on the behavior of similar steels, it would not be unreasonable to expect that the quenched and aged (425°C) specimens of AF1410 and HY180 blunt to vertices. These data are indicated with arrows in Figure 6a. It should also be noted that six out of nine data points in the former two figures (24,26) are plotted incorrectly with respect to the tabulated data in Reference 26 and the source references for these steels (22,24,25), although the correct positioning of the data points does not significantly alter the general trajectory of the trend lines.

$$\varepsilon_F = a[(R_V / R_I)|_{R_o}]^{\omega}, \qquad [13a]$$

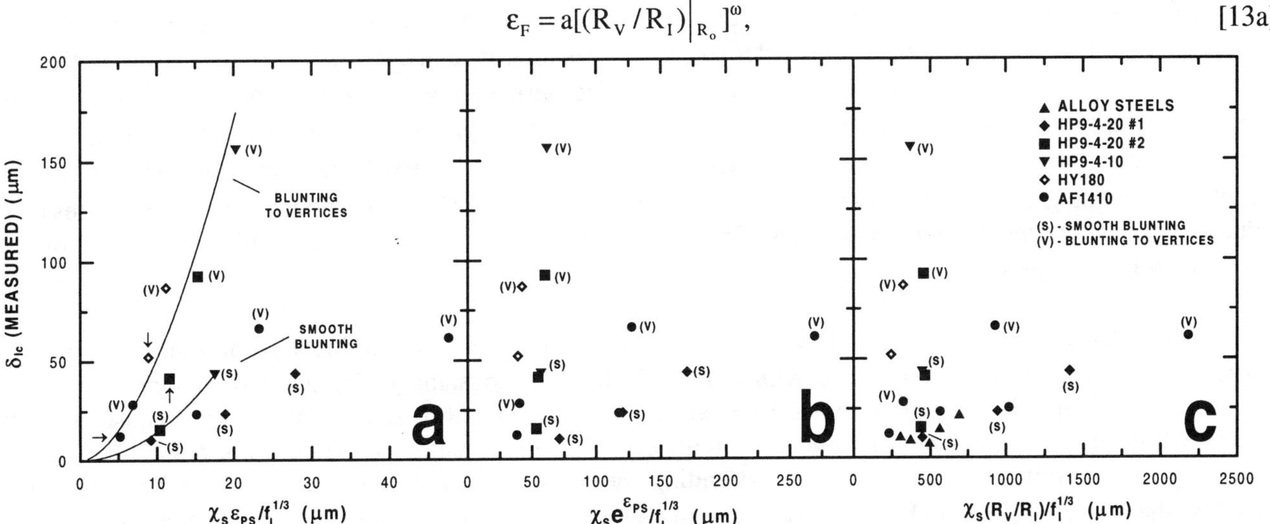

Figure 6. Measured values of δ_{Ic} for a variety of steels (3,8,11,13,22,24,25) are shown as a function of χ_S, f_I, and various definitions of matrix ductility parameter based on the general functional form of Equation 4. The matrix ductility parameter is defined in terms of (a) the plane strain tensile ductility (24,25), (b) the exponential of plane strain tensile ductility, and (c) the extent of primary microvoid growth, R_V/R_I.

and:

$$\ell_o = b\chi_S[(R_V / R_I)|_{R_o}]^{1-\omega} = b\chi_S\left[\frac{\varepsilon_F}{a}\right]^{\frac{1-\omega}{\omega}} \quad (b \geq 1), \qquad [13b]$$

where a and b are positive constants and ω is a partitioning constant ($0 < \omega \leq 1$). Unfortunately, this method of partitioning, while mathematically convenient, defines ε_F in a manner that is functionally inconsistent with local measures of strain on the fracture surface (29). The model also predicts characteristic distances that exceed the inclusion interparticle spacing (i.e., $\ell_o \geq \chi_S$) regardless of the relative resistance of the microstructure to the primary and secondary fracture processes.

The Rice-Johnson (6) model can be interpreted as predicting $\ell_o \leq \chi_S$ ($\chi_S = \chi_V$) since crack extension is defined as the separation of the ligament between the blunted crack tip and the first primary microvoid below the crack tip. However, since final fracture is arbitrarily defined in their mechanics analysis, it is not difficult to envision a material model as encompassing a substantial amount of primary microvoid growth at particles located more than one interparticle spacing below the blunted crack tip prior to the widespread onset of the secondary fracture process (i.e., $\ell_o \geq \chi_S$). A more reasonable description of characteristic microstructural distance can be obtained from an alternative definition of microstructural fracture strain, one that is functionally consistent with the definition of true strain:

$$\delta_{Ic} \propto \ell_o e^{\varepsilon_C}. \qquad [14]$$

Equations 7b and 14 can be combined to yield an expression for ℓ_o:

$$\ell_o \propto \left[f_I^{-\beta/3} e^{[C_2+(\kappa-1)\varepsilon_C]}\right]\chi_S, \qquad [15]$$

where C_2 and κ are the constants defined in Equation 7b for a given set of steels. This expression defines two regimes of fracture behavior with a characteristic microstructural distance that scales in direct proportion to χ_S. Negative values of β in combination with small values of ε_C define a regime in which primary microvoid growth is interrupted by the early onset of a secondary fracture process ($\ell_o \leq \chi_S$). Positive values of β in combination with large values of ε_C physically represent a ductile fracture process in which initiation toughness is primarily controlled by the growth of primary microvoids ($\ell_o \geq \chi_S$). In this case the matrix microstructure possesses enough ductility to allow a more extensive amount of primary microvoid growth prior to the widespread initiation of the secondary fracture process.

Calculated values of ℓ_o for a variety of steels (2,3,8,11,13,21-25) scale in direct proportion to χ_S for a given level of matrix ductility and increase with matrix ductility at constant χ_S for a constant mode of crack tip deformation, Figure 7. The latter comparisons are necessarily limited to steels that exhibit a constant mode of deformation since the constant of proportionality for Equation 14 is assumed to be unity for both smooth blunting behavior and blunting to vertices. Notwithstanding the utilization of a single constant for both deformation modes, the comparatively high values of ℓ_o for the as-quenched steels suggest that smooth blunting behavior favors the primary fracture process, although the development of toughness is limited by a low resistance to the secondary fracture process. Quenched and aged specimens of the ultrahigh strength steels, on the other hand, exhibit increases in δ_{Ic} and decreases in ℓ_o with an increase in aging temperature. The decrease in ℓ_o relative to χ_S exemplifies the increased dependence of δ_{Ic} on the secondary fracture process with an increase in aging temperature. Thus, the relative resistance of the microstructure to the primary and secondary fracture processes defines the magnitude of ℓ_o with respect to χ_S, and the development of initiation toughness is dependent on the factors that can significantly bias the degree of competition between the two fracture processes.

The power law material model, Equation 7b, provides additional support for this concept in that the volume fraction exponent η ($\eta = -\beta/3$) appears to define regimes of fracture behavior based on the mode of crack tip deformation. Steels that exhibit smooth blunting behavior typically possess relatively high strain hardening rates, and the development of toughness is associated with positive values of η ($\beta < 0$), Figure 8. High strain hardening rates promote both primary and secondary microvoid growth; however, the concurrent enhancement of secondary microvoid initiation, which represents a decrease in the resistance of the microstructure to the secondary fracture process, typically limits the development of toughness. In this connection, Equation 15 predicts that ℓ_o will increase relative to χ_S with increases in f_I when χ_S and ε_C are constant. Since increases in R_o, and hence R_V will be generally accompanied by decreases in the mean-free distance between primary microvoids, ℓ_o would be intuitively expected to increase relative to χ_S with decreases in the area fraction of intervoid material at the level of applied loading where secondary fracture dominates fracture surface formation. This type of microstructural change effectively decreases the relative importance of the secondary fracture process and would be expected to improve the initiation toughness of steels that exhibit smooth blunting behavior. Conversely, steels that blunt to vertices typically exhibit relatively low strain hardening rates, and the development of toughness is associated with negative values of η ($\beta > 0$), Figure 8. Low strain hardening rates concurrently limit the rate of change of microvoid growth with applied strain and the ability to initiate microvoids at secondary particles, such that toughening is promoted by primary microvoid growth in conjunction with a high resistance to the secondary fracture process. Equation 15 is also consistent with this type of behavior in that ℓ_o is predicted to decrease relative to χ_S with increases in f_I at constant χ_S and ε_C, but in this case a decrease in the area fraction of intervoid material (i.e., a decrease in the relative importance of the secondary fracture process) would act to degrade toughness.

In general, the toughness data for quenched and aged ultrahigh strength steels (8,21-25) supports the power law formalism, Equation 7, while the data for steels with lower levels of toughness (2,3,13,18-20) can be de-

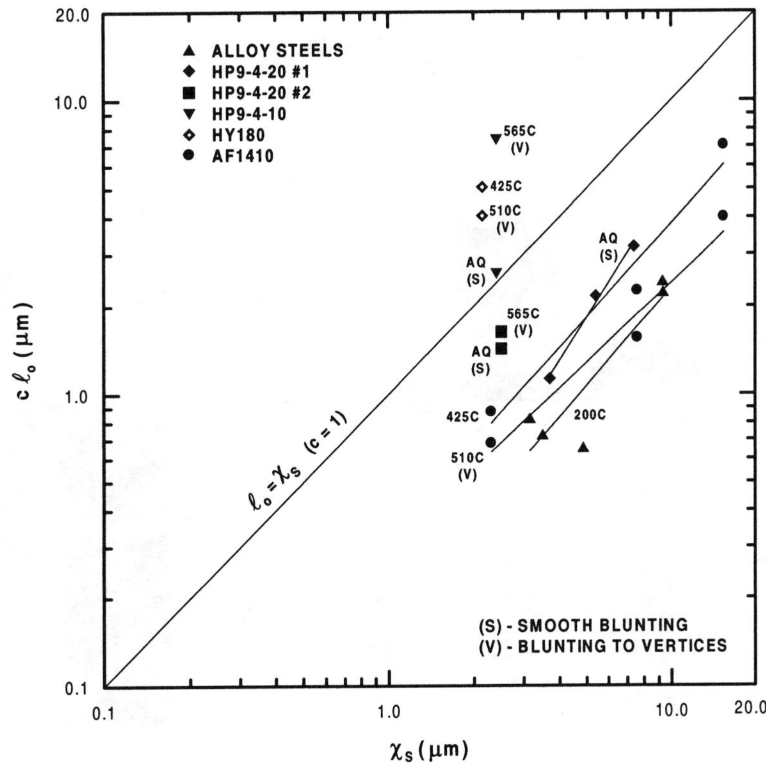

Figure 7. The characteristic microstructural distance, ℓ_o, is shown as a function of χ_S for a variety of steels (3,8,11,13,22,24,25) based on Equation 15. The parameter c, which is taken as unity in this figure, represents the geometric scaling constant in Equation 14 for different crack tip deformation modes.

scribed by either mathematical formulation. McMeeking's (28) calculations, on the other hand, are consistent with the linear relationship defined by Equation 1. These inconsistencies can be explained in terms of the toughness of the various steels relative to the limited ranges of δ_I/χ_S and R_V/R_I over which McMeeking evaluated the Rice-Johnson model, Table I. The ultrahigh strength steels evaluated by Garrison and coworkers, for example, exhibit values of δ_{Ic}/χ_S and R_V/R_I that far exceed the limiting values where δ_I/χ_S is directly proportional to R_V/R_I for either smooth blunting behavior or blunting to vertices. Conversely, the values of δ_{Ic}/χ_S for the lightly-tempered 0.35% C alloy steels (3,11), 0.45% C-Cr-Ni-Mo steels (18), and 0.2% C-12% Cr steels (20) are similar in magnitude to the limiting values of δ_I/χ_S. Thus, initiation toughness can be described by Equation 1 for steels with limited resistance to the secondary fracture process, although Equation 7 provides a better description of initiation toughness over a much broader range of competition between the primary and secondary fracture processes. The similar degree of correlation in materials with lower toughness results from the fact that a power law dependence of δ_I on χ_S can be approximated by a linear function over a limited range when the parameters are small in magnitude, Figures 3 and 4.

THE SECONDARY FRACTURE PROCESS

The formation of secondary microvoids contrasts with that of primary microvoids in that secondary microvoid nucleation typically requires a comparatively large amount of strain and final separation of the intervoid material is necessarily accompanied by the coalescence (i.e., complete impingement) of these features. If the ratio of secondary void size to particle size is taken as a measure of the local ductility of the intervoid material, the following approximation can be employed for the second-phase particles participating in the secondary fracture process:

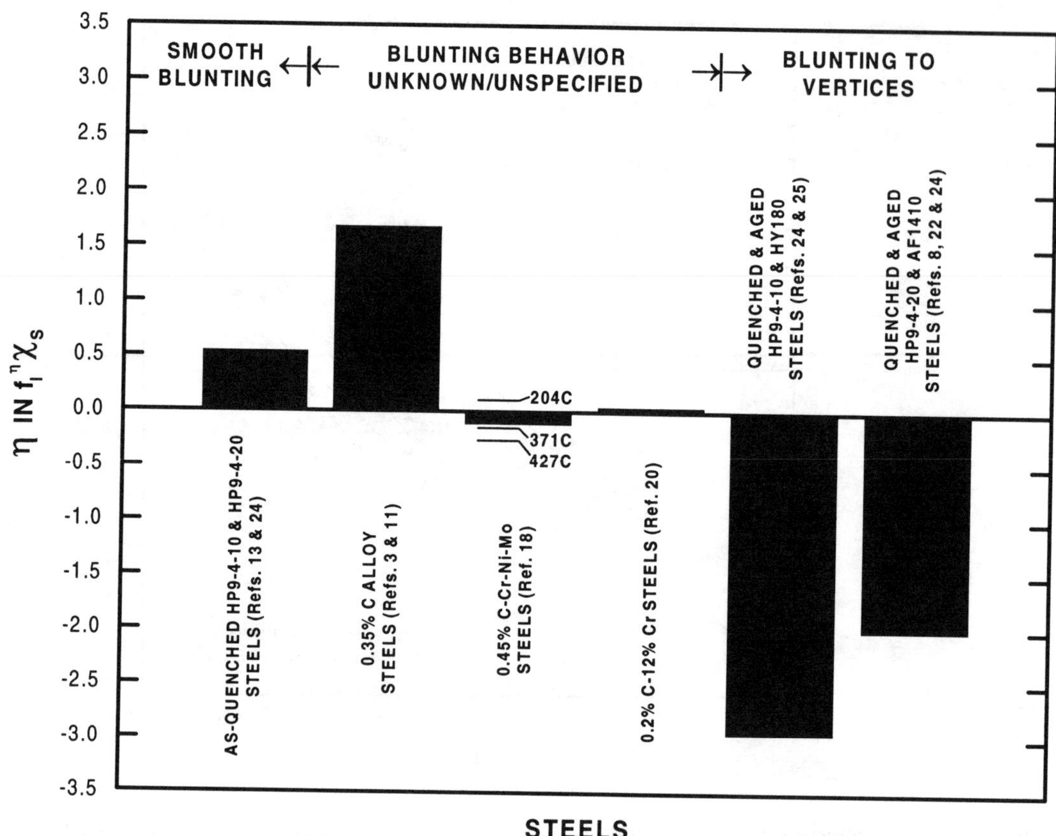

Figure 8. Differences in crack tip blunting behavior are defined in terms of the volume fraction exponent (η = -β/3) in Equation 7b for different sets of steels.

$$d_V / d_F \cong \frac{\chi_F}{d_F} = \frac{1}{N_F^{0.5} d_F}, \qquad [16]$$

where d_V/d_F = representative measure of the extent of secondary microvoid growth, d_V = secondary microvoid size, d_F = representative measure of the size of particles that nucleate secondary microvoids, N_F = areal density of small second-phase particles participating in the secondary fracture process, and χ_F = interparticle spacing of secondary particles on the fracture surface. The average microvoid growth strain in the fracture plane, ε_G, can be defined as:

$$\kappa\varepsilon_G = \ln\left[\frac{d_V}{d_F}\right]^{\kappa}, \qquad [17]$$

where κ = constant.

The applicability of this definition is evaluated with a series of low-sulfur steels melted to a nominal base composition of 0.33% C-0.8% Mn-0.5% Cr, Table II. The steels, which were processed to possess fine-grained austenite microstructures with different contents and dispersions of small second-phase particles, were evaluated in the fully hardened and lightly tempered condition (R_C 51). The processing and properties of the steels are summarized elsewhere (16). A value of ε_G for each steel/processing condition was evaluated from the dispersion of second-phase particles found on the fracture surface of a Charpy specimen tested at a temperature in the upper-shelf fracture regime (i.e., the small particles that directly participate in the ductile fracture process). Specifically,

Table I: Summary of Initiation Toughness and Deformation Mode Data for a Variety of Alloy Steels

Material	References	T_A (°C)/ T_T (°C)[1]	[%S]	Deformation Mode[2]	δ_I/χ_S	R_V/R_I	$\varepsilon_G = \ln(R_V/R_I)$
McMeeking's Analysis	28	N.A.	N.A.	S	**2.0**[3]	**4.0**[3]	----
				V	**0.8**	**2.3**	----
AF1410	21	816/510	0.001	V	8.7	9.2	2.22
HP9-4-20 (Heat #2)	24	840/AQ	0.002	S	6.2	10.6	2.36
		840/565		V	37	11.0	2.40
HP9-4-10	24	865/AQ	0.002	S	18	13.2	2.58
		865/565		V	65	10.9	2.39
HY180 (Heat #1)	25	816/510	0.005	V	40.7	11.3	2.42
0.35% C Alloy Steels	3,11	900/200	0.004-0.008	unknown	1.6-3.5	3.7-7.6	1.3-2.0
0.45% C-Cr-Ni-Mo	18	843/204	0.008-0.049	unknown	0.55-0.61	---	----
		843/316	0.025	unknown	0.73	---	----
		843/371	0.008-0.049	unknown	0.92-1.16	---	----
		843/427	0.008-0.049	unknown	1.18-1.72	---	----
		843/482	0.008-0.049	unknown	2.79-3.12	---	----
0.2% C-12% Cr	20	1020/670	0.004-0.035	unknown	0.60-1.04	---	----

[1] T_A = austenitizing temperature; T_T = tempering temperature.
[2] S = smooth blunting behavior; V = blunting to vertices.
[3] These values of δ_I/χ_S and R_V/R_I represent the limiting values associated with a relatively linear relationship between δ_I/χ_S and R_V/R_I for each type of blunting behavior.

an estimate of d_V/d_F was obtained from Equation 16 with the areal density and geometric mean size of particles for each field of view, and ε_G was then calculated as the geometric mean value of the distribution of d_V/d_F values for multiple fields of view.

Estimates of ε_G are summarized along with the secondary dispersion parameters in Table III for the 0.33% C-Mn-Cr steels. The values of ε_G range from 2.6 to 3.6 and generally scale in direct proportion to the longitudinal upper-shelf impact toughness of the lightly tempered steels. However, the values of microvoid growth strain based on the secondary particles, $\ln(d_V/d_F)$, are substantially greater than the corresponding measure of local strain based on primary microvoids, $\ln(R_V/R_I)$, in other high-strength steels. For example, the values of ε_G based on non-metallic inclusions range from 1.3 to 2.0 in the lightly-tempered 0.35% C alloy steels evaluated by Garrison (3,11). The ultrahigh strength steels evaluated by Garrison and coworkers (8,21,22,24,25) exhibit somewhat higher levels of $\ln(R_V/R_I)$, Table I, but the strain values are still substantially less than the values of $\ln(d_V/d_F)$ for the 0.33% C-Mn-Cr steels with comparatively low toughness. The large differences in the two measures of strain

Table II: 0.33% C-Mn-Cr Steel Compositions (weight percentages)

Steel	C	Mn	Si	Cr	Ni	Mo	S	P	Ti	Al	N	O
A1	0.32	0.79	0.24	0.48	0.12	0.06	0.002	0.009	-----	0.026	0.0086	0.0016
T1	0.33	0.78	0.24	0.49	0.12	0.03	0.001	0.001	0.041	0.026	0.0112	0.0022
N1	0.33	0.80	0.24	0.50	0.11	0.05	0.001	0.008	-----	0.004	0.0022	0.0024

Table III: Summary of Secondary Dispersion Parameters for the 0.33% C-Mn-Cr Steels

Steel	Precipitate Species	Processing Condition[2]	Bulk Material[1] d_B (nm)[3]	N_B/N_{REF}	Fracture Surface d_F (nm)	N_F/N_{REF}	d_V/d_F[4]	ε_G	E (J)[5]
A1	AlN	CP (800°C)	18 ± 11	≡ 1	29 ± 15	0.16	16.1 ± 1.5	2.78	17 (14-19)
		PR (800°C)	24 ± 9	0.92	31 ± 16	0.12	18.1 ± 1.4	2.90	17 (14-20)
		CP (900°C)	30 ± 14	0.58	84 ± 50	0.04	13.6 ± 1.6	2.61	17 (16-19)
		PR (900°C)	27 ± 11	0.60	33 ± 23	0.06	24.3 ± 1.3	3.19	26 (24-28)
T1	TiN	CP (900°C)	13 ± 7	0.41	22 ± 13	0.17	19.3 ± 1.6	2.96	23 (22-24)
N1	Fe-Based Carbides	CP (800°C)	----	----	22 ± 14	0.11	24.1 ± 1.2	3.18	26 (22-47)
		CP (900°C)	----	----	49 ± 28	0.01	38.2 ± 1.3	3.64	47

[1]The areal precipitate density values for both the bulk material and fracture surface measurements have been referenced to the bulk material estimate for the conventionally processed specimen of steel A1 after austenitization at 800°C.
[2]CP = conventional processing prior to hardening at the indicated temperature; PR = precipitate refining treatment prior to hardening at the indicated temperature.
[3]d = mean particle size and standard deviation; N = areal particle density.
[4]Geometric mean value and standard deviation.
[5]Mean longitudinal upper-shelf impact toughness and range.

can be rationalized by applying the mean value theorem to the integral expression for the mean strain associated with the uninhibited growth of an isolated void (29):

$$\varepsilon_G^{PP} = \ln\left[\frac{R_V}{R_I}\right] = 0.322 \int_{\varepsilon=\varepsilon_o^{PP}}^{\varepsilon=\varepsilon_P} e^{\left[\sqrt{3}\sigma/\tau_o\right]} d\varepsilon = C^{PP}[\varepsilon_P - \varepsilon_o^{PP}], \qquad [18a]$$

and:

$$\varepsilon_G^{SP} = \ln\left[\frac{d_V}{d_F}\right] = 0.322 \int_{\varepsilon=\varepsilon_o^{SP}}^{\varepsilon=\varepsilon_P} e^{\left[\sqrt{3}\sigma/\tau_o\right]} d\varepsilon = C^{SP}[\varepsilon_P - \varepsilon_o^{SP}], \qquad [18b]$$

where ε_P = plastic strain, ε_o^i = microvoid initiation strain, C^i = mean value of the integrand over the integrated range of plastic strain, and the superscripts PP and SP refer to the primary and secondary void-initiating particles, respectively. Solving the two expressions for ε_P and equating yields:

$$\ln\left[\frac{d_V}{d_F}\right] = \frac{C^{SP}}{C^{PP}} \ln\left[\frac{R_V}{R_I}\right] + C^{SP}[\varepsilon_o^{PP} - \varepsilon_o^{SP}]. \qquad [19]$$

Since the tensile deformation behavior of high-strength steels can be characterized by a continuous function in which stress increases with applied strain at a decreasing rate, $C^{SP} \geq C^{PP}$ and the critical initiation strain that defines the condition $d_V/d_F \geq R_V/R_I$ is given by:

$$\varepsilon_o^{SP} \leq \varepsilon_o^{PP} + \left[\frac{1}{C^{PP}} - \frac{1}{C^{SP}}\right] \ln\left[\frac{R_V}{R_I}\right]. \qquad [20]$$

Based on a comparison of two isolated microvoids, this analysis suggests that d_V/d_F may exceed R_V/R_I at any value of applied strain provided the difference between ε_o^{SP} and ε_o^{PP} is less than a critical fraction of $\ln(R_V/R_I)$. This explanation is an obvious oversimplification in that interaction, impingement, and/or the widespread initiation of a secondary fracture process limit the growth of primary microvoids and impingement necessarily limits the growth of secondary microvoids. The rationalization nevertheless provides a reasonable physical interpretation in which the nucleation, growth, and coalescence of primary and secondary microvoids are viewed as overlapping sequential processes that occur over a range of strain as a consequence of a large difference in the overall magnitude of microvoid initiation strain.

The secondary fracture process can be characterized by a representative statistical measure of the extent of void growth when the initiation strain is negligible since $\varepsilon_C = \varepsilon_G$. This is the tacit assumption that is made in the application of the linear material model to steels containing "weakly bound" MnS inclusions. The growth strain can also be substituted for ε_C when the microvoid initiation strain is constant for a given set of similar steels. This assumption is basically valid for materials containing a monodispersed system of particles or when initiation strain is independent of particle size. Since neither condition is typically satisfied for the primary or secondary void-initiating particles in high-strength steels, an appropriate definition of microvoid initiation strain is required in the application of both the linear and power law material models. Microvoid initiation strain could be defined in terms of a statistical measure representative of a dispersion of particles or a minimum value characteristic of initiation at the largest particle in a dispersion. However, these definitions fail to provide relevant independent measures of microvoid initiation that can be applied to a set of similar materials with different dispersions of second-phase particles. Initiation strain can be alternatively defined as a maximum value characteristic of the smallest particle that can nucleate a microvoid in the most highly stressed volume element of the fracture process zone. This latter measure of initiation strain, which is fundamentally related to the yielding and strain hardening behavior of the material, provides a single parameter characterization of microvoid initiation that can be applied to a set of "similar" materials in both the linear and power law material models. In other words, initiation toughness can be estimated without the need to explicitly specify microvoid initiation strain since this parameter can be treated as a constant in Equations 1 and 7 for a set of similar materials. The success of this approach is exemplified by the data in Figures 3-5. It should also be noted that this definition of initiation strain defines the fraction of the entire second-phase dispersion that can be potentially sampled in a given material, but it does not characterize any effects of sampling for particles larger than the minimum size.

SAMPLING OF SECOND-PHASE PARTICLES DURING DUCTILE FRACTURE

Primary Void-Initiating Particles

Most investigations (2,3,6,8,11,13,21,22,24,25) base the application of the linear material model on the implicit assumption that the entire dispersion of non-metallic inclusions participates in the ductile fracture process. The complete sampling of an inclusion population might be expected in steels with lower inclusion contents and/or when the barrier to microvoid nucleation is low relative to the level of applied loading associated with the widespread onset of the secondary fracture process. However, the preferential sampling of coarse inclusions may also be feasible under conditions in which the nucleation and/or growth of primary microvoids are appropriately constrained. Since this assumption has not been validated, MnS inclusion data for a variety of high-strength steels (3,8,11,13,22,24,25) have been analyzed in an attempt to determine if sampling occurs during the primary fracture process.

Sampling during the primary fracture process is evaluated from a comparison of χ_S as determined from stereological measurements of polished metallographic specimens and $2\overline{R}_V$, where $\overline{R}_V$ = average microvoid radius. The condition $\chi_S = 2\overline{R}_V$ represents the complete sampling of a monodispersed dispersion of particles that nucleates a monodispersed system of area-filling microvoids. For the more general case of a polydispersed system of particles, sampling can be defined in an average sense by the inequality $\chi_S < 2\overline{R}_V$ when $A_{PV} < 1$, where A_{PV} = area fraction of primary microvoids. Additional data (3,8,13,24,25) can be tested if $0.89R_o/f_I^{1/3}$ is substituted for χ_S and $\overline{R}_V/R_o$ is taken as a measure of the maximum value of R_V/R_I, $(R_V/R_I)_C$, associated with the complete sampling of an inclusion population (i.e., $\chi_S/2R_o \approx \overline{R}_V/R_o$).[1] The criterion for sampling then becomes $\overline{R_V/R_I} > (R_V/R_I)_C$, where $(R_V/R_I)_C = 0.45/f_I^{1/3}$. Based on these criteria, a majority of the steels (3,8,13,22,24, 25) listed in Table II of Reference 26 are predicted to exhibit a tendency towards sampling during the primary fracture process, Figure 9.

Sampling appears to occur in steels that contain relatively high densities of comparatively small, less widely spaced MnS inclusions. Steels in this regime that blunt to vertices are predicted to exhibit the largest degree of sampling as given in rough measure by the difference between $\overline{R_V/R_I}$ and $(R_V/R_I)_C$, which suggests that constrained deformation ahead of a crack tip tends to promote sampling of an inclusion population. High degrees of sampling are also predicted to occur in steels with high inherent toughness (i.e., low-carbon steels with high nickel contents), although a portion of the differences between $\overline{R_V/R_I}$ and $(R_V/R_I)_C$ may result from the localized coalescence of neighboring microvoids into macrovoids in as-quenched specimens that exhibit smooth blunting behavior (26). Notwithstanding any spatial effects (35) associated with deformation geometry and/or void interactions, the sampling of MnS inclusions is somewhat inconsistent with the traditional notion of a critical damage mechanism of ductile fracture in which the primary void-initiating particles are "weakly bound" to the matrix (36). However, HY180 steels containing MnS have been shown (25,37) to exhibit microvoid initiation

---

[1]This substitution is based on the assumption that $\overline{R}_V/R_o$ provides a representative measure of R_V/R_I. The difference between $\overline{R_V/R_I}$ and $\overline{R}_V/R_o$ is less than 12% for six of nine material conditions listed in Table II of Reference 26 (3,13,22,24), although the difference ranges from 17% to 34% for the other three steels (13,24) in this table.

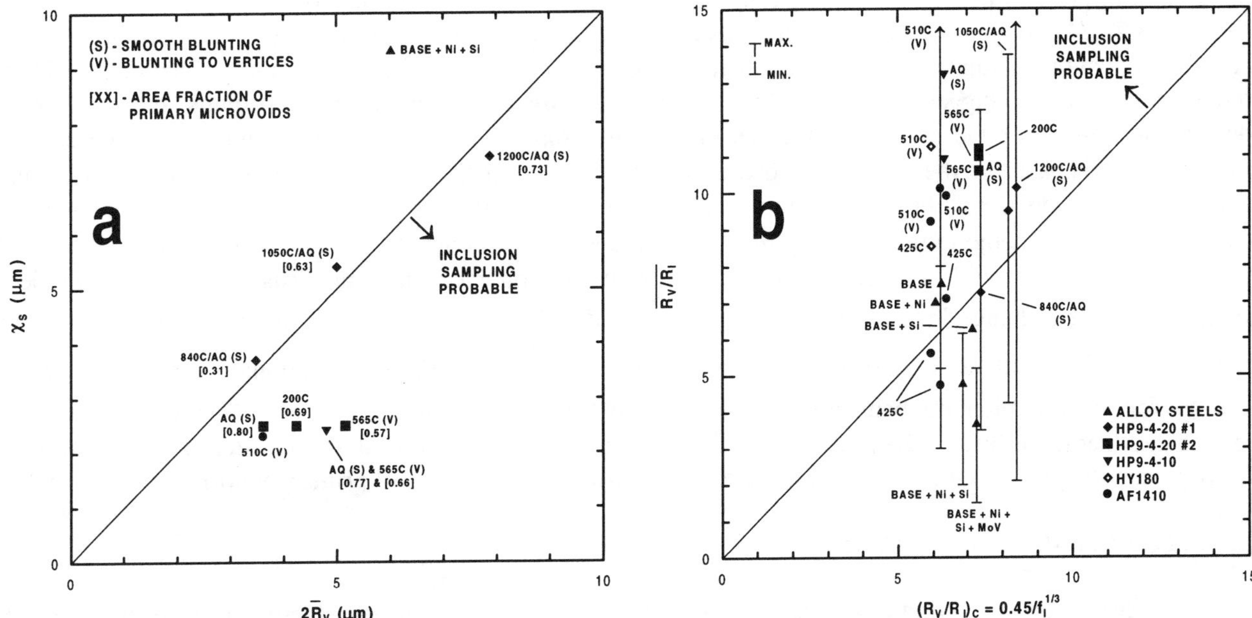

Figure 9. The propensity for sampling of MnS inclusions during ductile fracture is assessed for a variety of steels (3,8,11,13,22,24,25): (a) comparison of χ_S and average microvoid diameter and (b) comparison of the average extent of primary microvoid growth and $(R_V/R_I)_C$, where $(R_V/R_I)_C$ is defined in terms of the volume fraction of MnS inclusions, f_I. The symbols represent average values of the parameters reported in Table II of Reference 26, whereas the bars in (b) represent the range of R_V/R_I obtained from the original sources of data for several of the steels (3,8,11,13).

strains on the order of 0.25-0.55 as measured in the necked region of tensile specimens. Cox and Low (1) also determined that primary microvoid initiation is stress controlled in 4340 steel containing MnS inclusions, which in the general case should yield microvoid initiation strains greater than zero. This is consistent with their observations that microvoid initiation strain increases with decreasing sulfur content (i.e., generally decreasing MnS size in the 4340 steels) and microvoid initiation progresses from the largest inclusions in a dispersion to smaller inclusions with increases in applied strain. Thus, sampling during primary fracture appears viable in steels containing high densities of small, weakly bound primary particles such as MnS, although the effect of sampling on toughness is presumably small in comparison to that which occurs in steels containing more strongly bound primary particles such as $Ti_4C_2S_2$ (25,37).

The present analysis is indeterminate in the regime defined by the inequalities $\chi_S \geq 2\bar{R}_V$, Figure 9a, and $\overline{R_V/R_I} \leq (R_V/R_I)_C$, Figure 9b, since primary microvoid growth could be effectively terminated by a secondary fracture process either prior to or after the complete sampling of the inclusion population in a steel. However, the steels in this regime, which comprise two 0.35% C alloy steels (3,11) and an AF1410 steel tempered at 425°C (8), contain large, widely spaced inclusions relative to the other steels listed in Table II of Reference 26. Based on this distinguishing characteristic, it can be reasonably speculated that these steels may represent the case in which the sampling of inclusions is minimized. In support of this viewpoint, a presumably high strain hardening rate in the lightly-tempered (200°C) 0.35% C alloy steels would act to reduce sampling of the inclusion population by promoting microvoid initiation early in the primary fracture process. A similar argument can be made for AF1410 specimens tempered at 425°C and 510°C, where a difference in strain hardening rate is associated with a difference in the predicted response to sampling during primary fracture, Figure 9b.

The Rice-Johnson mechanics analysis is based on the premise that primary void-initiating particles are weakly bound to the matrix, and derived material models rely on the implicit assumption that the entire population of primary particles is sampled during ductile fracture. Most applications of the linear material model (3,8,11,13,22,24-26) employ this assumption as evidenced by the fact that the inclusion interparticle spacing is a calculated quantity obtained from measurements of R_o and f_I on polished metallographic specimens. However, the present analysis suggests that sampling is promoted by small interparticle spacings and/or constrained deformation ahead of the crack tip, and the effects of sampling could introduce potentially large errors into material models based on the value of χ_S that represents the entire inclusion population in a material. This may be a limiting factor in the ability of the linear material model to provide an adequate description of the ductile fracture process in some ultrahigh strength steels, Figure 5a. In contrast, $\ln(\delta_I/\chi_V)$ at fracture is dependent on $\beta\ln(\chi_V/R_i)$ in the power law interpretation of the Rice-Johnson mechanics analysis, and the magnitude of a sampling-induced change in δ_{Ic}/χ_S is defined in terms of $f_I^{-\beta/3}$ (i.e.,$(\chi_S/R_o)^\beta$) in the resulting material model, Equation 7. This method of analysis is powerful in that $f_I^{-\beta/3}$ also defines the functional relationship between ℓ_o and χ_S for smooth blunting behavior and blunting to vertices, Figures 7 and 8. It should be nonetheless recognized that the utility of the power law model is limited since β is treated as a constant that represents some average sampling effect for a set of similar steels with different dispersions of primary void-initiating particles.

Secondary Void-Initiating Particles

The dispersion of second-phase particles that directly participates in the ductile fracture of the intervoid material has been shown (16,17) to comprise a subset distribution of the entire dispersion of "small" particles in the microstructure of lightly-tempered high-strength steels. This observation is in general accord with the micromechanistic concept that the largest particles in a dispersion are preferentially sampled when there is a substantial barrier to microvoid initiation. However, large densities of smaller particles also provide sites for microvoid initiation in steels with broad precipitate size distributions, consistent with the possibility of nucleating microvoids at particles larger than a critical size that is inversely related to the magnitude of local stresses ahead of the crack tip. Secondary fracture can be viewed in this context as the progressive sampling of smaller particles with increases in applied stress, such that the development of toughness effectively depends on the degree of heterogeneity in size for a given content of particles.

Considering the wide range of particle size distributions that may exist in a steel, it would be reasonable to anticipate that the secondary fracture process varies between the extremes of nucleation-limited and growth-limited behavior as a result of sampling. Nucleation-limited behavior would be expected to result from a comparatively low content of particles with a broad size distribution, and secondary fracture resistance would be primarily limited by the initiation of a low density of microvoids at the largest particles in the dispersion. An AlN size distribution characteristic of nucleation-limited sampling is shown in Figure 10a for a conventionally-processed Charpy V-notch specimen of an air-melt 4340 steel (R_C 54) tested at an upper-shelf temperature (17). Although a broad size range of AlN precipitates is sampled during the secondary fracture process, the preferential sampling of coarse precipitates is exemplified by the size distribution of precipitates located on the fracture surface.[1] Growth-limited behavior, on the other hand, would be expected to result from microvoid nucleation at a comparatively high density of sites, and this mechanism would tend to favor refined dispersions, or more generally dispersions with a small range of particle size. An example more characteristic of growth-limited behavior is shown in Figure 10c for a Charpy V-notch specimen of a 0.33% C-Mn-Cr steel containing TiN (16), Tables II and III. A large fraction (41%) of the entire dispersion of fine TiN precipitates participated in the secondary fracture process, although the preferential sampling of coarser precipitates in the dispersion is manifested as a decreased density of 10-20 nm precipitates and a virtual absence of precipitates smaller than 10 nm on the fracture surface.

[1]The apparent inconsistency in the bulk and fracture surface distributions of precipitate size in Figure 10a result from difficulties associated with the detection of low densities of coarse precipitates on replicas obtained from polished metallographic specimens. Characterizing the entire dispersion of grain-refining precipitates from polished specimens requires an unreasonable amount of sampling. For example, based on the assumption that precipitates are uniformly dispersed throughout the microstructure, the complete sampling of precipitates larger than 100 nm would require the examination of ~300 additional fields of view from polished specimens.

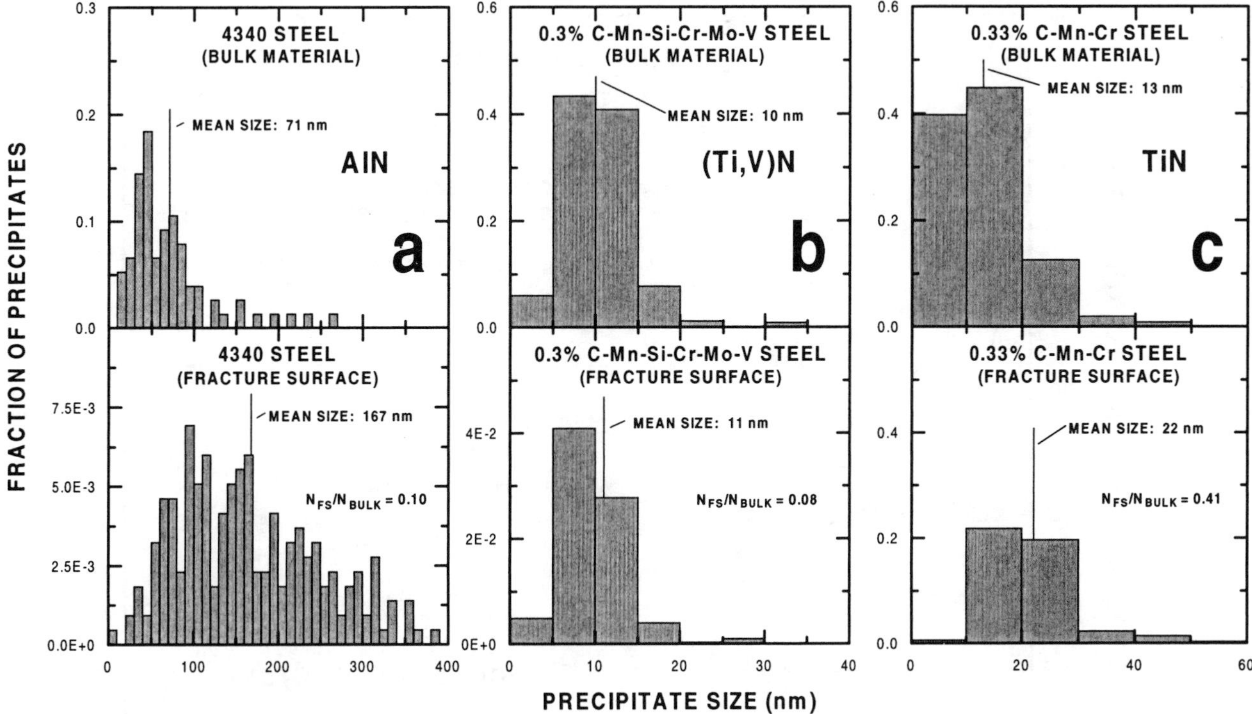

Figure 10. Precipitate size distributions are shown representing the transition from nucleation-limited sampling towards growth-limited sampling in several steels: (a) lightly-tempered 4340 steel containing AlN precipitates (17), (b) 0.3% C-Mn-Si-Cr-Mo-V steel containing (Ti,V)N precipitates (38), and (c) 0.33% C-Mn-Cr steel containing TiN precipitates, Table III (16). The fracture surface distributions were obtained from Charpy V-notch specimens tested at upper-shelf temperatures for the 4340 and 0.33% C-Mn-Cr steels and a compact tension specimen for the 0.3% C-Mn-Si-Cr-Mo-V steel.

An intermediate case that reflects a low density of uniformly sampled (Ti,V)N precipitates is shown in Figure 10b for a compact tension specimen of a 0.3% C-Mn-Si-Cr-Mo-V steel (38,39). As exemplified by these data, the transition from nucleation-limited to growth-limited behavior is characterized as the sampling of a progressively increasing fraction of the secondary particle dispersion; that is, the content and size distribution of particles on the fracture surface converge towards the dispersion parameters for the bulk material.

Differences in the state of precipitation and the resulting response of the microstructure to the secondary fracture process are not necessarily associated with the development of a unique level of toughness in a material, Table III. For example, steel A1 exhibits an equivalent level of upper-shelf impact toughness and drastically different fracture surface dispersion parameters in three of four material conditions, whereas the other steels exhibit concomitant changes in the dispersion parameters and upper-shelf toughness. These data suggest that the baseline level of toughness is dependent on the content of precipitates that participates in the secondary fracture process while changes in sampling behavior affect the development of toughness about the baseline. Based on the stereological correspondence between particle size, density, and volume fraction (27), general forms of Equations 16 and 17 can be combined to yield the expression:

$$\varepsilon_G = \frac{1}{\rho}\ln(1/f_F), \qquad [21]$$

where f_F = volume fraction of particles that participates in the secondary fracture process and ρ = dimensionality of the fracture process ($2 \le \rho \le 3$). The longitudinal upper-shelf impact toughness, E, is shown as a function of

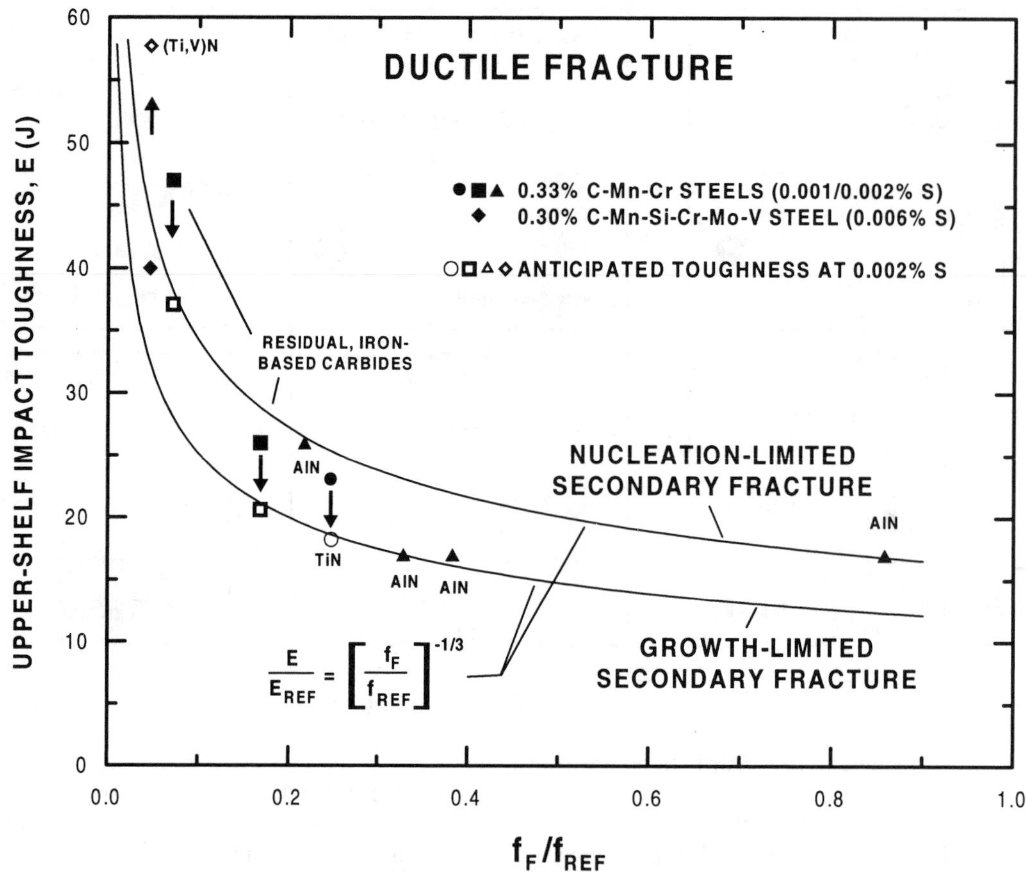

Figure 11. The longitudinal upper-shelf impact toughness, E, is shown as a function of the normalized volume fraction of secondary particles participating in the ductile fracture process, f_F/f_{REF}, for several 0.33% C-Mn-Cr steels (16) and a 0.3% C-Mn-Si-Cr-Mo-V steel (38) containing different species of particles. The reference value of volume fraction, f_{REF}, is taken as the bulk precipitate content in conventionally-processed steel A1 specimens austenitized at 800°C, Table III. In addition, the impact toughness data for several steels have been normalized to the 0.002% S level.

f_F/f_{REF} in Figure 11 for the 0.33% C-Mn-Cr steels (16) and the 0.30% C-Mn-Si-Cr-Mo-V steel (38), where f_F/f_{REF} is defined as (27):

$$\frac{f_F}{f_{REF}} = \left[\frac{N_F}{N_{REF}}\right]\left[\frac{d_F^2 + \sigma_F^2}{d_{REF}^2 + \sigma_{REF}^2}\right], \quad [22]$$

where σ_i = standard deviation of particle size. The estimates of f_F are referenced to the bulk material condition for the conventionally-processed steel A1 specimen austenitized at 800°C, Table III. The impact toughness data for steels T1 and N1 (0.001% S) and the 0.30% C-Mn-Si-Cr-Mo-V steel (0.006% S) have also been normalized to the 0.002% S level based on the assumption that E is proportional to $f_I^{-1/3}$, where f_I is evaluated from the solubility product for MnS in austenite (31).[1]

[1]This assumption can be reasonably applied since the 0.33% C-Mn-Cr steels (16) possess similar base compositions and were subjected to the same melting, forging, and hot-rolling procedures.

These data indeed demonstrate that upper-shelf impact toughness is primarily dependent on the content of second-phase particles that participates in the secondary fracture process, although the state of precipitation exerts a substantial second-order effect on toughness at any given level of f_F. The upper-shelf impact toughness is directly proportional to $f_F^{-1/3}$ in accord with the original premise that the fracture surface is ultimately formed by the complete coalescence of microvoids during the secondary fracture process (i.e., $d_V/d_F \propto \chi_F/d_F$).[1] Based on Equations 1, 7, and 17, δ_{Ic} should exhibit the following functional dependence on f_F:

$$\delta_{Ic} \propto \exp(\kappa\varepsilon_G)\Big|_{R_o,\chi_S} \propto \frac{1}{f_F^{1/\rho}}. \qquad [23]$$

However, generalization of this functional dependence to upper-shelf impact toughness requires initiation toughness to scale in direct proportion to propagation resistance for a ductile mode of fracture. Garwood and Turner (40) have shown that ductile propagation resistance defined in terms of the crack opening displacement near the crack tip, δ_A, is a constant fraction of the initiation toughness, δ_{Ic}, in a mild steel. Other investigators (41,42) have proposed that a direct proportionality between δ_A and δ_{Ic} is theoretically tractable since a constant level of crack-tip opening angle is required to maintain stable crack growth. The link between initiation and propagation resistance has also been established for different measures of toughness. In particular, the Rolfe-Novak-Barsom correlation (43,44) predicts a direct proportionality between K_{Ic}^2 and upper-shelf impact toughness, consistent with an energy interpretation of the J integral fracture criterion (45-47). Since δ_{Ic} is proportional to K_{Ic}^2, upper-shelf impact toughness should scale in direct proportion to δ_{Ic}, thereby providing a fundamental basis for the functional dependence of E on f_F, Figure 11.

SUMMARY

Linear and power law interpretations of the Rice-Johnson mechanics analysis provide good functional descriptions of the ductile fracture process in high-strength steels with relatively low toughness. However, several investigators (21,26) have noted that the linear material model fails to predict a decrease in the dependence of δ_{Ic} on χ_S above a critical interparticle spacing for a given level of matrix ductility (48). The power law interpretation of the model is attractive in that initiation toughness is predicted to vary continuously towards an upper bound with increases in χ_S if a material exhibits sufficiently high matrix ductility. The power law model also yields an operational definition for a characteristic microstructural distance, ℓ_o, that scales in direct proportion to χ_S for both smooth blunting behavior and blunting to vertices. The constant of proportionality between ℓ_o and χ_S defines the extent of competition between the primary and secondary fracture processes in terms of a function of the volume fraction of void-initiating particles, f_I, and microstructural fracture strain, ε_C. Based on a limited amount of data, it also appears that the mode of crack tip deformation can be characterized by the functional dependence of ℓ_o on f_I at constant χ_S and ε_C.

It is speculated that the extent of sampling during primary fracture may be minimized in steels containing low densities of widely spaced MnS inclusions, although a definitive conclusion cannot be reached from presently available data. However, experimental data from several investigations suggest that the sampling of primary void-initiating particles is the rule rather than the exception in steels containing higher densities of smaller, less widely spaced MnS inclusions. Sampling is prevalent during the secondary fracture process since the particles

[1]Secondary fracture obviously occurs as a volumetric process in the 0.33% C-Mn-Cr steels, although the application of Equation 16 to these data is based on areal parameters obtained from the fracture surfaces of Charpy specimens, Table III. The definition of d_V/d_F provides a measure of the extent of void growth in the fracture plane and avoids uncertainties associated with the estimation of volumetric stereological parameters from the areal parameters for non-spherical particles. The two approaches are compatible if the extent of secondary microvoid growth in the fracture plane is proportional to the average microvoid size.

that initiate microvoids are more strongly bound to the matrix, and sampling varies between nucleation-limited and growth-limited behavior depending on the size scale and heterogeneity in the size of particles that comprise the secondary dispersion. Despite variations in toughness with the state of precipitation, the content of small second-phase particles is the primary variable affecting the resistance to secondary fracture in tempered martensitic microstructures.

ACKNOWLEDGMENTS

The authors would like to thank The Timken Company for permission to publish the results of this research. We would also like to express our appreciation to J. C. Murza for reviewing the manuscript.

REFERENCES

1. T. B. Cox and J. R. Low, Jr., "An Investigation of the Plastic Fracture of AISI 4340 and 18 Nickel--200 Grade Maraging Steels," **Metallurgical Transactions**, 1974, vol. 5, pp. 1457-1470.

2. W. M. Garrison, Jr., "The Effects of Silicon and Nickel Additions on the Mechanical Properties of a 0.4 Carbon Low Alloy Steel," **Scripta Metallurgica**, 1982, vol. 16, pp. 877-880.

3. W. M. Garrison, Jr., "The Effect of Silicon and Nickel Additions on the Sulfide Spacing and Fracture Toughness of a 0.4 Carbon Low Alloy Steel," **Metallurgical Transactions A**, 1986, vol. 17A, pp. 669-678.

4. S. Lee, L. Majno, and R. J. Asaro, "Correlation of Microstructure and Fracture Toughness in Two 4340 Steels," **Metallurgical Transactions A**, 1985, vol. 16A, pp. 1633-1648.

5. R. O. Ritchie and R. M. Horn, "Further Considerations on the Inconsistency in Toughness Evaluation of AISI 4340 Steel Austenitized at Increasing Temperatures," **Metallurgical Transactions A**, 1978, vol. 9A, pp. 331-341.

6. J. R. Rice and M. A. Johnson, "The Role of Large Crack Tip Geometry Changes in Plane Strain Fracture," **Inelastic Behavior of Solids** (New York, NY: McGraw-Hill Book Company, 1970), pp. 641-670.

7. W. M. Garrison, Jr., "A Microstructural Interpretation of the Fracture Strain and Characteristic Fracture Distance," **Scripta Metallurgica**, 1984, vol. 18, pp. 583-586.

8. W. M. Garrison, Jr. and N. R. Moody, "The Influence of Inclusion Spacing and Microstructure on the Fracture Toughness of the Secondary Hardening Steel AF1410," **Metallurgical Transactions A,** 1987, vol. 18A, pp. 1257-1263.

9. C. L. Briant, S. K. Banerji, and A. M. Ritter, "The Role of Nitrogen in the Embrittlement of Steel," **Metallurgical Transactions A**, 1982, vol. 13A, pp. 1939-1950.

10. E. Chang, C. Y. Chang, and C. D. Liu, "The Effects of Double Austenitization on the Mechanical Properties of a 0.34C Containing Low-Alloy Ni-Cr-Mo-V Steel," **Metallurgical Transactions A,** 1994, vol. 25A, pp. 545-555.

11. W. M. Garrison, Jr., "A Micromechanistic Interpretation of the Influence of Undissolved Carbides on the Fracture Toughness of a Low Alloy Steel," **Scripta Metallurgica**, 1986, vol. 20, pp. 633-636.
12. M. J. Gore, G. B. Olson, and M. Cohen, "Grain-Refining Dispersions and Properties in Ultrahigh-Strength Steels," **Innovations in Ultrahigh-Strength Steel Technology, Proceedings of the 34th Sagamore**

Conference (Watertown, MA: U. S. Army Laboratory Command, Materials Technology Laboratory, 1987), pp. 425-441.

13. K. J. Handerhan and W. M. Garrison, Jr., "The Effect of Austenitizing Temperature on the Fracture Initiation Toughness of As-Quenched HP9-4-20 Steel," **Metallurgical Transactions A**, 1988, vol. 19A, pp. 2989-3003.

14. M. J. Leap and J. C. Wingert, "Recent Advances in the Technology of Toughening Grain-Refined, High-Strength Steels" (Warrendale, PA: SAE International, Technical Paper 961749, 1996).

15. M. J. Leap and J. C. Wingert, "Application of the AdvanTec™ Process for Improving the Toughness of Grain-Refined, High-Strength Steels," ***38th* Mechanical Working & Steel Processing Conference Proceedings** (Warrendale, PA: Iron and Steel Society, Inc., 1997), pp. 195-220.

16. M. J. Leap and J. C. Wingert, "Interrelationships Between Toughness and Grain-Refining Precipitates in Lightly-Tempered, High-Strength Steels," **39th Mechanical Working & Steel Processing Conference Proceedings** (Warrendale, PA: Iron and Steel Society, Inc., 1998), pp. 685-703.

17. M. J. Leap and J. C. Wingert, "The Effects of Grain-Refining Precipitates on the Development of Toughness in 4340 Steel," **Metallurgical Transactions A**, vol. 30A, 1999, pp. 93-114.

18. A. J. Birkle, R. P. Wei, and G. E. Pellissier, "Analysis of Plane-Strain Fracture in a Series of 0.45C-Ni-Cr-Mo Steels with Different Sulfur Contents," **Transactions of the American Society for Metals**, 1966, vol. 59, pp. 981-990.

19. W. A. Spitzig, "Correlations Between Fractographic Features and Plane-Strain Fracture Toughness in an Ultrahigh-Strength Steel," **Electron Microfractography, ASTM STP 453** (Philadelphia, PA: American Society for Testing and Materials, 1969), pp. 90-110.

20. V. P. Raghupathy, V. Srinivasan, H. Krishnan, and M. N. Chandrasekharaiah, "The Effect of Sulphide Inclusions on Fracture Toughness and Fatigue Crack Growth in 12 wt% Cr Steels," **Journal of Materials Science**, 1982, vol. 17, pp. 2112-2126.

21. W. M. Garrison, Jr. and K. J. Handerhan, "Fracture Toughness: Particle-Dispersion Correlations," **Innovations in Ultrahigh-Strength Steel Technology, Proceedings of the 34th Sagamore Conference** (Watertown, MA: U. S. Army Laboratory Command, Materials Technology Laboratory, 1987), pp. 443-466.

22. K. J. Handerhan, W. M. Garrison, Jr., and N. R. Moody, "A Comparison of the Fracture Behavior of Two Heats of the Secondary Hardening Steel AF1410," **Metallurgical Transactions A**, 1989, vol. 20A, pp. 105-123.

23. K. J. Handerhan and W. M. Garrison, Jr., "Observations on the Upper Shelf Crack Tip Blunting Behavior of Ultra High Strength Steels," **Scripta Metallurgica**, 1988, vol. 22, pp. 607-610.

24. K. J. Handerhan and W. M. Garrison, Jr., "A Study of Crack Tip Blunting and the Influence of Blunting Behavior on the Fracture Toughness of Ultra High Strength Steels," **Acta Metallurgica et Materialia**, 1992, vol. 40, pp. 1337-1355.

25. J. W. Bray, J. L. Maloney, K. S. Raghavan, and W. M. Garrison, Jr., "A Comparison of the Fracture Behavior of Two Commercially Produced Heats of HY180 Steel Differing in Sulfide Type," **Metallurgical Transactions A**, 1991, vol. 22A, pp. 2277-2285.

26. J. W. Bray, K. J. Handerhan, W. M. Garrison, Jr., and A. W. Thompson, "Fracture Toughness and the Extents of Primary Void Growth," **Metallurgical Transactions A**, 1992, vol. 23A, pp. 485-496.

27. M. F. Ashby and R. Ebeling, "On the Determination of the Number, Size, Spacing, and Volume Fraction of Spherical Second-Phase Particles from Extraction Replicas," **Transactions of the Metallurgical Society of AIME**, 1966, vol. 236, pp. 1396-1404.

28. R. M. McMeeking, "Blunting of a Plane Strain Crack Tip into a Shape with Vertices," **Transactions of the ASME, Journal of Engineering Materials and Technology**, 1977, vol. 99, pp. 290-297.

29. J. R. Rice and D. M. Tracey, "On the Ductile Enlargement of Voids in Triaxial Stress Fields," **Journal of the Mechanics and Physics of Solids**, 1969, vol. 17, pp. 201-217.

30. A. W. Thompson, "Modeling of Local Strains in Ductile Fracture," **Metallurgical Transactions A**, 1987, vol. 18A, pp. 1877-1886.

31. E. T. Turkdogan, S. Ignatowicz, and J. Pearson, "The Solubility of Sulphur in Iron and Iron-Manganese Alloys," **Journal of the Iron and Steel Institute**, 1955, vol. 180, pp. 349-354.

32. T. Lin, A. G. Evans, and R. O. Ritchie, "Stochastic Modeling of the Independent Roles of Particle Size and Grain Size in Transgranular Cleavage Fracture," **Metallurgical Transactions A**, 1987, vol. 18A, pp. 641-651.

33. M. F. Kanninen and C. H. Popelar, **Advanced Fracture Mechanics** (New York, NY: Oxford University Press, 1985), pp. 63-67.

34. A. A. Wells, "Application of Fracture Mechanics at and Beyond General Yielding," **British Welding Journal**, 1963, vol. 10, pp. 563-570.

35. W. A. Spitzig, J. F. Kelly, and O. Richmond, "Quantitative Characterization of Second-Phase Populations," **Metallography**, 1985, vol. 18, pp. 235-261.

36. D. Brooksbank and K. W. Andrews, "Thermal Expansion of Some Inclusions Found in Steels and Relation to Tessellated Stresses," **Journal of the Iron and Steel Institute**, 1968, vol. 206, pp. 595-599.

37. W. M. Garrison, Jr., J. L. Maloney, and A. L. Wojcieszynski, "Influence of Tempering and Second Phase Particle Distributions on the Fracture Behavior of HY180 Steel," **Gilbert R. Speich Symposium Proceedings** (Warrendale, PA: Iron and Steel Society, Inc., 1992), pp. 237-246.

38. M. J. Leap, unpublished research, The Timken Company, 1998.

39. J. E. McVicker, United States Patent 5,131,965, 1992.

40. S. J. Garwood and C. E. Turner, "Slow Stable Crack Growth in Structural Steel," **International Journal of Fracture**, 1978, vol. 14, pp. R195-R198.

41. R. O. Ritchie and A. W. Thompson, "On Macroscopic and Microscopic Analyses for Crack Initiation and Crack Growth Toughness in Ductile Alloys," **Metallurgical Transactions A**, 1985, vol. 16A, pp. 233-248.

42. W. M. Garrison, Jr. and A. W. Thompson, "Micromechanistic Expressions of Continuum Microscale Parameters for Stable Crack Growth," **Metallurgical Transactions A**, 1986, vol. 17A, pp. 2249-2253.

43. S. T. Rolfe and S. R. Novak, "Slow-Bend K_{Ic} Testing of Medium-Strength High-Toughness Steels," **Review of Developments in Plane Strain Fracture Toughness Testing, ASTM STP 463** (Philadelphia, PA: American Society for Testing and Materials, 1970), pp. 124-159.

44. J. M. Barsom and S. T. Rolfe, "Correlations Between K_{Ic} and Charpy V-Notch Test Results in the Transition-Temperature Range," **Impact Testing of Metals, ASTM STP 466** (Philadelphia, PA: American Society for Testing and Materials, 1970), pp. 281-302.

45. P. C. Paris, written discussion of the paper by J. A. Begley and J. D. Landes, **Fracture Toughness, ASTM STP 514** (Philadelphia, PA: American Society for Testing and Materials, 1972), pp. 21-22.

46. J. R. Rice, P. C. Paris, and J. G. Merkle, "Some Further Results of J-Integral Analysis and Estimates," **Progress in Flaw Growth and Fracture Toughness Testing, ASTM STP 536** (Philadelphia, PA: American Society for Testing and Materials, 1973), pp. 231-245.

47. J. G. Merkle, "Analytical Applications of the J-Integral," **Progress in Flaw Growth and Fracture Toughness Testing, ASTM STP 536** (Philadelphia, PA: American Society for Testing and Materials, 1973), pp. 264-280.

48. G. R. Speich and W. A. Spitzig, "Effect of Volume Fraction and Shape of Sulfide Inclusions on Through-Thickness Ductility and Impact Energy of High-Strength 4340 Plate Steels," **Metallurgical Transactions A**, 1982, vol. 13A, pp. 2239-2257.

Cleavage Fracture Stress of Plain Carbon Fully Pearlitic Steel Bars

G. M. Michal and T. D. Nixon*

Department of Materials Science & Engineering
Case School of Engineering
Case Western Reserve University
Cleveland, Ohio 44106-7204

*B. F. Goodrich Aerospace
9921 Brecksville Road
Brecksville, Ohio 44141

Key Words: Cleavage fracture stress, Pearlite, Lamellar spacing, Notched bend bar specimen, Heat treatment

INTRODUCTION

The two phase constituency of pearlitic steels makes them an excellent example of an in-situ composite material comprised of layers of ductile ferrite strengthened with cementite lamellae. A recent review has summarized the mechanical behavior of pearlitic steels and described the pearlite microstructure as an in-situ lamellar composite [1]. However, that review focused on only a single 1080 steel composition with pearlite that contained approximately 12 volume percent of lamellar cementite. The goal of the present work was to expand the scope of previous investigations of the fracture characteristics of pearlitic steels by evaluating steels with lamellar cementite contents ranging from 7 to 15 volume percent.

A key element of this study was the production of fully pearlitic microstructures in specimens of a large enough physical size to evaluate mechanical properties. This goal had to be achieved using steels with carbon contents ranging from 0.45 to 1.00 weight percent. The specimen geometry chosen for evaluation of the cleavage fracture stress was that of a notched bend bar with a square cross section 12.7mm on a side. A valid test demanded that the specimens had a uniform fully pearlitic microstructure throughout their cross section that was free of residual stresses. Those requirements could only be met if the test bars did not experience a significant temperature gradient during their heat treatment.

To assure that the heat treatment process never produced a temperature gradient in the test bars meant that conditions of newtonian cooling had to be maintained during cooling from the austenitization temperature. When the Biot number for a heat treatment process is less than 0.1, then the conditions are appropriate to achieve newtonian cooling [2]. That value for the Biot number established the maximum value of the convective heat transfer coefficient, h, of the cooling media that could be used during the heat treatment process. The maximum permissible value for h was approximately 630 Watts/m$^2\cdot$ °K. The corresponding maximum cooling rate at 700°C could not exceed approximately 1800°C/min. Such constraints required that the austenite to pearlite transformations be carried out in the test bars under conditions of continuous cooling [3]. The maximum cooling rate employed was marginally fast enough to avoid the formation of proeutectoid ferrite in the lowest carbon steel evaluated, i.e., 1045 and the formation of proeutectoid cementite in the highest carbon steel evaluated, i.e., 10100.

Utilizing continuous cooling procedures, four-point notched bend specimens were produced and tested to determine nominal bend fracture strengths at temperatures ranging from -125 to +250°C. Sub-scale tensile specimens were then machined from each fractured bend bar and used to determine its mechanical properties at the same temperature as was employed for its four point bend test. The ratio of the nominal fracture stress to the yield stress was then used in conjunction with a finite element analysis (FEA) by Griffiths and Owens [4] to determine the ratio of the maximum principal stress in the notched bend bar to its yield strength. That ratio immediately defined the value of the maximum principal stress at fracture which was taken to be the cleavage fracture stress of the pearlite. Microstructural and fractography analyses were also performed on the bend bar specimens and correlated with the values determined for the cleavage fracture stress of pearlitic microstructures that contained from 7 to 15 volume percent cementite.

EXPERIMENTAL PROCEDURES

Bars of commercially produced 1045, 1065, 1080 and 10100 steels 19 mm

Bars of commercially produced 1045, 1065, 1080 and 10100 steels 19 mm " \l 2} in diameter were sectioned into pieces 76 mm long, machined into blanks 13mm square and then heat treated. Chemical analyses of the four grades of plain carbon steel are listed in Table I. All blanks were austenitized at 1100°C for one hour. After austenitization the 1045 steel was quenched into oil. The 1065, 1080 and 10100 steels were quenched into molten salt. After heat treatment, final machining was performed, yielding the notched bend bar geometry illustrated in Figure 1. After four-point bend tests were conducted, sub-scale tensile specimens with a gage diameter of 3mm and a gage length of 15mm were machined from the ends of the fractured bend bars.

Table I: Chemical Analysis in Weight Percent for the Steels Employed

	1045	**1065**	**1080**	**10100**
Carbon	0.47	0.64	0.76	0.99
Manganese	0.73	0.87	0.74	0.41
Phosphorous	<0.003	0.009	0.012	0.007
Sulfur	0.003	0.029	0.029	0.013
Silicon	0.24	0.22	0.24	0.20
Chromium	0.06	0.14	0.11	0.06
Nickel	0.05	0.10	0.06	0.03
Molybdenum	<0.01	0.01	0.02	0.01
Copper	0.12	0.27	0.09	0.04
Vanadium	<0.005	<0.005	0.04	<0.005
Aluminum	0.02	<0.01	<0.01	0.02

Hardness measurements were made on the ends of all of the bend bars as a check of their heat treatments. Four-point bend tests were performed using an MTS 810 servo-hydraulic testing system equipped with a model 651 temperature chamber. Specimens were tested at -125, -75, 25, 100 and 250°C. A crosshead speed of 0.5mm/minute was employed for all tests. Tensile testing of the sub-scale specimens was conducted using an extensometer with a 12.7mm gage length.

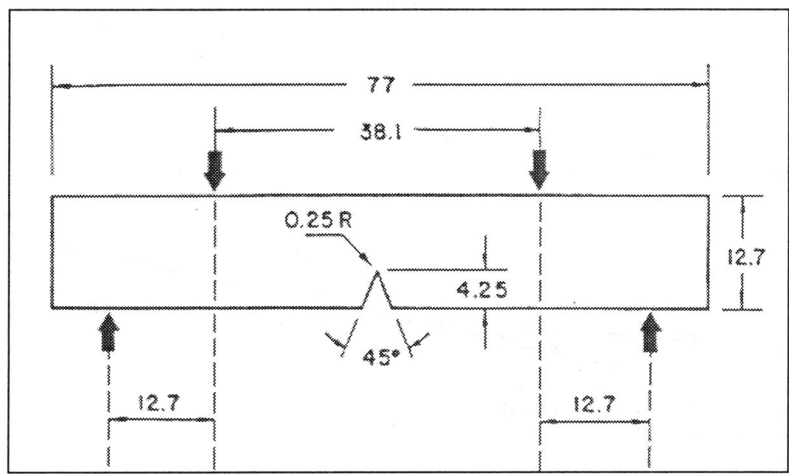

Figure 1: Notched bend bar specimen geometry. Dimensions given in millimeters.

End pieces of fractured bend bars were metallographically prepared and used to determine the prior austenite grain size of the four grades of steel using optical microscopy. Other polished and etched samples were viewed with a Hitachi S4500 FEG-SEM to establish the lamellar spacing of the pearlite. The fracture surfaces of select bend bar specimens were examined with the same SEM to define the locations of fracture initiation sites.

RESULTS

Hardness testing-The average hardness values for the four grades of steel after heat treatment are listed in Table II. Each average value was calculated based upon 13 hardness readings taken from each of 8 bend bars machined from each steel grade. The observed hardness levels are consistent with those expected for each of the steel grades in a fully pearlitic condition created by continuous cooling [3]. The relatively high hardness levels exhibited by each of the steel grades indicate that the continuous cooling heat treatment procedures did produce pearlite with very fine lamellar spacings.

Table II: Hardnesses of the Heat Treated Bend Bar Specimens.

Steel Grade	Avg. Rockwell Hardness (C-scale)	Standard Deviation (C-scale)
1045	17.5	2.6
1065	24.8	3.0
1080	33.6	5.5
10100	44.8	1.7

Notched 4-Point Bend Testing-The averages of two nominal bend fracture stress measurements are plotted as a function of test temperature in Figure 2 for each of the four grades of steel. For all of the steel grades, the nominal bend fracture stress was observed to increase with increasing test temperature. After evaluating the low-temperature test data for the 1045 and 1065 steel grades, extrapolation of that data indicated that the ratio of nominal bend fracture stress to the yield stress for tests conducted on those steels above 25°C would be greater than the ratio allowed for a valid FEA analysis of the cleavage fracture stress. As such, no elevated temperature testing was performed on those two steels. A result to note in Figure 2 was that for the tests run at

-125 and -75°C the lowest values for the nominal bend strength were observed for the 1065 steel.

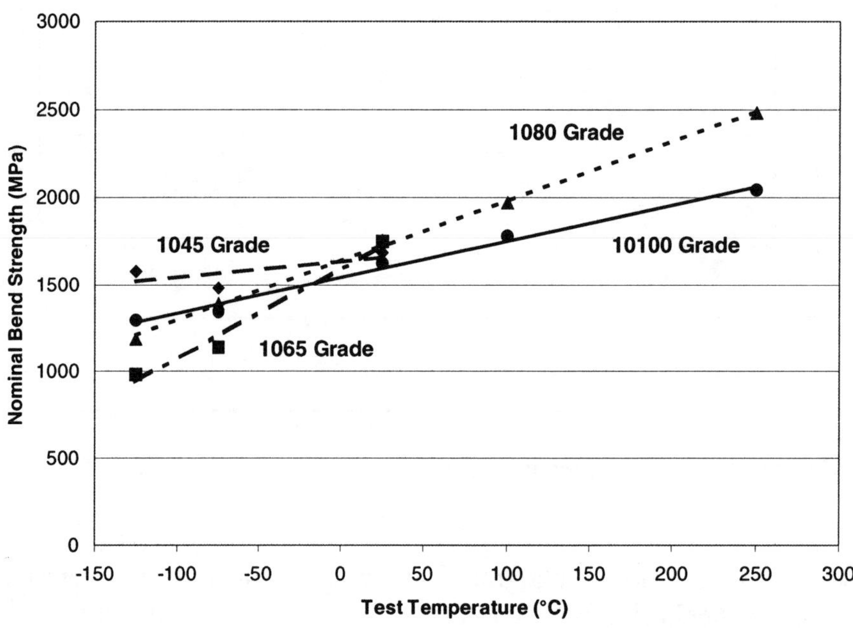

Figure 2: Average of two measurements of the nominal bend fracture stress as a function of test temperature.

Tensile Testing-The average of two 0.2% offset yield stress measurements are plotted as a function of test temperature in Figure 3 for each of the four grades of steel. All of the grades of steel exhibited a significant decrease in their yield stress with increasing test temperature over the range of temperatures evaluated. A result to note in Figure 3 was that the yield strengths of the 1080 and 10100 steels were quite similar over the temperature range evaluated even though the room temperature hardness levels of the two steels were quite different.

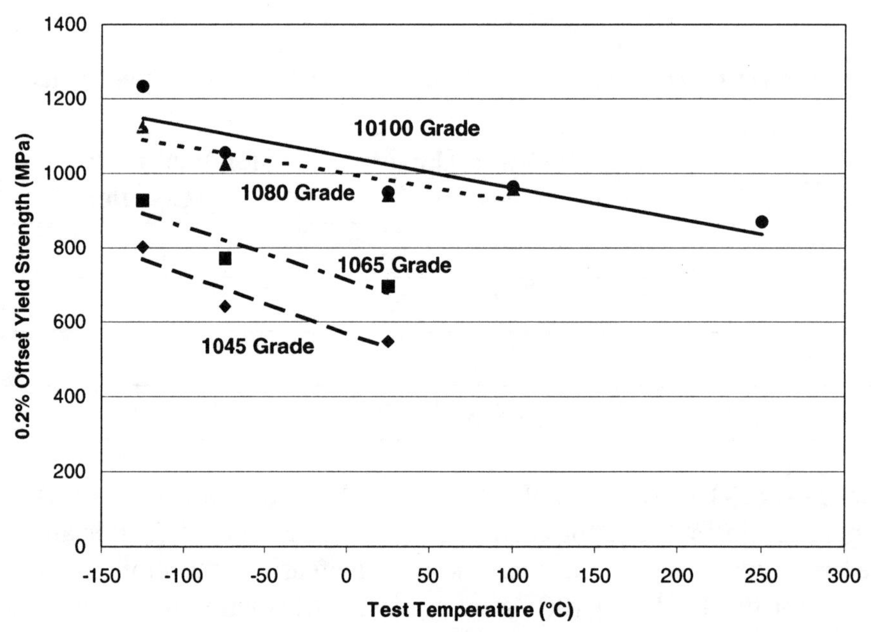

Figure 3: Average of two measurements of the 0.2% offset yield stress as a function of test temperature.

The reduction-in-area values measured for the four grades of steel are plotted as a function of test temperature as shown in Figure 4. Values plotted are the average of measurements from two specimens. Considerable scatter in the individual data points was observed. The small cross section of the tensile specimens certainly contributed to this scatter. The reduction-in-area values were observed to increase substantially with increasing test temperature for each of the four grades of steel. The reduction-in-area values were also very sensitive to the carbon content of the steels. The 1045 steel exhibited a large measure of macroscopic ductility prior to fracture even when tested at –125°C. An equivalent level of macroscopic ductility would only have occurred in the 10100 steel if it were tested above +200°C.

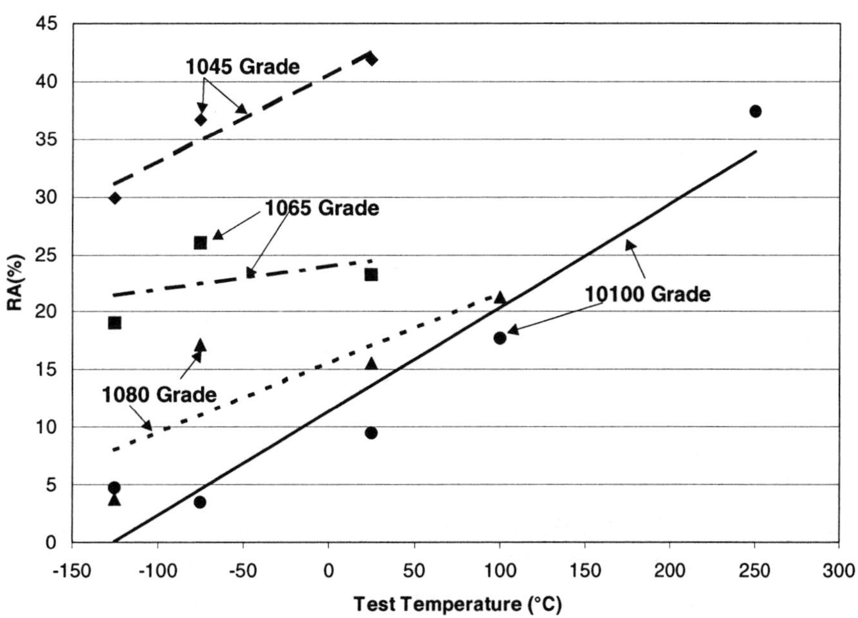

Figure 4: Average of two measurements of the reduction in area as a function of test temperature.

Cleavage Fracture Stress Determination-The ratios of the nominal bend fracture stress to the 0.2% offset yield stress for each test temperature/steel grade combination were used in conjunction with the FEA study of Griffiths and Owens to calculate values of the cleavage fracture stress [4]. Test conditions that produced ratios greater than 2.15 were not allowed based upon the FEA, and such tests could not be used to determine valid cleavage fracture stress values. The values of the cleavage fracture stress are plotted for each of the four grades of steel as a function of test temperature in Figure 5. The cleavage fracture stress for each of the grades of steel exhibited little to no temperature dependence. The lack of test temperature dependence for the cleavage fracture stress of a eutectoid composition steel with pearlitic microstructures was also observed by Lewandowski and Thompson [5]. The cleavage fracture stress values plotted in Figure 5 for the 1065 steel are significantly below those of the other steels. The low values for the 1065 steel are a direct result of the low nominal bend fracture stress values noted for the 1065 steel in Figure 2.

Microstructural Characterization-Optical metallographic analysis procedures were used to determine the prior austenite grain size of heat treated specimens. The prior austenite grain sizes produced by the 1100°C austenitization ranged from ASTM grain size number 4.2 (85μm) to 6.0 (45μm) for all four grades of steel. That analysis also revealed the presence of films of proeutectoid ferrite in the 1045 steel. The average volume percent of proeutectoid ferrite found in the 1045 steel was 7.5. Isolated regions of proeutectoid ferrite were discovered in the 1065 steel. The average volume percent of proeutectoid ferrite observed in the 1065 steel was 1.8. Bainitic regions were also found in some of the samples of the 1080 grade steel.

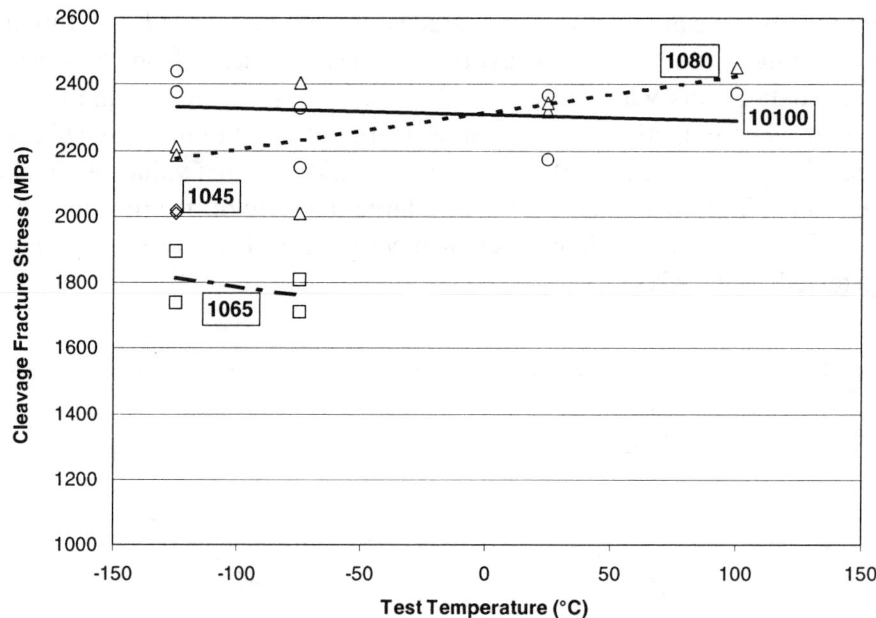

Figure 5: Cleavage fracture stress as a function of test temperature.

High magnification SEM images from replicate polished and etched samples of each of the four grades of steel were used to quantify the lamellar spacings of the pearlite. The average spacing from the two samples of each steel are shown in Figure 6. The pearlite spacing was observed to decrease as the carbon content of the steels increased. In all cases the spacing of the pearlite was quite fine, being on the order of 0.1μm, or less. Such spacings indicate that the degree of undercooling achieved prior to the start of pearlite formation was greater than 150°C [6].

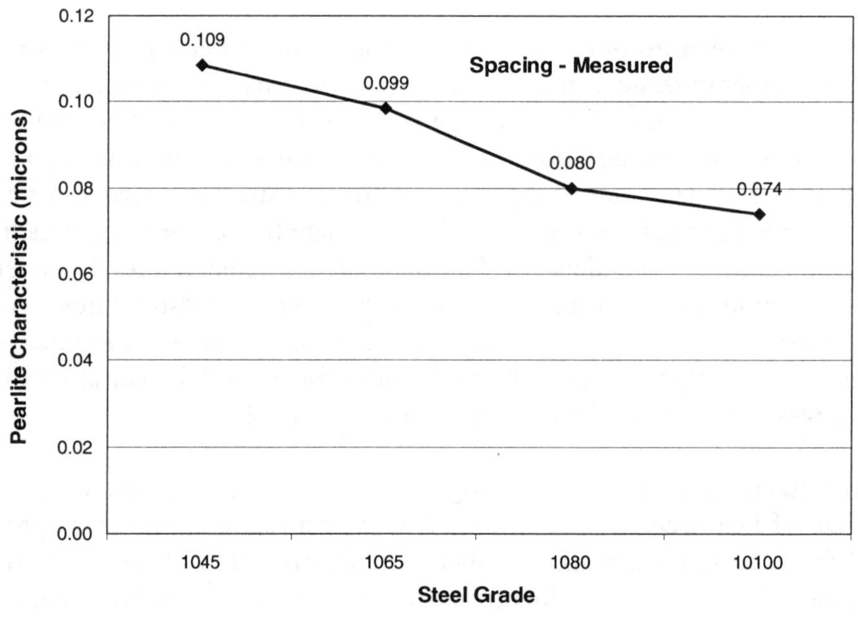

Figure 6: Average lamellar spacing of the pearlite.

The fracture surfaces of one notched bend specimen from each of the four grades of steel were characterized using an SEM. The aim of the work was to determine the location and cause of possible fracture origin sites and compare the locations with those predicted by the FEA to experience the maximum stress during a bend test. The FEA determined that the peak stress occurred 0.1 to 1.0mm below the bottom of the notch. All of the specimens examined were found to have three or more fracture origin sites. Sixteen of the 17 fracture origin sites found were within the FEA predicted peak stress location.

The large majority of the fracture origin sites were associated with the presence of a particle. Figure 7 shows a representative site where a fractured particle approximately 4μm in size was found. The EDS analysis of the particle indicated that it was a (Ti, V, Nb) C, N precipitate phase. Other particles commonly found at fracture initiation sites were manganese sulfides and magnesium silicate inclusions. The majority of the particles found at fracture origin sites were less than 5μm in size.

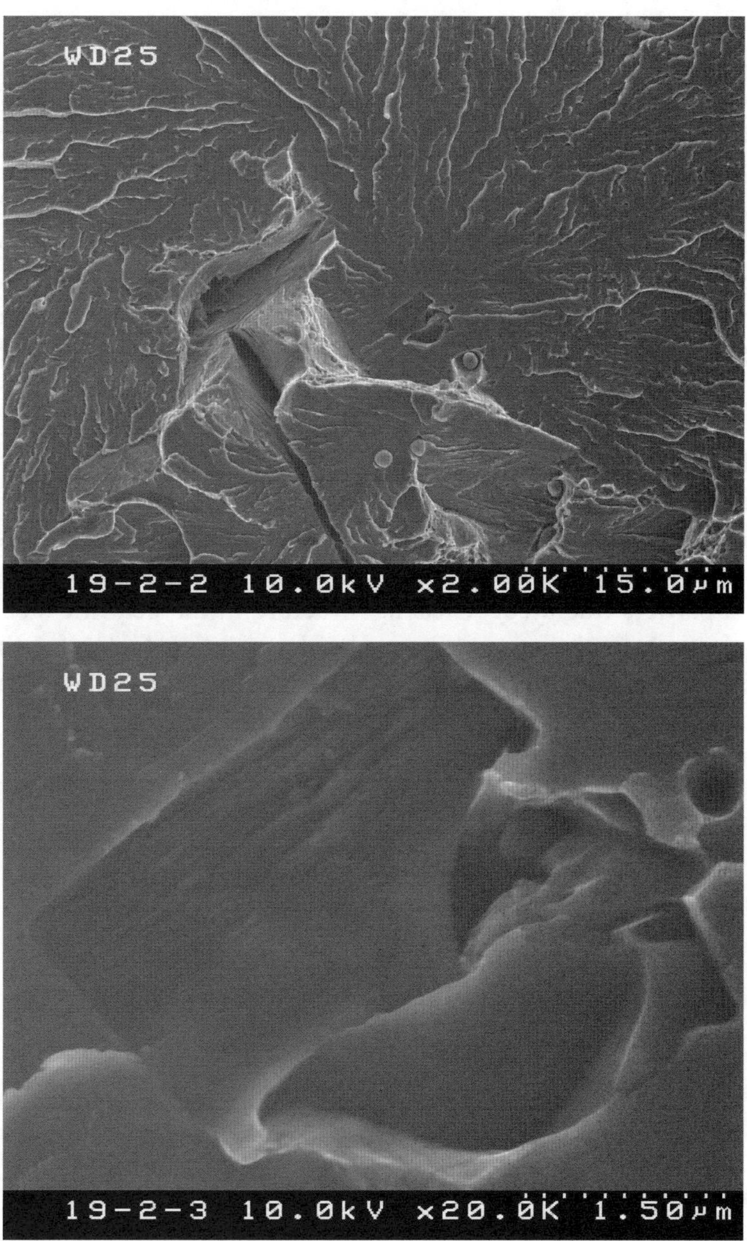

Figure 7: A candidate initiation site containing a (Ti, V, Nb)C, N particle found on a 1080 steel fracture surface.

DISCUSSION

A key finding of this investigation was that the cleavage fracture stress of the pearlitic steels increased when their pearlite spacings decreased. A secondary observation was a lack of temperature dependence by the cleavage fracture stress for a steel grade with a given pearlite spacing. The work by Lewandowski and Thompson on 1080 steels with pearlitic microstructures also found that the cleavage fracture stress increased when the pearlite spacing decreased and very little temperature dependence associated with the cleavage fracture stress of pearlite with a lamellar spacing of 0.11μm [3].

Cleavage fractures in the bend bar specimens were observed to have initiated at multiple sites. Particles were found at almost all of the fracture initiation sites. Since only one side of a fracture could be examined, it was possible that in the few cases where a particle was not detected an inclusion or precipitate was embedded in the other half of the fracture surface, i.e., the section used to make the sub-scale tensile specimens. The locations below the base of the notch of the fracture origin sites displayed good quantitative agreement with the peak stress location as predicted by the FEA.

In their study of 1080 steels Lewandowski and Thompson found multiple initiation sites and that the type of fracture initiator, i.e., precipitate or inclusion, had no influence on the value for the cleavage fracture stress of fine pearlitic microstructures [3]. The findings of this investigation are in good agreement with their observations. In both their work and the present study fracture initiation sites were observed with associated initiators having sizes ranging from 1 to 15μm. In comparison the thicknesses of the cementite lamellae in the steels used for this investigation were on the order of 0.01μm. The actual fracture initiators were at least two orders of magnitude larger than the thickness of the cementite found in the pearlite. As such cracking of individual cementite lamella was not found to be a potent fracture initiation site.

Based on the observation that each specimen contained multiple possible initiation sites as well as a variety of initiator types, the conclusion can be drawn that the magnitude of the cleavage fracture stress exceeded the level of stress required to fracture many particles in the steels. If the cleavage fracture stress was less than the stress required to fracture essentially all of the particles in the steel, catastrophic crack growth would then have occurred as soon as the first particle fractured. Cleavage fracture in the pearlitic steels under those conditions would be considered a nucleation controlled process. The determining factor for the characteristic cleavage fracture stress is the inherent resistance of the pearlitic microstructure to crack propagation. For the typical size of the crack initiators observed in the steels, i.e., less than 10μm, the cleavage fracture stress was larger than the fracture initiation stress, and catastrophic cleavage fracture did not occur immediately upon initiation. The cleavage fracture of pearlite is a propagation controlled process. This conclusion is in agreement with the findings of Lewandowski and Thompson [3].

The relationship between the cleavage fracture stress, σ_f, and the size of the crack, C_o, that initiates fracture is given by the following equation suggested by Smith [7].

$$\sigma_f = \left[\frac{4E\gamma_p}{\pi(1-v)c_0}\right]^{1/2} \tag{1}$$

In Equation 1, E is the elastic modules of the steel, v is Poison's ratio and γ_p is the total energy expended to create unit area of fracture surface. The value for γ_p for pearlite would be that for the cleavage fracture of ferrite modified by the presence of the cementite lamellae. It can be anticipated that some unit amount of energy, W_c, is required for a cleavage crack to propagate across a unit length of cementite in the pearlite. Under those conditions the energy required to create unit area of fracture surface would be proportional to the ratio of W_c and the lamellar spacing, S, of the pearlite as indicated in Equation 2.

$$\gamma_p \propto \frac{W_c}{S} \tag{2}$$

Inserting Equation 2 into Equation 1 establishes a prediction that the cleavage fracture stress of pearlite should be inversely proportional to the square root of the pearlite spacing. In Figure 8 the average cleavage fracture stress values determined for the four grades of steel are plotted as a function of the inverse square root of their measured pearlite spacings. A reasonable agreement with a linear relationship is observed for all but the 1065 steel. In addition the cleavage fracture stress values determined by Lewandowski and Thompson for 1080 steels with pearlite spacings of 0.11 and 0.25μm have been plotted on Figure 8 [3]. Their data fall remarkably close to the straight line best fit through the data generated in the present study.

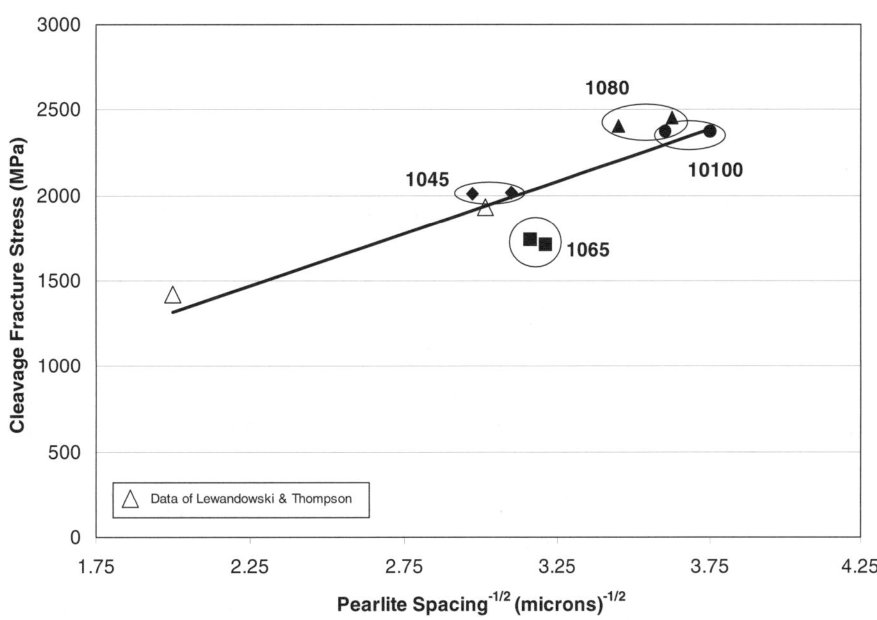

Figure 8: Cleavage fracture stress as a function of the inverse square root of the pearlite spacing.

Taking the cleavage fracture stress values determined in this study with the observation that initiation sites 10μm in size were present prior to cleavage fracture propagation allowed equation 1 to be used to estimate γ_p for the four steel grades. The resulting values for γ_p varied from 66 J/m^2 for the 1045 steel with a pearlite spacing of 0.109μm to 103 J/m^2 for the 10100 steel with a pearlite spacing of 0.074μm. Knott estimated that the value for γ_p for ferrite is 14 J/m^2 [8]. The fine pearlites appear to increase the energy needed to create unit area of fracture surface by a factor ranging from 5 to 7 times that of "unreinforced" ferrite. A number of possible mechanisms may contribute to the energy absorption found in the pearlite microstructures. It has been suggested and demonstrated that as cementite becomes finer in pearlitic steels there is a transition from brittle behavior to some localized ductility in the cementite [9]. Other possible mechanisms are the local blunting, and/or deflection of the crack at the cementite/ferrite interface resulting in the dissipation of an increased amount of energy.

One inconsistent result from this investigation were the outlying cleavage fracture stress values obtained for the 1065 steel. The poor agreement between the cleavage fracture stress for that steel compared to the others evaluated can be clearly seen in Figure 8. One possible microstructural feature that could have reduced the apparent cleavage fracture stress of the 1065 steel was the presence of discontinuous regions of proeutectoid ferrite. The proeutectoid ferrite observed in the 1065 steel was acicular in nature and found in isolated locations. The regions of highly constrained ferrite may have served as relatively large initiation sites for cleavage fracture. As such the proeutectoid ferrite would have effectively lowered the cleavage fracture stress of the 1065 steel.

CONCLUSIONS

Pearlitic microstructures with lamellar spacings of 0.1μm or less could be produced in bend bar specimens with 13mm size square cross sections from 1045, 1065, 1080 and 10100 steels using heat treatment procedures that maintained newtonian cooling conditions. The cleavage fracture stresses determined for the four steels were independent of test temperature over the range of -125 to +250°C. The cleavage fracture stress values were observed to be inversely proportional to the square root of the measured pearlite spacings.

Cleavage fracture of the pearlite microstructures produced in this investigation was a propagation controlled process. Multiple initiation sites were found on the fractured surfaces of all four grades of steel. The values determined for the energy required to create unit area of cleavage fracture surface ranged from approximately 60 to 100 J/m^2 based upon observed fractures initiating at particles 10μm in size.

ACKNOWLEDGEMENTS

Special thanks are extended to Jerry S. Lee, Larry D. Bentsen and S. K. Lau for helpful suggestions and guidance. The support for this project provided by B. F. Goodrich Aerospace is greatly appreciated.

REFERENCES

1. J. J. Lewandowski, "Mechanical Behavior of In-situ Composites," In-Situ Composites: Science and Technology, eds. M. Singh and D. Lewis, The Minerals, Metals & Materials Society, 1994, pp. 159-167.

2. J. R. Welty, C. E. Wicks and R. E. Wilson, "Fundamentals of Momentum, Heat and Mass Transfer, John Wiley & Sons, Inc., New York, 1969, pp. 282-284.

3. M. Atkins, Atlas of Continuous Cooling Transformation Diagrams for Engineering Steels, American Society for Metals, Metals Park, 1980, pp. 13-23.

4. J. R. Griffiths and D. R. J. Owne, "An Elastic-Plastic Stress analysis for a Notched Bar in Plane Strain Bending," J. Mech. Phys. Solids, vol. 19, 1971, pp. 419-431.

5. J. J. Lewandowski and A. W. Thompson, "Microstructural Effects on the Cleavage Fracture Stress of Fully Pearlitic Steel," Metallurgical Trans. A, vol. 17A, 1986, pp. 1769-1786.

6. J. C. Fisher, "Eutectoid Decomposition," Thermodynamics in Physical Metallurgy, American Society for Metals, Cleveland, 1950, pp. 201-241.

7. E. Smith, "The Nucleation and Grown of Cleavage Microcracks in Mild Steel," Proceedings of the Conference on the Physical Basis of Yield and Fracture, Int. Physics and Physical Society, Oxford, pp. 1966, pp. 36-46.

8. J. F. Knott, "Micromechanisms of Fracture and the Fracture Toughness of Engineering Alloys," Fracture 1977, Vol. 2, ed. D. M. R. Taplin, Univ. of Waterloo Press, Waterloo, ONT, 1977, pp. 61-91.

9. D. A. Porter, K. E. Easterling, and G. D. W. Smith, "Dynamic Studies of the Tensile Deformation and Fracture of Pearlite," Acta metall., vol. 26, 1978, pp. 1405-1422.

EXAMINATION OF FULLY BAINITIC STRUCTURES IN TRIP STEELS

M. Bouet, R. Fillipone, E. Essadiqi[1], J. Root[2] and S. Yue

Department of Mining and Metallurgical Engineering,
McGill University,
3610 University St.,
Montreal, Quebec, H3A 2B2, Canada
(514) 398-1378
steve@minmet.lan.mcgill.ca
[1]MTL-CANMET, Ottawa, Ontario, Canada
[2]National Research Council of Canada, Chalk River Laboratories
Chalk River, Ontario, Canada

Key Words: TRIP steel, retained austenite, martensite, bainite, intercritical deformation, hot deformation.

ABSTRACT

In Si-Mn TRIP steels, it is generally acknowledged that the optimum combination of strength and elongation is attained with a multiphase microstructure of retained austenite, bainite and polygonal ferrite. However, work at McGill university has indicated that ductilities comparable to this 'conventional' microstructure have been attained in microalloyed TRIP steels with fully bainitic structures. This work describes the production and characterization of these microstructures, and suggests mechanisms to explain this behaviour.

INTRODUCTION

The excellent combination of strength and ductility exhibited by TRIP steels is frequently considered to be mainly due to the transformation of retained austenite to martensite. However, in Si-Mn TRIP steels, the microstructure is comprised of not only retained austenite, but also bainite, polygonal ferrite and perhaps even martensite. The formation of considerable volume fractions of polygonal ferrite and bainite is necessary to retain austenite, by elevating C levels in austenite, as the austenite transformation progresses, until metastable austenite is retained to room temperature. Since typical retained austenite level in such steels is only about 10 to 15%, it is clear that the balance of the microstructural constituents will play a significant role in controlling the mechanical properties.

An obvious microstructural parameter to consider is the ratio of polygonal ferrite to bainite. Matsumura and his workers[(1)] concluded that an excellent combination of tensile strength (~1000 MPa) and elongation to fracture (~ 30%) is obtainable in a 0.4C - 1.5Si - 0.8Mn steel, when ferrite rather than bainite is the predominant phase in the microstructure. The reason being that; (i) ferrite is softer than bainite, and thus exhibits higher elongation to fracture and, (ii) ferrite easily accepts the volume expansion of the retained austenite accompanied by the strain-induced transformation, resulting in a better exhibition of the TRIP behaviour.

On the other hand, Sakuma et al.[2] showed that a high elongation to fracture can be obtained in TRIP steels with a significant amount of bainite. In this work, it was demonstrated that the high elongation to fracture (~30 % elongation to fracture), with about 35 vol.% bainite, was related to the increase in stable retained austenite formed by isothermal transformation. The increase in the stability of the retained austenite was due to the enrichment of carbon in the retained austenite during isothermal transformation. Nevertheless, polygonal ferrite remains the predominant constituent. As a consequence of these types of observations, a TRIP microstructure comprising of about 30% bainite 10-15% austenite, and the balance being polygonal ferrite is often the goal.

Because the target microstructure is usually as noted above, there is comparatively less information concerning fully bainitic TRIP steels. What little information there is points to increased strength, but, not surprisingly, a significantly decreased ductility. For example, a typical strength and ductility combination for a microstructure generated by intercritically annealing a cold rolled TRIP grade at 860 °C, is 750 MPa and 36% .[3] By contrast, a TRIP composition intercritically annealed at 950 °C, the strength and elongation to fracture was found to be 1305 MPa and 12%, respectively.[4] One of the main differences between these two microstructures, implied from the respective intercritical annealing temperatures, is the bainite volume fraction, which is presumably close to being fully bainitic in the case of the higher temperature anneal.

In this paper, microalloyed TRIP steels are subjected to thermomechanical treatments that generate bainitic or near bainitic structures. The mechanical properties are compared with a microstructure comprised of polygonal ferrite, bainite and retained austenite.

EXPERIMENTAL DETAILS

The composition of the steels examined are shown in Table I

Steel	C (wt%)	Si	Mn	Nb	Mo
A	0.16	1.66	1.4	0.025	
B	0.20	1.60	1.4	0.041	0.3
C	0.23	1.1	1.6	0.033	0.33

These are all essentially Si-Mn TRIP steels with carbon compositions around 0.2 wt%, which is commonly regarded as being the optimum carbon level for this type of TRIP composition. All steels also contain a microalloying level of Nb, which has been shown to be beneficial for steels used in the as-hot rolled condition.[5] The Mo was added mainly because of its ability to postpone carbide nitride precipitation. Thus, in steel B, the concept was to improve the effectiveness of Nb in solid solution. In steel C, Mo is used as a part replacement for Si. These steels were cast at MTL-CANMET and received in the as-hot rolled condition.

These steels were subjected to multi-step hot deformation using the torsion testing apparatus in the CSIRA hot deformation laboratory at McGill University. Two types of hot deformation schedule were used to generate fully recrystallised and pancaked structures, respectively. For the former, (Schedule A), 12 identical passes of 0.2 strain, delivered at a strain rate of 0.1 s^{-1}, were executed between 1165 and 1000 °C, following reheating to 1200 °C in order to dissolve any Nb(C,N). The interpass time was sufficient to allow full recrystallization to take place between each pass. In order to pancake the steels, the schedule differed in that 16 passes were executed between 1165 to 870 °C; the other details remain the same. In both cases, after the final deformation, the steels were subjected to three different cooling rates to room temperature: (i) air cooling (cooling rate at 700 °C ≅ 2 °Cs^{-1}); (ii) accelerated cooling 1 (F1 ≅ 9 °Cs^{-1}); (iii) accelerated cooling 2 (F2≅ 18 °Cs^{-1}).

For all microstructures, ultimate tensile strength and elongation to fracture were determined using the shear punch test(6), and the retained austenite characteristics were determined by neutron diffraction at National Research Council of Canada, Chalk River Laboratories, Canada.(7) The shear punch test involves punching out a disk of the material under examination. Thus, only a narrow annular ring of the microstructure is actually deformed. This is, therefore, an ideal test to investigate the microstructure generated by torsion testing, because there is a radial variation of microstructure corresponding to the strain distribution that occurs as a consequence of torsional loading. Thus, microstructures that were examined corresponded to the specimen 'ring' tested by the shear punch test. The neutron diffractometer at Chalk River also has the resolution capability needed to examine the annulus corresponding to the shear punch test region.

RESULTS AND DISCUSSION

Recrystallised specimens:

The UTS and ductilities for all conditions and grades are summarised in Figs 1 and 2. It should be noted that work at McGill has demonstrated that the strengths measured by the shear punch test compare well with those generated by tensile tests, although the absolute values of the ductilities may be questionable. However, all trends indicated by this test are reliable.

With regard to UTS values, in general, the Mo addition appears to compensate well for the reduction in Si, particularly at the faster cooling rates where the reduced Si steel actually exhibits the best UTS. On the other hand, a high Si level plus Mo does not lead to better strengths than the low Si version. For all steels, increasing the cooling rate increases the UTS.

The elongation to fracture (Fig.2) reveals that increasing the cooling rate also improves the ductility for all steels. A slightly different picture emerges with respect to the effect of chemical composition (compared to the effect on the UTS). The low Si, Mo 'compensated', steel exhibits the worse ductility of all the grades. However, the difference in ductilities between all the steels is reduced as the cooling rate increases.

With regard to the microstructures, the retained austenite measurements are shown in Fig. 3, and reveal that the retained austenite decreases drastically with increasing cooling rate. The behaviour of the low Si grade is particularly interesting. At the slowest cooling rate, it possesses the most retained austenite, but at the highest cooling rate, no retained austenite can be detected. The other two grades have quite similar retained austenite volume fractions.

The microstructures of all these specimens have no detectable polygonal ferrite, the microstructure being lamellar, probably a bainite (Fig. 4). There is no significant effect of cooling rate. However, assuming the structure shown in Fig. 4 is bainite, it can be assumed that the effect of increasing cooling rate reduces the amount of retained austenite by transforming it to martensite. However, more work is required to confirm this hypothesis.

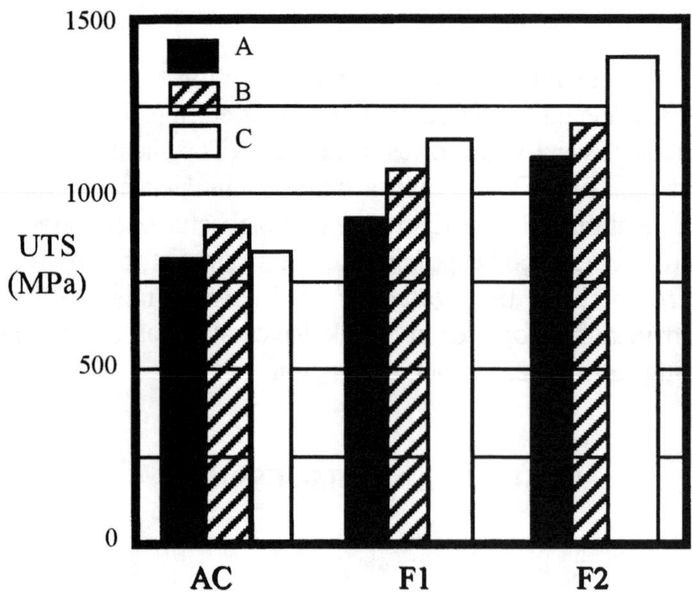

Fig 1 UTS of multi-pass, recrystallized specimens

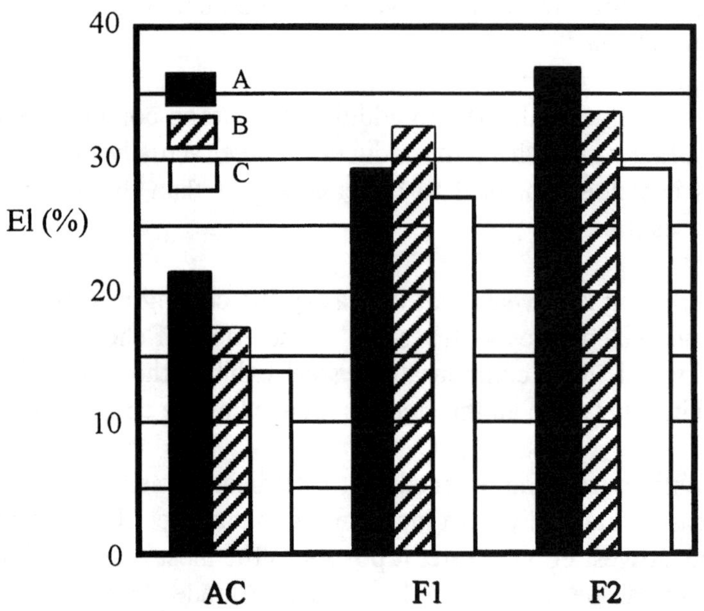

Fig 2 Elongation to fracture of multi-pass, recrystallized specimens

Pancaked specimens:

The UTS behaviour of the pancaked specimens is very similar to that of the recrystallized specimens. However, the elongation to fracture is somewhat different, as can be seen by comparing Fig. 5 to Fig. 2. The essential difference is that there is not as strong an effect of cooling rate as for the recrystallized specimens, with virtually no change in the ductility of specimen B. For the other two grades, there is still a tendency for the ductility to increase with faster cooling rates.

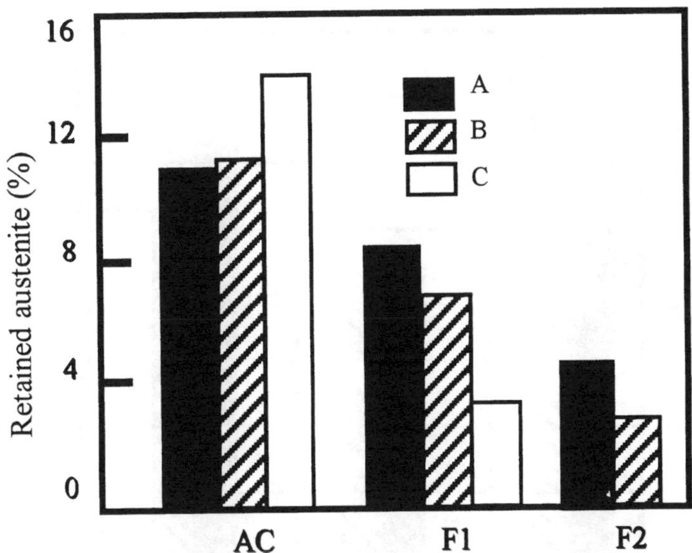

Fig.3 Retained austenite percentage, recrystallized specimens

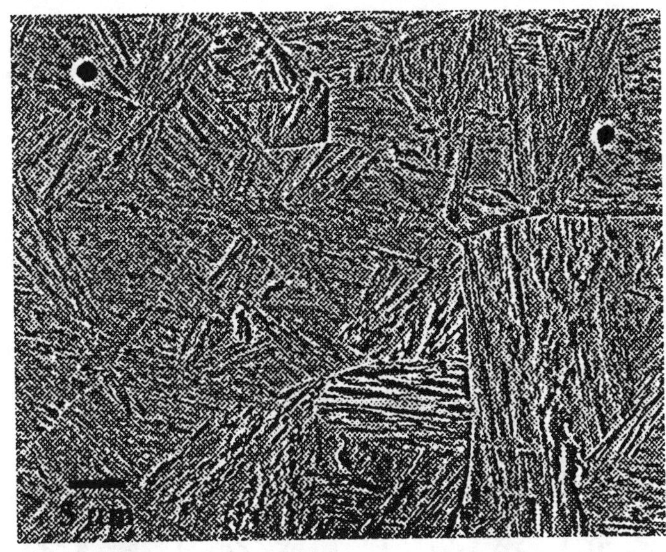

Fig.4 Grade C, recrystallised, F1 cooling rate, is typical of all the recrystallised

The retained austenite volume fraction, shown in Fig. 6, differs in several aspects from the behaviour exhibited by the recrystallized specimens. In particular, for grades A and B, the retained austenite volume fraction does not decrease with increasing cooling rate, but exhibits a maximum at F1. However, a striking similarity is that, for the low Si steel, there is no detectable retained austenite for the specimen subjected to the fastest cooling rate.

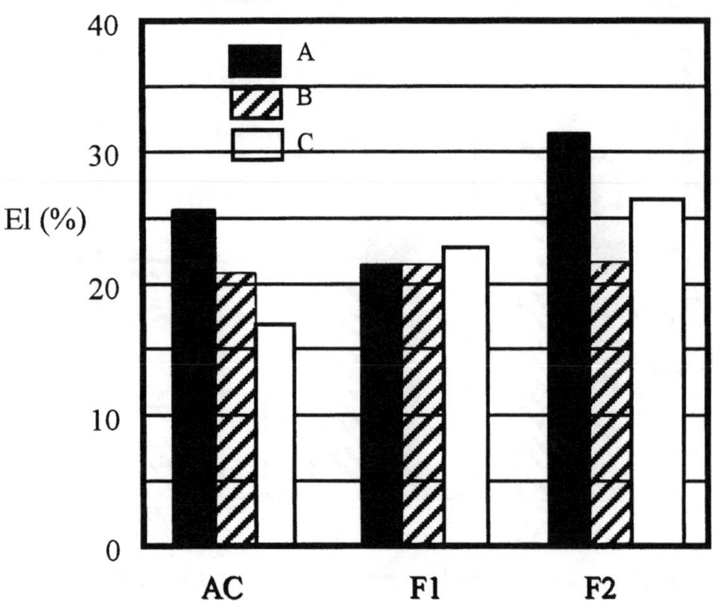

Fig 5 Elongation to fracture of multi-pass, pancaked specimens

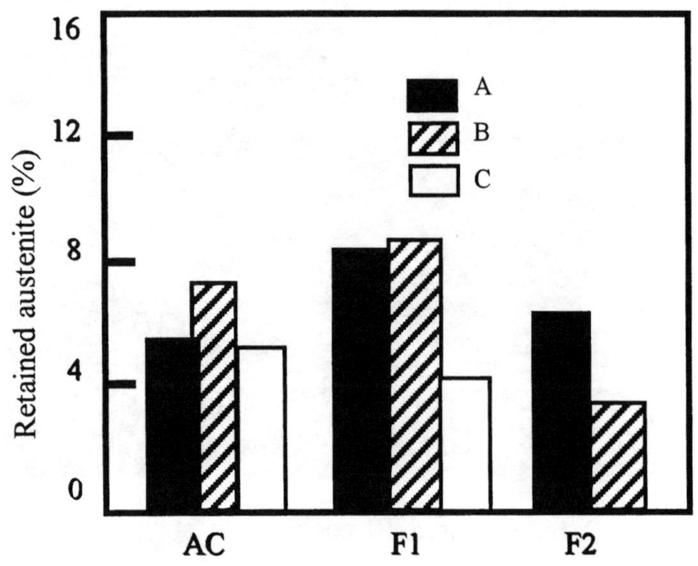

Fig.6 Retained austenite percentage, pancaked specimens

The air cooled microstructure of grade A, which did not contain any Mo, is comprised of about 60 or 70 % polygonal ferrite, i.e. close to the 'optimum' volume fraction of cold rolled and annealed Si-Mn TRIP steels. This indicates that the pancaking treatment was effective in accelerating the austenite transformation. However, all the other pancaked structures contained no resolvable polygonal ferrite, and were probably largely bainite. At magnifications of upto 2000X on the SEM, there was no particular difference between the bainites of the pancaked structure and the recrystallized structure.

It is notable that the structure containing polygonal ferrite, bainite and retained austenite possesses a superior ductility compared to the other grades in both the air cooled and F1 conditions. Thus, it appears that the these results are somewhat consistent with the findings from cold rolled and intercritically annealed TRIP steels, indicating that fully bainitic steels possess inferior ductilities to steels with polygonal ferrite. However, fully lamellar structures generated by 'high' cooling rates lead to ductilities comparable or better than the polygonal ferrite structure generated in this work. In fact, the ductilities of the recrystallized steels subjected to the faster cooling rates are higher than the corresponding pancaked specimens.

The microstructures of these steels are very complex, but the above observations seem to indicate that martensite may be a factor in dictating the properties of these steels. As mentioned above, an increasing cooling rate would tend to increase the martensite, as implied by the decrease in retained austenite. A recrystallized steel has more hardenability than a pancaked one, as exemplified by polygonal ferrite appearing in the air-cooled pancaked steel A. Therefore, the recrystallized steels would tend to have more martensite, for a given cooling rate.

It should be noted that none of these microstructures have been optimised for TRIP, since they have all been continuously cooled to room temperature, and have not been held isothernally (at about 400 °C) to form bainite, as is the usual approach for these steels. Therefore, although most if these steels contain retained austenite, it is not certain that this would impart an effective TRIP component to these microstructures. Indeed, the fact that structures with no detectable retained austenite exhibit comparable, or even superior properties, suggests that these microstructures may not rely on the TRIP effect to aid the mechanical properties. Nevertheless, even if the microstructure of Fig.7 was subjected to an appropriate bainite hold to maximize the TRIP effect, it would be difficult to surpass the UTS/ductility combination exhibited by the recrystallised steel C subjected to the fastest cooling rate.

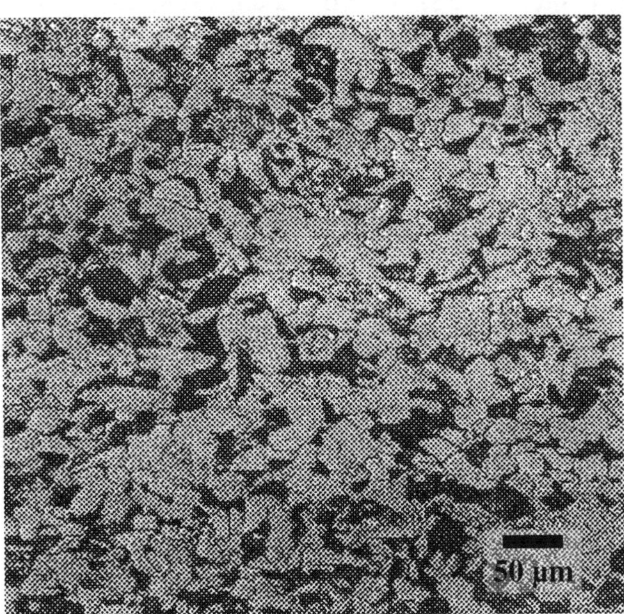

Fig.7 Grade A, pancaked and air cooled

SUMMARY AND CONCLUSIONS

Microalloyed TRIP steels were subjected to various thermomechanical treatments to generate both bainitic and polygonal ferrite/ bainite microstructures. The strength of the bainite structures were always superior to the polygonal ferrite bearing microstructure. Bainitic structures generated by 'accelerated' cooling rates exhibited superior ductilities to the polygonal ferrite bearing structure. Bainite structures having no retained austenite were generated in the low Si, Mo bearing grade. These structures exhibited excellent ductility and UTS properties, comparable to the correspondingly treated steels containing high Si levels and retained austenite.

ACKNOWLEDGEMENTS

The authors would like to thank the Canadian Steel Industry Research Association (CSIRA) and the Natural Science and Engineering Research Council (NSERC) of Canada for their financial support.

REFERENCES

1. O. Matsumura, Y. Sakuma and H.Takechi; ISIJ, 27,1987, 570.
2. Y. Sakuma, O. Matsumura and O. Akisue; ISIJ, 31, 1991, 1348.
3. T. Tanaka, M. Nishida, K. Hashiguchi and T. Kato; "Structure and Properties of Dual- Phase Steels", TMS/AIME, Warrendale, Pa, 1979, 221.
4. A.J. DeArdo; Proc. International Conf. on Physical Metallurgy of Thermomechanical Processing of Steel and Other Metals (Thermec- 88), Tokyo, Japan, 1988, 20.
5. A. Zarei-Hanzaki, P.D. Hodgson and S. Yue; Metall. Trans., 28A, A, (1997), pp. 2405-2414
6. M. Bouet, J. Root, E. Es-sadiqi and S. Yue: Mater. Sci. Forum, 284, (1998), pp. 319-326
7. A. Zarei-Hanzaki, J. Root, P.D. Hodgson and S. Yue; Acta. Metall., 43, (1995), pp. 569-578

Machinability of Bi-S and Pb-S-treated Free-Machining Steels

Samkyu Cho[1)] and Youngmoon Lee[2)]
1) Plate, Rod & Welding Group
POSCO Technical Research Laboratories
Pohang, P.O. Box 36, KOREA
Fax: 82-54-220-6911
Tel: 82-54-220-6344
E-mail: pc543027@posco.co.kr
2) Department of Mechanical Engineering
Kyungpook National University
1370, Sankyuk-dong, Buk-ku, Taegu
Fax: 82-53-950-6550
Tel: 82-53-950-5574
E-mail: ymlee@knu.ac.kr

Key Words: Bismuth-treated free-machining steel, Tool wear, Chip, Cutting energy, Cutting force, Chip-breaking cycle time, Surface roughness

ABSTRACT

The machinability of Bi-S and Pb-S-treated free machining steels was investigated in turning operations. The flank wear width(VB) of the single-point tool used to machine the Bi-S steel was narrower than that used to machine the Pb-S steel because of the soft matrix structure of the former. The specific cutting energy of the Bi-S steel was higher than that of the Pb-S steel. There were minimal differences in the chip-breaking cycle time and surface roughness between the two steels. The surface roughness was improved by controlling the hardness of the work-piece.

INTRODUCTION

The machinability of lead-treated free-machining steel has been well established since it was first developed in 1939. However, recent environmental concerns and problems related to lead contamination have resulted in certain regulations limiting the use of lead. Accordingly, Bismuth has been considered as an alternative since its physical and chemical properties are similar to those of lead. Although, there are many obstacles to commercializing bismuth-treated free-machining steel, including a high cost and high boiling point, many efforts continue to be made due to the issue of environmental protection.

However, there have been few reports on the machinability of bismuth-treated free-machining steel since the commercialization report by the Inland steel company in 1980s[1-3]. Therefore, this study investigated the machining properties of bismuth-treated free-machining steel, including tool wear, cutting energy, chip breaking, and machined surface roughness, and compared them with those of lead-treated free-machining steel.

EXPERIMENTS

Work-pieces for machining test

The chemical composition of the work-pieces is listed in Table 1. The lead and sulfur-treated work-piece was a commercially cold-drawn bar (ϕ30, now PbCD), whereas the Bismuth and sulfur-treated work-piece was continuously cast into a bloom, rolled into a billet, and then rolled into a rod (ϕ5.5~34mm) in a coil. The hot-rolled rod was finally cold-drawn into a bar using a 11.4%~22.6% reduction (now BiCD).

Table 1 Chemical composition of work-pieces (wt%)

Mark	C	Mn	P	S	Bi	Pb
Pb-S	.080	1.11	.079	.3	-	.28
Bi-S (A)	.071	1.33	.076	.3	.10	-
Bi-S (B)	.075	1.33	.077	.3	.12	-

Machining test

The tool-wear and chip-breaking properties were tested using a CNC lathe (MECCA-3). The cutting force was measured by a tool dynamometer (Kistler 9257A) in a general lathe. The composition of the tool included tungsten carbide, titanium carbide, tantalum carbide, and cobalt. The tool also had a groove-type chip former (CNMG120408, P20). The test was carried out at a cutting speed of 50~200 m/min, cutting depth of 1~3 mm, and feed rate of 0.05~0.54 mm/rev.

The flank wear width (VB)[4] was measured at regular intervals using optical microscopy and then compared. Fig. 1 shows a schematic diagram of the tool dynamometer and data processing procedures. The average value of each cutting force component was measured by the tool dynamometer and then transferred to a coordinate system to determine the shear plane and tool face. Thereafter, the shear force (Fs), shear stress (τ), shear strain (γ), and specific shear energy (u_s) were ascertained based on the shear plane, and the friction force (Fc), friction coefficient (μ), and specific friction energy (u_f) based on the tool face. The specific cutting energy (u), which is the cutting energy per unit volume of the work piece, was calculated based on the specific shear energy (u_s) and specific friction energy (u_f) [5]. There are several methods of evaluating the chip breakability including the chip-fracture strain [6], chip-breaking index [7], and chip-breaking cycle time [7]. In this study, the chip breakability was estimated using both the chip-breaking index and chip-breaking cycle time. For the purpose of quantitative comparisons and analysis it is useful to be able to express the roughness of machined surfaces in terms of a single factor or index. The most commonly used surface roughness index is known as the arithmetical mean value, Ra [8].

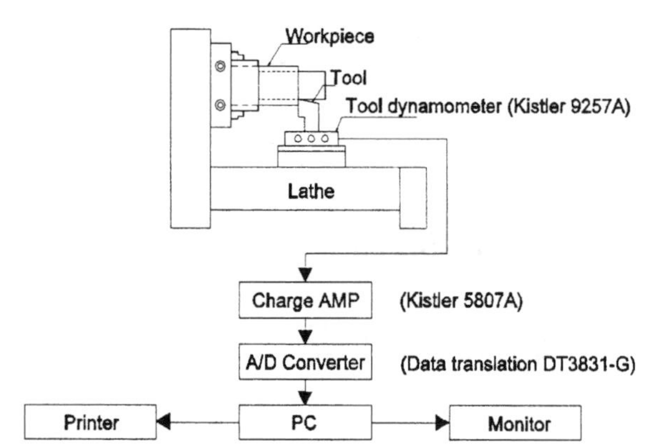

Fig. 1 Schematic diagram for tool dynamometer and data processing procedures

RESULTS AND DISCUSSION

Microstructure and hardness of work-pieces

The microstructure of the work-pieces consisted of ferrite, pearlite, plus inclusions of MnS and bismuth (or lead). The volume fractions of the ferrite, pearlite, and MnS and Bi (or Pb) inclusions were about 87.5%, 10%, and 2.5%, respectively. All the microstructure volume fractions were dependent on the chemical composition, including the carbon, sulfur, manganese, and bismuth (or lead).

Table 2 shows the microstructure hardness with a 5 gr load, 10 gr load, and 10 kg load, respectively. The hardness with a 10 kg load indicated the average deformation resistance of the ferrite, pearlite, and MnS and Bi (or Pb) inclusions. The hardness with a 5gr load indicated the deformation resistance of only the MnS inclusion, and that with a 10gr load indicated the deformation resistance of only the ferrite. The hardness of the PbCD was higher than that of the BiCD. The hardness of the MnS inclusions exhibited a wide deviation depending upon their morphologies.

Table 2 Hardness of work-pieces (Hv)

Mark	Microstructure			Dia.(Red., %)
	(Ferrite +Pearlite +MnS) (10kg)	MnS (5g)	Ferrite (10g)	
Pb-S	**187.8**	**170~200**	**187.0**	**ϕ34 →30 (22.2%)**
Bi-S (A)	**175.3**		**167.1**	**ϕ34 →31 (16.9%)**
Bi-S (B)	**160.9**		**146.8**	**ϕ34 →31 (16.9%)**

Tool flank wear

Fig. 2 shows the flank wear width (VB) of the sintered tungsten carbide tool (P20) used in the turning operation with BiCD and PbCD. The wear resistance of the tool flank to machining BiCD was stronger than that when machining PbCD. The factors that affect wear resistance are the hardness of the matrix structure, the volume fraction of the inclusions and pearlite, and the morphology of the inclusions. There were few differences in the volume fraction of the microstructure and the morphology of MnS between BiCD and PbCD. Therefore, it would appear that the wear resistance of the tool flank was mainly dependent on the hardness of the work-piece, accordingly, when the two types of work-piece had the same hardness there was little difference in the wear resistance of the tool flank.

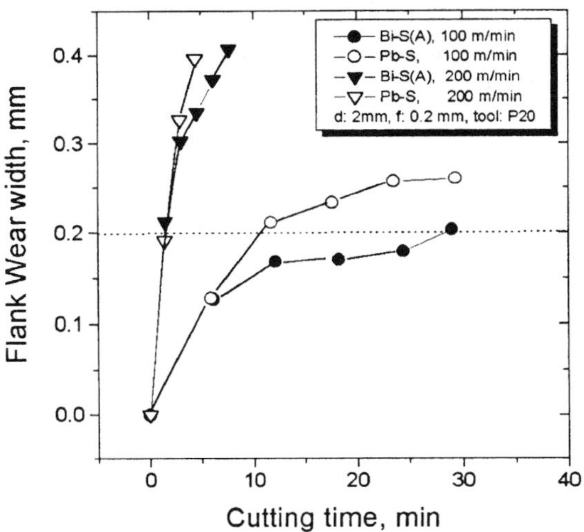

Fig. 2 Flank wear width relative to cutting time in turning operation

Specific shear energy on shear plane

Fig. 3 shows the shear force (Fs) of the two types of work-piece relative to the feed rates in the turning operation, where the shear force of the work-pieces increased relative to the feed rate. The shear force of the BiCD work-piece increased more steeply than that of the PbCD work-piece.

Fig. 4 illustrates the shear strain (γ) of the two work pieces with feed rates in turning operation. The shear strain of the BiCD is larger than that of the PbCD. That is due to the softer matrix of the BiCD. The larger shear strain means the smaller shear angle (ϕ) and the thicker chip in the constant feed rate. This means higher shear energy needs to cut the constant volume of the work piece.

Therefore, it is necessary to harden the matrix for reducing the shear strain and the shear energy. There are many methods to harden the matrix structure such as the adding interstitial elements and the increasing cold reduction ratio.

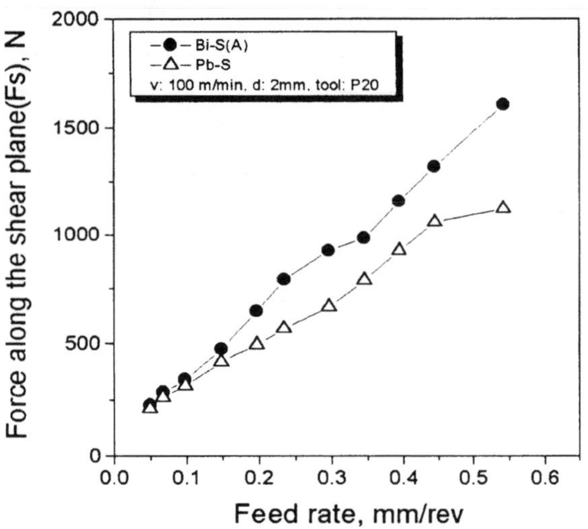

Fig. 3 Shear force (Fs) with feed rate in turning operation

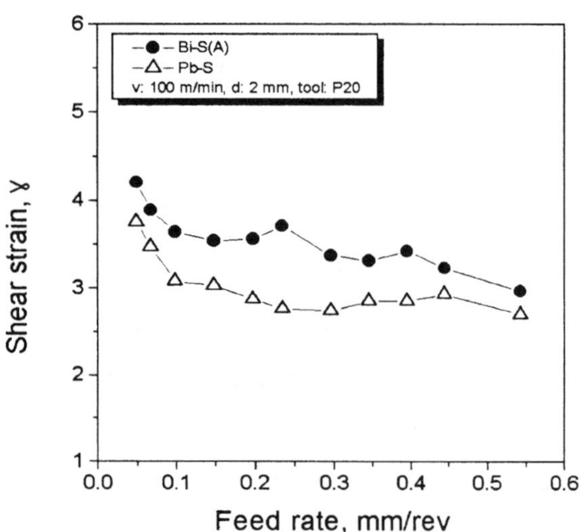

Fig. 4 Shear strain (γ) relative to feed rate in turning operation

Fig. 5 shows the shear stress (τ) of the two types of work-piece relative to the feed rate in the turning operation, which indicated little difference in the shear stress between the work pieces. It is still not exactly clear why there was hardly any difference in the shear stress between the two types of work-piece, despite the BiCD work-piece being softer than the PbCD work-piece. One possibility is that this could be attributed to the lower melting temperature and effective liquid metal embrittlement of bismuth.

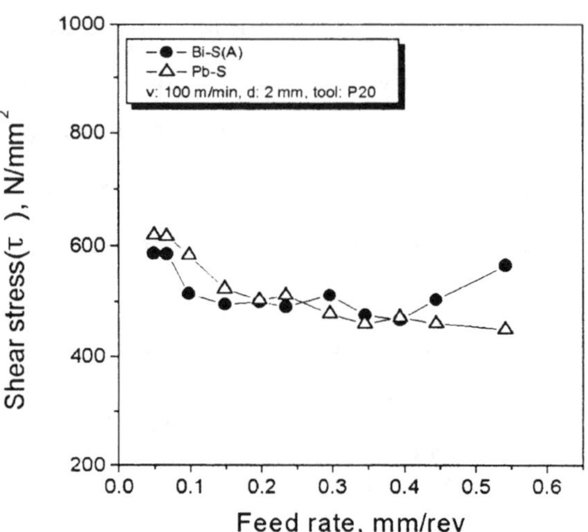

Fig. 5 Shear stress (τ) relative to feed rate in turning operation

Fig. 6 shows the specific shear energy (u_s) of the two types of work-piece relative to the feed rate in the turning operation. The specific shear energy of the BiCD work-piece was larger than that of the PbCD work-piece. The specific shear energy(u_s) is expressed as the product of the shear stress (τ) and the shear strain (γ), therefore, the larger shear energy of the BiCD work-piece was mainly due to its larger shear strain.

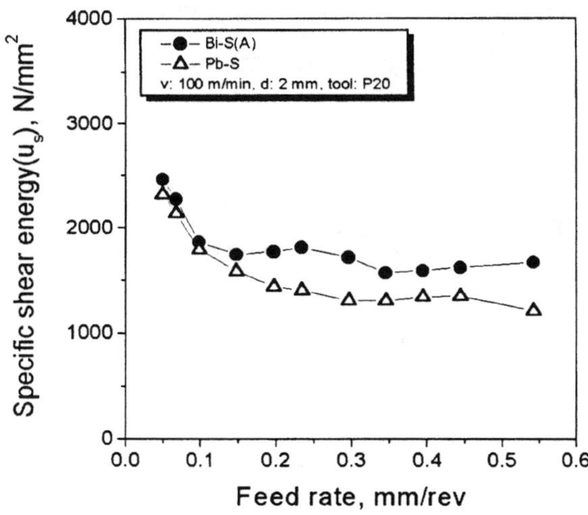

Fig. 6 Specific shear energy (u_s) relative to feed rate in turning operation

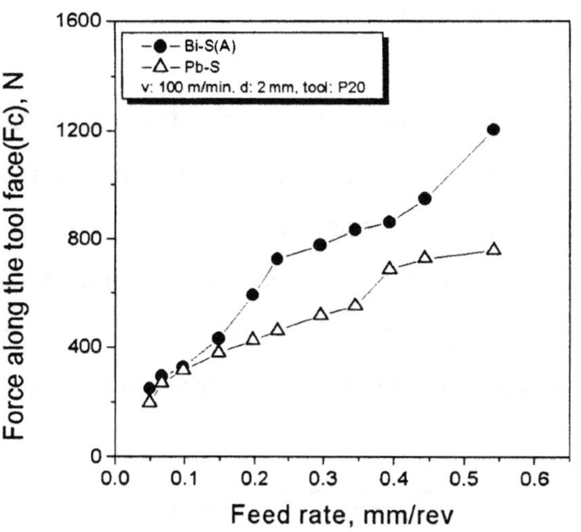

Fig. 7 Friction force (Fc) relative to feed rate in turning operation

Specific friction energy on tool face

Fig. 7 shows the friction force (Fc) along the tool face. The friction force increased with the feed rate. The friction force of the BiCD work-piece increased more steeply than that of the PbCD work-piece. The difference in the friction force between the two types of work-piece increased with an increased feed rate. It would appear that the increased feed rate generated sufficient deformation heat to melt the bismuth or lead inclusion in the work-piece. The friction force on the tool face was then decreased by the molten bismuth or lead. If it is assumed that there was little difference in the lubricant role between the two types of work-piece, the 0.3 wt% molten lead in the PbCD work-piece was evidently much more effective in reducing the friction force than the 0.1 wt% molten bismuth in the BiCD work-piece.

Fig. 8 shows the friction coefficient ($\mu = \tan\beta$) relative to the feed rate, and indicates that the friction coefficient decreased relative to the feed rate. The friction coefficient of the PbCD work-piece was smaller than that of the BiCD work-piece because of the larger shear angle (ϕ) in the latter caused by its smaller shear strain (γ) and thinner chip.

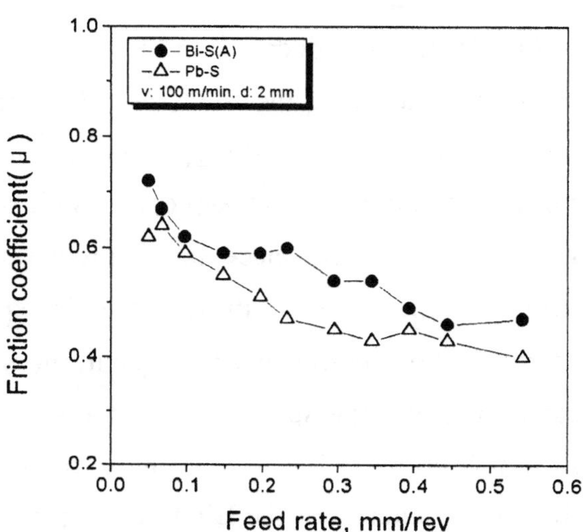

Fig. 8 Friction coefficient (μ) relative to feed rate in turning operation

Fig. 9 shows the specific friction energy (u_f) relative to the feed rate. There was basically no difference in the specific friction energy between the two types of work-piece. The data was scattered across a wide feed rate because of the inhomogeneous properties of the work pieces.

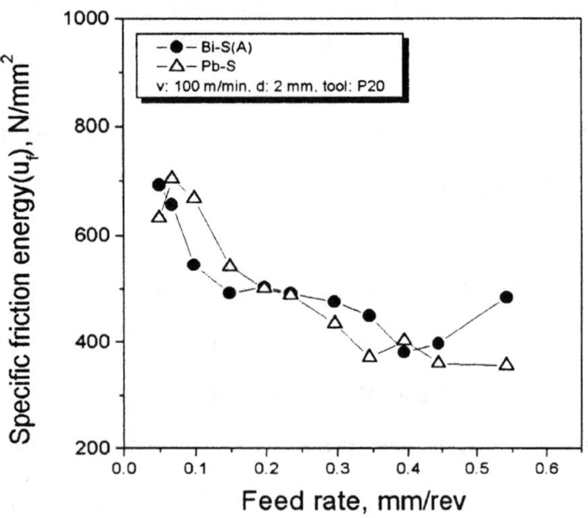

Fig. 9 Specific friction energy (u_f) relative to feed rate in turning operation

Specific cutting energy

Fig. 10 shows a decrease in the specific cutting energy (u) relative to the feed rate. This result is due to the size effect in cutting[9]. The specific cutt ing energy of the BiCD work-piece was higher than that of the PbCD work-piece above a feed rate of 0. 15 mm/rev. The specific cutting energy is expressed in terms of the specific shear energy in the primary shear region and the specific friction energy in the chip-tool friction region. Accordingly, the larger sp ecific cutting energy of the BiCD work-piece was d ue to its larger specific shear energy, which in turn was due to its larger shear strain and softer matrix. Therefore, the specific cutting energy of BiCD will decrease by hardening its matrix.

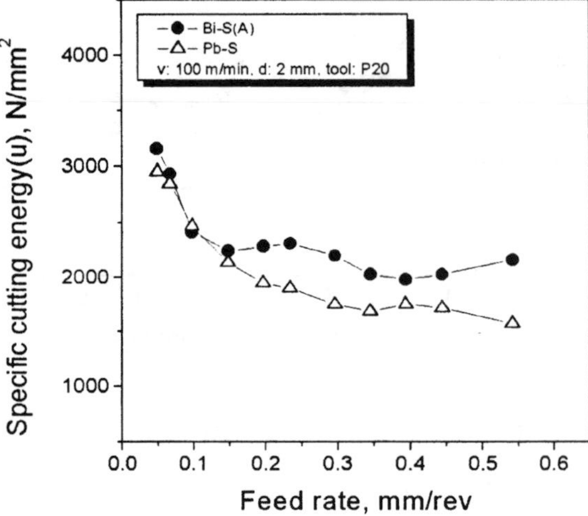

Fig. 10 Specific cutting energy (u) relative to feed r ate in turning operation

Chip-breaking cycle time

Fig. 11 shows the chip-breaking cycle time (T_B) relative to the feed rate at cutting speeds of 100 m/min and 200 m/min, respectively. With the slower feed rate, the chip breaking cycle time decreased drastically and an under-controlled chip was formed. In contrast, with the faster feed rate, the chip-breaking cycle time decreased smoothly and a properly-controlled chip was formed. Then the broken-chip type was transformed from a properly controlled chip type to an over-controlled chip type with an increased feed rate, however, the chip-breaking cycle time does not adequately explain the transformation. And as seen in the figure, the chip breaking cycle time with the higher cutting speed (200m/min) is always shorter than that with the lower cutting speed (100m/min). This indicates that the length of broken chip, at any given feed rate, hardly depends on the cutting speed. However there was little difference in the chip-breaking cycle time between the two types of work-piece with a feed rate above 0.24 mm/rev regardless of the cutting speed.

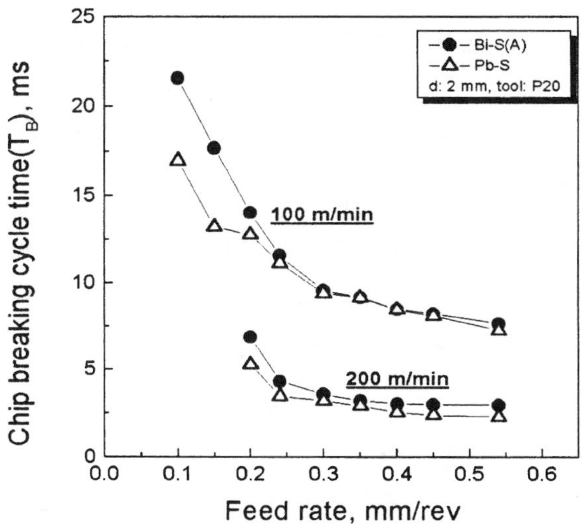

Fig. 11 Chip breaking cycle time (T_B) relative to feed rate in turning operation

Machined surface roughness

Fig. 12 shows the center-line mean roughness (Ra) relative to the feed rate at a cutting speed of (a) 50 m/min, (b) 100 m/min, and (c) 200 m/min in the turning operation. The minimum value of the center-line mean roughness appeared at a feed rate of 0.15 mm/rev regardless of the cutting speed or the addition of lead or bismuth. The Ra increased with a feed rate above 0.15 mm/rev, which matched well with the theoretical surface roughness [10] expressed in terms of the nose radius of the tool and the feed rate. The Ra slightly increased with a feed rate below 0.15 mm/rev, which was due to the minimum undeformed chip thickness [11].

At a cutting speed of 50 m/min, the Ra of the BiCD work-piece was larger than that of the PbCD work-piece. At a cutting speed of 100 to 200 m/min, there was little difference in the Ra between the two types of work-piece. It is common that the Ra of a softer work-piece will be larger than that of a harder work-piece at a lower cutting speed because of the tearing of the work-piece. With an increased cutting speed, the tearing phenomena of the work piece disappeared and the Ra approached that of the theoretical surface roughness, regardless of the addition of lead or bismuth.

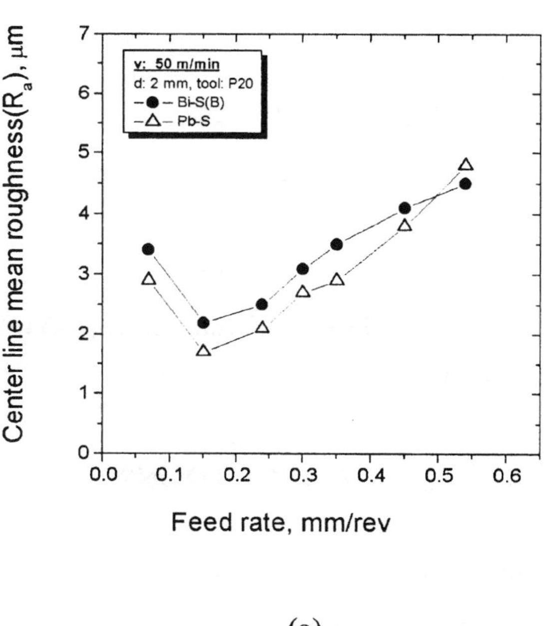

(a)

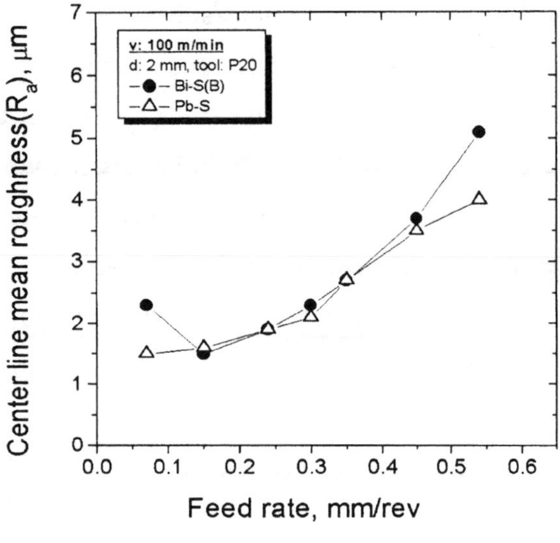

(b)

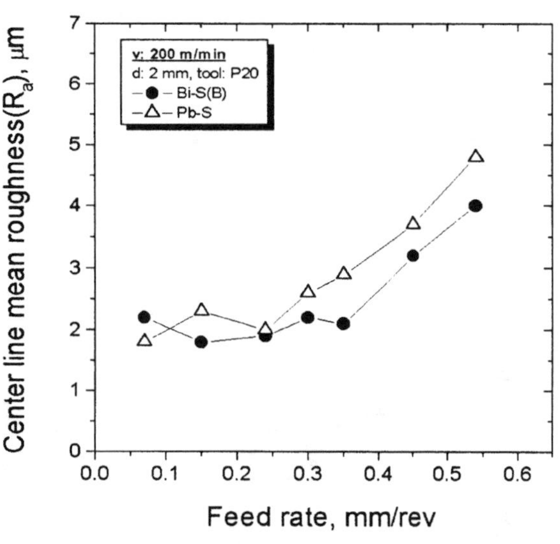

(c)

Fig. 12 Center-line mean roughness (Ra) relative to feed in turning operation

CONCLUSIONS

Flank wear width: The flank wear resistance to machining BiCD is stronger than that when machining PbCD. Accordingly, it would appear that the wear resistance to the work-piece mainly depends on the hardness of the microstructure. Therefore, if the hardness between two types of work-piece is the same there will be little difference in the wear resistance.

Specific cutting energy: The larger specific cutting energy of BiCD is due to its larger specific shear energy caused by its larger shear strain and softer matrix. Therefore, it would appear that the specific cutting energy of BiCD would be reduced if the matrix structure is hardened.

Chip breakability: The breakability of a BiCD chip is poorer than that of a PbCD chip in terms of the chip breaking index. However, there is little difference in the chip-breaking cycle time between the two types of work-piece above a feed rate of 0.24 mm/rev regardless of the cutting speed.

Machined surface roughness: At a cutting speed of 50 m/min, the Ra of BiCD is larger than that of PbCD. However, at cutting speeds from 100 to 200 m/min, there is little difference in the Ra between the two types of work-piece.

REFERENCES

[1] D. Bhattacharya : Mechanical Working and Steel Processing Proceedings, 18(1980) 153

[2] D. Bhattacharya : Iron and Steel Maker, vol.3, No.8(1981) 40

[3] D. Bhattacharya : The bulletin of the Bismuth Institute, 46(1985)

[4] Tool life testing with single-point turning tools, ISO, 5th Draft proposal, ISO/TC29/WG22 (Secretariat37) 91, March, (1972)

[5] Y.M. Lee, T.S. Song, T.J. Park, E.S. Jang : Journal of Korean Society of Precision Engineering, Vol.16, No.6(1999) 190

[6] K. Nakayama : Transaction of JSME, Vol.5/34, NO.7(1962) 142

[7] Y.M. Lee, W.S. Choi, S.L. Jang and I.H. Heo: Journal of Korean Society of Mechanical Engineering, Vol.A22, No.11(1998) 1989

[8] Korean Organization for Standardization : KS Handbook, Mechanics, KS B 0161(1988) 7

[9] G. Boothroyd : "Fundamentals of Metal Machining and Machine Tools", Scripta Book Company, Washington D.C., (1975) 70

[10] G. Boothroyd : "Fundamentals of Metal Machining and Machine Tools", Scripta Book Company, Washington D.C., (1975) 134

[11] M.C. Shaw: "Metal Cutting Principles", Clarendon Press, Oxford, (1984) 519

BAR PRODUCTS AND FORGINGS II

AN ANALYSIS OF TENSILE DEFORMATION BEHAVIOR AND MICROSTRUCTURAL EVOLUTION IN AN ARTIFICIALLY BANDED SAE 5140

Ted F. Majka, David K. Matlock, George Krauss, and Mark T. Lusk
Advanced Steel Processing and Products Research Center
Colorado School of Mines
Golden, Colorado 80401
Telephone: (303) 384-2238
Fax: (303) 273-3016

Key Words: Banding, Steel, Tensile Properties, Hardenability, SAE 5140

ABSTRACT

The effect of band thickness on the mechanical properties and microstructural evolution of an artificially banded SAE 5140 has been investigated. Results show little effect of Mn banding on mechanical properties when specimens are either quenched or very slowly cooled. Major differences in microstructure and mechanical properties develop as a function of band thickness when specimens are cooled at intermediate rates.

INTRODUCTION

Banding is the term used to describe a microstructure that consists of alternating layers of two different microconstituents. The source of banding is the segregation of alloying elements during dendritic solidification.[1-5] Alloying elements such as manganese are rejected into the liquid from solidifying dendrites resulting in low solute cores and high solute interdendritic regions. During subsequent hot rolling these regions are "pancaked" into long strands or bands. Differences in hardenability and A_3 temperature cause high solute bands to transform to different microconstituents than low solute bands and a banded microstructure forms.

Although the primary cause of banding is segregation during solidification, there are factors such as cooling rate and austenite grain size that influence the severity of banded microstructures. It is possible, under certain conditions, to form a uniform microstructure from material with compositional segregation. For example, if the cooling rate is fast enough, both high solute and low solute regions will transform to martensite.[6] Microstructural banding can also be suppressed when austenite grains are large enough to encompass multiple bands.[7]

Banded microstructures are readily observable during routine metallographic examination. As a consequence, questions arise about the effect of banding on the mechanical performance of steel. Conflicting results from several studies also add to the misunderstanding surrounding the subject. For example, several studies have shown that banding has a deleterious effect on distortion, machinability, and response to heat treatment.[8,9] Other studies, however, have shown the effect of banding on the yield and tensile strength of steel to be negligible.[10] Some investigations have even concluded the presence of banding to be beneficial to machining under certain conditions.[11,12] Even though the subject has been investigated for years, the results of these studies indicate there is still much to be learned about banding and its effects on the microstructure and properties of steel.

The purpose of this project is to explore the issue of banding in a systematic manner through the evaluation of artificially banded steel with a range of controlled band thicknesses. The effect of band thickness on tensile deformation behavior is examined along with the development of the banded microstructure at different cooling rates.

EXPERIMENTAL PROCEDURE

This section contains a description of the process that was used to create artificially banded steel for this project. The procedure is based on rolling together sheets of two steels differing only in manganese content. Rolling is performed at high temperature under the action of a vacuum. Surface area created as a consequence of deformation instantaneously bonds producing a solid plate of artificially banded material.

An SAE 5140 steel with 0.82 wt pct manganese and a SAE 5140M (modified) with 1.83 wt pct manganese were used to produce the artificially banded steel. Chemical compositions of the two alloys are shown in Table I. Compositions were chosen based on the results of a previous study which concluded manganese to be the major segregating element in banded structures.[13] Heats of the two alloys were vacuum induction melted, cast into 50 kg ingots and rolled into plates 500 mm x 165 mm x 16 mm. These plates were homogenized in a vacuum furnace for 12 hours at 1300° C to eliminate segregation produced during casting. They were then ground to a thickness of 12 mm to remove scale, decarburization, and surface irregularities produced during rolling and homogenizing. The ground 12 mm plates were then cut into 150 mm x 50 mm sections. These smaller sections were heated to 1150° C and hot rolled with a laboratory rolling mill into 2 mm thick sheets. The sheets were cleaned with 37% HCl solution at 70° C to remove scale and decarburization from hot rolling. The cleaned sheets were then cold rolled with a laboratory rolling mill to a thickness of 0.50 mm. The cold rolled sheets were sheared into 125 mm x 90 mm coupons and cleaned with ethanol and a commercial abrasive cleaning pad to remove oil and oxidation.

Table I: Steel chemistries (in wt%) of SAE 5140 and SAE 5140M used in this study.

Grade	C	Mn	P	S	Si	Cu	Ni	Cr	Mo	Al
5140	0.39	0.82	0.014	0.022	0.22	0.16	0.17	0.81	0.04	0.030
5140M	0.41	1.83	0.015	0.024	0.22	0.16	0.17	0.81	0.04	0.029

The cleaned coupons of the two alloys were alternately stacked into 200 sheet packets as shown in Figure 1a. These stacks were then TIG welded inside of boxes made of 18 gauge 304 stainless steel sheet as displayed in Figure 1b. A 750 mm long tube made of 18 gauge 304 stainless steel with a 6.35 mm outer diameter was also welded to the boxes. The purpose of the tube was to connect the box to an active vacuum system to prevent oxidation of the sheets. A vacuum of 23 mm Hg was maintained during furnace heating as well as the first few passes through the rolling mill. This was essential to ensure proper bonding of sheets during rolling. The boxes were placed in a furnace at 1150° C for 1 hour and hot-rolled to produce plates with controlled band thicknesses of 320, 160, 80, 40, and 20 μm. The starting thickness of each stainless steel box was approximately 50 mm so different reductions were required to produce the range of band thicknesses.

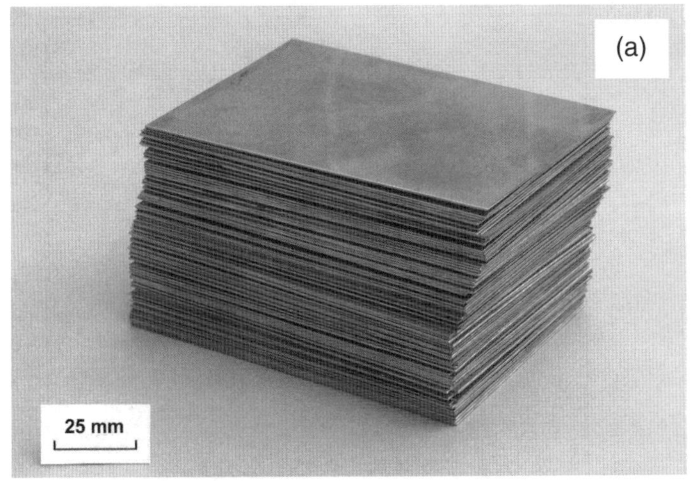

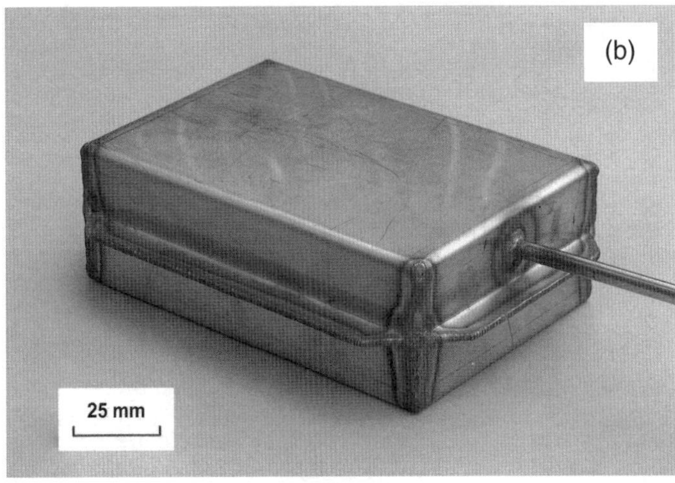

Fig. 1: Sheets of 5140 and 5140M steel were alternately stacked into 200 sheet packest as shown in (a). Packets were then sealed inside of 304 stainless steel boxes in preparation for hot rolling as displayed in (b).

Plates of artificially banded steel were cut into 80 mm x 12 mm strips with the 80 mm length transverse to the rolling direction. The strips were then ground into flat tensile specimens with a 31.75 mm reduced section and a 6.350 x 3.175 mm gage cross section. Tensile specimens were austenitized at 850° C for 20 minutes and cooled at six different rates (83° C/s, 5.1° C/s, 2.6° C/s, 0.6° C/s, 1° C/min, and 0.5° C/min) to produce a range of microstructures. The average cooling rates between 704° C and 538° C were used to calculate the specified rates. Specimens were tested in the untempered state.

Tensile samples were tested in accordance with ASTM E8 at room temperature and at a crosshead speed of 5.08 mm/min. Two valid tests were performed for each condition and results were put into a single-factor analysis of variance (ANOVA) with $\alpha=0.05$ to detect any statistically significant trends. Standard metallographic techniques were used for microstructural characterization of the tensile samples. Grip sections of the tensile samples were removed and mounted so that a longitudinal section was viewed in the transverse direction. All polished specimens were etched with 2% nital.

RESULTS

Tensile Tests

Representative engineering stress-strain tensile curves for the artificially banded 5140 steel cooled at 83° C/s, 5.1° C/s, 0.6° C/s, and 1.0° C/min, respectively, are shown in Figures 2 to 5. Each figure includes complete stress-strain curves in (a) and the initial yield region plotted on an expanded scale in (b). Data for the samples cooled at 2.6° C/s and 0.5° C/s are similar to the 0.6° C/s and 1.0° C/min samples, respectively, and are not considered further here. A complete analysis of the samples is presented elsewhere.[14] Average tensile properties derived from Figures 2 to 5 along with ANOVA p-values are summarized in Tables II-V.

At the highest cooling rate of 83° C/s, Figure 2 shows that all samples exhibited similar stress-strain curves with limited ductility and the data in Table II indicates that there was no statistically significant dependence of any property on layer thickness. With a decrease in cooling rate, Figures 3-5 show that the tensile properties depended on layer thickness. Figure 3 for the 5.1° C/s cooling rate shows that band thickness had a slight effect on tensile properties. Samples with 320 μm bands had an average UTS value of 1506 MPa while samples with 20 μm bands had an average value of 1350 MPa. Yield strength (YS), YS/UTS ratio, uniform elongation, and total elongation were not significantly affected as indicated by the high p-values for those properties listed in Table III.

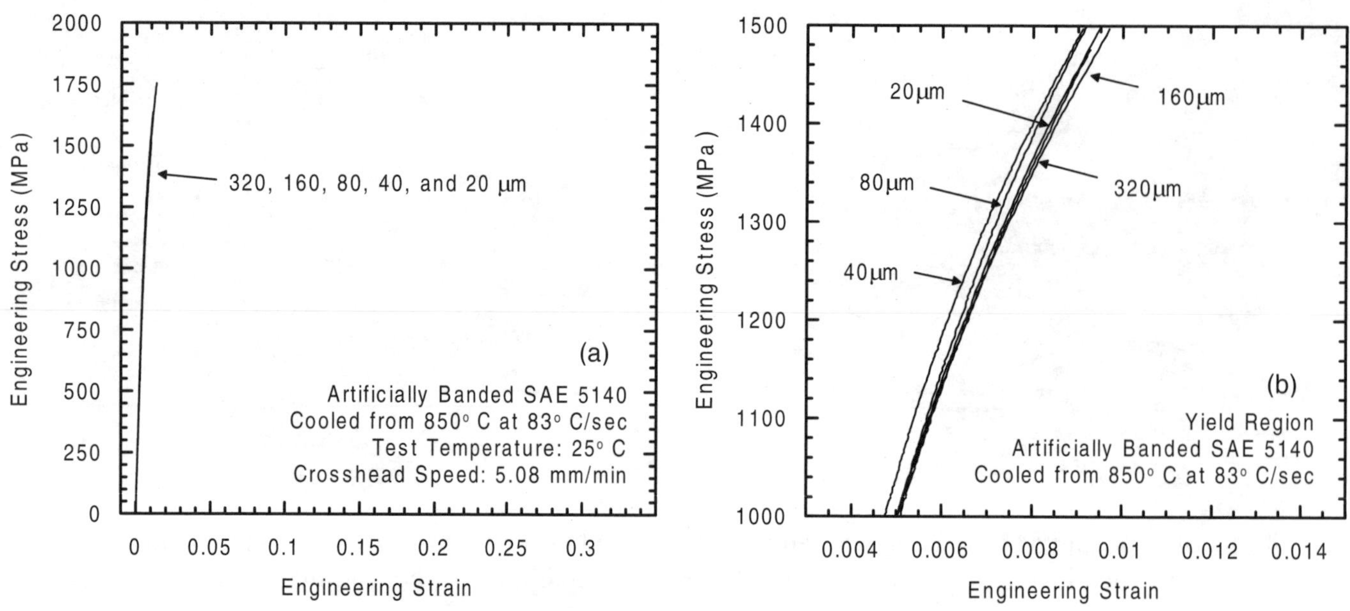

Fig. 2: Representative engineering stress-strain curves for samples cooled at 83° C/s are shown in (a). Yield region is shown on expanded scale in (b).

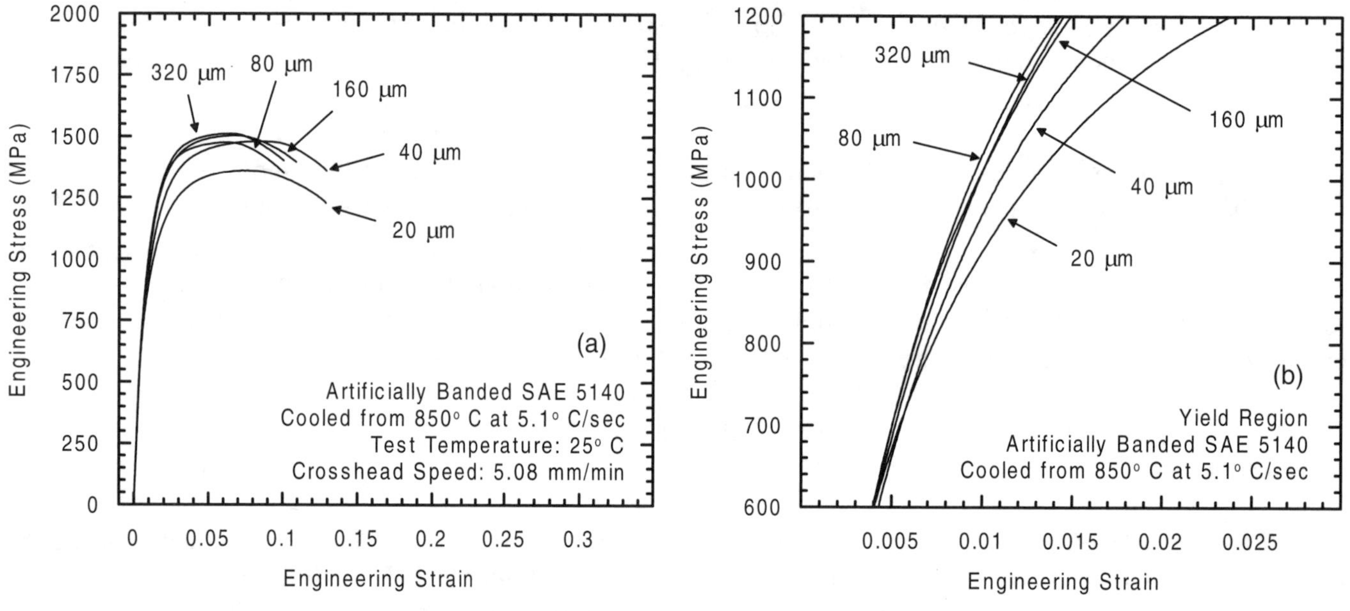

Fig. 3: Representative engineering stress-strain curves for samples cooled at 5.1° C/s are shown in (a). Yield region is shown on expanded scale in (b).

For samples cooled at 0.6° C/s, Figure 4 shows that band thickness had a pronounced effect on tensile properties. As shown in Table IV, average values of UTS dropped from 1025 MPa for samples with 320 μm thick bands to 756 MPa for those with 20 μm bands. YS/UTS ratio increased from 0.48 to 0.62 as band thickness decreased from 320 to 20 μm. Both uniform and total elongation also increased as band thickness decreased. Uniform elongation increased from 9.2% to 12.2% while total elongation increased from 14.3% to 25.7%. Results of the ANOVA indicated that 0.2% yield strength values were not significantly affected by band thickness. Yield behavior, however, shifted from continuous for samples with 320 μm to discontinuous for samples with 20 μm bands.

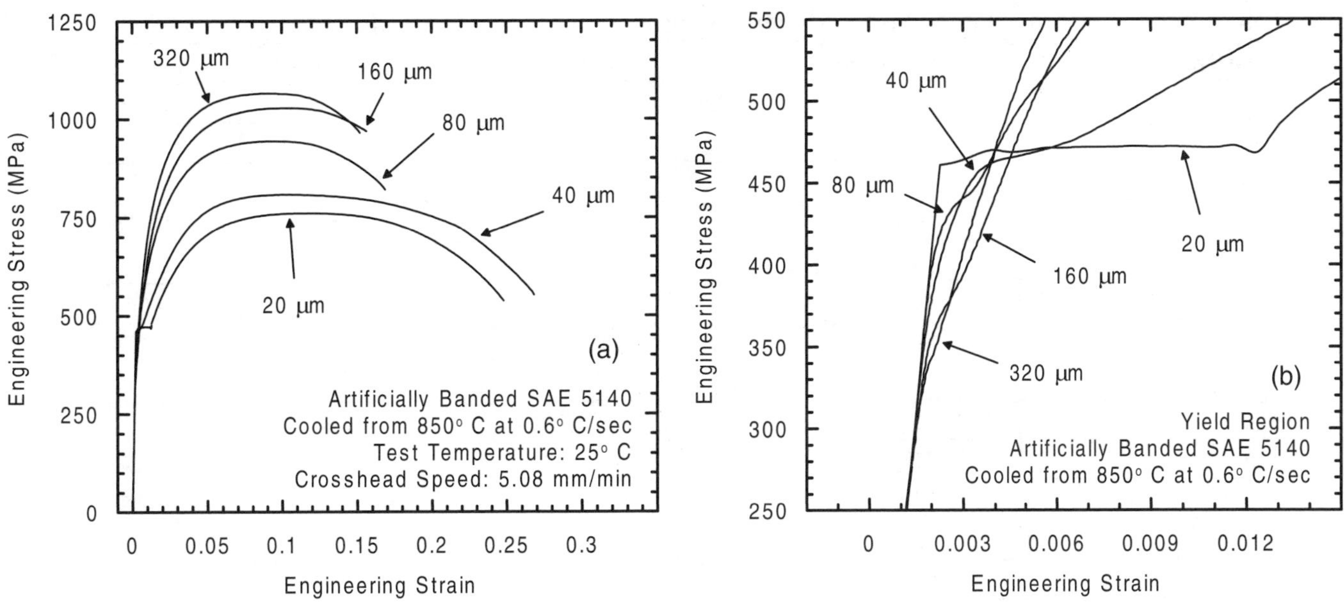

Fig. 4: Representative engineering stress-strain curves for samples cooled at 0.6° C/s are shown in (a). Yield region is shown on expanded scale in (b).

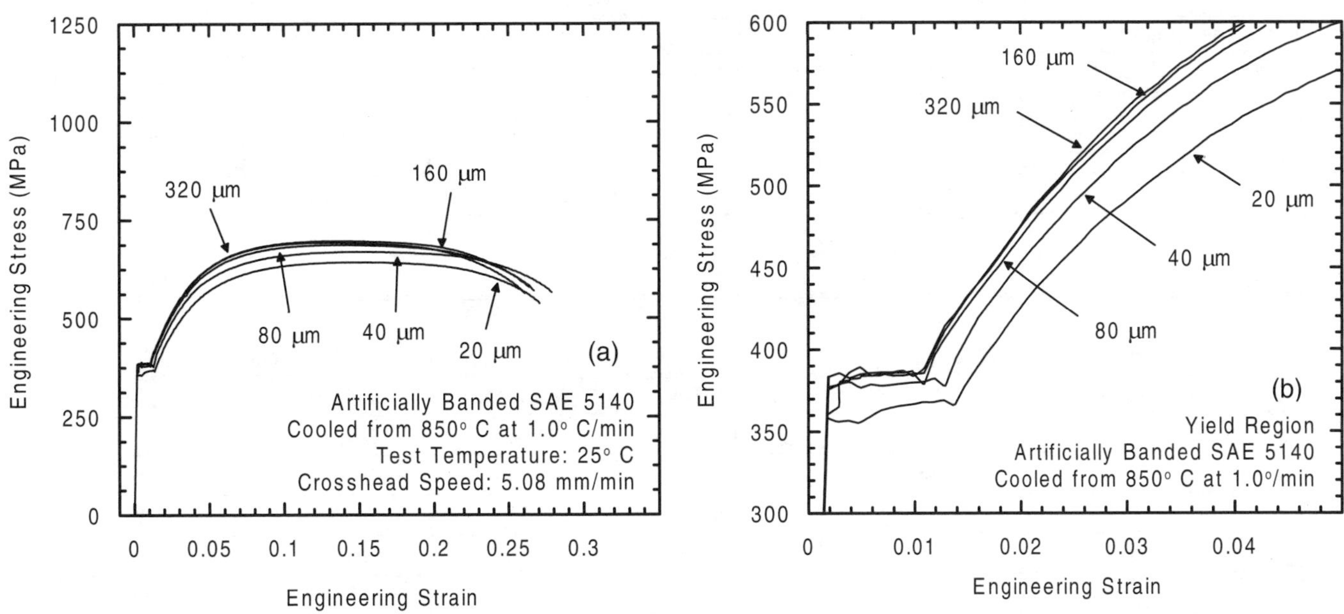

Fig. 5: Representative engineering stress-strain curves for samples cooled at 1.0° C/min are shown in (a). Yield region is shown on expanded scale in (b).

With a further decrease in cooling rate to 1.0° C/min, Figure 5 shows that all samples exhibited similar deformation behavior. Band thickness had a statistically significant effect on UTS as indicated by the low p-value. The magnitude of the difference, however, was much less than that displayed for samples cooled at 0.6° C/sec. Average values decreased by only 28 MPa. YS/UTS ratio, uniform elongation, and total elongation were not significantly affected by band thickness.

Table II – Average mechanical property values for artificially banded SAE 5140 cooled at 83° C/s.

Band Thickness (μm)	0.2% YS (MPa)	UTS (MPa)	YS/UTS	Uniform El	Total El
320	1420	1555	0.917	1.1%	1.1%
160	1420	1535	0.926	1.1%	1.1%
80	1480	1560	0.949	1.1%	1.1%
40	1490	1780	0.845	1.5%	1.5%
20	1420	1645	0.864	1.3%	1.3%
p-value (α=0.05)	0.60	0.38	0.69	0.64	0.64

Table III – Average mechanical property values for artificially banded SAE 5140 cooled at 5.1° C/s.

Band Thickness (μm)	0.2% YS (MPa)	UTS (MPa)	YS/UTS	Uniform El	Total El
320	723	1506	0.481	6.7%	10.6%
160	749	1515	0.495	6.1%	10.4%
80	723	1469	0.492	6.6%	10.7%
40	689	1474	0.463	7.5%	11.8%
20	713	1350	0.529	7.3%	12.2%
p-value (α=0.05)	0.25	0.000012	0.14	0.14	0.09

Table IV – Average mechanical property values for artificially banded SAE 5140 cooled at 0.6° C/s.

Band Thickness (μm)	0.2% YS (MPa)	UTS (MPa)	YS/UTS	Uniform El	Total El
320	493	1025	0.481	9.2%	14.3%
160	493	1012	0.487	9.8%	15.7%
80	484	929	0.521	9.9%	16.9%
40	466	816	0.571	11.2%	24.6%
20	467	756	0.618	12.2%	25.7%
p-value (α=0.05)	0.098	$4.2x10^{-7}$	$2.7x10^{-5}$	$1.2x10^{-3}$	$7.7x10^{-8}$

Table V – Average mechanical property values for artificially banded SAE 5140 cooled at 1.0° C/min.

Band Thickness (μm)	0.2% YS (MPa)	UTS (MPa)	YS/UTS	Uniform El	Total El
320	381	699	0.546	13.8%	27.0%
160	387	695	0.555	14.4%	26.9%
80	385	684	0.563	14.3%	26.9%
40	377	665	0.565	14.6%	27.6%
20	356	646	0.551	15.2%	28.1%
p-value (α=0.05)	$6.0x10^{-4}$	$5.1x10^{-7}$	0.12	0.25	0.30

Microstructural Evolution

Microstructures for samples cooled at 83° C/s are shown in Figure 6. The microstructure of material with 320 μm thick bands is displayed in Figure 6(a). At this high cooling rate both the 5140 and 5140M bands transformed to martensite. A slight difference in shading is the only distinguishable feature between the two alloys. The microstructure from samples with 80 and 20 μm bands are shown in Figures 6(b) and 6(c). As in the 320 μm band microstructure, a difference in etching response is the only distinguishable feature separating the two alloys. The interface between bands in the 320 μm microstructure is shown in Figure 6(d).

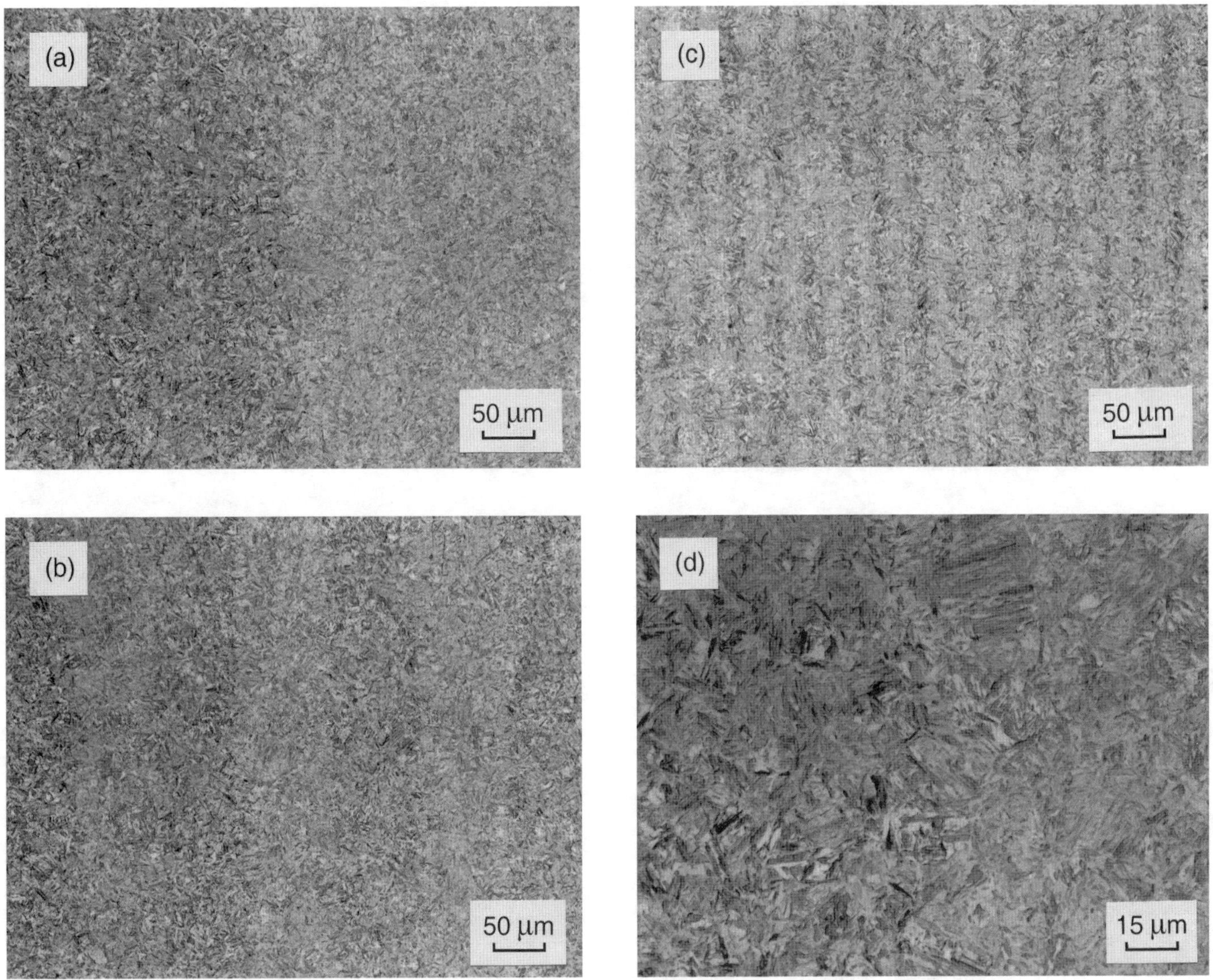

Fig. 6: Light optical micrographs for specimens cooled at 83° C/s. Material with 320, 80, and 20 μm bands are shown in (a), (b), and (c), respectively. Interface of 320 μm material is displayed at higher magnification in (d).

Microstructures for samples cooled at 5.1° C/s are shown in Figure 7. The microstructure of material with 160 μm bands is shown in Figure 7(a). The high manganese 5140M, the light band in the micrograph, transformed to martensite with a small amount of bainite. The low manganese 5140, the dark band in the micrograph, transformed to predominantly bainite with a small amount of martensite. The microstructure of the 20 μm band sample is shown in Figure 7(b). As in the previous microstructure, the 5140M bands are martensite with a small amount of bainite and the 5140 bands are bainite with a small amount of martensite. Figure 7(c) is a micrograph of the interface in between bands in the 160 μm specimen. Bainite laths extend from 5140 to 5140M bands. This effect becomes even more pronounced at finer band thicknesses as shown in Figure 7(d). In this microstructure, bainite laths extend completely across 5140M bands in some locations.

Microstructures for samples cooled at 0.6° C/s are shown in Figure 8. For this cooling rate, micrographs are displayed for each of the five band thicknesses. The microstructure of material with 320 μm bands is displayed in Figure 8(a). Bands of 5140M transformed to a mixture of bainite and martensite while bands of 5140 transformed to ferrite and pearlite. A discrete layer of pearlite also formed at the interface between bands.

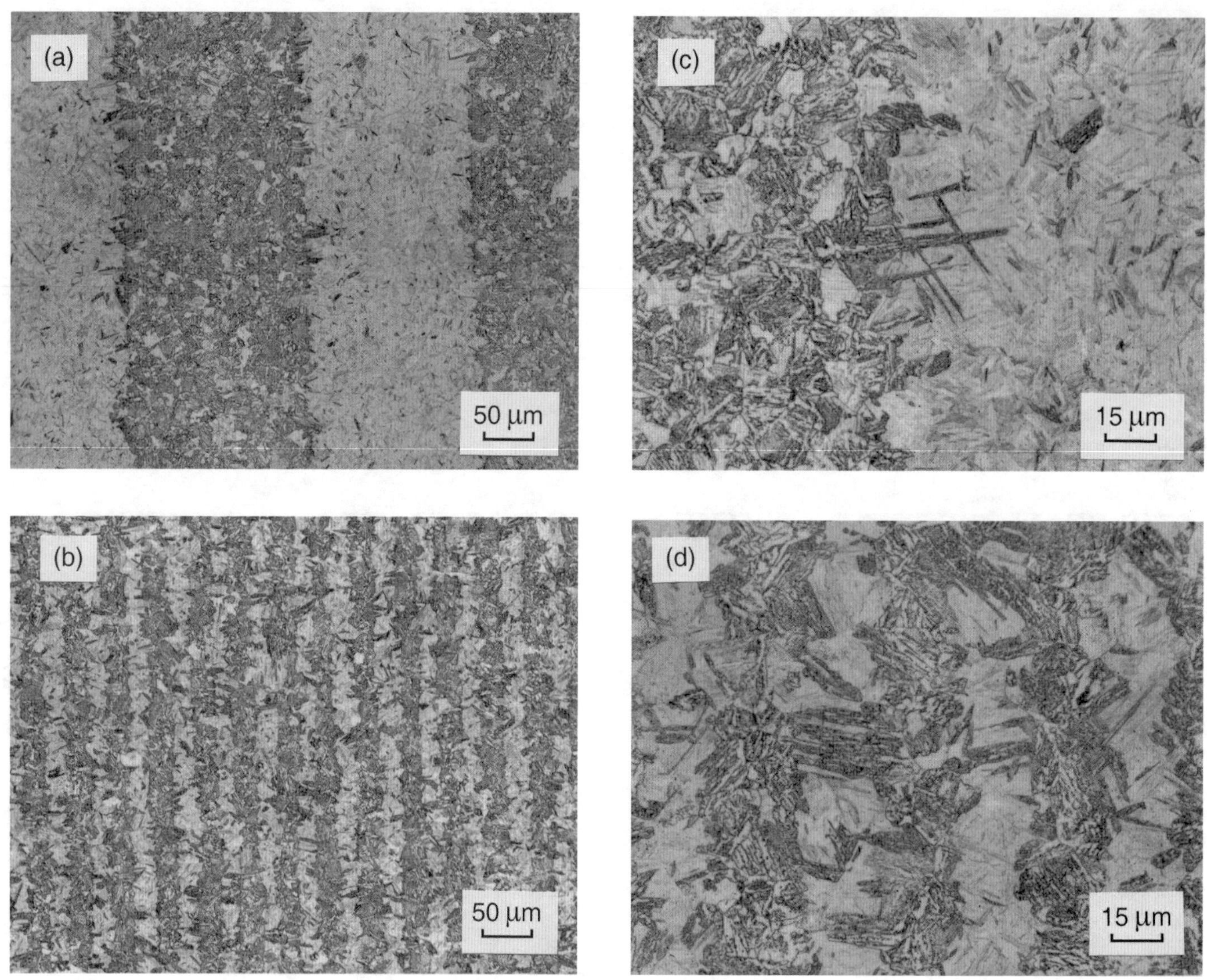

Fig. 7: Light optical micrographs for specimens cooled at 5.1° C/s. Material with 160 and 20 μm bands are shown in (a) and (b), respectively. Interface of 160 μm and 20 μm material is displayed at higher magnification in (c) and (d), respectively.

A similar microstructure developed in the 160 μm sample shown in Figure 8(b). A mixture of bainite and martensite formed in 5140M bands while proeutectoid ferrite and pearlite formed in 5140. An interfacial layer of pearlite is again present. A similar microstructure developed in the 80 μm material in Figure 8(c), except for a more pronounced effect associated with the interfacial band of pearlite. The interfacial layer of pearlite has grown at the expense of the 5140M bands. This effect is even more pronounced in the 40 μm material shown in Figure 8(d). As in the other samples, the 5140 bands transformed to proeutectoid ferrite and pearlite. However, at this band thickness, the interfacial pearlite layer has completely grown through the 5140M bands in most places. Patches of bainite and martensite still exist at the center of the bands but only in a few locations. The microstructure of material with 20 μm bands consists entirely of proeutectoid ferrite and pearlite as shown in Figure 8(e). The interfacial layer of pearlite is shown at a higher magnification in Figure 8(f).

Microstructures for samples cooled at 1.0° C/min are shown in Figure 9. The microstructure of material with 320 μm bands is shown in Figure 9(a). Bands of 5140M consist almost entirely of pearlite while bands of 5140 are proeutectoid ferrite and pearlite. Small patches of proetectoid ferrite are present in the center region of

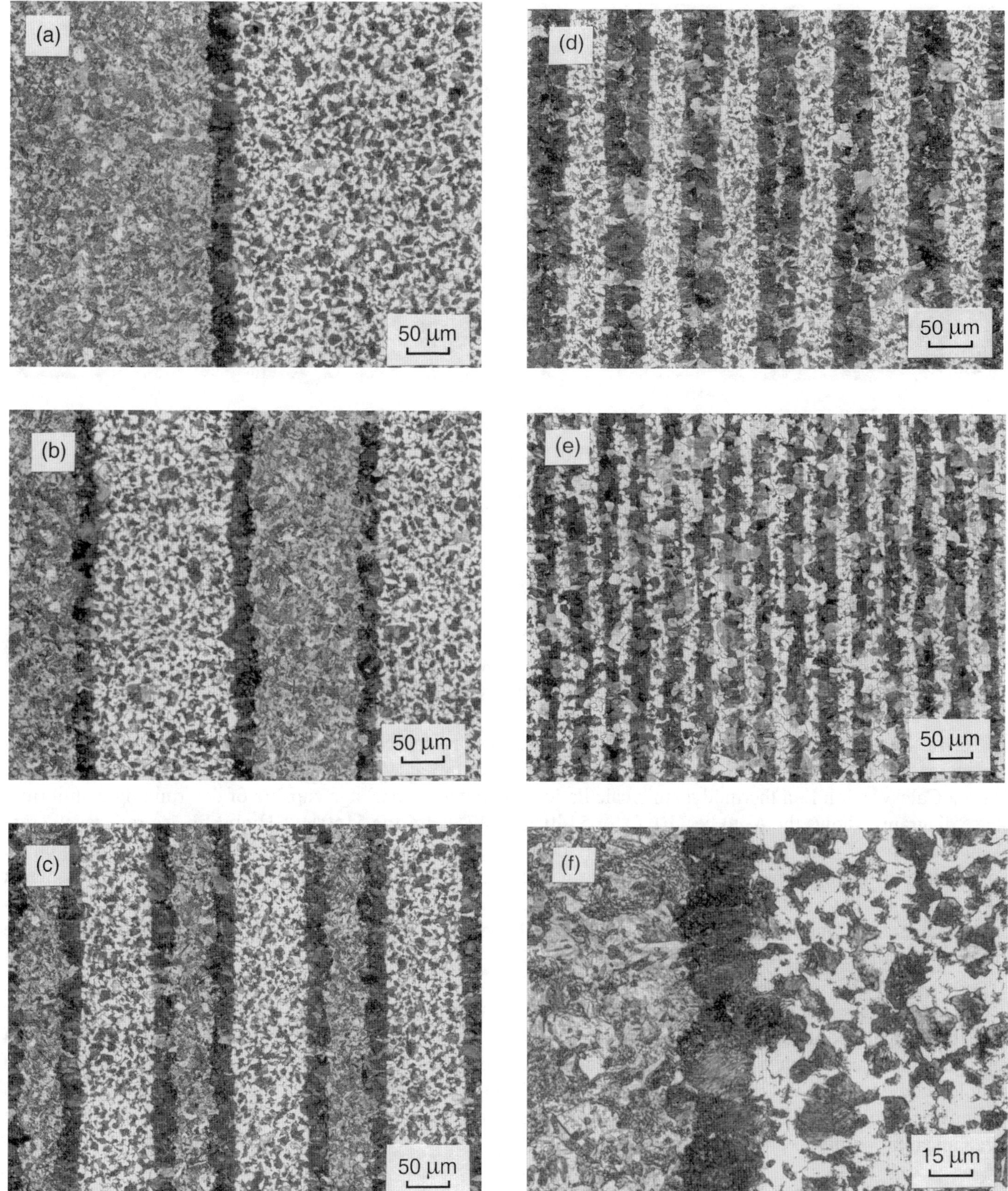

Fig. 8: Microstructures for specimens cooled at 0.6° C/s. Material with 320, 160, 80, 40 and 20 μm bands are shown in (a) to (e), respectively. Interface of 320 μm material is displayed at higher magnification in (f). Light optical micrographs. 2% nital etch.

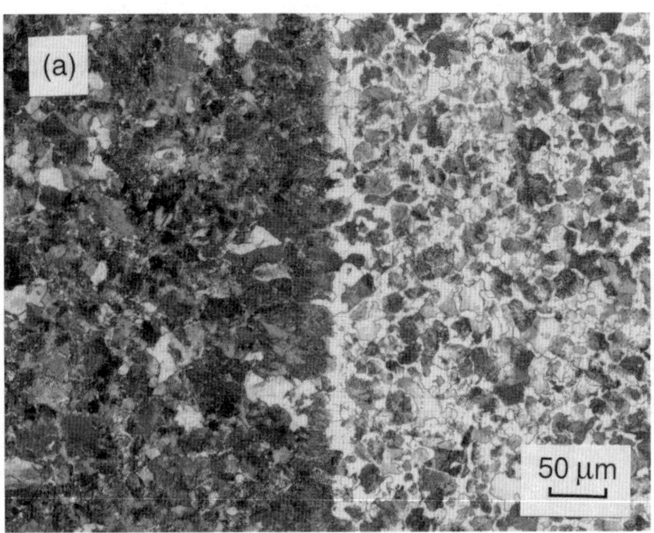

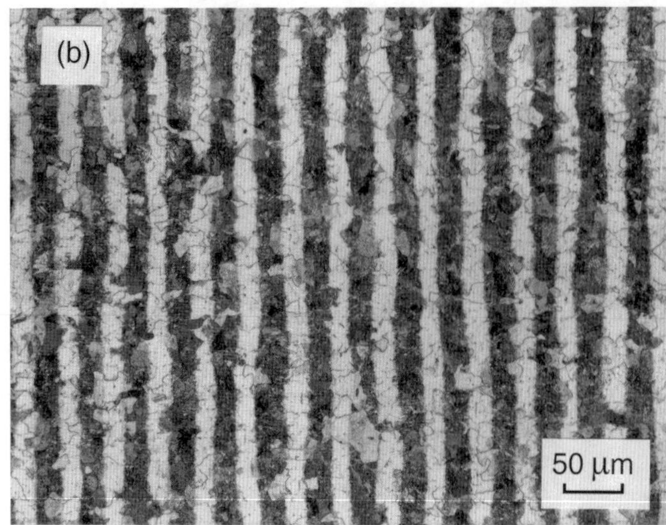

Fig. 9: Microstructures for specimens cooled at 1.0° C/min. 320 μm band material is shown in (a) along with microstructure of 20 μm band material in (b). Light optical micrograph. 2% nital etch.

5140M bands. The microstructure of a sample with 20 μm bands is shown in Figure 9(b). Bands of 5140M consist entirely of pearlite and bands of 5140 are almost entirely ferrite.

DISCUSSION

The above results show little effect of manganese banding on mechanical properties when specimens are either quenched or very slowly cooled. However, major differences in microstructures and mechanical properties develop as a function of band thickness when specimens are cooled at intermediate rates.

In the present work, all banding related differences are caused only by variations in Mn content. High Mn increases austenite stability and lowers the A_3 temperature at which ferrite begins to form. For example, ThermoCalc Version L, a thermodynamic calculation and database package capable of determining equilibrium phase diagrams, shows the A_3 to be 770° C for 5140 and 747° C for the 5140M.[15] High Mn also reduces the rate of austenite transformation to microconstituents depending on diffusion, i.e. high Mn increases hardenability. This is demonstrated by the diameter of bar that will contain 50% martensite at the center after an ideal quench (D_I): 80 mm for the 5140 and 180 mm for the 5140M.[16] Thus, Mn related stability factors, i.e. the effect of Mn on A_3 temperature, and kinetic factors, i.e. the effect of Mn on hardenability or phase transformations determine the extent of microstructural banding.

At high cooling rates, microstructural banding is suppressed altogether. When the cooling rate is faster than the critical cooling velocity required to form martensite in both high solute and low solute regions of a segregated structure then a uniform microstructure forms. This was demonstrated in the artificially banded samples cooled at 83° C/s. Although a 1 wt. pct. difference in manganese existed between bands, a uniform martensitic structure still formed. The interface does not appear to affect the microstructure and hence the mechanical properties as shown in Figure 2 and 6.

Microstructural banding due to hardenability differences appears at cooling rates where martensite forms in one band and diffusion controlled products form in the other. As cooling rates decrease, hardenabilty differences between bands become apparent as shown in Figure 7 for the samples cooled at 5.1° C/s. High manganese bands transformed to predominantly martensite while low manganese bands transformed to predominantly bainite. The effect of the interface between bands becomes apparent as well. Bainite laths extend from low manganese to high manganese bands and as the band thickness decreases these laths occupy a larger fraction of the high manganese bands. This interfacial effect manifests itself as a decrease in UTS with a decrease in band thickness as shown in Figure 3a. A_3 banding does not appear to be a factor at these higher cooling rates.

At slower cooling rates hardenability banding is still evident but the influence of A_3 banding is also apparent. This can be seen in Figure 8 for the samples cooled at 0.6^{o} C/s. Hardenability banding can be seen in the differences between bands: high Mn bands transformed to a mixture of martensite and bainite while low Mn bands transformed to proeutectoid ferrite and pearlite. The effect of A_3 banding can be seen at the interface between bands where a discrete layer of pearlite formed.

As explained by Kirkaldy et al[1], A_3 banding results from differences between segregated bands in the temperature at which ferrite forms upon cooling. Upon cooling from the austenite phase field, bands with high A_3 temperatures begin to form proeutectoid ferrite first. Ferrite nuclei at the interface reject carbon into adjacent low A_3 bands that are still austenite. The austenite in these bands is not bound by lever rule constraints at this temperature and can readily accept carbon. The interfacial band of pearlite is associated with the transfer of carbon to low A_3 bands and becomes more pronounced as cooling rate decreases and as A_3 temperature differences increase. For example, Figure 8 shows that the interfacial bands of pearlite grow into the high Mn bands instead of the low Mn bands. This type of microstructural banding is only observed at slow cooling rates due to its dependence on carbon diffusion and consequently, the A_3 temperature is a good approximation for the differences in continuous cooling behavior.

The interfacial layer of pearlite has a dramatic effect on the mechanical properties of artificially banded steel cooled at 0.6^{o} C/s. UTS, uniform elongation, total elongation and YS/UTS ratio are all significantly affected by changes in band thickness. More and more of the bainite and martensite in the high hardenability bands were replaced with the interfacial layer of pearlite as band thickness decreased. At the slowest cooling rate, the interfacial pearlite band occupies most of the high Mn bands even at the largest band thicknesses. Thus, all of the microstructure consist of the products of diffusion controlled transformations.

All banded materials in this study contained the same volume fraction (0.5) of 5140 and 5140M steel. If the interface did not affect the microstructure in each layer, then the rule of mixtures for composite strengthening predicts that for all cooling rates the tensile behavior would be independent of layer thickness. Deviations from the rule of mixtures prediction are then direct evidence that the interface affects transformation.

CONCLUSIONS

This study has investigated the effect of band thickness on the mechanical properties of an artificially banded SAE 5140 continuously cooled at different rates. It is hoped that the results will provide clarification on the issue of microstructural banding in low alloy steel. The following conclusions encompass mechanical properties and microstructural development as well as the effect of different types of banding on the properties of steel.

1. At cooling rates sufficiently high to form martensite in both alloys, the effect of band thickness on mechanical properties is negligible.

2. At intermediate cooling rates, where microstructural banding due to hardenability differences is the dominant mechanism and mixtures of martensite and bainite form, then band thickness has a slight effect on properties such as UTS.

3. At cooling rates where microstructural banding due to A_3 and hardenability differences coexist then band thickness has a significant effect on properties such as UTS, uniform elongation, total elongation, YS/UTS ratio, and yield behavior. This is the result of ferrite formation and the growth of an interfacial band of pearlite at the expense of the high hardenability band.

4. At slow cooling rates, typical of annealing and normalizing, where banding due to differences in A_3 temperatures between bands is the dominant mechanism then band thickness has only a slight influence on mechanical properties.

ACKNOWLEDGEMENTS

The support of the Advanced Steel Processing and Products Research Center, an Industry/University cooperative research center at the Colorado School of Mines. We extend special thanks to the Timken Company for providing the laboratory heats of experimental steel.

REFERENCES

1. J.S. Kirkaldy, J. von Destinon-Forstmann, and R.J. Brigham, "Simulation of Banding in Steel" Canadian Metallurgical Quarterly, Vol. 1, 1962, pp 59-81.

2. P. G. Bastien, "The Mechanism of Formation of Banded Structures", Journal of the Iron and Steel Institute, December 1957, pp.281-291.

3. J.D. Lavender and F.W. Jones, "An Investigation on Banding", Journal of the Iron and Steel Institute, September 1949, pp. 14-17.

4. C.F. Jatczak, D.J. Girardi, and E.S. Rowland, "On Banding in Steel", Transactions of the ASM, Vol. 48, 1956, pp. 279-305.

5. T.B. Smith, "Microsegregation in Low-Alloy Steels", Iron & Steel, November 1964, pp. 536-541.

6. G. Mukherjee, "Banding in Wrought Low Alloy Steels", Tool and Alloy Steels, Vol. 11, No. 2, February 1977, pp. 55-58.

7. S.W. Thompson and P.R. Howell, "Factors Influencing Ferrite/Pearlite Banding and Origin of Large Pearlite Nodules in a Hypoeutectoid Plate Steel", Materials Science and Technology, Vol. 8, September 1992, pp. 777-784.

8. H.S. Bavister and P.T. Clary, "Unwanted Variation in Steel Composition: Some Effects Upon Heat Treatment and Machining Practices", Journal-West of Scotland Iron and Steel Institute, Vol. 70, 1962-1963, pp. 63-82.

9. N.E. Woldman, "Good and Bad Structures in Machining Steel", Materials and Methods, Vol. 25, No. 2, February 1947, pp. 80-86.

10. R.A. Grange, "Effect of Microstructural Banding in Steel", Metallurgical Transactions, Vol. 2, February 1971, pp. 417-426.

11. J.N. Dutta and G.S. Patki, "Influence of Banded Structure on Machinability with Particular Reference to Case Hardening Steels", The Machinability of Engineering Materials, ASM International, Materials Park, OH, 1983, pp. 413-446

12. C.J. Moody and G. Felbaum, "Heat Treating for Better Machining", Metal Progress, October 1966, pp. 129-132.

13. J. Black, "Modeling the Effects of Chemical Segregation on Phase Transformations in Medium Carbon Bar Steels", M.S. Thesis MT-SRC-098-029, Colorado School of Mines, Golden, CO, December 1998.

14. T. Majka, "An Analysis of Tensile Deformation Behavior and Microstructural Evolution in an Artificially Banded SAE 5140", M.S. Thesis, Colorado School of Mines, Golden, CO, September 2000.

15. B. Sundman, B. Jansson, J.O. Andersson, "The Thermo-Calc Databank System", CALPHAD, 1985, pp. 153-190.

16. G. Krauss, "Hardness and Hardenability", Steels: Heat Treatment and Processing Principles, ASM International, Materials Park, OH, 1990, pp. 163-169.

Modeling Hot Bar Rolling to Predict Surface Quality

E. Buddy Damm, Praveen Pauskar, Jeffery Ives, Krich Sawamiphakdi, and Mark Conneely
The Timken Company
PO BOX 6930
Canton, OH 44706-0930
USA
330-471-2703
Damme@Timken.com

Key Words: surface quality, recrystallization, grain growth, flow stress, high temperature fracture, damage criteria, and modeling

INTRODUCTION

Forging quality bar steel requires excellent surface quality in order to ensure superior as-forged products. Steel producers seek to provide excellent surface quality in the as-rolled condition to eliminate rework or surface conditioning.

During hot bar rolling metallurgical phenomena such as recrystallization and grain growth control the state of the austenite microstructure. Factors that affect the evolution of the austenite microstructure include the thermal conditions, time, steel chemistry, and the deformation conditions. Microstructure in turn affects the metallurgical responses such as flow stress, and the ability to withstand deformation without high temperature fracture.

For this work, microstructure dependent material models for flow stress, recrystallization, grain growth, and fracture were developed and incorporated into a 3-dimensional finite element Hot Bar Mill Model which included a damage criterion to evaluate the potential for cracking.[1,2] Pass designs for The Timken Company's new Harrison Rolling Mill were evaluated prior to commissioning using the model. The initial evaluations of some of the pass designs indicated potential for both high temperature fracture, and for the formation of seams or laps due to overfill. Based on these results, adjustments were made to the pass design to eliminate surface defects.

Two steel grades were evaluated in the laboratory, and multiple pass designs were evaluated for each steel grade using computer simulation. The metallurgical models for each steel grade are described, and pass evaluation results are presented for a selected case.

EXPERIMENTAL PROCEDURES

The chemical compositions of the two steel grades selected for investigation are shown in Table I below. Thermomechanical experiments were conducted using a servo hydraulic thermomechanical process simulator to characterize, and model the grain growth, flow stress, recrystallization, and high temperature fracture behavior of both steels.

Table I Chemical compositions and designations used in this study.

Designation	C	Mn	S	Si	Cr	Ni	Mo	V	Al	N
4120	0.21	0.79	0.013	0.06	0.84	0.12	0.31	0.01	0.046	0.02
1538V	0.38	1.43	0.075	0.54	0.16	0.10	0.03	0.10	0.019	0.02

Thermomechanical process simulation experiments

Prior to sample manufacturing, sample blanks were solutionized at 1250°C for 1 hour and quenched in oil, followed by tempering at 400°C for 2 hours. Fig. 1 below summarizes the experimental procedures schematically. All tests were carried out in the temperature range of 950°C and 1250°C. Uniaxial compression tests on 10mm diameter by 15mm tall samples were used to characterize the flow stress and recrystallization behavior. Fraction softening, as shown schematically in Fig. 2, was used as a measure of recrystallization kinetics. Deformation conditions for the flow stress and recrystallization studies were as follows: axial compression strain, 0.05 to 1.0 mm/mm, strain rate, 0.1 to 10 per second, and initial austenite grain sizes, 30 to 200 micrometers.

Samples for grain growth were heated to selected temperatures, and held for between 0 and 600 seconds, followed by either rapid quenching, or by controlled cooling in order to provide microstructures that aided in the identification of the prior austenite grain boundaries. Recrystallized grain size was characterized by deforming, and then holding at temperature for a period of time, which corresponded to 95% recrystallization for the condition studied, followed again by either quenching, or controlled cooling to reveal the prior austenite grain size.

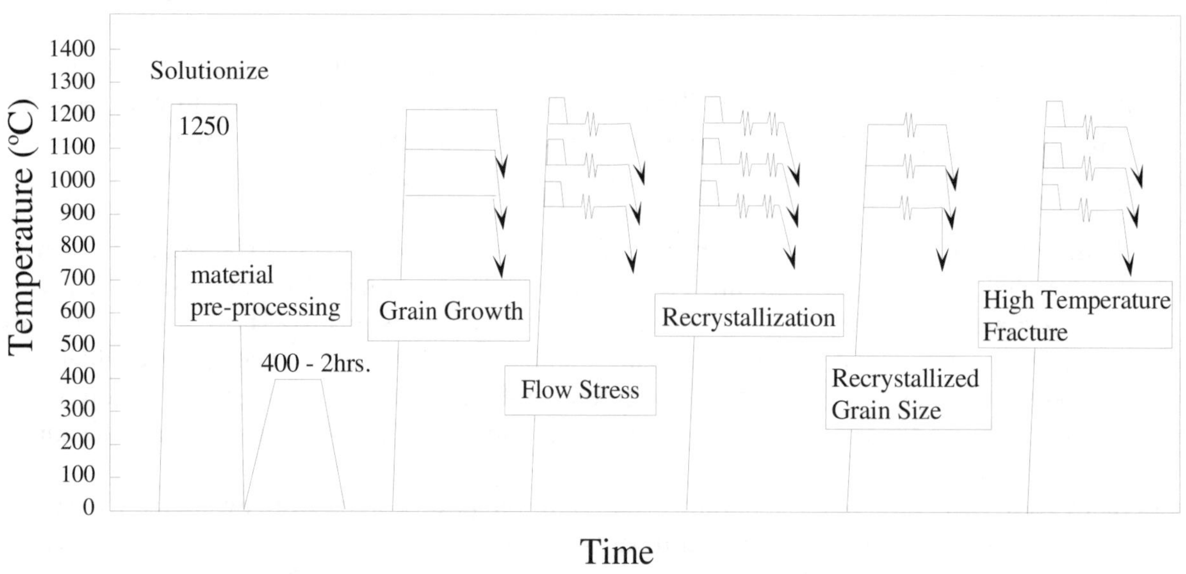

Fig. 1. Schematic representation of experimental procedures.

High temperature fracture was characterized using a constrained rod sample 10mm in diameter by 83.5mm long as shown in Fig. 3. For these tests the sample was held using copper grips, and heated using direct electrical resistance heating. A control thermocouple was spot welded to the sample at the mid-point location. With this configuration, the resulting thermal gradient coupled with the constrained sample ends in the grip provided a high degree of non-uniform deformation during compression, which generated secondary tensile strains at the sample surface.

The hoop strains, and compressive strains were measured using a diameteral extensometer, and a high temperature optical system. For each condition studied, the strain at which fracture was first detected was recorded. Deformation conditions for the high temperature fracture study were as follows: hoop strains between 0.1 and 0.8 mm/mm, hoop strain rate, 0.1 to 10 per second, grain sizes, 30 to 200 micrometers, and temperatures between 850°C and 1150°C.

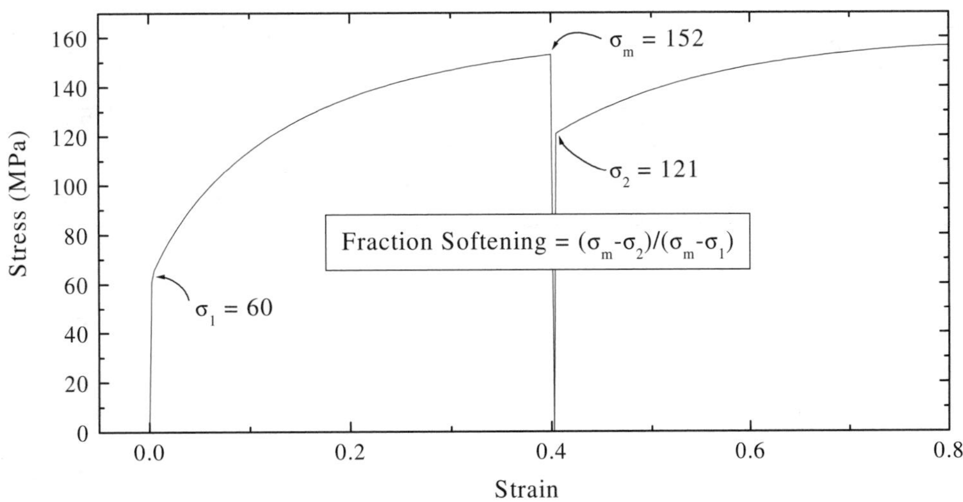

Fig. 2. Schematic representation of fraction softening measurements for double hit uniaxial compression tests used to characterize the recrystallization kinetics.

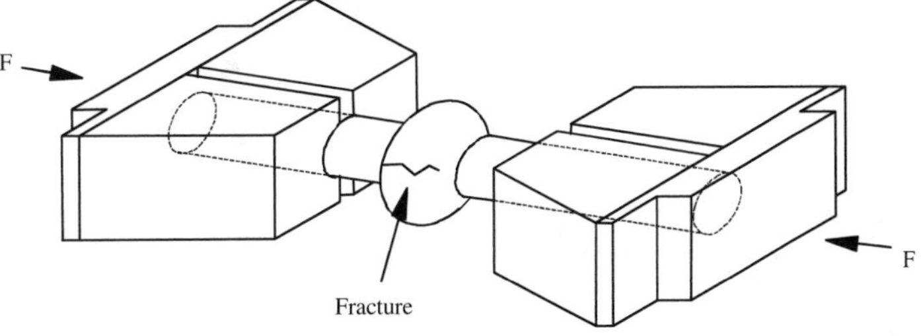

Fig. 3. Schematic of test used to characterize high temperature fracture behavior.

RESULTS & DISCUSSION

Grain growth

The grain coarsening kinetics of the 4120 steel relative to the 1538V steel were significantly more rapid (Fig. 4). This is believed to be due to the increased carbon content, and/or the higher silicon content, and/or the elevated sulfur level in the 1538V steel. At lower temperatures un-dissolved V(CN) and/or AlN precipitates can pin grain boundaries. Under mill processing conditions, initial soaking temperature are high enough to dissolve these precipitates. In order to simulate mill conditions, samples were solutionized to dissolve the V(CN) and AlN followed by quenching and tempering prior to specimen manufacture. During testing, samples were heated rapidly in order to avoid V(CN) and AlN re-precipitation[3,4]. At temperatures above the AlN and V(CN) precipitation dissolution temperature, MnS precipitates, and solid solution elements reduce grain boundary mobility.[5,6] Similarly, carbon concentrates at the grain boundaries, and is believed to reduce the grain boundary energy, and hence is believed to reduce the grain boundary mobility. Equation 1 and Table II show the grain growth models for each steel grade.

$$d^m = d_0^m + kt\exp\left[\frac{-Q}{RT}\right] \qquad [1]$$

where,
d = grain size (μm), d_o = initial grain size, t = time (seconds), T = temperature (Kelvin) and R = 8.314.

Table II Coefficients for grain growth models

Steel	Condition	m	k	Q
4120	All evaluated	4.0	3E30	675,000

Steel	Condition	m	k	Q
1538V	t<30 sec, and T>1100°C	7	5.4E38	724,000
	t<30 sec, and T<1100°C	5	8.9E38	829,580
	t>30 sec (where d_o=d_{30})	5.5	2.9E39	870,475

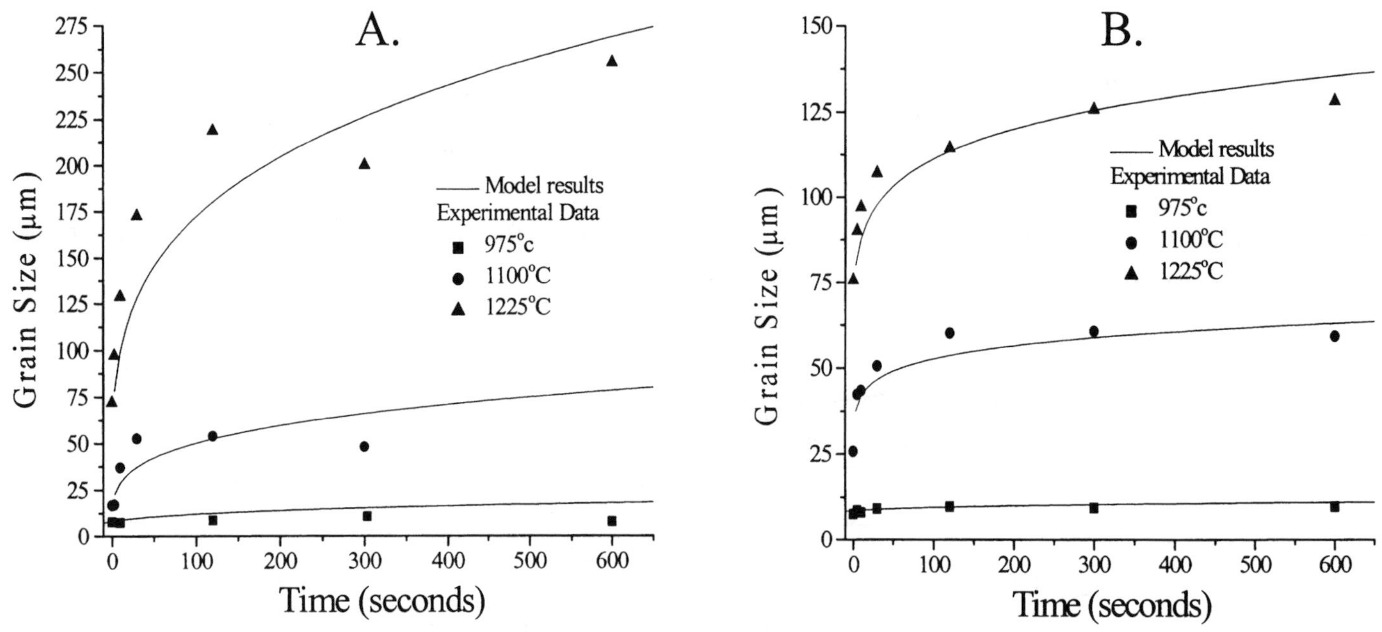

Fig. 4. Grain growth behavior, and model results for A. 4120 steel, and B.1538V steel.

Flow stress

The flow stress of each material was characterized in terms of the high temperature yield strength, work hardening rate, steady state flow stress (under dynamic recovery, and dynamic recrystallization conditions), strain to initiate dynamic recrystallization, and rate of softening upon the initiation of dynamic recrystallization. In each case, temperature, strain, strain rate, and initial grain size were incorporated as appropriate[7,8,9]. The flow stress model equations (eqn. 2-10), and coefficient values (Table III) are shown below.

The 4120 steel exhibited higher flow stresses, at a given strain, than the 1538V steel in all cases, but the percentage increase in flow stress cannot be simply stated. Fig. 5 shows selected examples of the flow stress value as measured, and as modeled. Each material exhibited different behaviors for yield, work hardening, and peak strain behavior as a function of strain rate, temperature, and initial grain size. In general, the 1538V steel exhibited significantly lower flow stress. 1538V in particular exhibited lower strains to initiate dynamic recrystallization, and significantly more rapid strain softening once dynamic conditions had been reached. Furthermore, 1538V tended toward finer grain size, which in turn reduces the strain needed to initiate dynamic recrystallization, resulting in additional reduction in the flow stress.

Higher carbon levels, and reduced nickel, chrome, and molybdenum additions in the 1538V steel are believed to be responsible for the reduction in flow stress relative to the 4120 steel. Carbon has been observed by other investigators to have a complex effect on the flow stress. At higher temperatures and lower strain rates, carbon tends to decrease flow stress, while at lower temperatures, and higher strain rates, carbon increases flow stress[10,11]. Similarly, Medina et. al. evaluated the impact of carbon, and some solid solution, and microalloy elements. They found that carbon reduces the flow stress slightly and also reduces the peak strain, while alloy additions increase the flow stress, and peak strain[12-15].

For the current work it is difficult to know what factor(s) are controlling the flow stress behavior of the 2 steels studied due to the complex chemistries inherent in alloy steel systems. Systematic variations in alloy contents were not conducted. It is reasonable to assume that the reductions in solid solution alloy elements in the 1538V cause the reduction in the flow stress behavior, and that the increased carbon contributes to the decrease in peak strain observed. The presence of vanadium in the 1538V steel further confounds this difficulty either as a solid solution element, or as a V(C,N) precipitate.

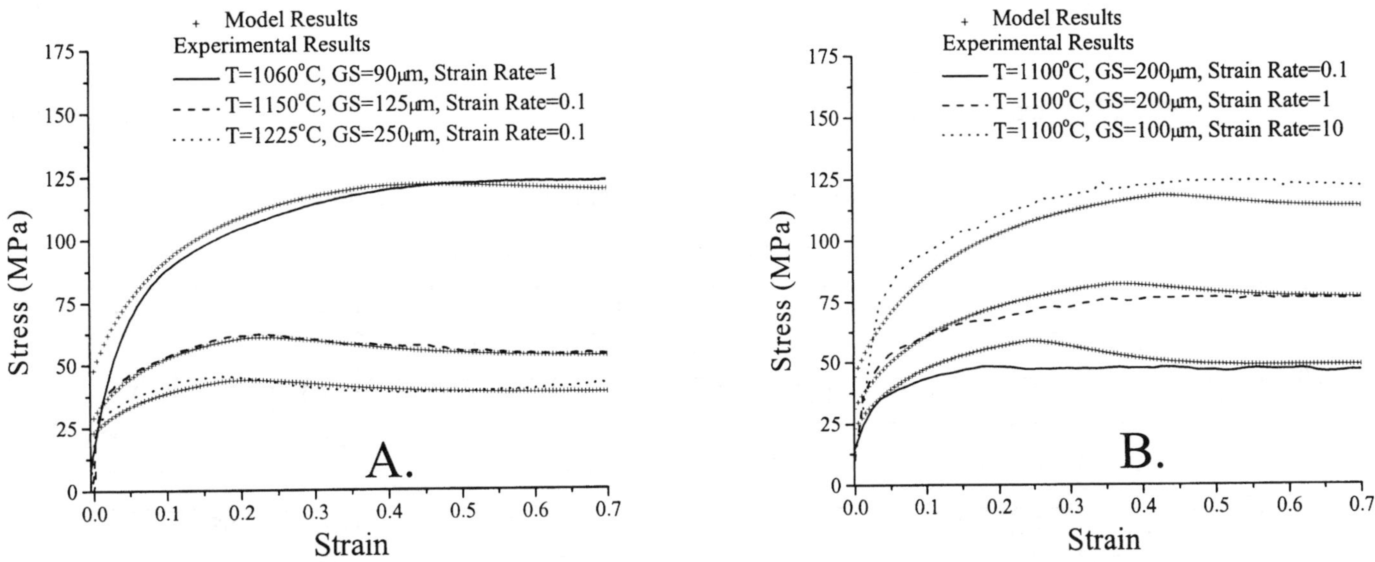

Fig. 5. Flow stress behavior, and model results for A. 4120 steel, and B.1538V steel.

Critical strain: $\varepsilon_c = 0.83 * \varepsilon_p$. [2]

Where

For $\varepsilon < \varepsilon_c$

$$\sigma^{drec} = \left[\sigma_{ss}^{*\,2} + \left(\sigma_0^{\,2} - \sigma_{ss}^{*\,2}\right) \cdot e^{-\Omega\varepsilon}\right]^{0.5} \quad [3]$$

For $\varepsilon > \varepsilon_c$

$$\sigma = \sigma^{drec} - \left[\sigma_{ss}^{*} - \sigma_{ss}^{drex}\right] \bullet \left[1 - \exp\left(-0.693\left(\frac{\varepsilon - \varepsilon_c}{\varepsilon_{0.5} - \varepsilon_c}\right)^n\right)\right] \quad [4]$$

where, ε_c = critical strain to initiate dynamic recrystallization, and ε_p = the peak strain

σ^{drec} = flow stress (dynamic recovery conditions)

σ^{*}_{ss} = flow stress (saturation stress)

σ_0 = flow stress (high temperature yield)

Ω = recovery rate term

σ^{drex}_{ss} = saturation flow stress (dynamic recrystallization conditions)

$\varepsilon_{0.5}$ = Strain to reach 50% dynamic recrystallization

where:
$$\varepsilon_p = A1 \cdot d_0^{b1} \cdot \dot{\varepsilon}^{b2} \cdot \exp\left(\frac{P1}{RT}\right) \quad [5]$$

$$\sigma_0 = A2 \cdot \dot{\varepsilon}^{b3} d_0^{-b4} \exp\left(\frac{P2}{T}\right) \quad [6]$$

$$\Omega = A3 \cdot \dot{\varepsilon}^{-b5} d_0^{-b6} \exp\left(\frac{-P3}{RT}\right) \quad [7]$$

$$\sigma^{*}_{ss} = A4\left(\frac{d_0}{C1}\right)^{-b7} \sinh^{-1}\left[B1\left[\dot{\varepsilon} \cdot \exp\left(\frac{P4}{RT}\right)\right]^{b8}\right] \quad [8]$$

$$\sigma^{drex}_{ss} = A5 \cdot \sinh^{-1}\left[B2 \cdot \left[\dot{\varepsilon} \cdot \exp\left(\frac{P5}{RT}\right)\right]^{b9}\right] \quad [9]$$

$$\varepsilon_{0.5} = A6 \cdot d_0^{b10}\left[\dot{\varepsilon} \cdot \exp\left(\frac{P6}{RT}\right)\right]^{b11} \quad [10]$$

Where d_o = initial grain size, T = temperature (Kelvin), ε = strain, $\dot{\varepsilon}$ = strain rate, and R = 8.314.

Table III Coefficients for flow stress models

	n	A1	A2	A3	A4	A5	A6	B1	B2	b1	b2	b3	b4	
4120	1.42	3.50E-03	2.96	129	100	90.91	9.95E-03	3.01E-03	1.18E-03	0.15	0.17	0.11	0.11	
	b5	**b6**	**b7**	**b8**	**b9**	**b10**	**b11**	**C1**	**P1**	**P2**	**P3**	**P4**	**P5**	**P6**
	0.09	0.20	0.07	0.175	0.238	0.1	0.145	125	46326	36300	24346	396800	338000	278772

	n	A1	A2	A3	A4	A5	A6	B1	B2	b1	b2	b3	b4	
1538V	1.46	6.05E-04	1.44	513	100	90.91	1.06E-03	4.38E-03	1.40E-03	0.32	0.17	0.13	0.07	
	b5	**b6**	**b7**	**b8**	**b9**	**b10**	**b11**	**C1**	**P1**	**P2**	**P3**	**P4**	**P5**	**P6**
	0.09	0.29	0.1	0.162	0.229	0.32	0.144	100	55530	4733	35617	390000	325513	360700

Recrystallization

The recrystallization kinetics of each material were evaluated and modeled using the Johnson-Mehl-Avrami type (JMA) kinetics equation. Equations 11-17 and Table IV show the model format and coefficients for each material respectively[17,18]. Recrystallized grain size under static and metadynamic recrystallization conditions was also modeled.

Selected cases of measured versus modeled recrystallization kinetics are shown in Fig. 6. As is expected, for 4120 steel strain rate increases caused on increase in recrystallization kinetics (Fig. 6a). Similarly, for 1538V (Fig. 6b), increased strain, and reduced grain size (diamond versus square symbols) cause an increase in recrystallization kinetics. Increased temperature (diamond versus circle symbols) also resulted in an increase in recrystallization kinetics.

Under most of the conditions studies, the time required for 50% recrystallization ($t_{0.5}$) for the 1538V steel was significantly shorter under static recrystallization conditions when compared to 4120 steel. Similarly, under metadynamic recrystallization conditions, the 1538V steel exhibited more rapid recrystallization kinetics however the difference was much less significant when compared to static recrystallization results.

Various authors have investigated or reviewed the effect of alloying elements on recrystallization kinetics[18,19]. For the most part, these investigations have focused on HSLA steels with Ti and Nb additions, or on simple C-Mn steels. Therefore, it is difficult to conclusively identify the impact of chemistry on recrystallization kinetics for the current work.

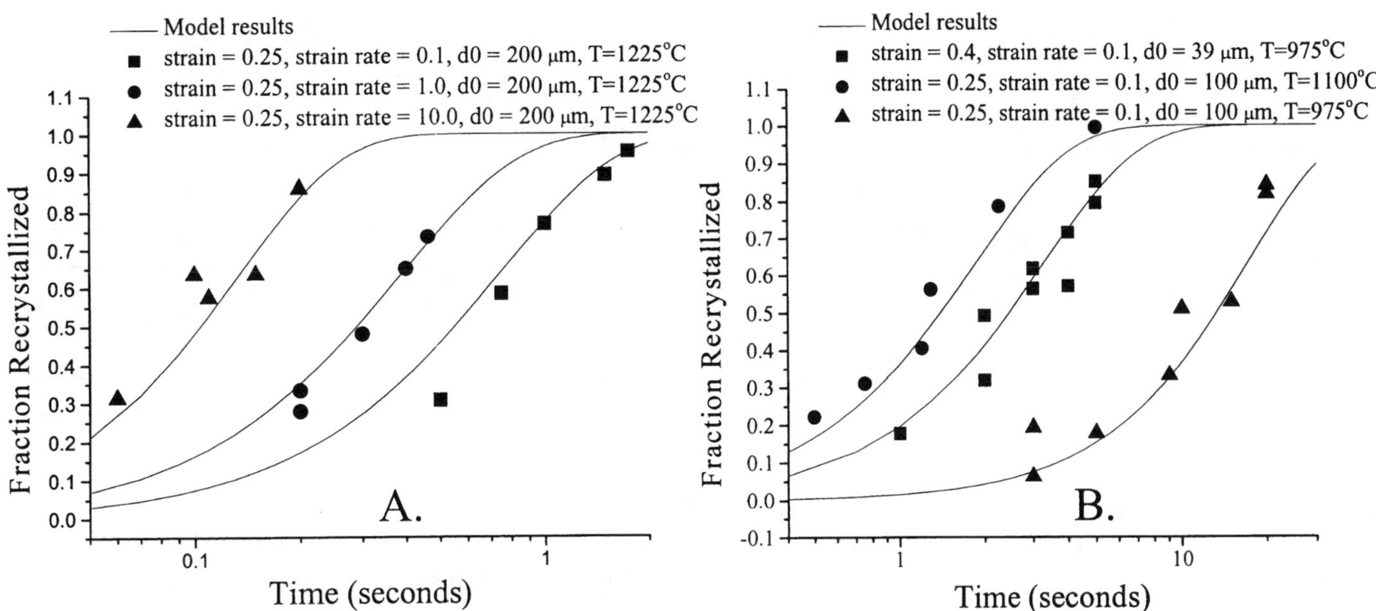

Fig. 6. Recrystallization behavior, and model results for A. 4120 steel, and B.1538V steel.

$$X = 1 - \exp\left(-0.693\left[\frac{t}{t_{0.5}}\right]^k\right)$$, Fraction recrystallized [11]

where,

$$\varepsilon_c = C\left(V1 \cdot d_0^{g1} \cdot Z^{g2}\right)$$, Critical strain for metadynamic conditions [12]

$$Z = \dot{\varepsilon} \cdot \exp\left(\frac{Q_{app}}{RT}\right)$$, Zener-Holloman parameter [13]

For $\varepsilon < \varepsilon_c$, Static Recrystallization,

$$t_{0.5} = V2 \cdot \dot{\varepsilon}^{-g3} d_0^{g4} \varepsilon^{-g5} \exp\left(\frac{L1}{RT}\right)$$, Time for 50% static recrystallization [14]

$$d_{Srxn} = V3 \cdot d_0^{g6} \varepsilon^{-g7} \exp\left(\frac{-L2}{RT}\right)$$, Static recrystallized grain size [15]

For $\varepsilon > \varepsilon_c$, Metadynamic Recrystallization;

$$t_{0.5} = V4 \cdot Z^{-g8} \exp\left(\frac{L3}{RT}\right) \text{(4120 steel); and}$$

$$t_{0.5} = V4 \cdot \dot{\varepsilon}^{-n1} d_0^{n2} \varepsilon^{-n3} \exp\left(\frac{L3}{RT}\right) \text{ (1538V steel), Time for 50\% metadynamic recrystallization} \quad [16]$$

$$d_{Drxn} = V5 \cdot Z^{-g9} \text{, Metadynamic recrystallized grain size} \quad [17]$$

Where t = time (seconds), d_o = initial grain size (μm), T = temperature (Kelvin), ε = strain, $\dot{\varepsilon}$ = strain rate and R=8.314

Table IV Coefficients for the recrystallization models

	C	V1	V2	V3	V4	V5	g1	g2	g3	g4	
4120	0.83	1.50E-03	1.00E-11	1.3	3.00E-05	2900	0.22	0.16	0.24	0.20	
	g5	**g6**	**g7**	**g8**	**g9**	**L1**	**L2**	**L3**	**Qapp**	**k_{stat}**	**k_{dyn}**
	2.36	0.46	1.4	0.5	0.16	241000	7200	270000	320000	1.0	1.0

	C	V1	V2	V3	V4	V5	g1	g2	g3	g4	
1538V	0.8	3.92E-03	1.60E-04	2.14	5.75E-08	1.22E+04	0.18	0.13	0.28	0.26	
	g5	**g6**	**g7**	**g8**	**g9**	**L1**	**L2**	**L3**	**Qapp**	**k_{stat}**	**k_{dyn}**
	1.17	0.81	0.76	na	0.193	84362	21000	154614	330000	1.46	1.28
	n1	**n2**	**n3**								
	0.28	0.26	1.17								

High temperature fracture

A statistically based model was developed for the fracture limit of both steels. Equations 18-20 and Fig. 7 below show the models for the damage criteria for both steel grades. Both steels exhibited a decreased fracture limit with increasing grain size. This is consistent with expectations since reduced grain size is known to increase room temperature toughness, but it is worth noting that this trend is also true at higher temperatures.

The 4120 steel showed greater sensitivity to temperature and strain rate compared to 1538V, particularly at medium to coarse grain sizes, but the overall ability to withstand fracture was greater for 4120. This is attributed primarily to the higher sulfur content of the 1538V steel. The minimum observed fracture limit for each steel grade was at low temperatures, low strain rates, and coarse grain sizes. Over the range of conditions studied no ductility trough as a function of temperature was observed.

The applied damage value is determined as follows:

$$\sum \frac{\sigma \cdot \Delta\bar{\varepsilon}}{\bar{\sigma}} = C_{applied} \quad [18]$$

where σ - largest principle tensile stress;

$\bar{\varepsilon}$ - Equivalent strain;

$\bar{\sigma}$ - Equivalent stress;

terms are summed over increments of $\bar{\varepsilon}$.

When $C_{applied} > C_{frac.}$ cracking is initiated

$$C_{frac.,1538} = 1/(4.96 - 4.322x10^{-3} \cdot T + 0.94 \cdot Log[d_0] + 0.83 \cdot Log[\dot{\varepsilon}] + 6.258x10^{-6} \cdot T^2 + 1.22 \cdot (Log[d_0])^2 - 0.19 \cdot (Log[\dot{\varepsilon}])^2 - 5.513x10^{-3} \cdot T \cdot Log[d_0] - 0.53 \cdot Log[d_0] \cdot Log[\dot{\varepsilon}]) \quad [19]$$

$$C_{frac.,4120} = 0.37 + 1.53x10^{-3} \cdot T - 0.19 \cdot Log[d_0] - 0.47 \cdot Log[\dot{\varepsilon}] + 1.785x10^{-4} \cdot T \cdot Log[d_0] + 4.669x10^{-4} \cdot T \cdot Log[\dot{\varepsilon}] + 0.46 \cdot Log[d_0] \cdot Log[\dot{\varepsilon}] - 4.268x10-4 \cdot T \cdot Log[d_0] \cdot Log[\dot{\varepsilon}] \quad [20]$$

Where d_o = initial grain size, T = temperature (°C), ε = strain, $\dot{\varepsilon}$ = strain rate

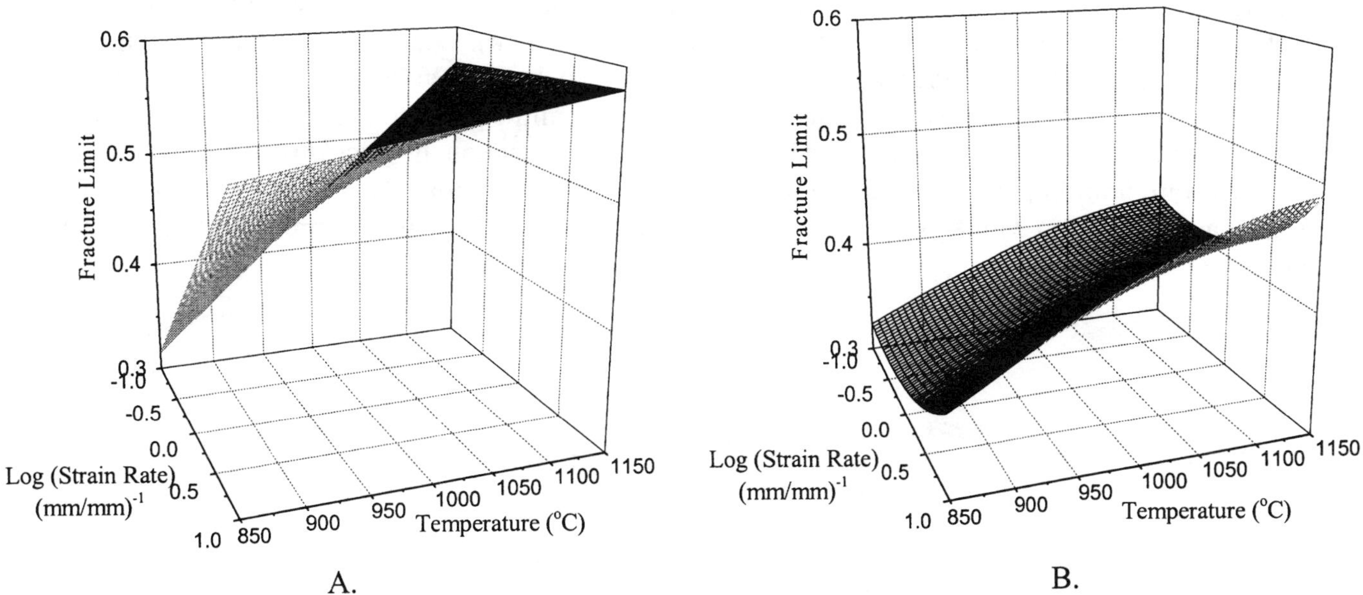

Fig. 7. High Temperature fracture model results for A. 4120 steel, and B.1538V steel.

Roll pass simulation

Initial pass designs for the new Timken - Harrison Avenue Rolling Mill were evaluated using the Hot Bar Mill Model to identify, and correct potential unacceptable surface quality events associated with high temperature fracture, and lap/seam formation. A pass sequence evaluated for the production of 4 inch round bars from 8 inch round corner square billets indicated potential for both types of defects. This pass sequence consisted of alternating horizontal, and vertical passes as follows: box, box, round box, round, round box, round, round box, round.

Fig. 8 shows the results of the computer simulations. The results presented are for the top center and side center locations of the initial billet since these locations were most susceptible to high temperature fracture. During the first pass, the fracture limit was exceeded at the top center location. Prior to the first pass, the austenite grain size was relatively coarse. The coarse grain size coupled with a relatively aggressive deformation schedule for the first pass was the most likely cause for the indicated surface fracture. Similarly, during the sixth pass, the fracture ratio for the top center location was relatively high due, in part, to the incomplete recrystallization that occurred between passes five, and six. Fig. 9a shows the fracture ratio throughout the billet cross section during pass number one. Fig. 9b shows the profile of the billet after the first pass. The billet exhibits some bulging on the sides as a result of pass overfill. The resulting bulging due to overfill can be rolled over in later passes, resulting in a lap, or seam.

Based on these results, a series of new pass designs were conceived, and evaluated using the FEM model to eliminate the potential for poor surface quality. A new pass design was then selected, and implemented for the reduction of 8" RCS billets to 4" rounds. Because this change was made prior to the commissioning of the new bar mill, it is not possible to compare the initially proposed pass design to the new pass design. However, one year after commissioning, there has been a 50% reduction in scrap associated with surface quality. It is reasonable to assume that the new pass design significantly improved surface quality, and aided in a smoother start up of the new mill.

SUMMARY

During hot bar rolling, there are complex interactions between the high temperature microstructural evolution (grain growth, recrystallization) the thermomechanical process variables (temperature, time, and deformation) and the metallurgical responses (flow stress, fracture limit). Coupling of metallurgical models, thermal models, and mechanical models through finite element analysis provides a tool to evaluate, and improve thermomechanical processing schedules without risking mill equipment, or reducing productivity.

For this work, the steel chemistries studied had a significant impact on the microstructural evolution, flow stress, and fracture limit. In order to evaluate the impact of chemistry thoroughly, it is necessary to perform a more systematic evaluation with a range of designed steel chemistries. Similarly, improved modeling methods than those used for this work may need to be developed or incorporated.

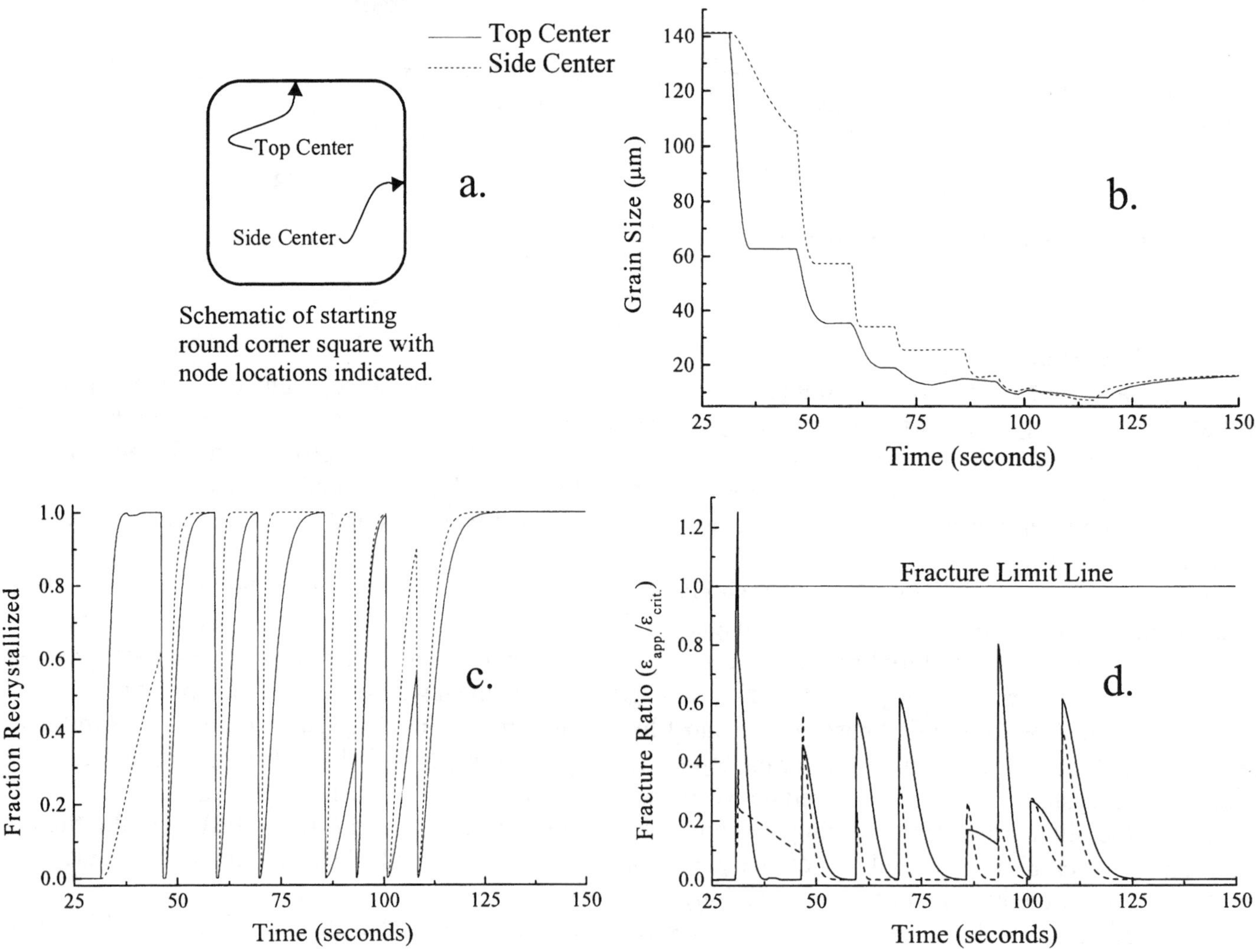

Fig. 8. Computer simulation results for production of a 4 inch round from a 8 inch round corner square. a. Location of simulation points, b. Austenite grain size evolution, c. Austenite recrystallization kinetics, and d. fracture ratio (indicating fracture at Top Center location, 1st pass).

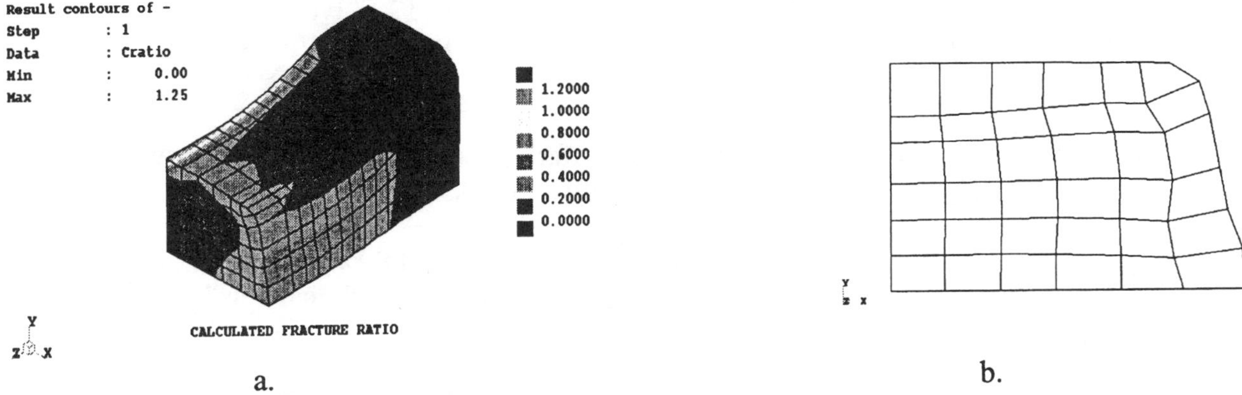

Fig. 9. Computer simulation results for pass number one showing; a. the distribution of fracture ratio in the roll bite, and b. the bulging of the billet at the side due to pass overfill.

ACKNOWLEDGEMENTS

The authors would like to thank The Timken Company for permission to publish work, and for supporting its development. We would also like to acknowledge the efforts of Ohio State University, and Prof. R. Shivpuri and his students for their efforts in rolling simulation development.

REFERENCES

1. P.M. Pauskar; "An Integrated System for Analysis of Metal Flow and Microstructural Evolution in Hot Rolling", Phd. Dissertation, The Ohio State University, 1998.

2. K. Sawamiphakdi, J. E. Ives, E. B. Damm, R. Shivpuri, and P. Pauskar; "The Evolution and Application of Computer Modeling of Hot Bar Rolling at the Timken Company", Proceedings of the7th international conference on steel rolling, Tokyo, Japan, 1998, pp. 557-562.

3. M. Militzer, A. Giumelli, E. B. Hawbolt, and T.R. Meadocroft; "Austenite Grain Growth Kinetics in Al-Killed Plain Carbon Steels", Met. Trans. A, Vol. 27A, pp. 3399-3409.

4. M.J. Leap and J.C. Wingert; "Application of the Advantec Process of Improving the Toughness of Grain-Refined, High-Strength Steels", 38th MWSP Conf. Proc., ISS, Vol. XXXIV, 1997, pp. 195-220.

5. G. Cai, and J. D. Boyd, "Austenite Grain Growth and Precipitate Dissolution in Medium Carbon Microalloy Steels", Grain Growth in Polycrystalline Materials III. TMS, 1998, pp. 579-585.

6. T. Gladman, The Physical Metallurgy of Microalloyed Steels, Institute of Materials , London, 1997

7. Laasraoui, A., and Jonas, J.J., "Prediction of flow stresses at high temperatures and strain rates", Metallurgical Transactions A, July 1991, pp. 1545-1558.

8. S.B. Davenport, N.J. Silk, C.N. Sparks, and C.M. Sellars, "Development of Constitutive equations for Modeling of Hot Rolling", Material Science and Technology, Vol. 16, May, 2000. Pp. 539-546.

9. P. Pauskar, E. B. Damm, R. Shivpuri, S. Phadke, K. Sawamiphakdi, and J.E. Ives; "A Microstructure Dependent Flow Stress Model for Steel Rolling", 39th MWSP Conf. Proc., ISS Vol. XXXV, 1998. Pp. 973-982.

10. D.C. Collinson, P.D. Hodgson, and C.H.J. Davies, "The Effect of Carbon on the Hot Deformation and Recrystallization of Austenite", THERMEC '97, TMS, 1997. PP. 483-489.

11. J. Jaipal, C.H.J. Davies, B.P. Wynne, D.C. Collinson, A. Brownrigg, and P.D. Hodgson, "Effect of Carbon on the Hot Flow Stress and Dynamic Recrystallization Behavior of Plain Carbon Steels", THERMEC '97, TMS, 1997. PP. 539-545.

12. S.F. Medina, C.A. Hernandez; "General Expression of the Zener-Hollomon Parameter as a Function of the Chemical Composition of Low Alloy and Microalloyed Steels", Acta. Mat. Vol. 44, 1996. PP 137-148.

13. S.F. Medina, C.A. Hernandez; "The Influence of Chemical Composition on Peak Strain of Deformed Austenite in Low Alloy and Microalloyed Steels", Acta. Mat. Vol. 44, 1996. PP 149-154.

14. S.F. Medina, C.A. Hernandez; "Modelling Austenite Flow Curves in Low Alloy and Microalloyed Steels", Acta. Mat. Vol. 44, 1996. PP 155-163.

15. S.F. Medina, C.A. Hernandez; "Modelling of the Dynamic Recrystallization of Austenite in Low Alloy and Microalloyed Steels", Acta. Mat. Vol. 44, 1996. PP 165-171.

16. Laasraoui, A. and Jonas, J. J.; "Recrystallization of austenite after deformation at high temperatures and strain rates - analysis and modeling", Metallurgical Transactions A, Volume 22A, January 1991, pp. 151-160.

17. C.M. Sellars; "Modeling Microstructural Development During Hot Rolling", Mat. Sci. & Tech., Vol. 6, Nov., 1990. Pp 1072-1081.

18. P.D. Hodgson; "Mathematical Modelling of Recrystallization Processes During the Hot Rolling of Steel", Ph.D. Thesis, Univ. Queensland, Australia, 1993.

19. S.F. Medina, J.E. Mancilla; "Influence of Alloying Elements in Solution on Static Recrystallization Kinetics of Hot Deformed Steels", ISIJ Intl.Vol. 36, No. 8. Pp. 1063-1069.

Physical and Mathematical Simulation of Phase Transformation during Accelerated Cooling of Eutectoid Steel Rods

Roman Kuziak[1] and Maciej Pietrzyk[2]
[1]The Institute for Ferrous Metallurgy
44-100 Gliwice, Poland
+48 32 234 52 41
rkuziak@steel.imz.gliwice.pl
[2]Akademia Gorniczo-Hutnicza
30-059 Krakow, Poland
+48 12 617 29 21
pietrzyk@metal.agh.edu.pl

Key Words: physical modeling, mathematical modeling, finite-element method, rods, eutectoid steel, phase transformation, air-cooling, accelerated cooling, mechanical properties.

INTRODUCTION

Various researchers have developed models describing kinetics of phase transformation for rod steels during cooling after hot rolling, see for example [1-4]. These models predict such characteristics of the austenite-ferrite transformation as the transformation start temperature, nucleation and growth and site saturation. Some of the models are extended and include austenite-pearlite, austenite-bainite and austenite-martensite transformations. Earlier Authors' works in this field focused on an inclusion of the phase transformation models into the finite element codes, which simulate temperature fields in products during rolling and cooling after rolling. The solutions for cooling of rods in TEMPCORE system[5] and for heat treatment of rail's head[6] were obtained. The objective of the present work was development of the model, which describes pearlite transformation during accelerated air-cooling of eutectoid steel rods. Constant cooling rate dilatometric experiments and thermomechanical tests on the Gleeble 3800 simulator were used for obtaining the experimental data. An inverse analysis was applied for determination of the constants in the model. Developed model was implemented into the finite element code, which simulates rolling and air-cooling of rods. Heat transfer coefficients for various air-cooling methods were determined on the basis of temperature measurements during cooling of rods under industrial conditions.

THERMOMECHANICAL MODEL

The model, which is used in the present work, is an extension of the thermal-mechanical-microstructural finite element solution developed at AGH in Krakow[7] for shape rolling processes. This solution uses generalized plane strain approach, which allows effective simulation of the metal flow and the heat transfer during shape rolling. In this approach the cross section of the shape moving with the actual metal velocity is considered and non-stationary solution of mechanical and thermal equations is performed in this section. In consequence, three-dimensional results are obtained from two-dimensional finite element model. Huber-Mises yield criterion associated with Levy-Mises flow rule and is used in the mechanical component of the model. Combination of this model with the closed form equations describing recrystallization and grain growth in steels enabled prediction of austenite microstructure at the exit from the rolling mill. This solution for rolling of rods is described in paper[8]. Thermal component of this model is used in the present work for simulation of accelerated cooling of rods after rolling. It is based on a numerical solution of the heat transport equation:

$$\nabla^T(k\nabla T)+Q-\rho c_p\left(\mathbf{v}^T\nabla T-\frac{\partial T}{\partial t}\right)=0 \tag{1}$$

where: k - conductivity, T - temperature, Q – heat generated due to plastic work or due to transformation, ρ - density, c_p – specific heat, t – time, $\mathbf{v}$ – vector of velocities, which represents convection and which is zero when cooling after rolling is simulated.

The solution of equation (1) is performed in a typical finite element manner[9]. Galerkin scheme is applied in time integration of equations. Details of this solution with the relevant boundary conditions at the surface are given in publication[10]. The boundary conditions are represented by the heat flux through the outer surface of the rod, which is calculated as:

$$k\frac{\partial T}{\partial \mathbf{n}}=\alpha(T_0-T) \tag{2}$$

where: $\mathbf{n}$ – vector normal to the rod's surface, T_0 - temperature of the surrounding medium, α - heat transfer coefficient.

For free air-cooling the heat transfer coefficient α is determined from typical radiation and convection equations[11]. Evaluation of α for various types of accelerated air-cooling, including fan cooling and air blowing under various pressures, was performed on the basis of in situ temperature measurements described below.

The experiment, aiming at an evaluation of the heat transfer coefficient α was performed under the laboratory conditions. The rods with the thermocouples inserted at the center and close to the upper and lower surfaces, were subjected to various types of forced air-cooling. The air velocity, which varied between value close to zero (free air-cooling) and 3.2 m/s, distinguished the cooling methods. Some of the experiments were carried out in the building, with a surrounding temperature of 15-20°C, some outside with a surrounding temperature of 0-10°C. A comparison between measured and predicted temperatures at two locations, obtained for free air-cooling, is shown in Fig.1. It is seen in this figure that the model underestimated the heating due to the phase transformation. However, an excellent agreement between measurements and predictions was obtained beyond the phase transformation range of temperatures. Similar results were obtained for various types of forced air-cooling, see for example Fig.2, where the results for central thermocouple for two air velocities are shown. On the basis of all experimental data the following formulae, describing heat transfer coefficient as a function of air velocity, was suggested:

$$\alpha=(1+0.09v)\varepsilon\sigma(T_K+T_0)(T_K^2+T_0^2) \tag{3}$$

where: v - velocity of the air [m/s], ε - emissivity, σ - Boltzman constant, T_0 - temperature of the surrounding air [K], T_K - temperature of the rod surface [K].

Equation (3) was tested and is valid for air velocities below 4.5 m/s, which were used in the present work. An analysis of all experimental data allows the conclusion that the finite element model with boundary conditions based on equation (3) predicts the temperatures during forced air-cooling with reasonable accuracy. In further sections of the paper this solution is used for a simulation of cooling of eutectoid steel rods under industrial conditions.

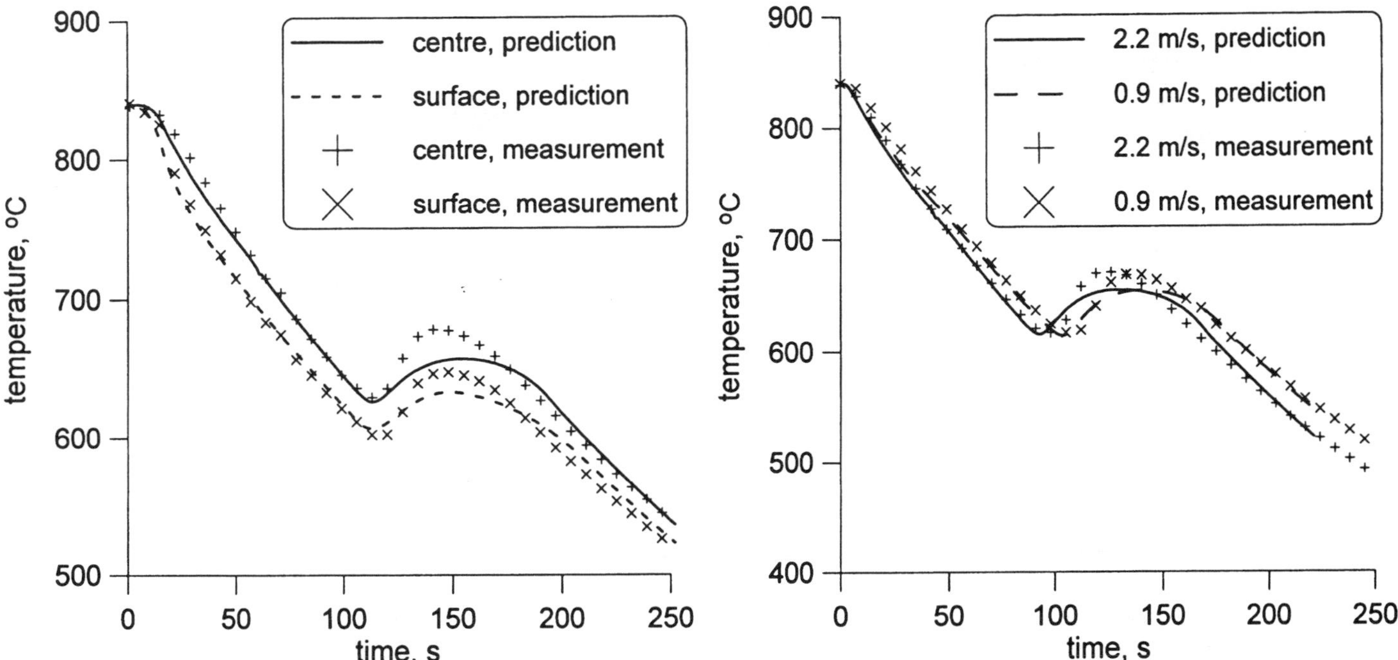

Fig.1. Measured and calculated time temperature profiles for free air-cooling

Fig.2. Measured and calculated time temperature profiles for forced air-cooling with various air velocities.

PHASE TRANSFORMATION MODEL

Transformation kinetics

The phase transformation model for eutectoid steels is based on equations, which describe volume fraction of the considered phase as a function of time and temperature. Incubation time for pearlitic and bainitic transformations is accounted for. The main equations of the model are given below[7,12].

- Temperature Ae_1:

$$Ae_1[^\circ C] = 723 - 10.7[Mn] - 16.9[Ni] + 29.1[Si] + 16.9[Cr] \tag{4}$$

- Incubation time for pearlitic transformation:

$$\tau_P = \frac{(x_1 C_r)^{x_{11}} k_P T_K^{0.5}}{(Ae_1 - T)^{x_3}} \exp\left(\frac{x_2 \times 10^3}{R T_K}\right) \tag{5}$$

- Volume fraction of pearlite:

$$X_p = 1 - \exp\left[-\frac{b_p}{D_\gamma^{x_5}}(t - \tau_P)^{x_6}\right] \tag{6}$$

where:

$$b_p = x_4 \times 10^{-34} \exp\left(0.28T - 0.00026T^2\right)$$

- Bainitic transformation start temperature:

$$T_b[^\circ C] = x_7 - 270[C] - 90[Mn] - 37[Ni] - 70[Cr] - 83[Mo] \tag{7}$$

- Incubation time for bainitic transformation:

$$\tau_b = \frac{k_b}{(T_{b0} - T)^3} \exp\left(\frac{Q_b}{RT_K}\right) \tag{8}$$

- Volume fraction of bainite:

$$X_b = 1 - \exp\left[-b_b (t - \tau_b)^{n_b}\right] \tag{9}$$

where, for $T \geq 350^\circ C$

$k_b = x_8$; Q_b =88000 J/mol; $T_{b0} = 923^\circ C$; $n_b = 1.2$

$$b_b = x_9 \exp\left[-127.923 + 0.3113 T_K - 0.000193 T_K^2\right]$$

and for $T < 350^\circ C$

$k_b = 20.8$; $Q_b = 82000$ J/mol; $T_{b0} = 812^\circ C$; $n_b = 1.25$

$$b_b = \exp\left[-51.89 + 0.1385 T_K - 0.0001039 T_K^2\right]$$

- Martensitic transformation start temperature:

$$T_m[^\circ C] = x_{10} - 474[C] - 33[Mn] - 17[Ni] - 17[Cr] - 21[Mo] \tag{10}$$

- Fraction of austenite which transforms into martensite (model of Koistinen and Marburger given also in[13]:

$$X_M = (1 - F_f - F_p - F_b)\{1 - \exp[-0.011(T_M - T)]\} \tag{11}$$

where: [Mn], [Ni], [Si], [Cr] - contents of manganese, nickel, silicon and chromium in steel, C_r - cooling rate, T - temperature in °C, T_K - absolute temperature in K, D_γ - austenite grain size, t - time, τ_p - time at the beginning of pearlitic transformation, τ_b - time at the beginning of bainitic transformation, F_f, F_p, F_b – fraction of ferrite, pearlite and bainite with respect to the whole volume of material.

The model described above allows simulation of microstructural phenomena, which take place during cooling of products after hot deformation. Model is capable of accounting for varying conditions of cooling. It is shown in[9] that implementation of this model into the finite element code reveals its interesting predictive capabilities. Simulations performed for compression and cooling of axisymmetrical samples[14] confirmed advantages of this approach. Local volume fractions of various phases could be predicted.

Mechanical properties

Simulation includes mechanical properties of products in room temperature. The equations used for prediction of these properties for the investigated steel are based on an analysis presented in[7] and they are repeated briefly below. The interlamellar spacing in pearlite, in m, is calculated as:

$$\frac{1}{S_o} = (129.28 - 54.373[Mn] - 4.378[Cr] - 17.5[Si]) - (0.1783 - 0.0723[Mn] - 0.0121[Cr] - 0.0274[Si])T_p \tag{12}$$

where: T_p - temperature of pearlitic transformation in °C.
Since the temperature varies during cooling, the interlamellar spacing is calculated as weighted average during the transformation, according to the formula:

$$\frac{1}{S_0} = \frac{\int \frac{dX_p}{S_0}}{\int dX_p} \qquad (13)$$

Pearlite nodule size and pearlite colony size are two additional parameters, which characterize microstructure of this phase. They are calculated from the following equations:

pearlite nodule size

$$D_n = \frac{6500\left[1 - \exp\left(-0.016 D_\gamma\right)\right]^{0.6}}{Ae_1 - T_p}$$

pearlite colony size

$$D_c = \left(0.857 - 0.00119 T_p\right)^{-1} \qquad (14)$$

Yield stress and tensile strength for eutectoid steel are calculated as:

$$\mathrm{R}_e(\mathrm{MPa}) = R_1 + k_1\chi^{-1} \qquad \mathrm{R}_m(\mathrm{MPa}) = R_2 + k_2\chi^{-1} + 72[\mathrm{Si}] \qquad (15)$$

where: $\chi(\mathrm{mm}) = 2(S_o - t)$ is the mean free path for dislocation glide in pearlitic ferrite, t - thickness of the cementite plate calculated as $t(\mathrm{mm}) = 0.15\,[\mathrm{C}]\,S_o$. Constants in equations (15) are:

$R_1 = 308$ $k_1 = 0.070$
$R_2 = 706$ $k_2 = 0.072$ for $S_o > 0.15$ m

$R_1 = 259$ $k_1 = 0.087$
$R_2 = 773$ $k_2 = 0.058$ for $S_o < 0.15$ m

Hardness (Vickers) is calculated with an assumption that three phases (pearlite, bainite and martensite) may appear in steel. The rule of mixture is applied. However, since an experimental validation showed that the model overestimates hardness for cooling rates 4 - 6°C/s, when martensite appears (see for example publication[2] or results presented in the next section), weighting coefficient q was introduced for martensite:

$$\mathrm{HV} = \frac{F_p\,\mathrm{HV}_p + F_b\,\mathrm{HV}_b + F_m^q\,\mathrm{HV}_m}{F_p + F_b + F_m^q} \qquad (16)$$

where: F_P, F_b, F_m - volume fraction of pearlite, bainite and martensite, respectively, HV_P, HV_b, HV_m - hardness of pearlite, bainite and martensite, respectively, calculated as:

$$\mathrm{HV}_P = 615 + 640[\mathrm{C}] - \left(0.57 + 0.4[\mathrm{C}]\right)\left(T_P + 273\right)$$
$$\mathrm{HV}_b = \mathrm{HV}_p \qquad \mathrm{HV}_m \cong 830 \qquad (17)$$

Identification of parameters

The material constants in the model depend on a chemical composition of steel. Determination of these constants for particular steel often presents difficulties. Identification technique based on optimization was used for this purpose in the present work. This approach is described in the publication[15]. The experiment included cooling of the rod steel samples with various cooling rates. The chemical composition of the investigated steel is given in Table I. The austenite grain size after hot deformation was 35 μm. The measurements included start and end temperatures of transformations as well as volume fractions of phases in final products and hardness of final products. The selected constants in the model were grouped in the vector of unknown coefficients (**x**). These coefficients for eutectoid steel were determined from a condition of minimum of the cost function defined as a distance between measured and calculated values of transformation start and end temperatures and volume fractions of phases, as shown in[15].

Table I Chemical composition of the steel used in this study.

C	Mn	Si	Ni	Cr	P	S
0.74	0.97	0.25	0.15	0.68	0.02	0.02

Optimization yielded the values of the coefficients **x** given in Table II. Comparison of the predicted CCT diagram for the optimal coefficients **x** with the experimental one is shown in Fig. 3. It is seen in this Figure that the model predicts properly start of all transformations involved. Some discrepancies appear between prediction and observation of pearlite transformation end temperature. This is probable due to the fact that model determines this temperature when volume fraction of pearlite achieves 0.95 while an effect of transformation in the dilatometric test disappears earlier. Comparison of measured and calculated volume fraction of phases for various cooling rates is shown in Fig. 4.

Table II Coefficients obtained from optimization.

x_1	x_2	x_3	x_4	x_5	x_6	x_7	x_8	x_9	x_{10}	x_{11}
10	106	4.89	3	1.52	1.96	855	22.5	0.5	581	2

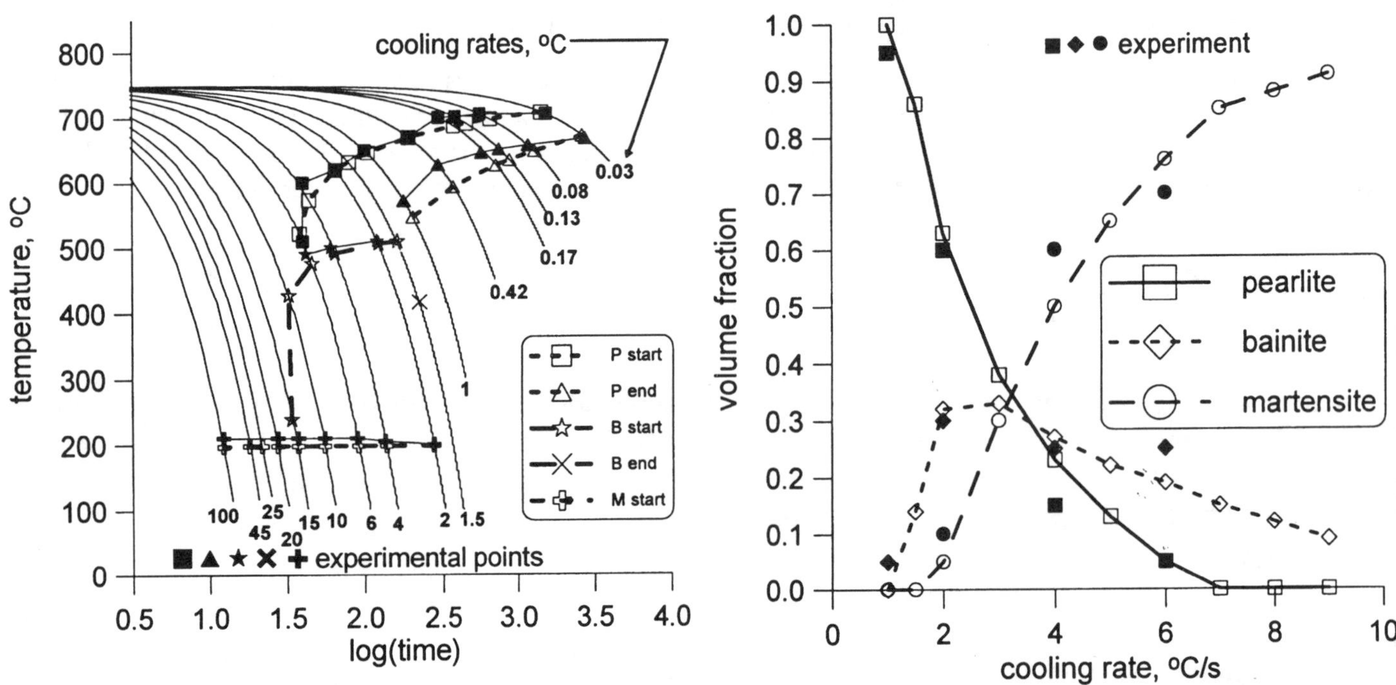

Fig. 3. Comparison of measured and predicted CCT diagram.

Fig. 4. Comparison of measured and predicted volume fraction of phases for various cooling rates.

Fig. 5 shows a comparison between measured and predicted hardness of the samples after the tests. An analysis of the results presented in Figs 3, 4 and 5 confirms, in general, good predictive capabilities of the model. Some discrepancies between predicted and measured volume fractions of phases and hardness appear for cooling rates around 4 - 6°C/s. In spite of using weighting coefficient q = 3/2 in equation (17), the model overestimated the hardness for these cooling rates. This problem requires further attention. Since, however, practical cooling rates in the air-cooling processes considered in the present work do not exceed 4°C/s, this problem was not furthered here.

Results of calculations of the yield stress and tensile strength for various cooling rates are shown in Fig. 6. These results were not validated experimentally in the present work. However, equations (15) were validated in publication[7] for steels with a similar chemical composition.

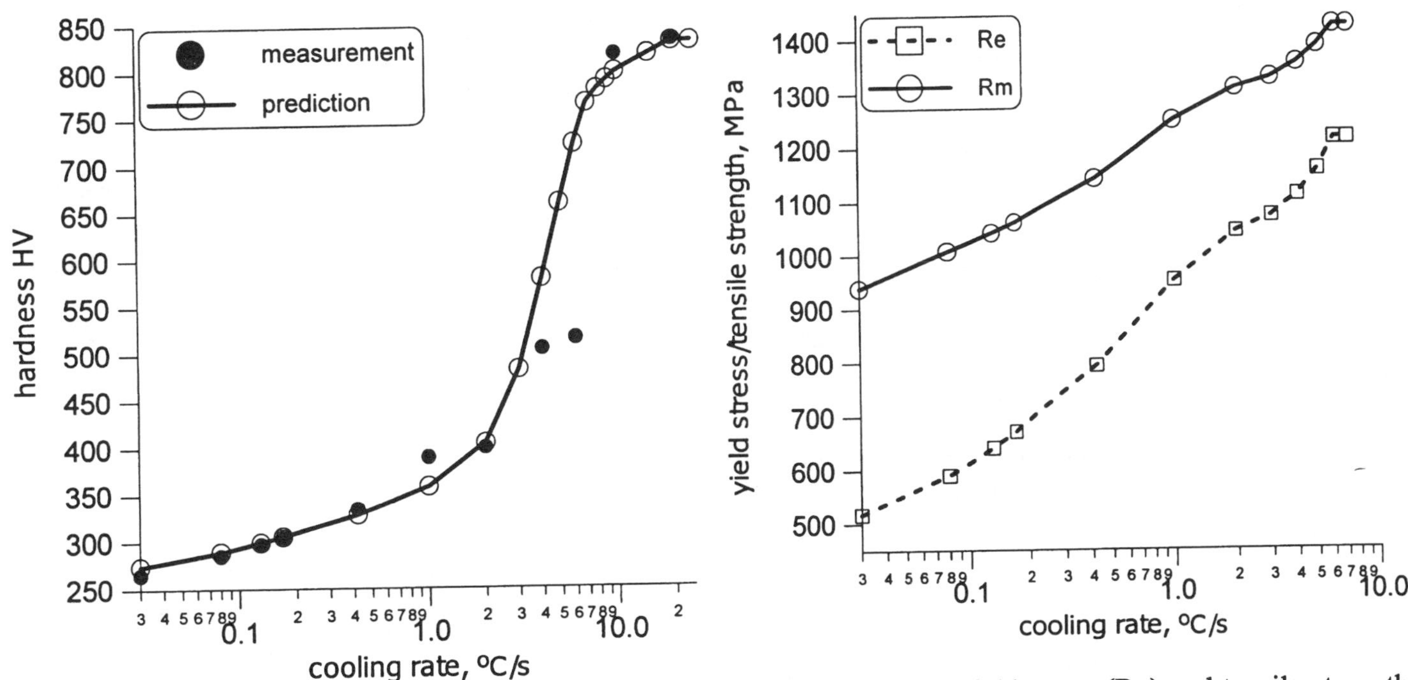

Fig. 5. Comparison of measured and predicted hardness of the samples after the tests

Fig. 6. Calculated yield stress (Re) and tensile strength (Rm) for various cooling rates.

SIMULATIONS

Simulations were performed for various processes of air-cooling of eutectoid steel rods. The results for the rods with a diameter of 15 mm are presented below. Predicted distributions of hardness along the rod radius for various cooling intensities are shown in Fig. 7a. It is seen in this figure that for low cooling intensity hardness distribution is reasonably uniform. When cooling intensity is increased, an increase of hardness in the surface layers is observed. Fig. 7b shows yield stress, tensile strength and hardness as a function of air velocity during cooling. Analysis of these results shows that hardness remains almost constant for lower air velocities, below about 2 m/s. An increase of the velocity above this value results in more significant increase of hardness. Contrary, yield stress and tensile strength are more sensitive to velocity increase for lower cooling intensities, when air velocity is below 3 m/s. Increasing the velocity above this value does not have significant influence on yield stress and tensile strength.

Predicted values of microstructural parameters, including interlamellar spacing in pearlite, pearlite nodule size and pearlite colony size for various cooling intensities, are presented in Fig. 8. It is seen that the values of all these parameters decrease with increasing air velocity during cooling. These values are shown for informative purposes only and they were not validated experimentally.

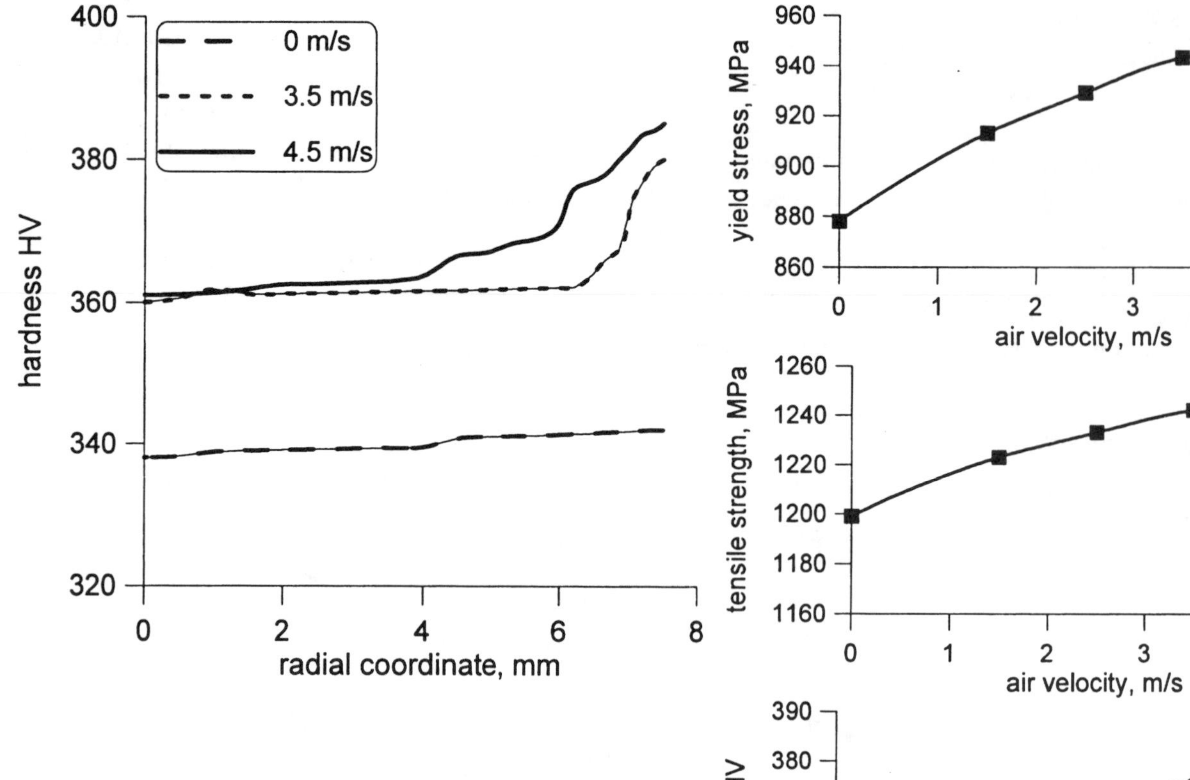

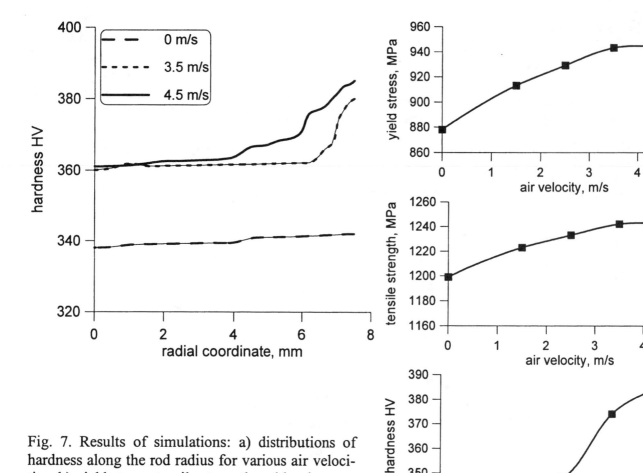

Fig. 7. Results of simulations: a) distributions of hardness along the rod radius for various air velocities, b) yield stress, tensile strength and hardness as a function of air velocity.

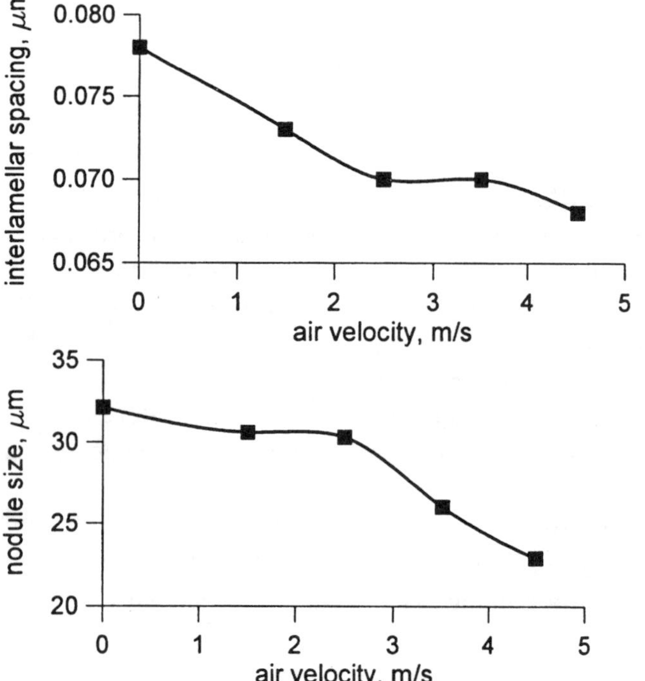

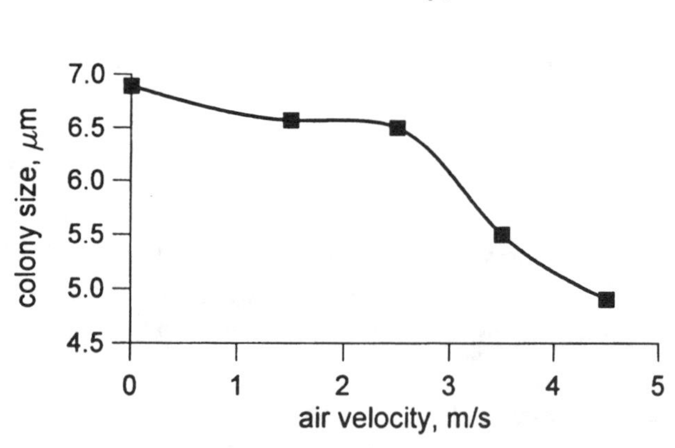

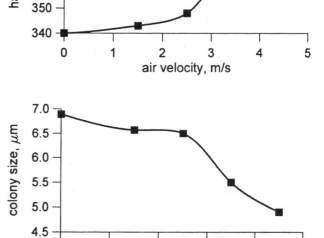

Fig. 8. Predicted values of the interlamellar spacing in pearlite, pearlite nodule size and pearlite colony size for various cooling intensities.

SUMMARY

The model, which predicts phase transformations during air-cooling of eutectoid steel reds, is presented in the paper. The material parameters in the model were determined using identification procedure based on optimization techniques. Validation of the model confirmed its good predictive capability. Some discrepancies between calculations and measurements appear for the cooling rates in the range 4 - 6°C/s and this problem requires further attention. It should be emphasized, however, that extending the model to higher cooling rates and including martensitic transformation was done to obtain complete CCT diagram. These higher cooling rates do not appear in the rod cooling processes, which were investigated.

The developed phase transformation model was implemented into the finite element code predicting temperatures during cooling of rods and simulations of industrial air-cooling processes were performed. Influence of cooling conditions on microstructure and properties of final products was investigated. It can be concluded from all the results of simulations that an increase of air velocity during cooling results in an increase of mechanical properties and in decrease of parameters characterizing microstructure of pearlite. It is seen in Figs 7 and 8 that sensitivity of various parameters to changes of the air velocity is different.

ACKNOWLEDGEMENTS

Financial assistance of KBN (AGH project no. 11.11.110.222 and IMZ project no. 2466/C.T08-7/99) is gratefully acknowledged.

REFERENCES

1. P.C. Campbell, E.B. Hawbolt and J.K. Brimacombe, Microstructural Engineering Applied to the Controlled Cooling of Steel Wire Rod: Part III. Mathematical Model - Formulation and Predictions, Metall. Trans. A, 22A, 1991, pp. 2791-2796.

2. E. Anelli, "Application of Mathematical Modelling to Hot Rolling and Controlled Cooling of Wire Rods and Bars", ISIJ Int. Vol. 32, 1992, pp. 440-449.

3. R.M. Davila, D.J. Vigueras and M.T. Velazques, Improving the Quality of High and Low-Carbon Steel Rods Employed in Wire Drawing Operations, Steel res., 66, 1995, 55-62.

4. A. Kumar, L.K. Singhai and S.K. Sarna, "Mathematical Model for Predicting the Thermal and Mechanical Behaviour of Rebar during Quenching and Self Tempering", Steel res., 66, 1995, pp. 476-481.

5. M. Pietrzyk and R. Kuziak, Finite Element Modelling of Accelerated Cooling of Rods after Hot Rolling, Proc. 41st MWSP Conf., Baltimore, 1999, pp. 405-413.

6. M. Pietrzyk and R. Kuziak, Modelling of Controlled Cooling of Rails after Hot Rolling, Proc. Conf. Rolling'2000, Vasteras, 2000 (CD ROM).

7. M. Glowacki, Termomechaniczno-mikrostrukturalny model walcowania w wykrojach ksztaltowych, Rozprawy Monografie AGH no. 76, Krakow, 1998.

8. M. Glowacki, R. Kuziak and M. Pietrzyk, “Modelling of Plastic Flow, Heat Transfer and Microstructural Evolution during Rolling of Vanadium Steel Rods”, Proc. Conf. Thermomechanical Processing in Theory, Modelling and Practice, eds, B. Hutchinson, M. Andersson, G. Engberg, B. Karlsson and T. Siwecki, Stockholm, 1996, pp. 258-265.

9. O.C. Zienkiewicz, Finite Element Method, McGraw Hill, 1977.

10. J.G. Lenard, M. Pietrzyk and L. Cser, Mathematical and Physical Simulation of the Properties of Hot Rolled Products, Elsevier, Amsterdam, 1999.

11. J.G. Lenard and M. Pietrzyk, "Thermal Mechanical Modelling of a Hot Strip Mill", Advanced Technology of Plasticity, Proc. 3rd ICTP, Kyoto, 1990, vol.2, pp. 601-606.

12. R. Kuziak, Y.W. Cheng, M. Glowacki, M. Pietrzyk, Modelling of the Microstructure and Mechanical Properties of Steels during Thermomechanical Processing, NIST Technical Note 1393, Boulder, 1997.

13. M. Umemoto, Mathematical Model of Phase Transformation from Work-Hardened Austenite, Symp. Math. Mod. of Hot Rolling of Steel, ed., S. Yue, Hamilton, 1990, pp. 404-422.

14. M. Pietrzyk and R. Kuziak, "Coupling the Thermal-Mechanical Finite-Element Approach with Phase Transformation Model for Low Carbon Steels", 2nd ESAFORM Conf. on Metal Forming, ed., J. Covas, Guimaraes, 1999, pp. 525-528.

15. M. Pietrzyk T. Kondek, J. Majta and A. Zurek, Method of Identification of the Phase Transformation Model for Steels, Proc. COM 2000, Ottawa, 2000, (in press).

A Systems Approach to Simultaneous Design of Roll Passes and Microstructure in Bar Rolling

Satish Kini and Rajiv Shivpuri
The Ohio State University
210, Baker Systems, 1971, Neil Avenue
Columbus, OH 43210
Tel: (614) 292-7874, Fax: (614) 292-7852
Shivpuri.1@osu.edu

Key Words: Bar rolling, Roll pass design, Microstructural evolution, FEM Simulation, Optimization

INTRODUCTION

The rolling process occupies the most important position of all the deformation processes. Over 90% of all metals that are ever deformed are subjected to rolling. During the hot rolling of rods defects are formed. These include flow-related defects such as fins, laps, seams etc. and material related defects such as segregations, inclusions, banding etc. in rolled products. These defects in rolled rods can often lead to failure of the parts formed by cold forging. The outstanding performance of steel as an engineering material is due to the wide range of microstructures and mechanical properties that are possible when it is subjected to controlled processing and heat treatment. The properties of rolled steel products depend upon the composition of the material, deformation history in the rolling mill and the heat treatment after rolling. Due to the large number of variables involved, designing a rolling process sequence often requires a large number of trials in order to produce a product with the desired tolerances and properties. Increased competitiveness among steel companies has accentuated the need to design roll pass sequences that give the desired product properties in the first production run. The objective is the "production of special bar quality (SBQ) rods at the same price as commercial grade (CG) rods" by controlling the rolling process

Many researchers have worked on the bar and flat rolling process and have developed empirical models. With the development of computer systems, numerical analysis has become a leading method for analyzing deformation in rolling processes. 3-D finite element methods to simulate the shape rolling process were developed by Kiuchi et al.[1] and Kim et al.[2]. Pauskar et al.[3] modified and improved Kim's software. Computer aided roll pass design systems have also been developed in the last decade. Mauk et al.[4] developed a computer aided roll design system for relieving the designer of tiresome routine work while leaving critical decisions to the roll pass design expert. Perotti et al.[5] introduced a new approach for computer aided roll pass design by combining empirical formulae and iterative schemes. Alberti et al.[6] proposed an integrated approach based on knowledge based systems and FEM methods. Shin[7] has reviewed some of the important work relevant to computer aided roll pass design.

Process modeling is one of the tools that can eliminate the trial and error process involved in designing a rolling process. Accurate prediction of dimensions and material properties in the hot rolled steel product requires modeling of the mechanical behavior of the material during rolling, microstructural evolution in the roll bite and in the interstand region, and the phase transformation during controlled cooling after rolling. Development of a comprehensive hot rolling model therefore requires an integrated approach to modeling metal flow, temperature distribution, microstructural evolution and phase transformation. With significant progress being made in recent years in modeling of microstructural evolution[8-13], mathematical models are gaining acceptance as powerful tools for conducting off-line prediction of metal flow and metallurgical changes during rolling. Fig. 1 shows the processes involved in the forming of a forged part, starting from casting, and ending with forging. Rolling and table cooling are the intermediate processes. Software that can be used for modeling these processes is shown in the figure. PROCAST and DEFORM are commercial software while ROLPAS and ROBROLL have been developed at The Ohio State University.

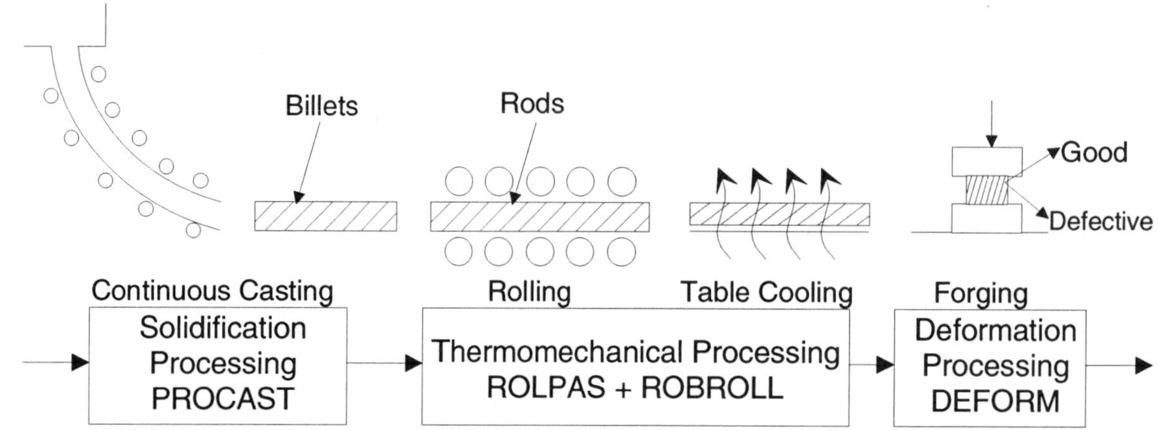

Fig. 1: Modeling of billet from casting to forging

In today's highly competitive and fast paced market, steel companies are very interested in developing offline analytical tools that would assist in the design of a rolling process by reducing or perhaps even eliminating the trial and error process. There is a need for mill design, roll pass design and microstructural design at the rolling mill. Each of these designs can be done separately but since they are dependant on each other there is a need for an integrated tool which can simultaneously and quickly design the roll pass sequence to take care of all these concerns. With the increasing use of computers in manufacturing, different kinds of modeling tools, such as the upper bound method, expert systems, optimization methods and the finite element method (FEM), have been developed. The integrated tool proposed in this paper will optimize the roll pass design and its various parameters so as to ensure a sound final product within geometrical tolerance and with no defects. It will use some of the newly developed modeling tools to achieve this purpose. The proposed optimization system has been incorporated with the current rolling model to rapidly give us the best roll pass design.

PROPOSED OPTIMIZATION SYSTEM

As mentioned above there are three main people who affect the quality of the final product. They are the roll pass designer, the mill operator, and the physical metallurgist. The roll pass designer specifies the intermediate passes trying to meet the desired geometry specifications and tolerance limits. The design should provide for highest possible production capacity and minimum cost of rolled stock. Earlier roll pass designs were mainly based on empirical formulae developed over the years, which varied in their functional character and applicability. Trial-and-error methods, which slightly modify one of the intermediate passes, take a lot of time, and costs are high. Hence the experience of the roll designers plays a major role in roll pass design. The mill operator is mainly concerned with the operation of the rolling mill and all related issues. He has to make sure that all parameters have been set correctly and that roll loads and torques are not above the permissible limits. The rolling speeds and the tension in the billets have to be controlled so as to avoid cobbles and defects. Finally the physical metallurgist looks at the microstructural characteristics of the final product. The final properties of the product are related to the austenite grain size due to microstructural evolution, the finishing temperature of rolling and the phase transformations during cooling after rolling. The metallurgist can suggest values of these parameters so as to make sure the final properties are within specifications. For a good sound final product it is required that all three people are taken into account. The current rolling model at The Ohio State University consists of an system developed for modeling metal flow and microstructural evolution during shape rolling. This is extremely useful for predicting thermal, mechanical and microstructural changes during rolling in steel bars. However it cannot optimize the rolling parameters so as to ensure that the final product falls within specifications.

A simple block diagram showing the proposed optimization system can be seen in figure 2. This system will be very useful in ensuring that all required product specifications are met and that the final product is defect free and of very high quality. There are five main steps in this strategy for roll pass design. The first step is the assessment of the initial pass design and the rolling setup. This is followed by a mathematical analysis of the initial pass design using finite element simulation software. This gives us detailed outputs of all important factors such as temperatures, loads, grain size, mechanical properties etc. The next step is the estimation of the errors in the initial design. Here we compare the results of the analysis with desired output specifications and find out what we need to improve and optimize. The next step is the macro optimization of the roll pass sequence. Here we use a very quick analytical solution method along with an optimization technique to optimize the roll pass design. However since we make certain assumptions the results are not as accurate as that of the finite element solution. This is an iterative procedure and continues until we get the best solution. The final step is the micro optimization of the results of the previous steps. Here we again use finite element software and optimization to fine-tune the parameters so as to get the final optimum solution. Each of these steps is now described in more detail.

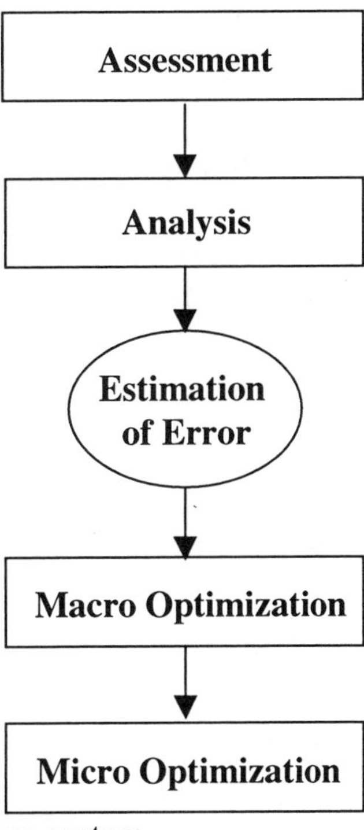

Fig. 2: Proposed roll pass design optimization system

Assessment

In this step the initial or current design is thoroughly assessed by getting information from the roll pass designer and his colleagues. The first thing done is to gather all the inputs required to analyze the roll pass design using simulation software. This information is gathered from the roll pass designer, the mill operator and the physical metallurgist. It may be obtained by using a questionnaire or by interviews. The inputs required for the analysis of the initial pass design include the following:

- Billet material and properties
- Roll material and properties
- Roll pass sequence and individual caliber shapes
- Roll diameters, roll gaps and roll speeds
- Interstand distances and interpass conditions
- Cooling conditions
- Final shape and tolerances required
- Ranges of final hardness and mechanical properties required

The following inputs may not be known for a new sequence but are known for a current sequence. These help later in error estimation and defect diagnosis.

- Austenite grain size
- Material defects
- Shape defects such as seams and cobbles
- Measured shape and properties
- Mill loads, finishing temperature of rolling

From this information we can assess if the current situation is under control. We also get an idea of the final properties and dimensions that are out of specification.

Analysis

After the assessment of the roll pass sequence to be optimized is completed, the next step is the analysis of this sequence. The Manufacturing Research Group (MRG) at The Ohio State University has developed software to predict the quality of the rolled bar. An integrated system developed for modeling metal flow (ROLPAS) and microstructural evolution (MICON) during shape rolling has been developed. The system includes modules for modeling of phase transformation (AUSTRANS) during table cooling and the prediction of mechanical properties of the rolled product. The central feature of the current integrated hot rolling model is a 3-D finite element program ROLPAS for simulating multi-pass shape rolling (Fig. 3). This FEM code predicts the distribution of temperature, strain, strain rate, the material flow as well as the roll separating forces and torques. The code is being used by a number of steel companies for metal flow simulations and has yielded very good results. Details on the formulation of the FEM model can be found in Pauskar[14].

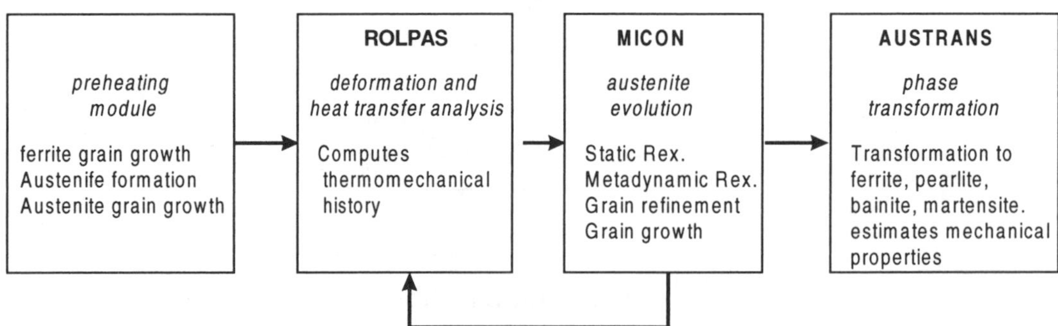

Fig. 3: Flow diagram of the integrated hot rolling model

A microstructural evolution module MICON (Fig. 3) has been integrated into ROLPAS to enable modeling of austenite evolution[9]. MICON uses the thermomechanical history computed by the FEM program in conjunction with empirical microstructure models to predict the evolution of austenite during hot rolling. The microstructural changes occurring in bar rolling are primarily due to static recrystallization and grain growth that occur in the interstand region. In roll passes where accumulated strain is large meta-dynamic recrystallization is modeled in the interstand region following Sellars[15]. The ability of the program to model the evolution of austenite during rolling with reasonable accuracy was demonstrated by Pauskar et al.[16]. ROLPAS uses the microstructural evolution data to model subsequent deformation and to compute the thermomechanical history. This thermomechanical history data is in turn used to model microstructural evolution in the next interstand.

The last segment in the system is a phase transformation modeling module AUSTRANS (Fig. 3). It uses the temperature history after rolling (computed by ROLPAS) and the isothermal transformation data to model the transformation of austenite to ferrite, pearlite, bainite and martensite. This model also uses structure-property relationships to predict the mechanical properties of the rolled product. Phase transformation occurs during cooling from the hot austenite phase to room temperature. The kinetics of diffusional transformation are completely described by the isothermal transformation (IT) diagram which includes the ferrite start curve, pearlite start and end curves, bainite start and end curves and precipitation of second phases or inclusions. Some form of digital representation of the various curves in the IT diagram is required to model transformation and to predict properties of the final part. An intelligent technique was used to adapt the isothermal Johnson-Mehl-Avrami equation to the non-isothermal transformation processes. AUSTRANS uses the temperature history and microstructural details from the last interstand to model transformation of austenite.

Hence in this step by using the rolling model developed at OSU we are able to predict the final properties of the billet as well as the loads and torques at individual passes. State variables such as the geometry of the billet and its microstructure (austenite grain size and different phases) are also predicted in this analysis step. Even though this step is relatively quick due to the advancements in computer technology and improvements in the FEM software, it is not instantaneous and takes a few hours depending on the number and type of passes.

Estimation of Error

The aim of this step is to estimate the error in the initial pass design so that we know how far we are from the final product specifications. To do this, the results of the analysis in the second step are compared with the final specifications originally assessed in the first step. A scheme will be designed so that both the magnitude and the distribution of the error from the nominal are estimated. This task of estimating the error is not as trivial as it seems since there are various specifications that have to be met. Examples of different specifications include mechanical properties such as % elongation, tensile strength, yield strength etc. and the different dimensions of the final product. Hence we have to not only find out which specifications are within the acceptable range but also how much weight each error should be given in the final error function. If certain specifications are within limits then the initial weights for these specifications in the initial error function are initialized to zero. During the optimization procedure in the next step the weights will change depending on which specifications are met.

Macro Optimization System

This step is the most important step in the whole optimization process. Figure 4 shows a block diagram of the macro optimization system. The first block shows the input into the system that basically consists of the data from the first three steps. This includes the data from the assessment and analysis of the initial roll pass design and the error estimated in the previous step. The next step is to perturb the roll pass design parameters so as to minimize the error and bring all the specifications within the acceptable ranges. To do this we need to know the effects of different roll pass parameters on the final properties and the final shape of the product. To do this we need a lot of information. A number of different tools will be used in order to decide the parameters to be perturbed and the magnitude of the perturbation. They include roll pass design rules and formulae from literature, influence coefficients developed using the FE software ROLPAS, the modules MICON and AUSTRANS and the software ROBROLL.

A lot of researchers have worked in the field of bar rolling and a great amount of information relating final outputs with the different parameters has been developed. Experienced roll pass designers have also developed rules for roll pass design. Influence coefficients tell us the effect of design parameters on the final output. Running various simulations using FE software and finding the effect of the different roll pass parameters on the various outputs will develop these coefficients. Design of experiments will be used to reduce the number of simulations to develop the influence coefficients. MICON and AUSTRANS will also be used to decide the parameters to be perturbed to get the final properties within specifications. AUSTRANS will be used to determine the best conditions needed after rolling so as to get the required properties since it relates the final properties to conditions after rolling and the cooling rate. MICON will be used to determine the thermomechanical history that the billet needs to experience to get a particular microstructure.

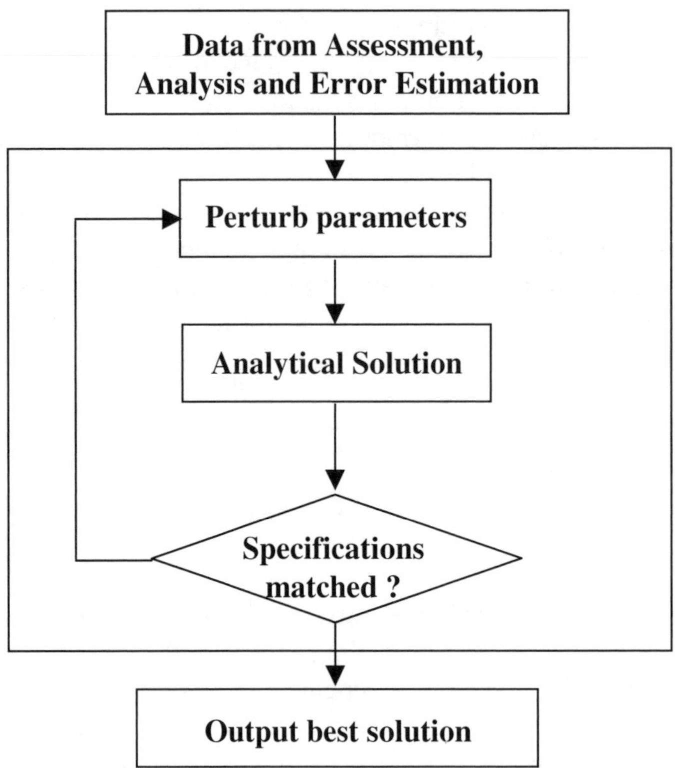

Fig. 4: Macro optimization system

Defect diagnosis will be carried out to relate the various defects to the different roll pass design parameters. The aim will be to be able to develop a strategy to change values of the rolling parameters so as to prevent the occurrence of defects. During rolling, we get variation in the output of the roll pass sequence or the fill of the final pass. This variance in filling (overfill and underfill) is related to the noise factors that exist in the rolling site, such as variance in material components, billet geometry, roll diameter, friction between rolls and material, roll wear, rolling temperature and so forth. The variance in spread often causes surface flaws (laps, seams, fins, etc.) in the product geometry that need to be minimized. ROBROLL (ROBust Design of ROd RoLLing Pass)[17] is a computer-aided system that takes care of these input variations and designs a robust roll pass sequence. This system integrates well-established empirical relations, Taguchi methods of experimental design and the finite element method of process simulation to arrive at robust roll pass designs. These designs are then incorporated into software for simultaneous consideration of metal flow and rolling mill operations in bar rolling plants. The software can be used to evaluate and improve existing passes in bar mills by reducing the variation in the dimensions of the final rolling product and minimizing the occurrence of flow related defects.

The parameters that can be perturbed so as to effect the final product and its properties include all the roll pass sequence parameters and the interstand conditions. Other parameters include the preheat temperature and the rolling speeds. Since there are a lot of parameters which can be changed we need an optimization technique to be used along with the rules and modules used for perturbing the parameters. An optimization strategy will be chosen for this purpose. We will not obtain the final result in one go but will have to run a number of iterations to get our final optimal result. In every iteration the parameters will be perturbed and the roll pass sequence will then be analyzed to see the effects on the final properties. Here the FEM analysis cannot be used since it would cause the optimization to take a prohibitively long period of time. This would lead to this method being ineffective in terms of time and cost.

Hence after the parameters are perturbed we use an analytical solution to analyze the roll pass sequence. The flow diagram of the analytical solution is shown in figure 5. Over the years various mathematical models have been developed to simulate rod rolling. By using these models we can analytically simulate the roll pass sequence and get macro results. Macro results do not give us cross sectional distributions but instead give us average values of the different paramters. This does not help us in determining defects but does give a very good idea of the final shape and mean properties of the product. Formulae and equations from literature will be used for each block in the flow chart (figure 5).

After analyzing the new sequence a check is made to see if the properties match the specifications (figure 4). If not the parameters are perturbed again and the sequence analyzed once more. The iterative procedure goes on until all the properties match the specifications. We then go to the final step in the roll pass optimization system.

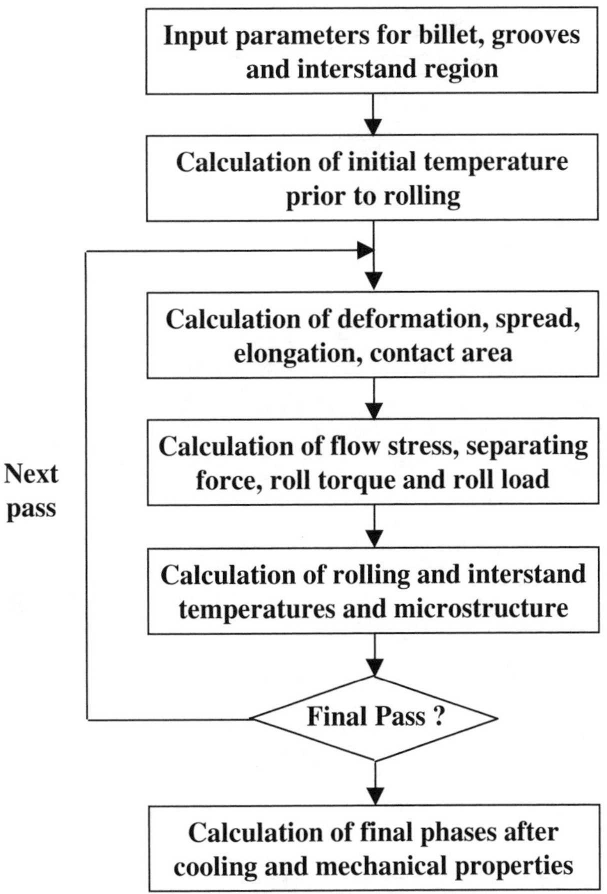

Fig. 5: Analytical solution

Micro Optimization

This is the final step of the proposed roll pass optimization system. Now that we have got a new roll pass design we need to check it to make sure that our analytical solution gave us the correct results. Even though the analytical solution has been calibrated with the more accurate FEM analysis method, due to the assumptions made, the solution will not be exact. Hence minor adjustments will be needed so as to micro optimize the new roll pass design. Here we use the FE software to check for defects and final properties and make minor changes to the roll pass parameters so that all properties are within specifications. We also use the rules and influence coefficients to guide the slight tweaking of the roll pass parameters that is required. A final FEM analysis is carried out to verify that the new roll pass design ensures that all properties and dimensions are within specifications. One advantage of this system is that it allows the user to conduct a quick sensitivity analysis to see the effect of varying process parameters on the final properties. Example contours showing distribution of austenite grain size are illustrated in figure 6. Figure 7 shows progress of phase transformation in the final product on cooling after rolling. Figure 8 shows the variation in the mechanical properties along the radial direction for three different reheat temperatures. Similar results can be obtained easily by changing other parameters such as the roll pass sequence, interstand cooling, and post rolling cooling conditions.

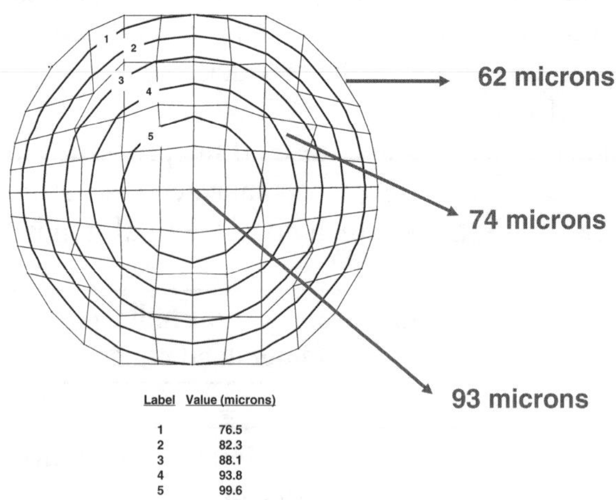

Fig. 6: Contours showing distribution of austenite grain size

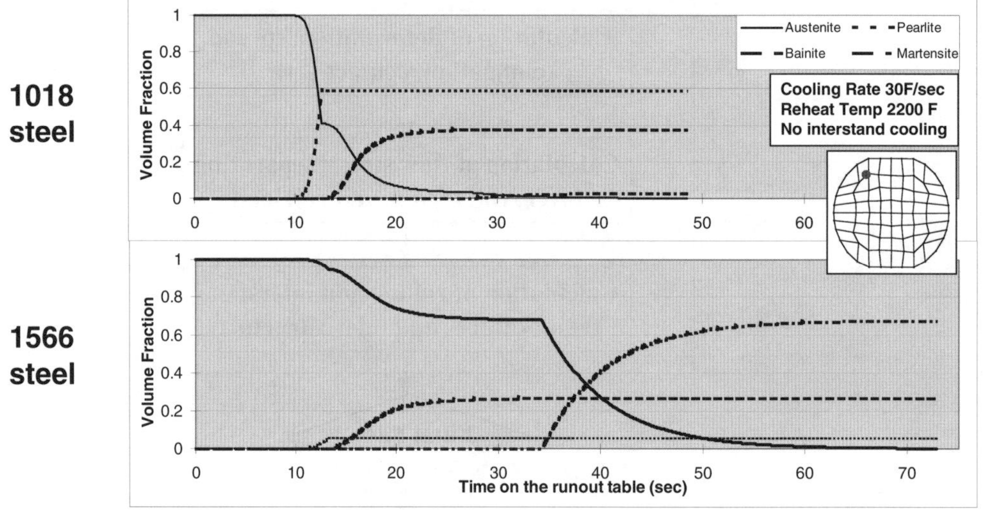

Fig. 7: Progress of phase transformation

CONCLUSIONS AND FUTURE WORK

During rolling defects are formed in the final product if the process is not properly controlled. Typical defects include seams, fins, segregations, banding, inclusions etc. These defects lead to failures such as internal and external cracks when the part is cold forged. Rolling thus has a great effect on the forgeability of the rods and bars. The aim is to reduce defects and improve product quality by controlling the hot rolling process. Optimizing the roll pass sequence and all its parameters so as to make all properties within specifications and preventing defects does this. For this purpose an optimization system has been proposed. The system has been described in the paper and uses software for rolling simulation developed by the Manufacturing Research Group at The Ohio State University. The software developed for performing off-line computer analysis of the roll pass sequences was used to model metal flow, microstructural evolution and phase transformation during hot rolling. An analytical solution along with an optimization technique is proposed for macro optimization of the roll pass sequence. The final micro optimization is then carried out. This optimization system is currently being implemented at The Ohio State University.

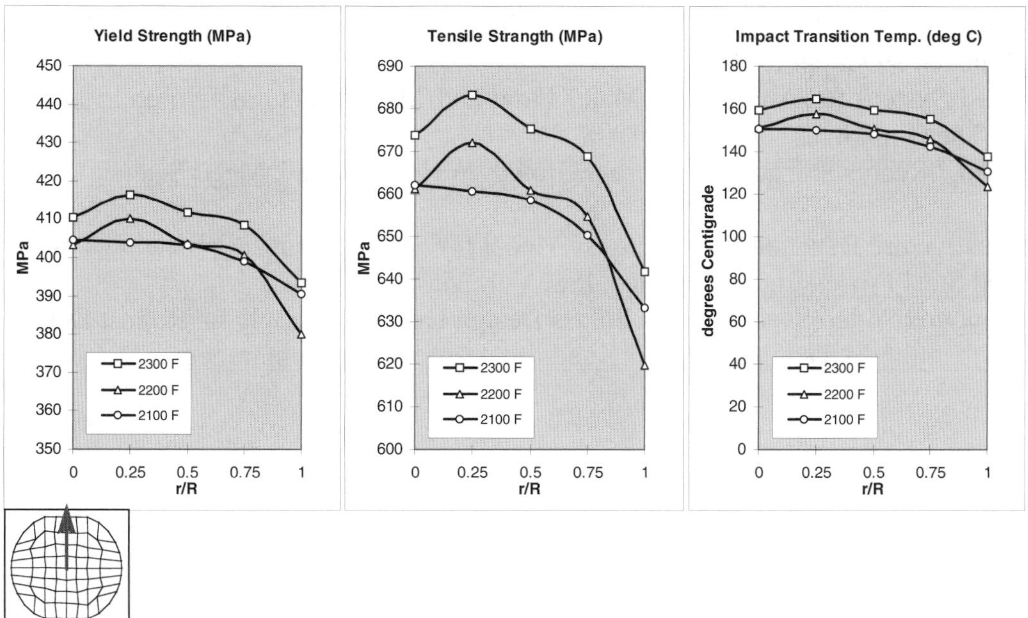

Fig. 8: Variation of mechanical properties along the radial direction for three different initial temperatures

ACKNOWLEDGMENT

The authors gratefully acknowledge the financial and technical support provided by the Consortium for Advancement of Rolling Technology (Members: Chaparral Steel, Inland Steel and Timken Steel).

REFERENCES

1. M. Kiuchi, and J. Yanagimoto, "Computer Aided Simulation of Shape Rolling Processes" Proceedings of 16th NAMRC, September 1987, pp. 34-40.
2. N. Kim, S. Kobayashi, and T. Altan "Three-dimensional Analysis and Computer Simulation of Shape Rolling by the Finite Element and Slab Element Method." Int. J. of Machine Tools and Manufacturing, Vol.31, No.4, 1991, pp. 553-563.
3. P. Pauskar, R. Shivpuri, H. Cho, and N. Kim, "Microstructure Prediction in Multipass Bar Rolling with Interstand Cooling", 38th Mechanical Working and Steel Processing Conference Proceedings, Vol. 34, 1996, pp. 137-146.
4. P.J. Mauk and R. Kopp, “Computer-aided Roll Pass Design”, Der Kalibreur, Vol. 37, 1982, pp.89-93.
5. G. Perotti, and N. Kapaj, “Roll Pass Design for Round Bars”, Annals of the CIRP, Vol. 39(1), 1990, pp. 283-286.
6. N. Alberti, L. Cannizaro, and F. Micari, “Knowledge-based Systems and F.E. Simulations in Metal-forming Processes Design: An Integrated Approach”, Annals of the CIRP, Vol. 40(1), 1991, pp. 295-298.
7. W. Shin, “Development of Techniques for Pass Design and Optimization in the Rolling of Shapes”, Ph.D. Dissertation, OSU, 1995.
8. P. Choquet et al.; Mathematical Modeling of Hot Rolling of Steel, edited by S. Yue, Canadian Institute of Mining and Metallurgy, Montreal, Canada, 1990, pp. 34-43.
9. Pauskar, P. et al., Proceedings of the 38th Mechanical Working and Steel Processing Conference, ISS, Vol. 34, 1997, pp.137-146.

10. Y. Saito, and C. Shiga, ISIJ International, 1992, pp. 414-422.
11. I.V. Samarasekhara, D.Q. Jin, and J.K. Brimacombe, 38th Mechanical Working and Steel Processing Conference Proceedings, ISS, Vol. 34, 1996, pp. 313-327.
12. C.M. Sellars and J.H. Benyon, ISIJ International, Vol. 32, No. 3, 1992, pp. 359-367.
13. H. Yada, Mathematical Modeling of Hot Rolling of Steel, edited by S. Yue, Canadian Institute of Mining and Metallurgy, Montreal, Canada, 1990, pp. 2-36.
14. P. M. Pauskar, "An Integrated System for Analysis of Metal Flow and Microstructural Evolution in Hot Rolling", Ph.D. Dissertation, The Ohio State University, 1998.
15. C.M. Sellars, and A. Whiteman, "Recrystallization and Grain Growth in Hot Rolling", Metal Science, Mar.-Apr. 1979, pp. 187-194.
16. P.M. Pauskar et al., "A Microstructure Dependant Flow Stress Model for Steel Rolling", Proceedings of the 39th Mechanical Working and Steel Processing Conference, ISS, Vol. 35, 1997.
17. S. Kini and R. Shivpuri, "Computer Aided System for Roll Pass Design Evaluation applying Robust Design Techniques", Proceedings of the 40th Mechanical Working and Steel Processing Conference, ISS, Vol. 36, 1998.

The Effects of Aluminum Nitride Precipitation on the Development of Toughness in 9313 Steel

M. J. Leap and J. C. Wingert
Materials Science Department
The Timken Company
Canton, Ohio 44706

Key Words: impact toughness, aluminum nitride, tempered martensite, solidification, precipitate dissolution, precipitate coarsening

INTRODUCTION

The toughness of grain-refined, high-strength steels is highly dependent on the content and dispersion of various types of second-phase particles in the microstructure. Coarse AlN precipitates, for example, degrade the toughness of tempered martensite over a broad range of test temperature by providing preferential sites in the microstructure for the formation of transgranular cleavage facets, quasicleavage facets, and secondary microvoids (1-5). Although high-strength steels containing 0.2-0.4% C typically contain AlN precipitates with sizes up to about 400 nm (4,5), the precipitate dispersions can be effectively refined during the reheating operation that precedes finish rolling or forging (6).

It has been more recently found that air-melt 9313 steel may contain low densities of extremely coarse (~1 μm) AlN precipitates after hot working, forging, and heat treating. The sporadic occurrence of coarse AlN precipitates degrades the fracture resistance and contributes to a high level of heat-to-heat variability in the toughness of forged and heat-treated components. However, the magnitude of this effect is unknown relative to the degradation in toughness that typically results from the presence of much higher densities of intermediate-sized (100-400 nm) AlN precipitates in high-strength steels (1-6). Since low-carbon alloy steels such as 9313 are utilized in a variety of applications that have demanding toughness requirements, the potential to develop low levels of toughness is unacceptable from the standpoint of product quality and consistency.

The purpose of this investigation is to evaluate the processing factors that promote the development of low toughness in air-melt 9313 steel. A primary goal of this study is to phenomenologically evaluate the effects of processing on toughness in terms of the evolution of AlN precipitates. A second objective is to theoretically identify the most likely mechanism(s) associated with the formation and retention of extremely coarse AlN precipitates in the microstructure and determine if the development of coarse AlN is inherent to this steel composition. Finally, the interrelated effects of intermediate-sized and coarse AlN precipitates are considered in the context of optimizing material toughness and minimizing variability in toughness.

EXPERIMENTAL PROCEDURE

Materials and Processing

Product chemistries for two 9313 steels are listed in Table I. Both ingot cast heats possess aluminum, nitrogen, and residual titanium contents characteristic of aluminum-killed EAF steel, although steel A exhibits a significantly higher sulfur content than steel B.

Table I: Steel Chemistries (weight percentages)

Steel	C	Mn	Si	Cr	Ni	Mo	S	P	Ti	Al	N (ppm)	O (ppm)	E (J)[1]
A	0.14	0.69	0.22	1.45	3.23	0.10	0.006	0.009	0.002	0.023	92	8	61
B	0.15	0.70	0.24	1.43	3.25	0.12	0.002	0.009	0.003	0.026	96	6	14

[1]E = room-temperature impact toughness obtained from forged and heat-treated components.

Material from steels A and B were obtained in the form of as-forged components (~40 kg). The manufacturing of the components typically comprises reheating sections of 140 mm bar for a minimum amount of time at a low-to-intermediate temperature (~1150-1175°C), upset forging the mults into the component shape, and bin cooling the forgings to room temperature. The axis of revolution of the conical-shaped component is maintained parallel to the axis of the hot-rolled bar during the forging operation such that the longitudinal orientation in the forged component generally parallels the longitudinal orientation in the hot-rolled bar. The forgings were subsequently carburized at 954°C for eight hours, slow cooled, reaustenitized at 829°C for two hours, oil quenched, and tempered at 204°C for four hours.

Material from steel B was also obtained in the form of 140 mm diameter wrought bar. The bar was sectioned into 102 mm lengths, and two, 38 mm thick hot-rolling specimens were extracted in the longitudinal orientation from positions straddling the centerline of each bar section, Figure 1. The hot-rolling specimens were subjected to several pretreatment schedules comprising reheating and hot rolling into 25 mm plate, Table II.

The longitudinal orientation of the bar was maintained and a limited amount of deformation was imposed during hot rolling to simulate the material flow that occurs during the forging of the components. Reheating temperatures were selected based on mid-range and upper-bound estimates of the AlN solution temperature for steel B, and specimens were reheated both above and below the AlN solution temperature prior to hot rolling.[1] The plates were either air cooled or oil quenched to room temperature after hot rolling, and one-half of the oil quenched plates were subcritically annealed at 700°C for two hours and air cooled to room temperature, Figure 2.

---

[1]The solubility products derived by Darken et al. (7) and Leslie et al. (8) yield AlN solution temperatures of 1060°C and 1188°C, respectively.

Table II: Summary of Processing Parameters for the Pretreatment Operation

Reheating Temperature (°C)	Reheating Time (hours)	Number of Passes	Hot-Rolling Reduction (%)
975	1.0	2	15 & 17
1050	1.0	1	33
1225	2.0	1	33

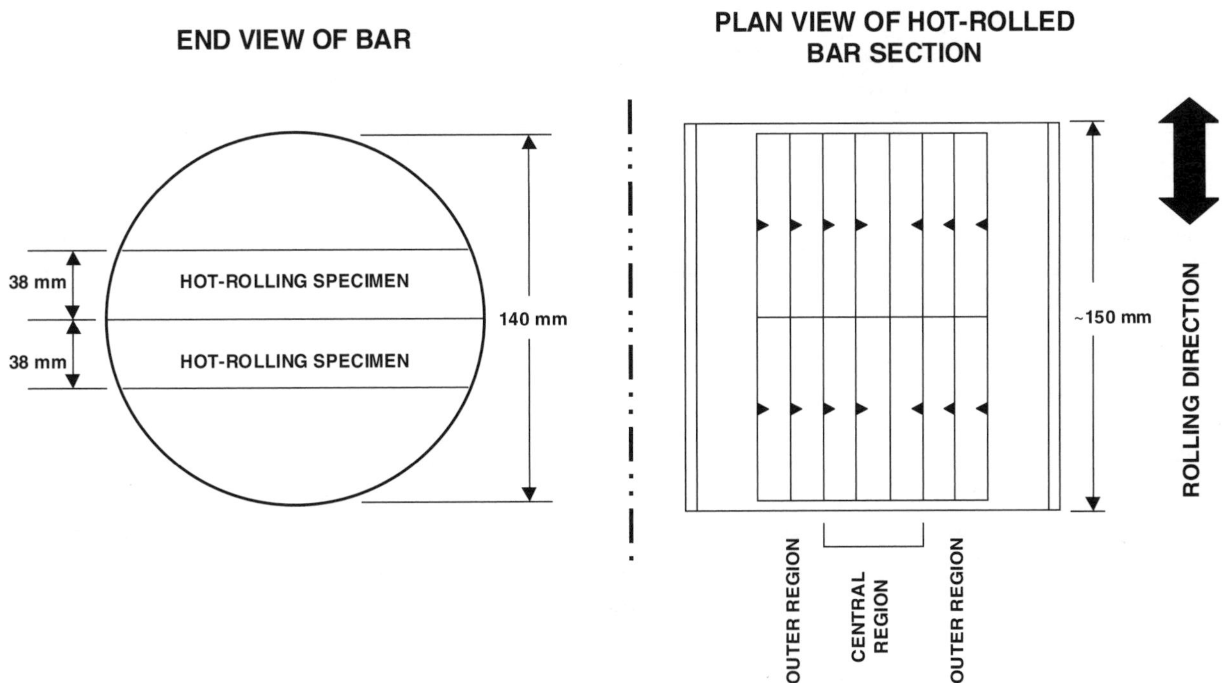

Figure 1. Schematic illustration of the hot-rolling specimen along with the sectioning plan for the Charpy V-notch specimens.

Plates subjected to each pretreatment condition were pseudocarburized at 954°C for eight hours in a fluid bed furnace with a nitrogen atmosphere and air cooled to room temperature, Figure 2a. Plates hot rolled at 975°C and air cooled were also pseudocarburized at 954°C for 16 hours and 1000°C for eight hours to evaluate the sensitivity of impact toughness to variations in carburizing parameters. All plates were subsequently austenitized at 829°C for two hours in a fluid bed furnace with a nitrogen atmosphere, oil quenched, and tempered at 204°C for four hours.

Plates of the air cooled and oil quenched/subcritically annealed steels were also austenitized after hot rolling without the intermediate pseudocarburizing operation, Figure 2b. These plates were austenitized at 829°C for two hours in a fluid bed furnace with a nitrogen atmosphere, oil quenched, and tempered at 204°C for four hours.

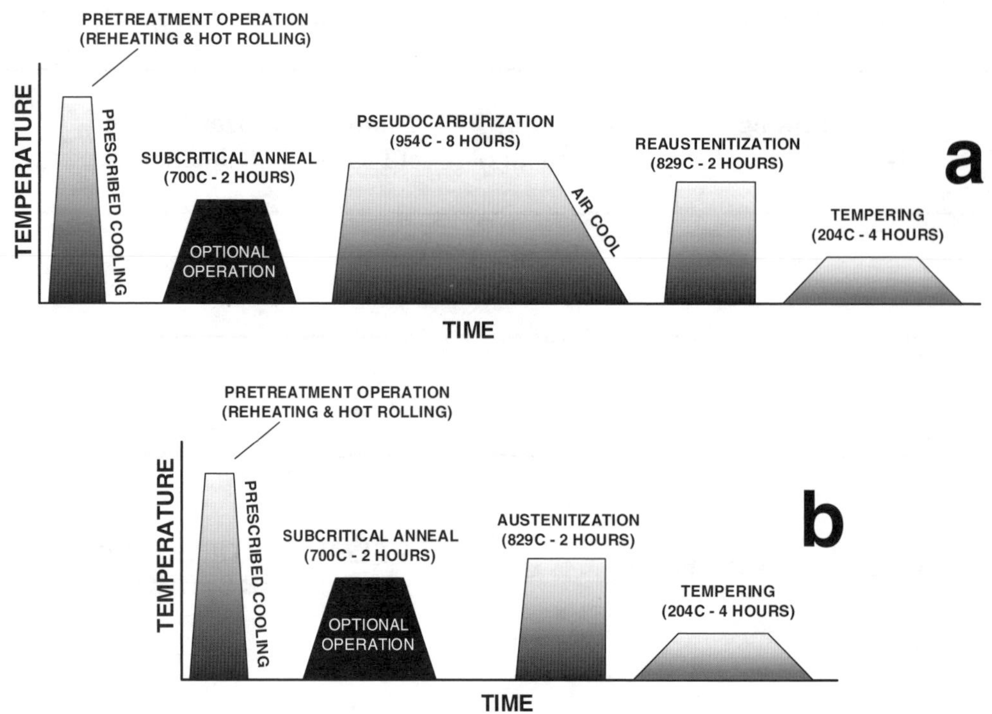

Figure 2. Schematic illustration of the processing history for hot-rolling specimens extracted from the wrought bars: (a) processes including both pseudocarburizing and hardening operations and (b) processes including a hardening operation.

Mechanical Testing and Metallography

Standard Charpy V-notch specimens were extracted from the *outer core region* of the forged and heat-treated components in the LR orientation.[1] Standard Charpy V-notch specimens were also extracted from the mid-plane of the heat-treated plates in the LT orientation, Figure 1. The general position of specimens in the plates was tracked to delineate differences in toughness between the central and outer regions of the plates. Charpy specimens were tested at room temperature and 120°C in accordance with ASTM E23. The latter test temperature was utilized in an attempt to obtain estimates of upper-shelf toughness for each material condition.

The austenite grain structures exhibited by steel B were qualitatively evaluated by light microscopy. Charpy specimens from the hot-rolled plates were sectioned on a plane perpendicular to the rolling direction, tempered at ~480°C for 24 hours, prepared for examination, and etched in a saturated picric acid solution containing sodium tridecylbenzene sulfonate as a wetting agent.

Fractography was conducted on Charpy specimens with an Amray 3300 scanning electron microscope operating at 20 kV.

The state of AlN precipitation on the fracture surfaces of Charpy specimens was evaluated by transmission electron microscopy. The Charpy specimens were extracted from forged and heat-treated components and tested at room temperature. Single-stage carbon extraction replicas were prepared by evaporating carbon onto the central, unstable fracture region of the specimens. The replicas were lifted in 10% nital, mounted on copper grids,

[1]The toughness of the core region is the limiting factor in the design of the forged and heat-treated components. Therefore, all impact toughness data in this investigation represent the (core) properties of the base steel composition in hot-worked and heat-treated specimens.

and viewed in a Philips CM-200 operating at 200 kV. Precipitates located on the fracture surfaces were primarily characterized in terms of morphology, although the compositional classification of many precipitates was verified by energy dispersive spectroscopy. Quantitative estimations of the state of precipitation were obtained following the general computational procedures outlined by Ashby and Ebeling (9).

RESULTS

Fracture Behavior of Forged and Heat-Treated Components

Steel A possesses significantly greater room-temperature impact toughness than steel B, Table I. Charpy specimens of steel B exhibit fractures in which a limited amount of ductile crack extension from the notch root is followed by unstable fracture in a brittle manner, Figure 3. Charpy specimens of steel A also exhibit fractures

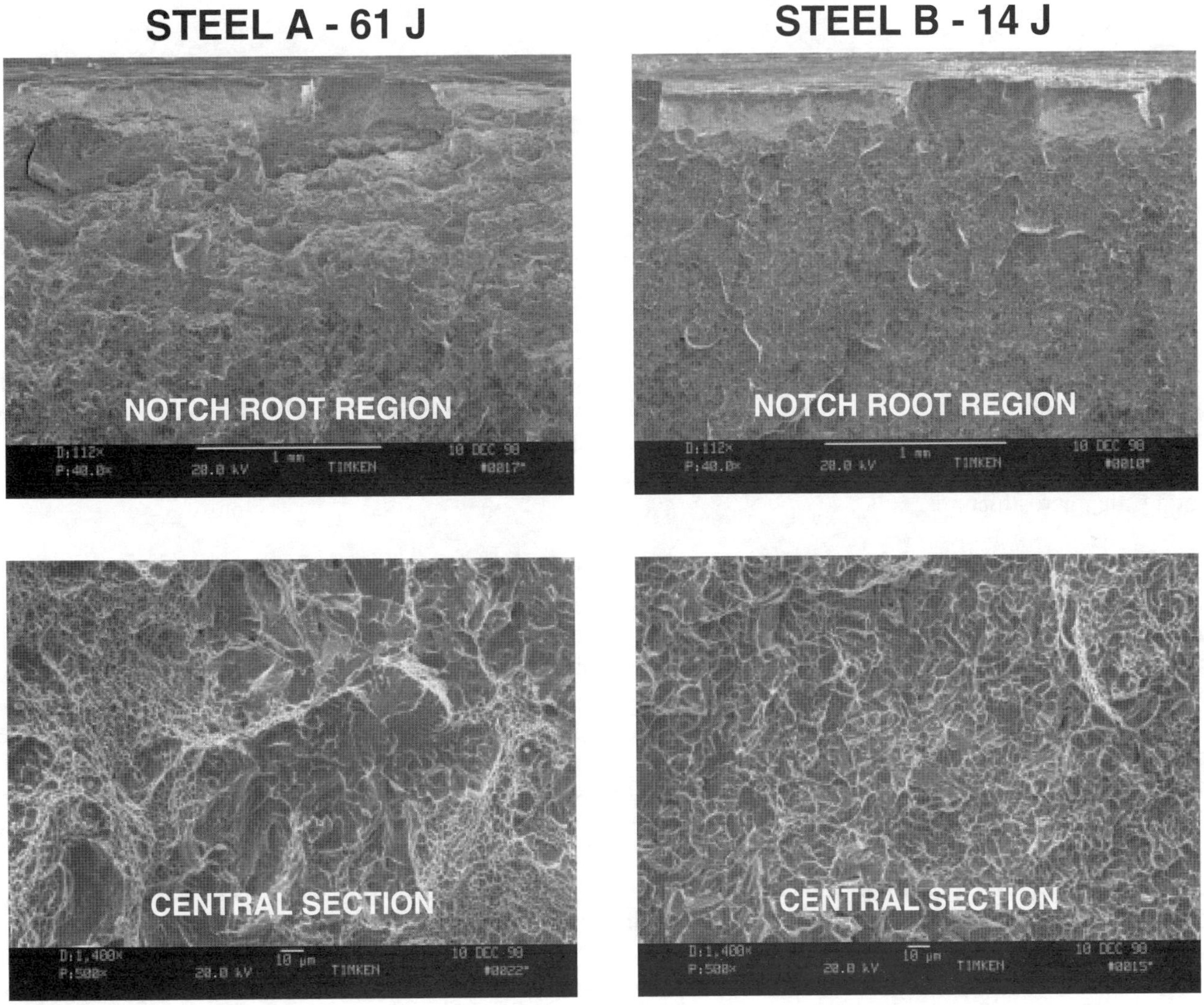

STEEL A - 61 J

STEEL B - 14 J

NOTCH ROOT REGION

NOTCH ROOT REGION

CENTRAL SECTION

CENTRAL SECTION

Figure 3. The morphology of fracture exhibited by Charpy V-notch specimens is shown for steels A and B. The Charpy specimens were extracted in the longitudinal orientation from the core region of forged and heat-treated components.

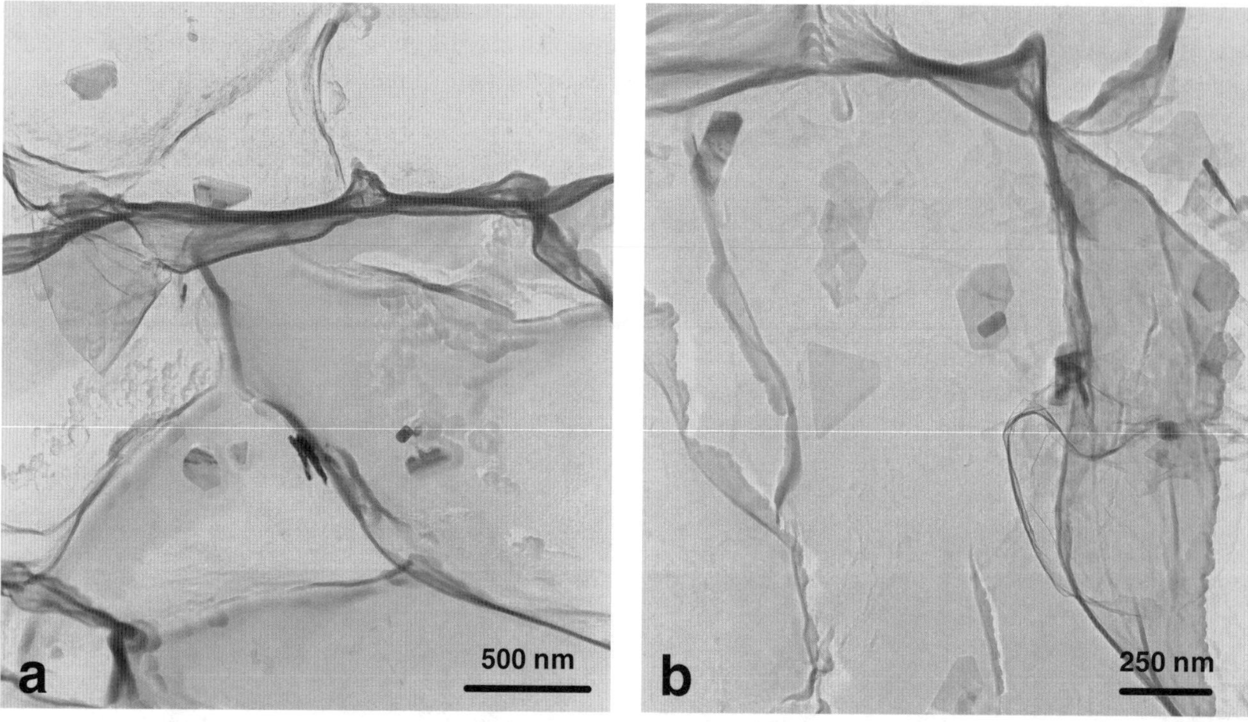

Figure 4. AlN precipitates on the fracture surface of Charpy V-notch specimens are shown for forged and heat-treated components of (a) steel A and (b) steel B. TEM fractographs of single-stage carbon extraction replicas.

characteristic of transition-temperature behavior, but in this case unstable fracture occurs via a mixed-mode mechanism. Notwithstanding the differences in unstable fracture morphology, the large difference in the amount of ductile crack extension between the two steels suggests the microstructural feature(s) that initiate unstable fracture (i.e., interrupt the ductile fracture process) limit the development of toughness in the lightly tempered martensitic microstructures.

Forged and heat-treated material from both heats exhibits coarse AlN precipitates on the fracture surfaces of Charpy V-notch specimens tested at room temperature, Figure 4. This observation indicates that coarse AlN precipitates are the microstructural feature responsible for the initiation of a majority of local fracture events that produce cleavage/quasicleavage facets in both steels. Coarse AlN precipitates as well as some TiN particles are located in the secondary microvoids that comprise void sheets on the fracture surface of the steel A specimen. A low density of smaller TiN precipitates is also present on the fracture surface of the Charpy specimen from steel B. However, the overwhelming presence of coarse AlN precipitates on the fracture surfaces suggests that the impact toughness of the steels is, for all practical purposes, relatively insensitive to the presence of a lower density of smaller TiN precipitates in the microstructure.

Size distributions of AlN precipitates located on the fracture surfaces of Charpy V-notch specimens are shown in Figures 5a and 5b for the two steels. The dispersion of AlN precipitates on the fracture surface of the Charpy specimen with comparatively high toughness (steel A) exhibits a mean size of 132 nm. A majority of the AlN precipitates in this material are less than 400 nm in size, although infrequent occurrences of precipitates as large as 625 nm are observed on the fracture surface. The fracture surface of the steel B Charpy specimen exhibits a somewhat coarser AlN dispersion with a mean size of 169 nm. While a majority of precipitates in this material are also less than 400 nm in size, a significant density of precipitates as large as 1 μm are found on the fracture surface. In addition, the steel B specimen exhibits a higher density of precipitates on the fracture surface than the steel A specimen.

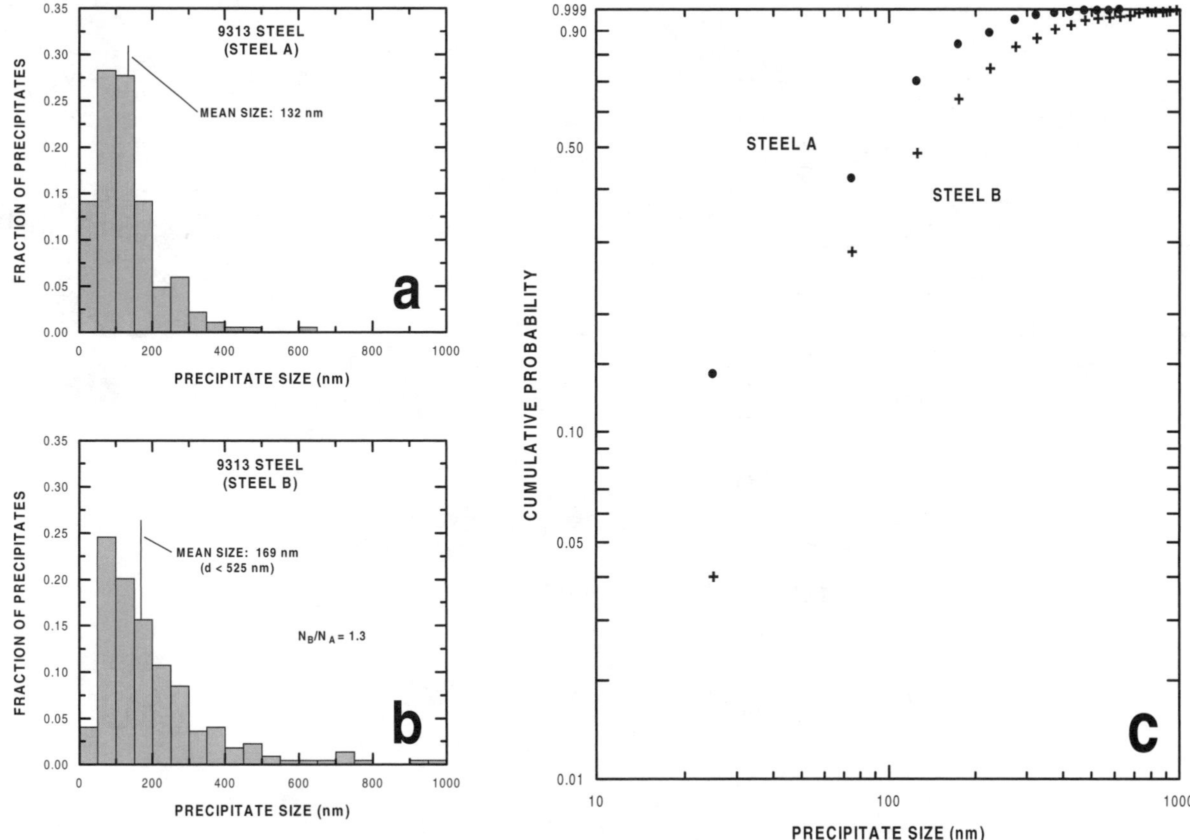

Figure 5. AlN size distributions are shown for precipitates located on the fracture surfaces of Charpy V-notch specimens obtained from forged and heat-treated components: (a) steel A and (b) steel B. The (c) cumulative distributions of precipitate size based on the grouped data in (a) and (b) are shown for Charpy specimens of steels A and B.

Effects of Processing on the Development of Austenite Grain Structures

Reheating at 975°C and hot rolling produces similar austenite grain structures in the hardened and pseudocarburized and hardened specimens for each pretreatment condition. The specimens generally exhibit relatively fine-grained structures, although limited amounts of abnormal grain coarsening are evident in the form of somewhat coarser grains dispersed throughout the microstructure, particularly in the air cooled plates. Subsequent to reheating at 1050°C and hot rolling, pseudocarburized and hardened specimens of both the air cooled and oil quenched/subcritically annealed plates exhibit regions of more extensive abnormal grain coarsening. However, the largest changes in grain structure appear to result from changes in the pseudocarburizing parameters after hot rolling at low temperatures, Figure 6. An increase in pseudocarburization time from eight to 16 hours at 954°C, for example, increases the extent and severity of abnormal grain growth in the air cooled plates, whereas similar grain structures result from pseudocarburizing for eight hours at 954°C and 1000°C.

Reheating at 1225°C and hot rolling produces a wide variety of austenite grain structures in steel B. Abnormal grain coarsening is evident in pseudocarburized and hardened specimens of both the air cooled and oil quenched/subcritically annealed plates, although the grain structure in the air cooled plate is coarser and exhibits regions containing extremely coarse grains, Figures 7a and 7b. In contrast, hardened specimens of both the air cooled and oil quenched/subcritically annealed plates exhibit extremely coarse austenite grain structures, Figures 7c and 7d.

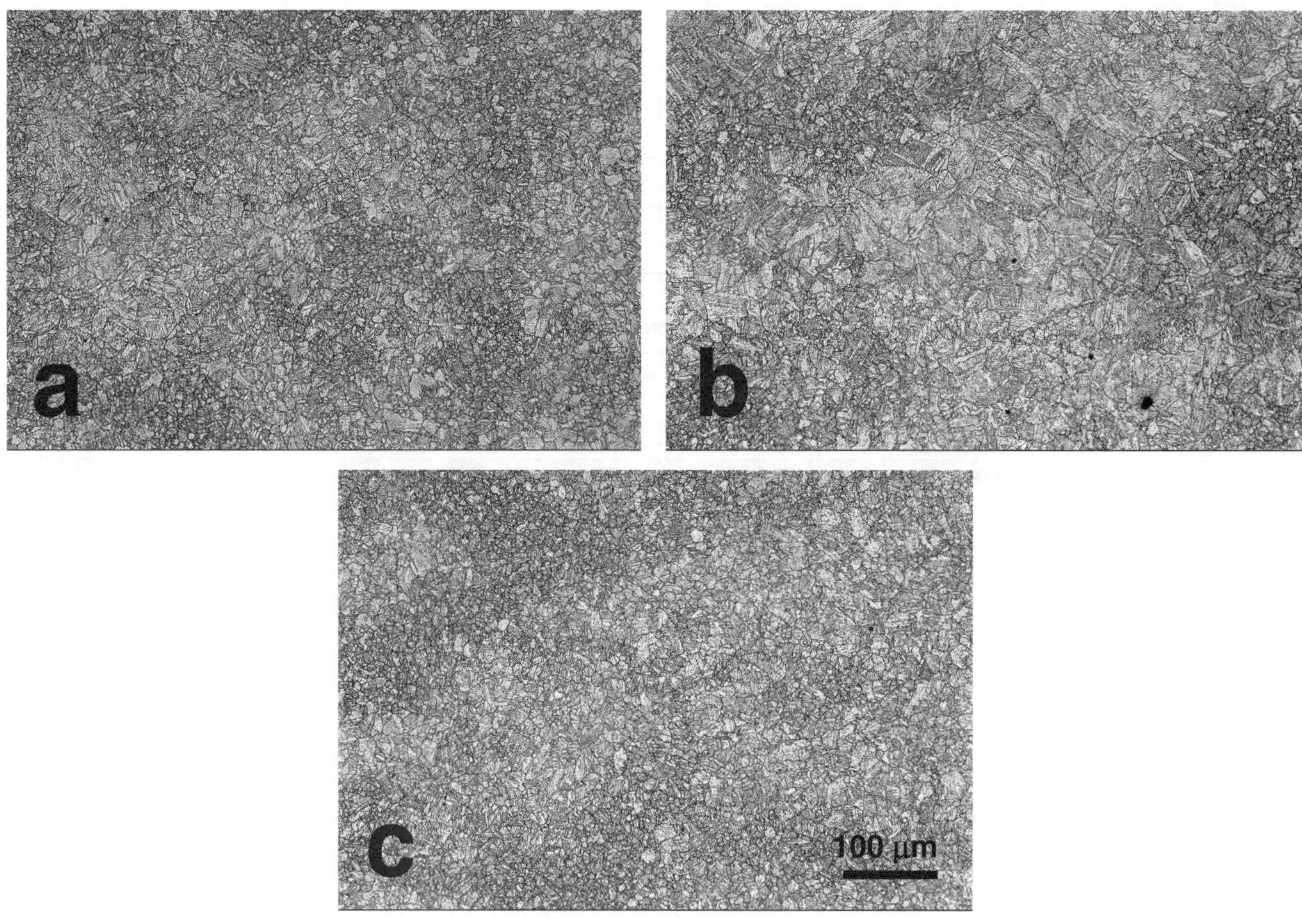

Figure 6. Austenite grain structures in steel B after reheating and hot rolling at 975°C, air cooling to room temperature, pseudocarburizing, reaustenitizing at 829°C for two hours, oil quenching, and tempering at 204°C for four hours: specimens pseudocarburized at (a) 954°C for eight hours, (b) 954°C for 16 hours, and (c) 1000°C for eight hours.

Effects of Processing on the Development of Impact Toughness

The effects of pretreatment conditions on the development of impact toughness in pseudocarburized and hardened specimens of steel B are summarized in Figure 8.[1] The oil quenched/subcritically annealed specimens exhibit the highest levels of toughness over the entire range of hot-rolling temperature, although the air cooled and oil quenched specimens exhibit similar levels of toughness after hot rolling in the 975-1050°C range and testing at 120°C. The impact toughness increases with hot-rolling temperature over the 975-1050°C range in a majority of instances, but the different material conditions exhibit drastically different responses with an increase in hot-rolling temperature to 1225°C. In particular, the toughness decreases to low levels in the air cooled and oil quenched specimens, whereas high toughness persists with increases in hot-rolling temperature to 1225°C in the oil quenched/subcritically annealed specimens. The data for each material condition also exhibit a consistent trend in which specimens extracted from the central region of the plates exhibit higher toughness than specimens extracted from the outer regions of the plates at both test temperatures.

[1]Since the number of specimens was limited and a substantial amount of variability is present in the toughness data for some of the material conditions, all data are presented in terms of the arithmetic mean and the observed minimum and maximum values of impact toughness.

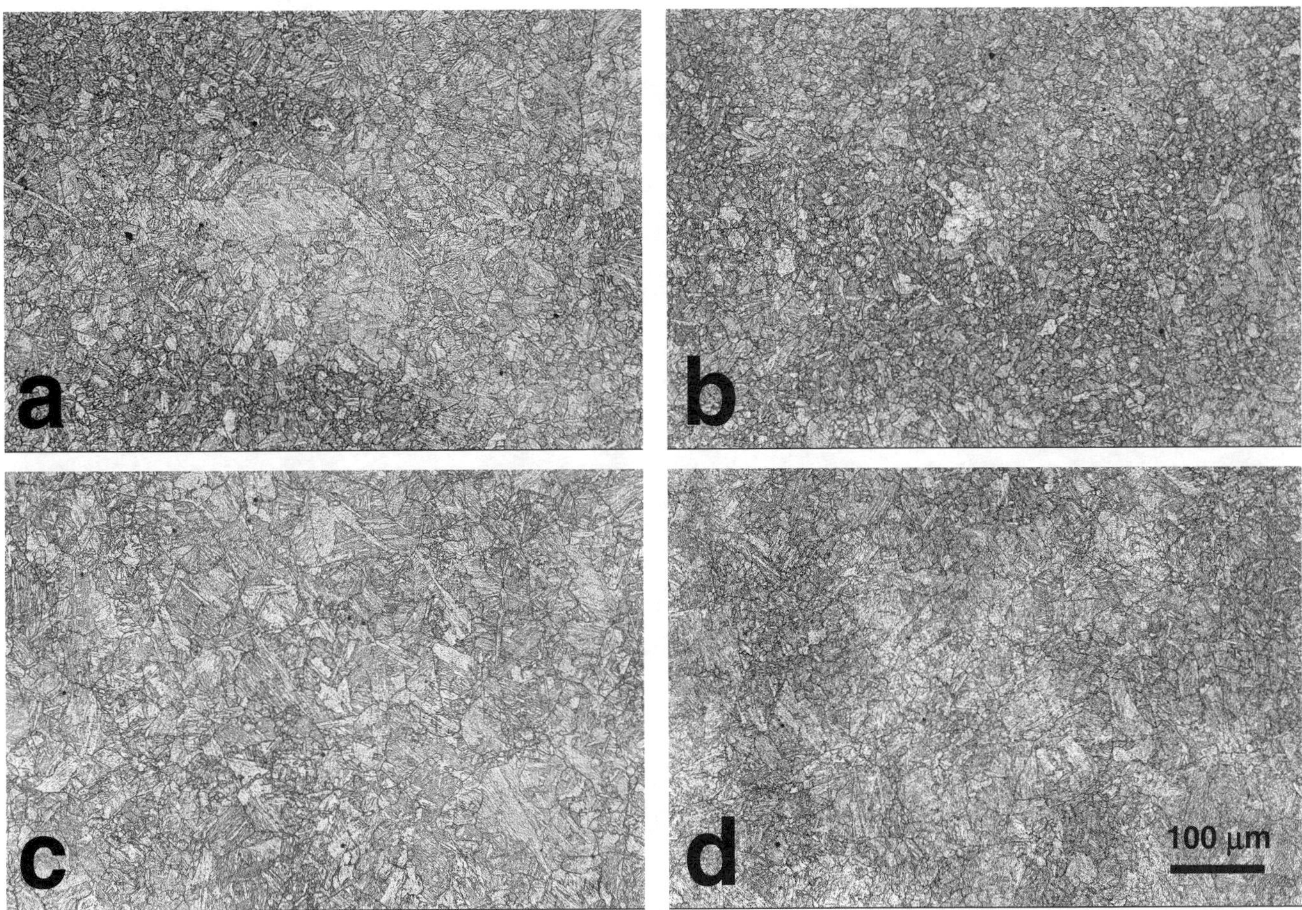

Figure 7. Austenite grain structures after reheating at 1225°C, hot rolling, and (a,c) air cooling to room temperature or (b,d) oil quenching to room temperature and subcritical annealing at 700°C: (a,b) pseudocarburized (954°C for eight hours) and hardened (829°C for two hours) specimens and (c,d) hardened (829°C for two hours) specimens.

The effects of pseudocarburizing on the development of toughness in air cooled and oil quenched/subcritically annealed specimens of steel B are summarized in Figure 9. Hardened specimens of the oil quenched/subcritically annealed steel exhibit lower toughness than the pseudocarburized and hardened specimens at both test temperatures, although the differences in toughness between the two material conditions diminish with increases in hot-rolling temperature. The trend in the toughness data for hardened specimens of the air cooled steel generally parallels that of the pseudocarburized and hardened specimens at both test temperatures. However, the toughness of the hardened specimens approaches and exceeds that of the pseudocarburized and hardened specimens with increases in hot-rolling temperature from 1050°C to 1225°C.

The sensitivity of impact toughness to changes in the pseudocarburizing parameters is summarized in Table III for steel B specimens hot rolled at 975°C and air cooled to room temperature. An increase in pseudocarburizing time at 954°C from eight hours to 16 hours produces either no change in toughness or a small increase in toughness at each test temperature. With the exception of specimens extracted from the outer regions of the plates and tested at room temperature, an increase in pseudocarburizing temperature from 954°C to 1000°C is associated with decreases in impact toughness.

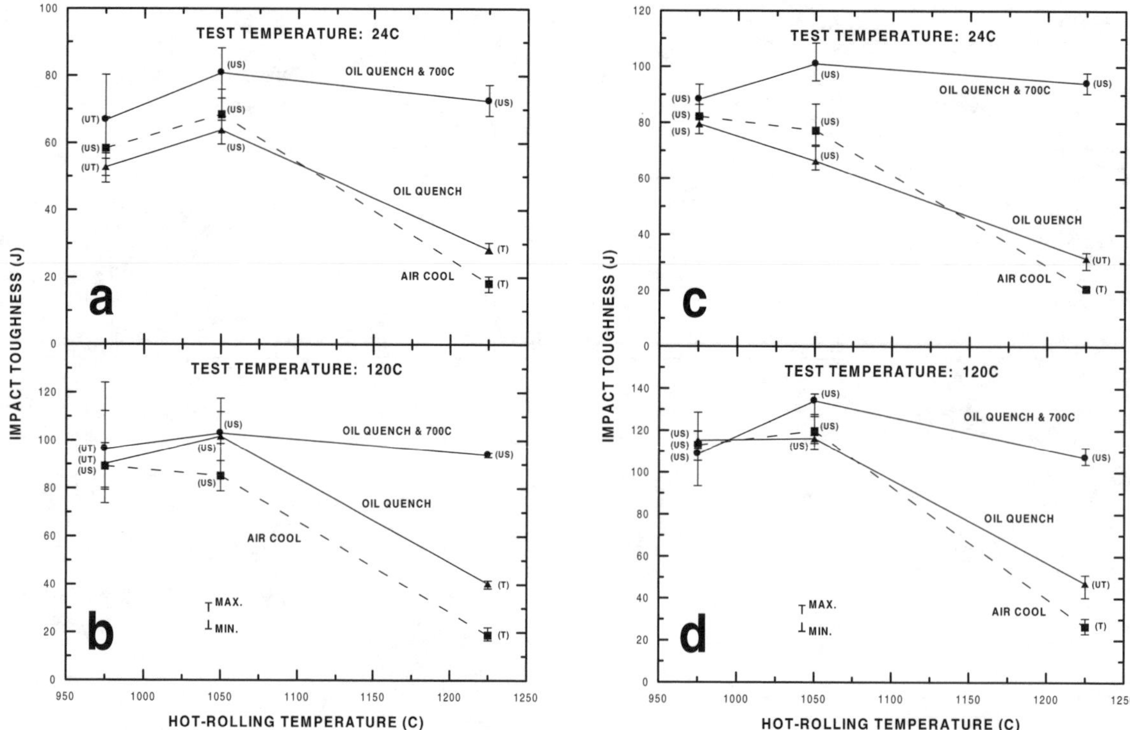

Figure 8. The longitudinal impact toughness of steel B is shown as a function of hot-rolling temperature and post-rolling conditioning for pseudocarburized and hardened specimens tested at (a,c) 24°C and (b,d) 120°C: (a,b) specimens extracted from the outer regions of the plates and (c,d) specimens extracted from the central region of the plates. The characters in parentheses indicate the macroscopic mode of fracture: US = upper-shelf behavior, T = transitional behavior, and UT = upper-transitional (near upper-shelf) behavior.

Table III: Summary of Impact Toughness Data for Different Pseudocarburizing Parameters

Pseudocarburizing Parameters[1]	Position in Hot-Rolled Plate	Mean Impact Toughness (J)[2] T_T = 24°C[3]	T_T = 120°C
954°C – 8 hours	Central Region	82.2 (82/83)	112.9 (106/120)
954°C – 16 hours	Central Region	82.2 (80/84)	120.3 (118/123)
1000°C – 8 hours	Central Region	68.7 (62/74)	102.7 (90/113)
954°C – 8 hours	Outer Regions	58.4 (48/68)	89.1 (74/112)
954°C – 16 hours	Outer Regions	62.3 (54/69)	93.5 (81/110)
1000°C – 8 hours	Outer Regions	61.4 (53/68)	79.3 (75/82)

[1]Subsequent to pseudocarburizing, all specimens were austenitized at 829°C for two hours, oil quenched, and tempered at 204°C for four hours.
[2]The values in parentheses represent the minimum and maximum observed values of impact toughness.
[3]T_T = test temperature.

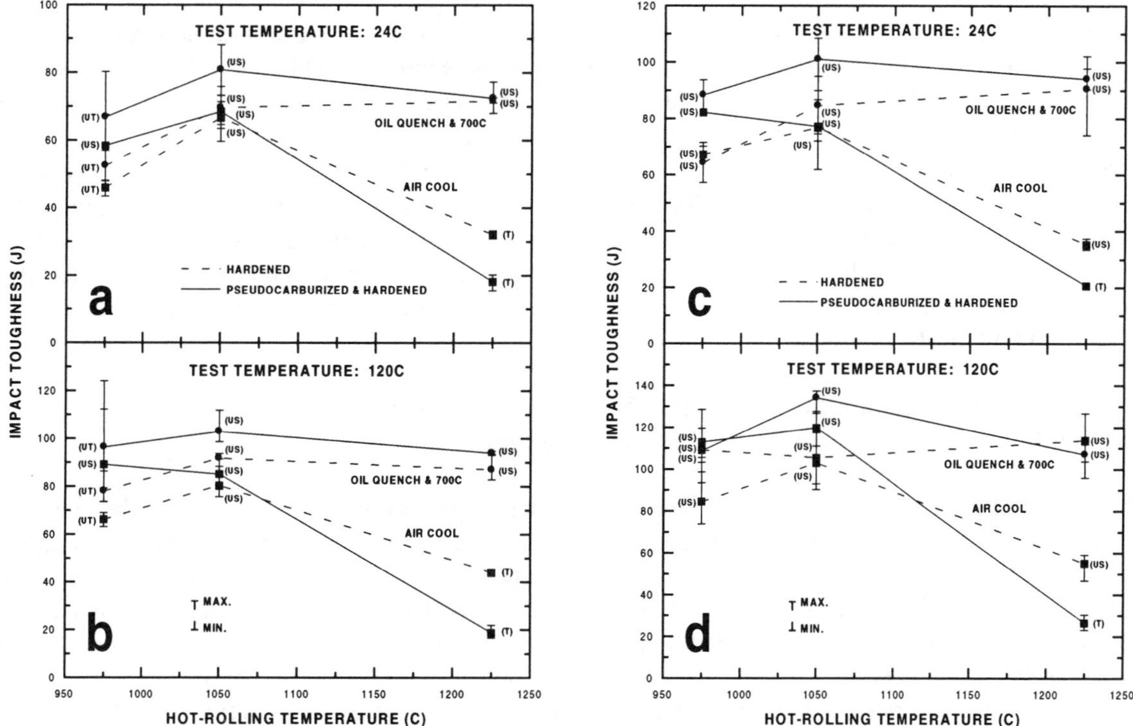

Figure 9. The longitudinal impact toughness of steel B is shown as a function of hot-rolling temperature and post-rolling conditioning for hardened specimens and pseudocarburized and hardened specimens tested at (a,c) 24°C and (b,d) 120°C: (a,b) specimens extracted from the outer regions of the plates and (c,d) specimens extracted from the central region of the plates. The characters in parentheses indicate the macroscopic mode of fracture: US = upper-shelf behavior, T = transitional behavior, and UT = upper-transitional (near upper-shelf) behavior.

DISCUSSION

Since room temperature coincides with the transition-temperature regime in material extracted from the forged components, the fracture of the Charpy V-notch specimens can be generally characterized as a three-stage process comprising the plastic deformation of the ligament, ductile crack initiation and extension from the notch root, and unstable fracture. The initiation of unstable fracture is associated with a change in the micromechanism of fracture from microvoid coalescence to transgranular cleavage intermixed with varying amounts of ductile rupture, and differences in toughness are primarily related to differences in the amount of ductile crack extension that precede unstable fracture. Based on this observation, there are three general reasons to discount MnS inclusions as the primary microstructural feature governing the toughness of the 9313 steel. First, the room-temperature toughness of the forged components scales in direct proportion to the sulfur content of the steels, Table I. Fiber toughening, which is typically observed in longitudinally-oriented Charpy specimens of resulfurized microalloyed forging steels, would only be expected at much higher sulfur contents (>>0.03%). Second, the differences in toughness between the central and outer regions of the hot-rolled plates are inconsistent with the expected segregation of MnS in the ingot during solidification; that is, lower toughness resulting from higher contents of MnS would be expected in the central section of the plates. Finally, relatively weak bonding between MnS inclusions and a ferritic matrix (10,11) promotes decohesion at the particle-matrix interface over broad ranges of temperature and particle size, thereby minimizing the possibility of propagating a microcrack in an inclusion through the particle-matrix interface (12).

Austenite grain structure can also be effectively eliminated as the primary variable affecting the impact toughness of the 9313 steel in both the transition-temperature and upper-shelf fracture regimes. Reheating at

1225°C and hot rolling, for example, produces similar levels of toughness in both hardened and pseudocarburized and hardened specimens of the oil quenched/subcritically annealed plates, Figure 9, although the austenite grain structures of the two material conditions are drastically different, Figure 7. This effect is further exemplified by the data for the air cooled plates, where the coarse-grained structure of the hardened specimens possesses significantly higher levels of impact toughness than the comparatively finer structure exhibited by the pseudocarburized and hardened specimens. Moreover, pseudocarburized and hardened specimens with significant differences in grain structure exhibit similar levels of impact toughness after hot rolling at 975°C and air cooling to room temperature, Table III and Figure 6. These data suggest AlN precipitates are the microstructural feature primarily responsible for the development of impact toughness in the air-melt 9313 steel, consistent with the preponderance of coarse AlN precipitates observed on the fracture surfaces of Charpy specimens from both heats, Figure 4.

Phenomenological Explanation for the Effects of Processing on the State of AlN Precipitation and the Development of Impact Toughness

The effects of processing on the development of impact toughness in the 9313 steel can be characterized in terms of the effects of processing on the state of precipitation prior to pseudocarburizing and the evolution of AlN precipitates during the pseudocarburizing and hardening operations. An increase in reheating temperature at levels below the AlN solution temperature is expected to decrease the content of intermediate-sized precipitates retained through hot rolling and cooling, such that precipitate ripening occurs in a less heterogeneous dispersion during subsequent austenitization. Since a significant decrease in the content of intermediate-sized precipitates will tend to decrease the potential for ripening, an increase in hot-rolling temperature would be expected to increase the overall level of toughness and decrease the difference in the toughness of the hardened specimens and the pseudocarburized and hardened specimens. The toughness data for the different material conditions is generally consistent with this line of reasoning; however, some of the air cooled and oil quenched specimens exhibit decreases in impact energy after hot rolling in the 975-1050°C range, Figure 8. The latter type of behavior suggests that any potential improvements in toughness resulting from AlN dissolution during reheating may be offset by increased heterogeneity in the dispersion of precipitates retained through hot rolling and cooling in these material conditions.[1]

An increase in hot-rolling temperature at levels above the equilibrium AlN solution temperature is associated with different responses in the air cooled, oil quenched, and oil quenched/subcritically annealed steel. Specimens of the air cooled and oil quenched steel exhibit substantial decreases in toughness with increases in hot-rolling temperature to 1225°C, Figure 8. The degradation in toughness presumably reflects the formation and coarsening of AlN precipitates during the pseudocarburizing and hardening operations. The improvements in the toughness of the oil quenched steel relative to the air cooled steel after hot rolling at 1225°C further suggest that the degradation in toughness is exacerbated by differences in the extent of AlN precipitation and growth during post-deformation cooling (1,2,5). The oil quenched/subcritically annealed steel, on the other hand, exhibits high levels of toughness over the entire range of hot-rolling temperature. The retention of toughness results from extensive AlN dissolution prior to hot rolling, the inhibition of AlN precipitation during post-deformation cooling, and the precipitation of a dense dispersion of fine AlN particles during subcritical annealing. In other words, the development of good toughness results from the formation of a homogeneous dispersion of precipitates prior to the pseudocarburizing and hardening operations.

Once hot rolling and post-rolling conditioning establish heterogeneity in the AlN dispersion, the development of impact toughness is dependent on the evolution of AlN precipitates during the pseudocarburizing and hardening operations. Isothermal ripening will provide an increment of toughening if decreases in the size and volume density of intermediate-sized precipitates are large enough to offset increases in the size of the

---

[1]Heterogeneity in the present context primarily refers to variations in the size of precipitates in a dispersion (i.e., the broadening or narrowing of a size distribution about a mean value of precipitate size).

coarsest precipitates in a dispersion. In this connection, a high precipitate coarsening potential at 954°C (i.e., high temperature and soluble aluminum content) would act to decrease heterogeneity in an initially heterogeneous dispersion and increase heterogeneity in a more homogeneous dispersion. Precipitate ripening also occurs during subsequent austenitization at 829°C, but since the process is retarded by a high volume fraction of AlN (i.e., low soluble aluminum content), the development of toughness will be primarily dependent on changes in the AlN dispersion that occur during pseudocarburizing. In contrast, any initial heterogeneity in the AlN dispersion may be largely inherited when hot rolling and post-rolling conditioning are followed by austenitization at 829°C for two hours. Thus, heterogeneity initially present in a dispersion of AlN would be expected to increase in the pseudocarburized and hardened specimens and decrease in the hardened specimens with increases in hot-rolling temperature. These types of microstructural changes would yield the general variation in impact toughness with hot-rolling temperature that is exhibited by the air cooled steel, Figure 9.[1]

The dependence of toughness on pseudocarburizing parameters after hot rolling at 975°C and air cooling is generally consistent with the foregoing explanation. An increase in pseudocarburizing time at 954°C will promote ripening in the dispersion of AlN, where steady state ripening is characterized by decreases in the density of small and intermediate-sized precipitates and increases in the size of the coarsest precipitates in a dispersion. The small changes in impact toughness that accompany the increase in pseudocarburizing time suggest that the increase in toughness resulting from decreases in the density of intermediate-sized precipitates is large enough to offset the decrease in toughness associated with increases in the size of the coarsest precipitates in the dispersion. Conversely, an increase in pseudocarburizing temperature to a level above the hot-rolling temperature (i.e., 1000°C) significantly increases the coarsening potential at a lower precipitate volume fraction, thereby exacerbating heterogeneity in the AlN dispersion. This heterogeneity in the dispersion, which is inherited and possibly amplified during subsequent austenitization at 829°C, acts to degrade the toughness of the steel.

Although the pseudocarburizing and hardening operations affect toughness in a manner stochastically consistent with the expected evolution of AlN precipitates in austenite, the largest changes in toughness result from the processing parameters that define the hot-rolling operation. Low levels of toughness, similar in magnitude to the impact toughness of the forged and heat-treated components, develop as a result of hot rolling above the AlN solution temperature and either air cooling or oil quenching to room temperature, Figure 8. However, these results are somewhat incongruent with the standard practice of reheating at low-to-intermediate temperatures for a minimum amount of time prior to forging the components. This inconsistency along with a position dependence of toughness in the hot-rolled plates suggests that the processing of the steel prior to hot rolling and/or forging affects the formation and retention of coarse AlN precipitates in the microstructure.

Factors Affecting the Formation and Retention of Coarse AlN Precipitates

The formation and retention of extremely coarse AlN precipitates in the microstructure would necessarily require nucleation along with a substantial amount of growth at high temperatures. Based on various estimates for the solubility product of AlN (13), the nominal composition for steel B yields AlN solution temperatures of ~1140°C and ~1285°C for the liquid and δ ferrite phases, respectively. The calculated AlN solution temperatures are well below the ranges of temperature associated with the stability of either phase. In contrast, the range of calculated (7,8) AlN solution temperature in austenite (1060-1188°C) is in general agreement with the dependence of impact toughness on hot-rolling temperature in that the steel can be effectively solution treated at reasonable temperatures in the austenite phase field. Accordingly, the formation of extremely coarse AlN precipitates must result from an enrichment phenomenon during the solidification of the steel. Since the composition of the 9313 steel is well within the peritectic regime, significant amounts of solute partitioning

[1]Unlike the air cooled steel, the toughness of hardened specimens does not exceed the toughness of pseudocarburized and hardened specimens with increases in hot-rolling temperature above the equilibrium AlN solution temperature in the oil quenched/subcritically annealed steel, Figure 9. This difference in behavior suggests that the dispersion of AlN resulting from the application of a solution pretreatment at 1225°C and a precipitation anneal at 700°C is extremely homogeneous, such that the pseudocarburizing operation has a minimal effect on proliferating heterogeneity in the dispersion.

between δ ferrite and the liquid steel would be expected during solidification. In order to investigate if solute enrichment in the interdendritic liquid provides a viable mechanism for AlN precipitation at high temperatures, a thermodynamic analysis was conducted to determine the potential for AlN precipitation under conditions that are expected to represent the range of behavior exhibited by the steel during solidification.

For the equilibrium partitioning of elements between δ ferrite and the liquid steel, the thermodynamic system can be defined in terms of a mass balance for each element:

$$\chi_k^T = \left[1 - \sum \phi_{MN}\right]\left[\eta \chi_k^\delta + (1-\eta)\chi_k^\ell\right] + 0.5 \sum \phi_{MN}^k, \quad [1]$$

where η = atomic fraction of δ ferrite with respect to the combined mole fractions of δ ferrite and liquid in the system, χ_k^i = mole fraction of the kth element in the ith phase, ϕ_{MN} = mole fraction of an MN precipitate phase, ϕ_{MN}^k = mole fraction of the kth element in an MN precipitate phase, and the superscript T refers to the total concentration of an element in the steel. Substitution of the equilibrium partition ratio, $\beta_k^{\delta\ell}$, in Equation 1 yields a relationship for the concentration of the kth element in the liquid phase:

$$\chi_k^\ell(\psi) = \frac{\chi_k^T - 0.5\sum \phi_{MN}^k}{\left[1 - \sum \phi_{MN}\right] - \left[1 - \beta_k^{\delta\ell}\right]\psi}, \quad [2]$$

where the local atomic fraction of δ ferrite in a volume element of material, ψ, is defined as:

$$\psi = \left[1 - \sum \phi_{MN}\right]\eta, \quad [3]$$

and η is treated as a prescribed independent variable.

Solute enrichment in the liquid can also be evaluated for conditions in which local equilibrium is maintained at the δ ferrite-liquid interface during steady-state solidification. An approximate mass balance can be written for a system comprised of δ ferrite, liquid, and any number of precipitate phases in the liquid:

$$\left[\chi_k^\ell - \chi_k^\delta\right] d\eta \cong [1-\eta] d\chi_k^\ell, \quad [4]$$

where the left-hand and right-hand sides of the expression represent the mass transfer of the kth element out of the δ ferrite and the concentration change in the liquid that accompanies an infinitesimal increase in η, respectively. The mass balance, which is based on the substitution of $d\eta$ for $d\psi$, neglects the formation of any MN precipitate phase(s) in the enriched liquid. However, the approximation is appropriate when $\Sigma\phi_{MN} << 1 - \eta$ since $d\eta \cong d\psi$ and $d\left\{\sum \phi_{MN}^k\right\}/d\eta << d\chi_k^\ell/d\eta$. Within the limitations of this constraint, Equation 4 can be integrated without the necessity of defining an analytical expression between χ_k^ℓ, ϕ_{MN}^k, and η.

Substituting $\beta_k^{\delta\ell}$ into this expression and integrating between appropriate limits yields the Scheil equation for the enrichment of the kth element in the liquid phase:

$$\chi_k^\ell(\eta) \cong \chi_k^R(\eta)[1-\eta]^{\beta_k^{\delta\ell}-1}, \quad [5]$$

where the lower limit of integration, $\chi_k^R(\eta)$, represents the total concentration of the kth element for the reduced δ ferrite-liquid system in which the liquid contains an equilibrium amount of MN precipitate phase(s) at η:

$$\chi_k^R(\eta) = \chi_k^T - 0.5\sum \phi_{MN}^k . \qquad [6]$$

χ_k^R is treated as an explicit variable in Equation 5 since the partitioning of elements at the δ ferrite-liquid interface is based on path-independent state functions and thermodynamic equilibrium is consistently defined for a system in which MN precipitate phases are present in the liquid:

$$\beta_k^{\delta\ell} = \left[\frac{\chi_k^\delta}{\chi_k^\ell}\right]_{\delta-\ell} = \left[\frac{\chi_k^\delta}{\chi_k^\ell}\right]_{\delta-\ell-MN} . \qquad [7]$$

Equations 2 and 5 degenerate into standard forms of the enrichment equations (14) as the solubility limit for the most thermodynamically stable species of precipitate is approached; that is, $\Sigma\phi_{MN} \to 0$ and $\chi_k^R(\eta) \to \chi_k^T$.

Assuming no barrier to nucleation exists, the precipitation of AlN or TiN in a volume element of liquid is possible when solute enrichment exceeds a critical level:[1]

$$\chi_M^\ell(\eta_c)\chi_N^\ell(\eta_c) = \frac{\Gamma_{MN}^\ell}{\gamma_M^\ell(\eta_c)\gamma_N^\ell(\eta_c)}, \qquad [8]$$

where γ_k^ℓ = activity coefficient for the kth element and η_c = critical fraction of solidification associated with the equilibrium solubility limit of MN in the solute-enriched interdendritic liquid. The parameter Γ_{MN}^ℓ represents the solubility product of a binary nitride defined on a mole fraction basis:

$$\log\Gamma_{MN}^\ell = \log K_{MN}^\ell + \log\left[\frac{\Omega_\ell^2 10^{-4}}{\Omega_M\Omega_N}\right], \qquad [9]$$

where Ω_i = atomic weight of the indicated element, Ω_ℓ = molecular weight of the liquid steel, and K_{MN}^ℓ = standard solubility product for MN in liquid steel. In addition, the activity coefficient for each element is given by the approximate relationship:

$$\ln\gamma_k^\ell(\eta) \cong \ln\gamma_k^\circ + \sum \varepsilon_k^m \chi_m^\ell(\eta), \qquad [10]$$

where ε_k^m = first-order interaction parameter and γ_k° = activity coefficient at infinite dilution.

Thermodynamic equilibrium between TiN, AlN, and the solute-enriched liquid is evaluated for two conditions that represent the expected range of behavior of the steel during solidification:

- Case 1: No back diffusion of any substitutional solutes, Equation 5, and complete diffusion of carbon and nitrogen, Equation 2, in δ ferrite; and
- Case 2: No back diffusion of any elements in δ ferrite during solidification, Equation 5.

---

[1]This assumption is reasonable since dendrite surfaces would be expected to provide potent sites for the heterogeneous nucleation of AlN.

The analysis is conducted utilizing the equilibrium partition ratio data compiled by Battle and Pehlke (15). However, $\beta_{Mo}^{\delta\ell}$ is taken as unity since this parameter is not readily accessible from the literature. The solubility of AlN and TiN in liquid iron is estimated from the data of Evans and Pehlke (16) and Morita and Kunisada (17), respectively. First-order interaction parameters, obtained from the data tabulated by Sigworth and Elliott (18), were converted from a weight percentage basis to a mole fraction basis (19,20) to maintain consistency with the partition ratio data of Battle and Pehlke (15). Temperature-dependent estimates for the interaction parameters were employed where available, but in the absence of these data the interaction parameters for elements in liquid iron at 1600°C were utilized for estimating the effects of solute enrichment on the activity of titanium, aluminum, and nitrogen.

The analysis is based on several assumptions. First, it is assumed that the phase boundary positions for the δ ferrite and liquid are unaffected by solute enrichment such that the temperature-dependent functions for the equilibrium partition ratios (15) apply throughout solidification. This alleviates the problem of obtaining reliable estimates of activity coefficients and interaction parameters for elements in δ ferrite. Second, it is assumed that solidification is isothermally completed at the peritectic temperature (1495°C), although it is recognized that carbon enrichment past a critical level will progressively lower the solidification temperature of residual liquid in interdendritic regions. Third, the molecular weight of the liquid, Ω_ℓ, is treated as a constant equivalent to the molecular weight of the alloy in Equation 9. This assumption is conservative with respect to predicting AlN formation since Ω_ℓ decreases with increasing amounts of solute enrichment.[1] Finally, TiN is assumed to precipitate in preference to Ti(C,N) during solidification. This assumption, which greatly simplifies the numerical calculations, is reasonable considering the large difference in the solubility of TiN and TiC in liquid iron (i.e., $K_{TiN}/K_{TiC} \cong 0.001$) (13). Based on these assumptions, the elemental concentrations, $\chi_k^\ell(\eta)$, activity coefficients, $\gamma_k^\ell(\eta)$, mass balances, and equilibrium expressions yield a system of 17 equations. Thermodynamic equilibrium was estimated by numerically iterating the system of equations with a modified Newton-Raphson procedure to a convergence of less than 0.5% for each variable.

The results of calculations for the steel B composition indicate that AlN precipitation during solidification is not viable when a significant amount of interstitial back diffusion occurs in δ ferrite (case 1). In contrast, the existence of both TiN and AlN is thermodynamically feasible under conditions in which local equilibrium is maintained at the δ ferrite-liquid interface during isothermal solidification at 1495°C (case 2), Figure 10. These results suggest that sufficiently rapid solidification (e.g., in the near-surface regions of a cast section) will promote the precipitation of AlN during solidification, whereas a comparatively low solidification rate will suppress the precipitation reaction. Moreover, bulk compositional modifications will not provide an effective means of inhibiting AlN precipitation during solidification since ϕ_{AlN} is relatively insensitive to titanium content within the range that would be considered a residual level (≤ 0.005%). Finally, this analysis also provides a feasible explanation for the position dependence of impact toughness in the hot-rolled plates of steel B, where the prediction of AlN precipitation in the interdendritic liquid correlates with the development of consistently lower toughness in material extracted from the columnar zone of the ingots. This differential in toughness is present regardless of the hot-working and heat-treating schedules applied to the steel, Figures 8 and 9 and Table III.

Although the appropriate mode of solidification provides a necessary condition for the formation of coarse AlN precipitates in the microstructure, a significant fraction of post-solidification cooling occurs in austenite at temperatures well above the equilibrium AlN solution temperature. Thus, once a dispersion of coarse AlN precipitates is established in the as-solidified material, sufficient conditions for the existence of these precipitates in the wrought steel necessarily rely on the processing factors that promote the retention of the precipitates

---

[1]This assumption was verified for the data shown in Figure 10.

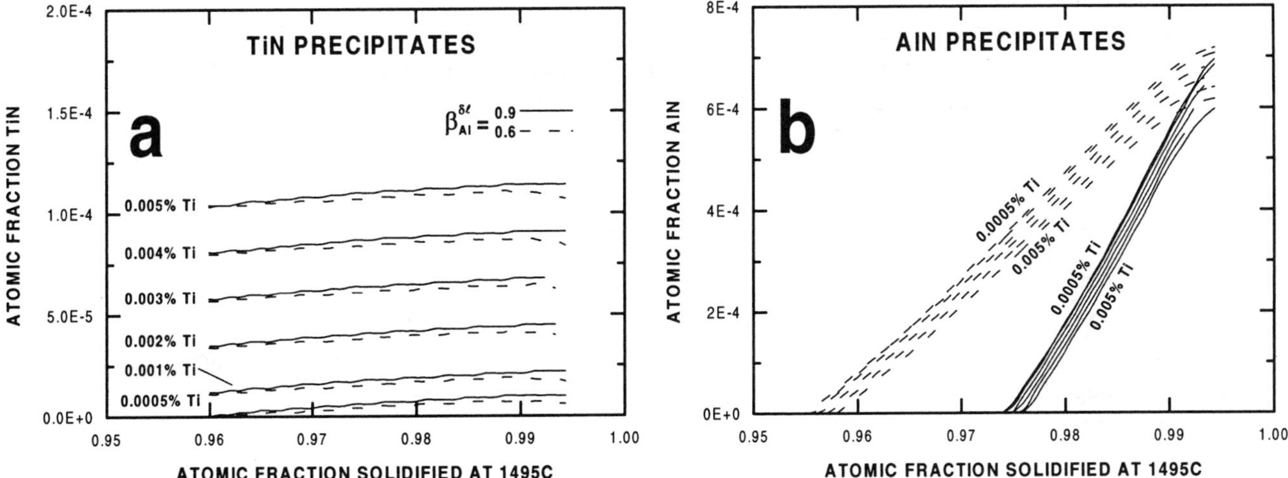

through post-solidification cooling. An indication of the stability of coarse AlN precipitates during post-solidification cooling can be obtained from the general kinetic model of Cheng, Hawbolt, and Meadowcroft (21). Based on a

Figure 10. Calculated estimates of the content of (a) TiN and (b) AlN are shown as a function of the local atomic fraction solidified and titanium content for the steel B base composition. The calculations are based on the assumptions that local equilibrium is established and maintained at the δ ferrite-liquid interface for both substitutional and interstitial elements and solidification is completed at 1495°C. In addition, precipitate contents are evaluated for the bounding values of aluminum partition ratio reported by Battle and Pehlke (15).

stationary interface approximation for a spherically symmetrical system, the solution of Fick's first and second laws yields:

$$a_f = a_o - \int_0^{t_D} \left[\frac{C_I - C_M}{C_P - C_I}\right]\left[\frac{D}{a} + \left(\frac{D}{\pi t}\right)^{1/2}\right] dt, \qquad [11]$$

where a_f = final precipitate radius, a_o = initial precipitate radius, D = diffusivity of the rate limiting element (i.e., aluminum), C_M = bulk matrix concentration of the rate limiting element, C_P = concentration of the rate limiting element in the precipitate phase, C_I = concentration of the rate limiting element at the precipitate-matrix interface, t = time, and t_D = time associated with AlN dissolution. Time can be eliminated as an explicit variable in Equation 11 by introducing cooling rate and integrating precipitate size over a range of temperature:

$$a_f \cong a_o - \int_{1768K}^{T} \left[\frac{C_I - C_M}{C_P - C_I}\right]\left[\frac{D}{a} + \left(\frac{D\omega}{\pi \Delta T}\right)^{1/2}\right] \frac{1}{\omega} dT, \qquad [12]$$

where ω = post-solidification cooling rate (constant) and ΔT = 1768 – T for cooling rates defined as positive values.

Since the coarse AlN precipitates exist as platelets with a high aspect ratio, the size of AlN precipitates is approximated as the diameter of a disk with an equivalent volume to that of a spherical precipitate:

$$d_{AlN} \approx \left[\frac{16\zeta}{3}\right]^{1/3} a, \qquad [13]$$

where d_{AlN} = AlN precipitate size and ζ = aspect ratio of the AlN platelets. A single-valued indication of the chemical driving force for dissolution is obtained from the radius of curvature, α, of the disk-shaped precipitate, and the elemental concentration at the precipitate-matrix interface, C_I, is estimated from the Gibbs-Thompson equation:

$$C_I \cong C_M e^{\left[\frac{2\psi_{AlN}\rho_{AlN}}{\Omega_{AlN}RT\alpha}\right]}, \qquad [14]$$

where ψ_{AlN} = interphase interfacial energy, ρ_{AlN} = density of AlN, R = universal gas constant, and T = absolute temperature. Values of ψ_{AlN} were estimated utilizing the method outlined in Reference 21, and Equation 12 was numerically integrated for various values of d_{AlN} and ω over the range of temperature from 1495°C to the vicinity of the equilibrium AlN solution temperature.

This analysis is based on the implicit assumption that the interparticle spacing is significantly larger than the size of the coarse precipitates, whereby the effective diffusion volume for each precipitate is modeled as a sphere at the radial position where the aluminum concentration approaches the bulk matrix concentration, C_M. It is also assumed that C_M is independent of time and equivalent to the bulk concentration of aluminum in the steel. Based on these assumptions, the analysis is equivalent to that for the dissolution of a single, isolated precipitate in a dilute matrix, and the numerical integration of Equation 12 is conducted without the need to specify stereological parameters for the AlN dispersion. The assumptions yield lower-bound estimates for the extent of precipitate dissolution if the far-field concentration of aluminum increases towards the bulk concentration of aluminum as AlN precipitates dissolve. If, on the other hand, the far-field concentration of aluminum is initially greater than the bulk concentration, the assumption of constant C_M will overestimate the extent of precipitate dissolution. The latter condition could result from precipitate clustering and/or inherited concentration gradients around precipitates in the as-solidified steel. The analysis is also based on the assumption of a constant precipitate aspect ratio (ζ = 10), which is not necessarily realistic for a polydispersed system of precipitates subjected to a dissolution process over a broad range of temperature. Nevertheless, the analysis should provide an intuitively correct indication of the changes in the system that would be expected during post-solidification cooling.

The minimum size of AlN precipitate retained through post-solidification cooling, d_{MIN}, is predicted to increase with decreases in both temperature and cooling rate, and the value of d_{MIN} for each cooling rate approaches a constant value, d_C, with decreases in temperature, Figure 11. The increases in d_C with decreases in ω represent increases in the extent of precipitate dissolution during post-solidification cooling, consistent with the fact that increases in residence time at temperatures above the equilibrium solution temperature will promote simultaneous reductions in the size, content, and volume density of AlN precipitates. However, particles initially larger than d_C will be retained through post-solidification cooling, and the extent of dissolution will scale as some inverse function of the difference between the initial precipitate size and d_C. A dissolution process in this context would be expected to amplify heterogeneity in the size of AlN precipitates during post-solidification cooling, thereby promoting the retention of a comparatively high density of intermediate-sized precipitates in combination with a much lower density of extremely coarse precipitates. The model predictions are in general accord with the presence of substantially different densities of intermediate-sized and coarse AlN precipitates on Charpy specimen fracture surfaces. In particular, differences in the relative density of the two general size groups, which result from the preferential evolution of smaller precipitates during both post-solidification cooling and subsequent processing, are large enough to be manifested as bilinear cumulative size distributions for the steels, Figure 5c.

Relative Effects of Intermediate-Sized and Coarse Precipitates on Toughness

Large differences in the toughness of forged and heat-treated components suggest that variability in toughness is dependent on differences in the size and content of both intermediate-sized and coarse precipitates in

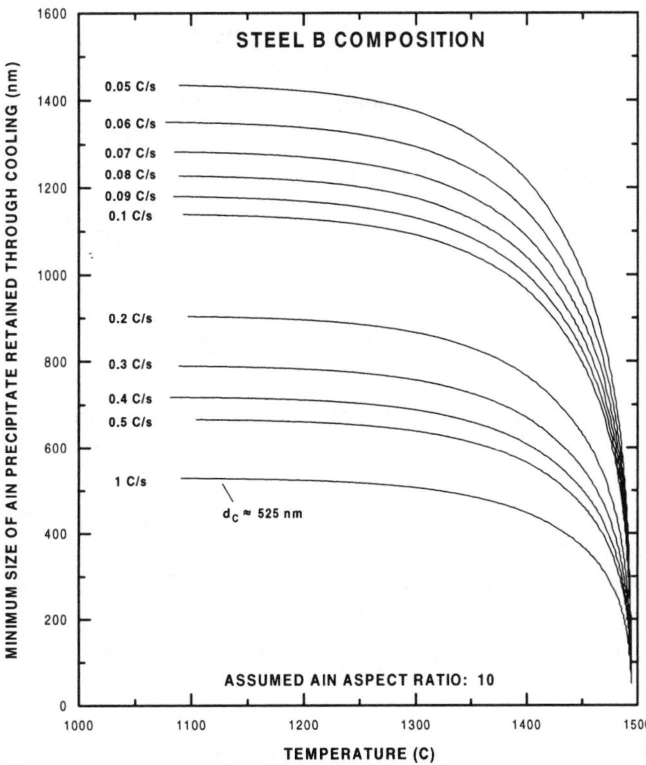

the microstructure. Coarse AlN precipitates that form under appropriate conditions during solidification are retained through processing and degrade the toughness of material corresponding to the columnar zone of an ingot. The sporadic occurrence of these precipitates contribute to heat-to-heat variability in toughness, toughness variations

Figure 11. Calculated estimates of the minimum size of AlN precipitate retained through cooling, d_{MIN}, are shown as a function of temperature and post-solidification cooling rate for the steel B composition.

within a heat, and toughness variations through the cross-section of the final product. For the case of steel B, the position dependence of toughness in the hot-rolled plates yields differences ranging from 2.5 J to 26.7 J in the mean level of room-temperature impact toughness exhibited by pseudocarburized and hardened specimens of the different material conditions. Conversely, kinetic calculations support the viability of retaining both intermediate-sized and coarse precipitates through post-solidification cooling and the breakdown of ingots into 140 mm bar. The evolution of these precipitates during subsequent hot rolling and post-rolling conditioning yields material with room-temperature toughness ranging from ~15 J to ~100 J, Figure 8. Thus, while the coarsest precipitates retained through post-solidification cooling may exert a significant effect on toughness, the largest contribution to the variability in toughness is expected to result from the evolution of intermediate-sized precipitates that are retained through the last hot-working operation, regardless of the origin of the intermediate-sized precipitates.

The kinetic model predicts that isothermal reheating at 975°C and 1050°C for one hour will completely dissolve precipitates as large as ~65 nm and ~110 nm, respectively. Reheating at 1225°C for two hours, on the other hand, is predicted to completely dissolve particles up to ~330 nm in size, corresponding to between 90% and 99% of the density of precipitates observed on the fracture surfaces of the Charpy specimens from forged and heat-treated components, Figure 5c. However, reheating at temperatures above 1050°C is associated with the development of low toughness in air cooled plates of steel B, Figure 8, and the potential to develop low toughness in forged components, Figure 3. Two options can be explored within these constraints for improving the toughness of 9313 steel. The first option comprises lowering the forging temperature to about 1050°C, although this type of change may very well have a negative impact on production rates and tooling costs. More importantly, since variability in toughness is dependent on the evolution of AlN precipitates retained through both post-solidification cooling and the last hot-working operation, lowering the forging temperature will act to increase variability in the toughness of the final component. A second more viable option comprises reheating at temperatures above the AlN solution temperature, forging, and downstream processing (i.e., accelerated cooling followed by a subcritical anneal) to provide a uniform dispersion of fine AlN precipitates prior to heat treatment.[1]

CONCLUSIONS

1. Intermediate-sized and coarse AlN precipitates are responsible for the development of low toughness in air-melt 9313 steel.

2. Thermodynamic calculations indicate that AlN precipitation in the solute-enriched interdendritic liquid is viable when local equilibrium is maintained at the δ ferrite-liquid interface during the latter stages of solidification. Low-carbon steels that contain substantial amounts of δ ferrite during solidification are susceptible to embrittlement by coarse AlN precipitates since the precipitation reaction only requires aluminum and nitrogen in concentrations representative of EAF steelmaking practices for fine-grained, aluminum-killed steels.

3. Kinetic calculations suggest that coarse AlN precipitates may be retained through post-solidification cooling at rates commensurate with the severity of cooling in the near-surface regions of large sections.

4. The formation and retention of coarse AlN precipitates can be controlled by reducing the extent of microsegregation during solidification and subsequently allowing the precipitates to dissolve at high temperatures in the austenite phase field. These objectives may be realized by processing to minimize both the solidification rate and the post-solidification cooling rate.

5. The general position dependence of toughness in hot-rolled plates is consistent with a difference in the propensity to form and retain coarse AlN precipitates in the columnar and equiaxed zones of cast sections.

6. The variation in impact toughness with the processing parameters for hot rolling, pseudocarburization, and final austenitization is stochastically consistent with the effects of time and temperature on the evolution of AlN precipitates in steel.

7. The results of this investigation suggest that impact toughness can be substantially improved through the combined application of post-solidification processing and reheating at sufficiently high temperatures for the hot-rolling and forging operations. However, the development of optimum toughness will also require downstream processing to produce a uniform dispersion of fine AlN precipitates prior to the carburizing and hardening heat treatments.

ACKNOWLEDGMENTS

The writers would like to thank The Timken Company for permission to publish the results of this work. The writers would also like to acknowledge G. M. Waid for providing the fractographs in Figure 3 and T. L.

---

[1]Accelerated cooling was achieved by oil quenching plates of steel B in the present investigation. However, the imposition of lower cooling rates and the application of a subcritical anneal prior to carburizing have provided substantial improvements in the toughness of forged and heat-treated components manufactured from the 9313 steel. This latter result is consistent with the dependence of toughness on post-deformation cooling rate (2) in other high-strength steels (1,5,6).

Davis for providing transmission electron fractographs of the Charpy V-notch specimens. Finally, the writers would also like to express their appreciation to M. S. Galehouse for assistance with the hot rolling of the steels, J. A. Anderson and C. L. Newhouse for metallographic support, and Ken Casner of Quality Circle Machine for machining the test specimens for this research.

REFERENCES

1. M. J. Leap and J. C. Wingert, "Recent Advances in the Technology of Toughening Grain-Refined, High-Strength Steels" (Warrendale, PA: SAE International, Paper 961749, 1996).

2. M. J. Leap and J. C. Wingert, "Application of the AdvanTec Process for Improving the Toughness of Grain-Refined, High-Strength Steels," **38th Mechanical Working & Steel Processing Conference Proceedings** (Warrendale, PA: Iron and Steel Society, Inc., 1997), pp. 195-220.

3. M. J. Leap, J. C. Wingert, and C. A. Mozden, "Development of a Process for Toughening Grain-Refined, High-Strength Steels," **Steel Forgings: Second Volume, ASTM STP 1259**, eds. E. G. Nisbett and A. S. Melilli (West Conshohocken, PA: American Society for Testing and Materials, 1997), pp. 160-195.

4. M. J. Leap and J. C. Wingert, "Interrelationships Between Toughness and Grain-Refining Precipitates in Lightly-Tempered, High-Strength Steels," **39th Mechanical Working & Steel Processing Conference Proceedings** (Warrendale, PA: Iron and Steel Society, Inc., 1998), pp. 685-703.

5. M. J. Leap and J. C. Wingert, "The Effects of Grain-Refining Precipitates on the Development of Toughness in 4340 Steel," **Metallurgical Transactions A**, vol. 30A, 1999, pp. 93-114.

6. M. J. Leap, United States Patent 5,409,554, 1995.

7. L. S. Darken, R. P. Smith, and E. W. Filer, "Solubility of Gaseous Nitrogen in Gamma Iron and the Effect of Alloying Constituents-Aluminum Nitride Precipitation," **Transactions of the Metallurgical Society of AIME**, vol. 191, 1951, pp. 1174-1179.

8. W. C. Leslie, R. L. Rickett, C. L. Dotson, and C. S. Walton, "Solution and Precipitation of Aluminum Nitride in Relation to the Structure of Low Carbon Steels," **Transactions of the American Society for Metals**, vol. 46, 1954, pp. 1470-1497.

9. M. F. Ashby and R. Ebeling, "On the Determination of the Number, Size, Spacing, and Volume Fraction of Spherical Second-Phase Particles from Extraction Replicas," **Transactions of the Metallurgical Society of AIME**, vol. 236, 1966, pp. 1396-1404.

10. D. Brooksbank and K. W. Andrews, "Thermal Expansion of Some Inclusions Found in Steels and Relation to Tessellated Stresses," **Journal of the Iron and Steel Institute**, vol. 206, 1968, pp. 595-599.

11. J. W. Bray, J. L. Maloney, K. S. Raghavan, and W. M. Garrison, Jr., "A Comparison of the Fracture Behavior of Two Commercially Produced Heats of HY180 Steel Differing in Sulfide Type," **Metallurgical Transactions A**, vol. 22A, 1991, pp. 2277-2285.

12. T. Lin, A. G. Evans, and R. O. Ritchie, "Stochastic Modeling of the Independent Roles of Particle Size and Grain Size in Transgranular Cleavage Fracture," **Metallurgical Transactions A**, vol. 18A, 1987, pp. 641-651.

13. E. T. Turkdogan, "Causes and Effects of Nitride and Carbonitride Precipitation During Continuous Casting," **Transactions of the Iron and Steel Society**, 1989, pp. 61-75.

14. M. C. Flemings, **Solidification Processing** (New York, NY: McGraw-Hill Book Company, 1974), pp. 31-36.

15. T. P. Battle and R. D. Pehlke, "Equilibrium Partition Coefficients in Iron-Based Alloys," **Metallurgical Transactions B**, vol. 20B, 1989, pp. 149-160.

16. D. B. Evans and R. D. Pehlke, "The Aluminum-Nitrogen Equilibrium in Liquid Iron," **Transactions of the Metallurgical Society of AIME**, vol. 230, 1964, pp. 1651-1656.

17. Z. Morita and K. Kunisada, "Solubility of Nitrogen and Equilibrium of Titanium Nitride Forming Reaction in Liquid Fe-Ti Alloys," **Transactions of the Iron and Steel Institute of Japan**, vol. 18, 1978, pp. 648-654.

18. G. K. Sigworth and J. F. Elliott, "The Thermodynamics of Liquid Dilute Iron Alloys," **Metal Science**, vol. 8, 1974, pp. 298-310.

19. C. H. P. Lupis and J. F. Elliott, "Generalized Interaction Coefficients, Part I: Definitions," **Acta Metallurgica**, vol. 14, 1966, pp. 529-538.

20. C. H. P. Lupis and J. F. Elliott, "The Relationship Between the Interaction Coefficients ε and e," **Transactions of the Metallurgical Society of AIME**, vol. 233, 1965, pp. 257-258.

21. L. M. Cheng, E. B. Hawbolt, and T. R. Meadowcroft, "Study of the Kinetics of Dissolution, Coarsening and Growth of Aluminum Nitride in Low Carbon Steels," **40th Mechanical Working & Steel Processing Conference Proceedings** (Warrendale, PA: Iron and Steel Society, Inc., 1998), pp. 947-957.

BAR PRODUCTS AND FORGINGS III

Thermal Desorption of Hydrogen from 4340 Steel

Peter Ganeff
Illinois Institute of Technology
10 West 32nd Street
Chicago, IL 60616-3793
Phone Number: (312) 567-3203
Fax Number: (312) 567-8875

Robert Foley
Illinois Institute of Technology
10 West 32nd Street
Chicago, IL 60616-3793
Phone Number: (312) 567-3052
Fax Number: (312) 567-8875

Key Words: Thermal Desorption Spectroscopy, Hydrogen Embrittlement, Hydrogen Detection, 4340 Steel, Microstructural Traps

INTRODUCTION

Hydrogen embrittlement is a general problem that can occur in the production and use of higher-strength steels. The embrittlement is associated with various types of hydrogen damage, most prevalent of which is hydrogen induced cracking (HIC), also known as hydrogen assisted cracking (HAC). Several physical conditions must all be present for HIC to occur. These include a critical level of diffusible hydrogen content in the microstructure, a restraint stress, a minimum hardness level, a susceptible microstructure, and the temperature of the piece of iron or steel that undergoes HIC must fall within a range of –100 to +100°C [1]. Consequently, HIC is preventable by keeping the steel under consideration from attaining any one of the aforementioned attributes. One method used for decreasing the diffusible hydrogen content is to add metallurgical hydrogen traps [2].

Hydrogen traps include imperfections, dislocations, dislocation debris, point defects, grain boundaries, microvoids, and second phase particles at room temperature. Most of the hydrogen within a given matrix exists inside of microstructural traps because hydrogen gas is nearly insoluble in steel [2]. The strength with which a trap holds hydrogen is called its binding energy. Traps with high binding energies, called irreversible traps, hold hydrogen atoms tightly, essentially removing them from the microstructure [3]. Traps with low binding energies, called reversible traps, do not hold onto hydrogen as tightly, and the hydrogen atoms inside such traps maintain an equilibrium concentration with hydrogen atoms present in normal interstitial lattice sites. Adding irreversible traps can reduce the diffusible hydrogen content and the susceptibility to HIC, so it is important to try to identify and characterize them within the steel microstructure.

Many different methods are used to study the attributes of hydrogen traps. Examples include the enhanced solubility method, hydrogen evolution from pre-charged iron, magnetic relaxation, internal friction measurements, the electrochemical permeation technique, and thermal desorption spectroscopy [4]. The most frequently used of these processes are the electrochemical permeation technique and thermal desorption spectroscopy.

The electrochemical permeation technique was used in the studies of Kumnick and Johnson, and it involves measuring the amount of time it takes for hydrogen to enter into, move through, and outgas from a steel membrane [5]. The hydrogen enters the steel via gas phase charging through a palladium coating, and it travels to an electrochemical cell where it can be detected by a change in potential. The time it takes hydrogen to pass through the steel membrane depends upon the apparent diffusivity, which, in turn, is affected by the number and strength of traps present in the microstructure [5]. Advantages of the electrochemical permeation technique are the development of a uniform hydrogen concentration in the microstructure and the direct control of input hydrogen fugacity [6]. Reversible and irreversible traps can be distinguished with repeated runs [7]. Disadvantages of the technique are that it is not suitable for low-temperature measurements, surface impedance reactions may not prevent hydrogen from outgassing into the electrochemical cell, and it usually only detects one type of microstructural trap [8]. Lee and Lee further point out that the technique measures the permeation of hydrogen out of the steel membrane rather than the hydrogen coming out of the traps directly [9]. Currently, there are no universally accepted methods for calculating permeation rates [8].

Thermal desorption spectroscopy is another method for studying the characteristics of microstructural traps for hydrogen in metals [7,9,10]. The theory upon which it is based centers on the fact that hydrogen leaves trap sites upon heating. As the microstructure is heated, atoms gain kinetic energy to the point where hydrogen atoms can make the jump from trap sites to normal lattice interstitial sites. The hydrogen atoms saturate the matrix, diffuse to the sample surface, and outgas into the surrounding atmosphere. There they can then be detected within a closed system by a pressure gauge, a gas chromatograph, or a mass spectrometer. If the tested metal is heated at a constant rate, the gas evolution rate from the specimen will reach a maximum at a specific temperature for each type of trap. This temperature is called the peak temperature for the trap, and it is related to the trap binding energy. Higher heating rates cause the peak temperature to increase by an amount that is related to the trap binding energy [7].

There are many advantages to studying the characteristics of microstructural trap sites using thermal desorption spectroscopy. The technique provides a wealth of information. It can be used to determine the binding energy of a trap site, the amount of hydrogen inside of a trap site, the density of a particular type of trap site, and the nature of hydrogen to trap site bonding [7]. Importantly, it can characterize more than one type of trap within a microstructure.

There are also disadvantages to using thermal desorption spectroscopy. The greatest drawback is the fact that it is a thermally destructive test [2]. Thermal desorption spectroscopy involves heating up a sample of steel to very high temperatures, often past the austenite transformation temperature. Recovery and recrystallization can take place upon heating thus altering or eliminating trap sites from the microstructure. The hydrogen in these traps is expelled into the lattice and gives a peak in the evolution rate versus temperature graph. Such effects make it difficult to distinguish between evolution rate peaks that arise from thermally desorbed hydrogen and peaks that arise from hydrogen kicked out into the matrix upon elimination of trap sites.

Thermal desorption spectroscopy was the method used in the research conducted for this paper. It was an ideal fit because there are many different types of trap sites within the microstructure of commercial high-strength bar and forging steels. The goal of this work is to characterize the hydrogen thermal desorption spectra of hot-rolled and quenched-and-tempered 4340 steel. Issues concerning design, construction, and operation of the thermal desorption spectrometer are also presented [10].

MATERIALS AND EXPERIMENTAL PROCEDURES

Equipment Though many different pieces of equipment were used to carry out the research for this paper, the most notable and prominent was the thermal desorption spectrometer which was designed and constructed as part of this study [10]. Figure 1 shows a diagram of the essential components of the apparatus. The apparatus is basically an instrumented vacuum chamber attached to a quartz tube that is surrounded by a three-zone furnace. The sample is placed into the quartz tube through a sample portal and positioned directly above a thermocouple probe. The steel samples can have most any bulk geometry though the samples used in the present study were two-inch long cylinders. Once the sample portal is closed, the system is pumped down to vacuum by a mechanical and a diffusion pump. The entire apparatus is allowed to outgas for a specified period of time following which the preprogrammed furnace controller is activated to heat the sample. The system detects hydrogen through changes in pressure and thermal conductivity of gases that occur with outgassing upon heating. The chamber is equipped with a penning gauge, a thermocouple gauge, and a capacitance diaphragm gauge. The electrical output of the penning and thermocouple transducers varies with the pressure and the composition of gases, so that lighter gasses produce a higher apparent pressure. The capacitance diaphragm gauge relays an output of absolute pressure, a pressure value independent of the type of gas present inside the chamber. This difference in transducer operational principles allows for a quick, though indirect, confirmation of whether a specified pressure increase is due to hydrogen outgassing from the sample. Though direct measurement of the gas composition can be obtained with a mass spectrometer, for example, the indirect method described here offers simplification without loss of results.

The two other important pieces of equipment used in this research were a Dell Dimension XPS D233 computer with a National Instruments 16-bit analog-to-digital data acquisition board and an EG&G potentiostat. The computer was used to transmit, record, and manipulate the voltage data taken from each desorption test. Voltage readings were recorded from a thermocouple probe in the furnace near the sample, a thermocouple vacuum gauge, and the capacitance diaphragm vacuum gauge. National Instruments Labview for Windows program was used to record the voltage outputs to a computer file, and the conversions from voltages to pressures and temperatures were made with Microsoft Excel 97. The EG&G potentiostat, together with a graphite anode and charging solution, was used to charge the steel samples with hydrogen.

Sample Preparation: The 4340 steel used in this work was obtained in the form of 3/4-inch diameter, ten-foot long hot-rolled bars. The as-received condition is designated as AR in the following paragraphs. The bars were sectioned in to four-inch lengths, normalized for thirty minutes at 900°C, and furnace cooled. Samples were austenitized at 870°C for thirty minutes and quenched into ice water held between 0 and 5°C. The as-quenched samples were tempered at either 350 or 600°C for four hours at temperature and allowed to furnace cool. The quenched-and-tempered conditions are designated as either QT-350 or QT-600 in the following paragraphs.

Metallographic examination revealed that AR bars had a uniform ferrite-pearlite microstructure while the QT bars had uniform tempered martensite microstructures. The martensitic laths appeared to be slightly coarser for QT-600 samples than for QT-350. Figure 2 shows microhardness values obtained from test traces across the sample diameters. The data reveal that the hardness values in the central section of the bars was uniform and varied as expected from changes in the base microstructure due to the applied heat treatments [11]. The QT-350 sample is the hardest of all samples because it received the lowest tempering temperature after quenching. The hardness appears to stay constant across the diameter of the bar, but it drops off dramatically within 0.1 mm of the outside diameter. The QT-600 sample showed a similar profile, but with a lower base hardness because of the higher-temperature tempering it received. The as-received samples have the lowest base hardness of all, and its hardness distribution was comparatively uniform. The drop in hardness values near the edges of the QT-350 and QT-600 samples is a result of surface decarburization that occurred due to reactions of carbon in the steel with oxygen in the atmosphere of the laboratory box furnace used to carry out the heat treatments.

Lengths near the center of the four-inch bars were turned to produce 5/8-inch diameter cylinders. These cylinders were designed to be immersed into the hydrogen charging solution and later outgassed in the thermal spectrometer. One end of the 3/4-inch section was retained and cleaned to provide electrical contact in the hydrogen charging procedure. The cathodic polarization curve for 4340 steel published by Shim was used to determine an appropriate current density range for hydrogen charging [12]. The charging solution was mixture of 1 liter of 1 N H_2SO_4 with 5 g of As_2O_3, and preliminary tests indicated that a current density of 0.1 mA/cm^2 would produce a pressure rise in the apparatus appropriate for the transducer ranges while minimizing possible changes in the microstructure. Ono and Meshii point out that the peak temperatures of evolution rate curves (in this case change in pressure curves) are pushed higher with more severe cathodic charging conditions [4]. The samples were cleaned, using alternate applications of soap-and-water, acetone, and 400 grit carbimet-polishing paper, and dried. Charging was carried out for six hours at 30°C after which the turned, two-inch section was separated from the bar using a cutoff wheel. Each sample was cleaned before being loaded into the thermal desorption spectrometer.

Experimental Procedures The procedure for the process of hydrogen charging, outgassing, and ramping is described in detail. The procedure was identical for every sample. A sample is cleaned and polished before cathodic charging. It is then suspended in a beaker and connected to the aforementioned potentiostat with alligator clamps. The potentiostat is subsequently hooked up to a graphite electrode as the anode and the sample as the cathode. An appropriate current density is set on the potentiostat to provide a charging density of 0.1 mA/cm^2, the potentiostat is turned on, and the beaker is filled with the hydrogen charging solution. A timer is started as soon as the level of the charging solution reaches the ridge between the two-inch, specified-diameter, cylindrical section, and the 3/4-inch bar section. The beaker is placed on a hot plate, and the hot plate warms the charging solution until it reaches a temperature of 30°C. The sample charges for six hours, and it is then removed, cleaned, polished, and placed into the preconditioned thermal desorption spectrometer through the sample portal. The process of transferring the sample from the charging apparatus to the thermal analysis apparatus takes approximately forty-five minutes due to the cleaning and sectioning procedures. The sample is placed directly over the furnace thermocouple probe, and the portal is resealed. The entire apparatus is pumped down and the tube-furnace is set to 30°C. Once the thermocouple beneath the sample registers 30°C, a timer is started, so that the sample could be outgassed for a consistent period of fourteen hours in order to remove the diffusible hydrogen from the sample. The computer with the A/D board is set up to measure the voltages from the thermocouple under the sample, the thermocouple vacuum gauge, and the capacitance diaphragm gauge. The A/D program is initialized to collect data once every five seconds. Upon completion of the outgas period, the pressure in the system is less than 0.01 mTorr. The furnace controller program is started, a ball valve is closed to isolate the system, and the computer program is started. For tests reported here, the furnace controller ramps the furnace temperature at 5°C/min to 900°C whereupon it stabilizes the furnace at that temperature for about an hour.

Samples from the as-received bars were machined, charged, and tested following the procedures used to test the quenched-and-tempered samples. However, the AR tests were performed in direct sequence without preconditioning or intermediate thermal cycles. Baseline tests were performed periodically to quantify the outgassing rate of the system as a whole as well as to check for leaks. These tests were performed with a sample that had been tested but which had not been removed from the apparatus.

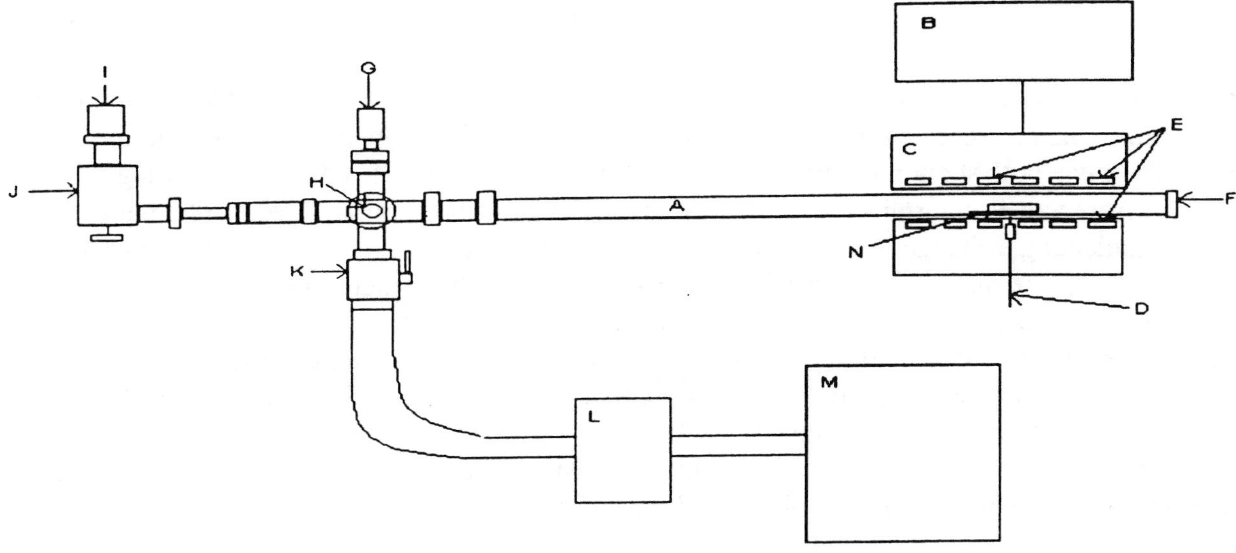

Figure 1 - Thermal Desorption Spectrometer. A: Quartz Tube, B: Furnace Controller, C: Furnace, D: Thermocouple Probe, E: Heating Elements, F: Sample Portal, G: Penning Gauge, H: Thermocouple Gauge, I: Capacitance Diaphragm Gauge, J: Right Angle Vacuum Valve, K: Ball Valve, L: Diffusion Pump, M: Mechanical Pump, N: Sample

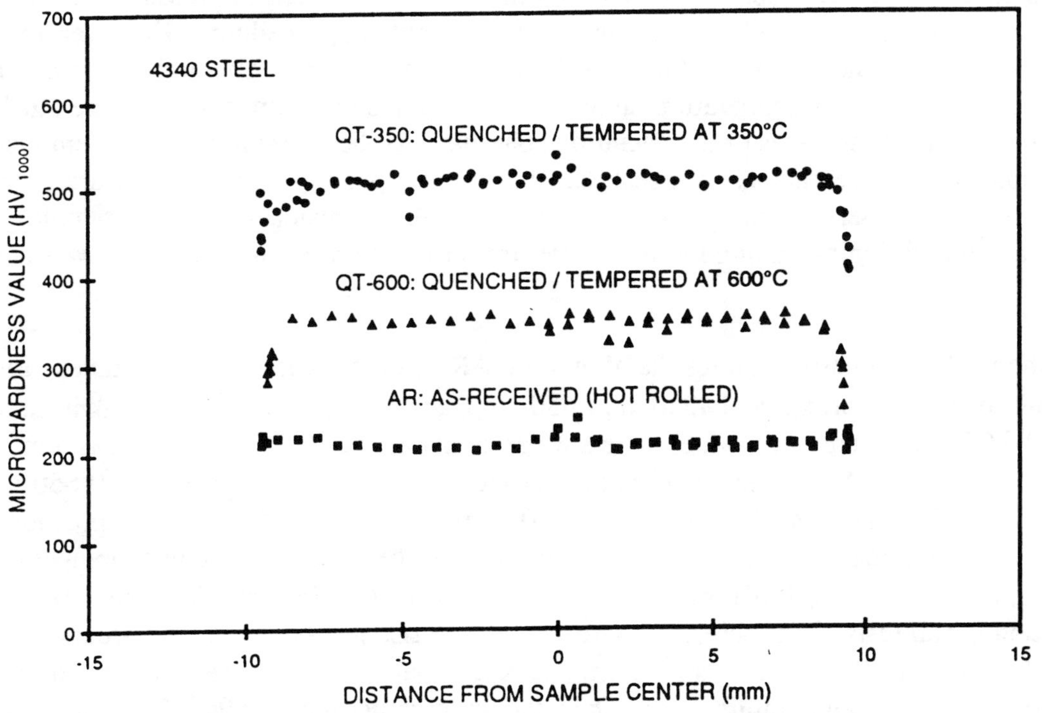

Figure 2 - Microhardness profiles through the samples AR, QT-350, and QT-600.

RESULTS

Figure 3 shows pressure rise and furnace temperature data obtained from an AR sample test and from a blank sample test. The pressure determined from the thermocouple (TC) gauge as well as from the capacitance diaphragm (CD) gauge both show pressure increasing monotonically with periods of accelerating pressure rise. Initially, the pressure values determined from the TC and CD gauges are similar. Eventually, the pressure values deviate, indicating a change in the composition of the system gases. Because a hydrogen-rich atmosphere has higher thermal conductivity than a nitrogen-oxygen atmosphere, the pressure value that is determined from the TC gauge is higher than the actual pressure value that is determined from the CD gauge. The pressure value determined from the CD gauge is insensitive to composition, and therefore this value indicates the true system pressure rise from system outgassing, from leaks, or from the sample. While the major difference between AR#4 and the blank samples is the amount of hydrogen each releases, comparison of the results of the tests show that the pressure increase due to the system leak rate is small and distinguishable from the pressure increase from hydrogen evolution out of the test sample.

Although the plot shown in Figure 3 can be used to characterize hydrogen desorption behavior, it is common to present the results of a thermal desorption test as a plot of hydrogen evolution rate versus temperature or time. The advantage of this approach lies in the fact that evolution rate curves can indicate the amount of hydrogen present in each trap site as well as the total amount of trapped hydrogen in the sample. And though the difference in TC and CD gauge pressure values is indicative of hydrogen outgassing, the time derivative of the TC gauge pressure can also give information about the overall system performance. Such a treatment does not change the location of the peaks, but the curves are exaggerated above their true values.

Initial test results from the AR samples revealed that the apparatus must be thoroughly outgassed before repeatable results could be obtained. Figure 4 shows the results as change-in-pressure versus temperature curves. All four tests in this set were carried out following identical procedures. However, the results of the first sample test are different from the results of subsequent sample tests. Although a fourteen-hour degassing cycle was applied prior to each temperature ramp, the different results from sample #1 are attributed to outgassing from the inside surfaces of the vacuum apparatus. As the pressure in the system increases to 1 atm with each sample change, the data also suggest that short time exposures to 1 atm pressure is not sufficient to reproduce the behavior observed after prolonged exposures. As the results of tests on samples #2-#4 varied only slightly, a preconditioning baking thermal cycle, from room temperature to 900°C, was added to the test procedure.

Figure 5 shows thermal desorption results of tests on AR, QT-350, and QT-600 samples. Figures 4 and 5 show that peaks occur at approximately 250, 600-650, 750, and 900°C. The 250°C peak is not due to hydrogen because TC and CD gauge pressure values were similar below about 450°C. Similar reasoning indicated that the other peaks are attributable, at least in part, to hydrogen desorption. The peak at 610-650°C is largest for the as-received samples, and it is slightly larger for QT-350 than it is for QT-600. The position of this peak depends on the prior thermal processing and microstructure of the samples. The peak temperature is 610°C for the AR sample, 625°C for the QT-600 sample, and 650°C for the QT-350 sample. The 750°C peak is similar in size and position for all tests. The increase in rate of pressure rise just below 900°C occurs due to hydrogen outgassing. However, the 900°C peak is not a true peak associated with release of hydrogen from a trap, but a reflection of the end of the temperature ramp. Once the temperature reached 900°C, the test proceeded isothermally for about an hour.

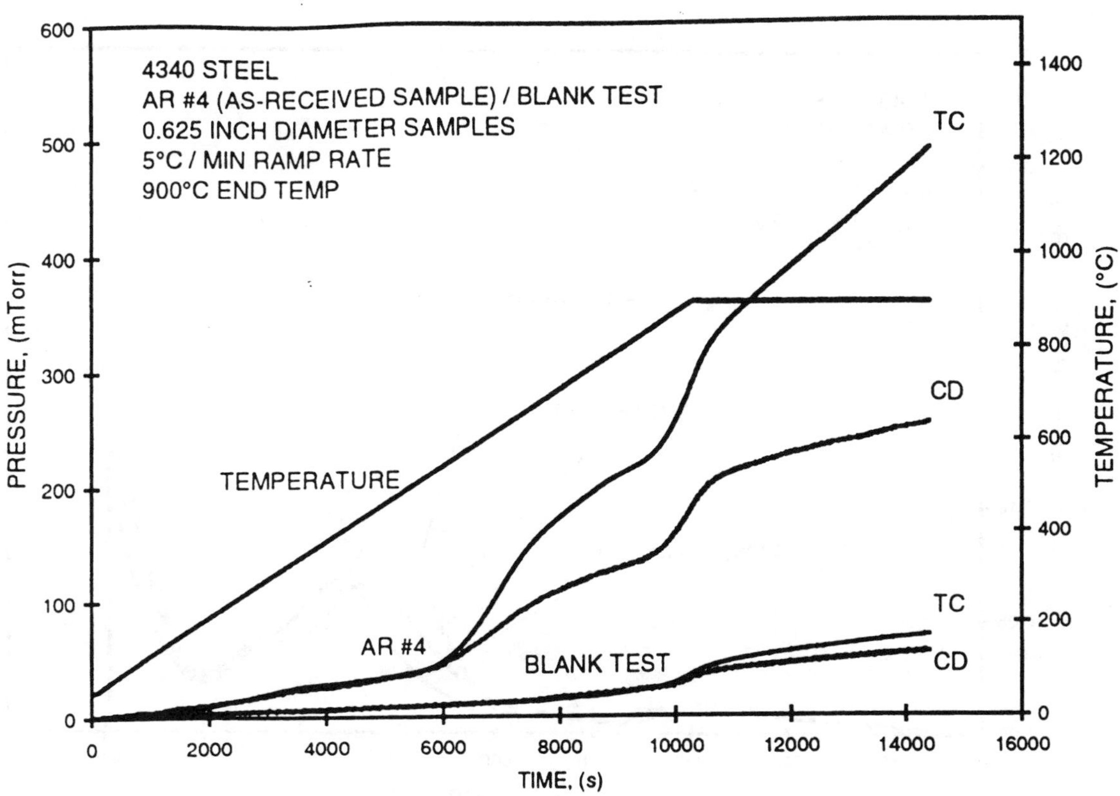

Figure 3 – Pressure and temperature data from thermal desorption tests.

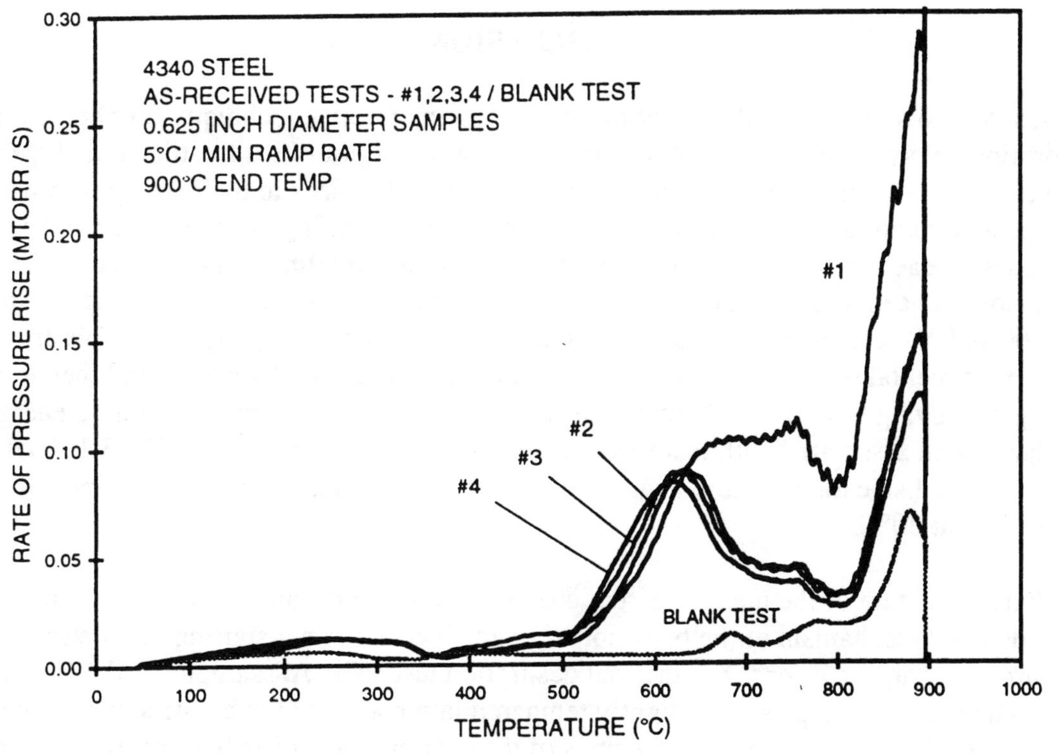

Figure 4 – Thermal desorption results from multiple tests of as-received samples of hot-rolled 4340 steel.

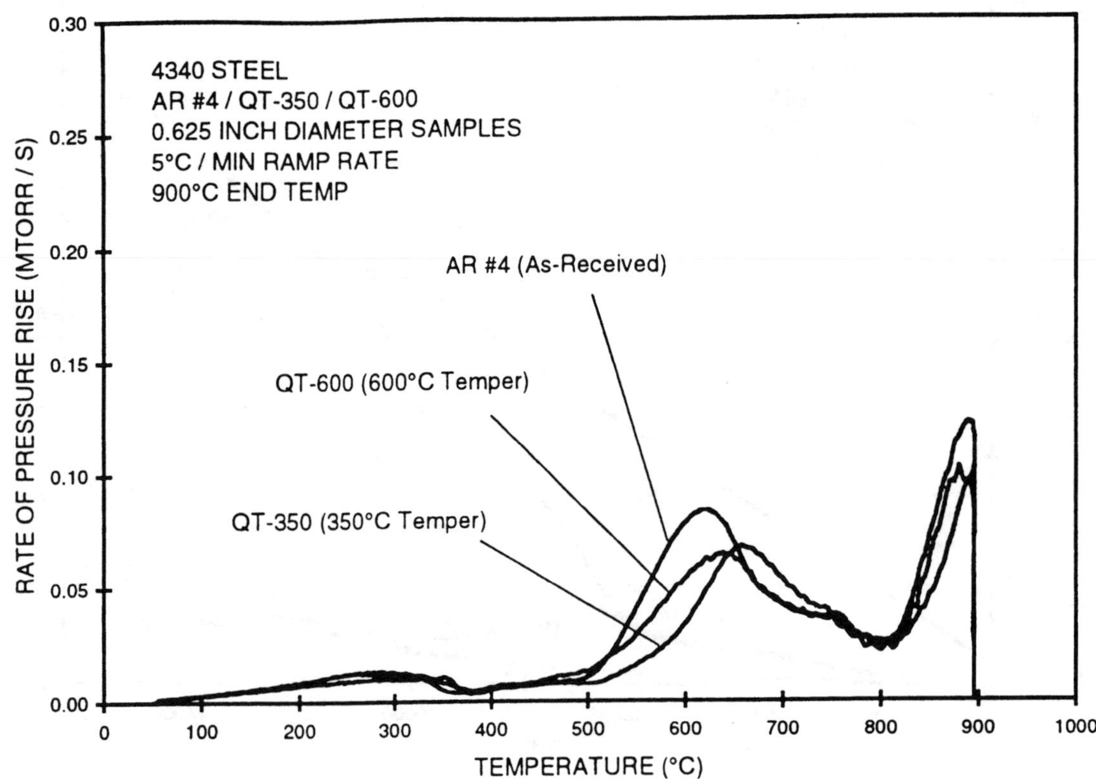

Figure 5 – Thermal desorption test results from AR, QT-350, and QT-600 4340 steel samples.

DISCUSSION

The above results show that thermal desorption analysis is an effective technique for characterizing hydrogen desorption behavior of steel samples. Figure 4 shows that prior heat treatment, and the accompanying microstructure changes, give rise to measurable, though subtle, differences in hydrogen desorption behavior. The similar shape and magnitude of the peaks indicates that similar binding energies and number density of hydrogen traps exist in each material. The major effect of prior heat treatment is exhibited by differences in the desorption behavior between 450 and 750°C. For the combination of sample size and temperature ramp used in this study, the 600-650°C desorption peak is associated only with very strong traps. This observation is consistent with experimental procedures designed to saturate the traps with hydrogen and then remove the diffusible hydrogen, leaving only tightly bound hydrogen. Such strong traps have a binding energy over 30-50 kJ/mol and have been associated with features such as microvoids or precipitates [7,13,14]. Hydrogen in weaker traps such as in dislocations or low-angle boundaries are expected to have been removed by the outgassing procedure at 30°C.

Due to the dynamic nature of the microstructure changes occurring during the test itself, only a qualitative description of a detailed mechanism of the behavior is given. Three distinct starting microstructures were subjected to hydrogen charging prior to the thermal desorption test. The AR samples had a fine ferrite-pearlite microstructure. The QT-350 samples had a lightly tempered lath martensite microstructure which consisted of a combination of lath packets containing high densities of dislocations, fine distributions of iron carbides, and a small volume fraction of inter-lath retained austenite [11]. The QT-600 samples also had a tempered martensite microstructure, but as the hardness values in Figure 2 indicate, the four-hour 600°C tempering treatment produced a much coarser microstructure consisting of iron carbides, alloy carbides, fewer defects and dislocations within a polygonallized ferrite base microstructure [11,15]. The different microstructures produced prior to hydrogen charging and thermal degassing each had a very different stability with respect to possible changes that might be incurred upon heating. The ferrite-pearlite microstructure, formed by diffusion decomposition of austenite after normalizing, is the most stable of the three microstructures while the QT-350 microstructure is the least stable.

Of the possible irreversible trap sites, voids, microcracks, precipitates, and precipitate-matrix interfaces were found to have the highest binding energies [7,13,14]. The results in Figure 5 support the idea that the 610-650°C peak is a result of degassing from traps with binding energies in excess of 50 kJ/mol. Gibala and Demiglio characterized the various traps present in 4340 steel and identified the incoherent Fe_3C and alloy carbides as the deepest hydrogen traps [14]. Pressouyre and Bernstein identified TiC precipitates that acted as hydrogen traps with binding energies close to 95 kJ/mol, and Turnbull confirmed that thermal desorption of these traps, in otherwise unalloyed iron, appears at approximately 690°C [7,16]. This peak temperature is close to the peak temperature of 610-650°C observed in Figure 5. As there are many other differences between the iron alloy and 4340 steel microstructures, it is likely the case that the peaks observed in the present study represent hydrogen desorption from precipitates as well as inclusions. Manganese sulfide inclusions can also act as irreversible traps in 4340 [13]. Notably, the desorption rates at 600°C rank according to microstructure coarseness. At 600°C, the rate of outgassing is highest for the AR samples, and lowest for the QT-350 sample. However, while this implies that the finer microstructure has higher binding energy hydrogen traps, the number density of these traps is similar in all materials as demonstrated by the consistent size of the three peaks. In 4340 steel, the most prevalent alloy carbides are those containing Cr and Mo, while the prevailing iron carbide is Fe_3C [11]. Gibala and Demiglio found that the primary, irreversible traps for hydrogen in 4340 steel, quenched-and-tempered at temperatures of 400°C and above, were microcracks and incoherent Fe_3C [14]. The supposition is supported by the fact that carbide particles and sulfide inclusions are common to all samples, although the particle distributions differ significantly between initial microstructures. The Fe_3C in the AR microstructure exists mostly in the lamellae of the pearlite. The QT-350 microstructure has a finer Fe_3C distribution because the dislocations within the laths of the martensite microstructure, the lath boundaries, the lath packets boundaries, and prior austenite boundaries present in the as-quenched microstructure provide nucleation sites for iron carbide precipitates [11]. Though coarsening of the Fe_3C precipitates occurs with increasing tempering temperature, higher tempering temperatures also promote the precipitation of alloy carbide precipitates. However, while such variations in initial microstructure largely control the mechanical behavior of the steel, the results imply that microstructure modifications through heat treatment do not significantly affect the strength or number density of deep hydrogen traps in 4340 steel.

Two more peaks were observed on the spectra of the experimental and as-received samples: a small peak at 750°C and a large peak at 900°C. The 750°C peak may correspond to the end of the austenite phase transformation. As hydrogen is more soluble in austenite than in ferrite, new austenite should act as a hydrogen trap and the rate of outgassing should decrease as hydrogen dissolved in the ferrite sinks into the austenite. The Ac_1 and Ac_3 temperatures of 4340 steel are 705 and 725°C. Indeed, all tests generally showed a decrease in the rate of outgassing between about 700 and 800°C. Above 750°C, no ferrite should remain and the process of hydrogen re-equilibration and desorption from traps should begin anew with austenite as the base microstructure.

SUMMARY

A major task involved in this research was the design, assembly, and the development of procedures for operation of a thermal desorption spectrometer. The goal of this work was to identify and characterize the deep irreversible trap sites for hydrogen within the microstructure of 4340 steel. Hot-rolled and quenched-and-tempered 4340 steel samples were charged with hydrogen and degassed in the apparatus. The results revealed generally similar degassing behavior, and peaks were observed at approximately 250, 610-650, 750, and 900°C. Although the 250°C peak was attributed to desorption of gases from the walls of the apparatus itself, the other peaks were associated with hydrogen evolution from the samples. The prior heat treatment and microstructure most significantly affect the position and shape of the 610-650°C peak with the trend that coarser microstructures bind hydrogen less effectively than do finer microstructures. The hydrogen desorption in this temperature range occurs only from very strong traps and was attributed to carbide particles and inclusions in the steel.

AKNOWLEDGEMENTS

The authors gratefully acknowledge the many helpful discussions with Mrs. Craig Johnson, John Swartz, and Ron Mashek of the E1 machine shop for their helpful suggestions concerning the design and construction of the apparatus. This work was supported though a graduate research fellowship provided by A. Finkl and Sons Company of Chicago, Illinois.

REFERENCES

1. N. Bailey, F.R. Coe, T.G. Gooch, P.M. Hart, N. Jenkins, and R.J. Pargeter, Welding Steel without Hydrogen Cracking, 2nd Edition, ASM International, Materials Park, OH, 1993.

2. M. Nagumo, K. Takai, and K. Okuda, "Nature of Hydrogen Trapping Sites in Steels induced by Plastic Deformation," Journal of Alloys and Compounds, Vol. 293-295, 1999, pp. 310-316.

3. J.-Y. Lee and J.L. Lee, "A Trapping Theory of Hydrogen in Pure Iron," Philosophical Magazine A, Vol. 56, No. 3, 1987, pp. 293-309.

4. K. Ono and M. Meshii, "Hydrogen Detrapping from Grain Boundaries and Dislocations in High Purity Iron," Acta Metallurgica et. Materiala, Vol. 40, No. 6, 1992, pp. 1357-1364.

5. A.J. Kumnick and H.H. Johnson, "Deep Trapping States for Hydrogen in Deformed Iron," Acta Metallurgica, Vol. 28, 1980, pp. 33-39.

6. P.L. Chang and W.D.G. Bennett, Journal of the Iron and Steel Institute, 1952, pp. 205-213.

7. A. Turnbull, R.B. Hutchings, and D.H. Ferriss, "Modeling of Thermal Desorption of Hydrogen from Metals," Materials Science and Engineering, Vol. A238, 1997, pp. 317-328.

8. R.W. Lin and H.H. Johnson, "Advanced Techniques for Characterizing Hydrogen in Metals," edited by N.F Fiore and B.J. Berkowitz, TMS-AIME, Kentucky, 1981, pp. 105-118.

9. J.-L. Lee and J.-Y. Lee, "Hydrogen Retrapping after Thermal Charging of Hydrogen in Iron Single Crystal," Metallurgical Transactions A, Vol. 20A, September 1989, pp. 1793-1802.

10. W.Y. Choo and J.-Y. Lee, "Thermal Analysis of Trapped Hydrogen in Pure Iron," Metallurgical Transactions A, Vol. 13A, January 1982, pp. 135-140.

11. G. Krauss, "Tempering of Steel," Steels: Heat Treatment and Processing Principles, ASM International, Materials Park, OH, 1990, pp. 205-241.

12. I.O. Shim and J.G. Byrne, "A Study of Hydrogen Embrittlement in 4340 Steel I: Mechanical Aspects," Materials Science and Engineering, Vol. A123, 1990, pp. 169-180.

13. B.G. Pound, "Hydrogen Trapping in High Strength Steels," Acta Materiala, Vol. 46, No. 16, 1998, pp. 5733-5743.

14. R. Gibala and D.S. DeMiglio, "Hydrogen in Iron and Steels: Interactions, Traps, and Crack Paths," Conf. Proc. Hydrogen Effects in Metals, edited by I.M. Bernstein and A.W. Thompson, The Metallurgical Society of AIME, Warrendale, PA, 1981, pp.113-122.

15. William C. Leslie, "Carbon Steels," The Physical Metallurgy of Steel, TechBooks, Herndon, VA, 1981.

16. G.M. Pressouyre and I.M. Bernstein, "A Quantitative Analysis of Hydrogen Trapping," Metallurgical Transactions A, Vol. 9A, November 1978, pp. 1571-1579.

EFFECT OF THERMOMECHANICAL HISTORY ON THE HIGH TEMPERATURE MECHANICAL PROPERTIES OF A MICROALLOYED STEEL AND A LOW CARBON STEEL

Siamak Akhlaghi, Farid Hassani and Steve Yue
Department of Mining and Metallurgical Engineering, McGill University
3610 University Street
Montreal, Quebec, H3A 2B2
Canada
(514) 398-4755 ext.1378
steve@minmet.lan.mcgill.ca

Key Words: Hot ductility, Microalloyed steel, Low carbon steel, Thermomechanical processing, Isothermal hot ductility curve

ABSTRACT

The low ductility of steels at elevated temperatures is one of the main causes of surface cracking in continuous casting. Much work has been conducted to study the hot ductility behavior of various steels using simple isothermal tests in laboratories. However, there are several considerable differences between the thermomechanical history of such tests and the complex history that occurs during continuous casting. Previous work at McGill has shown that the thermal history can strongly control the hot ductility of steel. The present work concentrates on the effect of strains imposed during thermal cycles representative of continuous casting. In the casting machine, such strains could be generated by friction, ferrostatic pressure or thermal gradients. It is observed that imposing such strains markedly improves or deteriorates hot ductility at the unbending point of the thermal history in a Nb-Ti microalloyed steel, depending largely on the temperature at which the strains are executed. On the other hand, applying identical thermomechanical histories to a low carbon steel has little affect on the hot ductility. The metallurgy and consequences of these results are discussed in this paper.

INTRODUCTION

Transverse cracking in strands produced by continuous casting process is an ongoing problem.[1-3] There have been many hot ductility studies to determine the metallurgical reasons for this problem. In the majority of these efforts, the specimens have been subjected to a simple thermal profile immediately prior to tensile testing to fracture. However, thermal analyses of the casting process, indicates that the thermal history that the strands are actually experiencing during the continuous casting process is entirely different from these simple laboratory tests.[4-5] In addition to this, there is some evidence that the surface of the strands are also being subjected to stresses generated thermally or mechanically.[2,6-7]

In the present work, the simultaneous effects of thermal history and mechanical deformation (i.e. the thermomechanical history) in a Nb-Ti microalloyed steel and a low carbon steel, have been studied. The intriguing results are explained with the help of microstructural observations.

EXPERIMENTAL PROCEDURE

The hot ductility of specimens subjected to different thermomechanical histories, was studied in an in-situ melting and solidification tensile test. Details of the test setup have been reported elsewhere.[8]

Two steels having close carbon contents were examined. One was a Nb-Ti microalloyed steel with carbon content of 0.06 % and the other was simple low carbon steel with carbon content of 0.08 % and no microalloyed elements. It should also be noted that the Mn and Si contents of these steels were different. The full compositions of the steels are given in Table I.

Table I Chemical composition of studied materials (wt %)

	C	Mn	Si	P	S	Nb	Ti
Nb-Ti	0.058	1.210	0.159	0.008	0.011	0.037	0.017
Low carbon	0.079	0.426	0.062	0.009	0.012	0.002	0.001

Conventional isothermal hot ductility tests were conducted on both steels. In these experiments, specimens were heated up to right above their melting points and then cooled down to the test temperature at a cooling rate of 10 °C/s and finally pulled to fracture at a rate of 5×10^{-3} s^{-1}. The effect of deformation on isothermal hot ductility tests was measured by applying 10 % compressive deformation at the rate of 8×10^{-3} s^{-1} in the proximity of melting point during cooling down from melting.

In another set of experiments, tensile specimens were subjected to two different time/temperature schedules, Table II and Fig. 1, which were obtained from mathematical model simulations of a billet continuous casting process. In these schedules, the temperature drops to T_{min} after melting, because the surface of the strand is in contact with the mould at this stage. On exiting the mould, the temperature immediately increases due to outward heat diffusion from the core of the strand to the surface. Finally the temperature decreases slightly to the point of unbending.

Table II Key temperatures of the thermal histories studied.

Thermal History	Melting Temp.	Cooling rate after Solidification	T_{min} Temp.	T_{max} Temp.	$T_{unbending}$ Temp.
No. 1	1450 °C	10 °C /s	780 °C	1200 °C	1020 °C
No. 2	1450 °C	10 °C/s	800 °C	1180 °C	1020 °C

Different levels of deformation, applied at a rate of 8×10^{-3} s^{-1}, were superimposed at different stages of these thermal histories. The effect of these deformations on the hot ductility was measured by tensile testing at the point of the thermal history, representing the unbending point, (Fig. 1) at a rate of 5×10^{-3} s^{-1}.

RESULTS

Conventional isothermal hot ductility tests

The hot ductility versus test temperature curves of both steels are shown in Fig. 2. As can be seen, the hot ductility at 1050 °C drops down precipitously in the Nb-Ti microalloyed steel, whereas only a very shallow trough is seen in the low carbon steel. This behavior is in agreement with previous reports for both steels.[1]

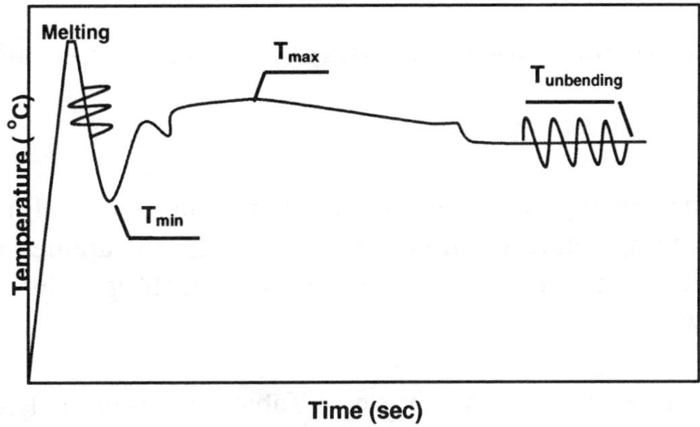

Fig. 1: Schematic diagram of thermal history used in experiments.

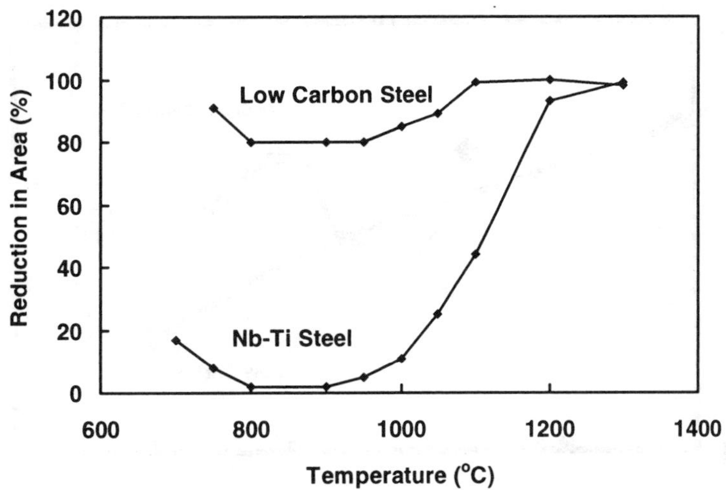

Fig. 2: Hot ductility curves for both steels.

Effect of deformation on conventional isothermal hot ductility curve of Nb-Ti microalloyed steel

In order to study the effect of deformation on the hot ductility as measured by conventional isothermal hot ductility tests, a compressive deformation of 10 % was applied beginning at 1450 °C, during the cooling segment of the thermal profile. This temperature was chosen due to the significant changes observed in the case of billet simulation thermal histories following such a deformation at this temperature, as will be shown later. The stress-temperature curve obtained from applying such a deformation at 1450 °C exhibits a hump, as shown in Fig. 3. This suggests that a change in microstructure is occurring during deformation. Zhou et al. have reported that in a steel similar to this composition, the transformation of δ-ferrite to austenite takes place over this temperature range.[9] If this is the case, then the hump shape can be explained as follows: the stress first decreases because strain free austenite replaces work-hardened ferrite, but the stress increases again because austenite is work hardened. As is illustrated, the work hardening rate of austenite is higher that ferrite, which is as expected.

The hot ductility trough of the Nb-Ti microalloyed specimens subjected to deformation during the conventional isothermal hot ductility test, is presented in Fig. 4. A comparison between the hot ductility curves with deformation, and without applied deformation, indicates that applying deformation does not have a strong influence on the hot ductility trough, although the minimum ductility occurs at a slightly lower temperature after deformation has been executed.

Effect of superimposed deformation on the hot ductility of the Nb-Ti microalloyed steel in thermal history experiments

Thermal History 1

Single-step deformation. The first region studied was in the proximity of melting point. It was found that the hot ductility at the unbending temperature when no prior deformation is applied is 10 %. After imposing 10 to 20 % compressive deformation at the thermal region close to the melting point, the hot ductility at unbending point improves to about 100 %.

The other region studied was at 900 °C. As shown in Table III, applying 1 to 20 % tensile deformation at this temperature leads to a decrease in hot ductility at unbending from 10 to 3 %.

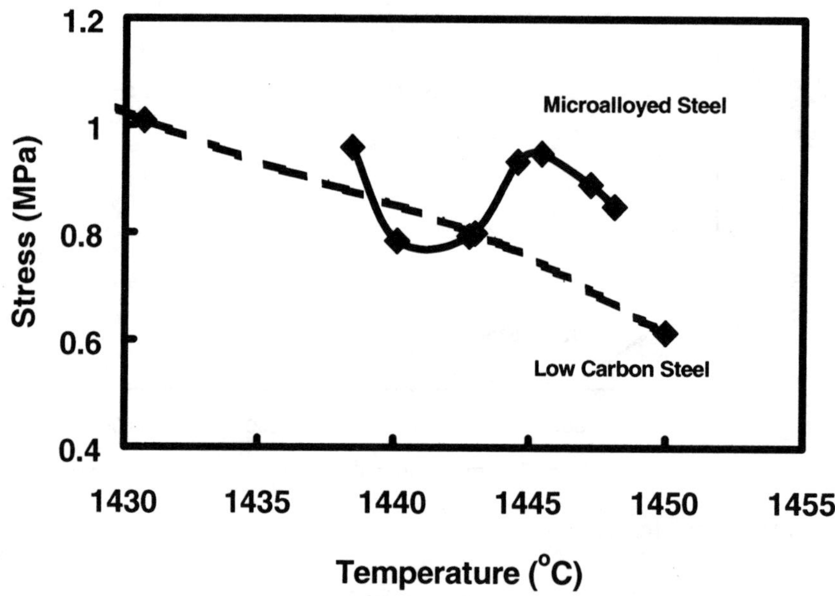

Fig. 3: The stress-strain curve of the Nb-Ti microalloyed steel and low carbon steel, obtained after applying 10 % compressive deformation in the proximity of melting point.

Multi-step deformation. Real thermal profiles of strands cooled during the continuous casting process show that the surfaces of strands may experience cyclic thermal changes at the secondary cooling zone during continuous casting. This region corresponds to the last cooling segment of the thermal history illustrated in Fig. 1, encompassing a temperature range from 1200 to 1020 °C. In order to consider this effect, cyclic deformation in the thermal range of 1200 to 1020 °C was imposed. As can be seen in Table IV, superimposing tension-compression cyclic deformation with 2 % deformation at each cycle raises the hot ductility at the unbending point to 15 %, as compared to only 3 % after deformation at 900 °C.

Also, since it was found that the ductility at unbending point changes with the temperature at which deformation is applied, i.e. in the proximity of melting point or at 900 °C, the effect of combinations of these deformations on the hot ductility at the unbending point was studied. As shown in Table IV, applying 20 % compressive deformation in the proximity of melting point, following by 1 % tensile deformation at 900 °C leads to 97 % hot ductility at unbending point.

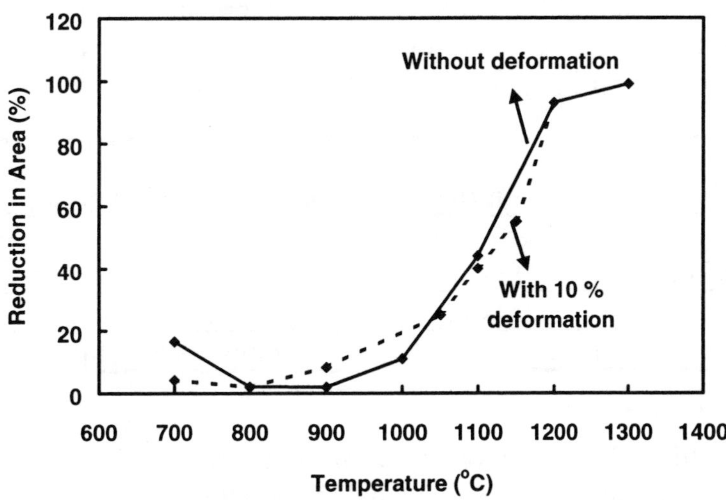

Fig. 4: Effect of deformation on hot ductility trough. 10 % compressive deformation is applied in the proximity of the melting point, at 1450 °C.

Table III Effect of single step deformation on the hot ductility of Nb-Ti microalloyed steel

Condition	Deformation Start Temp.	Deformation Percent	Mode of Deformation	Reduction in Area
No Strain	N/A	N/A	N/A	10 %
20 % Com.	1450 °C	20 %	Compression	98.5 %
10 % Com.	1450 °C	10 %	Compression	98 %
1 % Ten.	900 °C	1 %	Tension	1 %
20 % Ten.	900 °C	20 %	Tension	3 %

The last multi-step deformation, which was studied, was composed of 20 % compressive deformation at the proximity of melting point, 1% tensile deformation at 900 °C and cyclic deformations in the temperature range of 1200 to 1020 °C. As can be seen in Table IV, the hot ductility at the unbending point of this schedule is 97 %.

Effect of thermal history changes on the hot ductility of Nb-Ti microalloyed steel

The effect of the thermal history changes, which were shown in Table II, on the hot ductility of Nb-Ti microalloyed steel are shown in Table V. It was observed that increasing the T_{min}, or decreasing the T_{max} in the thermal history, improves the hot ductility at unbending from 10 %, in thermal history 1, to 60 %, in thermal history 2. Also it is shown in Table V that superimposing 5 % compressive deformation at the proximity of the melting point in thermal history 2, improves the hot ductility at the unbending point from 60 % to 90 %.

Table IV Effect of multi-step deformation on the hot ductility of Nb-Ti microalloyed steel

Condition	Deformation Start Temp.	Deformation Percent	Mode of Deformation	Reduction in Area
1 % Ten.	900 °C	1 %	Tension	1 %
1 % Ten. + Cyclic deform.	900 °C + 1200-1020 °C	1 % + 2 % per each cycle	Tension + Tension-Compression	15 %
20 % Comp. + 1 % Tension	1450 °C + 900 °C	20 % + 1 %	Compression + Tension	97 %
20 % Comp. + 1 % Tension + Cyclic deform.	1450 °C + 900 °C + 1200-1020 °C	20 % + 1 % + 2 % per each cycle	Compression + Tension + Tension-Compression	97 %

Table V Effect of thermal history on the hot ductility of Nb-Ti microalloyed steel

Condition	Deformation Start Temp.	Deformation Percent	Mode of Deformation	Reduction in Area
No Strain	N/A	N/A	N/A	60 %
5 % Com.	1450 °C	5 %	Compression	90 %

Effect of deformation on hot ductility of low carbon steel, thermal history 1

Deformation studies similar to those performed on the Nb-Ti microalloyed steel were executed on the low C steel. In the absence of any deformation, the hot ductility at the unbending point was 100 %. By contrast with what was observed in the Nb-Ti microalloyed steel, applying 10 to 20 % compressive deformation at the proximity of melting point, leads to a decrease in the hot ductility from the 100 % to less than 74 %. However, it was found that, regardless of either the deformation temperature, the mode of deformation or the extent of deformation, applying deformation in any other thermal region did not change the hot ductility.

DISCUSSION

Conventional isothermal hot ductility tests

The hot ductility curve shown in Fig. 2 indicated that the hot ductility of low carbon steel is higher than that of the Nb-Ti microalloyed steel at all temperatures below 1200 °C. This is because of the absence of precipitates in the low carbon steel. Precipitates can cause void formation, which, by coalescence, leads to failure.[1,3]

Table VI Effect of deformation on the hot ductility of low carbon steel

Condition	Deformation Start Temp.	Deformation Percent	Mode of Deformation	Reduction in Area
No Strain	N/A	N/A	N/A	100 %
20 % Com.	1450 °C	20 %	Compression	74 %
10 % Com.	1450 °C	10 %	Compression	70 %
20 % Com.	1420 °C	20 %	Compression	100 %
10 % Com.	1420 °C	10 %	Compression	100 %
6 % Ten.	950 °C	6 %	Tension	100 %
40 % Ten.	950 °C	40 %	Tension	100 %

Effect of deformation on the isothermal hot ductility curve of Nb-Ti microalloyed steel

It was shown that imposing deformation in the proximity of melting point, 1450 °C, leads to a slight widening of the hot ductility trough, to lower temperatures, but no change at high temperatures, Fig. 4.

It was illustrated in Fig. 3 that the deformation in the proximity of melting point is applied during the transformation of δ-ferrite to austenite. The microstructure of specimens subjected to this deformation, as a result, showed that grain size is refined leading to a more continuous ferrite network on cooling to 780 °C.[10]

Thus, in the conventional isothermal tests, when this structure is subjected to tensile deformation, the ferrite, which is the softer phase relative to austenite due to its higher stacking fault energy and thus higher recovery rate, is subjected to considerably more deformation.[1,11-12] Therefore void linkage occurs along the thin layers of ferrite at grain boundaries, and because of the more continuous ferrite network, the ductility trough widens at low temperatures.

Effect of deformation on the hot ductility of Nb-Ti microalloyed steel subjected to thermal histories

Single-step deformation. Results given in Table III indicated that hot ductility at the unbending point due to thermal history 1 alone is 10 %. It can be argued that during solidification most of the alloying elements, including Nb, segregate to grain boundaries. Thus after decreasing the temperature to below the solubility temperatures of the species during the thermal history, these elements will precipitate out.

Furthermore, it was shown in Fig. 4 that the stress-temperature curve obtained from applying such a deformation shows a hump that suggests a change in microstructure occurring during deformation. This change in microstructure corresponds to the reported δ-ferrite to austenite transformation temperature for the present steel.[9] Therefore it could be speculated that deformation in the proximity of melting point was applied in two regions: L+δ and δ+γ phase regions. This causes accumulation of strain in the δ phase and consequently finer austenite grains form due to deformation-induced transformation. The finer austenite grains provide more preferential sites for the subsequent nucleation of α-ferrite at lower temperatures. Therefore, the α to γ transformation starts at higher temperatures. This in turn leads to a higher volume fraction of ferrite at T_{min}.[10] The reason why this improves the ductility can be suggested as follows. In the absence of deformation, the as-solidified grain boundaries would contain high concentrations of segregated elements, such as Nb. This would encourage grain boundary precipitation as the temperature falls below the solubility temperatures of the precipitating species. However, if deformation induced transformation takes place, the austenite boundaries move, leaving behind the segregated regions. Hence fewer precipitates would be present at the unbending temperature. Also it is worth noting that long holding times at high temperature help the homogenization of segregated areas that are left behind as the grain boundaries move away.

At this point, it should be recalled that such deformation did not improve the ductilities of the specimens that were subjected to conventional isothermal hot ductility testing. For these specimens, the deformation at 1450 °C would also have caused grain boundary migration. However, because of the short time interval between this migration and the tensile test, it may be that the segregated regions remained segregated and still led to relatively unchanged numbers of precipitates. In addition, when ferrite forms, precipitation will still be favored at interphase boundaries. This, plus the presence of the thin ferrite film will still embrittle the structure, even if the numbers of precipitates are reduced.

Implementing deformation at temperature of 900 °C encourages deformation induced transformation, and hence intensive precipitation takes place at unbending temperature.[13] This is the reason when the ductility is impaired due to such a deformation.

Multi-step deformation. Applying cyclic deformation in the thermal range of 1200 to 1020 °C coarsens the precipitates, which had already been formed by the deformation induced coarsening process.[14] This leads to coarser precipitates at the unbending point, and therefore the hot ductility improves.

Applying deformation in the proximity of melting point, as discussed above, moves grain boundaries away from the segregated areas. This appears to be enough to counteract the tendency of further deformation at lower temperatures to accelerate precipitation.

Effect of thermal history changes on the hot ductility of Nb-Ti microalloyed steel

It was found that the transformation temperature of austenite to ferrite in Nb-Ti steel is around 800 °C.[10] Therefore decreasing the T_{min} to below 800 °C, (thermal history 1), leads to the formation of a thin layer of ferrite along grain boundaries of austenite. Since the solubility of carbonitride precipitates in ferrite is lower than in austenite, more precipitates form at T_{min} and subsequently at the $T_{unbending}$ point in thermal history 1, thereby decreasing the hot ductility.

The reason for the improvement of hot ductility by applying deformation in the proximity of the melting point in thermal history 2 is as was mentioned for thermal history 1.

Effect of deformation on hot ductility of low carbon steel

It was shown that imposing a compressive deformation in the proximity of the melting point leads to a noticeable decrease in hot ductility from 100 % to less than 74 %, Table VI.

The flow stress behavior of both steels subjected to 10 % compressive deformation in the proximity of melting point is shown in Fig. 3. As can be seen, a hump appears in the flow stress curve of the microalloyed steel, as was discussed earlier, whereas there is no such behavior in the low carbon steel. This suggests that a transformation took place during deformation in microalloyed steel, but not in the low carbon steel. Thus, it may be concluded that deformation in low carbon steel has been applied at single δ-ferrite phase during compression in the proximity of melting point.

Applying deformation to the δ-ferrite phase leads to the formation of deformation-induced vacancies in the microstructure. These excess vacancies form sulfur-vacancy complexes that, due to non-equilibrium segregation, migrate to grain boundaries.[15] Also the comparatively low ratio of Mn to S in the low carbon steel assists the segregation of sulfur. Sulfur segregation to grain boundaries has been suggested as a possible cause of poor ductility.[16] It has been demonstrated [17] that segregated S exerts an attractive force on the electrons associated with the bonding of Fe atoms, reducing the strength of the boundaries in this way. Also, it is reported that sulfur may precipitate with iron and manganese and thus reduce the hot ductility at unbending point.[16] Such a rationale may explain the deleterious effect of deformation in this steel.

CONCLUSION

The effects of thermomechanical histories on a Nb-Ti microalloyed steel and a low carbon steel were studied. It was observed that:

1- The conventional isothermal hot ductility curve of the Nb-Ti microalloyed steel exhibits a deep trough, while only a very shallow trough is exhibited by the low carbon steel.

2- Applying compressive deformation in the proximity of melting point during the conventional isothermal hot ductility tests widens the ductility trough to lower temperatures in the Nb-Ti microalloyed steel.

3- For the thermal histories examined, the application of deformation at temperatures close to melting point of the Nb-Ti microalloyed steel significantly improves the hot ductility at the unbending point, regardless of any subsequent deformation. Such a deformation, however impairs the hot ductility in the low carbon steel.

4- A severe decrease in hot ductility occurs when deformation is applied at 900 ^{o}C in the Nb-Ti microalloyed steel, whereas no such change occurs in the low carbon steel.

5- Imposing cyclic deformation following the implementation of deformation at 900 ^{o}C improves the hot ductility slightly.

ACKNOWLEDGMENTS

The authors acknowledge the Canadian Steel Industry Research Association (CSIRA) and the Natural Science and Engineering Research Council of Canada (NSERC) for providing financial support. They also wish to thank IPSCO Inc. Regina, Canada and Ivaco Rolling Mills. L'Original, Canada for providing the steels.

REFERENCES

1- B. Mintz, S. Yue and J.J. Jonas, "Hot ductility of steels and its relationship to the problem of transverse cracking during continuous casting", International Materials Review, Vol. 36, No. 5, 1991, pp. 187-227.

2- W. T. Lankford, "Some considerations of strength and ductility in the continuous-casting process", Metallurgical Transactions, vol. 3, Jun. 1972, pp. 1331-1357.

3- H.G. Suzuki, S. Nishimura, J. Imamura and Y. Nakamura, "Embrittlement of steels occurring in the temperature range from 1000 to 600 deg C, Transactions ISIJ, vol. 24, 1984, pp. 169-177.

4- W.R. Irving, Continuous Casting of Steel, The Institute of Materials, London, 1993.

5- M. El-Bealy, N. Leskinen and H. Fredriksson, "Simulation of cooling conditions in secondary cooling zone in continuous casting process", Ironmaking and Steelmaking, vol. 22, No. 3, 1995, pp. 246-255.

6- B. Barber, B.A. Lewis and B.M. Leckenby, "Finite-element analysis of strand deformation and strain distribution in solidifying shell during continuous slab casting", Ironmaking and Steelmaking, vol. 12, No. 4, 1985, pp. 171-175.

7- J.E. Kelly, K.P. Michalek, T.G. O'Connor, B.G. Thomas and J.A. Dantzig, "Initial development of thermal and stress fields in continuously cast steel billets", Metallurgical Transactions A, vol. 19A, Oct. 1988, pp. 2589-2602.

8- S. Akhlaghi, F. Hassani and S. Yue, "Effect of thermomechanical history on the hot ductility of a Nb-Ti microalloyed steel", 40th Mechanical Working and Steel Processing Conference Proceedings. vol. 36 (USA), Oct. 1998, pp. 699-705.

9- C. Zhou and R. Priestner, "The evolution of precipitates in Nb-Ti microalloyed steels during solidification and post-solidification cooling", ISIJ International, vol. 36, No. 11, 1996, pp. 1397-1405.

10- S. Akhlaghi, F. Hassani and S. Yue, "Effect of deformation on the hot ductility and microstructural evolution of a Nb-Ti microalloyed steel", 41st Mechanical Working and Steel Processing Conference Proceedings, vol. 37 (USA), Oct. 1999, pp. 125-132.

11- J. Lewis, J.J. Jonas, and B. Mintz, "The formation of deformation induced ferrite during mechanical testing", ISIJ International, vol. 38, No.3, 1998, pp. 300-309.

12- B. Mintz and J. J. Jonas, "Influence of strain rate on production of deformation induced ferrite and hot ductility of steels", Materials Science and Technology, vol. 10, August 1994, pp. 721-727.

13- B. Dutta and C. M. Sellars, "Effect of composition and process variables on Nb(C,N) precipitation in niobium microalloyed austenite", Materials Science and Technology, Mar. 1987, vol. 3, pp. 197-206.

14- I. Weiss, PhD Thesis, McGill University, 1978.

15- M. Militzer, W.P. Sun and J.J. Jonas, "Modelling the effect of deformation-induced vacancies on segregation and precipitation", Acta metall. mater., vol. 42, No. 1, 1994, pp. 133-141.

16- G.A. Wilber, R. Batra, W.F. Savage and W.J. Childs, "Effects of thermal history and composition on the hot ductility of low C steels", Metallurgical Transactions A, vol. 6A, Sept., 1975, pp. 1727-1735.

17- R.P. Messmer and C.L. Briant, "The role of chemical bonding in grain boundary embrittlement", Acta metall., vol. 30, no. 2, 1982, pp. 457-467.

The Effect of Electro-Slag Remelting (ESR) Process on the Forgeability Characteristics of DIN 17350 (AISI L6) Tool Steel

Ali M. Shahhosseini[*]
Mohammad H. Shahhosseini[**]
Ali Aghamohammadi[**]

[*] Mechanical Engineering Dept.
University of Louisville
Louisville, KY 40292
Tel: 502-852-6334
E-mail: amshah01@gwise.louisville.edu

[**] Metallurgical Engineering Dept.
Tehran University
Tehran, Iran

Key Words: Radial Forging, Electro-Slag Remelting, Tool Steel

INTRODUCTION

In hot working of metals, apart from the change of shape, modification of the cast structure required to break down the undesirable cast structure of any metal or alloy is highly important. Radial Forging may be utilized to refine the dendrite structure and to obtain remarkable properties. Radial Forging is a new way of Open-Die Forging. Many industrial products, such as axels with different cross sections are produced by this method. Forging breaks down the unwanted casting microstructure of metals. In the same way if we have a specific purified microstructure after casting, less break-down deformation would be necessary during forging. Electro-slag Remelting (ESR) is a good process to achieve such clean steel. Forgeability of AISI L6 low alloy tool steel ingots produced by both conventional and ESR methods is investigated in the present work.

In this research, the mechanical properties of the conventional products (AISI L6), after producing them in a conventional way of casting and Radial Forging, are compared with ESR products. In latter products, ESR method is applied as an intermediate process before Radial Forging. Tensile and impact energy tests were conducted to study the effect of strain on the mechanical properties and microstructure of the steel under investigation. The results obtained are indicative of the improvement of the mechanical properties of the steel by increasing the amount of deformation applied. Moreover, the minimal ratios of area reduction necessary to break down the casting structure in conventional and ESR ingots happened to be 5/1 and 3/1, respectively.

EXPERIMENTAL PROCEDURE

AISI L6 Tool Steel- L6 tool steel could be considered as a carbon steel with a small amount of alloying elements, such as nickel, chromium, and molybdenum (Table I). These elements are added to increase hardenability, wear resistance, and toughness of the steel. [1,2,3,4] L6 tool steel is applied as a plastic mold steel. According to ASTM A370 standard, L6 tool steel is produced as annealed and based on its application could be subject to quench and tempering process. [2]

Table I: Chemical Composition of DIN 17350 (AISI L6) Tool Steel [5]

Element	C	Si	S	P	Mn	Ni	Cr	Mo	V
wt. %	0.5-0.6	0.1-0.4	≤ 0.030	≤ 0.030	0.65-0.95	1.5-1.8	1.0-1.2	0.45-0.55	0.07-0.12

Steelmaking and Forging- A 12 tons electrical arc furnace is utilized to produce L6 steel. The molten steel is cast as ingots with a uniform cross section. One of the ingots is Radial Forged after conventional casting and the other is also Radial Forged but after getting purified by ESR method. Radial Forging specification of the ingots produced by conventional and ESR methods are listed in Table II and Table III, respectively.

Table II: Radial Forging Specification of the Ingots Produced by Conventional Method [5]

Cross Section No.	Cross Section Diameter (mm)	Reduction of Area (%)	Amount of Strain	A_o/A (approximately)
1	384	51	0.71	2/1
2	324	65	1.05	3/1
3	283	74	1.35	4/1
4	182	89	2.23	9/1

Table III: Radial Forging Specification of the Ingots Produced by ESR Method [5]

Cross Section No.	Cross Section Diameter (mm)	Reduction of Area (%)	Amount of Strain	A_o/A (approximately)
1	221	52	0.75	2/1
2	180	68	1.15	3/1
3	160	75	1.40	4/1

Axial Mechanical Properties- In order to obtain the mechanical properties of the forged products, tensile and Charpy specimens were made according to ASTM E8 and ASTM E23 standards, respectively. Fig. 1 shows the axial mechanical properties of the Radial Forged ingots produced by conventional and ESR methods at different amounts of strain. As shown in this figure, Ultimate Tensile Strength (UTS) of ESR products varies with a high rate in low amount of strain during forging but becomes less sensitive to strains higher than 1.5. In contrast, UTS of conventional products does not become steady at higher amounts of strain and the steel shows brittle characteristic. The variation of Elongation and Impact Energy with strain shows the expected results. In general, the mechanical properties improve with higher levels of strain. The reason for this improvement is mainly because of the break-down of cast structure and the refinement of primary austenite grain size as a result of recrystalization. [5] By having finer austenite and increasing the grain boundaries, there is more area for pearlite nucleation. Therefore, the steel microstructure will be finer and mechanical properties improve as a result of such fine microstructure. [6,7] By the same token, it may be stated that with the same amount of plastic deformation, mechanical properties of ESR products are better than the conventional ones. The reason can be the more directional solidification, less segregation, and cleaner ESR ingots. [5]

Radial Mechanical Properties- Fig. 2 shows the radial mechanical properties of the conventional and ESR forged steels in different amounts of strain. As shown in this figure, there is also better mechanical properties in higher amount of strain in radial specimens. As a matter of fact, there is not too much change in the mechanical properties of the conventional products, in comparison with ESR ones. At the highest amount of strain, the

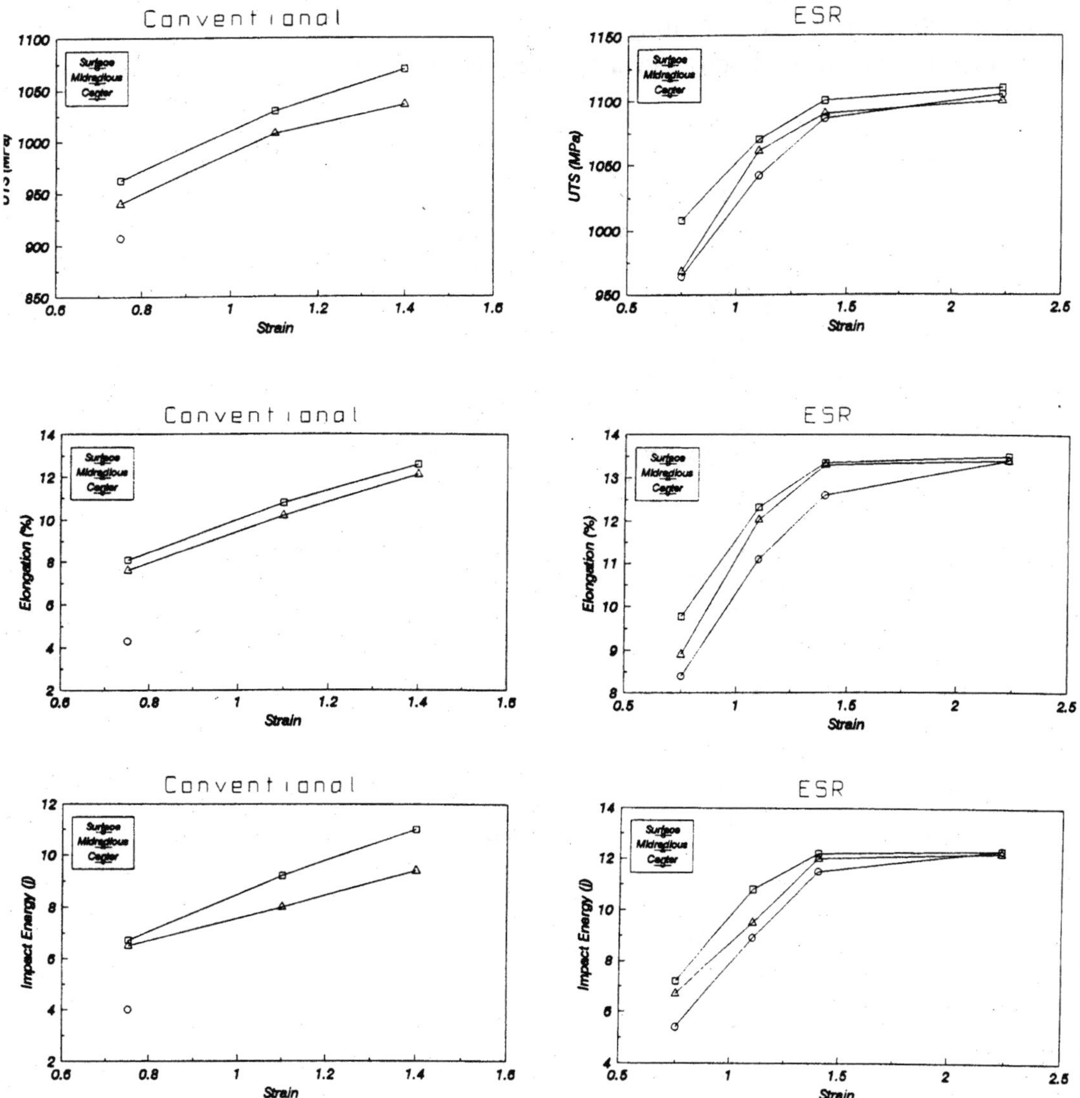

Fig. 1: Axial mechanical properties of the Radial Forged ingots produced by conventional and ESR methods in different amount of strain.[5]

□ surface

△ mid-radius

○ center

anisotropy ratio of the conventional specimens is 1/3, whereas the ratio of the ESR specimens is 1/2. Therefore, one of the significant differences between the conventional and ESR products is the difference in anisotropy.[5]

Element Segregation- Segregation ratio of different elements in conventional and ESR products, from surface to center, is shown in Fig. 3. The difference in chemical composition in micro and macro scales dictates the forgeability and mechanical properties of steel in different locations. In Fig. 3, the horizontal axes show the different locations of the cross section from surface (No. of sample = 1) to center (No. of sample = 9). The vertical axes shows macrosegregation ratio:

$S_{macro} = [\ (C - C_\circ)/\ C_\circ]\times 100$

C: element percentage measured after forging

$C_\circ$: element percentage measured before casting (ladle composition)

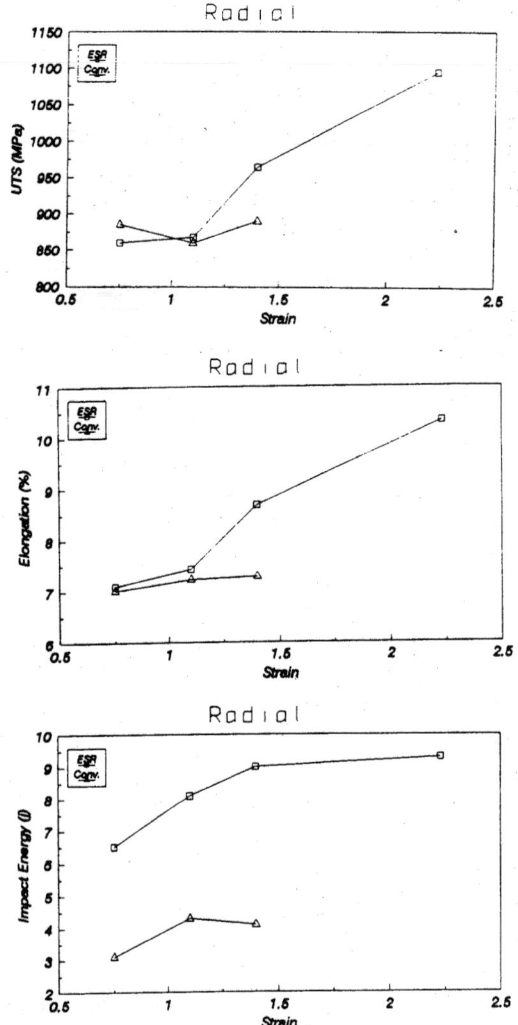

Fig. 2: Radial mechanical properties of the conventional and ESR forged steels in different amount of strain. [5]
□ ESR
△ Conventional

For ESR ingots, it is assumed that the steel composition does not change during ESR process. In general, Fig. 3 confirms the positive effect of ESR process on the chemical uniformity of the steel. In this case, segregation ratio of different elements does not change significantly, in comparison with the conventional products. It can be concluded that macrosegregation is almost eliminated in ESR ingots. [5] In contrast, macrosegregation is significant in the conventional ingots. On the surface of the ingots there is some positive macrosegregation and in the center an intensive negative macrosegregation is visible. [5] This difference in segregation can be justified by the solidification process of the conventional ingots. During solidification, directional dendrites grow toward the center of the ingots from the surface and in this growth alloying elements are rejected to the space between the dendrites. In this case, at a certain distance from the surface, the inter-dendrite space is very small and does not cause a remarkable non-uniformity in steel. When the solidification front progresses more toward the center of the ingot, the liquid steel streams in touch with the front breaks down the dendrites.

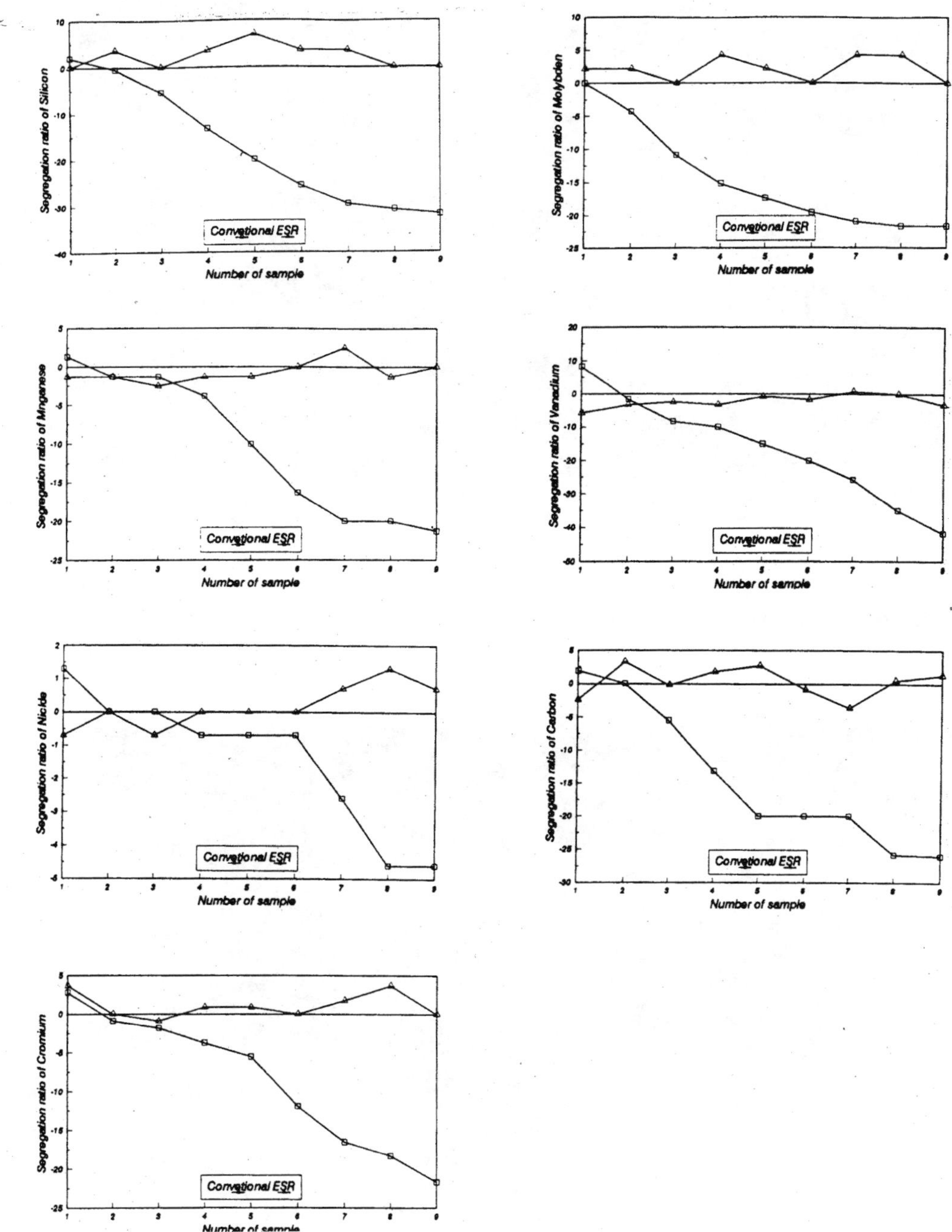

Fig. 3: Segregation ratio of different elements in conventional and ESR products at different locations of the cross section from surface (No. of sample = 1) to center (No. of sample = 9).[5]
△ ESR
□ Conventional

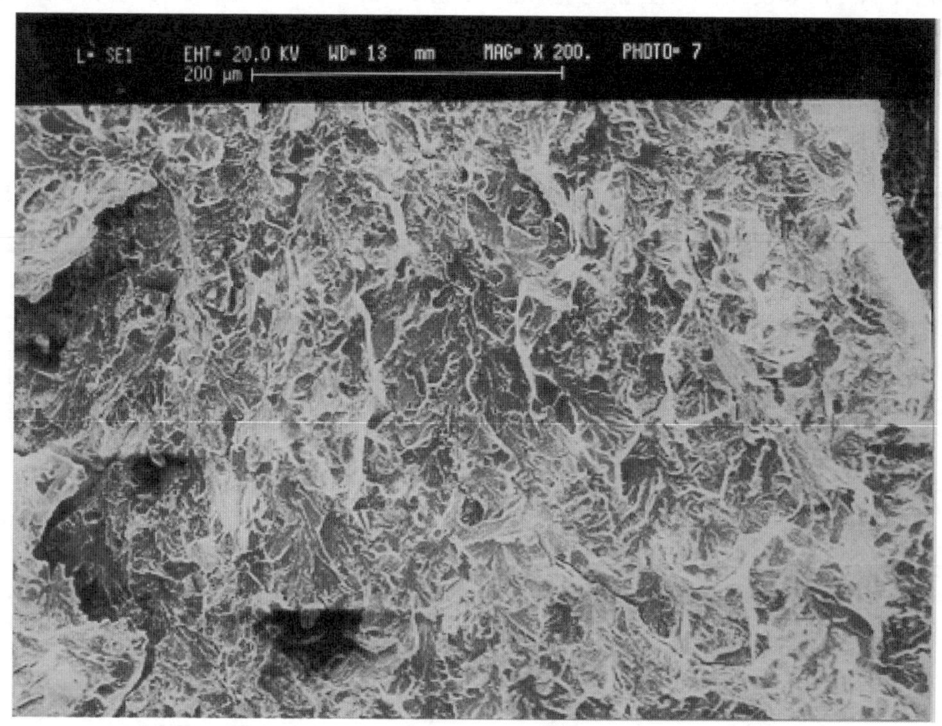

Fig. 4: Fracture surface of a ESR specimen (Mag. = 200 and A∘/A = 2/1). [5]

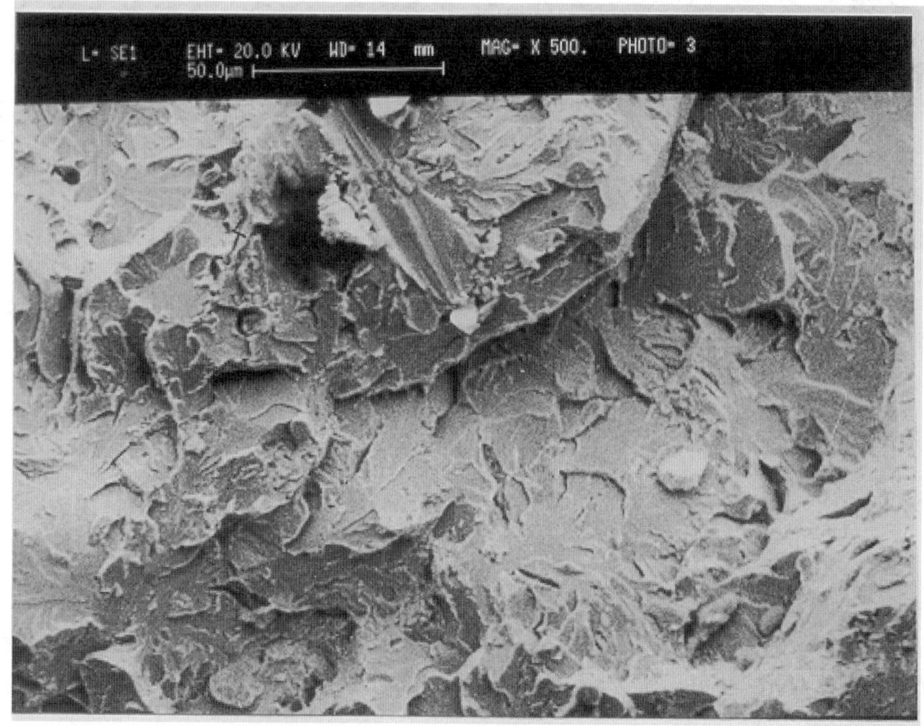

Fig. 5: Fracture surface of a ESR specimen (Mag. = 500 and A∘/A = 2/1). [5]

The stack of such dendrites on the bottom of the ingot creates the intensive negative segregation.[8,9] The rate of negative segregation depends on the uniformity of the dendrite core. By having more microsegregation, more alloying elements are rejected from the dendrites and the amount of macrosegregation becomes higher. Lack of alloying elements in the center of ingots, because of macrosegregation, degrades the mechanical properties of steel in that area. It can be predicted that the rate of negative segregation at the bottom of ingots is as high as positive segregation at the top of the ingots. The difference in element composition affects the mechanical properties and forgeability of conventional ingots tremendously, but even in ESR ingots with a large cross section, less non-uniformity in element composition is visible. In general, the ESR process almost eliminates macrosegregation and decreases the amount of microsegregation in steels.[5]

SEM Observation of Fracture Surfaces of ESR Specimens- Fracture surfaces of impact specimens are investigated using a Scanning Electron Microscope (SEM), Cambridge Model-S360. As shown in Fig. 4 and Fig. 5, the fracture surface of ESR specimens is a combination of intergranular and transgranular fractures.

SUMMARY

From the results obtained the following conclusions can be drawn:

1. Radial Forging increases the mechanical properties of AISI L6 tool steel at different amounts of strain. Higher strain gives better mechanical properties but after a certain strain level, the change in such properties is not significant.
2. In all cross sections, the mechanical properties of ESR products are higher than the conventional ones.
3. The mechanical properties are different at different depths from the surface of products in both ESR and conventional products. In general, the mechanical properties are better toward the surface. For smaller cross sections, this difference in the properties degrades and in a certain strain level, the mechanical properties will be almost uniform in the whole cross section.
4. Macrosegregation exists in both ESR and conventional ingots, but the amount of macrosegregation is much higher in the conventional ones.
5. The results from this research demonstrate that, in Radial Forging, ESR products have much better mechanical properties than the conventional ones.

ACKNOWLEDGEMENTS

The authors would like to express their sincere appreciation to Arak Machine Manufacturing Plant, Arak, Iran as well as Isfarayen Steel Complex, Isfarayen, Iran. The authors also wish to thank A. Amadeh, M.H. Parsa, A. Zaraie, and E.G. Brehob who generously gave a great deal of their time throughout this research.

REFERENCES

1. G.A. Roberts, "Tool Steels," American Society for Metals, 1988, pp. 11-50.

2. Metals Handbook, 10th ed., Vol. 1, American Society for Metals, 1993, pp. 117-143.

3. J. Geller, "Classification of Tool Steel by Properties and Application," Mir Pub., Moscow, 1978.

4. Metals Handbook, 10th ed., Vol. 3, American Society for Metals, 1993, pp. 430-440.

5. A. Aghamohammadi, M.S. Thesis, Metallurgical Engineering Dept., Tehran University, Iran, 1997.

6. K.A. Jackson, "Defect Formation, Micro segregation, and Crystal Growth Morphology," American Society for Metals, 1986.

7. K. Kudrin, "Steel making," Mir Pub., Moscow, 1985.

8. S.H. Avner, "Introduction to Physical Metallurgy," 2nd ed., McGraw-Hill Company, London, 1974.

9. G.E. Dieter, "Mechanical metallurgy," McGraw-Hill Company, London, 1989.

Hot Profile Ring Forming

James A. Laverick, Kevin P. Daiger and Kevin A. Raketich
The Timken Company
Mail Code: RES-13
1835 Dueber Avenue, S.W.
P.O. Box 6930
Canton, OH 44706-0930 U.S.A.
Phone: (330) 471-2294
Fax: (330) 471-7142
E-mail: laverick@timken.com

Key Words: Forming, Bearings, Forgings, Profile Forming

INTRODUCTION

Profile Ring Mill – A technology solution for high volume rings

Over the last two decades there has been a significant change in the manufacture of annular wrought steel components for automatic transmission and anti-friction wheel bearing applications at virtually every major producer in North America. This change is largely due to the aging of capital equipment and advances in manufacturing technology in component machining processes. The dominant trend has been removal of machining operations and a re-direction of capital to higher value-added capabilities considered core to the original equipment manufacturer's competitive position. Additionally, where the internal machining has been modernized, computer numerically controlled (CNC) turning has been chosen over the traditional screw machine or cam-operated turning processes.

These trends have changed the supplier base for both machined components and raw material for those components. First, the transfer of the manufacturing responsibility created numerous suppliers with the critical mass to focus investment in CNC turning capabilities that could be leveraged over several customer's requirements. Second, the proliferation of CNC turning at the OEM and supply base substantially increased the requirement for discrete components ready for chucking versus full-length bar or tubing.

This secondary requirement for discrete raw material was not as readily available as the CNC turning operations. Capital and technical investment in machinery that can effectively produce these components is significantly greater than turning. The volume, quality, and cost requirements of these blanks have limited wrought material manufacturing options to two primary choices:

- parted seamless mechanical tubing and
- high speed hot formed rings

Each of these options has the advantages detailed below:

Parted seamless mechanical tubing
Seamless tubing process does not require waste web or taper
Seamless tubing process gives wide size capability and high throughput
Cold parting methods are readily available and provide high flexibility

High speed hot forming
Closed die tooling can produce highly contoured products as well as duplex forgings
Part specific tooling increases size capability for contoured products
High speed hot separation gives product ready for CNC machining

As can be expected, product opportunities have gravitated to the most efficient process for producing each shape. For example, planetary ring gears whose shape requires a straight-walled blank have generally been produced using seamless mechanical tubing. Conversely, highly contoured sun gears and clutch hubs have been manufactured using the closed die high-speed hot formers.

The commercial problem occurred as the growth in component requirements increased in the "middle ground" where neither of the processes was optimal. There were still disadvantages of each process that required a specialization which did not allow cost efficient production of a wide product range. Those disadvantages are listed below:

Parted seamless mechanical tubing
Cold parting (saw) efficiency limited as wall thickness increased
Material utilization decreased as contouring increased
Cold parting (saw) efficiency limited with introduction of alternate gear and race materials

High speed hot forming
Forming efficiency limited as outer diameter increased
Material utilization decreased as inner diameter increased (waste web)
Limited size range on any one machine model

With an understanding that requirements for discrete input materials would continue, additional processing technology was investigated. The aim was a manufacturing method that would maximize the advantages of the current processes and create new value for the CNC turning operations. Specifically, the objective was a process that would capitalize on the efficiency and flexibility of the rotary forming (tube) process while introducing contouring and parting capability of the high-speed hot former. This combination has been achieved with the North American introduction of the Profile Ring making process.

Process Description

Round bars are received at the Tryon Peak Plant in Columbus, North Carolina from the Timken Company's main plant in Canton, Ohio. The bars ranging from 67 to 121mm OD x 1 to 2m in length arrive by truck in the sheared or sawn condition and are stored in an outdoor laydown yard until needed. The plant is designed to process low to high carbon steels, examples of the range include 8620, 1050 and 52100. As certain orders are processed, the required material is retrieved from the laydown yard and brought into the facility. The schematic of the process is shown in Figure 1.

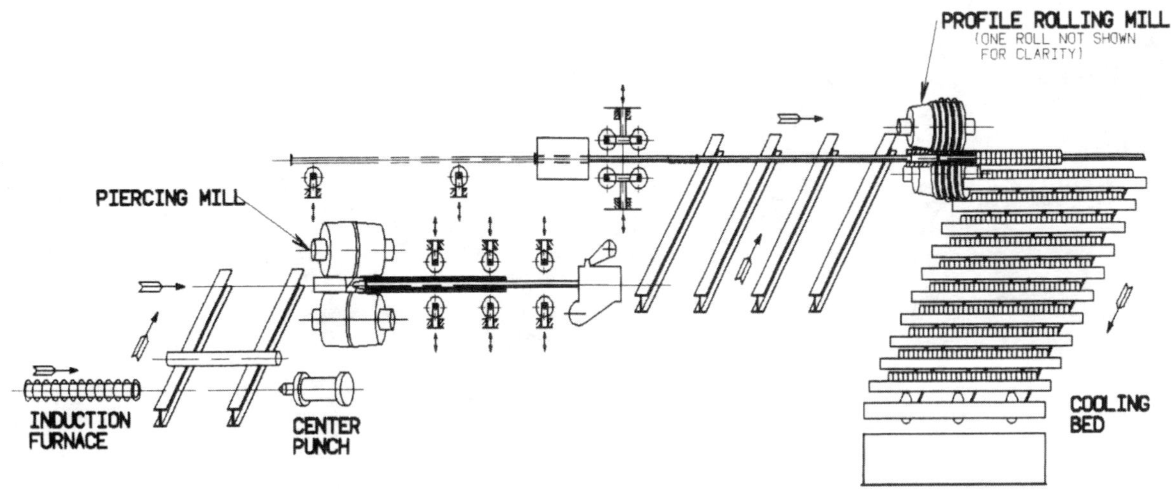

PROCESS SCHEMATIC OF PROFILE ROLLING PROCESS

Fig. 1: Overview of Profile Rolling Process.

The bars are brought to the incoming loading area, seen in Figure 2 and are unbundled and are ready to be introduced individually into the process, Figure 3. There are two parallel pusher type induction furnace lines, these can be used individually or sequenced together. These can be seen in Figures 4 and 5.

Fig. 2: Bars entering the building

Fig. 3: Bars entering the system

Fig. 4: Bars entering induction furnace

Fig. 5: Dual induction furnace lines

Properly heated bars, Figure 6, proceed through a dimensional vision gauge, Figure 7, which measures the OD and length of the bar, to the piercing mill while off temperature bars are discharged to the reject bucks. Piercing temperatures vary by material grade and are held to +/- 25 degrees C within and between bars. The bars are pierced on a two-roll mill with a piercing plug held on a mandrel, Figures 8 and 9.

Fig. 6: Heated bar exiting induction furnace

Fig. 7: Bar exiting through DVG

Fig. 8: Bar piercing into a tube shell

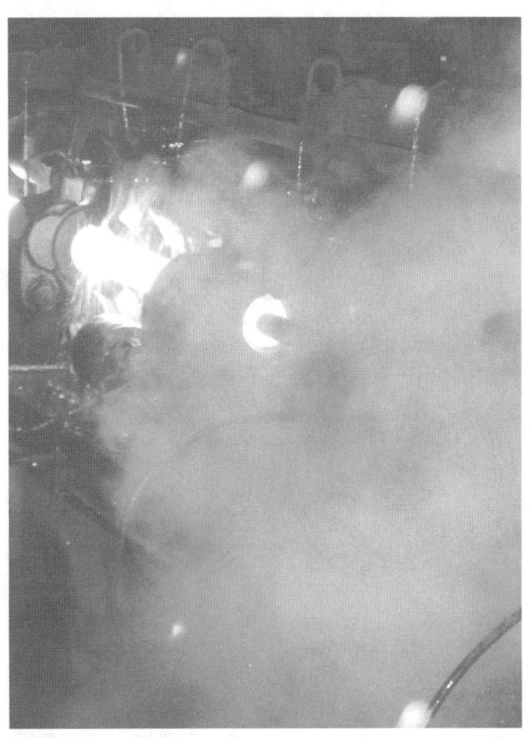

Fig. 9: Tube shell being made

The tube shells are stripped off the mandrel and pass through another dimensional vision gauge, which measures the OD and length of the shell. The wall thickness of the shell is calculated. The shell proceeds to a transfer table. This plant has the capability of sending the shells to a cooling table where these shells can be used or sold in this form, Figure 10. Shells continuing in the process are sent over the transfer table to the forming mill, Figure 11.

Fig. 10: Tube shells on the cooling bed

Fig. 11: Hot shell transferring to the forming mill

Once the shell arrives at the inlet table of the forming mill, a mandrel and mandrel bar are inserted into position in the forming mill through the shell ID. Figure 12 shows a schematic of this operation. The shell pusher advances the shell over the mandrel bar into an inlet guide and into the forming rolls where the OD can be profiled and the metal is parted into discrete preforms.

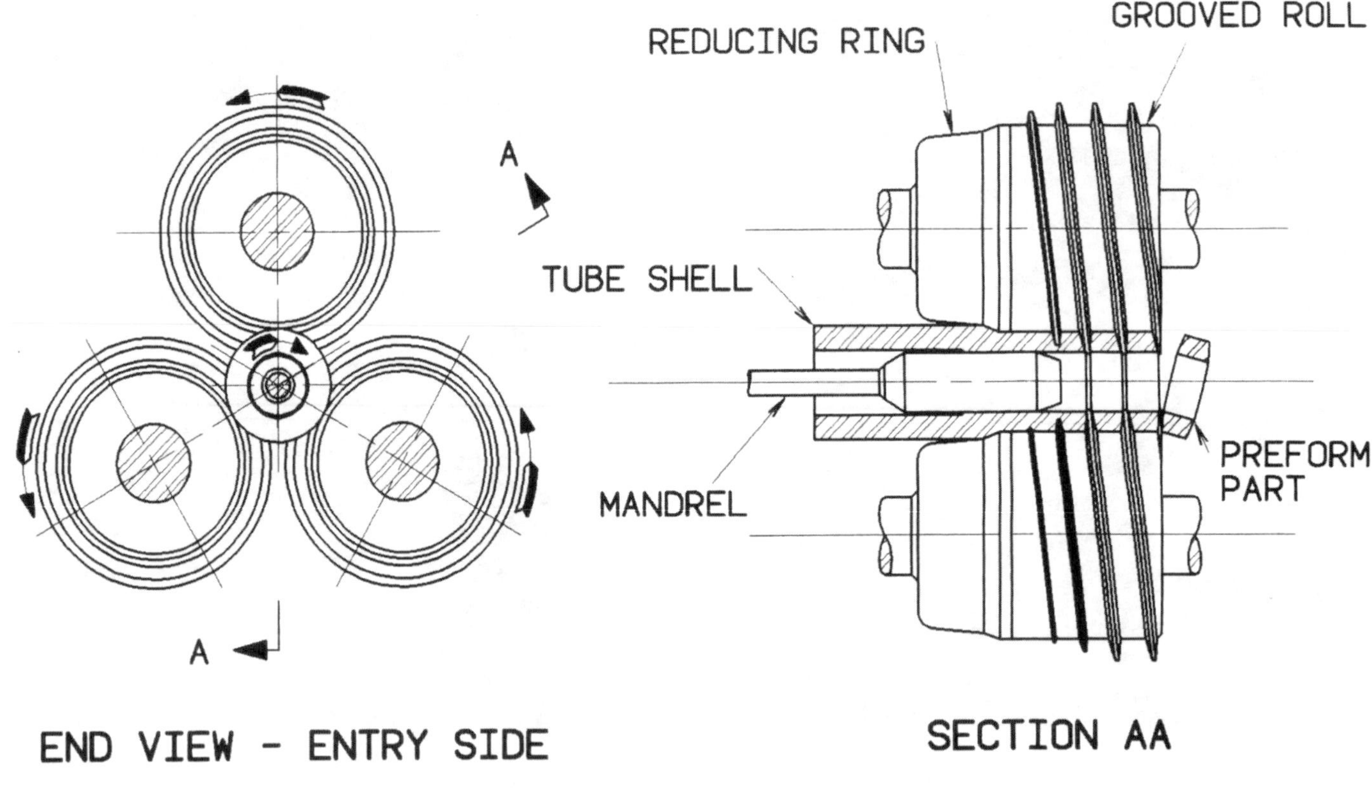

Fig. 12: Schematic of the forming mill

The forming mill rolls rotate at 165 to 280 RPM producing a part in each rotation, Figures 12 and 13. The parts are discharged from the forming mill and are transported to a cooling conveyor. The parts are spread uniformly on the mesh belt. The parts are control cooled depending on the metallurgical characteristics required using regulated airflow through four zones. Figures 14, 15 and 16 show the cooling conveyor with parts entering and exiting. The parts are then loaded into hoppers for movement to other processes. The plant also has deflash, heat treat and machining capabilities.

Fig. 13: Preforms exiting the forming mill.

Fig. 14: Four zone Cooling Conveyor

Fig. 15: Hot Preforms entering the Cooling Conveyor

Fig. 16: Control cooled Preforms exiting

Process Capabilities

Table I shows the process capabilities.

Location	High Carbon Preforms		Low & Medium Carbon Preforms	
	Minimum	Maximum	Minimum	Maximum
OD (mm)	55	107	55	112
ID (mm)	37	759	37	80
Length (mm)	15	70	15	100
OD/Wall Ratio	5	8.5	5	9

Process Advantages:
Profile Ring Mill vs. Upset forging

- Less stock
- Better tolerances
- No draft angle needed
- Less waste material in the flash
- Can have OD undercut profiles

Profile Ring Mill vs. high speed forging (e.g. Hatebur)

- Less tooling
- No waste material in the ID (waste web)
- No draft angles required
- Can have OD undercut profiles
- Shorter set up time

This process does not require additional equipment or steps for parting. Utilization of the controlled cooling capability can provide metallurgical advantages in the form of process-controlled microstructure and hardness with less or no subsequent heat treating.

Product Examples:
As seen in Figures 17 through 20, applications include bearing inner races, transmission parts and blanks for other automotive parts. The parts are shown with the darker shading representing the finish-machined part and the light shading representing the machining stock.

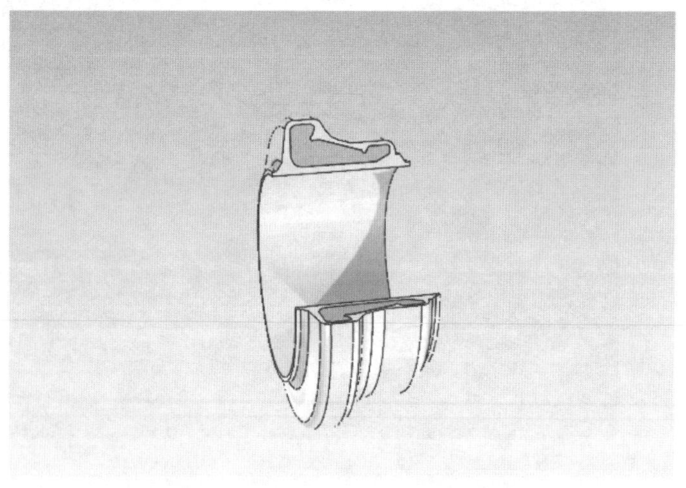

Fig. 17: Tapered bearing inner race

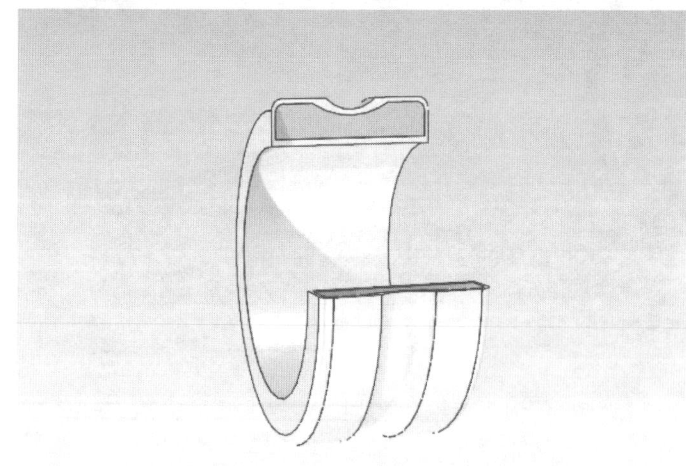

Fig. 18: Ball bearing inner race

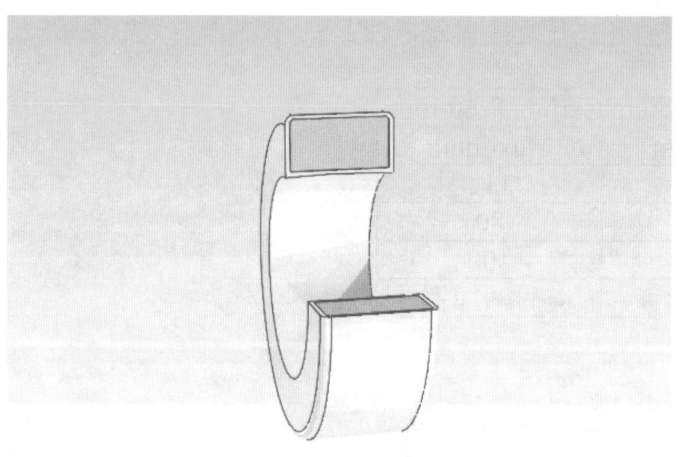

Fig. 19: C.V. Joint outer race

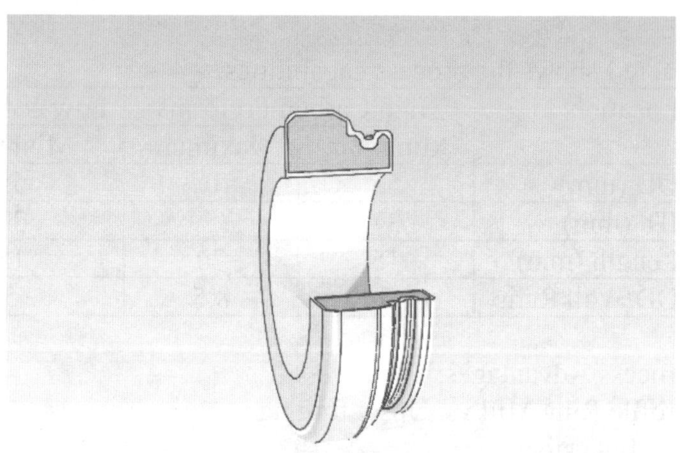

Fig. 20: Shift sleeve

Summary

As evidenced, the Profile Ring Mill is a unique process that encompasses the process advantages of seamless tube manufacturing with the contouring and parting capabilities traditional to high speed hot forming. This combination creates a new value offering for the power transmission industry, particularly in automotive transmission and wheel bearing applications. Applications such as thick-walled rectangular race blanks, OD contoured bearing and gear blanks and circumferentially grooved sleeves and races can now be produced in an efficient rotary forming process. Additional opportunities exist when this unique forming technology is considered in the power transmission design process.

ACKNOWLEDGEMENTS

The authors would like to thank The Timken Company for the opportunity to publish this paper.

The Centrifugal Casting of HSS Roll for Narrow Strip and Rod Mills

Zhang Xin*
Michael C. Perks**

ABSTRACT

This paper records the development and production of the latest HSS quality, horizontally centrifugally cast, with the core material in spheroidal graphite iron. The composition and subsequent heat treatment ensures the hardness of the working layer of the HSS material will reach 80/85 shore, with virtually no fall off in hardness throughout the working layer, while the structure of complex carbides in a martensitic matrix ensures uniform wear as well as high wear resistance. Heat treatment as well as producing a satisfactory wear resistant shell also allows the core material to maintain its high strength properties. By using the CC duplex HSS roll in the finishing stands of narrow strip mills, or in the pre-finishing stands of high speed wire rod mills, gives significant productivity gains both in roll life, and roll shop/mill operations.

INTRODUCTION

The introduction of high speed continuous mills together with increased customer demands in surface finish and shape, the rollmakers are faced with the dual challenge of producing roll qualities for increased productivity as well as end product quality. In recent years new roll qualities have been developed, as well as different manufacturing processes needed to successfully produce these rolls. The conventional AIC roll in some stands has been replaced by High Chrome alloys and more recently by HSS qualities. Quantum increases in performance have been recorded and yet still further development is continuing.

The foundry is especially designed for the production of small to medium sized centrifugally cast rolls. There are in line 8 horizontal centrifugal casting machines, each one capable of casting 3 rolls per hour of varying sizes. To feed these casting machines there are 5 in-line induction furnaces, medium frequency of varying capacity from 1 tonne to 5 tonnes. Liquid metal is weighed into the moulds by electronic crane mounted weighers, and metal composition is controlled by a. Quantavac. Metal temperatures are measured by standard thermocouples, while metal solidification is measured by black body radiation thermocouples. A heat treatment shop is computer controlled to very close temperature ranges ±5°C.

Current annual production for small rolls and rings amounts to approximately 8,000 t.p.a. The quality control system is independent from production, and produces certificates of assurance for dimensions, hardness, structure, and ultrasonics as well as monitoring the production systems.

* Ms. Zhang Xin, Deputy Gen. Manager, BRC-Lian Qiang Foundry, Tangshan, China
** Mr. Michael C. Perks, Technical Director, British Rollmakers (China) Ltd., Hong Kong, China

THE EFFECT OF ALLOYING ELEMENT ON THE STRUCTURE OF THE WORKING LAYER

Carbon
Carbon, Iron and Chromium when in specific ratios, combine to form complex carbides M7C3. At the same time carbon will also combine with Mo, V, Nb, Cr, and W, to form various high hardness carbides such as MC, M6C, and M2C. To obtain the most suitable wear resistant and castable structure, the carbon level should be between 1.6%-2.0%. Increasing the carbon above this level promotes the formation of more and continuous carbide, thus reducing the ductility, heat and crack resistance.

Silicon
Silicon together with carbon helps with the fluidity during casting, maintains ductility and strength, and should be at a level of 0.3-1.0%

Manganese
Manganese helps to keep the oxygen level of liquid HSS composition to a minimum, while at the same time combining with the residual Sulphur to form MnS, preventing grain boundary embrittlement. Manganese level should be 0.5-1.0%

Chromium
Cr, FE, V, Nb, Mo, W and Carbon combine together to form complex carbides during solidification and subsequent heat treatment. Some of the Chromium is dissolved in the matrix, and increases the wear resistance and hardenability. Corrosion resistance is also increased. To give the best wear resistance while at the same time maintaining the ductility, the Chromium level should be between 4.0-8.0%

Tungsten
Fe and C combine together with W, Cr, Mo, Nb and V to form duplex carbides which enhance the high temperature and tempering characteristics and subsequent hardness of these HSS qualities. W is usually present in the Fe4W2C form. Eutectic M6C carbides have a fish bone type shape and due to the low quenching temperature cannot be easily be broken down into the non continuous form. W2C type carbides have a theoretically high density of 17.2 which under centrifugal force can easily be changed.W level is set at 1.5-2.5% to get the best wear resistance, ductility, heat and crack resistance. It also helps to maintain the hot hardness during the rolling operation, and increases the corrosion resistance at elevated temperature.

Molybdenum
Molybdenum has a similar function when combined with Fe and C, as W. Eutectic Mo2C shape carbides form in the shape of rods/sticks or features, easily dispersed. The theoretical density of Mo2C is 9, 18 and can partially be substituted for W when the total amount of carbides and variance analysis is considered. It is not advisable to substitute all the W by Mo as some of the oxidation and corrosion properties will be diminished. The Mo level is ideally set at 4.0-6.0%

Vanadium and Niobium
Both V and Nb combine with Fe, Cr, Mo to form MC carbides, which have very high hardness levels and will precipitate as fine carbides, thus playing a major role in the rolling process, increasing the wear and crack resistance, while at the same time strengthening the ductility. To be effective the level of both alloys should be 4.0-7.0% but above this level segregation will occur during solidification using the centrifugal process.

Nickel
While Nickel is used extensively in steels to increase the hardenability, it is not used in HSS qualities because of its tendency to increase the residual austenite levels after transformation treatments have been carried out. Therefore to maintain the desired mechanical property levels, we suggest that the level of Nickel be 0.5-1.5%

Table I: The chemical compositions in the working layer of HSS rolls (%)

C	Si	Mn	W	Cr	Mo	V	Nb	Ni
1.6/2.0	0.3/1.0	0.3/1.0	1.5/2.5	4.0/8.0	4.0/6.0	4.0/6.0	0.5/1.5	0.5/1.5

HSS/Shell MICROSTRUCTURE

The high speed steel shell is closest to a D2 type analysis, the microstructure consisting of primary and eutectic carbides in a martensitic matrix. Strong carbide stabilising elements, such as Vanadium, Tungsten, Molybdenum and Chromium are added to improve the hardness and resistance to wear, oxidation and roughness. The type, morphology and volume fraction (see table II) of carbides as well as the characteristics of the matrix all effect the properties. Hardness is affected by the intercellular carbides and their distribution. Wear and surface quality are directly influenced by carbide morphology. High amounts of Vanadium give the best wear properties but compromise the surface roughness characteristic of the roll. The carbide stabilising elements are balanced to ensure the correct morphology and volume fraction whilst retaining adequate solid solution strengthening. The high spinning speeds necessary to produce small diameter rolls limits the vanadium content, due to potential problems of segregation. The transition or white iron zone is made up of complex Chromium and Vanadium carbides. The temperature profile across the interface has to be carefully controlled to avoid the formation of intercellular micro-porosity. The core is an SG cast iron, well spheroidised with a high nodule count. The excellent feeding characteristics of the SG ensures centreline soundness even in the thinnest sections.

Figures I and II Show the typical shell microstructure at x 500 and x 200.

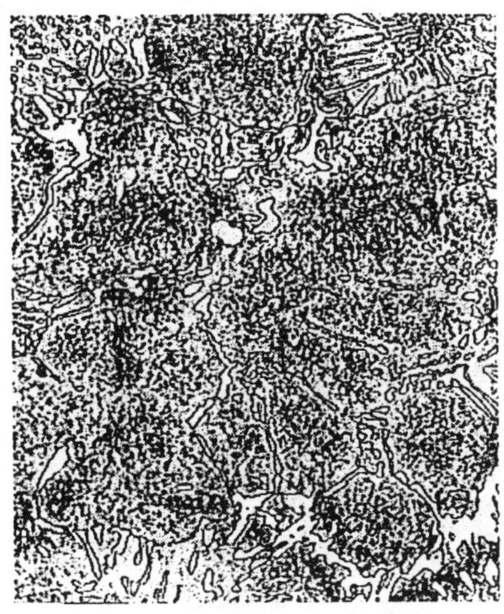

Figure I
Shell x 500

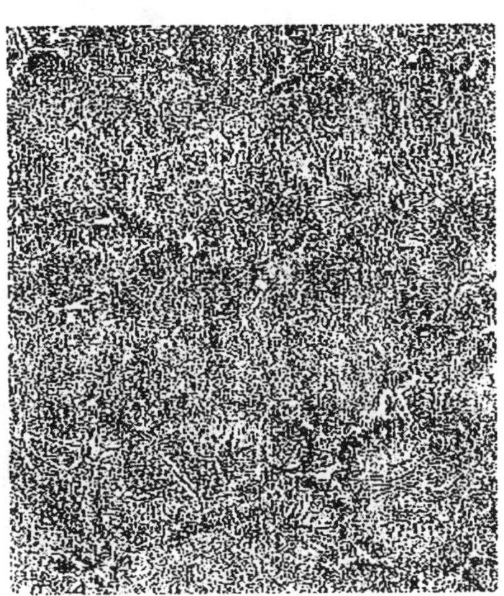

Figure II
Shell x 200

Table II: The carbide composition & mechanical properties of several quality of duplex rolls

Roll Type	Carbides			Tensile Strength (Mpa)	AV (HSC)	Carbide HV	Matrix HV	Wear Resis	Mod of Elasticity /x103 Mpa
	Form	Area (%)	Size (μm)						
HSS 1.5-2.4C	VC MC M2C M6C	7-15	<100	700-900	80-90	3000 2000 1800	650-750	4-5	230
HiCr 2.0-2.9C	M7C3	20-30	100-150	650-800	70-90	1200-1600	450-550	1-2	200
HiNiCr 3.0-3.4C	M3C	30-40	200-500	400-600	70-85	840-1100	430-460	1	170

Table III: The chemical composition of the core materials of CC duplex HSS rolls

Quality	C	Si	Mn	Ni	Cr	Mo	Mg
SGP Cast Iron	3.2/3.6	1.9/2.3	0.3/0.7	0.5/1.0	≤0.20	0.2/0.5	≥0.05

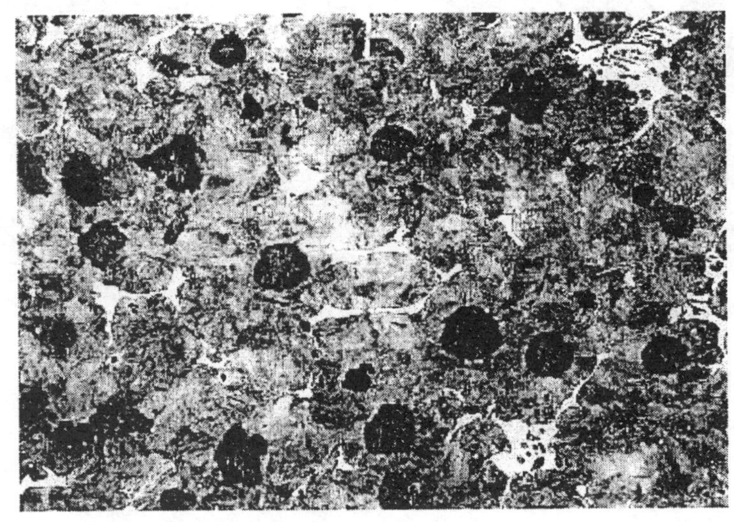

Figure III
Core x 100

CORE COMPOSITION & PROPERTIES

The tensile strength of the roll neck has to be high enough to withstand the rolling loads as well as the heat treatment. Chemical compositions are chosen very carefully according to the specific working conditions of the rolls and the strength required for the core and the necks, to withstand melting, casting conditions and heat treatment effect. The chemical compositions of the cast iron pearlitic core and CC duplex HSS roll core are listed in Table III while the mechanical properties are listed in Table IV.

The microstructure is shown in Figure III.

Table IV: The mechanical properties of the core materials of CC duplex HSS rolls

Quality	Tensile Strength (Mpa)	Elongation (%)	Bending Strength (Mpa)	Hardness (HSC)
SGP Cast Iron	500-600	0.7-1.5	1000-1200	40-50

SHELL CORE INTERFACE

To ensure successful roll life and avoid mechanical failure, a complete fusion between the shell and the core is vital. In manufacturing HSS rolls by horizontal CC duplex casting method, the working layer will be removed from the casting machine after solidification. The position of the working layer will change from horizontal to vertical. The positions of the chill and the upper and lower necks are matched, and casting of the core material can then proceed. To obtain an optimum working layer, various factors need to be considered.
(1) The temperature of the inner surface of the shell
(2) The protection from oxidation of the inner surface
(3) The casting temperature of the shell and core iron
(4) Time interval between casting shell & core
(5) Dressing thickness & Cooling rate.

It is essential to get a good fusion between the inner core and outer shell but avoiding excessive mixing the working layer and the core.

The temperature of the internal layer should be between 1,000 - 1,100°C measured by radiation thermocouple, while the core iron to be within 1,350 - 1,450°C dependent upon liquidus in order to obtain the required optimum conditions for fusion of the shell and core, but not to mix the working layer & core. In practice, when the inner surface of the shell is oxidized, fusion of the shell/core requires a higher temperature of core iron but this is not recommended. This oxidation will increase the difficulty of the fusion between the core and the inner surface of the working layer. Thus, in our CC casting technology, we are using a special kind of material to act as a flux protection layer under the high temperature of the inner surface of the shell. This protection layer carries the function of anti-oxidation of the inner surface of the shell and also preserving uniform temperature. This will enables the shell metal to have a unidirectional solidification from the surface to the inner layer. After aligning all the chills (barrel & necks) in position, core metal will be top poured or bottom poured into the mould. Due to the wash action of the high temperature core metal and the buoyant force, the flux protection layer will separate from the solidified working layer and float to the upper neck chills. A 3-10mm thick solidified working layer will be remelted and fused with the molten iron of the core. The final sodification of the shell and the core is a metallurgical bond.

One of the characteristics of CC duplex HSS roll is that the interface between the shell and the core is very distinct. The quality of the interface and the depth of the working layer is checked by ultrasonics. Indepth investigation was carried out to determine the integrity of the shell/core bond. We considered it necessary for this interface to be narrow, free from porosity, and of suitable structure to withstand the rolling loads.

Electron probe micro analysis together with a quantitative line scan was carried out by Swinden Technology Centre, Corus, across the interface and into the shell and core at either side. Figure IV show the alloy distribution and concentration in the area of scan. Electron scan microscopy quantitatively confirmed these conclusions that the interface was approximately 3mm wide and there is some minor chrome carbide concentration within this area, but it is not continuous. See Table VI.

From these results we can draw a number of conclusions : -

- The shell/core interface is narrow
- The interface does not contain any porosity and is sound
- There is little or no alloy migration into the core material & little shell solution into the core
- The core material remains in this case normal pearlitic nodular graphite. Tensile tests across this bond show levels of 600 MPa.

Table VI: % Carbide levels (Electron Scan)

Distance from Interface		VK	CRK	NiK	MoK	WK
	mm					
↑	2.00	4.74	5.56	2.27	2.93	2.40
	1.00	2.76	2.66	0.83	1.84	0.86
SHELL	0.50	0.96	1.79	1.09	1.38	0.99
CORE	0.50	0.63	1.06	0.53	0.24	1.62
	1.00	0.73	1.04	0.97	1.49	0.29
↓	2.00	0.45	0.74	0.90	0.72	0.85

HEAT TREATMENT

HSS rolls have high hardness, good wear resistance at high temperatures due to quenching, tempering and heat treatment. The quenching temperature for CC duplex HSS rolls is governed by the alloy composition. We are required to maintain a good working depth of the rolls as well as to maintain the quality of the core materials, especially when using alloy spheroidal nodular graphite. If the quenching temperature is too high, the core will be distorted or even melted. Thus, the foundry controls the alloy content and subsequent quenching temperature up to the point that the alloy carbides just transform to become austenite. Adopting forced air cooling, the material will be transformed into bainitic and martensitic form, which is then tempered. The Mo, W, V and Nb dissolved in the austenitic matrix then become stable component of carbides. This enacts the release of micro carbides and leads to the hardening effect. The outer shell is then formed with high hardness and good wear resistance at high temperatures. This will not affect the ductility of the inner layer while the tensile strength of the neck is greater than 550 Mpa.

A standard production HSS roll #A612-503 was used as a sample for heat treatment testing. Test pieces were cut from the barrel in the dimensions of 50mm x 30 mm x 10 mm. There are 10 testing pieces, the 10th one is as cast. The first to ninth testing pieces are separated into 3 groups and undergo quenching in 3 different temperature : 950°C, 1,050°C and 1,150°C. After quenching, tempering is also done in 3 different temperature : 500°C, 550°C and 600°C. The hardness after quenching and tempering are listed in Table VII.

From Table VII we can see that at quenching temperature 1,150°C and tempering temperature 550°C, the HSS roll has hardening effect. The hardness after tempering is 3°HSC higher than the hardness after quenching. For other temperature for quenching and tempering, the hardness drops, especially when tempering is done at 600°C, the hardness drop is the most radical.

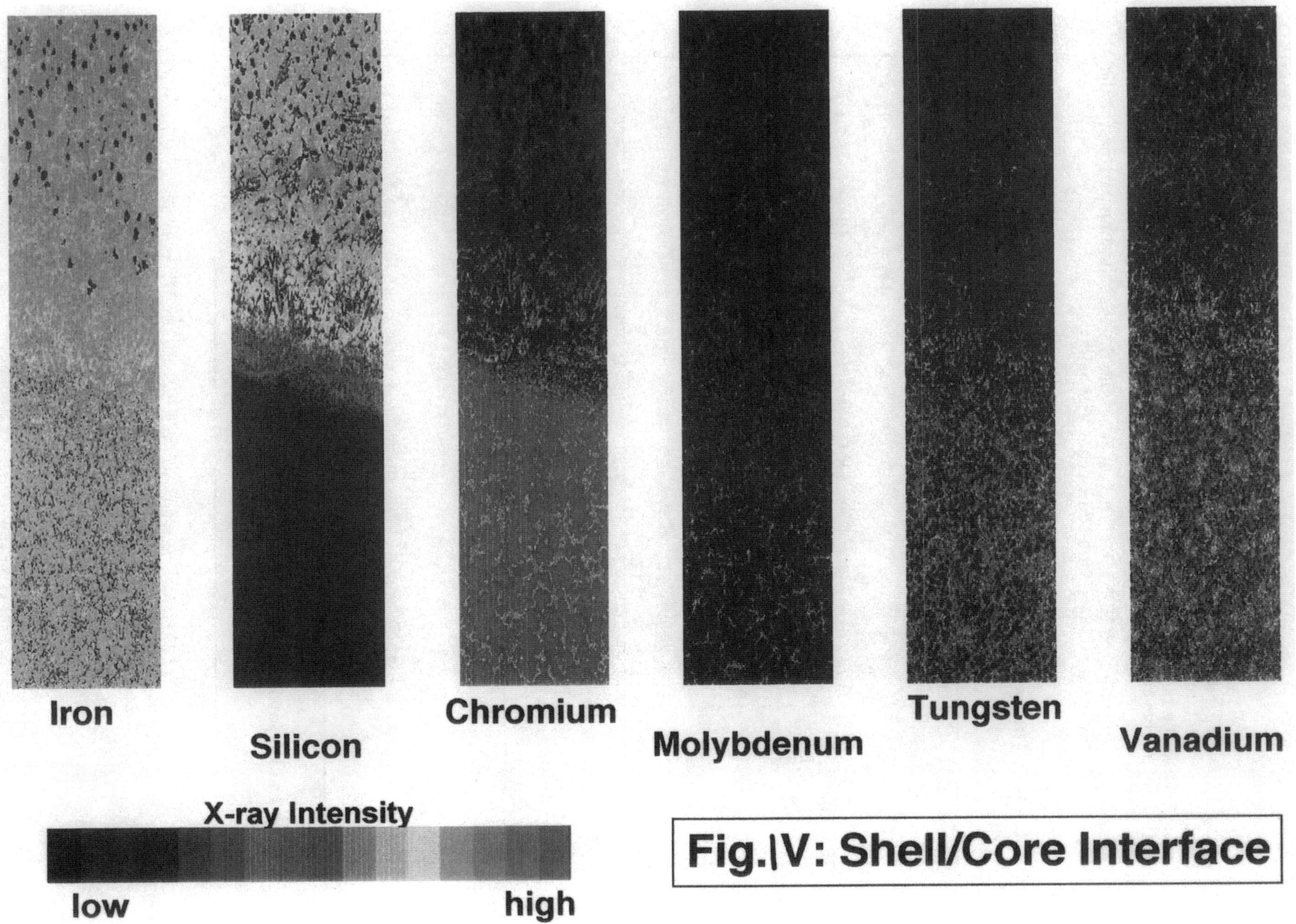

Fig.IV: Shell/Core Interface

Though at 950°C quenching + 500°C tempering there is also hardness drop, the final hardness can still attain 83°HSC. According to the findings from the above testing, there are two suitable handling for heat treatment : (1) 1,150°C quenching + 550°C tempering (2) 950°C quenching + 500°C tempering but (2) is recommended as being more practical. Subsequent tempering can be carried out to give correct structure.

After heat treatment of the HSS rolls in accordance with the above testing parameters, the hardness profile of the roll is illustrated in Table VIII below.

Table VII: The hardness change under different temperature for quenching and tempering

Sample	Quenching Temp °C	Hardness after Quench HRC	HSC	Tempering Temp °C	Hardness after Temper HRC	HSC
1	950	64	88.5	500	61	83
2	950	64	88.5	550	58	78.5
3	950	63	87	600	52	69.5
4	1050	61.5	84	500	60.5	82
5	1050	62.5	86	550	60.5	82
6	1050	62	85	600	57	77
7	1150	61	83	500	61	83
8	1150	60.5	82	550	62	85
9	1150	61	83	600	52	69.5
10	casting state	55.5	74			

Table VIII: Roll hardness testing results

Roll#	Barrel Hardness in Casting State				Barrel Hardness after Heat Treatment			
A75	A	B	C	Ave	A	B	C	Ave
313	68	69	69	69	83	82	84	83
315	70	70	71	70	83	82	83	83
317	72	70	71	71	79	81	80	80
319	76	73	74	74	83	85	85	84

Table IX: Performance Data

				AIC/SG		HSS		
Area	Size	Roll Type	Strands	Tonnes	Dressing	Tonnes	Dressing	Improvement
	312 x 500	Strip	1	180	2.2mm	475	0.55	x 10
China	310 x 150	Ring	1	980	1mm	3160	1mm	x 3.2
	380 x 600	Strip	1	80	1	160	0.4	x 5
	370 x 50	Ring	1	175	4	Carbide Replacement		
						500	0.5	
Ex-China	280 x 400	Rod	1	500	2.5	1600	2.5	x 3
	410 x 710	Rod	Multi	1000	0.37mm	1000	0.066	x 5.5

PRECAUTIONS IN USE

1. Good water supply both in volume and pressure, properly directed.
2. Good shape from previous stands. Directional control of guides critical.
3. Crack removal after redressing.
4. Regular Ultrasonic testing of interface.
5. Correct tension and drafting between stands.

BENEFITS WHEN USING HSS ROLLS

1. Longer rolling campaigns.
2. Less dressing per campaign.
3. Improved mill and roll shop productivity. (Less mill down time)
4. Better shape and size to the sizing mill.
5. Reduced roll cost/tonne rolled.
6. Lower roll inventory.

CONCLUSION

The use of HSS rolls can show quautum increases in benefits to both the mill and roll shops in productivity and quality. To make full use of these benefits, some changes in practices will be necessary particularly in inspection, records and rescheduling. Roll qualities in earlier stands will have to be reviewed to minimise mill downtime and even maintainance schedules may change.

ACKNOWLEDGEMENT

The authors wish to thank their colleagues at British Rollmakers China and Lian Qiang Foundry for their assistance towards this paper.

John Mees, Sheffield Forgemasters Rolls Technical Department

Development of High Carbon Wire Rod for Pre-Stressed Concrete Applications in TSSI

Arun K. Mukherji : Project Manager
David Jones : Quality Control Manager

Thai Special Steel Industry P.C.L.
26/56 TPI Tower Floor.19th, Chan Tat Mai Road ,Sathorn, Bangkok, Thailand 10120.
Tel:- 662-678-6500-4 Fax:- 662-678-6505

ABSTRACT

One of the most modern and sophisticated wire rod mill has been commissioned by Thai Special Steel Industry Co.,Ltd (TSSI) in Rayong province of Thailand in May 1998. The mill supplied by Morgan Construction company of U.S.A / Kocks GmbH Germany, consists of a Walking Beam Furnace (Stein Heurtey, France), 14 Horizontal / Vertical Stands; a five stand, 3 roll Kocks block, and 8 stand No Twist Mill. The mill is followed by a 96 meter long Stelmor conveyor with both accelerated blowing (MORAIR) & retarded stelmor features. The electric's have been supplied by ABB Sweden, which includes a higher level production control computer. Apart from the wire rod line (5.5 mm.∅ –20.0 mm ∅), Danieli has supplied the Bar in Coil Line, through which bigger size bars (18 mm – 40.0 mm ∅) can be produced.

In this article, authors wish to share their experiences gained during the development of high carbon wire rod in TSSI Wire Rod Mill for pre-stressed concrete application. The topics include, customer's plant visit, drawing and stranding operation charts, billet out sourcing specification, development of metallurgical process chart in wire rod mill, actual rolling operation and testing of wire rod at customer plants.

1. INTRODUCTION

Thai Special Steel Industry (TSSI) Bangkok, was set up in 1995. The aim of the company was to install a state of the art Wire Rod and Bar Mill to produce Special Steel wire rod & bar products for automobile sector, general engineering applications and P.C. wire for construction sector.

The order for WRM equipment was placed on Morgan Construction Company, Worcestor, USA in mid 1995. Later on a 5 Stand 3 Roll Reducing and Sizing Block was ordered on Kocks Germany, to be installed ahead of a 8 Stand Morgan V-Mill (NTM). The designed finishing speed for 5.5 mm. diameter wire rod is 120 meters / second.

A bar in coil line was added to the project to produce bars of diameter 18 mm. to 40 mm. This equipment has been supplied by Danieli (Italy) and consists of water box, shear, switch, two garret coilers, and a walking beam type coil conveyor system with both accelerated and retarded cooling arrangement. The land for the plant was aquired in an industrial park in Rayong Province situated in Southern Thailand. It is very near to sea cost of Gulf of Thailand and was started as a greenfield site project.

The equipment details have been shown in Table I Layout of the plant has been shown in Figure 1 and 2. The mill layout and equipment selection has been based on the following important features.

- Highest rod surface quality.
- Very stringent rod tolerances.
- Improved physical and metallurgical properties.
- On line quality assurance.
- Decreased decarburization.

2. CUSTOMER REQUIREMENTS

As part of the Product Development Programme, it was decided to produce high carbon wire rod for Pre-stressed concrete applications. This product had a large customer base in Thailand and till 1998, most of the wire rod had been imported from Japan, South Africa, Brazil and other west European countries. In order to ascertain the market requirement for various PC Wire applications, a team conisting of marketing, quality management and production personnel undertook visits to several large customer's factories and had discussions regarding their technical requirements.

TABLE-I : EQUIPMENT DETAILS FOR TSSI WIRE ROD MILL

EQUIPMENT	TYPE	CAPACITY/SIZE	SUPPLIER
FURNACE	WALKING BEAM	100 T/HR LIGHT FUEL OIL/ GAS/RECUPERATOR	STIEN HEURTEY FRANCE LEVEL-2 COMPUTER
ROUGHING	1 H 2 V 3 H 4 V 5 H 6 V	625 mm/ 400 kw 625 mm/ 400 kw 625 mm/ 400 kw 515 mm/ 400 kw 515 mm/ 500 kw 515 mm/ 400 kw	MORGAN/ ABB ALL MOTORS ARE A.C.SQ. CAGE INDUCTION MOTOR SUPPLIED THROUGH A.C. INVERTER
CROP & COBBLE SHEAR #1	CRANK TYPE	1145 mm	D.C. START/ STOP TYPE
INTER #1	7 H 8 V 9 H 10 V 11 H 12 V	425 mm/ 600 kw 425 mm/ 500 kw 425 mm/ 600 kw 425 mm/ 500 kw 350 mm/ 600 kw 350 mm/ 500 kw	SAME AS EARLIER MAIN DRIVE
CROP & COBBLE SHEAR #2	CRANK TYPE	1080 mm	D.C. START/ STOP TYPE
INTER #2	13 H 14 V	350 mm/ 600 kw 350 mm/ 500 kw	MORGAN/ ABB A.C. DRIVE
KOCKS RSB	15 (K1) 16,17,18,19 (K2)	300 mm/ 400 kw 300 mm/ 1200 kw	A.C. DRIVE
UPLOOPERS	THREE UPLOOPERS	BETWEEN	STAND #12,13,14 & KOCKS RSB.
WIRE ROD/BAR DIAMETER GAUGE	ROTARY TYPE	FOR DIAMETER UP TO 40 mm	ORBIS
SURFACE FLAW DETECTOR #1	EDDY CURRENT TYPE	-DO-	SOFRA TEST/ CASTEL
SIDE LOOPER	PNEUMATIC OPERATION	BEFORE NO TWIST MILL	MORGAN/ ABB
SWITCH	MANUAL	TO CHANGE BETWEEN WIRE ROD/ BAR LINE	DANIELI
CROP & DEVIDE SHEAR CHOPPING SHEAR	950 mm	FRONT/ TAIL/ DEVIDE/ SAMPLE CUT	D.C. DRIVE
WATER BOXES	WITH CLOSE LOOP MORGAN TEMPERATURE CONTROL COMPUTER	FIVE NUMBERS/BEFORE NTM-1 AFTER NTM-4	MORGAN WITH SPLIT NOZZLE
NO TWIST MILL STAND #20-24 STAND # 25-27	V-MILL 230 mm SIZE 160 mm SIZE	MOTOR 5000 kw A.C. LCI DRIVE	MORGAN/ ABB
ROLLERISED TURN DOWN	FOUR ROLLS INDIVIDUAL DRIVE	A.C. INVERTORS	ABB
DIAMETER ONLINE GAUGE	ROTARY TYPE	5.5 mm-20.0 mm DIAMETER	ORBIS GAUGE
SURFACE FLAW DETECTOR #2	EDDY CURRENT TYPE	-DO-	SOFRA TEST/ CASTEL
PINCH ROLL	250 mm DIA. ROLLS	D.C. DRIVE 150 kw 1750 RPM	MORGAN/ ABB
LAYING HEAD	WITH FRONT END ORIENTATION CONTROL	D.C. DRIVE 320 kw 1440 RPM	-DO-
STELMOR CONVEYOR	ROLLER TABLE WITH D.C. VARIABLE SPEED DRIVES	NO. OF SECTIONS-12 WITH HOODS FOR RETARDED AS WELL AS STANDARD OPERATION	-DO-
STELMOR BLOWERS	CENTRIFUGAL FANS WITH A.C. VARIABLE SPEED DRIVES	NO. OF BLOWERS-14 154,000 m^3/HR. CAPACITY FAN DISCHARGE HAS OPTIFLIX DISCHARGE SYSTEM. 280 KW., 1500 RPM., 380V., 3 PHASE., A.C. MOTORS	-DO-
REFORM TUB	WITH RING DITRIBUTOR SYSTEM	HYDRAULIC MOTOR DRIVE	MORGAN

FIGURE 1

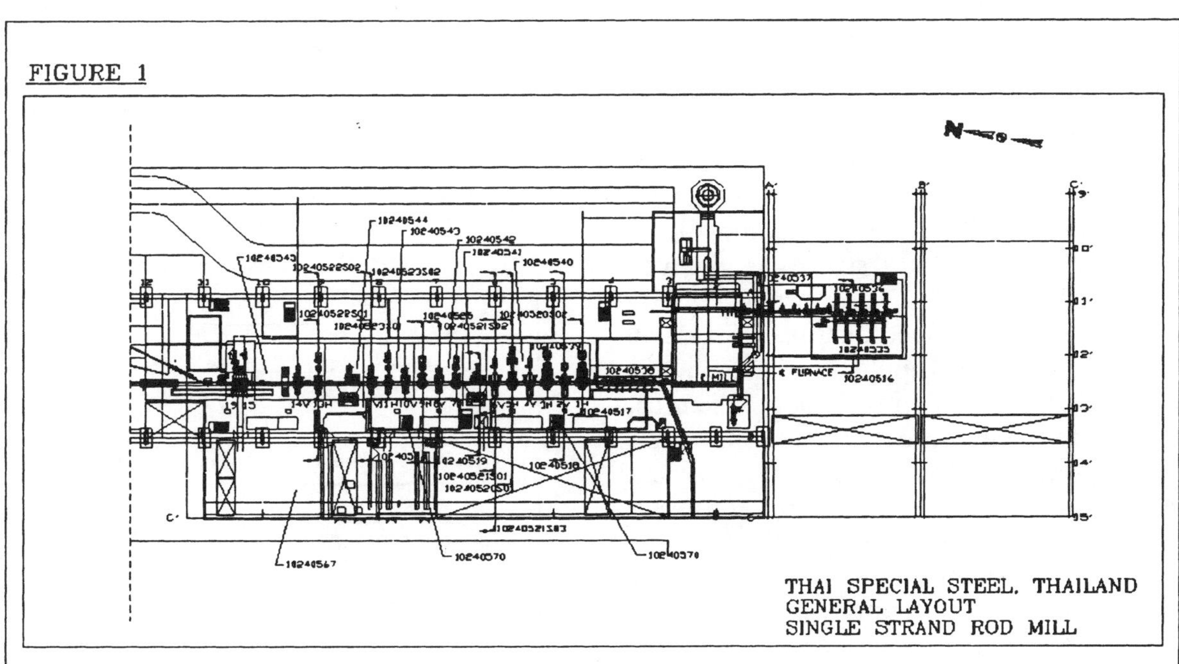

FIGURE 2

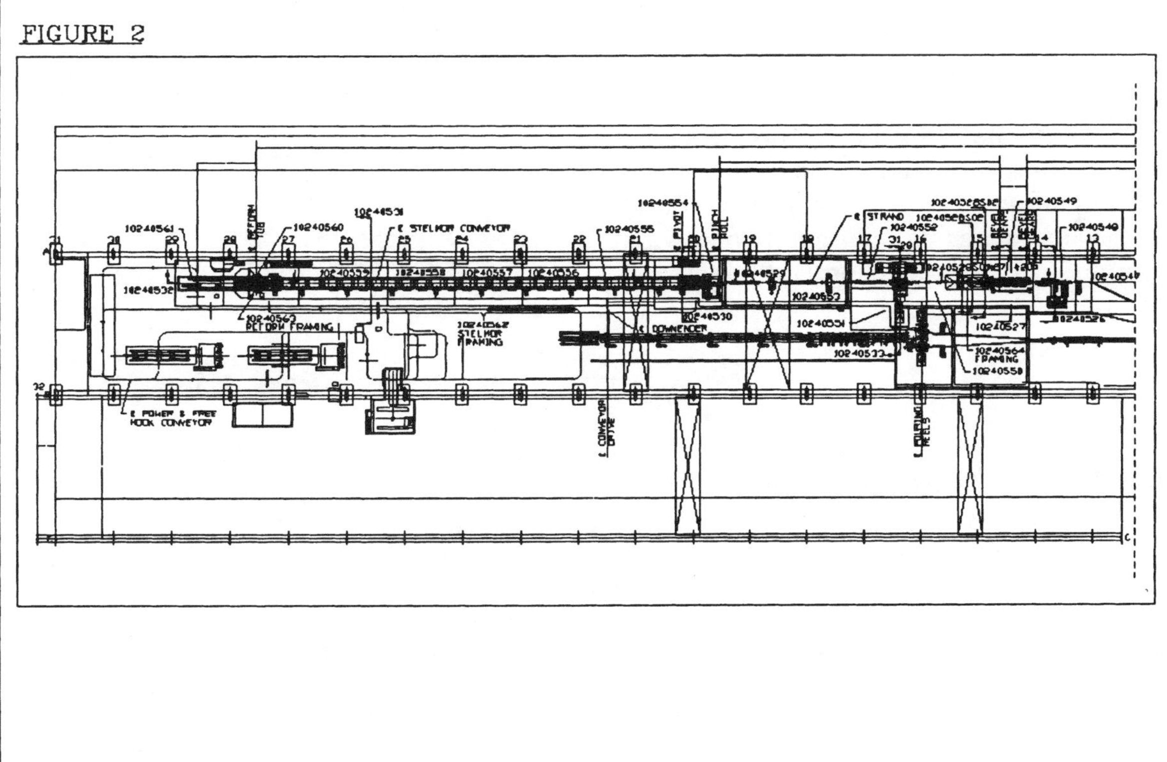

The team also observed the wire drawing and other processes of different customers. This helped to conceptualise the basic parameters for the production of wire rod for PC wire application.

The main items produced by Thai Wire Drawing Industry are summarised in Figure 3. Tables II and III give the Technical standard specifications for PC single wire and PC7 wire strands. The process followed for manufacturing the PC wire (single & multi strand) is given in Figure 4. Most of the Thai Wire producers buy wire rod with Japanese Specification JIS G3506 grade SWRH82B for PC application, which was the base for formulating the Chemical composition for billet procurement. However, depending upon the different tensile requirement of the "as rolled wire rod" by different customers, the chemical compositions had to be modified to achieve the end results. Table IV summaries the major categories of customers and their process parameters, including the drawing machine details and approximate drawing speeds being followed. The table also indicates the approximate reduction ratios being followed in the drawing process.

Table V gives the Tensile requirement for various customers. While analysing the customer requirement, it was generally found that the tensile ranges fall within two groups depending upon their processing conditions and final aim grade.

FIGURE-4 PC WIRE AND STRAND MANUFACTURING PROCESS

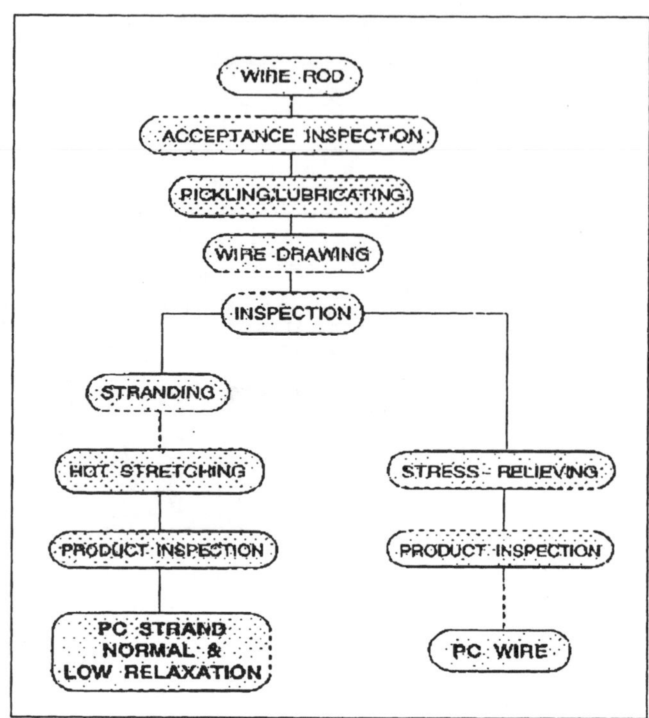

TABLE : V TENSILE REQUIREMENT

Customer Group	Rod Sizes	Tensile Strength Newton/mm2
Group A	9 mm , 11 mm , 13 mm	1100 - 1180
Group B	9 mm , 11 mm , 13 mm	1170 - 1230

TABLE : IV CUSTOMER PROCESS DETAILS

S.No.	Customer Group.	Descaling System.	No. of drawing holes per machine	Average reduction %	Approx drawing speed	Wire rod size and final drawn wire size
1	A	Chemical	6 Holes	24 - 25 %	5.0-6.0 M/S	13 mm ∅ drawn to 7.0 mm 11 mm ∅ drawn to 5.0 mm 9 mm ∅ drawn to 4.0 mm
2	B	Chemical	7 Holes	20 – 21 %	4.0-5.0 M/S	ditto
3	C	Chemical	8 Holes	18 – 20 %	3.5-4.5 M/S	ditto

TABLE II

PC WIRE SPECIFICATION

Standard	Nominal Diameter		Tolerance on Diameter		Nominal Area		Tensile Strength			Yield Strength			Elongation % (min.)	Coil Diameter	Nominal Weight	
	in.	mm.	in.	mm.	in^2.	mm^2.	psi (min.)	kgf/mm^2. (min.)	N/mm^2. (min.)	psi (min.)	kgf/mm^2. (min.)	N/mm^2. (min.)			lb/ 1,000ft.	kg./ 1,000m.
TIS 95-2534																
PC4	0.157	4	±0.002	±0.050	0.019	12.57	(248,900)	(175)	1,720	(213,300)	(150)	1,470	3.5(in 100mm)	1.2 m.	66.3	98.7
PC5	0.197	5	±0.002	±0.050	0.030	19.64	(248,900)	(175)	1,720	(213,300)	(150)	1,470	4.0(in 100mm)	1.2 m.	103.5	154.0
PC7	0.276	7	±0.002	±0.050	0.060	38.48	(227,900)	(160)	1,720	(192,000)	(135)	1,325	4.5(in 100mm)	1.8 m.	202.9	302.0
PC9	0.354	9	±0.002	±0.050	0.098	63.62	(206,200)	(145)	1,720	(177,800)	(125)	1,225	4.5(in 100mm)	1.8 m.	335.3	499.0
ASTM A421-80																
Type BA	0.196	4.98	±0.002	±0.050	0.030	19.48	(240,000)	(168.7)	1,655	(204,000)	(143.4)	1,407	4.0 (in 10")			
	0.250	6.35	±0.002	±0.050	0.049	31.67	(240,000)	(168.7)	1,655	(204,000)	(143.4)	1,407	4.0 (in 10")			
	0.276	7.01	±0.002	±0.050	0.060	38.59	(235,000)	(165.2)	1,620	(199,750)	(140.4)	1,377	4.0 (in 10")			
Type WA	0.192	4.88	±0.002	±0.050	0.029	18.70	(250,000)	(175.8)	1,725	(212,500)	(149.4)	1,465	4.0 (in 10")	Coil inside		
	0.196	4.98	±0.002	±0.050	0.030	19.48	(250,000)	(175.8)	1,725	(212,500)	(149.4)	1,465	4.0 (in 10")	dia		
	0.250	6.35	±0.002	±0.050	0.049	31.67	(240,000)	(168.7)	1,655	(204,000)	(143.4)	1,407	4.0 (in 10")	48" min.		
	0.276	7.01	±0.002	±0.050	0.060	38.59	(235,000)	(165.2)	1,620	(199,750)	(140.4)	1,377	4.0 (in 10")			
BS 5896 : 1980																
Normal	0.157	4	±0.0016	±0.04	0.020	12.6	(242,200)	(170)	1,670	(201,600)	(142)	1,390		1.25 m.	66.3	98.9
Relaxation	0.157	4	±0.0016	±0.04	0.020	12.6	(256,700)	(180)	1,770	(213,200)	(150)	1,470	For all wires	1.25 m.	66.3	98.9
	0.177	4.5	±0.002	±0.05	0.025	15.9	(235,000)	(165)	1,620	(195,800)	(138)	1,350	3.50%	1.25 m.	48.0	125
	0.197	5	±0.002	±0.05	0.030	19.6	(242,200)	(170)	1,670	(201,600)	(142)	1,390	in 200 mm.	1.5 m.	103.5	154
	0.197	5	±0.002	±0.05	0.030	19.6	(256,700)	(180)	1,770	(213,200)	(150)	1,470	at max. load	1.5 m.	103.5	154
	0.236	6	±0.002	±0.05	0.044	28.3	(242,200)	(170)	1,670	(201,600)	(141)	1,390		2.0 m.	149.2	222
	0.236	6	±0.002	±0.05	0.044	28.3	(256,700)	(180)	1,770	(213,200)	(150)	1,470		2.0 m.	149.2	222
	0.276	7	±0.002	±0.05	0.060	38.5	(227,700)	(160)	1,570	(188,500)	(133)	1,300		2.0 m.	202.9	302
	0.276	7	±0.002	±0.05	0.060	38.5	(242,200)	(170)	1,670	(201,600)	(142)	1,390		2.0 m.	202.9	302
BS 5896 : 1980 SWPD 1																
	0.114	2.9	±0.001	±0.030	0.010	6.61	12.7 kN (min)	(197)	1,300 kgf(min)	11.3 kN (min)	(174)	1,300 kgf(min)	3.5 (in 100mm)		34.8	51.8
	0.138	3.5	±0.0015	±0.040	0.015	9.62	16.2 kN (min)	(172)	1,650 kgf(min)	14.2 kN (min)	(151)	1,300 kgf(min)	3.5 (in 100mm)		50.7	75.5
	0.157	4	±0.0015	±0.040	0.019	12.57	21.1 kN (min)	(172)	2,150 kgf(min)	18.6 kN (min)	(151)	1,300 kgf(min)	3.5 (in 100mm)		66.3	98.7
	0.177	4.5	±0.002	±0.050	0.025	15.90	25.5 kN (min)	(164)	2,600 kgf(min)	22.6 kN (min)	(145)	1,300 kgf(min)	4.0 (in 100mm)		84	125
	0.197	5	±0.002	±0.050	0.030	19.64	31.9 kN (min)	(166)	3,250 kgf(min)	27.9 kN (min)	(145)	1,300 kgf(min)	4.0 (in 100mm)		103.5	154
	0.236	6	±0.002	±0.050	0.044	28.27	44.1 kN (min)	(159)	4,500 kgf(min)	38.7 kN (min)	(140)	1,300 kgf(min)	4.0 (in 100mm)		149.2	222
	0.276	7	±0.002	±0.050	0.060	38.48	58.3 kN (min)	(155)	5,950 kgf(min)	51.0 kN (min)	(135)	1,300 kgf(min)	4.5 (in 100mm)		202.9	302
	0.315	8	±0.002	±0.050	0.078	50.27	74.0 kN (min)	(150)	7,550 kgf(min)	64.2 kN (min)	(130)	1,300 kgf(min)	4.5 (in 100mm)		265.4	395
	0.354	9	±0.002	±0.050	0.098	63.62	90.2 kN (min)	(145)	9,200 kgf(min)	78.0 kN (min)	(125)	1,300 kgf(min)	4.5 (in 100mm)		335.3	499

TABLE III NORMAL RELAXATION PC STRAND SPECIFICATION

Standard		Nominal Diameter		Tolerance on Diameter		Nominal Area		Breaking Strength (Min)			Minimum load at 1% extension			Elongation (%) (min.) in 24 inches. (in 610 mm.)	Nominal Weight	
		in.	mm.	in.	mm.	in^2.	mm^2.	lbf.	kgf.	kN.	lbf.	kgf.	kN.		lb/ 1,000ft.	kg./ 1,000m.
TIS 95-2534																
Grade 1725	SPC 9A	3/8	9.53	±0.016	±0.40	0.080	51.61	(20,000)	(9,070)	89.0	17,000	(7,710)	75.6	3.5	272	405
	SPC 12A	1/2	12.70	±0.016	±0.40	1.440	92.90	(36,000)	(16,320)	160.1	30,600	(13,880)	136.2	3.5	490	730
	SPC 15A	0.6	15.24	±0.016	±0.40	0.216	139.35	(54,000)	(24,490)	240.2	45,900	(20,820)	204.2	3.5	737	1,094
Grade 1860	SPC 9B	3/8	9.53	+0.026 / -0.006	+0.65 / -0.15	0.085	54.840	(23,000)	(10,430)	102.3	19,550	(8,870)	87.0	3.5	290	432
	SPC 12B	1/2	12.70	+0.026 / -0.006	+0.65 / -0.15	0.153	98.710	(41,300)	(18,730)	183.7	35,100	(15,910)	156.1	3.5	520	775
	SPC 15B	0.6	15.24	+0.026 / -0.006	+0.65 / -0.15	0.217	140.00	(58,600)	(26,580)	260.7	49,800	(22,580)	221.5	3.5	740	1,102
ASTM A416-88b																
	Grade 250	1/4	6.35	±0.016	±0.41	0.036	23.22	9,000	(4,080)	(40.0)	7,650	(3,470)	(34.0)	3.5	122	182
		5/16	7.94	±0.016	±0.41	0.058	37.42	14,500	(6,580)	(64.5)	12,300	(5,580)	(54.7)	3.5	197	294
		3/8	9.53	±0.016	±0.41	0.080	51.61	20,000	(9,070)	(89.0)	17,000	(7,710)	(75.6)	3.5	272	405
		7/16	11.11	±0.016	±0.41	0.108	69.68	27,000	(12,240)	(120.1)	23,000	(10,430)	(102.3)	3.5	367	548
		1/2	12.70	±0.016	±0.41	0.144	92.90	36,000	(16,320)	(160.1)	30,600	(13,880)	(136.2)	3.5	490	730
		0.6	15.24	±0.016	±0.41	0.216	139.35	54,000	(24,490)	(240.2)	45,900	(20,820)	(204.2)	3.5	737	1,094
	Grade 270	3/8	9.53	+0.026 / -0.006	+0.66 / -0.15	0.085	54.84	23,000	(10,430)	(102.3)	19,550	(8,870)	(87.0)	3.5	290	432
		7/16	11.11	+0.026 / -0.006	+0.66 / -0.15	0.115	74.19	31,000	(18,730)	(137.9)	35,100	(15,910)	(117.2)	3.5	390	582
		1/2	12.70	+0.026 / -0.006	+0.66 / -0.15	0.156	98.71	41,000	(26,580)	(183.7)	35,100	(22,580)	(156.1)	3.5	520	775
		0.6	15.24	+0.026 / -0.006	+0.66 / -0.15	0.217	140.00	58,000	(26,580)	(260.7)	49,800	(22,580)	(221.5)	3.5	740	1,102
BS 5896-1980																
Normal Relaxation 7 - wire standard		3/8	9.3	+0.012 / -0.006	+0.3 / -0.15	0.081	52	20,682	9,381	92.0	18,209	8,260	(81.0)	3.5	274	408
		7/16	11.0	+0.012 / -0.006	+0.3 / -0.15	0.110	71	28,100	12,746	125.0	24,728	11,217	(110.0)	3.5	374	557
		1/2	12.50	+0.012 / -0.006	+0.4 / -0.2	0.146	93	36,867	16,723	164.0	32,371	14,684	(144.0)	3.5	491	730
		0.6	15.2	+0.012 / -0.006	+0.4 / -0.2	0.215	139	52,156	23,657	232.0	45,859	20,802	(204.0)	3.5	733	1,090
JIS G3536-1988																
	SWPR 7A	1/4	6.2	+0.016 / -0.008	+0.4 / -0.2	0.036	23.23	9,000	4,100	40.2	7,600	3,450	(33.8)	3.5	123	182
		5/16	7.9	+0.016 / -0.008	+0.4 / -0.2	0.058	37.42	14,540	6,600	64.7	12,300	5,600	(54.9)	3.5	197	293
		3/8	9.3	+0.016 / -0.008	+0.4 / -0.2	0.080	51.61	19,900	9,050	88.8	17,000	7,700	(75.5)	3.5	272	405
		7/16	10.8	+0.016 / -0.008	+0.4 / -0.2	0.108	69.68	26,900	12,200	120.0	23,000	10,400	(102.0)	3.5	368	546
		1/2	12.4	+0.016 / -0.008	+0.4 / -0.2	0.144	92.90	35,900	16,300	160.0	30,600	13,900	(136.0)	3.5	491	729
		0.6	15.2	+0.016 / -0.008	+0.4 / -0.2	0.215	138.7	5,400	24,500	240.0	45,800	20,800	(204.0)	3.5	740	1,101
	SWPR 7B	3/8	9.50	+0.016 / -0.008	+0.4 / -0.2	0.085	54.84	22,900	10,400	102.0	19,500	8,850	(86.8)	3.5	291	432
		7/16	11.1	+0.016 / -0.008	+0.4 / -0.2	0.115	74.19	31,000	14,100	138.0	26,500	12,000	(118.0)	3.5	391	580
		1/2	12.7	+0.016 / -0.008	+0.4 / -0.2	0.153	98.71	41,200	18,700	183.0	35,000	15,900	(156.0)	3.5	521	774
		0.6	15.2	+0.016 / -0.008	+0.4 / -0.2	0.215	138.7	58,600	26,600	261.0	49,800	22,600	(222.0)	3.5	740	1,101

FIGURE-3 STANDARD PC WIRE & PC STRAND PRODUCTS

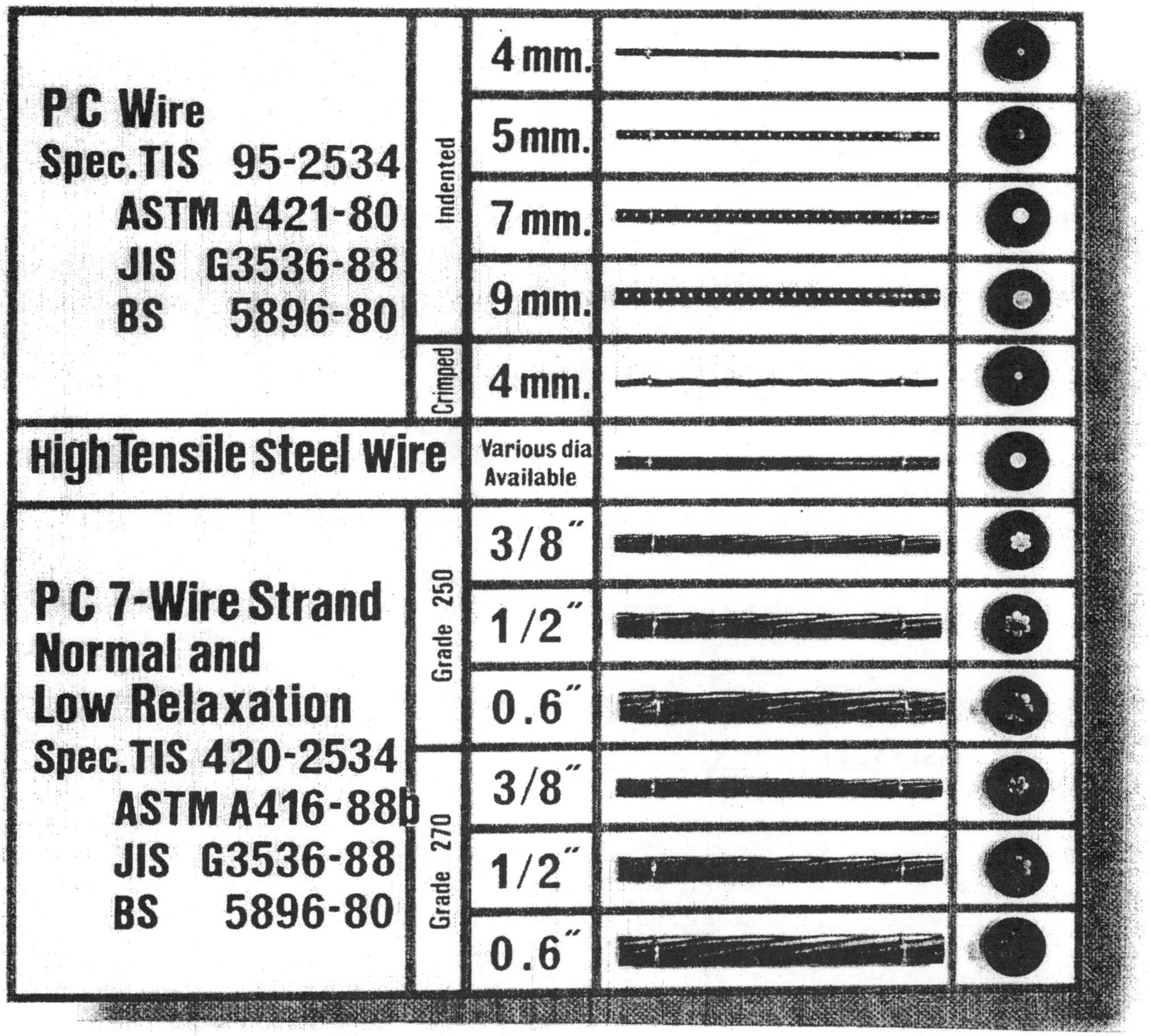

P C Wire Spec.TIS 95-2534 ASTM A421-80 JIS G3536-88 BS 5896-80	Indented	4 mm.
	Indented	5 mm.
	Indented	7 mm.
	Indented	9 mm.
	Crimped	4 mm.
High Tensile Steel Wire		Various dia Available
P C 7-Wire Strand Normal and Low Relaxation Spec.TIS 420-2534 ASTM A416-88b JIS G3536-88 BS 5896-80	Grade 250	3/8"
	Grade 250	1/2"
	Grade 250	0.6"
	Grade 270	3/8"
	Grade 270	1/2"
	Grade 270	0.6"

Other major requirement for PC Wire Rod, are as follows:

A. The wire rods should be free from fins, blowholes, rolled in scales, rust, roll marks and seams.
B. The Inclusion rating when tested as per ASTM-E-45 shall not exceed.

Maxm Inclusion Rating		
	Thin (T)	Heavy (H)
Type A	2.0	1.5
Type B	2.0	1.5
Type C	2.0	1.5
Type D	1.5	1.5

C. Surface Defects: allowable depth of surface defects should not exceed 0.1 mm.
D. Decarburisation: maximum depth should not exceed 0.07 mm.
E. Size tolerance:- diameter tolerance to be ±0.25 mm with ovality at 0.40 mm max.
F. Microstructure:- Grain size of ASTM6 or finer. Fine pearlitic structure with minimal ferrite and no secondary transformation products (bainite or martensite).
G. Scale:- suitable for acid descaling

3. CHEMICAL COMPOSITION FOR STEEL OUTSOURCING.

As per the JIS G3502 SWRH82B specification (or Thai Industrial Standard TIS349-2532) the chemical composition of steel is as follows.

Carbon – 0.79 to 0.86%
Silicon – 0.15 to 0.35%
Manganese - 0.60 to 0.90%
Phosphorus 0.030% Max.
Sulphur 0.030% Max.

However, as explained earlier, most of the wire rod producers micro-alloy the steel to achieve the tensile strength requirement stipulated by the PC Wire manufacturers. While surveying the international billet market, it was noticed that generally the European Steel plants added Vanadium (V) Whilst the South American & Japanese steel plants added Chromium (Cr) as single alloy. There were other Steel plants in South Africa, Italy and China which offered billets with both Vanadium (V) as well as Chromium (Cr) as micro-alloying elements. Since TSSI had to outsource the steel billets for producing Prestressed Concrete wire-rod, four different steel compositions were made as standards depending upon the source of billets and customer Tensile Strength requirements. The specifications are given in Table VI.

Table VI TSSI STEEL COMPOSITIONS

Base Composition (AIM SPECS)	Base Plus (Vanadium)	Base Plus (Chromium)	Base Plus (V + Cr)
Carbon – 0.80 – 0.84% Silicon – 0.15 – 0.30% Manganese – 0.65 – 0.80% Phosphorus – 0.025% max Sulphur – 0.025% max	Base Spec + Vanadium 0.05 – 0.07%	Base Spec + Chromium 0.20 – 0.30 %	Base Spec + Vanadium 0.03 – 0.05% Chromium 0.20 – .30%

4. PROCESSING OF HIGH CARBON WIRE RODS.

Wire used in pre-stressed concrete require a tensile strength of approximately 1760-1850 Newton / mm2. To achieve this strength level, wire rod must have fully pearlitic microstructure before wire drawing process. During hot rolling of steel, it may lead to different transformation products, depending on following factors:

A. Carbon content and Micro alloying element.
B. Temperature of Steel at various stages.
C. Austenite Grain size.
D. Final cooling rate at Stelmor conveyor.

In general, with the increase in carbon content and grain size, the hardenability increases. Grain size in turn depends greatly on the preheating temperature, rolling temperature and laying temperature. Ideal micro-structure for direct drawing of high carbon wire rods consists of fine unresolved pearlite and minimal amount of pro-eutectoid ferrite. Resolved pearlite has much influence on the wire rod properties and is always aimed to be lower than 5-10%

The hardenability of high carbon wire rods is significantly affected by the micro-alloys. Main objectives of micro-alloying are:

A. To increase the strength above the level typical of plain carbon steels.
B. It is of importance when continuously casting billets of small dimensions, to avoid centerline segregation and resulting cementite formation, which is harmful to drawability.

The increase in strength induced by micro-alloy addition relies mainly on two strengthening mechanisms.

A. Precipitation hardening, as a result of precipitation in ferrite during or after the transformation from austenite. (vanadium)
B. Increased hardenability by delaying the pearlite transformation on controlled cooling, lower transformation temperature and thus finer microstructure achieved than in plain C steel (Chromium)

Figure 5. Shows the components of tensile strength in high carbon steel due to its microstructure. Figure 6 shows the components of tensile strength for base composition of high carbon steel. Main tasks are to:

A. Increase Pearlite Content.
B. Reduce inter lamellar spacing.

For a fully pearlitic micro-structure, the carbon content must be close to the eutectoid composition: 0.82 – 0.86%. The interlamellar spacing is smaller, the lower is the temperature of pearlite formation. This is achieved in the process by increasing the cooling rate of the wire rod on the Stelmor, where major metallurgical transformation takes place.

FIGURE-5 COMPONENTS OF STRENGTH IN HIGH CARBON STEEL

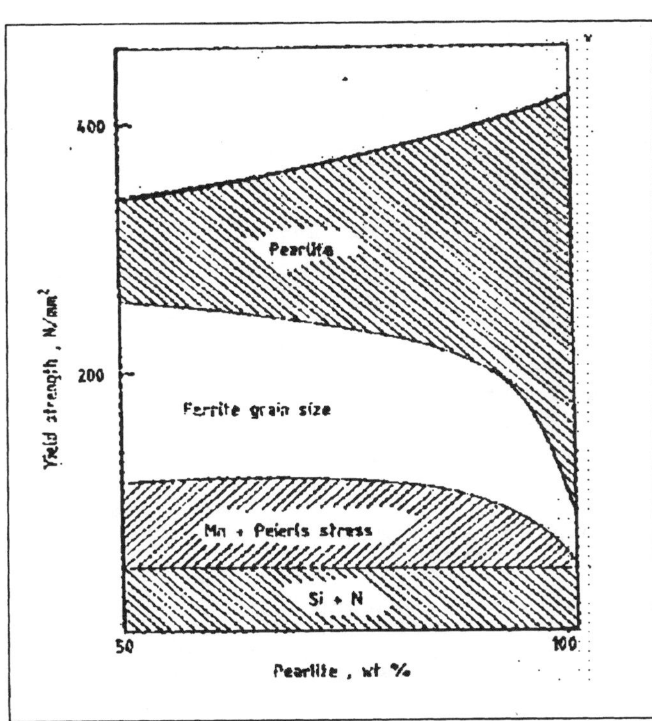

5. TSSI STELMOR DETAILS

One of the most important features of TSSI Wire Rod Mill is the Water Box Control system to have a close control of rod temperature and the Stelmor conveyor for the post rolling metallurgical treatment of the wire rod.

The equipment provided in the Stelmor Controled cooling system include the following components.

- Water Box before NO TWIST MILL (1 No.)
- Water Boxes between NO TWIST MILL and inclined Laying Head (4 Nos)
- Orbis gauge for online diameter measurement.
- Sofratest Eddy Current Testing Equipment for on line Crack detection.
- Pinch Roll
- Inclined Laying Head
- Stelmor controlled cooling conveyor along with 14 Blowers.
- Coil Forming Chamber including Ring Distributor system.

FIGURE-6 COMPONENTS OF TENSILE STRENGTH FOR BASE COMPOSITION OF HIGH CARBON STEEL

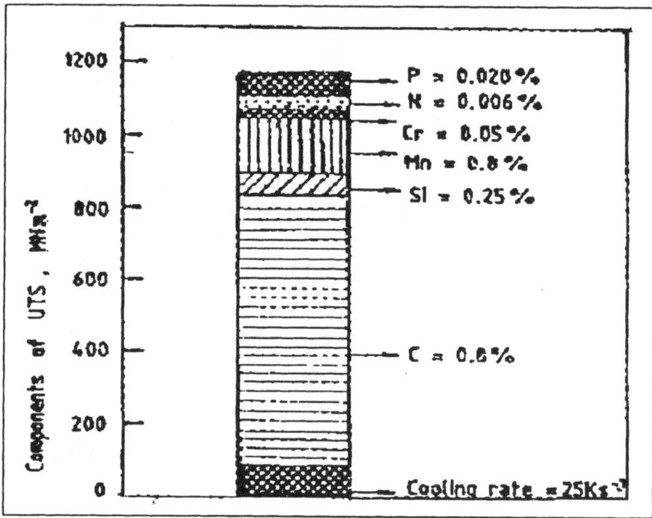

Water Box ahead of NTM

The purpose of this waterbox is to cool the stock entering the NTM, so that it can finish rolled at a temperature necessary to meet the specific property or metallurgical requirements for a particular grade. The size of the water box has been chosen for producing a temperature drop of approx. 100°C at the maximum design

Rolling rate which is 100 tonnes / hour for this mill. All functions of this box are controlled by Morgan Enhanced Temperature Controlled system (METCS).

Water Boxes after NTM

There are four water boxes installed after NO TWIST MILL. The purpose of these water boxes are to achieve the correct laying head temperature for the wire rod. Depending upon the grades being rolled, different combinations of waterboxes are used. Morgan Enhanced Temperature Controlled System control the pressure and water flow through these waterboxes, in order to achieve a temperature tolerance of ±10 °C for the laying head temperature.

Stelmor Conveyor and Air Blowers

Stelmor roller conveyors are driven by d.c. motors and can achieve a maximum linear speed of 1.5 meters per second for the wire rod rings. The entire Stelmor roller conveyor has been devided into 12 zones. First six conveyor zones have two blowers each with optiflex configuration. Zones 7 and 9 have one blower each. Zone 8 does not have any blowers.

A. Blower capacity – 154,000 M3/HR (total number-14) at 325 mm water column pressure (MORAIR)
B. Motor Capacity 250 KW AC Variable speed with 1500 RPM.

Due to this large air blowing capacity of the Stelmor, 20-22°C / second cooling rate for the wire rod can be achieved with maximum conveyor speed.

6. PROCESS CHART FOR PC WIRE ROD

Keeping the above points in mind, the following temperature profiles were followed to roll PC wire rod of 13 mm, 11 mm and 9 mm diameters. Rolling speeds at the NTM exit were between 35 meters / sec upto 60 meters / sec respectively.

Furnace soaking Zone - 1025 – 1050°C
Billet Temperature at Stand #1 - 980° – 1000°C
Stock Temperature at Kocks exit -1000° – 1020°C
Stock Temperature at NTM entry - 900° – 920°C
Laying Head Temperature - 820° – 850°C

Stelmor Conveyor Speeds:
Zone 1 - 0.8 – 1.0 meters/ second
Zone 6 - 1.0 – 1.2 meters / sec.
Zone 10 - 0.75 – 0.80 meters / sec
Zone 11 - 0.65 meters / sec.

Stelmor Blowers Setting:
Blowers 1 to 12 - 100% speed of 1500 RPM
With optiflex position at 25%

7. ACTUAL ROLLING OPERATION

However, when different compositions of billets were rolled with the above Process Chart, it was found that tensile strengths achieved for billets with both chromium and vanadium were too high. During the trials at certain customers plants, who followed 6 hole drawing practice, the tensiles achieved in the PC wire was too high. This was due to the fact that TSSI Stelmor blowing capacity was very high because of MORAIR configuration.

Further trials were conducted for different steel compositions with modified process parameters. After wire drawing operation at the customer plants, process chart was modified to accommodate two different groups of customers as well as different steel compositions.

Table VII. gives the two different process parameters which were adopted for rolling P.C. Wire rod with single alloy addition (Chromium or Vanadium) and with both alloying elements (Chromium and Vanadium)

Table VIII. gives the final matrix, with customer Grouping / Process chart (standard or modified) on Y-AXIS and rod sizes 9 mm,11mm,13mm) on the X-AXIS. Three different chemical compositions of steel has been categorised as
BASE - Plain Carbon Chemistry
+1 - One microalloys (Cr or V)
+2 - Two microalloys (Cr +V)

Table VII. Mill Process Conditions.

Process Route	Lay Head Temp °C	Air Blowers 1 - 12	Conveyor Speed, m/sec *		
			Zone 1	Zone 6	Reform
Standard	820 - 850	100%	.8– 1.0	1.0-1.2	.65
Modified	780 - 800	50%	.8– 1.0	1.0-1.2	.65

* Conveyor speeds, gradual increment from Zone 1 to Zone 6, then reduce on the drop sections to produce satisfactory coil reformation.

Table VIII. Steel Compositions and process routes for different applications.

Customer group	Process Route	Rod Size, mm		
		9	11	13
A	Standard	Base	Base	+1
	Modified		+1	+2
B	Standard	+1	+1	+2
	Modified	+1	+2	

Base = Base Chemistry
+1 = + one Microalloy (Cr or V)
+2 = + two Microalloys (Cr + V)

8. ROD PROPERTIES ACHIEVED

Table IX summarises the tensile strength ranges achieved in the various diameter wirerods with different chemistries and stelmor process parameters. It can be seen from this chart, how the tensiles varied with three different chemistry material rolled with standard stelmor parameters and modified stelmor parameters.

As TSSI is a local re-roller in Thailand without having steel making facility, the suitability of available imported billet does not always ideally match customer's requirements, thus requiring modifications to the stelmor process charts to achieve target strengths.

Table IX. Typical rod tensile strengths.

Steel Composition	Process Route	Tensile strength, N/mm² at Rod Size Range		
		9	11	13
BASE	Stardard	1120-1170	1100-1150	
+V or + Cr	Stardard	1190-1230	1160-1200	1140-1180
+V or +Cr	Modified	1160-1200		
+Cr + V	Stardard		1200-1250	1160-1220
+Cr + V	Modified	1190-1250	1160-1220	

9. CONCLUSIONS

By a combination of steel chemistry selection and modified stelmor process routes, the wide range of customer requirements have been successfully met. As a result TSSI has been accepted as the first local producer to supply PC Grade wire rods into the local market.

FLAT ROLL PRODUCTS I/ ROLL TECHNOLOGY II (Joint)

THE WEAR OF TOOL STEEL ROLLS DURING HOT ROLLING OF LOW CARBON STEELS

J.J. Fitzpatrick and J.G. Lenard*
Department of Mechanical Engineering
University of Waterloo
Waterloo, Ontario N2L 3G1 Canada
*Tel.: (519) 888 4567, ext. 3114
*Fax: (519) 888 6197
*e-mail: jglenard@mecheng1.uwaterloo.ca

Key Words: Wear, Tool Steel, Rolls, Hot Rolling

INTRODUCTION

Three general phenomena control the wear process between two contacting surfaces, regardless of what mechanisms are causing the wear [(1)]:

- *Chemical and physical interactions of the surface with lubricants and other constituents of the environment*
- *Transmission of forces at the interface through asperities and loose wear particles*
- *The response of a given pair of solid materials to the forces at the surface*

The two dominant modes of wear, observed on the work rolls during hot rolling of steels are adhesive and abrasive wear.

Adhesive wear is the result of relative sliding between two surfaces under load. The two surfaces form atomic bonds between contacting asperities, which may be stronger than the softer parent material, provided their chemical affinity for one-another is high. The relative sliding action may break the bond or tear out the asperity of the softer material in which case the resultant particle adheres to the harder surface. While this is the fundamental theory of adhesive wear, it is often disputed in the literature.

First, for any particle transfer to take place, the junction between the asperities must be stronger than either parent material. This does not seem reasonable under normal circumstances due to defects, contaminants or oxides present at the surface of the asperity. Any one of these would inhibit or prevent a strong molecular bond from forming. Thus, complete adhesion of asperities is quite rare.

In addition, according to the proposed process, there would be no wear on the surface of the harder material which is contradictory to published experimental data. Rabinowicz[(2)] attempts to modify the adhesion theory by proposing that the harder surface may contain some weak spots along the asperities which can fracture instead of the softer material. According to Jahanmir[3] this explanation is overly simplified. If the wear particles are indeed generated through asperity-asperity junctions, then the adhesive wear mechanism should be a direct function of surface roughness, but experiments have shown that this is not the case. Microstructure and mechanical properties of the material also affect the wear rate, which is another factor ignored in the overall adhesion theory.

The basic theory also overlooks one important and obvious consequence of adhesive wear; the generation of loose wear particles. According to the basic concept of adhesion, all particles that separate from the softer material should adhere and remain bonded to the harder material. That is, the adhesion theory only considers the formation of adherent particles[3]. Rabinowicz[2] suggests a modification that accounts for particle generation based on an adhesion energy versus elastic energy concept, but this only accounts for particles that are at least 1 µm in size. Wear particles are usually much smaller than that.

Schey [4] further defines the specific mechanisms involved in adhesive wear, citing observations taken from dies in metalworking. He agrees that the junction between asperities must be stronger than the softer material, but indicates that the junction may in fact be as strong or stronger than the harder material. This results in a separation that adheres either to the die or work piece. Also noted in Schey's work is the observation of wear on the die resulting from diffusion at elevated temperatures, a concept ignored by other adhesion models. It is possible for the alloying elements in the die, which give the die its strength, to diffuse into the work piece after some adhesion has occurred, resulting in material loss. This process is mainly noted in machining but is present in rolling and drawing, as well[4] .

If the adhesion junction is stronger than the work piece only, then separation results in die pick-up of the loose material. Schey[4] notes several further possible scenarios. The adhered particle can grow by accumulating more work piece material and may strain harden due to the applied contact loads. Separation of this particle can then occur through a fatigue process, resulting in a loose particulate. In contrast to Rabinowicz's energy theory, Schey places no limit on the minimum particle size that can be produced. Schey also suggests that the separated particle can transfer back and forth between the tool and work piece a number of times until the particle eventually leaves the deformation zone, bonded to the work piece.

Abrasive wear occurs where hard particles come into contact with a loaded surface. The particles could be loosely held between two surfaces (3-body wear) or be a part of the second surface (2-body wear), such as a protruding inclusion or hard asperity. The material is removed through a process of "micro-cutting" where the hard particles act as cutting tools and remove chips from the surface [3]. A secondary mechanism may also occur, similar to the delamination process. The hard particles may do some subsurface damage that aids in the removal of material through micro-cutting. The subsurface cracking is an accelerant only, and is not the primary mode of surface failure. For abrasion to take place, the Vickers hardness of the abrasive particle should be at least 1.5 times that of the softer material. Particles such as oxides and the intermetallics of precipitation-hardened alloys satisfy this criterion, and are of interest in metalworking since they may be as hard or harder than the constituents in the tool material. Therefore, abrasion becomes a significant factor when the lubrication between the die and work piece fails to provide proper separation. If the particles are embedded in the softer metal then the effect becomes particularly damaging to the tool if the matrix of the softer material is strong, and the embedded particles are not pulled loose under load [4].

The geometry of the abrasive particle and of the contact zone also plays a significant role in determining the wear. Particle size and distribution of particles, as well as the relative orientation to the abraded surface are of key importance. If the particulate is spherical in nature and comes into contact at a small attack angle, there will be little surface penetration. Thus, only elastic deformation of the surface will occur. This will have some damaging effects, as the contact creates local heating and possible lubrication breakdown, but there will be no wear, at least not by abrasion. If the penetration increases, "plowing" can occur, where large sections of material are forced out by a digging action from the abrasive particle. The end result of the plowing depends on the ductility of the deformed material. If the material has low ductility or is susceptible to strain hardening, the fragments that separate themselves from the surface stay in the contact zone, becoming third body wear particles. If the material is ductile, the separated material can be forced into other wear grooves, and may be removed by subsequent contact.

If the attack angle of the abrasive particle becomes steep, the mode of material removal is through a cutting action, and the material is instantly removed from the surface. Abrasive wear can be reduced mainly by increasing the hardness of the wearing material. This can be accomplished by employing a heat treatment method on alloyed metals, where the matrix hardness is increased in conjunction with the forming of coherent precipitates. Batchelor and Stachowiak [(5)] discuss the mechanisms of friction and wear. Ploughing between the contacting asperities causes abrasive wear. Erosive wear is the result of impact of particles. Repeated contact causes fatigue wear and liquid droplet erosion causes cavitational wear. These phenomena are not independent and often occur simultaneously. Studies show that roll wear rates are highest at temperatures of 850 - 950°C, which are precisely the temperatures used in the last of the finishing stands of hot strip mills.

The fairly recent introduction of tool steel work rolls in some of the stands of hot strip mills [(6)] requires the establishment of the dependence of the mechanisms of the wear process as well as the rate of wear on the process parameters. As indicated, use of HSS rolls reduced the wear depth of work rolls by 75% over high chrome cast iron [(6)]. The reduced wear when using HSS rolls is also indicated by Webber [(7)] who wrote that "...in the normal operating zone for the hot mill, HSS rolls had 2 to 4 times less wear...". Hashimoto et al. [(8)] also indicate improved productivity and product quality with the use of HSS rolls.

The objectives of the present study are to identify the effect of the process parameters on the wear of tool steel rolls during hot rolling of low carbon steels. Following Lundberg [(9)] and Savage et al. [(10)], a wear simulator was developed that attempts to duplicate, as close as possible, the conditions existing in a hot strip mill. The process parameters considered are the temperature, the rolling speed and the normal load, each of which is varied from a low to a high value.

EQUIPMENT, MATERIAL AND PROCEDURE

Equipment - The schematic diagram of the wear rig used is shown in Figure 1.

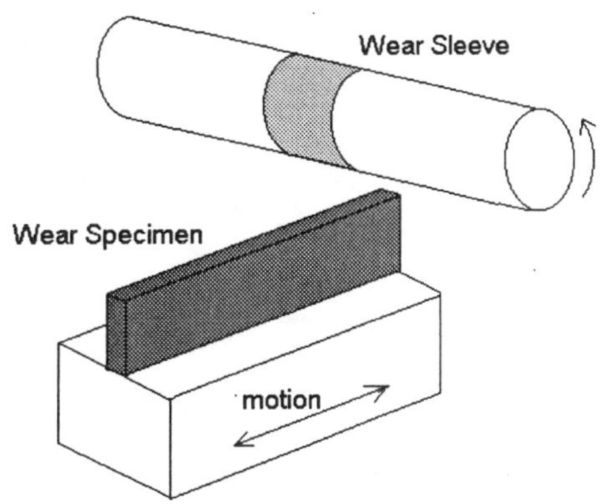

Figure 1: The schematic diagram of the wear apparatus

The wear apparatus is built on a Stanat laboratory rolling mill. Using the four-high configuration, the bottom work roll and back up roll are removed and are replaced by the wear specimen in its holder, as shown in Figure 1. The work roll is a sleeve, held in place by the upper back up roll. The sleeve, of D2 tool steel, surface hardness of Rc = 52 and a roughness of 1.5 μm (Ra), is fitted onto the work roll spindle and the sample is mounted. The outside diameter of the sleeve is 40 mm. The surface temperature of the specimen is monitored by two optical pyrometers, on each side of the work roll sleeve.

Materials and Procedure - A strip of hot rolled AISI 1018 steel, 38 mm wide by 6.4 mm thick is cut into 100 mm lengths. The initial weight of both the sample and the work roll sleeve are measured. Two torches, located on either side of the work roll sleeve, begin heating the sample. The specimen is moved against the rotating sleeve while its surface is heated by two torches, one on either side of the work roll.

Once the sample is at the appropriate, consistent temperature, the data acquisition system is started, collecting data on the sample's surface temperature, the vertical load on the sample and the friction load. The rolls are lowered until they come into contact with the sample and are forced further down until the appropriate vertical load is reached. The vertical and the friction forces and the temperature are continuously monitored. The data - the forces, temperatures, the roll and sample speed - are collected using a digital data acquisition system. The work rolls are continuously cooled using water, during the tests.

The roll and sample are worn for 7.5 minutes, removed and weighed. On reassembly the sample and roll are placed back in the exact position that they were removed from to ensure that the contact geometry remains the same. The same sample and roll are used in the next test, under the same circumstances, but the test is run for 10 minutes. After the second weighing, the sample and roll are re-installed and run for a final 10 minutes. The reduced length of time for the first run aids in identifying any run-in characteristics that may surface.

RESULTS AND DISCUSSION

Roll wear - The dependence of the roll wear on the temperature is shown in Fig. 2 and the dependence of the roll wear rate on the normal load is shown in Fig. 3. The effects of roll speed are given in Fig. 4.

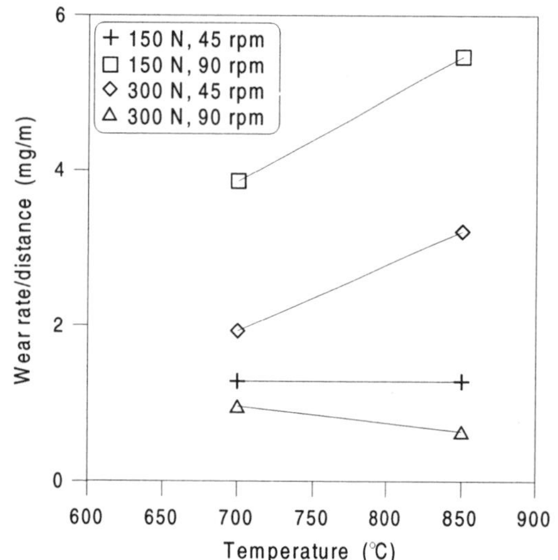

Figure 2: The effect of the temperature on the amount of wear at various speeds and temperatures

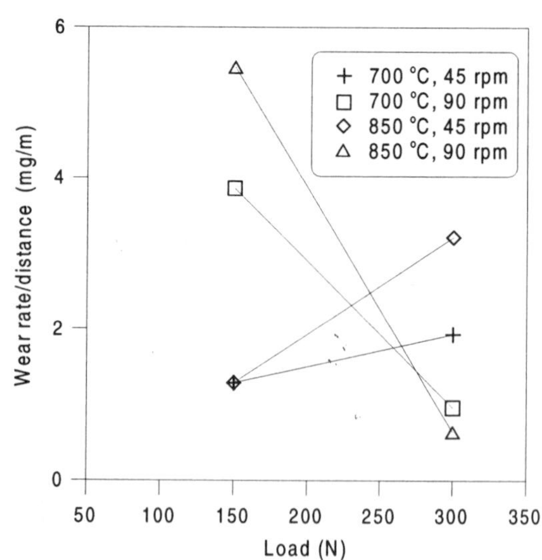

Figure 3: The effect of the normal load on the amount of wear at various speeds and normal loads

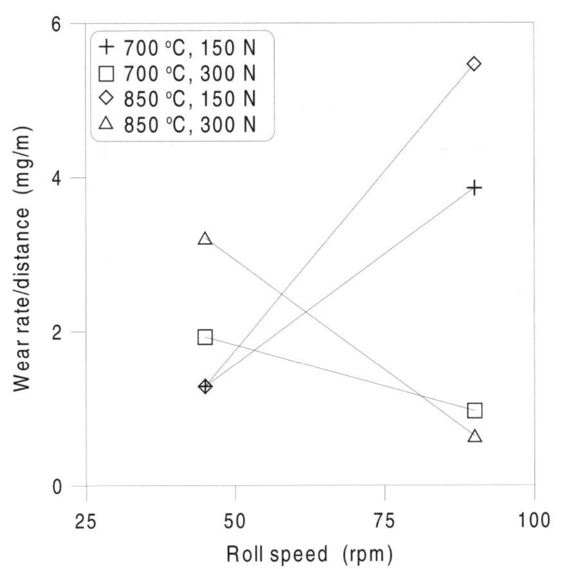

Figure 4: The dependence of the rate of wear on the roll speed (Note that 100 rpm corresponds to 0.21 m/s surface velocity).

The plots shown in Figures 2, 3 and 4, give the roll wear rates in mg/m of rolled distance versus the temperature, vertical load and the roll speed, respectively. Note that the roll wear is in terms of mg/m of sliding distance. This is done to remove the effect of the increased sliding distance at the faster speeds.

From Figure 2 it is observed that in the test performed at 150 N load and 90 rpm roll speed, the wear on the roll increased for an increase in temperature. While there was a slight increase in wear for the sample at 300 N, 45 rpm, this was not as significant as in the case of the previous sample.

The two other process conditions showed little and virtually constant amount of wear, essentially irrespective of the sample's surface temperature.

The results in Figure 3 indicate that at higher relative velocities increasing loads reduce the amount of metal loss of the work roll. At lower speeds the amount of wear increases. The same conclusions may be drawn when the data of Figure 4 are observed.

The small changes in the amount of wear are most likely due to the narrow range in temperature used for these experiments. The flow stress of the sample material only drops by less than 20% for an increase in temperature from 700 to 850°C. Thus, the contact zone changes very little with respect to geometry and interfacial interaction as the temperature is increased. As a result, it is observed that the temperature has very little effect on the wear of the rolls for these tests, agreeing with the conclusions of the recent overview of Spuzic et al. (11).

These phenomena can be partially explained using lubrication theory. Although all of the experiments were performed without any lubricants or additives, it is possible for the oxide layer on the sample to behave like a lubricant. Schey explains: "Solid lubricants…separate the two surfaces with a layer of low shear strength. They lubricate even when sliding speeds are low or temperatures are high" (4). If the low shear strength scale formed on the surface of the sample were to remain in the contact zone during the rolling process, it is quite possible for it to lubricate the surfaces. This becomes prevalent at the high loads, as the high vertical force, the tensile and shear stresses break up the thin layer of scale, providing the lubricating action. As well, the increasing lubricating effect of the scale with the increase of the relative velocity of the contacting surfaces parallels the observations made when lubricated flat rolling experiments at low temperatures are conducted. In those, higher speeds invariably lead to lower coefficients of friction.

The Stribeck Curve - The idea of an increase in load decreasing the coefficient of friction is not new. This concept is prevalent in journal bearing analysis to which rolling has a similar contact geometry. The Stribeck curve, originally developed to represent the behaviour of lubricated journal bearings, may be used to study the active mechanisms. The curve demonstrates three distinct regimes of contact lubrication.

When no lubricant is introduced, dry rolling occurs, and the coefficient of friction is high. Region I, following dry rolling, is the regime of boundary lubrication. Here there is definite metal-to-metal contact of the two surfaces, thus an increase in load on the contact will cause an increase in the coefficient of friction. This is also the behaviour demonstrated by dry contacts. Region II represents mixed lubrication where there is still some metal to metal contact, but the exact nature of the interfacial interaction is highly dependent on the contact parameters: the viscosity, relative speed and the pressure (η, ω and p, respectively). The third regime indicates hydrodynamic lubrication. Here, the two surfaces are fully separated by the lubricating agent and there is no metal-to-metal contact. In this area, any increase in the load will cause a decrease in the coefficient of friction. In general, when interfacial friction is high, so is the rate of wear, as indicated by the empirical relation of Roberts [(12)], formulated for the conditions of hot rolling of steels.

Calculating the Sommerfeld number - The Sommerfeld number is defined as

$$S = \frac{\eta\omega}{p}$$

but η, the lubricant viscosity, is very difficult to quantify in the present context. A solid substance does not have a real viscosity in its normal state, so an "effective viscosity" for the solid must be estimated. This is done using the fundamental definition of viscosity, for Newtonian fluids.

The shear stress of the surface scale can be estimated as half of its yield strength, and the velocity gradient is found by relating the thickness of the surface scale to the relative velocity between roll and sample. The velocity gradient is estimated using the relative velocity between the rolls and the moving sample. The height of the gradient is calculated by approximating the distance between the roll and the sample as the thickness of the scale layer. This is difficult to measure physically but Munther [(13)] provides graphical solutions of oxide growth kinetics at elevated temperatures that are useful for this approximation. The yield strength of scale is estimated using the data of Tiley, Zhang and Lenard [(14)], where high temperature compression tests were performed on cylindrical samples of scale, formed on the surface of low carbon steels.

The last component needed is the average contact pressure. This is a difficult number to quantify, since the contact is dynamic in nature, and involves both elastic and plastic deformation. Williams [(15)] provides a solution for a nominal line contact, where the surface is elastic/plastic. In this equation, P is the hydrostatic pressure, Y and E are the yield strength and Young's modulus of the compressed material, respectively, ϕ is the value of the angle between the local tangent and the surface at the edge of contact and D is a small numerical term which accounts for the variation of the surface angle during the deforming process.

$$\frac{P}{Y} \approx \frac{1}{\sqrt{3}}\left\{1 + \ln\left(\frac{4E\tan\phi}{3\pi Y}\right)\right\} + D$$

The yield strength of the deforming material is found using the equations by Shida[(16)], which account for the material's softening at elevated temperatures. The angle between the edge of contact and the local tangent is dependent on the vertical load and is estimated to be less than 10 degrees. Since the D value is considered small in comparison to the other terms, it is neglected for the purposes of this estimation.

Figure 5 is a plot of the calculated Sommerfeld number for all of the tests versus the amount of roll wear per meter of sliding distance. Since the temperature was found to have little effect on the amount of roll wear, the numbers shown are the average wear rates for the same tests performed at the two temperatures.

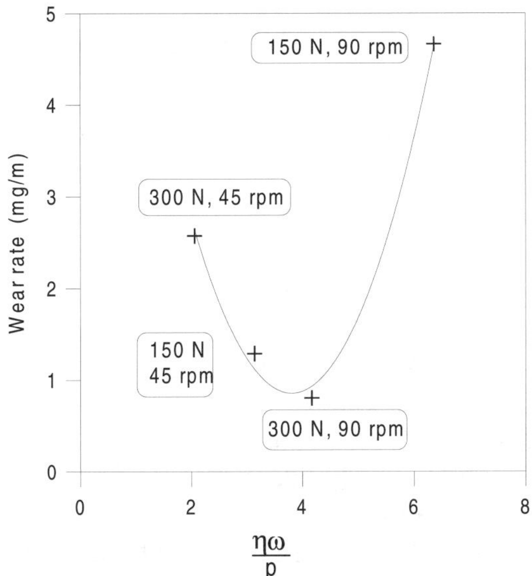

Figure 5: The dependence of the rate of wear on the Sommerfeld number.

The Sommerfeld number vs the roll wear rate shows a parabolic relation. The far right data point, which experienced the most amount of wear is the most curious. The first three points follow the Stribeck analogy quite well, where the first point is near the Region I/Region II boundary and the next two points are well into Region II and are approaching hydrodynamic conditions.

One would then expect that the last point would follow with a slightly increased coefficient of friction and the same or slightly increased amount of wear. However, the fourth point showed the greatest amount of wear for all tests. This is an indication that complete surface separation did not occur here, and that there are other factors that contribute to the surface wear. It is hypothesized that the nature of the wear mechanism changed from adhesive to abrasive wear.

The basis for the lubrication analogy was the presence of a coherent scale layer that can separate the two contacting bodies. However, if under certain circumstances the scale layer breaks up, and is ejected from the contact area, there is no basis for any kind of lubrication to occur. In fact, the break up of the scale will only accelerate the wear, as the particles become abrasives that cut and dig at the two contacting bodies, as in 3-body abrasion. If the broken pieces of scale are not ejected, they may aid in the lubrication process. Full hydrodynamic lubrication would only occur if the oxide layer remains intact.

CONCLUSIONS

While there are many types of wear, abrasion and adhesion dominate the causes of material removal during hot rolling of steels. Using a wear simulator, designed to duplicate conditions as close to industrial situations as possible in the laboratory, the loss of material from tool steel rolls was monitored as a function of the speed, load and the temperature. While the temperatures and the interfacial pressures were close to those under industrial conditions, the relative velocities were substantially lower. The amount of wear varied as a function of the process parameters.

Increasing relative velocity and increasing interfacial normal loads resulted in reduced wear. Changes in the surface temperature of the rolled sample did not create significant variations of the amount of material loss.

A wear map, similar to the Stribeck curve, used to study lubrication regimes, was plotted, after assuming that the rolled metal's apparent viscosity may be taken to equal its shear strength. The plot shows that the rate of wear depends on the relative speed and the normal load. When the junction is subjected to low loads, increasing the speed caused a significant increase in the rate of wear. At high loads the opposite observation was noted, indicating that the wear mode changed from adhesive to abrasive, responding to the altered process conditions.

ACKNOWLEDGMENTS

The financial assistance of the Natural Sciences and Engineering Research Council of Canada is gratefully acknowledged.

REFERENCES

1. N.P. Suh, "Wear Mechanisms: An Assessment of the State of Knowledge", in Fundamentals of Tribology, N.P. Suh and N. Saka, eds., MIT press, Cambridge Massachusetts, 1978.
2. E. Rabinowicz, Friction and Wear of Materials, 2nd ed, John Wiley and Sons, Toronto, 1995.
3. S. Jahanmir, "On the Wear Mechanisms and the Wear Equations", in Fundamentals of Tribology, N.P. Suh and N. Saka, eds., MIT press, Cambridge Massachusetts, 1978.
4. J.A. Schey, Tribology in Metalworking, American Society for Metals, Metals Park, Ohio, 1983.
5. A.W. Batchelor and G.W. Stachowiak, "Tribology in Materials Processing", J. Mat. Proc. Techn., Vol. 48, 1995, pp. 503-515.
6. H. Takigawa, T. Tanaka, S. Ohmoto and M. Hashimoto, "Development of High-Speed Tool Steel Rolls and Their Application to Hot Rolling Mills", Nippon Steel Technical Reports, No. 74,1997, pp. 77-83.
7. R. Webber, "The Performance of High Speed Steel Rolls at Dofasco", Proc. 37th MWSP Conference, 1995, Hamilton, pp. 267-269.
8. M. Hashimoto, H. Takigawa and T. Kawakami, "Development and Application of High-Speed Tool Steel (HSS) Rolls in Hot Strip Rolling", Proc. 37th MWSP Conference, 1995, Hamilton, pp. 275-282.
9. S-E Lundberg, Evaluation of Deterioration Mechanisms and Roll Life of Different Roll Materials", steel research, 1993, pp. 597-603.
10. G. Savage, R. Boelen, A. Horti, H. Morikawa and Y. Tsujimoto, "Hot Wear Testing of Roll Alloys", Proc. 37th MWSP Conference, 1995, Hamilton, pp. 333-337.
11. S. Spuzic, K.N. Strafford, C. Subramanian and G. Savage, "Wear of Hot Rolling Mill Rolls: an Overview", Wear, Vol. 176, 1994, pp. 261-271.
12. W.L. Roberts, Cold Rolling of Steel, Marcel Dekker, Inc. New York, 1978.
13. P.A. Munther, "The Effect of Material and Process Parameters on the Frictional Conditions in Hot Flat Rolling of Steels", Ph.D. Thesis, University of Waterloo, 1997.
14. J. Tiley, J., Y. Zhang and J.G. Lenard, "Hot Compression Testing of Low Carbon Steel Industrial Reheat Furnace Scale", steel research, Vol. 70, 1999, pp. 437-440.
15. J.A. Williams, Engineering Tribology, Oxford University Press, New York, 1994.
16. S. Shida, "Empirical Formula of Flow Stress of Carbon Steels", Hitachi Ltd., Tokyo, Japan, Internal paper, 1974.

ROLL BITE DEFORMATION OF THE THIN SCALE LAYER ON A PLAIN CARBON STEEL DURING HOT ROLLING

John B. Tiley*, John G Lenard**, Yangchun Yu**
*Dofasco Inc.,1330 Burlington Street E.
Hamilton/Ontario L8N 3J5/ Canada
Tel.: 905-544-3761
E-mail: John_Tiley@dofasco.ca
** University of Waterloo
Waterloo, Ontario, Canada

Key Words: Oxide Scale, Deformation, Hot Rolling, Roll Bite

INTRODUCTION

The deformation of the scale layer, that is always present during hot rolling of steels, is the subject of many investigations [1-10]. Rolled in scale defects are always a concern for meeting demanding product applications. The literature explains that this occurs by suggesting that:

[1] the scale cracks and separates in the roll bite[1] allowing the substrate material to flow into the vacated rift, or
[2] the scale rucks or delaminates at roll bite entry[3], allowing the scale fragments to be pressed into the surface.

In order to avoid this defect, the scale layer must be thin enough and plastic enough to be elongated in the same proportion as the metal substrate. Arguably, the most significant parameters affecting the development of the rolled-in-scale problem, are the strength and ductility of the scale and the strength and ductility of its adhesion to the surface of the parent material.

In this investigation, a two-high laboratory hot rolling mill and heating furnace were used to control the test parameters of scale thickness and rolling temperatures to be similar to the entry conditions of the finishing stands of a hot strip mill. Using oxygen-free nitrogen both during the heating and the cooling processes closely controlled the development of the scale layer.

The objective was to examine the process conditions under which the layer of scale, remaining on the surface of the rolled steel, remains continuous or becomes cracked.

EXPERIMENTAL PROCEDURE

Material, Sample Preparation, Procedure and Equipment

Material and Sample Preparation. Three steel grades were investigated in the laboratory experiments. The samples were an AISI 1018 grade (12.7mm X 50.8mm X 304mm), an Ultra-low carbon grade (5.1mm X 50.8mm X 304mm); and a low carbon grade (4.7mm X38mm X304mm).
Their chemical compositions are shown in Table I.

Table I. The Steel Chemical Composition (by wt%)

Grade	C	Mn	P	S	Si	Cr	Mo	V	N
AISI 1018	0.18	0.71				0.21			
Low Carbon	0.14	0.76	0.009	.021	0.24	0.073	0.008		0.01
Ultra-low C	0.002	0.176	0.005	0.004	0.008			0.004	0.002

The AISI 1018 and low carbon steel samples had a cold machined surface finish. The ultra-low carbon steel samples had an as hot rolled mill finish; a thin, smooth scale layer was already present.

The goal of the experimental procedure was to obtain a thin monolayer of scale, similar to that after descaling in a hot strip mill operating environment.

Procedure. In each case, the samples were pre-heated (soaked) for 20 minutes in the laboratory furnace, whose set point was held above the required rolling temperature. Two sets of tests were done with the 1018 grade; purging the furnace with oxygen-free nitrogen and no purging. The reheating furnace was purged with nitrogen for all of the remaining samples. Each sample was extracted(with tongs) from the reheating furnace and rolled immediately (within 5-10 seconds). The samples were reduced over a range of 10% to 50% (nominally). The low carbon steel and ultra-low carbon steel samples were cooled in a nitrogen purged box, installed beside the rolling mill, until they reached room temperature.

Because the AISI 1018 grade experiments resulted in a very thick scale layer(after rolling and cooling), two of the pieces* (see Table **II**) were cut into 150 mm long sections, after the first reduction, with the as-produced scale layer. These were reheated to 1010°C in the nitrogen atmosphere and re-rolled with a total of 50% reduction, in order to roll in the scale, so that a comparison could be made with commercial defect samples. The reduction was taken in four successive passes, with the samples being reheated to above the rolling temperature in the protective atmosphere, between each pass.

Each sample was examined using an optical microscope in order to view the surface and check it for cracks (plane view). Selected samples were also cross-sectioned and mounted in epoxy in order to metallographically measure the scale thickness. These samples were compared to samples of material hot rolled at Dofasco Inc, with and without rolled in scale defects.

Equipment. The rolling mills used in the experiments were 2-high STANAT mills, with 150 mm diameter work rolls, both at Dofasco Inc. (for the re-rolling simulations), and at the University of Waterloo(for the main experiments). The Waterloo mill is highly instrumented and has tool steel rolls with a roll surface hardness of Rc = 54 and the roll surface roughness of 1 μm. It is powered by a constant torque DC motor of 11 kW power. The rolling speed is continuously adjustable, up to 1.4 m/s. The Dofasco mill is similar with no instrumentation and it has cast steel rolls.

EXPERIMENTAL RESULTS

AISI 1018 Grade Steel

The AISI 1018 steel samples were processed under conditions shown in Table II. In all tests the rate of oxidation after removal from the reheating furnace was excessive. The samples blistered before rolling. Each sample was stopped in the roll bite in order to capture this region for metallography. After rolling all showed a heavy oxide finish, which is more typical of a thick slab rather than a finished hot band coil.

The micrographs (Example: Figure # 1) revealed that 2 or 3 layers of scale existed on the surface, which are typical of blistering and previously published experiments [6,7,8]. Since the oxidation rate of the chosen grade was excessive, the procedure was abandoned. The scale layer was too thick to represent finishing mill hot rolling scale thickness conditions.

Table II. Experiments Group 1 and 2 with 12.7mm thick samples (AISI 1018 grade)

Experiment Group#1 No purge	Soak Temp. (ºC)	Reduction (%)	Scaling Time: seconds before rolling	Observations: after rolling
1	1000	32	5 seconds	Thick scale
2 and 10	1150	32	5	Thick scale
3	1117	32	10	Thick scale
4	1070	32	5	Thick scale
5	1045	32	10	Thick scale
6	1057	58	5	Air cool
*7(reheated)	1058	58	10	Thick scale
8	1119	58	5	Double scale?
*9(reheated))	1090	58	10	Large tiger stripe

Experiment Group#2 nitrogen purge	Soak Temp. (ºC)	Reduction %	Scaling Time: seconds before rolling	Comments after rolling
1	1055	50	5 seconds	Thick scale
2	1035	50	12	Thick scale
3	1080	50	5	Thick scale
4	1000	50	15	Thick scale
5	1080	50	5	Thick scale

The thickness and the make-up of the scale, shortly after entry into the roll gap, are observable in Figure 1, below. The location of the magnified region is also shown.

Stopping the rolling process while the strip was still being rolled produced the curved shape, visible in the figure. As a result, some thinning of the scale layer is also noticed. As well, several layers of the scale are present.

Figure 1. Entry into the Roll bite example; Scale thickness ~ 100 microns

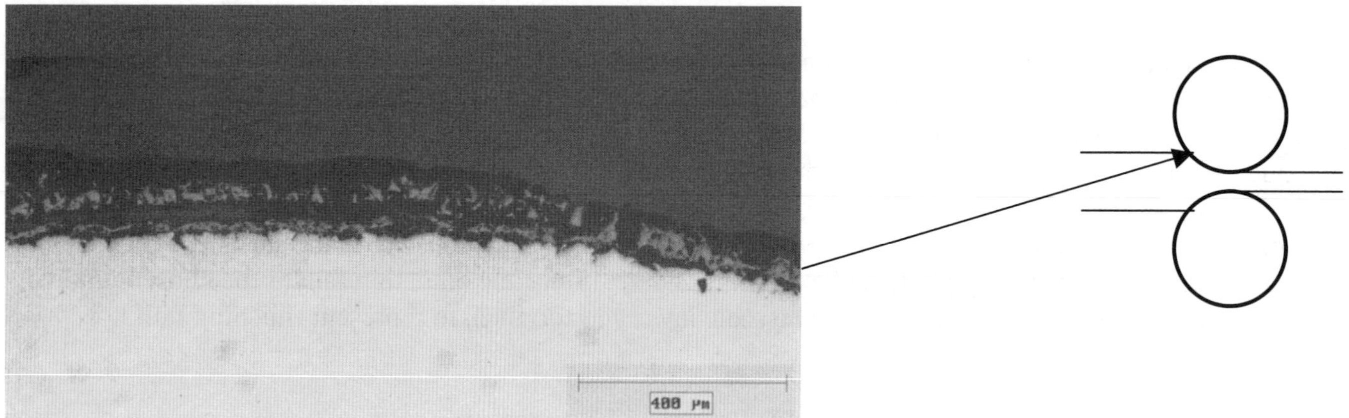

Low Carbon Steel

Rolling the thinner samples of lower carbon steel, with the controlled furnace atmosphere and oxygen-free nitrogen cooling, and inspecting the surfaces and the cross-sections of the scale layers proved more fruitful. The low carbon steel samples were processed under conditions shown in Table III. In order to evaluate the effect of the rolling pass on the thickness of the scale layer, selected samples were heated and cooled, following a procedure similar to those that were rolled. These samples showed the same scale thickness, less the reduction, as after rolling. After rolling, each was inspected for cracks at a magnification of 50X – see Figure # 2, showing sample yu29, heated to 1019 °C and rolled at 877 °C entry temperature, at a roll speed of 70 mm/s, to a reduction of 38.7%. The sample's surface appears smooth with no visible cracks. This may be compared to the surface of sample yu4, heated and rolled at somewhat lower temperatures, at a much higher speed and to a significantly lower reduction. The resulting surface shows cracks of 40 – 60 µm thickness. Cross-sections allowed the measurement of the scale thickness, shown in Figure # 3. The first sample, yu26, rolled at 874 °C, at 169.8 m/s and reduced by 31.6%, indicates fairly thin scales, of 9.23 µm average thickness. The second sample, yu28, rolled at a higher speed and temperature, produced a much thicker scale, of 52.08 µm average thickness.

Table III. Experiment Group 3 with 4.7 mm thick samples (low carbon) **[c]racked**; **[s]mooth**

Sample No.	Temperature (°C)		Rolling speed	Reduction	Scale Thickness
	Furnace	Rolling	(mm/s)	(%)	(µm)
yu4* [c]	976	842	728.4	10.3	63.89
yu14 [s]	978	848	160.6	39.6	17.39
yu25 [s]	1019	868	73.4	31.0	8.04
yu26** [s]	1019	874	169.8	31.6	9.23
yu28*** [c]	1019	910	612.5	31.6	52.08
yu29****[s]	1019	877	70.0	38.7	10.75
yu31 [s]	1019	852	294.9	36.6	13.75
yu38 [c]	1080	930	182.8	20.8	18.59

Figure 2. Plane view micrographs of the surface, which was the technique used to inspect for cracks.

**** Sample yu29: X50 - smooth surface * Sample yu4: X50 - cracked surface

The smooth surface texture of the un-cracked sample (on the left) is similar to the sand blasted roll texture of the tool steel rolls in the STANAT mill.

Figure 3. Micrograph examples of cross-sections taken to measure the scale thickness.

** Sample yu26: X1000. *** Sample yu28: X1000

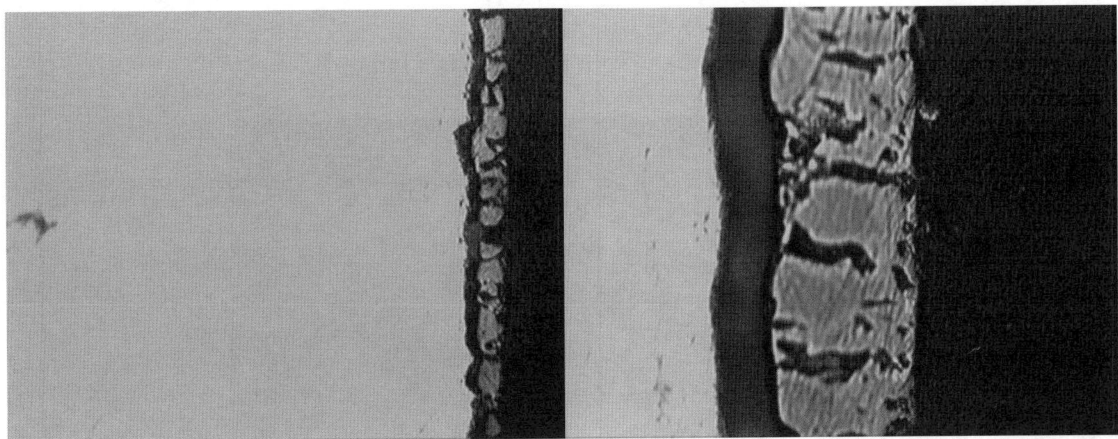

The scale (dark) is on the right hand side of each cross-section. The metal substrate (light in contrast) is on the left-hand side.

In these experiments the main process parameters that were varies were extraction and rolling temperature, to obtain a range of scale thickness; rolling speed (varied by a factor of 10) to vary the strain rate; and reduction from 10%-40% to vary the roll bite deformation pressure. The experiments were designed to explore conditions that might cause cracking, in the roll bite, of a thin (mono) scale layer.

Dofasco Ultra-low Carbon Grade As-Rolled Hot Band

The ultra-low carbon steel samples were processed under conditions shown in Table IV. Each was inspected for cracks at a magnification of 50X, as shown in Figure # 4. Cross-sections were made for the measurement of the scale thickness.

Table IV. Experiment Group 4 with 5.03 mm thick samples (ultra-low carbon grade) **[c]racked; [s]mooth**

Sample No.		Processing		Rolling speed	Reduction	Scale thickness: after rolling
		Furnace (°C)	Scaling time	(mm/s)	(%)	(μm)
jt1	**[c]**	1035	10 s	380	8.5	30
jt2	**[c]**	1050	15 s	‘’	8.5	30
jt3	**[c]**	1055	20 s	‘’	8.5	40
jt4	**[c]**	1055	5 s	‘’	8.5	30
jt5*	**[c]**	1062	25 s	‘’	8.5	50 (blistered)
jt6**	**[s]**	n/a	n/a	n/a	0.0	33

The technique preserved the initial scale thickness (33μ) until it started to blister, however, all scales cracked during the rolling experiments. Note that sample #jt6 was from the Dofasco as rolled ultra-low carbon hot band used for the experiments.

Figure 4. Plane view micrographs of the surface, which was the technique used to inspect for cracks.

Sample jt5*: X50 Sample jt6** from as rolled hot band: X50

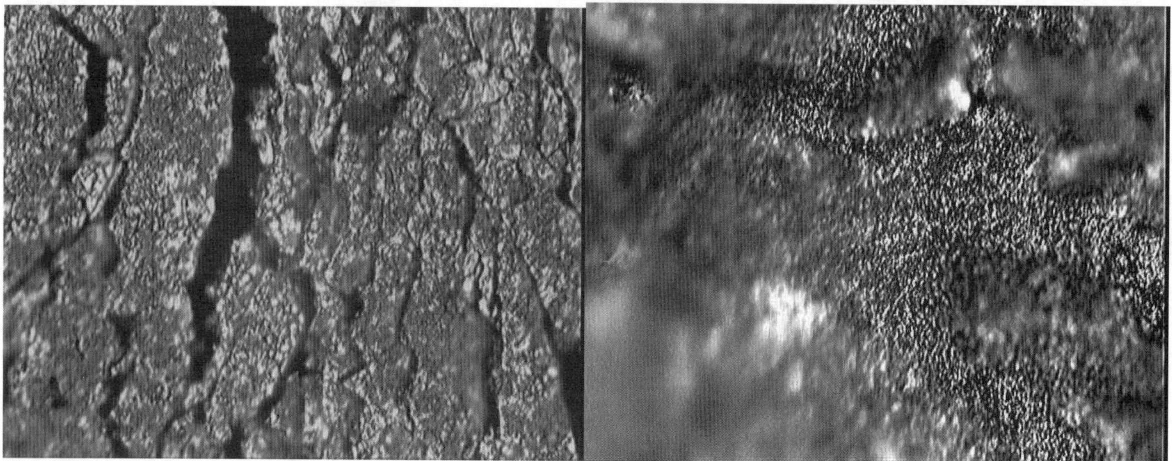

The smooth surface texture of the as rolled sample (on the right) is similar to the roll texture of the last rolling stand in the Dofasco hot mill.

DISCUSSION

The results of these experiments can be compared to defects of the 'salt and pepper' or 'wave' scale type, found in Dofasco's hot band products. Cracking is associated with the defective area, as shown in the examples below.

Figure # 5a is a typical Dofasco rolled in scale defect. Figure # 5b is the experimentally produced defect. This defect was created by further hot reducing the heavily cracked thick scale by 50%. As the cracked areas are spread further apart, during the subsequent rolling, the oxygen in the surrounding air penetrates and continues to oxidize the parent metal. The cracked scale particles can now be pressed into the surface on the next rolling pass. In a seven-stand hot mill finishing train (like Dofasco's) there will be six further reductions after the steel passes through the entry stand.

Figure 5a. Scale defects in commercial hot band: X50

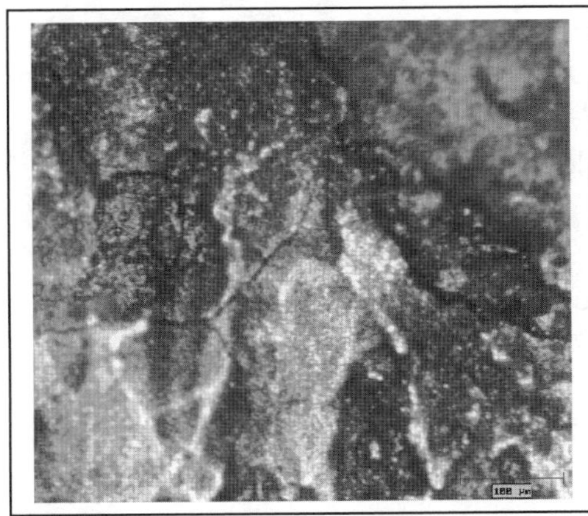

Figure 5b. Experimentally rolled in defect: X50

From the experimental data, it is observed that a relationship exists in the extraction temperature and scale thickness domain, which causes a cracked or smooth surface, when the sample is hot rolled. The values are plotted from the data in Tables III and IV, and shown in Figure # 6.

Figure 6. Scale Thickness in μm versus Furnace Temperature ^{o}C

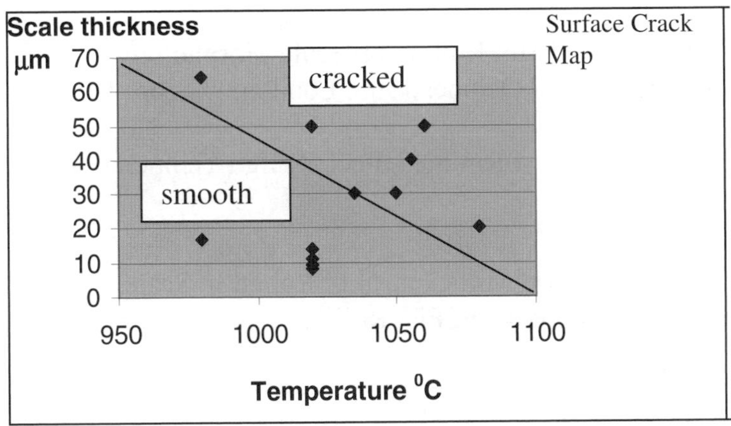

The samples were cracked above the diagonal line in Figure # 6 and not cracked below it. Oxidation is rapid and can blister the samples at high furnace extraction temperature. It is important to maintain a thin scale layer, if cracking is to be avoided. A mono-layer less than 20 microns thick did not crack. As in previous work [5,6] the surface is rough and cracked when the residual scale layer is thicker than ~ 20 microns.

Other relationships such as the tendency to crack versus rolling speed or reduction % are not evident within the range of these experiments. These may be worth investigating in future. Further work is required to relate the conditions of scale thickness and its tendency to crack, to the roll bite friction coefficient; particularly with and without the lubricating emulsions that are becoming more commonly used in industrial hot rolling mills.

ACKNOWLEDGEMENTS

The authors express their gratitude for the support of Dofasco Inc. for providing samples of steel for the experiments. The financial assistance of the Natural Sciences and Engineering Research Council of Canada and NATO is gratefully acknowledged.

REFERENCES

1. Blazevic, David T., Rolled in Scale - The Consistent Problem, 4th International Steel Rolling Conference, The Science and Technology of Flat Rolling Deauville - France ,June 1-3, 1987 pp. A.38.1-A.38.13.

2. Blazevic, David T., Rough Surface: The Major Problem on Thin Cast Hot Strip Mills, presented at the A.I.S.E. Spring Convention, Salt Lake City , May 8,9,10, 1995.

3. Luyckx, L. and Lorang, F., Formation of Fleck Scale During Hot Rolling of Wide Strip, C.N.R.M. internal report, Jan.1965, TRANSLATED from the German by B.I.S.I. Translation Service.

4. Seki H., Prevention of Scale Defects in Hot Strip Rolling Using High Speed Steel Rolls CAMP-ISIJ, Vol.9, 1996, p.972.

5. Li, Y.H. and Sellars, C.M., Modelling Deformation Behaviour of Oxide Scales and Their Effects on Interfacial Heat Transfer and friction During Hot Steel Rolling, 2nd Int. Conf. on Modelling of Metal Rolling Processes, London, UK, 9-11 December, 1996, pp. 192-201.

6. Boelin, R., Curcio, P., Assefpour, M., Laboratory Simulation of Hot Rolling on Scale Morphology, Hydraulic Descaling in Rolling Mills, Institute of Materials, 9-10 October 1995, London, UK. pp. 1-9.

7. Matsumo, F., Blistering and Hydraulic Removal of Scale Films of Rimmed Steel at High Temperature Transactions. I.S.I.J, Vol.20, 198, pp. 413-421.

8. Grigg, C., R., Seaton, B., G., Matteson, L., E., 1420 mm Hot Strip Mill Control of Rolled In Scale, 4th International Steel Rolling Conference, The Science and Technology of Flat Rolling Deauville- France, June 1-3, 1987, pp. A.39.1-A.39.8.

9. Torres, M., and Colas, R., Growth and Breakage of the Oxide Layer During Hot Rolling of Low Carbon Steels, in '2nd International Conference on Modeling of Metal Rolling Processes', The Institute of Materials, London 1966, pp. 629-636.

10. Fletcher, J., D., and Beynon, J., H., Relating the Small Scale Thermo-TriboMechanics in Hot Strip Rolling to the Global Deformation Behaviour, 2nd Int. Conf. on Modelling of Metal Rolling Processes, London, UK, 9-11 December, 1996, pp. 202-212.

CORRELATION OF ROLLED-IN DEFECTS IN IF STEEL AND THE THERMAL FATIGUE PROPERTIES OF THE HOT STRIP MILL WORK ROLLS

Jae-Hwa Ryu and Han-Bin Ryu
Technical Research Labs., Pohang Iron & Steel Company, Ltd.
Pohang PO Box 36, 1 Koedong-dong, Nam-ku,
Pohang-shi, Kyungbuk, 790-785, Korea
Tel.: 82-54-220-6151
e-mail: jaeryu@posco.co.kr

Key Words: Rolled-in Defect, IF Steel, Hot Strip Work Roll, Thermal Fatigue

INTRODUCTION

The production of ultra-low carbon steel (IF steel) has greatly increased owing to high demands for good formability steels. IF steel strips require severe surface quality as well as excellent formability because most of them are being used for the outer panels of automobile and home appliance. However, the tiny rolled-in defects interrupted further increase in production of IF steel.

Therefore, in this work, the cause of surface defects was investigated by tracking the original location of the defects and analyzing correlation between the defects and surface peeling of hot strip mill work rolls. Trials for preventing the defects were conducted by selecting the roll material with high thermal fatigue resistance through thermal fatigue testing, and then applying it to the finishing stand in trouble.

EXPERIEMNTS

The shape of the rolled-in defect and its chemical composition were examined in the cold-rolled IF steel by using EPMA(Electron Probe Micro Analyzer).

The original site causing the defects was tracked by measuring the elongated ratio (length/width) of the defects in the cold-rolled strips, and then matching it with the reduction schedule from hot-rolling to cold-rolling. From these investigations, it was found that the surface defects originated from surface peeling of the fourth finishing stand work rolls (F4 rolls) in hot strip mill.

In order to identify effect of hot-rolling conditions on surface peeling of F4 rolls and subsequent surface defects of the strips, their grades were ranked on the base of the peeled area of F4 rolls and length of surface defects as listed in Table I and Table II, respectively.

Table I. The grades of surface peeling in F4 rolls.

Grade	Appearance of roll surface
1	Beginning of surface peeling
2	Peeled area/total working zone < 10%
3	10% ≤ peeled area/total working zone <40%
4	40% ≤ peeled area/total working zone

Table II. The grades of surface defects in cold-rolled IF steels.

Grade	Appearance of strip surface
0	No defect
1	Defect length < 2mm
2	2mm ≤ defect length <5mm
3	5mm ≤ defect length

The roll surface peeling is known to be caused by the thermal fatigue damage because hot strip mill work rolls are being operated under the repeated thermal cycling.[1~3] The amount of surface defects considerably decreased through several kinds of trials reducing thermal fatigue damage of F4 rolls such as intermediate roll change and increase in roll coolant and lubricant, but it was difficult to prevent the surface defects completely due to a higher hot-rolling temperature. Therefore, the roll material with high thermal resistance was desired.

Thermal fatigue testing was conducted for two roll materials of an existing nickel-grain iron roll and a newly developed high speed steel roll (Table III) by using thermal fatigue tester [4].

Table III. Chemical compositions of roll materials used for thermal fatigue testing.

Roll material	Chemical composition (wt%)							
	C	Si	Mn	Cr	Ni	Mo	V	W
Nickel-grain iron	2.8/3.5	0.5/1.5	0.5/1.0	1/3	3/5	0.1/0.5	-	-
High speed steel	0.5/1.5	0.5/1.6	0.5/1.5	4/7	1/4	1/4	3/7	0/9

The maximum temperature was varied in the ranges of 400℃ to 600℃ with a fixed minimum temperature of 200℃. The change in the gauge length with thermal cycling was detected by an extensometer, and the load was controlled accordingly to keep the gauge section at the complete constraint state because the thin surface layer of the work roll which experiences the thermal fatigue is in a complete constraint by the roll interior. [3,4] The specimen in the fatigue test was put under repeated compressive and tensile stresses until it eventually failed, and then the number of cycles to failure was counted. In order to interpret the thermal fatigue test results, the microstructures and the crack initiation sites were examined by sectioning the fractured specimens longitudinally, and the tensile strength and thermal expansion coefficient for two roll materials were also measured.

RESULTS & DISCUSSIONS

Origin of surface defects

Fig.1 shows the shape of surface defect in the cold-rolled IF steel and its elemental distribution. The defect is placed on the surface like a rolled-in scab and it contains nickel, silicon, and chromium which are not added in the IF steel at all. It indicated that the defect was not rolled-in scale defect often observed in cold-rolled IF steels but it came from outside source. Therefore, the work roll materials directly contacting with strip were reviewed from hot-rolling to cold-rolling, and then it was found that nickel-grain iron roll which were being used for the work rolls in the rear finishing stands of hot strip mills contained the alloying elements close to those of the defect (Table III). In POSCO's No.2 hot strip mill in trouble, high speed steel rolls and nickel-grain iron rolls were being used in the front (F1~F3 stands) and rear (F4~F7 stands) finishing stands, respectively.

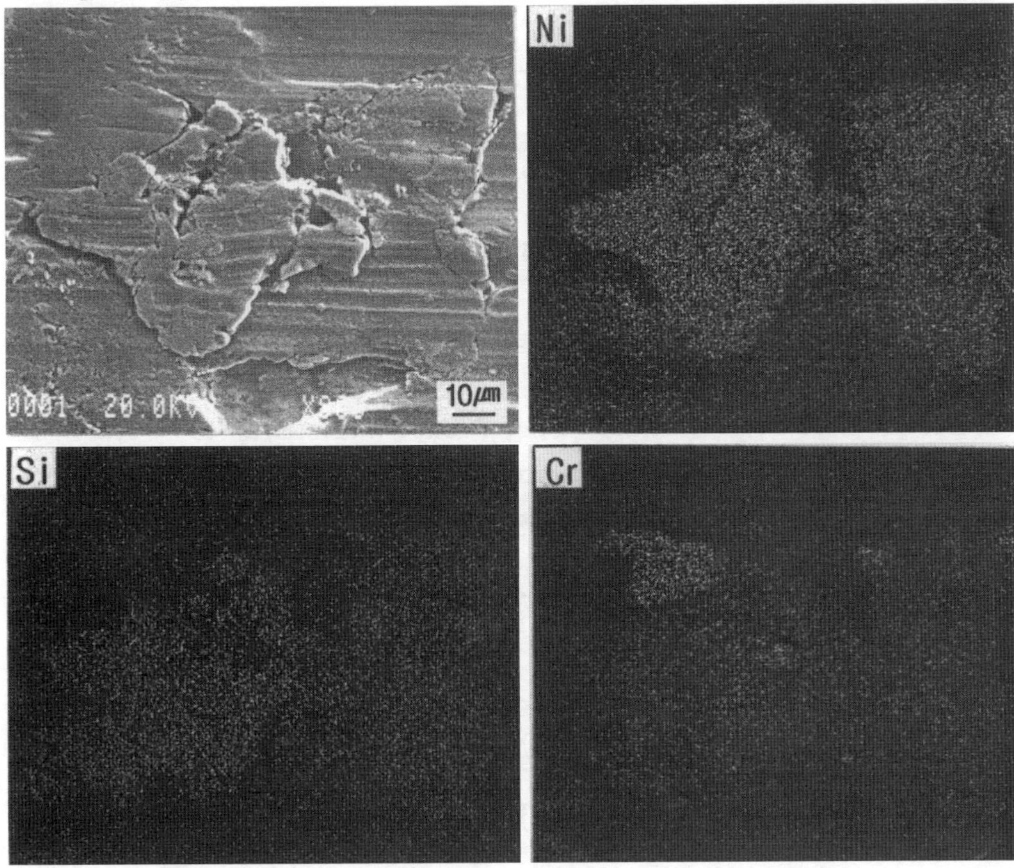

Fig.1: The surface defect and its elemental mapping in the cold-rolled IF steel.

The elongated ratio of the defects in the cold-rolled strips with 0.75mm thickness was measured to be 8.2~11.6. When it was matched with hot-rolling and cold-rolling reduction schedule, it was corresponded to No.3 and No.4 finishing stands in hot strip mill. Moreover, roll surface of F4 rolls was observed to be severely peeled when IF steels were rolled. Therefore, it was clear that the surface defects were caused by the surface peeling of F4 rolls.

Correlation of hot-rolling condition with roll surface peeling and strip defects

Fig.2 shows relation between hot-rolling sequence and the grades of surface defects in cold-rolled IF steel strips. The amount of defective strips and their grades greatly increased in the latter part of a rolling campaign, which is thought to be attributed to easy peeling of roll surface due to accumulated thermal fatigue damage.

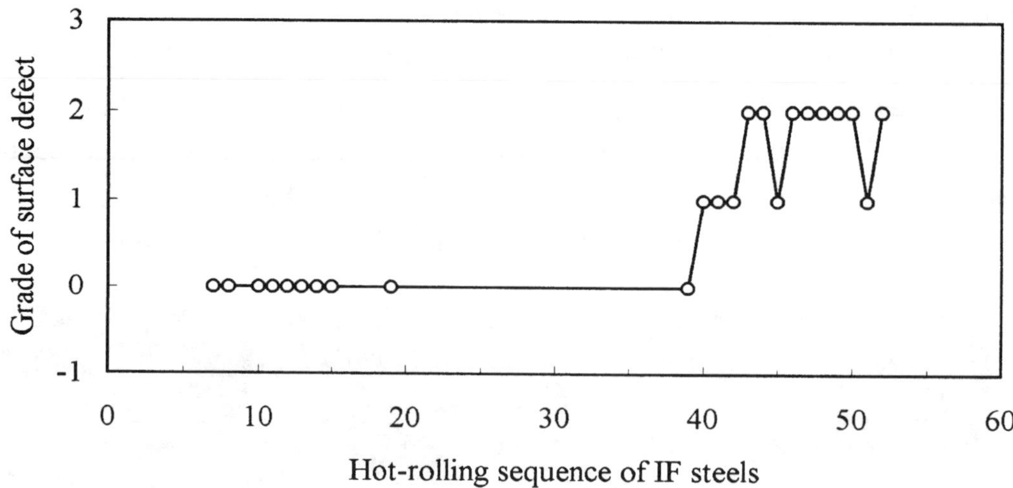

Fig.2: Relation between hot-rolling sequence of IF steels and grade of surface defects.

Fig.3 shows relation between grade of surface peeling of F4 rolls and the numbers of hot-rolled IF steel strips in a rolling campaign. The grades of surface peeling increased as the numbers of strips increased. In order to make this effect more clearly, the hot rolling practice was interrupted by every 10 strips of IF steel, and then roll surface was observed. It was confirmed that F4 roll surface had started to peel after rolling 10 strips of IF steel and had been peeled almost completely after 40 strips.

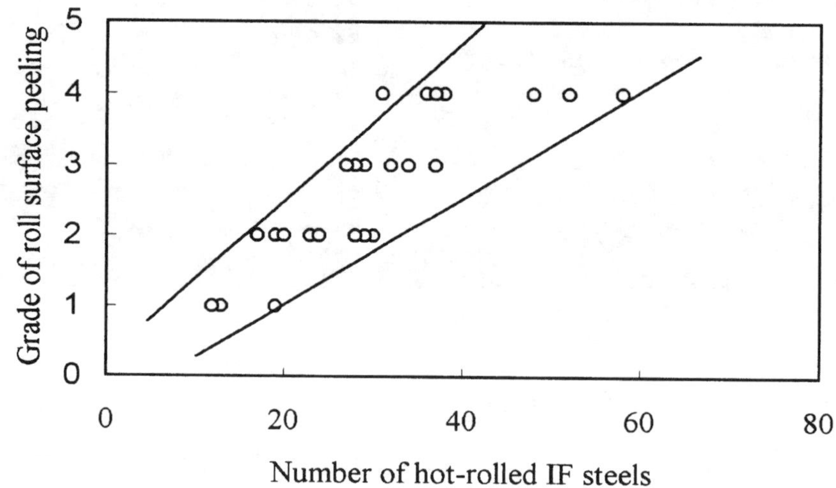

Fig.3: Relation between roll surface peeling of F4 rolls and numbers of hot-rolled IF steels in a campaign.

Fig.4 shows relation between the grade of roll surface peeling and the amount and grade of surface defects. When the grade of roll surface peeling was 2, the grade of strip defects is only 1 and the percent of defective strips is 12%. However, the grade of strip defects got worse to 2 and 3, the percent of defective strips considerably increased to 45% when the grade of roll surface peeling was 4.

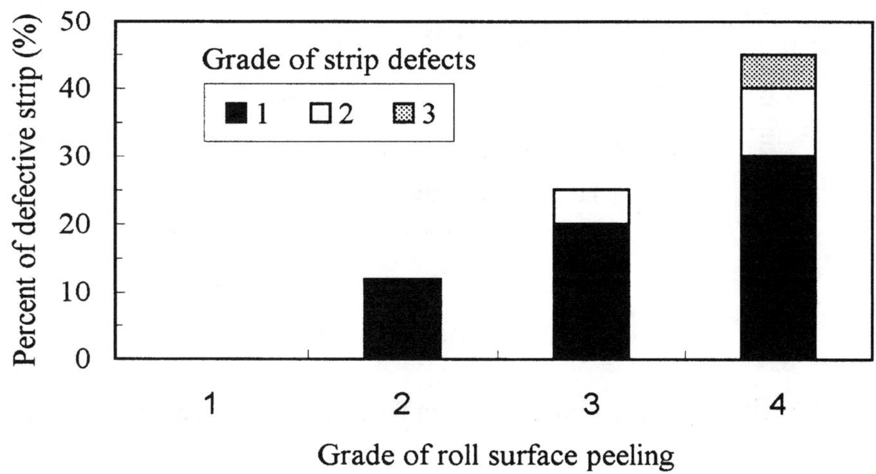

Fig.4: Relation between the grade of roll surface peeling and the amount and grade of surface defects.

Trials for preventing the surface defects

In order to produce the defect-free IF steel strips, it is essential to prevent the surface peeling of F4 rolls which is due to an increased thermal fatigue damage because IF steels are being hot-rolled at a higher temperature. The amount of defective strips was reduced from 45% to 35% by increasing roll coolant and lubrication of 20% that suppress the surface peeling. Further reduction was obtained by replacing F4 rolls with newly ground rolls in the middle of a rolling campaign. However, it seemed to be difficult prevent the defects completely by using nickel-grain iron roll. Therefore, the roll material with a higher thermal fatigue resistance was requested in the F4 stand.

Thermal fatigue properties of roll materials

High speed steel roll which was reported to have excellent thermal resistance [5)] was chosen to compare the thermal fatigue property with an existing nickel-grain iron roll. Fig.5 shows the results of the thermal fatigue tests obtained by varying the maximum temperature with a minimum temperature set at 200℃ under total strain of 0. At a given maximum temperature, high speed steel roll shows much longer fatigue life than nickel-grain iron roll. The fatigue lives of both rolls are greatly shortened with increasing the maximum temperature due to an increased thermal stress. This result indicated that nickel-grain iron roll had lower thermal fatigue resistance, consequently, it would be easily peeled when the rolling temperature like IF steel was higher.

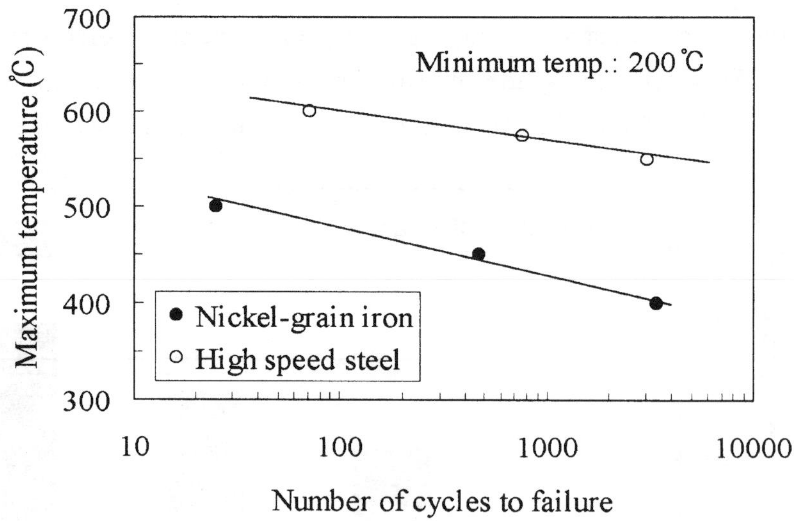

Fig.5: Variation of thermal fatigue life with maximum temperature for two roll materials.

Fig.6 shows the microstrucures of high speed steel and nickel-grain iron rolls, repectively. The microstructures of both rolls are composed of a tempered martensitic matrix (grey region) and the carbides (white region) except nickel-grain roll has a few percent of graphites (black region). High speed steel roll contains less volume fraction of carbides than nickel-grain iron roll because of the lower carbon content (Table III), but it has finer carbides. [4)]

Fig.6: Optical microstructures of high speed steel and nickel-grain iron rolls.

Fig.7 shows the sectioned surface of thermal fatigued specimens. The coarse carbides existing on the specimen surface acted as the preferred crack initiation sites because they have low ductility as well as a large difference in thermal expansion coefficient with that of the matrix.

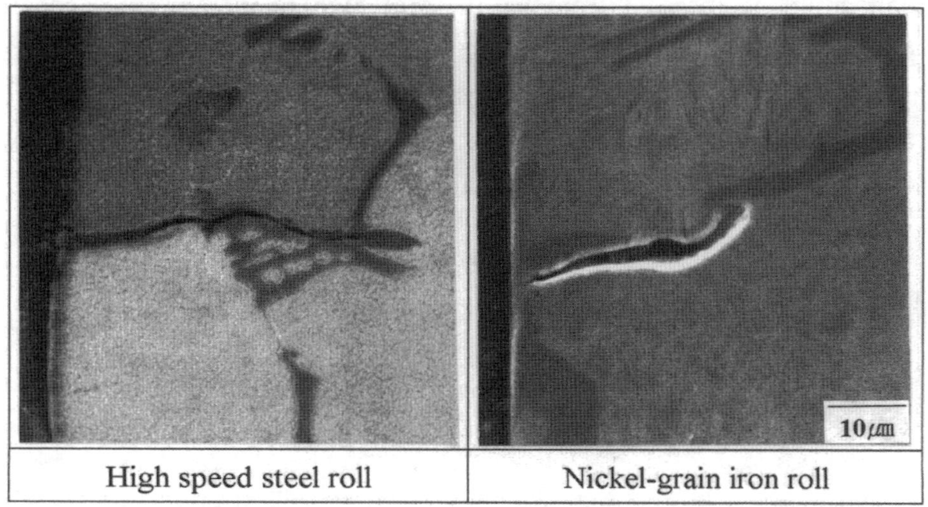

Fig.7: Surface cracks on the surfaces of thermal fatigued specimens.

Fig.8 shows the density of fatigue cracks measured on the surface of thermal fatigued specimens. High speed steel roll has lower crack density than nickel-grain iron roll because of low volume fraction of carbides. The density of cracks increases with increasing the maximum temperature because a higher tensile stress would be applied to the specimen as the temperature rises.

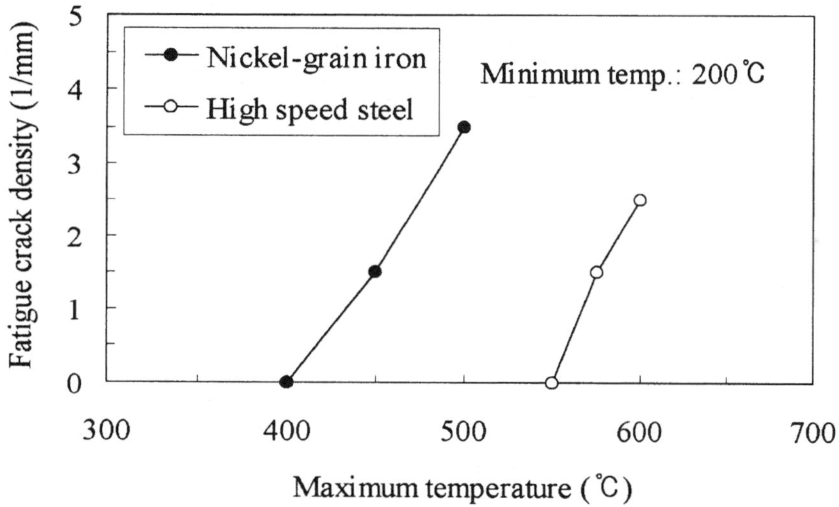

Fig.8: Surface crack density in the thermal fatigued specimens.

Thermal fatigue properties of roll materials were determined by collective effect of their microstructures, mechanical and physical properties. A collective understanding of those factors is necessary because they affect the thermal fatigue property in an inter-related and complicated manner. First, when consideration is taken into microstructure, high speed steel roll has lower volume fraction of coarse carbides which acted as the initiation sites of fatigue cracks as shown in Fig.7. Second, mechanical properties to affect thermal fatigue property are tensile strength and compressive strength at high temperatures. High speed steel roll has higher tensile strength (Table IV) and compressive strength at high temperatures which could be inferred from hardness at high temperatures [6], accordingly, it has better thermal fatigue property. The lower tensile strength of nickel-grain iron roll is related with existence of flake-type graphites that act as notches under tensile load.[7] Third, physical properties to affect thermal fatigue property are thermal expansion coefficient, thermal conductivity and Young's modulus.

Their collective effects including mechanical and physical properties can be evaluated by using Eichelberg quality factor [8];

$$\text{Eichelberg quality factor} = \frac{(1-\nu)\cdot k\cdot \sigma_u}{E\cdot \alpha} \qquad (1)$$

Where, ν: Poisson's ratio, k : thermal conductivity, σ_u : tensile strength at an ambient temperature, E : Young's modulus, α : thermal expansion coefficient.

Thermal fatigue properties of rolls would be better as the value of Eichelberg quality factor becomes higher. The calculated values of Eichelberg quality factor for two rolls are listed in Table IV including their physical and mechanical properties. High speed steel roll has much higher Eichelberg quality factor than nickel-grain iron roll mainly due to higher tensile strength and higher thermal conductivity.

Table IV Mechanical and physical properties, and calculated Eichelberg quality factor of the roll materials.

Roll material	ν^*	k^* (W/m·k)	σ_u (kg/mm^2)	E^* (kg/mm^2)	α (10^{-6}/k)	Eichelberg quality factor
High speed steel	0.27	25.5	98	23500	14.0	5545
Nickel-grain iron	0.27	23.5	52	17500	14.5	3515

* : Reference 6

Therefore, high speed steel roll has much better thermal fatigue property than nickel-grain iron roll as shown in Fig.5.

On the base of thermal fatigue test results, it was decided to introduce high speed steel roll into F4 stand. As a result, there was no surface peeling of F4 rolls even after rolling a lots of IF steels, subsequently, complete prevention of the surface defects was achieved.

CONCLUSIONS

(1) The surface defects of cold-rolled IF steel strips originated from surface peeling of the fourth finishing work rolls in hot strip mill which was caused by the severe thermal fatigue damage due to a higher rolling temperature.

(2) Complete prevention was achieved by replacing nickel-grain iron rolls with high speed steel rolls that have superior thermal fatigue property.

REFERENCES

1. R.V.Williams and G.M.Boxall, "Roll Surface Deterioration in Hot Strip Mills," Journal of The Iron and Steel Institute, Vol. 203, April 1965, pp.369-377.

2. J.J.deBarbadillo and Trozzi, "Mechanism of Banding in Hot Strip Mill Work Rolls," Iron and Steel Engineer, Vol.58, No.1, January 1981, pp.63-72.

3. J.H.Ryu, O.Kwon, P.J.Lee, and Y.M.Kim, "Evaluation of the Finishing Roll Surface Deterioration at Hot Strip Mill," The Iron and Steel Institute of Japan International, Vol.32, No.11, 1992, pp.1221-1223.

4. S.Lee, D.H.Kim, J.H.Ryu, and K.Shin, "Correlation of Microstructure and Thermal Fatigue Property of Three Work Rolls," Metallurgical and Materials Transactions A, Vol.28A, December 1997, pp.2595-2608.

5. H.Yamamoto, S.Uchida, S.Araya, K.Nakajima, M.Hashimoto, and K.Kimura, "Characteristcs of High-Speed Toll Steel as Materials of Work Roll in Hot Rolling," METEC Congress 94, 6th International Rolling Conference, Dusseldorf Gemany, June 20-22, 1994, Proceedings Vol.2, pp.59-64.

6. M.Hashimoto, S.Otomo, K.Yoshida, K.Kimura, R.Kurahashi, T.Kawakami, and T.Kouga, "Development of High-Performance Roll by Continuous Pouring Process for Cladding," The Iron and Steel Institute of Japan International, Vol.32, No.11, 1992, pp.1202-1210.

7. W.L.Bradley and M.N.Srinivasan, "Fracture and Fracture Toughness of Cast Irons," International Materials Reviews, Vol.35, No.3, 1990, pp.129-161.

8. J.C.Radon, D.J.Burns, and P.P.Benham, "Push-Pull Low-Endurance Fatigue of Cast irons and Steels," Journal of The Iron and Steel Institute, Vol.204, September 1966, pp.928-935.

INFLUENCE OF THE ROLL BITE CONDITIONS ON THE SURFACE QUALITY OF HOT ROLLED COIL

Frédéric GEFFRAYE, Vincenzo LANTERI, Pascal GRATACOS, Sabine DAUPHIN *
Irsid, R&D Usinor Group, Voie Romaine, BP30320, F-57283 Maizières-Lès-Metz, Cedex, France
Tel.: + 33 3 8770 4000, Fax: + 33 3 8770 4711
E-mail: frederic.geffraye@irsid.usinor.com, vincenzo.lanteri@irsid.usinor.com, pascal.gratacos@irsid.usinor.com
* Sollac, Usinor Group, Rue du Comte Jean - Grande Synthe, BP 2508, F- 59381 Dunkerque, Cedex, France
Tel.: + 33 3 2829 3513, E-mail: sabine.dauphin@sollac.usinor.com

Key Words: Hot rolling, Roll Surface Degradation, Surface Quality, Friction, Finite Element Modeling

INTRODUCTION

The production of hot rolled strips with a high surface quality in the hot strip mill (HSM) is hampered by high operating costs in order to preserve the integrity of the work rolls. The ever-increasing productivity of the HSM, the struggle for limiting costs, along with the ever-rising customer requirements for high surface quality, make roll wear control (especially in the front stands of the finishing mill) the prime concern of HSM operators.

Surface roll degradation results from the complex interaction between strip parameters (oxide characteristics, steel grades), rolls parameters (oxide types, carbide phases, hardness), process parameters (stock temperature, reduction rate, rolling speed, rolling time) and tribological factors (friction, wear, lubricant) [1-4]. As the mechanisms of roll surface degradation are still today imperfectly quantified, the control of the roll wear behavior imposes costly constraints. These include: specific rolling schedules for high quality grades, use of lubrication, use of optimal roll cooling, use of surface strip cooling, shortening of the rolling campaign, restrictions on rolling temperature and on the rate of reduction in the first stands.

The major difficulty in studying roll surface degradation in the HSM is that the work roll surface can be fully characterized only at the end of the rolling schedule. Although this analysis can give valuable information about roll wear during a specific rolling schedule, it is not sufficient to understand the evolution of the roll surface during a rolling campaign. Several research projects are currently underway with focus on the direct observation of the work roll through a CCD camera inspection system installed in the rolling stand [5-6]. Up to now, promising results shed light on the mechanisms of roll degradation especially in the presence of lubrication with the new generation of HSS rolls, and attempts are made to correlate these data with process parameters [7].

Another interesting approach is to monitor the evolution of the friction coefficient during rolling. For example, Uijtdebroeks, et al., established relations between the evolution of the roll gap friction and roll surface degradation [6]. It has been shown that severe roll degradation (i.e., banding) can lead to an increase of the friction coefficient. However, in the case of moderate or localized roll degradation, the surface modification is often not sufficient to induce an increase of the friction coefficient.

In this paper, we present an alternative strategy in order to evaluate roll degradation based on the estimation of the roll bite conditions. The method is based on the determination of the friction coefficient with a roll gap FEM-based model from the measure of the forward slip. Then roll bite conditions are computed using the same model. For practical purposes, mainly aimed at reducing calculation time, the FEM results were modeled using wavelet networks. Finally, a few examples of the practical use of this work are given based on actual industrial data.

Evaluating the roll gap friction in the finishing mill has been the subject of many research projects from a theoretical point of view as well as from a practical standpoint.

Munther and Lenard reviewed various methods for the measurement of the friction coefficient during hot rolling [8]. Most of these are based on models that have been derived from the equilibrium equation under assumptions on geometry, friction and material behavior. The results of these models are generally used to estimate process parameters, such as roll force, roll torque, power or forward slip. However, the authors pointed out that they are not accurate enough to deal with sensitive computations such as inverse calculations of the friction coefficient using industrially measured rolling forces, and if used, would yield inaccurate data. To overcome this difficulty, Munther and Lenard highlighted the interest in using a roll gap model based on the finite element method (FEM) in order to identify the friction coefficient.

Measuring friction coefficient with a roll gap FEM-based model

A procedure was developed in order to identify the friction coefficient in hot rolling using LAM3, a steady-state thermo-elastoviscoplastic FEM-based model of rolling with coupled thermo-elastic roll deformation using a steady-state approach [9].

The friction law used is a Coulomb friction type law. Roughly, the shear stress is written as:

$$\tau = \mathrm{Min}\left(\mu.\sigma_N.V_{gcar}^{p}, \overline{m}.\frac{\sigma_0}{\sqrt{3}}.V_{gcar}^{p} \right) \tag{1}$$

where μ is the Coulomb friction coefficient, σ_N is the normal stress, $\overline{m}$ is the Tresca friction factor, σ_0 is the yield stress and V_{gcar} is the characteristic relative velocity. In application, the relative velocity sensitivity coefficient p is taken equal to 0.01, the Tresca friction factor $\overline{m}$ is fixed to 1, and the characteristic relative velocity V_{gcar} is estimated to 10% of the roll speed. This friction law can thus be seen as a Coulomb friction law with a small sensitivity to the relative velocity and the normal stress limited by the maximal shear stress defined from the Von Mises yield criterion.

The rheological law is taken from the work done by Thonus, et al., where the yield stress of the steel depends on the carbon and manganese contents, strip temperature, grain size and strain rate [10].

In figure 1, we plotted the influence of the Coulomb friction coefficient on the forward slip and on the linear roll separating force (force per unit width of strip) in conditions close to industrial practice (F2 stand). The forward slip ratio is defined by:

$$\Theta_{forward} = \frac{V_s - V_r}{V_s} \tag{2}$$

where V_s is the speed of the strip at the exit of the roll gap and V_r is the linear roll speed. The measurement of the forward slip in the HSM is based on the computation of the time of the strip head to go from one stand to the other. The curves in figure 1 are obtained by a set of finite element calculations, each point corresponding to a value of the Coulomb friction coefficient. The values of the process parameters measured on the HSM make it then possible to identify a friction coefficient. In both cases, the measured values are located in a region of the curves where the process parameters are sensitive to friction. We can also notice that the value of the friction coefficient is different according to whether the coefficient is identified from the rolling force or from the forward slip. In the selected example, the measured forward slip ratio is $\Theta_{forward} = 0.058$ and the linear rolling force is F = 12.054 MN/m, giving values of $\mu = 0.155$ and $\mu = 0.130$, respectively.

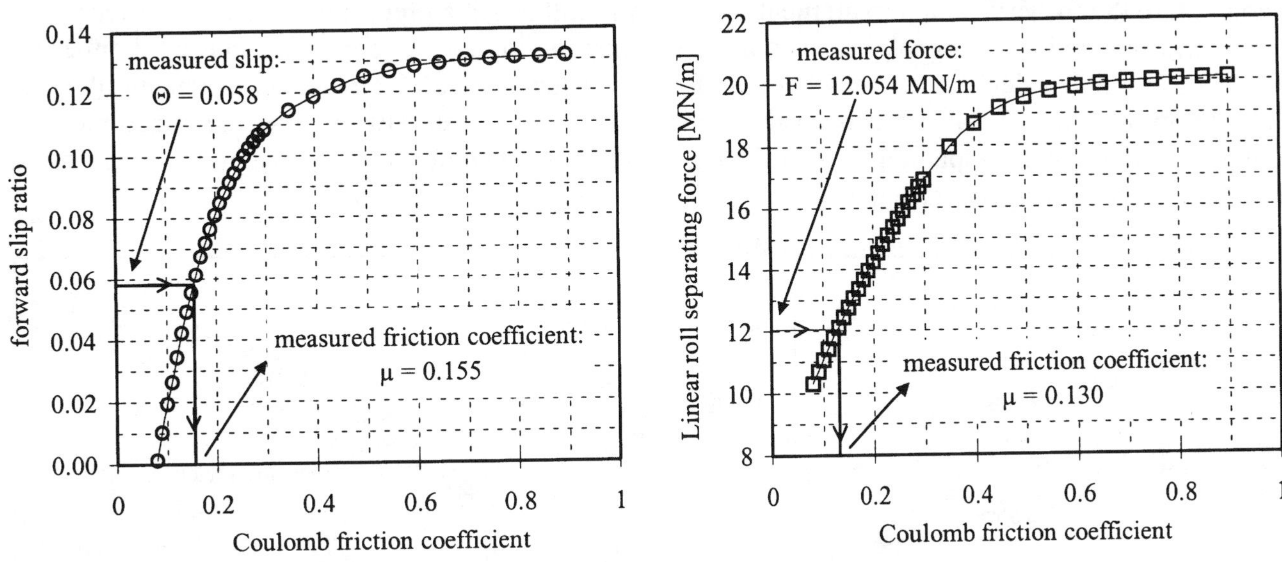

Fig. 1: Influence of the Coulomb friction coefficient on the forward slip and linear roll separating force calculated by the FEM-based model (h_{entry} = 21.16 mm, h_{exit} = 12.50 mm, V_{roll} = 142.98 m/min, Φ_{roll} = 692.6 mm, T_{strip} = 944 °C, backward tensile stress = 4.41 MPa).

As Munther and Lenard underlined, the identification of the friction coefficient is possible starting from various process parameters: rolling force, rolling torque, power, forward slip, etc. Martin, et al., proposed another approach combining the forward slip and the roll separating force [11]. Their method consists in minimizing a residual function constituted by the square root of the sum of the squared errors of the predicted forward slip and roll separating force. But, as mentioned by the authors, the model inversion process may require several solutions of the roll gap model in conjunction with a root finding or minimization algorithm to solve the friction coefficient for a given case. In case of identification on just one of the process parameter, it is thus a question of retaining the most relevant one. The parameter selected must obey two criteria: it must be measured accurately on the HSM, and its computation by the model must not be very sensitive to uncertainty in the input data.

Taking into account former results, we were interested only in the roll separating force and in the forward slip. Power and roll torque are, for various reasons, process parameters which are either difficult to measure on the HSM or difficult to compare with the results of simulation. To determine which parameter is the least sensitive to uncertainties in the input data, a sensitivity study to the various parameters of the model was carried out. It appears that the forward slip is not very sensitive to the input parameters that have an impact on the yield stress (strip temperature, metal hardness, roll speed), which is obviously not the case for the rolling force. In the same fashion, the characteristics of the working roll (diameter, Young's modulus) only slightly affect the forward slip. These characteristics, on the other hand, have more impact on the roll force, mostly because of the change in the arc of contact length. Thus, the sensitivity study led us to identify friction from the forward slip rather than from the rolling force, the latter being more sensitive to many process parameters. Moreover, we noticed for many different cases that, when identified from forward slip, the error made on the roll separating force (i.e., the relative difference between the measured and the predicted force) is much less than the error made on the forward slip when identified from the roll separating force.

The identification procedure of the friction coefficient with the FEM-based model is carried out, for each strip of the campaign, with the secant method described in [12]. Two or three increments are generally sufficient to converge towards the selected measured process parameter (forward slip or roll force).

We should also point out the influence of the strip tensile stress on the forward slip and on the roll separating force .We can indeed see in figure 2, that the backward tensile stress has a significant influence both on the forward slip and on the roll force. This means that the level of tension taken into account in the input data of the FEM-based model strongly influences the value of the measured friction coefficient. This feature is illustrated in figure 3 where the values of friction determined for various levels of tensile stress are indicated. We should notice that backward tensile stress only is of concern as long as there is no forward tension before the forward slip measurement. As it is difficult to get a precise measurement of tension at the time of strip-in-stand (SIS) signal, because of the transient period of the looper, we thus used either the set point value or a null value in the case where the F1 stand is concerned. A precise measurement of tension is necessary to obtain a "real" friction value. Nevertheless, we can always use the measured friction coefficients for relative comparisons, in particular to follow its trend during a campaign, considering that the backward tension should not vary much.

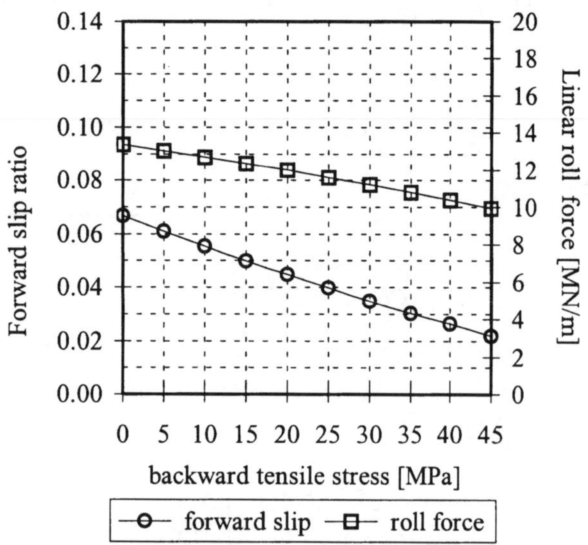

Fig. 2: Influence of the backward tensile stress on the forward slip and linear roll separating force calculated with the FEM-based model ($h_{entry} = 21.16$ mm, $h_{exit} = 12.50$ mm, $V_{roll} = 142.98$ m/mn, $\Phi_{roll} = 692.6$ mm, $T_{strip} = 944$ °C, $\mu = 0.16$).

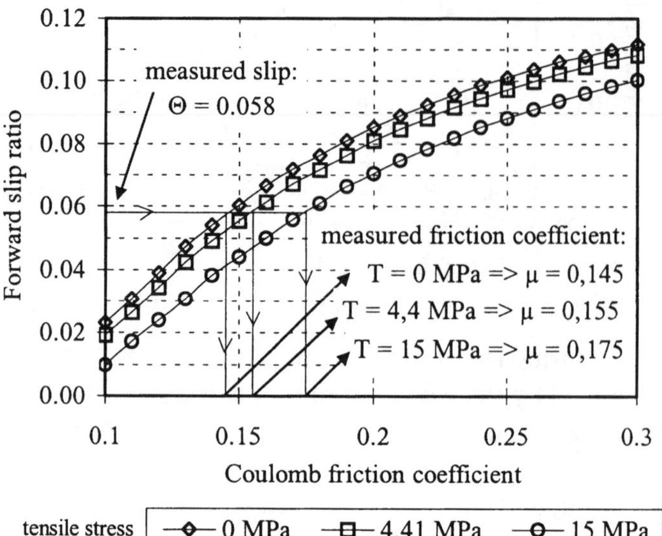

Fig. 3: Influence of the backward tensile stress on the value of the measured friction coefficient calculated with the FEM-based model ($h_{entry} = 21.16$ mm, $h_{exit} = 12.50$ mm, $V_{roll} = 142.98$ m/mn, $\Phi_{roll} = 692.6$ mm, $T_{strip} = 944$ °C).

Measuring friction coefficient with a roll gap adapted Sims model [13]

In order to avoid inconveniences due to the long computational time of the FEM-based method, the results were compared to an analytical expression of the Tresca friction factor adapted from the Sims theory [14]. With respect to the original model, where the shear stress is equal to the maximal shear stress defined from the Von Mises yield criterion (i.e., sticking friction), we used instead a Tresca formulation with:

$$\tau = \overline{m}.\frac{\sigma_0}{\sqrt{3}} \tag{3}$$

If we do not consider strip tension, the analytical resolution of the equilibrium equations gives the following expression:

$$\overline{m} = \frac{\dfrac{\pi}{8.\sqrt{\dfrac{R'}{h_{exit}}}}.Ln\left(\dfrac{h_{exit}}{h_{entry}}\right)}{Arc\tan\sqrt{\Theta_{forward}} - \dfrac{1}{2}.Arc\tan\left[\sqrt{\dfrac{R'}{h_{exit}}}.Arc\cos\left(1 - \dfrac{h_{entry} - h_{exit}}{2.R'}\right)\right]} \tag{4}$$

where h_{entry} and h_{exit} are, respectively, the strip thickness at the entry and the exit of the roll bite and R' is the deformed roll radius calculated with the Hitchcock formula.

Evolution of the friction coefficient through a rolling campaign

We compare in figure 4 the evolution of the Tresca friction factor (identified from equation (4)) and the Coulomb friction coefficient (identified with the FEM-based model) during a rolling schedule. This figure shows that the trends are similar throughout a rolling campaign, although the ranges of both friction coefficient values are different. This observation has been made with various campaigns and with different front stands of the finishing mill (up to F4) on at least three different Usinor HSMs. Note that, the strip tensile stress was not taken into account. Therefore, we may underestimate the real values of the friction coefficient.

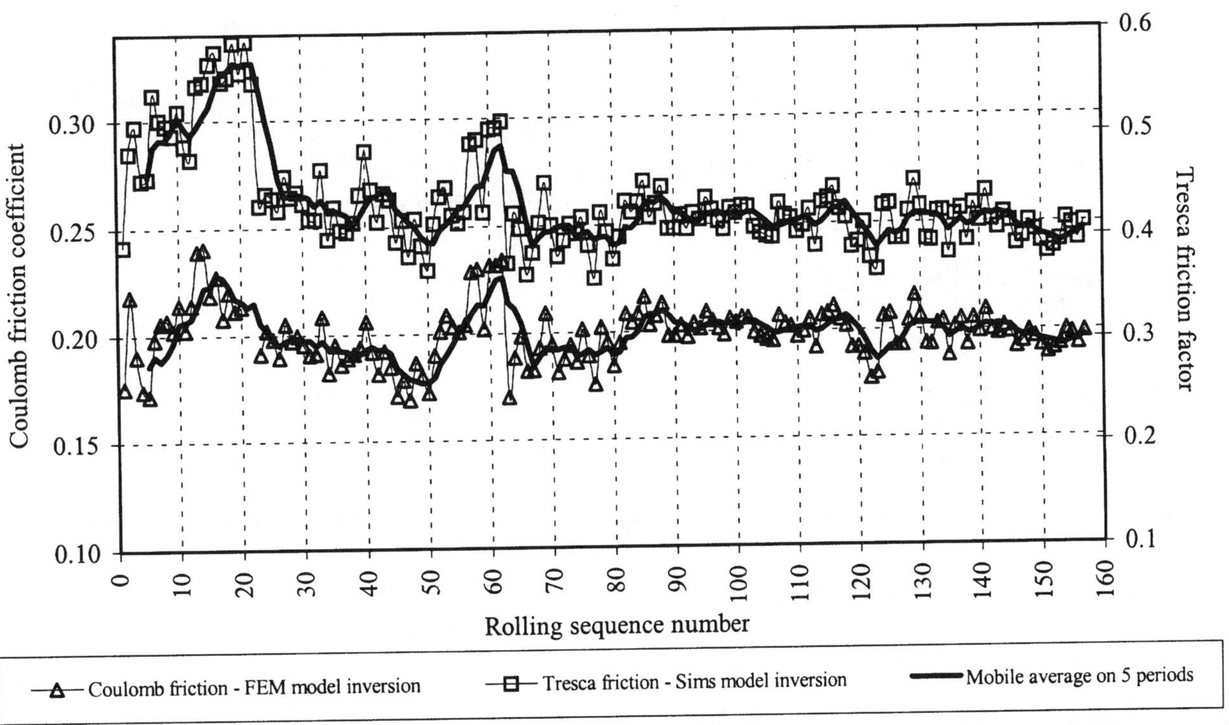

Fig. 4: Comparison of the trends of Coulomb and Tresca friction coefficients throughout a rolling campaign for the first finishing stand.

The conclusion of the friction analysis in the front stands of the finishing mill is that two approaches can be used to monitor the evolution of friction during a rolling campaign. One is based on a modified Sims model and gives an analytical expression of the Tresca friction factor from the forward slip ratio (equation (4)). The second is based on a FEM-based model of the roll bite coupled with an inverse method. In the conditions of the study (i.e., the four first stands), these two methods give the same friction trends throughout the rolling campaign.

Moreover, from a large number of rolling campaigns, we have noticed that the evolution of friction can be linked to tribological-related phenomena such as: formation of the black oxide film at the beginning of the campaign, efficiency of lubrication, higher friction coefficient when using the new generation HSS work rolls. However, we only found a weak correlation between the friction coefficient values and the occurrence of catastrophic surface degradation leading to defects on the strips.

In order to study the roll surface degradation, we assume that it is necessary to take into account the mechanical conditions endured by the work roll surface during a rolling campaign. For this purpose, the FEM-based model is suitable to compute the mechanical stresses in the roll bite.

EVOLUTION OF THE ROLL BITE CONDITIONS DURING ROLLING

As the roll surface degradation is considered to be an aggravating cause of poor strip surface quality, it is interesting to study the influence of the friction coefficient on the mechanical stresses generated at the roll/strip interface in the roll bite. Calculations of the roll bite conditions are made with the FEM-based model used for the friction coefficient identification. Note that, in the calculations, we took simplified friction and heat transfer boundary conditions since the presence of an oxide scale at the strip surface is not taken into account. The influence of the scale on the roll bite conditions is being currently studied by Beynon, et al., [15].

In figure 5, we plotted the distribution of the normal and shear stresses in a particular roll bite (close to an industrial F2 stand) using different Coulomb friction coefficients. As expected, an increase of the friction coefficient leads to a higher friction hill and a higher magnitude of the shear stresses. These results are, by all means, in agreement with published theories. Thus, one can expect that an increase of the friction coefficient is detrimental in terms of mechanical stresses generated at both the strip and roll surfaces. Another example of the influence of friction on the mechanical conditions that occur at the interface is given in figure 6. The relative velocity is the absolute difference between the strip and the roll speed, and we arbitrarily define the sticking zone as the portion of the roll bite where the relative velocity is lower than 1% of the linear roll speed. The higher the friction coefficient, the larger the sticking zone. Moreover, the position of the sticking zone is shifted toward the roll bite entry. We presume also that the friction-dissipated power could be a relevant factor for the quantification of the roll surface degradation. The friction-dissipated power is calculated in the FEM-based model from the following expression:

$$P_{friction} = \int_{contact} \tau . |V_r - V_s| . dS \quad (5)$$

where τ is the shear stress.

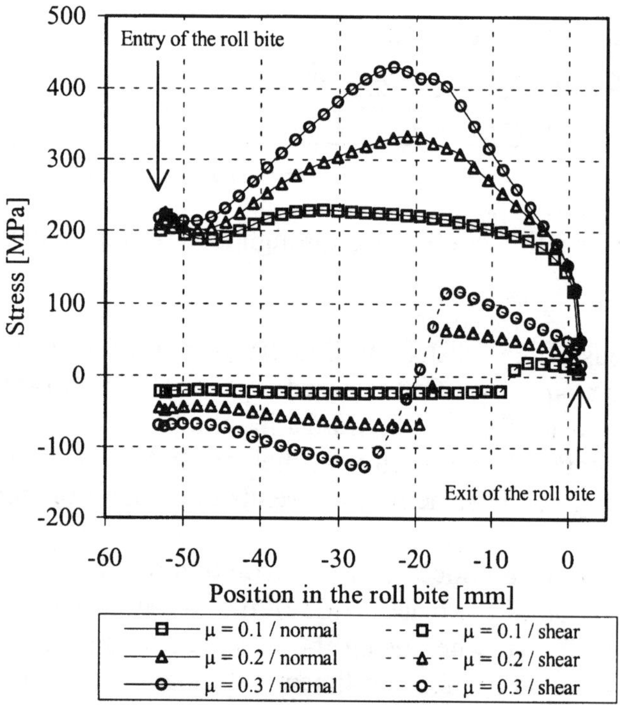

Fig. 5: Influence of the Coulomb friction coefficient on the normal and shear stresses in the roll bite calculated with the FEM-based model (h_{entry} = 21.16 mm, h_{exit} = 12.50 mm, V_{roll} = 142.98 m/mn, Φ_{roll} = 692.6 mm, T_{strip} = 944 °C).

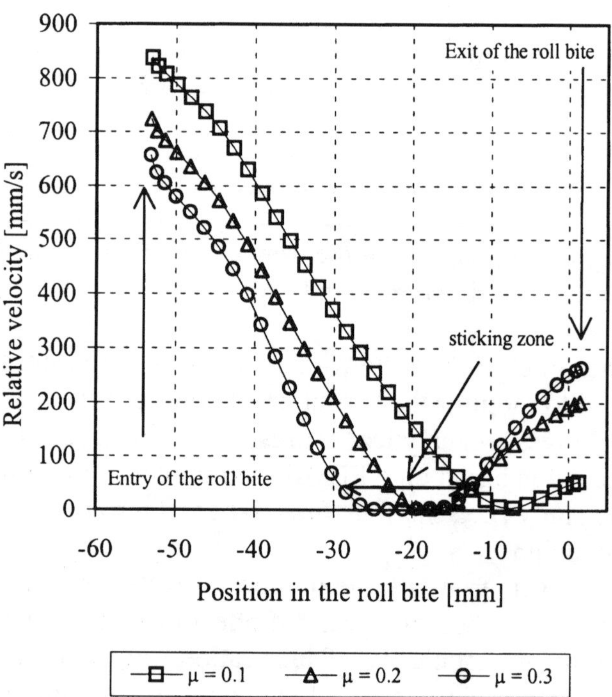

Fig. 6: Influence of the Coulomb friction coefficient on the relative velocity in the roll bite calculated with the FEM-based model (h_{entry} = 21.16 mm, h_{exit} = 12.50 mm, V_{roll} = 142.98 m/mn, Φ_{roll} = 692.6 mm, T_{strip} = 944 °C).

The mechanical conditions in the roll bite do not exclusively depend on the friction coefficient, but also on other process parameters (reduction, temperature, rolling speed, metal hardness, strip thickness, etc.), and each process parameter has a distinctive influence.

For an industrial use of the roll bite conditions, it is thus necessary to build functions suitable for making an on-line estimation. For this purpose, we conducted an exhaustive numerical parametric study of the roll gap using the FEM-based model. Six input process parameters were chosen for their influence on the roll bite conditions: forward slip, reduction, strip temperature and thickness, roll speed and metal hardness. For each input parameter, we defined their range of variation in order to cover the front stands of the finishing mill set points. Four or five values were taken in this interval. Each combination of parameters was then simulated with the FEM model, leading to 1340 computations of roll gap cases. From this set of numerical results, an estimation function was built for each roll bite condition (maximal normal stress, maximal shear stress, friction-dissipated power, width of the sticking zone) using a wavelet networks method as described in [16]. An example of the accuracy of the estimation function of the roll bite conditions is given in figure 7.

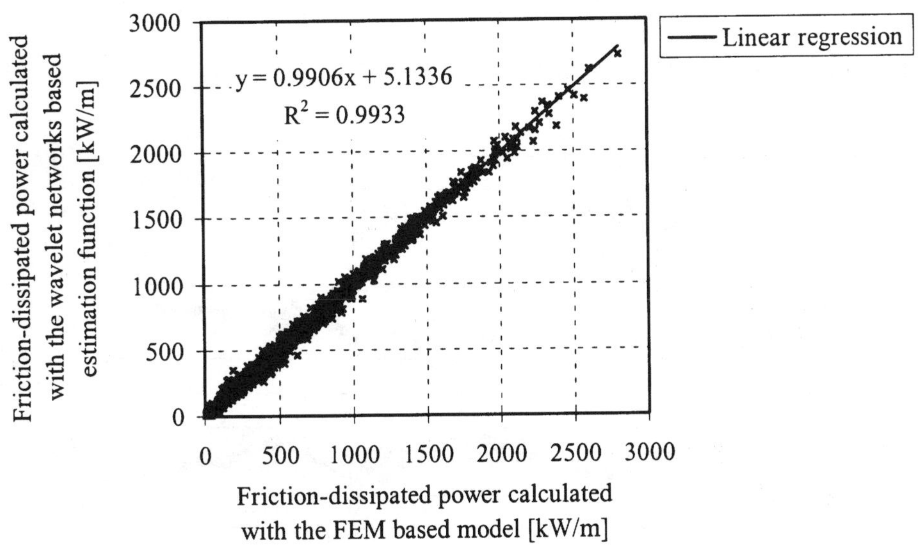

Fig. 7: Precision of the wavelet networks estimation function for one of the roll bite conditions.

Figure 8 presents plots of the evolution of the friction coefficient and of the maximum normal stress calculated with the FEM-based model during a rolling campaign. The trends of both parameters are different: a weak friction in the roll bite can be concomitant with a significant normal stress (i.e., when simultaneously the temperature decreases and the reduction increases). The roll bite conditions monitoring can also be used to evaluate the mechanical stresses applied on the work roll surface. As an example, figure 9 shows the evolution of friction coefficient and the friction-dissipated power as a function of rolling sequence number. Once again, one can observe that severe conditions can be imposed on the work roll even with a low level of friction.

Within the scope of the study on the contribution of roll bite conditions to the prediction of work roll degradation and subsequent surface defects on the strips, we analyzed a database built with 207 campaigns collecting process parameters from the rolling of more than 26,000 low-carbon strips. Roll bite conditions have been determined with the wavelet networks estimation function for the F1 to F4 rolling stands. The general trend observed on these campaigns is that, the higher the friction-dissipated energy endured by the work rolls, the higher the probability to observe roll degradation and therefore surface defects on the strip. This point is illustrated in figure 10 where two types of rolling schedule are shown. Schedules with a lower average friction-dissipated power level (type "A") are less sensitive to surface defects that schedules with higher friction-dissipated power values (type "B").

At this time, a better correlation has to be established between the observed surface defects and the roll bite conditions. Indeed, it appears that a surface defect prediction function should at least take into account a combination of different roll bite conditions to be more predictive. Studies are currently underway on this topic.

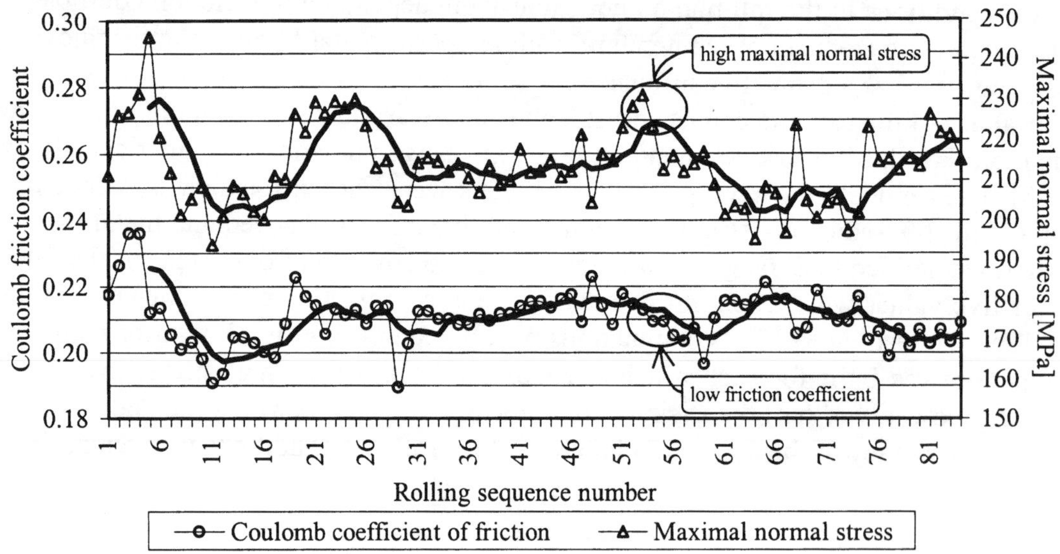

Fig. 8: Trends of Coulomb friction coefficient and maximal normal stress during a rolling campaign for the first finishing stand.

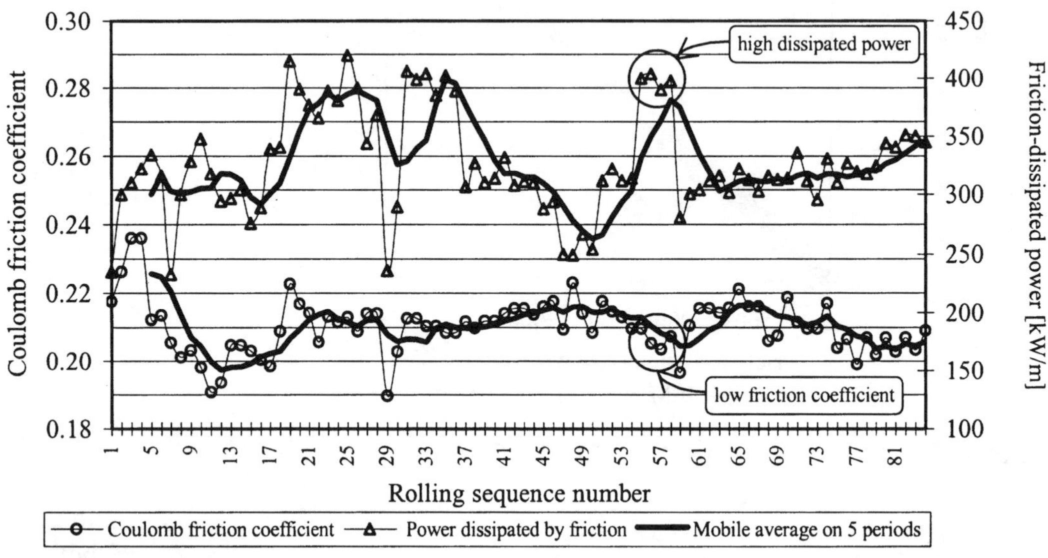

Fig. 9: Trends of Coulomb friction coefficient and friction-dissipated power during a rolling campaign for the first finishing stand.

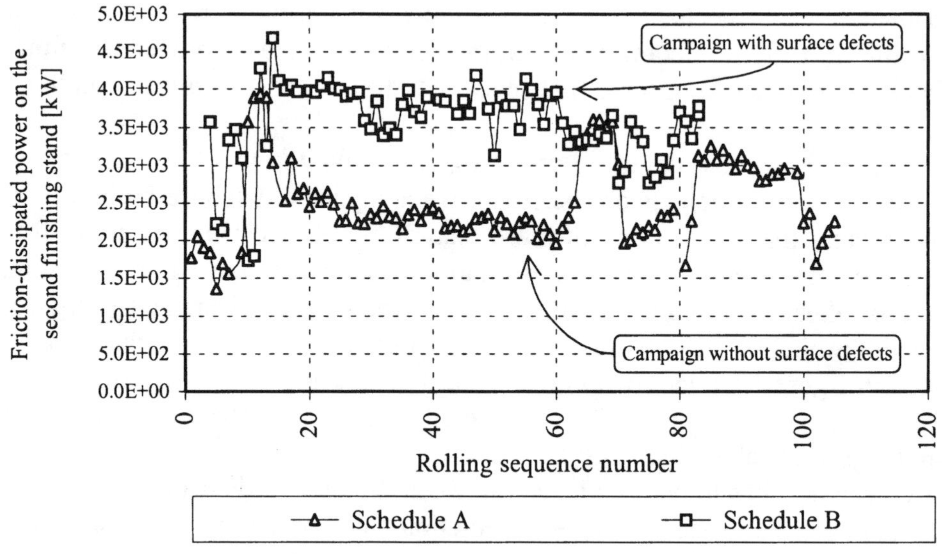

Fig. 10: Influence of the rolling schedule on surface defect occurrence.

CONCLUSIONS

We investigated a method for determining roll gap friction in the finishing mill using a FEM-based model. Forward slip is the best-suited process parameter to identify friction, as it is less sensitive to the various process parameters than the roll separating force. An alternative method, based on an adapted Sims model taking into account Tresca sliding friction, gives similar trends in the evolution of the friction coefficient throughout the rolling campaign, at less computing time expense.
The former method exhibits advantages when a quantification of the mechanical roll bite conditions is required. This is the case when studying mechanisms of roll surface degradation and subsequent defects generated on the strip surface. From a numerical parametric study of the roll gap in the front stands of the finishing mill, estimation functions of the roll bite conditions were established for on-line use. The first results obtained on an industrial database showed promising correlation between roll bite conditions and roll degradation. Further work includes the introduction of scale at the surface of the steel strip in the roll gap FEM-based model.

ACKNOWLEDGEMENTS

The authors gratefully thank Mr. G. Mathis, Sollac, Usinor Group, for his assistance in the use of wavelet networks. We also would like to thank R. Dalmayrac for the process database from Sollac Atlantique, Dunkerque. This work received partial support from the European Community of Steel and Coal.

REFERENCES

1. P. Carless, "Present and future hot strip mill finishing train work rolls", Rolls 2000+ Conference Proceedings, 12-14 April 1999, Birmingham, UK

2. L. Caithness, S. Cox, S. Emery, "Surface behaviour of HSS in hot strip mills", Rolls 2000+ Conference Proceedings, 12-14 April 1999, Birmingham, UK

3. V. Lanteri, C. Thomas, J. Bocquet, H. Yamamoto, S. Araya, "Black oxide film generation on work rolls and its effects on hot-rolling tribological characteristics", Steel Rolling'98 Conference Proceedings, ISIJ, November 9-11, 1998, Chiba, Japan

4. H. Uijtdebroeks, R. Franssen, J.J. de Roo, G. Sonck, D. Steinier, "Surface evolution of HiCr work rolls during hot rolling: on-line observations", Rolls 2000+ Conference Proceedings, 12-14 April 1999, Birmingham UK

5. T. Yasui, J. Azuma, K. Dejima, M. Fukuda, T. Nishimura, K. Yoshitake, "Surface observations of work roll in finishing stand of hot strip mill", 33rd MWSP Conference Proceedings, ISS-AIME, Vol. XXIX, 1992, pp. 115-118

6. H. Uitjdebroeks, R. Fransen, G. Sonck, A. Van Schooten, "On-line analysis of the surface work roll deterioration in a hot strip mill", La Revue de Métallurgie-CIT N°6 - Juin 1998 - pp. 789-799

7. J.C. Herman, D. Steinier, H. Uijtdebroeks, J. Lacroix, G. Sonck, D. Liquet, "Improvement of the steel rolling by the use of lubrication and HSS rolls", Steel Rolling'98 Conference Proceedings, ISIJ, November 9-11, 1998, Chiba, Japan

8. P. A. Munther, J. G. Lenard, "A study of friction during hot rolling of steels", Scandinavian Journal of Metallurgy, Vol. 26, pp. 231-240, 1997.

9. A. Hacquin, P. Montmitonnet, J.-P. Guillerault, "A steady-state thermo-elastoviscoplastic finite element model of rolling with coupled thermo-elastic roll deformation", Journal of Materials Processing Technology, 60, pp. 109-116, 1996.

10. P. Thonus, J.C. Herman, J.P. Breyer, M. Sinnaeve, A. Charlier, D. Liquet, R. Marquet "Off-line analysis of the HSS roll behaviour in the hot-strip mill by the use of a rolling load prediction model", 38th MWSP Conference Proceedings, ISS, Vol. XXXIV, pp. 43-49, 1997.

11. D.C. Martin, P. Mäntylä, P.D. Hodgson "An analysis of friction in finishing rolling of hot steel strip", Modelling of Metal Rolling Processes 3 Conference Proceedings, 13-15 December, 1999, London, UK

12. W.H. Press, S.A. Teukolsky, S.A. Vetterling, B. P. Flannery, "Numerical Recipes in Fortran. The Art of Scientific Computing", 2nd ed., Cambridge University Press

13. J.-P. Rinck, M. de Conti, "Calcul de la force et du couple de laminage à l'aide du modèle de Sims. Introduction d'un frottement de Tresca", internal report, IRSID, May 1980.

14. Sims RB, "Calculation of roll force and torque in hot rolling mills", Proc. Inst. Mech. Engrs, 168, pp. 191-219, 1954.

15. J.H. Beynon, M. Krzyzanowski, "Finite-element model of steel oxide failure during flat hot rolling process", Modelling of Metal Rolling Processes 3 Conference Proceedings, 13-15 December, 1999, London, UK

16. Q. Zhang, "Using wavelet networks in nonparametric estimation", Internal publication, n° 833, IRISA, Rennes, France, June 1994

A Mechanism for Generating Aluminum Debris in the Roll Bite and Its Partitioning between the Surface of the Work Roll and the Surface of the Sheet as Smudge©

RONALD A. REICH, JUNE M. EPP and DAVID E. GANTZER*
Surface Technical Division
ALCOA Research Center
ALCOA Center, Pennsylvania 15069

A series of tests of cold rolling aluminum on a laboratory rolling mill showed that the amount of debris generated in the roll bite increased with an increase in the amount of reduction. It also showed at light reductions, the debris generated distributed evenly between the surface of the sheet exiting the mill as smudge and the surface of the work roll where it was washed off. The rate of generating debris at light reductions also followed the Archard equation. At higher reductions, however, it was shown that much more debris adhered to the work roll and was washed off than that which remained on the surface as smudge. The rate of generating debris also deviated from the Archard equation. To explain this deviation, it was proposed that the mechanism for generating debris in the roll bite was due to abrasive wear at light reductions but at higher reductions, the mechanism was due to adhesive wear.

KEY WORDS

Aluminum Industry Tribology, Rolling, Wear and Failure

INTRODUCTION

Smudge is defined as debris that remains on the surface of aluminum sheet after rolling. It is an aesthetic defect since it makes the surface of a sheet appear dark. It can also build up on the surface of a work tool in a post-rolling deformation process of the sheet, decreasing the efficiency of the tool due to high friction.

Smudge is generated in the roll bite due to the tribological interaction between the work roll, the sheet and the lubricant. Normally, the asperities on the surface of the work roll push the aluminum metal to the side in the roll bite, forming troughs and mounds which repeat in a pattern across the sheet. However, there are a few spots on the work roll surface in which the geometry of the asperity, Fig. 1, forces the surface to abrade, forming debris particles rather than a trough. The optical micrograph in Fig. 2 shows several places where these asperities have cut paths on the surface of the sheet, leaving behind trails of metallic debris.

To cut a surface, one surface must rub against another. In the roll bite, this rubbing takes place due to the mismatch between the incoming speed of the metal and the speed of the work roll. Up to the neutral point, the surface of the work roll rubs against the surface of the sheet. After the neutral point, the sheet's surface rubs against the surface of the work roll. The excess length which the sheet and the work roll rub against each other is defined as sliding distance, and a visual representation of it is shown in Fig. 3.

If it is assumed for now that debris is formed in the roll bite due to abrasive wear, then the rate of debris formation should be a function of sliding distance and the area of contact between the metal and the work roll asperity according to the Archard equation (*1*):

$$V/SD = KW/H \qquad [1]$$

where V is the volumetric wear, SD is sliding distance, H is the hardness of aluminum, W is the load on the work rolls, K is the coefficient of wear and W/H is equal to the area of contact.

To test this idea, cleaned (smudge free) 5182 aluminum alloy was cold rolled on a laboratory rolling mill using a proprietary lubricant. Several reductions were taken to vary the sliding distance and load (Table 1). Sliding distance for a point on the surface of metal moving through the roll bite was calculated from slip and reduction values, using Eq. [2] (*2*):

$$\text{Sliding Distance} = \left(\frac{\sqrt{a^*t_i}}{3^*(1+s)}\right)^*\left\{\frac{\beta^{1.5}}{1-\beta} + 4s(s(1-\beta))^{0.5} - 3s\beta^{0.5}\right\} \qquad [2]$$

where a is equal to the work roll radius, t_i is equal to the thickness of the metal at the entry of the roll bite, s is equal to slip and β is equal to the metal reduction ratio.

*Currently with Calgon Corporation, Pittsburgh, PA.

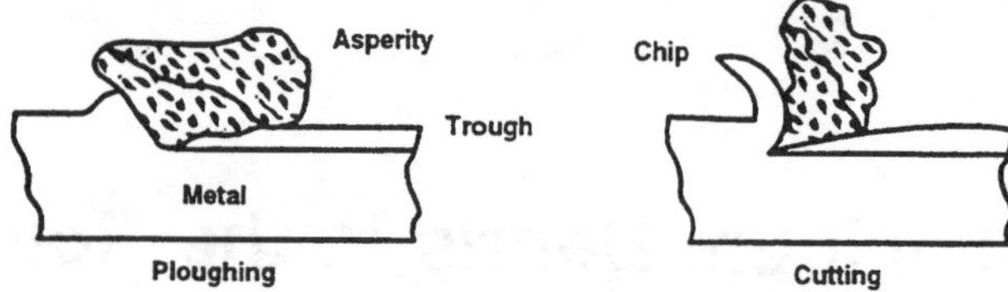

Fig. 1—Geometry of work roll asperity and its effects on cutting.

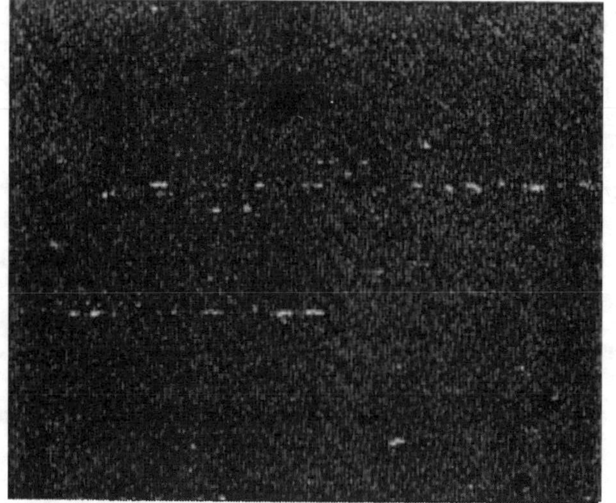

Fig. 2—Smudge debris on the surface of Al sheet (100X). Fines were removed by tape and then photographed.

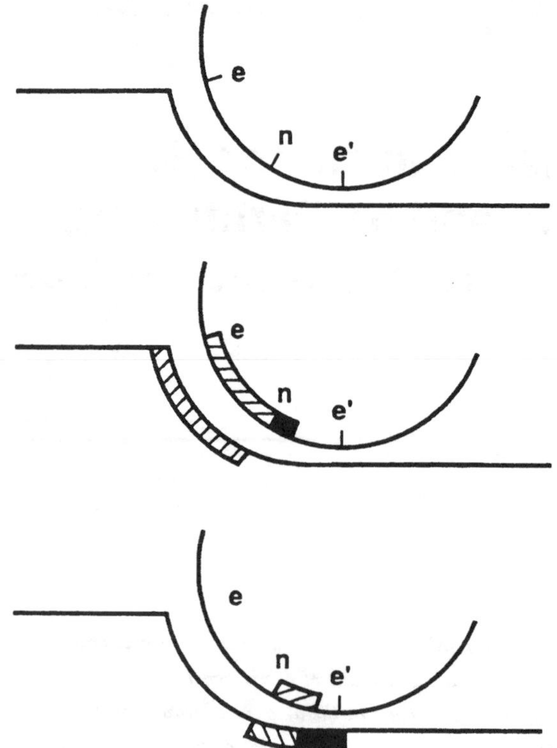

Vr = Velocity of work roll
T = Time
$\widehat{EN}$ = Arc length from e to n
$\widehat{NE'}$ = Arc length from n to e'

Sliding Distance = (Vr•Ten-$\widehat{EN}$) + ($\widehat{NE'}$-Vr•Tne')
= Black Area

Fig. 3—Schematic of the work roll and visualization of sliding distance.

An average sliding velocity (SD/t) was then calculated by dividing the sliding distance by the time it takes a point on the surface of the metal to pass through the roll bite. This period of time was calculated by integrating Eq. [3] to obtain Eq. [7]:

$$t = \int_0^{\sqrt{(l_i - l_e)a}} \frac{dx}{V_x} \qquad [3]$$

where l_e is equal to the thickness of the metal at the exit of the roll bite, V_x is the velocity of the metal at position x in the roll bite and $\sqrt{(l_i - l_e)a}$ is the length of the contact arc in the roll bite.

To integrate Eq. [3], it was assumed that the deformation throughout the roll bite was homogeneous which required that the product of metal velocity and thickness was constant throughout the roll bite, as shown in Eq. [4]:

$$l_i V_i = l_x V_x = l_e V_e \qquad [4]$$

where l_x is equal to the thickness of the metal at position x in the roll bite, V_i is equal to the velocity of the metal at the entry of the roll bite and V_e is equal to the velocity of the metal at the exit of the roll bite.

It has also been shown that the thickness of the metal at position x in the roll bite (l_x) can be calculated from Eq. [5], (2).

$$l_x = l_e + \frac{x^2}{a} \qquad [5]$$

Equations [4] and [5] can now be combined to derive Eq. [6] which upon integration gives Eq. [7]

$$t = \int_0^{\sqrt{(l_i - l_e)a}} \left(\frac{x^2 + l_e a}{V_e l_e a}\right) dx \qquad [6]$$

$$t = \frac{((l_i - l_e)a)^{3/2}}{3 V_e l_e a} + \frac{\sqrt{(l_i - l_e)a}}{V_e} \qquad [7]$$

Knowing the average sliding velocity, the total sliding distance for a specified period of time could then be calculated by multiplying the average sliding velocity by this time.

To measure the amount of debris generated in the roll bite for a specified amount of time, it was necessary to quantitatively measure the amount of debris which remained on the sheet as smudge and the amount of debris which initially adhered to the work roll and then was washed off by the lubricant. This was done by buffing the debris off the surface of the sheet using cheesecloth and an orbital sander. The cheesecloth was soaked in HCl to dissolve the metallic fines,

TABLE 1—SLIDING DISTANCE CALCULATION					
REDUCTION RATIO[1]	DISTANCE (m)[2]	FORWARD SLIP RATIO[3]	SLIDING DISTANCE FOR A POINT ON THE SURFACE (m)[4]	AVERAGE SLIDING SPEED (m/sec)[5]	SLIDING DISTANCE/m^2 OF METAL ROLLED[6]
0.07	2.3693e-1	0.011	3.48e-5	0.0387	0.094
0.15	2.4171e-1	0.032	1.14e-4	0.0838	0.203
0.25	2.4526e-1	0.047	2.82e-4	0.1537	0.371
0.35	2.4844e-1	0.060	5.49e-4	0.237	0.575
0.45	2.5123e-1	0.072	9.68e-4	0.341	0.829
0.50	2.5479e-1	0.087	1.23e-3	0.393	0.953
0.55	2.5717e-1	0.098	1.58e-3	0.456	1.107

[1]The reduction ratio is calculated by the following equation:

Reduction Ratio = (Entry metal thickness − Exit metal thickness)/Entry metal thickness

[2]This is the distance between the marked lines on the aluminum sheet. This value represents the average of six different measurements.

[3]The forward slip ratio is calculated using the following equation:

$$\text{Forward Slip Ratio} = \frac{\text{Distance measured between lines on sheet} - \text{Distance between lines on the roll}}{\text{Distance between lines on the roll}}$$

[4]The sliding distance for a point on the surface of the metal passing through the roll bite was calculated using the following equations and parameters:

$$\text{Sliding Distance} = \left(\frac{\sqrt{a * t_i}}{3*(1+s)}\right) * \left\{\frac{\beta^{1.5}}{1-\beta} + 4s(s(1-\beta))^{0.5} - 3s\beta^{0.5}\right\}$$

Only four parameters are needed to determine the sliding distance:

a = work roll radius = 7.457e-2 m
t_i = entry metal thickness = 6.096e-4 m
s = forward slip ratio
β = metal reduction ratio = 0.07, 0.15, 0.25, 0.35, 0.45, 0.50 and 0.55

[5]Average sliding speed was calculated by dividing the sliding distance for a point passing through the roll bite by the time required for that point to pass through the roll bite (see Eq. [7]).

[6]Sliding distance for rolling one square meter of metal.

and atomic absorption spectroscopy (AA) was used to measure how much aluminum was in the HCl solution. The amount of debris in the lubricant was quantitatively measured by capturing the lubricant after it cascaded down from the work rolls and then mixing it with HCl to dissolve the fines and once again using atomic absorption spectroscopy to measure how much aluminum was in the HCl solution. Debris particles were also analyzed to determine their composition and size distribution. The exact details of the tests procedures are found in the next section.

EXPERIMENTAL

Laboratory Rolling Mill Setup

Parameters and mill settings used during the test:

- Lubricant: Proprietary
- Lubricant Flow Rate: 10,000 mL/min
- Lubrication Mode: Mineral Oil Application
- Alloy: Cleaned 5182-O (zero smudge), 6.096e^{-4} m thick, 0.102 m wide
- Rolling Conditions: 122 mpm exit speed
- Work Roll Setup: Bare 52100 Steel, 0.46 micron roughness, 0.149 m diameter, 2 high configuration

Seven coils were rolled taking the following reductions: 7 percent, 15 percent, 25 percent, 35 percent, 45 percent, 50 percent and 55 percent. Above a 50 percent reduction, the roll coating became excessive, and a herringbone defect was present on the metal.

Mill Operation

The mill operator threaded the mill at 61 mpm until the desired reduction was obtained. After threading, the exit speed of the metal was increased to 122 mpm. Thirty seconds after the metal reduction and exit speed were obtained, a marker was placed in the coil. A lubricant sample was obtained using a 1000 mL tall beaker to collect the dirty oil washing down from the work rolls and the mill housing. Approximately 900 mL of lubricant was collected. The lubricant was not recycled to prevent a buildup of debris in the lubricant. The lubricant was collected again at 60 and 90 seconds. Markers were placed in the coil to designate these times.

Procedure for Measuring Debris in Lubricants

The three samples of lubricant taken at 30, 60 and 90 seconds of run time were combined to obtain approximately 2700 mL of total volume. Before combining, each sample of lubricant was thoroughly stirred with a magnetic stir bar to insure that all the debris was in suspension.

1000 mL of the combined lubricants were placed into a 1000 mL beaker, and 100 mL of 1:1 HCl/H_2O solution were added. This mixture was allowed to react overnight. A small aliquot (~10 mL) of the reacted solution was taken out of each container and analyzed for aluminum by AA. From the AA value and the value of the flow rate of the lubricant (10 L/min), it was possible to calculate the amount of debris washed off the surface of the work rolls per square meter of metal rolled.

Prior to dissolving the fines with the HCl solution, a visible spectrum of a small oil sample (1 mL) from each reduction was taken using a UV-visible spectrophotometer. The relative differences in spectra at 525 nm were due to the scattering of light by the debris particles and, therefore, represented differences in turbidity.

Rayleigh scattering measurements of the oil samples were taken using a commercial instrument to measure the size distributions of the debris.

Smudge Analysis

The procedure developed to measure the smudge is found in the appendix. Each smudge value represents the average of three separate smudge buffings at 30, 60 and 90 seconds of run time.

Surface Analysis

The composition and morphology of the smudge particles were analyzed using Auger electron spectroscopy. Auger spectra and secondary electron (SEM) images were obtained with a PHI Model 670 Scanning Auger Nanoprobe System which is equipped with a Schottky Field Emission source, a coaxially mounted Cylindrical Mirror Analyzer and a multichannel detector. The samples were analyzed using a primary electron beam energy of 10 keV, a specimen current of 10 to 20 nA and an 8 mil aperture. The diameter of the electron beam under those operating conditions was 500 Å (50 nm). The smudge particles were removed directly from the metal surface using conductive carbon tape and mounted on a stainless steel specimen holder for analysis. Auger depth profiles were obtained by rastering a 4 keV argon ion beam over a 6 × 6 mm^2 area. The sputtering rate was calibrated by depth profiling a 200 Å Al_2O_3/Al barrier oxide standard. The sputter rate used for these analyses was 20 Å/min. Oxide thicknesses were calculated from the sputtering time needed to reduce the O KLL Auger signal to half its maximum intensity. The Auger profile data were analyzed using the PHI Matlab Linear Least Squares software, separating the Al LMM Auger peak into its oxide and metal components. Smudge particles were analyzed from the 7 percent and 50 percent reduction experiments. The conductive carbon tape was analyzed prior to its use to insure that no contaminants were introduced onto the surface of the particles from the tape. The Auger spectrum obtained for the surface of the carbon tape showed the only surface components to be carbon and oxygen.

RESULTS

Table 2 lists the measured values for smudge, debris washed off the work roll and total debris (which is the sum of the first two). The values are reported in mg of Al/m^2 of aluminum exiting the roll bite. It should be noted that both the bottom and top surfaces of the metal are accounted for.

The visible spectra of the oil phase and its debris for each reduction are shown in Fig. 4. Comparing the relative transmittance at 525 nm it can be seen that there is a gradual decrease in transmittance up to a reduction of 50 percent. The transmittance increases again for the sample taken at 55 percent reduction. These differences in values represent turbidity differences and are due to differences in the number of suspended fines or debris particles. The relative differences also correspond with the relative differences in the amount of debris in the oil phase as measured in units of mg of Al/m^2 of rolled metal (Table 2). This indicates that the results of the experiments are reliable.

The size distribution of the fines in the oil samples from the Rayleigh scattering measurements are shown in Table 3. Quite surprisingly, this data shows that the average size of the debris particles (fines) in the oil phase are relatively the same for all of the different reductions taken, but the reliability of the data at the lower reductions is questionable due to a small number of particles in solution. This data also supports the conclusion that the relative differences in the visible spectra are a relative measure of the number of fines or debris suspended in solution since the size does not change. (Turbidity is a function of both number and size of suspended particles.)

SEM micrographs for the smudge debris particles removed directly from the rolled sheet for the 7 percent and 50 percent reductions are shown in Figs. 5 and 6, respectively. The micrographs taken at low magnifications (<1000×) show the arrangement/distribution and directionality of the smudge particles; the debris was oriented in the rolling direction. The micrographs also show the variability in the size of the smudge particles. The "length" of the smudge particles was observed to be as large as 50 μm to as small as <5 μm. The "width" of the particles was usually 10 μm. Particles that appear in long "strings," ~50 μm in length, are actually composed of many smaller pieces of debris. The morphology of the particles can be seen in the higher magnification SEMs. Many of the debris particles appeared to be grooved (also in the rolling direction), and many have several grooves indicating that the particles were probably rolled over several times. The morphology and size distribution of the particles did not appear to be different for either reduction; neither the 50 percent or 7 percent reduction favored the formation of one particular size of particle. In other words, the debris removed from the metal surface for the 7 percent reduction appeared the same as that removed from the 50 percent reduction metal.

Auger survey spectra taken from the surfaces of several smudge particles are shown in Fig. 7. The spectra indicated that the major surface components were Al, O, C and Mg, as would be expected. Major amounts of Si were also detected on the particle surfaces, and its origin is not known at this time; Si was not introduced from the C tape (see Experimental section). The surface composition of the smudge particles was uniform, and the average surface composition is summarized in Table 4. Auger spectra taken of the sheet surface,

Table 2—Smudge and Debris vs. Reduction					
Reduction	Smudge (mg)[1]	Total Smudge per Area (mg/m^2)[2]	Debris (mg)[3]	Debris per Area (mg/m^2)[3]	Load (W) per Reduction (kN)
7	2.20	11.84	65.2	13.13	65.0
15	2.62	14.10	68.2	13.78	112.8
25	3.04	16.36	74.9	15.07	146.6
35	3.24	17.44	99.2	20.02	176.9
45	3.08	16.58	116.1	23.36	208.4
50	2.76	14.85	129.0	26.05	225.8
55	2.68	14.42	116.4	23.47	255.3

[1]Smudge is defined in this chart as the aluminum fines which are found attached to the surface of the metal sheet. This is the sum from the top and bottom of a 0.91 meter long by 0.10 meter wide sheet.

[2]The exit speed for all reductions was 121.92 meter per min. The sample size was 0.91 m by 0.10 m. This length of sheet represents 1.82e-1 m^2 of area. This is determined by the following equation:

$$\text{Area } (m^2) = \text{Length of sample } (m) \times \text{Width } (m) \times 2 = 0.91 \text{ m} \times 0.10 \text{ m} \times 2 = 0.182 \text{ m}^2$$

The total smudge per area was then determined by the following equation:

$$\text{Total smudge per area } (mg/m^2) = \text{Total smudge } (mg) \div \text{Area } (m^2) = \text{Total smudge } (mg) \div 0.182 \text{ m}^2$$

[3]Debris is defined as the amount of aluminum fines in the lubrication after rolling. To determine the amount of debris in the oil, 1000 mL of oil was captured and mixed with 100 mL of 1 : 1 HCl/H_2O. The acid was sent to analytical for AA analysis for aluminum. The flow rate of the oil was 3000 mL per 18 sec. Therefore, 2000 mL represents 12 sec of run time. The area of the metal which exits the Fenn mill in 12 sec is:

$$\text{Area of metal } (m^2) = \text{Time of sampling (sec)} \times \frac{1 \text{ min}}{60 \text{ sec}} \times \text{Exit speed} \left(\frac{m}{min}\right) \times \text{Width } (m) \times 2$$

$$= 12 \text{ sec} \times \frac{1 \text{ min}}{60 \text{ sec}} \times \frac{121.9 \text{ m}}{\text{min}} \times 10.16e\text{-}2 \text{ m} \times 2 = 4.95 \text{ m}^2$$

The debris per area (mg/ft^2) is equal to:

$$\text{Debris per area } (mg/m^2) = \frac{\text{Total debris (mg)}}{\text{Area of metal } (m^2)} = \frac{\text{Total debris (mg)}}{4.95 \text{ m}^2}$$

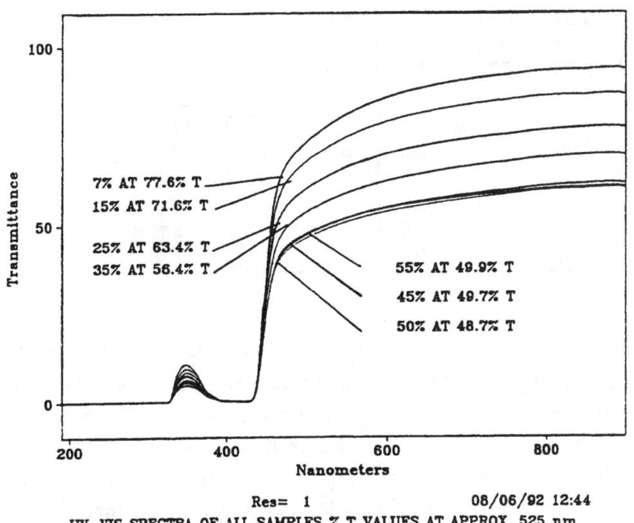

Fig. 4—UV-Vis spectra of dirty oil samples.

as-rolled and after a 10-min sputter cleaning are shown in Fig. 8. Also included in that same figure is the Auger survey spectrum of a sputtered smudge particle. The surface compositions are indicated in Fig. 9. These spectra show that the surface composition of the smudge particles is very similar to the sheet surface except for the presence of Si, which was only observed on the smudge particles. Mg, present in the alloy at 0.8–1.3 percent, is almost always observed on sheet surfaces (of Mg containing Al alloys) after rolling and was also seen to be present on the surface of the smudge particles. One of the main driving forces for the cause of the migration of Mg to the metal surface is heat (*3*), (*4*), and the observation that Mg was also present on the surface of the smudge particle shows that the oxide formation is similar to the sheet surface, and the particles probably see the same temperatures as the sheet during rolling. After removal of the oxide surface by sputtering, the spectra indicate that the surface compositions were nearly identical, and the smudge particle had the same composition as the metal sheet.

Auger depth profiles were obtained for the smudge particles to determine oxide thicknesses. A representative depth profile is shown in Fig. 9. The depth profile shows the compositional variation with depth of the surface oxide components (C, O, metallic Al, Al oxide and Mg oxide). The depth profile shows that the oxide thickness was <100 Å, and the oxide was a mixture of Al and Mg oxides. The depth profile is very similar to that observed for cold-rolled Al sheet surfaces (also shown in Fig. 9). These results indicate that each of the smudge particles present on the rolled Al surface is a metallic Al particle covered with a thin oxide layer.

DISCUSSION

According to the Archard equation, the amount of debris

TABLE 3—RAYLEIGH SCATTERING: SIZE DISTRIBUTION OF DEBRIS AT VARIOUS REDUCTIONS

#2 7% Time = 100 sec Samp. Avg. of 3 Runs		#2 15% Time = 100 sec Samp. Avg. of 3 Runs		#2 25% Time = 100 sec Samp. Avg. of 3 Runs		#2 35% Time = 100 sec Samp. Avg. of 3 Runs		#2 45% Time = 100 sec Samp. Avg. of 3 Runs	
DV	0.0019	DV	0.0019	DV	0.0089	DV	0.0071	DV	0.0100
%10	0.29	%10	0.33	%10	0.29	%10	0.35	%10	0.35
%50	0.46	%50	0.70	%50	0.90	%50	0.78	%50	0.89
%90	0.70	%90	1.23	%90	2.52	%90	1.36	%90	2.39
MV	0.48	MV	0.73	MV	1.12	MV	0.78	MV	1.15
CS	14.946	CS	11.627	CS	10.132	CS	9.922	CS	8.832

Percent Passing			Percent Passing			Percent Passing			Percent Passing			Percent Passing		
Chan.	Cum.	Vol.	Chan.	Cum.	Vol.	Chan.	Cum.	Vol.	Chan.	Cum.	Vol.	Chan.	Cum.	Vol.
42.21	100.0	0.0	42.21	100.0	0.0	42.21	100.0	0.0	42.21	100.0	0.0	42.21	100.0	0.0
29.85	100.0	0.0	29.85	100.0	0.0	29.85	100.0	0.0	29.85	100.0	0.0	29.85	100.0	0.0
21.10	100.0	0.0	21.10	100.0	0.0	21.10	100.0	0.0	21.10	100.0	0.0	21.10	100.0	0.0
14.92	100.0	0.0	14.92	100.0	0.0	14.92	100.0	0.0	14.92	100.0	0.0	14.92	100.0	0.0
10.55	100.0	0.0	10.55	100.0	0.0	10.55	100.0	0.0	10.55	100.0	0.0	10.55	100.0	0.0
7.46	100.0	0.0	7.46	100.0	0.0	7.46	100.0	0.0	7.46	100.0	0.0	7.46	100.0	0.0
5.27	100.0	0.0	5.27	100.0	0.0	5.27	100.0	0.0	5.27	100.0	0.0	5.27	100.0	0.0
3.73	100.0	0.0	3.73	100.0	0.0	3.73	100.0	11.7	3.73	100.0	0.0	3.73	100.0	4.2
2.63	100.0	0.0	2.63	100.0	2.7	2.63	88.3	7.6	2.63	100.0	0.0	2.63	95.8	21.7
1.69	100.0	0.0	1.69	97.3	17.3	1.69	80.8	20.4	1.69	100.0	25.4	1.69	74.0	15.4
1.01	100.0	11.1	1.01	80.0	28.7	1.01	60.4	23.7	1.01	74.6	33.4	1.01	58.6	22.6
0.66	88.9	45.8	0.66	51.3	35.3	0.66	36.7	8.7	0.66	41.1	19.5	0.66	36.0	15.9
0.43	43.1	20.3	0.43	26.0	8.8	0.43	27.9	11.4	0.43	21.6	12.7	0.43	20.1	11.0
0.34	22.9	12.0	0.34	17.2	9.2	0.34	16.5	9.1	0.34	8.9	5.8	0.34	9.2	5.2
0.24	10.8	7.4	0.24	7.9	5.4	0.24	7.5	5.2	0.24	3.1	2.5	0.24	3.9	2.8
0.17	3.5	3.5	0.17	2.5	2.5	0.17	2.3	2.3	0.17	0.7	0.7	0.17	1.1	1.1

#2 50% Time = 100 sec Samp. Avg. of 3 Runs		#2 55% Time = 100 sec Samp. Avg. of 3 Runs	
DV	0.0143	DV	0.0072
%10	0.21	%10	0.32
%50	0.86	%50	0.67
%90	2.35	%90	1.77
MV	1.04	MV	0.89
CS	12.975	CS	10.676

Percent Passing			Percent Passing		
Chan.	Cum.	Vol.	Chan.	Cum.	Vol.
42.21	100.0	0.0	42.21	100.0	0.0
29.85	100.0	0.0	29.85	100.0	0.0
21.10	100.0	0.0	21.10	100.0	0.0
14.92	100.0	0.0	14.92	100.0	0.0
10.55	100.0	0.0	10.55	100.0	0.0
7.46	100.0	0.0	7.46	100.0	0.0
5.27	100.0	0.0	5.27	100.0	0.0
3.73	100.0	4.2	3.73	100.0	0.0
2.63	95.8	16.5	2.63	100.0	11.1
1.69	79.3	29.8	1.69	88.9	25.9
1.01	49.4	3.8	1.01	63.0	12.3
0.66	45.6	0.0	0.66	50.7	20.0
0.43	45.6	9.7	0.43	30.7	15.6
0.34	35.8	19.8	0.34	15.1	9.6
0.24	16.1	11.0	0.24	5.5	4.6
0.17	5.1	5.1	0.17	1.0	1.0

generated in the roll bite should increase with an increase in sliding distance and load. Figure 10 shows that the total amount of debris does increase with an increase in load and sliding distance, but the rate of increase is not linear, plus there is a slight decrease in the amount of debris produced at 55 percent reduction (Table 2). This slight decrease is probably a result of lubricant failure because a herringbone defect was seen on the metal, and there was also a heavy buildup of roll coating on the work rolls.

It does appear, however, the Archard equation is followed at light reductions (7–30 percent) where the amount of debris produced increases linearly with the product of load and sliding distance before it deviates at higher reductions. Interestingly, when accounting for the total amount of debris exiting the roll bite as to how much remains on the sheet as smudge and how much adheres to the work roll and is washed off, it is seen at reductions less than 30 percent that the partition ratio is equal to one and at higher reductions it is less than one. In fact, at 45 percent reduction, so much more debris adheres to the work roll that the amount of smudge actually decreases even though the total amount of debris generated in the roll bite continues to increase with an increase in reduction. This suggests that the deviation from the Archard equation is due to a change in the mechanism of generating debris in the roll bite and that this change occurs when the partition ratio deviates from one.

A closer examination of the Archard equation may help to explain this. According to Eq. [1], the amount of debris generated in the roll bite should increase with an increase in reduction because the area of contact between the asperities on the work roll and the aluminum sheet increases. (As reduction increases, the asperities dig deeper into the surface of the aluminum metal and should remove more material.) However, an analysis of the size and composition of the smudge particles by Auger or the size distribution of particles in the oil phase or on the surface of the sheet show that there is very little change either in composition or size distribution with a change in reduction. In other words, it does not appear that the average asperity on the surface of the work roll

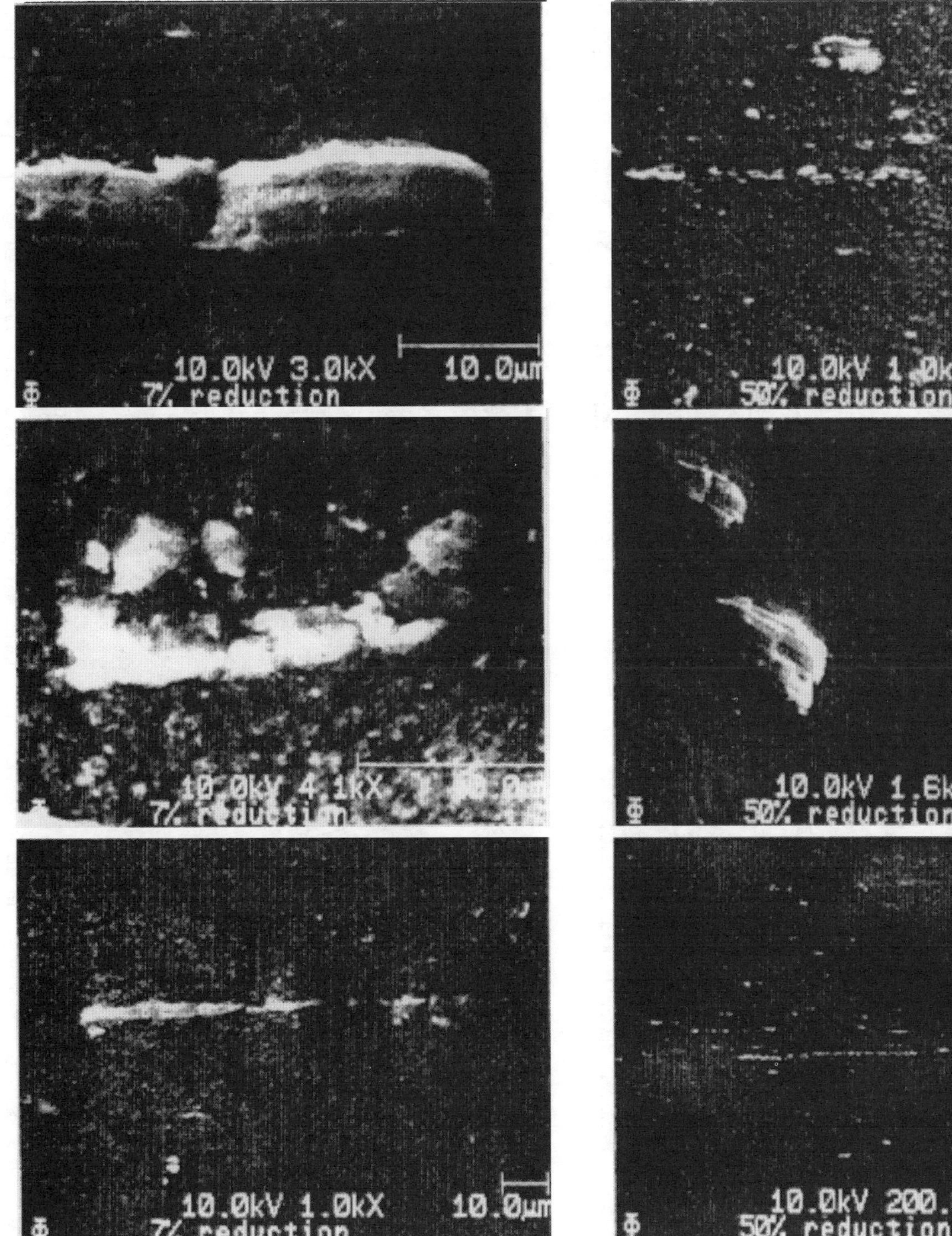

Fig. 5—Smudge debris from seven percent reduction.

Fig. 6—Smudge debris from 50 percent reduction.

cuts the surface to form debris by digging into the surface, but rather metal flows around it to form a trough.

Furthermore, Wilson and Sheu have shown that the area of contact between an asperity on a workpiece and tool during plastic deformation is not equal to W/H (5) which is true for elastic deformation only. Equation [1], therefore, must be modified to calculate the real area of contact from the apparent area of contact, giving Eq. [8]:

$$\frac{V}{SD} = KfA \qquad [8]$$

where A is equal to the apparent area of contact and f is equal to the fractional area of contact.

It is not possible to calculate the real area of contact for this series of reductions. However, if it is assumed the fractional area of contact (f) is equal to one, which requires the lubrication regime to be boundary lubrication (5), then it is found that the shape in Fig. 10 does not change when milligrams of aluminum is plotted against the product of sliding distance (SD) and apparent area of contact (A) (Fig. 11). This suggests that increasing reduction simply increases the number of cutting surfaces (i.e., an irregular shaped asperity)

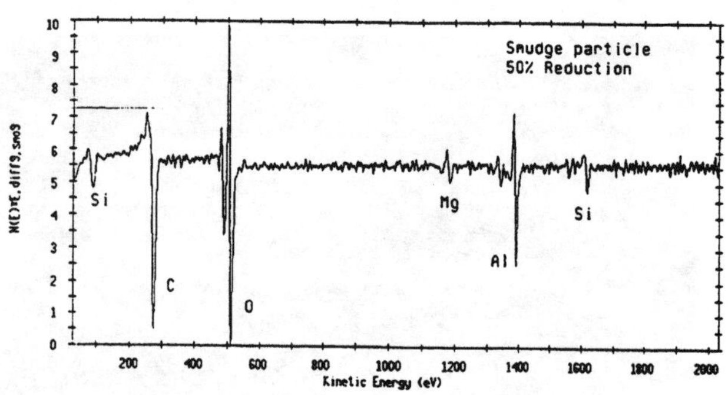

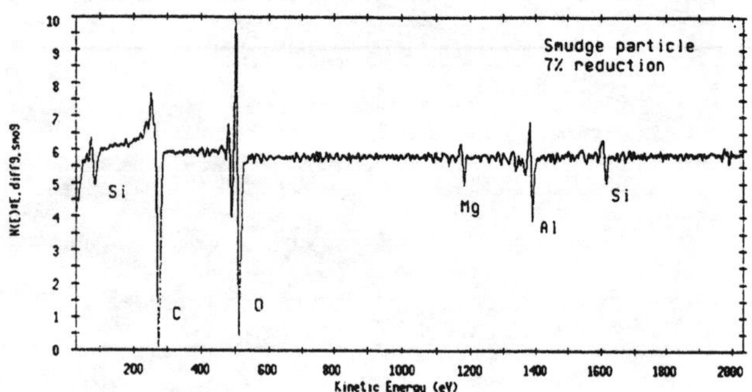

Fig. 7—Auger spectra of smudge particles.

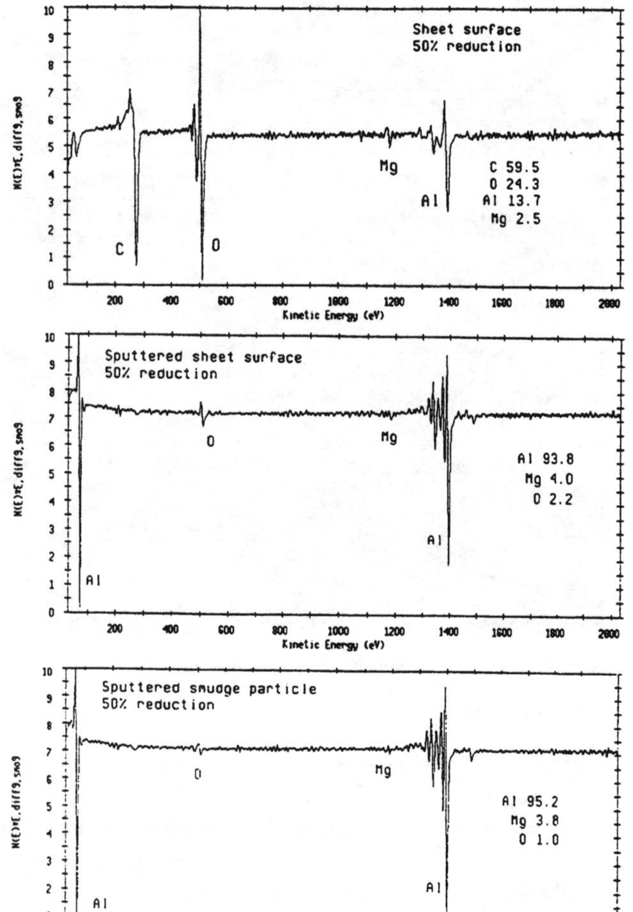

Fig. 8—Auger survey spectra of sheet surface (top), sputtered sheet surface (middle) and sputtered smudge particle (bottom).

TABLE 4—SURFACE COMPOSITION OF SMUDGE PARTICLES DETERMINED BY AUGER (ATOMIC %)					
SAMPLE	C	O	Al	Mg	Si
50% reduction	53.8	21.8	13.8	3.4	7.2
7% reduction	66.9	16.8	8.2	2.8	5.3

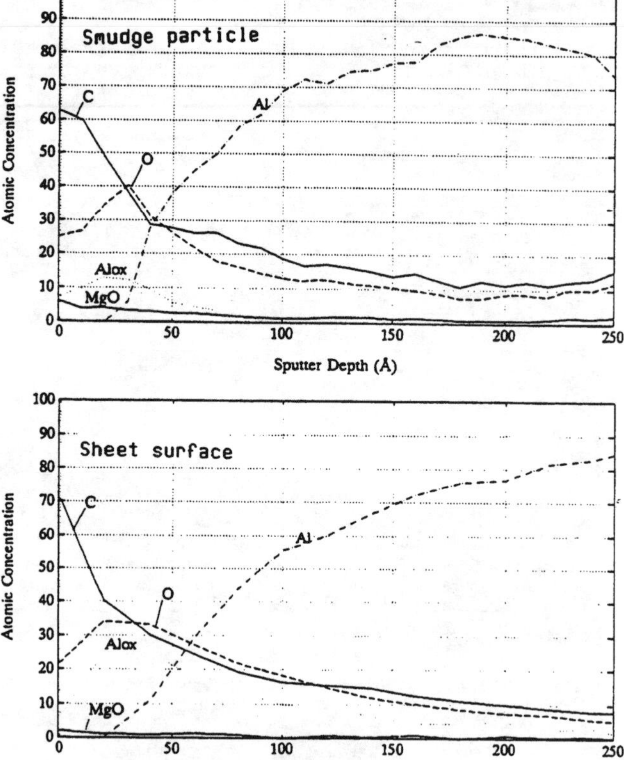

Fig. 9—Auger depth profile of a smudge particle (top) and sheet surface (bottom).

in contact with the surface of the aluminum sheet rather than the increase in the area of contact of each asperity.

Unfortunately, this does not explain why the wear rate deviates from the Archard equation at higher reductions or why the partition ratio is one at reductions less than 30 percent and then deviates from one at the higher reductions. A third factor changes, however, when changing reduction which should also affect how much debris is generated in the roll bite, namely the lubricant film.

According to the Wilson-Walowit equation (*6*), represented by Eq. [9], increasing reduction decreases the amount of lubricant in the roll bite. In fact, a +55 percent reduction, there is not enough lubricant to prevent metal from grossly adhering to the surface of the work roll, causing a herringbone defect. Hence, it is proposed that at light reductions, there is a sufficient amount of lubricant in the roll bite such that the debris which is generated becomes completely coated with boundary additives and, therefore, the debris does not favor adhering to the work roll surface or the sheet's surface. Consequently, the debris is evenly divided between the two. As the reduction increases, the amount of lubricant

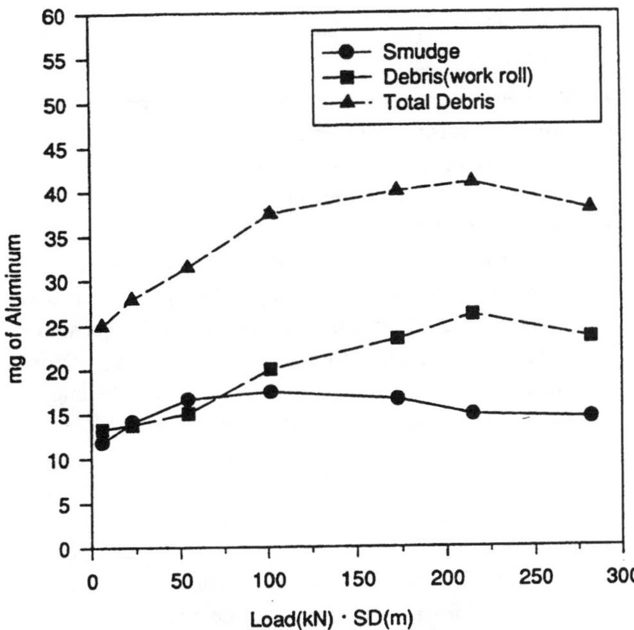

Fig. 10—Smudge, debris on the work roll, and total debris.

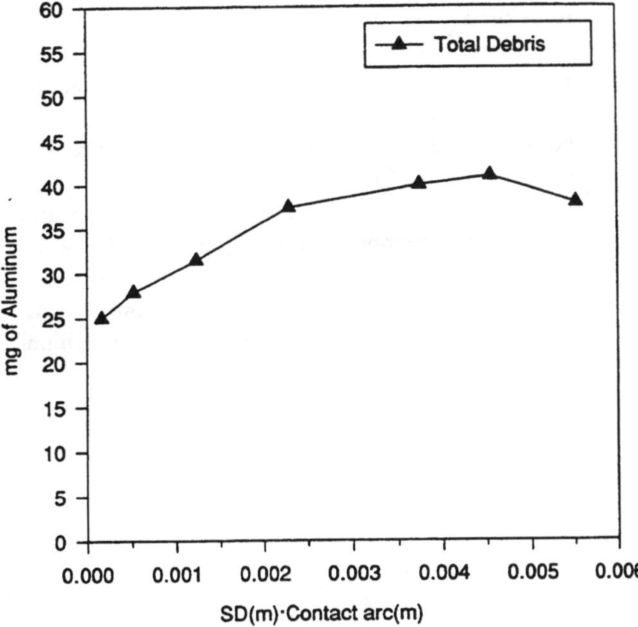

Fig. 11—Total debris as a function of sliding distance and apparent area of contact.

in the roll bite decreases, increasing the amount of adhesion of metal to the work roll and the amount of adhesive wear due to a lack of protection from the adsorption of boundary additives. This also causes the debris to adhere more to the work roll rather than the sheet. It is proposed, therefore, at light reductions the mechanism of forming debris in the roll bite is due to abrasive wear, and at very high reductions the mechanism is due to adhesive wear. It should be emphasized, this is only a hypothesis which satisfactorily explains the test results.

$$h = \frac{3\eta\infty(V_1 + V_2)}{1 - e^{-\infty(Y - S)}}\left(\frac{R}{G_1 - G_2}\right)^{1/2} \quad [9]$$

In Eq. [9], h = film thickness, η = viscosity, ∞ = viscosity pressure coefficient, V_1 = velocity of metal at the entry to the roll bite, V_2 = velocity of the work roll, Y = yield stress of the metal, R = work roll radius, G_1 = entry gauge, G_2 = exit gauge and S = tension.

SUMMARY

The amount of debris generated in the roll bite increases with an increase in reduction. At light reductions (and under these process conditions), half of the debris remains on the sheet as smudge and the other half adheres to the work roll where it is washed off. However, at heavier reductions, more debris adheres to the work roll than that which adheres to the sheet such that the amount of smudge actually decreases even though the amount of debris generated in the roll bite has increased. Also, at light reductions, the wear rate appears to follow the Archard equation but deviates from it at heavier reductions. It is proposed this deviation is due to the wear mechanism changing from abrasive wear at light reductions to adhesive wear at heavier reductions. It's also proposed that the change in wear mechanism is due to a decrease in the amount of lubricant going through the roll bite at heavier reductions.

ACKNOWLEDGMENT

The authors would like to thank Dr. Simon Sheu for his helpful comments and valuable suggestions.

REFERENCES

(1) Archard, J. F., "Contact and Rubbing of Flat Surfaces," *Jour. of Appl. Phys.*, **24**, pp 981–988, (1953).

(2) Hector, L. G. and Sheu, S., "Focused Energy Beam Work Roll Surface Texturing Science and Technology," *Jour. of Materials Processing and Manufacturing Sci.*, **2**, 1, pp 63–117, (1993).

(3) Wakefield, G. R., "The Composition of Oxides Formed on Al-Mg Alloys," *Appl. Surf. Sci.*, **51**, pp 95–102, (1991).

(4) Goldstein, B. and Dresner, J., "Growth of MgO Films with High Secondary Electron Emission on Al-Mg Alloys," *Surf. Sci.*, **71**, pp 15–26, (1978).

(5) Wilson, W. R. and Sheu, S., "Real Area of Contact and Boundary Friction in Metal Forming," *Int'l. Jour. of Mech. Sci.*, **30**, 7, pp 475–489, (1988).

(6) Wilson, W. R. D. and Walowit, J. A., "An Isothermal Hydrodynamic Lubrication Theory for Strip Rolling with Front and Back Tension," *Inst. Trib. Conv.*, I. Mech. E., London, pp 169–172, (1971).

APPENDIX

Standard Operating Procedure to Measure Smudge

1. Equipment and Material

The following equipment and materials were used: pneumatic orbital/straight line pad sander; ~(23 cm × 23 cm) square cheesecloth; N-butanol; concentrated HCl; goggles; latex, rubber and cotton gloves; glass syringe; stopwatch; 4 oz. plastic bottles; 20 mL plastic vial; burettes; brown wrapping or Kraft paper; and hearing protection.

2. Preparation Steps

2.1 (Wear cotton gloves when cutting the metal.) Prepare metal sample from the plant stocks or the laboratory mill. The sample made from the laboratory mill should be a 96.5 cm long strip with as-received width (usually 15.2 cm).

2.2 Cut two ~(23 cm × 23 cm) square cheesecloths in half for each test. Need a minimum of three halves (11.5 cm × 23 cm).

2.3 Obtain a hood area with a large work space. Place masking tape down on the hood countertop 91.44 cm apart. Place brown wrapping paper down on the hood countertop to cover the entire working area beneath the metal sample being tested.

2.4 Prepare a 20% volume acid solution, in a hood, using concentrated hydrocyloric acid and DI water. (Example: 80 mL DI and 20 mL HCl.)

3. Start-Up Steps

3.1 Using the masking tape marks on the hood countertop, as a guide, tape ~2.54 cm on each end of the metal strip leaving 91.44 cm of length exposed. The top side of the metal should be facing upwards.

4. Run Test

4.1 Buffing Steps

4.1.1 (Wear latex gloves and hearing protection.) Secure a new cheesecloth to the orbital sander. Draw 1 mL of N-butanol into the glass syringe and apply to the metal surface. Start buffing immediately for two minutes. Apply no pressure to the sander.

4.1.2 Apply an additional 1 mL of N-butanol. Continue to buff, for two minutes, with the orbital sander.

4.1.3 Repeat Step 4.1.2.

4.2 Remove cheesecloth and place the cheesecloth to the side until completely done buffing.

4.3 Repeat buffing Steps 4.1.1 and 4.1.2 using a second cheesecloth (second buffing).

4.4 Repeat Step 4.1 ONLY! (third buffing).

4.5 Visually check the appearance of the cheesecloth.

4.6 If the cloth is still dark, remove the cheesecloth and repeat Step 4.1.1 until the cheesecloth no longer shows any sign of smudge (fourth buffing). [Note: A more "quantitative" method of smudge removal efficiency is the Tape Test. On the metal surface apply a ~2.54 cm piece of double sided tape. Smooth out all air bubbles between the metal and the tape. With tweezers remove the tape and place onto a glass microscope slide. Smooth out all air bubbles and view the slide under a low magnification microscope. If smudge is present it will be visible as dark specks on the tape.]

4.7 Place all the cheesecloths into a 4 oz. plastic bottle. If more than three cheesecloths were used in the smudge buffing, then a plastic bottle larger than 4 oz. will have to be used.

4.8 (Wear proper gloves and goggles.)
For one cheesecloth, place 50 mL of 20% acid solution into the bottle.
For two–three cheesecloths, place 100 mL of 20% acid solution into the bottle.
For three cheesecloths or more, place 50 mL of acid solution into the bottle for EACH cheesecloth used.

4.9 After soaking a minimum of 10 hours, stir the cheesecloth around for about one minute. Remove ~15 mL of acid solution and send for AA analysis of aluminum only.

FLAT ROLL PRODUCTS II

GALVANNEALING BEHAVIOR AND COATING PERFORMANCE OF Ti-IF AND Nb-Ti-IF STEELS

Chann C. Cheng
Ispat Inland Research Laboratories
3001 E. Columbus Drive
East Chicago, IN 46312, USA

Key Words: Galvanneal Coating, Ti-IF, Nb-Ti-IF, Hot Dip Simulator, Structure, Double Olsen Test

ABSTRACT

The Hot Dip Simulator (HDS) at Ispat Inland Research Laboratories was used to simulate production line galvannealing conditions and produce galvannealed Ti-IF and Nb-Ti-IF samples for comparison. Results show that, by using the same heating cycle, Zn pot conditions, and atmosphere environments, the average galvannealing rates of Ti-IF and Nb-Ti-IF steels are comparable. However, the galvannealing response of Ti-IF steel is not as consistent as that of Nb-Ti-IF steel. With the similar Fe% in the coating, the two substrates exhibit different phase distributions within the galvanneal coating as well as different Double Olsen test results.

INTRODUCTION

Interstitial-Free(IF) steel is one of the preferred substrates for automotive exposed galvannealed products. In North America, between two kinds of galvannealed IF steel, galvanneal manufacturers prefer Nb-Ti-IF steel than Ti-IF steel for automotive exposed applications. The preference is based on observations by galvanizing line operators that galvannealing conditions are more difficult to optimize for Ti-IF steel than that for Nb-Ti-IF steel[1,2]. In addition, during the stamping process, galvannealed Ti-IF steel generally possesses a worse powdering resistance than galvannealed Nb-Ti-IF steel[2,3]. However, galvannealed Ti-IF steel offers increased productivity advantages in both steelmaking and hot rolling as well as reduced raw alloy costs without mechanical property deterioration.

The general perception of production difficulties and poor powdering resistance of galvannealed Ti-IF steel was based on data obtained from commercial hot dip galvanizing lines. Most laboratory studies on galvannealed IF steels considered Ti-IF and Nb-Ti-IF steels as the same steel substrate. To date, there has been no direct evidence showing a difference in galvannealing behavior and coating performance between Ti-IF steel and Nb-Ti-IF steel that were processed under the same conditions. One possible explanation is that the investigative tools and methodologies used were not capable of detecting the subtle differences between the two subject IF steels. For instance, most galvannealing research methodology in galvannealing uses pre-galvanized steel that is heat-treated by rapid heating in a Gleeble Furnace. The methodology of re-melting pre-coated steel substrates cannot accurately simulate the dynamic conditions of the actual hot dipping and galvannealing processes in which Zn remains in the liquid state until the galvannealing process is near completion. Another significant deviation of the Gleeble methodology from galvanized steels is that the Al content in the coating of pre-galvanized steels is much higher than that in the actual galvannealed steels. This may result in a much thicker Fe_2Al_5 layer at the coating/steel interface during the hot dipping stage, thus eliminating or minimizing the already subtle differences between the Ti-IF and Nb-Ti-IF steels.

The Hot Dip Simulator (HDS) at Ispat Inland Research Laboratories was used to simulate production line galvannealing conditions and produce galvannealed Ti-IF and Nb-Ti-IF samples for comparison[4]. Results show that, by using the same heating cycle, Zn pot conditions, and atmosphere environments, the average galvannealing rates of Ti-IF and Nb-Ti-IF steels are comparable. However, the galvannealing response of Ti-IF steel is not as consistent as that of Nb-Ti-IF steel. With the similar Fe% in the coating, the two substrates exhibit different phase distributions within the galvanneal coating as well as different Double Olsen test results.

EXPERIMENTS

Materials

The two laboratory IF steels used in this study are essentially ultra low carbon IF steels, one stabilized with Ti only and one stabilized with Nb and Ti. They were processed in the laboratory to the same final thickness of 1.0 mm.

Hot Dip Simulator

A schematic drawing of the Ispat Inland HDS is shown in Figure 1. It was designed and built by Ispat Inland researchers[4]. The HDS consists of six major components: 1) Glove Box, 2) Infrared Furnace, 3) Induction Furnace, 4) Melt Pot, 5) Coating Weight Control Knives, and 6) Hydraulic Transfer Rod. The Control and Data Analysis System of the HDS automatically controls and integrates all the major functions. The panel size for the HDS is 102 mm x 306 mm. Two sets of samples were processed with the HDS. The first set consisted of 6 panels of each steel and the second set had 5 panels of each. During the simulation, sample panels were randomized within each set. The Zn spelter was obtained from the Zn bath of the Inland's commercial hot dip galvanizing line. At the beginning of each simulation, the Al content in the Zn bath was determined and changed, if necessary, to 0.14 wt%.

Figure 2 shows the heating cycle used in this study. The cycle includes an annealing step (controlled heating and holding) at 800°C, a controlled cooling step to the immersion temperature (460°C), dipping, coating knife wiping, galvannealing, and rapid cooling. All steps were controlled and monitored by the control system (microVax), with a thermocouple on the sample panel.

Galvanneal Coating Structure and Performance Test

Cross-sectional micros were prepared for phase evaluation. Rowland's reagent[5] was used to etch the micros to reveal galvanneal coating phases. Optical Microscopy was used to observe ς phase by the color contrast and SEM was utilized to measure the thickness of Γ phase at the steel/coating interface. The Double Olsen test (by weight loss) was conducted on a 5 cm diameter disc for the coating performance.

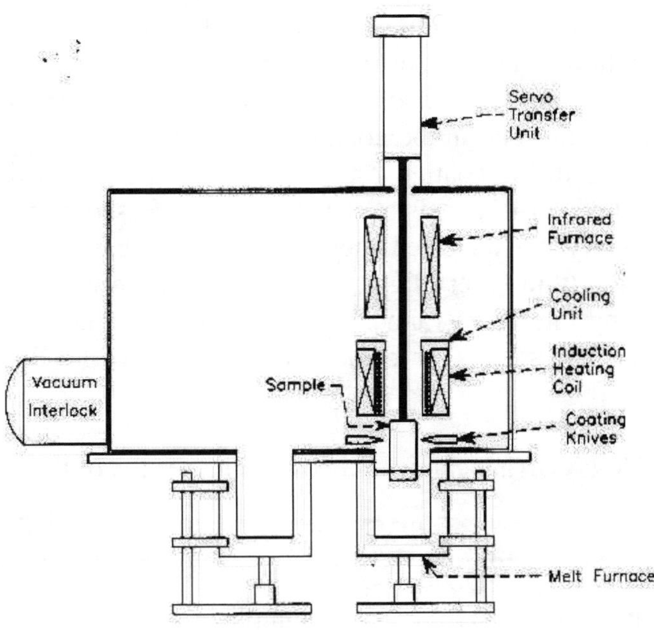

Fig. 1: Schematic drawing of the Ispat Inland Hot Dip Simulator.

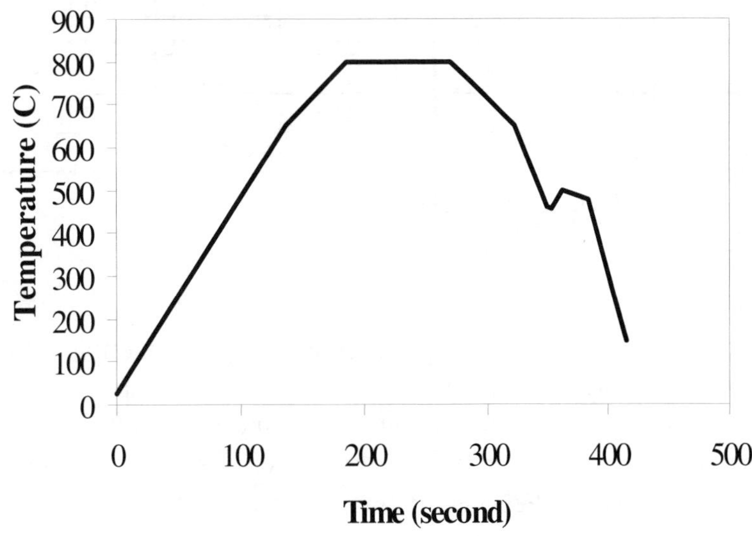

Fig.2: The Hot Dip Simulator heating cycle used in this study.

RESULTS

Galvannealing Behavior

Two sets of steel panels were processed by the HDS. The heating cycle of the first set was designed to produce a galvanneal coating with a coating weight aim of 40 g/m^2 and 11% Fe in the coating. The aim for the second set was 45 g/m^2 coating weight and 11% Fe in the coating. Table I shows the results for the two sets. The coating weight and %Fe for Nb-Ti-IF steel are quite close to the aims. The coating weight is consistent for Ti-IF steel in both sets. However, the %Fe in the coating is more variable than that of Nb-Ti-IF steel. Figure 3 shows the variation of the %Fe in the coating for both steel substrates in the two sets. Figure 4 shows the galvannealed Ti-IF panels and Nb-Ti-IF panels of the second set. Based on the appearance of the galvannealed panels, Nb-Ti-IF steel apparently galvannealed more consistently than Ti-IF steel. Within the same set, the coating weight was controlled within a very tight range, therefore Fe% in the coating can be considered as a good indicator of galvannealing rate or galvannealing behavior. In other words, by using a set condition, the galvannealing behavior of Nb-Ti-IF steel is more consistent than that of Ti- IF steel.

Table I: Galvannealing and powdering results of the two sets.

Set 1

Sample ID	Steel Spec.	Coating Weight (g/m^2)	Fe (%)	Double Olsen (mg)
73	Ti-IF	40.7	9.8	2.0
74	Ti-IF	41.0	10.8	3.8
75	Ti-IF	39.7	11.1	3.9
76	Ti-IF	38.7	9.4	2.3
77	Ti-IF	38.3	11.9	4.7
78	Ti-IF	42.5	10.5	3.0
82	Nb-Ti-IF	41.3	11.3	3.7
84	Nb-Ti-IF	42.1	11.3	3.9
85	Nb-Ti-IF	41.4	11.2	3.6
86	Nb-Ti-IF	38.5	11.8	4.2
87	Nb-Ti-IF	38.2	10.7	1.7
88	Nb-Ti-IF	38.8	11.6	2.7

Set 2

Sample ID	Steel Spec.	Coating Weight (g/m^2)	Fe (%)	Double Olsen (mg)
79	Ti-IF	44.0	10.6	3.7
710	Ti-IF	46.4	8.1	1.4
711	Ti-IF	43.8	11.3	4.4
712	Ti-IF	44.2	8.4	2.1
713	Ti-IF	43.8	10.0	2.8
89	Nb-Ti-IF	44.6	11.5	3.5
810	Nb-Ti-IF	44.5	11.1	2.9
811	Nb-Ti-IF	44.2	11.4	4.0
812	Nb-Ti-IF	44.4	10.4	2.7
813	Nb-Ti-IF	44.2	11.1	3.0

Galvanneal Coating Structure

Figure 5a is an optical micrograph showing the ς phase morphology of Sample 74 (Ti-IF steel with 10.8% Fe) near the coating surface. Comparably, Figure 5b shows the ς phase morphology of Sample 87 (Nb-Ti-IF with 10.7%Fe). It was observed that, with a similar coating weight and Fe% in the coating, galvannealed Ti-IF steel contained more ς crystals at the surface than galvannealed Nb-Ti-IF steel. Figures 6a and 6b are SEM photos of the two samples showing the Γ phase at the steel/coating interface. The thickness of Γ phase was estimated to be about 0.5 μm to 0.8 μm for all samples in both sets. There was no obvious difference in Γ phase thickness between the two galvannealed steels. However, as seen in Figures 5 and 6, the steel/coating interfacial structure is different between the two galvannealed steels. The steel/coating interface in galvannealed Ti-IF steel was found to be much rougher than that in Nb-Ti-IF steel. Further, the galvanneal coating surface of Ti-IF steel was also seen, Figure 5a, to be rougher than that of Nb-Ti-IF steel.

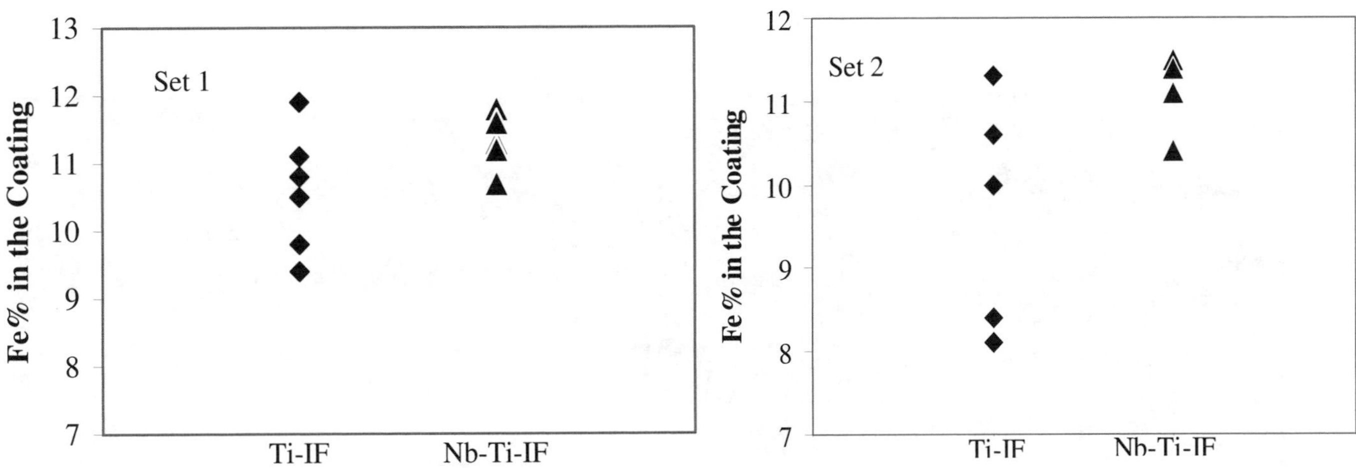

Figure 3: Galvannealing results of the two sets.

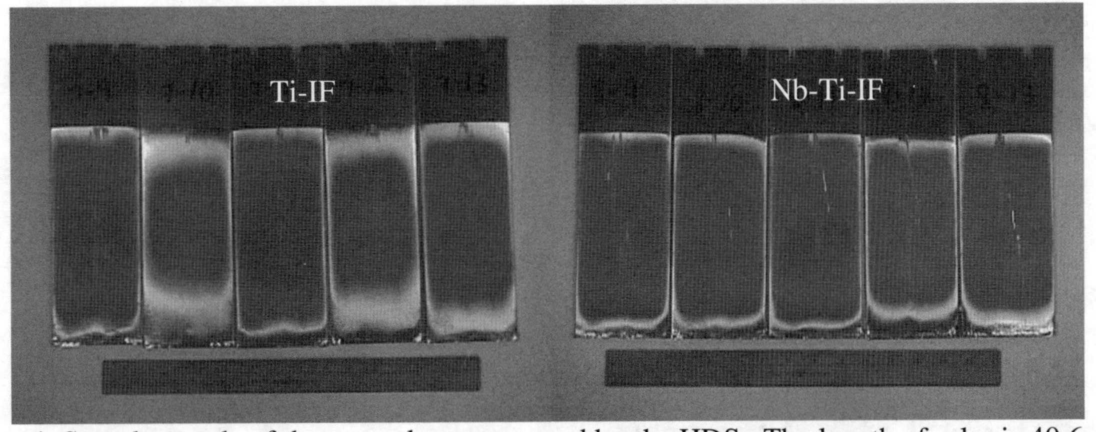

Figure 4: Sample panels of the second set processed by the HDS. The length of ruler is 40.6 cm.

Coating Performance

Coating performance was determined by weight loss using the Double Olsen method. Figure 7 is a plot of weight loss (mg) versus Fe% in the coating for both sets. It is clear that, with a similar coating weight and Fe content, galvannealed Ti-IF steel exhibited greater coating loss than galvannealed Nb-Ti-IF steel.

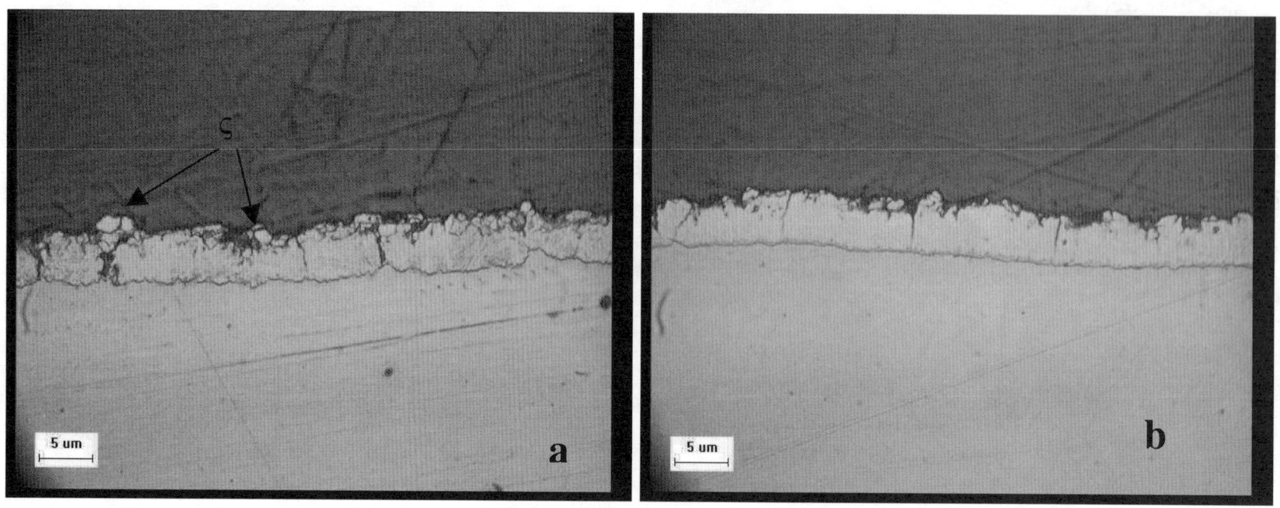

Figure 5: Optical micrographs of galvanneal coating. (a) Ti-IF steel, Sample ID 74, and (b) Nb-Ti-IF steel, Sample ID 87.

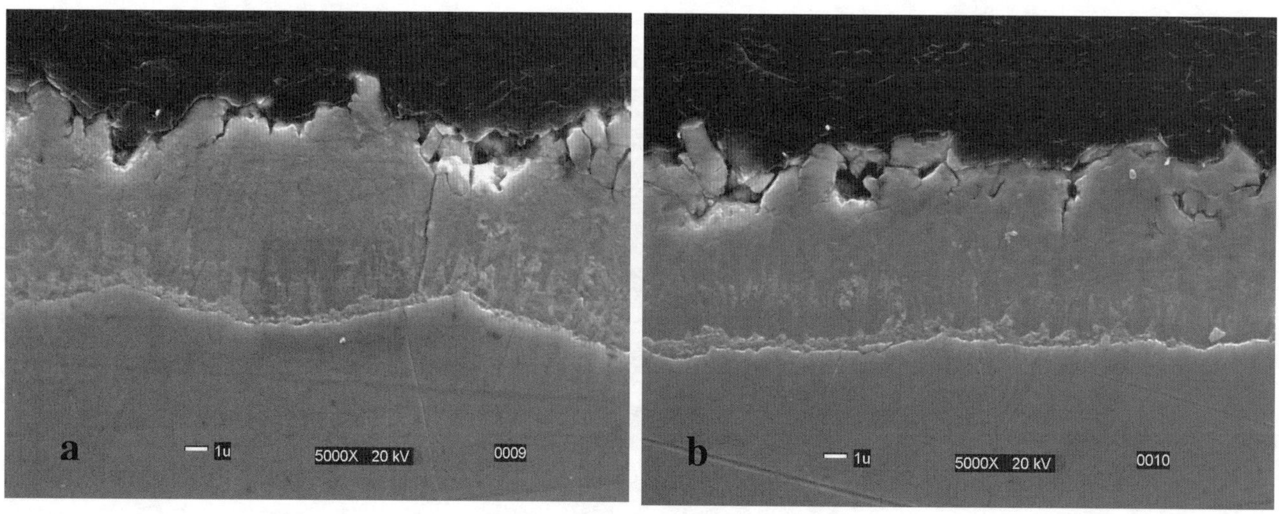

Figure 6: SEM micrographs of galvanneal coating. (a) Ti-IF steel, Sample ID 74, and (b) Nb-Ti-IF steel, Sample ID 87.

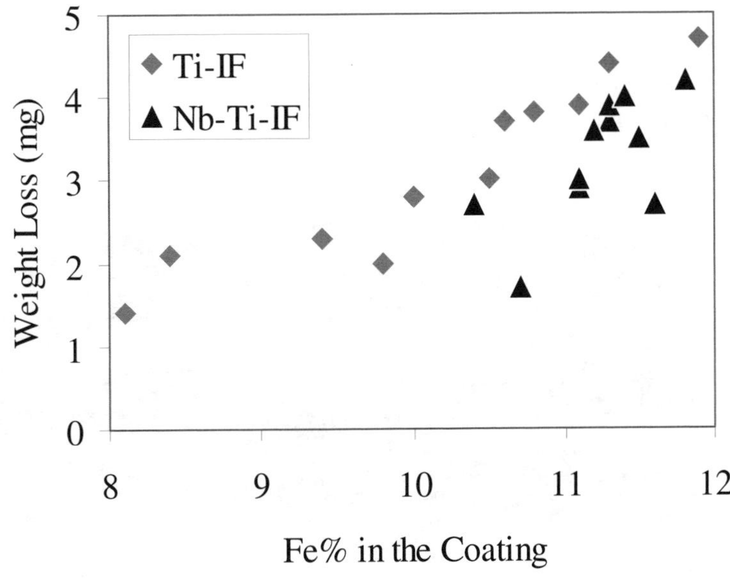

Figure 7: Double Olsen results of the two galvannealed steels.

DISCUSSION

A general conception about the galvannealing of Ti-IF steel is that it galvanneals faster than Nb-Ti-IF steel. The results of this study, Table I and Figure 3, suggest that this is not true. Rather, Nb-Ti-IF steel galvanneals at a comparable and more consistent rate. Moreover, Ti-IF steel sometimes (4 panels out of 11 panels) galvanneals at a slower galvannealing rate. The reason for the slow galvannealing is not clear. One possible reason is due to the differences in steel grain and sub-grain structures. It is calculated that, after stabilization with N, S, and C, Nb-Ti-IF steel has about 10 to 20% of its Nb left in the Fe matrix[6,7]. Ti-IF steel is estimated to have about 30 to 50% of its Ti in the Fe matrix[6,7]. It was demonstrated that free Nb atoms, since they are larger than Fe atoms, tend to segregate to steel grain and sub-grain boundaries[8]. Free Ti atoms, on the other hand, are more soluble in the Fe matrix and tend to stay within the steel grains[8]. Therefore, grain boundaries in Ti-IF steel are cleaner than that in Nb-Ti-IF steel.

It is generally believed that δ phase outburst at IF steel grain and sub-grain boundaries is the main mechanism for the galvannealing process[9,10]. Therefore, steel grain and sub-grain structure and their chemical conditions are the two major factors in determining the galvannealing process of IF steel. The difference in galvannealing between Ti-IF and Nb-Ti-IF steels seems to be in the preferred outburst sites. Nb atoms tend to segregate to grain and sub-grain boundaries, and therefore significantly reduce the magnitude of outburst at steel grain boundaries. Segregation of Nb to grain and sub-grain boundaries makes grain and sub-grain boundaries equally favorable as outburst sites and hence increases the total number of outburst sites in Nb-Ti-IF steel. On the other hand, grain boundaries in Ti-IF steel are very clean and hence are the preferred outburst sites over sub-grain boundaries. The magnitude of outburst at grain boundaries of Ti-IF steel is considered to be much greater than that in Nb-Ti-IF steel due to the cleanness of grain boundaries. However, the density of outburst site is considered much lower, because outburst is more likely to occur only at grain boundaries for Ti-IF steel. A large outburst structure with low outburst site density makes the coating/steel interface very rough as shown in Figures 5a and 6a for Ti-IF steel. A small outburst structure with high outburst site density leads to a flatter coating/steel interface, Figures 5b and 6b for Nb-Ti-IF steel. Similarly, larger outburst and low site density (Ti-

IF steel) tend to make the coating surface much rougher[11]. Moreover, ς crystals are always left at the concave locations between the two large outburst sites due to the capillary effect[11] of liquid Zn in partially alloyed galvannealed Ti-IF steel as shown in Figure 8. This is the reason that galvannealed Ti-IF steel tends to have more ς crystals at the coating surface than galvannealed Nb-Ti-IF steel. The presence of ς phase at the coating surface is believed to increase the friction coefficient of the galvannealed Ti-IF steel and degrade the powdering resistance.

Figure 8: Large outburst structure in a partially alloyed galvannealed IF steel. ς phase is seen at the concave locations between the two outburst structures.

The presence of ς phase at the coating surface of Ti-IF steel makes the surface color of galvannealed steel lighter or brighter. In commercial CGLs, there are on-line devices to determine the completion of galvannealing. Generally, those devices measure the strip surface emissivity or reflectivity. Excess ς phase at the surface trends to mislead the on-line devices to increase the galvannealing temperature and hence over-alloy the product. Without checking the Fe content in the coating, one may mistakenly conclude that Ti-IF steel is galvannealed faster and has worse powdering resistance as CGL operators generally believe.

SUMMARY

A Hot Dip Simulator was used to study the galvannealing behavior of Ti-IF and Nb-Ti-IF steels. The galvannealing rate of the two steels was found to be comparable. Nb-Ti-IF steel is galvannealed at a more consistent rate than Ti-IF steel. About 30% of Ti-IF steels were galvannealed at a considerably slower rate. Galvannealed Ti-IF steel, with a similar Fe% in the coating, has slightly more ς crystals than galvannealed Nb-Ti-IF steel. Tested by the Double Olsen method, galvannealed Ti-IF steel also has worse powdering resistance. The differences in coating structure and powdering resistance seem to be due to the differences in the structure and chemical condition of steel grain and sub-grain boundary, which leads to different outburst mechanisms.

ACKNOWLEDGEMENTS

The author gratefully acknowledges critical review of the manuscript by Oscar Lanzi III and Don Jordan. The author is indebted to Tom Michniewicz and Rento Bejasa for their intelligent work in HDS simulation and metallographic preparation. Finally, the author would like to thank Ispat Inland management for their permission to publish this paper.

REFERENCES

1. B. M. Gray and C. Belleau, "Development of Hot-Dip Alloy Coated Steels for Exposed Automotive Applications," 1989 Mechanical Working and Steel Processing Conference, pp. 3-15.
2. N. Kino, M. Yamada, Y. Tokunaga, and H. Tsuchiya, "Production of Nb-Ti-Added Ultra-Low-Carbon Steel for Galvannealed Application," Metallurgy of Vacuum-Degassed Steel Products, edited by R. Pradhan, TMS, Warrendale, PA, 1990, pp. 197-213.
3. Y. Tokunaga and H. Kato, "Application of Intersitial-Free (IF) Steel Sheets to Automobile Parts," Metallurgy of Vacuum-Degassed Steel Products, edited by R. Pradhan, TMS, Warrendale, PA, 1990, pp. 91-108.
4. C. C. Cheng, L. L. Franks, and D. A. White, "Design and Application of a Hot Dip Simulator," SAE Technical Paper Series #930024, SAE, Detroit, 1993.
5. Rowland, D. H., "Metallography of Hot-dipeed Galvanized Coatings," Am. Society for Metals Trans. 40, 983-1011(1948).
6. M. Hua, C. I. Garcia, and A. J. DeArdo, "Precipitation Behavior in Ultra-Low-Carbon Steels Containing Titanium and Niobium," Met. Trans. 28A, 1769 (1997).
7. O. Lanzi III, Ispat Inland Research Laboratories, private communications.
8. M. Hua, C. I. Garcia, and A. J. DeArdo, "The Grain Boundary Segregation of P, B, C and Nb in ULC Steels," 40th Mechanical Working and Steel Processing Conference Proceedings, ISS, Warrendale, PA, 1998, pp. 877-881.
9. Y. Hisamatsu, "Science and Technology of Zinc and Zinc Alloy Coated Sheet Steel," GALVATECH'89, ISIJ, Tokyo, Japan, 1989, pp. 3-15.
10. A. Nishimoto, J. Inagaki, and K. Nakaoka, "Effects of Surface Microstructure and Chemical Composition of Steels on Formation of Fe-Zn Compounds During Continuous Galvanizing, " Trans. ISIJ, 26,807(1986).
11. M. Guttmann, "Diffusive Phase Transformations in Hot Dip Galvanizing," Material Science Forum, 155-156, 527(1994).

INFLUENCE OF PRECIPITATES IN IF STEEL ON GALVANNEAL COATINGS STRUCTURE

Yoshio Shindo
Ispat Inland Research Laboratory
3001 E. Columbus Drive
East Chicago, Indiana 46312
U.S.A
219-399-7032
Yxshin@Ispat.com

Galvanneal, IF steel, Precipitates, MnS, CuS, Inclusions

INTRODUCTION

Galvannealed sheet steels are widely used as automobile body panels. Good press formability is important for galvannealed sheet steel. The influence of galvannealing conditions and steel chemistry on the structure of galvanneal coating has been studied intensively by many researchers to improve the press formability of IF steel. However, there are still problems with press forming under certain conditions and there needs to be improvement. This paper is the summary of an investigation conducted on the effect of precipitates in IF steel on the galvanneal coating structure. TiN, NbC and alumina particles were found in galvanneal coating and had no effect on coating structure. Cu_2S and MnS were not found in galvanneal coatings. MnS and Cu_2S were probably decomposed by the molten zinc. Mn and Cu diffused into the bulk of molten Zn coaing. However S diffused slowly and stayed in or around a certain $\delta 1$ crystal. S affected the $\delta 1$ Zn-Fe crystal growth. If the MnS in the steel was close to the coating, molten zinc penetrated into the MnS location and formed a Zn-Fe alloy pocket beneath the substrate surface.

EXPERIMENTS

Galvanneal sheet steel, full hard sheet steel and annealed cold rolled sheet steel specimens were made by splitting a full hard experimental coil. The steel was TiNb IF steel and specimens were cut from the center of the coil. The investigation was conducted on the side which came from the bottom side of the hot rolled coil. The chemical composition was listed in Table 1. Cross sectional and surface observations by SEM were applied to detect the precipitates and also inclusions from the substrate steel. Each specimen was electrochemically etched to develop its coating structure. EDAX was used to analyze the coating layer. Inclusions and precipitates in the coating and in the steel surface layer equivalent to the steel consumed by galvannealing were collected using modified SPEED method and counted. All precipitates and inclusions collected from the specimens were investigated by TEM and SEM with EDAX.

1. Sampling Procedure

A full hard experimental coil was produced under standard processing conditions. Slab is produced using CC and scarfed to eliminate any kinds of defects on it. The slab was hot rolled to 3mm and shipped cold roll mill to make a 0.75mm full hard coil. The full hard coil was spliced at the middle of it to make two small coils for galvannealing and cold rolled sheet annealing. Annealing temperature was set to the experimental annealing temperature of recrystallizing temperature+70°C to obtain well developed precipitates in the specimens.
A 1 meter length of full hard steel sheet was cut from the split edge of the small coil for galvannealing to make full hard specimens. The galvannealed sheet steel and annealed cold rolled sheet steel specimens were cut from each small coil after eliminating about 1.5m in length from the end.

Table 1. Basic Al killed steel chemistry for the experimental IF steel coil.

Element	C	N	S	Mn	Cu	Al
	36 ppm	28 ppm	120 ppm	0.45%	0.06%	0.044%

Ti and Nb are added equivalent to the C, N and S

2. Electrochemical Extraction and Etching Method (Modified SPEED Method) [1, 2, 3]

The SPEED method was developed for investigation of precipitates in the steel. However, when applied galvannealed steel, the method should be modified to prevent the over etching the galvanneal coating because of the large electrode potential difference between the zinc alloy and the substrate steel. The key point for controlling the dissolution condition is controlling of the Cl ion in the solution. The electrode potential used for the dissolution and etching is also an important part of the modification. If the electrode potential has exceeded 200 mV from the reversible electrode potential of the specimen, uneven etching starts at once, which makes the SEM observation difficult. Therefore, in this experiment, the Cl ion in the etching solution was controlled using silver wire auxiliary electrodes for delivering the Ag ion into the solution to automatically eliminate the excess Cl ion. The solution used for the experiment was 5%Acetylacetone + 02%EDTA + 0.5%methyl-anmonium chloride + 50ppm Cl ion + 200 ppm Fe ion + 100 ppm Zn ion in methanol which was developed by Ispat Inland R & D. The electrode potential was set to natural electrode potential + 150mV using the potentiostat.

3. Counting Precipitates and Inclusions

Precipitates and inclusions were collected separately according to their size using different pore size filters (pore size ranged from 50 μm to 0.1μm). The particles larger than 0.5 μm were counted visually under SEM. The particles less than 0.5μm were estimated from the particles' population on the screen-collector for TEM. If the particles were composed of more than one oxide, the particle was classified by its main oxide component. For example, if the inclusion was estimated as 60%Al_2O_3 + 40%SiO_2 from EDAX analysis, the particle was counted as an alumina inclusion. However, for the TiN and NbC, if a particle was composed of a TiN particle and a NbC particle which stuck together, the particle was counted twice. One count is for TiN and the other is for NbC. MnS and CuS particles were directly counted on the filter to eliminate the breakage through handling.

RESULTS

1. Precipitates and Inclusions Found Beneath the Surface of the Substrate IF steel

The galvanneal coating was about 8μm. Therefore, the steel sheet surface consumed by the galvanneal coating was estimated to be 0.8μm because the galvanneal coating contained about 10% Fe. The number of alumina inclusions and precipitates from the surface to a 0.8μm depth was measured for comparison with galvanneal coatings. The measurement results were shown in Table 2, Table 3 and Table 4. The full hard sheet steel has a large number of tiny MnS particles compared to annealed sheet steel. The tiny MnS particles were probably made from large MnS particles in hot rolled sheet steel which were broken down by cold rolling. There was no significant difference in the numbers of alumina inclusions and precipitates between the full hard sheet steel and the annealed sheet steel. The size distribution for annealed sheet steel is summarized in Table 4. SiO_2, CaO-MgO, Fe_2O_3 were also found in the annealed steel sheet specimens. There were 57 particles of SiO_2, 44 particles of CaO-MgO, 28 particles of Fe_2O_3 and 56 (Al,Si,Ca, Na, K)-oxides with a total number of 185; these were listed in the column called "Others" in Table 2. MnS was the main precipitate of sulfide and TiN was the main precipitate of nitride. NbC and TiC were found on the TiN. Photo 1 is an example of the particles collected on a 0.1 μm filter.

Table 2. Condition of substrate steel from surface to 0.8μm depth point (average of 10 specimens)

	Inclusions		Precipitates				
Particles	Alumina	Others	MnS	Cu2S	TiS	TiN	NbC &TiC
Full hard	34.6 /cm2	21.0 /cm2	>44.2/cm2	11.3 /cm2	1.5 /cm2	172.8 /cm2	7.8 /cm2
Annealed	38.9 /cm2	18.5 /cm2	48.2 /cm2	7.5 /cm2	2.2 /cm2	168.3 /cm2	11.3 /cm2

Table 3. Size distributions of alumina and MnS in full hard sheet steel (total numbers of 10 specimens=Numbers for 10cm2)

Size(μm)	0.1-0.49	0.5-0.99.	1.0-4.99	5-9.99	10-20	20-50	50-
Alumina	59	97	141	40	9	0	0
MnS	Many	217	131	94	0	0	0

A large number of broken MnS particles were observed.

Table 4. Size distributions of alumina and MnS in annealed sheet steel (total numbers of 10 specimens=Numbers for 10cm2)

Size(μm)	0.1-0.49	0.5-0.99.	1.0-4.99	5-9.99	10-20	20-50	50-
Alumina	69	112	156	45	7	0	0
MnS	0	32*	109*	199	105	34	3

* The number may contain the number of broken particles from large MnS.

2. Precipitates and Inclusions Found in Galvanneal Coatings

The galvanneal coatings were measured three ways:

1. Extraction of precipitates and inclusions from galvanneal coatings by electrochemical method.
2. Surface measurement after electrochemically dissolving ζ layer and δ1-palisade layer.
3. Cross sectional measurement of the coating layer with and without electrochemical etching.

2.1. Particles Extracted from Galvannealed Coatings

The extraction results were summarized in Tables 5 and 6. 50.4% of the alumina coming from substrate steel remained in the galvannealed coating. Large alumina inclusions were lost through the galvanizing and galvannealing process. 49.7% of inclusions other than alumina, coming from the substrate remained in the galvanneal coating. The alumina inclusions from the galvannealed coatings were tiny spherical particles which were different from the larger irregularly shaped alumina particles found in the annealed sheet steel (compare Photo 2 to Photo 1)

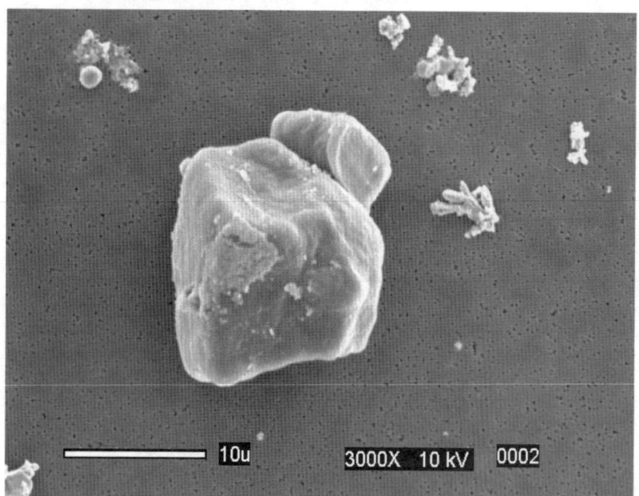

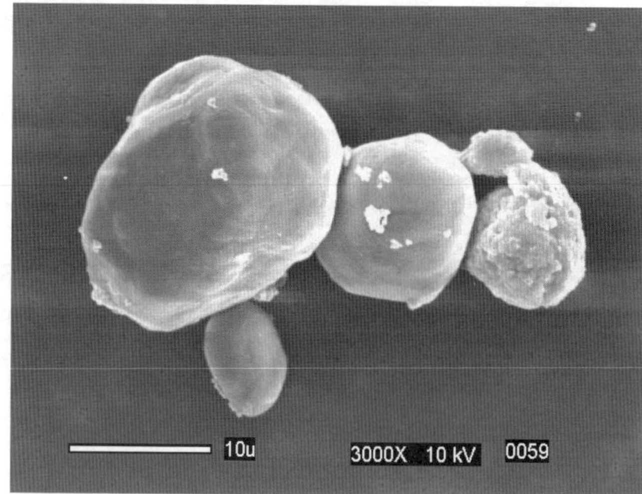

Photo 1. Particles extracted from annealed sheet steel (Right: alumina inclusions)
Tiny particles are mixtures of alumina, MnS,Cu_2S,TiN, TiN with NbC, TiC and MnS

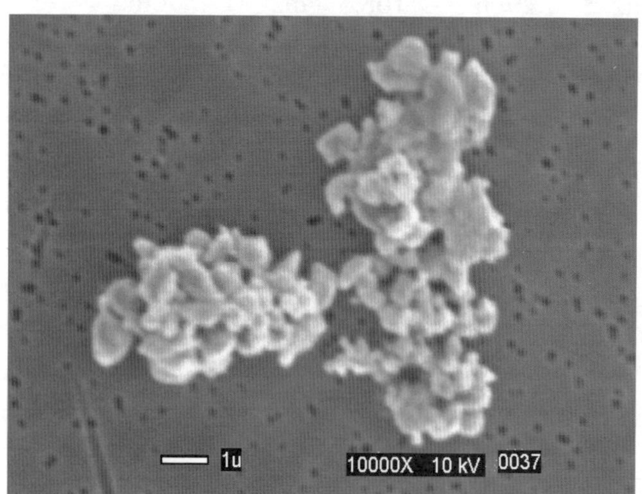

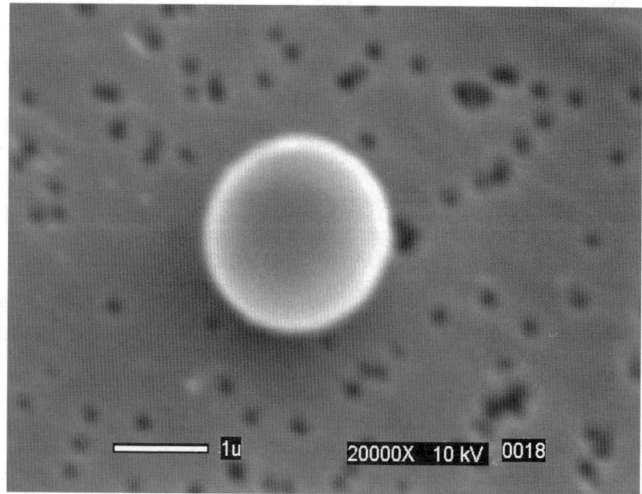

Photo 2. Particles extracted from galvanneal coatings (Right: alumina inclusions)
Tiny particles are mixtures of TiN, alumina and other particles.

The sulfide precipitates were not found in the galvanneal coating. However, 57.4% of TiN precipitates were found in the coatings, but only 5.8% of NbC precipitates were found. TiN obtained from the galvannealed coating and from the surface layer (0.8μm depth from the surface) of the full hard sheet steel were shown in Photos 3 and 4. Most of the TiN obtained from the galvannealed coating was clean cubic.

Table 5. Precipitates and inclusions extracted from coatings

	Inclusions		Precipitates				
Particles	Alumina	Others	MnS	Cu2S	TiO2	TiN	NbC
Numbers	19.6 /cm2	9.2 /cm2	1.1 /cm2	0 /cm2	0.8 /cm2	108 /cm2	2.4 /cm2
Retain ratio	50.4 %	49.7 %	2.3 %	0 %	-	57.4 %	5.8 %

Table 6. Size distribution of alumina inclusions from coatings and retain ratio
(Retain ratio=Number of particles in coating layer / Number of particles at the surface of cold rolled steel)

Size (μm)	0.1-0.49	0.5-0.99.	1.0-4.99	5-9.99	10-20	20-50	10 over
Alumina	39	91	65	1	0		
Retain ratio	56.5 %	81.2 %	41.7 %	2.2 %	0 %		

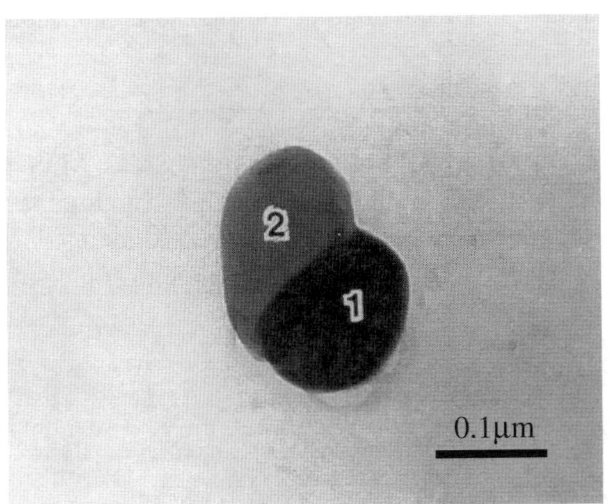

Photo 3. TEM picture of TiN with (Ti,Nb)C extracted from full hard sheet steel. (Left photo/ 1:TiC, 2:TiN)
Similar figures of the co-precipitated particles were obtained from the annealed cold rolled sheet steel.
The precipitates were mostly TiN with NbC, TiN with TiC, TiN with MnS or TiN with (Mn,Cu)S.

Photo 4. TEM picture of TiN extracted from galvanneal coatings
Most of TiN obtained from the galvanneal coating had clean cubic figure. Co-precipitates such as MnS, Cu_2S, TiC, NbC were washed out from the surface of TiN.

Possible reasons for the disappearance of MnS and Cu2S are:

a. Dissolution of MnS and Cu2S in to the coating layer,
b. Decomposition of MnS and Cu2S by molten zinc,
c. Dispersion of small size MnS into the galvannealing layer and oxidation at the coating surface, or
d. Prevention of precipitate growth by formation of Fe-Zn alloy.

Analysis of the annealed sheet suggests that a certain number of the tiny MnS particles will be formed into larger particles by the annealing process in the galvanizing and galvanealing line. Therefore, dissolution or decomposition of MnS is the reason for its disappearance from the galvanneal coating. Further study is needed.

2.2 Particles Found on the Half Dissolved Coating Surface

The particles detected on the surface by SEM were alumina (Photo 5), silica(Photo 6) and iron oxides. MnS and Cu2S were not found on the half dissolved galvannealed coating's surface. Both photos suggest that alumina and silica did not affect the Zn-Fe alloy structure surrounding them. The most frequently observed inclusion sizes were 1 to 5μm. The inclusions less than 1μm were hard to observe because of the similar appearance of the ZnFe alloy crystals.

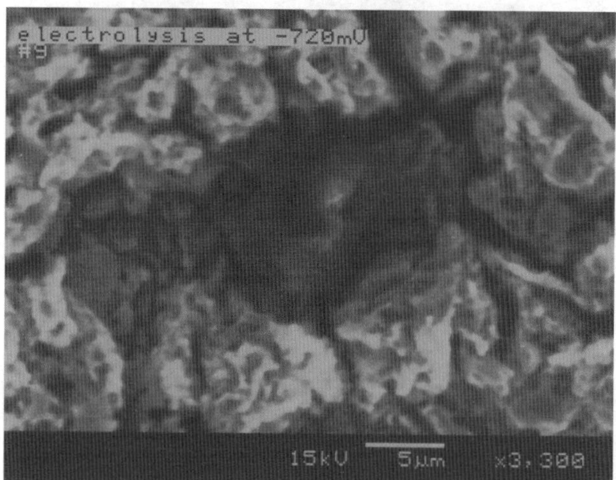

Photo 5. Alumina found in galvanneal coating
The ZnFe alloy layer was dissolved about 4μm under –720mV vs. Ag/AgCl reference electrode in the solution.

Photo 6. Silica found in galvanneal coating
The ZnFe alloy layer was dissolved about 4μm under –720mV. vs. Ag/AgCl reference electrode In the solution

2.3 Particles Found on the Cross Sectioned Coating Layer and Substrate

To verify the results of analysis on the particles from the coating, 5 specimens were cross-sectioned and SEM observations were made. Alumina inclusions were found on all five sectioned surfaces. Photo 7 is an example of the alumina inclusions found inside the coatings. Particles found in the coating were summarized in Table 7. Alumina inclusions were mainly found in the bottom half and bottom boundary of the galvanneal coatings. No MnS and Cu2S were found in the entire region of the cross-sectioned surface (investigated area size: 10mm x 8 μm). This observation result agreed with the result of the extraction from the galvanneal coating layer.

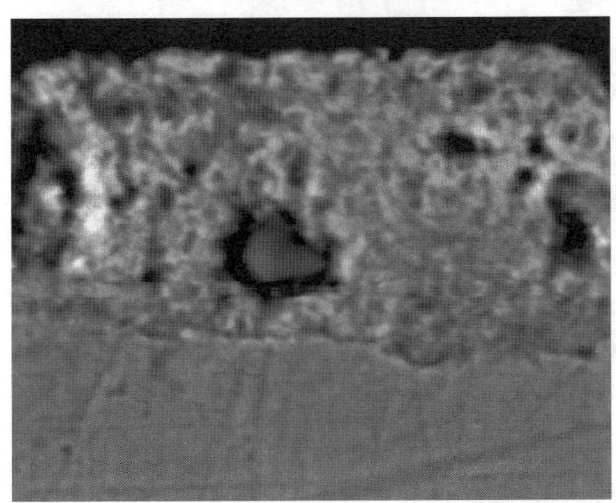

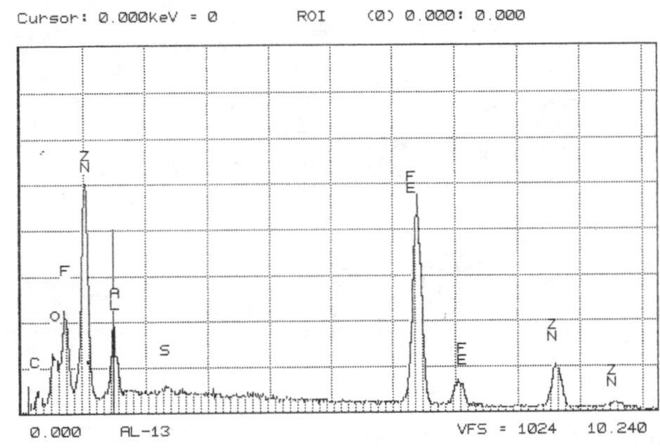

Photo 7. Alumina inclusion observed inside the galvanneal coating layer and its EDAX analytical data.

Table 7. Alumina inclusions found on the cross sectioned surface (investigated area size: 10mm x 8 μm).

Specimen Number .		GA-CS-1	GA-CS-2	GA-CS-3	GA-CS-4	GA-CS-5
Location inside the GA coating layer	Surface	0	0	1	1	0
	Upper half	0	0	1	0	0
	Bottom half	4	3	2	7	4
	Bottom boundary	0	2	1	0	2

3. Investigation of MnS and Cu2S Disappearance Inside Galvanneal Coating.

MnS and Cu2S were not obtained from the coatings by extraction analysis and cross sectional observation. However, they had to be found somewhere inside the coating, whether they were dispersed or decomposed. To trace the location of the sulfides, the cross sectioned specimens were investigated before and after the electrochemical etching.

First, two specimens, which were only polished with and without buffing, were measured visually using SEM to find the MnS. Photos 7 and 8 are examples of the polished surface. Only five alumina inclusions and a silica inclusion were found in the coating. Then, electrochemical etching was applied on these specimens for about 3 minutes under –650mV vs. Ag/AgCl reference electrode. Photos 9 and 10 are the typical coating structure obtained from the cross sectioned specimens after the electrochemical etching.

The one structure shown in Photo 9–Left coincided with the established concept of δ1 phase crystal layers [4, 5]. The upper layer was composed of the palisade δ1crystals and the lower layer was composed of the compact δ1 crystals.

Unfortunately, cutting the specific location of the etched crystal for TEM was not reliable, thus the author investigated the structure of each layer separately and verified both layers. Detail will be presented at another time.

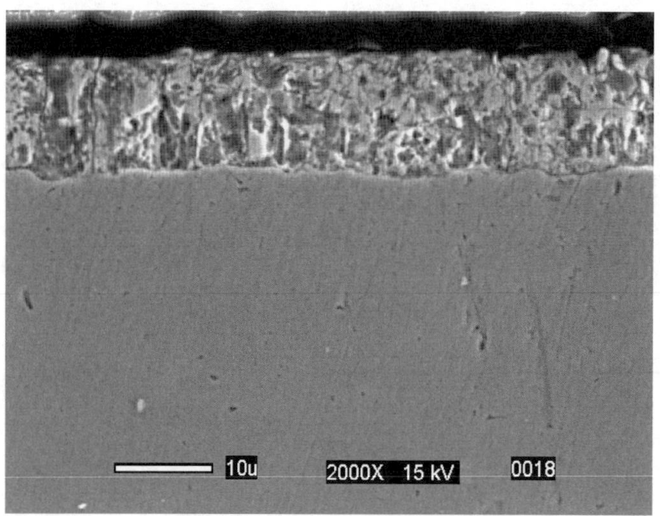

Photo 7. As polished without buffing
Coating layer was observed as a fractured surface of polycrystal layer composed of tiny Zn-Fe crystals.

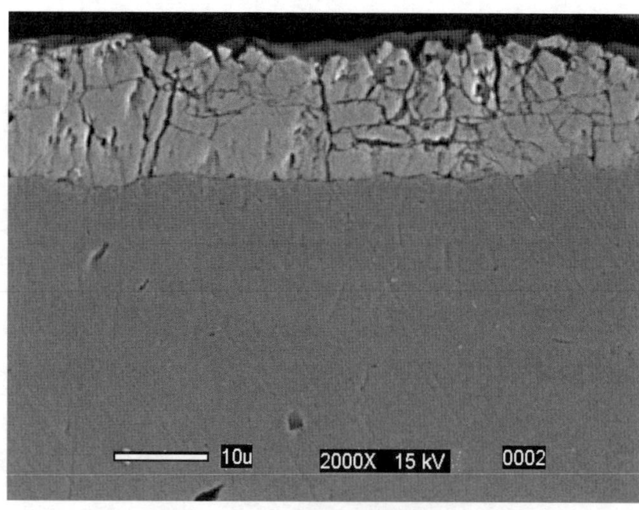

Photo 8. As polished with buffing
Coating layer showed fractures inside the layer. The coating layer was composed of large Zn-Fe crystals.

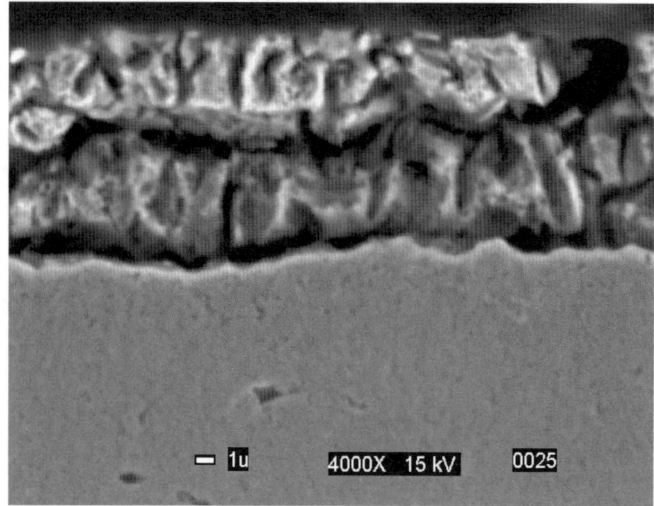

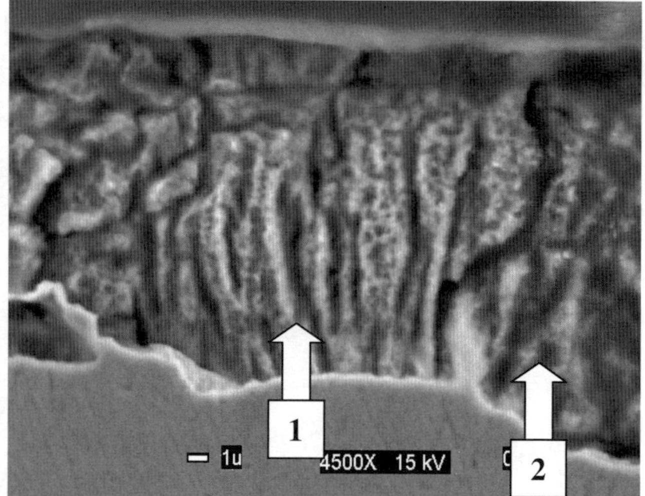

Photo 9. After electrochemical etching (Left: 4000x, Right: 4500x)

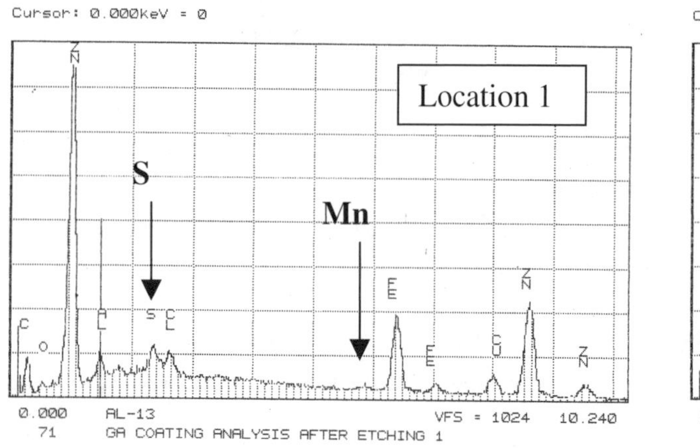

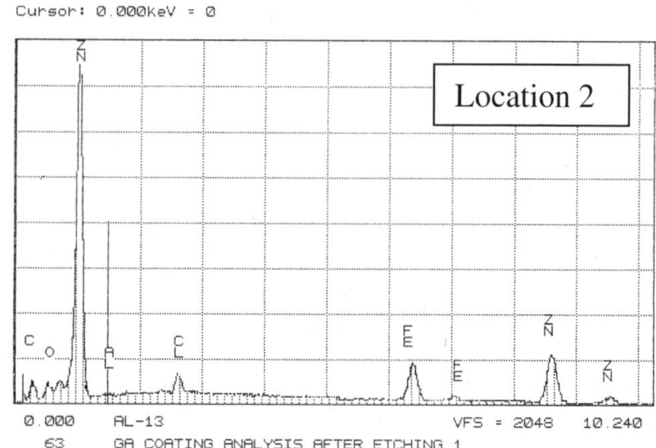

Figure 1. EDAX analysis on columnar crystals (Left: location 1, Right: location 2 in photo 9)

The other typical coating structure obtained from the electrochemical etching was columnar as shown in Photo 9-Right. EDAX analysis picked up small peaks of S and Mn at location 1. However, it was very difficult to tell the precise location of S and Mn. Therefore, weaker and longer electrochemical etching (-750mV vs. Ag/AgCl reference electrode) was applied on the specimen. Every 5 minutes, the specimens were analyzed using EDAX. Photo 10 was the cross sectioned coating surface appearance after etching.

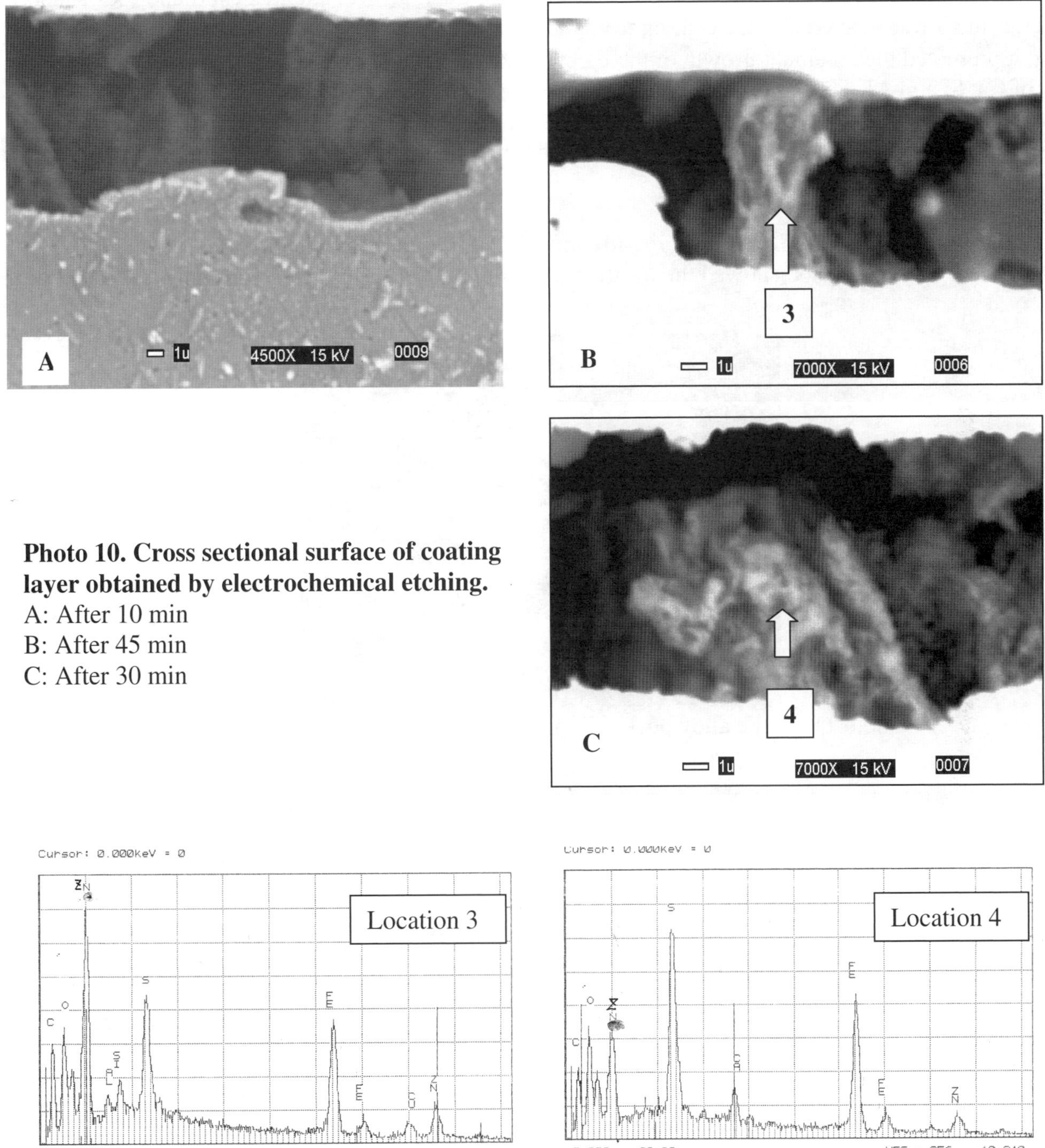

Photo 10. Cross sectional surface of coating layer obtained by electrochemical etching.
A: After 10 min
B: After 45 min
C: After 30 min

Figure 2. EDAX analysis of the residual crystals shown in photo 10B and 10C. Strong S peak was appeared on the EDAX charts. However, Mn peak was not observed on both charts.

After 10min of etching, the coating showed a deeply etched surface. However, there was no significant difference among the etched crystals. When the etching time reached at 30 minutes, a small number of crystals were left on the steel substrate. Their shape was like a dog bone or a round rock. Element analysis by EDAX on these crystals showed the strong S peak without the Mn peak. On the dog bone shaped crystal, the S peak was strong and all over the crystal. However, on the round rock crystal, the S peak was strong at the top of the crystal. The results suggested that Mn was diffused into the bulk of coatings and only S was left in some Zn-Fe crystals. The effect of S on the δ1 phase Zn-Fe alloy crystals was different when it was delivered from MnS. If S was delivered before δ1 crystal formed, it prevented the columnar growth and made a round rock shape. However, if the S was delivered to the coating from the substrate at the time when the δ1 crystals started forming, it promoted the columnar growth of the δ1 phase Zn-Fe alloy. The EDAX chart showed also a small amount of Ca, Si and Al. The oxides of those elements were normally found in MnS and Cu_2S. The data also suggest the decomposition of sulfide precipitates through the Zn-Fe alloying reaction. In this investigation, however, the author could not find the any segregation or enrichment of Mn inside the coating.

4. Investigation of MnS at the Boundary of Galvanneal Coating and Substrate Steel.

Electrochemical etching method showed the location of S enriched Zn-Fe alloy crystals. However, the location

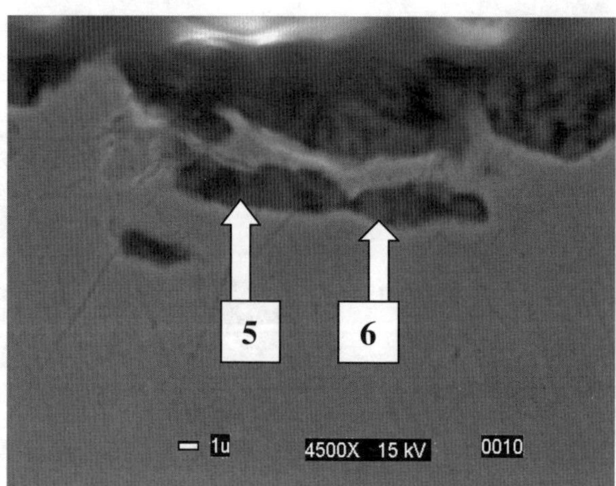

Photo 11. Zn-Fe alloy pocket formed under the surface of substrate steel

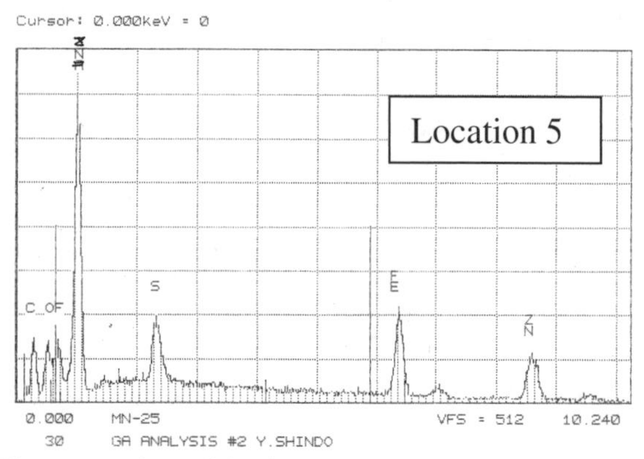

Figure 3. Elements found inside the pocket by EDAX.

of Mn delivered from the MnS was still not clear. A possible location where MnS would be found was the boundary of the galvanneal coating and the substrate steel. Therefore, further investigation was done on the boudary. In this experiment, when the etching had progressed, four zinc penetrated places (Zn-Fe alloy pockets)

appeared in the specimen. EDAX analysis showed a S peak without a Mn peak. However, when further electrochemical etching was applied on the specimen, one of the pockets became like Photo 11. At location 5 in Photo 11, EDAX chart showed a strong S peak without a Mn peak as shown in Figure 3-Left.
However, at location 6 in Photo 11, the Mn peak was appeared with the S peak on the EDAX chart. Figure 4 was the schematics of the etching results. It had a Zn-Fe alloy pocket and an opening on the left hand side. Alumina was also detected at **D** above the opening **A.** The results suggest that the pocket was originally occupied by MnS developed on an alumina particle as shown in figure 5. When the galvannealing started, molten zinc penetrated into that place and decomposed the MnS particle. In the case of Photo 11, some part of the MnS particle survived without decomposition. On this specimen, the number of MnS with alumina particles was 7. Some of them were found very close to the boundary of coating and substrate steel.

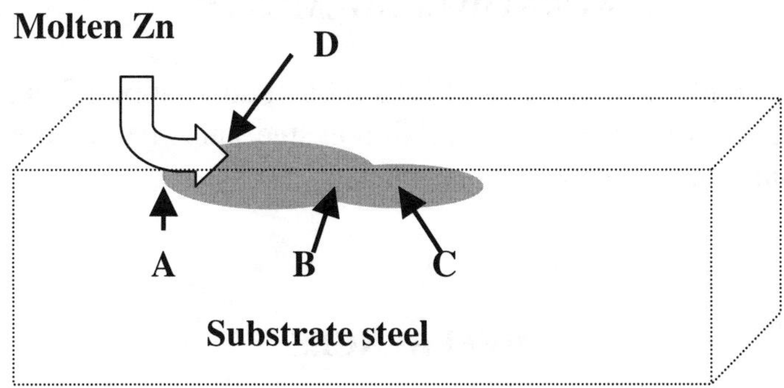

Figure 4. Schematics of the Zn-Fe alloy pocket formed under the substrate surface. Zinc was penetrated from **A** and diffused to **B**, then to **C**. Alumina was found at **D.**

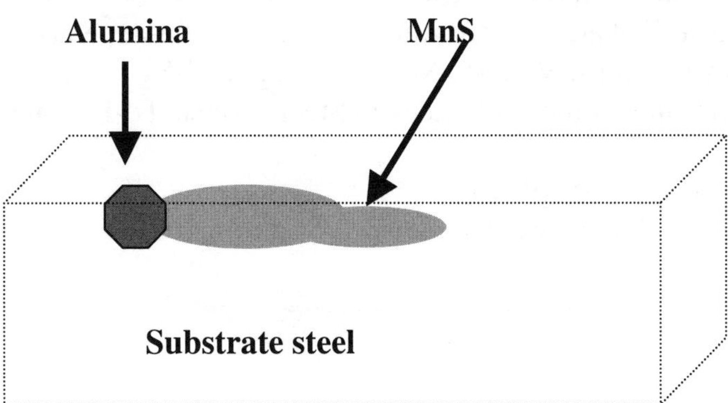

Figure 5. Estimated precondition at the Zn-Fe alloy pocket. That pocket was originally occupied by MnS developed on an alumina particle. Molten zinc penetrated into the place and decomposed the MnS particle. In the case of Photo 11, still some part of the MnS particle survived without decomposition.

SUMMARY

The effect of precipitates and inclusions in the sheet steel on galvannealed coatings was investigated using a trial coil. Inclusions such as alumina, CaO, SiO_2, etc. made the discontinuous spot in the coating but did not affect the galvanneal coating structure. Precipitates produced by Ti and Nb were less than 1 μm and did not affect the coating structure. Only MnS and CuS affected the galvanneal coating structure. MnS was thought to be decomposed by diffused molten zinc. When MnS was decomposed by zinc, Mn diffused to the Zinc or Zn-Fe alloy layer and left sulfur. This sulfur segregated in the nearest Zn-Fe alloy crystal and affected the coating structure. Sulfur was found around the columnar crystals and round rock crystals. S probably made the Zn-Fe crystal discontinuous and made the crystal columnar if the S diffused when δ1 alloy growth occurred. Some Zn-Fe crystals showed a dog bone shape by the effect of S. However, if S diffusion was finished before δ1 crystal started growing, it prevented the columnar crystal growth inside the coating and made the crystal a round rock shape. If the MnS decomposed beneath the substrate surface, a Zn-Fe alloy pocket was formed by penetrated Zn, which kept not only S but also Mn inside of it. These conclusions were based on only EDAX analysis and no TEM work. Therefore, further study is required for verification.

ACKNOWLEDGEMENTS

This experiment was conducted with the help of Ispat Inland Steel and Nippon Steel Co. Mr. Kenji Takahashi helped the sampling of spacimens. Mr. Yoshimi Kata (Nippon steel) also helped the analysis of inclusion and precipitates. I would like to thank both engineers very much.

REFERENCES

1. F. Kurosawa, I.Taguchi, and M. Tanino, "An application of the SPEED Method to the Study of Ferrous Materials" , *Kaihou*. Japan Inst. Metal, Vol.20, No.5,1981, pp. 377-384.
2. F. Kurosawa, I.Taguchi, M. Tanino, H. Suzuki and R. Matsumoto, "Observation and Analysis of Precipitates Formed in Steels at Elevated Temperatures using the Non-aqueous Electrolyte-Potentiostatic Etching Method", J.Japan, Inst. Metals, Vol. 45, No. 1, 1981, pp. 72-81
3. K. Narita, "Isolation and Determination of Sulphides in Steel", Tetsu-To-Hagane, Vol. 73, No.1,1987 pp.67-83.
4. W. Katz, Stahl und Eisen, Vol. 75 1955, p1101-.
5. Thaddeus B. Massalski "Binary alloy phase diagrams" 2nd edition 1991

WETTING OF ZINC ALLOYS ON STEELS CONTAINING SILICON

Yongsug Chung, Matthew X. Yao*, Jian Wang and James M. Toguri
Department of Metallurgy and Materials Science, University of Toronto,
184 College Street, Toronto, Ontario M5S 3E4 Canada
Tel.: 416-978-1584 Fax: 416-978-4155
*Deloro Stellite Inc., P.O. Box 5300, Belleville, Ontario, K8N 5C4
Tel: 613-968-3481 Fax: 613-966-8269

Key Words: Galvanizing Process, Silicon-Bearing Steel, Surface Tension, Contact Angle, Work of Adhesion.

INTRODUCTION

Zinc galvanizing is a well-developed practical surface treatment technology of steel structural materials. The types, properties and relative merits of Zn coatings are extensively discussed in the literature. Six commercial methods of applying Zn coatings to iron and steel are hot dip galvanizing, Zn plating, mechanical coating, Zn spraying and coatings incorporating Zn dust or flakes. There are a number of problems associated with galvanizing. For example, excessive coating thickness, a gray surface appearance and poor adhesion are frequently observed when galvanizing silicon killed and semi-killed steels. Particularly, in steels with 0.03-0.12 mass % silicon, there is a marked increase in the coating thickness, this being the Sandelin effect [1]. When the silicon content surpasses 0.3 mass %, a thick coating again appears.

Studies have been carried out to overcome the detrimental effects of silicon on the abnormal growth of the zinc-iron intermetallic compounds. The problem is generally satisfactorily resolved by adding aluminum [2] and nickel [3] to the galvanizing bath and by increasing the galvanizing temperature [4]. However, the industrial application of aluminum additions has been limited because of the rather complicated surface preparation required prior to immersion in the galvanizing bath. High temperature galvanizing increases the corrosion of the zinc containing pots and associated hardware and also leads to excessive dross formation.

Information on the surface tension of zinc and its alloys together with information on the wettability between molten zinc/zinc alloys and steels, and plating equipment are of fundamental importance to the better understand the galvanizing process. Consequently, a study on the effect of the addition of alloying elements (Al, Ni and Si) to zinc on the surface tension and the wetting properties between Zn/Zn-alloys and Si-bearing steels has been initiated. In this paper, the measurement of the surface tension of zinc and the wetting behavior between zinc and silicon bearing steel are described.

EXPERMENTS

Sessile drop technique combined with X-ray- There are several techniques to study the surface tension and contact angle. Among them, the sessile drop method combined with X-ray radiography is widely used, especially at high temperatures. The classical sessile drop technique can be briefly described as follows: A sample of the solid metal is placed upon a substrate. This arrangement is then placed into a furnace, which has the capability for imaging and photographing. Once the furnace temperature reaches the melting point of the metal sample, the sample melts and takes its droplet shape and forms an angle with the substrate. The main advantage of this technique is that it provides the surface tension and the contact angle, simultaneously.

The apparatus used in this project is shown in Fig. 1. It includes the X-ray source, the furnace, the image capturing and recording system, and the vacuum and gas purging system. The entire set up is located in two rooms with the X-ray unit and the furnace being set in a lead lined chamber for radiation control and with the units for controlling both the X-ray and the furnace situated in a room adjacent to the radiation chamber. To obtain images generated from the irradiation, either X-ray films or X-ray fluoroscope can be used. A fluoroscope converts the invisible X-ray wave into the visible range. To see the image produced by the fluoroscope, a video camera is used. Both the fluoroscope and the camera are mounted on a movable stand for mobility. A video cassette recorder that is connected to a high-resolution video monitor receives the signal from the video camera. This arrangement allows the X-ray images, generated as a result of the X-ray radiation, to be observed on the video monitor and recorded on tape as desired. Selected images are then recorded on the X-ray film. To allow the passage of X-rays through the sample with minimal attenuation, holes are made on the radiation shields along the X-ray path. To create a controlled environment, a reaction tube made from quartz is used (see Fig. 2 (a)). The tube is inserted into the furnace through an opening located on the top furnace plate. A control thermocouple is inserted through an opening on the bottom of the furnace into the hot zone to regulate the temperature. During the experiment, the sample assembly is loaded into the reaction tube. The tube is sealed with O-rings. The tube is repeatedly evacuated, to the maximum vacuum achievable by a mechanical vacuum pump, and then back filled with high purity argon gas prior to experimentation. A protection gas (high purity Argon and hydrogen) always flows during the experiment. To remove trace amount of oxygen in the argon gas, an oxygen getter (Copper turning) is used.

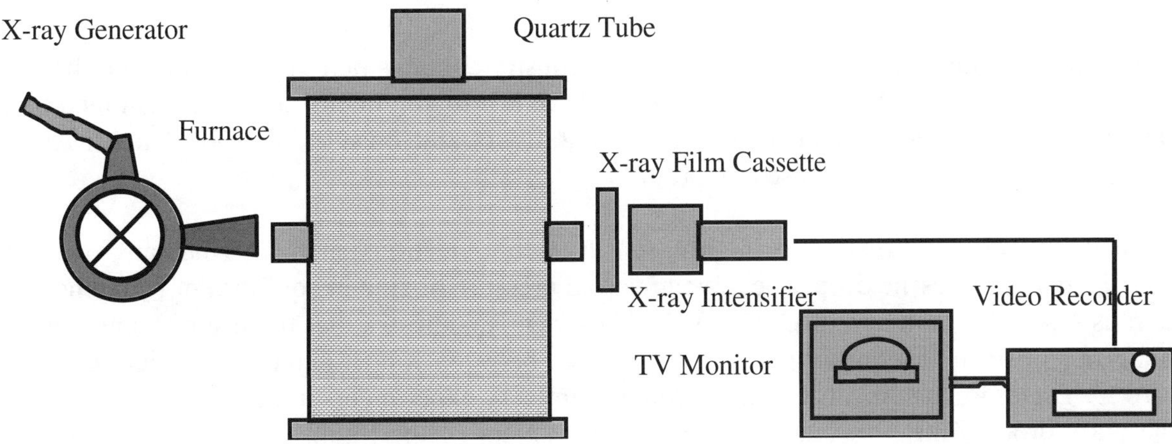

Figure 1. Apparatus of SiC electric resistance furnace combined with X-ray system used for surface tension and contact angle measurements.

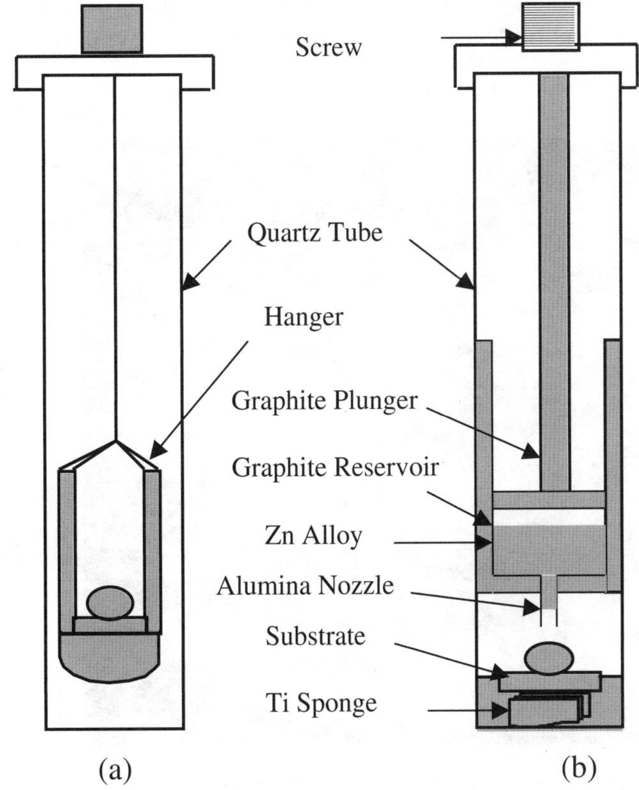

Figure 2. Reaction tube setups of hanger-type (a) and of plunger-type (b) used in the wetting study.

Zn and associated special concerns- The zinc used in industry contains ppm levels of Fe and Pb. The addition of the reactive element Al and the transition elements Ni and Si to zinc melts are to be examined to determine whether they lower surface tension and enhance the wettability of zinc on steel. The pure Zn, Zn-Al, Zn-Ni, Zn-Si, Zn-Al-Fe are being studied in this research program.

When the metal under investigation is highly sensitive to oxidation, such as zinc, an oxide skin often encapsulates the resulting sessile drop. The oxygen partial pressure of zinc in equilibrium with zinc oxide (ZnO) is calculated as 6.64×10^{-16} atm at 470°C by FACT. This clearly indicates that the use of copper as the oxygen getter is not efficient since the equilibrium oxygen pressure between copper and copper oxide (Cu_2O) is 8.5×10^{-17} atm at 470°C. Therefore, the results obtained from preliminary experiments indicated the presence of an oxide film on the alloy drops. This film hindered the free movement of the liquid within the drop and resulted in distorted drop shapes. To alleviate this problem, a plunger-type device was used to form the sessile drop for all subsequent experiments. In this setup, the sample was put into the furnace inside a quartz reaction tube. The schematic configuration is shown in Fig. 2 (b). As illustrated in Fig. 2 (b), the sample is loaded in the graphite reservoir and extruded on to the substrate to form a droplet. Figure 3 shows a successful image obtained by this method. A clean zinc droplet is seen on the alumina substrate. The residual liquid zinc in the upper part of the picture has an irregular shape, which is the result of the initial oxidation of the zinc. Titanium sponge, which is not seen in this figure, is placed in the bottom of the reaction tube to prevent the oxidation of the droplet.

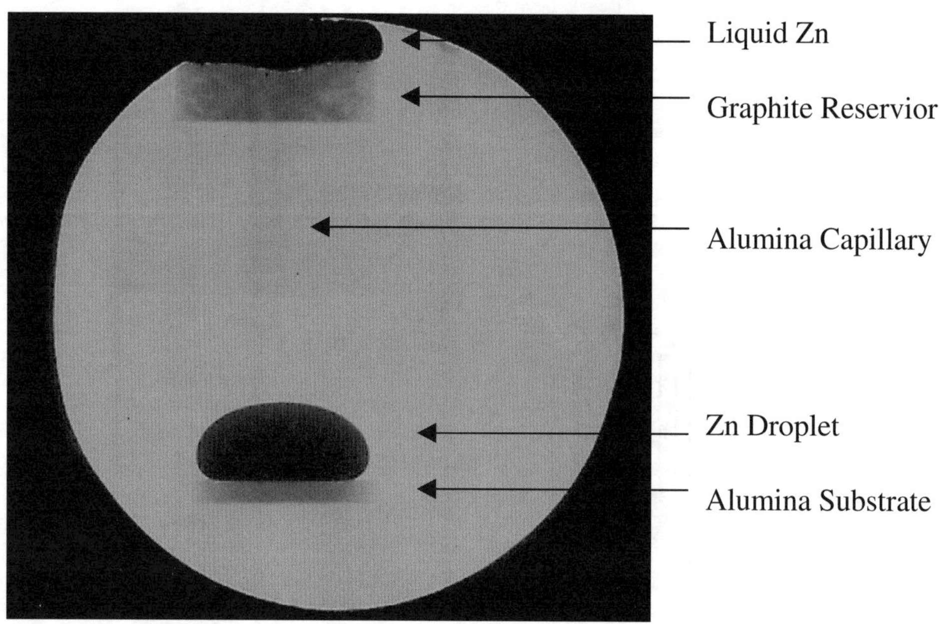

Figure 3. A scanned X-ray picture of zinc droplet sitting on alumina

Steel specimen- Contact angle measurements are planned for steels containing different levels of silicon. Table I shows the chemical composition of the steels used in the experimentation. These cold-rolled steels were supplied from Dofasco Inc. Each plate was cut to 2 by 2 centimeter in size and polished with 3μ alumina abrasive. The plate was kept in acetone to avoid oxidation prior to the experiment.

Table I. Chemical composition of the steels containing different silicon content, supplied by Dofasco Inc.

Steel	**CC007B**	**CC040**	**440P**	**440W**
Si	**0.009**	**0.012**	**0.15**	**0.225**
C	0.0031	0.05	0.0035	0.08
Mn	0.18	0.21	1.50	1.00
Cu	0.01	0.02	-	-
Al	0.032	0.035	0.05	0.05
Ti	0.055	0.002	0.02	-
Ni (Nb)	0.01	0.01	- (0.03)	-
V(N)	0.004	0.002	- (0.004)	- (0.006)
Cr (Cb)	0.02 (0.008)	0.02 (0.001)	-	-
Ca	0.0002	0.0001	-	0.0033
P	0.005	0.004	0.09	0.01
S	0.004	0.007	0.006	0.006
Cr (Sn)	0.005 (0.004)	0.02 (0.005)	-	-

Calculation of surface tension and contact angle- The calculation of the surface tension and the determination of contact angles are based on the sessile drop technique combined with x-ray radiography and a computer enhancing method. There are four steps involved in this droplet image processing. (1) capturing images from x-ray film by using Adobe Photoshop, (2) making a binary image and size/pixel relationship by applying NIH Image, (3) digitizing the processed picture to 41 data set through HyperCard and (4) running a computer program, which is a numerical solution of the Laplace equation, written in Fortran 77 to determine surface tension, contact angle and density. The input data are weight of the sample, digitized data sets, density of sample and size relation. This computer-enhanced method was primarily developed at the Carnegie Mellon University and details can be found in elsewhere [5].

RESULTS AND DISSCUSION

Surface tensions of pure zinc - Even though the plunger-type of the apparatus created a clean liquid zinc droplet a low oxygen potential near the sample was required to prevent oxidation, which may cause an error in the surface tension measurement. Many precautions were, therefore, taken such as: (1) reaction tube was evacuated by vacuum pump and was reflushed by Ar gas. This operation was repited at least 3 times. (2) reaction tube was sealed, titanium sponge was placed in the bottom of the crucible to serve as an oxygen trap and (3) a mixture of 5% hydrogen and 95% argon gases was used in order to remove the oxygen within the reaction tube.

For the surface tension measurement, the droplet sitting on the substrate must be the equilibrium shape over a long period of time. However, a reaction occurred between the steel substrate and the zinc droplet. For this reason an alumina substrate (poly-crystal, 99.8% in purity) was substituted for steel. Under this condition, the surface tension of pure zinc (99.999% in purity) was measured and is listed in Table II

Table II A comparison of surface tension, density and contact angle of zinc on an alumina substrate (* Estimated value based on $d\gamma/dT$= -0.17 mN/m°C)

Temperature [°C]	Surface Tension [mN/m]	Desity [g/cm^3]	Ref	Contact Angle [deg]
470	771	6.462	Ar-5%H_2, Sessile Drop	154
470	775 [6)]	6.545 [6)]	Vacuum, Max. Drop Pressure	-
600 470	791 [7)] 813*	-	Vacuum, Sessile Drop	155 [7)]

As can be seen in Table II, the surface tension was found to be 771 mN/m at 470°C, which is in a good agreement with previous studies. The surface tension [6] was reported as 775 mN/m in vacuum condition by employing the maximum drop pressure technique, which has the advantage in avoiding oxidation but has the disadvantage not to providing the contact angle. By using the sessile drop technique, Shinozaki et al. [7], recently, obtained value of 791mN/m in vacuum at 600°C. The temperature coefficient of the surface tension of pure zinc of –0.17 is widely accepted. The surface tension can be estimated as 813 mN/m at 470°C by adopting this coefficient.

The density and the contact angle were calculated simultaneously with the surface tension. The density (6.462 g/cm^3) found in the present study was in a reasonable agreement with the one from Lucas (6.545 g/cm^3). The contact angle, is explained in the next paragraph in detail, between zinc and the alumina was found to be 154 degrees in the present study. This is also in a good agreement with the one (155 degrees) reported by Shinozaki et al [7].

Contact angles between pure zinc and Si-bearing steels- It was in 1805 that T. Young suggested the famous equation which defines the contact angle of a liquid resting on a solid assuming mechanical equilibrium among the three interfacial tensions: γ_{SV} at the interface between the solid and the vapor, γ_{SL} at the surface between the solid and liquid, γ_{LV} at the surface between the liquid and the vapor. This equation is called Young's equation and is given by

$$\cos\theta = \frac{\gamma_{SV} - \gamma_{SL}}{\gamma_{LV}} \qquad (1)$$

When θ > 90°, the liquid is non-wetting with respect to the particular solid. Inversely, for systems in which θ < 90°, the solid is wetted by the liquid. θ = 90° is the transition angle. Both non-wetting and wetting systems are respectively described in Fig. 4.

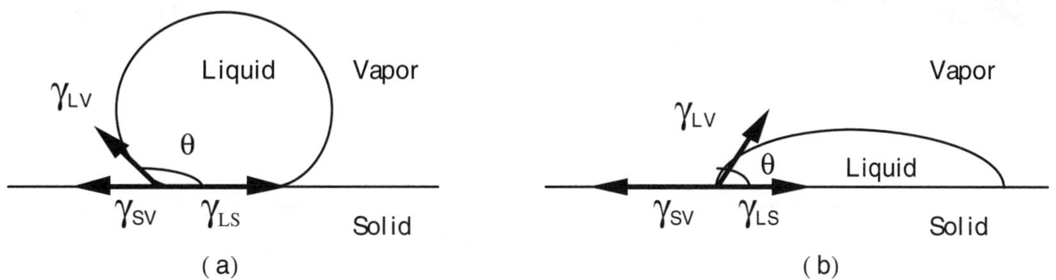

Figure 4. The relation of surface tension of solid-vapor (γS), surface tension of liquid-vapor (γLG) and interfacial tension between liquid-solid (γLS) with contact angles at non-wetting (a) and wetting (b).

The contact angle between pure zinc and 440P steel was determined as a function of time at 470°C. Figure 5 shows the appearance of the zinc droplet sitting on the steel (a) and the measured value of cosθ (b) varied with time. Figure 5 (a) clearly shows that the droplet progressed from non-wetting to wetting as time increased. The zinc no longer existed as a droplet and fully coated the steel surface in 24 minute.

Work of adhesion between pure zinc and Si-bearing steel- From knowledge of the surface tension and the contact angle, one can calculate the work of adhesion. This thermodynamic wetting property is defined as the free energy change (ΔG) of the work per unit area necessary to separate a liquid from a solid as expressed in Eq. (2)

$$\Delta G = W_{SL} = \gamma_{SV} + \gamma_{LV} - \gamma_{SL} \quad (2)$$

where γ_{SV}, is the surface tension of the solid, γ_{LV} is the surface tension of the liquid and γ_{SL} is the interfacial tension between the solid and the liquid. Combining Eq. (2) with Eq. (1) results in the Young-Dupre equation.

$$W_{SL} = \gamma_{LV}(1 + \cos\theta) \quad (3)$$

The work of adhesion is a measure of the wettablity between the two phases. The value varies from 0 to $2\gamma_{LV}$. If the value is close to 0, there is a weak bonding between the liquid and solid phases while if the value is close to $2\gamma_{LV}$, there is a strong bonding between the two phases.

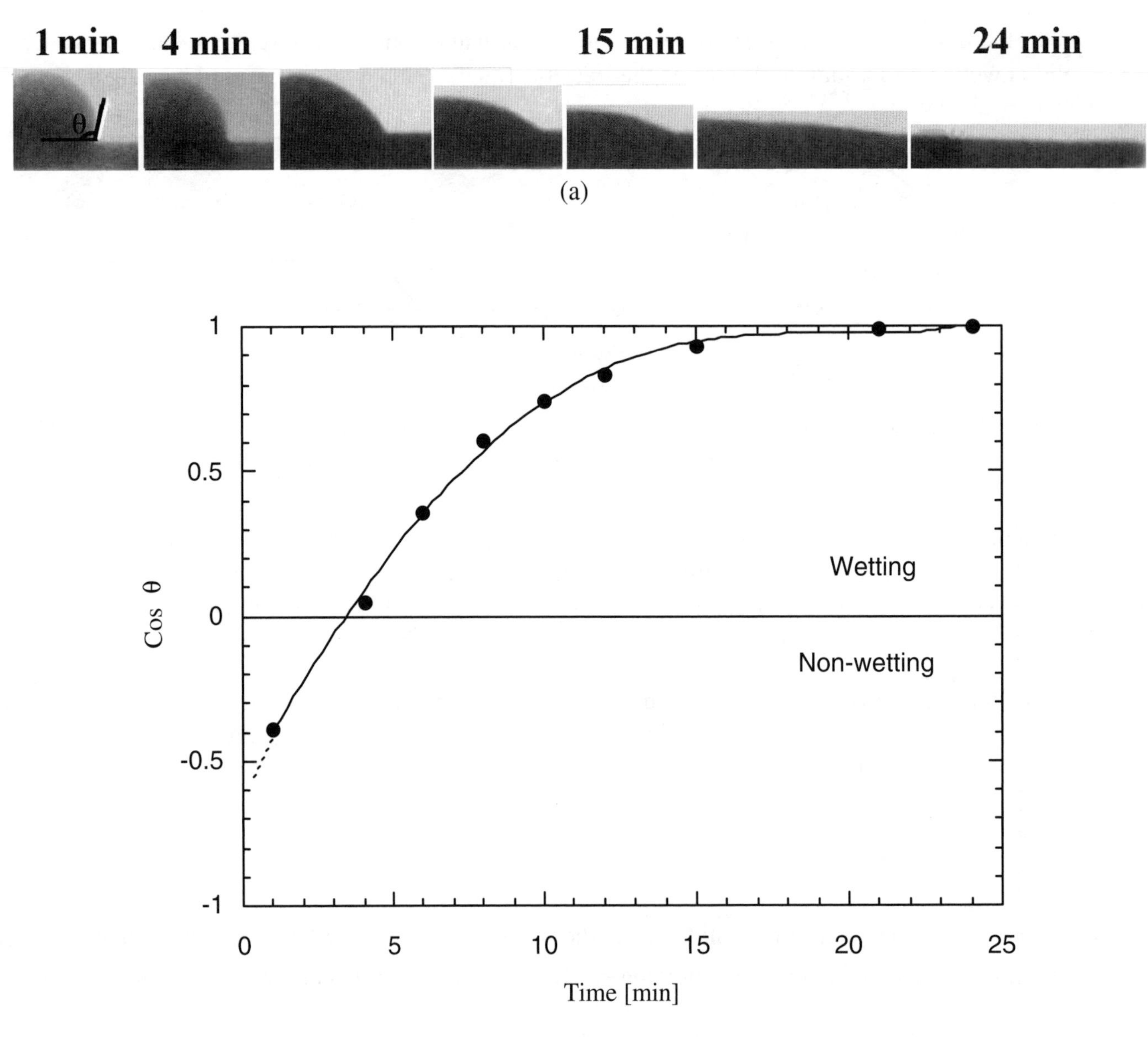

Figure 5. Appearance of the zinc droplet in contact with 440P steel (a) and the variation of cosθ (b) as a function of time at 470°C.

Figure 6 shows the work of adhesion between pure zinc and Si-bearing steel as a function of time at 470°C. The work of adhesion increased abruptly with increasing time and reached a maximum value ($2\gamma_{LV}$) in about 20 minute. This indicates that there is a strong reaction between zinc and the steel. The variation of the work of adhesion between the zinc and the alumina is also shown in Fig. 6. The work of adhesion was obtained as 78 mN/m, which is very close to zero and indicates that there is a weak force between the zinc and the alumina The work of adhesion also remained unchanged with time. This implies that there is almost no reaction during the period of time. The first data point of this experiment was obtained at 1 minute after the droplet sat on the substrates. As can be seen in Fig. 6, the work of adhesion of the zinc-steel is larger than the one of the zinc-alumina. It implies that the reaction between the zinc and the steel may occur as soon as the droplet was in contact with the substrate.

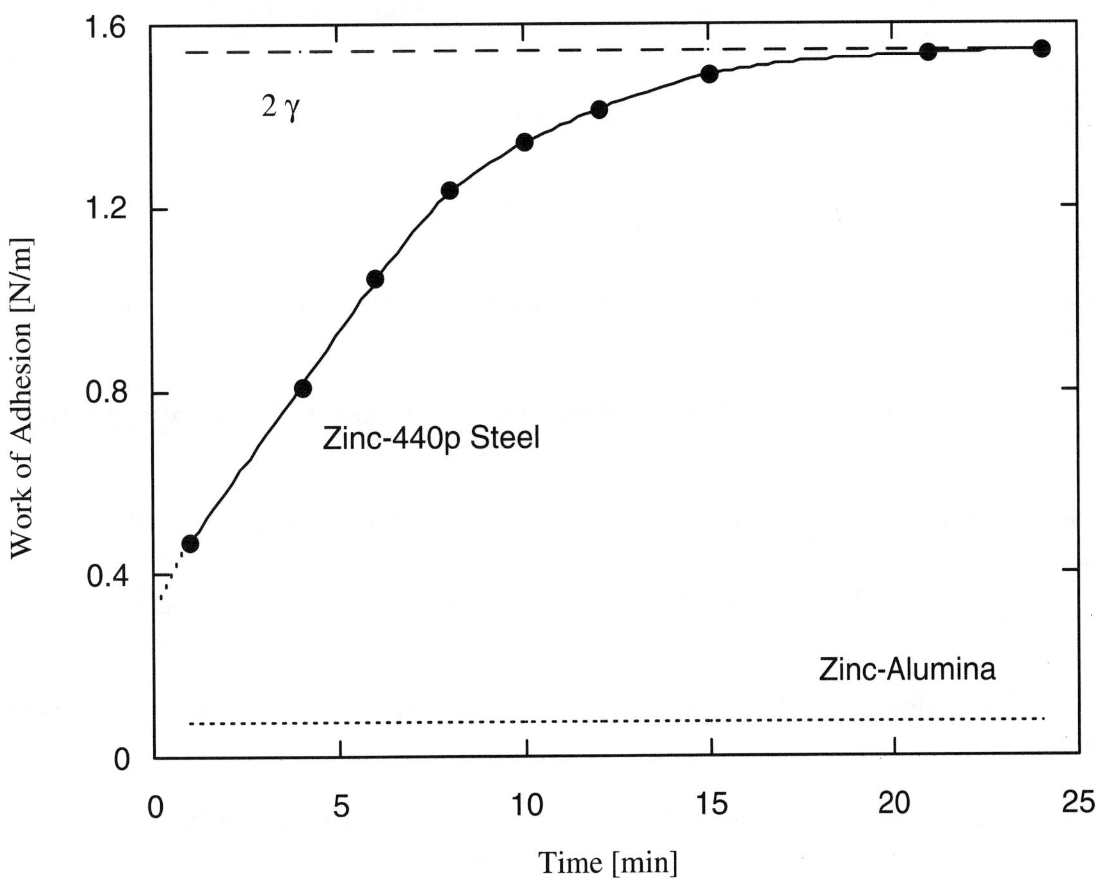

Figure 6. shows the work of adhesion between pure zinc and Si-bearing steels as a function of time at 470°C.

CONCLUSIONS

Sessile drop technique combined with X-ray radiography was successfully carried out for measurement of the surface tension of zinc and contact angle between zinc and steel by using a plunger-type device. The surface tension of pure zinc was found to be 771 mN/m at 470°C, which is a good agreement with previous studies. The contact angle and the work of adhesion between pure zinc and the steel (440P) abruptly increased with time. The zinc completely coated the steel in 20 minutes.

ACKNOWLEDGEMENTS

The authors are grateful to Ontario-Singapore Joint Research Program for its financial support of this project.

REFERENCES

1. R. W. Sandelin, Wire and Wire Products, 11 (1940) 655.

2. J. Mackowiak and N. R. Short, "Metallurgy of Galvanized Coatings," International Metals Review, Vol.24, 1979, No.1., Pages 1-19.

3. G. Reumont, P. Perrot and J. Foct, "Thermodynamic study of the galvanizing process in a Zn-0.1%Ni bath," Journal of Materials Science, Vol. 33, 1998, Pages 4759-4768.

4. G. Reumont, P. Perrot and J. Foct, Proceedings 17th Galvanizing Conference, Paris, 1994.

5. , I. Jimbo and A. W. Cramb, "Computer Aided Interfacial Measurements," ISIJ International, Vol. 32, 1992, No. 1, Pages. 26-35

6. B. C. Allen, Liquid Metals: Chemistry and Physics (ed. Beer S.Z.). Marcel Dekker, New York, NY, 1972, Page 186.

7. N. Shinozaki, M. Suenaga and K. Mukai, "Wettability of Zirconia and Alumina Ceramics by Molten Zinc," Materials Transactions of JIM, Vol. 40, 1999, No. 1, Pages 52-56.

Effects of double zinc coating on steel sheets

Hyun Tae Kim
Technical Research Labs., POSCO
Pohang Iron & Steel Company
Pohang P.O. Box 36
1, Koedong-Dong
Pohang-shi Kyungbuk 790-985
Korea
82-54-220-6234
pc543346@posco.co.kr

Key Words : zinc coating, sulfate bath, chloride bath, whiteness, brightness, preferred orientation

INTRODUCTION

Zinc coated steel sheets have been used as the automobile body and home appliance panels due to its superior corrosion resistance. When the coated steel sheets are used without any treatment or with post-treatment, their surface appearance affects the products quality[1]. Therefore, in order to improve surface appearance of coated steel sheets, various studies have been carried out. Originally, this property depends on crystal structure as well as bath types, electrolyte compositions, electroplating condition. Due to its hexagonal close-packed(hcp) crystal structure, zinc exhibits anisotropic properties and strong preferred crystallographic orientation[2~3]. These texture properties have effects on various aspects of performance, and they would be desirable to change the surface appearance. Besides, in the continuous electrogalvanized line, coating appearance also depends on equipment's condition and type. Contaminated surface of plate is generally appeared as band mark, stain and striation. These contaminations are almost generated by the contact of roll. Unfortunately, there are lots of contacts between steel strip and roll in the carosel cell comparative to other type cell.

In the present paper, the variation of surface appearance of zinc electrogalvanized steel sheets was investigated, mainly focusing on the effect of chloride bath and sulfate bath on the whiteness and brightness. And then, the effect of double coating was tested ; the first zinc coating was plated at carosel cell of chloride bath and the second was continuously coated at vertical cell of sulfate bath.

EXPERIMENTAL PROCEDURE

Zinc coating was carried out in the experimental apparatus which is mainly composed of rectangular coating cell with the size of 100(w)x250(l)x20(gap)mm and electrolyte circulation tank of 45 liter connected to pump. The cold-rolled steel sheets were used as the cathode, and zinc plate as the anode. The electrolyte was prepared with the industrial grade chemicals and distilled water. The additive used in this study was the polyethyleneglycol compound which was added to the electrolyte with the concentration of 0.2ml/l. The electrolyte and coating conditions are summarized in table•. Both electrolytes were tested at each optimum condition. The coating weight was controlled between 5 and 30g/m^2.

In the case of test in the plant, the first zinc layer was coated at eight carosel cells and the second at four vertical cells. The coating weight of the former was 15g/m^2 and the latter was varied from 3.0g/m^2 to 10g/m^2. The electrolyte and coating conditions summarized at table•are also used. The coating layer was stripped in a 1:3 HCl solution, and the coating weight was measured. Surface morphology was observed by a scanning electron microscope, and the preferred orientations were measured using X-ray diffraction tester. The whiteness and brightness were evaluated with a color difference meter and a glossmeter, respectively.

Table •Experimental conditions of zinc coated specimen (g/L)

Bath	Zn($ZnSO_4.7H_2O$)	Zn($ZnCl_2$)	Na_2SO_4	$(NH_4)_2SO_4$	Cl(KCl)	Temp.(°C)	pH	C/D(A/dm^2)
Chloride	-	100	-	-	245	60	4.5	60
Sulfate	120	-	35	40	-	70	3	60

RESULTS and DISCUSSION

1) Surface appearance of zinc coating with bath type

Fig.1 and Fig.2 show the typical morphology and preferred orientation of zinc coatings obtained under a sulfate bath and chloride bath without or with additive. In the case of addition–free chloride bath, the zinc coating consists of large and non-uniform rectangular platelets. These shapes result in the poor formability and surface leveling.

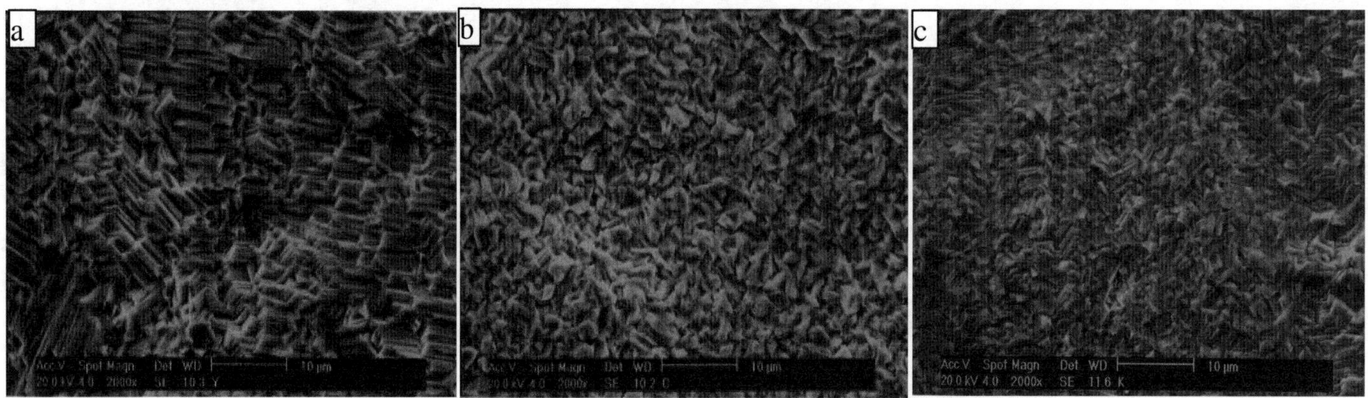

Fig.1 SEM morphology of zinc coatings with bath type (a:addition-free chloride bath, b:addition-additive chloride bath, c:sulfate bath)

The crystallographic orientation was highly preferred as the order of (002), (103), (102) at addition-free chloride bath. This shows the typical basal orientation. However, the fine and dense grains were obtained, and the preferred orientation was changed as the order of (101),(100),(102) by the addition of additive to chloride electrolyte. The basal plane intensity was decreased. At the sulfate bath, the reduced size rectangular-shaped grains were observed, and the preferred coating orientations with (002), (101), (103) were obtained. Especially the preferred orientation and grain morphologies were greatly changed with addition additive in chloride bath. Such results would be connected with the surface appearance as compared Fig.3.

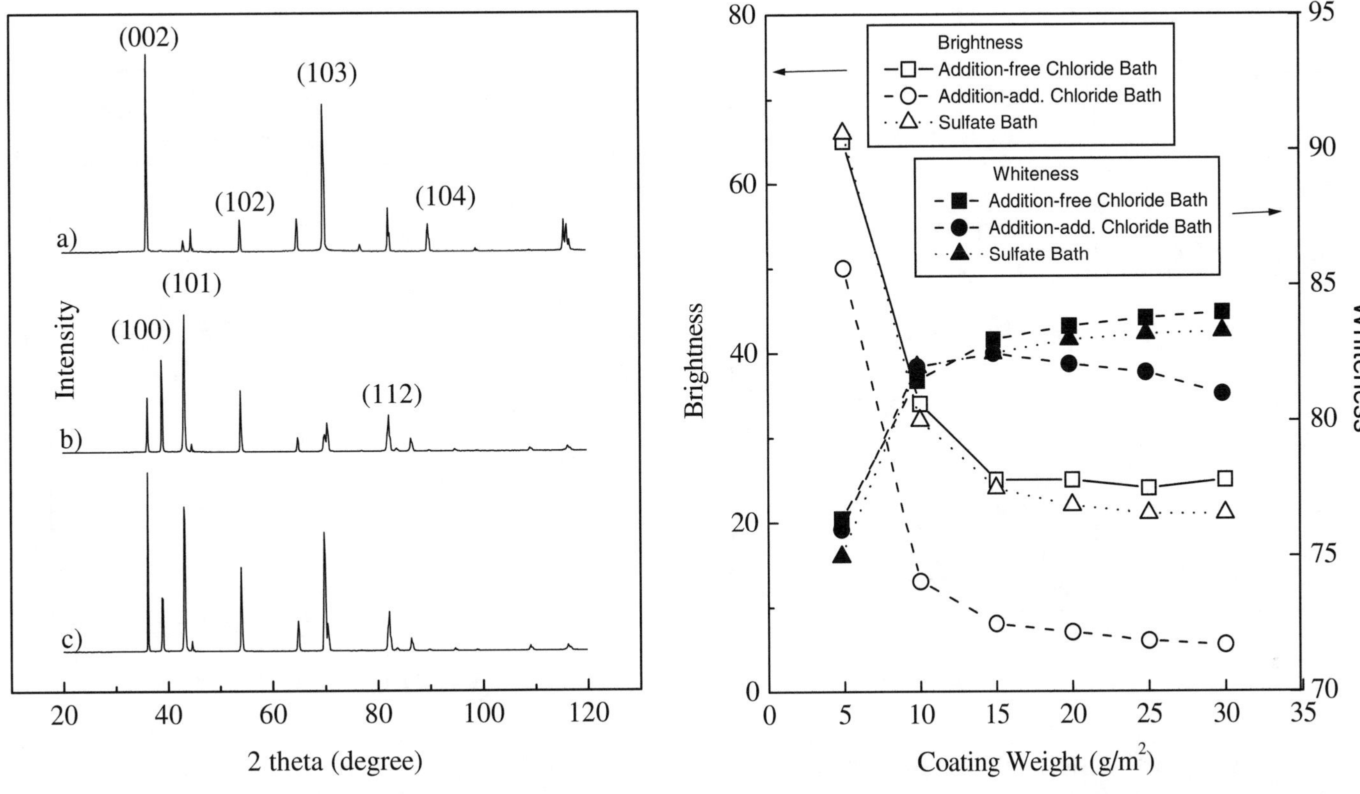

Fig.2 X-ray diffraction patterns of zinc coating layer with bath type (a:addition-free chloride bath b:addition-additive chloride bath, c:sulfate bath)

Fig 3. Variation of surface appearance of zinc coating layer with bath type

The surface appearance of coating layer was compared in terms of the whiteness and brightness(Fig.3). The whiteness and brightness of layers coated at sulfate bath and addition-free chloride bath represent higher than at addition-additive chloride bath. From these results, it can be assumed that the improvement of surface appearance related with the increasing of the basal orientation (002) intensity and rectangular grains. With the increase of coating weight, the brightness was decreased at all kind of bath, while the whiteness was increased at the sulfate bath and obtained critical point at 15 g/m^2 at the addition-additive chloride bath. Fig.4 shows the preferred coating orientation with coating weight at addition-additive chloride bath. At coating weight less than 15g/m^2, those depended on coating weight, but not over it. From the above results, It can be assumed that the growth at initial stage of coating is associated with epitaxial growth. In this condition, it occurred by 15g/m^2 of coating weight. Fig.5 shows the variation of the whiteness and brightness with coating weight and electrolyte bath after 15g/m^2 coating at the addition-additive chloride bath. The increase of coating weight lowers the whiteness and brightness at addition-additive chloride bath, while adverse at sulfate bath. From this consequence, the sulfate bath was believed to be effective in obtaining good surface appearance of coating layer

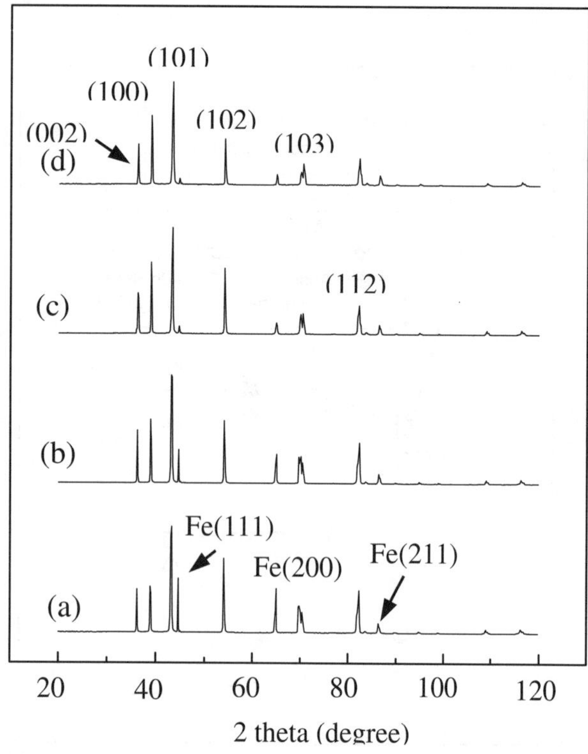

Fig.4 X-ray diffraction patterns of zinc coating deposited at addition-additive chloride bath (a:5, b:10, c:15, d:20g/m^2)

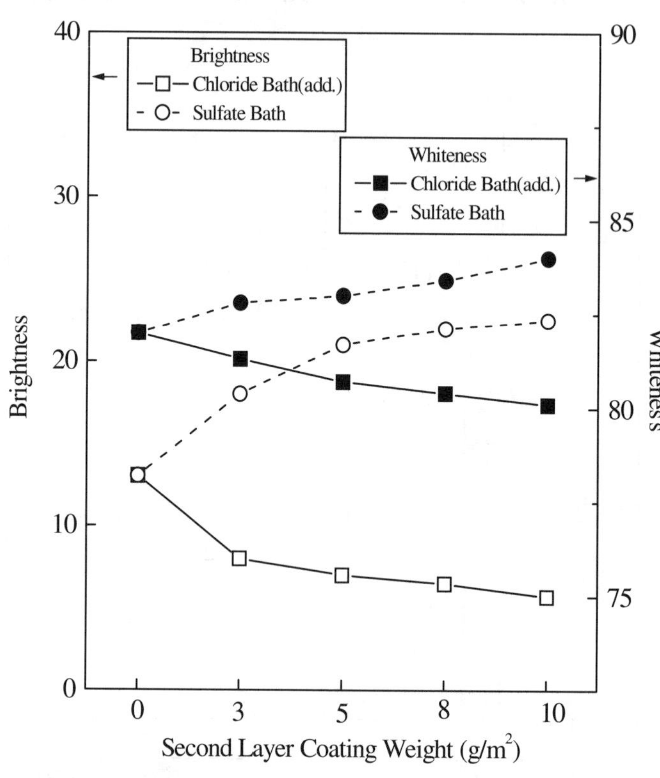

Fig.5 Variation of surface appearance with second layer coating weight (after 15g/m^2 coated at addition-additive chloride bath)

2) Double coating test at working line

The overall layout of the POSCO Pohang EGL is shown in Fig.6. This line consists of entry section, post-treatment section, and coating section which is composed of the carosel and the vertical cell. The carosel cell consists of a sink roll attached conductor band, a conductor band cleaning block, 13 segmented soluble anodes, counter-flow electrolyte injection headers and wringer rolls, and the vertical cell consists of a sink roll, a conduct roll and 4 soluble anodes. The zinc coating is usually conducted at carosel cell of addition-aditive chloride bath. However, the quality of coating surface was poor comparative to other cell type, because of lots contacts between steel strip and roll. Contaminated surface of coating is generally appeared as band mark, stain

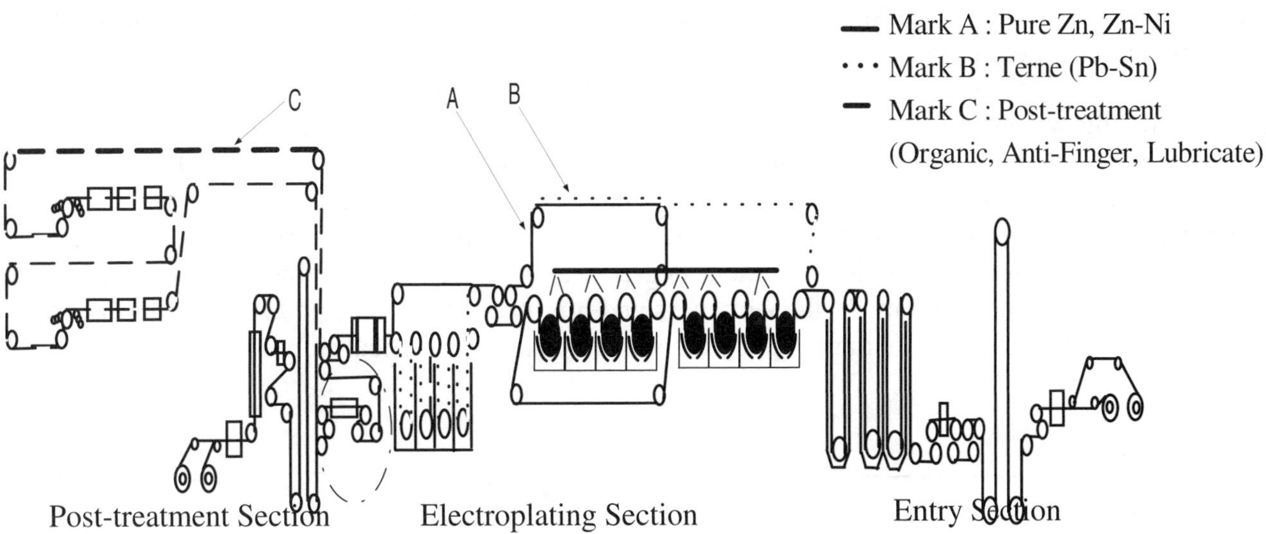

Fig. 6 Schematic diagram of POSCO Pohang EGL

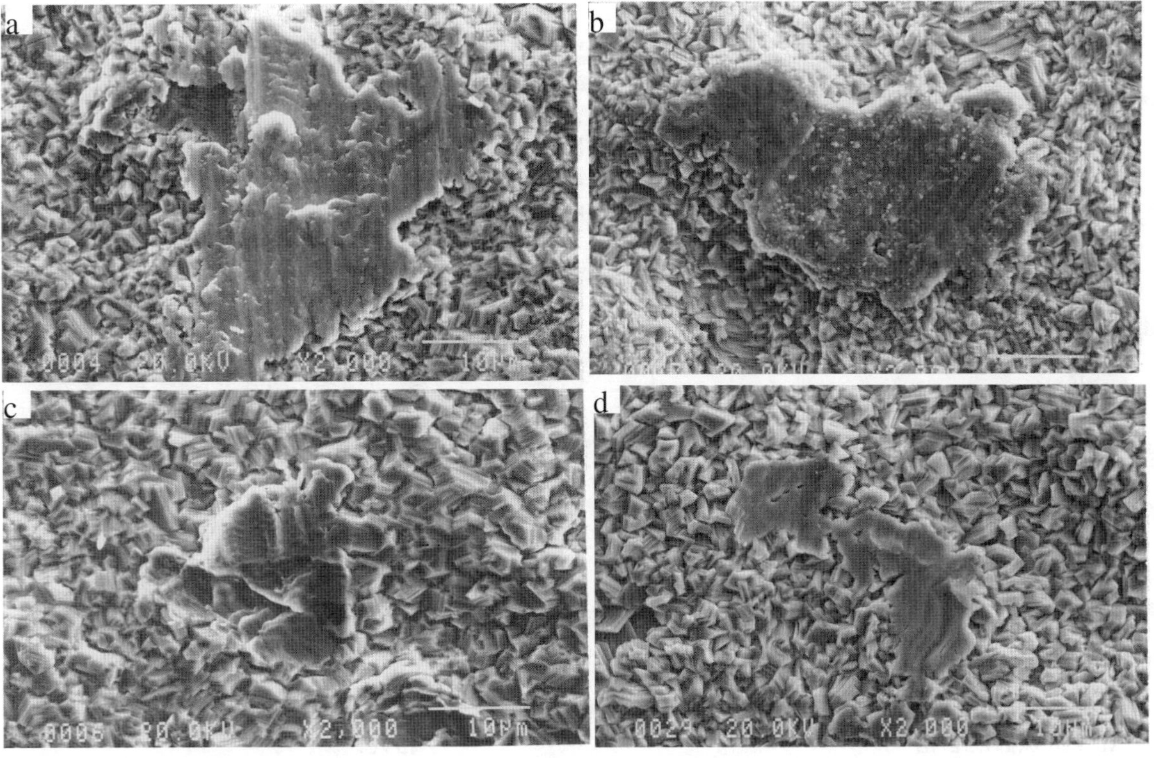

Fig.7 SEM morphology of band-mark area with coating weight of second layer {(second layer coating weight : a(0.0g/m^2), b(3.0g/m^2), c(5.0g/m^2), d(8.0g/m^2)}

and striation. Fig. 7a shows the SEM morphology at band mark area of 15g/m^2 coated steel sheets. The band mark is clearly appeared on the coating surface and can be observed with line by the naked eye. However, when the first coated steel sheet was continuously coated at vertical cell of sulfate bath, the band mark is relaxed with second layer coating weight (Fig.7b~d). At above 5g/m^2 coating weight of the second coating, band mark is almost disappeared. The preferred orientations changed with coating weight from (101), (100), (002) to (002), (101), (103) are shown in Fig.8. The increase of second coating weight, preferentially increased the (002), (103) orientation, whereas the (101) and (100) orientation decreased. The whiteness and brightness were improved by second coating (Fig.9). However, its improvement degree was reduced with coating weight. From the above results, it can be said that the band mark, the whiteness and brightness of zinc-coated steel sheets at carosel cell of addition-additive chloride bath are improved by the second coating at vertical cell of sulfate bath.

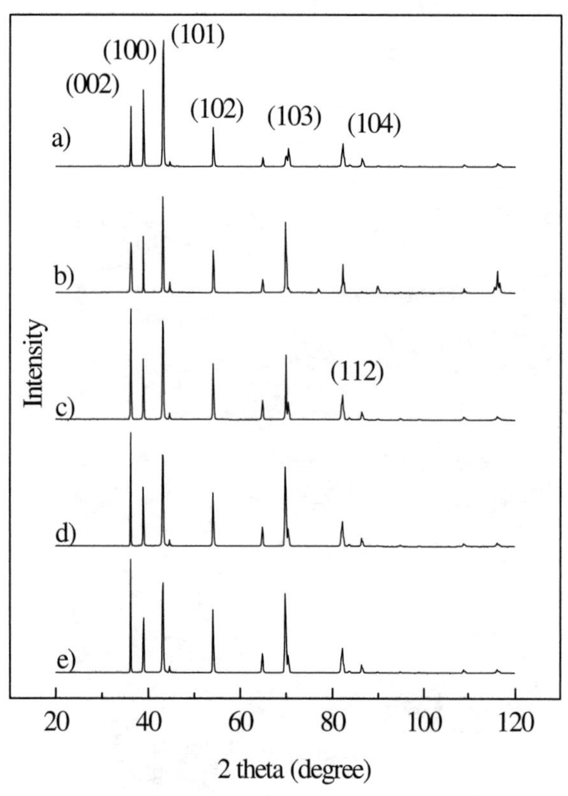

Fig.8 X-ray diffraction patterns with coating weight of second layer (a:0.0, b:3.0, c:5.0 d:8.0, e:10.0g/m^2)

Fig.9 Variation of surface appearance of zinc coating with coating weight of second layer

SUMMARY

The types of bath to improve the surface appearance of zinc coating were investigated, and then the effects of coating equipment were compared. They are summarized as follows:

1) The surface appearance of coating is variously changed with the electrolyte bath type. The sulfate bath represents the better effect on surface appearance than the addition-additive chloride bath.

2) The effects of bath on surface appearance come out of the variation of coating structure and preferred orientation, which seems to be related the increasing of the basal orientation (002) intensity and rectangular grains.

3) With the increase of coating weight, the critical point for the surface appearance was observed at coating weight 15 g/m^2. At coating weight less than 15g/m^2, those depended on coating weight, but not over it. It can be assumed that the growth at initial stage of coating is associated with epitaxial growth.

4) The surface appearance of layer plated at carosel cell comparative to the other cell is poor. However, it can be improved by coating weight over 5g/m^2 at vertical cell of sulfate bath.

REFERENCE

1) K. Ishii, M,Kimoto and A.Yakawa , “Effect of crystal orientation on properties of zinc electroplated steel sheets,” Galvatech`98, p547.

2) R.Sato, “Crystal growth of electrodeposited zinc,” J. Electrochem. Soc., v.106, No.3, (1959) p206

3) I. Tomov,”Factors influencing the preferential orientations in zinc coating electrodeposited from chloride baths,” J. Appl. Electrochemistry, v.19. (1989) p377

Effect of Chromate on the Corrosion and Fuel Resistance of Resin Coated Steel Sheets for Automobile Fuel Tank

Soohyoun Cho, Jaeryung Lee, Sangkeol Noh
Technical Research Lab., Pohang Iron & steel Co.
1, Koedong-Dong, Nam-ku
Pohang, 790-985
Korea
Tel: 82-54-220-6193
Fax: 82-54-220-6913
E-mail: happyman@posco.co.kr

Key Words: corrosion resistance, fuel resistance, fuel tank, chromate, thermal stability, resin

INTRODUCTION

In these days, the service life of automobile was more extended and these trends propelled the development of highly anti-corrosive materials. Especially, the fuel tank of automobile which is an important safety part must have good corrosion and fuel resistance, as well as strength to endure the inside gas pressure and the vibration during driving. Therefore, materials used for the fuel tank need to be suitable for those properties. In aspect of fabrication, the materials must have good formability, weldability, and paintability for economical car manufacturing.[1]

On the purpose of the fuel tank application, the terne(Pb-Sn) sheet has been widely used as a material for the fuel tanks of motor vehicles. Plastic is also used for many cars but weak against the gasoline penetration. It is urgently demanded that fuel tank materials have higher corrosion resistance and environmental harmlessness. However, the terne sheet has an excellent corrosion resistance, this disadvantageous because of environmental problem of lead usage.

Recently, the environmental regulation became the major issue in the automobile industry. Japan has lowered the permissible lead content in the shredder dust from 3 ppm to 0.3 ppm since 1996. In Europe, the automobile containing harmful elements such as heavy metal will not be allowed to dispose on the earth after 2003.[2] Therefore, the terne sheet required to be substituted for some other lead free coat steels such as Zn-Ni resin coat steel, Al coat steel, and Ni flash coat steel.[3-5]

After considering above many factors, we have developed PF-COAT steel with the special chromate and resin layer on the electroplated Zn-Ni alloy. It did not contain any lead to meet environmental regulation, which possessed good fuel and cosmetic corrosion resistance. However, the fuel tank materials need not only good formability, paintability, corrosion resistance but also welding property. The welding property is directly related to what kinds of surface treated layers on the steel and the contamination of electrodes used in welding process.[6] Especially, if non-conductive resin is coated on the steel, it is impossible to get good welding property. It is known that the welding property is related to the resin thickness. Although the addition of various metal powders was conducted in order to gain the conductivity for the welding property, the thickness of resin coating seriously affects welding properties. Therefore, the chromate which is between electroplated layer and resin layer must have excellent corrosion resistance in considering of thin resin layer.

The chromate was classified into three groups which are roll on type, reaction type, and electro deposit type one. In our study, the roll on type chromate was used, and it has high corrosion resistance than the others. Also, this type chromate can reduce environmental pollution by little solution loss.

In the present work, the effect of chromate and resin coating on the corrosion and fuel resistance was investigated in considering the manufacturing condition for newly developed PF-COAT. Also, it was compared with other materials for fuel tank application.

EXPERIMENTS

(1) Test Material

The materials used in the work are in detail in Table I. The substrate of test material is cold rolled steel sheet which contains low carbon, and it has low carbon equivalent level. The test materials whose top is Zn-Ni electroplated layer were investigated according to the variation of chromate condition, as well as effect of resin coating. Other materials such as electro galvanized and terne sheet were used for comparing with the test materials.

Table I. Test materials.

Product	Steel Thickness (mm)	Coating Weight (g/m^2)	Coating Composition	Chromate Weight (mg/m^2)	Resin Thickness (μm)	Remarks
PF-COAT	0.8	30	Zn-12% Ni	100	2.0	Electrolytic
Terne (E)	0.8	60	Pb-8% Sn	15	No	Electrolytic
Terne (H)	0.8	60	Pb-8% Sn	No	No	Hot-dip
EG	0.8	40	Pure Zn	80	No	Electrolytic

(2) Test Procedure

The developed PF-COAT was initially prepared by Zn-Ni electroplating with 30 g/m^2, and followed by coating of the roll on type chromate with 100 mg/m^2. The curing after chromating was performed at the 120, 160, 200, and 250 °C temperature to investigate corrosion resistance according to curing condition. Finally, special resin was coated by roll coater with 2.0 μm thickness, and the second curing was done. For the comparing of Ni flash coating effect, Ni with 0~300 mg/m^2 was coated on the substrate before the Zn-Ni electroplating.

The external corrosion resistance was measured for the flat panel and the drawn cup by the salt spray testing according to ASTM B 117 after 600 or 1000 hours. The cyclic corrosion test of test materials were carried out in 50 cycles. A cycle consists of salt spraying for 6 hours, drying at 50 °C in 35% humidity for 3 hours and holding at 50 °C in 95% humidity for 15 hours.

The gasoline resistance test was conducted in two kinds of fuel systems; deteriorated gasoline consists of 70% regular gasoline and 30% 2g/l NaCl, and gasohol was made up of 84% regular gasoline, 15% methanol and 1% water with formic acid. This fuel was filled in the drawn cup of each material and the fuel corrosion resistance test was carried out for three or six months in "fill and shake" method. The drawn cups filled with the deteriorated gasoline and gasohol were shaken for 8 hours and preserved in stationary state for 16 hours. The contaminated gasoline was changed every month.

RESULTS

(1) Ni flash coating

Ni flash coating with thin coating weight was treated between substrate and Zn-Ni alloy coating for the improvement of corrosion resistance of Zn-Ni electroplating. Fig. 1 shows the effect of Ni flash coating weight on the corrosion performance of chromated steel sheet. When Ni was not coated, red rust and white rust occurred after salt spray test for 1000 hours. However, as Ni flash coating weight increased, the area of white rust decreased. It revealed that the corrosion resistance was improved with Ni flash coating weight increase. Moreover, Ni flash coating diminished the crack caused by Zn-Ni electroplating, and led to the improvement of corrosion resistance after forming.

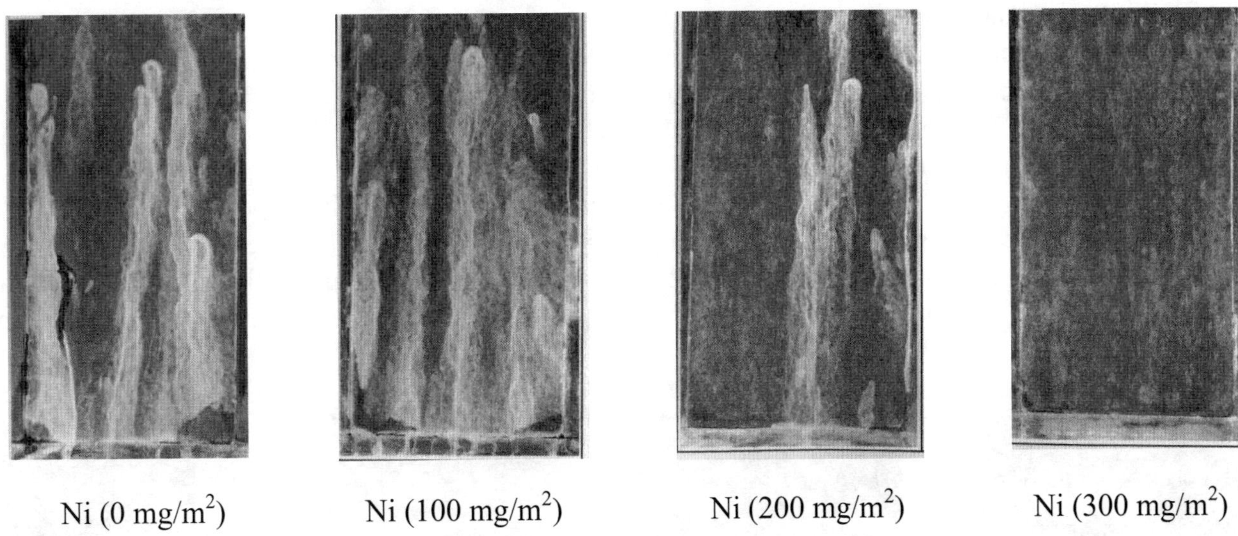

Ni (0 mg/m^2) Ni (100 mg/m^2) Ni (200 mg/m^2) Ni (300 mg/m^2)

Fig. 1. Effect of Ni flash coating weight on the corrosion performance after salt spray test for 1000 hours.

(2) Curing effect

Fig. 2 shows the surface appearance of chromated steel sheets after the salt spray test for 1000 hours according to the peak metal temperature. The developed chromate solution contained silane coupling agent to connect organic resin and inorganic chromate. If sufficient thermal energy was supplied for cross linking reaction, this coupling agent led to inhibit chromate eruption under wet condition due to forming the strong chromate layer. However, the white rust was obviously appeared at the chromated steel sheets at 120 °C, because curing reaction did not fully occurr in the chromated layer.

Also, corrosion resistance was decreased in case of performing curing reaction under excessively high temperature condition. Therefore, it was estimated that corrosion resistance of chromate layer was affected by film properties according to curing condition, and appropriate thermal energy was demanded for optimum properties.

Fig. 2. Surface appearance of chromated steel sheets after the exposure of salt spray test for 1000 hours

Fig. 3 shows the surface appearance of chromated cylindrical cups after the salt spray test for 600 hours according to the peak metal temperature. In order to simulate the corrosion resistance for the press formed fuel tank, flat panels were drawn as cylindrical cups. At low temperature, curing reaction of chromated layer was not enough and white rust occurred in spot shape at the soluble hexavalent chromium position. The chromated sample of 160 °C curing temperature did not show any white rust. On the contrary, at high temperature above 200 °C, the corrosion resistance was dramatically decreased. The deterioration of corrosion resistance becomes more serious at the high temperature with increasing experimental temperature. From the EPMA analysis, it was confirmed that formed part had higher zinc, nickel, and iron atomic contents than flat part one. Therefore, it was estimated that the deterioration of corrosion resistance of formed part was caused by exposure of substrate at the crack of chromate layer after forming.

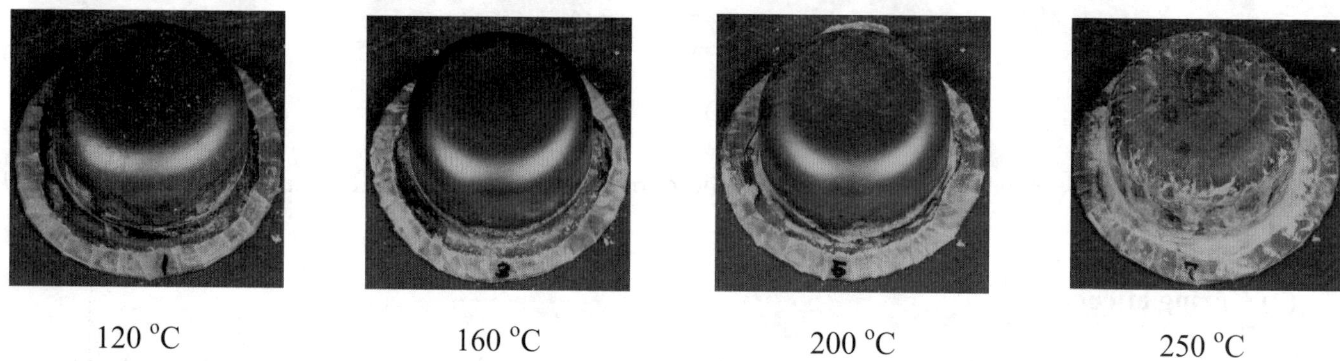

Fig. 3. Surface appearance of chromated cylindrical cups after the exposure of salt spray test for 600 hours.

Fig. 4 shows Auger electron peak intensity ratio for zinc to chromium in the chromate layer. As curing temperature increases, zinc to chromium ratio becomes higher. Therefore, the permeation of zinc into chromate layer is more serious at high temperature, which is considered as one of main factors for deterioration of corrosion resistance. The zinc permeation was more severe when surface was at the bulgy point, and it was confirmed by Zn Auger mapping for the surface layer. Furthermore, if chromate coating weight was lower than sufficient ones, corrosion resistance was more seriously deteriorated due to zinc permeation at the high temperature. Generally, corrosion resistance is probably improved according to the increase of chromate coating weight. However, excessive chromate coating brings about declination of corrosion resistance because of crack in chromate layer after forming. It was found that the optimum chromate coating for the PF-COAT was in the range of 100~120 mg/m^2.

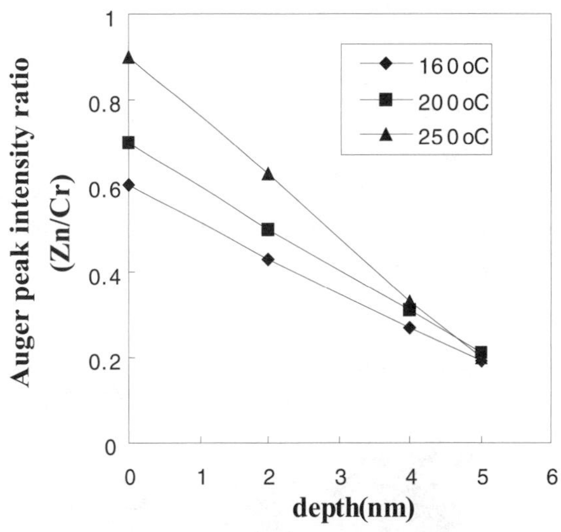

Fig. 4. Auger electron peak intensity ratio for Zn/Cr in the chromate layer.

(3) Corrosion behavior

Fig. 5 shows scanning electron microscopic image of white rust part of chromated zinc-nickel alloy steel sheet after salt spray test. It was detected that crack was caused by zinc-nickel alloy layer, and this crack grew bigger after forming. From the WDS analysis, it was assured that chromate layer was not in this crack. Therefore, the substrate was exposed at this crack and corrosion reaction firstly occurred at the crack of formed part. In the figure, it was detected that the needle-shaped corrosion product appeared at the crack of formed part and was changed into the globular shape corrosion product as time. This needle-shaped corrosion product was composed of zinc and chloride, and the globular shape one was zinc oxide from EPMA and XRD.

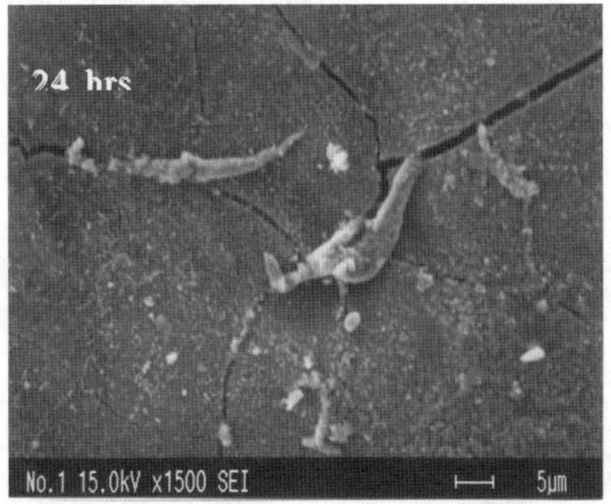

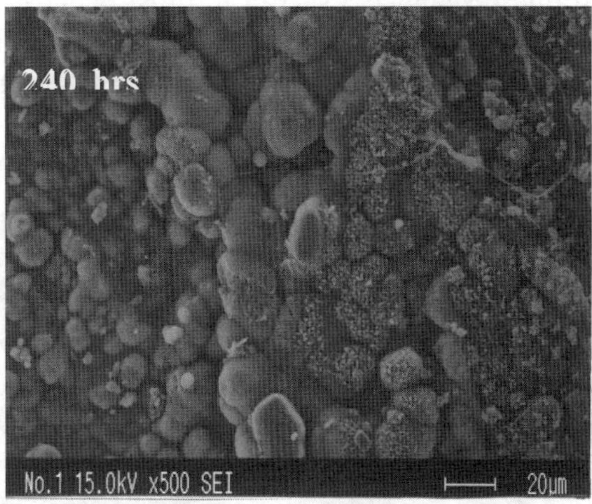

Fig. 5. Scanning electron microscopic image of white rust part at cylindrical cup sample with exposure time after salt spray test.

(4) External corrosion resistance

Fig. 6 shows the corrosion performance of test material with other test materials after the cyclic corrosion test for 50 cycles. In electrogalvanized steel and terne sheets, red rust occurred at the whole part. However, once it was treated by resin on the chromated Zn-Ni alloy plate, the corrosion resistance property was significantly improved. The resin coated steel doesn't reveal red rust but the terne sheet does. It is assured that the PF-COAT steel chromated on Zn-Ni alloy coating layer is able to have a galvanic corrosion protection to block the corrosion of the substrate under such a salty environment.

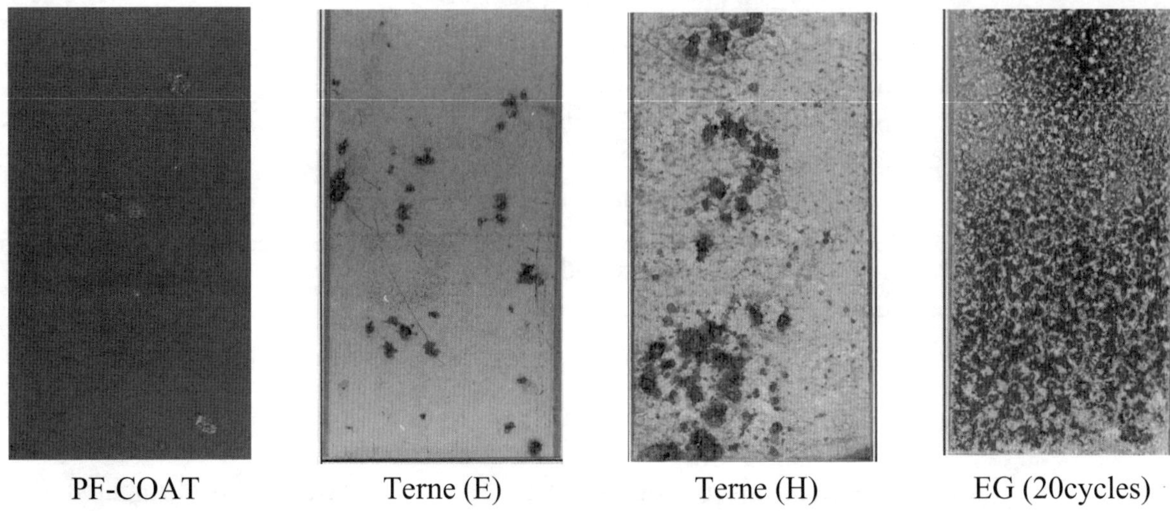

PF-COAT　　Terne (E)　　Terne (H)　　EG (20cycles)

Fig. 6. Surface appearance of compared steel sheets after cyclic corrosion test for 50 cycles.

(5) Internal fuel corrosion resistance

Fig. 7 and 8 show the surface appearance after the fuel corrosion tests for 90 days in the deteriorated gasoline and for 180 days in gasohol. The PF-COAT had little white rust in deteriorated gasoline resistance test. But red rust occurred at the electrogalvanized and terne sheets for this test. In the gasohol, white rust was not remarkably observed at the all tested materials for 180 days. From the results, it can be concluded that resin coating on the thicker chromated layer is very effective for fuel corrosion resistance.

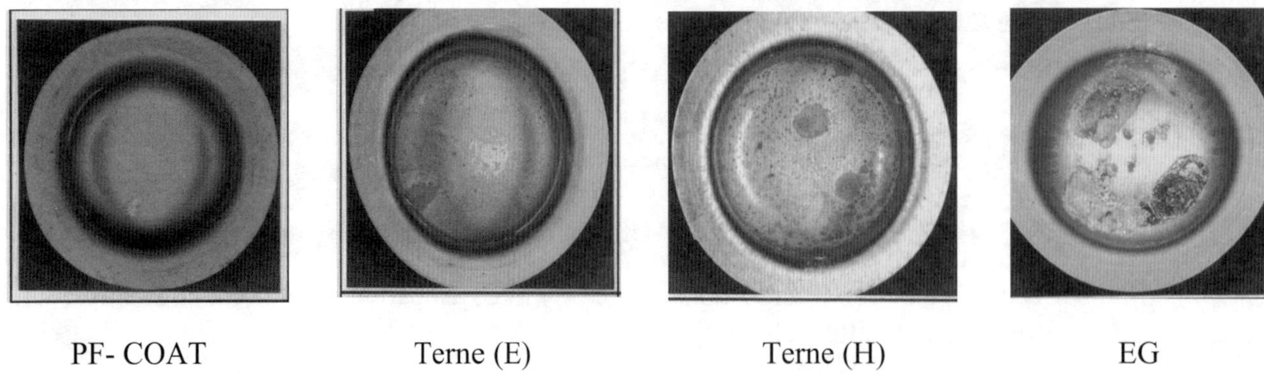

PF- COAT　　Terne (E)　　Terne (H)　　EG

Fig. 7. Surface appearance of compared steel sheets after holding for 90 days in deteriorated gasoline.

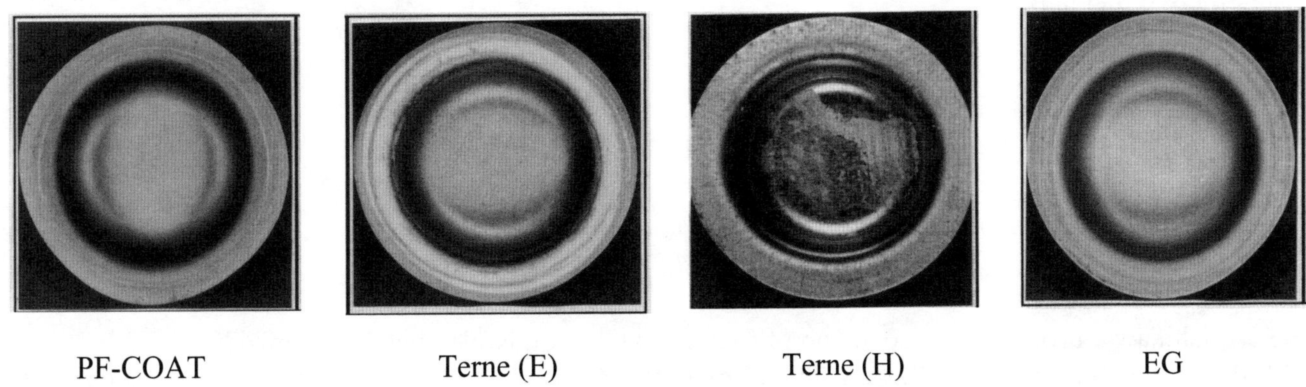

Fig. 8. Surface appearance of compared steel sheets after holding for 180 days in gasohol.

Table II shows amount of metal ion in deteriorated gasoline solution after internal fuel resistance evaluation for PF-COAT and other tested materials. In the PF-COAT, more metal ions of Zn-Ni coating layer were detected as time passes, and iron and chromium ion were identified after three months. For the terne sheets, more iron ions were detected than PF-COAT, and it revealed that corrosion was severely progressed at the metal substrate. It is assured that the PF-COAT steel is able to have a galvanic corrosion protection to block the corrosion of the substrate under such a wet fuel environment.

Table II. Amount of metal ion in detreriorated gasoline solution after fuel resistance evaluation for the tested materials. (unit: μg/piece)

Classification		Zn (mg/piece)	Ni	Fe	Cr	Sn	Pb
PF-COAT	30 days	0.722	3.4	Tr	Tr	-	-
	90 days	0.894	5.0	65	4.0	-	-
Terne (E)	30 days	-	-	44.0	-	Tr	389.0
	90 days	-	-	6660	-	81.0	286.0

CONCLUSION

The chromate plays an important roll on the corrosion resistance, but the curing temperature must be appropriately controlled to get good corrosion property. Also, if sufficient Ni flash coating was applied, Ni flash coating could conceal the crack caused by Zn-Ni electroplating. Therefore, it led to improvement of corrosion resistance after forming for chromate layer. From the present work, it was confirmed that PF-COAT steel with the special chromate and resin layer had excellent external corrosion resistance and internal fuel corrosion resistance. It can be concluded from our work that newly developed steel sheet which has no environmentally harmful lead can be substituted for terne sheet of automobile fuel tank.

REFERENCES

1. T. Bessho, "Surface treatments to correspond to environment of Automotive Industry", The Journal of the Surface Finishing Society of Japan, Vol. 50, No. 12, 1999, pp. 1085.

2. H. Ishikawa, Proc. Japan association of Corrosion Control, November 1999, pp. 420.
3. P. J. Alvarado, "Steel vs. Plastics: The Competition for Light-Vehicle Fuel Tanks", JOM, July 1996, pp. 22.

4. H. Kashiwagi, S. I. Tsuchiya, H. Nagai, K. Fukui, and N. Kimiwada, "Development of Zn-Ni Coated Steel Sheet for Fuel Tank", GALVATECH'98, 1998, pp. 558.

5. H. N. Hahn, S. G. Fountoulakis, and T. Ogonowski, "Corrosion Performance and Fuel Compatibility of Prepainted Zinc-Nickel Coated Steel for Fuel Tank Applications", SAE Technical Paper Series 971006, 1997, pp. 163.

6. S. Thoru, Y. Nobutaka, and G. Masahisa, Metal & Technology, Vol. 65, No. 10, 1995, pp. 909.

FLAT ROLL PRODUCTS III

RECIPROCAL EFFECTS OF ROLL SURFACE CONDITIONS AND WORKING TEMPERATURE ON THE NUMERICAL COMPUTATION OF ROLLING FORCE OF HSLA STEEL

Taher Ahmed El-Bitar
Central Metallurgical R&D Institute (CMRDI),
Metal Working Department
P O Box 87 Helwan, Cairo - Egypt.
Tel.: 202 5010640, Fax.: 202 5010639
E. mail: rucmrdi@rusys.eg.net

Key Words: Numerical Modeling, Hot Rolling, Friction, Working Temperature, Rolling Force, HSLA Steel.

INTRODUCTION

One of the major parameters controlling roll pass design process is the calculation of rolling force and torque to achieve microstructure and mechanical property control without the need for expensive and time consuming trial and error technique. Furthermore, rolling force and torque are main parameters in the advanced and efficient finite element method (FEM), which allows prediction of metal flow during the forming operations, establishing the limits of formability, and predicting forces and stresses needed to execute the forming operations so that tooling and equipment can be designed[1].

Due to the computational proposes, a numerical model has been constructed to calculate force and torque during hot rolling[2]. Like numerical models were enhanced by including a description of the material behavior, mill limitations, roll surface condition (friction coefficient), and thermal conditions[3-6]. The numerical model[2], depends mainly on the behavior of workpiece during deformation at different temperatures. This behavior can be described by the flow stress - true strain relationship (flow curves). These flow curves are mathematically formulated. Many investigators developed formulas[7-12], to characterize these flow curves. The most suitable ones is that developed by Halal and Kaftanoglu[11] of the form; $(\sigma = Y_o (1+B\varphi)^n - A\varphi^m)$, where σ and φ are the flow stress and true strain respectively. Y_o, B, A, n, and m are constants. This form is applicable for both hardening and softening phenomenon, which occurred during hot rolling. Flow stress is varied with strain, strain rate, and temperature. These parameters are clearly reflected on the value of rolling force and torque. Working temperature has a pronounced effect, where it is inversely proportional to the value of rolling force and torque[2,13]. However, the friction between rolls and workpiece is influenced by their temperatures[5,14]. Moreover, most of the numerical models for hot rolling, consider the coefficient of friction as in a direct proportional with the value of rolling force and torque[2,6].

This work is a part of a series partially published previously[2,13], and aims to clarify the opposite effects of both rolling temperature and friction between the working rolls and the work piece.

EXPERIMENTAL WORK

Y-blocks of V-HSLA steel is cast with a chemical composition as that followed in table I;

Table I. Steel Chemistry, wt. %.

C	Si	Mn	S	P	Al	V	Fe
0.25	1.00	0.99	0.025	0.033	0.068	0.093	Rest

Cylindrical compression test specimens were machined so that the height was two times the diameter. Heating of test specimens was in a muffle furnace without using protective gas. Series of numerous axisymmetric hot compression tests were conducted under a universal testing machine with an automatic controlled variable cross head speed, at temperatures 900, 950, 1000, and 1050°C. The cross head speed of the test machine was maintained at 50 mm/min. The strain rate of the experiments could not be exactly computed due to very short time of compression, however, lower cross head speeds could not be used to avoid high loss of specimens temperature. The adiabatic heating during testing can be compensated with heat dissipated due to conduction and radiation[15].

The roll mill data used in the numerical model is considered as 2-high reversing experimental mill with 320x320 mm diameter rolls and 450 mm barrel length. The average thickness and temperature of workpiece after and before each pass are as shown in table II;

Table II Pass schedule of the Rolling Process

Parameter	Pass 1	Pass 2	Pass 3	Pass 4
h_o, mm	80.0	50.0	35.0	27.0
h_i , mm	50.0	35.0	27.0	23.0
Temp., °C	1050	1000	950	900

The coefficient of friction between the workpiece and working rolls is changed from 0.2 up to 0.4. Forward and backward strip tensions are considered to be zero.

RESULTS AND DISCUSSIONS

Flow stress and corresponding true strain are initiated from data of compressive yield loads and displacements. A multi-dimensional matrix having the flow stress, true strain and temperature as coordinates is established. Fig. 1 represents the experimental compressive flow curves of the steel at 900, 950, 1000, and 1050 °C. For both temperatures 900 and 950°C, the flow stress increases continuously with strain indicating that the steel raises its resistance to deformation as it has neither sufficient amount of deformation nor high enough temperature to reach to either the level of dynamic recrystallization or steady state flow[16]. However, in case of the tests at both 1000 and 1050°C, the flow curves reach a plateau indicating the initiation of dynamic restoration followed by a steady state flow[17].

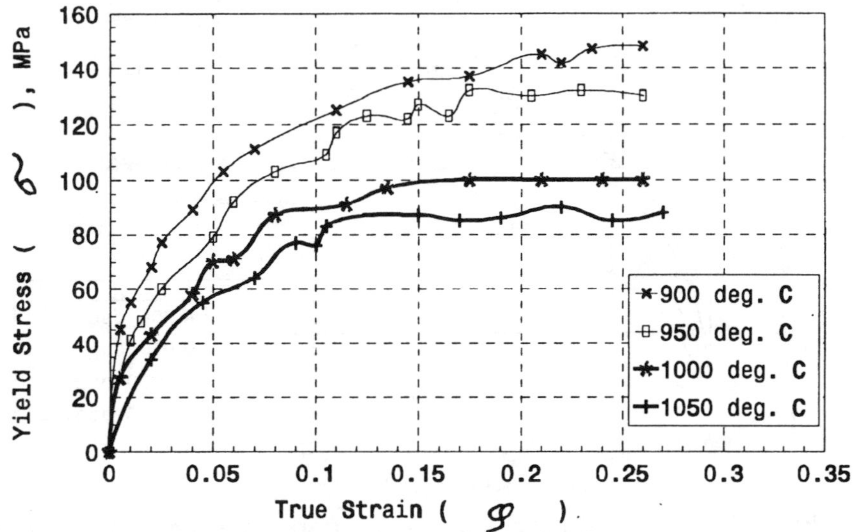

Fig. 1. Experimental Flow Curves at different testing temperatures.

Steels usually have three identified regions on the flow curves when they subjected to hot deformation. These include the initial hardening region followed first by softening and then either by steady state flow or cyclic hardening/softening. In case of hot rolling, steel reaches to the softening region only[17]. Hence, it is evident to examine the suitability of the experimental data with the eq. (1), which was developed by Halal, and Kaftanoglu[11] of the form:-

$$\sigma = Y_o\,(1+B\varphi)^n - A\varphi^m \qquad (1)$$

The process of curve fitting of the experimental data with eq. (1) defines the identifying constants, Y_o, B, n, A, and m. Table III represents the identifying constants of the flow curves at different testing temperatures.

Table III. Constants of the flow curves equation at different temperatures.

Temperature °C	Y_o MPa	B	N	A MPa	m
900	58.0	107.0	.510	500.5	0.80
950	54.0	97.0	.525	505.5	0.82
1000	40.0	95.0	.560	478.0	0.89
1050	30.0	93.0	.590	420.0	0.98

The calculated flow curves at different testing temperatures are presented in Fig. 2.

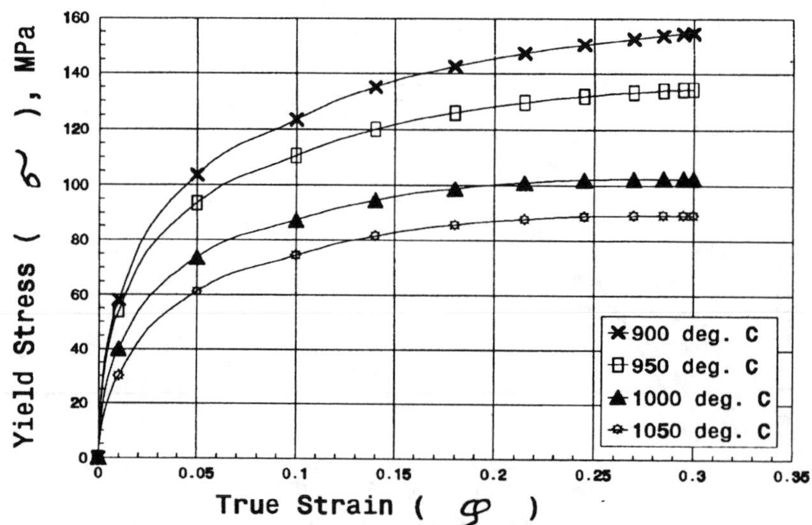

Fig. 2. Calculated Flow curves at different testing temperatures.

The corrugation of the experimental curves in figure 1 due to experimental uncertainties is corrected by the calculated ones in figure 2.

Fig. 3 shows a relation between the specific rolling force (rolling force per 1 m sheet width) / amount of true strain (φ) and working temperature. Obviously, the value of rolling force needed to roll the steel raises appreciably as the workpiece loses some of its temperature at the last finishing passes where high working temperature lowers the resistance of steel to deformation[15].

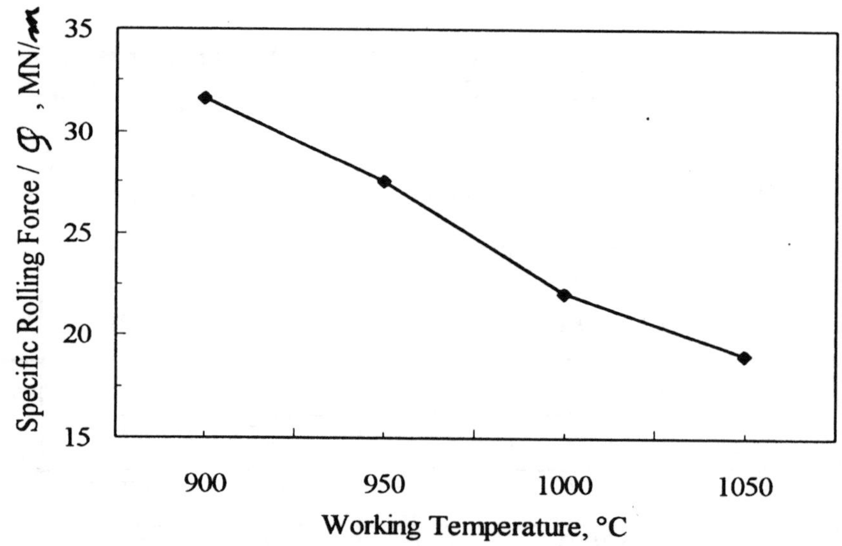

Fig. 3 Relation between specific rolling force per unit strain and working Temperature, at 0.3 Coefficient of Friction.

Friction has a reciprocal effect on the rolling force compared with the effect of the working temperature. Fig. 4 represents the effect of friction coefficient between working rolls and workpiece on the specific rolling force / amount of true strain (φ). Really, the specific rolling force increases with the increase of the friction coefficient (μ), where μ is always a positive term during computation in the numerical model[2]. However, the effect of increasing friction in hot rolling is not so seriously effective as in the case of cold rolling[18]. This may lead to the conclusion that the theory of regions of restricted deformation is not applied to the case of hot rolling.

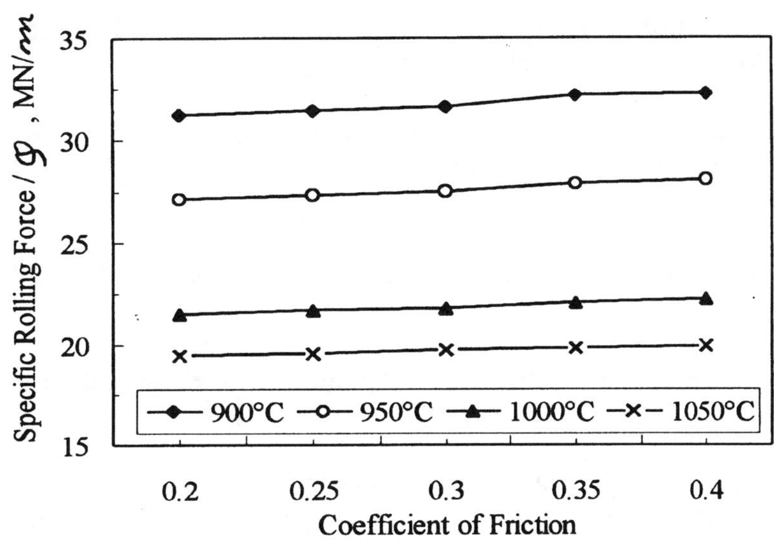

Fig. 4 Effect of Coefficient of Friction on the Specific Rolling Force.

CONCLUSIONS

1- The flow stress for curves of 900 and 950 °C increases continuously with strain, as the steel does not reach the level of dynamic recrystallization. However, at both 1000 and 1050 °C, the flow curves reach a plateau indicating the initiation of dynamic restoration.

2- Curve fitting of experimental data by the numerical model identifies the constants of the flow curves at different temperatures.

3- Rolling force raises appreciably as the workpiece losses some of its temperature.

4- The effect of increasing friction in hot rolling is not seriously effective as in case of cold rolling.

SUMMARY

Hot compressive tests are carried out at temperatures between 900 and 1050°C on HSLA steel specimens. Flow curves at different testing temperatures are initiated and mathematically curve fitted according to the formula; $\sigma = Y_o (1+B\varphi)^n - A\varphi^m$, that takes into consideration hardening and softening phenomenon, which occurred during hot rolling. The flow curve equations are fed into a numerical model, which is constructed on the bases of a two-dimensional plastic deformation and Orwan's theory for calculation of roll pressure in hot rolling. A four passes rolling process is designed to reduce 80.0 mm thickness steel strip at 1050 °C to 27.0 mm at 900 °C with different friction coefficients. Due to computational proposes, the entry and exit thickness at each pass, roll diameter, strip wide and value of front and back tensions are fed into the model together with the working temperature and friction coefficient between working rolls and rolled strip. The vertical roll force, for each pass, is then calculated and correlated to both temperature and friction coefficient. From the rolling force point of view, it is concluded that working temperatures have a more pronounced effect than the friction between working rolls and strip.

REFERENCES

1- S. Kobayashi, Metal Forming and the Finite Element Method, Oxford Univ. press, New York, 1989.

2- T. A. El-Bitar, "Constitutive Modeling and Analysis of Hot Flat Rolling", *1st. European Rolling Conf (HUNGAROLLING' 96)*, Balatonszeplak, Hungary, vol. I, 4-6 Sep. 1996.

3- B. Richter, "The exactness of the roll force- precalculation for cold strip rolling and temper rolling or different mill stands" , *1st. European Rolling Conf. (HUNGAROLLING' 96)*, Balatonszeplak, Hungary, vol I, 4-6 Sep. 1996.

4- B. Richter, et al., "Calculation and improvement of the rolling schedule of cold strip mill", Stahl und Eisen, vol. 115, No. 2, Feb. 1995, pp: 53-59.

5- W. Yuen, Y. Popelianski, and M. Prouten, "Variations of friction in the roll bite and their effects on cold strip rolling", Iron & Steelmaker, vol. 23, No. 2, Feb. 1996, pp: 33-39.

6- N. Silk and Y. Li,, "Friction Measurement in a Tandem Finishing Mill by the Forward Slip Method", Proc. Modeling of Metal Rolling Processes, London, UK, 13-15 Dec. 1999, 400-409.

7- T. Altan and F. Boulger, "Flow stress of metals and its application in Metal Forming Analysis", ASME Journal of Engineering for Industry, vol. 95, 1973, pp: 1009-1019.

8- Siltari, T., "Ermittlung von Flußkurven beim Kaltwalzen an hand gemessener walzparameter", Steel research, 63, No. 5, 1992, pp: 212-218.

9- S. Shida, "Empirical Formula of Flow stress of Low Carbon Steels", Hitachi Research Report, 1974, pp: 1-8.

10- U. Saeed and J. Lenard, J. Eng. Mat. and Tech. , vol. 102 , Apr. 1980, pp: 223-228.

11- A. S. Halal and B. Kaftanoglu, "Computer-Aided Modeling of Hot and Cold Rolling of Flat Strip", *Int. Conf. on Computers in Engineering*, Las Vegas, Nevada , 1984, pp: 120-126.

12- D. Farrugia, M. Zhou and P. Ingham, "Development of Accurate and Reliable Constitutive Models for Flow Stress Predictions during the Hot Rolling of Steel", Proc. Modeling of Metal Rolling Processes, London, UK, 13-15 Dec. 1999, pp: 133-139.

13- T. A. El-Bitar, "Role of Compressive Flow Curves in the Evolutionary Computing Model for Hot Flat Rolling", La Metallurgia Italiana, No. 11, 1998, pp: 17-20.

14- M. Kiuchi, J. Yanagimoto and E. Wakamatsu, "Thermal Analysis of Hot Plate/Sheet Rolling", Proc. 7th Intr. Conf. Steel Rolling'98, Chiba, Japan, Nov. 1998, pp: 227-230.

15- T. A. El-Bitar, "Hot and Cold Deformability of Si-steel Electrical Sheets", Ph.D. Thesis, Cairo Uni., Egypt, 1989.

16- F. Siciliano, "Mathematical Modeling of the Hot strip Rolling of Nb Microalloyed Steels", Ph. D. Thesis, Mc Gill University, Montreal, Canada, Feb. 1999.

17- L. D'orazio, A. Mitchell and J. Lenard, "On the response of Nb bearing HSLA steels to single and multistage compression", Proc. HSLA steels'85 Conference, Beijing, China, 4-7 Nov. 1985, pp: 1-10.

18- T. A. El-Bitar, T., "Investigation of some Factors Controlling Roll Pass Design during Cold Rolling of Aluminum", Industrial Development Journal, No. 3, Nov. 1994, pp: 32-36.

A Study on the Edge Cracking of Low Carbon Steel Sheets Manufactured by Mini-Mill Process

Jai-Hyun Kwak, Jin-Hwan Chung and K.Mox Cho
Kwangyang Rolling Products Research Group
POSCO Technical Research Laboratories, 699 Kumho-dong
Kwangyang City, Cheonnam, 545-090
KOREA
Tel; 82-61-790-8679, Fax; 82-61-790-8799
jhkwak@posco.co.kr

Key Words: Edge cracking, Mini-mill process, Low carbon steel, Mn/S ratio, precipitation, MnS

INTRODUCTION

Thin slab-direct hot rolling, called mini-mill process, is expected to become a future process for production of low carbon steel sheets. Mini-mill has some strong points compared to integrated mill in terms of low capital costs as well as low energy and labor costs per unit of output. The facilities of mini-mills have been improved in terms of better quality and productivity since the first generation mini-mill for hot strip production in 1989.[1] Recently, mini-mill processes have adopted not only a thicker slab caster to increase the productivity and reduce surface defects of slabs, but also a roughing mill for a thinner gauge of hot strip production.[2] All of these efforts culminate into a competitive advantage over the integrated hot strip mill. However, there are several problems to be solved in the processing, one of which is edge cracking in low carbon steel sheet.[3,4]

The edge cracking of low carbon steels is regarded as a kind of embrittlement at high temperature. Even though there are several different opinions about the causes of embrittlement in the temperature range of 900 to 1200℃, it is well known that the super cooling and low deformation temperature close to 900℃ causes ductility of plain low carbon steel to deteriorate[4-7]. Since the thickness of cast slab produced in mini-mill processes is thinner than that in integrated mill, the cooling rate at the surface of slab edge must be higher. Compared to integrated mill, these characteristics put mini-mill at a disadvantage. Even though some mini-mill processes attach tunnel furnaces or reheating furnaces prior to roughing, the major functions of these furnaces are to obtain the rolling temperature, to act as a buffer for thin slabs, and to homogenize the segregated elements.[1,2] If optimum conditions of preheat treatment to prevent the edge cracking in mini-mill process were obtained, the furnace could be applied as an edge cracking inhibitor prior to roughing.

The present study tackles the metallurgical subjects involving the thin slab-direct hot rolling process, i.e. mini-mill process. In order to clarify the effect of chemical composition of steel and MnS precipitation behavior on the development of edge cracking during hot rolling, the content of manganese and sulfur in low carbon steel was varied and the isothermal treatment was applied prior to roughing. In addition, this study proposes the concept of an edge crack index for more accurate estimation of the total amount of edge cracking as a function of Mn/S ratio.

EXPERIMENTS

Material

The steels with a basic composition of 0.04 wt.% carbon were vacuum melted. The chemical compositions of steels investigated are given in Table 1. To study the effect of Mn and S content on the edge cracking of low carbon steel in mini-mill process, Mn and S were varied from 0.1 to 0.3 and 0.003 to 0.013 wt.% respectively. The specimens described as 'DR' were for investigating the effect of Mn and S content as well as preheat temperature on the edge cracking. On the other hand, 'CCR' specimens were for studying the effect of preheat treatment time prior to roughing on the edge cracking.

Table 1. Chemical compositions of steels (in weight %)

	C	Mn	P	S	Si	Al	Cu	Cr	Ni	N
DR	0.033~ 0.048	**0.093~ 0.305**	0.014~ 0.016	**0.003~ 0.013**	0.006~ 0.016	0.014~ 0.037	0.076~ 0.082	0.039~ 0.043	0.031~ 0.043	0.004~ 0.005
CCR	0.023~ 0.045	**0.196~ 0.32**	0.011~ 0.017	**0.007~ 0.0095**	0.007~ 0.012	0.021~ 0.043	0.079~ 0.082	0.039~ 0.043	0.029~ 0.034	0.003~ 0.0077

DR; Direct Rolling, CCR; Cold Charge Rolling

Roughing simulation of mini-mill process

The cast slabs, 60mm thick, were hot charged in the moving furnace for roughing simulations. The thermal history from casting to roughing used in this study was based on the calculated result of surface heat transfer analysis of a slab in ISP (Inline Strip Production). After repeated measurements of temperature changes at a 2mm depth 10mm from the slab corner, similar thermal history to heat transfer analysis of slab surface could be obtained by the adjustment of extruding time of slab from steel mould after casting. The preheat temperatures of the moving furnace, set prior to vacuum melting, were set at 1130 and 1230℃ for the slab to reach temperatures of 1050 and 1150℃ respectively. Fig. 1 shows the schematic time-temperature diagram prior to the roughing simulation of mini-mill process. The roughing was simulated in a laboratory simulator to the final thickness of 11.2mm by three passes in a laboratory simulator. The roughing reduction rates were 31.7, 48.8 and 46.7%.

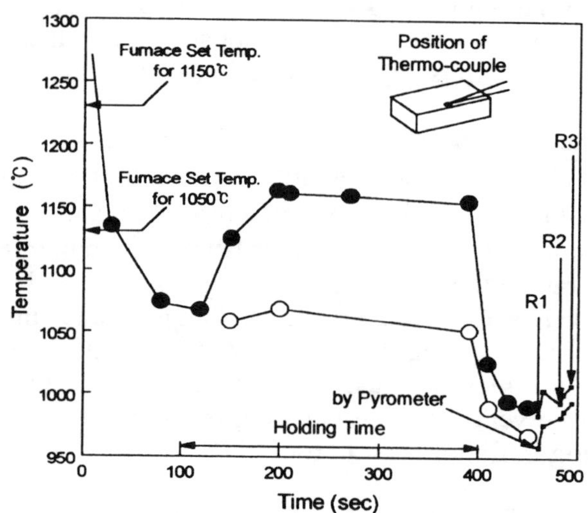

Fig. 1. Schematic time-temperature diagram prior to roughing mill simulation in mini-mill process.

On the other hand, specimens 'CCR' were reheated at 1400℃ for 30 minutes followed by air cooling for 100 seconds. Then preheat treatments were done same way as for the 'DR' specimens. But as-cast specimens

(specimens without preheat treatment) were air cooled for 140 and 120 seconds to get similar hot rolling temperature (roughing) corresponding to 1050 and 1150℃ preheat treatment respectively. The entry hot rolling temperature of 1050℃ preheat treatment was about 960℃ and that of 1150℃ one was 980℃.

Estimation of edge cracking

For the purpose of estimating the edge cracking severity, the edge of hot rolled specimens were ground down to half of their thickness and etched in HCl solution. The length of the edge crack was measured to the accuracy of 1/100mm using a digital caliper.

In the present study, the concept of an edge crack index was proposed for more accurate estimation of the total amount of cracking as a function of Mn/S ratio. The edge crack index proposed in this study is the summation of crack length per unit edge length of specimens (cm), which is defined as equation (1).

$$\text{Crack Index} = \overline{L_C} \cdot \overline{D_C} = \sum L_C / \sum L_{SE} \qquad (1)$$

Where $\overline{L_C}$ is the average crack length (mm), $\overline{D_C}$ is the number of cracks per unit length of specimen edge (1/cm), L_C is the individual crack length, L_{SE} is the edge length of the specimen. When the micro crack with high density and the deep crack with low density appeared in the steel edge, the average crack length of the latter is longer than that of the former even though edge cracking contributes same amount of edge deformation during hot rolling in both cases. But the crack index, which denotes the total length of edge crack per unit edge length of hot rolled specimens, takes into account average crack length as well as the density of edge cracks.

RESULTS

Effect of Mn/S ratio and preheat treatment temperature

The average crack length and crack index of low carbon steels as a function of Mn/S ratio are shown in Fig.2 and 3. The applied preheat treatment temperature in Fig 2 and 3 was 1050 and 1150℃ respectively. The crack

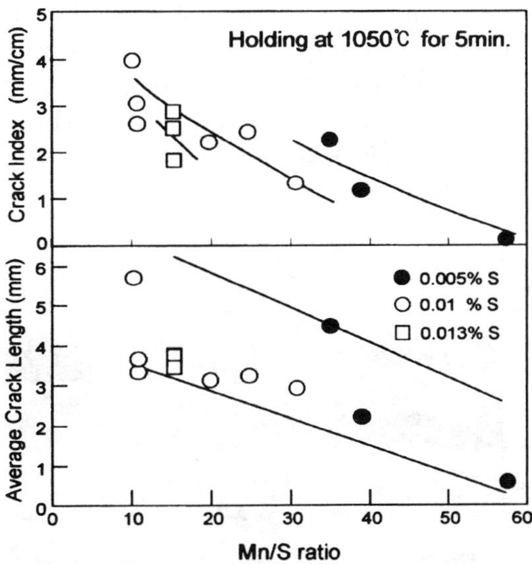

Fig. 2. The effect of Mn/S ratio on the average edge crack length and edge crack index of low carbon steel held at 1050℃.

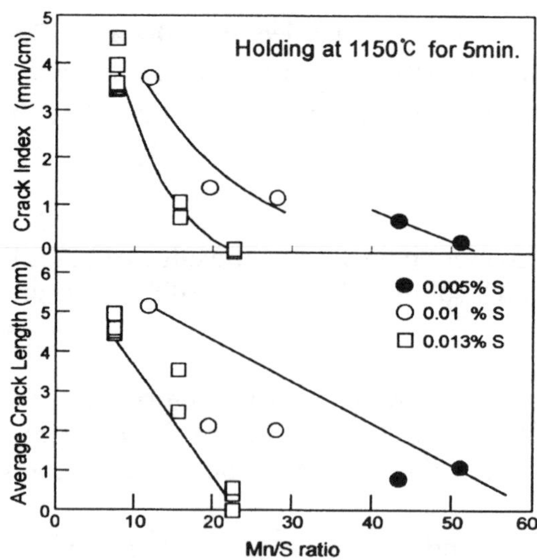

Fig. 3. The effect of Mn/S ratio on the average edge crack length and edge crack index of low carbon steel held at 1150℃.

index shows stronger correlation to the Mn/S ratio in both cases than the average crack length does. Higher Mn/S ratios corresponded with lower average crack lengths and crack indices of both of the preheat treated specimens. The edge crack index decreased as the Mn/S ratio increased, and this tendency was more dominant in low carbon steels containing 0.01~0.013% S. If the low carbon steel was heat treated at 1050℃ for 5 minutes prior to roughing in the mini-mill process, the required Mn/S ratio to prevent edge cracking was above 65 and 50 for S levels of 0.005% and 0.01% respectively (Fig.2). In the case of 1150℃, Mn/S ratio required was at least 60, 40 and 25 for S levels of 0.005%, 0.01% and 0.013% respectively. So It is considered that the 1150℃ preheat treatment is more effective to reduce the edge cracking of low carbon steel than 1050℃ preheat treatment, and the minimum Mn content is around 0.3~0.4% in commercial mini-mill processes if the heating furnace is adopted before the roughing-mill.

Effect of holding time

As shown in 'CCR' of Table 1, low carbon steels containing around 0.008% S were used to investigate the effect of preheat treatment time on the edge cracking. The Mn/S ratios in steels were 23~28, 28~31 and 34~43 as the Mn content in steels was varied at levels of 0.2, 0.25 and 0.3% respectively. The edge cracking of as-cast specimens (without preheating) was more severe than that of isothermal preheat treated specimens as shown in Fig. 4. The edge cracking during roughing of low carbon steel containing 0.2% Mn and 0.008% S could be prevented by preheat treatment at 1150℃ for 15 minutes. In low carbon steel containing 0.25% Mn and

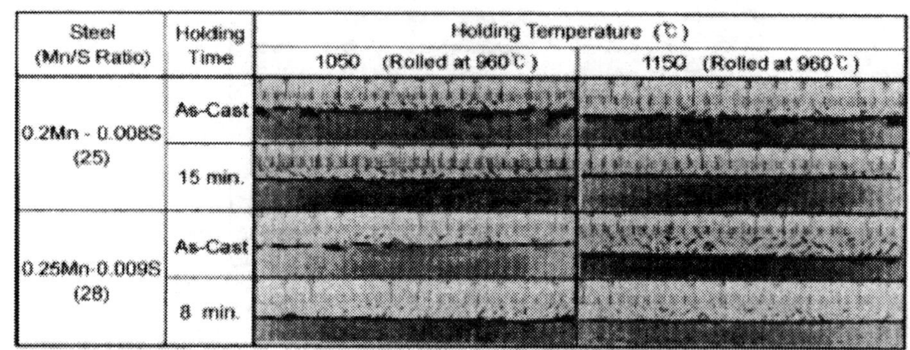

Fig. 4. The changes of edge cracking in low carbon steels with isothermal preheat treatment and holding time.

0.009% S, the edge cracking preventive condition of preheat treatment was at 1150℃ for 8 minutes. However, even though low carbon steels have the same compositions, edge cracking was not thoroughly prevented at the preheat treatment temperature of 1050℃. The edge cracking was remarkably reduced and preheat treatment time for edge cracking prevention was shortened to half as Mn content increased from 0.2 to 0.25%. Fig. 5 and 6 show the effect of preheat treatment time on edge cracking of low carbon steels held at 1050℃ and 1150℃ respectively. In Fig. 5, there was no edge cracking regardless of holding time in low carbon steel containing 0.3% Mn and the edge cracking was reduced as holding time increased in low carbon steel containing 0.2 and

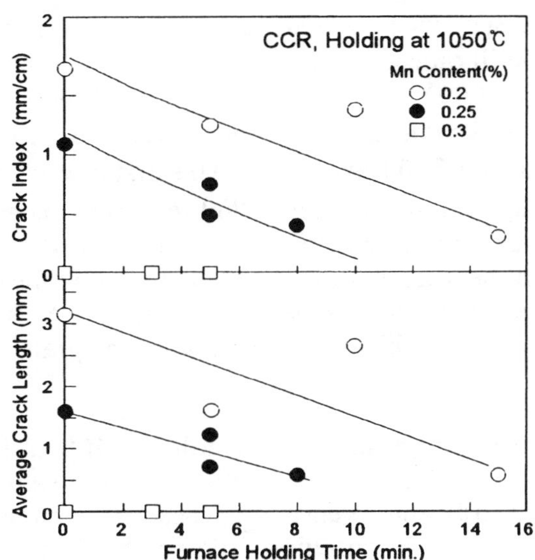

Fig. 5. The effect of preheat treatment time on the average edge crack length and edge crack index of low carbon steel held at 1050℃.

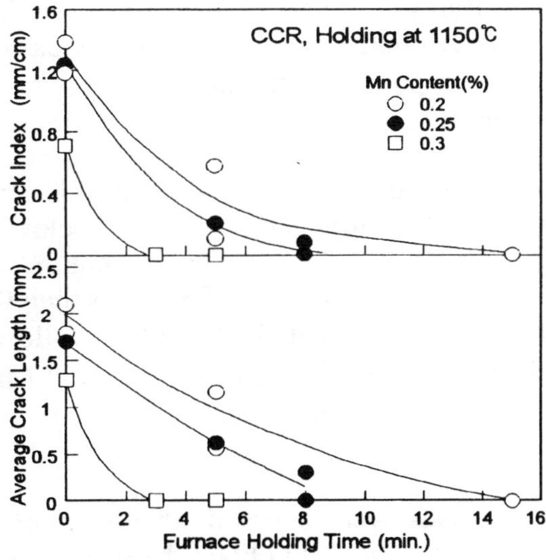

Fig. 6. The effect of preheat treatment time on the average edge crack length and edge crack index of low carbon steel plate held at 1150℃.

0.25% Mn. The edge cracking tendency with respect to holding time at 1150℃ preheat treatment was similar to Fig. 5, but the edge cracking could be prevented within a shorter time than with a preheat treatment temperature of 1050℃.

DISCUSSION

Effect of Mn/S ratio

The hot rolling temperature of the specimens was at least 950℃. Even though there are some descalers in roughing units in commercial mini-mill processes, thermal analysis by FEM has verified that the temperature of the slab surface has not fallen below 900℃. The edge cracking of low carbon steels is regarded as a kind of embrittlement of steel at high temperatures. There are several different opinions about the causes of embrittlement in the temperature range of 900 to 1200℃. The first one is segregation of impurity at austenitic grain boundary.[3,5,8] The low ductility region from about 900 to 1200℃ is attributed to interaction between the precipitates or inclusions and viscous liquidus such as FeS or segregated S and P at austenite grain boundary.[8] Since one of the major deformation mechanisms of low carbon steel is the sliding of grain boundaries at high temperatures, the strength of grain boundaries is very important. If S is segregated at the austenite grain boundary, the strength of grain the boundary is lower[8] than that of the grain, and this results in a grain boundary fracture.

The second opinion regarding causes of embrittlement is as follows. If S and O are concentrated at the austenite grain boundary due to temperature decreases, they exist either in super-saturated solid solution, or as very fine closely spaced precipitates of (Fe,Mn)S and (Fe,Mn)O. These precipitates inhibit movement of the grain boundary and become the source of stress concentration followed by forming cracks.[4-6,9] The last opinion deals with the effect of precipitates and PFZ (precipitate-free zone) adjacent to grain boundary.[7,10] The most preferable precipitation site is the grain boundary at high temperatures. Since PFZ is softer than grain, plastic deformation is concentrated in the PFZ. The precipitates become preferred sites for void initiation, fracture progress by void coalescence, and crack propagation along grain boundary.

The above arguments differ in their explanation of the mechanism of hot embrittlement, but there is a common point, that is, the fact that higher Mn/S ratios give better ductility of low carbon steel in the temperature range of 900 to 1200℃. So, It has long been believed that embrittlement of low carbon steel in this temperature range is very sensitive to the Mn/S ratio. Lankford pointed out that the lowest Mn/S ratio required is 40 in order to prevent embrittlement of low carbon steels.[5]

The results about effects of Mn/S ratio on edge cracking obtained in this study, as shown in Fig. 2 and 3, were similar to Lankford's. When the content of S in low carbon steels is below 0.013%, the minimum Mn content needed to prevent edge cracking was 0.3 to 0.4%. This implies that not only the Mn/S ratio but also the critical amounts of Mn and S are important. Suzuki has shown that if S is less than 0.006% or Mn is higher than 0.7%,[6] embrittlement of low carbon steels, which are melted at 1400℃ followed by cooling rate of 20℃/s to 900~1200℃, does not occur even when Mn/S ratio is small.[6] The critical Mn content of this study, however, is less than that of Suzuki's and it is considered that this difference comes from the different deformation mode or the from effect of preheat treatment.

In this study, two different rolling methods were adopted, that is, 'DR' and 'CCR'. 'DR' steels are for simulation mini-mill directly. However, it is difficult to obtain the exact effect of preheat treatment time within the same composition, because a simulation can be done only once per melting. If dendrite forms on the skin surface of the cast slab and results in segregation of S and Mn, the composition of S and Mn can be different from the average composition.10,11 Even though the precipitation temperature of MnS is below austenitic temperature in a certain composition, MnS can be formed at the temperature of delta-ferrite because of concentration of Mn and S in primary dendrite arm spacing.[11,12] In contrast, 'CCR', the solution treated steels, have uniform composition and MnS as dissolution. So, it was dubious that the results of the two different rolling methods show good agreement.

For the purpose of comparing the effect of Mn/S ratio on the edge crack index between the two different rolling methods with preheat treatment temperatures, the results of low carbon steel containing around 0.01% S in Fig. 2,3,5 and 6 are summarized in Fig.7. All the results of 'DR' steels were preheat treated for 5 minutes, but that of 'CCR' ones were not. The edge crack occurrence depended mainly on the Mn/S ratio, not on the rolling method used. So the most important factor to prevent edge cracking is considered the Mn/S ratio in low carbon steels manufactured by mini-mill process.

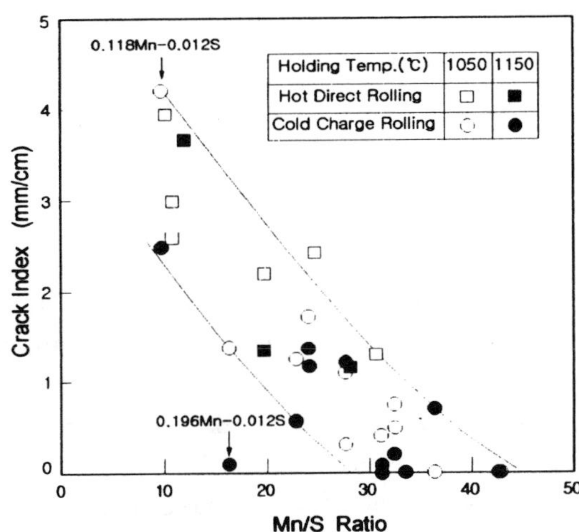

Fig. 7. Comparison the effect of Mn/S ratio on the edge crack index between two different rolling methods with preheat treatment temperatures.

Effect of preheat treatment

As shown in Fig. 2, 3 and 5, 6, the crack index of low carbon steels preheat treated at 1150℃ was less sensitive to the Mn/S ratio and preheat treatment time than those at 1050℃ were. One of reasons for this difference may come from the different rolling temperature, that is, the hot rolling temperature of steels preheat treated at 1050 and 1150℃ was from 960 to 1000℃ and 980 to 1100℃ respectively. If the severe embrittlement occurs around 900℃, the deformation temperature can affect edge cracking. It seems that such an effect is not so considerable, because the level of edge cracking occurrence in as-cast specimens is not so different between two holding temperatures(1050 and 1150℃) as shown in Fig. 4. In this study, the Mn/S ratio, holding temperature and time are considered the more influential factors on edge cracking than deformation temperature.

The effect of preheat treatment temperature on edge cracking of 'DR' is shown as a function of Mn/S ratio in Fig. 8. At temperature range of 900 to 1200℃, the higher holding temperature increases ductility of steels within a shorter holding time.[6,8] In the results of Suzuki et al[6], for example, the holding time required to prevent embrittlement of low carbon steel at least 500 seconds at the holding temperature of 900℃ , 200 seconds at that of 1150℃ and about 40 seconds at that of 1150℃.

The higher heat treatment temperature gives faster growth of MnS, and results in decreasing not only the density of MnS precipitates, but also the S segregation at austenite grain boundary.[3,5,8] As mentioned before, the sites of crack initiation decrease as the density of precipitates at grain boundaries decreases and grain boundary cohesion increases as the amount of S segregation decreases.

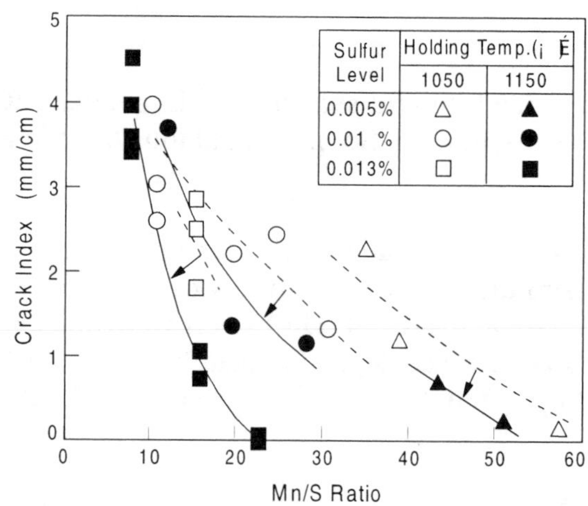

Fig. 8. The effect of Mn/S ratio and holding temperature on the edge crack index of low carbon steel in roughing mill.

Assumed cause of edge cracking

There are several proposed mechanisms on embrittlement of low carbon steels in the temperature range of 900 to 1200•. But there is a common point, that is, embrittlement can be prevented by high Mn/S ratios and preheat treatment of low carbon steel. In this study, as the content of S in low carbon steels increases, the required Mn/S ratio to prevent edge cracking decreases as shown in Fig. 8. When the content of S is below 0.013%, the minimum Mn content to prevent edge cracking was 0.3 to 0.4%. This implies that not only the Mn/S ratio but also the critical amount of Mn and S are important. If steels have different Mn and S contents but same Mn/S ratio, MnS begins to precipitate at higher temperature in steel containing a lot of S than that containing a little S in equilibrium state as shown in Fig. 9. As a result, a lower amount of S remains in steel.

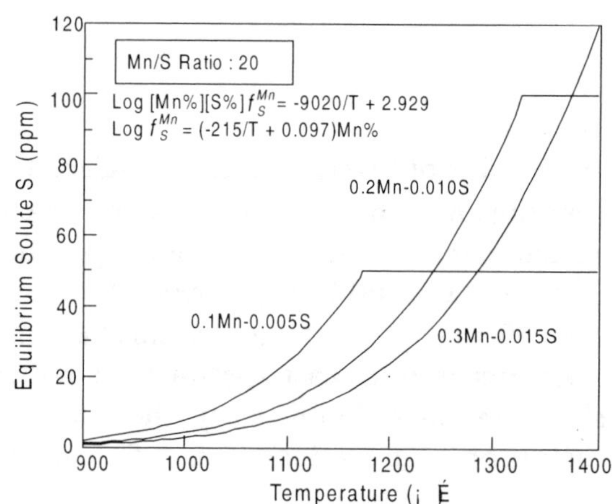

Fig. 9. Comparison of the equilibrium solute S with three different Mn and S compositions.

In case of high S levels, a lower Mn/S ratio is required than with low S levels because the amount of solute S decreased fast due to the active nucleation and growth of MnS. This fact can be confirmed by TEM observation of sulfides in Fig. 10. The sulfides of steel containing 0.1% Mn and 0.01% S in Fig. 10-a) were very fine and the most of them are (Fe,Mn)S. These precipitates are considered to be formed during deformation or at just above rolling temperature. In contrast, there were sparse MnS precipitates in steels containing 0.3% Mn and 0.01% S. In Fig. 10-b), the composition of MnS is considered to be stoichiometric, because the intensity of Mn and S were almost same in EDS analysis. So, the precipitation temperature 0.3Mn-0.01S steel was high enough to agglomerate during preheat treatment or cooling.

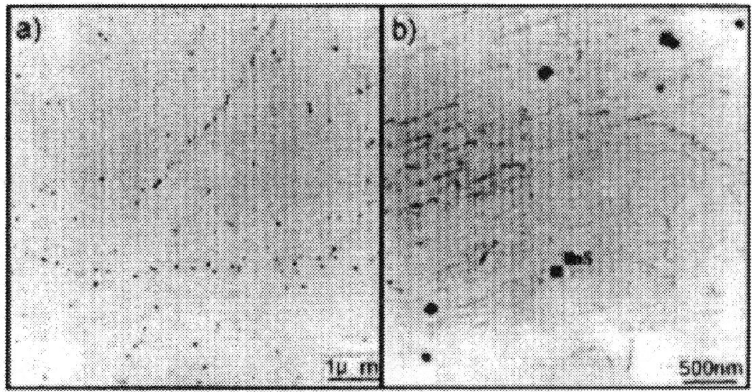

Fig. 10. T.E.M. observation of sulfides using extraction carbon replica in :
a) 0.1Mn-0.01S, held at 1150℃ and b) 0.3Mn-0.01S, held at 1150℃ for 5 minutes.

Matsubara has shown that the fine and disperse precipitations like FeS or (Fe,Mn)S can precipitate rapidly upon cooling below 1200℃ and result in a loss of ductility of steel containing more than 0.03% S.[9] He distinguished sulfides by their shape. One is a plate-like inclusion which has a widmanstätten structure and is precipitated on {100} plane of austenite grain. The other one is a very fine particle inclusion which is precipitated in grain boundaries and causes a loss of ductility by precipitation hardening of austenitic grain. This result implies that the precipitation rate of FeS or (Fe,Mn)S is very fast; on the other hand, precipitation of MnS requires a certain amount of time for a given temperature.[3]

Suzuki pointed out that MnS begins to precipitate at 1027℃ in low carbon steel containing 0.22% Mn and 0.007% S, and the fine MnS at grain boundary causes embrittlement.[12] On the other hand, if FeS or (Fe,Mn)S is regarded as a major factor causing edge cracking as Matsubara suggested,[9] it would be concluded that the precipitation temperature of FeS is higher than that of MnS in low carbon steel containing 0.2% Mn and 0.01% S. In TEM observation, some times the 'plate-like' FeS could be found along the micro cracks but only the fine FeS or (Fe,Mn)S could be found near the crack.

In Fe-S phase diagram which was calculated by Senuma using THERMO-CALC,[3] the solubility of S at 1200℃ is about 0.03% and at 1000℃ is about 0.01%. An abrupt decrease in the solubility of S occurs, however, with the addition of Mn because Mn decreases the solubility of S in steel at austenitic temperature. So, FeS or (Fe,Mn)S is considered to be precipitated during the cooling or deformation process.But in this study, FeS or (Fe,Mn)S were easily observed in steels where edge cracks had occurred.[5] MnS were frequently found in the steels without edge cracks, though they are very fine - as small as 60nm. From the results of this study, it is suggested that not only fine precipitates but also segregated S at grain boundaries are important factors in causing the edge cracking of low carbon steels.

CONCLUSIONS

With the increase in manganese content in low carbon steel, not only coarser MnS precipitated, but also smaller amounts of FeS or (Fe,Mn)S were found, both of which resulted in suppression of edge cracking. It is suggested that not only fine precipitates but also segregated S at grain boundaries are important factors in causing the edge cracking of low carbon steels. The influential factors in edge cracking of low carbon steels were, in order of importance, Mn/S ratio, preheat treatment time and temperature. Edge cracking during roughing in the hot-rolling process of mini-mill was effectively prevented by means of the isothermal treatment at 1150℃ for 5 minutes in the 0.3% manganese steel containing sulfur lower than 0.013%. The concept of an edge crack index was proposed for more accurate estimation of the total amount of edge cracking as a function of Mn/S ratio in this study.

REFERENCES

1. F.K. Iverson ans K. Busse, "A Review of First Year CSP Operations at Nucor Steel's New Thin Slab Casting Facility", Metallurgical Plant and Technology International, 1991, No. 1, pp. 40-51

2. M. Korchynsky, "High Strength Low Alloy Steels Produced by Thin Slab Casting", Iron & Steel Review, April 1997, pp. 1-7

3. T. Senuma, S. Sanagi, K. Kawasaki, S. Akamatsu, T. Hayashma and O. Akisue, "Precipitation Control of Cold Rolled Mild Steel Sheets in Thin Slab-Drect Hot Rolling Process", Tetsu-To-Hagane, Vol. 79, No. 2, Feb. 1993, pp. 194-200.

4. G.A. Wilber, R. Batra, F. Savage and W.J. Childs, "The Effect of Thermal History and Composition on Hot Ductility of Low Carbon Steels", Metallurgical Transactions A, Vol. 6A, No. 9, Sep. 1975, pp. 1727-1735

5. W.T. Lankford,Jr., "Some Considerations of Strength and Ductility in the Continuous-Casting Process", Metallurgical Transactions, Vo. 3, No. 6, June 1972, pp. 1331-1357

6. H.G. Suzuki, S.Nishimura and S.Yamaguchi, "Physical Simulation of the Continuous Casting of Steels", Proceedings of Intn. Symp. on Phys. Simulation of Welding, Hot Forming and Cont. Casting, Ottawa, Canada, 1988, pp. II-1 – II-14.

7. K.Yasumoto, M.Maehara, S.Ura and Y. Ohmori, "Effect of Sulphur on Hot Ductility of Low Carbon Steel Austenite", Mat. Sci. and Tech., Vol. 1, No. 2, Feb. 1985, pp. 111-116.

8. H. Kobayashi, "Hot-Ductility Recovery by Manganese Sulphide Precipitation in Low Manganese Mild Steel", ISIJ International, Vol. 31, No. 3, March 1991, pp.268-277.

9. K. Matsubara, "On the Behavior of the Precipitated Sulfide Inclusions in Solid Steel", Tetsu-To-Hagane, Vol. 51, No. 12, Dec. 1965, pp. 2220-2232.

10. S. Yue, J.J. Jonas and B. Mintz, "Relationship between Hot Ductility and Cracking During the Continuous Casting of Steel", Proceedings of 13th PTD Conf., 1995, pp. 45-52.

11. Y. Ueshima, N. Komatsu, S. Mizoguchi and H. Kajioka, "Effects of Elements on Interdendritic Microsegregation of Carbon Steel", Tetsu-To-Hagane, Vol. 73, No. 11, Nov. 1987, pp. 1551-1558.

12. M.Suzuki, C.H. Yu, H. Shibata and T. Emi, "Recovery of Hot Ductility by Improving Thermal Pattern of Continuously Cast Low Carbon and Ultra Low Carbon Steel Slabs for Hot Direct Rolling", ISIJ International, Vol. 37, No. 9, Sep. 1997, pp. 862-871.

Effect of Pickling Inhibitors on Surface Properties of Steel Sheets

Rhobum Park and Sug-Kyu Lee
Technical Research Laboratories
POSCO, 699 Kumhodong
Kwangyang City, Cheonnam, 545-090
KOREA
Tel.: 82-61-790-8734
E-mail: rhobum@posco.co.kr

Key words: Hot-rolling, scale, pickling, hydrochloric acid, inhibitor, steel sheet

INTRODUCTION

Acid solutions are generally used for the removal of undesirable scale and rust from metal surface in several industrial processes. Hydrochloric acid and sulfuric acid are widely adopted for the pickling processes of metals. In addition, inhibitors are applied for preventing metal dissolution and/or acid consumption. The protecting action of an inhibitor in metal corrosion is often associated with chemical or physical adsorption involving a variation in the charge of adsorbed substance and a transfer of charge from one phase to the other. Most of the efficient acid inhibitors are high molecular weight organic compounds which mainly contain nitrogen, sulfur or oxygen atoms in their structure[1-5]. In this study, effects of pickling inhibitors on surface properties of steel sheets in hydrochloric acid solution are investigated.

EXPERIMENTAL PROCEDURE

Inhibitors- Newly tried inhibitor in this study was composed of polycycloaminate $[(CH_2)_3N_2]_n$, n=1-8], phenylmethylpyrasol $[C_{10}H_{n1}N_{n2}O$, n1=10-12, n2=1-2], polypeptides $[(NHCH(R_1)CONHCH(R_2)CO)_n$, n=25-1,500, R_1 & R_2=H, alkyl or aryl], and fatty acylchloride $[CH(CH_2)_nCl$, n=12-22] with the ratio of 3.5:1.5:2.5:2.5. For comparison, commercially available inhibitor which was mainly composed of aminate compounds (a product of Buhmwoo Chemical Co, Korea) was used.

Specimens- Hot-rolled low carbon steel samples used in this study have the composition (wt %) as shown in Table 1. For scale removal rate test in acid pickling, samples 33mm in diameter with 2.5mm thick were cut from the center area of as-prepared sheet No. 1 in Table 1. On the other hand, for inhibition efficiency test, samples were cut from the same material with same size (33mm $\phi \times 2.5$ mm t). They were then polished, degreased in acetone, dried and weighed. Most of the samples for industrial trials (on-line tests) were directly taken from the processing line. Cold-rolled steel strip which had same composition as No.1 sample was used for anti-rust effect test.

Table 1. Chemical composition of the hot-rolled steel specimen used in this study

Sample No.	C	Mn	Si	P	S	Sol.Al	N
1	0.04	0.15	≤0.02	≤0.010	≤0.010	0.020~0.05	≤0.004
2	0.04	0.25	≤0.02	≤0.015	≤0.015	0.025~0.05	≤0.006
3	0.10	1.10	≤0.10	≤0.020	≤0.007	0.020~0.05	≤0.006
4	0.16	0.80	≤0.03	≤0.015	≤0.008	0.010~0.05	≤0.004

Inhibitor test- 1) Inhibition test in acid pickling process was carried out in 12% HCl solution at 80℃. Inhibitors were added at the concentration range from 0.025 to 0.8%. After 30 minutes immersion, inhibition efficiency and surface whiteness were measured. **2)** Scale removal rate test was carried out in the same solution. In this case, the immersion time was changed to 15 seconds. Removal rate was evaluated by measuring the surface whiteness. **3)** Anti-rust test was carried out in 5% HCl solution at 70℃. Inhibitors were added at concentrations of 0, 0.1 and 0.2%. Test was carried out as following procedure; 12 seconds pickling, 5 seconds in air, 2 seconds rinsing, 5 seconds in air, 12 seconds rinsing, and drying. **4)** The industrial trials were carried out on POSCO Kwangyang Steel Works No.1 POL (Pickling and Oiling Line) in 12% HCl solution at 80℃. Inhibitors were added at concentrations of 0.1% and immersion time was adjusted to about 45 seconds. Before surface analysis, in order to remove organic contaminant, steel sheets were cleaned by ultrasonication in trichloroethylene solution.

Surface properties analysis- To evaluate the surface properties of steel sheets in acid pickling, whiteness, yellowness, roughness and morphology were measured and observed using color difference meter, 3 dimensional roughness meter and SEM.

RESULTS AND DISCUSSION

The effectiveness of the inhibitor was studied by 3 different methods; the inhibition efficiency, scale removal rate and whiteness. Fig. 1 shows the results of the inhibition efficiency as a function of concentration of the newly developed (hereinafter called new) and commercially used (hereinafter called present) ones in 12% hydrochloric acid. It is found that inhibition efficiency and whiteness are improved by increasing the concentration of both new and present inhibitors. Note that the new inhibitor shows better results than that of the present one. For example, in the widely used concentration of 0.1% in the established pickling process, the former improved the inhibition efficiency (9.94%) and whiteness (5.45 point) compared to the later. Here the inhibition efficiency was evaluated according to weight loss method: that is, % inhibition was defined as

$$\frac{(\text{weight loss by uninhibited acid} - \text{weight loss by inhibited acid}) \times 100}{(\text{weight loss by uninhibited acid})}$$

Fig. 2 shows the results of the scale removal rate test as a function of whiteness with concentration of the new and present inhibitors. The scale removal rate is estimated from the whiteness after dipping the steel sheets in 12% HCl solution for 15 seconds. To simply measure scale removal rate, we considered the whiteness of the steel surface as a measure of the scale removal rate. This is a rationale; the more scale remained, the lower whiteness measured. It is found that whiteness is increased by increasing the new inhibitor concentration upto 0.2%, shows plateau to 0.4%, and then decreased above 0.4%. Interestingly, in the case of the present inhibitor, the whiteness is decreased continuously upon increasing the inhibitor concentration. These results suggest that scale still remained at the steel surface by removing the sample from the solution in the short time such as 15 seconds, and that addition of the new inhibitor to HCl solution is an effective way to accelerate the pickling rate. On the other hand, addition of the present inhibitor was not effective in improving the whiteness compared to that of uninhibited solution (0% addition).

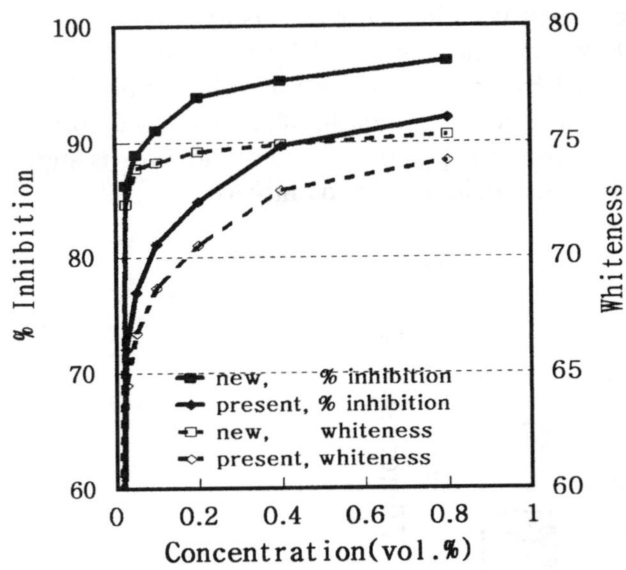

Fig. 1. Inhibition efficiency of additive concentration in 12% hydrochloric acid at 80℃. Immersion time was 30 min.

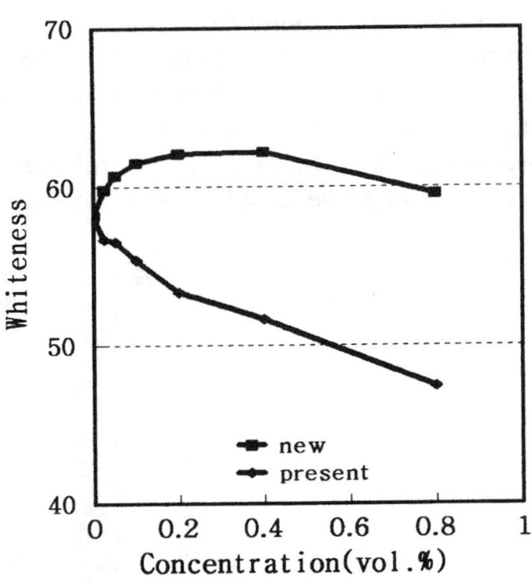

Fig. 2. Effect of additive concentration on scale removal rate in 12% hydrochloric acid at 80℃, immersion time 15 seconds.

Fig. 3 shows the SEM images of the steel surface after inhibition efficiency test (as shown in Fig. 1) by adding 0.2% of the (a) new or (b) present inhibitor in 12% HCl solution. It is clearly observed that surface corrosion is suppressed by adding the new inhibitor. While the present inhibitor give rise to the severe pitting due to the less efficiency in inhibition activity. Furthermore, average roughness (Ra) of the both surfaces is measured to be 1.27μm (new) and 3.29μm (present), respectively.

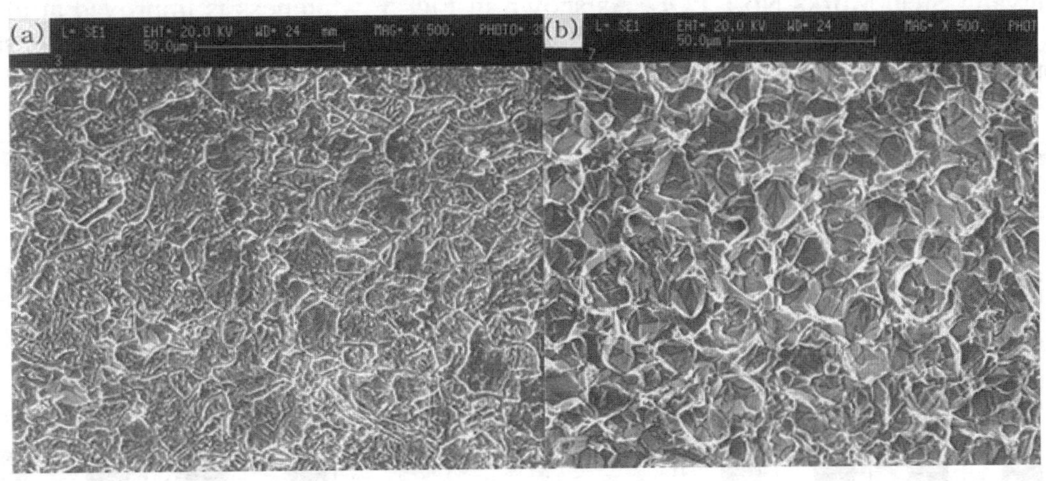

Fig. 3. SEM pictures of surface of the steel after inhibition test in 12% hydrochloric acid containing 0.2% new (a) and present (b) inhibitors at 80℃. Immersion time was 30 min.

It is generally known that inhibitor has various effects to reduce metal dissolution, acid consumption, and/or hydrogen absorption (especially, in high carbon steel), during acid pickling [6]. It is known that rinsing rust occurs when acid pickling treatment is performed on the some cold rolled steel sheets. In this study we also investigated tendency of anti-rust by adding two different inhibitors. In order to elucidate the effect of inhibitors to the tendency of anti-rust, 5 steps as described in experimental procedures are adopted. It is very similar to the procedures of acid pickling in CAL (continuous annealing line) with water cooling section, which is known as causing rust readily. The tendency of anti-rust is estimated by measuring the yellowness and whiteness. Anti-rust is increased by decreasing the level of yellowness and by increasing that of the whiteness. Fig. 4 shows the tendency of anti-rust is improved by adding the new inhibitor. Note that new inhibitor (B and C) provided a decreased yellowness and an increased whiteness of steel surface compared to those of the uninhibited. Present inhibitor also improved anti-rust capacity. But, new inhibitor is superior to present one in preventing rust. In both, higher concentration results in improved anti-rust.

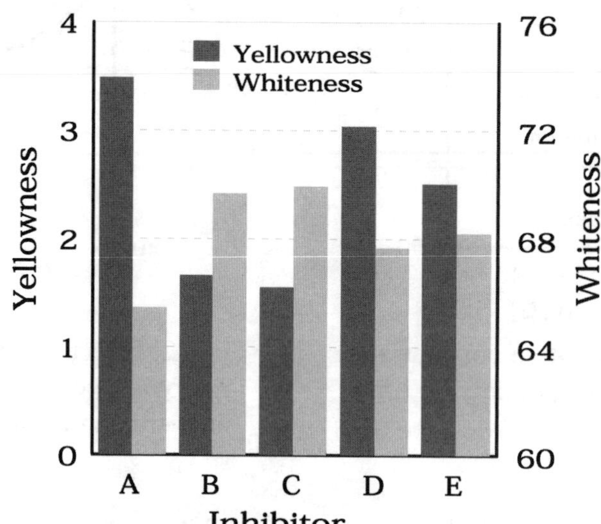

Fig. 4. Anti-rust effect of additives in 5% hydrochloric acid at 70℃. A: uninhibited, B: new inhibitor 0.1%, C: new inhibitor 0.2%, D: present inhibitor 0.1% and E: present inhibitor 0.2%

To apply the above results to the on-line process, the effect of new and present inhibitor was compared at POSCO Kwangyang Steel Works No.1 POL. As shown in Fig. 5, whiteness is improved at all the samples by adding new inhibitor compared with the present inhibitor. The difference in mean value is estimated to be 3.97. On the other hand, yellowness is decreased approximately to the mean value of 1.15 (Fig. 6).

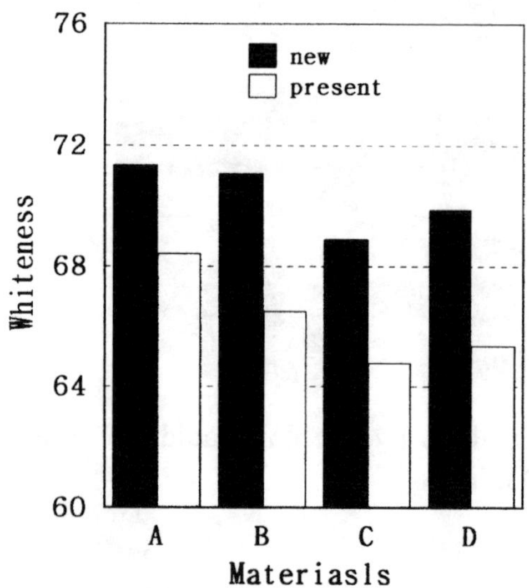

Fig. 5. Comparison of surface whiteness of on-line test samples after immersion for 45 seconds in 12% hydrochloric acid containing 0.1% inhibitor at 80℃. A, B, C, and D stand for sample number 1, 2, 3, and 4 in Table 1, respectively.

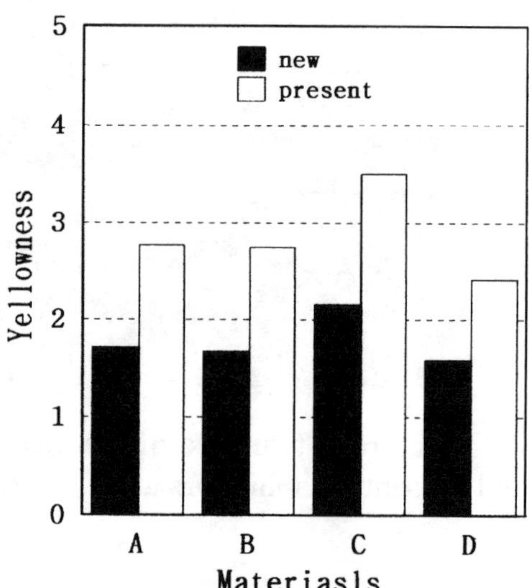

Fig. 6. Comparison of surface yellowness of on-line test samples after immersion for 45 seconds in 12% hydrochloric acid containing 0.1% inhibitor at 80℃. A, B, C, and D stand for sample number 1, 2, 3, and 4 in Table 1, respectively.

Fig. 7 shows the SEM images of the sample 1 (sample A in Fig. 5 and 6) after on-line test with addition of (a) new and (b) present inhibitor. It is roughly assumed that corrosion is less progressed by adding the new inhibitor than the present inhibitor. Average roughness (Ra) of the both surfaces was measured as 0.89 μm (new) and 1.09 μm (present). These results suggest that the inhibitor newly prepared in this study is better than the commercial one in improving the properties of steel surface.

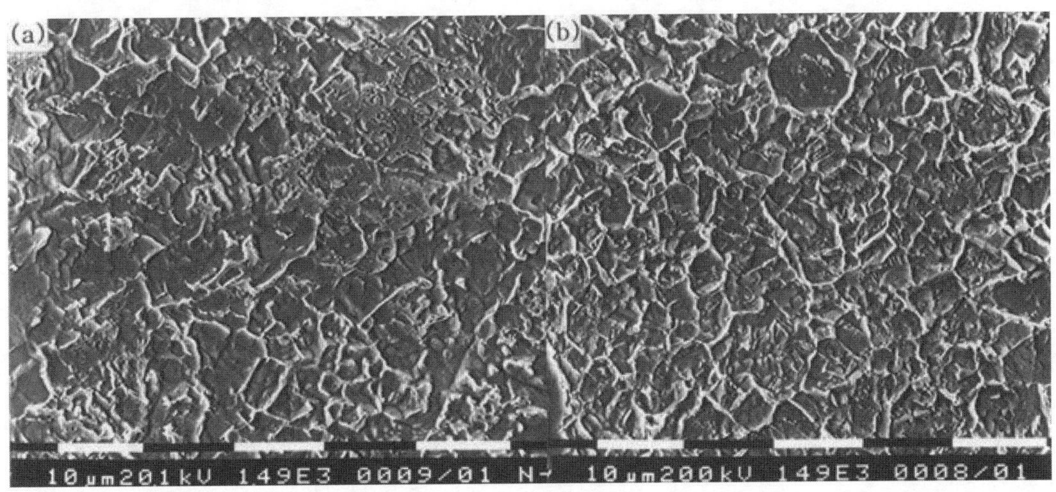

Fig. 7 SEM pictures of surface of the steel in 12% hydrochloric acid containing 0.08% new(a) and present(b) additives at 80℃, immersion time 45 sec.

CONCLUSIONS

1. Inhibition efficiency and scale removal rate are improved by adding newly developed inhibitor. At 0.1%, the new inhibitor improves the inhibition efficiency about 9.94% and surface whiteness about 5.45 point compared to commercially used one.

2. In the on-line test, new inhibitor is turned out to be effective to improve the surface lightness. The estimated whiteness increases to 3.97 and yellowness decreases to 1.15.

3. Higher performance of inhibitor results in improved quality of steel surface related to degree of etching, pitting, roughness and yellowness.

REFERENCES

1. L. L. Shreir, "Corrosion", 2nd Ed., Newnes-Butterworth, London, Vol. 2, 18.2-18.3, 1977
2. R. L. Every and O. L. Riggs, "Mater. Prot.", Vol. 3, 46, 1964
3. W. Machu, "Proc. 3rd European Symp. on Corrosion inhibitors", Ferrara Univ., 107, 1971
4. Y. Kumar, D. Dhirandra, B. Sanyal, and G. N. Pandey, "Met. Finish.", Vol.81, 45, 1983
5. P. Sharma and J. N. Gaur, "J. Electrochem. Soc.", India, Vol. 25, 4, 1976
6. I. Singh, A. K. Dey and V. A. Altekar, "NML Tech. J.", Vol. 24, 36, 1982

GENERAL FERROUS METALLURGY

A PRECIPITATION MODEL FOR PREDICTING HOT DUCTILITY BEHAVIOUR IN MICROALLOYED STEELS

Kevin Mark BANKS*, Alan Paul BENTLEY* and Andreas KOURSARIS**

* Industrial Metals and Minerals Research Institute, University of Pretoria, South Africa.
**University of the Witwatersrand, South Africa
Tel. +27 12 420 4552
Fax. +27 12 362 5304
e-mail. kbanks@postino.up.ac.za

Key Words: carbonitride precipitation, modelling, hot ductility, Nb steels, V-Nb steels

INTRODUCTION

Hot ductility in the austenite of microalloyed steels is associated with carbonitride precipitation at grain boundaries and dislocations within the matrix[1)]. Although investigations into the hot ductility behaviour of microalloyed steels are extensive[1),2),3),4)], precipitation models for predicting hot ductility during continuous casting are rare. Mathematical modelling has been used with some success[5),6),7)] in describing carbonitride precipitation in both undeformed and deformed austenite. However, most precipitation models have focused on simple alloying systems such as Nb-microalloyed steels. Precipitation start-time-temperature (PTT) diagrams for multi-component microalloying systems are scarce. In this work, the mutual solubility of complex carbonitrides in austenite, together with the classic nucleation theory, are used to establish relationships between precipitation and hot ductility in V-, Nb- and V-Nb microalloyed steels. The model is used to explain hot ductility behaviour during thin slab casting of low C-V-Nb steels.

PRECIPITATION MODEL

Solubility of carbonitrides in austenite - In this model, the solubility of NbV(C,N) precipitates, which have type B_1 (NaCl) structures, is estimated. Nb and V sites are on one sub-lattice (fcc) and C and N on another sub-lattice (fcc) in the complex carbonitride precipitate of the form $Nb_xV_{1-x}C_yN_{1-y}$. Subscripts x and y are the mole fractions of Nb and C in each sub-lattice, where $1 \geq x \geq 0$ and $1 \geq y \geq 0$. If only one substitutional element is present in the system *i.e.* $x = 1$, the second substitutional element simply falls away. This is the case for Nb(C,N) precipitates in Nb steels and V(C,N) in V steels. When the precipitates are considered to be aggregates of four kinds of binary compounds, namely NbC, VC, NbN and NbC, the mole fraction of the binary compounds in the precipitate, x_{NbC}, x_{VC}, x_{NbN} and x_{VN} are related to x and y by the following equations[8)]:

$$x = x_{NbC} + x_{NbN} \tag{1}$$

$$y = x_{NbC} + x_{VC} \tag{2}$$

$$1 = x_{NbC} + x_{VC} + x_{NbN} + x_{VN} \tag{3}$$

The solute in the austenite is assumed to obey Henry's Law since the solution is very dilute. The solubility products of the binary precipitates, converted from mass % to mole fraction, are:

$$[Nb][N] = x_{NbN}M^2_{Fe} \times 10^{-4}/(M_{Nb}M_N)10^{3.57-9660/T} \quad \text{ref[9]} \quad (4)$$

$$[Nb][C] = x_{NbC}M^2_{Fe} \times 10^{-4}/(M_{Nb}M_C)10^{3.28-8266/T + (983/T - 0.598)(\%Mn)} \quad \text{ref[10]} \quad (5)$$

$$[V][N] = x_{VN}M^2_{Fe} \times 10^{-4}/(M_V M_N)10^{3.1-8030/T} \quad \text{ref[9]} \quad (6)$$

$$[V][C] = x_{VC}M^2_{Fe} \times 10^{-4}/(M_V M_C)10^{6.73-8500/T} \quad \text{ref[9]} \quad (7)$$

where M_i is the atomic mass of element *i*. The equilibrium solubility product for NbC includes a term for Mn which increases its solubility. In addition, the following mass balances apply[8)]:

$$x = x_{NbC} + x_{NbN} = (Nb^0 - [Nb])/(Nb^0 - [Nb] + V^0 - [V]) \quad (8)$$

$$1\text{-}x = x_{VC} + x_{VN} = (V^0 - [V])/(Nb^0 - [Nb] + V^0 - [V]) \quad (9)$$

$$y = x_{NbC} + x_{VC} = (C^0 - [C])/(C^0 - [C] + N^0 - [N]) \quad (10)$$

$$1\text{-}y = x_{NbN} + x_{VN} = (V^0 - [V])/(C^0 - [C] + N^0 - [N]) \quad (11)$$

The superscript *0* refers to the total content. Substituting eqs. 4 to 7 into eqs. 8 to 11 results in four non-linear equations in four unknowns, which were solved using the Newton-Raphson numerical method. The dissolution temperature, T_{DISS}, the equilibrium composition of the second-phase particles and effective solute concentrations of the precipitate-forming elements were calculated. The mole fraction of precipitates, *f*, is:[9)]

$$f = Nb^0 - [Nb] + V^0 - [V] = C^0 - [C] + N^0 - [N] \quad (12)$$

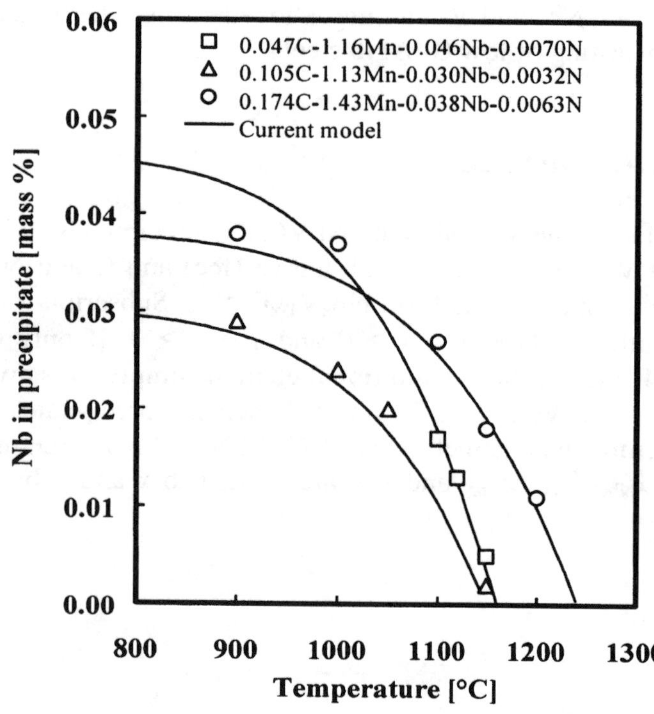

Fig. 1: Nb in precipitate as a function of temperature [12]

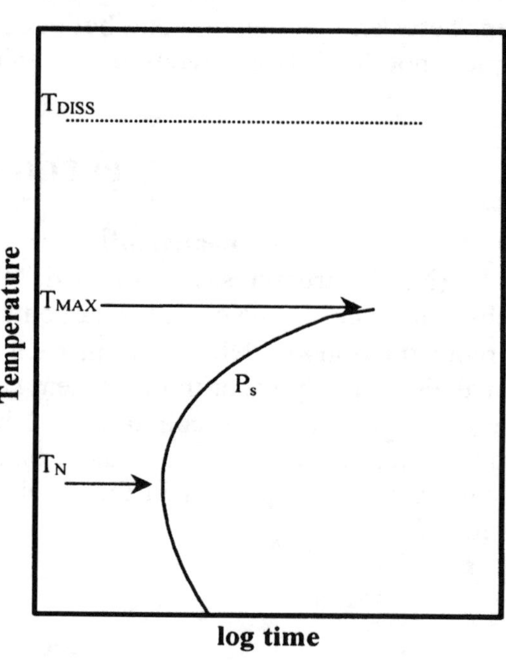

Fig. 2: Schematic of a PTT diagram

AlN precipitation in austenite was excluded from the solubility model because it has a different crystal structure (hcp), and therefore no mutual solubility with the carbides and nitrides of Nb, V and Ti, all of which have cubic crystal structures[11]. Also, the kinetics of AlN precipitation has been shown to be somewhat sluggish in austenite[4] and was ignored in the nucleation model that follows. The solubility model for Nb(C,N) was validated by comparing the measured[12] and calculated precipitate Nb concentrations in Nb-bearing steels in fig. 1. No measured precipitate fractions for V-Nb steels could be found for comparison.

Solid state nucleation at dislocations – Fig. 2 shows a schematic diagram of a PTT curve. The isothermal precipitation start time is defined as P_s and the nose temperature, T_N, corresponds to the maximum rate of precipitation. T_{MAX} is the asymptotic temperature at the upper limit of the PTT curve. Although hot ductility is influenced by precipitates at both the austenite grain boundaries and at dislocations[1], only precipitation at dislocations is considered here.

At an early stage of precipitation, when the particle fraction is sufficiently low, the incubation time is negligibly small and there is an absence of excess vacancies, a good approximation of P_s is given by[13]

$$P_s = Aa_M^2 [S]_{eff}^{-1} D_{eff}^{-1} \exp\left[\frac{16\pi\xi_{eff}^3 \gamma_{eff}^3}{3k_B\ T(\Delta G_{chem})^2}\right] = Aa_M^2 [S]_{eff}^{-1} D_{eff}^{-1} \exp[\Gamma] \qquad (13)$$

The lattice parameter of the matrix phase (austenite) is a_M (Table I) and k_B is the Bolzmann's constant. A is a constant fitting parameter equivalent to the ratio of the critical number of nuclei per unit volume which must be formed for precipitation to be detected, to the total amount of nucleation sites at dislocations. In reality, A is not constant, but a function of strain, strain rate, particle diameter and precipitate mole fraction at P_s.[7] A complete analysis of parameter A, which influences the kinetics considerably, was not performed and focus was placed on the temperature range over which precipitation occurs. It is, however, well known[14),15] that both Nb(C,N) and VN precipitate out rapidly at strain rates less than $10^{-1}s^{-1}$, and nucleation and growth were assumed to have taken place in all unbending (hot ductility) tests described later which were performed at strain rates between $1x10^{-3}s^{-1}$ and $4x10^{-3}s^{-1}$.

D_{eff} (m^2/s) is the effective diffusion coefficient in dilute substitutional solid solutions, and is considered to be independent of steel composition and a function only of absolute temperature, T. D_{eff} in V-Nb steels was calculated using the methodology employed by Okaguchi et al[8]:

$$D_{eff} = xD_{0,Nb} \exp\left[\frac{Q_{diff,Nb}}{RT}\right] + (1-x)D_{0,V} \exp\left[\frac{Q_{diff,V}}{RT}\right] \qquad (14)$$

According the diffusion coefficients in Table I, the bulk diffusivity of V at a given temperature in austenite is slower than that of Nb. D_{eff} influences the nucleation kinetics and, hence, the location of the PTT curve on the time axis. It must be noted that the diffusivity equations used for both Nb and V are applicable to diffusion in the austenite matrix, whereas diffusivity at dislocations is considerably faster[16] and more appropriate. The diffusivity of V at dislocations could not, however, be found in the literature. $[S]_{eff}$ is the effective solute concentration in mole fraction of the substitutional elements which are controlling the rate of the nucleation process. $[S]_{eff}$ for the complex precipitate $Nb_xV_{1-x}(C_yN_{1-y})$ can be expressed as: [8]

$$[S]_{eff} = x[Nb] + (1-x)[V] \qquad (15)$$

The chemical driving force, ΔG_{chem}(kJ/mol), is the difference in chemical free energy between the precipitate and the nucleus, which is composed of atoms that have the same chemical potential as those in austenite[8]. Therefore, ΔG_{chem} of the complex precipitate $Nb_xV_{1-x}(C_yN_{1-y})$ in V-Nb steels is given as a function of the supersaturation, k_s, as:

$$\Delta G_{chem} = -\frac{RT}{2V_P}\ln(k_S) = -\frac{RT}{2V_P}\left[\ln\frac{[Nb]^x[V]^{1-x}[C]^y[N]^{1-y}}{[Nb]_e^x[V]_e^{1-x}[C]_e^y[N]_e^{1-y}}\right] \quad (16)$$

where subscript *e* refers to the equilibrium concentration and *R* is the gas constant. V_P (m^3/mol) is the molar volume of the mixture of Nb(C,N) and V(C,N) precipitates[8]:

$$V_P = \frac{a_P^3}{8}N_A \quad (17)$$

where N_A is Avogadro's number. The lattice parameter of NbV(C,N), a_p, was estimated from the lattice parameters of Nb(C,N) and V(C,N) given in Table I:

$$a_P = xa_{Nb(C,N)} + (1-x)a_{V(C,N)} \quad (18)$$

The effective interfacial energy of the nucleus and matrix, γ_{eff} (J/m^2), was estimated from the γ values of NbC, NbN, VC and VN at dislocations (Table I) expressed in terms of absolute temperature[17]:

$$\gamma_{eff} = x_{VC}\gamma_{VC} + x_{VN}\gamma_{VN} + x_{NbC}\gamma_{NbC} + x_{NbN}\gamma_{NbN} \quad (19)$$

Table I. Constants used in the precipitation model

Parameter	Value	Units	Reference
$a_{Nb(C,N)}$	0.444	nm	18
$a_{V(C,N)}$	0.415	nm	19
a_M (austenite)	0.359	nm	20
$D_{0,Nb}$	1.40	m^2/s	21
$D_{0,V}$	3.65	m^2/s	22
$Q_{diff,Nb}$	270	kJ/mol	21
$Q_{diff,V}$	293	kJ/mol	22
γ_{NbC}	$1.0058 - 0.4493 \times 10^{-3}T$	J/m^2	17
γ_{NbN}	$0.9717 - 0.4340 \times 10^{-3}T$	J/m^2	17
γ_{VC}	$0.8469 - 0.3783 \times 10^{-3}T$	J/m^2	17
γ_{VN}	$0.8090 - 0.3614 \times 10^{-3}T$	J/m^2	17

ξ_{eff} is a modifier for the interfacial energy that arises from nucleation at dislocations or at grain boundaries because of their strain energy and surface energy and is unity for homogeneous nucleation and less than 1 for heterogeneous nucleation[5]. The uncertainties surrounding the accurate evaluation of ξ_{eff} makes it difficult, or even impossible, to predict the dominant nucleation mode without the benefit of a comprehensive physical simulation programme and microscopic investigations.[7] For this reason, ξ_{eff} is commonly used as a fitting parameter to match predicted precipitation start times with experimentally determined values.[5),6),16] Eqn. 20 predicts the upper limit of ξ at dislocations[23],ξ_{dis}, where the exponent, *m*, was given as 2/3. Although the quoted value of *m* gave good correlation between measured and calculated dynamic P_s times in a low C-V steel[10] *m* had to be modified to ½ to obtain good agreement in low C-Nb steel[10].

$$\xi_{dis} \le \left(\frac{-\Delta G_{chem}}{-\Delta G_{chem} + \frac{\mu}{4\pi^2}}\right)^m \quad (20)$$

where μ (MPa) is the shear modulus of the austenite matrix[24] :

$$\mu = 81000(1-0.91(T-300)/1810) \quad (21)$$

Following the law of mixtures used in eqs. 14, 15 and 18, ξ_{dis} for V-Nb steels was estimated from the mole fraction of NbC, NbN, VC and VN as follows:

$$m = 0.5x + 0.67(1-x) \quad (22)$$

If x approaches 1, ξ_{dis} is equal to that of Nb steel, whilst if $x = 0$, the material is essentially a V steel. Fig. 3 shows reasonable agreement between measured and predicted P_s values in low C-V-Nb, V and Nb steels[10] using fitted A values in eq. 13 of $3x10^{-3}$, $3x10^{-3}$ and $2.2x10^{-6}$ respectively.

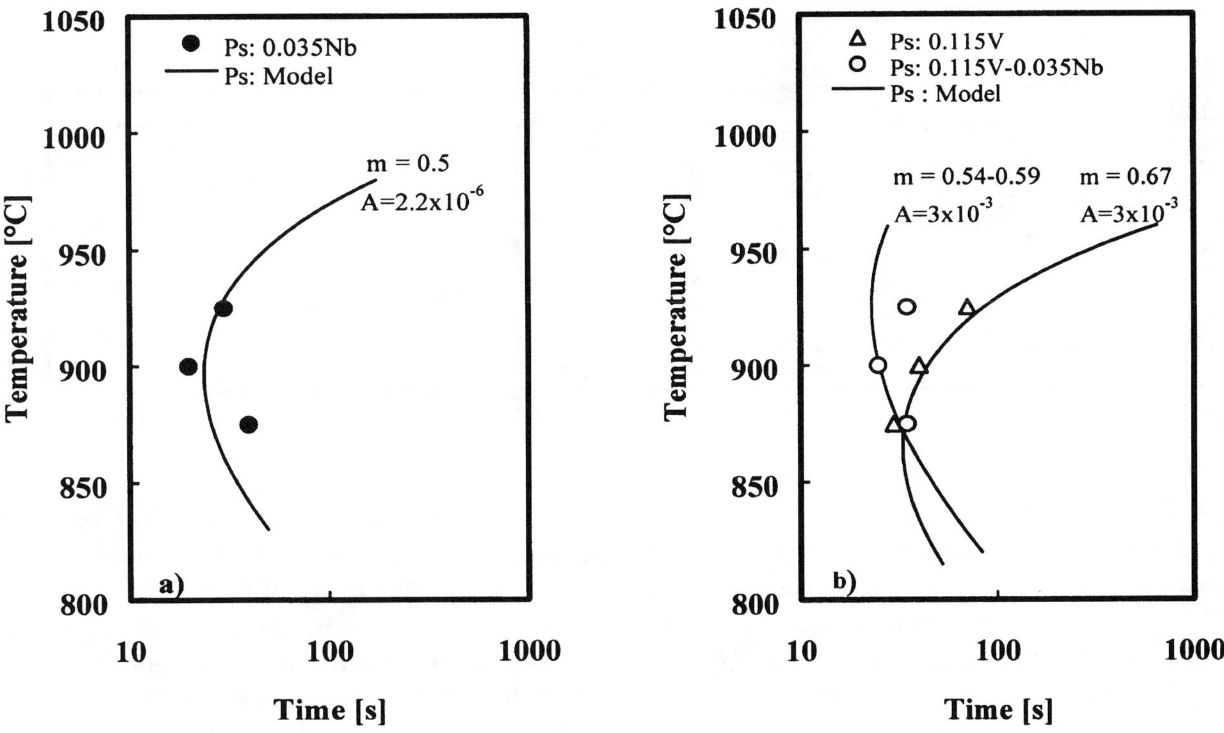

Fig. 3: Dynamic PTT diagrams for a)Nb steel and b) V and V-Nb steels [10]
(Base chemistry: 0.05C-1.2Mn-0.006N)

MODEL APPLICATIONS TO HOT DUCTILITY

1. Predicting Ductility Temperatures In HSLA Steels

An attempt was made to find relationships between critical ductility temperatures and the T_N temperature. The chemical composition of Nb, V and V-Nb steels from various sources are collected in Table II, together with the respective hot ductility testing conditions. The data considered was restricted to cooling rates and strain rates typically found in conventional continuous casting or thin slab casting. The holding time prior to testing was 0.5 or 5min. A values used previously for each steel type were employed in the calculations. The temperature at which ductility starts to fall below 90% reduction-in-area ($RA_{90\%}$) was defined as T_{HDH}. The temperature at $RA_{50\%}$ was defined as T_{TROUGH} and represented the start of the trough region at the high-temperature end of the ductility curve. Since reported critical RA values corresponding to cracking in continuously cast strands vary; 60%[25], 30-50%[26] and 40%[27], an RA value of 50% was assumed.

Table II Chemistries and ductility testing conditions of Nb, V and V-Nb steels

Steel	Ref.	C %	Mn %	V %	Nb %	N %	Cool rate °C/s	Hold time min	Strain rate $x10^{-3}s^{-1}$
Nb	**Steels**								
A*	37	0.12	1.40	-	0.038	0.005	3	0.5	2
B	28	0.16	1.24	-	0.023	0.009	1	5	3
C*	29	0.094	1.47	-	0.032	0.0065	1-3.3	5	2
D*	30	0.09	1.40	-	0.028	0.0052	0.83	5	3
E*	30	0.12	1.40	-	0.015	0.0042	0.83	5	3
F	28	0.014	1.48	-	0.028	0.007	1	5	3
G	31	0.08	1.46	-	0.015	0.002	NA	5	4
V	**Steels**								
H*	30	0.10	1.34	0.11	-	0.010	0.83	5	3
I*	30	0.10	1.40	0.05	-	0.0051	0.83	5	3
J*	37	0.05	1.35	0.085	-	0.012	3	0.5	2
K*	37	0.05	1.35	0.085	-	0.007	3	0.5	2
V-Nb	**Steels**								
L	31	0.14	1.45	0.075	0.035	0.005+	NA	5	4
M	32	0.05	1.30	0.056	0.048	0.008	0.4	5	1
N	32	0.08	1.30	0.056	0.031	0.008	0.4	5	1
P	30	0.11	1.42	0.110	0.030	0.0044	0.83	5	3
Q	30	0.11	1.43	0.053	0.030	0.0046	0.83	5	3
R	30	0.10	1.40	0.100	0.015	0.0057	0.83	5	3
S	30	0.10	1.40	0.048	0.015	0.0062	0.83	5	3
T*	37	0.05	1.35	0.085	0.015	0.012	3	0.5	2
U*	37	0.05	1.35	0.085	0.015	0.007	3	0.5	2

* *In-situ* melted, all other steels re-austenitized. + Assumed value.

Nb steels – Fig. 4 shows that T_N was approximately 230°C below the calculated dissolution temperature, indicating that a certain amount of supersaturation occurs before precipitation can commence. Generally, good correlation was found between the T_{HDH} temperature and T_N, with the onset of ductility loss located 55 to 100°C above T_N for both reheated and *in-situ* melted specimens. T_{HDH} was always below T_{MAX}, which differs somewhat from the suggestion by Mintz *et al*[33)] that the onset of ductility loss corresponds with T_{MAX}, the highest temperature at which dynamic precipitation can occur. Clearly, if the onset of ductility loss was chosen to correspond with $RA_{99\%}$ for example, T_{HDH} would be shifted upwards and closer to T_{MAX}. It is possible that, since T_{MAX} is associated with relatively long P_S times little, if any, dynamic precipitation takes place during deformation at this temperature. If dynamic precipitation does indeed occur above T_{HDH}, the high diffusivity and low supersaturation would produce coarse particles and small particle fractions, which do not impair ductility. For the given range of cooling rates and strain rates, the relationship between T_N and T_{HDH} does not appear to be overly sensitive to the testing method, *i.e. in-situ* melted or reheated tests.

The T_{TROUGH} temperatures in peritectic steels A-E, which contain typical N contents, were between 0 and 30°C above T_N. This is consistent with the idea that the onset of the ductility trough coincides with the maximum rate of precipitation[33)]. In the peritectic region, the influence of C content on ductility is small because of the overriding effect of Nb(C,N) precipitates[3)]. When the interstitial content is lowered considerably, as in steels F (0.014%C) and steel G (0.002%N), the T_{TROUGH} temperature was below T_N. The superior ductility in steel F was assumed to be due to lower precipitate fractions.[28)] However, the calculated equilibrium precipitate mole fraction in steel F at T_N was $1.5x10^{-4}$(fig. 5), almost double that of steel E which had a substantially higher T_{TROUGH} temperature. This implied that slower precipitation kinetics, rather than precipitate fraction, was the controlling factor for good ductility in steel F. The RA of steel G did not fall below 60% and T_{TROUGH} could not be defined. Due to low Nb and N contents in steel G, the precipitate mole fraction is small, and lower temperatures are necessary to provide sufficient supersaturation for precipitation to occur and, hence, impair ductility. At lower temperatures, diffusion of Nb is relatively slow, which shifts P_s to longer times and is beneficial for ductility.

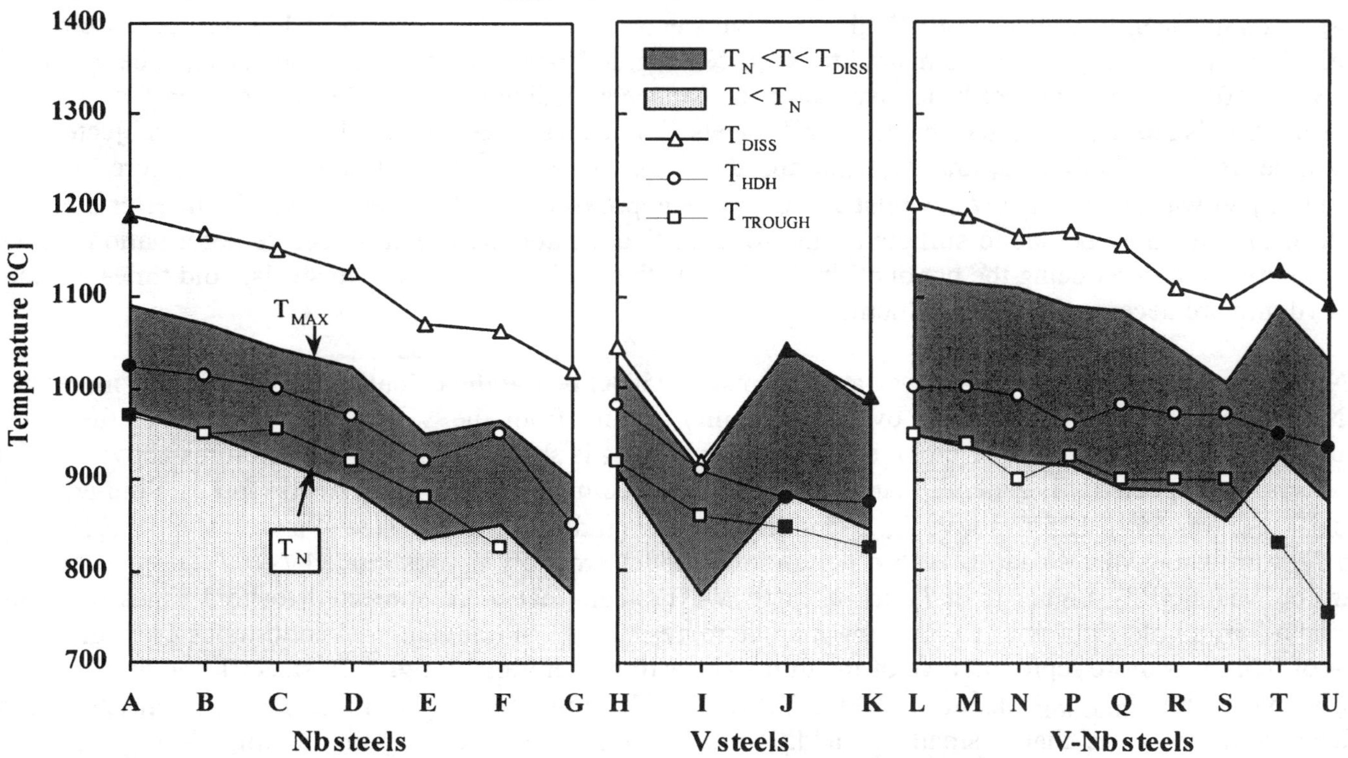

Fig. 4: Relationship between calculated precipitation temperatures and critical ductility temperatures
(Open symbols: 5 min hold time, closed symbols: 0.5 min. hold time.)

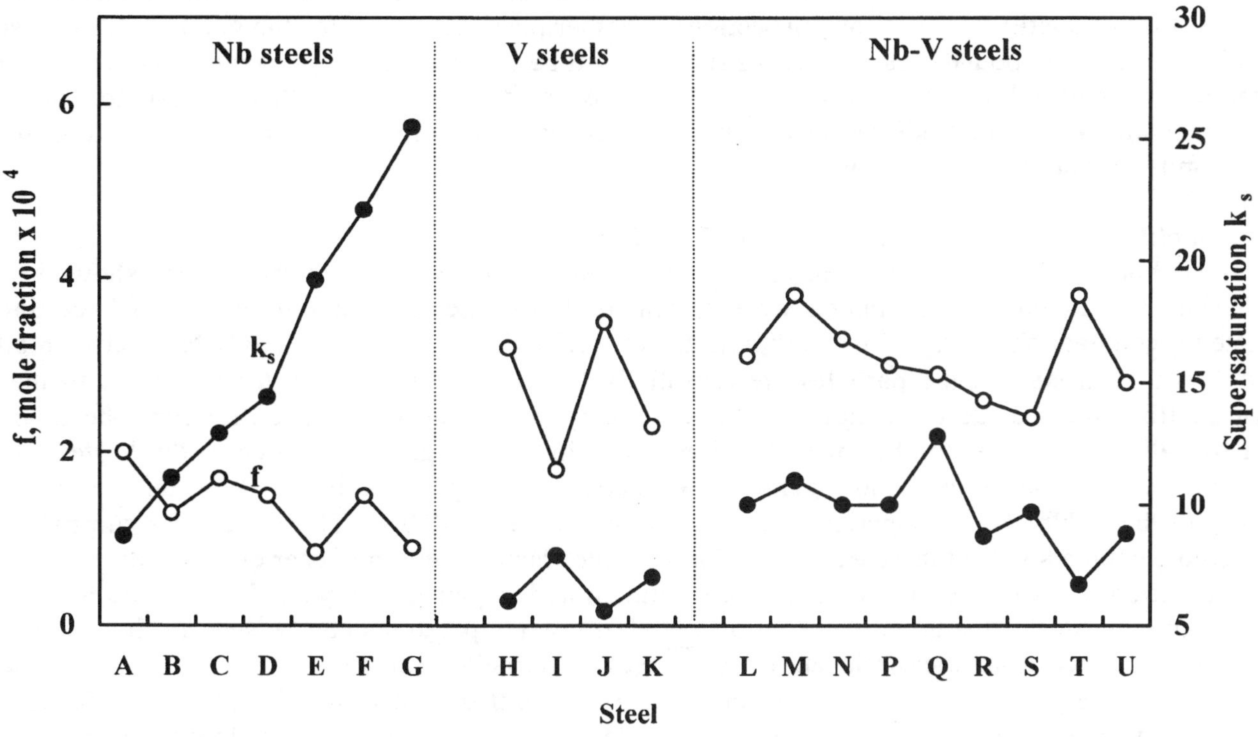

Fig. 5: Calculated precipitate mole fraction and supersaturation at T_N in Nb, V and V-Nb steels

V steels – Fig. 4 shows that T_{MAX} is just below T_{DISS}, and the average temperature difference between T_N and T_{DISS} was significantly smaller (140°C) than the Nb steels. The T_{DISS} temperature is lowered because of the relatively high solubility of VN compared to Nb(C,N). Fig. 5 shows that the calculated *f* at T_N is between 1.9 and 3.5 x 10^4, which is generally higher than *f* in Nb steels, concurring with the observation that a larger mole fraction is required to impair ductility in V steels.[34)] In as-cast steels H and I, which were subjected to a hold time of 5min, both T_{HDH} and T_{TROUGH} are above T_N. In as-cast steels J and K, which were held for 0.5min, T_{HDH} was at, or above T_N, whilst T_{TROUGH} was approximately 20°C below T_N. The inferior ductility in steels H and I may be due to sufficient time for significant static precipitation before deformation, which is more effective in reducing the hot ductility in V steels than in Nb steels[4)]. In Nb steels, hold times of more than 10 min are necessary to impair ductility[31)].

V-Nb steels – Fig. 4 shows that T_N was approximately 250°C below the calculated T_{DISS}, slightly more than in Nb steels due to the decrease in overall solubility arising from the V addition. Because of the lower solubility of NbV(C,N), the PTT curve is expanded and T_{MAX} is shifted to higher temperatures above T_N than Nb steels. As expected, *f* is generally larger in V-Nb steels compared to Nb-only steels (fig. 5). Similar to Nb steels, T_{HDH} was 55 to 125°C above T_N. In higher nitrogen grades H and J(V steels) and T(V-Nb steel), T_{HDH} and T_N were somewhat closer to each other and the model probably overestimated T_N. T_{TROUGH} coincided with, or was slightly higher than T_N in reheated V-Nb steels that were subjected to hold times of 5min. Whilst a 5 min hold time prior to straining may be representative of conventional continuous casting, a hold time of 0.5min is more representative of the thin slab casting operation[37)], fig. 6. In steels T and U, T_{HDH} was also above T_N but T_{TROUGH} was about 100°C below T_N. The good ductility in these steels is consistent with industrial observations that a small Nb addition to V steels does not impair ductility during thin slab casting[35)]. Significant improvement in ductility occurred despite the relatively fast cooling rate, which is well known to deteriorate ductility.[29,37)] As in V steels J and K, improved ductility in steels T and U compared to the other V-Nb steels may be partly due to the limited static precipitation prior to testing during the short hold time. Nb additions to V steel also delay precipitation at low temperatures in the austenite, which is beneficial for hot ductility (see following section). Another possible reason for the good ductility in steels T and U is that the grain boundaries in as-cast structures will be less effectively pinned by coarser particles and cavitation processes will take place less readily than in reheated material[36)]. Further, as-cast structures have a lower concentration of fine MnS particles compared to reheated specimens, which re-precipitate at austenite grain boundaries and impair ductility.

2. Hot Ductility In V-Nb Steels During Thin Slab Casting

The as-cast hot ductility curves of steels J, K, T and U, which are detailed elsewhere[37)], are shown in Figs. 7 and 8. The testing conditions applied were typical of those encountered during thin slab casting. The predicted precipitate chemistry of the carbonitrides in steel U is $Nb_{0.25\text{-}0.65}V_{0.75\text{-}0.35}C_{0.3}N_{0.7}$ between 800 and 1100°C, whilst in steel K the particles are essentially VN at all precipitation temperatures. In the V-Nb steels, ductility starts to decrease earlier at the high-temperature end of the ductility curve because of the lower overall solubility and higher mole fraction of precipitates (Fig. 9) compared to the V steel. The RA values in the V-Nb steels drop more gently with decreasing temperature below T_{HDH} until eventually the ductility in the V-Nb steel is superior to the V steel below about 850°C. This result was surprising, since larger solute contents should increase the precipitate mole fraction, and the faster effective diffusivity due to the Nb addition was expected to accelerate precipitation and impair ductility at all austenitic temperatures. The probable reason for the good ductility in steel U at low temperatures can be seen in fig. 10, where the PTT curve, obtained from stress relaxation tests[37)] on *in-situ* melted specimens, revealed a distinct delay in post-dynamic precipitation below 900°C when compared with that of steel K. The addition of 0.035%Nb to reheated V steel was found by Akben *et al*[10)] to delay dynamic precipitation at low temperatures in the austenite, fig. 3b. These authors have suggested that this is due to the increased solubility of VN in the presence of Nb. However, this effect was observed rather weakly in fig. 9, where the amount of precipitated V in steel U is slightly lower than steel K below 920°C. Because more N and C are consumed due to the Nb

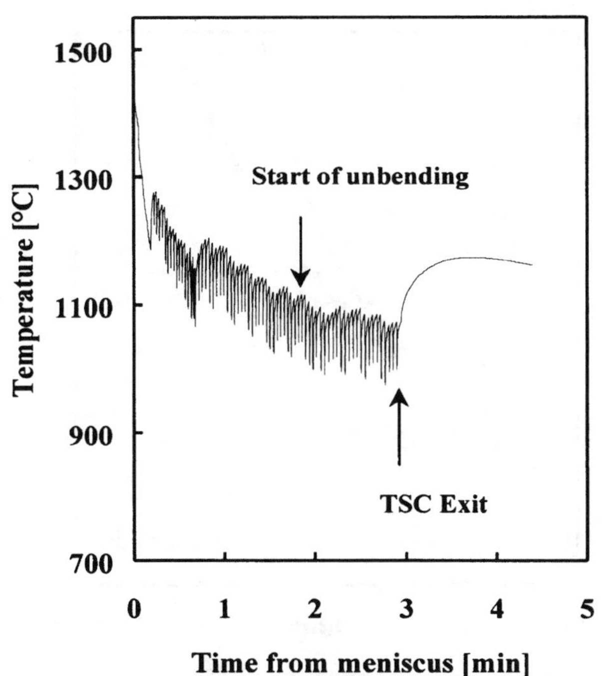

Fig. 6: Predicted strand surface temperature during thin slab casting[37] (mid-broad face, 90mm strand cast at 5m/min)

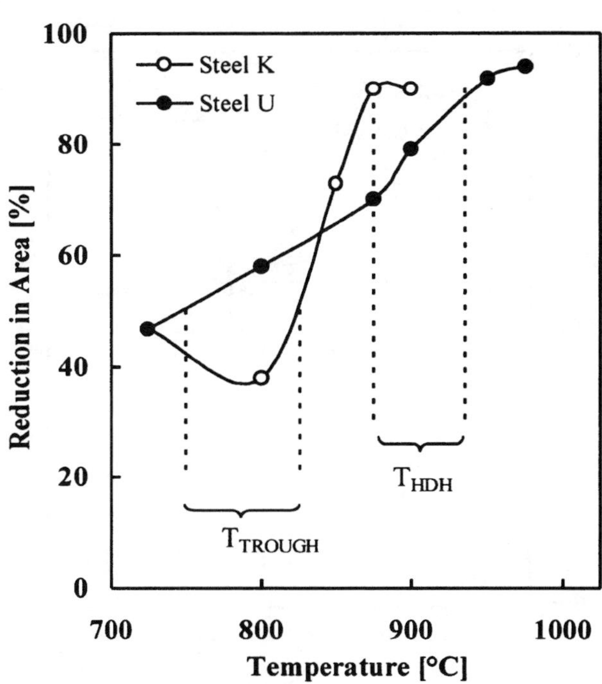

Fig. 7: As-cast hot ductility for low C-V steel K and low C-V-Nb steel U [37]

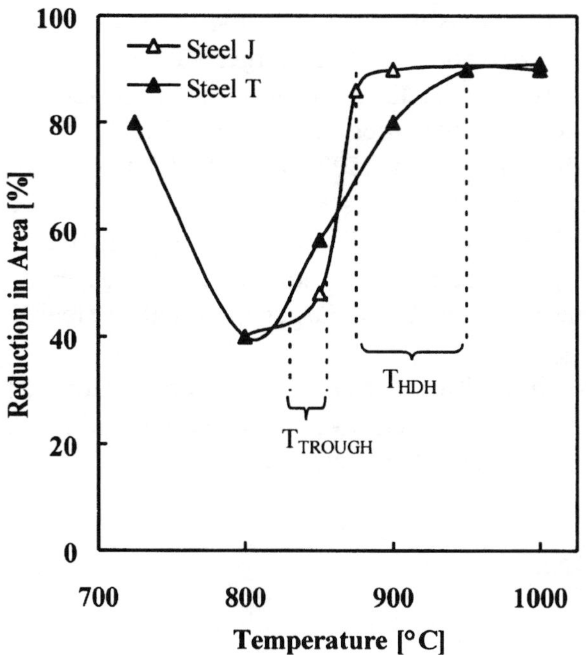

Fig. 8: As-cast ductility curves for high nitrogen steels J and T [37]

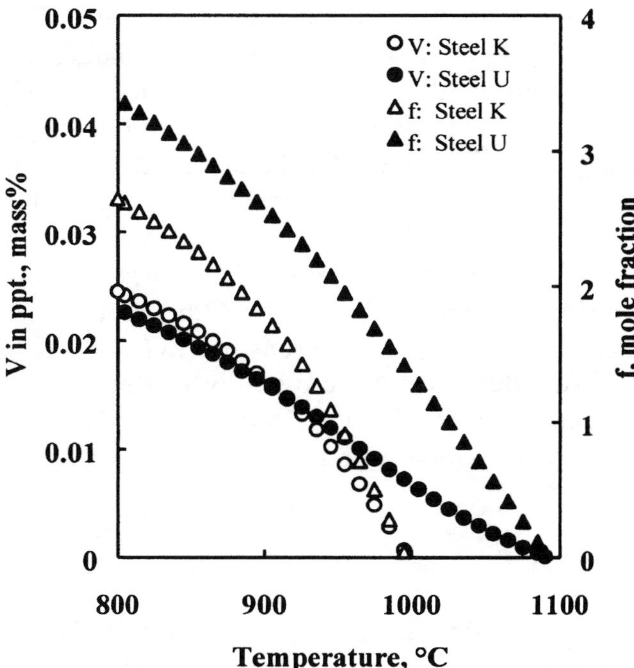

Fig. 9: Precipitate mole fraction, *f*, and precipitated V content in steels K and U

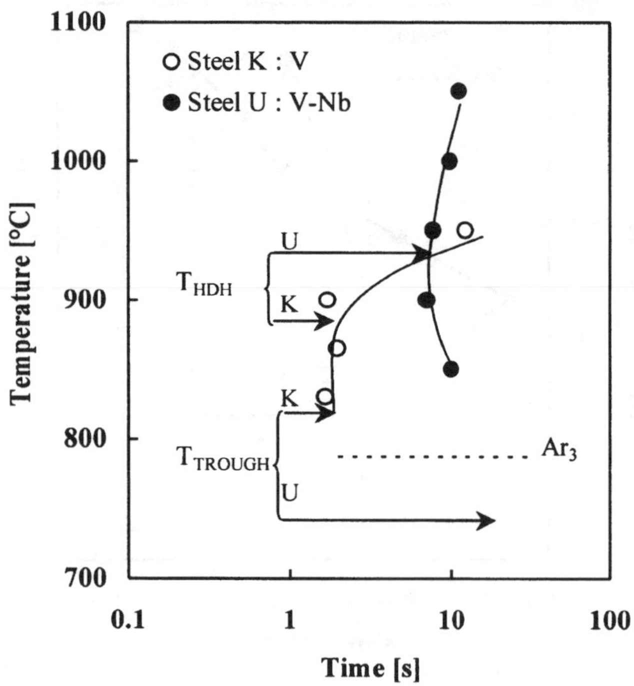

Fig. 10: Measured post-dynamic PTT curves for steels K and U [37]

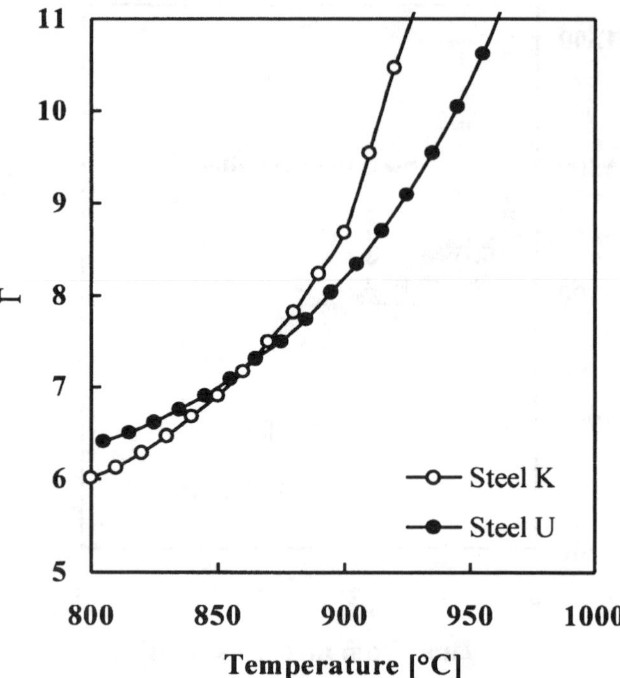

Fig. 11: Γ as a function of temperature in steels K and U

addition in steel U, the solubility of V is increased. Another contributing factor to the delay in precipitation in steel U at low temperatures is the thermodynamic parameter Γ in eq.13. Γ, which is proportional to P_s, is plotted against temperature in fig. 11 for steels K and U. At decreasing temperatures, the difference in Γ between the two steels gets progressively smaller until about 870°C, below which larger values are found for steel U. A similar effect (not shown) was found in high N steels J and T. This shift to larger Γ in steel U will, in part, contribute to the observed delay in precipitation at low temperatures in the austenite. This delay in precipitation may account for the observed improvement in ductility below 850°C in both V-Nb and V-Nb-N steels. Γ is a function of the interfacial energy, which according to Table I is larger for NbC and NbN at a given temperature than both VC and VN. Larger γ_{dis} values, which indicate increasing difficulty for nucleation, together with larger ξ_{dis} values in Nb-containing steels, result in smaller decreases in Γ with decreasing temperature in steel U. Although the chemical energy in steel U is larger than that of steel K at a given temperature, Γ is more sensitive to $(\gamma_{dis}\xi_{dis})^3$ than $(\Delta G_{chem})^2$.

An added benefit to ductility in steel U is that the Nb addition lowers the Ar_3 temperature[1)], (The calculated[38)] Ar_3 for 0.05C-1.35Mn steel is 787°C) which reduces the negative influence of strain-induced ferrite film formation at grain boundaries in the low temperature region of the austenite.

CONCLUSIONS

1. A mathematical model, based on mutual solubility and the classic nucleation theory, was developed to correlate precipitation and critical ductility temperatures in V, Nb and V-Nb steels.

2. Generally, the temperature at which ductility started to deteriorate in Nb, V and V-Nb steels was above the nose temperature, T_N, but below the highest precipitation temperature, T_{MAX}. For moderate to slow cooling rates, the onset of the ductility trough corresponded approximately to T_N for long hold times. Short hold times in low C-V and low C-V-Nb steels shift the onset of the ductility trough to below T_N.

3. The addition of Nb to V steel increases both the interfacial energy and the interfacial energy modifier between the matrix and the precipitates, which contribute to a delay in precipitation at low temperatures in the austenite. This delay in precipitation may be responsible for an improvement in hot ductility.

ACKNOWLEDGEMENTS

The authors would like to thank Saldanha Steel and Iscor Ltd. for financial assistance and permission to publish this paper.

REFERENCES

1. Y. Maehara, K. Yasumoto, H. Tomono, T. Nagamichi and Y. Ohmori, Materials Science and Technology, Vol. 6, Sept. 1990, pp. 793-806.

2. B Mintz and J.M. Arrowsmith, Metals Technology, Vol. 6, January 1979, pp. 24-32.

3. A. Guillet, S. Yue and M.G. Akben, ISIJ International, Vol. 33, No.3, 1993, pp. 413-419.

4. D.N. Crowther, Z. Mohamed and B. Mintz, Transactions ISIJ, Vol. 27, 1987, p. 366.

5. B. Dutta and C.M. Sellars, Material Science and Technology, Vol. 3, March 1987, pp.197-206.

6. S.H. Park and J.J. Jonas, Mathematical Modelling of Hot Rolling of Steel, Ed. S. Yue. Canadian Institute of Mining and Metallurgy, Hamilton, Ontario, Canada, 1990, pp. 446-456.

7. M. Militzer, W.P. Sun and J.J. Jonas, Proc. Int. Conf. Microstrostructure and properties of Microalloyed and Other Modern HSLA Steels, Ed. A.J. DeAardo, Pittsburgh, June 1991, pp. 271-279.

8. S. Okaguchi and T. Hashimoto, ISIJ International. Vol. 32, No. 3, 1992, pp. 283-290.

9. D.C.Houghton, Acta Metallurgica Materialia, Vol. 41, No. 10, 1993, pp. 283-290.

10. M.G. Akben, B. Bacroix and J.J. Jonas, Acta Metallurgica, Vol. 31, 1983, pp. 161-174.

11. H. Adrian, Microalloying '95 Conference Proceedings, pp. 285-305.

12. Ch. Perdrix, B. Chamont, E. Amoris and H. Biausser, Proceedings of the international Conference on Physical Metallurgy of Thermomechanical Processing of Steel and Other Metals, Thermec-88, Ed. I. Tanuma, Toko, Japan, Vol. 1, June 1988, pp. 807-814.

13. K.C. Russell, Advances in Interface and Colloid Science, Vol. 13, 1980, pp. 205-318.

14. M.G. Akben, I. Weiss and J.J. Jonas, Acta Metallurgica, Vol. 29, 1981, p. 111.

15. I. Weiss and J.J. Jonas, Metallurgical Transactions, Vol. 10A, 1979, p. 831.

16. W.J. Liu, Metallurgical and Materials Transactions, Vol. 26A, July 1995, pp. 1641-1657.

17. S. Fuyu and C. Wenxuan, "HSLA Steels: Processing, Properties and Applications", Eds. G. Tither and Z. Shuohua , The Min. Met. and Mat. Society, 1992, pp. 43-50.

18. X. Wenchung and S. Fuyu, Acta Metallurgica Sinica, Vol. 3, 1990, p. B211.

19. R.O. Elliot and C.P. Kempter, Journal of Physical Chemistry, Vol. 62, 1958, pp. 630-631.

20. H.J. Goldschmidt, Advances in X-Ray Analysis, Vol. 5, New York, Plenum Press, 1961, p. 191.

21. S. Kurokawa, J.E. Ruzzante, A.M.Hey and F. Dyment, 36th Annual Congress ABM, Recife, Brazil, July 1981, Vol. 1, pp. 47-63.

22. C.D. Morton, Ph.D Thesis, University of Sheffield, 1975.

23. W.P. Sun, Ph.D Thesis, McGill University, Montreal, 1991.

24. H.J. Frost H.J and M.F. Ashby, Deformation-Mechanism Maps, Pergamon Press, Oxford, 1982, p.21.

25. H.G. Suzuki, S. Nishimura and Y Nakamura, Transactions ISIJ, Vol. 24, 1984, p. 54.

26. B. Mintz and S. Yue, 34th Mechanical Working and Steel Processing Conference, ISS-AIME, Vol. XXX, 1993, p. 391.

27. N.E. Hannerz, Transactions ISIJ, Vol. 25, 1985, p. 149.

28. B. Mintz and Z. Mohamed, 7th International Conference on Fracture, Vol. 4, Houston, Pergamon Press, 1989, p. 2545.

29. R. Abushosha, S. Ayyad and B. Mintz, Material Science and Technology, Vol. 14, April 1998, pp. 346-351.

30. B. Mintz and R. Abushosha, Microalloying in Steels, Eds. Rodriquez-Ibabe, I Gutierrez and B. Lopez, Trans. Tech. Publications. Ltd, Switzerland, 1998, pp. 461-468.

31. C. Ouchi and K Matsumoto, Transactions ISIJ. Vol. 22, 1982, pp. 181-189.

32. F.A. Verdoorn, Unpublished work, University of Pretoria, 1999.

33. B. Mintz, S. Yue and J.J.Jonas, International Materials Review, Vol. 36, No. 5, 1991, pp. 187-217.

34. B. Mintz, ISIJ International, Vol. 39, No. 9, 1999, pp. 833-855.

35. P.J. Lubensky, S.L. Wigman and D.J. Johnson, Proceedings of Microalloying '95, 1995, pp. 225-233.

36. B. Mintz, J.R. Wilcox and D.N. Crowther, Materials Science and Technology, Vol. 2, 1986, p. 589.

37. K.M. Banks, Ph.D Thesis, University of the Witwatersrand, March 2000.

38. C. Ouchi, T. Sampei and I. Kozasu, Transactions ISIJ, Vol. 22, 1982, pp. 214-222.

PRECIPITATION AND REHEAT CRACKING SUSCEPTIBILITY OF A514-TYPE STEELS CONTAINING SULFUR AND BORON

C. Huang, J. S. Keske, D. K. Matlock
Advanced Steel Processing and Products Research Center
Colorado School of Mines
1500 Illinois St., Golden, CO 80401
Tel: (303) 384-2139, Fax: (604) 273-3016

ABSTRACT

Reheat Cracking susceptibility was evaluated in A514-type steels with different levels of sulfur and boron. Coarse-grained heat-affected zones were simulated in specimens that were later subjected to a constant load hot ductility test on a thermomechanical simulator. The removal of boron in these steels resulted in a significant reduction in susceptibility to reheat cracking. SEM observation on fracture surfaces and STEM examination on carbon replicas extracted from fracture and sectioned surfaces reveals that large particles contain Ce, S, Fe, etc., and most medium particles are of M_3C type. That BN was not detected in these steels suggests that the negative effect of boron on reheat cracking may result from either grain boundary segregation of boron or increasing stability of M_3C by boron partition to cementite.

Key Words: reheat cracking, boron partition, cementite, precipitation, fracture.

INTRODUCTION

In welded structures, reheat cracks at elevated temperature form because relaxation strains exceed the creep ductility of the coarse-grain heat-affected-zone (HAZ). The mechanisms of reheat cracking are often related to precipitation and segregation processes in steels.[1,2] One mechanism suggests that precipitation strengthening of grain interiors during post-weld-heat-treatment (PWHT) causes relaxation of residual stress or plastic deformation to be restricted in grain interiors and concentrated along weaker grain boundaries.[3] The resultant localized deformation leads to the formation of voids at steps or other discontinuities on grain boundary interfaces. Such cavities, when linked, form grain boundary cracks.

In addition to precipitate effects, the susceptibility of welds to reheat cracking is also affected by solute segregation. Various interactions among alloying and trace elements can exist and affect ductility, either by direct influence on the non-metallic inclusion content, its composition or shape, or by embrittlement caused by segregation of trace impurities to grain boundary interfaces at high temperatures. Impurity segregation can affect the rates of both cavity generation and cavity growth.[4,5,6]

While sulfur is believed by many investigators to be an element of primary concern, several researchers have noted that the presence of boron in steel increases reheat cracking tendencies.[4,7-11]

Nose and Katsube[8] studied HT-80 steels and claimed that the presence of even small amounts of boron and vanadium greatly increased the sensitivity to reheat cracking. McPherson and co-workers[9,10] confirmed these claims with commercial and laboratory heats which showed that boron levels as low as 3 ppm increased the susceptibility to reheat cracking. Intergranular fracture was most severe when boron was present with aluminum and vanadium. In a recent study[11], it was suggested that if boron nitride particles form within the matrix and diffusion conditions and free nitrogen concentrations are such that nitrogen and chromium do not co-segregate to grain boundaries, then sulfur atoms can readily diffuse to and occupy sites on grain boundaries to promote embrittlement.

This present work is an extension of a recent study by Keske[12,13]who evaluated the reheat cracking resistance of a series of laboratory alloys designed to assess the boron effect. The previous study showed that steels with boron additions exhibited an increased susceptibility to reheat cracking and identified the need for a electron microscopy study to evaluate the role of precipitates. Thus this current study was designed to clarify further the effects of boron through an electron microscopy (both scanning, SEM, and scanning-transmission, STEM) analysis of precipitate observed on fracture surface or extracted with carbon replica.

EXPERIMENTS

In the Keske study[12,13], ten steels were produced to investigate reheat cracking in an ASTM A514 grade F alloy that contains boron. The alloy base composition (in wt. pct.) used in this study was 0.15C-0.9Mn-0.25Si-0.01P-0.5Cr-0.5Mo-0.05V-0.027Ti-0.03Al-0.005N, and the controlled variations in Ni, S, B, Ca and/or Ce that are listed in Table 1 along with slight variations in carbon content. The alloy designation used in Table 1 refers, in ppm, to the boron and sulfur content (e.g. alloy 12B185S contain 12ppm boron and 185ppm sulfur). The alloys include three boron-free heats that do not meet the ASTM specification, due to the modified boron content. The steel heats were ingot-cast, hot-rolled, and cooled to room temperature. Subsequently, the steels were heat-treated to produce a tempered martensitic microstructure. After heat treating, cylindrical test specimens were machined from the plates, and then subjected to coarse-grained heat-affected zone simulations in a Gleeble 1500 thermomechanical test system. Following HAZ simulation, constant-load hot-ductility testing was carried out to estimate the reheat cracking susceptibility. Grain size and tempering experiments were conducted to determine the effect of grain size and the degree of tempering on the reheat-cracking susceptibility. The details of experimental procedures are summarized more completely elsewhere.[12,13]

Table 1 Alloy compositions for elements which were varied in the base alloys

	C	S ppm	Ni	B ppm	Ca ppm	Ce ppm
12B185S	0.153	185	0.74	12	<10	<20
12B101S	0.152	101	0.74	12	<10	<20
13B40S	0.160	40	0.74	13	<10	<20
14B49S	0.168	49	0.73	14	<10	<20
12B33S +Ca	0.155	33	0.71	12	29	<20
13B40S +Ce	0.150	40	0.73	13	<10	110
13B15S +Ce	0.133	15	0.72	13	<10	60
0B30S	0.159	30	0.93	<2	<10	<2
0B32S +Ca	0.160	32	0.94	2	26	30
0B14S +Ce	0.180	14	0.96	<2	<10	80

Some of the fractured specimens were examined with a JEOL 840A scanning electron microscope. In combination with the SEM, energy-disperse spectroscopy (EDS) was employed to determine the elements present within the identifiable particles on fracture facets.

Some of the fractured specimens were sectioned along the central plane. The sectioned specimens were mounted, polished and etched following standard metallographic procedures. A carbon film was deposited on the etched surface and then extracted in a 5% Nital solution. To extract particles from fracture surfaces, the specimens were heavily carbon coated at various angles. Then, the specimens were extracted in a 10% Nital electrolyte, freeing the carbon film from the surface of the fracture by applying a 10V DC power supply. Precipitates on carbon replicas were examined using scanning-transmission-electron-microscopy (STEM). A light element detector was employed during EDS analysis.

RESULTS

Constant-Load Hot-Ductility Testing

Table 2 shows a comparison of hot-ductility testing values (by %RA) at a constant temperature of 621°C and an initial stress of 410MPa. All boron alloys exhibited less than 4.7% RA while boron-free steels exhibited greater than 8.5% RA. For some steels two samples were tested and the results for both are shown.

Table 2 Comparison of %RA for Constant-Load, Hot-Ductility Testing at 621°C and an Initial Stress of 410MPa

Alloys	12B185 S	12B101 S	13B40S	14B49S	12B33S +Ca	13B40S +Ce	13B15S +Ce	0B30S	0B32S +Ca	0B14S +Ce
%RA	4.7	1.6	3.1	3.5	4.1	3.4	2.5	8.5	10.8	35.5
	-	2.5	-	1.6	-	-	3.5	11.5	8.4	17.4

In addition to measuring the %RA, stress *vs.* rupture-time data were collected for the steels. Fig. 1 shows the stress *vs.* rupture-time behavior for steels after testing at temperatures of 565°C and 621°C. Clearly, the boron-free alloys exhibit superior stress *vs.* rupture-time behavior since at a particular initial stress, the boron-free alloys require a longer time to rupture during high temperature testing. Alternatively, for a particular rupture time, the boron-free alloys can withstand higher initial stress levels before failure.

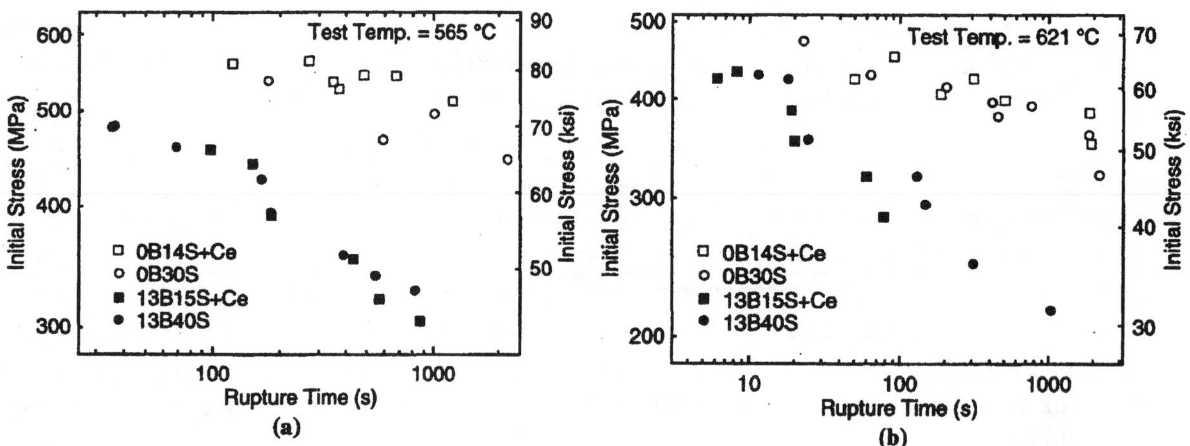

Fig. 1: Initial stress *vs.* rupture time for test temperature of (a) 565°C and (b) 621°C.

Precipitate Analysis

Fracture surfaces and carbon extraction replicas were evaluated on failed samples at the alloys listed in Table 1. Precipitate observations from several of samples were similar and the samples discussed below represent examples which characterize the observation.

Micrographs obtained from fracture surfaces showed 100% intergranular fracture and featureless prior austenite grain boundaries at comparatively low magnification. However, at higher magnification, cavities were revealed on many of the grain-boundary facets. As summarized previously[13], typical cavities appeared to be shallow with diameters of 1 to 10μm. Most of the particles within the cavities were small, sub-micron particles that could not be characterized with EDS. However, the deeper cavities at the prior austenite grain boundaries were associated with comparatively large particles or inclusions, as shown in Fig. 2. Correspondingly, large particles were always surrounded by comparatively ductile fracture feature areas with deeper cavities. Results from EDS analyses showed the presence of S, Cr, Mn, Al, Ti and Ce in the particles, depending on the alloy and the individual particle. Titanium was believed to be present in the form of a nitride as cubic particles were observed, and sometimes complex particles were shown to contain Ti, Mn, S, and Cr.

As presented in Fig. 3, the fracture surfaces of boron free and cerium treated alloys composed more ductile fracture feature area with more spherical sulfide particles which were shown to contain Ce. However, the fracture surface of boron-bearing alloys showed less ductility, with very few large particles.

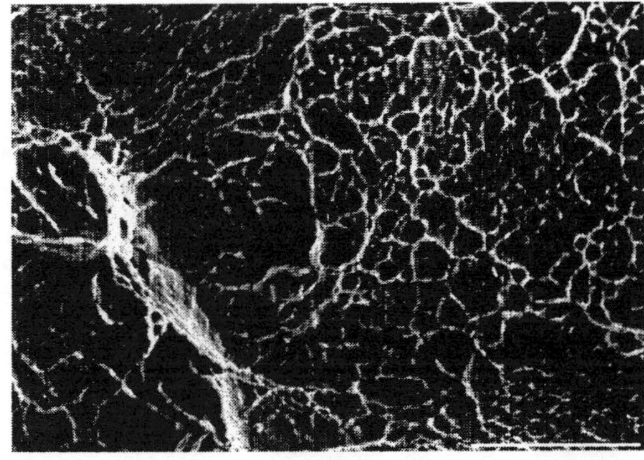

Fig. 2: grain boundary facets (SEM micrograph) in 0B14S+Ce showing ductile fracture around large particles.

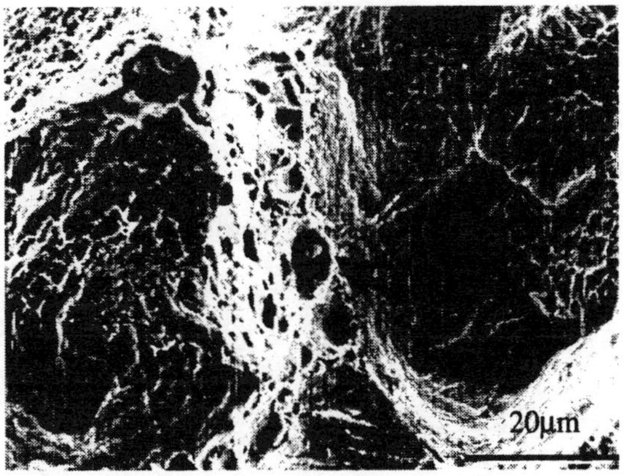

Fig. 3: Cerium sulfide particle (SEM micrograph) in alloy 0B14S+Ce.

STEM observations on carbon replicas extracted from HAZ simulated and polished specimens indicated that carbide precipitates were present before tempering. In the STEM, carbide particles with a variety at composition were identified by EDS analysis and the carbides contained some of the follows: C, Fe, Cr, Mo, Mn, Ti. Sulfides were rarely detected except that very few TiS particles were detected and cerium-bearing sulfides could not be found before tempering and/or constant-load hot-ductility testing.

Precipitates were also analyzed in the STEM on polished sections removed from constant-load hot-ductility test samples. Sample sections were taken close to the fracture surface and thus the precipitates that were examined represent the alloy behavior just prior to fracture. TiN particles were detected in all specimens examined. The presence of the TiN particles indicated that Ti gettered N to protect boron. Even though much effort was made to find BN during STEM examination, no BN particles were observed. Many specimens were examined and hundreds of particles were analyzed by EDS and diffraction patterns. EDS analysis indicated that most medium particles (about 100nm) contained Fe, Mn, Cr, and C and the smaller particles (10 to 50nm) were similar but some of them had higher Cr contents. Figure 4 shows one of these medium sized particles and the corresponding diffraction pattern analysis. Diffraction pattern analysis suggested that most of the carbide particles were a M_3C type instead of $M_{23}C_6$. Some extremely large particles were detected in Ce added alloys, as shown in Figure 5. EDS analysis indicated that they were sulfides that contained Ce, Fe, and/or Ti. It is worth mentioning here that no MnS particles were detected during STEM examination.

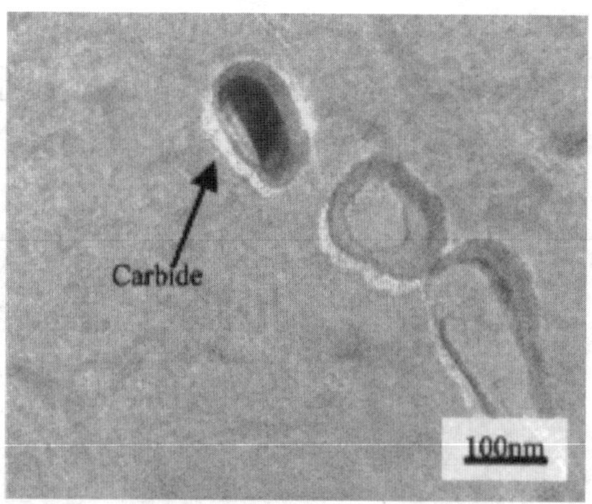

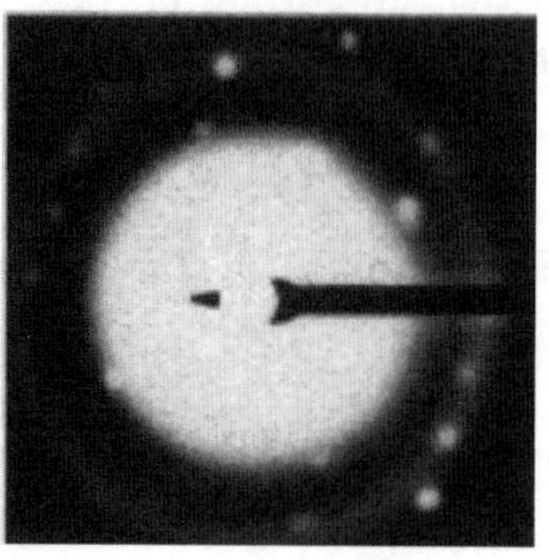

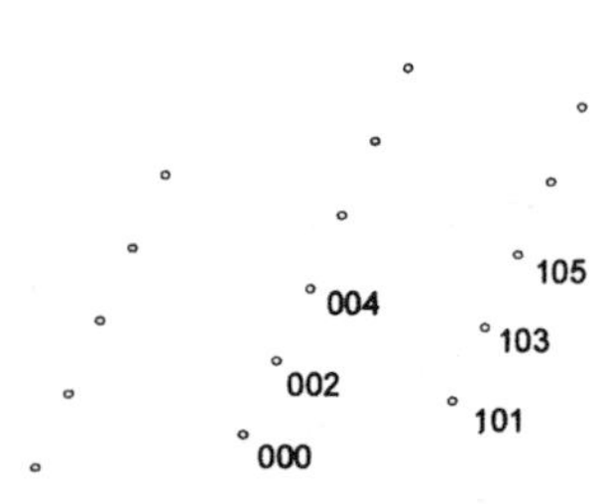

Fig. 4: STEM micrograph of carbides and diffraction pattern analysis in replica extracted from sectioned surface of alloy 13B40S.

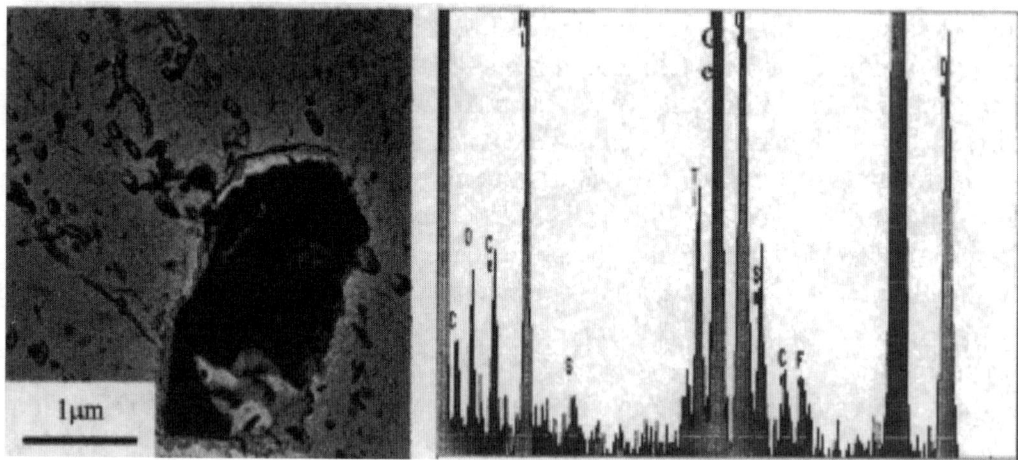

Fig. 5: STEM micrograph of sulfide and its EDS profile (intensity *vs* energy) in replica extracted from sectioned surface of alloy 13B40S+Ce.

Carbon replicas extracted directly from fracture surfaces contained particles that were difficult to evaluate because oxides were also extracted from fracture surface. These oxides formed at the test temperature and on cooling after fracture in the constant-load hot-ductility tests. The oxides obscured most of the particles present in the microstructure and associated with the fracture process. Nevertheless, some particles, particularly those at the edge of a crack in a carbon replica, could be characterized by EDS. Figure 6 shows one of these particles and the corresponding EDS analysis. The observed particles ranged in size from 50nm to 200nm and were mostly iron carbides that contained Mn and Cr. Large particles were also extracted from the fracture surface and they were detected to contain Ce, Ti, Mn, S and Al, as shown in Figure 7. They are believed to be the rare-earth sulfides that were also detected on fracture surfaces by SEM and on carbon replicas extracted from sectioned surfaces by STEM. Again, it should be pointed out that no BN or MnS particles were found in this examination.

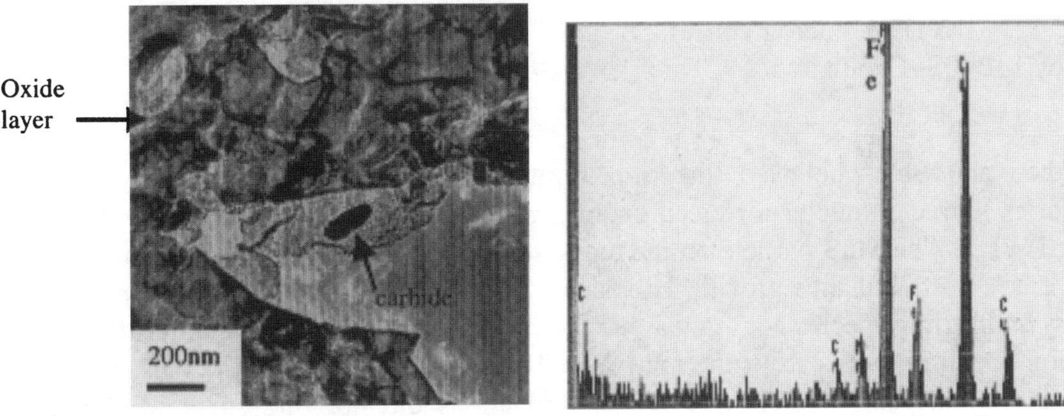

Fig. 6: STEM micrograph of carbide and its EDS profile (intensity *vs* energy) in replica extracted from a fracture surface of alloy 12B101S.

DISCUSSION

Precipitate analysis (SEM and STEM) indicates that most particles examined are carbides and very few MnS particles are detected. Large cubic feature particles can be found in all specimens and are identified as TiN instead of BN. The morphology of carbides is similar in carbon replicas extracted from HAZ simulated specimens and specimens sectioned from constant-load hot-ductility tested samples. EDS and diffraction pattern analysis indicates that these carbide particles are type M_3C. It is reasonable to assume that the M_3C carbides form during HAZ simulation process and persist during tempering and/or constant-load hot-ductility testing.

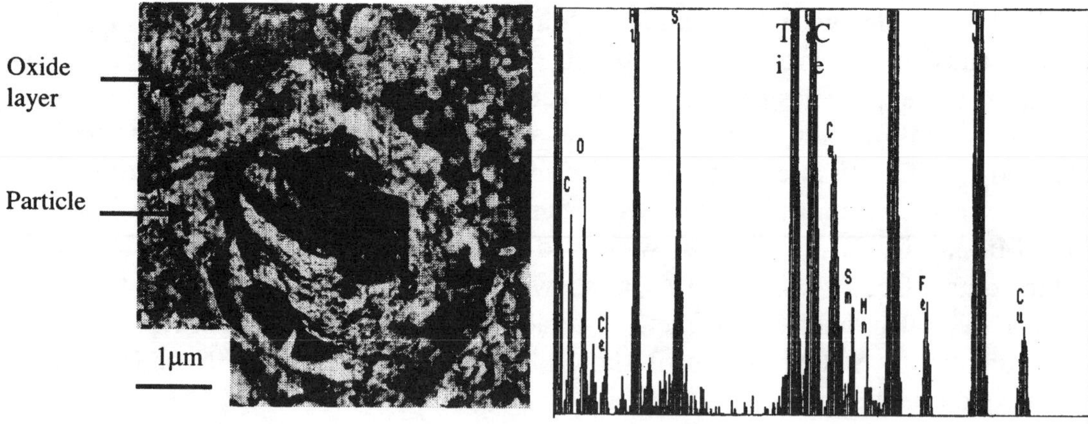

Fig. 7: STEM micrograph of cerium sulfide in replica extracted from fracture surface of alloy 13B40S+Ce.

McMahon and Shin[2] believed that re-precipitated fine MnS arrays along prior austenite grain boundaries served as nuclei for creep cavities which grown and linked to form boundary cracking during PWHT. The MnS particle on the crack surface dissolved and provided sulfur which diffused to the crack tip to enhance growth. Hippsley and co-workers[14] concluded the role of sulfur as the primary embrittling species but argued that prior austenite boundaries are embrittled by stress-driven diffusion of sulfur atoms from the matrix to the boundaries ahead of a crack, instead of from re-dissolving small MnS particles. Present examinations did not find small MnS either on carbon replicas extracted from HAZ specimens or on carbon replicas extracted from sections or from fracture surfaces. These observations suggest that small MnS particles in the steels examined here were dissolved during the high temperature excursion in the HAZ and remained in matrix during rapid cooling.

The detrimental effect of boron is clearly shown in Figure 1, where the hot ductility is consistently lower, the rupture times for a given initial stress are consistently shorter, and the fracture surfaces show little plasticity. Several theories exist to explain the detrimental effects of boron on reheat cracking sensitivity. The most well-defined theory deals with boron being a scavenger of nitrogen, where nitrogen is believed to be an element that will co-segregate with chromium to prior austenite grain boundaries and compete with sulfur for grain boundary sites.[11] Results from the present study suggest that boron must influence susceptibility to reheat cracking in a different manner. Excess titanium is added in all of the alloys tested to tie up the nitrogen. The ability of Ti to tie up nitrogen was confirmed as many TiN particles were found in alloys and BN precipitates were not detected. Hence, the significant effect of boron in the present study is unlikely to be associated with changes in solute nitrogen.

Boron is believed to precipitate as $Fe_{23}(C,B)_6$, $Fe_3(C,B)$, and Fe_2B in an Fe-C-B system.[15] In the current study, diffraction pattern analysis revealed that the precipitates are a M_3C type. Considering the large solubility of boron in cementite, it is suggested that significant amounts of boron tend to partition to cementite and most of the particles examined are $(Fe,Mn,Cr)_3(C,B)$. An atom probe field ion microscopy investigation suggests that the concentration of boron in M_3C phase is higher than that in the $M_{23}C_6$ phase in tempered martensitic steel.[16] C. D. Lundin and K. K. Khan[3] concluded that the M_3C type carbide persisted longer in reheat cracking sensitive steels than in resistant steels. Some of specimens examined have been tempered at 621°C for one hour before constant-load hot-ductility testing. The carbide precipitates observed on carbon replicas extracted from sectioned surface transverse to the fracture surface were still type M_3C instead of $M_{23}C_6$. Therefore, the addition of boron in these steels seems to increase the stability of M_3C by partitioning to cementite. Thus, the effect of boron on reheat cracking could be associated with the effect of boron on the decohesion of the cementite/matrix interface. The segregation of boron at prior austenite grain boundaries could be another factor that needs to be investigated further.

CONCLUSIONS

The principal conclusions that can be drawn from the present studies are as follows:

(1) The removal of boron in A514-type steels resulted in a significant reduction in susceptibility to reheat cracking.

(2) That no MnS particles were detected on replicas extracted from either sectioned or fracture surfaces suggests that the sulfur that goes into solution during weld cycle is response for minor effect of sulfur on reheat cracking in current alloys.

(3) Carbide precipitates detected in the HAZ simulated and/or constant-load hot-ductility tested specimens are type M_3C.

(4) Boron partitioning to cementite could be one of the detrimental effects of boron on reheat cracking.

ACKNOWLEGEMENTS

This projected was supported by the Advanced Steel Processing and Products Research Center, a NSF Industry/University Cooperative Research Center at the Colorado School of Mines. The authors wish to thank Dr. S. W. Thompson for his advice and discussion.

REFERENCES

1. A. Dhooge and Vinckier: "Reheat Cracking-Review of Recent Studies (1984-1990)", Welding in the World, vol. 30, No.3/4, 1992, pp.44-71.

2. J. Shin and C. J. McMahon Jr., "Mechanisms of Stress-Relief Cracking in a Ferritic Steel", Acta Metall., vol. 32, 1984, pp.1535-1552.

3. C. D. Lundin, and K. K. Khan: "Fundamental Studies of the Metallurgical Causes and Mitigation of Reheat Cracking in 11/4Cr-1/2Mo and 21/4Cr-1Mo Steels", WRC Bulletin 409, Feb., 1996, pp. 1-117.

4. J. Sun, R. Zaiss, M. Menyhard and C. J. McMahon, "Impurity Effects in Stress-Relief Cracking of Mn-Cr-Mo-Ni Steel", Matls. Sci. and Eng., vol. 3, 1987, pp. 139-145.

5. J. J. Lewandowski, C. A. Hippsley, M. B. D. Ellis and J. F. Knott, "Effects of Impurity Segregation on Sustained Load Cracking of 21/4Cr-1Mo Steels-I. Crack Initiation", Acta Metall., vol. 35, No.3, 1987, pp. 593-608.

6. J. J. Lewandowski, C. A. Hippsley, and J. F. Knott, "Effects of Impurity Segregation on Sustained Load Cracking of 21/4Cr-1Mo Steels-II. Crack Propagation" Acta Metall., vol. 35, No.8, 1987, pp. 2081-2090.

7. T. Fujii, K. Yamamoto, and Ueno, "Effects of Microalloying on Susceptibility to Intergranular Fracture at about 600°C in Heat-Affected Zone of High Strength Low Alloy Steel", Testsu-to-Hagane, vol.67, 1981, pp.1523.

8. J. Nose and C. Katsube, "The Effect of Chemical Composition on Stress-Relief Cracking in HT-80 Steels", IIW Doc. X-617-71, 1971.

9. R. I. Presser and R. McPherson, "Boron Segregation and Elevated Temperature Embrittlement of Ferritic Steel", Scripta Metall., vol.11, 1977, pp.745-749.

10. R. McPherson, "Compositional Effects on Reheat Cracking of Low-Alloy Ferritic Steels", Metals Forum, vol.3, 1980, pp. 175-186.

11. S. Ishikawa, J. A. Pfaendtner, and C. J. McMahon Jr., "The Effect of Boron on Stress-Relief Cracking of Alloy Steels", Matls. Sci. and Eng., vol. A272, 1999, pp. 16-23.

12. J. S. Keske, "Reheat-Cracking Sensitivity in ASTM A514 Steels as Influenced by Sulfur and Boron", M. S. Thesis, Colorado School of Mines, July, 1999.

13. J. S. Keske et al., "Reheat-Cracking Susceptibility of ASTM A514-Type Steels Containing Different Levels of Sulfur and Boron", Proc. of Intern. Symp. on Steel for Fabricated Structures, ASM International, Materials Parks, Ohio, 1999, pp. 214-222.

14. C. A. Hippsley, J. F. Knott, B. C. Edwards, "A Study of Stress-Relief Cracking in 21/4Cr-1Mo Steels", Acta Metall., vol. 28, 1980, pp.869-85.

15. Y. Ohmor, "The Isothermal Decomposition of an Fe-C-B Austenite", Transactions ISIJ, vol. 11, 1971, pp. 339-348.

16. P. Hofer M. K. Miller, S. S. Babu, S. A. David, and H. Cerjak, "Atom Probe Ion Microscopy Investigation of Boron Containing Martensitic 9% Chromium Steel", Metallurgical & Materials Trans A, vol. 31A, March 2000, pp. 975-985.

Effect of Carbon and Niobium Content on HAZ Toughness of HT590 Steel

Hitoshi Furuya Ryuji Uemori Yukio Tomita Shuji Aihara Yukito Hagiwara
Nippon Steel Corp./ Steel Research Laboratories
20-1, Shintomi, Futtsu, Chiba-pref., Japan
Tel.: +81-439-80-3095
Fax.: +81-439-80-2744
E-mail: furuya@atsuita.re.nec.co.jp

Key Words: HT590 steel, HAZ toughness, Carbon, Niobium, Martensitic island, Heterogeneity

SUMMARY

A study was performed to clarify the effect of alloying elements such as carbon and niobium on heat-affected zone (HAZ) toughness of HT590 steel. With increasing carbon and niobium content, HAZ toughness tends to deteriorate. It was found that high number density of martensitic island (M*) with increasing content of carbon and niobium plays an important role on the deterioration of HAZ toughness, in addition to the effect of volume fraction, size and aspect ratio of the M*. Particularly, good correlation between fracture appearance transition temperature and maximum value of area fraction of localized M* was confirmed by quantitative analysis on the distribution of the M*. Carbon and niobium both affect the distribution of M* due to the increased hardenability of untransformed austenite adjacent to grain boundary ferrite allotriomorph and to microsegregation of niobium and manganese.

INTRODUCTION

For the safety against brittle fracture of welded steel structures, it is important to use steels possessing enough weld heat-affected zone (HAZ) toughness. Since higher heat input has recently been applied for fabricating structures efficiently, it is necessary to maintain a certain level of toughness at HAZ of high heat input welding. Refining HAZ microstructure by e.g. nitride or oxide particles through intragranular ferrite transformation and austenite grain refinement is essentially important for improving the HAZ toughness. However, it is not necessarily sufficient, especially at higher strength steels, because another factor also affects the HAZ toughness. It is well known that the presence of brittle phase such as martensitic island (M*) brings about the loss of HAZ toughness[1)].

The M* is often formed during welding. On the way of cooling stage of weld thermal cycle, transformation from austenite to ferrite or pearlite or bainite occurs. The area of untransformed austenite decreases with a drop in temperature. This area becomes enriched in carbon. Some part of this area dissolves into ferrite and carbide, but the other transforms into martensite at lower temperature, leading to the M*. The formation of M* depends on the content of alloying elements and cooling rate.

The M* is known to reduce HAZ toughness because it acts as a brittle fracture initiation site[2)]. It was found that debonding at the interface between M* and matrix or cracking of M* itself, because of its high hardness, is the cause of brittle fracture[2)]. Many researches have been conducted to clarify the controlling factor by which the M* reduces HAZ toughness. At first, volume fraction of the M* was correlated to HAZ toughness[3)]. On the other hand, the effect of M* morphology was examined and it was revealed that size or aspect ratio has a significant influence on the initiation of brittle fracture[4)]. However, relationship between heterogeneity of M* distribution and HAZ toughness is not clear. Furthermore, in carbon steels containing niobium, which factor of M* plays the most significant role on the

HAZ toughness with the variation of alloying elements has not been clarified.

In the present study, controlling factor for the loss of HAZ toughness of high heat input welding (20kJ/mm) was examined for steels of 590MPa class tensile strength from a viewpoint of carbon and niobium content. Especially, relationship between HAZ toughness and heterogeneity of M* distribution in addition to area fraction, size and aspect ratio of the M* was discussed.

EXPERIMENTAL PROCEDURE

Nine steels were prepared by laboratory vacuum melting. The chemical compositions are given in Table I. Steel A is the base steel of 470MPa yield strength, 590MPa tensile strength. In steel B to I, carbon and niobium contents were varied to provide different microstructures, especially different volume fractions, sizes, morphologies and distributions of M* particles at HAZ. Alloying elements such as copper and nickel were added to increase the strength of steels. The steels were reheated at 1423K and hot-rolled to 15mm in the thickness.

Table I. Chemical compositions of the steels in mass percent.

	C	Si	Mn	P	S	Nb	Ti	Al	N	Ceq.	Pcm
A	0.10	0.20	1.48	0.009	0.0042	**0.012**	0.012	0.027	0.0042	0.37	0.21
B	0.10	0.20	1.49	0.008	0.0037	**0.020**	0.012	0.026	0.0040	0.37	0.21
C	0.10	0.20	1.48	0.008	0.0038	**0.031**	0.012	0.027	0.0042	0.37	0.21
D	0.10	0.21	1.50	0.009	0.0038	**0.038**	0.013	0.027	0.0047	0.37	0.21
E	0.10	0.21	1.49	0.009	0.0039	**<0.005**	0.012	0.028	0.0049	0.37	0.21
F	**0.05**	0.20	1.49	0.010	0.0038	**0.011**	0.012	0.029	0.0042	0.32	0.16
G	**0.05**	0.20	1.49	0.009	0.0036	**0.020**	0.013	0.030	0.0042	0.32	0.16
H	**0.05**	0.20	1.48	0.009	0.0039	**0.040**	0.014	0.030	0.0043	0.32	0.16
I	**0.05**	0.20	1.47	0.009	0.0039	**0.059**	0.014	0.030	0.0044	0.32	0.16

Ceq.=C+Mn/6+Si/24+Mo/4+Cr/5+Ni/40+V/14
Pcm=C+Si/30+(Mn+Cu+Cr)/20+Ni/60+Mo/15+V/10+5B

Simulated HAZ specimens of 12x12x120mm in size were machined from the plate in the rolling direction and subjected to simulated heat cycle. The simulated heat cycle, which corresponds to that of welding with heat input of 20kJ/mm, is shown in Fig.1. 20 to 30 simulated HAZ specimens were made for each steel, and used for various experiments mentioned below.

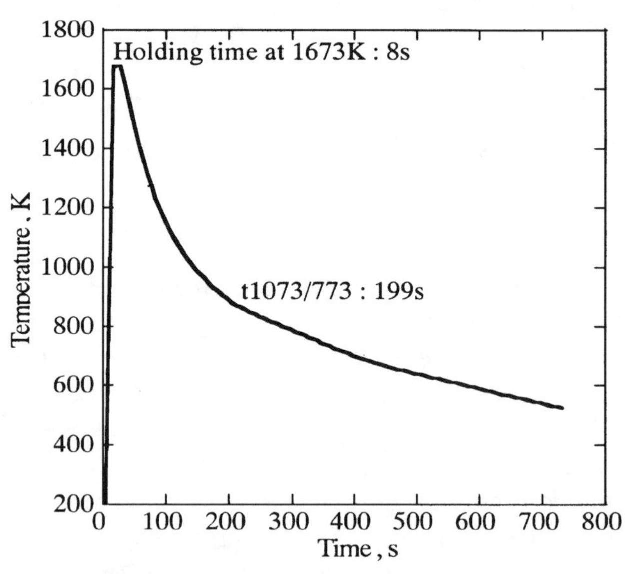

Fig.1. Heat cycle profile which corresponds to heat input of 20kJ/mm.

(1) Microstructure and hardness of simulated HAZ-A simulated HAZ specimen was subjected to microstructural observation and Vickers hardness measurement. The specimen was polished and etched with 2% nital, and microstructure was subsequently observed by an optical microscope(OM). The observed surface was perpendicular to the rolling direction. The hardness measurements were performed at 5kgf indentation loading.

(2) Charpy impact test-Charpy impact test specimens were machined from the simulated HAZ specimens and subjected to the tests at temperature of 193 to 313K. Absorbed energy (vE) and crystallinity were measured and subsequently the temperature at which crystallinity is 50% (vTrs) was determined.

(3) Fracture surface observation of Charpy impact test specimens-Fracture surfaces of Charpy specimens were observed by scanning electron microscope (SEM) to clarify a brittle fracture initiation site and the surrounding microstructure. In case a particle was found at the brittle fracture initiation site, Electron Dispersive X-ray Spectroscopy (EDS) analysis was performed to identify the particle. To observe the microstructure surrounding the fracture initiation site, the specimen was sectioned perpendicular to the fracture surface and coinciding with the initiation site, polished and etched with 2% nital, and subsequently microstructure at the initiation site was observed by SEM.

(4) Quantitative analysis of M*-After polished and etched using LePera etchant[5], M* was observed by OM with x500 magnification. 25 images were obtained for all steels and quantified by a computer with image analysis software. Total area fraction of M*, area of each M* and aspect ratio of each M* were measured. The measurement was limited to the M*s with the area of greater than $0.8\mu m^2$, because the M* with the area of smaller than $0.8\mu m^2$ is considered to have less effect on HAZ toughness and it is hard to be identified and quantified by OM. It is noted that the aspect ratio is defined as the ratio of the longest length to the shortest length.

(5) Low speed bend test-To observe microcracks prior to brittle fracture initiation, low speed bend tests were conducted. Specimens with the same configuration as the Charpy impact test specimens were prepared, and three-point-bent at 233K and with crosshead displacement rate of 0.1mm/s. A specimen was fractured by monotonic loading and fracture initiation load was determined. Subsequently, another specimen was tested, but in this case, the loading was interrupted just before attaining the brittle fracture initiation load and then unloaded. Load and the displacement of the load point were measured and the absorbed energy was calculated by integrating the load-displacement curve. The tested specimen was cut at the mid-section, polished and etched with 2% nital. Microcracks and structures around the microcracks were observed by SEM.

RESULTS

(1) Microstructure and hardness of simulated HAZ -Microstructures of the steel A to E (0.1%C), "Group1", and the steel F to I (0.05%C), "Group2", are shown in Fig.2 and Fig.3, respectively. Vickers hardness values are also shown. The microstructure consists of grain boundary ferrite allotriomorph (FA), grain boundary polygonal ferrite (GPF), ferrite side plate (FSP), intragranular ferrite (IGF) and upper bainite (UB). The steels of Group1 exhibit lower area fraction of FA and GPF because of higher hardenability associated with higher carbon content. Furthermore, the area fraction of FA and GPF decreases with the increase in niobium content, but the variation is not significant. In the steels of Group2, the area fraction of FA and GPF is higher due to the lower hardenability, and decreases markedly with the increase in niobium content. On the other hand, the increase in niobium content leads to the increase in area fraction of FSP and UB for the both groups, because the hardenability increases with the increase in niobium content.

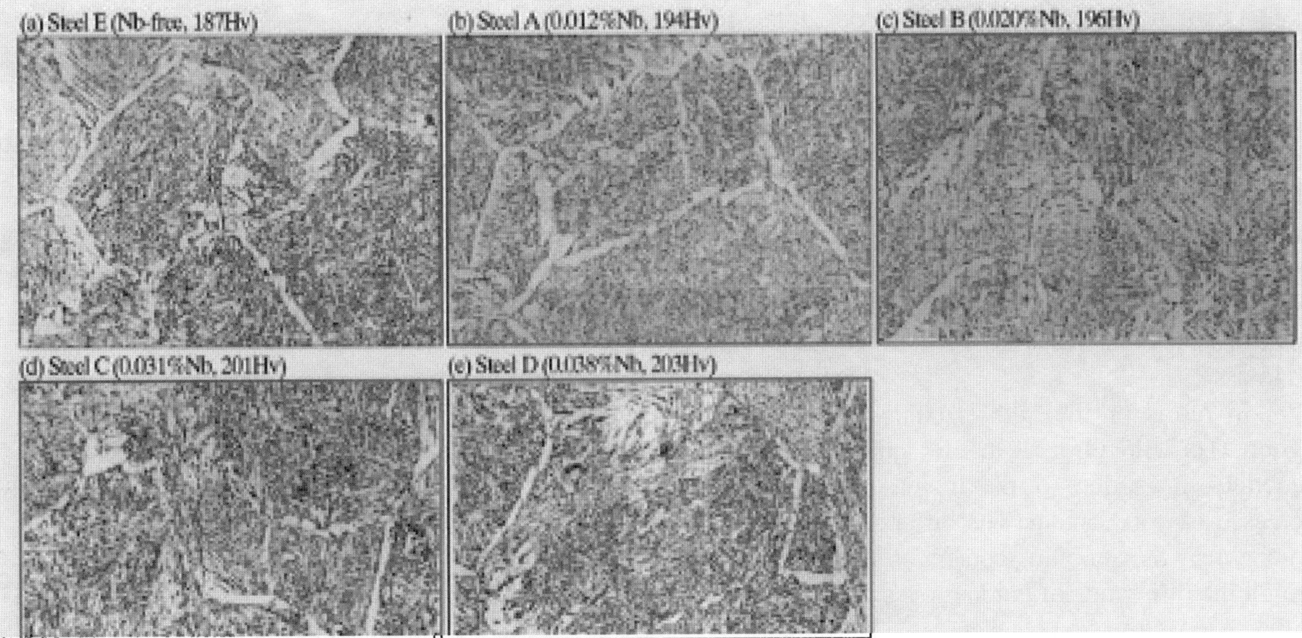

Fig.2. Optical micrographs of the steel A to E (0.1%C), "Group1", hardness data is indicated in each micrograph.

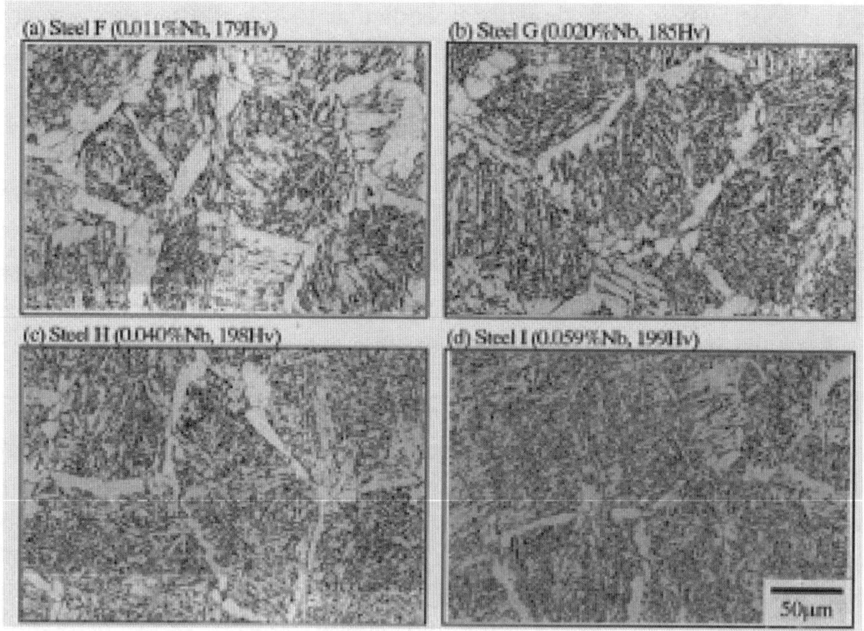

Fig.3. Optical micrographs of the steel F to I (0.05%C), "Group2", hardness data is indicated in each micrograph.

Hardness of the simulated HAZ increases with increasing carbon and niobium content. Because the change in hardness corresponds well to the change of microstructure, e.g. area fraction of FA, GPF and FSP etc., it is expected that the hardening by the precipitation of niobium carbide or nitride during the cooling stage of simulated HAZ heat cycle has a negligible effect on the hardness, if any.

If compared at the same niobium content, difference of hardness between Group1 and Group2 is large when the niobium content is low. For example, at around 0.01mass% niobium content, the difference of hardness is 15Hv (Steel A and Steel F). It decreases when the niobium content is higher; e.g. compare steel D and steel H (0.04%Nb), the hardness difference is 4Hv. It may be attributed to the variation of microstructure based on hardenability. In case of lower niobium (0.01%), since hardenability is relatively low, the area fraction of FA and GPF in the steel F (0.05%C) become larger, while the area fraction is smaller because of the increase in hardenability in the steel A (0.1%C). Therefore the difference in hardness is large. On the other hand, in case of higher niobium (0.04%), since hardenability is relatively high, the area fraction of FA and GPF is smaller than in case of lower niobium content regardless of carbon content, leading to almost the same hardness.

(2) Charpy impact test-Figure4 shows the relationship between vTrs and niobium content. For the both groups, vTrs tends to increase with increasing niobium content. In Group1, vTrs reveals a marked rise with increasing niobium content, but except for steel E. In Group2, vTrs increases with niobium content up to 0.04mass%. However, the increase of vTrs is saturated over 0.04mass%.

In terms of the carbon content, Group1 exhibits higher vTrs than Group2 in the same niobium content. Generally, it is due to the increase of hardness with increased carbon content. The difference in vTrs become larger when the niobium content is around 0.04mass%, though the difference of hardness is small. This point will be discussed in the next chapter.

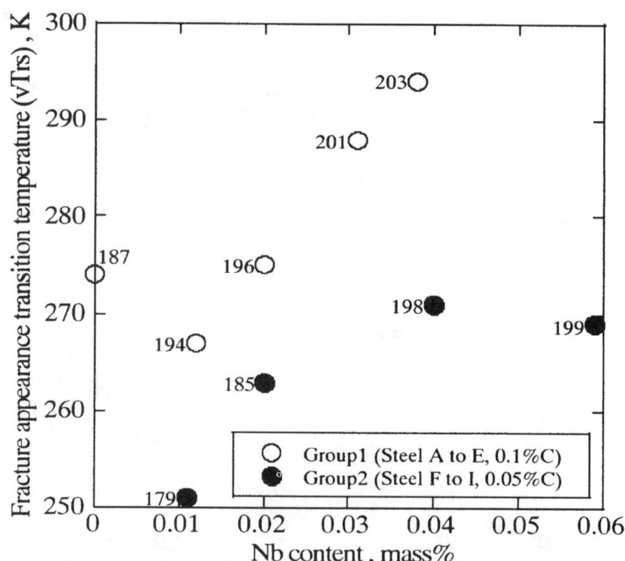

Fig.4. Relationship between vTrs and Nb content. (Hardness data is indicated for each plot.)

(3) Fracture surface observation of Charpy impact test specimens -Examples of the fracture surface surrounding the brittle fracture initiation site are shown in Fig.5 for the steel E, D, F and I. In some of these specimens, some particles were observed precisely coinciding with the fracture initiation point (Fig.6 and Fig.7). From the EDS analysis of the particles, only the peak of Fe was detected in most cases. From this analysis together with the morphology of the particle, the particle is presumably M* and acts as a brittle fracture initiation site (Fig.6). Occasionally (less than 10% of observation), peaks of Ti and Nb were detected from similar particle, which is considered to correspond to the nitride or carbonitride of Ti and Nb (~μm) (Fig7). Note that there was no remarkable difference in facet size between the steels.

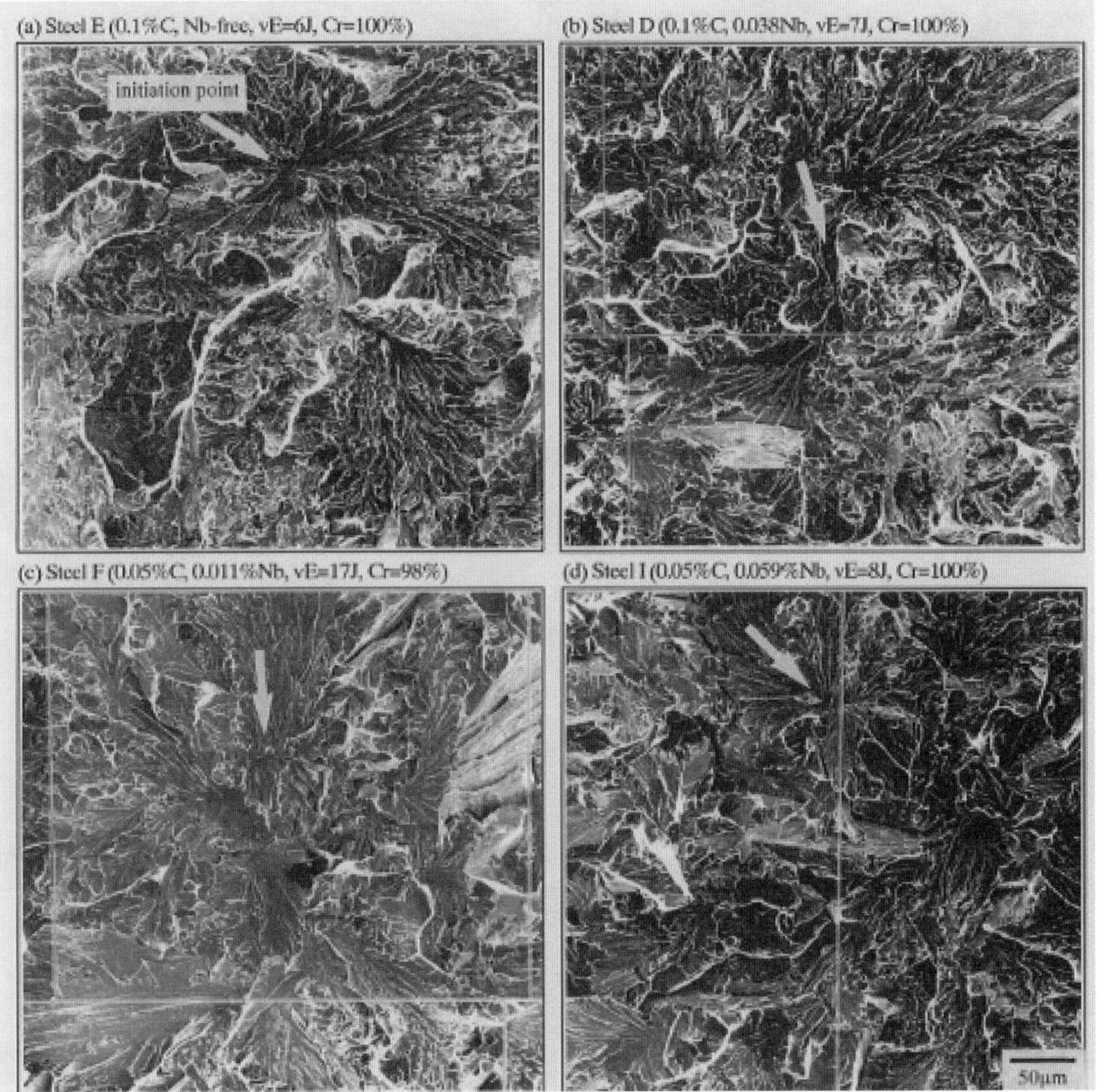

Fig.5. SEM micrographs of fracture surface near the brittle fracture initiation site.
(Test temperature is 193K for each specimen,
Absorbed energy (vE) and Crystallinity (Cr) are indicated for each micrograph.)

Examples of the microstructure near the initiation site, which were derived from the observation of the section perpendicular to the fracture surface, are shown in Fig.8. In Fig.8(a) and (b), initiation site corresponds to the M* because only the peak of Fe appeared in EDS, while Fig.8(c) and (d) is the case where initiation site was Ti(Nb) carbonitride. These microstructures have common characteristics; regardless of a kind of the initiation site, microstructure surrounding the brittle fracture initiation site is FSP or other microstructures adjacent to FA and GPF. Furthermore, a number of segregated M*s are found in the region of the initiation site. In other word, the existence of M* has the most important role on the brittle fracture initiation. Especially, the high number density of M* in the localized region is expected to promote the fracture initiation.

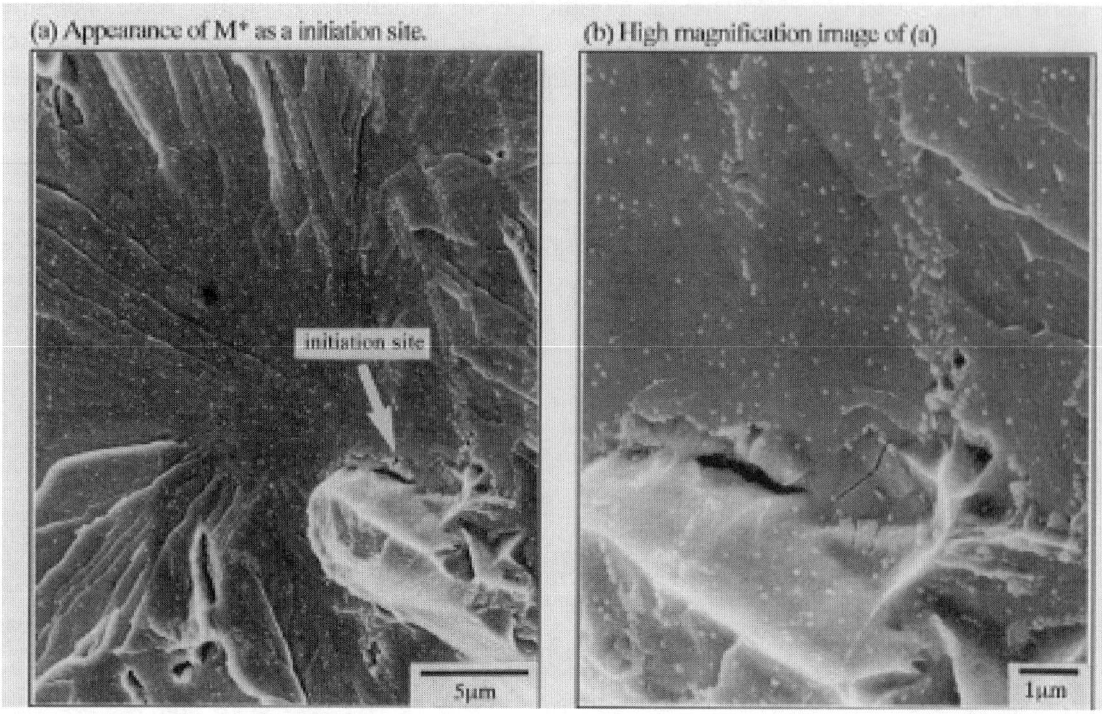

Fig.6. Appearance of fracture surface surrounding the brittle fracture initiation site. (Steel D, 0.1%C, 0.038%Nb, Test temperature is 233K., vE=14J, Cr=90%).

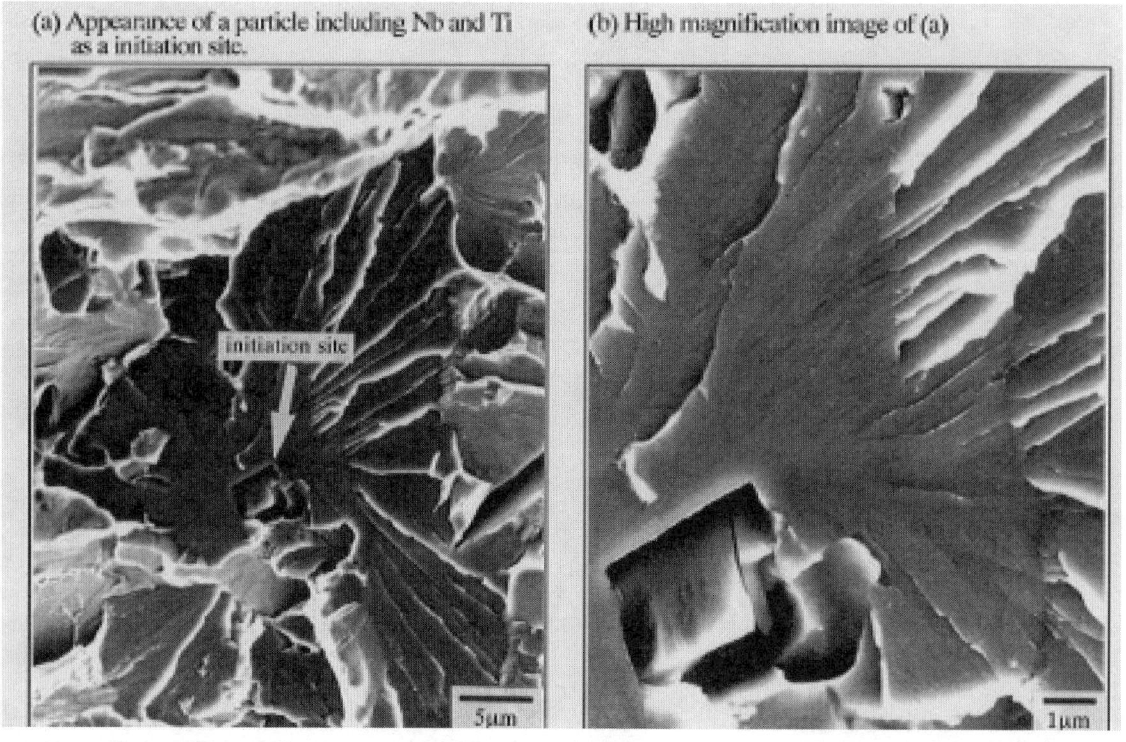

Fig.7. Appearance of fracture surface surrounding the brittle fracture initiation site. (Steel B, 0.1%C, 0.020%Nb, Test temperature is 193K., vE=11J, Cr=100%).

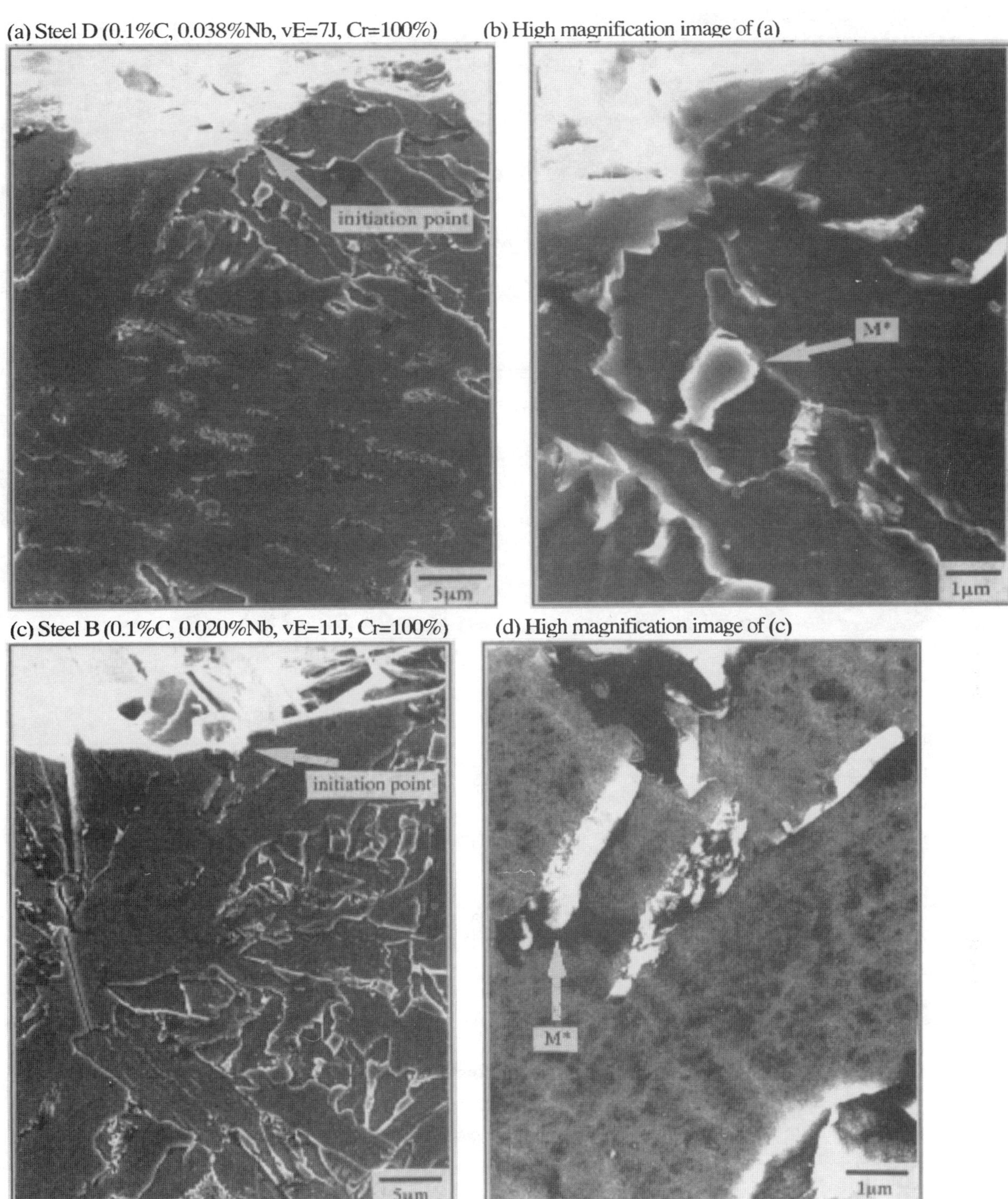

Fig.8. Appearance of the microstructure near the brittle fracture initiation site.
((a),(b):Steel D, (c),(d):Steel B, Test temperature of each specimen is 193K.)

(4) Quantitative Analysis of M*-Examples of observed M* (indicated by arrows) are shown in Fig.9 for the steels E, D, F and I. It appears that the content of M* markedly increases with increasing niobium content in both carbon content levels. And also in the steel D (0.1%C, 0.038%Nb), most of M*s exist in FSP or UB. Examples of M* area and aspect ratio histograms are shown in Fig.10 and Fig.11, respectively. These results correspond to the tendency of M* area fraction in optical micrographs.

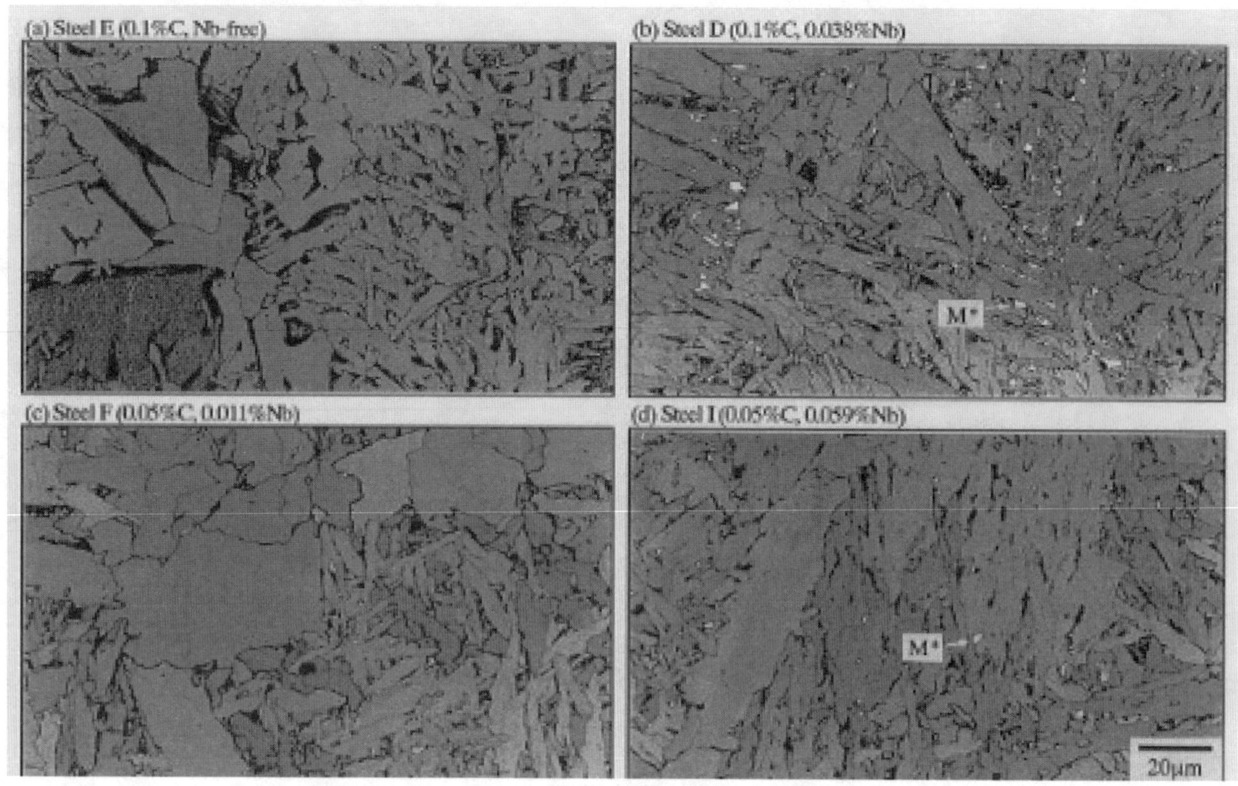

Fig.9. Appearance of M* with optical microstructure.

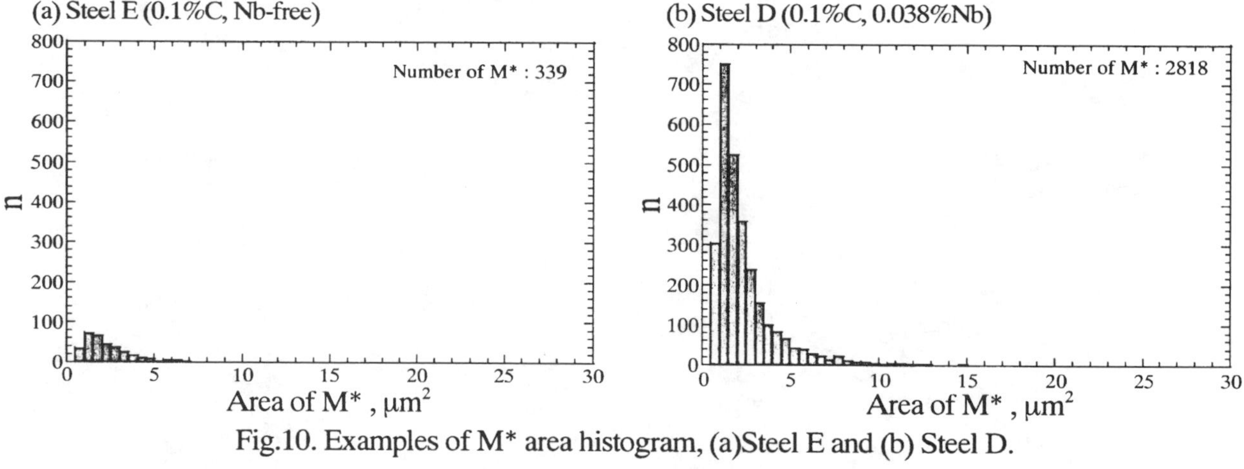

Fig.10. Examples of M* area histogram, (a)Steel E and (b) Steel D.

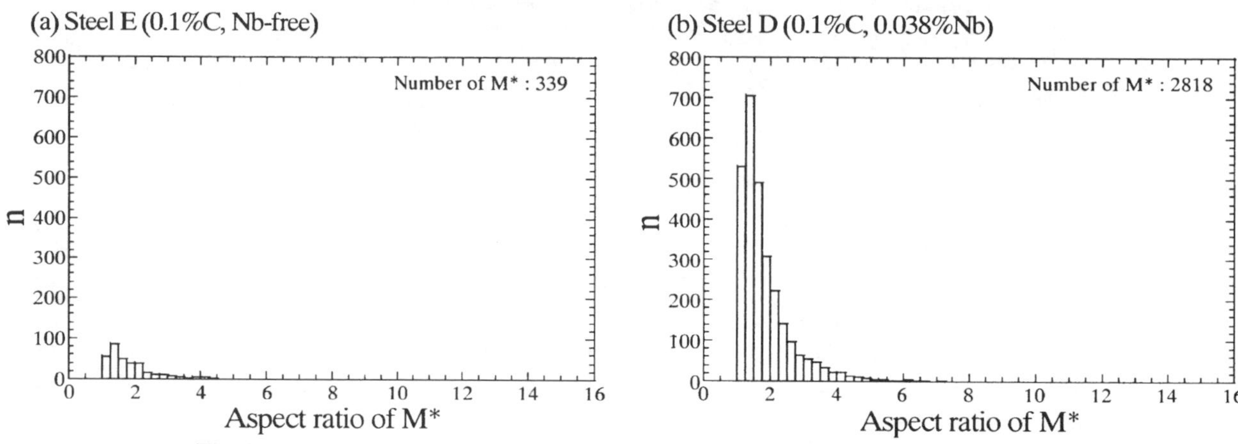

Fig.11. Examples of M* aspect ratio histogram, (a)Steel E and (b) Steel D.

Figure12 shows the change in total M* area fraction as a function of niobium content. The total M* area fraction increases with increasing niobium content for the both carbon content levels. But the rate of increase is not the same. For Group1 (0.1%C), M* increases steeply with increasing niobium content, while the increase of M* area fraction is saturated for Group2 (0.05%C) around 0.04mass% niobium content.

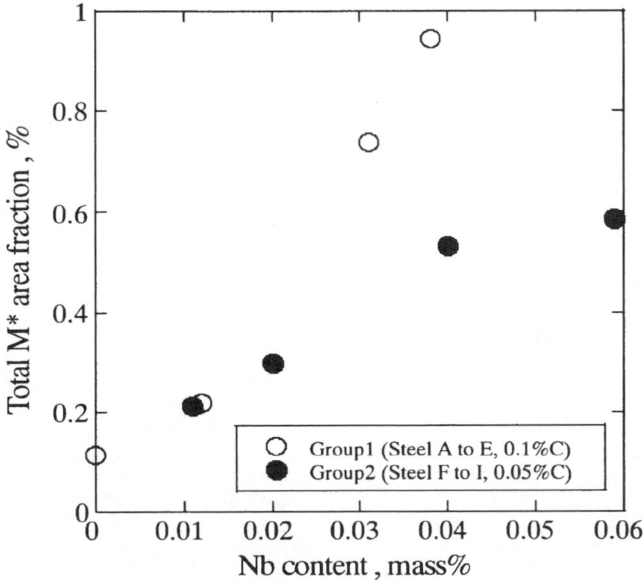

Fig.12. Relationship between total M* area fraction and Nb content.

Figure13 and Fig.14 shows maximum value, average value and minimum value of M* area and M* aspect ratio, respectively, as a function of niobium content. For the both figures, the maximum value is defined as the average of the upper 10 M* particles, and the minimum value is defined as the average of the lower 10 M* particles. In Group1 (0.1%C), only maximum value of M* area or aspect ratio increases with the increase in niobium content. In Group2 (0.05%C), the maximum value is also almost constant.

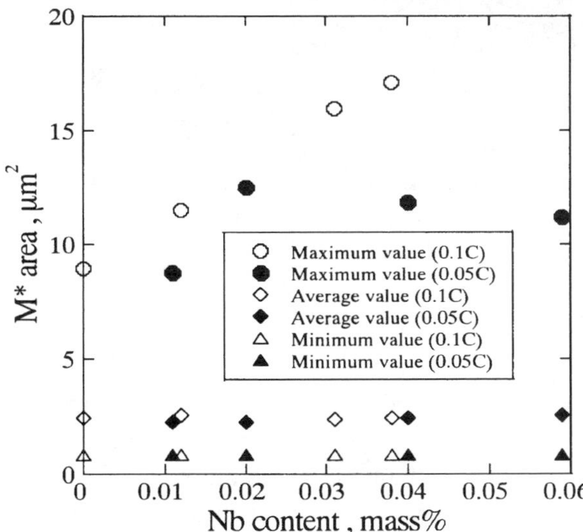

Fig.13. Relationship between M* area and Nb content.

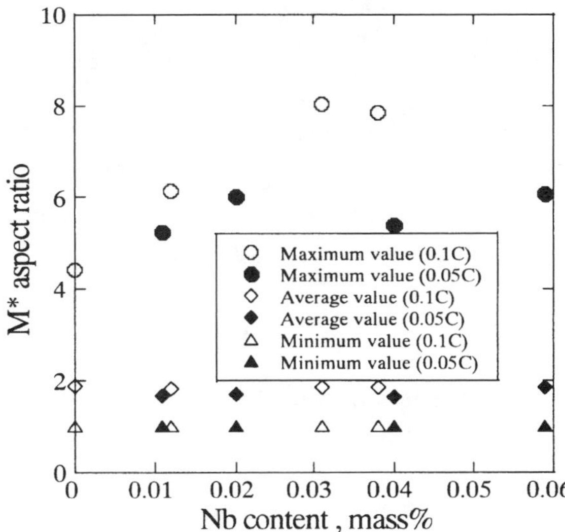

Fig.14. Relationship between M* aspect ratio and Nb content.

(5) Low speed bend test-Figure15 shows SEM images of the etched mid-section at the notch root region of the interrupted-unloaded slow bend specimen of steel D and I. Some microcracks were found to initiate at the interface between M* and ferrite matrix and others were found inside of the M*. In both cases, these microcracks were found to initiate in the region of the segregated M*s. Furthermore, the microcracks were initiated from the M* with relatively large size or aspect ratio.

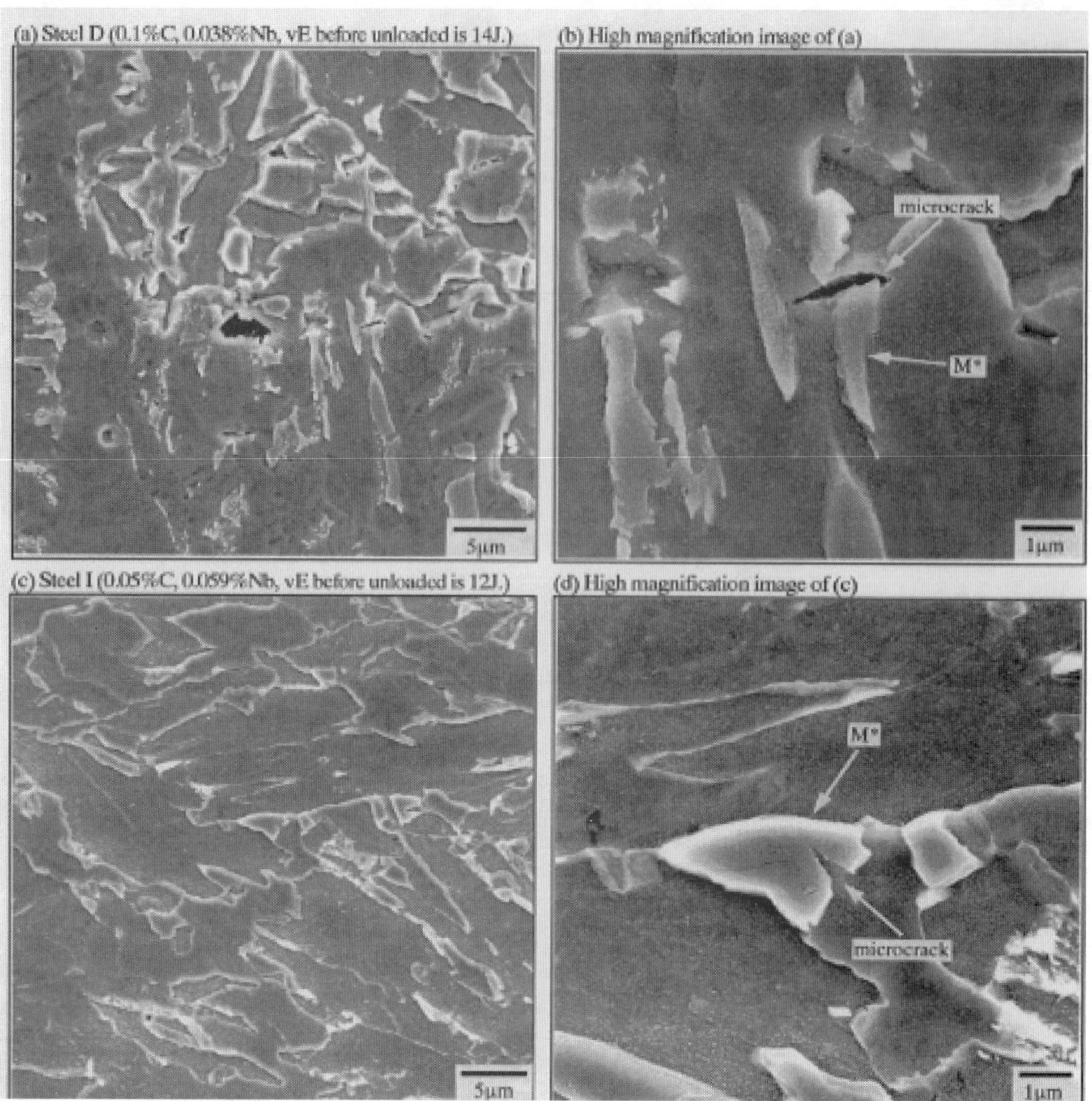

Fig.15. Microcrack initiation appearance at M* of (a),(b)Steel D and (c),(d)Steel I. (Test temperature is 233K.)

DISCUSSION

(1) Effect of carbon and niobium content on HAZ toughness-From the relationship between vTrs and niobium content shown in Fig.4, the effect of carbon and niobium content on HAZ toughness under large heat input (20kJ/mm) became clear. In the case of 0.1mass% carbon, the increase of niobium content led to the marked increase of vTrs, while it resulted in the gradual increase of vTrs and the saturation above 0.04mass% niobium in the case of 0.05mass% carbon. In general, increase of carbon content tended to increase vTrs. Difference of vTrs at the same niobium content tended to be larger with increasing niobium content.

On the other hand, it was clarified from the various kinds of observation that M* has the important and direct influence on the brittle fracture initiation. Hence, relation between HAZ toughness and M* was sought. Fig.12 shows variation of M* area fraction as a function of niobium and carbon content. The overall trend is very similar to the relationship between vTrs and niobium and carbon content. By the observation of the microstructure near the brittle fracture initiation site of Charpy impact test specimens and the sectioned plane of low speed bend test specimens, the localized and high number density of M* was found to have strong influence on the initiation of brittle fracture.

To examine the heterogeneity of the M* distribution, area fraction of M* was measured in the localized areas of 50x50μm and the maximum value was selected. Fig.16 shows the maximum value of local M* area fraction (f_{M*max}) as a function of niobium content. The tendency of the f_{M*max} is different between Group1 (0.1%C) and Group2 (0.05%C). In Group1, f_{M*max} steeply increases and amounts to 9% at 0.04mass% niobium content, while the increase is less marked in Group2. In other words, the localization of M* occurs extremely in higher carbon steels.

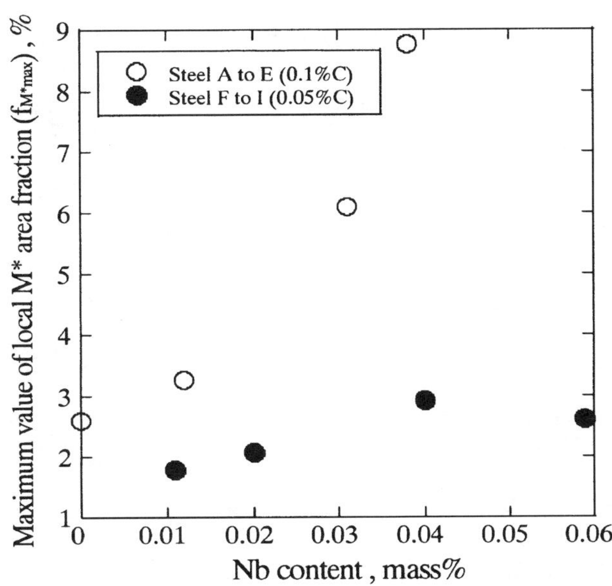

Fig.16. Relationship between maximum value of local M* area fraction (f_{M*max}) and Nb content.

When the niobium content is around 0.01mass%, vTrs increases with increasing carbon content (Fig.4). Since the total M* area fraction is almost the same (Fig.12), the effect of M* area fraction on HAZ toughness seems to be little. But f_{M*max} increases with the increase of carbon content (Fig.16). Therefore, the f_{M*max} is considered to be one of the major factors of deterioration in HAZ toughness in addition to hardness. At 0.04% niobium content, rapid increase in f_{M*max} corresponds to the larger difference of vTrs between 0.1% carbon and 0.05% carbon steels.

Figures 17 to 20 shows the relationships between vTrs and the various parameters related to the M*. Each figure includes the hardness data on each plot. According to Fig.17, although the increase of the total M* area fraction leads to deterioration in HAZ toughness, the tendency is divided by the carbon content. In addition, Steel I and Steel C have the almost same hardness and the average M* area fraction, but they reveal much different HAZ toughness. Therefore, the total M* area fraction is not expected to be an important factor on HAZ toughness. Fig.18 reveals an excellent correlation between vTrs and f_{M*max}. Regardless of the carbon content, the relation is represented as a single curve. This curve also includes the effect of hardness. Since the solid line consists of the steels of almost the same hardness (around 190~200Hv), the effect of M* localization is much large. The dotted line consists of the steels whose hardness varies significantly with niobium content. In these data, change in the f_{M*max} is relatively small (around 2~3%). So the effect of f_{M*max} is not so large, but the effect of the hardness or the microstructure is considered to be large. According to Fig.19 and Fig.20, a certain effect of maximum M* area or aspect ratio is expected.

(2) Formation of high number density of M* region-The brittle region containing high number density of M* is produced in the case of high niobium and carbon content. When austenite is transformed into FA or GPF from austenite grain boundary, the untransformed austenite near the transformed FA and GPF is enriched in carbon because of lower solubility of carbon in ferrite. From equilibrium transformation diagram of Fe-Carbon system, although carbon content is the same at austenite/ferrite interface at same temperature, the area with relatively high carbon is larger in higher carbon steel. On the other hand, microsegregation of manganese, niobium etc. inevitably takes place at solidification stage. Microsegregation of these elements is enhanced at higher carbon and niobium steel. The superimposition of these two effects is expected to produce higher hardenability at localized region and result in the high number density of M* (local concentration of M*) especially in high carbon and high niobium steels. Therefore,

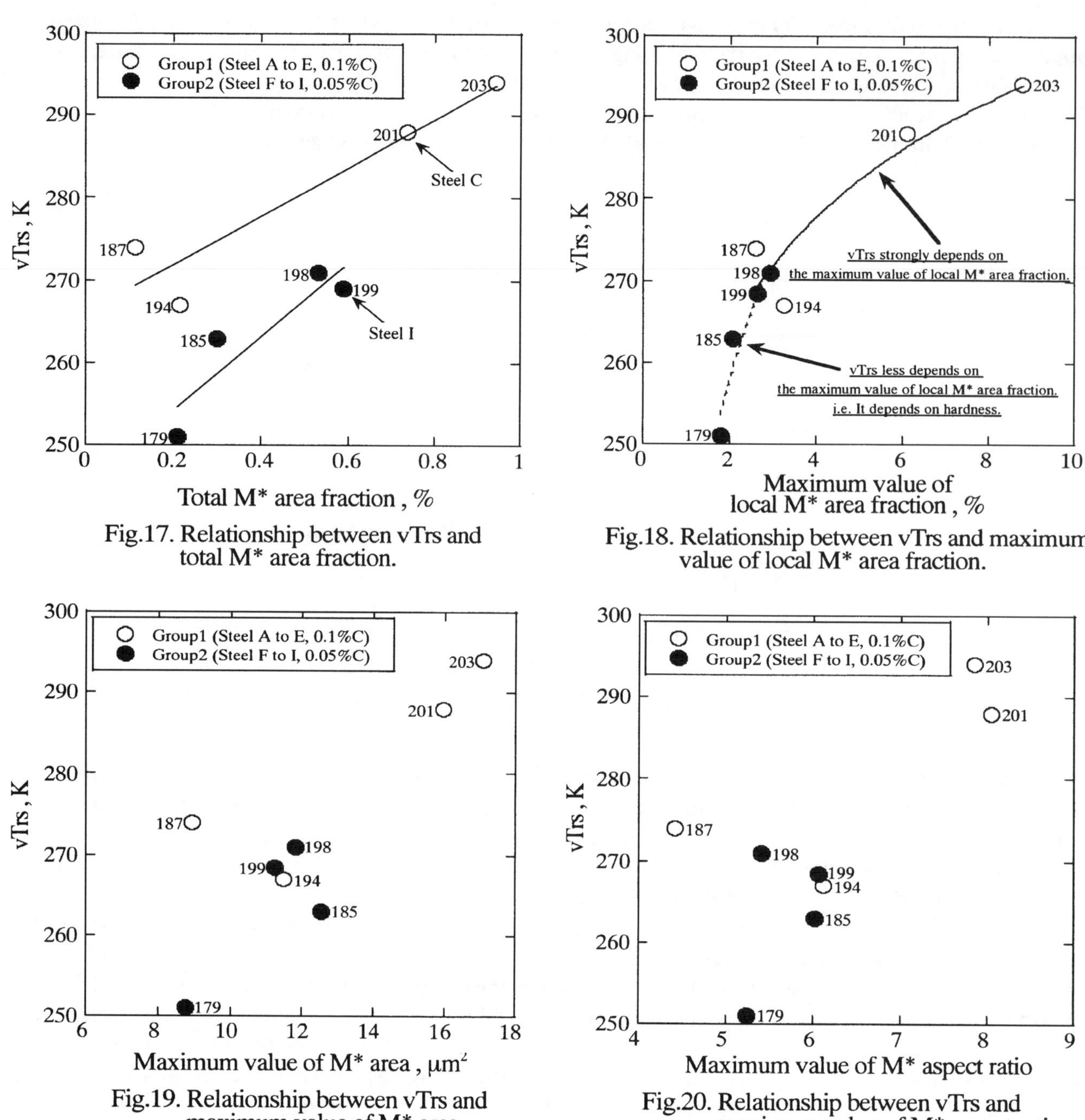

Fig.17. Relationship between vTrs and total M* area fraction.

Fig.18. Relationship between vTrs and maximum value of local M* area fraction.

Fig.19. Relationship between vTrs and maximum value of M* area.

Fig.20. Relationship between vTrs and maximum value of M* aspect ratio.

FSP or UB neighboring FA and GPF has much high number density of M*. Schematic illustration of this mechanism is shown in Fig.21. It should be noted that f_{M^*max} increased with niobium especially at higher carbon steel. This can be explained by the fact that manganese and niobium are enriched more significantly at microsegregation region in higher carbon steel.

(3) Relationship between the region of high number density of M* and brittle fracture-As shown in Fig.15, microcracks were observed in the region of high number density of M*. Furthermore, these regions were found to correspond to FSP or UB. These microcracks are initiated at the interface between M* and matrix or at the interior of M*. The microcrack initiation may be attributed to the stress concentration in such an area. When the steel is deformed, the plastic deformation of FSP is restricted because of higher hardness of FSP than FA and GPF; the stress in FSP may be increased. Inside of the FSP, heterogeneity of strength between M* and matrix leads to further stress concentration in M* and in the highly strained region near the interface. These effects may be superimposed to promote stress concentration at M* leading to the microcrack initiation at the interface or interior of the M*. Consequently, the region of high number density of M* is expected to promote the brittle fracture initiation.

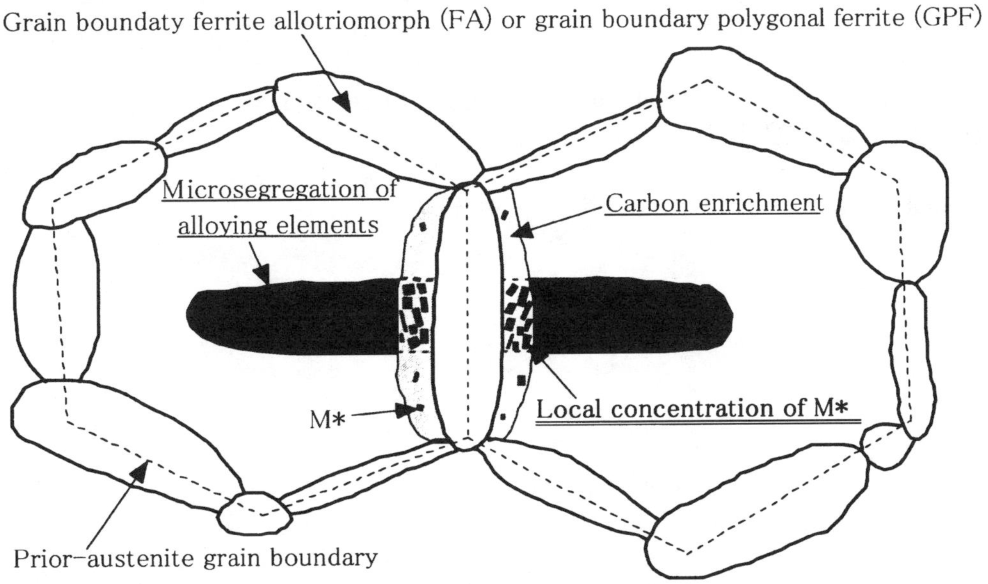

Fig.21. Schematic illustration appearing formation of local concentration of M*.

CONCLUSION

(1) Simulated HAZ toughness of HT590 deteriorates with increasing carbon and niobium content. Especially, the effect of niobium becomes larger in higher carbon steel.

(2) Brittle fracture was found to initiate from the localized region of concentrated martensitic island (M*). The local concentration of M* has the most important effect on HAZ toughness. Since the local concentration of M* is enhanced with increasing niobium content in higher carbon steels, HAZ toughness markedly deteriorates. On the other hand in lower carbon steels, the local concentration of M* is not enhanced by niobium significantly. Therefore the cause of HAZ toughness degradation in lower carbon steels is not so much local concentration of M* as hardness.

(3) Local concentration of M* is formed because of the superimposition of carbon enrichment in untransformed austenite near grain boundary ferrite allotriomorph (FA) or grain boundary polygonal ferrite (GPF) and of microsegregation of alloying elements such as manganese and niobium. In higher carbon and niobium steels, carbon enriched region in austenite is larger and also microsegregation of manganese and niobium is more promoted. Consequently, local concentration of M* is more frequently formed resulting in lower HAZ toughness.

(4) When the steel containing FSP with high number density of M* is deformed, the plastic deformation of FSP is restricted because of its higher hardness. Therefore, the stress in FSP may be increased. Within the FSP, heterogeneity of strength between M* and matrix leads to further stress concentration in M*. These two effects may be superimposed to promote stress concentration at M* leading to microcrack initiation at the interface or interior of the M*. Consequently, the local concentration of M* is expected to promote the brittle fracture initiation.

REFERENCES

1. H.Mimura, M.Iino, H.Haga, N.Nomura, Kazuo Aoki and Koichi Aoki, "On the Toughness and Micro-Structure in Low Carbon Steels Subjected to Weld Thermal Cycles," Transactions of Japan Welding Society, April, 1970, pp.28-34.

2. S.Aihara and T.Haze, "Influence of High-Carbon Martensitic Island on Crack-Tip Opening Displacement Value of

Weld Heat-Affected Zone in HSLA Steels," TMS Annual Meeting, 1988.

3. F.Matsuda, Z.Li, P.Bernasovsky, K.Ishihara and H.Okada, "An investigation on the behaviour of the M-A constituent in simulated HAZ of HSLA steels," Welding in the World, vol.29, No.9/10, 1991, pp.24-30.

4. F.Minami, H.Jing, M.Toyoda, F.Kawabata and K.Amano, "Effect of Local Hard Zone on Fracture Initiation of Weld HAZ," IIW Doc.X-1254-92, 1992, pp.23-34.

5. F.S.LePera, Journal of Metals, vol.32, No.3, 1980, pp.38-39.

Influence of Batch Annealing Soak Temperature on the Structure and Properties of Fully Stabilized and High Strength, Fully Stabilized Steels

D. P. Hoydick
U. S. Steel
4000 Tech Center Drive
Monroeville, PA 15146
USA
Tel.: 412-825-2919

Key Words: Fully Stabilized Steels, High Strength Fully Stabilized Steels, Batch Annealing, Mechanical Properties, Abnormal Grain Growth, r_m-value, n-value

INTRODUCTION

Over the past several decades, research on the physical metallurgy of fully stabilized (i.e. interstitially free) steels has focussed on the influence of composition and processing variables on the resultant mechanical properties and formability parameters [1-7]. Studies abound on the influence of stabilization mechanism, precipitation behavior, hot and cold mill processing, and annealing parameters on the structure and properties of these materials. In terms of the influence of annealing parameters, much of the work has involved the influence of continuous annealing variables on resultant properties. Generally, for continuously annealed fully stabilized steels, an improvement in r_m-value is observed along with a small change in n-value as the annealing temperature is increased up until the practical temperature limit of 1600°F. For conventional processing, a r_m-value of about 2.0 is observed. As a large portion of the commercial market exists for continuously annealed FS steel, this emphasis on continuously annealed FS material is not surprising. However, a substantial market exists for batch annealed fully stabilized steel product. A review of the literature shows little information with respect to the influence of batch annealing (BA) cycle on the properties of fully stabilized steel.

One research group, however, did investigate the influence of soaking time and temperature, although not quite in the batch annealing regime, on the r_m-values of Ti-stabilized FS steel [8]. In this study, substantial r_m-value improvements were observed for samples annealed at the highest temperatures and times. For example, a r_m-value of 2.67 was observed for a Ti-stabilized FS steel annealed for 30 minutes at 900°C (1650°F). As this cycle is not representative of what is typically done commercially for batch annealing, it was of interest to determine if these types of r_m-value improvements could be attained within commercial BA limitations, i.e. utilizing BA soak temperatures in the 1350 to 1500°F range. In addition to r_m-value, changes in n-value as a function of BA cycles were also of interest from a formability improvement standpoint. From a metallurgical perspective, as batch annealing temperatures increase, structural stability becomes an issue due to the threat of abnormal grain growth and must be considered if higher temperature batch annealing cycles are to be an option.

Even less information exists on the influence of batch annealing soak temperature on the resultant structure and properties of high strength, fully stabilized (HSFS) steels. These steels rely on the addition of strengthening elements such as P, Mn, and B to produce fully stabilized steels with elevated strength levels [9-11]. Once again, as in the FS steel case, much of the recent study has focussed on continuous annealing processing. It was also of interest to determine the properties of these steels through batch annealing as a function of batch annealing soak temperature.

In this study, the influence of batch annealing soak temperature on the microstructures and mechanical properties of both fully stabilized (FS) and high-strength, fully stabilized (HSFS) steels was investigated. From a practical standpoint, the study was undertaken because of the addition of hydrogen annealing facilities at U. S. Steel Gary Works. These bases have higher annealing temperature capabilities than the bases previously available and should allow for BA-cycle temperatures to reach soak temperatures in the vicinity of 1500°F.

EXPERIMENTAL PROCEDURE

In order to progress with the annealing study, samples of several conventionally processed FS and HSFS grades of steel were procured from Gary Works after cold rolling, i.e. in the full hard state. The following grades were obtained: a Ti FS grade (steel #1), a Ti/Nb FS grade (steel #2), a Ti/Nb HSFS grade with a P addition (steel #3), a Ti/Nb HSFS grade with P and B additions (steel #4), and a Ti/Nb HSFS steel with Mn, P, and B additions (steel #5). The composition of the materials used in the investigation is shown in Table I. Each steel was

Table I. Compositions of steels used in the investigation (in wt. %)

Element	Steel #1	Steel #2	Steel #3	Steel #4	Steel #5
C	0.0026	0.0034	0.0044	0.0047	0.0046
Mn	0.17	0.16	0.2	0.3	1.25
P	0.009	0.009	0.035	0.044	0.11
S	0.0081	0.0074	0.0078	0.0083	0.0065
Ti	0.069	0.03	0.039	0.021	0.065
N	0.003	0.0033	0.0038	0.0022	0.004
B	<0.0002	<0.0002	<0.0002	0.0009	0.001
Nb	<0.002	0.035	0.035	0.036	0.036

subsequently laboratory batch annealed using three different annealing cycles: one to simulate a conventional commercial cycle (1350°F soak) and two utilizing higher soak temperatures (1425 and 1500°F, respectively). The higher temperature cycles were designed to remain within the commercial capabilities of the hydrogen annealing facility. Computer simulations along with thermocouple production data were utilized in the design of the annealing cycles. The temperature-time profiles of the annealing cycles are shown in Figure 1.

RESULTS

Microstructural Analysis

After laboratory batch annealing, each of the materials was examined metallographically via optical microscopy. At the initiation of the study, one of the concerns of utilizing higher batch annealing temperatures was the possibility of abnormal grain growth leading to unstable microstructures and the production of both variable and undesirable properties. Thus, the microstructures of the steels were observed after annealing to investigate the stability of the structures with the higher annealing temperatures of particular interest. The microstructures are illustrated in Figure 2. Figures 2(a), 2(b), and 2(c) show the resultant ferrite structures for

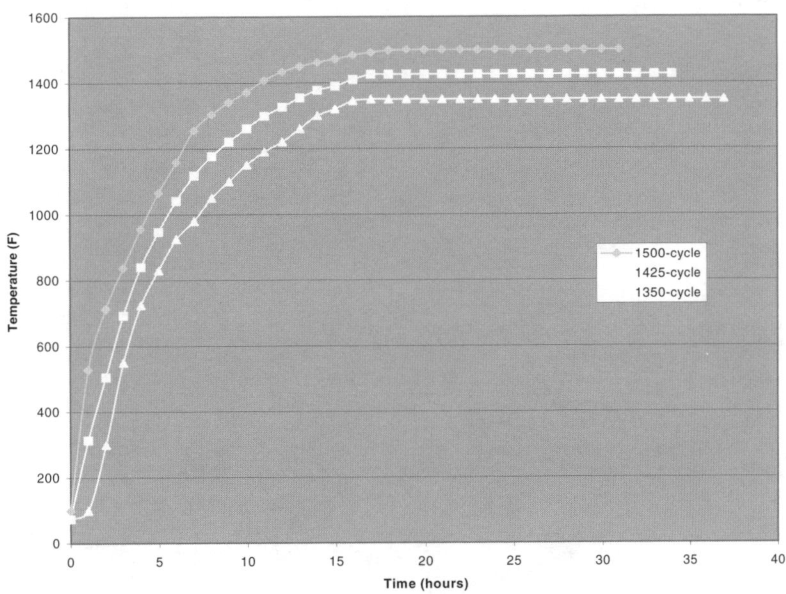

Figure 1. Batch annealing cycles utilized in investigation.

steel #1 after annealing using the 1350°F, 1425°F, and 1500°F cycles, respectively. The corresponding optical images for steels 2, 3, 4, and 5 through each cycle are illustrated in Figures 2(d) through 2(o). A summary of the microstructural observations are shown in Table II below. In terms of grain sizes for the individual steels, the results of the 1350°F cycle are consistent with commercial experience. For the higher temperature cycles, the grain size of steels 1 and 2 increase as the annealing temperature is increased. For example, the grain size of steel 1, the Ti-stabilized steel, goes from ASTM 8.0 (approximately 20 μm grain diameter) to ASTM 7.0 (28 μm) to ASTM 6.0 (40 μm) as the soak temperature increases from 1350 to 1425 to 1500°F. It should be noted that, in terms of structural stability, the grain structures for steels 1 and 2 remained equiaxed and homogeneous independent of the annealing temperature.

Table II. Summary of microstructural observations of as-batch annealed steels

Steel	1350°F	Cycle	1425°F	Cycle	1500°F	Cycle
	Grain Size	Comment	Grain Size	Comment	Grain Size	Comment
1	8.0	equiaxed, homogeneous	7.0	equiaxed, homogeneous	6.0	equiaxed, homogeneous
2	8.5	equiaxed, homogeneous	8.0	equiaxed, homogeneous	7.5	equiaxed, homogeneous
3	9.5	equiaxed, homogeneous	9.5	equiaxed, homogeneous	NA	regions of abnormal grain growth
4	10.0	equiaxed, homogeneous	10.0	equiaxed, homogeneous	NA	regions of abnormal grain growth
5	10.5	bimodal grain distribution	10.5	bimodal grain distribution	10.5	equiaxed, homogeneous

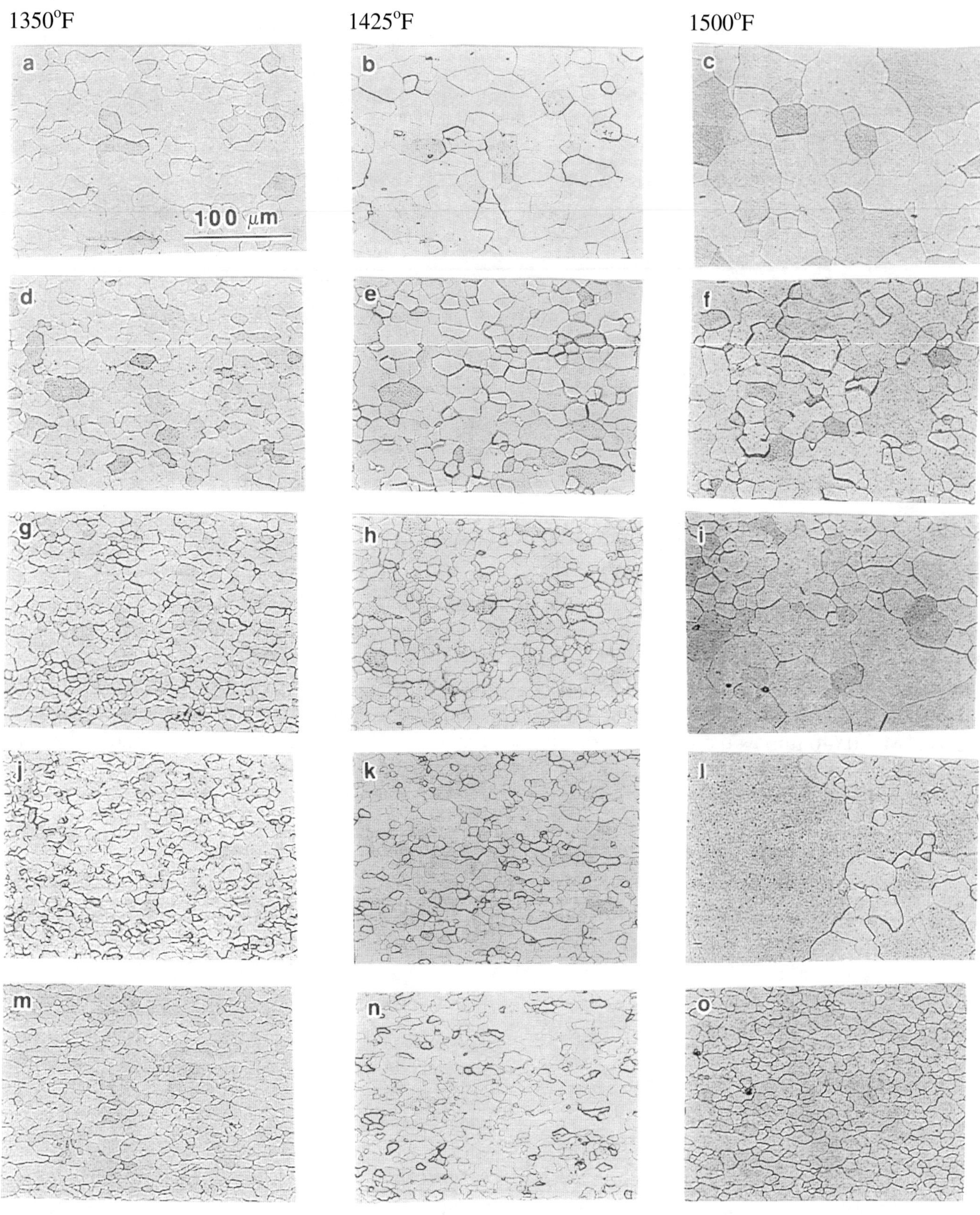

Figure 2. Microstructures of each steel through the 1350, 1425, and 1500°F batch annealing cycles: (a)-(c) steel 1, (d) – (f) steel 2, (g)-(i) steel 3, (j)-(l) steel 4, (m)-(o) steel 5.

For the HSFS steels, different behavior was observed. For steels 3 and 4, little grain size increase occurred when going from the 1350 to the 1425°F cycle. Additionally, abnormal grain growth occurred in steels 3 and 4 for the 1500°F cycle, making the commercial application of this cycle to these two steel grades impractical. For steel 5, the resultant grain sizes were rather independent of the annealing soak temperature. There was a difference in grain morphology, however, as a more homogenous grain structure was observed at the 1500°F annealing temperature as compared to the 1350 and 1425°F cycles.

Mechanical Properties

Mechanical testing was performed on each of the steels annealed at the 1350 and 1500°F isotherms, the results of which are shown in Table III. For each steel, the table shows the as-annealed mechanical properties, i.e. no temper rolling was undertaken. Keep in mind that the results for steels 3 and 4 for the 1500°F cycle represents testing on material exhibiting abnormal grain growth and have very little practical significance. Generally, for all of the steels, as the annealing temperature was increased, a softening of the steels occurred (i.e. lower yield and UTS) as well as an increase in formability parameters (n-value and r-value). In terms of n-value (measured from 5% strain to strain at UTS), only a modest increase is observed for steels 1 and 2 when going from the lower temperature cycle to the higher temperature cycle. An increase of 0.004 and 0.002 is observed for grades steels 1 and 2, respectively. This is a rather insignificant increase, particularly if one considers the fact that a slight softening of the steel also occurs. Thus, in terms of n-value improvement, going to a higher temperature batch annealing cycle does not appear to be advantageous for steels 1 and 2. A much more significant increase in n-value is observed for steels 3, 4, and 5. As discussed previously, the abnormal grain growth present at the higher annealing cycle for steels 3 and 4 precludes the utility of this property improvement for these grades. For grade 5, the n-value improves from 0.225 to 0.248 when going from a conventional cycle to the higher temperature cycle.

Table III. Mechanical properties of the steels annealed at the 1350 and 1500°F isotherms

Steel	Anneal Cycle	Mechanical Properties				
		Yield Strength (MPa)	UTS (MPa)	TE	N-value (5/U)	r-value (long)
1	1350°F	88.2	280	50.3	0.291	2.14
	1500°F	77.9	268	50.2	0.295	2.50
2	1350°F	104	300	46.6	0.280	1.87
	1500°F	95.8	294	49.5	0.282	1.92
3	1350°F	129	334	42.9	0.265	1.61
	1500°F	107	310	38.5	0.283	2.17
4	1350°F	158	351	41.2	0.263	1.09
	1500°F	127	305	24.2	0.280	1.09
5	1350°F	291	434	34.6	0.225	1.33
	1500°F	263	406	33.1	0.248	1.39

In terms of r-value (longitudinal), the r-value was found to increase with annealing temperature in all of the steels except for steel 4, where the r-value remained unchanged. For steels 2 and 5, only a slight increase in r-value was observed whereas for steels 1 and 3, a significant increase in r-value was observed. Once again, the higher r-value for steel 3 at the higher temperature cycle has no practical significance due to the abnormal grain growth phenomenon. For steel 1, the r-value (longitudinal) increased from 2.14 to 2.50 when going from the 1350°F cycle to the 1500°F cycle. Because of this significant increase, L, T, and D tensile samples of this steel annealed at 1500°F were prepared so that R_m could be determined. R_m measurements taken on the steel resulted in a value of 2.8, significantly higher than the value of about 2.0 generally observed for Ti-stabilized FS steel.

DISCUSSION

Batch Annealing Behavior and Structural Stability

The results of the microstructural investigation as a function of batch annealing soak temperature indicate that the degree of structural stability and grain coarsening behavior may be different between FS and HSFS type materials and thus highly dependant upon alloying additions. The grain sizes of the FS grades, steels 1 and 2, increase gradually with annealing temperature in the 1350 to 1500°F range and remain equiaxed and homogeneous regardless of batch annealing soak temperature. Thus, normal grain growth behavior appears to be occurring for the FS steels. The HSFS grades, however, do not appear to exhibit the normal grain growth behavior one might expect upon raising the BA soak temperature from 1350 to 1425°F as the grain size of these grades remain virtually the same for these two annealing cycles. Furthermore, two of the HSFS grades, steels 3 and 4, exhibit abnormal grain growth at the 1500°F cycle. As both normal and abnormal grain growth are complex phenomena that can be affected by a number of metallurgical factors such as solute content, precipitation behavior, and texture effects [12], a detailed understanding of the annealing behavioral differences between the different grades was not sought after here. A comparison of the compositions of the HSFS with the FS grades, however, show P (steel 3), P+B (steel 4), and P+B+Mn additions to the HSFS grades. Thus, as little grain coarsening occurred in these steels when going from the 1350 to 1425°F cycle, the presence of these alloying elements, either in solute and/or precipitate form, was able to exhibit a pinning effect on ferrite grain boundaries in some manner as to not allow grain growth. As P and B have been shown to create solute drag effects during recrystallization, it is not surprising that these elements could be influential in grain coarsening behavior, also [9]. Additionally, P has been shown to potentially exist as FeTiP particles in these types of steels which could also influence annealing behavior [13]. Last, but not least, B and P have been shown to alter final product r-values and thus texture, which could certainly influence the propensity to induce abnormal grain coarsening [14]. Whatever the mechanism, the pinning effect lessens at the 1500°F isotherm for steels 3 and 4, making abnormal grain growth a reality for these steels. The pinning effect remains for steel 5, however, probably due to the high level of alloying in this steel. A detailed study involving precipitate behavior in these steels would be necessary to provide a clearer fundamental understanding of the grain growth observations seen here.

Mechanical Properties

For low carbon steels, the improvement of r-value with grain size increase during annealing has been pretty well established. For Ti-stabilized IF steels, Osawa has shown that a similar phenomenon should be expected [8]. This present work agrees with the observation that r-value improvement scales with grain size in Ti-only stabilized steels. For examples, an exceptional r_m-value of 2.8 was achieved for a steel exhibiting a grain size of about 40 μm that was produced utilizing high temperature batch annealing. Keep in mind that no other composition and/or processing optimization was done in this study and that further r_m improvements may be a possibility using this route. This r_m value is on par with r_m-values presently available only via specialized processing such as lubricated ferrite hot rolling [15,16]. In other words, a low cost alternative to ultra-high r_m-values in Ti-only FS steel may be via higher temperature batch annealing as demonstrated in this study. One thing to consider, though, is that the steels exhibiting the very high r_m-values in this work also exhibit grain sizes on the order of ASTM 6.0, getting into the realm where surface irregularities or "orange peeling" may become a factor. No surface irregularities were observed on the high r_m-value steels during tensile testing in this study, however.

The other steels showed no significant change in r-value (longitudinal) as a function of annealing temperature from a practical standpoint. A slight increase in r-longitudinal was observed for the Ti-Nb FS steel (steel 2). Whether or not this small r_m-value increment is a result of primarily a grain size effect would require further study, however, to determine if a similar relationship between r-value and grain size exists in the Ti-Nb FS case as well as the Ti-only FS case. For the HSFS steels 3 and 4, no conclusions can be drawn with respect to the relationship between r-value and grain sizes as the different annealing processing resulted in abnormal grain coarsening at the highest annealing temperature. For steel 5, no grain size variation was observed between the 1350 and 1500°F annealing cycles, and thus, once again, no r-value/grain size relationships could be determined.

For the case of n-values, Takechi [17] has shown that n-values of FS steels tend to be relatively independent of grain size. His experimental work shows n-values to range from 0.28 to 0.30 over a range of grain size from 20 to 60 μm. The present work agrees with this conclusion as n-values of about 0.295 are found for both the ASTM 8.0 and ASTM 6.0 grain sizes found for the 1350 and 1500°F annealing cycles, respectively. The present work also suggests that the same situation may also be true for Ti-Nb as well as Ti-only stabilized steels. The results on steel 2 indicate similar n-values for samples annealed at both the 1350 and 1500°F cycles. For steels 3 and 4, little grain size differences exist for the samples measured and thus no conclusions can be drawn as to the relationship between n-value and grain size in these HSFS steels. For steel 5, however, a significant rise in n-value from 0.225 to 0.248 is observed when going from the 1350 to 1500°F annealing temperature even though the grain sizes are about the same. A different grain morphology is found, however, indicating that a different annealing mechanism may have become active at the higher annealing temperature for this steel that results in an improved strain hardening behavior.

CONCLUSIONS

In this study, laboratory batch annealing (BA) simulations using both a conventional BA cycle (1350°F soak) and higher temperature BA cycles (1425 and 1500°F soaks) were performed on both fully stabilized (FS) and high-strength, fully stabilized steels (HSFS) in an effort to determine if the formability of the steels could be improved with higher temperature batch annealing. The results of the investigation reveal the following:

1) Microstructural analysis revealed homogeneous, equiaxed grain structures for all grades undergoing the 1350°F cycle with grain sizes in general agreement with commercial experience. For samples undergoing the 1425°F cycle, slightly larger equiaxed grain structures were observed for FS grades whereas the HSFS steel grain sizes remained similar as for the 1350°F cycle. For the 1500°F cycle, the grain structures of the FS steels remained equiaxed and homogeneous albeit larger in size. However, two of the HSFS grades, steels 3 and 4, underwent abnormal grain growth when put through this cycle, making the practical application of this cycle to these steels unfeasible. These differences in annealing and grain coarsening behavior between the FS and HSFS grades no doubt have to do with the influence of alloying additions such as P, B, and Mn on annealing response and grain boundary mobility.
2) Mechanical testing revealed n-value improvements upon going from the conventional cycle (1350°F soak) to the higher temperature cycle (1500°F soak) to be nominal in a practical sense for all grades except for steel 5. For this grade, an HSFS steel with a high degree of P, B, and Mn alloying, the n-value jumped from 0.225 to 0.248.
3) The results of r-value (longitudinal) testing revealed little r-value improvement for the majority of the grades. For steel 1, the Ti-stabilized FS steel, however, an r_m-value of 2.8 was achieved, significantly higher than the value of about 2.0 expected from conventional processing.

ACKNOWLEDGEMENTS

The author wishes to thank J. H. Gallenstein and U. S. Steel for permission to publish this work. Additionally, Brandon Hance should be acknowledged for helpful discussions. Also, Rob Ives and Eric Pohlmann of Gary Works are acknowledged for their assistance in acquiring the samples and the thermocouple test data used in this study.

REFERENCES

1. H. Takechi, "Metallurgical Aspects on Interstitial Free Sheet Sheet from Industrial Viewpoints", ISIJ International, Volume 34, No. 1, pp.1-8.

2. H. Takechi, "Research on Metallurgical Behavior and Application of Modern LC and ULC Steels in Japan", Institute of Ferrous Met. Auchen, 1998, pp. 133-144.

3. T. Asamura, "Recent Developments of Modern LC and ULC Sheet Steels in Japan", Institute of Ferrous Met. Auchen, 1998, pp. 1-14.

4. I. Gupta, T. Parayil, and L-T. Shiang, "Effect of Processing Parameters on the Properties of Cold Rolled Interstitial Free Steels", Hot and Cold Rolled Sheet Steels, TMS, 1988, pp. 139-153.

5. M. Hua, C. I. Garcia, and A. J. DeArdo, "Precipitation Behavior in Ultra-low Carbon Steels Containing Titanium and Niobium", Metallurgical and Materials Transactions A, Volume 28A, pp. 1769-1780.

6. I. Gupta and D. Battacharya, "Metallurgy of Formable Vacuum Degassed Interstitial-Free Steels", Metallurgy of Vacuum-Degassed Steel Products, TMS, 1990, pp. 43-71.

7. G. Krauss, D.O. Wilshynsky, and D.K. Matlock, "Processing and Properties of Interstitial Free Steels", Interstitial Free Steel Sheet: Processing, Fabrication, and Properties, Canadian Institute of Mining, Metallurgy and Petroleum, 1991, pp. 1-15.

8. K. Osawa, S. Satoh, T. Obara, H. Abe, and K. Tsunoyama, "Recrystalization Behavior of Extra-Low C Cold-rolled Sheet Steels During Continuous Annealing", Metallurgy of Vacuum Degassed Steel Products, TMS, 1990, pp, 181-195.

9. D. P. Hoydick and T.M. Osman, "Influence of the Phosphorus Content on the Recrystallization Behavior and Mechanical Properties of Boron-bearing Ti-Nb Fully Stabilized Steels", 40th MWSP Proceedings, 1998, pp. 195-204.

10. S. Luo, B. Liu, D. Zhong, X. Wang, Yl Kang, and S. Jiao, "Effect of Composition and Processing Parameters on Microstructure and Formability of P Bearing IF Steels", Institute of Ferrous Met. Auchen, 1998, pp. 509-520.

11. H. Chen, and Y. Hwang, "Effects of Phosphorus and Boron on the Mechanical Properties of High-Strength IF Sheet Steels", China Steel Technical Report, No. 9, 1995, pp. 44-53.

12. R. E. Reed-Hill, Physical Metallurgy Principles, D. Van Nostrand Company, New York, 1973, pp. 298-321.

13. O. Homart and S. Lanteri, "Influence of Mn and P on Precipitation in High Strength IF-Ti Steels", 38th MWSP Proceedings, ISS, Volume 34, 1997, pp. 431-441.

14. C. Brun, P. Patou, and P. Parniere, "Influence of Phosphorus and Manganese on the Recrystallization Texture Development During Continuous Annealing in Ti-IF Sheets", Metallurgy of Continuously Annealed Sheet Steels, AIME, 1982, pp. 173-197.

15. S. Matsuoka, K. Sakata, D. Furukimi, and T. Obara, "Development of Super Deep Drawable Sheet Steels by Lubricant Hot Rolling in Ferrite Region", Institute of Ferrous Met. Auchen, 1998, pp. 85-96.

16. T. Senuma, "Physical Metallurgy for Producing Super Formable Deep Drawing Steel Sheets", Institute of Ferrous Met. Auchen, 1998, pp. 157-168.

17. H. Takechi, "Application of IF and IF Based Sheet Steels in Japan", IF Steels 2000 Proceedings, ISS, 2000, pp. 1-12.

Product Physical Metallurgy Sessions

Strengthening Mechanisms in Steel Products

Dedicated in Memory of Paul E. Repas

Paul Repas became an AIME member in 1961 while still a student at then Case Institute of Technology, and later was a founding member of the Product Physical Metallurgy Committee of the Mechanical Working and Steel Processing Division of ISS. Upon completing his Ph.D. thesis at Case in 1965, Paul joined U. S. Steel where he began his long association with steel development and the application of strengthening mechanisms to achieve desired product performance.

A great deal of the progress in the development and evolution of HSLA steel technology occurred during Paul's 33 year career at U. S. Steel, and he played an important role in that progress. He devoted a significant portion of his career to develop a thorough understanding of the strengthening mechanisms available in steel. His contributions included organizing seven conferences aimed at understanding the various strengthening mechanisms in steel. Paul was perhaps best known for his work on high-strength, low-alloy steels and their processing during hot rolling, on which he published extensively. Overall, he contributed over 20 papers on various strengthening topics, as varied as phase transformations in uranium alloys to dual phase steels to microalloyed plate steels. He was an excellent synthesizer of HSLA technology with demonstrated capability to creatively use it to develop useful new products.

It is with great honor that the Product Physical Metallurgy committee dedicates this year's programming in memory of Paul Repas.

On behalf of the Product Physical Metallurgy Committee:
Joseph Defilippi and Matthew Merwin
U.S. Steel Division of USX Corp.

PRODUCT PHYSICAL METALLURGY I

A Tribute to Paul E. Repas

Joseph D. Defilippi and Matthew J. Merwin
U. S. Steel Research and Technology Center
4000 Tech Center Drive
Monroeville, PA 15146
USA
(412) 825-2000 / (412) 825-2223 (fax)
jdefilippi@uss.com, mmerwin@uss.com

INTRODUCTION

The origin of high-strength low-alloy (HSLA) steels is generally attributed to Williams[1], who in 1900 called attention to the increased corrosion resistance of a copper containing steel when alternately wet and dried. However, it was not until 1933 when U. S. Steel introduced COR-TEN steel that the added bonus of high strength in combination with markedly improved atmospheric corrosion resistance was achieved. From these humble beginnings, the continued development and application HSLA steels evolved in part because of the increasing technology base associated with these steels as well as the constantly increasing sophistication in the equipment and techniques used to process them. A great deal of the progress occurred during Paul E. Repas' 33 year career at U. S. Steel, and he played an important role in that progress. In reviewing his works, it is clear that it is easy to achieve strength alone in steel, but it is a challenge to do so while simultaneously achieving a desirable combination of other properties. To help meet this challenge, Paul devoted a significant portion of his career to develop a thorough understanding of the strengthening mechanisms available in steel, as a necessary prelude to the development of new steel products. His contributions included organizing conferences aimed at understanding the various strengthening mechanisms in steel, preparing papers on various strengthening topics, and creatively using the technology to achieve useful new steel products. In this keynote paper, his role in the evolution of HSLA technology is reviewed, utilizing both his published and unpublished works.

THE CONFERENCES

As the recognized corporate expert for U. S. Steel on the physical metallurgy of hot-rolled products, Paul understood that he would not be able to achieve the level of understanding needed in the field through his individual work alone. He therefore embraced activities in conferences and professional societies, including being a charter member of the Product Physical Metallurgy (PPM) committee of the Mechanical Working and Steel Processing Division (MWSP) of Iron and Steel Society (ISS). He also helped to organize or edit several conferences and proceedings including: *HSLA Steels: Metallurgy and Applications,*[2] *Microalloyed HSLA Steels* (Microalloying '88),[3] *Low-Carbon Steels for the 90's,*[4] *Microalloying '95,*[5] and the *Gilbert R. Speich Symposium.*[6] Similar to this ISS tribute to Paul, the Gilbert R. Speich Symposium was a memorial one comprising the PPM programming of ISS's MWSP conference 1992. Each of these conferences contains works reflecting the research of interest at the time. Table I presents some of the focus areas for each of these conferences.

Table I. Conferences Paul Helped Organize.

Conference	Focus Areas
HSLA Steels: Metallurgy and Applications[2]	Physical Metallurgy Steelmaking Pipe and Tubular Metallurgy Plate Metallurgy and Weldability Dual Phase Steel End User Requirements Sheet Steels
Microalloyed HSLA Steels (Microalloying '88)[3]	Products for the Oil and Gas Industry Products for Offshore Installations Marine, Military and Pressure Vessels Automotive and Off-Highway Machinery, Engineering and Construction
Low-Carbon Steels for the 90's[4]	Steel Processing Precipitation and Computer Modeling Plate and Structural Steels Sheet Steels Interstitial-Free Steels Welding
Microalloying '95[5]	Line Pipe Steels Thermodynamics of Microalloy Precipitation and Microstructure
Gilbert R. Speich Symposium[6]	Tempering and Aging of Martensite and Bainite High Strength Commercial and Military Steels

In these conferences, the fundamentals of the physical metallurgy of HSLA and high strength steels were uncovered and discussed by international experts. Once understood, these fundamentals, such as the contributions of grain size, solid solutions, dislocation hardening, phase transformations, and precipitation to the performance of steels could then be applied to improve current products and develop new ones. Consequently, in providing the service of helping to organize the conferences and symposia, Paul also helped himself in gaining new information that could be applied in his development duties.

INDIVIDUAL WORK

However, one does not achieve the level of respect garnered by Paul simply by helping to organize symposia. One must also contribute to the available knowledge, which he did through authoring over twenty papers in the open literature. Paul participated in the landmark Microalloying '75 conference through contributing a paper on microalloyed plate steels for use in arctic-grade line pipe.[7] He continued to contribute publications until his final paper was published in 1996, again in the area of microalloyed steels, but this time on their application to hot-strip-mill products.[8]

Over his career Paul achieved a level of understanding of metallurgy of hot-rolled steels which allowed him to write overview articles describing the state of the art. One example of such a review paper is "Metallurgical

Fundamentals of HSLA Steels".[9] In this paper Paul illustrates how the various strengthening mechanisms are combined in HSLA steels using a modified Hall-Petch relationship as shown below.

$$\sigma = \sigma_0 + \sigma_{SSS} + \sigma_{ISS} + \sigma_{PPT} + \sigma_{DSL} + \sigma_{SUB} + \sigma_{SPH} + k_y d^{-1/2} \quad (1)$$

Where σ_0 is the intrinsic matrix hardening, the other sigma (σ) terms are the strengthening increments due to substitutional solid solution (SSS), interstitial solid solution (ISS), precipitation (PPT), dislocations (DSL), substructure (SUB), and second phase (SPH). K_y is of course the Hall–Petch constant, and d is ferrite grain size. The paper details the degree to which these mechanisms contribute to the strength of hot-rolled steels, and consequently impact other properties, such as ductility and toughness. It was particularly well demonstrated that increased strength comes at a price, such as in figures depicting the loss of ductility accompanying increased strength regardless of strengthening mechanism, and the decrease in toughness with increased strength through all mechanisms excepting grain refinement (see Figure 1).[9]

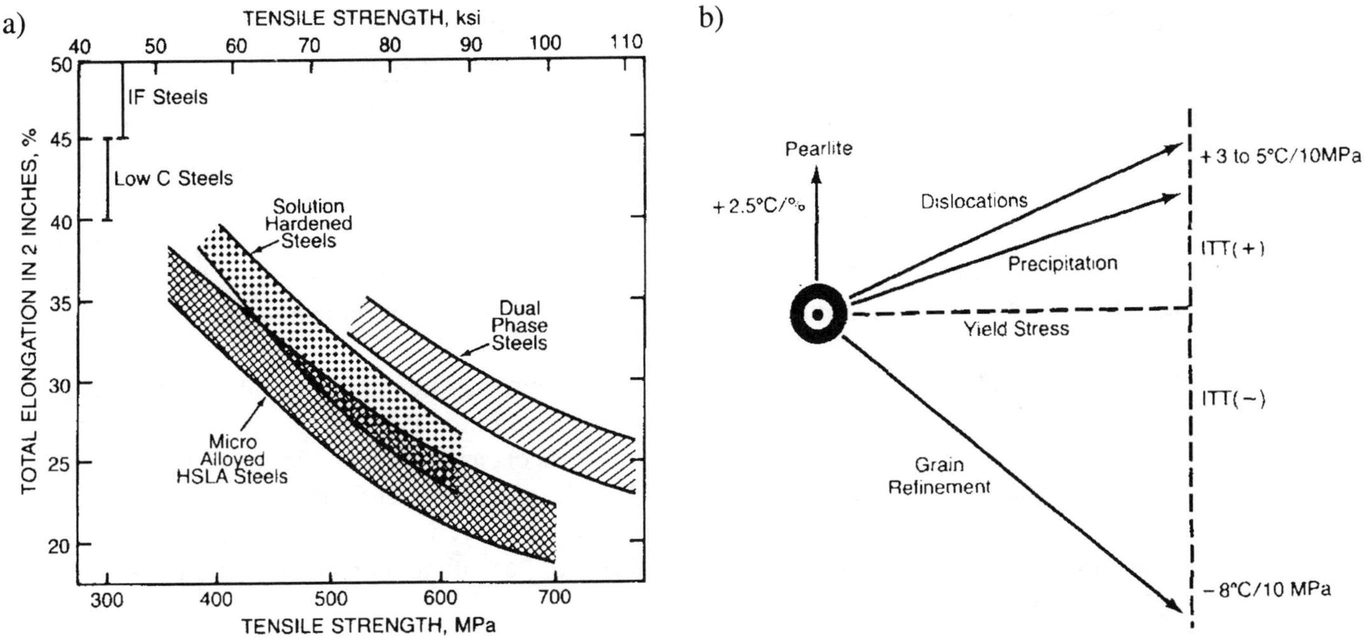

Figure 1. a) Demonstration of the effect of strength on ductility for a variety of strengthening mechanisms. b) Effect of strengthening mechanisms on the Charpy V-Notch impact transition temperature (after Pickering[10]).

Paul was particularly adept at applying his basic understanding of physical metallurgy phenomena to the real-world environment of commercial production. This quality was well demonstrated through the Gilbert R. Speich Lecture Paul presented to the Chicago chapter of ASM International, entitled "The Hot Strip Mill as Thermomechanical Processing Tool."[11] In a time when most work concerning thermomechanical processing was geared toward application in plate production, Paul reminded us that the hot strip mill was an ideal environment for TMCP with the controlled rolling temperatures of the finishing deformation and the interrupted-accelerated cooling accomplished on the runout table. The relative effects of the strengthening mechanisms were depicted in diagrams such as those presented in Figure 2, which illustrate how one can expect the mechanisms to contribute theoretically (Figure 2a), and their consequences in actual materials (Figure 2b).

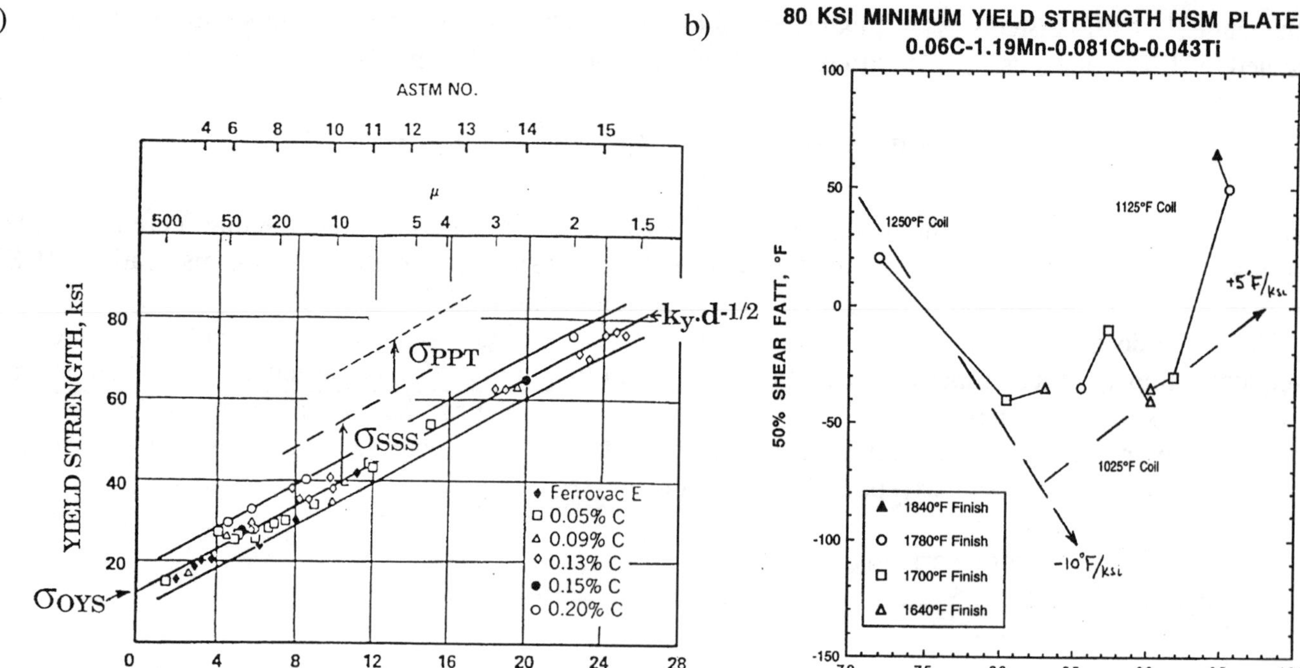

Figure 2. a) Magnitude of strengthening increment through grain refinement, solid solution strengthening, and precipitation strengthening. b) Application of Pickering-type vector notation for the effect of strengthening mechanisms on strength and toughness in a hot-strip-mill plate grade.

While the previous examples of papers presented by Paul demonstrate his knowledge of microalloyed HSLA steels, particularly with respect to their processing on hot-strip mills, Paul's expertise was not limited to that arena. Paul also was involved in examination of dual phase steels as the interest in these grades increased in the late '70's. In 1979, he wrote a paper for the ISS MWSP conference, entitled "Physical Metallurgy of Dual-Phase Steels."[12] In this paper, an overview of the technology involved in dual phase steel production was presented, as well as the results from work examining the effect of varying silicon content and cooling rate on the properties of continuously annealed steels.

He showed that dual-phase microstructures could be developed in a wide variety of steels by appropriate combination of steel composition and processing cycles. The challenge, particularly for as-hot-rolled sheet, was to select a composition with sufficient robustness to transform from austenite into the desired fine-grained polygonal ferrite plus a dispersion of martensite under relatively complex thermomechanical processing sequence. Generally, relatively rich alloys are required for processes that involve slow cooling rates such as for an as-hot-rolled coil or a box-annealed one. Relatively lean alloys can be used when cooling rates are high such as for accelerated-cooled or continuously annealed sheet. Figure 3 presents some of the results discussed in the paper.

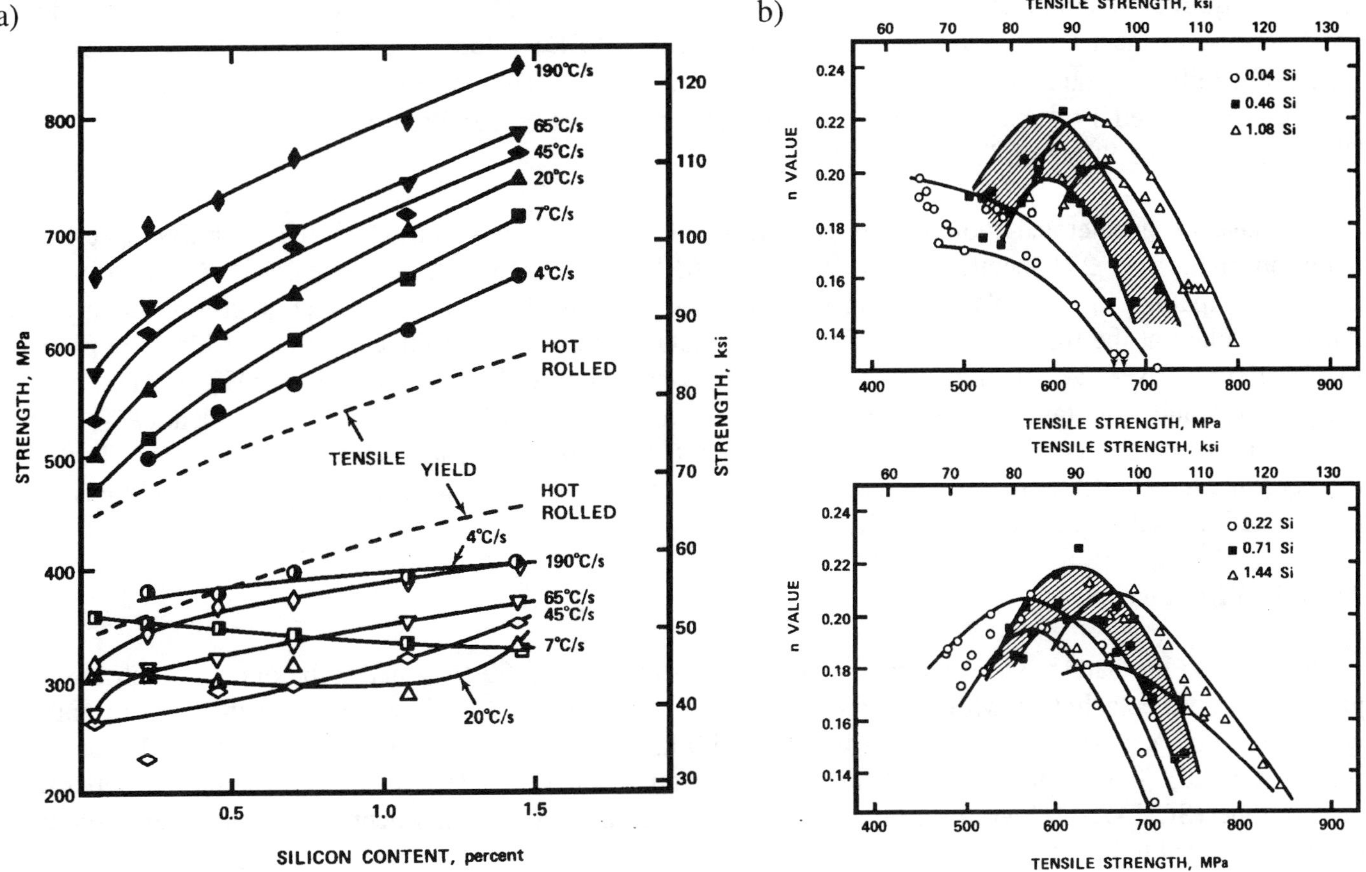

Figure 3. a) Effect of silicon content and cooling rate on strength on intercritically annealed steel sheet. b) Effect of silicon on n-value, as a function of ultimate tensile strength, for intercritically annealed steel sheet.

The last of Paul's work to be discussed in this paper is his contribution to the development of a grade of steel for application in offshore oil platforms. The importance of this development to U. S. Steel and the North American steel industry is such that it warrants special attention, and is a graphic example of his creative ability to utilize his knowledge of strengthening mechanisms to develop useful steel products.

Development of API 2Y 50 Steel

Background

During the early 1980's, low heat-affected-zone (HAZ) toughness values were reported in microalloyed offshore platform steels by companies working in the North Sea. These results were attributed to small areas of limited cleavage resistance within the coarse-grained HAZ that were termed local brittle zones (LBZ).[13] While the LBZ's were not associated with any platform failures at the time, the possibility that failures could occur spurred development of a new class of steel with improved HAZ properties to ameliorate concerns of the platform owners.

In the United States, the American Petroleum Institute (API) was the responsible specification body dealing with this issue. When considering replacement of API 2H, a nominal 0.18C-1.35Mn normalized grade, that had many years of proven service in platform applications, the API not only specified the new properties required, but a mandatory plate production process to achieve them. In the sole specification initially considered, API

2W, the required yield and tensile strengths for the 50T grade were 50- and 70- ksi, respectively.[14] In addition, the steel had to meet Crack Tip Opening Displacement (CTOD) values, used to assess the toughness of the HAZ, of 4 mils minimum at 14° F in critical areas of the HAZ. The prescribed test methods were set forth in API Recommended Practice 2Z (RP2Z), "Recommended Practice for Preproduction Qualification for Steel Plates for Offshore Structures".[15]

Because an accelerated-cooled or direct-quenched process was specified in the API 2W specification, adoption of the API 2W specification alone by the API effectively would legislate North America producers from participating in new platform construction inasmuch as there were no such facilities in any North American plant at the time. When a U. S. Steel member of the API committee argued that identical properties to those of an API 2W steel could be achieved by a reheat-quench-and-temper (RQT) process, the API reconsidered and simultaneously adopted a separate specification, API 2Y.[16] API 2Y was identical to API 2W in all respects except the plate production process. All these API specifications were issued on May 1, 1987. It then became the challenge for Paul and one of his colleagues to develop an API 2Y steel that would pass all the preproduction qualification tests in RP 2Z.

Technical Requirements

Stripped to the barest essentials, the composition of an API 2Y 50 steel had to be one in which the tensile properties could be achieved while minimizing HAZ deterioration in three key regions of the HAZ in multipass welds prepared at heat inputs ranging from 1.5 to 5.0 kJ/mm. The key areas were the unaltered coarse-grain zone (CGHAZ), the intercritically reheated coarse-grain zone (IRCGHAZ), and the unaltered subcritically heated zone (SCHAZ). Obviously, grain growth, formation of MA constituent, and adverse grain-boundary precipitation had to be avoided.

Alloy Design

The composition devised for the candidate API 2Y-50 steel is shown in the following table, and the rationale for the design displays Paul's ability to creatively choose among the various strengthening mechanisms to develop a new steel product.

Table II - Composition of Electric Furnace Heat No. 2R1990

C	Mn	P	S	Si	Cu	Ni	Cr	Mo	Cb	Ti	Al	N	V	B
0.087	1.38	0.014	0.003	0.26	0.20	0.36	0.05	0.01	0.01	0.008	0.025	0.009	0.005	Res.

Of all the elements in the steel, carbon has the most profound affect on transformation to martensite. Accordingly, reduction to levels significantly below the typical level of 0.18 percent for API 2H steel was required. The target of 0.09 percent was chosen in part to meet a restrictive carbon equivalent maximum (Pcm) of 0.23. To compensate for the loss of strength from carbon, manganese was targeted at 1.35 percent.[17] At this level, manganese exerts complex beneficial effects including transformation strengthening, grain refinement, and reduction of grain-boundary carbide thickness. All are important for improving toughness.

Because the strength of a 0.09C, 1.35Mn steel is lower than required in the API 2Y specification, solid-solution-strengthening was considered next, and, in the process, precipitation strengthening was ruled out. In the First and Second Long Period of Elements, the size factors for vanadium, chromium, cobalt, nickel, and molybdenum are favorable for formation of solid solutions with iron.[18] Most of these elements, however, exert some undesirable effects that make them unacceptable for the alloy design. Vanadium forms very soluble carbides that precipitate during multipass welding to create a very high hardness HAZ. Chromium exerts a high hardenability effect in steel and also forms a series of carbides that readily dissolve and re-precipitate on HAZ boundaries in multipass welds.[19] This is particularly undesirable in a HAZ of a structure that must be stress-relieved after welding. Molybdenum exerts a high hardenablity effect in steel. Cobalt is technically acceptable, but is too expensive. Among these series, only nickel is acceptable. It exerts a weaker hardenability effect in steel than some of the other aforementioned elements, does not form undesirable carbides, and promotes cross slip at low temperatures. Copper, which lies just outside the First Long Period, has some of the same characteristics as nickel. It, also, was included in the steel design.

Control of grain growth in the HAZ is an important feature, which must be accomplished through pinning of grain boundaries during multiple excursions to temperatures reaching the melting point. A dispersion of very small precipitates having as high a temperature stability as possible is needed. However, large primary particles must be avoided inasmuch as they are ineffective in pinning boundaries and are also potential initiation sites for cleavage fracture. Titanium and columbium can create the desired dispersion of fine precipitates when combining with carbon and nitrogen in the solid state at temperatures significantly below the melting point. Accordingly, the levels of titanium and columbium were set less than stoichiometric values to avoid precipitation of large primary precipitates at temperatures near the melting point.

Finally, steelmaking practices were used to minimize the inclusion and hydrogen content of the steel. Calcium additions were made to spheroidize the sulfides. The levels of chromium, molybdenum, vanadium, boron, and phosphorus, although a little higher than desired, were all residual ones, typical for electric furnace steel at the time.

Processing

A 200-ton electric-furnace heat (No. 2R1990) was melted at the USS South Works plant in 1987. The steel was vacuum degassed, and calcium wire was injected for sulfide shape control. The heat was teemed into big-end-up ingots, slabbed at South Works, and subsequently rolled into 1-, 3-, and 4-inch-thick plates at USS Gary Works.

Results

Selective data from the successful qualification of the plates according to the provisions of API 2Z are shown in the following tables. As shown in Table III, the tensile and impact properties of the base metal exceeded the requirements of API 2Y, and upper shelf toughness behavior was exhibited at – 76° F.

Table III Tensile and Impact Properties of API 2Y-50 Steel Plates

Plate Thickness, inch	**Yield Strength, ksi**	**Tensile Strength, ksi**	**Elong. in 2 inches, %**	**Reduction of Area, %**	**Charpy V-Notch Energy, ft-lb**		**Through Thickness Reduction of Area, %**
					-76°F	**-40°F**	
1	61.5	76.4	30.5	78.0	156,165,239	239,240,240	60.0, 70.0
3	53.1	71.5	34.5	80.0	239,239,240	239,240,240	53.0, 74.0
4	52.4	72.5	33.0	78.7	239,239,239	238,239,239	48.0, 53.0
API 2Y (Minimums)	50*	65	23	-	25 (30 average)	25 (30 average)	30

Shown in Table IV are CTOD values for a 3-inch-thick plate welded at a heat input of 127 kJ/in. Note that the values are excellent for both as-welded and post-weld heat-treated (stress relieved) plate. The excellent stress relieved results are attributed in part to the absence of chromium in the alloy design.

Table IV. Coarse-Grain HAZ CTOD Results for 3-Inch-Thick Weldments

Heat Input, kJ/inch	Weldment Condition	Specimen	CTOD, mils	Notch Placement		Rating
				%HAZ	%CG	
127	As Welded	81-04	46.9	80	16	Pass
		83-08	31.0	71	15	Pass
		81-05	15.3	68	15	Pass
		86-02	15.6	55	24	Pass
		86-03	81.4	52	15	Pass
		86-05	43.9	81	24	Pass
	PWHT	2-04	65.4	73	30	Pass
		2-05	54.7	58	32	Pass
		2-06	63.0	82	26	Pass
		2-07	62.7	85	36	Pass
		2-08	61.2	79	25	Pass
		97-04	86.0	55	52	Pass
		97-05	90.9	79	57	Pass
		97-06	44.5	66	46	Pass
		97-07	10.1	50	40	Pass

Applications

In 1988, U. S. Steel became the first company to qualify a steel to the API 2Y specification, and designated the new grade as O-TEN 50 steel. In 1989, British Petroleum became the first company to begin construction of a platform for installation in the Gulf of Mexico using either API 2W steel or API 2Y steel. The platform, designated Mississippi Canyon 109, was constructed from O-TEN 50 steel and installed in 1990. Shown in Figure 4 is the upper section of the 22,000-ton platform being towed to sea.

Additional structures built in the 1990's with O-TEN steel include the Auger, Mars, Lobster, and Petronius platforms. The development of O-TEN steel saved the offshore platform business in the Gulf of Mexico for North American plate producers. Paul's alloy design worked the first time, and is a lasting tribute to his ability to choose among a variety of strengthening mechanisms to optimize other steel attributes while achieving strength.

Figure 4 – Upper section of Mississippi Canyon 109 Platform.

EPILOGUE

We at U. S. Steel are extremely grateful that the Iron and Steel Society has chosen to honor one of our most distinguished colleagues, Paul E. Repas, at this symposium. As we sought appropriate words to express our feelings and admiration for him, we came across those spoken by Morris Cohen at a symposium to honor another of our distinguished U. S. Steel colleagues, Gilbert Speich, who succumbed at an early age to the same disease as Paul.

To paraphrase Morris' elegant eulogy,

> "Sadly, physical metallurgy lost one of those rare engineering scientists who was capable of bridging the broad spectrum of theory, experiment, and practical purpose. Paul Repas epitomized all this with exquisite care, quiet perseverance, and scholarly understanding. As a metallurgist and human being, he touched us all in many wholesome ways, and we are profoundly grateful for it".

Yes, Paul was all this, but more simply, he was a genuinely nice person. He was a generous man, generous with his time, advice, mentoring skills, and patience. Unquestionably, he strove to be the best he could be. He shared his immense talents for the good of his profession at every opportunity. He worked hard, but was wise enough to smell the roses along the way with his beloved wife, Ginny, and his family. What a wonderful legacy and example he left for those who carry on.

REFERENCES

1. F. H. Williams, "Influence of Copper in Retarding Corrosion of Soft Steel and Wrought Iron," Proceedings Engineering Society of Western Pennsylvania, Vol. 16, 1900, pp. 231-233.

2. HSLA Steels: Metallurgy and Applications, edited by J. M. Gray, T. Ko, Z. Shouhua, W. Baorong, and X. Xishan, ASM International, 1986.

3. Microalloyed HSLA Steels, Proceedings of Microalloying '88, ASM International, 1988.

4. International Symposium on Low-Carbon Steels for the 90's, edited by R. Asfahani, and G. Tither, TMS, Warrendale, PA, 1993.

5. Microalloying '95, Iron and Steel Society, Warrendale, PA 1995.

6. Gilbert Speich Symposium Proceedings: Fundamentals of Aging and Tempering in Bainitic and Martensitic Steel Products, edited by G. Krauss and P. E. Repas, Iron and Steel Society, Warrendale, PA, 1992.

7. P. E. Repas, "Control of Strength and Toughness in Hot-Rolled Low-Carbon Manganese-Molybdenum-Columbium-Vanadium Steel," Microalloying '75, Union Carbide, New York, NY, 1975, pp. 387 – 398.

8. P. E. Repas, G. A. Dries, and D. J. Wincko, "Property Control in High-Strength Heavy-Gage Bands," Thermo-Mechanical Processing in Theory, Modelling & Practice [TMP]^2, edited by B. Hutchinson, M. Andersson, G. Engberg, B. Karlsson, and T. Siwecki, The Swedish Society for Materials Technology, 1997, pp. 329 – 341.

9. P. E. Repas, "Metallurgical Fundamentals for HSLA Steels," Microalloyed HSLA Steels, Proceedings of Microalloying '88, ASM International, 1988, pp. 3 –14.

10. F. B. Pickering, "High-Strength, Low-Alloy Steels – A Decade of Progress," Microalloying '75, Union Carbide, New York, NY, 1975, pp. 3 – 30.

11. P. E. Repas, "The Hot Strip Mill as a Thermomechanical Processing Tool," Gilbert R. Speich Lecture to the Chicago Chapter of ASM International, Chicago, IL, April 9, 1996.

12. P. E. Repas, "Physical Metallurgy of Dual-Phase Steels," Mechanical Working and Steel Processing XVII, AIME, Warrendale, PA, 1979, pp. 277 – 305.

13. D. P. Fairchild, "Fracture-Toughness Testing of Weld Heat-Affected Zones in Structural Steel," Symposium on Fatigue and Fracture Testing of Weldments, ASTM, 1988.

14. API Specification 2W, "Specification for Steel Plates for Offshore Structures, Produced by Thermo-Mechanical Control Processing (TMCP)," American Petroleum Institute, May 1, 1987.

15. API Specification 2Z, "Recommended Practice for Preproduction Qualification for Steel Plates for Offshore Structures," American Petroleum Institute, May 1, 1987.

16. API Specification 2Y, "Specification for Steel Plates, Quenched and Tempered, for Offshore Structures," American Petroleum Institute, May 1, 1987.

17. B. Mintz, "Influence of Silicon and Nitrogen on the Impact Properties of As-Rolled Mild and Carbon-Manganese Steels," JISI, 1973, 213, p.238.

18. W. Hume-Rothery andG. V. Raynor, The Structure of Metals and Alloys, Institute of Metals, London, 1956, pp. 243-245.

19. K. J. Irvine, D. J. Crowe, and F. B. Pickering, "The Physical Metallurgy of 12% Chromium Steels," JISI, 1960, 195, pp. 399-400.

Role of Structure and Microstructure in the Enhancement of Strength and Fracture Resistance of Ultra-High Strength Hot Rolled Steels

R.D.K. Misra, G.C. Weatherly*, J.E. Hartmann**, and A.J. Boucek
LTV Steel Technology Center
6801 Brecksville Road
Independence, OH 44131, USA
Phone: 216-642-7236 / Fax: 216-642-7225

*Department of Materials Science and Engineering
McMaster University
Hamilton, Ontario, L8S 4M1, CANADA

**LTV Steel Indiana Harbor Works
3001 Dickey Road
East Chicago, IN 46312, USA

Key Words: Hot Rolled Steels, Yield Strength, Toughness, Ferrite-Bainite Microstructure

INTRODUCTION

There is presently a strong interest in using 'as-hot rolled' steels in automotive and mechanical applications in place of the quenched and tempered (Q&T) steels because of the inherent metallurgical disadvantages associated with the latter, namely, retained austenite, residual stresses, quench-cracking, and distortion. Additionally hot rolled steels are likely to be cost effective and energy efficient, in that the additional processing through the conventional heat treatment system is eliminated. LTV Steel has recently pioneered the development of ultra-high strength hot rolled microalloyed steels with a minimum yield strength of 758 MPa (110 ksi)[1]. In general, these steels have 0.05 wt.% C and 1.5 wt.% Mn, and utilize microalloying concepts to attain a fine-grained ferrite-bainite microstructure possessing ultra-high strength and good toughness. The mechanical behavior of the newly developed steels, in terms of toughness, fatigue and formability, is comparable to or better than the equivalent strength Q&T steels. These steels have found use in applications such as truck bodies, car brackets, crane booms, structural tubings etc., and as a viable substitute for lower strength 550 MPa (80 ksi) HSLA steel, with weight savings proportional to increased strength.

The scope of this paper is restricted to ultra-high strength hot rolled steels of yield strength 689 MPa (100 ksi) and 758 MPa (110 ksi) which are currently being commercially processed on a continuous hot strip mill and coiled in thicknesses up to about 9.5 mm (0.375 inches). To be useful as a structural material, it is important that the steels possess a good strength-toughness combination, in addition to formability, weldability and fatigue resistance. The present paper is an attempt to correlate the part played by solute additions, grain size, precipitates and microstructural constituents in obtaining strength and fracture resistance, and consolidate the

the understanding that has been implemented in the development of hot rolled microalloyed steels, together with the discussion of the relevant published work. The influence of solutes, Mn, and the simultaneous addition of Cu and Ni on the yield strength is also outlined.

In the ultra-high strength microalloyed steels presented here the contributions to strength can be described in terms of the following:

(i) grain size,
(ii) solid solution strengthening from substitutional elements (σ_{SS}), such as Mn, Si etc., and additional interstitial hardening from C and B (σ_{int}),
(iii) precipitation hardening effect from the carbide particles (σ_{ppt}),
(iv) transformation hardening (volume fraction of bainite, slip band length which includes both packet and lath size) (σ_{transf}), and
(v) dislocation hardening (σ_{disl}).

The material used in the study described here was commercially produced at the LTV Steel. Continuously cast slabs of ~ 25.4 cm thickness were hot rolled down to a minimum of ~ 3 mm gauge.

I. CONTRIBUTIONS TO STRENGTHENING

(a) Grain Size

Maximum grain refinement constitutes one of the primary metallurgical objectives of the hot rolling process in the effort to obtain high strength and toughness in controlled-rolled microalloyed steels, and the Hall-Petch relationship (equation 1) between yield strength (σ_y) and mean grain diameter (d) is well known:

$$\sigma_y = \sigma_o + k_y d^{-1/2} \qquad [1]$$

σ_o is a measure of friction stress resisting dislocation movement or resistance of the grain interior, and k_y is a measure of grain boundary strength. A fine polygonal ferrite grain size of ~ 3-8 microns (depending on the final gauge of the processed steel) (Fig. 1) was achieved through the addition of nitride and carbide forming elements, Ti and Nb, and also through controlled thermo-mechanical processing. Titanium and Nb invariably form nitrides or carbonitrides either in the liquid steel, during solidification or during reheating at the soaking temperature assisting in the grain refinement process during the early and also later stages of thermo-mechanical processing. These particles are usually coarse, cuboidal-shaped particles, and remain undissolved during slab reheating. They also tend to crack during hot rolling of the steel. One such cracked particle is presented in Fig. 2, together with the electron diffraction pattern and EDX analysis. The analysis confirmed that they were Ti-rich nitrides having the approximate composition of $Ti_{0.9}Nb_{0.1}N$. In Fig. 2a, the edges of the precipitates are aligned along the <100> direction, although the diffraction pattern (Fig. 2b) shows some radial streaking and spots from more than one zone axis. The asterism of the diffraction spots arises from the slight misorientations introduced by cracking. The term particle fragmentation[2] has been used to describe the formation of precipitates with radially streaked diffraction patterns or diffraction patterns that come from more than one zone axis. A random distribution of these particles of size ranging from 0.3 microns to as large as 3 microns in edge length were observed. Their size and thermo-mechanical stability precludes any further role in the precipitation sequence for these particles. It is, however, interesting to note that the volume of a one micron cube particle is equivalent to 10^6 precipitates 10 nm in diameter, making any quantitative estimate of the effective volume fraction of small particles available for grain size control difficult.

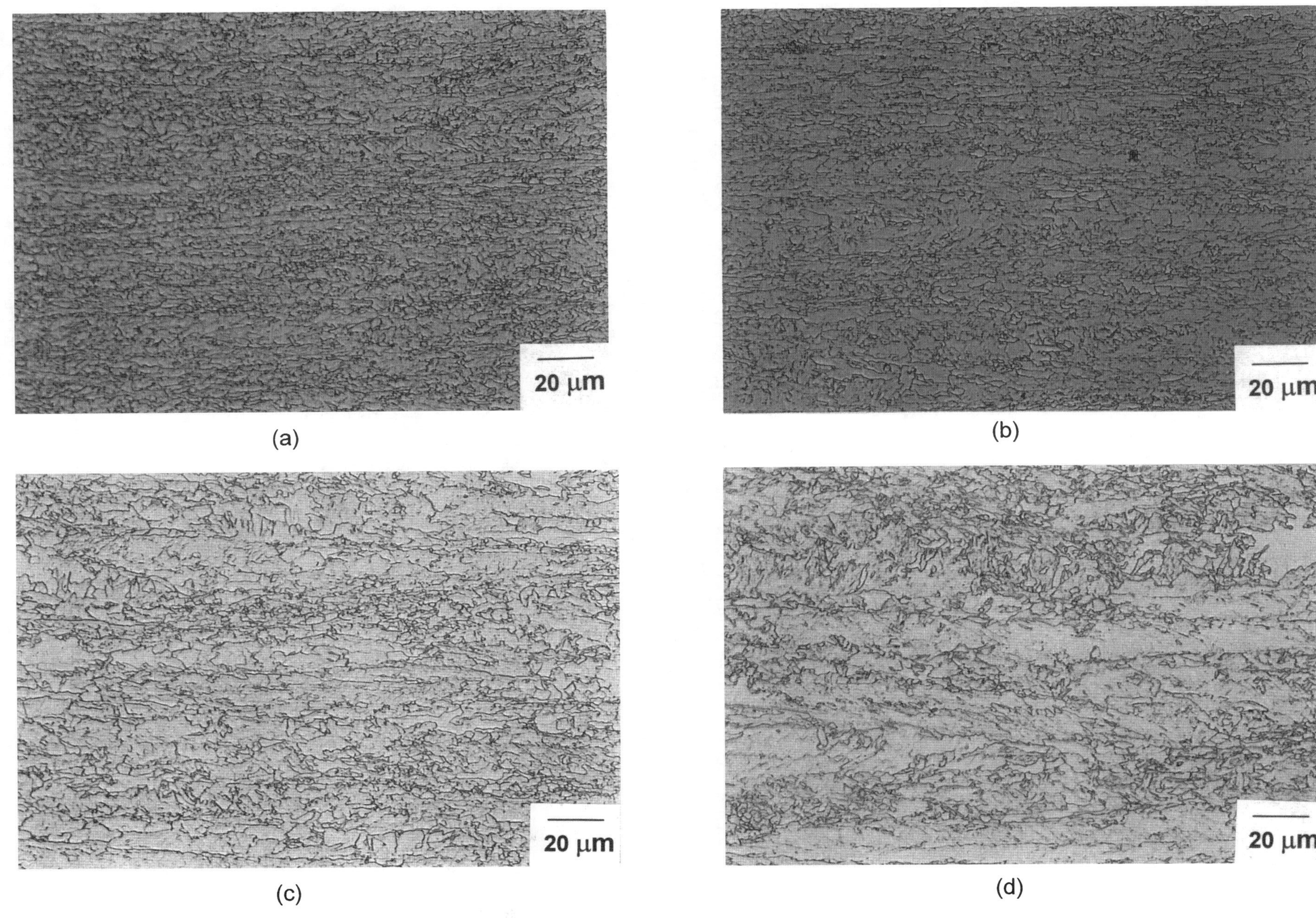

Fig.1. Microstructures of ultra-high strength hot rolled steels showing variation in grain size in coils of varying gauge. (a) 3.1 mm, (b) 6.3 mm, (c) 7.8 mm, and (d) 9.5 mm.

The processing parameters of relevance to the final grain size in controlled rolling were finish temperature, extent and rate of deformation, time between deformation and beginning of transformation, cooling rate on the run-out-table, and coiling temperature. These will not be discussed here as it is beyond the scope of the paper.

In controlled thermo-mechanical processing the structure of the deformed austenite at the onset of transformation reflects the kinetics of high temperature deformation processes namely, recovery, recrystallization and grain growth. The rate of these processes, the microstructure of the deformed austenite, and the resulting grain size depend, particularly on the carbide forming elements Nb and Ti as discussed below.

It may be inferred from the microstructures in Fig. 1 that during transformation of Nb and Ti-containing steels, the austenite grains retain their elongated shape, and this grain morphology and possibly incomplete recovery (see section 1(f) on dislocation hardening), influenced ferrite nucleation, resulting in finer polygonal ferrite. The resulting microstructure formed through nucleation and growth is, however, dependent on the % deformation of the parent austenite prior to transformation. In thermo-mechanically processed Ti and Nb-containing steels deformation takes place at temperatures at which the kinetics of recovery and recrystallization are normally expected to occur rapidly. However, the carbide-forming elements (Ti, Nb) make the steel amenable to thermo-mechanical treatment by increasing the recrystallization temperature and retarding the recrystallization of austenite by several orders of magnitude. In addition to this, strain-induced precipitation of Ti and Nb carbides at lattice defects such as dislocations and vacancies effectively pin the grain boundaries and retard the recrystallization of austenite. The net result is that the elongated shape of the deformed austenite grains was retained in the transformed ferrite (Fig. 1), which influenced the size of the ferrite grains because of aspect ratio considerations. Another role of the precipitation is that of acting as additional ferrite nucleation sites during transformation which further reduces the size of the ferrite grains. Apart from this, they act as barriers to the migration of ferrite boundaries during cooling and cause further grain refinement. Thus, the observed fine grain size was a cumulative contribution of a number of factors outlined above. The size, nature and distribution of precipitates in the steels is described below in section 1(c).

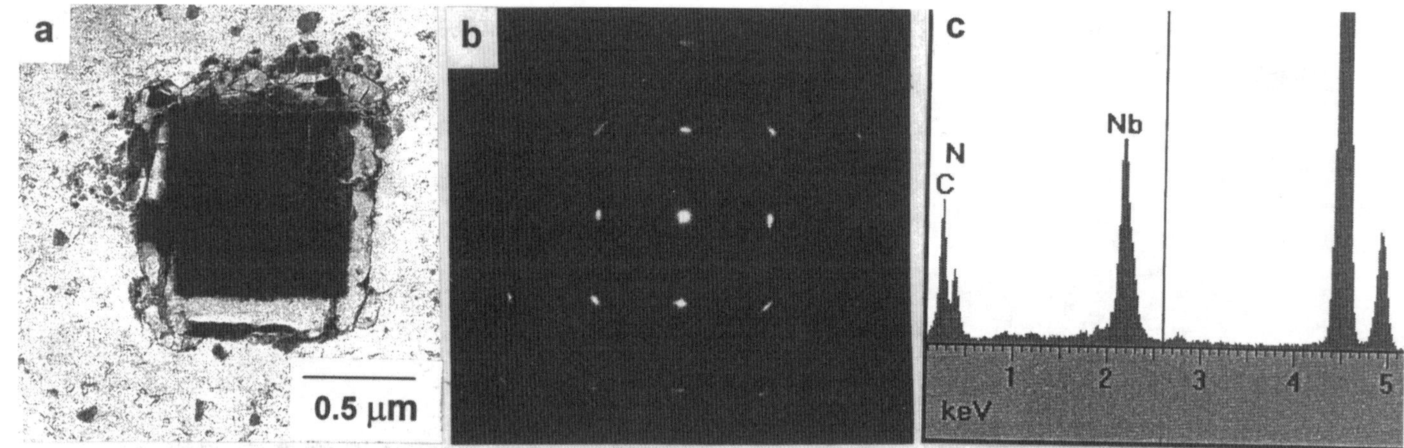

Fig. 2 (a) Cracked TiN particle, (b) <001> diffraction pattern; the asterism of the diffraction spots arises from the slight misorientation (s) introduced by the cracking, (c) EDX spectrum showing Ti-K, Nb-L and N-K peaks from the particle (the carbon signal comes from the extraction replica).

(b) Substitutional and Interstitial Strengthening

The effect of substitutional elements is well documented in the literature[3,4]. The substitutional elements considered here for the attainment of strength were Si, Mn, and Mo. Simultaneous additions of Cu and Ni were also considered for further enhancement in the yield strength of the steels. Manganese and Mo not only strengthen the steel through solid solution hardening but also through grain refinement. They influence the kinetics of transformation by delaying the onset of proeuctectoid ferrite nucleation (lowering of austenite to ferrite transformation temperature)[5, 6]. If there was any appreciable substitutional solute strengthening by Mo, it would not be simple to quantify because of interactions with interstitials, C and N, in conjunction with the carbide/nitride forming elements, Ti and Nb (see section 1c on precipitation hardening).

In our recent effort towards further development of ultra-high strength steels, simultaneous additions of Cu and Ni were made. Alloying with 0.3 wt.% Cu and 0.15 wt.% Ni to the base steel led to an increase of yield strength by about 32.4 MPa (4.7 ksi) (Table I). Copper is known to increase the tensile strength of steels through precipitation of Cu-rich particles and also improves hardenability[5,7]. Ni additions were made to prevent the hot shortness normally associated with Cu-containing steels.

In regard to the hardening it is known that the interstitial carbon has a dramatic solid solution strengthening effect in iron, but the equilibrium carbon content of ferrite is negligible at room temperature. Considering that the alloy chemistry was designed to encourage bainitic transformation in the ultra-high strength steels discussed

Table I: Effect of Cu and Ni addition, and decrease in Mn-content of 689 MPa (100 ksi) steels.

	Increase in Yield Strength	Increase in Tensile Strength
Base steel + 0.3Cu/0.15Ni (wt.%)	32.4 MPa (4.7 ksi)	34.5 MPa (5.0 ksi)
Base Steel – 0.3 wt.% Mn	21.4 MPa (3.1 ksi)	14.5 MPa (2.1 ksi)

here, any supersaturation of carbon is expected to be quickly reduced by partitioning into either the surrounding untransformed austenite or as inter/intralath carbides because of the high mobility of carbon at the baintic transformation temperature. An approximate estimate for these processes at the bainite transformation temperature of 600 °C is only a few milliseconds. There is also substantial evidence in the literature to suggest that the residual supersaturation of carbon atoms bound to dislocations in bainitic ferrite makes a significant contribution to strength[8-10]. Also, it may be noted that the C-content of the steel was in excess of the stoichiometric composition required to precipitate Ti and Nb as carbides/nitrides. Thus, the effect stated above is most likely to occur, and contribute significantly to the observed high strength of steels.

(c) Precipitation Strengthening

Microalloying additions, Ti and Nb, were used for both grain size control and precipitation strengthening. The precipitation of Ti and Nb carbides and carbonitirides can yield a range of strength and toughness, depending on their precipitation behavior. The size, morphology, and chemical composition of precipitates was determined using standard TEM and STEM methods, involving both carbon extraction replicas and thin foil samples prepared by conventional techniques. The TEM samples were studied in an FEG JEOL 2010F at 200kV and a Philips CM12 at 120kV. The former microscope was equipped with a Link EDX system for chemical analysis. The carbon extraction replicas were examined in a Philips CM12 TEM/STEM electron microscope with parallel detection electron energy loss spectrometer. It may also be noted that it is difficult to detect carbides and carbonitrides particles by direct transmission electron microscopy of thin foils, because in steels containing low temperature transformation product (e.g. bainite) the precipitates are incoherent, and also the precipitates are so small that it is very difficult to obtain a diffraction pattern from which to orient a darkfield examination. Consequently, most of the electron microscopy was done using carbon extraction replicas.

The particles, found in all areas of the extraction replicas, ranged in size from about 3 nm up to several 100 nm in diameter. These spheroidal particles always contained Ti and Nb, and were identified as (Ti, Nb) carbides. The distribution of the particles was measured directly from C-extraction replicas. Figure 3 shows a typical example of the distribution of the particles in a sample, and Fig. 4 summarizes the data obtained by measuring approximately 200-300 particles from two representative areas of each replica. The large size cuboidal particles in Fig. 3 are Ti-rich nitrides. (The areas used for distribution analysis did not contain some of the largest particles detected in steels, and the omission of these particles from the statistical analysis had no impact on the average particle size because of their low density). The different samples analyzed had the same mean particle size, approximately 9 nm, however, there were significant differences in the distribution of coarser particles (20 nm or larger) between the samples. This was due to unintentional variations in the processing parameters of the steels. Occasionally spheroidal particles up to about 500 nm in size were observed.

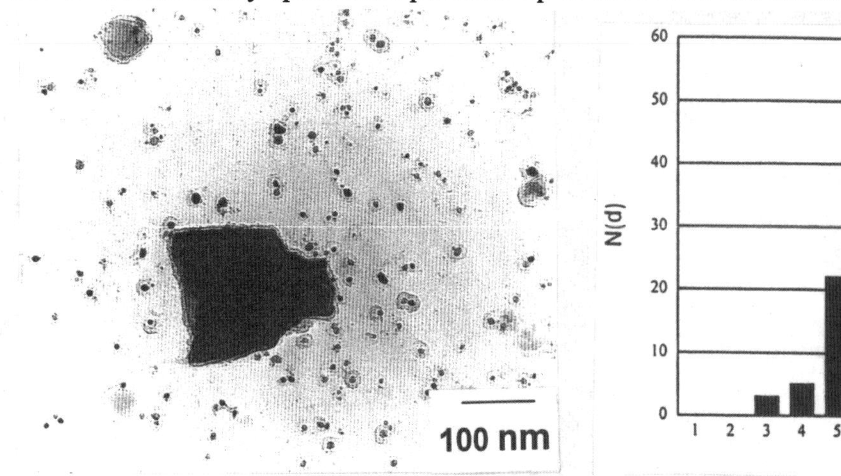

Fig. 3. Distribution of particles observed in the extraction replicas.

Fig. 4. Spheroidal particle size distribution as measured from the extraction replicas.

The composition of the spheroidal particles was determined by the EDX method, using the thin film approximation to convert the raw data (after background subtraction) to compositions. Many of the spectra showed significant Mo peaks (see Fig. 5) and this led to problems in quantification because the L-peaks of Mo overlap those of Nb. However the overlap problem was circumvented by using the K-peaks, with a "k-factor" of 0.27 to convert from intensity data to composition (in wt%):

$$C_{Ti} / C_{Nb} = 0.27 (I_{Ti})/(I_{Nb}) \qquad [2]$$

where I_{Ti} and I_{Nb} are the Ti-K and Nb-K intensities after background subtraction, respectively. The results of the analysis of the compositions (expressed as atomic %) of the analyzed particles in the different samples is summarized in Table II. In general, particles of three different stoichiometries were observed (Table II).

Table II: The composition of spheroidal particles as analyzed by EDX.

Mean Composition $(Ti_xNb_{1-x})C$**	Maximum	Minimum
$Ti_{0.39}Nb_{0.61}$ (32 readings)	$Ti_{0.5}Nb_{0.5}$	$Ti_{0.27}Nb_{0.73}$
$Ti_{0.4}Nb_{0.6}$ (14 readings)	$Ti_{0.5}Nb_{0.5}$	$Ti_{0.33}Nb_{0.67}$
$Ti_{0.48}Nb_{0.52}$ (26 readings)	$Ti_{0.72}Nb_{0.28}$	$Ti_{0.37}Nb_{0.63}$

**The analysis in Table II assumes that the precipitates do not contain any Mo (see comments below).

Generally steels exhibiting the narrowest size distribution of precipitates (Fig. 4), contained particles that were richer in Nb as compared to steels with wider distribution. In the steels examined the larger particles (> 100 nm) were found to be richer in Ti with average Ti content being approximately 20% higher than the mean value indicated in Table II. This behavior of different precipitates is expected in a commercial production because of variations in processing parameters.

As noted earlier the EDX spectra from many of the particles, particularly those in the smaller size range, showed the presence of significant quantities of Mo (Fig. 5), up to 25% of the value of Nb. There are no reports in the literature to suggest that Mo can dissolve in (Ti, Nb) carbides. The presence of Mo could come from a small Mo carbide particle, heterogeneously nucleated on the (Ti, Nb) carbide, or it may be thermodynamically possible for Mo to substitute for Nb in the particles at low temperatures in this class of steel. Further information is required to substantiate either of these possibilities.

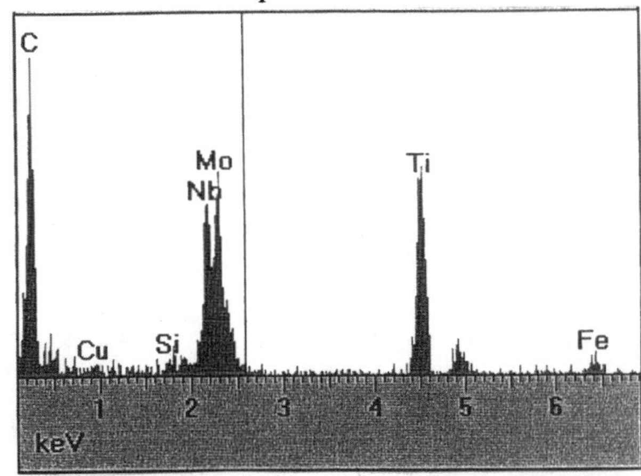

Fig.5. Typical EDX spectrum from 10 nm diameter (Ti, Nb) carbide particle showing pronounced Mo peak.

Size range of 5 to 20 nm is the approximate size of the particles for precipitation hardening since the interparticle spacing is optimal for hardening. As can be seen from the electron micrograph (Fig. 3), the matrix precipitates are fairly uniformly distributed. One may conclude from this observation that the precipitation hardening was homogeneous in the steels. It should, however, be noted that the extraction replica reveals most, but not all the precipitates[11]. In microalloyed steels the yield strength depends on the volume fractions (and strengths) of the ferrite grains and bainitic structures in the steel. The strength of the ferrite grains will be determined primarily by the spacing of the (Ti, Nb) particles which act as Orowan obstacles, and the dislocation density. On the basis of the size distribution and volume fraction of the particles found from the extraction replica study, it can be concluded that precipitates contribute significantly to strength.

The observed precipitates are expected to be formed at different stages of processing of the steel, namely, (a) relatively coarse precipitates (100 nm or greater) formed at high temperatures in the austenite region, (b) fine precipitates ($> \sim$ 20-50 nm) produced by strain-induced precipitation after rolling in the non-recrystallized region, (c) fine precipitates (20-50 nm) formed in the high-temperature ferrite region, and (d) very fine precipitates ($\sim$ 5-20 nm) formed during coiling after rolling. Precipitates (b) and (c) are expected to nucleate at dislocations and austenite-ferrite interfaces, thereby losing coherency with the matrix. Type (d) precipitates should be coherent with the matrix and their precipitation is expected to cause significant hardening. No effort was made to examine these aspects since the precipitation behavior is dealt with in detail in the literature.

It may now be appropriate to briefly summarize the sequence of precipitation of carbides with progress in thermo-mechanical processing. As expected in controlled rolled steels at soaking temperatures most of the Ti and Nb are taken into solution, according to limitations imposed by the solubility products. For carbide (also nitride) forming elements the solubility in austenite at any given temperature depends on the C and N-content of the steel. If we assume that equilibrium solute solubility was reached during the reheating process, then as the slab is cooled during the processing in the roughing stands and finshing stands, the solubility product is lowered, and an unstable situation exists where austenite is supersaturated with the microalloyed solute. Thus, with the lowering in temperature during various stages of hot rolling, saturation of solute increases and precipitation of solute begins when the kinetic factors are favorable. First, the large pre-existing cuboids and austenite boundaries are apparent sites for early nucleation for carbide precipitation. Deformation of austenite during rolling introduces lattice defects such as dislocations and vacancies assisting diffusional processes controlling the kinetics of precipitation. As a consequence, 'strain induced precipitation' occurs at the prior austenite boundaries or defects, which besides contributing to strengthening effectively pin the grain boundaries and refines the microstructure. Even though this precipitation is helpful to achieve thermo-mechanical response during hot rolling, the magnitude of precipitation strengthening of the ferrite phase is reduced because of the precipitation of solute in austenite that depletes the total solute concentration.

Subsequently, after the final finishing pass in the strip mill, the strip is cooled through the transformation range, such that on coiling transformation is almost completed. Ferrite has significantly reduced solubility for carbon and solute elements compared with the austenite phase, and thus the precipitation of carbides takes place during slow cooling of the coil to the room temperature.

(d) Low-Temperature-Transformation (Bainite) Hardening and Carbide Strengthening in Bainite

The significance of bainite as a transformation product lies in its effect on mechanical properties and the extent to which the characteristic features of the transformation can be exploited in the design of commercial steels. It was intended to produce a ferrite-bainite microstructure in the steels. Baintic microstructures in the steels were obtained through appropriate design of the alloy composition and by suitable control of processing.

Scanning electron microscopy of the steels prepared for the TEM examination revealed the structure shown in Fig. 6. This figure shows regions with distinct 'patches' of lath or plate-like carbides that are characteristic of upper bainite. These observations were confirmed by the thin foil study of samples by TEM. A region of upper bainite containing discrete carbide particles is shown in Fig. 7. A close look at the TEM bainitic microstructure also shows fine distribution of Ti-rich carbides in the bainite thus the strength of bainite may also be derived from both the distribution of cementite and Ti-rich carbides. Presumably much of the precipitation of these Ti-rich carbides occurs in the austenite prior to ferrite transformation. It may be noted that some authors have observed the propensity towards the formation of an irregular, non-lamellar ferrite/cementite aggregate when austenite contains a comparatively low carbon content[12], similar to that present in the steels discussed here.

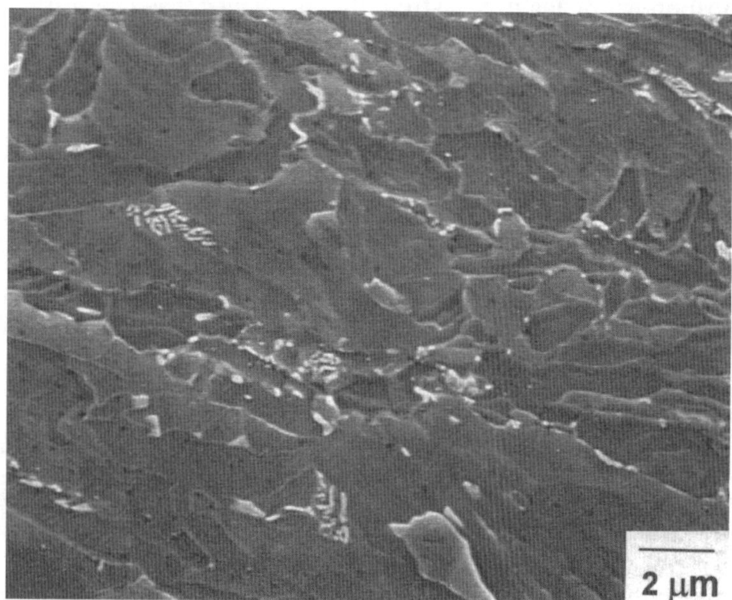

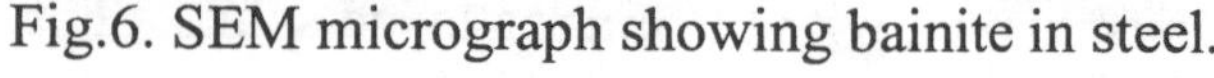

Fig.6. SEM micrograph showing bainite in steel.

Fig.7. TEM micrograph showing upper bainite.

The coarse cementite particles formed at lath boundaries in upper bainite (Figs. 6 and 7) are not expected to greatly affect the strength of the microstructure via dispersion hardening. However, they may influence the ability of the lath boundaries to resist the movement of dislocations, with the result that slip is confined within the laths, and the contribution of lath size to flow stress is increased. This behavior is in contrast to lower bainite, where the more refined dispersion of intralath carbides contributes to strengthening, such that the contribution increases with decreasing transformation temperature. Limited TEM work did not convincingly show that lower bainite of significant content was present in steels, even though some was expected.

Upper bainite begins with the nucleation of ferrite of paraequilibrium concentration, and since partitioning of microalloying elements are unlikely (especially at cooling rates used), the only significant compositional difference between proeuctoid ferrite and bainite are the carbides[10,13]. (Paraequilibrium transformation refers to the situation in which the substitutional alloying elements are unable to partition during the time scale of processing, although carbon, which is fast diffusing interstitial element, redistributes between the phases and attains equilibrium subject to this constraint). A simple mass balance is useful to estimate the amount of carbon in the enriched austenite[13]. The relevant equation is:

$$\%C_{alloy} = f_v(\alpha)\ \%C_{\alpha} + f_v(LTTP)\ \%C_{LTTP} \qquad [3]$$

where the weight percent carbon (%C) in the alloy, in the proeutectoid ferrite, and in the low-temperature-transformation product are indicated with appropriate subscripts, and the volume fraction (f_v) of each microconstituent is indicated in parentheses. Based on the alloy composition, volume fraction measurements, and an assumed carbon level of 0.02 wt. pct. in the proeutectoid ferrite, the carbon content of the LTTP was calculated to be 0.32. This carbon is present as cementite particles (Fig. 7), and is also expected to be segregated to dislocations. Since partitioning of substitutional alloying elements is unlikely (especially at the cooling rates employed), the only significant compositional difference between proeutectoid ferrite and the LTTP in steels is the carbon content.

Mn and Mo have a significant effect on strengthening through modification of the transformed microstructure[6], and also in influencing the transformation start temperature of the low temperature transformation product, bainite. The effect of decreasing Mn-content by 0.3 wt.%, led to an increase in the strength of the base steel by 21.4 MPa (3.1 ksi, Table II). This was observed to be due to an increase in the volume fraction of bainite by about 10% by volume.

(e) Lath strengthening

Earlier studies have correlated the change in strength (and also toughness) of bainitic steels with packet size and lath size (width)[14-18] as schematically illustrated in Fig. 8, where BI, BII and BIII, refer to the different types of bainite. In Fig. 8 only refinement of scale in lath and packet size is illustrated; size and distribution of carbides is not depicted. The nature of bainite depends on the temperature of transformation, BIII is formed at the lowest temperature. As we go from BI through BIII, the steel is expected to exhibit increased strength in association with good toughness. Most of the studies that have used lath size as the criteria showed that the flow stress varies with the reciprocal of some characteristic lath dimension[14-17], and one of the possible ways used to rationalize lath strengthening was to consider the orientation of the active slip systems in bainite[8]. Such considerations have shown that 75% of all the possible slip systems are oriented at 55° to the lath axis. Thus the slip across the lath axis suggests that the flow stress of the material can be controlled by the lath width, but there may be some contribution from the lath size and packet size (Fig. 8), where the packet is defined as a group of laths with a specific orientation. Thus the slip band length is expected to control the strength. The slip systems have not been determined for the steels under investigation, however, the lath size as determined from Fig. 7 was of the order of 0.5 microns, and was of the size expected for bainitic steels.

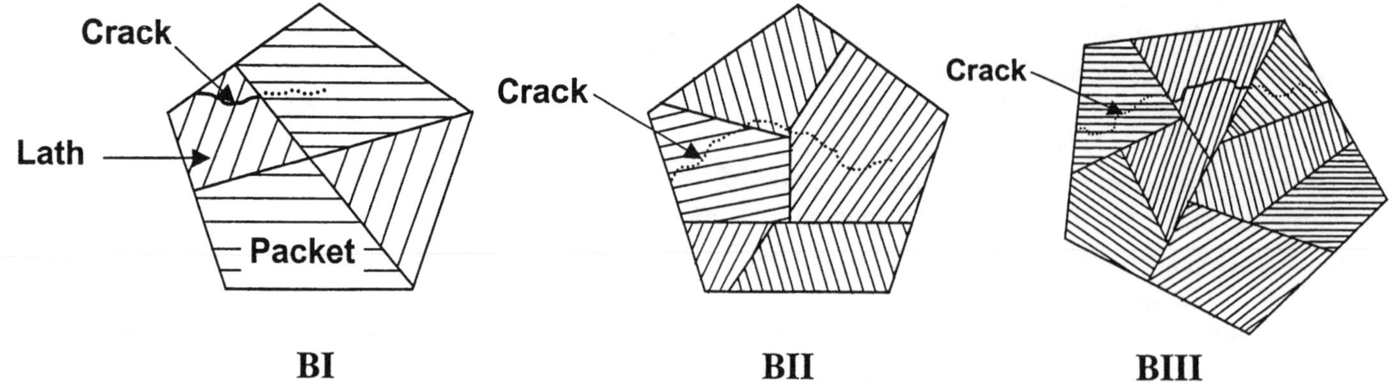

Fig. 8. Schematic illustration of three types of bainite, BI, BII, and BIII.

(f) Dislocation hardening

Bainitic ferrite is expected to contain a higher dislocation density than ferrite formed with polygonal or Widmanstatten morphology[8]. It has been suggested that the dislocation density and details of the substructure can be modified by thermomechanical processing, and part of the dislocation substructure of the parent austenite can be inherited by bainite leading to some additional strengthening[8,19]. Examples of the nature of dislocation substructure in proeutectoid ferrite and bainite are presented in Figs. 9 and 7, respectively. The substructure, particularly in Fig. 9 consists of relatively high and uniform dislocation density, with little or no recovery. No cell structure is evident within the proeutectoid grains, implying the near absence of recovery. The irregular nature and density of dislocations are typical of low temperature transformation product (bainitic ferrite, acicular ferrite) formed by continuous cooling from the austenite phase field. Another feature of relevance to strength to be noted in Fig. 9 is the fine distribution of (Ti, Nb) carbide particles. The observed substructure of upper bainite is one aspect of this relatively high dislocation density that has generally been assumed to arise as a result of the amalagamation of separate subunits formed by a displacive transformation mechanism[20].

Since all the factors described above control the strength of the final product, it is appropriate to express the relationship between strength and microstructure by equation [3] (notations defined in the introduction section):

$$\sigma_y = \sigma_o + \sigma_{ss} + \sigma_{int.} + \sigma_{ppt} + \sigma_{transf} + \sigma_{disl} + k_y d^{-1/2} \qquad [4]$$

In equation [4] it is important to realize that the strength will be some function of the relative volume fraction of various constituents (ferrite and bainite contents), and thus it may be appropriate to predict strength on the basis of an expression of the type

$$\sigma_y = \sum_{i=1}^{N} V_i \sigma_i \qquad [5]$$

where V_i is the volume fraction of constituent 'i' and σ_i is the flow stress of constituent 'i', which in principle can be estimated from the concepts outlined above. In the present steels, $\sigma_o + \sigma_{ss} + \sigma_{int.} + \sigma_{ppt} + k_y d^{-1/2}$, all contribute to ferrite matrix strengthening, and σ_{transf} (+ some degree of σ_{ppt} from cementite particles and fine Ti-rich carbides) contributes to bainite hardening.

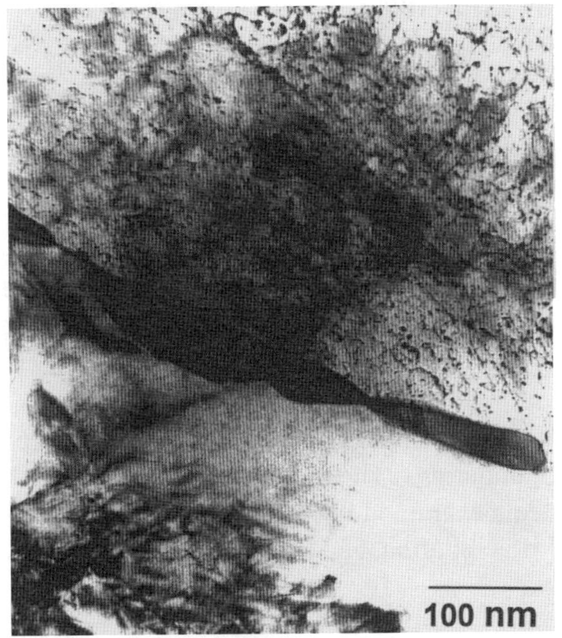

Fig. 9. Region of ferrite microstructure showing uniform dislocation density in the ferrite grains, Fe_3C precipitation at the grain boundary, and fine-scale (Ti, Nb) precipitates in the grains.

Fig.10. Schematic representation of the fracture process in a steel with high density of grain boundary precipitation of TiC.

II. FRACTURE RESISTANCE

(a) Effect of Grain Size and Precipitates on Impact Toughness

The strength and toughness obtained by controlled-rolling in microalloyed steels largely depends on the grain refinement and precipitation hardening, although these properties can be widely changed by controlled rolling parameters and the basic composition of the steel.

The effect of grain size on Charpy V-notch impact toughness and ductile-to-brittle transition temperature (DBTT) is listed in Table III. The DBTT relates to the temperature at which 50% shear was observed in the impact fractured sample. As expected the impact toughness was found to increase with decrease in grain size, accompanied by a decrease in DBTT. This increase of toughness occurred at a similar yield strength of ~ 689 MPa (100 ksi) (Table III). It may be noted that the finer grain size was a result of greater % reduction in the roughing and finishing stands (and hence differences in the final gauge of steels). The effect shown in Table III is an exaggeration because of an associated effect of section size (Table IV). It was observed that when the section size of the Charpy specimen was reduced from 1/2 to1/4, the DBTT decreased by about 50% (Table IV). However, the influence of grain size on impact toughness and DBTT is apparent even after taking into consideration the section size effect.

Table III: Effect of grain size on impact toughness and DBTT of 689 MPa (100 ksi) yield strength steels.

Gauge	Average Grain Size	Impact Toughness at - 40 ^{0}F (full-size equivalent)	DBTT
3.175 mm (0.125 in.)	3 microns	135.6 Nm (100 ft-lbs)	< - 100 ^{0}F
6.35 mm (0.250 in.)	5 microns	131.5 Nm (97 ft-lbs)	- 75 ^{0}F
9.525 mm (0.375 in.)	8 microns	101.7 Nm (75 ft-lbs)	- 50 ^{0}F

Table IV: Effect of section size on DBTT of 689 MPa (100 ksi) yield strength steels.

Section Size of Charpy Specimen	DBTT
1/2	- 60 °F
1/3	- 85 °F
1/4	- 120 °F

It is also important to realize that in spite of the fine grain size of the material, toughness of ultra-high strength steels can be significantly reduced by grain boundary precipitation of carbides[21]. This is applicable if strengthening is achieved by titanium carbides. Grain boundary precipitation of titanium carbides has an adverse effect on fracture toughness[21], where the effect of grain boundary precipitation is to provide a macroscopic continuous fracture path, while on the microscopic scale the fracture is of mixed-mode type consisting of quasi-cleavage and dimple rupture. This situation is most likely to occur if relatively coarse titanium carbides of 100-500 nm are formed at the grain boundaries. This adverse effect of grain boundary precipitation is schematically illustrated in Fig. 10. In the steels examined here grain boundary precipitation of Ti-rich carbides was not observed, only a few instances of precipitation of Mn-containing cementite particles was noted $(Fe, Mn)_3C$ (Fig. 11).

As described above the ultra-high strength steels typically consisted of a polygonal ferrite-bainite microstructure, and embedded within this structure were various type of particles, namely, large Ti-rich nitrides and fine carbides precipitated during hot rolling. The size of these particles depends on the thermo-processing conditions. In discussing the role of microstructure in determining the fracture resistance of the material, it is helpful to separate the microstructure into three different categories, primary particles, secondary particles and the fine scale microstructure. Under appropriate conditions any of three types of particles can nucleate voids[22].

The primary particles are those at which voids first nucleate. These are inclusions such as MnS and Al_2O_3. They most strongly influence the fracture initiation toughness by creating pre-existing voids. Void nucleation can take place by either decohesion of the particle-matrix interface or by fracture of the particle. As the size of the particle increases the void nucleation becomes easier resulting in lower toughness. The steels were Ca-treated to minimize MnS inclusions which act as a source of crack initiation leading to decohesion of the matrix/inclusion interface and lower toughness. Calcium-treatment of steels resulted in round spinel-type inclusions that are less prone to crack initiation and propagation.

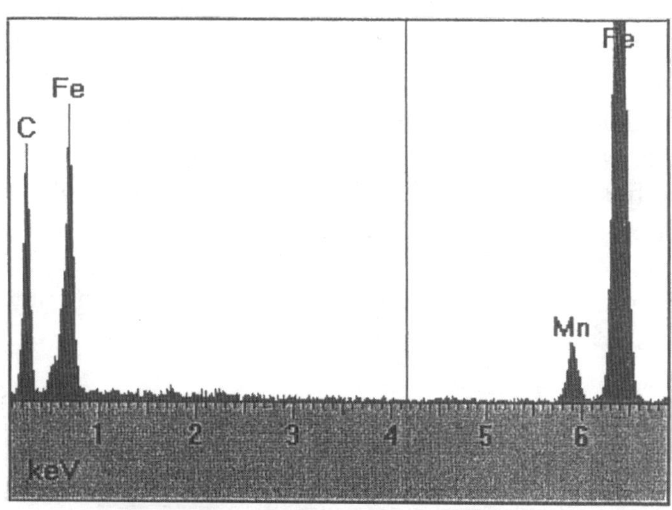

Fig.11. EDX confirming Mn-containing Fe_3C grain boundary precipitate in Fig. 9.

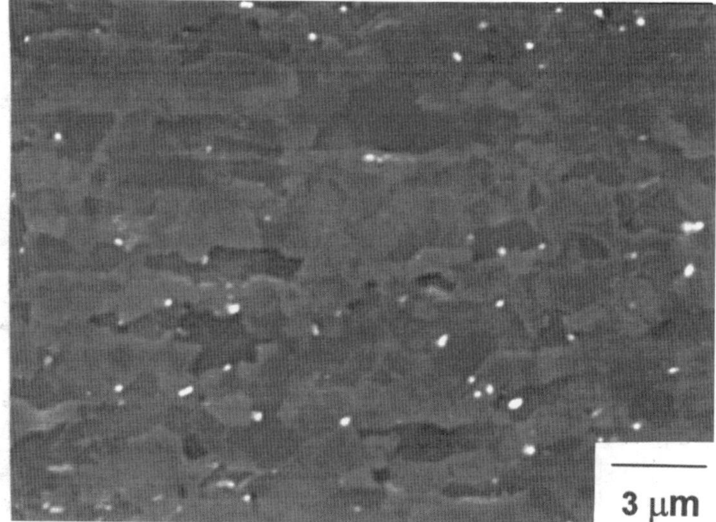

Fig. 12. Distribution of calcium aluminate inclusions in calcium-treated steels.

The rationale for the division between the primary and secondary particles is that the void nucleation will be much more difficult at the secondary particles and they will be either activated later or probably not at all. The voids nucleated at the primary particles can grow either until they impinge or until they coalesce via the formation of a void sheet. The secondary particles are undissolved nitrides or carbonitrides. The secondary particles can directly influence the fracture process if they actually nucleate voids and they can indirectly influence the fracture process by influencing the prior austenite grain size. As discussed in section 1(a), Ti and Nb were added to obtain grain refinement, in addition to precipitation strengthening.

Microstructural features other than primary and secondary particles are regarded as the fine scale microstructure. The fine scale microstructure can directly influence the fracture process if the fine particles precipitated during thermo-mechanical processing actually nucleate voids. These fine precipitates can also indirectly influence the fracture process through their role in determining the flow properties of the material. The flow properties can influence void nucleation and growth and the nature of crack tip blunting. Other microstructural features that are expected to indirectly influence the fracture process include bainite packet and lath sizes, their amounts, morphology and dislocation structure.

In the Ca-treated steels, the primary particles were fine spinel-type calcium inclusions ($CaO.Al_2O_3$) of size 100-200 nm fairly uniformly dispersed (Fig. 12) in the matrix. The evidence of secondary TiN particles (size range 0.3-3 microns) being involved in the fracture was difficult to observe. It is expected that if an increasing number of Ti-rich particles are involved in the fracture process, a decrease in toughness would be observed. It is interesting to note that the average primary dimple size was less than about 3 microns (Fig. 13). While the a majority of the voids nucleated at the Ca particles, a number of small voids nucleated at TiC particles, and a few large ones at Ti-rich nitride particles. For all the fracture surfaces examined it appeared that void initiation at the TiC particles occurred next to voids nucleated at primary Ca-inclusions. The fracture process in the present case can be visualized as follows. Voids are first nucleated at the primary Ca inclusions, and subsequently, these primary voids grow and at some point in time voids are nucleated at the TiC particles in the vicinity of the primary Ca-inclusions. Fracture occurred when primary and secondary voids coalesce. Voids at TiN particles (the large void in the left-hand top corner in Fig. 13) probably nucleated independently. The foregoing discussion emphasizes the importance of different particles and their size in determining the fracture resistance, and is not unique for the steels described here.

Another contribution to toughness comes from the bainite constituent. It is known that bainite, in general, provides significantly increased resistance to fracture[10]. The extent of fracture resistance, however, depends on the type of bainite, BI, BII, and BIII, as schematically illustrated in Fig. 8. The fine structure of the bainite assists in obstructing the initiated cracks by deflecting the crack at packet and lath boudaries. In the present study evidence for only upper bainite could be unambiguously confirmed. It is known that upper bainite (BII) has lower toughness compared to lower bainite (BIII). However, it is also believed that polygonal ferrite obtained by thermo-mechanical processing played a very important role in influencing toughness in this type of steels presumably by making the crack path small, and yielded low DBTT. This discussion is in accord with the earlier studies on low carbon commercial grade high strength steels where crack length was correlated with Charpy impact transition temperature[23].

(b) Texture

Besides the microstructure, alloying elements and precipitates, the ultimate properties of the steel are also dictated by the crystallographic texture of the material. During the controlled rolling of steels, three types of processes, namely, deformation, recrystallization, and transformation give rise to textures and induce texture changes[24]. The {332}<113> is the most beneficial among the transformation texture components from the point of view of achieving good combination of strength and toughness[24-26].

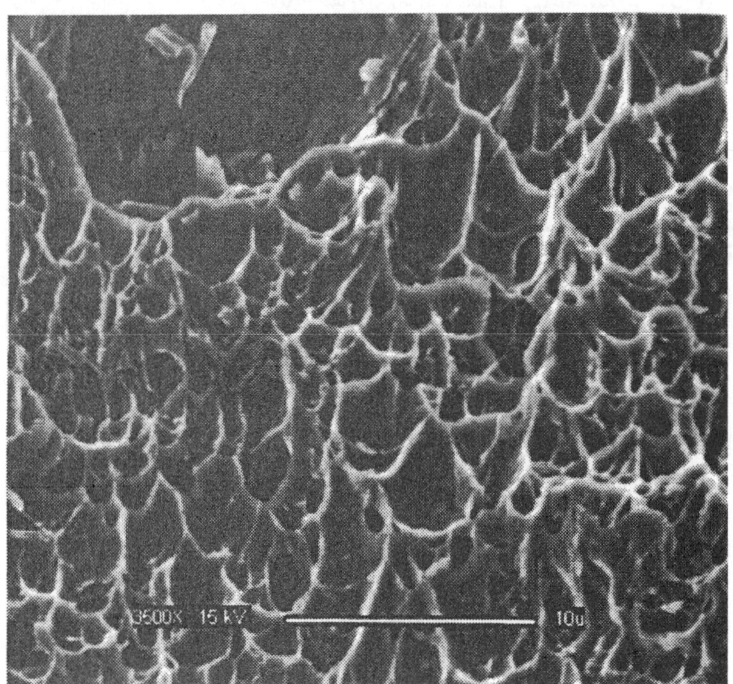

Fig. 13. SEM micrograph of the fracture surface of ultra-high strength steels. The large void in the top left-hand corner is nucleated by TiN.

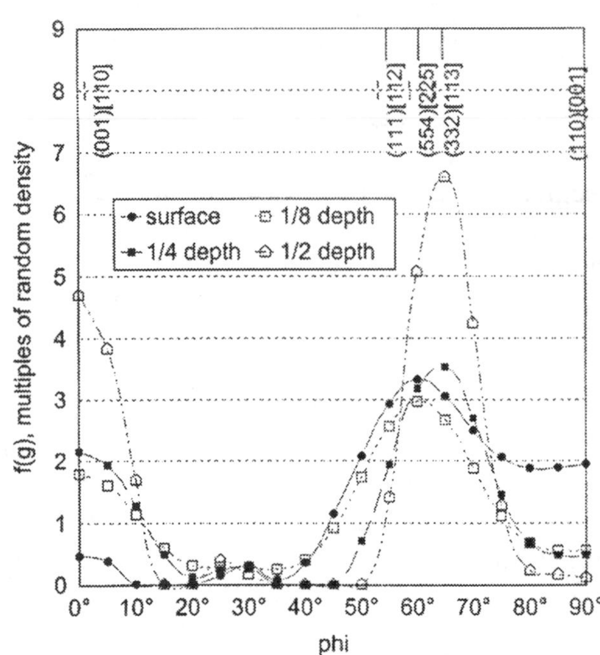

Fig. 14. Skeletal ε-fiber plot for the indicated depths-below the surface in ultra-high strength hot rolled steels.

Specimens for texture analysis were prepared by chemically attacking the hot mill top surface in a solution containing HF and H_2O_2. Depths selected for examination included the original surface and 1/8, 1/4 and 1/2-thickness positions.Orientation distribution functions (ODF) were calculated from the {110}, {211} and {222} x-ray diffraction pole figures themselves measured using a Scintag diffractometer. The data, corrected for background and defocusing, were collected up to an 80° tilt angle using Cu-Kα radiation and a Ge detector to eliminate Fe fluorescence counts. The pole figure outer rim estimates and ODF calculations (spherical harmonic, l=22 maximum) were accomplished using the popLA software package[27] after the method of Roe[28,29] and Bunge[30]. Samples were also examined as thinned from the bottom hot mill surface to observe any physical asymmetry, as reported by Hook[31] in hot rolled material. In our case, the pole figures measured with respect to both surfaces were comparable, indicative of negligible physical asymmetry in the materials. As is typical for rolled cubic materials, the pertinent textures in this study existed on the $\varphi_2 = 45°$ ODF planes (Bunge notation). Moreover, the textures present were part of one or more common fiber textures: α-fiber (<110> || RD), γ-fiber (<111> || ND) and ε-fiber (<110> || TD). In Fig. 14, an ε-fiber plot of relevance to toughness is presented. Important ideal {hkl}<uvw> orientations annotate the ε-fiber plot.

Fig. 14 shows, the center and 1/4-depth possesses considerable {332}<113>, which is the desired texture for toughness and may offset the undesired {100}<011>. This may well be the reason for good strength and impact toughness combination of the steels (Table III) in association with good edge formability[1].

The observed good toughness in steels can be attributed to (a) reduction in the size of the TiN particles (large particles can cause cleavage), (b) greater difficulty in nucleating voids at the primary spinel-type Ca-inclusions, (c) greater spacing between the primary particles, (d) a fine-grained and ferrite-bainite microstructure that is inherently more ductile, and (e) the high intensity of the desired {332}<113> texture.

REFERENCES

1. R.D.K. Misra, J.E. Hartmann and A.J. Boucek, Proceedings of 41st Mechanical Working and Steel Processing Conference, AISI, Pittsburgh, 1999, **37**, pp.509-514.
2. M. Prikryl, A. Kroupa, G.C. Weatherly, and S.V. Subramanian, Metall. Trans., 1996, **27A**, pp. 1149-1165.
3. C.E. Lacy and M. Gensamer, Trans. ASM, 1944, **32**, pp.88-95.
4. R.W.K. Honeycombe: Structure and Strength of Alloy Steels, Climax Molybdenum Co., Ann Arbor, MI, 1974.
5. F.B. Pickering, High Strength Low Alloy Steels – A Decade of Progress, Microalloying 75, New York, 1977, pp. 3-24.
6. S.A. Khan and H.K.D.H. Bhadeshia, Metall. Trans., 1990, **21A**, pp. 859-875.
7. R.D.K. Misra, C.Y. Prasad, T.V. Balasubramanian, and P. Rama Rao, Scripta Met., 1986, **20**, pp. 713-716.
8. D.V. Edmonds and R.C. Cochrane, Metall. Trans., 1990, **21A**, pp. 1527-1540.
9. H.K.D.H. Bhadeshia and J.W. Christian, Metall. Trans., 1990, **21A**, pp. 767-797.
10. H.K.D.H. Bhadeshia and A.R. Waugh, Acta Metall., 1982, **30**, pp. 775-784.
11. M.F. Ashby and R. Eberling, TMS-AIME, 1966, **236**, pp. 1396-1412.
12. D. Cheetham and N. Ridley, Met. Sci., 1975, **9**, pp. 411-414.
13. R.D.K. Misra, S.W. Thompson, T.A. Hylton, and A.J. Boucek, Metall. Trans., 2000 (in press).
14. P. Brozzo, G. Buzzichelli, A. Masconzoni, and M. Mirabile, Met. Sci., 1977, **11**, pp. 123-129.
15. T. Kunitake, Trans. Iron Steel Inst. Jpn., 1967, **7**, pp. 254-262.
16. M.J. Roberts, Metall. Trans., 1970, **1**, pp. 3287-3294.
17. J.P. Taylor and R. Blondeau, Metall. Trans., 1976, **7A**, pp. 891-894.
18. R.D.K. Misra and P. Rama Rao, Mater. Sci. Technol., 1997, **13**, pp. 277-288.
19. R.H. Edwards and N.F. Kennon, Metall. Trans., 1978, **9A**, pp. 1801-1809.
20. J.W. Christian and D.V. Edmonds: in Phase Transformations in Ferrous Alloys, Eds. A.R. Marder and J.I. Goldstein, AIME, New York, NY, 1984, pp. 293-325.
21. R.D.K. Misra, T.V. Balasubramanian, and P. Rama Rao, J. Mater. Sci., 1987, **6**, pp. 125-130.
22. W.M. Garrison, Jr., in Maraging Steels: Recent Developments and Applications, Eds. Richard K. Wilson, TMS, 1998, pp.177-209.
23. H. Ohtani, S. Okaguchi, Y. Fujishoro and Y. Ohmoni, Met. Trans., 1990, **21A** , pp. 877-888.
24. R.K. Ray and J.J. Jonas, Int. Met. Rev., 1990, **35**, pp.1-36.
25. PH. Lequeu and J.J. Jonas, Metall. Trans., 1988, **19A**, p.105-108.
26. H. Inagaki, K. Kurihara, and I. Kozasu, Trans. Iron. Steel Inst. Jpn., 1977, **17**, pp. 75-83.
27. J.S. Kallend, U. F. Kocks, A. D. Rollet and H.-R. Wend, Mater. Sci. Eng., 1991, **A132**, pp.1-10.
28. R.-J. Roe, J. Appl. Phys., 1965, **36**, pp. 2024-2031.
29. R.-J. Roe, J. Appl. Phys., 1966, **37**, pp. 2069-2072.
30. H.J. Bunge, Z. Metallknde., 1965, **56**, pp. 872-874.
31. R.E. Hook, Metall. Trans., 1993, **24A**, pp. 2009-2019.

An Integrated Model for Microstructural Evolution and its Application to the Properties Predictions of Plate Steels

Se Don Choo, Wung Yong Choo
POSCO/Tech. Res. Labs.
Plate,Rod & Welding Res. Gr.
1, Koedong-dong, Nam-gu
Pohang, Kyungbuk
South Korea
82-054-220-6079
sedonju@posco.co.kr

Key Words: Phase Transformation, Metallurgical Modeling, Mechanical Properties, Plate Steels

INTRODUCTION

Recently intensive works[1~3] have been conducted for the mathematical model to predict the final mechanical properties of hot rolled steel products. In view of the processing route, the model consists of three submodels for austenite structure, phase transformation and the mechanical properties, respectively. These submodels include numerous metallurgical relations which clarify the microstructural evolution and the structure-property relationships. As regarding the phase transformation submodel, the characteristics of the entire transformation process are not yet fully understood although phase transformations in steel have been investigated extensively since the last three decades. And semiemperical relationships are proposed to quantify the effect of chemistry and austenite grain size on the transformation kinetics. Most of these relationships are based on the so called Avrami type equation[4] which is powerful in describing the kinetic behaviors of isothermal phase transformation. For the Avrami type equations to be used practically, additive rule[5] should be applied since the recrystallization or transformation reaction take place under continuously cooled conditions in the industrial situation. However, the additive rule is still in doubt to be valid or not in the transformation from austenite to ferrite[6]. Besides the validity of the additive rule, the experimentally determined kinetic parameters in the isothermal transformation may have their own errors.

In this investigation, we have derived a new kinetic equation for austenite to ferrite transformation under continuous cooling condition. It was somewhat different from the conventional Avrami type equation. In this paper, the features of our model, focused on phase transformation from austenite to ferrite, are described. Combined with the temperature prediction model, this transformation submodel has been integrated to the full property prediction model. The temperature prediction model can calculate all the thermal histories of plate during reheating, rolling and cooling with the event data gathered through factory Local Area Network (LAN) in real time. For the property prediction submodel, a regression between the mechanical properties and the microstructure of hot rolled plate was carried out. To evaluate the developed model the predicted tensile strength were compared to the measured ones for about 5000 plates produced at POSCO plate mill.

PHASE TRANSFORMATION MODEL

Isothermally Cooled Transformation

The Avrami type equations are widely used for the prediction of nucleation and growth phenomena such as recrystallization, precipitation of carbonitrides and phase transformation. In the case of phase transformation of steels, the variation of fraction transformed, X_t, with time ,t , is described as :

$$X_t = 1\text{-exp}(-A\ t^n) \text{ -------------------------- } (1)$$

where A and n are constants which are dependent on steel chemistry and temperature. These constants are determined by experimental data. The Avrami type equations are valid for isothermal phenomena if the values of constants in equations are properly determined. However, the Avrami type equation can not be applied directly to the field condition where the temperature is continuously varied. Equation (1) was simply derived as follows. The transformation velocity, dX_t/dt, for all the diffusion controlled process is proportional to both non-transformed fraction, $(1-X_t)$, and diffusivity which has the Boltzmann type equation. So the transformation velocity can be expressed as:

$$dX_t/dt = A1\times (1-X_t) \times \text{exp}(-Q/RT) \text{ ------------ } (2)$$

where A1 is proportional constant, Q is activation energy for transformation and R is gas constant.
For the isothermal reaction , equation (2) can be rewritten as:

$$dX_t/dt = A2\times (1-X_t) \text{ -------------------------- } (3)$$

Solving the differential equation (3), we can obtain the equation (4) which is exactly same as the equation (1) if the exponent n is considered to be unity.

$$X_t = 1\text{-exp}(-A2\cdot t) \text{ ------------------------- } (4)$$

Experimentally, the n values for ferrite and pearlite transformation were found to range from 0.90 to 1.50.

Continuously Cooled Transformation

We have derived a new kinetic equation describing the transformation from austenite to ferrite under continuous cooling with the various cooling rate. The starting point was same as for the isothermal transformation. However, different from the isothermal transformation, the temperature related term in equation

(2), exp(-Q/RT), is not constant during the continuous cooling transformation. And for the austenite to ferrite transformation of low carbon steel the activation energy, Q, is given by[7]:

$$Q = (C_\gamma - C_\alpha)/\, C_\gamma \cdot RT \cdot \ln((1-C_0)/(1-C_\gamma)) \quad \text{---------------} \quad (5)$$

where C_γ is the carbon contents in austenite during transformation, C_α is the carbon contents in ferrite during transformation and C_0 is the matrix carbon contents.

For the low carbon steel, carbon contents C_0 and C_γ are extremely lower than unity. Using the logarithmic series, $\ln(1-X) \approx X$ when $X << 1$, we have:

$$\ln((1-C_0)/(1-C_\gamma)) = \ln(1-C_0) - (1-C_\gamma) = C_\gamma - C_0 = \Delta C \quad \text{---------------} \quad (6)$$

Fig.1 shows the schematic Fe-C diagram, where both the phase boundaries between $\alpha/\alpha+\gamma$ and between $\gamma/\alpha+\gamma$ are assumed to be straight. As transformation proceeds with decreasing temperature, the differences of carbon contents, ΔC, is proportional to the supercooling, ΔT, as shown in Fig. 1. ΔT is also proportional to the time after beginning of transformation, t, because the cooling rate was set to be constant during the transformation.

$$\Delta C = k\,\Delta T = k\,(CR \times t) \quad \text{-------------------------} \quad (7)$$

where CR is constant cooling rate and k is proportional constant.

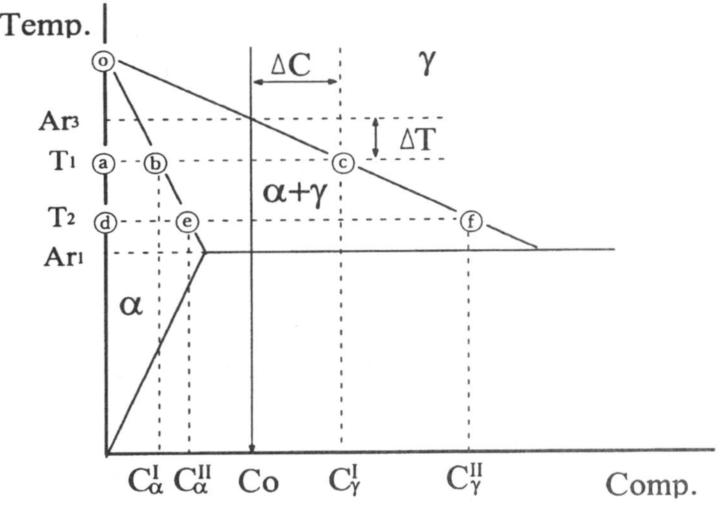

Fig.1 Schematic Fe-C phase diagram

We can also see in fig.1 that the triangle OBC is similar to the triangle OEF and the triangle OAC is similar to the triangle ODF. So, the content ratio, $(C_\gamma-C_\alpha)/ C_\gamma$ at an arbitrary temperature T1 is same to $(C_\gamma-C_\alpha)/ C_\gamma$ at another arbitrary temperature T2. It means that the content ratio, $(C_\gamma-C_\alpha)/ C_\gamma$, in equation (5) is constant during the transformation.

$$(C_\gamma^{I}-C_\alpha^{I})/ C_\gamma^{I}= bc/ac = ef/df =(C_\gamma^{II}-C_\alpha^{II})/ C_\gamma^{II}= B \text{ (constant)} \text{ ------ } (8)$$

Substituting equation (7) and (8) into the equation (5) we can obtain equation (9).

$$\exp(-Q/RT) = \exp(-B\times k\times CR\times t) \text{ ------------------------------- } (9)$$

Because the values of B and k are constant during transformation, product of them can be replaced another constant β, which is not dependent on time but dependent on composition and CR. Substituting equation (9) to equation (2), we have:

$$dX_t/dt = A1\times (1-X_t) \times\exp(\beta\times CR\times t) \text{ ----------- } (9)$$

Equation (9) is merely a homogeneous linear differential equation, which will be solved easily. Solving it with the assumption that the transformation begins with the point where 5% transformation proceeds, we have the final equation:

$$\ln(-\ln(1-X_t)) = \beta\times CR\times t -2.97 \text{ ---------------------- } (10)$$

Here the time, t, means the duration after 5% transformation. Equation (10) is a governing equation predicting the transformed fraction at any instant after the start of transformation. It is for the prediction of transformation behaviors occurred under continuous cooling with a given cooling rate. We have constructed a new property prediction model based on the equation (10) and we have evaluated the new model compared to the traditional model based on the Avrami type equation. It will be described in the next sections

Comparisons Between Avrami Type Equation and New Equation

Although the Avrami type equation is the most commonly used tool for tracking volume fraction kinetics of the isothermal phase transition, in authors' opinions, it has several weak points of itself. At first, there may be some experimental errors in determining both the start point and the ending point of the isothermal transformation where the holding temperature is apart from the nose temperature. At these temperatures some inevitable electrical noises may cause the transformation curves disturbed because the driving force of transformation is low and constant with time. On the other hand, in the continuous cooling experiments where the driving force is increasing with time, the transformation is pronouncedly accelerated with time. So we can obtain the starting point of transformation with higher confidence. Secondly, for an isothermal experiment of some low carbon steels whose nose time is too short, transformation may be already proceed even though the specimen is quenched to the aiming temperature as fast as possible. In this case, the transformation curves are incorrectly obtained. Under the continuous cooling, we can obtain the full transformation curves with no limit of composition ranges. Finally, different from the Avrami type equation, the new equation can be applied to the practical situation without the aid of the additive rule. With these reasons, we think that the new equation is more useful in predicting the transformation behaviors of steel plate.

ORGANIZATION OF THE INTEGRATED MODEL

Based on the new kinetic equations, the integrated property prediction model was established. The organization of the model is shown in Fig.2

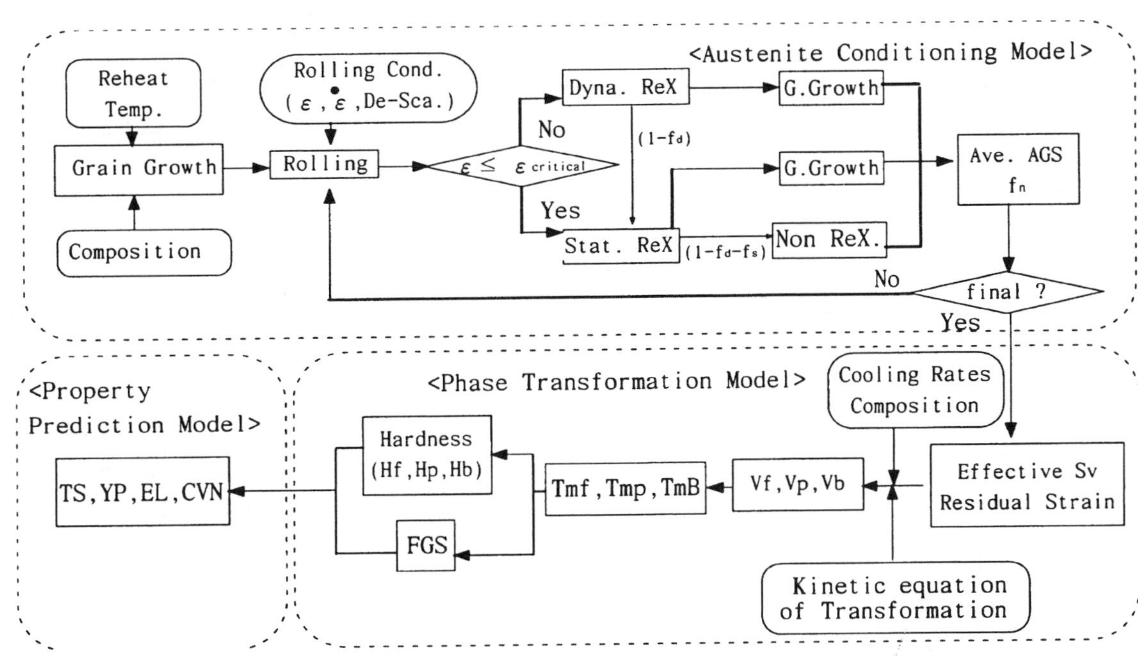

Fig. 2 Three submodels for the prediction of microstructute and mechanical properties

Data Gathering System

A specially designed data gathering system was developed and installed in plate mills of POSCO. Database of mainframe is composed of process data such as composition, reheating, strain, strain rate, interval time of rolling and cooling pattern for each plate. And all the data are gathered through LAN in real time and used for the temperature prediction model which calculates the thermomechanical histories of a plate. All the data needed for three submodels in Fig.2 are supplied by the data base and the temperature prediction model.

Austenite Structure Model

The microstructural change during hot rolling are grain growth in the reheating furnace, work hardening during plastic deformation, recrystallization during inter-pass time and grain growth after full recrystallization. These phase transition behaviors were described in the austenite conditioning model. All the equations used in the austenite conditioning model for C-Mn steels are listed in Table 1.

Table 2. Equations in the austenite conditioning model

Reheating grain size	$D_0 = 2.93\times10^5 t^{0.194} \exp(-94000/RT)$
Critical strain for dynamic recrystallization	$\varepsilon_c = 0.8\times1.573\times10^{-4}\times D_\gamma^{0.5}\times Z^{0.17}$ Z = (strain rate) $\times\exp(Q_d/RT)$ where Q_d = 375kJ/mole
Dynamic recrystallized grain size	$d_{mrx} = 2.6\times10^4 Z^{-0.23}$,where Qd=300kJ/mole,
Time constant for 50% dynamic recrystallization	$t_{50\%} = 1.1\times Z^{-0.8}\exp(230,000/RT)$,where Qd=300kJ/mole
Grain growth after dynamic recrystallization	$D_{mrx}{}^7 = d_{mrx}{}^7 + 8.2\times10^{25}(t_{ip}-t_{95\%})\exp(-400kJ/RT)$ where t_{ip} : interpass time between rolling pass
Static recrystallized grain size	$dsrx = 0.5\times\varepsilon^{-1}\times D_o^{0.37}$
Time constant for 50% static recrystallization	$t_{50\%} = 2.5\times10^{19}\varepsilon^{-1} D_o^{0.37}\exp(Q_{sr}/RT)$ where Q_{sr}=327kJ/mole at $T > 1277K$ Q_{sr}=780kJ/mole at $1277K > T > 1164K$ Q_{sr}=130kJ/mole at $1164 < T$
Grain growth after static recrystallization	$D_{srx}{}^7 = d_{srx}{}^7 + 1.5\times10^{15}(t_{ip}-t_{95\%})\exp(-400kJ/RT)$

The time constant for 50% recrystallization in table 1 is obtained from the isothermal recrystallization experiments and the kinetic equation is based on the Avrami type equation. However, in plate mill where the interpass time is approximately 10sec, temperature drop between the rolling pass is considerable. So the additive rule should be applied for the Avrami type equation to be used in plate mill. We have derived another kinetic equation for the recrystallization occurred under continuous cooled transformation. The derivation process will be described elsewhere in detail. The final equation was expressed as equation (14);

$$t_{50\%} = 2.5\times10^{19}\varepsilon^{-1} D_o^{0.37}\exp(Q_{sr}/RT_o) \text{ ---------------- } (11)$$

where T_o : temperature just after rolling

$$k = CR*Q_{sr}/T_o^2 \text{ -- } (12)$$

where CR : cooling rate during interpass time

$$A = \ln(2)/(1-\exp(-kt_{50\%})) \text{--------------- } (13)$$

And

$$Xt = 1-\exp(-A\times (1-\exp(-kt))) \text{ ------------------------ } (14)$$

With equation (14), we can predict the recrystallized fraction of austenite during the interpass time for each pass

Important outputs of the austenite conditioning model are the austenite grain size and the accumulated strain retained after hot rolling. These properties of austenite play important roles on subsequent austenite-ferrite transformation during cooling.

Phase Transformation Model

Hot rolled austenite transforms to ferrite, pearlite or bainite. These phase transformation are described by the transformation submodel. Both the microstructure of austenite and the cooling condition affect transformation products, and they determine the type of transformation products, the volume fraction of each phases and the ferrite grain size. For the transformation model derived here to be used, the kinetic constant, β, in equation (10) should be known. The transformation start temperature and volume fraction of 2nd phases according to the chemistry and cooling rate are also necessary. We measured the continuous cooling transformation kinetics using a dilatometer for 16 kinds of plain steels whose compositions range C: 0.05%~0.2%, Mn:0.5%~2.0%. The cooling rate was varied between 0.2°C/sec and 30 °C/sec. The experimental data were statically treated and the kinetic parameter, β, was revealed to be dependent on the chemical composition and cooling rate.

$$\beta = 0.114+0.268\times(\text{wt \%C})-0.264\times(\text{wt \%C})^{0.5}-0.01356\times(\text{wt \%Mn})^{0.5}-0.005\times(CR)^{0.5} \text{ -------- } (15)$$

The transformation start temperatures , Ar_3, for various conditions were also statically treated and we obtained an regression equation as:

$$Ar_3 = 925.95-494.7\times(\text{wt \%C})-64.8\times(\text{wt \%Mn})-10\times(CR)^{0.5} \text{ ----------------------------- } (16)$$

Using equations (15) and (16), we can predict the transformation curves for various cooling rates. For an example, the predicted CCT diagram of 0.1C-1.0Mn steel is shown in Fig.3. The temperatures where the transformed volume fraction are 10%, 50% and 90%, are well correspondent to the measured ones.

The microstructures after transformation were investigated and the volume fraction of second phases were measured for every condition. The ferrite grain size was also measured. Statistically treated, we could obtain the following equations:

$$V_p = -0.00548+1.31\times(\text{wt \%C})+0.0289\times(\text{wt \%Mn})-0.01842\times\ln(CR) \text{ ------------- } (17\text{-a})$$
$$V_b = -0.18+1.2\times(\text{wt \%C})+0.09\times(\text{wt \%Mn})+0.0544\times\ln(CR) \text{ ---------------------- } (17\text{-b})$$
$$V_f = 1-(Vp+Vb) \text{ --- } (17\text{-c})$$

where V_f, V_p, V_b are the volume fractions of ferrite, pearlite and bainite respectively.

And the ferrite grain size

$$D_\alpha = 7.32\times10^{-17} \times D_\gamma^{0.442} T_{mf}^{5.858} \times V_f^{0.65} \quad \text{------} \quad (18)$$

where D_γ is austenite grain size before transformation and T_{mf} is the 50% transformation temperature.

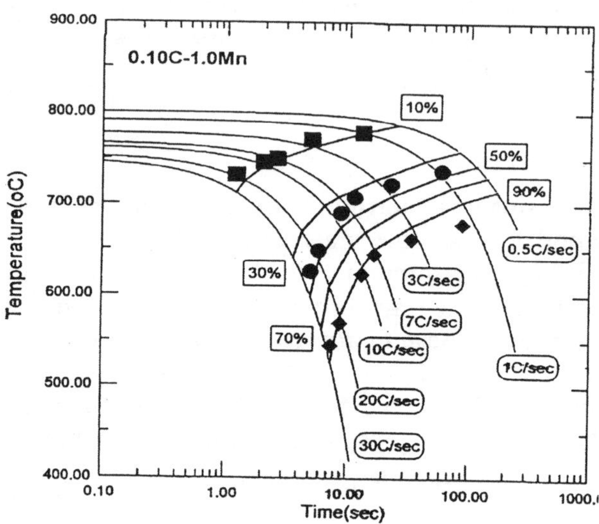

Fig. 3 Comparison between the calculated and observed results of CCT diagram

Property Prediction Model

One of the most important microstructural parameters is the ferrite grain size, for which the relation with strength is expressed by the well known Hall-Petch equation[8~9]. In describing the mechanical properties of composite aggregates such as steel, the mixture law[10] was found to be useful. In this investigation, we also used the following equation to predict the tensile strength

$$TS = a \times H_f V_f + b\times H_p V_p + c\times H_b V_b + d\times Vf/(FGS)^{0.5} + e \quad \text{------} \quad (19)$$

The volume fraction of each phase and ferrite grain size are calculated from the phase transformation model and the hardness of each phase transformed was given as:

$$H_f = 361-0.357\times T_{mf}+50\times (\%Si) \quad \text{------} \quad (20\text{-a})$$
$$H_p = 305.6-0.1832\times T_{mp}+84.92\times (\%Si)+2.203\times (\%Mn) \quad \text{------} \quad (20\text{-b})$$
$$H_b = 508-0.588\times T_{mb}+50\times (\%Si) \quad \text{------} \quad (20\text{-c})$$

The T_{mi} are the mean transformation temperature of i phase and calculated in the phase transformation model. The coefficients a~e in equation (19) was determined by a multiple regression method

EVALUATION

The proposed prediction model was evaluated compared to the conventional models which were based on the Avrami type equations. Fig. 4 through Fig.6 show predicted versus measured values of tensile strength for about 5,000 plain carbon steel plates. The measured results were from the production data in POSCO plate mill .

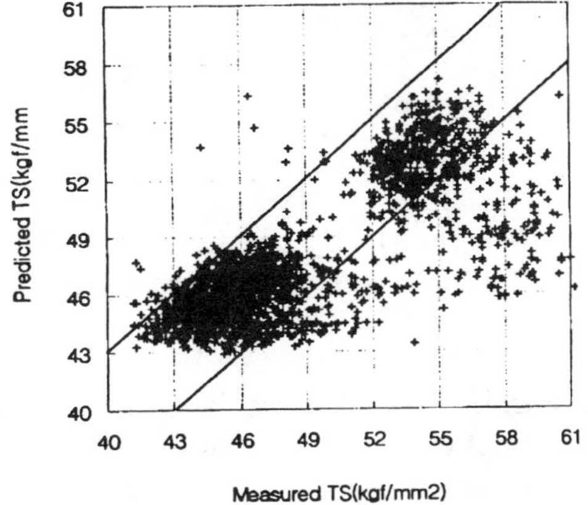

Fig. 4 Comparisons of calculated tensile strength by prediction model with the measured ones. Both of recrystallization and transformation behaviors were predicted by the isothermal model

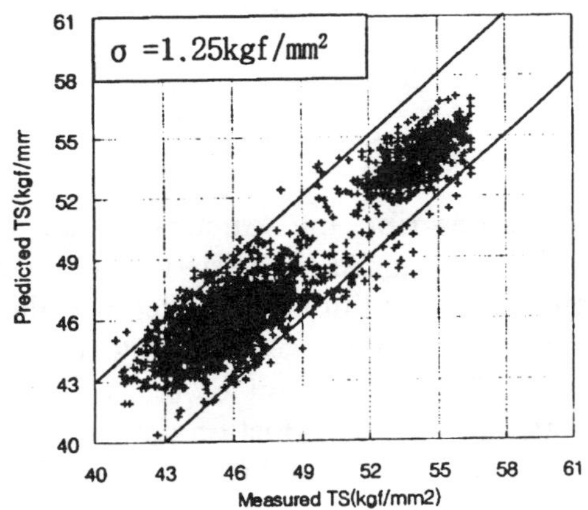

Fig. 5 Comparisons of calculated tensile strength by prediction model with the measured ones. Transformation behaviors were predicted by the continuously cooling model while recrystallization behaviors were predicted by the isothermal model

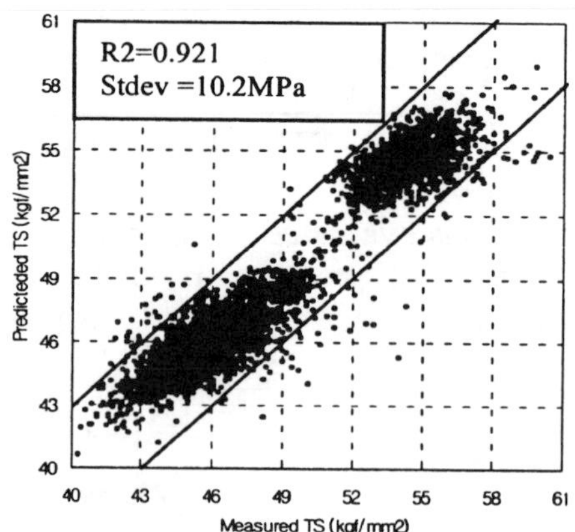

Fig. 6 Comparisons of calculated tensile strength by prediction model with the measured ones. Both of recrystallization and transformation behaviors were predicted by continuously cooling models which are proposed in this work.

Three kind of models were evaluated. Fig.4. shows the results of prediction in which both of recrystallization and transformation models are based on the Avrami type equations. And fig.5 shows the results of prediction in which transformation behaviors were predicted by the continuously cooling model while recrystallization behaviors were predicted by the isothermal model. Finally, fig. 6 shows the results of prediction in which both of recrystallization and transformation models which are based on the proposed models in this work. Among the models, the last one shows the greatest accuracy and the standard deviation of the residuals is about 10MPa. In the authors' opinions, this prediction error is also the level of accuracy that is reproducible with the available accuracy of measured data.

SUMMARY

A new model was suggested to describe the kinetics of austenite to ferrite transformation under continuous cooling condition. It was derived on the simplified Fe-C phase diagram and some thermodynamic theories were considered. For the austenite conditioning model we also derived an equation describing the recrystallization behaviors during the interpass time of plate rolling. These submodels were integrated into a full microstructural model which were applied to prediction of tensile strength. The full model developed was evaluated for the plain carbon steel plates with the production data in POSCO plate mill. Comparisons with the conventional property prediction models were also conducted. Among the models, the prediction accuracy of the model developed in this investigation was highest. The standard deviation of the residuals is about 10MPa, which is enough to be used.

REFERENCES

1. J. Andorfer, D.Auzinger, M. Hirsch G.Hubmer and R Picher, "Controlling the Mechanical Properties of Hot Rolled Strip" MPT International, No.5 1997, pp.104-110
2. M. Piette and C. Petrix, " An Integrated Model for Microstructural Evolution in the Hot Strip Mill and Tensile Properties Prediction of Plain and Microalloyed C-Mn Hot Strip", Material Science Forum, Vol.284-286, 1998, pp.361-368
3. G.Wang, Z.Liu and W.Gao, "Visual Modelling of Thermal and Microstructural Evolution during Hot Strip Rolling", THERMEC 97 1997, pp.2141-2147
4. M. Avrami, " Kinetics of Phase Change. I; General Theory", Journal of Chemical Physics, Vol.7, December 1939, pp.1103-1112
5. E.Sheil, Arch. Einsenhuttenw., Vol 12, 1935, p.565
6. M. Lusk and H.Jou, " On the Rule of Additivity in Phase Transformation Kinetics", Metallurgical and Materuals Transactions A, Vo 28A, Feb. 1997, pp.287-291.
7. R.Doherty, "Diffusive Phase Transformation in the Solid State", Physical Metallurgy, North-Holland Physics Publishing, Amsterdam,1983, pp.937-1030
8. E.Hall, Proc.Phy.Soc.,London, B64, 1951, p747
9. N.Petch, J.Iron and Steel Inst., 174 , 1953, p25
10. Y. Tomota, K.Kuroki and I.Tamura, Tetsu-to-Hagane, Vol.61, 1975, p107

MICROSTRUCTURAL STUDY OF GRADE 100 STEEL

Siamak Akhlaghi, Douglas G. Ivey and Laurie Collins*
Department of Chemical & Materials Engineering,
University of Alberta, Edmonton, Canada T6G 2G6
*IPSCO Inc., Regina, Saskatchewan, Canada
email: siamak@ualberta.ca

Key Words. Microalloyed steel, Bainitic ferrite, Allotriomorphic ferrite, (Ti,Nb)(CN) precipitates.

INTRODUCTION

Considerable work has been done to improve the properties of microalloyed steels by controlling the evolution of the microstructure.[1] It is well established that refinement of the microstructure promotes the improvement of the mechanical properties.[1-6] Grain refinement can be achieved by thermomechanical processing and/or changing the composition. However, thermomechanically rolled conventional steels usually exhibit ferritic-pearlitic structural components with a defined minimum grain size. The resulting limited toughness properties put restrictions on the usability of these kind of steels and further grain refinement is only possible by changing the microstructure to ferritic-bainitic via changing the composition.[5] The development of bainitic ferrite microstructures enhances both the toughness and the strength. Bainitic ferrite is a phase acicular in form and individual ferrite laths have a width that is typically of the order of 1 μm.[4] The results obtained from different researchers show that controlling the composition, using a themomechanical processing schedule like the one is shown in Fig. 1, and accelerated cooling from the finishing temperature assists in the formation of bainitic ferrite in microalloyed steels. In the shown thermomechanical schedule, the slab is first reheated to elevated temperatures for dissolution of the precipitated carbonitrides and then rough rolled to produce fine, polygonal austenitic grains by means of recrystallization. This is followed by finish rolling below the no recrystallization temperature for austenite to generate elongated, highly deformed austenite grains. Also, it has been reported that in order to develop bainitic ferrite in the microstructure, it is necessary to suppress the formation of high temperature transformation products. This may be accomplished through additions of elements like Mn and Mo.[2-3]

The extent to which strength increases by refining the microstructure in microalloyed steels is limited and any further increase is only possible by introducing precipitates. Microalloying elements like Ti, Nb and V form carbonitrides that decrease the mobility of dislocations and consequently increase the strength of the steels.[7]

Recently, a new grade of structural steel, Grade 100, has been developed to achieve the high strength needed without jeopardizing the toughness. Table I shows the mechanical properties of this newly developed steel.[8] Mechanical properties of two previously developed microalloyed steels are shown as well for comparison.[9] As is illustrated, changing the composition and adding more alloying elements, specifically Nb and Mo [9], gives rise to a significant increase in strength. In the present work, the microstructural characterization of this newly developed microalloyed steel, containing 0.08%C, 1.8%Mn and alloy additions of titanium, niobium, vanadium and molybdenum will be exhibited. Microstructures are studied by means of optical and electron microscopy and results are compared with reports of other investigators.

EXPERIMENTAL PROCEDURE

The steel plate studied in the present work was reheated to about 1250ºC in a reheating furnace, rough rolled at temperatures above 1000°C, finish rolled in the no recrystallization region, followed by accelerated cooling at moderate cooling rates and finally coiled at 500-600°C.

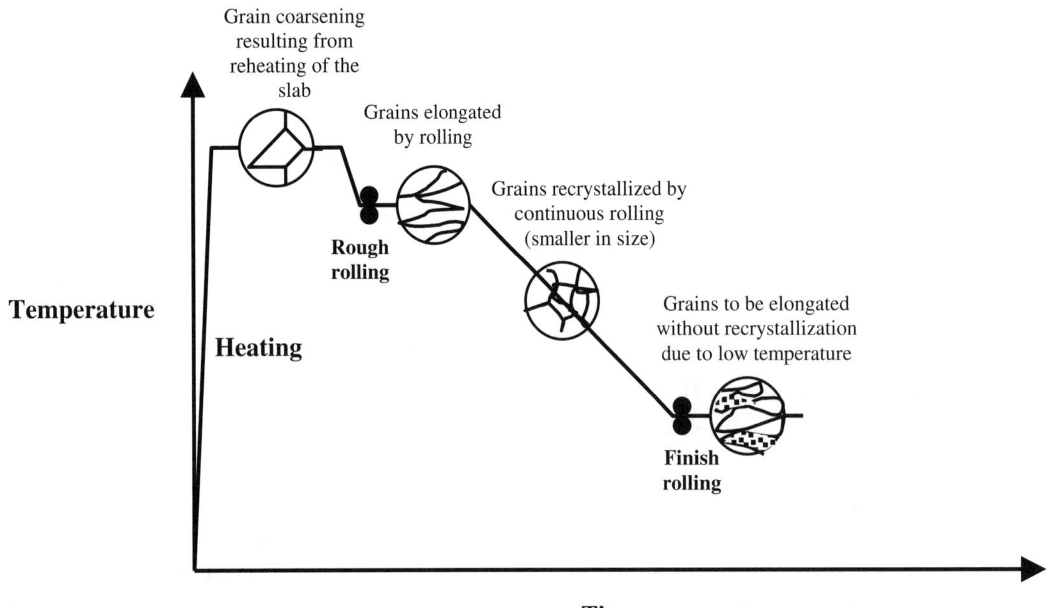

Fig. 1: Schematic illustration of thermomechanical processing.[5]

Table I Mechanical properties of Grade 100 steel, as well as two other microalloyed steels

Properties	X70	X80	Grade 100
Yield Strength (MPa)	545	568	723
UTS (MPa)	642	668.5	764

Sections of the plate in directions parallel and perpendicular to rolling were investigated. Initial microstructural examination was performed using an optical microscope and a Hitachi S-2700 scanning electron microscope (SEM) operated at 20 kV, equipped with an ultra thin window (UTW) Ge x-ray detector. Different microstructural features were brought out by etching polished specimens with a mixture of 10% nital and picral solution.

Grain diameters were determined by measuring the average intercept length. Microhardness measurements, performed with a 25 g load, were made of the different constituents of the microstructure. Inclusion distribution was determined from polished, unetched, specimens.

Several specimens were selected for more detailed examination in a JEOL 2010 transmission electron microscope (TEM). Two types of specimens were examined, i.e., thin foils and carbon extraction replicas. Thin foils were prepared by cutting thick slices (250-350 μm) from the side parallel to rolling direction. The slices were mechanically abraded to about 100 μm thickness. Discs of 3 mm diameter were punched out and thinned to perforation in a double jet electropolisher with acid-glycerin-ethanol electrolyte at -20 to -30°C, followed by a sputtering (<30 min) in an ion mill.

To prepare carbon extraction replicas, for precipitation analysis, mechanically polished specimens were lightly etched in 2% nital and carbon coated in a vacuum evaporator. Coated specimens were etched in 10% nital. Individual carbon replicas were floated in ethanol and picked up on 150 mesh Cu grids.

RESULTS

Phase characterization

The various phases identified are shown in Figs. 2-4. The optical micrograph (Fig. 2), taken from the section perpendicular to rolling direction, shows a microstructure that is composed of mostly bainitic ferrite (denoted as BF) and to the lesser extent, allotriomorphic ferrite (denoted as AF). At higher magnifications, it is clear that allotriomorphic ferrite forms at the prior austenite grain boundaries, along with Widmanstatten ferrite (denoted as WF) which forms at the same location (Fig. 3). This figure was taken from a section parallel to the rolling direction. Elongation of the parent austenite grains in the direction of rolling is the other feature revealed in Fig. 3. Fig. 4 illustrates the presence of pearlite (denoted as P) in the microstructure. Fig. 5 shows the microstructure at the top surface of a cross section. The microstructure is finer relative to the bulk microstructure and is entirely bainitic ferrite. Table II shows the average intercept length of the microstructure in the bulk and at the surface of the plate.

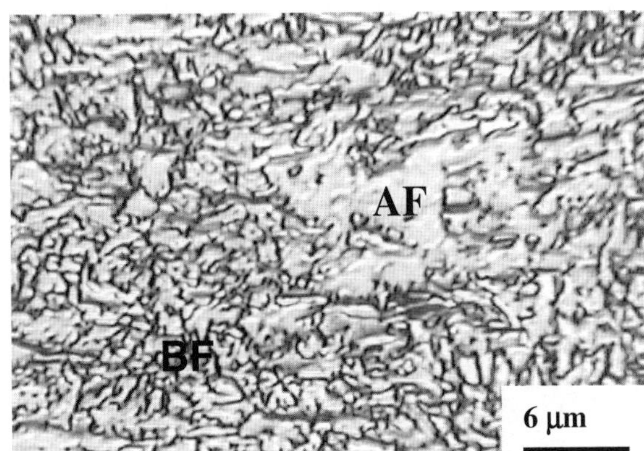

Fig. 2: Optical micrograph showing the presence of bainitic ferrite (BF) and allotriomorphic ferrite (AF). The sample is taken perpendicular to the rolling direction.

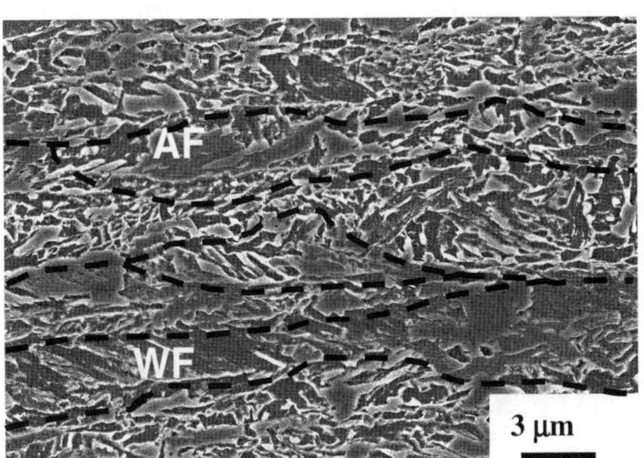

Fig. 3: SEM secondary electron image showing the presence of allotriomorphic ferrite (AF) and Widmanstatten ferrite (WF), which have formed along the prior austenite grain boundaries. The sample is taken parallel to the rolling direction.

A TEM bright field image taken from a thin foil specimen is shown in Fig. 6a. The microstructure consists of polygonal ferrite (Grain A) and laths of bainitic ferrite (Grains B and C). The corresponding selected area diffraction (SAD) patterns of grains A, B and C are shown in Figs. 6b, 6c and 6d, respectively. The SAD patterns of Grain B and Grain C (Figs. 6c and 6d) were taken at the same electron beam-specimen orientation. All diffraction patterns match with the b.c.c. α-ferrite crystal structure. There is a small misorientation between Grains B and C, which can be measured from the SAD patterns as ≈3°.

Microhardness of different phases

Microhardness measurements for allotriomorphic ferrite, Widmanstatten ferrite and bainitic ferrite are listed in Table III. Allotriomorphic ferrite has the lowest hardness, while Widmanstatten ferrite is the hardest constituent. The microhardness of bainitic ferrite at the surface and center is shown as well. As can be seen, in spite of the lower average intercept length at the surface relative to the interior, the microhardness is lower at the surface.

Distribution of inclusions in microstructure

Energy dispersive x-ray (EDX) analysis, in the SEM, of the inclusions showed that the most frequently observed inclusion types were oxy-sulfides of Ca and Al, Fig. 7. They are most likely complex inclusions of Al_2O_3-CaO-CaS with CaS as the major constituent.[10] The Fe and Mn peaks in the EDX spectra come from the ferritic matrix.

The distribution of inclusions in samples taken perpendicular to the rolling direction is illustrated in Fig. 8. As shown, the precipitates are generally less than 8 μm in size, however, a few inclusions were a few hundred in μm size. Also, the average areal fraction of the surface covered with inclusions was about 0.5%. This indicates that the inclusion severity level, according to ASTM Standard E-45, is fairly low.[11]

Precipitates

The precipitates could be classified into four types, based on their size and composition. The first type was coarse, cuboidal shaped TiN particles, several μm in size, which were formed alone or at the rim of inclusions (Fig. 9a). The second type of precipitates was intermediate-size Ti-rich carbonitrides, 100 to 300 nm in size (Fig. 9b). Nb-rich precipitates, slightly smaller than the previous ones, i.e., 30 to 150 nm, were also observed, either attached to the sides of Ti-rich precipitates (Fig. 9b) or formed separately in the microstructure. The last type of precipitates was very fine Nb-rich carbonitrides, less than 10 nm in size, formed within ferritic grains at sub-grain boundaries or dislocations developed by deformation during rolling at temperatures below the no recrystallization temperature (T_{nr}) (Fig. 9c). Details of the precipitate behavior are discussed elsewhere.[12]

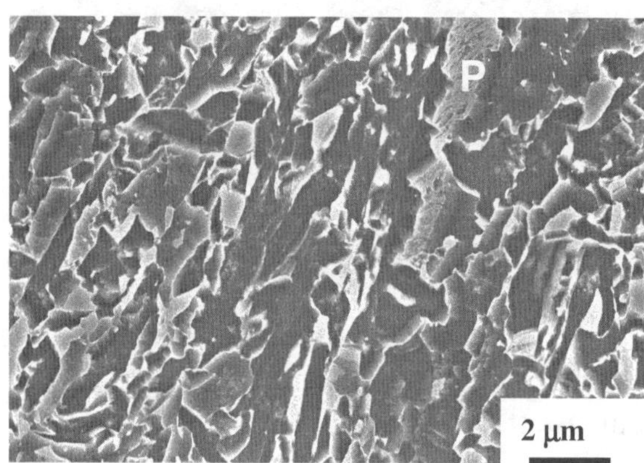

Fig. 4: SEM secondary electron image showing the presence of pearlite (P) in the microstructure.

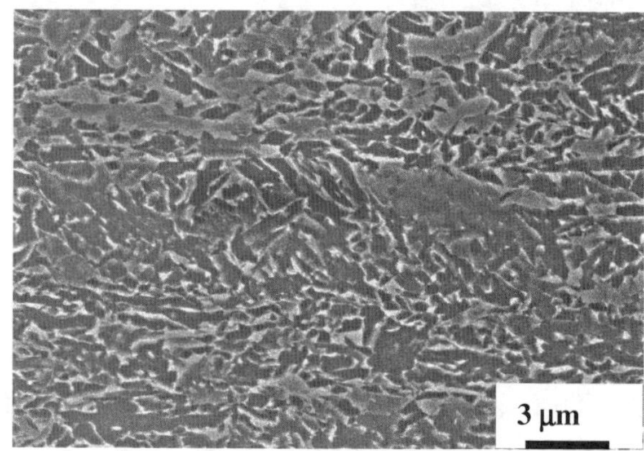

Fig. 5: SEM secondary electron image showing refinement of the microstructure at the plate surface. The sample is taken parallel to the rolling direction.

Table II Average intercept length of Grade 100 steel

Region	Parallel to rolling direction	Perpendicular to rolling direction
Surface	2.78 μm	2.36 μm
Center	3.68 μm	3.30 μm

Table III Microhardness (VH_{25g}) of different constituents and locations in the microstructure

	Allotriomorphic ferrite	Widmanstatten ferrite	Bainitic ferrite at interior	Bainitic ferrite at surface
Microhardness	259 VH ±2	309 VH ±3	298 VH ±12.2	269 VH ±3.5

DISCUSSION

Phase Formation

A continuous cooling transformation (CCT) curve of the present steel, in the absence of deformation [8], is shown in Fig. 10. The dashed line shows the cooling rate that the plate surface experiences during accelerated cooling. This graph shows that the microstructure, in the absence of deformation, is entirely bainitic. However, the thermomechanical history of the plate prior to reaching the accelerated cooling stage causes a considerable change in austenite grain size and segregation pattern, leading to microstructural differences.

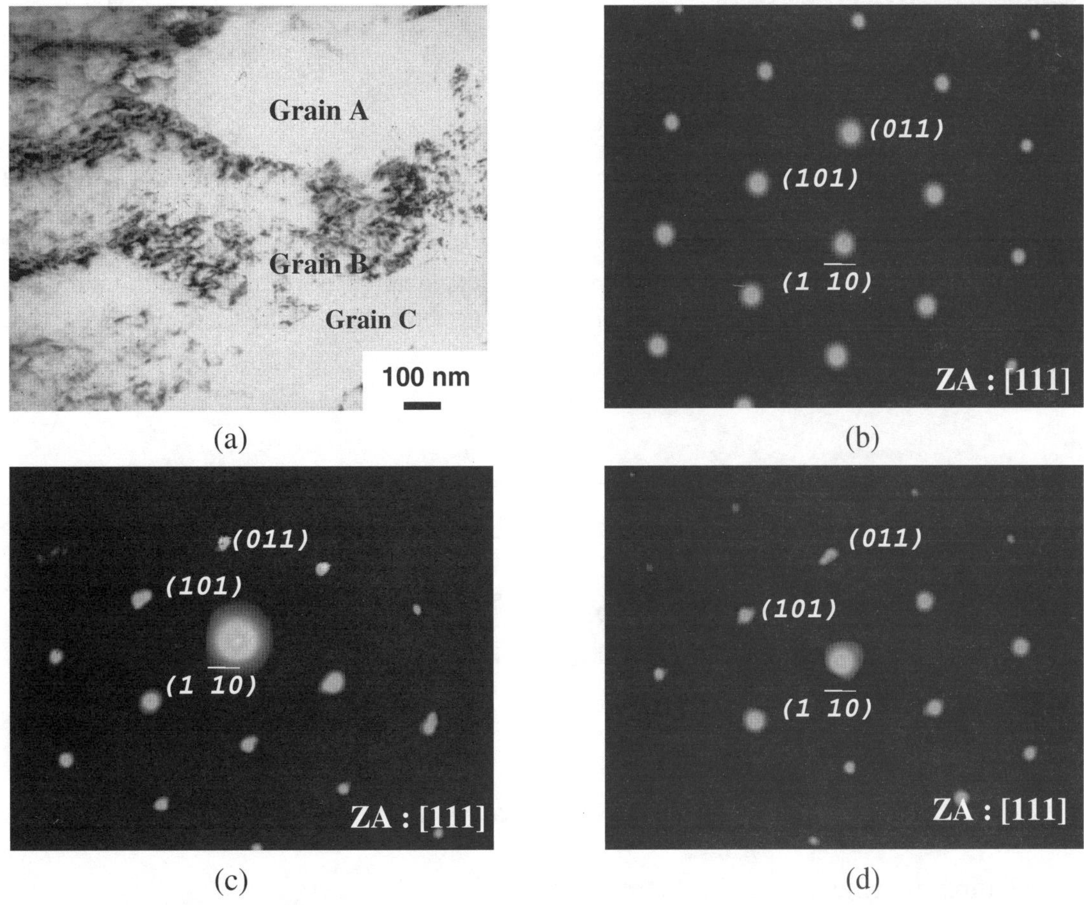

Fig. 6: a) TEM bright field image of the microstructure taken from the side parallel to rolling direction; b) SAD pattern of Grain A; c) SAD pattern of grain B; d) SAD pattern of Grain C. The zone axis for all three patterns is [111].

Addition of high levels of microalloy elements enhances the start temperature of precipitation in this steel and consequently the no recrystallization temperature increases to around 1010°C, according to the equation proposed by Barbosa et al.[13] Finish rolling is carried out at temperatures below 1000°C. This implies that the entire finish rolling process has been carried out within no recrystallization region and therefore the microstructure is heavily deformed before transformation begins. Refinement of the austenite grain size, owing to recrystallization during rough rolling, in addition to pancaking of grains, during finish rolling, provides additional nucleation sites when the transformation starts and this accordingly leads to a shift of the CCT curve in Fig. 10 towards shorter times. In this new condition, the first phase that forms during cooling is allotriomorphic ferrite. This phase, along with Widmanstatten ferrite, covers the prior austenite grain boundaries.

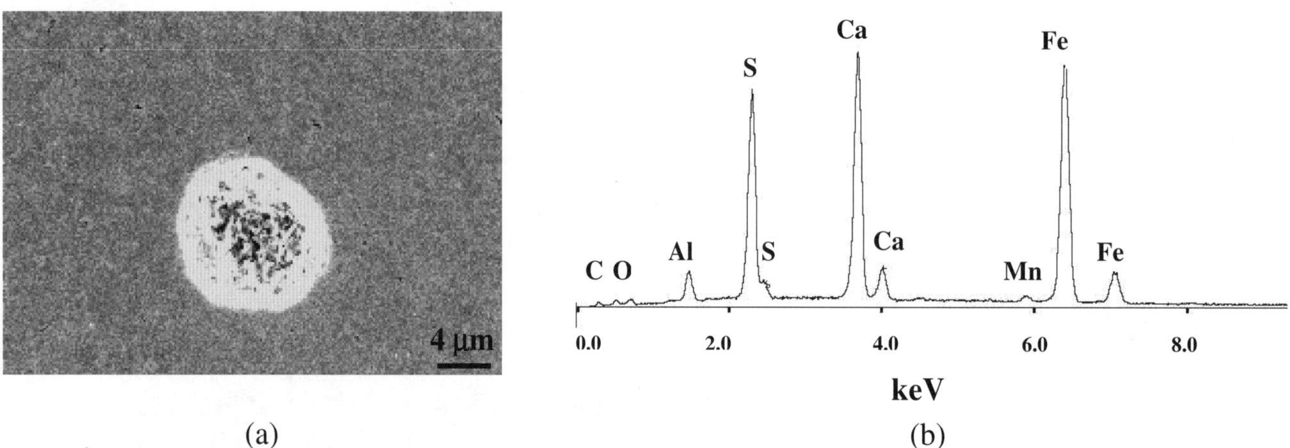

Fig. 7: a) SEM secondary electron image of an inclusion in the microstructure; b) EDX spectrum of the inclusion shown in Fig. 7a.

Formation of ferrite is normally accompanied by the rejection of carbon to the neighboring austenite grains at the interface of austenite and ferrite. Enrichment of austenite from the carbon paves the way for the formation of pearlite.

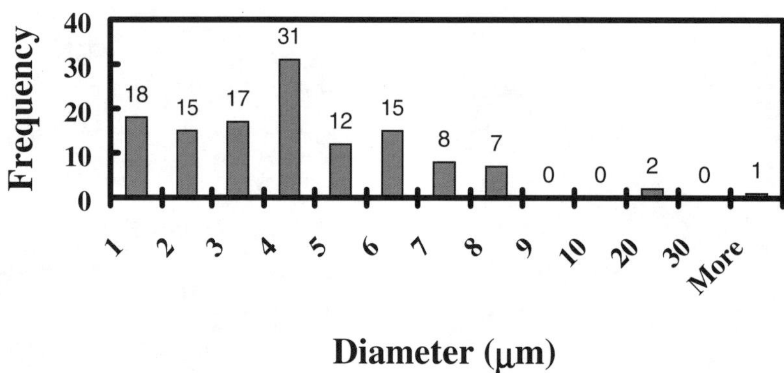

Fig. 8: Size distribution of inclusions in the microstructure.

The presence of Mn and Mo in the microstructure suppresses the austenite to ferrite transformation temperature to some extent and promotes the formation of low temperature microstructures such as bainitic ferrite.[3,14] The relatively fast cooling associated with accelerated cooling, in addition to the presence of Mn and Mo, prevents the formation of a high volume fraction of allotriomorphic ferrite and encourages instead development of bainitic ferrite which is more stable phase under the processing conditions.

The SAD patterns obtained from TEM confirmed the observations made by optical and scanning electron microscopy. The lath shape of the grains in Fig. 6a, and the presence of a low angle boundary between neighboring grains (Figs. 6c and 6d) indicate that the dominant constituent of the microstructure is bainitic ferrite.

Finer microstructure at the surface relative to bulk could be attributed to the higher cooling rates that the surface experiences, in addition to normal segregation of alloying elements from the surface towards the center. The higher cooling rate that the surface experience leads to the formation of finer products during transformation and the absence of alloying elements at the surface affects the type of products generated during transformation. There are also reports that show that deformation is concentrated at the surface compared with the center, thereby producing a finer microstructure at the surface.[15] Lower alloy content at the surface, relative to the plate interior due to segregation, causes a decrease in the hardness of this region. The higher cooling rate and segregation counteract one another, but presumably the depletion of alloying elements due to segregation is the stronger one in dictating the hardness.

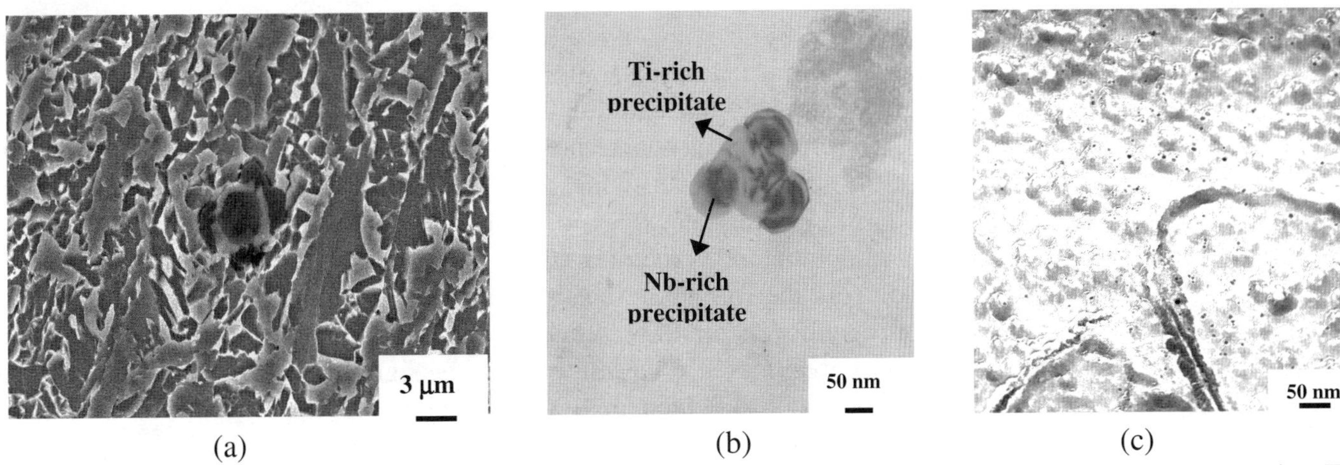

Fig. 9: Various precipitates observed in Grade 100 steel: a) Coarse TiN precipitates; b) intermediate-size Ti- and Nb-rich precipitates; and c) fine Nb-rich precipitates.

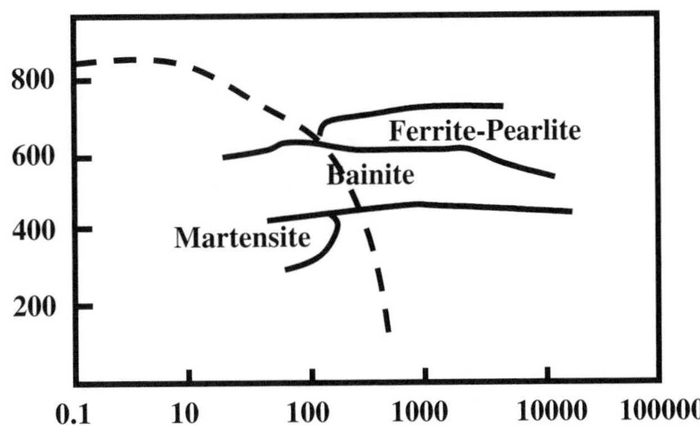

Fig. 10: CCT curve of the Grade 100 steel in the absence of deformation. [8]

It is of interest to note the relatively high hardness of the different microconstituents, particularly the ferrite phase. Hardness of ferrite in the present steel is increased to 260 HV as opposed to 180 HV in a plain carbon steel with similar carbon and manganese content.[16] This could be linked to presence of precipitation or dislocations in matrix. Dislocations may have developed during rolling below the no recrytallization temperature or after accelerated cooling.[17]

Inclusions in microstructure

Chemical analysis of the inclusions shows that they are probably complex inclusions of CaS-CaO-Al_2O_3. Application of calcium to the melt is a general recipe for producing quality steel, today. It is known, though, that the motive for calcium addition is based on modification of pure alumina inclusions with a high melting point to complex clacium-aluminate inclusions with a low melting point, as well as reducing the level of sulfur available for reaction with manganese or iron.[18] The solubility of CaS in the slag decreases as the temperature is decreased during solidification, CaS then precipitates on existing CaO-Al_2O_3 particles, forming complex inclusions of CaS-CaO-Al_2O_3.[10] The high intensity of sulfur and calcium relative to aluminium and oxygen in EDX spectrum, Figure 6b, suggests that a significant amount of calcium has been added to the melt and thus a high fraction of CaS inclusions is formed.[18] The inclusions are trapped in the microstructure during solidification of the rest of melt. These inclusions provide nucleation sites for the precipitation of TiN (Fig. 8a) and also promote formation of bainitic ferrite.[1] The overall distribution and volume fraction of inclusions suggest that the steel can be classified as a clean steel, according to ASTM E-45.

Precipitation

The large TiN precipitates, around 3 μm in size, are likely formed during solidification or at the elevated temperatures soon after solidification. The formation of some TiN precipitates on the refractory inclusions confirms that these particles have heterogeneously nucleated on the inclusions during solidification. The high titanium and nitrogen levels lead to the formation of coarse precipitates at elevated temperatures. The large cuboidal TiN precipitates are not useful in inhibiting austenite grain growth at reheating temperatures despite their stability, because they are too large, greater than 0.5 μm in size, rather than the 0.1 μm required.[19] Furthermore, they remove titanium, niobium and vanadium from solution, thereby reducing precipitation hardening. It is therefore desirable to avoid the formation of large cuboidal TiN precipitated in the liquid, and to reduce their size. Titanium and nitrogen levels could be reduced, but decreasing the nitrogen content is not suggested because the volume fraction of precipitates in austenite would be significantly reduced and thus is not desirable. However, it would probably be better to reduce the titanium content only.

The Ti- and Nb-rich intermediate size precipitates, 30 to 300 nm in size, are microalloy carbonitrides formed in the austenite region at temperatures below 1150 °C. They have precipitated by a deformation induced precipitation process during rolling.[20] Formation of these precipitates increases the no recrystallization temperature in steel and retards the recrystallization process.[13]

The various microalloyed carbides and nitrides, TiC, TiN, NbC, NbN, VC and VN, all form with the same NaCl-type crystal structure and similar lattice parameters. They are therefore mutually soluble and form complex (Ti,Nb,V)(C,N) precipitates.[11] The similarity in the crystal structure facilitates the nucleation of Nb-rich precipitates on existing Ti-rich precipitates, which provide thermodynamically favorable heterogeneous nucleation sites (Fig. 9b).[20]

The fine Nb-rich particles (<10nm) have been reported to precipitate in low carbon steels at temperatures between 500 and 700°C.[21] It has been argued that niobium carbides form by an aging process in the ferrite matrix. This temperature range coincides well with the coiling temperature of the experimental material in the present work. The arrangement of fine precipitates in microstructure generally displays a cellular pattern, which suggests that precipitates may have formed on sub-grain boundaries or dislocations developed during deformation in the no recrystallization region of austenite and/or in ferrite during finishing rolling.[22]

CONCLUSIONS

Microstructural characterization of a rolled, accelerated cooled and coiled Grade 100, Nb-Ti-V microalloyed steel has been performed. The microstructure was primarily composed of bainitic ferrite, along with allotriomorphic ferrite and Widmanstatten ferrite formed at prior austenite grain boundaries. Some regions of pearlite were also observed. Analysis of inclusions showed that they were most likely CaS-CaO-Al_2O_3, with a uniform and relatively fine distribution.

Four types of precipitates formed during processing, i.e., coarse TiN precipitates, which formed during solidification or soon after, Ti and Nb-rich carbonitrides, 30 to 300 nm in size, which formed in the austenite during rolling, and fine niobium carbonitrides, which formed at sub-grain boundaries and on dislocations during the final coiling stage.

ACKNOWLEDGEMENTS

The authors would like to thank the Natural Sciences and Engineering Research Council (NSERC) of Canada and IPSCO Inc. for providing financial support. Also provision of experimental material by IPSCO Inc. is gratefully acknowledged.

REFERENCES

1- T. Gladman, The Physical Metallurgy of Microalloyed Steels, The Institute of Materials, London, UK, 1997.

2- B. Dogan, L.E. Collins, and J.D. Boyd, “Effects of Thermomechanical Processing on the Microstructure and Mechanical Properties of a Ti-V-N Steel”, Metallurgical Transactions A, Vol. 19A, No. 5, May 1988, p. 1221.

3- L.E. Collins, M.J. Godden and J.D. Boyd, ”Microstructures of Linepipe Steels”, Canadian. Metallurgical Quarterly, Vol. 22, No. 2, Apr.-June 1983, p. 169.

4- J.M. Barnes, B.C. Muddle and P.D. Hodgson, ”Microstructure and Properties of Continuously Cooled Low-Carbon Bainitic Steels”, Phase Transformations During the Thermal/Mechanical Processing of Steel, Vancouver, British Columbia, Canada, 20-24 Aug. 1995, p. 161.

5- M.K. Graf, H.G. Hillenbrand, and P.A. Peters, “Accelerating Cooling of Plate for High Strength Large Diameter Pipe”, Conference Accelerated Cooling of Steel, Pittsburgh, Pennsylvania, USA, 19-21 Aug. 1985, p. 165.

6- M. Diaz-Fuentes, I. Madariaga, and I. Gutierrez, “Accicular Ferrite Microstructures and Mechanical Properties in a Low Carbon Wrought Steel”, Materials Science Forum, Vols. 284-286, 1998, p. 245.

7- B.S. Nelson, F. Zhang, and J.D. Boyd, “Evolution of Precipitate Distributions during Thermomechanical Processing of Steels”, THERMEC'97, Int. Conf. On Thermomechanical Pro. of Steels & other Materials, Ed. T. Chandra & T. Sakai, 1997, p. 491.

8- Private communication, L.E. Collins, IPSCO Inc., Canada.

9- U. Sharma and D.G. Ivey, "Microstructure of Microalloyed Linepipe Steels", to be presented at IPC 2000, ASME International, Oct. 1-5, 2000, Calgary, AB, Canada.

10- T. Takenouchi and K. Suzuki, "Influence of Ca-Al Deoxidizer on the Morphology of Inclusions", Transcations ISIJ, Vol. 18, 1978, p. 344.

11- Annual Book of ASTM Standard, E-45, Vol. 03.01, American Society for Testing and Materials, West Conshohocken, PA, USA, 2000.

12- S. Akhlaghi and D.G. Ivey, "Precipitation Behavior of a Grade 100 Structural Steel", to be published.

13- R.Barbosa, F. Boratto, S. Yue and J.J. Jonas, "The Influence of Chemical Composition on the Recrystallization Behavior of Microalloyed Steels", Processing Microstructure and Properties of HSLA Steels, Ed. By A.J. DeArdo, The Minerals, Metals and Materials Society, 1988.

14- R.E. Reed-Hill, R. Abbaschian, Physical Metallurgy Principles, 3rd Ed., PWS Publishing Company, Boston, MA, 1994.

15- B. Chin, S. Yue, J.A. Nemes and G.E. Ruddle, "Development of Strain Through the Cross-Section of a Steel Rod during Hot Rolling", 38th Mechanical Working and Steel Processing Conference Proceedings. Vol. XXXIV, Cleveland, Ohio, USA, 13-16 Oct. 1996, p. 147.

16- Metals Handbook, Desk Edition, Ed. by H. E. Boyer and T. L. Gall, ASM, 1985.

17- R.L. Bodnar and S.S. Hansen, "Effects of Deformation Below the Ar_3 Temperature on the Microstructure and Mechanical Properties of Structural Steels", 36th Mechanical Working and Steel Processing Conference Proceedings, ISS-AIME, vol. XXXII, 1995, p. 503.

18- V. Presern, B. Korousic and J.W. Hastie, "Thermodynamic Conditions for Inclusion Modification in Calcium Treated Steel", Steel Research, Vol. 62, No. 7, 1991, p. 289.

19- H. Gondoh, H. Nakasagi, H. Matsuda, H. Tamehiro and H. Chino, Nippon Steel Tech. Rep., 1979, 14, p. 55.

20- W.J. Liu and J.J. Jonas, "Nucleation Kinetics of Titanium Carbonitride in Microalloyed Austenite", Metallurgical Transactions A, Vol. 20A, No. 4, 1989, p. 689.

21- Y. Ohmori, "Precipitation of Fine Niobium Carbide Particles in Low Carbon Steels", Transactions ISIJ, Vol. 15, 1975, p. 194.

22- W.C. Leslie, The Relation between the Structure and Mechanical Properties of Metals, Her Majesty's Stationary Office, 1963, p. 334.

STENGTHENING MECHANISM IN DUAL-PHASE ACICULAR FERRITE + M/A MICROSTRUCTURES

I.A. Yakubtsov*, J.D. Boyd*, W.J. Liu** and E. Essadiqi**
* Materials and Metallurgical Engineering Department,
Queen's University, Kingston, ON K7L 3N6, CANADA,
Tel.: 613-533-6000 ext.77264
** Materials Technology Laboratory, CANMET,
568 Booth St., Ottawa, ON K1A 0G1, CANADA

Key Words: Strengthening mechanisms, bainite, acicular ferrite, M/A constituent, precipitates, dislocations, grain boundaries.

INTRODUCTION

Low carbon microalloyed plate steels for Grade 550 ($\sigma_Y = 550\,\text{MPa}$) linepipe have a wide range of microstructures from a mixture of ferrite + bainite to completely bainite. The latter can be conventional bainite CB (nucleated at previous austenite boundaries) or acicular ferrite AF (intragranular nucleated bainitic ferrite). The best combination of strength, toughness and weldability for linepipe steels is achieved with a microstructure comprising dual phase AF+M/A (martensite/austenite constituent) [1-7].

The effects of thermomechanical processing (TMP) and on-line accelerated cooling (OLAC) parameters on the final microstructures have been studied extensively for these types of steels. It has been shown that to develop Grade 690 ($\sigma_Y = 690\,\text{MPa}$) properties requires the formation of a mixture of fine AF laths and small islands of M/A constituent. It has been shown experimentally that varying TMP and OLAC parameters (austenite deformation below the non-recrystallization temperature (T_{NR}), accelerated cooling rate ($\dot{T}$) and interrupt temperature (T_I)) changes the final microstructure from a mixture of CB+M/A, through CB+AF+M/A, to a completely dual phase AF+M/A [8]. These steels contain various microstructural features which contribute to the overall strength: i.e, boundaries of CB and/or AF laths, dislocations, solid solution alloying, microalloy precipitates and fine islands of M/A constituent. Each microstructural feature produces obstacles which have different scales, densities and magnitudes of strengthening.

In this paper a quantitative description of each of these microstructural features in a dual phase AF+M/A is presented, the contribution of each to the yield strength is estimated and the appropriate approach for assessment of overall yield strength is discussed.

EXPERIMENTAL

The chemical composition of the commercial Grade 550 steel used in this investigation is given in Table I.

Table I Chemical composition of steel (wt. %)

C	Mn	Mo	Ni	Cr	Cu	Nb
0.042	1.78	0.31	0.32	0.04	0.27	0.08
Ti	V	N	Si	P	S	Al
0.012	0.006	0.006	0.372	0.011	0.003	0.025

Laboratory simulations of TMP and OLAC typical of an industrial plate processing schedule were carried out on an MMC quench-deformation dilatometer and a Gleeble apparatus. The detailed descriptions of the equipment, sample preparation and choices of TMP parameters are given elsewhere [8].

The basic laboratory TMP + OLAC schedules are shown in Fig.1. Each sample was held at 1200°C for 900 s. Single-deformation schedules were used in the dilatometer tests (Fig. 1a). In this case, samples were cooled at 1°C/s from the reheat temperature to the deformation temperature T_D (800°C), held 10 s, deformed in uniaxial compression at 1 s^{-1} to a selected true strain (ε), held 10 s, accelerated cooled at a selected rate ($\dot{T}$) to a selected accelerated cooling interrupt temperature (T_I) and cooled at 1°C/s to room temperature. In the Gleeble tests (Fig. 1b), the same basic schedule was followed, except there were two deformation steps, at 950°C and 800°C. These two values of T_D were selected to be above and below T_{NR}, respectively. A series of processing schedules were used to determine the effects of varying ε (0, 0.2, 0.5 and 0.7), $\dot{T}$ (5 - 35°C/s) and T_I (200 and 450°C).

Details of the various microstructures were obtained by TEM. Thin slices were cut parallel to the deformation axis, 3 mm discs were cut by EDM and the discs were thinned in a TENUPOL 3 electropolishing system. The electrolyte was 7% perchloric acid – glacial acetic acid, and the electropolishing conditions were 25 V at room temperature. Carbon extraction replicas were prepared from each quenched sample by light etching in 2% Nital solution, carbon coating, extracting in 10% Nital solution and collecting on Cu grids. TEM studies were carried out using a Philips CM 20 electron microscope at 200 kV with an ultra thin window (UTW) EDS system. For each quenched condition about 300-700 precipitates were measured and quantitative analysis of the size distributions was carried out using small-particle statistics [9].

Quantitative measurements of the volume percent of constituents were made on SEM micrographs by a point count method. For each condition, at least 30 fields were measured with a grid of 400 points, giving a typical standard deviation of ± 0.5 vol. %.

Dislocation densities in the final microstructures were determined by measurements of broadening of X-ray diffraction lines using the (222) reflection [10,11].

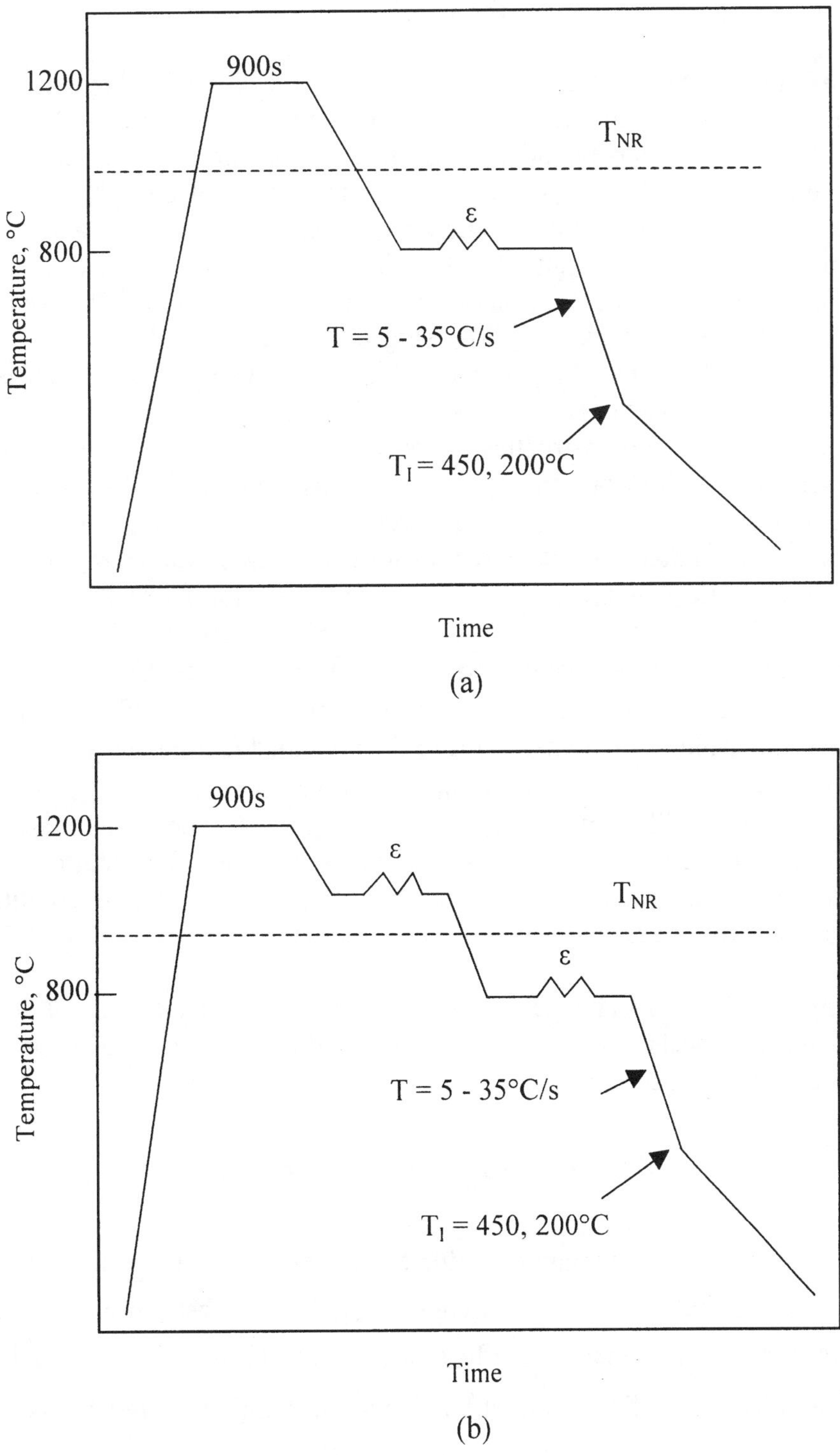

Fig.1 Schematic diagrams of laboratory TMP schedules: a) single deformation (dilatometer); b) double deformation (Gleeble).

RESULTS

Evolution of Final Microstructure For all of the TMP schedules, the final microstructures were one of two distinct types, or a mixture of the two types. The first type was CB+M/A, i.e., grain boundary nucleated ferrite laths with elongated interlath M/A constituent. This structure is illustrated in Figs. 2a and 2c. The bainitic ferrite packets grew until they impinged either the opposite grain boundary or another growing packet. The packets comprised parallel ferrite laths, having low angle misorientations and containing a high, uniform density of dislocations (Fig. 2c). The interlath phase appeared light in SEM – SE [8] and dark in TEM (Fig. 2c). The detailed structure had a high dislocation density and alternate bright and dark contrast areas, which is consistent with "blocky" martensite observed in low – carbon steels [12].

The second type of microstructure observed was AF+M/A, i.e., intragranular nucleated ferrite laths with dispersed M/A islands. This structure is illustrated in Figs. 2b and 2d. The intragranular ferrite generally grew as individual laths, impinging other growing laths, and the remaining austenite transformed to the equiaxed regions of M/A [8]. The bainitic ferrite contained a high, uniform dislocation density (Fig. 2d). The carbon-rich regions again appeared light in SEM – SE [8] and dark in TEM (Fig. 2d). These regions were generally equiaxed and $\leqslant 0.5$ μm in diameter. Carbon enrichment of these regions relative to the ferrite matrix was confirmed by EDS [8]. The detailed structures were either blocky martensite (Fig. 3a) or twinned martensite (Fig. 3b).

The effects of TMP parameters ε (austenite deformation below T_{NR}), $\dot{T}$ and T_I on the final microstructure are illustrated schematically in Fig. 4. The dual phase AF+M/A microstructure is obtained with a combination of high austenite deformation below T_{NR}, high accelerated cooling rate, and low interrupt temperature. These conditions also produce the highest volume fraction of M/A constituent ($f_{M/A}$) which is about 0.1 [8].

Strengthening Mechanism The general expression for yield strength of low carbon microalloyed bainitic steels, which assumes linear additivity of individual strengthening contributions (linear approach), is as follows:

$$\sigma_Y^L = \sigma_o + \sigma_{SS} + \sigma_{gb} + \sigma_d + \sigma_p + \sigma_{M/A} \tag{1}$$

where σ_o is the iron lattice strengthening due to the Peierls - Nabarro force and σ_o = 70 MPa [15]; σ_{SS} is the solid solution strengthening due to substitutional and interstitial atoms; σ_{gb} is strengthening due to the effective grain size (D_B) of conventional bainite or acicular ferrite laths; σ_d is dislocation strengthening of bainitic ferrite; σ_p is microalloy precipitation strengthening; $\sigma_{M/A}$ is strengthening due to M/A constituent.

The solid solution strengthening is given by:

$$\sigma_{SS} = \sum_i k_i c_i \tag{2}$$

Fig.2: Conventional bainite (CB+M/A): a) OM, c) TEM; acicular ferrite (AF+M/A): b) OM, d) TEM.

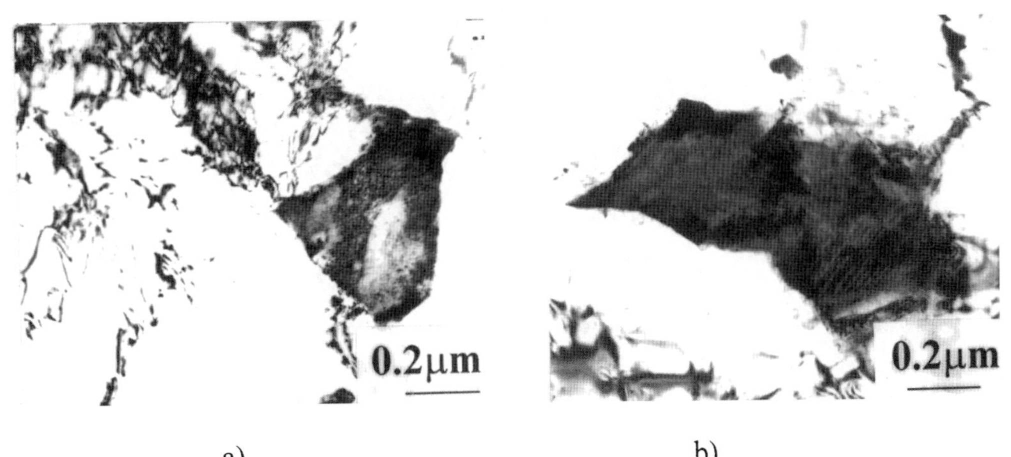

Fig. 3: M/A constituent in dual- phase AF+M/A: blocky (a) and twinned (b) martensites.

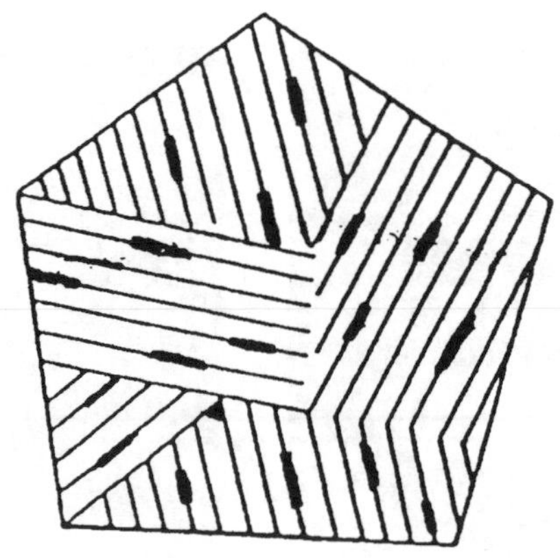

a) CB+M/A: $\varepsilon = 0$, $\dot{T} = 5\text{-}35$ °C/s, $T_I = 450, 200$ °C.

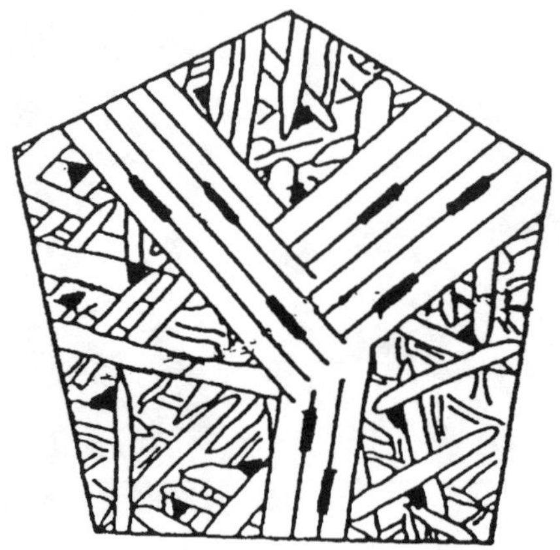

b) CB+AF+M/A: $\varepsilon < 0.5$, $\dot{T} = 5\text{-}35$ °C/s, $T_I = 450, 200$ °C.

c) AF+M/A: $\varepsilon \geq 0.5$, $\dot{T} \geq 30$ °C/s, $T_I = 200$ °C.

Fig. 4. Schematic illustration of the effects of TMP parameters on the final microstructure (black areas are M/A constituent): ε is austenite deformation below T_{NR}, $\dot{T}$ is accelerated cooling rate, T_I is interruption temperature. (Models extended from [13,14].)

where c_i are the concentrations of alloying elements in wt.% and k_i are the strengthening coefficients for each solute element in BCC iron [16]. The calculation of σ_{SS} for our steel composition is given in Table II. It is assumed that interstitial elements C and N are combined as microalloy precipitates during austenite deformation below T_{NR} and their solid solution strengthening effect in bainitic ferrite is weak.

Table II Solid solution strengthening

Alloying element	k_i in MPa per wt.%	Wt. %	$\Delta\sigma_i$, MPa	$\sigma_{SS} = \sum_i \Delta\sigma_i$, MPa
Mn	37	1.78	65.9	
Si	83	0.372	30.9	
Ni	33	0.328	10.8	120.4
Cr	-33	0.034	-1.1	
Cu	38	0.275	10.5	
Mo	11	0.307	3.4	

According to the model for martensitic and bainitic microstructures of low carbon low alloyed steels [17,18] the actual dimensions of the slip planes in an elongated lath are intermediate between its transverse thickness (w) and its length (L_α). A random distribution of slip orientations with respect to lath axis during deformation is assumed. In that case, the effective grain size (D_B) for a bainitic or martensitic lath microstructure is:

$$D_B \approx w\left(1 - 0.65\ln\left(\frac{w}{L_\alpha}\right)\right) \quad (3)$$

The strengthening due to lath boundaries can be described by the Hall-Petch relation for $D_B > 1$ μm:

$$\sigma_{gb} = k_{gb} D_B^{-1/2} \quad (4)$$

where k_{gb} is a constant and $k_{gb} = 18.6\,\text{MPa} \times \text{mm}^{1/2}$. The experimental data give an effective grain size D_B in an acicular ferrite structure of about 3 μm and the calculated strengthening magnitude due to lath boundaries is $\sigma_{gb} = 342\,\text{MPa}$.

The dislocation strengthening can be expressed as [19]:

$$\sigma_d = \alpha G b \rho^{1/2} \quad (5)$$

where the constant α for polycrystalline iron is 0.38 [20]; G is the shear modulus and G = 83 GPa for pure iron; b is the Burgers vector in BCC iron lattice and b = 0.248 nm; ρ is dislocation density in bainitic ferrite. Measurements of X – ray diffraction line broadening give $\rho_{exp}^{AF} = 6.13 \times 10^{15}\ m/m^3$ for acicular ferrite and the estimated dislocation strengthening is $\sigma_d = 612\,MPa$.

The precipitation strengthening component following the Ashby-Orowan approach is [21]:

$$\sigma_p = 0.538Gb\frac{f^{1/2}}{X}\ln\left(\frac{X}{2b}\right) \tag{6}$$

where f is the volume fraction of precipitates, and X the real (spatial) diameter of precipitates. Experimental measurements give the mean size of microalloy precipitates in acicular ferrite microstructure as X_{exp} = 40 nm. It is assumed that all Nb is consumed in the formation of carbo-nitride precipitates and their volume fraction can be estimated by $f_{max} = 1.1 \times 10^{-2} \times$ (wt.% Nb) [22]. In that case, the calculated precipitation strengthening is $\sigma_p = 30\,MPa$.

It has been shown experimentally that in low carbon low alloy microalloyed steels the yield and tensile strengths increase linearly with volume fraction of M/A constituent. For granular bainite microstructure the yield strength is given by $\sigma_Y = 360 + 900 f_{M/A}$, [MPa] [23], and for a mixture of primary ferrite and acicular ferrite $\sigma_Y = 324.4 + 1310 f_{M/A}$, [MPa] [5]. We have used an average of these two expressions to determine the strengthening effect of M/A:

$$\sigma_{M/A} = 1100 f_{M/A}, [MPa] \tag{7}$$

where $f_{M/A}$ is the volume fraction of M/A constituent in the final microstructure. The measured M/A volume fraction of 0.1 then gives $\sigma_{M/A} = 110\,MPa$.

The linear combination of individual strengthening contributions is usually used when the microstructure has different types of obstacles which have large differences in the microstructural size and magnitude of strengthening. In that case the small and weak obstacles can be considered as smooth to the large and strong microstructural features, and the total strengthening is equal to the arithmetic sum for strong and weak strengthening effects (Eqn. 1) [24,25].

When the microstructure is characterized by different types of obstacles with approximately the same order of magnitude strengthening for each, but with different densities, the total strengthening effect can be described by a root mean square approach [25]:

$$\sigma_Y^{RMS} = \sqrt{\sum_i \sigma_i^2} \tag{8}$$

where σ_i is the strengthening contribution of each microstructural feature.

When the microstructure has obstacles with differences in strengthening contributions, scale and densities, the overall strengthening can be described by a combination of the linear and root mean square approaches. Under these conditions, the various types of obstacles can be grouped with some of them by the linear approach, and then all groups are described with the root mean square summation. For example, in the case of the dual phase AF + M/A microstructure, the yield strength was determined from the expression:

$$\sigma_Y^{L+RMS} = \sqrt{\sigma_A^2 + \sigma_B^2 + \sigma_C^2} \tag{9}$$

where $\sigma_A = \sigma_o + \sigma_{SS} + \sigma_p + \sigma_{M/A}$, $\sigma_B = \sigma_{gb}$ and $\sigma_C = \sigma_d$.

The results of the calculation of the yield strength for the dual phase AF + M/A steel are given in Table III for the linear (σ_Y^L), root mean square (σ_Y^{RMS}) and combination of linear and root mean square (σ_Y^{L+RMS}) approaches.

Table III Theoretical estimations of the strengthening contribution for each microstructural component and the overall yield strength, [MPa]

σ_o	σ_{SS}	σ_p	σ_{gb}	σ_d	$\sigma_{M/A}$	σ_Y^L	σ_Y^{RMS}	σ_Y^{L+RMS}
70	120	30	340	600	110	1270	713	764

It can be seen from Table III, that the largest contributions to the strengthening of the AF+M/A microstructure is due to dislocations and boundaries of acicular ferrite laths. The dislocation strengthening is determined by the accelerated cooling rate under continuous cooling, and lath boundary strengthening can be governed by both the total austenite deformation below T_{NR} and the cooling rate which determine nucleation density and size of acicular ferrite laths in the final microstructure. The results in Table III show that the linear approach overestimates the yield strength and both the root mean square and a combination of the linear and root mean square approaches give reasonable values of the yield strength.

CONCLUSIONS

- The dual phase AF+M/A microstructure in low carbon plate steel can be obtained by a combination of high austenite deformation below T_{NR}, high accelerated cooling rate and low interruption temperature under continuous cooling.

- The dual phase AF+M/A is characterized by microstructural features which have large differences in their scale, densities and magnitudes of strengthening. The largest contributions to the yield strength of dual phase AF+M/A microstructure are due to the dislocation substructure and the boundaries of AF laths.

- In calculating the yield strength of dual phase AF+M/A, the linear approach overestimates the yield strength. Both the root mean square approach and the combination of linear and root mean square approaches give reasonable values for the yield strength.

ACKNOWLEGMENTS

The authors thank Dr. Tibor Turi (Stelco Inc.) for providing steel samples, Dr. Brad Diak (Materials and Metallurgical Engineering Department, Queen's University) for assistance with X - ray diffraction and Murray Letts (CANMET) for doing the Gleeble experiments. This research is a part of a collaborative project supported by Stelco Inc., Materials and Manufacturing Ontario, and CANMET.

REFERENCES

1. F. Kawabata, M. Okatsu, K. Amano and Y. Nakano, "Metallurgical and Mechanical Features of X 100 Linepipe Steel", Pipeline Technology, Vol.II, Ed. R. Denys, Elsevier, 1995, pp. 263-271.
2. H.G. Hillebrand, E. Amoris, K.A. Niedorhott, C. Perdix, A. Sreisselberger and U. Zeislmair, " Manufacturability of Linepipe in Grades up to X 100 from TM Processed Plate ", ibid, pp.273-285.
3. M. Okatsu, F. Kawabata and K. Amano, "Metallurgical and Mechanical Features of X100 Linepipe Steel", Materials Engineering, Vol. III, ASME, 1997, pp. 119-124.
4. Y. Terada, M. Yamashita, T. Hara, H. Tamehiro and N. Ayukawa, "Development of API X100 UOE Line Pipe", Nippon Steel Techn. Report, #72, Jan. 1997, pp. 47-52.
5. B.P. Wynne, P. Cizek, C.H.J. Davies, B.C. Muddle and P.D. Hodgson, " The Effects of Processing Parameters on the Mechanical Properties of Low Carbon Microalloyed Steels", THERMEC 97, The Minerals, Metals & Materials Society, 1997, pp. 837-843.
6. J.R. Yang, C.S. Chiou and C.Y. Huang, "The Effect of Prior Deformation of Austenite on Toughness Property in Ultra-Low Carbon Bainitic Steel", ibid, pp. 435-441.
7. P.S. Mitchell and F. Grimpe, "The Development of Vanadium-Containing AP15LX-100 Linepipe Steel", Materials for Resources Recovery and Transport, The Metallurgical Society of CIM, 1998, pp. 219-235.
8. I.A. Yakubtsov, J.D. Boyd, W.J. Liu and E. Es-Sadiqi, "Formation of Dual Phase Acicular Ferrite + M/A Microstructures for High Strength Plate Steels", 41-st MWSP Conf., Vol. XXXVII, ISS, 1999, pp. 801-810.
9. G. Herdman: "Small Particle Statistics", 2nd edn, 418, 1960, London, Butterworth & Co.
10. A.C. Vermeulen, R. Delhez, Th.H. de Keijser and E.J. Mittemeijer, "Changes in the Densities of Dislocations on Distinct Slip Systems During Stress Relaxation in Thin Aluminum Layers: The Interpretation of X-ray Diffraction Line Broadening and Line Shift", J. Appl. Phys., Vol. 77, 1995, No10, pp. 5026-5049.
11. T. Ungar, I. Dragomir, A. Revez and A. Borbely, " The Contrast Factors of Dislocations in Cubic Crystals: The Dislocation Model of Strain Anisotropy in Practice", J. Appl. Cryst., Vol. 32, 1999, pp. 992-1002.
12. G. Krauss, A.R. Marder, "The Morphology of Martensite in Iron Alloys", Met. Trans. A, Vol. 2, 1971, pp. 2343-2357.

13. D.V. Edmonds and R.C. Cochrane, "Structure – Property Relationship in Bainitic Steels ", Met. Trans. A, Vol.21, 1990, pp. 1527-1540.
14. J.R. Yang, C.Y. Huang, and C.F. Huang, " Influence of Acicular Ferrite and Bainite Microstructures on Toughness for an Ultra – Low - Carbon Alloy Steel Weld Metal", J. Mat. Sci. Letters, Vol.12, 1993, pp. 1290-1293.
15. W.B. Morrison, "The Effect of Grain Size on the Stress – Strain Relationship in Low-Carbon Steel", Trans. Am. Soc. Met., Vol. 59, 1966, pp. 824-846.
16. T. Gladman and F.B. Pickering, "Metallurgical Development in Carbon Steels", Spec. Rep. #81, The Iron and Steel Institute, London, 1963, pp. 10-21.
17. J.P. Naylor, "The Influence of the Lath Morphology on the Yield Stress and Transition Temperature of Martensitic - Bainitic Steels", Met. Trans. A, Vol.10, 1979, pp. 861-873.
18. J. Daigne, M. Guttman and J.P. Naylor, "The Influence of Lath Boundaries and Carbide Distribution on the Yield Strength of 0.4% C Tempered Martensitic Steels", Mat. Sci. and Eng., Vol. 56, 1982, pp. 1-10.
19. G.I. Taylor, "The Mechanism of Plastic deformation of Crystals. Part I.-Theoretical", Proc. Roy. Soc., Vol. A145, 1934, pp. 362-388.
20. A.S. Keh, " Direct Observation of Imperfections in Crystals", Wiley-Interscience, New York, 1962, pp. 213-233.
21. T. Gladman, "Precipitation Hardening in Metals", Mat. Sci. & Techn., Vol.15, 1999, pp. 30-36.
22. T. Gladman, "Properties and Applications", The Physical Metallurgy of Microalloyed Steels, The Institue of Materials, London, 1997, p. 363.
23. M.E. Bush and P.M. Kelly, "Strengthening Mechanisms in Bainitic Steels", Acta Metal., Vol.19, 1971, pp. 1363-1371.
24. U.F. Kocks, A.S. Argon and M.F. Ashby, "Thermodynamics and Kinetics of Slip", Prog. Mater. Sci., Vol.19, 1975, pp. 224-229.
25. U.F. Kocks, "Superposition of Alloy Hardening, Strain Hardening and Dynamic Recovery", Strength of Metals and Alloys, Vol.3, ICSMA 5, Pergamon Press Ltd., Oxford, 1980, pp.1661-1680.

Effect of V and N on Processing and Properties of HSLA Strip Steels Produced by Thin Slab Casting

Robert J. Glodowski
STRATCOR
4955 Steubenville Pike
Pittsburgh, PA 15205-9604
USA
Tel: 412-787-5055
Fax: 412-787-5030
bob.glodowski@stratcor.com

Key Words: HSLA, Vanadium, Nitrogen, Thin Slab Cast Strip, Strain Aging

INTRODUCTION

The production of microalloyed HSLA hot rolled sheet steel with thin slab casters using direct charging has presented several metallurgical challenges. Vanadium and nitrogen microalloying has provided some solutions to those difficulties, proving to be less prone to cracking during casting, more soluble in the reheat or holding furnaces, and easier to roll at temperatures similar to plain carbon steel. Using vanadium allows taking advantage of the higher levels of nitrogen typical of EAF steels for improving the effectiveness of precipitation strengthening.

Because of the ease of processing, vanadium plus nitrogen microalloying frequently has become the preferred choice in thin slab cast rolling mills for producing low carbon High Strength Low Alloy (HSLA) strip. Low carbon (less than 0.07% C) hot-rolled strip can be produced with yield strength levels from 350 to 550 MPa using casting and hot-rolling practices similar to that used for plain carbon steels.

The role of V and N in production lots of low carbon EAF steels cast in a thin slab caster and direct charged into a hot strip mill has been evaluated. Mechanical properties, ferrite grain size, and strain aging characteristics were investigated. The effects of chemistry and gauge on yield strength have been explored by means of regression models. The ability of vanadium to assist in refining the as-rolled ferrite grain size is supported by significant improvements compared to carbon steels, competitive with alternate microalloying systems. The completion of the V(C,N) precipitation strengthening process occurs in the coil after leaving the runout table. Strain aging was eliminated in all vanadium and vanadium-nitrogen steels when a normal coiling practice was either implemented or simulated.

Thin-Slab Casting Process

In the 1990's, there was a major expansion in the production of hot rolled strip steels by the thin slab casting process. In the USA and Canada, over 18 million metric tons of new steelmaking capacity has been added using this new technology. The major driving force for this dramatic change is the improved economics available from in-line conversion of liquid steel to a marketable final product.

The steel, made predominately in electric arc furnaces, is continuously cast into thin slabs from 50 to 90 mm in thickness. These slabs are immediately charged into tunnel furnaces to equalize slab temperatures for rolling, then charged into a hot strip mill. For the thinner slabs, the steel is rolled directly to final gauge, usually in five to seven rolling stands. The heavier slabs may be rolled through one or two roughing mills, coiled hot in a coil box system, then rolled to final gauge in the finish stands. Laminar flow cooling equipment capable of rapid quenching to coiling temperatures completes the processing from liquid steel to hot rolled strip.

Metallurgical features that differentiate this process from classical integrated steel manufacturing are many. First, the melting of the steel is typically done in electric furnaces. The iron units for this process tend to be from a high percentage of scrap, causing the metallic residuals to be higher than normally seen in blast furnace and BOF steels. Also, the nitrogen levels are higher in the EAF steels. Next, the rapid solidification rate, necessary in thin slab casting machines, produces smaller and well dispersed non-metallic inclusions reducing the anisotropy of properties in the longitudinal and transverse directions. Any microalloy precipitates, such as TiN, that may occur during solidification also tend to be finer than in slower cooling heavy slabs or ingots.

The slab enters the tunnel or "equalization" furnace while still austenitic. This process prevents the possibility of an austenite to ferrite to austenite transformation cycle prior to the rolling process. Any austenitic grain boundary embrittlement mechanism that may have developed during solidification and does not dissipate in the tunnel furnace will carry over to the rolling process. By necessity, due to the large length of the tunnel furnace, the reheat temperatures are relatively low compared to those normally reached in conventional slab reheat operations. Maximum reheat temperatures of 1150 °C are quite common. Any precipitates that need to be in solution prior to rolling must be soluble below this temperature.

The rolling of thin slabs (50mm) is usually continuous without the additional time between passes normally seen in a conventional strip mill with separate roughing stands. The short time interval between passes may be insufficient to produce the desired recrystallization of the austenite. To compensate for the short time interval, heavy reductions per pass are necessary to refine the austenitic grains.

Microalloy Choices for Thin Slab Cast Steels

From the previous discussion, one can deduce that the preferred microalloying system to produce HSLA grades in thin slab casting operations should have the following characteristics:

- Compatibility with electric furnace steels, including higher nitrogen levels.
- Minimal precipitation during solidification to reduce cracking problems during casting.
- Low solution temperature (high solubility) to insure the microalloy is in solid solution prior to rolling.
- Precipitation strengthening should occur after finish rolling to minimize roll force requirements.
- Development of a fine ferrite grain size should be achievable using high finish mill exit temperatures similar to carbon steel.
- Optimum coiling temperatures should be compatible with standard runout table cooling equipment.

Vanadium microalloying provides virtually all of these desirable characteristics. Higher nitrogen levels are advantageous because nitrogen is the preferred element for precipitation with vanadium.[1,2] Vanadium has been shown to be less prone to develop surface cracks than niobium steel under conditions simulating the straightening of continuously cast thin slabs.[3,4] V(C,N) has the lowest solubility temperatures compared to Cb (C,N) or TiN.[5] With the fast cooling rate on the runout table, most of the V(C,N) precipitation will occur as random precipitation in the ferrite after transformation. This preferred precipitation occurs at temperatures that are typical for coiling strip (600°C), well below the finish rolling temperature. Therefore the precipitation strengthening will not contribute to increased rolling forces.(6,7) Fine as rolled ferrite grain size can be achieved, first utilizing recrystallization controlled rolling for austenite grain refinement. The rapid cooling rate on the runout table along with the beneficial synergy of vanadium and nitrogen will maximize the grain refinement ratio ($D\gamma / D\alpha$). Final ferrite grain sizes below 5μ are readily easily produced using V microalloying with finish mill exit temperatures around 900°C.

MATERIAL EVALUATION PLAN

Test Material

To provide assistance in utilizing the advantages of vanadium microalloying in the thin slab cast steelmaking process, this investigation was initiated to quantify the strengthening effects of vanadium and nitrogen in commercially produced hot rolled strip steels produced in a thin slab cast and direct charge operation. Microstructures (ferrite grain size) and strain aging characteristics were also to be evaluated.

Test material was acquired from carbon and microalloyed hot rolled steel produced from a 50 mm slab casting process, direct charged through a tunnel furnace to a hot strip mill rolling process. Sheet gauge (thickness) varied from 2 mm to 15 mm. Three different steel mills contributed samples for this evaluation. Each steel source is differentiated by the prefix A, B, or C in the sample codes. There is likely to be some variations in processing, both due to equipment and practices. However, in all cases, the tunnel furnace temperatures were reported to be in the 1100 to 1150°C range. The aim finish rolling temperatures were reported to be 870 to 900°C, and the aim coiling temperatures were reported to be from 600 to 620°C. While in some cases the processing data was well documented, in others the samples were taken from production runs where only the standard practice information was available. Two of the mills (A and B) collected test samples from the tail of the strip on the runout table prior to coiling. Sampling location of mill C was not documented, but test results indicated that they were taken out of the coil well after the rolling process.

Steel types included 5 heats of standard carbon steels (Type C-Mn), 5 heats of vanadium steel grades without nitrogen enhancement (Type V), and 11 heats of vanadium grades with nitrogen enhancement (Type V-N). For additional comparison, two heats each of niobium grade (Type Nb), vanadium-niobium-nitrogen grade (Type V-Nb-N), and niobium-titanium grade (Type Nb-Ti) steels were evaluated using the same test procedures. All of the steels were Al killed, containing 0.02 to 0.05% aluminum.

Test Procedure

Testing of the collected material included a complete product chemistry check analysis of each sample. The ferrite grain size was determined on each sample using the intercept count method. Duplicate tensile tests were performed, both on the as-received material and also after a simulated coil cooling treatment. Strain aging index tests were performed on additional duplicate samples. This test involved straining the tensile specimen to 7.5%, stopping the test, aging the sample for 1 hr. at 100°C, then completing the test in the tensile machine.

The increase in flow stress after aging (initial flow stress on retest after aging minus the final flow stress at 7.5% strain prior to aging) was identified as the strain aging index.[8]

Because some of the samples were obtained prior to the coiling operation, a simulated coil cooling cycle was run on companion samples from each material. Test samples were reheated to 600°C in a laboratory furnace for 1 hour, then furnace cooled for 48 hr. The 600°C temperature was chosen as most representative of typical aim coiling temperatures. The time at the coiling temperature can be important in precipitation mechanisms, and evaluating different samples would not be comparative without all having experienced the coil cooling cycle.

TEST RESULTS

Chemistry

Alloys - Table 1 includes the key test results from the investigation. The carbon content was essentially constant at 0.05%, reflecting the desire of most thin slab casting operations to avoid the complications of casting the peritectic carbon range. Manganese and silicon contents varied depending on the grade requirements and final applications. All samples were Al killed, with total Al contents from 0.02% to 0.05%. Microalloy contents varied, again depending on the grade and application requirements. Vanadium additions up to 0.13% were included. Typical niobium additions were from 0.03% to 0.04%. The limited time and temperature in the tunnel reheat furnaces, along with the nitrogen levels typically present in these electric furnace steels, make it difficult to keep higher niobium contents in solution prior to the rolling process.[9] The titanium additions were limited to 0.014% maximum, typical of additions recommended for optimum grain coarsening control.[5] Nitrogen levels varied, starting from the 0.007% to 0.010% residual range typical of these electric furnace steels, to 0.020% for the higher levels of vanadium addition.

Residuals – Phosphorous and sulfur levels were consistently low for all steels, running 0.015% max for both. Metallic residuals were higher than normally seen in BOF type steels, but were quite low for electric furnace steels. Copper levels were all under 0.15%, with an average value of 0.11%. Ni, Cr and Mo levels averaged 0.05%, 0.04%, and 0.01% respectively. Residual levels were assumed to be constant in the analysis of the results of this investigation.

Microstructure

Since all the test steels had a carbon content less than 0.06%, the microstructures were predominately polygonal ferrite with little pearlite. The only measurable microstructural results reported are the average ferrite grain size determined for each test sample. Standard procedures using a line intercept count technique were used on all samples. Results are given in Table 1.

Mechanical Properties

Table 1 includes the values of yield strength (YS), ultimate tensile strength (UTS), and % elongation (El) in a 50 mm gauge length for each sample of material in the as-received condition. Only samples A-1 through A-15 were tested for UTS and % El after the simulated coil cooling treatment, since that required an additional duplicate set of samples. Yield strength (YS) and strain aging index (A.I.) values were determined from duplicate samples for both the as-received and simulated coil cooled conditions of each sample material.

Table 1. Chemistry, grain size, and mechanical properties of hot rolled samples from strip steels produced by the thin slab / direct charging process.

			Product Chemistry									As-Received				600 C, Furnace Cool			
Sample Code	Steel Type	Gauge, mm	C wt.%	Mn wt.%	Si wt.%	Al wt.%	V wt.%	Nb wt.%	Ti wt.%	N wt.%	G.S. μ	YS MPa	UTS MPa	El. %	A.I. MPa	YS MPa	UTS MPa	El. %	A.I. MPa
A-1	C-Mn	3.3	0.050	0.42	0.06	0.038	<.01	<.01	<.01	0.0068	7.71	337	391	38	70	303	366	42	31
A-2	C-Mn	2.4	0.056	0.28	0.04	0.029	<.01	<.01	<.01	0.0078	7.61	305	390	37	69	322	368	39	44
A-3	C-Mn	12.7	0.056	0.45	0.04	0.035	<.01	<.01	<.01	0.0102	9.02	288	402	43	61	274	387	43	69
A-4	C-Mn	4.7	0.048	0.28	0.02	0.039	<.01	<.01	<.01	0.0076	8.32	294	368	38	57	293	349	44	55
A-5	C-Mn	6.4	0.054	0.44	0.01	0.033	<.01	<.01	<.01	0.0110	9.20	293	401	40	63	324	378	43	-3
A-6	V	15.3	0.050	0.83	0.01	0.046	0.085	<.01	<.01	0.0090	7.91	397	493	32	63	431	470	38	2
A-7	V	3.0	0.059	0.76	0.02	0.036	0.054	<.01	<.01	0.0098	5.72	382	462	30	46	404	474	29	2
A-8	V	6.3	0.056	0.79	0.04	0.039	0.057	<.01	<.01	0.0086	6.90	401	459	31	5	423	468	33	-6
A-9	V	4.7	0.054	1.10	0.02	0.043	0.066	<.01	<.01	0.0105	5.44	382	491	32	43	383	518	31	-6
A-10	V	2.1	0.051	0.83	0.03	0.034	0.057	<.01	<.01	0.0084	5.00	375	462	32	26	391	453	29	2
A-11	V-N	12.6	0.064	1.01	0.01	0.039	0.065	<.01	<.01	0.0140	9.19	425	505	33	62	467	502	36	2
A-12	V-N	7.8	0.053	0.99	0.01	0.036	0.067	<.01	<.01	0.0156	5.29	506	529	31	23	525	550	31	12
A-13	V-N	2.1	0.056	0.78	0.03	0.037	0.058	<.01	<.01	0.0136	4.30	415	485	32	34	430	512	34	3
A-14	V-N	3.8	0.058	0.86	0.02	0.037	0.086	<.01	<.01	0.0163	5.43	444	555	28	35	473	575	25	-2
A-15	V-N	6.3	0.050	0.80	0.01	0.050	0.057	<.01	<.01	0.0142	3.39	395	496	29	4	364	502	32	-3
B-1	V-N	2.0	0.054	0.99	0.19	0.024	0.050	<.01	<.01	0.0150	5.68	442	517	28	30	463	na	na	1
B-2	V-N	7.6	0.051	0.97	0.20	0.022	0.046	<.01	<.01	0.0140	6.38	399	506	33	53	430	na	na	3
C-1	V-N	5.7	0.053	1.62	0.35	0.026	0.130	<.01	<.01	0.0200	3.35	560	650	27	6	567	na	na	8
C-2	V-N	5.8	0.054	1.57	0.32	0.022	0.130	<.01	<.01	0.0200	4.35	558	645	29	8	572	na	na	5
C-3	V-N	5.7	0.052	1.58	0.36	0.026	0.120	<.01	<.01	0.0200	3.78	536	627	30	6	548	na	na	-1
C-4	V-N	5.7	0.057	1.62	0.35	0.025	0.130	<.01	<.01	0.0200	3.73	567	658	27	5	568	na	na	1
A-16	Nb	12.5	0.043	0.93	0.25	0.039	0.016	0.045	<.01	0.0100	5.46	396	490	36	52	406	na	na	2
B-3	Nb	6.2	0.038	1.17	0.16	0.020	<.01	0.030	<.01	0.0070	4.94	403	463	37	47	390	na	na	-1
A-17	V-Nb-N	4.5	0.056	1.19	0.30	0.034	0.120	0.044	<.01	0.0190	3.36	546	614	26	17	575	na	na	3
A-18	V-Nb-N	6.2	0.069	1.20	0.03	0.033	0.130	0.043	<.01	0.0160	3.63	525	597	29	13	572	na	na	7
A-19	Nb-Ti	12.6	0.061	0.91	0.30	0.043	<.01	0.042	0.014	0.0100	5.69	367	491	36	20	338	na	na	5
B-4	Nb-Ti	9.3	0.031	1.17	0.19	0.020	<.01	0.029	0.012	0.0070	4.40	408	470	38	40	385	na	na	1

DISCUSSION

Effect of V and N on Strengthening

Estimating the effect of alloying elements on strength is required when setting chemistry specifications for producing minimum strength grades in the hot-rolled condition. Although the as-rolled strength is a function of both process and chemistry, it is often necessary to estimate the effects of chemistry on strength prior to rolling the product. To help with this estimation, it can be useful to develop an empirical model of chemistry vs. strength. The limitations of that model include the assumption that the process will be controlled to duplicate the process conditions used to generate the data from which the model was derived.

Given that the data available were from three different processes, it is not likely that all process conditions were similar. As such, there is some risk in generating a strength model from the combined data. Yet, developing a model is a practical way to estimate the specific effects of specific alloying elements. This is particularly true with a production database obtained without a specific experimental design.

A linear regression analysis was run on the V and V-N grades from all three suppliers. This included 16 samples from A-6 to C-4. The yield strength used was from simulated coil cooled (600°C, furnace cooled) samples. The independent variables used were gauge (mm), Mn, Si, V, N, and Al. Equation (I) was derived, with an R square statistic of 0.93.

$$\text{Predicted YS (MPa)} = 427 + 3.3(\text{mm}) + 34.5(\%\text{Mn}) - 343(\%\text{Si}) + 1370(\%\text{V}) + 6995(\%\text{N}) - 5461(\%\text{Al}) \qquad \text{(I)}$$

Some of the coefficients are suspect, particularly the positive size coefficient and the large negative Si and Al coefficients. Yet the vanadium coefficient and the nitrogen coefficient agree with previous models(10,11). Analysis of this type can be very useful and provide more accuracy when large amounts of production data are accumulated under controlled conditions. Even using this model, one can estimate that the strengthening effect of nitrogen is about 7 MPa for every 10 ppm N (0.001%) as long as the V:N ratio is greater than 4:1.

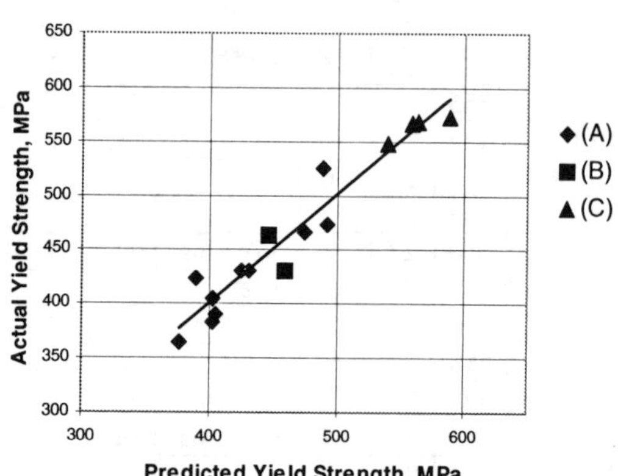

Figure 1. Regression Results (Eq. I).

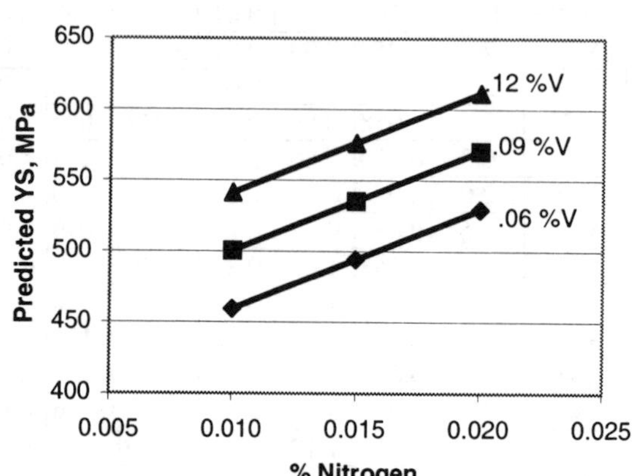

Figure 2. Effect of Nitrogen on Yield Strength.

Figure 1 is a graph of the predicted yield strength (PYS) versus the actual yield strength (AYS), in this case the yield strength after the simulated coil cooling. The material from supplier "C" comprised all of the 550 MPa grades, dominating the high strength side of the equation. Any processing differences that may exist between suppliers would tend to affect the model because of this choice of samples. Further substantiation would have to be done with a more uniform range of variables to improve confidence in the equation. However, generating this model was a viable way to provide some quantification of V and N effects from this data set.

These results support the well established observations that vanadium provides more strengthening when nitrogen is available up to the stoichiometric levels of V:N. The implication for thin slab casting operations is that the nitrogen levels of electric furnace steels can be used to an advantage to optimize the cost effectiveness of the vanadium alloy addition.

Figure 2 graphically depicts the strengthening effect of nitrogen in these grades. While the effects may not be truly linear as depicted by this simple linear regression model, the concept is that the strengthening effects of vanadium and nitrogen are reasonably linear. Therefore, vanadium and nitrogen are excellent "control" additions to make necessary strength adjustments. This is particularly true if the "control" additions contain a fixed ratio of each element, maintaining the efficiency of the vanadium addition while providing control over the final strength.

Ferrite Grain Size

Because of the nature of the data set available, a direct relationship between vanadium microalloying and as-rolled ferrite grain size could not be established. The data do demonstrate that fine ferrite grains can be readily achieved using the V-N alloying along with processing typical of the thin slab processing lines. The key features of the process include relatively high finishing temperatures of 850 to 900°C, allowing refinement of the austenite grains during rolling by repeated recrystallization. Additionally, lowering the austenite to ferrite transformation temperature provides additional refinement by increasing the grain refinement ratio, $D\gamma/DA$. Vanadium and nitrogen, along with manganese, have been shown to play a strong role in maximizing this grain refinement ratio.[12]

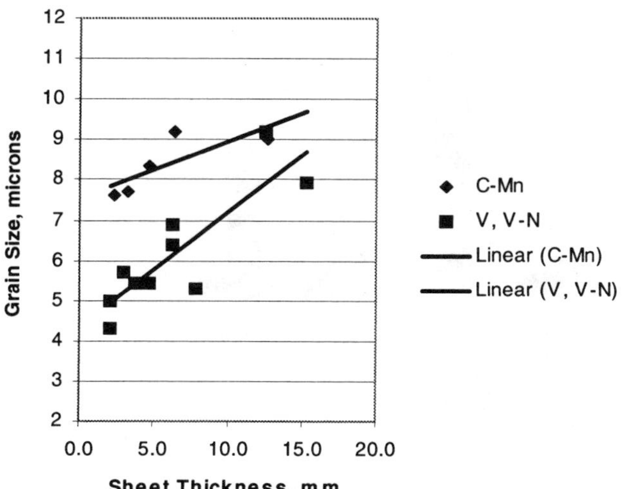

Figure 3. Ferrite grain size vs. gauge for C-Mn, V & V-N grades from mill "A".

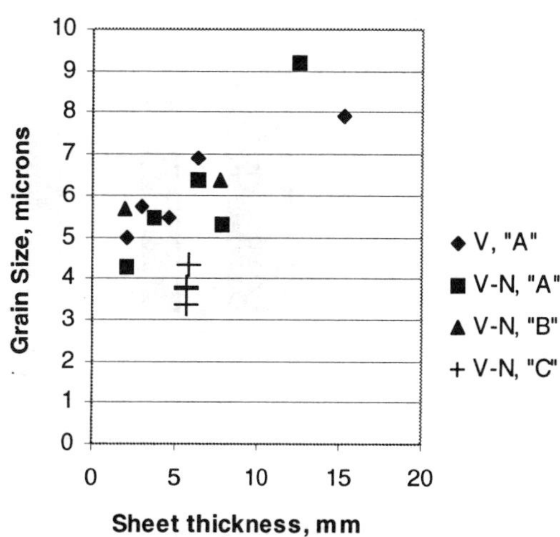

Figure 4. Ferrite grain size vs. gauge for all V-N material. "C" grades have highest V, N & Mn.

As shown in Figure 3, vanadium bearing grades show significant ferrite grain refinement over the carbon steel grades produced on the same mill with similar processing. Both the vanadium-nitrogen and the manganese levels play a role in this refinement. The faster cooling rates of the thinner gauges help minimize austenite grain growth after recrystallization from the last rolling pass, and lower the austenite to ferrite transformation temperatures.

Figure 4 shows the grain sizes observed in all V bearing steels. The steels from mill "C" had the highest alloying levels of vanadium (0.13%), nitrogen (0.020%), and manganese (1.6%). These alloy levels resulted in particularly fine ferrite grains as shown.

Strain Aging

One of the concerns often expressed about steels with enhanced nitrogen levels is the susceptibility to excessive strain aging. To evaluate this possibility, the strain aging tests were included as part of this evaluation. As explained previously, the strain aging index is defined as the amount of increase in flow stress after aging a prestrained sample at 100°C for 1 hour[8]. The amount of prestrain chosen for these tests was 7.5%. Since the samples obtained from some of the mills were taken before coiling, aging index values were determined before and after the samples were given a simulated coil cooling cycle. The cycle chosen to simulate the coil cooling process was to reheat to the samples to 600°C in a laboratory furnace, then slow cool in the furnace for 48 hours to ambient temperature.

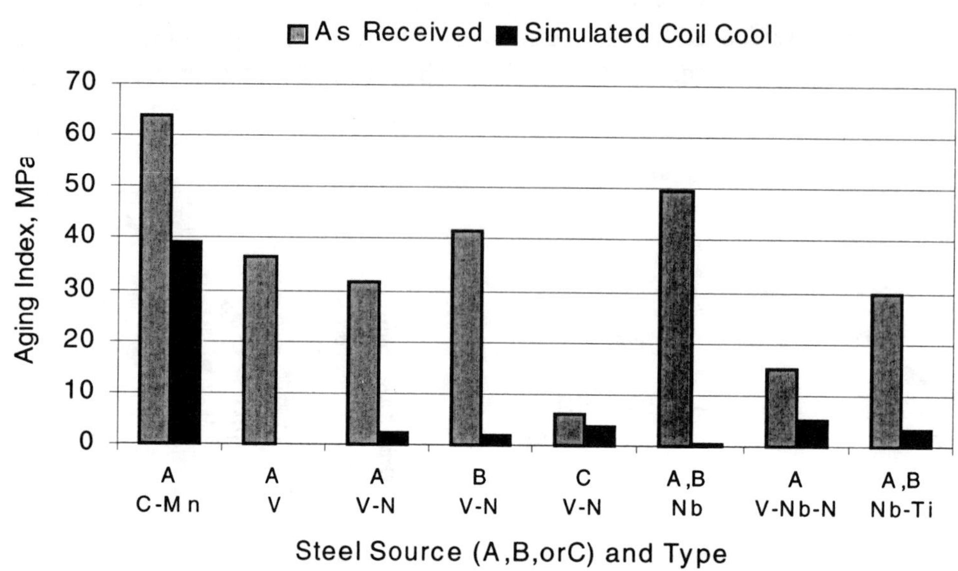

Figure 5. Average aging index values for each grade and steel mill source.

The results of the strain aging index study are shown in Figure 5. Each bar represents the average index value for the two to five samples in each group. The as-received results are adjacent to the simulated coil cooled results. A number of observations are of interest.

Samples from mill A were taken prior to coiling at the hot strip mill. Both the C-Mn grade and the V grade had the same average nitrogen levels, about 90 ppm. While the as-received samples of the V grade exhibited some strain aging, it was only about half that of the carbon steel. After simulated coil cooling, the V grade had no aging, while the C-Mn still had a substantial amount of aging. Both the V and V-N grades from mill A had the same amount of aging prior to the simulated coil cooling, and both had no aging after simulated coil cooling. The V-N grade from mill B had nearly identical performance to the V and V-N grades from mill A. This would seem to confirm that the samples from mill B were also taken off the runout table before the coil cooling process. The Nb and Nb-Ti grades from mill B had similar performance to the V-N grade.

Samples from mill C, on the other hand, showed virtually no aging either in the as-received or the coil cooled condition. These samples from mill C had the highest nitrogen levels of any of the steels. These results give a strong indication that the samples from mill C were taken out of the coil well after the coiling process. The coil cooling time was apparently sufficient to complete V(C,N) precipitation, resulting in no strain aging even without the simulated coil cooling treatment.

The elimination of strain aging susceptibility after coil cooling, either simulated or actual, in the V, V-N, and V-Nb-N grades was expected based on the anticipated precipitation of the uncombined nitrogen with the V in solution. What was not expected was the elimination of aging in the Nb and Nb-Ti grades. The Ti level was low in the Nb-Ti grades, and although the Ti should be fully precipitated as Ti-N there would still be ample free N left for aging. This was assuming the Nb would be primarily precipitated as NbC, or at least predominately C in Nb[C,N]. The elimination of aging in these grades after simulated coil cooling, unlike the C-Mn steels which still had aging, suggests that the nitrogen was tied up by either the Nb or the Al, which is present in all the steels.

If the free nitrogen in Nb steels was precipitated as Nb(C,N), that precipitation would normally be expected to provide some strengthening. If, however, the steel was not strengthened even after eliminating the aging caused by free nitrogen, then the nitrogen was most likely precipitated as AlN.

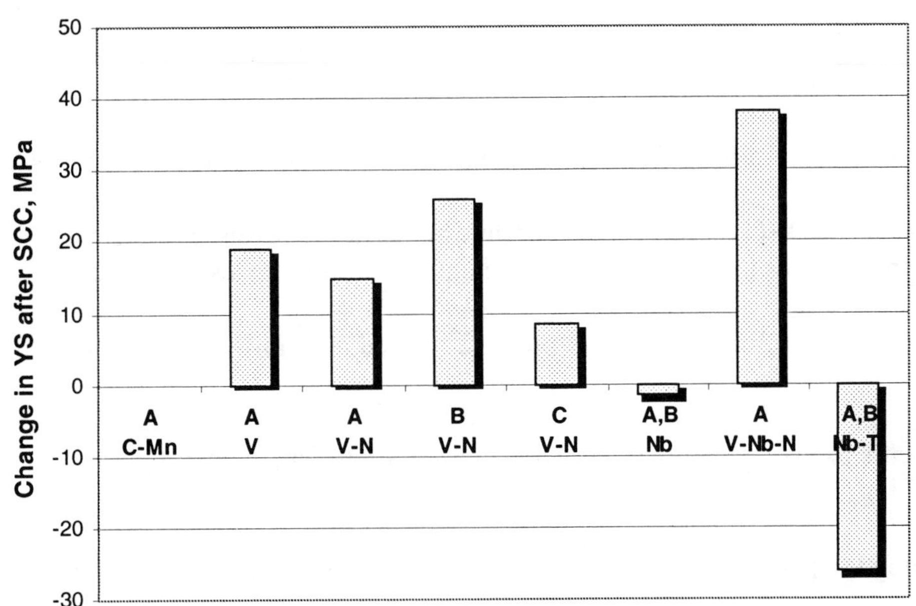

Figure 6. Average change (coil cooled YS – as-received YS) in yield strength after simulated coil cooling.

Figure 6 shows the average change of yield strength after the coil cooling simulation. A positive change means the yield strength increased as a result of the simulation. The C-Mn grade showed no change in yield strength, even though there was some reduction in the aging tendency. The primary nitride-forming element present in the C-Mn steel is aluminum. AlN is not known to be a strengthener, and therefore it is presumed that the coil cooling simulation with the C-Mn steels precipitated some, but not all, of the free nitrogen in the steel as AlN.

All of the V bearing steels exhibited strengthening after the coil cooling simulation, although the steel from mill C had the least amount. From the aging index results from "C" samples, it would be expected that there would be little or no free nitrogen left in the as-received condition. Without free N there would not be any significant additional precipitation of VN, therefore no significant strengthening. Even with the high nitrogen levels in the "C" steels, the V:N ratio was 6.5:1 compared to the stoichiometric ratio of 3.6:1. This would mean that there might have been some additional V in solution, which could have precipitated as a VC providing some marginal strengthening during the simulated coil cooling.

In the Nb and Nb-Ti grades, there was no indication of a strengthening reaction during the simulated coil cooling even though the strain aging susceptibility was reduced. In fact, in the Nb-Ti steels there was a significant reduction in strength from the 600 °C and furnace cool treatment. If there were some Nb(C,N) precipitation, one would normally expect strengthening from the process. Possible explanations would include that the nitrogen precipitation was as AlN, or that any strengthening from Nb(C,N) precipitation was offset by a loss in the dislocation strengthening that might be expected in these alloys. In any case, there is a risk of a strength loss in the body of the coil in these grades.

SUMMARY AND CONCLUSIONS

Vanadium has proven to be a popular choice as a microalloy for strip steels produced by the thin slab cast direct charging process. Primary reasons for the compatibility of vanadium with thin slab casting are as follows:

- Vanadium utilizes nitrogen as part of the alloy system. Nitrogen levels common to EAF steelmaking processes often associated with thin slab casting can be used to improve alloy cost effectiveness.
- Castability problems are minimized with vanadium compared to some other microalloy approaches.
- Higher solubility of the V(C,N) precipitates permit higher alloy levels in the steel even when restricted to the low reheat temperatures available in the tunnel equalization furnaces.
- Recrystallization control rolling techniques produce refined austenitic grains at normal carbon steel rolling temperatures.
- Precipitation occurs primarily during or after transformation to ferrite, thereby not contributing to roll forces during finish rolling.
- Reasonable coiling temperatures (580-620 °C) are compatible with normal mill equipment.

Data obtained in the course of this investigation has led to the following observations:

- The V and N strengthening effect in sheet material from a thin slab cast direct charging production process is quantitatively consistent with previous studies.
- Nitrogen enhances the strengthening effect of vanadium, improving the cost effectiveness of the vanadium alloy system.
- Hot rolled ferrite grain size can be refined using the V-N alloy system, achieving levels competitive with other alloy systems.
- The post-coiling slow cool thermal cycle is important to maximizing the precipitation of V(C,N) even in Al killed steels. With proper coiling temperatures (near 600 °C), yield strength is maximized and no strain aging is evident.
- Similar post-coiling slow cooling cycles on non-vanadium Al killed steels can reduce strain aging susceptibility, but may also lose yield strength.

ACKNOWLEDGEMENTS

A special thanks to Andrzej Wojcieszynski for his careful supervision of the metallurgical testing, and to Mike Korchynsky for his continuing support, guidance, and most of all, patience.

REFERENCES

1. M. Korchynsky, "Cost effectiveness of Microalloyed HSLA Steels," Conf. Proceedings, International Symposium on Steel for Fabricated Structures , ASM International, Cincinnati OH, Nov. 1999, pp. 139-145.

2. T. Siwecki, A. Sandberg, W. Roberts, R. Lagneborg, "The Influence of Processing Route and Nitrogen Content on Microstructure Development and Precipitation Hardening in V-microalloyed HSLA Steels," Thermomechanical Processing of Microalloyed Austenite, (Ed. By DeArdo, Ratz, Wray) TMS-AIME, Warrendale, USA, 1982, pp. 163-192.

3. P.S. Mitchell, D.N. Crowther and M.J.W. Green, "The Manufacture of High Strength, Vanadium-Containing Steels by Thin Slab Casting," Proc. of the 41st Mechanical Working and Steel Processing Conference, Baltimore, USA, 1999, pp. 459-470.

4. P.L. Lubensky, S.L. Wigman, D.J. Johnson, "High Strength Steel Processing via Direct Charging using Thin Slab Technology," Microalloying '95, Iron and Steel Society Inc., Pittsburgh, PA, 1995, pp. 225-233.

5. R. Lagneborg, T. Siwecki, S. Zajac, and B. Hutchinson, "The Role of Vanadium in Microalloyed Steels," Scandinavian Journal of Metallurgy, Vol. 28, issue 5, October 1999.

6. T. Kimura, F. Kawabata, K. Amano, A. Ohmori, M. Okatsu, K. Uchida, "Heavy Gauge H-Shapes with Excellent Seismic-Resistance for Building Structures Produced by the Third Generation TMCP," Proc. of International Symposium on Steel for Fabricated Structures, Cincinnati, USA, 1999, pp. 165-171.

7. T. Siwecki & S. Zajac, "Recrystallization Controlled Rolling and Accelerated Cooling of Ti-V-(Nb)-N Microalloyed Steels," Proc. of the 32nd Mechanical Working and Steel Processing Conference, Cincinnati, USA, 1990, pp. 441-451.

8. J.D. Baird, "Strain Aging of Steel – a critical review," Iron & Steel, May, 1963, pp. 186-192.

9. M. Korchynsky, "The Growing Role for Vanadium in HSLA Steels for Thin-Slab Casting," 33 METALPRODUCING, Penton Publications, Nov. 1998

10. R.J. Glodowski, "Vanadium-Nitrogen Microalloyed HSLA Strip Steels", The 4th International Conference on High Strength Low Alloy Steels, Xi'an, China, Oct. 2000.

11. Y. Zhang, C. Yang, S. Liu, "Study of V-N Microalloyed Steels for Reinforcing Bars," Symposium on the Promotion of New Grade III Reinforcing Bar and V-N Microalloying Technology, Beijing China, Feb. 2000.

12. W.B. Hutchinson, "Microstructure development during cooling of hot rolled steels" Proc. of Thermomechanical Processing of Steels, Vol. 1, May 2000, pp. 233-244.

PRODUCT PHYSICAL METALLURGY II

Metallurgical Review of Processes for Obtaining Strength and r-value in Galvannealed Sheet Steels

Robert P. Foley
Illinois Institute of Technology
10 West 32nd Street
Chicago, IL 60616
312-567-3052
foley@iit.edu

David K. Matlock
Colorado School of Mines
1500 Illinois Street
Golden, CO 80401
303-173-3775
dmatlock@mines.edu

George Krauss
Colorado School of Mines
1500 Illinois Street
Golden, CO 80401
303-273-3774
gkrauss@mines.edu

Key Words: Review, Strengthening Mechanisms, r-value, Galvannealed Sheet Steel, Ultra-Low Carbon Steels, Interstitial Free Steels

INTRODUCTION

Although electrogalvanized sheet has been used primarily for exposed automotive applications, galvannealed sheet has gained favor among some companies [1]. For exposed applications, weight reductions have been realized through the development of high strength steels and bake-hardenable steels such that panel stiffness is now a limiting consideration for setting the minimum thickness [2]. To be considered for this application, hot-dip galvannealed steels should obtain similar strength levels, mean r-values, and surface quality as existing electrogalvanized steels. Another approach to weight reduction is one-piece stamping [2,3]. However potential steels for this application require lower strength and greater formable (higher r-values) than present extra-deep drawing quality (EDDQ) steels (average r-values > 2.3). In contrast, weight savings through strength increases may still be obtained for unexposed applications [2,4]. In either exposed or unexposed applications, good drawability (high mean r-value) is considered a basic material requirement.

Figure 1 summarizes combinations of obtainable tensile strengths and average r-values as reported by one sheet manufacturer [5]. Candidate steels for hot-dip galvannealed automotive application are steels with similar surface quality and mechanical properties as existing electrogalvanized steels, steels with higher strength than existing ultra-low carbon galvannealed steels (approximately 300 MPa yield strength and 450 MPa tensile strength), steels with formability in excess of EDDQ level, and steels with good combinations of strength and average r-value. With these objectives in mind, thermal constraints of the hot-dip galvannealing line will be examined and used to identify metallurgical approaches for obtaining desired property combinations. Baseline yield and tensile strengths for ULC steels are approximately 130 and 300 MPa and the reported methods for increasing these low strengths are the subject of this work.

Figure 2 shows the various stages involved in the hot-dip galvanizing process [6]. The total process time is much less than five minutes. While post hot-dip batch annealing can be used to remove carbon from solution [7,8], use of interstitial-free steels avoids this issue altogether. In addition, retention of some unstabilized C in the final product is in fact beneficial for obtaining strengthening by bake-hardening mechanisms.

DESIGN PHILOSOPHY FOR ULTRA-LOW CARBON STEEL PROCESSING

The driving force for producing ultra-low carbon (<0.005 wt-pct C) steels is the beneficial effect which very low C and N contents have on microstructure and formability. As dissolved interstitial levels decrease, the cold rolling and recrystallization-annealing process produces a greater fraction of grains crystallographically oriented for good deep drawing performance (high average r-value). While a generally accepted mechanism for this improvement has yet to emerge [9], it is clear that the texture is influenced by the amount of dissolved interstitial elements present during cold- or warm-rolling [10] as well as during the recrystallization [11,12]. Presumably, the low grain boundary strength (small Hall-Petch term) associated with low interstitial levels [13-15] promotes development of homogeneous deformation structures with few shear bands, which act as nucleation sites for the undesirable texture components [9,16]. At recrystallization temperatures, unstabilized C or N can form dipoles with substitutional atoms such as Mn; the presence of the complexes likely reduces the growth rate of grains with the favorable {111}<uvw> texture [17,18].

Steelmaking in oxygen converters, although capable of reducing N to levels near 25 wt-ppm, typically can reduce C only to 100 to 200 wt-ppm. Ultra-low carbon steels are vacuum degassed steels. With vacuum degassing, C levels can be reduced to the 30 to 50 wt-ppm level consistently. Stabilization of N and C prior to cold rolling and recrystallization annealing is essential for obtaining the {111}<uvw> texture necessary for good formability. Nitrogen is effectively removed from solution by AlN or TiN formation. Major attention is therefore focused on stabilization or removal of carbon residual in interstitial solid solution by additions of Ti and Nb. Beyond the reduction in C obtained by the advances in steelmaking technology, alloying additions of Ti and/or Nb remove C from solution through the formation of carbon-containing precipitates. The best properties are obtained when C is removed from solution prior to cold rolling and recrystallization annealing. In practice, coiling is performed at relatively high temperatures (>680°C) to promote coarse carbide and nitride formation. Coarse precipitates, being harder to dissolve than fine precipitates, restrict the level of dissolved C or N that can be reached before recrystallization starts in subsequent processing. The phases which determine the degree of N and C stabilization are TiN, TiC, TiS, $Ti_4C_2S_2$, FeTiP, AlN, MnS, and NbC. It is also well established [19] that higher coiling temperatures and excess stabilizing elements are beneficial for increasing r-value. However, high coiling temperatures reportedly reduce pickling efficiency and may lead to handling problems [20, 21]. Therefore, reduction in total C content remains a basic steelmaking objective [3].

In ferrite recrystallization, nucleation of grains with the preferred orientation {111}<uvw> (usually {111}<110> and {111}<112>) occurs at or near prior ferrite grain boundaries, and production paths that increase the grain boundary area per unit volume are desirable for improving the deep-drawing performance. Such measures include: (1) using a low slab reheat temperature to inhibit austenite grain growth [9], (2) using Nb additions to inhibit austenite recrystallization, (3) alloying or cooling rate modification to enhance the austenite-to-ferrite transformation ratio [22], and (4) maximizing the cold-rolling reduction to increase surface area-to-volume ratio of the ferrite grains [10,23].

Higher r-values are realized with greater amounts of homogeneous rolling reduction [7,23]. While cold-rolling reduction is limited by thickness of the hot band and the target gage [24], combined warm and cold rolling practices can be used to increase the obtainable cold-reduction [25]. Both types of reduction can lead to the favorable texture formation if IF conditions are met during rolling [10]. Formability also increases with increasing annealing temperature [26]. This behavior is due to the grain growth of the favorable texture at the expense of the expense of other texture components. Consequently, elimination of second-phase pinning particles and use of prolonged annealing conditions are encouraged. Developing production capabilities allow C and N levels to be reduced to approximately 20 and 30 wt-ppm with consistency [27]. As discussed below, the very low C content can be removed from solution by coiling at lower temperatures.

As noted, Ti forms precipitate phases containing N, C, and S. TiN is the most stable of these precipitates and is even more stable than AlN. Likely formation of TiS, $Ti_4C_2S_2$, TiC, and MnS will depend on the soaking temperature and the amounts of each element in the steel. $Ti_4C_2S_2$ is favored at low soaking temperatures (<1050°C) [28] and low Mn content [29], whereas TiS is favored at higher soaking temperatures (1250°C) and at higher S levels [30,31]. The TiC phase has lower stability compared to the other phases. Consequently, the amount of Ti that needs to be added to produce a stabilized product depends on the amount of Ti that is tied up in other precipitates. For completely stabilized IF steels, the Ti addition is usually in excess of that required to completely combine with N, S, and C [32] as:

$$\text{excess Ti wt-pct} = \text{Ti wt-pct} - \{48/14\}(\text{N wt-pct}) - \{48/32\}(\text{S wt-pct}) - \{48/12\}(\text{C wt-pct}). \quad (1)$$

While the best ductility and drawability are obtained at excess Ti levels near 0.05 wt-pct [19,32,33], excess Ti deteriorates the galvannealed surface quality. Ti accelerates the galvannealing process making it more difficult to control [27] and is a contributor to a cosmetic streaking phenomena [7,8,34]. With excess Ti, very little solute C exists at grain boundaries and nucleation of outbursts can occur at these boundaries. In addition, stable Ti surface- and subsurface-oxide particles [8,35] can cause rolled-in surface defects related to slivers and may also act as nucleation sites for outburst phenomena in the galvannealing reaction [36]. Hashiguchi [7] has shown that streaky surface defects associated with Ti are rare below 0.03 wt-pct Ti.

In contrast to Ti, C control by Nb is more straightforward. The available Nb for forming NbC is simply the total Nb addition. However, NbC is less stable than TiC and may partially dissolve during high temperature annealing at temperatures in excess of 830°C. Excess Nb is generally undesirable from a processing point of view due to its effect of increasing both the ferrite and the austenite recrystallization temperatures [37-39]. Reportedly, the high finishing and coiling temperatures needed in the production of Nb-bearing steels can lead to unsuitable property variations along the length of the coil [40]. Despite this drawback, Nb in excess of that needed to stabilize C has benefits to both base mechanical properties and coating performance. In particular, excess Nb has a similar effect as excess Ti for producing highly formable steels. In addition, Nb reduces Δr values [26,39,41,42], improves powdering resistance [40,43], and reduces cold-work embrittlement associated with high P concentrations [33,31].

For exposed galvannealed applications, Ti+Nb steels can offer an advantage over straight Ti or Nb stabilized steels. For complete stabilization, Ti additions are made to combine with N and S such that Ti is not in excess, and Nb completely stabilizes C [32], according to the following relationships:

$$\{48/14\}(\text{N wt-pct}) < \text{Ti wt-pct} < \{48/14\}(\text{N wt-pct}) + \{48/32\}(\text{S wt-pct}), \quad (2)$$

$$\text{Nb wt-pct} > \{93/12\}(\text{C wt-pct}). \quad (3)$$

Keeping the Ti level low avoids surface appearance problems; Nb stabilizes C and improves powdering resistance. Kino [40] has shown that Ti+Nb steels, with coating weights as high as 60 g/m^2, perform as well as Ti steels with coating weight of only 30 g/m^2. Reduction in the N and C contents by steelmaking is still desired to improve formability and to keep Ti and Nb additions to a minimum. While stabilization of C and N by Ti and/or Nb additions promotes drawability, Ti can lead to unfavorable surface appearance and Nb can act to increase the recrystallization temperature.

STRENGTHENING MECHANISMS IN SHEET STEELS

This section presents literature results on strengthening mechanisms in sheet steel and their reported or likely effect on the galvannealing behavior and performance of the coating. With respect to the basic issues related to yield or tensile strength, many studies have attempted to summarize and review the practical contributions of different strengthening mechanisms to structure-property relationships [44-47]. For low-carbon or mild steels, the relative importance of different strengthening mechanisms are given by a modified Hall-Petch relationship:

$$S_{YS} = S_O + S_{SS} + S_{PPTN} + S_D + k\,d^{-1/2} \quad . \qquad (4)$$

where, S_{YS}, the yield stress, is expressed as a linear combination of different strengthening mechanisms. Here, S_O is the Peierls stress or intrinsic barrier to dislocation motion, S_{SS} is the strengthening increment from solid-solution hardening, S_{PPTN} is the contribution from precipitation hardening, S_D is substructure strengthening from dislocations, and $kd^{-1/2}$ is the Hall-Petch term. The numeric coefficients that enter Eq. 4 and similar equations are often obtained from regression analysis of mill or laboratory data.

Equation 4 is applicable to steels with a homogeneous microstructure and only a small volume fraction of second-phase particles. For steels with mixed-microstructures, e.g. ferrite-martensite, ferrite-bainite, etc., the constitutive behavior of each "phase", the extent of connectedness or contiguity, and the extent of deformation compatibility between the phases are additional microstructure variables affecting strength. For these microstructures, the relationship between strength and microstructure is complex and has been represented, for example in dual-phase steels, as a linear function of the volume fraction of martensite, V_m, (or stronger constituent) [48]:

$$S_{UTS} = K_1 + K_2\,V_m, \qquad (5)$$

where, S_{UTS} is the ultimate tensile stress, K_1 is the ultimate stress of the ferrite in the absence of any martensite and is itself a sum of effects reflected in Eq. 4. The term K_2V_m represents the increment of strength due to the presence of a unit volume fraction of martensite. The transformation of austenite to martensite represents another mechanism that can strengthen steel. Appreciable r-values in dual-phase microstructures are beginning to be obtained in production and understood metallurgically [22].

Solid Solution Strengthening A number of studies report the use of solid solution strengthening alloying elements to obtain high-strength ULC or ULC-IF steels [27,33,40]. Such elements typically include P, Mn, Si, and Cr. Figure 3 summarizes the strengthening effects of a number of different elements. Of these, P, Mn, and Si are the most common. Ushioda *et al.* showed data for both yield and tensile strength for different alloy content after applying a correction for grain size differences and illustrate that the strength contribution to yield strength per wt-pct addition is different than the contribution to ultimate strength [18]. These authors suggested that the effect results from different work hardening characteristics as related to changes in stacking fault energy. As discussed below, this effect has been exploited to increase work hardening rate and the formability of ULC steels.

In low-carbon steels the effect of P is closely tied to the strength of grain boundaries. For normal IF or ULC steels, the general strengthening by P is very effective at 7-12 MPa / 0.01 wt-pct P. In ULC steels, the strengthening increment can also depend on the type of stabilizing element present [27]. Meyer and coworkers have presented data to show that P strengthening is slightly lower in Ti-stabilized steels compared to Nb-stabilized steels. This difference is may be related to a change in grain size as well as to formation of the FeTiP phase which, when present, reduces the solute P available for strengthening. If the FeTiP phase is formed in the hot-band or during batch annealing, the recrystallization texture favorable for deep drawability is also suppressed [33,49]. In addition, loss of Ti to FeTiP clearly lowers the effective Ti content. Low coiling temperatures (500°C), higher cooling rates after hot rolling, and fast heating rates (10°C/s) can suppress FeTiP precipitation and favorable textures can be obtained [4,50,51]. In continuous annealing of Nb stabilized IF steels, P additions have little deleterious effect on mean r-value [33].

In steels that are overaged after hot-dip galvanizing (non-IF or non-ULC) to reduce dissolved carbon, the apparent strengthening increment from P can be up to 20 MPa / 0.01 wt-pct P. The origin of the additional strengthening has been related to precipitation hardening by fine epsilon carbide particles, which form because of a special interaction between P and C [52].

Like C, P tends to segregate to grain boundaries and to free surfaces. At grain boundaries, segregated P slows the alloying rate in the galvannealing process [36,53,54], thereby reducing the number of outbursts. The slower alloying rates associated with higher P content can be increased, if necessary, by lowering the Al content of the zinc bath [27,36,55]. At levels greater than approximately 0.03 wt-pct, P can cause unacceptable secondary cold-work embrittlement behavior in IF steels [41]. In stabilized IF steels, P segregates to grain boundaries and decreases boundary resistance to fracture. Embrittlement in P-containing steels can be alleviated somewhat by the addition of B at levels of 15 wt-ppm or less [30,56]. Analysis has shown that B blocks P at grain boundaries and, like C, has a cohesive effect on the boundaries [57,58]. With increased aging time, such as in batch annealing, P replaces B and higher ductile-to-brittle transition temperatures result [58,59]. Because BN forms rapidly, the effective B content can be reduced if N levels are not controlled by a suitable Ti addition. High coiling temperatures, in excess of 700°C, are also useful to promote AlN or TiN precipitation in preference to BN [57,60].

While the effect of B on the galvanizing reaction is not clear, it is likely that B acts similar to P, as both segregate to grain boundaries. In addition to its basic role in P-containing steels, B also contributes to solid solution strengthening. However, the strength increment is small at approximately 1 MPa / 10 ppm addition. Other process-related issues concerning B address changes in the ferrite recrystallization temperature and the Ar_3. With respect to recrystallization in the galvanizing line, B has a dramatic effect on raising the ferrite recrystallization temperature [32]. With respect to hot rolling, B also increases the mill loads in the finishing stands [21]. Consequently, B-containing steels must be finished at higher temperatures than B-free steels.

Tokunaga *et al.* [33] indicated that Nb and Nb+Ti steels can be less susceptible to cold-work embrittlement than straight Ti steels even when B is not present. This improvement, although small, likely reflects the lower stability of NbC compared to TiC [57]. In the Nb steels, partial dissolution of NbC is possible. Like B, free C enhances the grain boundary cohesive strength. Smaller grain size in the Nb steels may also contribute to the better ductile-to-brittle transition behavior.

Developed high strength P-containing steels for galvannealing applications are reported by Thyssen [38,61], NKK [62], Nippon Steel [40], Kawasaki [64] and Sumitomo [4,51]. The highest strength steels obtain YS and UTS values of approximately 300 and 450 MPa, with mean r-values of 1.95 [62,63]. The maximum P content in these steels is approximately 0.10 wt-pct, which contributes approximately 80-120 MPa to the base strength.

Additions of Mn, Si, and Cr have also been considered for solid solution strengthening. While the strengthening increment obtained from such elements is lower than that of P, the elements do not contribute to cold-work embrittlement phenomena. The elements Mn, Cr, and Si can oxidize in preference to Fe. As oxides can deteriorate the structure and performance of the galvanneal coating [36,64]. It has been suggested that the Si content be limited to 0.5 wt-pct [63]. The upper limit of the Mn is, in part, determined by oxide formation, in part, by it effect on hardenability. For Mn, an upper limit in the range between 1.5 and 2.0 wt-pct has been suggested [62-64]. The number of oxide particles can be reduced by controlling the sources of oxygen in the annealing atmosphere or by limiting the amount of oxide-forming alloying elements present in the steel.

Manganese is generally added to remove S from solution as MnS. In addition, Mn can also be beneficial to deep drawability as it stabilizes austenite and decreases the Ar_3 temperature. However, in low-carbon steels or ULC steels that are not processed to the IF condition, the Mn-C complex forms and {111}[uvw] texture deteriorates [17,51]. In strict IF steels however, the Mn addition can produce both a benefit to strength and to deep-drawing texture [49,50,62]. Figure 4 shows plots of ultimate tensile strength versus mean r-value for two IF steels [62]. As Mn increases, the ferrite start temperature decreases, super-saturation increases, and ferrite grain size is refined, promoting favorable texture formation upon subsequent cold rolling and annealing. At the same time, additional Mn provides solid solution strengthening so that r-value and UTS increase simultaneously.

After P, Si has the highest elemental hardening effect (approximately 85 MPa increase in tensile strength per wt-pct Si). However, Si is a ferrite stabilizer and it raises the Ar_3 temperature. Alone, this effect is generally undesirable as higher finishing temperatures are required to maintain austenite rolling conditions and large ferrite grain sizes are obtained [51]. However, in combination with Mn, Si additions have been used to balance the transformation temperature so to prevent substantial austenite grain coarsening after finishing [63]. The SiO_2 phase can be deleterious to coating performance.

Ushioda and coworkers [18] report that P and Si influence yield and tensile strength differently than do Mn and Cr. The authors show benefits obtained by taking advantage alloys with different work-hardening rates. With the Mn and Cr alloyed steels, lower yield strengths are obtained at UTS levels comparable to those of traditional steels. Low yield strength without loss in tensile strength benefits formability (shape fixability) and performance [2,18].

Bake-Hardening (Strain-Aging) Strengthening Bake-hardening (BH) is a strain-aging phenomena whereby dissolved interstitial C atoms (or less often, N atoms) diffuse to and strengthen grain boundaries as well as pin otherwise mobile dislocations [28,65-68]. The increment of strength is obtained in the paint-bake cycle and is characterized by the magnitude of the stress difference between the 2% offset tensile flow stress and the yield stress of a tensile sample strained 2% unloaded, aged at 170°C for 20 minutes and then restrained. Alloy design and processing parameters that contribute to retaining solute C or N and slowing room temperature aging will contribute to increase the strength increment from bake-hardening. Figure 5 shows the relationship between solute C, N, and BH [69]. The lower portion of the figure indicates that an unsuitable level of yield point elongation occurs when the dissolved C and N reach level greater than 10-20 wt-ppm. A number of approaches exist for controlling dissolved C and N levels to obtain good BH in ULC and ULC-IF steels.

Bleck has considered the range of C contents and BH values for steels produced by hot-dip galvanizing, continuous annealing and batch annealing [61]. Figure 6 summaries his findings. Without a post-annealing step, the total C levels should be kept near the target C content desired to produce the desired BH level. One approach is to lower the stabilization and let some carbides redissolve during high temperature annealing. Sakata *et al.* showed the effect of the reheat temperature on the average r-value and obtainable BH [sakata94]. Both high r-values and BH values are obtained. During recrystallization, C is stabilized in either NbC or TiC and during the high temperature annealing process, precipitates dissociate according to the stability of the alloy. If the alloy chemistry is very stabilized, the amount of dissolved C will vary with annealing temperature, leading to variable BH values. In addition to dissolving C, retention of C to room temperature is favored by high cooling rates after annealing. For galvannealing applications, Sakata *et al.* [70] also showed that recrystallize annealing followed by galvannealing can produce a steel with good BH values if the annealing temperature is sufficiently high to permit dissolution of some NbC and if the cooling rates from the annealing temperature and the galvannealing temperature are fast enough to avoid NbC and Fe_3C precipitation.

Alternatively, control of the dissolved C has also been attempted by inhibiting TiC nucleation. Mizui and coworkers [28,71] proposed a process whereby the solute C for bake hardening is equal to the total C. To accomplish this, the following constraints are needed: (1) the total C is controlled between 15 and 25 wt-ppm, (2) the Ti content is set between {48/14} (N wt-pct) and {48/14} (N wt-pct) + {48/32} (S wt-pct), (3) the Mn concentration is low enough to prevent MnS precipitation in preference to TiS, and (4) the slab reheat is high enough to prevent $Ti_4C_2S_2$ precipitation. Figure 7 schematically illustrates this and other scenarios. In case (c) of the figure, Ti is completely tied up with N and S; low Mn is used so that free S first combines with the remaining Ti as TiS; the remaining S combines with Mn as MnS; and the total C content is available to give a good BH value. In cases (a) (b) (d) and (e), the C stabilized in either TiC or $Ti_4C_2S_2$ is not available for providing a BH increment. As mentioned, the target C level is approximately 20 wt-ppm for the desired BH and aging resistance. Consequently, great care must be expended during steelmaking to obtain the precise C content.

Tsunoyama *et al.* [30] suggest lowering the S content to 30 ppm or less to reduce the number of TiS particles. As TiS acts as a heterogeneous nucleation site for TiC, TiC precipitation is inhibited by lack of TiS particles and C remains in solution for bake hardening. These authors indicate that bake-hardening can be obtained even at excess Ti/C ratios greater than one. While low S is essential for this procedure, C control is critical to insure that the total dissolved C content does not exceed the target level needed for good room temperature aging resistance.

The approaches to obtaining BH in ULC continuous annealed steels mentioned so far suffer from either residual C being present during recrystallization or the need for high annealing temperatures to dissolve the NbC or TiC precipitates. Kitamura and coworkers [72,73] suggest a method for avoiding these problems. First, the recrystallization texture favorable for deep drawing is produced from a cold-rolled Ti stabilized microstructure. Then, excess C and bake-hardening properties are imparted by substituting the normal furnace atmosphere with a carburizing atmosphere. Under these conditions, C diffuses in to the sheet, the excess Ti/C ratio decreases, and IF behavior is lost. Consequently, the sheet obtains BH properties and improved resistance to cold word embrittlement. The authors have shown that good combinations of mean r-value and BH are obtained by this process (up to 50 MPa BH and r-value of 2.2) [72]. This process has only been applied in the laboratory. Application to industrial processing would require major modifications of existing equipment.

At room temperature, segregation of dissolved C to grain boundaries and dislocations and the accompanying return of the yield point is largely unavoidable. However, the discontinuous yielding can be masked to some extent by temper rolling. Furthermore, the masking can be optimized to prolong the return of discontinuous yielding. Figure 8 illustrates this point and shows that compressive straining (with high loads), rather than tensile straining, improves the resistance yield point return [4]. The mechanism of the improvement is related to the inhomogeneous residual stress patterns introduced in the sheet [74,75]. Similar benefits from strain pattern control can also be obtained from roll texturing [27] or by the introduction of a dual-phase microstructure, in which, the residual stress patterning arises from the diffusionless transformation of austenite.

Substructure and Transformation Strengthening Dislocation substructure can originate from many processes. If recrystallization of cold-rolled sheet is not accomplished in the annealing furnace, then the sheet may be only partially recrystallized or recovery annealed. In this case, full recrystallization texture is not obtained in the final product and the drawability may suffer. Figure 9 shows the relationship between annealing temperature and UTS for a series of solution strengthened alloys [61]. As the annealing temperature decreases below 800°C, UTS values increase as a consequence of incomplete recrystallization. At 800 and 850°C, UTS

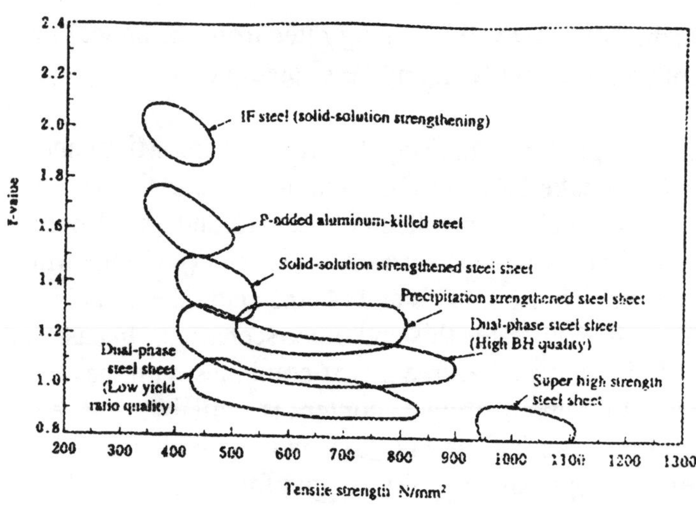

Figure 1 - Typical tradeoff in r-value and UTS [5].

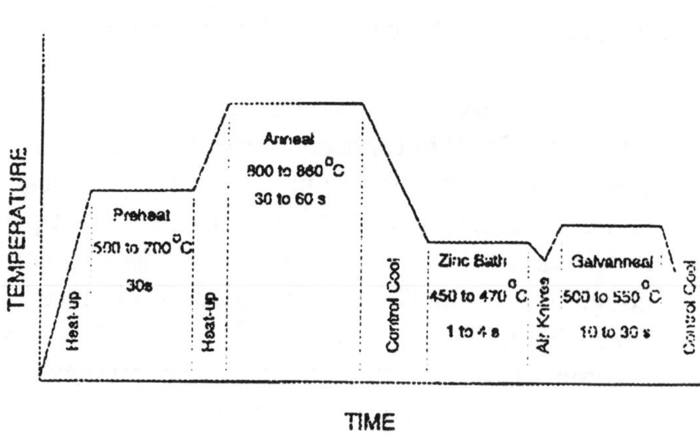

Figure 2 - Galvanneal cycle thermal path [6].

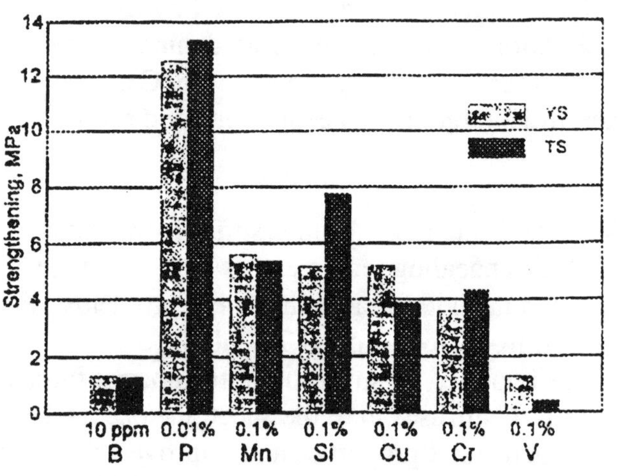

Figure 3 - Strengthening per alloying addition [27].

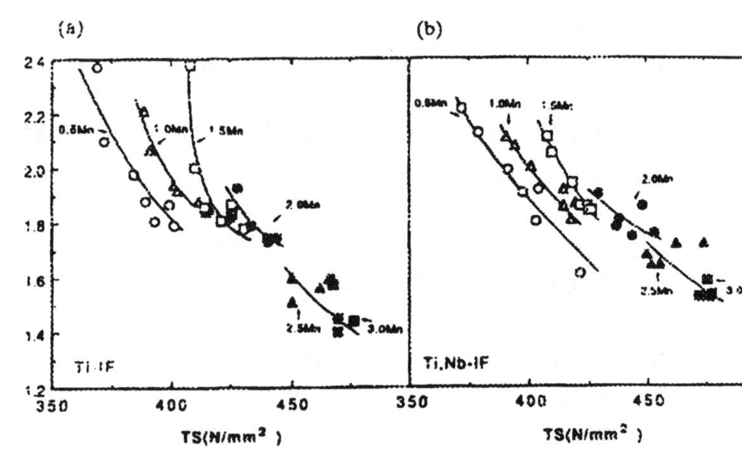

Figure 4 - Mn Effect on r-value-UTS balance [62].

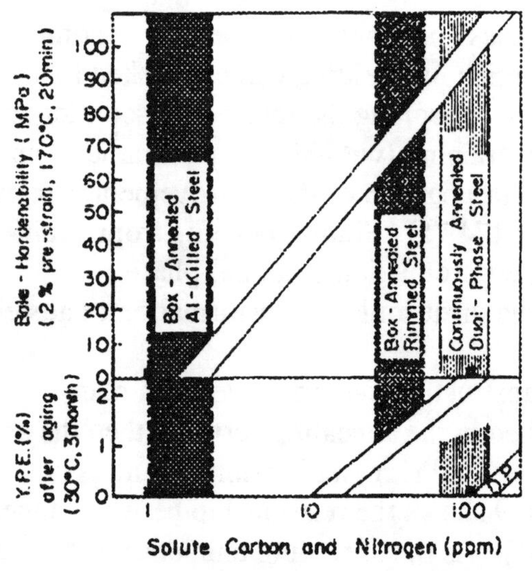

Figure 5 - Increase of BH with solute C+N [69].

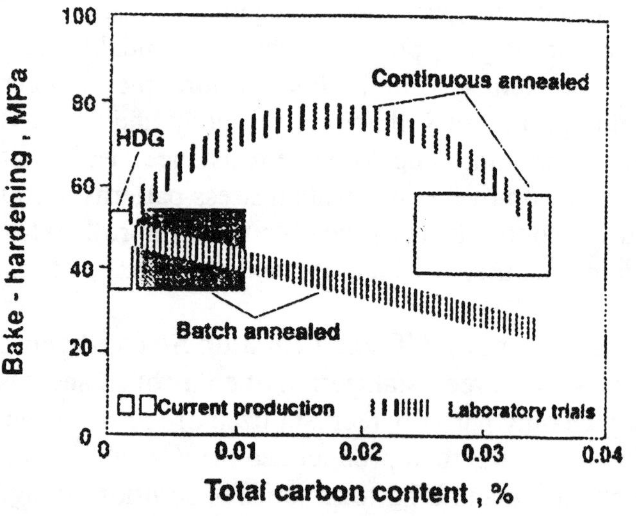

Figure 6 - BH vs total C and process [61].

values are nearly constant and obtain a maximum value of approximately 450 MPa that has been traditionally identified as the upper limit to the ultimate strength attainable in the recrystallized state [41]. For the best combination of properties, the alloying additions, line speeds, and annealing temperature need to be controlled to give a fully recrystallized structure and the highest possible r-values.

Alternative methods of substructure strengthening add dislocations after the recrystallization texture has been obtained. This may be accomplished either by alloy design and annealing routes that produce textured dual-phase type microstructures or by crafted temper rolling procedures. Studies from Nippon Steel [18,22] and Kawasaki Steel [70] have indicated that austenite decomposition of alloyed ULC steels can be accomplished to produce both favorable deep-drawing properties and benefits from transformation substructure. A similar approach by Makimattila [76] led to the successful development of the Zinquench process.

Figure 10 compares the behavior of two steels that were intercritically annealed or completely austenitized during annealing [22]. The graph shows that steel MP maintains nearly consistent mean r-values as annealing temperature is increased. Steel NT represents the traditional ULC behavior in that complete austenitization leads to texture randomizing and loss of high r-value. The authors presented pole figures obtained at different stages of annealing. Initially, both steels have similar textures, however after austenitizing and cooling, steel NT has nearly random texture while the texture of steel MP very closely resembles the preanneal texture. Retention of the deep-drawing texture in steel MP was related to the stresses introduced during the reverse transformation (ferrite-to-austenite) and the subsequent effect these stresses have on variant selection during subsequent austenite decomposition. Precise control of the decomposition is required to obtain the good texture. In addition to producing a higher strength dual-phase type microstructure, the developed steels may also take advantage of the superior BH and aging resistance offered by dual-phase steels.

After recrystallization, substructure can also be added by temper rolling without impairing the drawing texture. Temper rolling imparts shape control and eliminates discontinuous yielding in low-carbon and bake-hardenable steels. In low-carbon or bake-hardenable steels, the yield strength initially decreases with temper strain due to a change in yielding mechanism. For these types of steels, the minimum yield strength occurs before the temper rolling deformation contributes appreciable strengthening. In stabilized IF steels however, yielding is continuous in the as-annealed condition. Consequently, the temper rolling strain work hardens the steel immediately. As IF steels can be susceptible to cold work embrittlement, cold work strain from temper rolling is expected to reduce resistance to brittle fracture.

A number of published studies have examined strengthening by temper rolling [62,77,78]. Figure 11 shows the change in r-value versus temper elongation [77]. As mentioned, this strengthening mechanism is appropriate when yield rather than ultimate strength is of principal concern. Although total elongation decreases with temper elongation, the deep drawing performance, as determined by mean r-value, is unaffected until very high strains are attained. However, no studies address the effects of high temper rolling elongation for strengthening galvannealed steels. Limits of this method with respect to such issues as coating adhesion and cold work embrittlement need to be quantified with respect to the type and magnitude of the imparted strain pattern.

Grain Size and Precipitation Strengthening As with solid solution and substructure strengthening, independent control of grain size and precipitation hardening is not straightforward for Ti and Nb-microalloyed ULC steels used for hot-dip galvanizing. Both: chemistry and texture influence grain size strengthening. A small Hall-Petch slope, ky, is realized if either the level of the dissolved C and/or N (P and B, too) decreases or if the degree of preferred orientation increases [13,79]. However, both of these factors are desired for the production of stabilized ULC steels with deep drawing properties. In addition, dispersion strengthening is a function of interparticle spacing and volume fraction or particle size. In ULC steels, a low precipitate volume fraction results from the desire to reduce C and N contents. Furthermore, particles coarsening occurs during processing with high coiling temperatures and during recrystallization anneal.

For completely stabilized continuously annealed IF steels, the range of recrystallized grain sizes typically varies between approximately 5 and 20 μm [32]. Over this range, a 40-50 MPa increment of strength can be realized if the Hall-Petch slope has the 5-7 MPa $mm^{-1/2}$ values typical of IF steels. Notably, finer grain sizes a obtained in Nb-containing steels in comparison to straight Ti-stabilized steels. However, generally lower r-values result in Nb-containing steels under similar processing conditions due to the pinning effect that NbC particles have on grain growth during recrystallization. Consequently, high r-values and small grain sizes can be promoted by enhancing the nucleation of grains with favorable texture in preference to other orientations ar then by limiting the growth of such grains. Efforts applied to increase the ferrite surface area per volume in th cold-rolled condition naturally benefit grain size strengthening.

Precipitation of NbC during hot rolling is particularly effective for maintaining a fine austenite grain size a providing some precipitation strengthening [20]. The amount of strengthening increases with Nb content. However, Nb has a detrimental effect on increasing the recrystallization temperature of the cold-rolled sheet. Bleck has shown the relationship between Nb content, annealing temperature, and recrystallization condition f continuous annealed sheet steel [38]. In stabilized IF steels, the level of C or N is stoichiometrically lower tha the alloying addition used to tie up such elements. Consequently, rapid precipitate coarsening can occur. As the annealing temperature increases, so does the coarsening rate of NbC precipitates. Consequently, a process window exists that defines the limits for Nb content, strengthening and annealing temperature.

By decreasing the degree of stabilization, lower Ti/C and Nb/C ratios, a larger grain size strengthening increment can be realized if the interstitial solute segregates to the ferrite grains. In this case, the aging resistance is lowered and the good surface appearance of stamped parts can be lost if the steel is not used befo discontinuous yielding reappears. Consequently, the use of such strengthening can only be employed after stamping, *i.e.* during the bake-hardening process. Wilson [13] and Hanai *et al.* [65] clarified that C segregatio to grain boundaries produces higher Hall-Petch (ky) values. Figure 12 shows the effect of aging time at 90°C on the ky value of a low-carbon steel quenched from 700°C. For example, starting with a grain size of 10 μm, an increase in ky from the IF value of 6 MPa $mm^{-1/2}$ to the well aged value of 25 MPa $mm^{-1/2}$ would have the calculated effect of increasing yield strength by 190 MPa.

To summarize, many general steps can be taken which should improve general surface quality and performance as well as the overall formability of ULC galvannealed steels. One of the most important improvements has to be the reduction of oxide and other surface defects. All efforts should be made to minimize oxygen contact with the steel. Possible procedures include using: (a) caster shrouding, (b) low soaking temperatures for hot-rolling, (c) oxygen removal from annealing atmosphere, (d) reduction in oxide-forming elements Ti, Si, Mn, etc, when possible (Max: Ti 0.03 wt-pct and Si 0.5 wt-pct). Processing routes th develop microstructures with good deep drawing performance, high average r-values and low Δr-values, are to be promoted. Such routes rely on the absence of solute C during cold rolling and recrystallization, increased ferrite grain boundary surface area per unit volume before recrystallize annealing, and easy growth of grains with preferred orientation. The processing routes can depend on the degree of stabilization: (a) fine austenite grain size (Nb addition, low finishing temperatures with high rolling reductions), (b) rapid cooling rate after hc rolling, (c) lower Ar3 by Mn (IF only) or B additions, (d) ferrite-rolling in hot-mill (IF only), (e) high cold-rolling reductions, (f) easy grain growth after recrystallization (Nb removal, higher annealing temperatures).

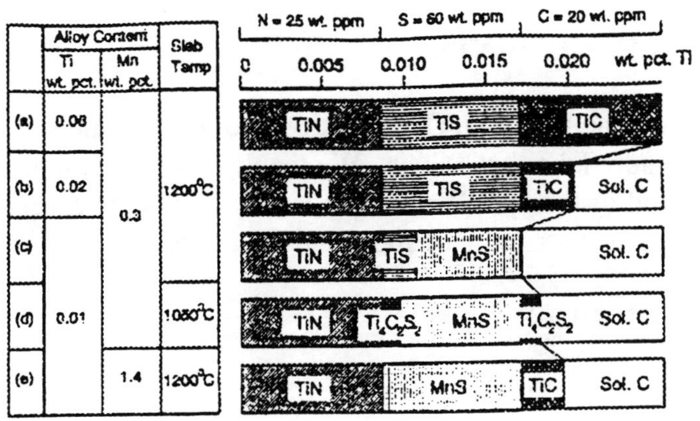

Figure 7 - A design for Bake Hardening [71].

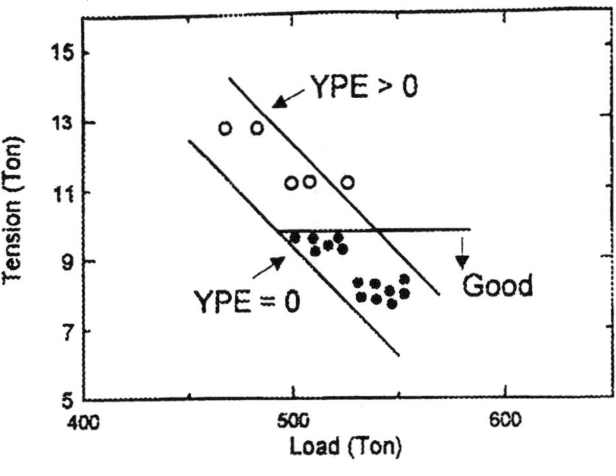

Figure 8 - Temper rolling for aging resistance [4].

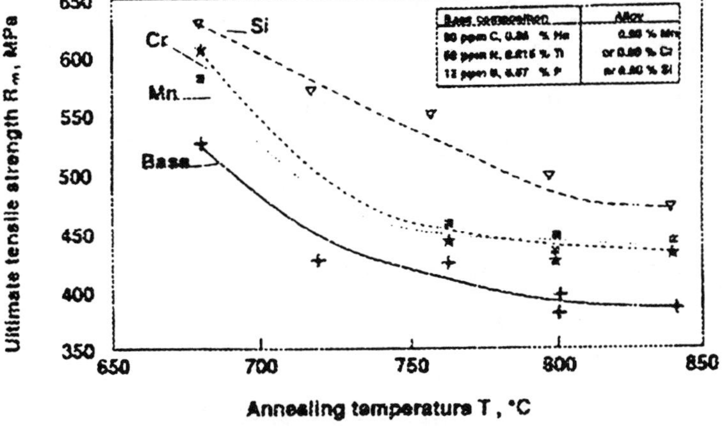

Figure 9 - Annealing response of CR Strip [61].

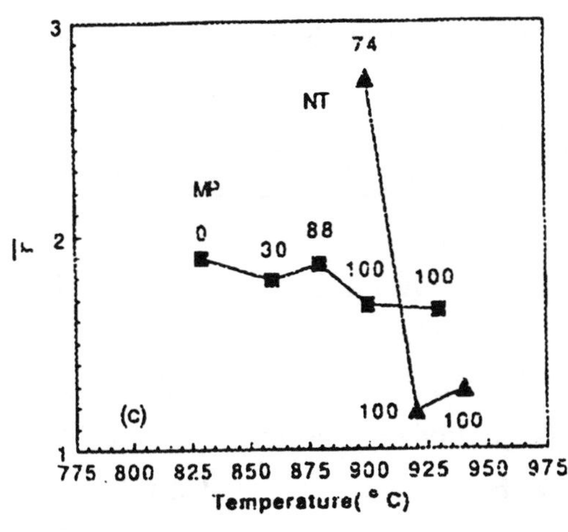

Figure 10 - R-value by high γ reheat [22].

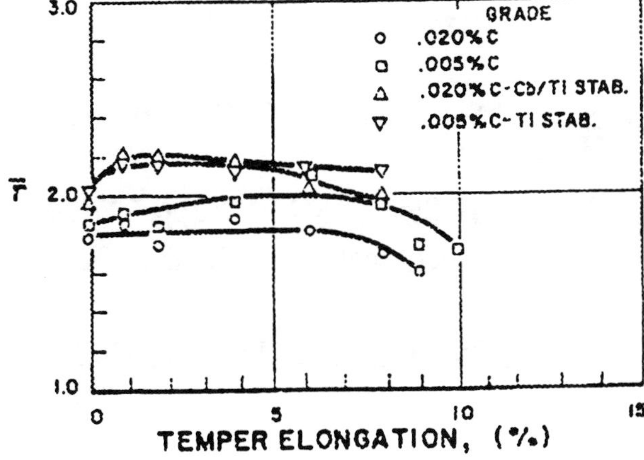

Figure 11 - R-value versus temper elongation [77].

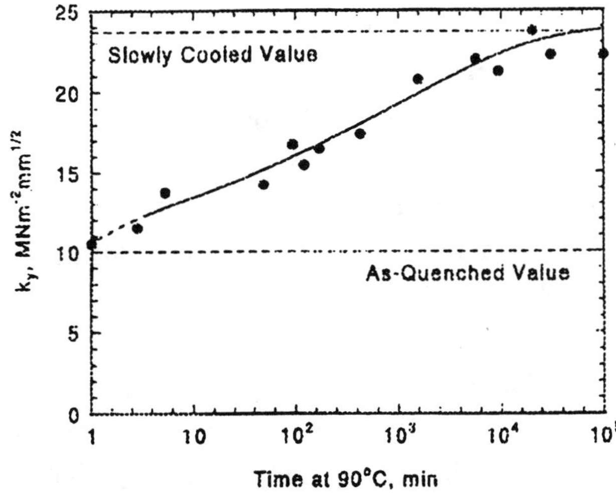

Figure 12 - Return of grain boundary strength [13].

ACKNOWLEDGEMENTS

This work was supported by the International Lead-Zinc Research Organization and the Advanced Steel Processing and Products Research Center at the Colorado School of Mines. We thank the cited authors for the adaptation and/or reproduction of their original data and/or figures. We acknowledge, with deep respect, the contributions of Paul Repas to the physical metallurgy understanding of sheet and plates steels.

REFERENCES

1. S. Robertson, New Steel, June 1994, pp. 16-19.
2. O. Akisue and M. Usuda, Nippon Steel Technical Report, Nippon Steel Company, Japan, No. 57, April 1993, pp. 11-15.
3. Itami et al., SAE Technical Paper Series, Paper No. 930783, SAE, Warrendale, PA 1993.
4. N. Mizui et al., SAE Technical Paper Series, Paper No. 920248, SAE Warrendale, PA, 1992.
5. NKK Technical Bulletin, Tec. No. 243-116-01, February 1995.
6. R.P. Foley et al., ILZRO Publication, ZC02, Research Triangle Park, NC, November 1995.
7. K. Hashiguchi et al., La Revue de Metallurgie - CIT, March 1990, pp. 277-283.
8. J.F. Butler, Proc of the Galvanizers Association 1991, vol. 83/84, January 1993, pp. 1-33.
9. B. Hutchinson, Materials Science Forum, vol. 157-162, 1994, pp. 1917-1928.
10. S. Hashimoto et al., Conf Proc Eighth Intl Conf on Textures of Materials, eds J.S. Kallend and G. Gottstein, TMS-AIME, Warrendale, PA, 1988, pp. 673-678.
11. M. Fukuda, Tetsu-to-Hagane, vol. 53, 1967, pp. 559-561.
12. W.B. Hutchinson et al., Conf Proc Technology of Continuously Annealed Sheet Steel, ed. R. Pradhan, TMS-AIME, Warrendale, PA, 1985, pp. 109-125.
13. D.V. Wilson, Metal Science, vol. 1., 1967, pp. 40-47.
14. W.C. Leslie, The Physical Metallurgy of Steels, Hemisphere Pub., New York, 1981, p. 126.
15. B. Mintz, Metals Technology, vol. 11, 1984, pp. 265-272.
16. Y. Hosoya and Y. Nagataki, Conf Proc 37th MWSP Intl Symp Recovery and Recrystallization, ISS-AIME, Warrendale, PA, 1995, pp. 915-925.
17. H. Abe, Conf Proc Eight Intl Conf on Textures of Materials, eds J.S. Kallend and G. Gottstein, TMS-AIME, Warrendale, PA, 1988, pp. 661-666.
18. K. Ushioda et al., Conf Proc Intl Forum on Physical Metallurgy of IF Steels, ISIJ, 1994, pp. 227-244.
19. H. Katoh et al., Conf Proc Technology of Continuously Annealed Sheet Steel, ed. R. Pradhan, TMS-AIME, Warrendale, PA, 1985, pp. 37-60.
20. W. Bleck et al., Conf Proc Metallurgy of Vacuum Degassed Steel Products, ed. R. Pradhan, TMS-AIME, 1990, pp. 73-90.
21. F. Rana, Inland Steel, private communication, August 1995.
22. N. Yoshinaga et al., Conf Proc High Strength Sheet Steels for the Automotive Industry, ed. R. Pradhan, ISS-AIME, Warrendale, PA, 1994, pp. 149-158.
23. I. Gupta and D. Bhattacharya, Conf Proc Metallurgy of Vacuum Degassed Steel Products, ed. R. Pradhan, TMS-AIME, Warrendale, PA, 1990, pp. 43-72.
24. I. Gupta, Inland Steel Company, private communication, 1992.

25. S. Hashimoto and T. Kashima, Kobelco Technology Review, vol. 13, 1992, p. 51
26. I. Gupta et al., Conf Proc Hot- and Cold-Rolled Sheet Steels, eds. R. Pradhan and G. Ludkovsky, TMS-AIME, Warrendale, PA, 1988, pp. 139-153.
27. L. Meyer et al., Conf Proc Intl Forum on Physical Metallurgy of IF Steels, ISIJ, 1994, pp. 203-222.
28. T. Tanioku et al., SAE Technical Paper Series, Paper No. 910293, Warrendale, PA, 1991.
29. K. Kawasaki et al., Proc Intl Conf on Processing, Microstructure and Properties of Microalloyed and Other Modern High Strength Low Alloy Steels, ed. A.J. DeArdo, ISS-AIME, 1992, pp. 137-144.
30. K. Tsunoyama et al., Conf Proc Hot- and Cold-Rolled Sheet Steels, eds. R. Pradhan and G. Ludkovsky, TMS-AIME, Warrendale, PA, 1988, pp. 155-164.
31. G. Tither et al., Proc Intl Conf on Physical Metallurgy of IF Steels, ISIJ, 1994, pp. 293-322.
32. R. Pradhan, Conf Proc Intl Forum on Physical Metallurgy of IF Steels, ISIJ, Tokyo, Japan,1994,pp.165-178.
33. Y. Tokunaga and H. Kato, Conf Proc Metallurgy of Vacuum Degassed Steel Products, ed. R. Pradhan, TMS-AIME, Warrendale, PA, 1990, pp. 91-108.
34. R.E. Miner et al., Conf Proc The Use and Manufacture of Zinc and Zinc Alloy Coated Sheet Steel Products into the 21st Century - Galvatech '95, ISS-AIME, Warrendale, PA, 1995, pp. 407-412.
35. R. Chatterjee-Fischer, Metallurgical Transaction A, vol. 9A, 1978, pp. 1553-1560.
36. M. Guttmann, Materials Science Forum, vol. 155-156, 1994, pp. 527-548.
37. D.O. Wilshynsky-Dresler et al., Conf Proc Developments in the Annealing of Sheet Steels, eds. R. Pradhan and I. Gupta, TMS-AIME, Warrendale, PA, 1992, pp. 189-218.
38. W. Bleck, et al., Conf Proc Microalloying'88, ASM, Materials Park, OH, 1988, pp. 337-344.
39. K. Tsunoyama et al., Conf Proc Metallurgy of Vacuum Degassed Steel Products, ed. R. Pradhan, TMS-AIME, 1990, pp. 127-141.
40. N. Kino et al., Conf Proc Metallurgy of Vacuum-Degassed Steel Products, ed. R. Pradhan, TMS-AIME, Warrendale, PA, 1990, pp. 197-213.
41. J. Hartmann, LTV Steel, private communication, August 1995.
42. S. Satoh et al., Trans. ISIJ, vol. 24, 1984, pp. 838.
43. E. Gallo et al., Conf Proc The Use and Manufacture of Zinc and Zinc Alloy Coated Sheet Steel Products into the 21st Century - Galvatech '95, ISS-AIME, Warrendale, PA, 1995, pp. 739-748.
44. P. Repas, Conf Proc Microalloying'88, ASM, Materials Park, OH, 1988, pp. 3-14.
45. D.T. Gawne and G.M.H. Lewis, Materials Science and Technology, vol. 1., March, 1985, pp. 183-191.
46. L. Parilak et al., Conf Proc Microalloying '88, ASM, Materials Park, OH, 1988, pp. 559-569.
47. F.B. Pickering, ISIJ, JIM, and Climax Molybdenum, Japan, 1971, pp. 9-31.
48. R. Pradhan, Conf Proc Technology of Continuously Annealed Cold-Rolled Sheet Steel, TMS-AIME, Warrendale, PA, 1985, pp. 297-317.
49. A. Okamoto and N. Mizui, Conf Proc Metallurgy of Vacuum-Degassed Sheet Steel, ed. R. Pradhan, TMS-AIME, Warrendale, PA, 1990, pp. 161-180.
50. A. Okamoto and N. Mizui, Tetsu-to-Hagane, vol. 76, No. 3, 1990, pp. 422-429.
51. N. Mizui and N. Kojima, ISIJ International, vol. 34, No. 1, 1994, pp. 123-131.
52. W. Bleck et al., Conf Proc 18th IDDRG, 1989, pp. 1-14.
53. M. Urai, CAMP ISIJ, vol. 5, 1992, pp. 1645.

54. C. Coffin and S.W. Thompson, Conf Proc The Use and Manufacture of Zinc and Zinc Alloy Coated Sheet Steel Products into the 21st Century - Glvatech 95, ISS-AIME, Warrendale, PA, 1995, pp. 121-131.
55. Y. Shindo, Inland Steel, private communication, September, 1995.
56. S.P. Bhat et al., Conf Proc Symp on High-Strength Steels for the Automotive Industry, ed. R. Pradhan, ISS-AIME, Warrendale, PA, 1994, pp. 209-222.
57. M. Yamada et al., Tetsu-to-Hagane, vol. 73, No. 8, 1987, pp. 1049-1056.
58. E. Yasuhara et al., Trans. ISIJ, vol. 34, No. 1, 1994, pp. 99-107.
59. G. Themelis, LTV Steel, private communication, August 1995.
60. N. Takahashi et al., Conf Proc Metallurgy of Continuous Annealed Sheet Steels, eds. B.L. Bramfitt and P.G Mangonon, TMS-AIME, Warrendale, PA, 1982, pp. 133-151.
61. W. Bleck et al., Conf Proc Symp. on High-Strength Steels for the Automotive Industry, ed. R. Pradhan, ISS-AIME, Warrendale, PA, 1994, pp. 141-148.
62. Y. Hosoya et al., Conf Proc Interstitial-Free Steel Sheet: Processing, Fabrication and Properties, eds L.E. Collins and D.L. Baragar, CIM, Montreal, Canada, 1991, pp. 107-111.
63. K. Seto et al., Conf Proc Symp on High-Strength Steels for the Automotive Industry, ed. R. Pradhan, ISS-AIME, Warrendale, PA, 1994, pp. 201-208.
64. L. Zhang and T.R. Bensinger, Conf Proc The Use and Manufacture of Zinc and Zinc Alloy Coated Sheet Steel Products into the 21st Century - Galvatech '95, ISS-AIME, Warrendale, PA, 1995, pp. 115-120.
65. S. Hanai et al., Trans. Iron Steel Inst. Japan, 24, 1984, pp. 17-23.
66. A. Okamoto et al., Sumitomo Search, Simitomo Metals, Japan, No. 39, September 1989, pp. 183-194.
67. T. Matsuoka et al., Conf Proc HSLA Steels Metallurgy and Applications, eds. J.M. Gray et al., ASM International, Materials Park, OH, 1986, pp. 969-976.
68. R.P. Foley et al., Conf Proc 39th MWSP, ISS-AIME, Warrendale, PA, 1998, pp. 653-666.
69. A. Okamoto et al., SAE Technical Paper Series, Paper No. 820018, SAE, Warrendale, PA, 1982.
70. K. Sakata et al., Conf Proc Intl Forum on Physical Metallurgy of IF Steels, ISIJ, Tokyo, Japan, 1994, pp. 279-288.
71. N. Kojima et al., Sumitomo Search, Sumitomo Metals, Japan, vol. 45, No. 5, 1993, pp. 12-19.
72. M. Kitamura et al., ISIJ International, vol. 34, No 1, 1994, pp. 115-122.
73. M. Kitamura et al., Conf Proc Developments in the Annealing of Sheet Steels, eds. R. Pradhan and I. Gupta TMS-AIME, Warrendale, PA, 1992, pp. 159-176.
74. R.D. Butler and D.V. Wilson, JISI, January 1963, pp. 16-33.
75. F.D. Bailey et al., Conf Proc Symp. on High-Strength Steels for the Automotive Industry, ed. R. Pradhan, ISS-AIME, Warrendale, PA, 1994, pp. 16-19.
76. S.J. Makimattila and P. Sippola, Conf Proc Hot- and Cold-Rolled Sheet Steels, eds. R. Pradhan and G. Ludkovsky, TMS-AIME, Warrendale, PA, 1988, pp. 255-270.
77. A.J. Boucek and J.F. Butler, Conf Proc Hot- and Cold-Rolled Sheet Steels, eds. R. Pradhan and G. Ludkovsky, TMS-AIME, Warrendale, PA, 1988, pp. 177-195.
78. R.J. Schepis et al., Conf Proc Intl Symp. on Low Carbon Steels for the 90's, eds. R. Asfahani and G. Tither, TMS-AIME, Warrendale, PA, 1993, pp. 435-444.
79. J.S. Kallend, Illinois Institute of Technology, private communication, July 1995.

PRECIPITATION HARDENING OF COPPER-CONTAINING IF STEELS

Jeremy Frimer, Duncan Meade, Matthias Militzer and Warren Poole
The Centre for Metallurgical Process Engineering
The University of British Columbia
Vancouver, BC
Canada V6T 1Z4
Tel.: (604) 822-3676

Key Words: Interstitial-Free Steel, Copper Precipitation, Ageing, Recrystallization

INTRODUCTION

In the past, the steel industry has successfully responded to the drive for more fuel efficient lightweight vehicles by developing higher strength steels. These steels have been the material class with the highest increase of any other new material in automotive applications over the last decade [1]. To further meet the challenges posted by the demands of the automotive industry, a new generation of steels has to be explored. Currently, multi-phase steels, in particular those showing the effect of transformation-induced plasticity (TRIP), are being discussed as new grades which promise a superior combination of strength and formability. However, the complexity of automotive parts suggests a variety of new steel grades to be used in future automobiles. In this regard, post heat treatment (PHT) steels may provide an interesting alternative [2]. These steels would show similar formability characteristics in the cold-rolled-annealed state as regular interstitial-free (IF) steels. After the automotive part has been stamped, it would be subjected to a heat treatment leading to an increase in strength due to precipitation hardening. Excellent candidates for PHT steels are copper-bearing IF steels.

Cu is currently being employed as alloying element in the 1 wt% range for high-strength low-alloy (HSLA) steels used in natural gas pipelines, shipbuilding and offshore platforms with benefits for strength, toughness, weldability and corrosion resistance [3]. The precipitation hardening effect of Cu is significant and may increase yield strengths by 200 MPa or more. However, heat treatment has usually to be performed for several hours at temperatures in the range 450 – 600 °C to realize this strengthening potential [4]. Further, Cu is known to create hot shortness during hot rolling and may significantly decrease the surface quality of steel sheets. Despite, these potential drawbacks it is important to assess in more detail the viability of Cu-bearing IF steels.

Previous work to study Cu precipitation was primarily focused on Fe-Cu model alloys [5-7]. It is now generally accepted that the following complex sequence is characteristic of precipitation in this system. Initially, metastable bcc precipitates form which are fully coherent with the iron matrix. At a critical size with a radius of approximately 3 nm, the precipitates undergo a martensitic transformation to the 9R structure which has a fcc lattice with a high density of twins. Finally, the twins disappear such that the precipitates attain the fcc equilibrium structure known as ε phase. Liu et al. [8] quantified precipitation hardening in IF steels with a variety of Cu levels ranging from 0.25 to 1.7 wt%. Precipitation hardening is observed for Cu levels above 0.5 wt% and increases with Cu content. Recently, the effect of Cu alloying on microstructure evolution was also investigated in low-carbon steels [9]. Interestingly, some differences in the precipitation hardening curves can be registered when comparing the data for these steels with those of Fe-Cu alloys having the same Cu content. This indicates that the overall steel chemistry may affect details of the Cu precipitation process.

In the present work, precipitation behavior of Cu and its effect on recrystallization is investigated for laboratory IF steels. Ageing curves of solutionized material are measured to quantify Cu precipitation hardening kinetics. Consistent with the aforementioned precipitation sequence, a complex model appears to be required to predict the ageing behavior. Here, a simplified model approach is suggested based on the experimental data. Further, the effect of coiling temperature on subsequent recovery and recrystallization during annealing of cold-rolled material is investigated.

EXPERIMENTAL WORK

Materials - Two laboratory IF steels with different levels of Cu alloying; i.e. 0.7 and 1.4 wt%, have been included in this study. The steel compositions are given in Table I. Both steels were provided by Dofasco Inc. as forged bars. In addition, material was obtained which was rolled at the pilot mill of CANMET. The rolling schedule consisted of hot rolling with a finishing temperature of 920 °C followed by a coiling simulation and subsequent cold rolling with a reduction of 80%. To simulate coiling, the hot rolled steel was held for one hour at designated coiling temperatures of 750 and 450 °C and then furnace cooled at a rate of 30 °C/hour.

Table I Chemical Compositions of the Steels (in wt%)

Steel	C	Mn	P	S	Si	Cu	Ni	Al	Nb	B	Ti	N
0.7Cu	0.0037	0.155	0.005	0.006	0.005	1.41	0.50	0.060	0.009	0.0008	0.058	0.0042
1.4Cu	0.0028	0.152	0.005	0.006	0.005	0.71	0.43	0.025	0.009	0.0008	0.059	0.0036

Procedures - Samples for ageing tests were taken from the as-received bars. Small blocks were cut from the bars and cold rolled to a reduction of 90%. Then, a solution treatment was performed in a vacuum furnace for 24 hours at 820°C. Samples were heat treated at temperatures in the range 450 – 600 °C. A salt bath was used for ageing times of up to 4 hours and a vacuum furnace for longer ageing times. Recrystallization studies were conducted on the as-received cold rolled material. Samples were cut from the sheets and aged for up to two hours in a salt bath at temperatures in the range 650 – 800 °C. To characterize precipitation and softening behavior, the hardness of all processed samples was determined. For this purpose, a microhardness indenter was used with a 1kg load. In addition, standard tensile tests were performed for selected treatment conditions.

PRECIPITATION HARDENING

Experimental Results - Ageing curves for the 1.4 Cu Steel are shown in Fig. 1 in terms of hardness increase from the initial hardness of approximately 110 HV. In the 600°C to 500°C range, two noticeable phenomena occur. First, peak ageing times increase with decreasing temperature from 15 minutes at 600 °C to approximately 6 hours at 500 °C. Second, peak strengths increase with decreasing temperature from 90 HV at 600 °C to 115 HV at 500 °C. This can be attributed to decreasing copper solubility with decreasing temperature. However, below 500 °C while the peak time increases further, the peak strength remains constant. This would suggest that Cu precipitation reaches a saturation volume fraction below 500 °C. From these results, it can be concluded that the Cu solubility limit at 500 °C is below 0.1 wt%, although solubility data reported in the literature [10-12] are controversial. At temperatures of 500 °C or lower, at least 93% of the Cu in the 1.4 Cu steel is precipitated.

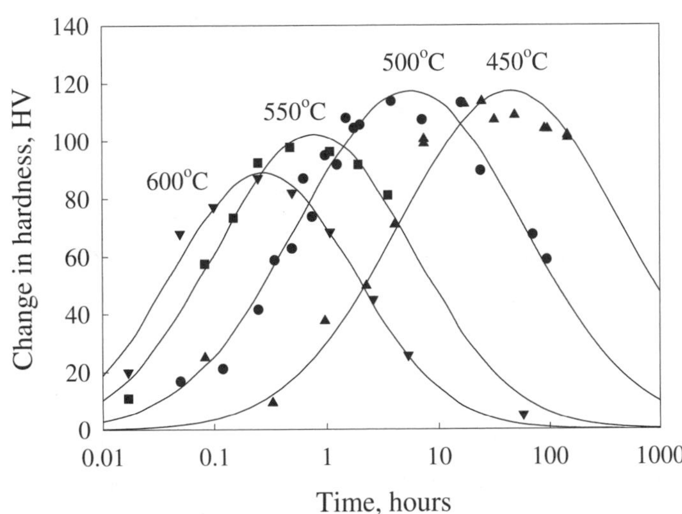

Fig. 1: Experimental ageing curves for the 1.4 Cu steel (note: solid solution hardness = 110 HV)

The substantial hardening effect of copper precipitation in this steel is confirmed with tensile tests. After solution treating, the yield strength is approximately 180 MPa and failure occurs at a total elongation of 23%. By ageing the material at 500 °C for 23 minutes, the yield strength increases considerably to nearly 400 MPa with only a minor loss in elongation to failure of about 1%. The material peak-aged at 500 °C shows a further increase in yield strength to 530 MPa; i.e. 350 MPa higher then before ageing, but incurs a significant loss in total elongation to approximately 18%. Peak aging at 600 °C shows a significant increase in yield strength (i.e. approximately 500 MPa) although it is slightly lower than the peak strength at 500 °C.

Copper content has a large effect on precipitation hardening behavior. For 500 °C, the ageing curves for both the 1.4 Cu and 0.7 Cu steels are compared in Fig. 2. The initial hardness of the 0.7 Cu steel is with 85 HV somewhat lower than that of the 1.4 Cu steel. The maximum peak strength increases with Cu alloying and the time to peak decreases. The increase of peak strength is simply related to the fact that more Cu is available to precipitate. At 500 °C, the peak strength maximum in the 0.7 Cu steel is approximately 45% of that in the 1.4 Cu steel. Further, the time required to peak harden the material increases by an order of magnitude when decreasing the copper content by a factor of 2. This can likely be attributed to a long incubation period being necessary for the slower nucleation caused by the smaller amount of copper.

For temperatures higher than 500 °C, the hardening peak decreases quickly with temperature in the 0.7 Cu steel and amounts to only approximately 25 HV at 550 °C. At lower temperatures, excessive holding times are required to attain peak hardness. Therefore, modeling analysis of Cu precipitation strengthening kinetics is restricted to the 1.4 Cu steel.

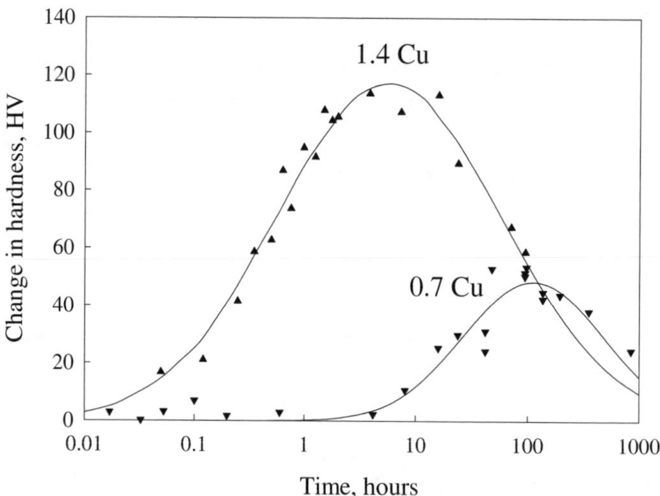

Fig. 2: Comparison of ageing curves in the 1.4 Cu and 0.7 Cu steels at 500 °C (note: Solid solution hardness of 0.7 and 1.4 Cu steels are 85 and 110 HV, respectively.)

Modelling - A new model is developed to predict the precipitation behavior of the 1.4 Cu steel. As discussed elsewhere [13], the Shercliff-Ashby analysis, which had been successfully adopted for precipitation hardening in HSLA steels [14, 15], cannot be applied to Cu precipitation strengthening. However, the temperature dependence of the peak strength amount can be incorporated into the model in a way similar to that proposed by Shercliff and Ashby [14]. In a first approximation, particle growth as opposed to particle coarsening may be assumed to be rate controlling. Based on classical parabolic particle growth, the normalized temperature-corrected time takes the form

$$\Lambda^* = \frac{t\exp(-\Theta/RT)}{\Lambda_p} \qquad (1)$$

From the experimental data for peak time as a function of temperature, Θ=175 kJ/mol and Λ_p=3.3×10^{-8}s have been determined for the 1.4 Cu steel. Further, the model has to take into account the evolution of volume fraction and particle strength as a function of aging time. As delineated elsewhere in detail [16], these evolution laws can be expressed in terms of the Λ^* defined by equation (1). The precipitation strengthening kinetics may be described by [13]

$$\frac{\Delta\sigma}{\Delta\sigma_{peak}} = \frac{2((11\Lambda^{*3/2})^{-2}+1)^{-1/4}(1-\exp(-\sqrt{\Lambda^*}))^{3/2}}{\sqrt{\Lambda^*}} \qquad (2)$$

where $\Delta\sigma$ is the precipitation strength increment and $\Delta\sigma_{peak}$ denotes the precipitation strength increment at peak. Using the above relation, a master curve can be constructed to describe precipitation hardening kinetics in the 1.4 Cu steel. As shown in Fig. 3, this new model accurately describes the experimental data.

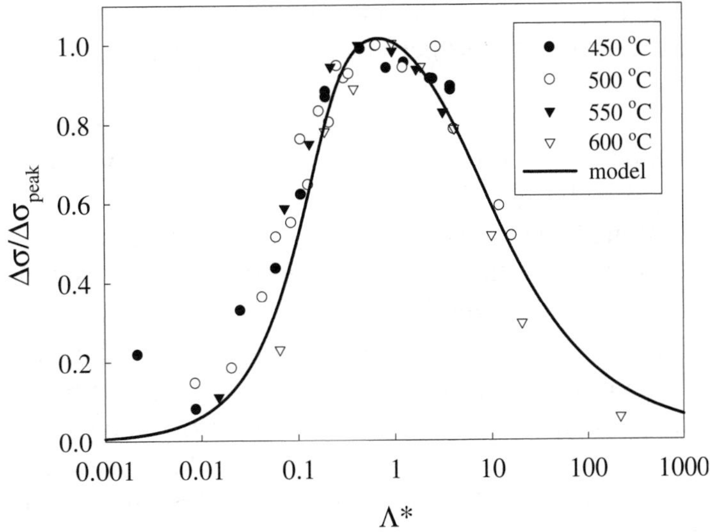

Fig. 3: Model for precipitation hardening kinetics in 1.4 Cu steel

RECRYSTALLIZATION BEHAVIOR

In addition to precipitation hardening, the softening behavior has been investigated for the high and low Cu content steels with 80% cold-reduction. Fig. 4 compares the softening behavior of both steels observed at 700 °C. Clearly, the 0.7 Cu steel shows much faster softening rates. Complete recrystallization is observed in the 0.7 Cu steel after 10 minutes whereas the 1.4 Cu steel is not fully recrystallized even after more than 1000 minutes of holding. This result confirms the retardation of recrystallization by Cu, as reported by Diligent-Berveiller et al. [17].

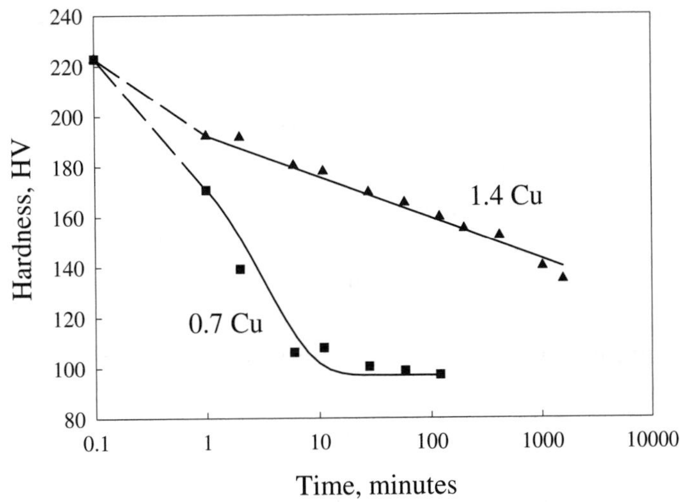

Fig. 4: Effect of Cu content on softening behavior at 700 °C for material coiled at 750 °C; values shown at 0.1 minutes indicate as-received hardness

The results shown in Fig. 4 are obtained on material which is coiled at 750 °C. Based on the precipitation results, it can be expected that these steels are in an overaged state with no precipitation strength. This suggestion is supported by the fact that the initial hardness in both steels is the same. Some of the precipitated Cu may go into solution during the softening heat treatment. The effect of Cu on recrystallization would then primarily be one of solute drag. However, further studies are required to clarify the mechanisms of Cu on recrystallization kinetics.

To further evaluate the potential interaction of precipitation and recrystallization, softening kinetics are compared for 1.4 Cu steel coiled at 450 and 750 °C. A one hour holding at 450 °C results in an underaged material with a precipitation strength increment of approximately 40 HV (see. Fig. 1). This appears to be consistent with the as-received strength of the cold rolled materials, which is approximately 30 HV higher for coiling at 450 °C than for coiling at 750 °C. Performing a heat treatment at 650 °C leads to the development of further precipitation strength in the material coiled at 450 °C but no such effect is observed after coiling at 750 °C, as illustrated in Figure 5. The peak at 650 °C is observed at the shortest holding time of 1 minute whereas the proposed precipitation model suggests a peak time of approximately 5 minutes. The observation of a somewhat shorter peak time can be expected since the as-cold rolled material had already acquired approximately half the precipitation strength attainable at the holding temperature and the presence of dislocations may accelerate the precipitation reaction. In addition, recovery processes occur simultaneously and may shift the apparent peak further to shorter times. Precipitation strength decreases due to overaging during the softening treatment. Based on the precipitation hardening model it can be expected that residual precipitation strength is basically eliminated at holding times which are approximately two orders-of-magnitude larger than the peak time (see Fig. 3); i.e., 500 minutes in the given case. Indeed, after 1000 minutes of holding, hardness values in the material coiled at 450 °C are reduced to similar values as observed for coiling at 750 °C.

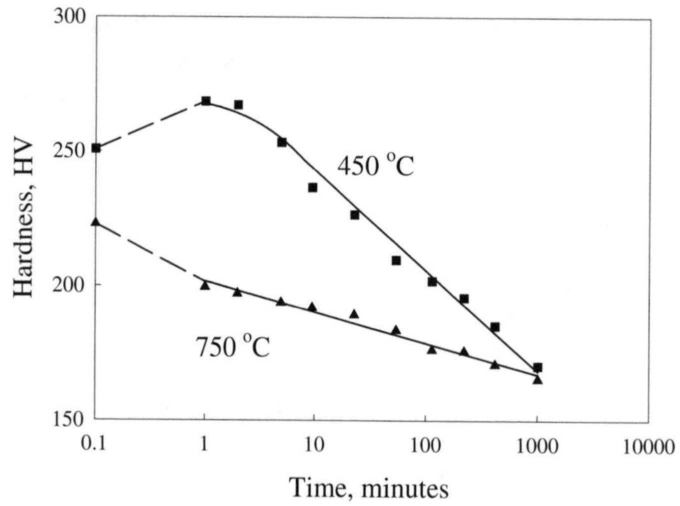

Fig. 5: Effect of coiling temperature on the softening behavior in the 1.4 Cu steel at 650 °C; values shown at 0.1 minutes indicate as-received hardness

If the material is aged at 750 °C, there is no observation of a precipitation peak for either the low or high coiling temperature, as illustrate in Fig. 6. The material coiled at 450 °C has no additional precipitation potential at this holding temperature and presumably overages rapidly within the first few minutes of holding; the softening behavior is then essentially independent of coiling temperature. Both materials are completely recrystallized after approximately 100 minutes of holding.

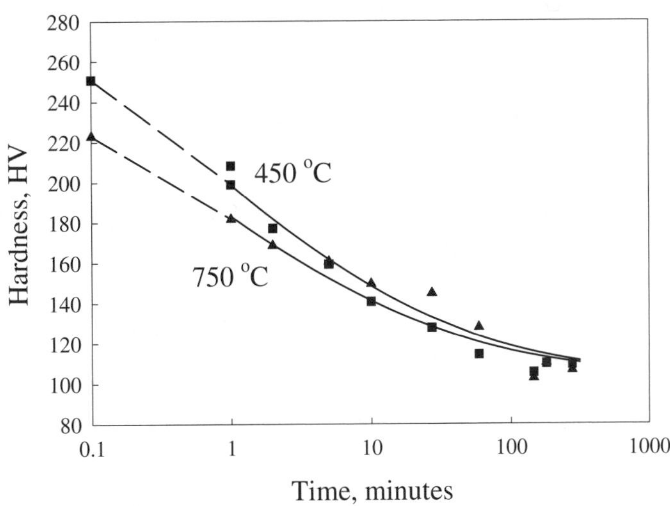

Fig. 6: Effect of coiling temperature on the softening behavior in the 1.4 Cu steel at 750 °C; values shown at 0.1 minutes indicate as-received hardness

CONCLUSIONS

Precipitation hardening and its effect on recrystallization behavior have been investigated in two Cu-bearing IF steels containing 0.7 wt% Cu and 1.4 wt% Cu, respectively. The strengthening potential of Cu precipitation increases markedly with Cu content in the investigated alloying range. A maximum yield strength increase of 350 MPa has been observed in the 1.4 Cu steel when aged for approximately 6 hours at 500 °C. A new model has been proposed to describe precipitation hardening in the 1.4 Cu steel.

While Cu levels in the range of 1.4 wt% promise a maximum strengthening potential, lower levels of Cu alloying are of interest to mitigate potential surface quality problems. Further, holding times of several hours at temperatures in the order of 500 °C appear excessive for a potential industrial application. Pre-deformation may accelerate precipitation rates thereby reducing required holding times and temperatures. Thus, additional studies have been initiated to quantify the effect of Cu contents and pre-deformation on the precipitation kinetics with the goal to establish an improved model as predictive tool incorporating these two variables.

ACKNOWLEDGEMENTS

This work was conducted as part of the Dofasco Chair program in Advanced Steel Processing. The authors would like to thank Dofasco Inc. for the material and support provided to conduct this study. Further, financial support from the National Sciences and Engineering Research Council of Canada (NSERC) is gratefully acknowledged.

REFERENCES

1. R.A. Heimbuch, "Going Common in a Global Market", 40th Mechanical Working and Steel Processing Conf. Proc., ISS, Warrendale, PA, Vol. XXXVI, 1998, pp. 3-10.

2. H. Shirasawa, "Recent Development of Cold Formable High Strength Hot Rolled Steels in Japan", High-Strength Sheet Steels for the Automotive Industry, R. Pradhan, ed., ISS, Warrendale, PA, 1994, pp. 3-10.

3. T.W. Montemarano, B.P. Sack, J.P. Gudas, M.G. Vassilaros and H.H. Vanderveldt, "High Strength Low Alloy Steels in Naval Construction", J. Ship Production, Vol. 2, 1986, pp. 145-162.

4. S.S. Ghasemi Banadkouki, D. Yu and D.P. Dunne, "Age Hardening in a Cu-Bearing High Strength Low Alloy Steel", ISIJ International, Vol. 36, 1996, pp. 61-67.

5. E. Hornbogen and R.C. Glenn, "A Metallographic Study of Precipitation of Copper from Alpha Iron", Trans. TMS-AIME, Vol. 218, 1960, pp. 1064-1070.

6. K.C. Russell and L.M. Brown, "A Dispersion Strengthening Model Based on Differing Elastic Moduli Applied to the Iron-Copper System", Acta Metall., Vol. 20, 1972, pp. 969-974.

7. P.J. Othen, M.L. Jenkins, G.D.W. Smith and W.J. Phythian, "Transmission Electron Microscope Investigations of the Structure of Copper Precipitates in Thermally-Aged Fe-Cu and Fe-Cu-Ni", Phil. Mag. Lett., Vol. 64, 1991, pp. 383-391.

8. J. Liu, D. Xie, D. Wang, L. Gao, M. Wen, S. Luo, P. Li, J. Gui and R. Wang, "Effects of Cu Content on the Ageing Kinetics of Cu-Added Interstitial-Free Steels", ISIJ International, Vol. 39, 1999, pp. 614-616.

9. M. Militzer, A. Deschamps, S. Dilney and D. Meade, "Effect of Cu on Microstructure in Low-Carbon Steels", 41st Mechanical Working and Steel Processing Conf. Proc., ISS, Warrendale, PA, Vol. XXXVII, 1999, pp. 735-742.

10. H.A. Wriedt and L.S. Darken, "The Solubility of Copper in Ferrite", Trans. Metall. Soc. AIME, Vol. 218, 1960, pp. 30-36.

11. G. Salje and M. Feller-Kniepmeier, "The Diffusion and Solubility of Copper in Iron", J. Appl. Phys., Vol. 48, 1977, pp. 1833-1839.

12. M.K. Miller, K.F. Russell, P. Pareige, M.J. Starink and R.C. Thomson, "Low Temperature Copper Solubilities in Fe-Cu-Ni", Mat. Sci. Eng., Vol. A250, 1998, pp. 49-54.

13. M. Militzer and W.J. Poole, "Modelling Precipitation Strengthening in HSLA Steels", HSLA Steels'2000, Xian, China, in press.

14. H.R. Shercliff and M.F. Ashby, "A Process Model for Age Hardening of Aluminum Alloys – I. The Model", Acta Metall. Mater., Vol. 38, 1990, pp. 1789-1802.

15. M. Militzer, W.J. Poole and W.P. Sun, "Precipitation Hardening of HSLA Steels", Steel Research, Vol. 69, 1999, pp. 279-285.

16. J. Frimer, M. Militzer and W.J. Poole, "Modelling of Precipitation Strengthening in a 1.4 wt.% Cu IF steel", to be published.

17. S. Diligent-Berveiller, E. Gautier, T. Iung, H. Régle and S. Chabanet, "Influence of Copper on Recrystallization of Ti-IF Sheet Steels", The 4th International Conference on Recrystallization and Related Phenomena, T. Sakai and H.G. Suzuki, eds., The Japan Institute of Metals, 1999, pp. 757-762.

An Investigation of Precipitation Reactions in 12Cr / 12Co / 5Mo Nickel Modified Martensitic Precipitation Strengthened Stainless Steel

Aytekin Hitit
Warren M. Garrison Jr
Department of Materials Science and Engineering,
Carnegie Mellon University,Pittsburgh,PA.15213
Phone:412-268-2699 Fax:412-268-7596

Key Words: martensitic, stainless steel, precipitation strengthening, retained austenite, reverted austenite, R-phase, σ-phase

INTRODUCTION

Martensitic precipitation strengthened stainless steels are strengthened by intermetallic compounds which precipitate on tempering. The types of precipitates depend on composition, tempering temperature, and tempering time. The main precipitates found in stainless maraging steels are Fe_2Mo Laves phase[1-3], Ni_3Ti[4-6],NiAl[7,8], χ-phase[1,2,9-13], Ni_3Be[14], G-$Ni_{16}Ti_6Si_7$[4,5], σ–phase[2,4,5,15,16], and R-phase[10-12,15-17]. Also, precipitation of quasicrystalline R′ has been reported[12]. Moreover, depending on the initial composition, some of austenite from the solution treatment temperature might be retained. Also, during tempering, precipitation of austenite might occur.

The objective of this work was to investigate particle precipitation and austenite formation as a function of aging temperature and nickel content in a 12Cr / 12Co / 5Mo stainless maraging steel.

EXPERIMENTAL PROCEDURES

In this work, four experimental alloys in which nickel content was varied from 4.5 to 6.0 wt.% were evaluated. Carbon levels of the alloys are low and quite constant. The compositions of the alloys are given in Table I.

The standard heat treatment was solution treatment at 1050°C for one hour, oil quenching and refrigeration in liquid nitrogen for four hours. Then tempering was carried out for 3.16 hours in neutral salt baths followed by water quenching. Room temperature tensile properties were determined in accordance with ASTM-E8[18]. Thin foils were electrochemically polished by using an electrolyte of 60g sodium chromate in 500ml acetic acid at 35V and room temperature. For the preparation of extraction replicas, polished specimens were etched with 10% nital, then coated with a thin layer of carbon. The replicas were stripped from the specimen by etching in 5% bromine in water free methanol. Copper grids were used to support the carbon replicas. Thin foils were studied by using Philips EM-420 transmission electron microscope operating at 120kV. Composition of precipitates was determined by using the carbon replicas with the EDX unit in Philips CM-200 transmission electron microscope. Austenite measurements were performed by X-ray with Mo K_α radiation using the method described by Miller[19].

Table I. Compositions of the Alloys in Weight Per-Cent*

Alloy	C	Cr	Co	Mo	Ni	Mn	Si	S	P	O_2	N_2
WD59	0.005	11.9	11.4	5.0	4.5	0.01	0.01	28	10	94	1
WD60	0.006	11.9	11.4	5.0	4.9	0.01	0.01	27	10	100	2
WD61	0.006	11.8	11.5	5.1	5.5	0.01	0.01	20	10	97	1
WD62	0.005	12.0	11.5	5.1	6.0	0.01	0.01	30	10	98	1

*Sulfur, phosphorus, oxygen, and nitrogen are given in weights parts per million.

RESULTS

Tensile Properties

The aging response of the alloys are shown in Figure 1. Age hardening starts at around 400°C for all the alloys. The yield strength increases with increasing the nickel content up to 5.5 wt.%. For the first three alloys the peak strength is achieved at about 550°C, and the peak yield strength increases with increasing nickel. However, further increase in nickel content results in lower peak yield strength, and rapid overaging.

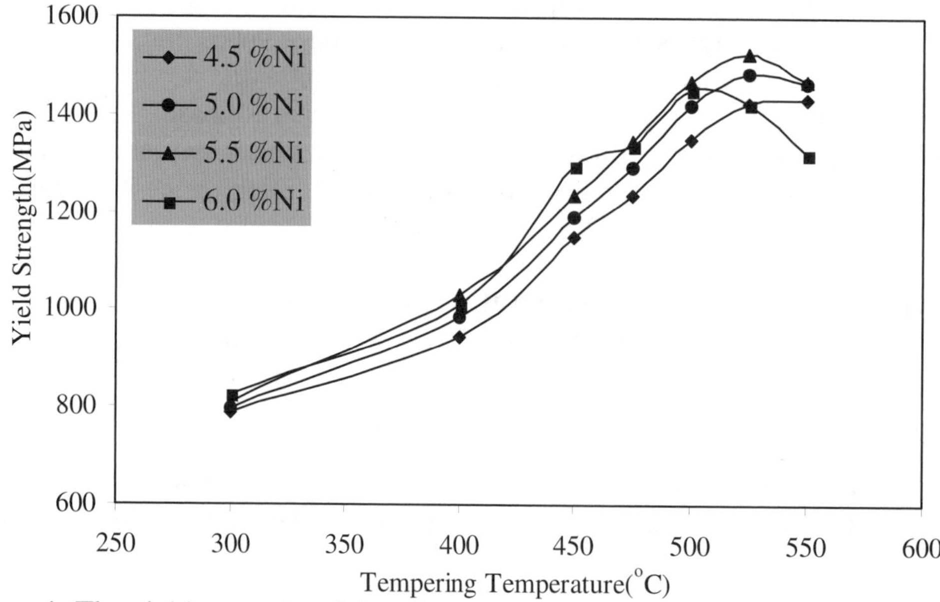

Figure 1: The yield strengths of the alloys as a function of tempering temperature.

Microstructure

4.5 %Ni alloy- X-ray diffraction showed that the unaged material is essentially martensite. After tempering at 525°C, two types of precipitates were observed. The size of the larger precipitates is about 15nm whereas the size of smaller precipitates is about 5nm. Electron diffraction patterns obtained from the smaller precipitates indicate that the precipitate is σ-phase. A selected area diffraction pattern, and the corresponding dark field image are given in Figure 2 for this precipitate type.

a) $[012]_M || [310]_\sigma$

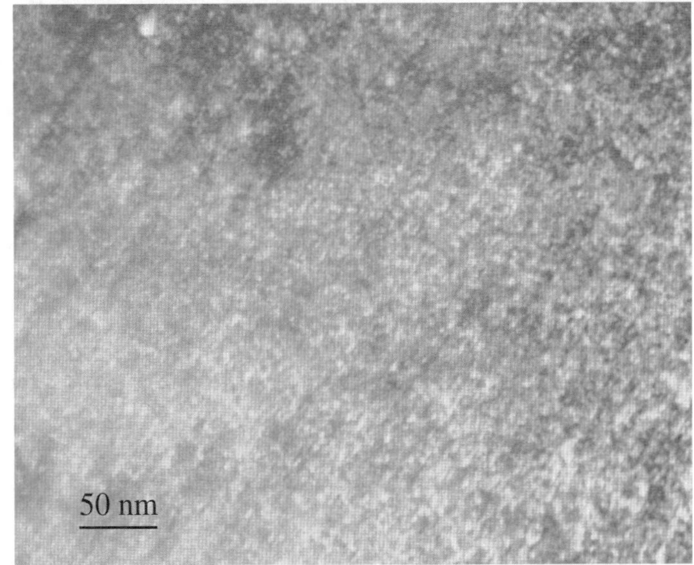

b) Dark field image obtained with the spot A

Figure 2 : Transmission electron micrographs of 4.5 wt.% nickel alloy tempered at 525°C

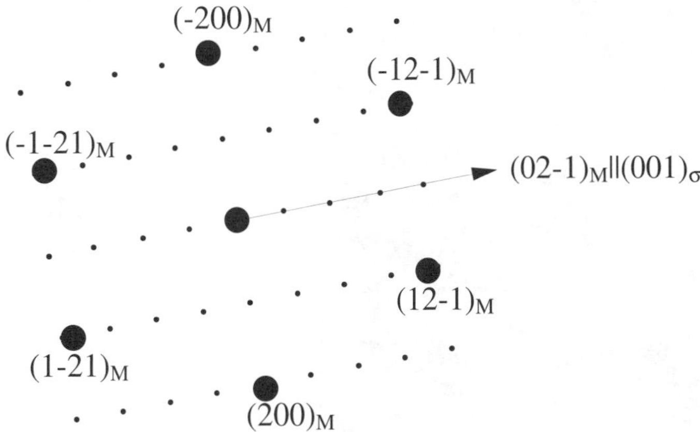

Figure 3 : Solution of the pattern given in Fig. 2-a

The electron diffraction pattern in Figure 2-a gives the following orientation relationship between the bcc-martensite matrix and σ-phase: $(02\text{-}1)_M \parallel (001)_\sigma$ and $[012]_M \parallel [310]_\sigma$.

EDX results showed that σ-phase particles are rich in chromium and iron whereas the larger precipitates are rich in iron, chromium, and molybdenum (Fig. 4).

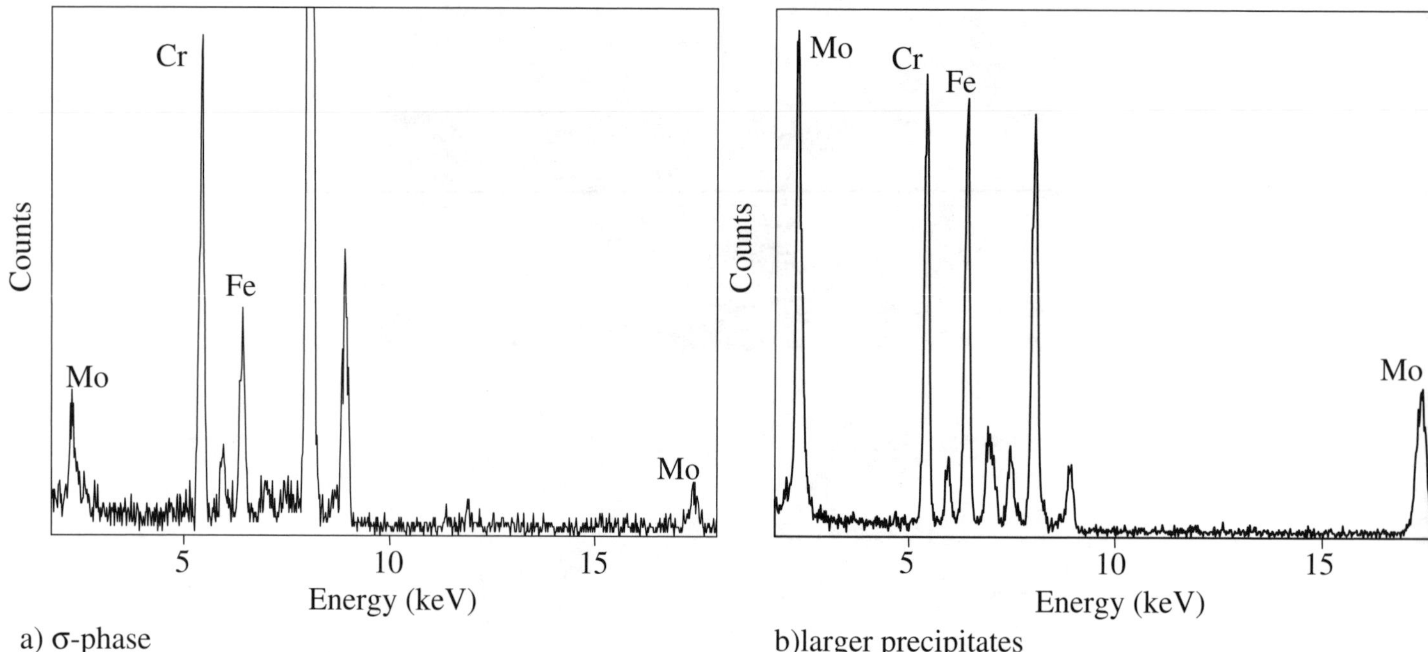

a) σ-phase b)larger precipitates

Figure 4 : EDX spectra of the precipitates obtained from the material tempered at 525°C

At the peak aging temperature, 550°C, particle size of the dominant precipitate is about 20-25nm, and the smaller 5nm size precipitate is also present. Chemical composition of the σ-phase and the larger precipitates were almost same as the compositions obtained at 525°C.

In order to determine the identity of the larger precipitates unambiguously, an overaged specimen tempered at 700°C for 3.16 hours was used.

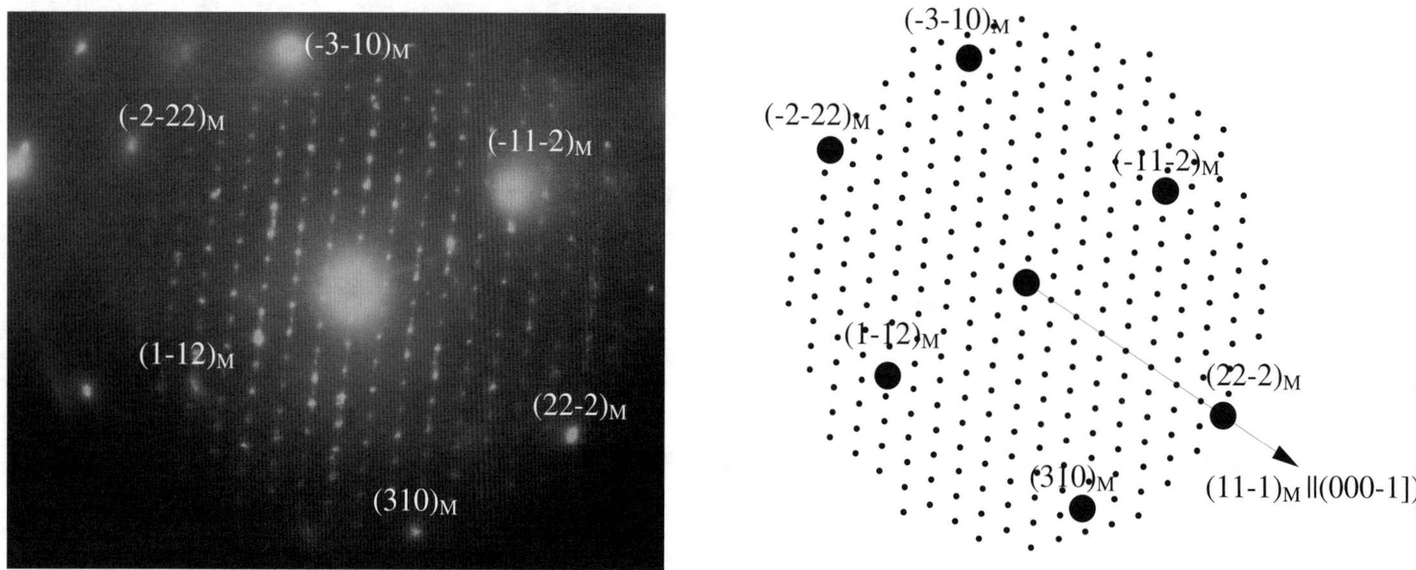

a) $[-132]_M \parallel [11\text{-}20]_R$ b) Solution to the pattern

Figure 5: An electron diffraction pattern obtained from the specimen tempered at 700°C and its solution

This single crystal electron diffraction pattern showed that the larger precipitates are R-phase with a trigonal structure (Table II). The following orientation relationship between the bcc-martensitic matrix and R-phase can be derived which was reported previously[17]: $(11\text{-}1)_M \parallel (0001)_R$ and $[\text{-}132]_M \parallel [11\text{-}20]_R$

5.5 %Ni alloy- X-ray diffraction measurements showed the presence of some retained austenite in the as quenched condition. This is due to lowering effect of nickel on the martensite start temperature. Also, reverted austenite was observed after tempering at 550°C and higher. After tempering at 550°C, σ-phase and R-phase were observed.

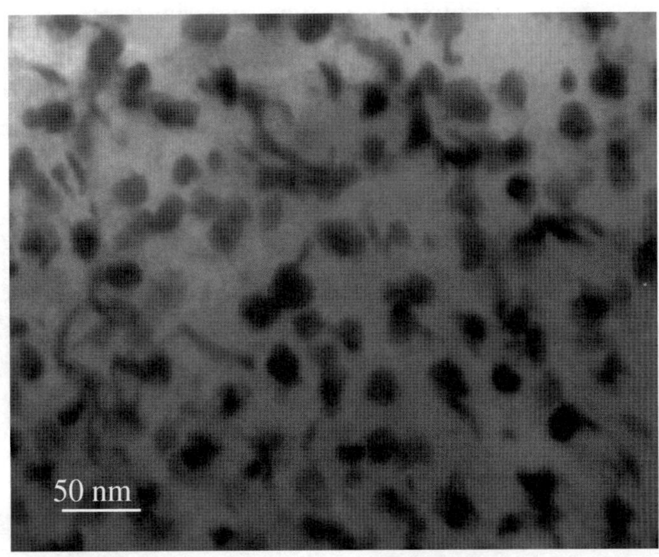

Figure 6: Bright field image of the specimen tempered at 550°C

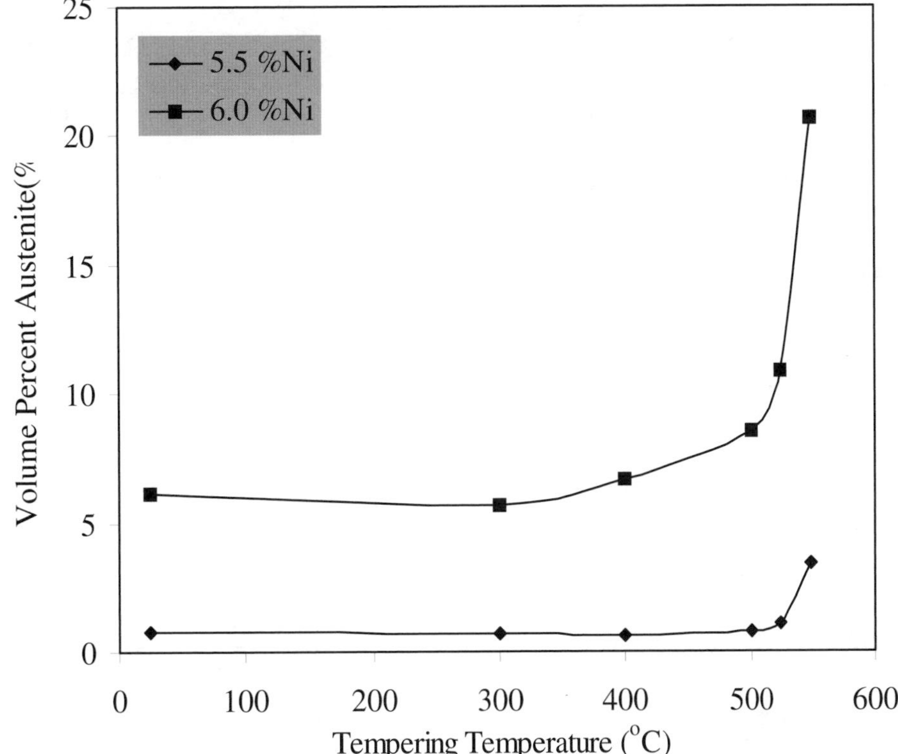

Figure 7: Volume fraction of austenite as a function of tempering temperature

Table II. Crystallographic data for the observed phases

Phase	Space group	Lattice Parameters (nm) a	b	c	Reference
α	Im-3m	0.2843			Present work
γ	Fm3m	0.3558			Present work
R	R-3	1.0903		1.9342	20
σ	$P4_2/mnm$	0.8797		0.4559	21

6.0%Ni alloy- Similar to the 5.5% nickel alloy, unaged material contains some retained austenite. It has a much higher austenite content than do the other three alloys. As seen in Figure 7, in addition to retained austenite, there is also some reversion on tempering which begins on tempering at about 550°C. Retained austenite as interlath films, and reverted austenite can be clearly seen in the micrographs below with R-phase particles discussed earlier. As for the 4.5 and 5.5 wt.% nickel alloys, R-phase and σ–phase were observed after tempering at 550°C.

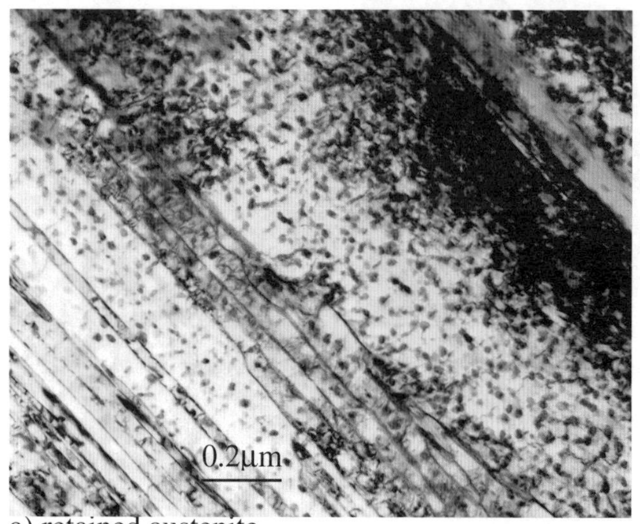

a) retained austenite

b)reverted austenite

Figure 8: Transmission electron micrographs of 6.0 wt.% nickel alloy tempered at 550°C

Table III: Chemical compositions of the phases obtained by EDX (wt.%)*

Alloy	Phase	Temp.(°C)	Cr	Fe	Mo	Co	Ni
WD59	R	525	27.17	23.82	35.10	7.11	6.77
		550	28.12	26.89	30.14	8.43	6.39
	σ	525	51.34	23.93	12.16	6.07	6.47
		550	55.70	23.58	5.98	7.43	7.30
WD61	R	550	24.07	25.11	35.44	8.59	6.78
	σ		52.75	22.01	8.31	7.35	9.54
WD62	R	550	22.77	27.27	34.55	8.58	6.82
	σ	550	49.17	26.42	11.23	6.95	6.21

*The results are average of at least 5 separate measurements.

DISCUSSION

Increasing the nickel content up to 5.5 wt% improves the yield strength. One explanation would be solid solution hardening effect by nickel. However, after tempering at 300°C, all the alloys have same yield strength which would not be expected, if the solute solution hardening effect of nickel was the main reason for the improvement of the strength. In addition, nickel content of the precipitates is quite constant (Table III). Therefore, nickel must have an indirect effect on the precipitation processes. It is believed that the peak yield strength of the 6.0wt.% nickel alloy would have been greater than that of 5.5wt.% nickel alloy, only if the substantial austenite reversion, observed in the 6.0wt.% nickel alloy, could have been avoided. Obviously, the rapid overaging of 6.0wt.% alloy has resulted from a high austenite content due to austenite reversion on tempering at 525°C and higher.

Two types of precipitates were observed. It is found that precipitation of σ–phase starts at about 500°C. The size of σ–phase particles is about 5nm, and it is the majority phase at 500°C. The chemical composition of σ–phase particles implies that it is CrFe and half of the iron sites seem to be filled by molybdenum, cobalt, and nickel. R-phase precipitation was observed at and above 525°C. Single crystal diffraction patterns obtained from the overaged specimen for various zone axes clearly showed that it is in fact R-phase and is coherent with the matrix. At 525°C, both R-phase and σ–phase are present. At 550°C, R-phase becomes the majority phase.

CONCLUSIONS

1. Increasing the nickel content up to 5.5 wt.% improves the yield strength.

2. The major precipitates are R-phase with the chemical composition of 35%Mo-25%Fe-25%Cr- 8%Co-7%Ni and σ-phase with the chemical composition of 52%Cr-24%Fe-8%Mo-8%Co-8%Ni.

3. σ–phase precipitation begins on tempering at 500°C

4.R-phase precipitation commences on tempering at 525°C, and it becomes the majority phase on tempering at 550°C

5. More rapid overaging of 6.0wt.%Ni alloy results from its substantially higher austenite content after tempering at 525°C and higher.

REFERENCES

1. A. Kasak, V.K. Chandhok and E.J. Dulis, "Development of Precipitation Hardening Cr-Mo-Co Stainless Steels," Trans. ASM., Vol.56, 1963, pp.455-467.

2. D. Coutsouradis, J.M. Drapier, E. Diderrich and L. Habraken, "Precipitation Hardening in High-Strength Stainless Steels," Cobalt, Vol.36, 1967, pp.144-156.

3. T. Kosa and T.A. DeBold, "Effect of Heat Treatment and Microstructure on the Mechanical and Corrosion Properties of a Precipitation Hardeneble Stainless Steel," , 1979, pp.367-381.

4. K. Hoshine and T. Utsunomiya, "Effects of Ti and Si on Precipitation Behavior of Martensitic Stainless Steel," Tetsu-to-Hagane, Vol.72, 1986, pp.77-84.

5. A. Gamperle, J. Gemperlova, W. Sha and G.D.W. Smith, "Aging Behaviour of Cobalt Free Chromium Containing Maraging Steels," The Institute of Materials, 1992, pp.546-554.

6. V.M. Schastlivtsev, Y.V. Kaletina, A.Y. Kaletin and I.L. Yakovleva, "The Influence of Heat Treatment on Mechanical and Fatigue Properties of Maraging Steels," Phys.Met.Metall, Vol.73, 1992, pp.83-89.

7. J.M. Haudin and F. Montheillet, "Etude par Microscopie Electronique des Precipitations Durcissantes dans la Ferrite d'un Acier Inoxydable (15%Cr, 7%Ni, 2%Mo). Mise en Evidence de la Precipitation d'une Phase R," Metallography, Vol.11, 1978, pp.391-428.

8. V. Seetharaman, M. Sundararaman and R. Krishnan, "Precipitation Hardening in a PH13-8 Mo Stainless Steel," Mater. Sci. Eng., Vol.47, 1981, pp.1-11.

9. C.M. Hammond, "The Development of Maraging Steels Containing Cobalt," Cobalt, Vol.25, 1964, pp.195-202.

10. F.A. Thompson and D.R.F. West, "Intermetallic Compound Precipitation in An Fe-10%Cr-13%Co-5%Mo Alloy," JISI, 1972, pp.691-697.

11. M.Y. Poptsov and N.V. Zvigintsev, "Influence of Segregation in the Martensite and Formation of R-Phase on the Stability of the Mechanical Properties of Steel 03KH11N10M2T1," Phys.Met.Metall., Vol.67, 1989, pp.164-168.

12. J.O. Nilsson, A.H. Stigenberg and P. Liu, "Isothermal Formation of Quasicrystalline Precipitates and Their Effect on Strength in a 12Cr-9Ni-4Mo Maraging Stainless Steel," Metall. Trans., Vol.25A, 1994, pp.2225-2233.

13. S.V. Grachev Proceeding of the 16th ASM Heat Treating Society Conference and Exposition, 1996.

14. J.Smolinski, "Properties of New Stainless Maraging Steels Containig Beryllium," JISI, Vol.204, 1966, pp.57-58.

15. N. Lambert, J.M. Drapier and D. Coutsouradis, "Precipitation-Hardening Cobalt-Copper-Molybdenum Stainless Steels," Cobalt, Vol.47, 1970, pp.68-74.

16. Y. Asayama, "The Effect of Aging on the Notch Toughness of High Strength Stainless Steels," Japan Institute of Metal Journal, Vol.40, 1976, pp.973-981.

17. D.J. Dyson and S.R. Keown, "A Study of Precipitation in a 12%Cr-Co-Mo Steel," Acta. Met., Vol.17, 1969, pp.1095-1107.

18. ASTM E8-81 ,1983 Annual Book of ASTM Standards, ASTM: Philadelphia, PA, 1983, p.197.

19. R.L. Miller, "A Rapid X-Ray Method for the Determination of Retained Austenite," Trans. A.I.M.E., Vol.57, 1964, pp.892-899.

20. Y. Komura, W.G. Sly and D.P. Shoemaker, "The Crystal Structure of the R-Phase, Mo-Co-Cr," Acta. Cryst., Vol.13, 1960, pp.575-585.

21. G.Bergman and D.P. Shoemaker, "The Determination of Crystal Structure of the Sigma Phase in the Iron-Chromium and Iron Molybdenum Systems", Acta Cryst., Vol.7, 1954, pp.857-865

PRODUCT PHYSICAL METALLURGY III

Effect of Deformation Schedule on the Microstructure and Mechanical Properties of Thermomechanically Processed C-Mn-Si TRIP Steel

I.B. Timokhina,[1] P.D. Hodgson[1] and E.V. Pereloma[2]
[1] School of Engineering and Technology, Deakin University, Geelong, VIC3217, Australia
Tel.: (03) 9905-1910;
Tel.: (03) 5227-1251;
Facsimile: (03) 5227-1103.
[2] Department of Materials Engineering, Monash University, Clayton,
VIC3800, Australia
Tel.: (03) 9905-4916
Facsimile: (03) 9905-4940

Key Words: TRIP steel, Thermomechanical processing, Retained Austenite, Deformation, Recrystallised Austenite condition, Non-Recrystallised Austenite Conditions, Bainite Morphology.

INTRODUCTION

From the mid-1980s attention have been drawn to enhancing the safety level of automobiles while reducing vehicle weight. The benefits of weight savings are essentially cost related. One of the important factors contributing to the use of steel in the automotive industry is the higher strengths and higher residual ductility at any given strength level.[1] TRIP (TRansformation Induced Plasticity) steels have been introduced as a fundamentally different type of higher strength steel.[2,3] The microstructure of these steels, with a typical composition of Fe-0.2C-1.7Mn-1.5Si (%wt), consists of the polygonal ferrite, bainite and a significant amount of retained austenite (RA).[4] The higher level of ductility is achieved by the transformation of metastable retained austenite to martensite under straining. The amount and stability of RA controls the mechanical properties of TRIP steels.[5] Stability of RA depends on the carbon content in the remaining austenite lattice, size, distribution and bainitic structure morphology.[6,7,8] For example, acicular ferrite has been shown to result on a significant increase in the quantity and stability of the retained austenite, which then improved the room temperature mechanical properties in TRIP steels.[8] The thermomechanical processing (TMP) schedule can affect the quality and quantity of RA.

The main objective in conventional TMP is to refine the ferrite grain size.[9] However, it has also been observed that the volume fraction of RA is related to the grain diameter of the polygonal ferrite.[10] In steels with a bainitic microstructure, the other aim of TMP is to refine the bainite by the deformation in the non-recrystallised austenite region; this can increase the stability of RA in TRIP steels. The bainitic microstructure can be significantly improved by deformation of more than 50% in the non-recrystallised region.[11] Furthermore, TMP influences the transformation behaviour leading to different morphologies of the second bainitic phase. The effect of morphology is substantial for the stability of RA.

The aim of the current investigation was to clarify the effect of different prior austenite conditions on the transformation kinetics and mechanical properties of TRIP steels.

EXPERIMENTAL

The composition of investigated steel is given in Table 1. The steel was received as hot rolled plate from BHP Research-Melbourne Laboratories.

Table I. Chemical Composition of Steels, wt %

C	Mn	Si	Ni	Mo	Cu	Nb	Ti	N
0.21	1.55	1.55	0.009	≤0.008	0.003	0.005	≤0.003	0.0035

The experiments were performed using a Hot Torsion Machine with a strain rate of 2.95 s^{-1}. Torsion allows generate higher amount of deformation in the recrystallised and non-recrystallised austenite region than compression or laboratory rolling.[12] Torsion specimens with a gauge length of 17 mm and diameter of 7.5 mm were machined from the plates with the longitudinal axis parallel to the rolling direction.

The processing schedule is shown in Fig. 1. After austenitising at 1250°C for 120 s the first sample was subjected to the double deformation, equivalent tensile strain (ε) (hereafter called "strain") for torsion: $\varepsilon_1 =$ 0.25 in the austenite recrystallisation region and $\varepsilon_2 = 0.47$ in the non-recrystallisation region. After the second deformation, specimens were cooled at 1 Ks^{-1} to 670°C, to form ~50-60% of ferrite. This was followed by accelerated cooling at 20 Ks^{-1} to the temperature 50°C above the simulated coiling temperature (T_C = 450°C) and further slow cooling at 5 Ks^{-1} to avoid undershoot. Samples were quenched to ambient temperature after a 600 s hold which allowed the isothermal bainitic transformation to take place.

To study the effect of the deformation in non-recrystallised austenite region, ε_2 was increased to 1.2 while the deformation in recrystallised region was kept constant (0.25). To clarify the effect of heavy deformation (ε_1) in the recrystallised region, various amount of deformation (0.25, 0.5 and 0.75) were used with a strain of 1.2 in the non-recrystallised region. The accelerated cooling start temperature (T_A) was varied from 670 to 690°C to maintain the polygonal ferrite content between 50-60%.

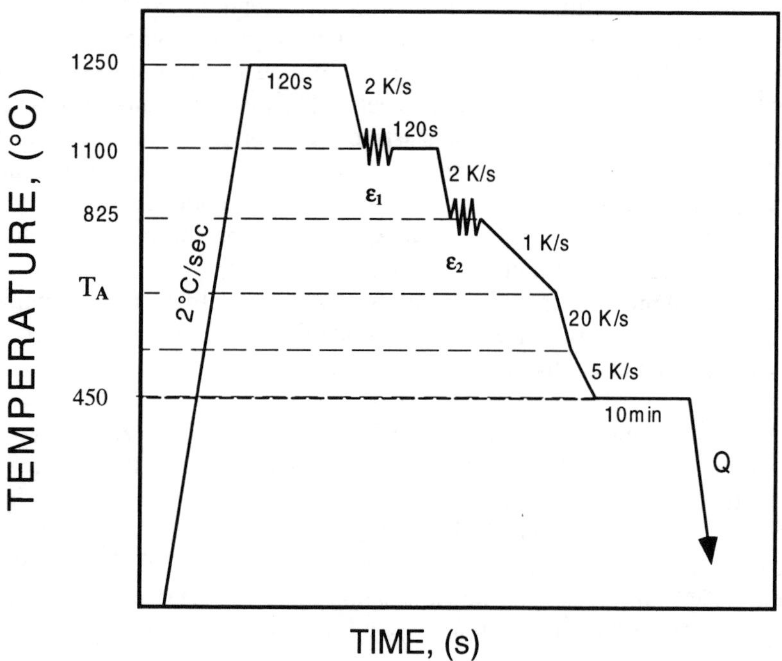

Fig. 1: TMP simulation schedule

The specimen preparation for optical examination involved standard metallographic procedures. These samples were etched with 2% Nital to reveal the microstructure. Micrographs were taken along the torsion axis. Selected samples were investigated using a field emission scanning electron microscope (JEOL JSM 630), operating at 15 kV accelerating voltage. The volume fraction of RA was measured by a tint etching technique. The color etchant was a mixture of three ingredients (4% HNO_3 + 7%$(NO_2)_3C_6H_2OH$ + saturated $Na_2S_2O_3$).[13] The prior austenite grain size was revealed using a solution of 2 g picric acid and 0.5 g copper(II) chloride in 100ml of water with 2 ml wetting agent. The quantitative analysis of RA and the measurements of the prior austenite grains were carried out using UTHSCSA image analysis software.

The microhardness of the second phase was measured utilising a Matsuzawa Microhardness Tester with 25-g load. Room temperature mechanical properties were determined by a shear punch technique. This technique has been developed for the cases when only a small amount of material is available. The manufactured in-house rig was attached to a Instron 4500 Tensile Tester. A 5 kN load cell with 0.25 mm/min. crosshead speed and a 3 mm-diameter punch was used in the tests. The samples were taken from the torsion specimens, cutting parallel to the torsion axis in the range of 300 to 350 μm thickness. The test was instrumented to provide punch load-displacement data to determine tensile strength using the following empirical equation:[14]

$$\sigma_{eff} = \frac{P-F}{2\pi rt} = C\sigma \qquad (1),$$

where P = maximum load (N), σ = corresponding uniaxial stress (MPa), F = friction load (N), r = punch radius (mm), t = specimen thickness (mm), C = empirical correlation coefficient.

From the previous work,[15] the correlation coefficient between the shear punch load-displacement curves and uniaxial tensile data was determined to be C = 0.58.

The total elongation was calculated using the empirical equation given by:[14]

$$El_{Total} = \frac{d_f}{t} \qquad (2),$$

where El_{Total} is total elongation and d_f is displacement at failure.

RESULTS AND DISCUSSION

The effect of the refinement of the austenite and subsequently the ferrite grain size on the mechanical properties of TRIP steel was studied using different levels of deformation in the austenite recrystallisation region while the effect of deformation on the further ferrite grain refinement, morphology of bainite and mechanical properties was determined from the different deformations in the non-recrystallisation region.

Effect of deformation in the non-recrystallised austenite region on the microstructure and mechanical properties

Austenite pancaking is achieved by deforming at temperatures below the non-recrystallisation temperature which alters the substructure of the austenite just prior to transformation.[16] The samples were deformed to 0.47 and 1.2-strain in the non-recrystallisation region, while the strain in the recrystallised region was 0.25. A representative microstructure after a strain of 0.47 is shown in Fig. 2. The microstructure consisted of 50-55% polygonal ferrite (PF), 30-35% non-carbide bainitic phases, such as granular bainite (GB) and acicular ferrite (AF), 8-10% martensite (M) and 6-10% retained austenite (RA). The RA appeared as coarse, blocky islands in PF, between the PF grains and bainitic ferrite laths. The average microhardness of the bainitic phase was 284 VHN. This microstructure showed an improved combination of the ultimate tensile strength (UTS) ~ 924 MPa and elongation (%El) ~ 49 % (Fig. 2) due to the balance between the ductile ferritic matrix, hard martensite and RA phase in bainitic ferrite in addition to the TRIP effect.

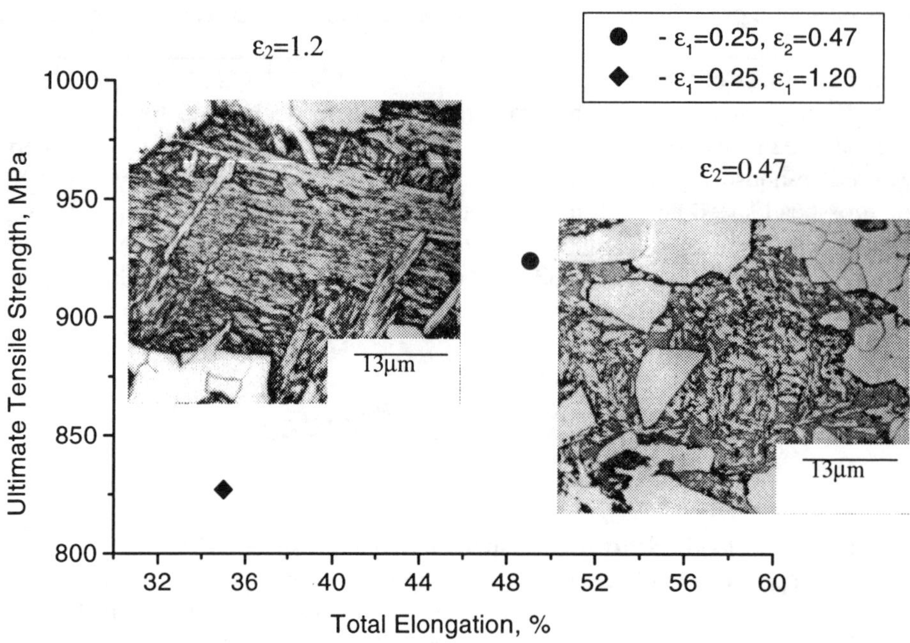

Fig. 2: Mechanical properties and microstructures of the samples after 0.47 and 1.2 strain in the non-recrystallised austenite region

After increasing the deformation in the non-recrystallised region ($\varepsilon_2 = 1.2$), the "pancaked" austenite grain thickness decreased from ~ 45 (0.47 of strain) to 25μm. Furthermore, further refinement of the ferrite grain size was observed: from ~ 15 ($\varepsilon_2 = 0.47$) to 9μm ($\varepsilon_2 = 1.2$). Ferrite grain refinement after deformation in the non-recrystallisation region was due to the formation of additional nucleation sites inside the austenite grains (deformation bands, twin bands and also dislocation arrays).[18] Based on other work,[10] the reduction of austenite and ferrite grain size should enhance the quantity of RA, because the volume fraction of RA (V_{RA}) has been related to the proportion V_{PF}/d_{PF}, where V_{PF} – volume fraction of PF, d_{PF} – grain diameter of PF. Thus, decreasing the ferrite grain size would lead to an increase of V_{RA} for a similar amount of PF. Moreover, the refinement of the bainite packets obtained by increasing the density of nucleation sites within "pancaked" austenite grains should increase the quality of RA, because the stability of RA depends on the RA size (the smaller RA is more stable). These two reasons are important in order to enhance the TRIP effect.

However, the contribution of surrounding phases to the stabilisation of RA is critical. The presence of carbides in the bainitic structure can dramatically reduce the carbon content in RA leading to transformation to martensite during quenching. This is the reason that the volume fraction of RA decreased to 4% after 1.2 strain. The microstructure after increasing the deformation in the non-recrystallised region was PF, upper bainite (UB), low bainite (LB), M and RA (Fig. 2), because the bainite transformation was retarded by the deformation debris in the non-recrystallised austenite.[11] The average microhardness of the bainitic phase was ~ 250 VHN. Microhardness reduced because of the presence of UB and a decrease the RA content. RA appeared as small particles, uniformly distributed between the bainitic ferrite laths. However, in spite of a decrease in V_{RA} the mechanical properties of the sample still exhibited a good combination of UTS (827Mpa) and %El (35%) (Fig. 2), because even a small amount of stable RA can provide a TRIP effect although significantly lower than in case when GB and AF were present. Thus, the contribution of the grain refinement in the mechanical properties of investigated steel competes with the effect of the stability of RA. The influence of the bainitic phase morphology on the mechanical properties has been identified as a critical aspect.

Effect of the recrystallised austenite conditions on the final microstructure and mechanical properties

The resulting austenite grain sizes after 0.25, 0.5 and 0.75-strain in the recrystallisation region, with the strain in the non-recrystallised region of 1.2, were respectively: ~ 25μm, 20μm and 15μm in thickness. The ferrite grain size also decreased from ~ 9 (0.25) to 7.5μm (0.50) to 7μm (0.75) because the deformed austenite grains provided an increase in the effective nucleation area and consequently the additional nucleation sites for the ferrite formation. The microstructure after 0.5-strain in the recrystallisation region consisted of UB, LB, and a small amount of RA (less than 2%) in addition to PF (Fig. 4). The microstructure was similar to the microstructure after 0.25-strain, but the bainitic microstructure was finer with a higher average microhardness (287VHN) due to the bainitic packet refinement. The formation of the UB and LB structures with a higher density of the fine carbides between the bainitic ferrite laths reduced the stability of RA and consequently the volume fraction of RA. Hence, the effect of the second phase on the quality and quantity of RA is more critical than a reduction in the ferrite grain size.

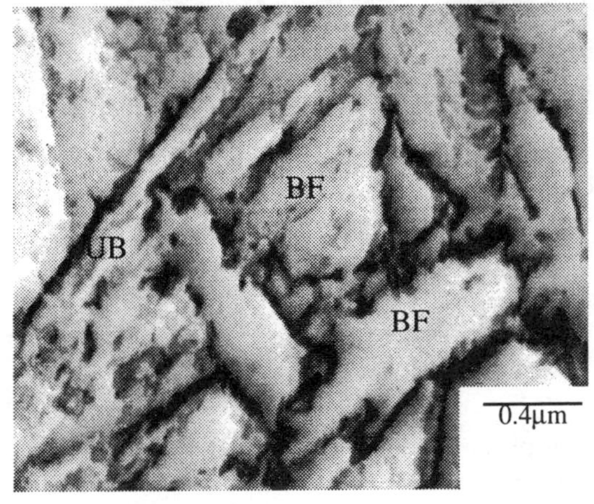

a

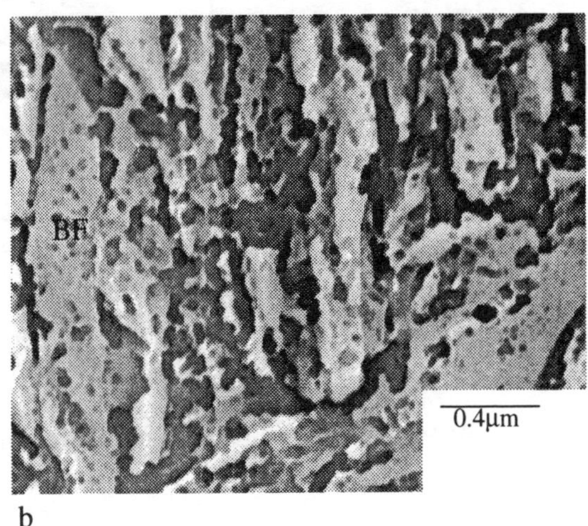

b

Fig. 3: SEM micrographs after 0.25 (a) and 0.50-strain (b)

Increasing the amount of deformation in the recrystallised austenite region led to the refinement of the bainitic packets compared with 0.25-strain (Fig. 3). Furthermore, the 0.50-strain significantly changed the morphology of bainitic ferrite (BF): i) increased the number of the crystallographic orientations of the bainitic ferrite laths because a heavy deformation enhances the nucleation of bainitic ferrite within austenite grains;[11] ii) significant refinement of the bainitic ferrite laths due to the interruption of their growth at the boundaries of cell-structures (Fig. 3).[11]

0.75-strain stimulated the formation of the carbide free GB structure with ~ 6% RA, ~ 9% M, UB and pearlite (P) in addition to 50% PF (Fig. 4). The average microhardness of the bainitic phase was 300 VHN, due to the bainitic phase morphology. Thus, an increase of strain up to $\varepsilon_1 = 0.75$ in recrystallised austenite region accelerated the bainite transformation and significantly improved the bainitic structure. RA was in the form of the coarse blocks between PF and layers between the bainitic ferrite laths. The increase in V_{RA} was due to the both reduction of the grain size and morphology of the bainitic structure.

The mechanical properties of steel after 0.25, 0.50, 0.75-strain in the recrystallised austenite region are displayed in Fig. 4. The UTS increased with decreasing the austenite and ferrite grain size when the samples obtained the similar bainitic structure. The reduction of the grain size did not lead to an increase in the elongation (Fig. 4). The V_{RA} played an important role in enhancing the elongation. The maximum UTS (883Mpa) and elongation (43%) was observed after 0.75-strain corresponding to the minimum ferrite grain size and maximum volume fraction of RA and the carbide-free bainitic structures, such as GB and AF. Hence, the morphology of the bainitic structure had a more pronounced effect on the mechanical properties than the grain size reduction.

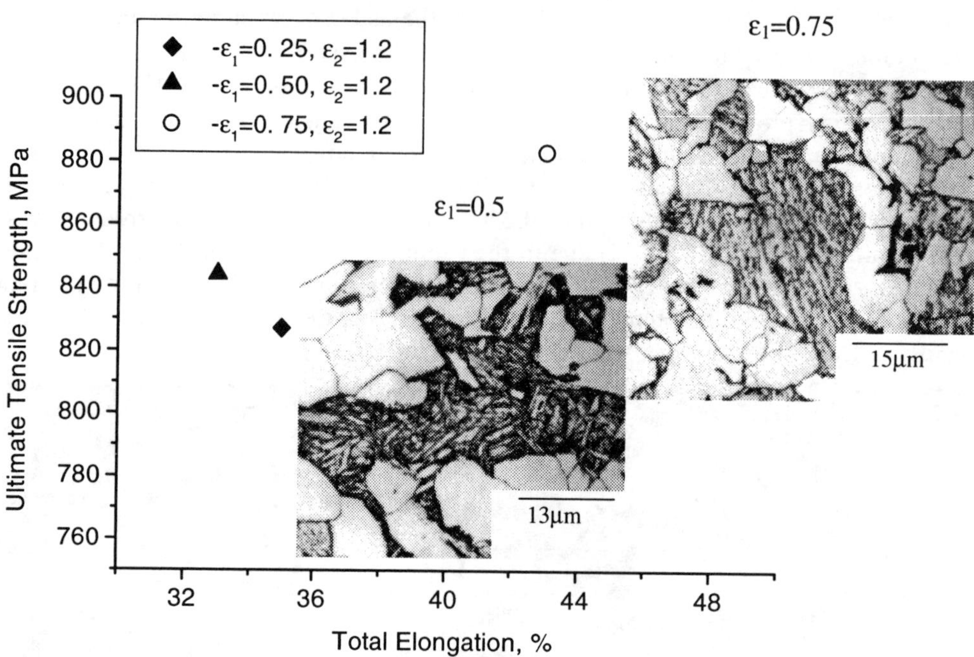

Fig. 4: Mechanical properties and microstructures of the samples after 0.25, 0.50 and 0.75-strain in the recrystallised austenite region

A common relationship between the strength and the austenite grain size is the Hall-Petch equation:

$$\sigma_y = \sigma_i + k_y d^{-1/2} \quad (3),$$

where σ_y is the lower yield stress, σ_I is the friction stress, k is the strengthening coefficient and d is the ferrite 'grain size'.[1] However, it was found that the yield strength in the TRIP steel depended not only on the grain size, but also on the bainitic morphology and quality and quantity of RA. The reduction of grain size only led to an increase in the yield strength when the contribution of RA to the mechanical properties was negligible (Fig. 5).

The relatively high V_{RA} obtained after 0.70-strain in the recrystallised austenite region stimulated a decrease in the yield strength at the good combination of UTS and %El. These results show that the Hall-Petch dependence is probably fortuitous for TRIP steel that relate to the results reported in other publications.[20]

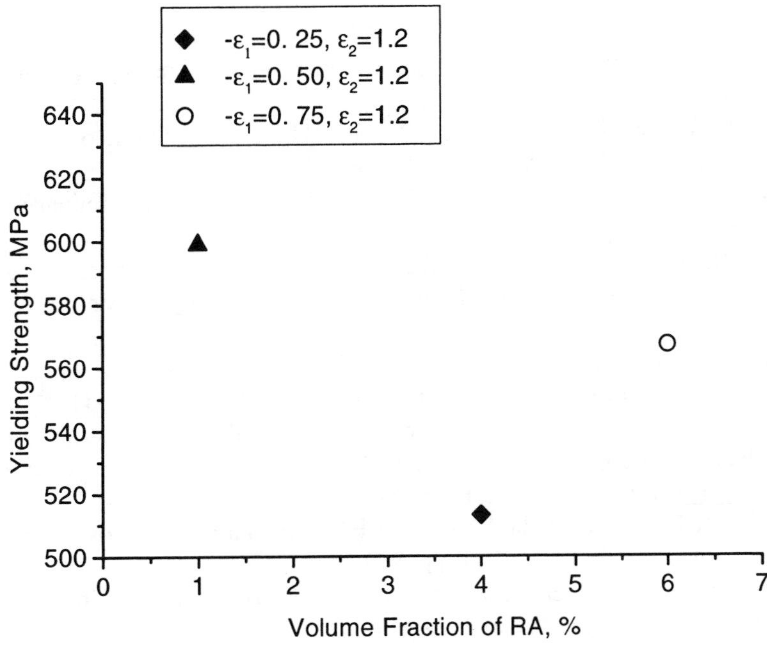

Fig. 5: The effect of V_{RA} on the Yielding Strength

Thus, the reduction in RA content led to an increase in yield strength. However, the effect of the ferrite grain size on the mechanical properties is still significant.

SUMMARY

The effect of the heavy deformation in the recrystallised and non-recrystallised austenite region on the transformation kinetics and mechanical properties of TRIP steel was studied. It was found that:

1. ferrite grain refinement does not always lead to an increase in V_{RA} because of the role of the bainite morphology;
2. the contribution of the grain refinement to the mechanical properties of TRIP steel is less than the effect of the bainite morphology and stability of RA;
3. deformation in recrystallised region decreases the bainitic packet size and changes the morphology of the bainitic ferrite that may enhance the RA quality;
4. even a small amount of RA can affect the mechanical properties more than ferrite grain refinement.

ACKNOWLEDGEMENTS

The authors would like to express their gratitude to BHP Research for providing experimental steels. They also would like to acknowledge the assistance in hot torsion testing from Mr. J. Whale. IT is grateful to the Deakin University for provision of a scholarship.

REFERENCES

1. T.Gladman, "The Physical Metallurgy of Microalloyed Steels", The Institute of Materials, Cambridge, 1997, UK, pp. 1-17.
2. V.F. Zackay, E.R. Parker, D. Fahr and R. Busch, "The Enhancement of Ductility in High-Strength Steels", Transactions of the ASM, Vol. 60, 1967, pp. 252-259.
3. W.W. Gerberich, P.L. Hemmings, M.D. Merz and V.F. Zackay, "Preliminary Toughness Results on TRIP Steel", Transactions Technical Notes, Vol. 61, 1968, pp. 843-847.
4. S.K. Liu and J. Zhang, "The Influence of the Si and Mn Concentation on the Kinetics of the Bainite Transformation in Fe-C-Si-Mn Alloys", Metallurgical Transactions A, Vol. 21A, June 1990, pp. 1517-1525.
5. O. Matsumura, Y. Sakuma, Y. Ishii and J. Zhao, "Effect of Retained Austenite on Formability of High Strength Sheet Steels", ISIJ International, Vol. 32, No. 10, 1992, pp. 1110-1116.
6. Y. Sakuma, O. Matsumura and H. Takechi, "Mechanical Properties and Retained Austenite in Intercritically Heat-Treated Bainite-Transformed Steel and Their Variation with Si and Mn Additions", Metallurgical Transactions A, Vol. 22A, February 1991, pp. 489-498.
7. L.C. Chang and H.K.D.H. Bhadeshia, "Austenite Films in Bainitic Microstructures", Materials Science and Technology, September 1995, Vol. 11, pp. 874-881.
8. D.Q. Bai, A. DiChiro and S. Yue, "Stability of Retained Austenite in Nb Microalloyed Mn-Si TRIP Steel", Materials Science Forum, 284-286, 1998, pp. 253-260.
9. F.B. Pickering, "Physical Metallurgy and Design of Steels", Applied Science Publishers Ltd., Barking, Essex, UK, 1978, p. 64.
10. O. Kawano,J. Wakita, K. Esaka and H. Abe, "Formation of Retained Austenite and Effect of Retained Austenite on Elongation in Low Carbon Hot-rolled High Strength Steels", ISIJ, Vol. 82, 1996, pp. 232-244.
11. K. Fujiwara and S. Okaguchi, "Morphology and Mechanical Properties of Bainitic Steels Deformed in Unrecrystallised Austenite Region", Materials Science Forum, Vols. 284-286, 1998, pp. 271-278.
12. P.D. Hodgson, D.C. Collinson and B. Perret, "The Use of Hot Torsion to Simulate the Thermomechanical Processing of Steel", Proceedings of 7th International Symposium on Physical Simulation of Casting, Hot Rolling and Welding, 21-23 January 1997, pp. 219-229.
13. A. Zarei-Hanzaki, "Transformation Characteristics of Si -Mn TRIP Steels After Thermomechanical Processing", PhD Thesis, McGill University, 1994, Montreal, Canada, pp. 44-45.
14. G.E. Lucas, J.W. Sheckherd and G.R. Obette, "Shear Punch and Microhardness Tests for Strength and Ductility Measurements", ASTM STP 888, ASTM, Philadelphia, PA, 1986, pp. 112-140.
15. E.V. Pereloma, I.B. Timokhina and P.D. Hodgson, "Microstructure and Mechanical Properties of Thermomechanically Processed C-Mn-Si Steels", Proc. Int. Conf., "Thermomechanical Processing of Steels", May 2000, London, IOM Communications LTD, pp. 418-427.
16. S.Yue, A. DiChiro and A. Zarei-Hanzaki, "Thermomechanical Processing Effects on C-Mn-Si TRIP Steels", JOM, September 1997, pp. 59-61.
17. P.H. Shipway and H.K.D.H. Bhadeshia, "Mechanical Stabilisation of Bainite", Materials Science and Technology, Vol. 11, November 1995, pp. 1116-1128.
18. R. Bengochea, B. Lopez and I. Gutierrez, "Microstructural Evolution during the Austenite-to-Ferrite Transformation from Deformed Austenite", Metallurgical and Materials Transactions A, Vol. 29A, February 1998, pp. 417-426.
19. T. Kvackaj and I. Mamuzic, "A Quantitative Characterisation of Austenite Microstructure after Deformation in Nonrecrystallisation Region and Its Influence on Ferrite Microstructure After Transformation", ISIJ International, Vol. 38, No 11, 1998, pp. 1270-1276.
20. H.K.D.H. Bhadeshia, "Bainite in Steel", The Institute of Materials, Cambridge University Press, Great Britain, 1992, pp. 292-308.

GRAIN REFINEMENT BY DANYMIC RECRYSTALLIZATION DURING WARM DEFORMATION IN ($\gamma+\alpha$) REGION OF LOW CARBON MICROALLOYED STEELS

Huoran HOU*, Huai GAO*, Qingyou LIU*, Han DONG*, Yuqing WENG**
* Central Iron & Steel Research Institute
** The Chinese Society for Metals
No.76, Xueyuan Nanlu, Beijing, 100081,P.R.China
Tel: +86-10-62182765
E-mail: hhr.cisri@263.net

Key Words: Grain Refinement, Dynamic recrystallization, Warm deformation, Microalloyed steel

INTRODUCTION

Warm rolling (especially ferrite rolling) of ultra low carbon steels is a new rolling schedule to produce thin hot-strip steels. In the direct application of warm rolling, a major metallurgical process influencing on the mechanical properties is recrystallization [1-7]. But it is generally accepted that the hot deformation of ferrite is controlled by dynamic recovery process alone. The reports of dynamic recrystallization in the ferrite range have been restricted to its occurrence in high purity materials such as zone-refined or vacuum-melted iron [8, 9]. Dynamic recrystallization of ferrite in IF steels containing Nb, Ti microalloying elements could occur during multi-passes deformation especially when strain rates are less than $0.1s^{-1}$. The purity of steels has significant effects on the dynamic recrystallization of ferrite [10, 11]. IF steels have very high purity and the interstitial atom concentrations in the steels such as C, N reach sufficient low levels, these atoms are easy to react with Nb,Ti so as to form Nb(C,N), Ti(C, N), *etc.* The relative absence of interstitial atoms permits the dynamic recrystallization of ferrite to occur. It is of interest that ultra-fine grain ferrite is produced by the dynamic transformation of austenite to ferrite as well as the dynamic recrystallization of ferrite when plain C-Mn steels are warm deformed in the intercritical region (*i.e.*, in the austenite-plus-ferrite, two-phase region) by Yada *et al* [12].

In the present study, the microstructural changes taking place during simulated warm deformation both on a Gleeble machine and a laboratory rolling mill were studied on two low carbon microalloyed steels.

The grain refinement features of the dynamic recrystallization of ferrite were also studied, and mechanical properties of ultra-fine grain specimens were measured.

EXPERIMENTALS

Materials The experimental steels are super-clean low carbon microalloyed steels A and B. The chemical compositions are given in Table I . The steels were melted in the vacuum induction furnace of 200kg in capacity. Prior to testing, the steel was austenitized at 1323K for 30 minutes and normalized by cooling to room temperature in the air. The measured critical transformation temperatures of the steels are shown in Table II.

Table I Chemical compositions of the experimental steels (in wt.%)

Specimens	C	Si	Mn	S	P	O	N	Nb	Ti	Als
A	0.04	0.18	1.20	0.0018	0.0012	0.0021	0.0038	0.058	0.024	0.055
B	0.11	0.17	1.42	0.0013	0.0012	0.0022	0.0044	0.048	0.019	0.044

Table II Transformation temperatures of the steels (K)

Specimens	Ac_1	Ac_3	Ar_1	Ar_3
A	1003	1173	1038	1118
B	998	1213	948	1078

Plain Strain Compression Hot compression tests were conducted on a Gleeble-2000 thermal simulator to study the dynamic recovery and the dynamic recrystallization behaviors. Specimens for plain strain compression were machined as 18(L) x 10 (W) x 12 (H) mm. The specimens were reheated to 1423K for 180s and then cooled down to 973K at cooling rate 10K/s, and subsequently with isothermal treatment for 600s to ensure obtaining equilibrium ferrite microstructures as much as possible. After being deformed with the reduction of 80% and the strain rates ranging from $0.01s^{-1}$ to $5s^{-1}$, the specimens were quenched into the water in about 2s. After compression, the specimen was cut in the midplane perpendicular to the compression plane and parallel to the compression direction. Since the upper and lower anvils were not coaxial due to the experimental error, shear stress was effected on the specimens.

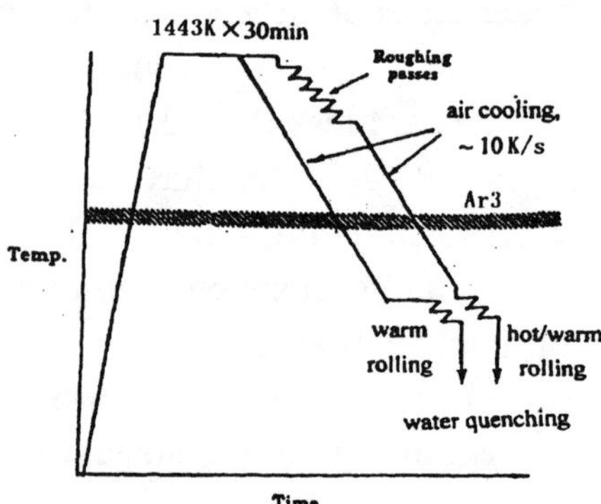

Fig.1 Two rolling procedures of the steels on the laboratory mill

Rolling Two different rolling procedures were employed on a laboratory rolling mill shown as Fig.1. In both cases, the finishing passes were applied in the warm rolling range (ferrite + few austenite). These methods are designed as warm and hot-warm rolling. The thickness of the slabs was 20-30mm. In warm rolling procedure, the slabs were heated to 1443K for 30min in the furnace, then cooled down to 973K in the air with cooling rate about 10K/s. Without rough rolling at austenite region, the steels were deformed at 973K without/with isothermal treatment for 10min with reduction of 80% and the specimens were quenched into water within 3s. In hot/warm rolling procedure, after heating at the same temperature, two rough rolling passes were applied on the austenite with the reduction of 30%, respectively. Then cooled down to the warm rolling temperature at 973K with 10K/s cooling rate. Single and two finishing passes were employed with the total reduction of 80% and 100% respectively.

Specimens for optical observation were prepared by polishing and etching with 4% nital solution. A Leica 500S microscope was employed to obtain more details of the microstructures. Further microstructural analysis was performed by using TEM technique. Ferrite grain size was determined by standard linear intercept methods and the volume fraction of recrystallized ferrite by point counting. For the latter, a grid of 100 points was employed, and counting was performed on 8 different fields of each microstructure. An MVK-E hardness tester was used to investigate microhardness of the microstructures with the applied load of 0.1kg.

RESULTS AND DISCUSSION

Plain Strain Compression After heated to 1423K for 180s and cooled down to 973K at cooling rate 10K /s and then subsequently with isothermal treatment for 600s, the specimens were directly quenched into the water quickly so that the microstructures and the features of the specimens before compression could be obtained shown as Fig.2. For steel A, there were almost total ferrite except for a little austenite (~5%), the hardness was 244HV0.1. For steel B, ferrite in the microstructures decreased to 70%, and hardness was 224HV0.1.

Stress-strain behaviors-Fig.3 shows the stress-strain curves of the experimental steels of A and B at strain rates of $0.01s^{-1}$, $0.1s^{-1}$, $1s^{-1}$ and $5s^{-1}$ during plain strain compression. For steel A, the compression could be assumed in the ferrite range since the microstructure consists of 95% ferrite before deformation. Shown as Fig.3 (a), the forms of the stress-strain curves for steel A are similar. At lower strain there is a distinct initial peak in stress. The maximum stress is about 250MPa when the strain rate reaches at $5\ s^{-1}$. If the strain

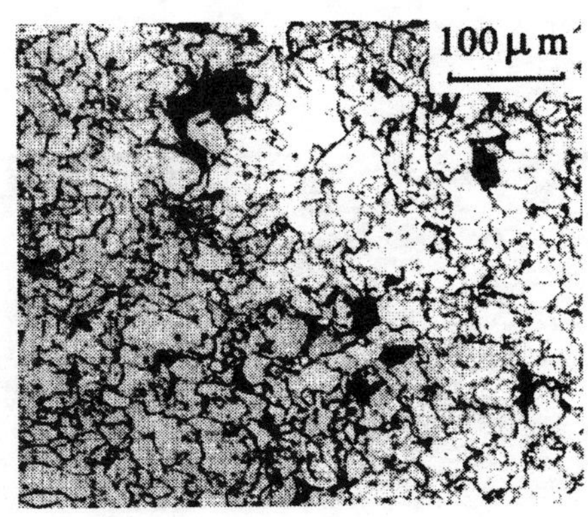

(a) Steel A

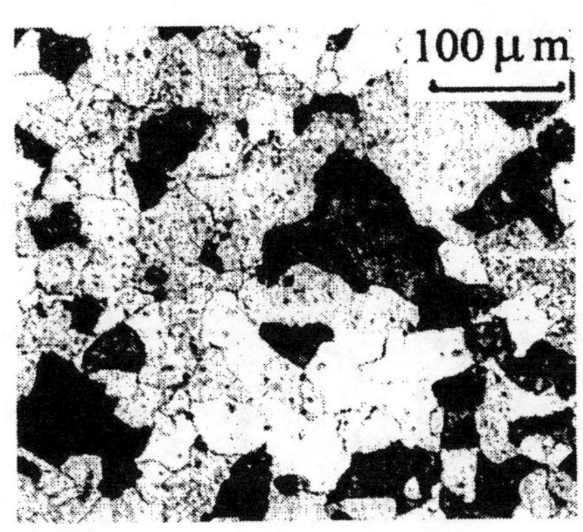

(b) Steel B

Fig.2 Microstructures of steels A and B before deformation

is further increased to permit the dislocation density to attain an appreciable level and stresses decrease, dynamic recrystallization is initiated during deformation. This dynamic softening mechanism can be recognized by the existence of a peak on the stress-strain curve as subsequent softening with increasing strain. Continuous dynamic recrystallization occurs in steel A. For steel B (Fig.3 (b)), the effect of the strain rates on the stress-strain curves was obvious. In the case of higher strain rate, i.e., $5s^{-1}$ and $1s^{-1}$, the curves show dynamic recovery and work hardening. But for lower strain rates, i.e., $0.1s^{-1}$ and $0.01s^{-1}$, the stress decreases after reaching to a peak stress, so the dynamic recrystallization occurs. It could be concluded that because of the increasing of the strain rates, time for dynamic recrystallization is diminished and the nucleation rate decreases so that inhibiting the dynamic recrystallizatin.

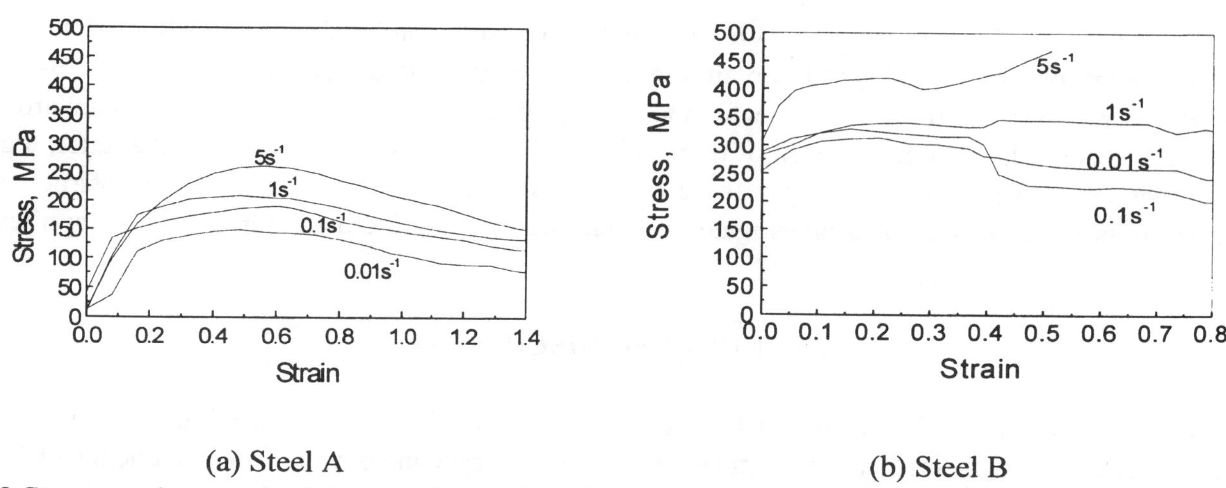

(a) Steel A (b) Steel B

Fig.3 Stress-strain curves of the experimental steels with varied strain rates from $0.01s^{-1}$-$5s^{-1}$

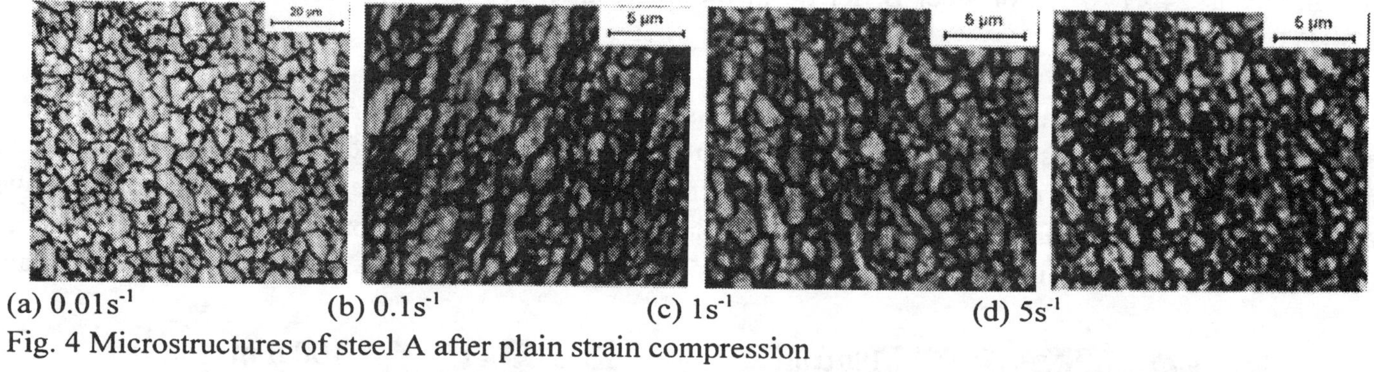

(a) $0.01s^{-1}$ (b) $0.1s^{-1}$ (c) $1s^{-1}$ (d) $5s^{-1}$

Fig. 4 Microstructures of steel A after plain strain compression

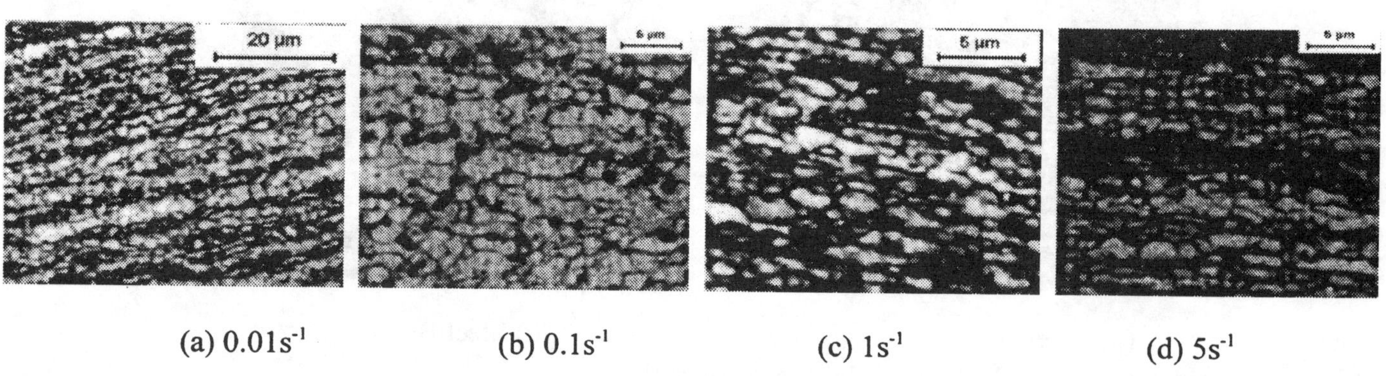

(a) $0.01s^{-1}$ (b) $0.1s^{-1}$ (c) $1s^{-1}$ (d) $5s^{-1}$

Fig.5 Microstructures of steel B after plain strain compression

Microstructures-The microstructures of the deformed specimens are shown as Fig.4 and Fig. 5 respectively. For steel A, it is illustrated in Fig.4 that when the strain rate was $0.01s^{-1}$ (Fig.4 (a)), an average ferrite grain size of 3.92 μ m was produced. When the strain rate was $0.1s^{-1}$, $1s^{-1}$ and $5s^{-1}$, the grain size decreased to 1.30 μ m (Fig. 4(b)), 1.12 μ m (Fig. 4(c)) and 0.96 μ m (Fig. 4(d)) respectively. The ferrite grain size decreases with the strain rate increasing. It is evident from the figure that the ferrite grain boundaries are fairly irregular and that the grain size is not uniform. It seems that these microstructural features are similar to those associated with the dynamic recrystallization of austenite. For steel B, the microstructures shown as in Fig.5 are different from those of steel A. The deformation structures consist of more elongated ferrite grains containing sub-boundaries that revealing subgrain clusters and some fine equiaxed grains. These fine ferrite grains should be formed by the partial dynamic recrystallization. As the strain rates decrease to less than $0.1s^{-1}$, more fine recrystallized ferrite grains were produced.

The microstructures and associated stress-strain curves presented above show that dynamic recrystallization occurs in the ferrite of low carbon microalloyed steels during plain strain deformation. But it is interesting that the characteristics of the dynamic recrystallization of ferrite are different in the two experimental steels with varied content of carbon. In steel A with lower carbon content (0.04wt%), continuous dynamic recrystallization of ferrite occurs under the conditions of all the experimental strain rates; but for higher carbon content (0.11wt%), only partial dynamic recrystallization occurs when the strain rates are of low values. A possible explanation is that the presence of high carbon levels leads to sufficient strain-induced carbide precipitation for the inhibition of dynamic recrystallization. By contrast, when the carbon concentration reaches at sufficiently low level, the relative absence of precipitation permits the dynamic recrystallization of ferrite to occur[10].

Besides, the state of stress also has great effect on the dynamic recrystallization. When cylinder compression test is employed, the deformation manner and state of stress are different from those of plain strain compression, only partial dynamic recrystallization occurs when strain rates are low enough. More details are discussed elsewhere [13].

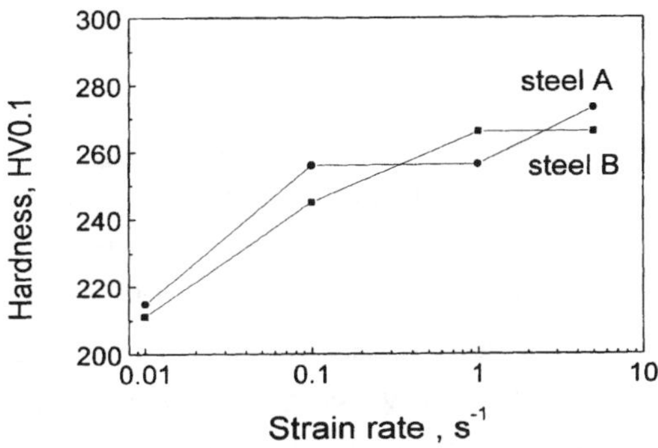

Fig.6 Hardness of the steel A and B

Microhardness of the steel A and B after compression is shown as Fig.6. The microhardness HV0.1 of steel A had the highest value of 273 at $5s^{-1}$. When the strain rate decreases to $1s^{-1}$, hardness decreases; hardness maintains the same level when strain rate is $0.1s^{-1}$. The hardness sharply decreases to 214 as the strain rate further decreases to $0.01s^{-1}$. From the curve of steel B, hardness scarcely decreases at higher strain rates, i.e., $5s^{-1}$ and $1s^{-1}$. Only if the strain rates are lower to $0.1s^{-1}$ and $0.01s^{-1}$, the hardness decreases sharply at lower strain rates. For the changes of the hardness, it is believed that dynamic recrystallization of ferrite occurs since the hardness decreases sharply. Otherwise, slight decrease of hardness shows dynamic recovery.

Further, the ferrite grain size becomes more and more small when the strain rates increase. The grain size of steel A is smaller than that of steel B when deformed at the same conditions, so the hardness of steel A are higher than that of steel B though the carbon content of steel B is higher.

Behavior of dynamic recrystallization of ferrite-TEM micrographs of the steel A and B are shown in Fig.7. TEM features of steel A compressed at strain rate $5s^{-1}$ are shown in Fig.7 (a) and those of steel B compressed at strain rate $0.1s^{-1}$ are shown in Fig.7 (b) and (c). In steel A, there are both ferrite structures containing subgrains and dislocation tangles in ferrite grains. Fig.7 (b) shows a fine ferrite grain nucleated at the intersection of prior boundaries of three ferrite grains. There were some precipitations and amount of dislocation net in it. Fig.7 (c) shows a typical example of interphase precipitation. The diameter of these carbonitride precipitations is about 8 nm. It is also observed that dislocation tangles exist with these carbonitride particles in ferrite. In thin foil specimens for TEM, interphase precipitation is easily recognized when the incident electron beam passes parallel to a previously moving γ/α interface. In such a case, sheets of carbonitride particles that formed along the advancing interface are observed and appear as parallel rows on the electron microscope screen [14].

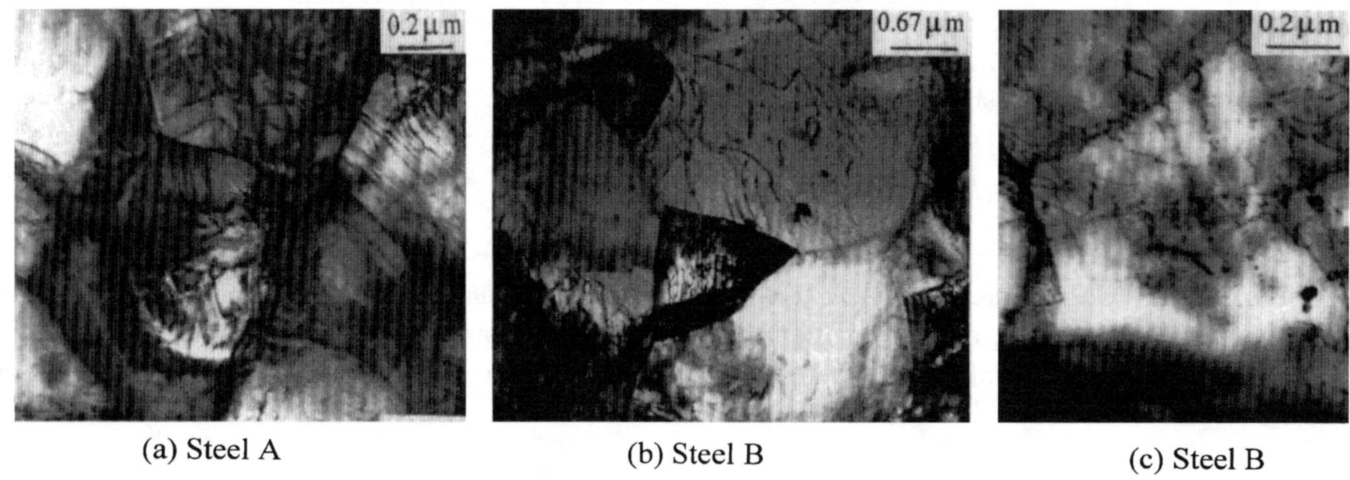

(a) Steel A (b) Steel B (c) Steel B

Fig.7 TEM micrographs of steel A and B

It is well known that work hardening and dynamic softening behaves contradictorily during working of materials. Dislocation generation and intersections of dislocations during plastic deformation leads to work hardening, while softening is mainly caused by dynamic recovery and/or dynamic recrystallization, which results from coalescence, annihilation and reconstruction of dislocations by cross-slip and climb with the aid of applied stress and thermal activation. Metallic materials should be divided into two kinds such as dynamic recovery and dynamic recrystallization by stacking fault energy in view of traditional theory. Dislocations in the metals with *bcc* structure and high stacking-fault energy (such as Al, α-Fe, ferrite steel, *etc.*) are easy to cross-slip and climb, so only dynamic recovery occurs in the metals. In the metals with *fcc* structure and low stacking-fault energy, dynamic recovery is difficult to occur because dislocations are difficult to cross-slip, so partial dislocation density is cumulative high enough so as to form big angle grain boundary, thus dynamic recrystallization occurs.

It is also generally believed that dynamic recrystallization nucleating manners are mainly: homogeneous nucleation, grain boundary nucleation, sub-grain nucleation, dislocation cluster nucleation, *etc.*. In present study, shown in Fig.7, there are interphase carbonitride precipitations and large amount of dislocation net in it. It appears that the large amount of carbonitride particles hinders the motion of dislocations and recrystallization nucleii form when dislocations cluster at obstacles formed by other dislocation or at small submicrometre size particles producing tangles of dislocations. The subgrain structure with low angle boundaries is formed during the early stage of deformation at the dislocation cluster site. With further deformation, the low angle boundaries are gradually transformed into high angle ones by means of absorbing dislocation or/and subgrain rotation, which indicates that continuous dynamic recrystallization occurs. The behaviors of dynamic recrystallization of ferrite are similar to that of Al [15].

Rolling Microstructures of steel B rolled with warm rolling procedure are shown in Fig.8. Fig. 8(a) shows the micrograph without isothermal treatment, approximately equiaxed ferrite grains are refined to about 1.30 μ m. Fig.8 (b) shows that with isothermal treatment for 10 min, many fine ferrite grains are formed with

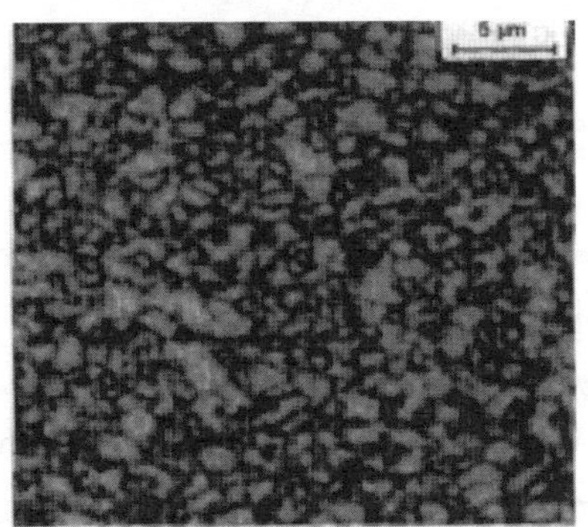

(a) Without isothermal treatment

(b) Isothermal treatment for 10min

Fig.8 Microstructures of steel B after warm rolling

(a) Steel A, single pass

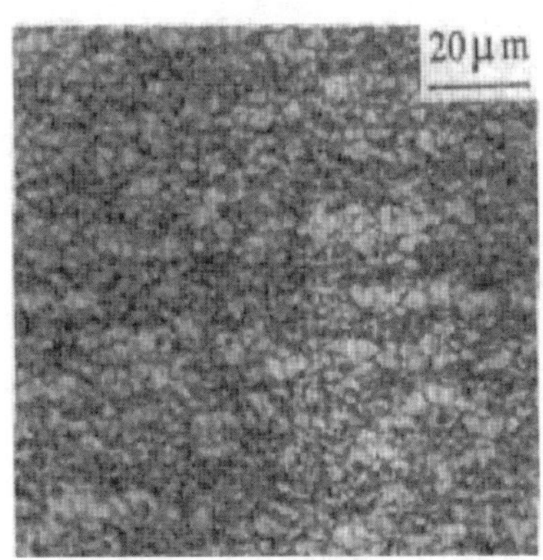

(b) Steel A, two passes

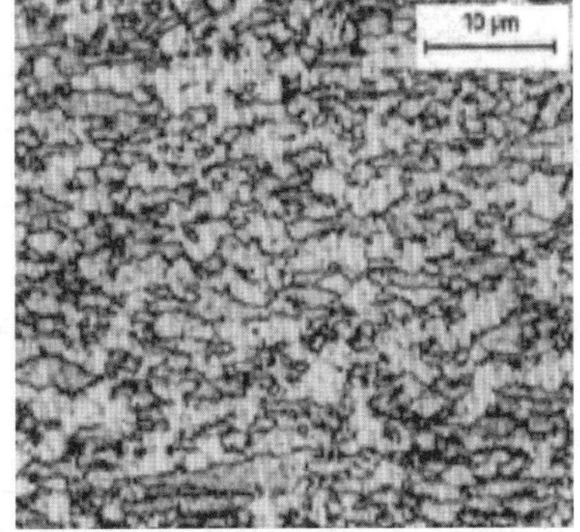

(c) Steel B, single pass

(d) Steel B, two pass

Fig.9 Microstructures of steel A and B after hot/warm rolling

the average grain size 1.10 μm. The tensile strength σ_b of these two specimens are of 905MPa and 930MPa respectively. The yield strength $\sigma_{0.2}$ are of 630MPa and 605MPa respectively. The strength of the steels increases significantly by grain refinement.

Microstrutures of hot/warm rolling are shown in Fig.9. Compared with warm rolling (Fig.8 and Fig.9 (c-d)), the grain size of hot/warm rolling is larger than that of warm rolling and the structures are more inhomogeneous in hot/warm rolling procedure. Shown as in Fig.9 are microstructures of steel A and B, when two passes rolling is applied, elongated grains with substructures increase significantly and grain size are larger than that of single pass. The tensile strength σ_b of these specimens (in the order from Fig.9 (a) to (c)) are 635MPa, 605MPa and 825MPa respectively. The yield strength $\sigma_{0.2}$ are of 445MPa, 545MPa and 650MPa respectively.

CONCLUSIONS

1. Dynamic recrystallization of ferrite occurs during warm deformation in ferrite/austenite+ferrite range through heavy reduction in low carbon microalloyed steels.
2. Ultra-fine microstructure of low carbon microalloyed steels with ferrite grain size refined to about 1.00 μm can be obtained through dynamic recrystallization. Submicron microstructure of 0.96 μm in size can be achieved in a Nb-Ti microalloyed steel with 0.04 percentage carbon content during warm deformation.
3. The yield strength of the microalloyed steels through warm rolling is over 600MPa through grain refinement.

ACKNOWLEDGMENTS

The authors thank the Ministry of Science and Technology of China for finacial support to this project (G1998061502).

REFERENCES

1. A.Najafi-Zadeh, J.J.Jonas and S.Yue, "Effect of Dynamic Recrystallization on Grain Refinement of IF Steels", Material Science Forum, Vols.113-115, 1993, pp. 441-446.

2. B.Feng, D.Dunne and T.Chandra, "The Effect of Ti-Nb Additions on The Restoration of α in HSLA Steels After Rolling in The ($\gamma+\alpha$) Two-Phase Region", Recrystallization'90, The Minerals, Metals & Materials Society, 1990, pp. 319-324.

3. E.A.Simielli, S.Yue and J.J.Jonas, "Recrystallization Kinetics of Microalloyed Steels Deformed in The Intercritical Region", Metallurgical Transactions A, Vol.23A, February 1992, pp. 597-608.

4. P.R.Cetlin, S.yue and J.J.Jonas, "Warm Working of Ferrite in The Simulated Rod Rolling of IF Steels", Material Science Forum, Vols.113-115, 1993, pp. 405-410.

5. M.Schmickl, D.Yu, C.Killmore, D.Langley and T.Chandra, "Prediction of Ferrite Grain Size after Warm Deformation of Low Carbon Steel", ISIJ International, Vol. 36, No.10, 1996, pp. 1279-1285.

6. R.Z.Wang and T.C.Lei, "Dynamic Recrystallization of Ferrite in A Low Carbon Steel during Hot Rolling in The ($\alpha+\gamma$) Two-Phase Range", Scripta Metallurgica *et* Materialia, Vol.31, No.9, 1994,

pp. 1193-1196.

7. I.A.Rauf and J.D.Boyd, "Microstructural Evolution during Thermomechanical Processing of A Ti-Nb Interstitial-Free Steel just below The Ar_3 Temperature", Metallurgical. and Material Transactions A ,Vol.28A, July 1997, pp. 1437-1443.

8. G.Glover and C.M.Sellars, "Recovery and Recrystallization during High Temperature Deformation of α-Iron", Metallurgical Transactions, Vol.4, March 1973, pp. 765-775.

9. G.Glover and C.M.Sellars, "Static recrystallization after hot deformation of α iron", Metallurgical Transactions, Vol.3, August 1972, pp. 2271-2280.

10. A.Najafi-Zadeh, J.J.Jonas and S.Yue, "Grain Refinement by Dynamic Recrystallization during the Simulated Warm-Rolling of Interstitial Free Steels", Metallurgical Transactions A, Vol.23A, September 1992, pp. 2607-2617.

11. N.Tsuji, Y.Mstsubara and Y.Saito, "Dynamic Recrystallization of Ferrite in Interstitial Free Steel", Scripta Material, Vol.37, April 1997, pp. 477-484.

12. H.Yada, Y.Matsumura and K.Nakajima, United States Patent, No. 4, 446,842 (1984).

13. Huoran Hou, Qingyou Liu and Han Dong, "Dynamic Recovery and Dynamic Recrystallization of Low Carbon Microalloyed Steels during Warm Deformation", The 4th International Conference on High Strength Low alloy Steels, Xian China, October 2000 (to be published).

14. H.-J.Kestenbach, "Dispersion Hardening by Niobium Carbonitride Precipitation in Ferrite", Material Science and Technology, Vol.13, September 1997, pp. 731-739.

15. T.Sheppard, N.C.Parson, and M.A. Zaidi, "Dynamic Recrystallization in Al-7Mg Alloy", Metal Science, Vol.17, October 1983, pp. 481- 490.

FORMATION OF ULTRA-FINE FERRITE GRAINS IN COMMON CARBON STEELS THROUGH LOW TEMPERATURE HEAVY DEFORMATION

Zhongmin YANG • Yan ZHAO • Qian CHEN, Huai GAO, Ruizheng WANG • Yanwen MA
Division of Structural Materials Central Iron & Steel Research Institute
No.76, Xueyuan Nanlu, Bejing 100081
Beijing
P. R. China
Tel. • +86-10-62182763
E-mail • yzm@public.east.net.cn

Key Words: Plain Carbon Steel, Ultra-fine Grains, Deformation Induced Ferrite • Dynamic Recrystallization

INTRODUCTION

Ferrite grain in microalloying steel plate and strip can be refined to 5μm and yield strength can be increased by about 200MPa, by controlled rolling and controlled cooling technology developed in 1970's. In the last decade or so, the research works on achieving ultra-fine grain structure in steel were carried out, in order to effectively improve the strength and toughness of steels. In recent years, an emphasis was focused on the process mechanism to obtain ultra-fine ferrite grains finer than 3μm for C-Mn steel[1-6].

In this paper, the focus on formation of ultra-fine grains in plain carbon steel by heavy deformation at low temperature. The isothermal compression were carried out on a hot deformation simulator (Gleeble 2000) with the methods of water cooling during deformation and decreasing strain. The analysis shows that the formation of ultra-fine grains must be the result from the common effect of dynamic recrystallization (DRX) and deformation induced ferrite (DIF). In this study, the roles of strain, strain rate, and temperature on the formation of ferrite during deformation have been examined in detail. The influence of cooling rate on the ferrite microstructure was also established.

EXPERIMENTAL PROCEDURES

The compositions of plain carbon steel Q235 are shown in Table I, that was supplied by Capital Iron and Steel Company in the square form of a continuous casting billet of 130 mm×130 mm in section size. The steel was hot forged into small round bars and normalized at 920 ℃ for 1 h. And then the bars were machined into specimens with size of Φ10 mm×12 mm for simulating compression test. The grain size of the specimens after normalization is around 25 μm.

The temperature region of 900~600 ℃, the reduction region of 10~80% and strain rate region of 30~350 s^{-1} were selected to study the influence of processing parameters on ferrite grains size. The experimental conditions are as follows: reheating at 900 ℃ for 2 min and followed by cooling at 10℃/s to

Table I. Chemical composition of experimental steel. wt%

C	Si	Mn	P	S	Al	Ti	O	N
0.18	0.21	0.60	0.016	0.020	0.0082	<0.005	0.017	0.0059

test temperature, and water quenching during deformation or after deformation. The cooling rate was given 3 – 20 ℃/s. The ultra-fine grain specimens for the tensile examination were machined into small specimens with size of Φ5 mm × 7 mm. The austenite to ferrite transformation start temperature (A_{r3}), the stop transformation temperature (A_{r1}), the eutectoid temperature of pearlite (T_s) and equilibrium temperature (A_{e3}) of steel Q235 are 780 ℃, 600 ℃, 700 ℃ and 840 ℃ respectively. The equilibrium temperature was calculated by using program Thermo-Calc, and others were measured.

A first series of tests was conducted at a strain rate of 30 s^{-1} in the temperature range from 600 ℃ to 870 ℃. The deformations were given in one or two stages with intervals between 0.5 and 1.5 s. The strain of the deformation of one stage was 0.8. In order to the possibility of the static recovery and static recrystallization (SRX) of ferrite were eliminated, the specimens were water quenched during deformation. And the strain of each deformation of two stages was 0.5. At the end of each test, the sample was immediately quenched.

The effect of the amount of deformation on the characteristics of the DIF transformation was investigated in a second series of tests. In this case, the specimens were deformed at 830 ℃ and 760 ℃ at the stain rate 30 s^{-1}. The amount of deformation was ranging between 0.1 and 0.8. The specimens were water quenched during deformation of one stage.

In a third series of tests, the effect of stain rate on the DIF transformation was determined. For this purpose, four different strain rates (30, 60, 100, 350 s^{-1}) were employed. The specimens were deformed by 0.8 at 870 ℃ and water quenched during deformation of one stage.

The effect of others factors on DIF transformation (dynamic recrystallization of austenite, reversal transformation of ferrite, cooling rate) also was indicated in detail.

Optical metallography was carried out on longitudinal sections of the specimens after polishing and etching with a nital. Microscopic observations were made on a specific area of the lateral section of a specimen where the strain is equal to the nominal strain. Foils for transmission electron microscope (TEM) were prepared from surface section of specimens. Austenite and ferrite grain sizes were obtained for all the specimens before or after deformation, using the intercept counting method.

RESULTS AND DISCUSSION

Deformation Temperature Making use of mechanisms of deformation induced ferrite transformation and dynamic recrystallization of ferrite, Yada et.al[1] had achieved ultra-fine grain ferrite with a fraction exceeding a half in the microstructure by means of heavy deformation in the temperature region of A_{e3}~A_{r3}[1]. Mintz et.al [2]thought that ferrite could be induced in the temperature range of A_{e3}~A_{r3}-20 ℃. The latest study reported that gamma (γ) → alpha (α) transformation could occur during deformation at temperature above A_{e3}[7].

There deformation temperature regions of 870~ 600 ℃ were selected. Figure 1 illustrates the microstructure of specimens of the single pass deformation. The specimens were water quenched during deformation. The possibility of the static recrystallization of ferrite was eliminated by water quenching during deformation. Figure 1 shows the microstructure of specimens deformed by this process. Fig.1 shows that 90%~95% ultra-fine ferrite of size 2-3 μm is obtained, when single pass deformation between 870~760 ℃ (A_{e3}+30 ℃~A_{r3}-20 ℃). Experimental result indicate the temperature has little influence on the volume fraction and the grain size of DIF when the deformation was given over than 80% at temperature range A_{e3}+30 ℃~A_{r3}-20 ℃.

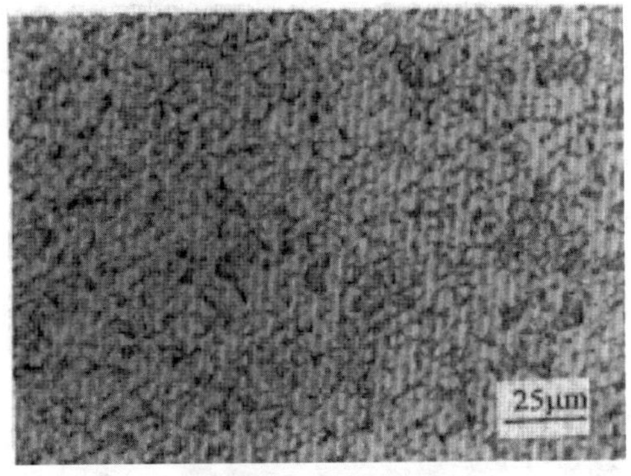

A At 850 ℃

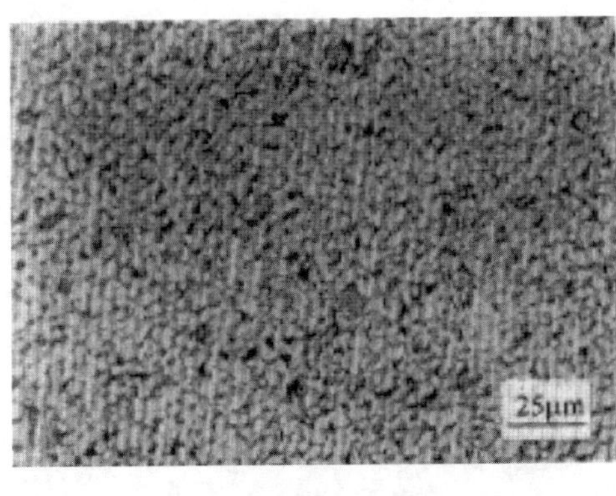

B At 760 ℃

Fig.1 Microstructures of the specimen, deformed by 80% once at given temperature, and water cooled during deformation

Figures (2, 3) show transmission electron micrograph and stress-strain curve of specimen, as shows as in Fig.1A, respectively. Transmission electron micrograph (Fig.2) of a typical grain of ultra-fine ferrite

Fig.2 TEM micrograph of DRX of DIF of specimen at deformed temperature 850 ℃

showed a low density of dislocations. Thin films of cementite are observed around the boundaries of grain of ultra-fine ferrite and pearlite is not. This is feature of the ultra-fine ferrite produced by this process in plain carbon steels. Figure 2 illustrates that the ultra-fine ferrite has an equiaxed morphology and suggests that the must be recrystallization of ferrite has occurred during deformation. The deformation stress nearly reaches its maximum value at strain of 1.3 and then rapidly decreases, as shown as in Fig.3.

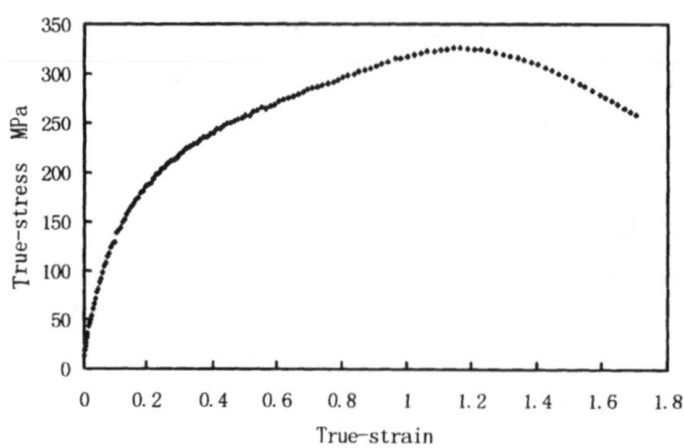

Fig.3 Stress-strain curves of specimen at deformed temperature 850 ℃

Figure 4 illustrates the microstructure of specimens of the two-pass deformation. Fig.4c shows that the deformed ferrite and the deformed pearlite still remain when the deformation temperature is lower than T_s, which results in a strip microstructure. When deformation is carried out in the range of T_s-T_s+ (30~50) ℃, the deformation bands appear in some ferrite grains and DIF shows cell structure with heredity of deformed austenite, as shown in Fig.4b. When deformation is carried out in A_{e3}+30 ℃~A_{r3}-20 ℃ (870~760 ℃), homogeneous and equiaxed ferrite grains can be obtained. The deformation with one pass shows the same tendency as that of two-pass deformation, as shown in Fig.4a. Therefore, the deformation temperature has influence on the microstructure feature of DIF.

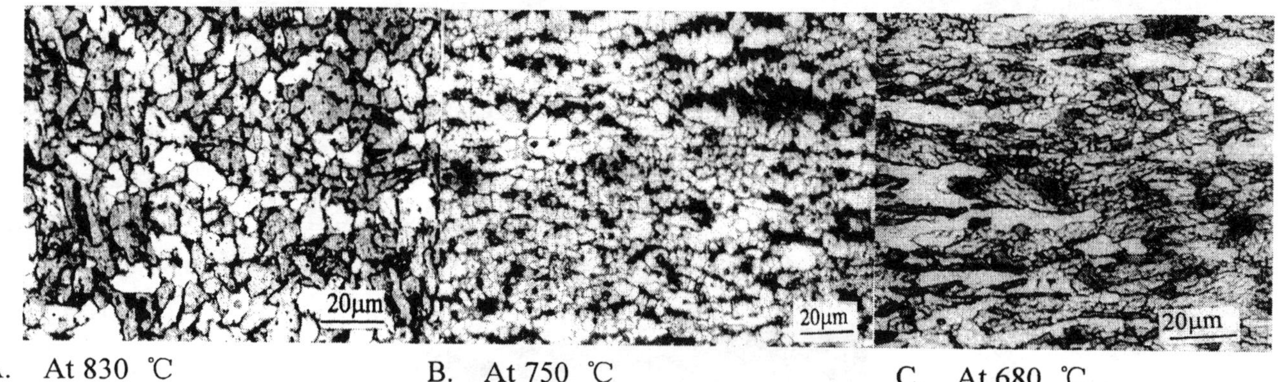

A. At 830 ℃ B. At 750 ℃ C. At 680 ℃

Fig 4. Ferrite structure of specimens deformed 50% twice at different temperature (A), (B) DIF and DRX. (C) elongated ferrite.

Amount of Deformation The results of this work show that the interpass time has effect on microstructural constituents and grain size. It can be seen that the ferrite grain size reduces with the interpass time decreasing. For steel Q235, according to a single pass deformation experiment, the reduction of 70% is necessary to obtain DIF with a fraction more than 50% in the microstructure. Also, the different amount of deformation results in different volume fraction and grain size of DIF, as shown in Fig.5. The fraction volume of ferrite increases and its grain size decreases with amount of deformation increasing. At the same time, the morphology of DIF also changes. Dynamic recrystallization of ferrite does not occur when amount of deformation is less than 40%, as shown in Fig.6. The possibility of the dynamic recrystallization of ferrite was eliminated by the method of reducing amount of deformation. However, the condition of generating dynamic recrystallization of DIF during deformation is not very clear.

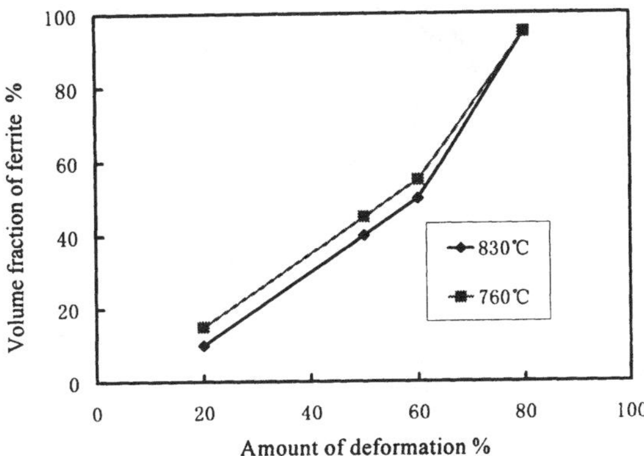

Fig.5 Volume fraction of DIF and reduction

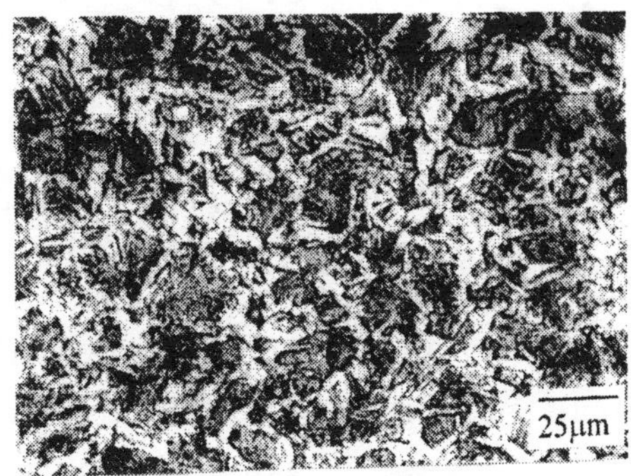

Fig.6 Microstructure of specimen deformed by 40% at 830 ℃

Figure 5 shows that, the fraction volume of ferrite is not change when the deformation was given over 80%. Also, when the deformation was given less than 20%, the DIF does not occur, as shown as Fig.7. A small amount of ferrite has precipitated during water quenching, as shown as in Fig.(7). Figure 8 shows stress-strain curve at deformation with 20% and 40%. The deformation stress not reaches its maximum value at 20% deformation that indicates the softer ferrite not occurs at same deformation temperature.

So this 20% of deformation is “critical amount of deformation” of DIF by this process. The minimum amount of deformation is required to the DIF. And the 80% of deformation should be maximum amount of deformation.

Fig.7 Microstructure of specimen deformed by 20% at 830 ℃

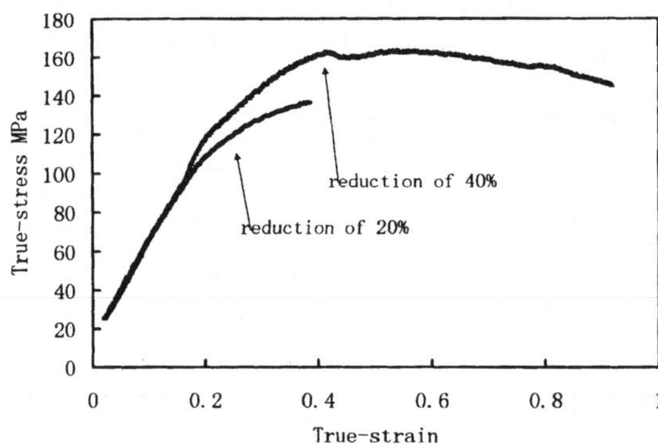

Fig.8 True stress-strain curves of specimen deformed at 830 ℃

Dynamic Recrystallization of Austenite The generally accepted fact is that in low carbon steels ferrite is considerably softer than austenite compared at same temperature. Because of this, peak value has occured in the stress-strain curve when the $\gamma \rightarrow \alpha$ transformation happened during deformation, as shown as in Fig.(3,8). Also, dynamic recrystallization of austenite has effect as such as $\gamma \rightarrow \alpha$ transformation.

For this problem, in the present study, the experiments were carried out on the Thermecmastor-z. The experimental condition were adopted as followed: heated to 1100 ℃ and held for 5min at same temperature, cooled to deformation temperature at 10 ℃/s and held 30 s at same temperature, deformed with 65% at different strain rate, water quenching after deformation. The experimental results show that dynamic recrystallization of austenite has very difficult occured during deformation of specimen by this process, as shown as in Fig.9. It would not contribute to deformation stress substantially.

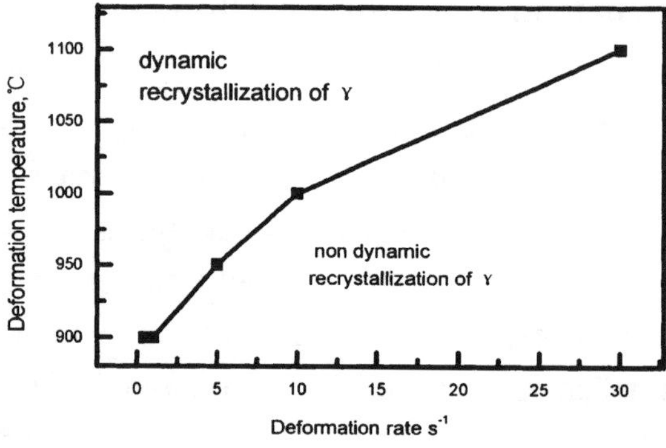

Fig.9 Schematic for the Q235 steel of dynamic recrystallization of austenite

Strain Rate Increasing strain rate will enhance the amount of deformation required to achieve DIF. Figure10 shows that in the experiment of the one pass deformation, the volume fraction of ferrite decreases with strain rate increasing when strain rates $\dot{\varepsilon}$ vary from 30 to 350s^{-1}. Morphology of ferrite changes from homogeneous and equiaxed to acicular, massive, as shown in Fig.11, which shows the feature of dynamic process of DIF. Increasing strain rate is equivalent to reducing amount of deformation and is not helpful to induced ferrite precipitation and grain refinement.

Feature of Ferrite In order to know whether water quenching has influence on ferrite formed after deformation, the test of water quenching was performed. The specimens holding for the same deformation time 0.05 s at same deformation temperature which as was adopted as shown in Fig.1, then water quenching.

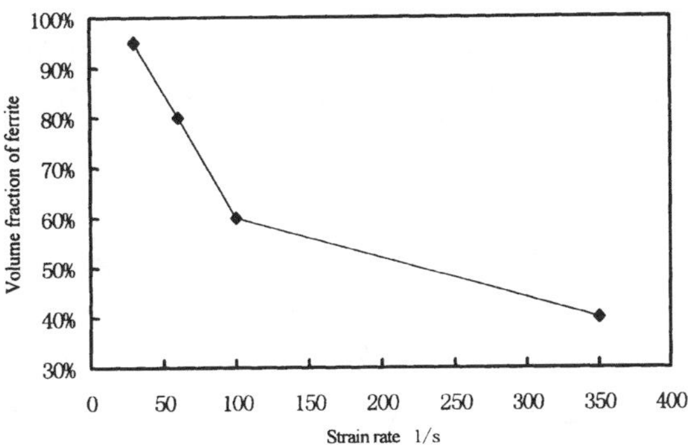

Fig.10 Strain rate and volume fraction of ferrite

The deformation was no exerted on specimens. The time 0.05 s equal same deformation time as adopted as shown in Fig.1. Figure 12 shows the microstructure of non-deformation specimens of water quenching. A small amount of massive and acicular ferrite has observed (see Fig.12). This transformation, however, is usually observed only in iron alloys with very low carbon content, e.g. The morphology of massive ferrite also has observed, as shown as in Fig. (6, 11b). This leads us to the reasoning that short-range diffusion of carbon during deformation. The other fact is that the pearlite is not in microstructure (see Fig.2) that has

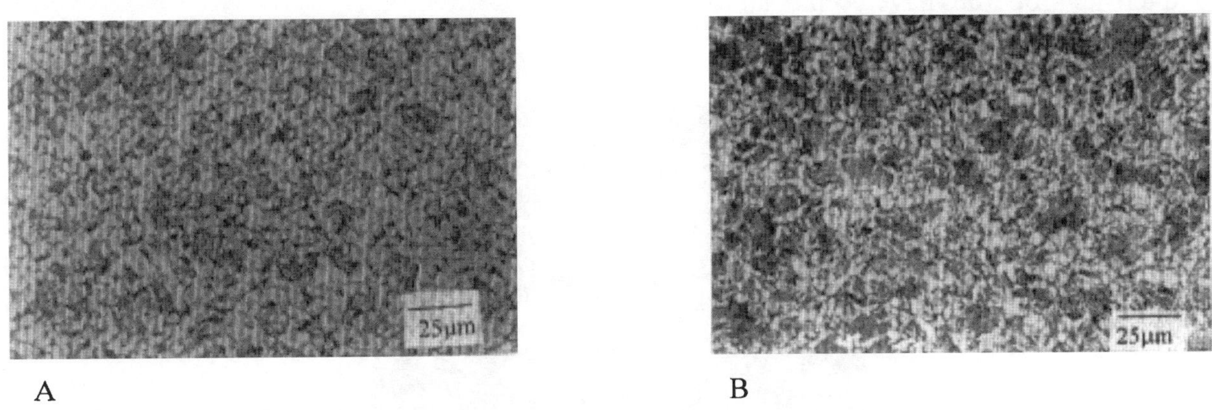

A B

Fig.11. Microstructure after deformed by 80% at 870 ℃, (A) $\dot{\varepsilon} = 30$ 1/s, (B) $\dot{\varepsilon} = 350$ 1/s

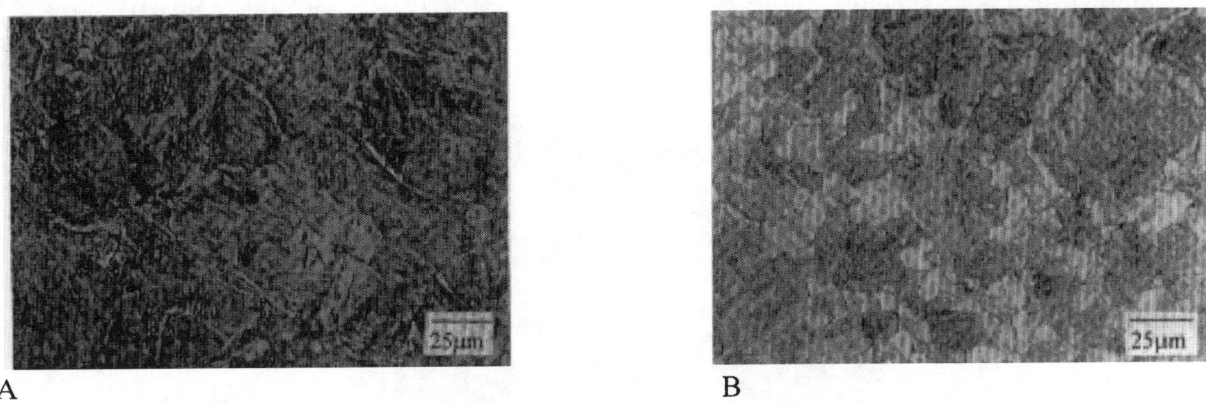

A B

Fig.12 Microstructure of specimens held by 0.05 s at different temperature and quenched by water. (A) at 850 ℃ (B) at 760 ℃

been substituted by cementite. Furthermore, the long-range diffusion of carbon usually accompanies $\gamma \rightarrow \alpha$ transformation. Hence, the above results show that the DIF is precipitated not by the long-range diffusion of carbon and that deformation does not enhance the eutectoid transformation temperature of pearlite. The effect of deformation is only enlarging the temperature range where ferrite is precipitated by deformation. In addition, the experimental results show that even if over 95% of ferrite content in the microstructure that is obtained. And morphology of ferrite is homogeneous and equiaxed (see Fig.1). If the ferrite is formed during cooling, its form should be massive form or acicular form. Because of this, the fine ferrite could be formed during deformation.

Reversal Transformation of Ferrite The DIF is usually obtained in austenite temperature zone. After the deformation, the ferrite would have automatically reversal transformation to austenite. To get the experimental evidence, the specimens were held for 0.5 s after deformation with 80% and then quenched by water (see Fig.13). Figure 14 shows the microstructure of specimen deformed by 80% at same temperature 810 ℃ and direct quenched by water during deformation. Their strain rate was given 30 1/s. Increasing the holding time led to decreasing amounts of ferrite and increasing size of ferrite, as can be see from Figs 13 and 14. The result shows the ferrite was unstable above A_{r3} and stable in the range $A_{r3} - T_s$.

Cooling Rate Controlled cooling after rolling is one of the most effective ways to control ferrite grain size for steel Q235. The samples are heated at 900 ℃ for 30 s, cooled to deformation temperature, reduction of 34% and strain rate 10 s^{-1}, and cooled at different cooling rate after deformation. The result is shown in Table II. It can be seen that the grain size has been reduced to ~4 μm after accelerating cooling at 20 ℃/s and good tensile properties have been obtained.

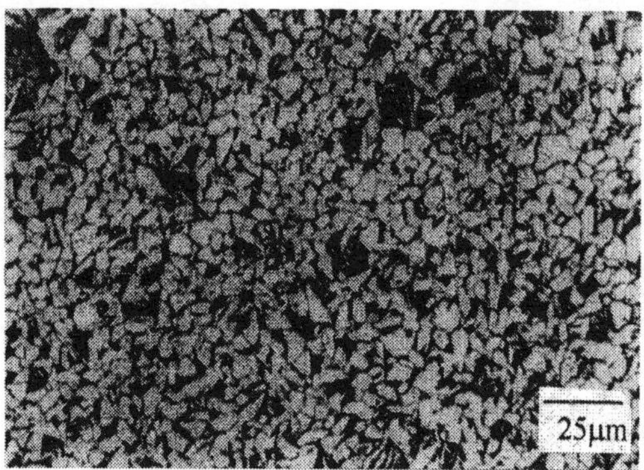

Fig.13 Microstructure of specimen after deformed by 80% at 810 ℃, hold 0.5 s at 810 ℃ and then water cooled

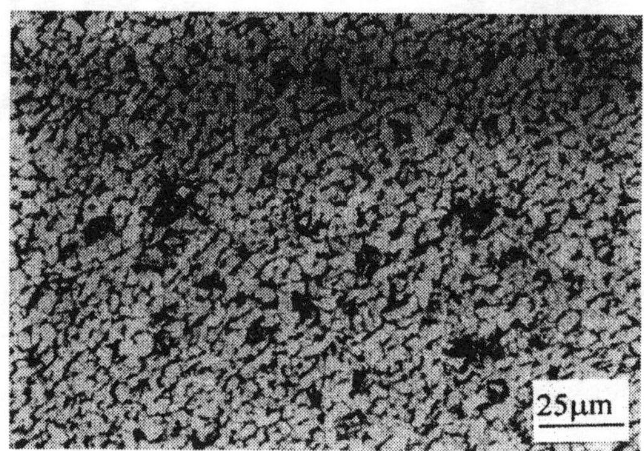

Fig.14 Microstructure of specimen deformed by 80% at 810 ℃ and water cooled during deformation

TableII. The effect of cooling rate on grain size and tensile property

№	Deformation temp ℃	Grain size μm	Cooling rate ℃/s	Microstructure	σ_s MPa	σ_b MPa	σ_s/σ_b	ψ %
1	850	16	3	F+P	340	495	0.69	60
2	850	8	5	F+P	355	510	0.70	58
3	850	8	10	F+P	370	525	0.70	57
4	850	4	20	F+P	430	545	0.79	64

CONCLUSIONS

Volume fraction of ferrite and recrystallization are dependent on reduction. Strain rate has influence on critical amount of deformation for obtaining DIF. Cooling process plays little effect on the volume fraction of ferrite, but it acts as a controlling fact on grain growth rate.

Homogeneous and equiaxed ferrite with 2μm for the common carbon steel Q235 can be obtained by controlling deformation parameters in the temperature range of $30+A_{e3}$~$A_{r3}-30$℃. The result is regarded as the joint effects of deformation induced ferrite transformation and dynamic recrystallization of ferrite.

The ferrite grain size for steel Q235 can be refined to 4.5~6μm and its yield strength reaches to 400MPa by means of large amounts of deformation near to A_{r3} and accelerated cooling after deformation.

ACKNOWLEDGMENTS

The authors thank the Ministry of Science and Technology of China for financial support to this project (G1998061501).

REFERENCES

1. Yada H. Matsumura Y., " A New Thermomechanical Heat Treatment For Grain Refining In Low Carbon Steel," Proc. Int. Conf. 'Physical metallurgy of thermomechanical processing of steel and other materials'. Tokyo, ed., Tamura I, Iron and Steel Institute of Japan, 1988, pp.200~207

2. B.Mintz, R.Abu-Shosha, M.Shaker. "Influence of Deformation Induced Ferrite, Grain Boundary Sliding, and Dynamic Recrystallisation on Hot Ductility of 0.1-0.75%C Steels,"Materials Science and Technology October 1993, Vol.9, pp. 907-914

3. Wung Yong Choo,Dong-Han Seo,Sang Woo Lee. CAMP-ISIJ 1999,Vol.12, pp.369-372.

4. P.J.Hurley, P.D.Hodgson and B.C. Muddle, "Analysis and Characterisation of Ultra-fine Ferrite Produced During a New Steel Strip Rolling Process," Scripta Materilia, 1999,Vol.40,No.4, pp.433-438

5. Matsumura Y. and H. Yada, "Evolution of Ultrafine-grained Frrite in Hot Successive Deformation," Transactions ISIJ, 1987, 27, pp.492-498

6. B.Mintz, J.Lewis, and J.J.Jonas, "Importance of Deformation Induced Ferrite and Factors Which Control Its Formation," Materials Science and Technology, May 1997, Vol.13, pp.379-388

7. Li C-M, H.Yada, H.Yamagata, "In situ observation of gamma →algha transformation during hot deformation in an Fe-Ni alloy by an x-ray diffraction method," Scripta Materilia, Sept. 1998, 39, (7), 4 pp.963-967

ULTRAFINE FERRITE PRODUCTION BY STRAIN-INDUCED TRANSFORMATION

G.L. Kelly and P.D. Hodgson
School of Engineering and Technology, Deakin University
Pigdons Rd.
Geelong, Vic. 3217
Australia
Tel.: 61 3 5227 1251
phodgson@deakin.edu.au

Ultrafine Ferrite, Strain-induced Transformation, Hot Deformation, Torsion Testing

INTRODUCTION

The production of ultra-fine grained (1-3μm) steels is the subject of considerable international research. The appeal of ultra-fine ferrite (UFF) is the remarkable increase in strength and toughness that can obtained by decreasing the grain size. There are a number of possible mechanisms by which UFF can form, strain-induced transformation of austenite and dynamic recrystallisation of ferrite are the two which are most commonly investigated.

Previous work by one of the current authors has focussed on the production of UFF by the strain-induced transformation rolling (SITR) process.[1,2,3] Careful characterisation of the UFF surface layer of rolled strip included texture analysis and TEM.[3] The UFF grains were equiaxed and free of dislocations with carbides uniformly distributed at triple points and grain boundaries. This suggested that strain-induced transformation was the mechanism by which the UFF formed. Texture analysis showed that the ferrite grains had transformed from the austenite by the conventional Kurdjumov-Sachs relationship.[3] These results were compared to the deformation structures and texture of a Ni-Fe model alloy.[4] The deformation structure of the Ni-Fe alloy closely corresponded with the scale and features of the UFF. This pointed strongly to strain-induced transformation, not dynamic recrystallisation of ferrite, being the dominant mechanism by which UFF formed in this process.

UFF has been produced in laboratory tests including hot compression[5], hot torsion and rolling, although these methods usually use extreme conditions which are not easily translated to commercial processes. The work carried out using the SITR process identified the critical parameters as high undercooling, high strain rate and high shear strain.[1,6] Hot compression results also suggest that high undercooling and high strains are required.[5] The processing window of the production of UFF is not well defined. The present study uses hot torsion testing of a plain carbon steel to examine the effects of strain rate and deformation temperature on the production of UFF by strain-induced transformation.

EXPERIMENTAL

Hot torsion tests of 1020 grade plain carbon steel were performed at three different strain rates ($0.3s^{-1}$, $3.0s^{-1}$ and $10.0s^{-1}$) and temperatures between 600°C and 900°C. The samples had a gauge length of 20mm and a gauge diameter of 6.7mm. The strains used were as close as possible to the failure strains for each condition. Due to the range of strain rates and temperatures, these strains varied between 2 and 6. The schedule used in the tests is shown in Figure 1. Prior to the deformation, the samples were heated to 1200°C for three minutes to increase the austenite grain size to ~250μm. Immediately after deformation the samples were quenched by a water spray. An argon atmosphere was used during the heating and deformation to minimise decarburisation of the sample surfaces. The Ar_3 of this steel is approximately 660°C for the starting grain size and cooling rate used.

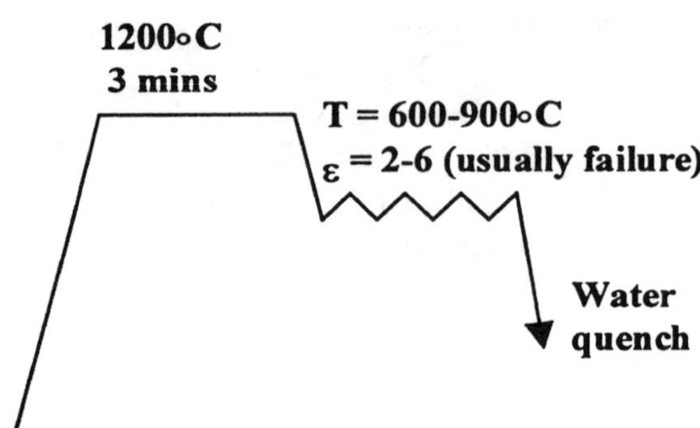

Figure 1: schedule of torsion tests

After quenching, the samples were sectioned and prepared for metallographic examination by polishing followed by etching in nital.

RESULTS

Flow curves

Comparing the curves obtained from the different strain rates, it can be seen that both the peak stress and the steady state flow stress increase as the strain rate increases (Figure 2). This is typical of flow curves of most materials.

The curves also show a more interesting effect. If we consider the flow curves obtained at a strain rate of $0.3s^{-1}$ (Figure 2a), it can be seen that at 900°C the curve is typical of deformation of austenite. This curve shows that the material slowly work hardened to the peak stress, then softened slightly before steady state flow was achieved. As the temperature was decreased, the peak stresses and strains of the curves increased as expected until the deformation temperature was close to the Ar_3. The curve from 650°C has a different shape, more typical of the deformation of ferrite, with initial high work hardening and softening. This result is not surprising since 650°C is just below the Ar_3 and some ferrite should have been present at that temperature to influence the shape of the flow curve.

As the strain rate increased to $3.0s^{-1}$ (Figure 2b), the flow curves at intermediate temperatures (750 and 800°C) began to display two distinct peaks. This "double peak" flow curve shape is even more strongly apparent when the strain rate was increased to $10.0s^{-1}$ (Figure 2c). At $10s^{-1}$, the majority of curves had either two peaks or an extremely broad, flattened peak.

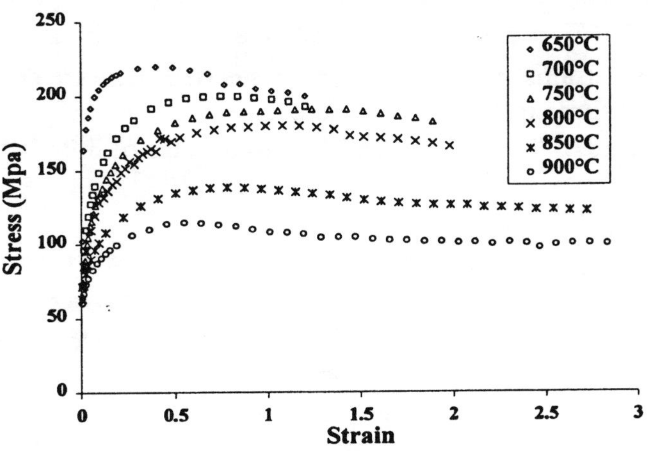

(a) strain rate = $0.3s^{-1}$

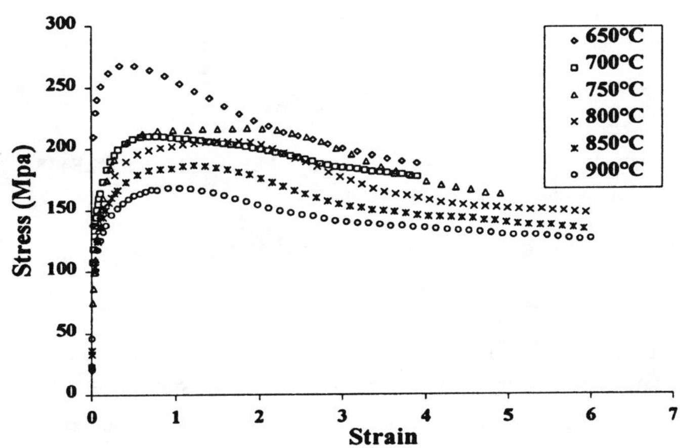

(b) strain rate = $3.0s^{-1}$

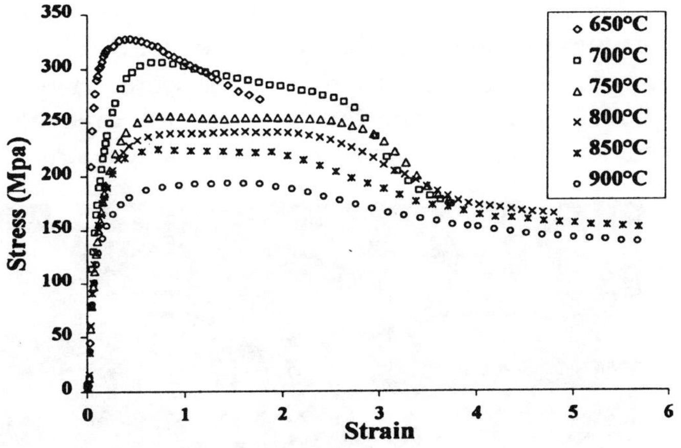

(c) strain rate =$10.0s^{-1}$

Figure 3: Flow curves from hot torsion deformation at various temperatures.

The data from the flow curves can be compared by plotting peak strain against the Zener-Hollomon parameter $Z = \dot{\varepsilon}\exp(Q_{def} / RT)$ (Figure 4). Q_{def} was taken as 300kJ/mol. It can be seen that at low values of Z (high temperatures), the peak strains increase as Z increases (temperature decreases). The data for all strain rates fall on a straight line. At some critical value the peak strain suddenly begins to decrease. This critical value is $6\times10^{14}s^{-1}$ for a strain rate of $0.3s^{-1}$, $5\times10^{15}s^{-1}$ for strain rates of $3.0s^{-1}$ and $10.0s^{-1}$.

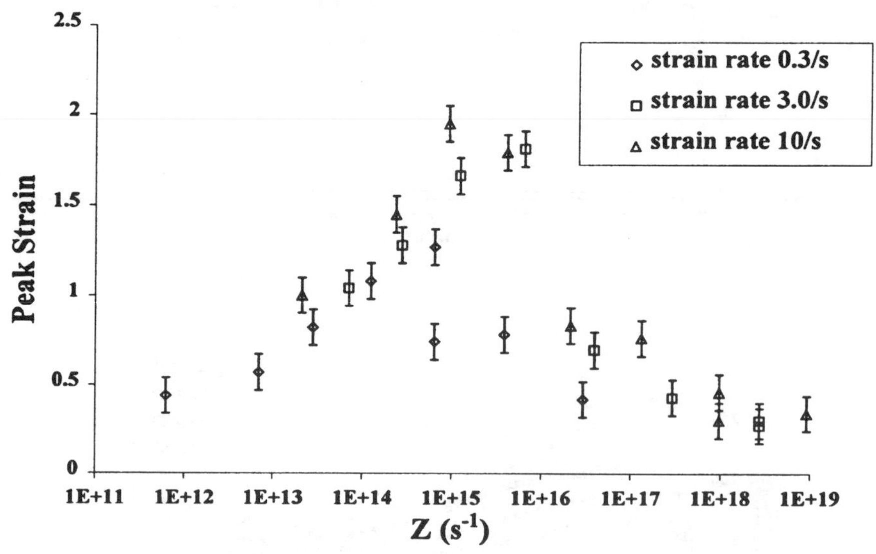

Figure 4: Peak strain vs Zener-Hollomon parameter for all strain rates.

Microstructures

The microstructures give more insight into the processes at work, particularly the formation of strain-induced UFF. UFF was not formed in any of the tests using a strain rate of $0.3s^{-1}$. Given the undercooling requirement for the formation of UFF, it was most likely to form at deformation temperatures close to Ar_3. At the lower deformation temperatures, the final microstructures of the samples strained at $0.3s^{-1}$ consisted largely of warm-worked pro-eutectoid and fine, acicular ferrite with some martensite.

As the strain rate was increased to $3.0s^{-1}$, UFF appeared in the sample deformed at 650°C, just below the Ar_3 temperature (Figure 5). Under these conditions, the volume of UFF formed was quite high and extended almost to the centre of the sample. The equiaxed ferrite grains were generally 1-3μm in size with carbides uniformly distributed throughout the ferrite. The ferrite grains are arranged in lines (running vertically in Figure 5). No UFF was formed at any other temperatures tested using a strain rate of $3.0s^{-1}$.

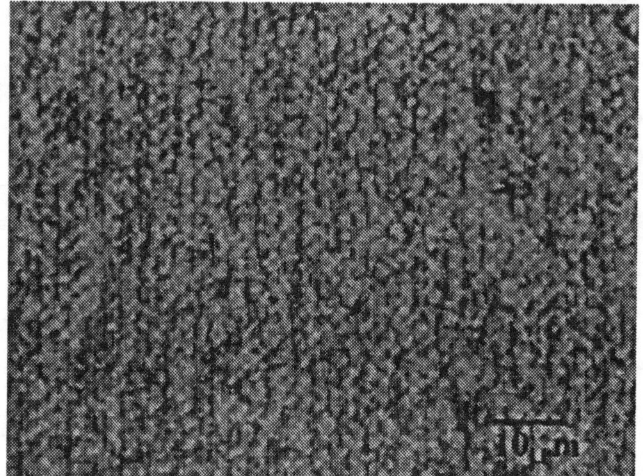

Figure 5: UFF formed in sample deformed at 650°C at a strain rate of $3.0s^{-1}$ and a final strain of 4.

At the highest strain rate tested, $10.0s^{-1}$, the temperature window for the formation of UFF widened and areas of UFF were seen in samples deformed at 650 and 700°C. For both temperatures the volume of UFF was much lower than in the sample deformed at $3.0s^{-1}$. The microstructures of the samples deformed at $10.0s^{-1}$ consist of patchy UFF mixed with other phases. The areas of UFF do not penetrate to near the centre of the sample. Figure 6 shows one area near the surface of the sample deformed at 650°C to a strain of 2 at two different magnifications. These microstructures consist of warm-worked pro-eutectoid ferrite (A) that formed prior to, or at an early stage of, the deformation. There is some acicular ferrite and there are also patches of fine, equiaxed ferrite grains (UFF) approximately 3μm in size (B). It appears that the strain-induced transformation to UFF has not proceeded to completion. The boundary of the patch of UFF (C) is adjacent to areas of bainite and/or martensite which would have formed from the austenite present during the quench.

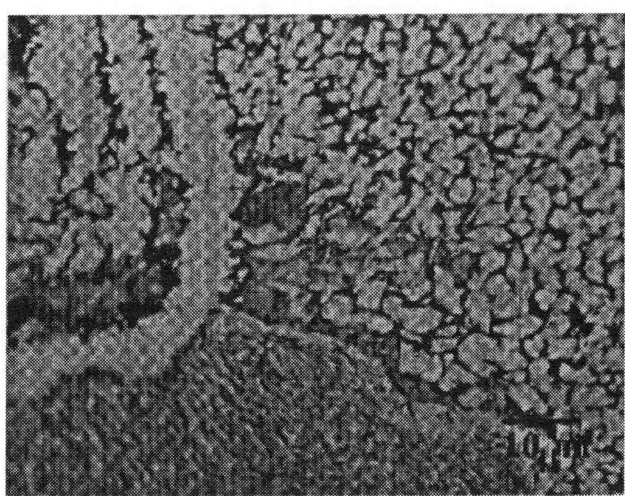

Figure 6a: Complex microstructure of sample deformed at 650°C and a strain rate of $10.0s^{-1}$ to a strain of 2.

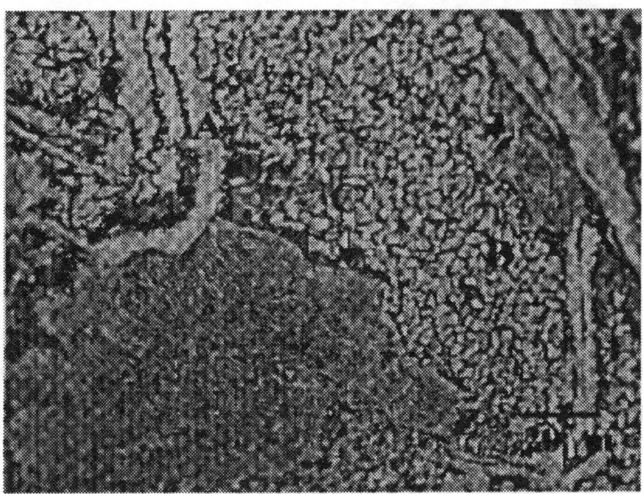

Figure 6b: Same area as in Figure 5a. A: warm-worked pro-eutectoid ferrite, B: UFF, C: transformation front.

DISCUSSION

Increasing the strain rate had two observable effects on both the microstructures and flow curves. Higher strain rates made it possible to form UFF at higher temperatures. The unusual flow curves with two peaks occurred over a wider temperature range and the two peaks were most clear at the highest strain rate. This prompts the question of whether the two effects are related.

The peak strains increased as temperature decreased (Z increased) for the high temperature deformations (Figure 4). The data from all strain rates fell on the same straight line. This trend changed at a critical value of Z which depended on the strain rate. This change in behaviour corresponded to the maximum stress of first peak exceeding that of the second. The "double peak" behaviour has been attributed to the presence of two phases in the material. The peak at low strains being explained by the transformation to ferrite or ferrite being present and that at higher strains being due to the austenite.[7]

While the two peaks at low temperatures are explained simply by the presence of two phases, it is not so clear why they are present at temperatures as high as 850°C (200°C above Ar_3) when the strain rate is increased to $10.0 s^{-1}$. It could be thought that the higher strain rate was driving the strain-induced transformation from austenite to ferrite at higher temperatures. This would result in UFF being present well above the Ar_3 and thus explain the two peaks. The microstructural examination showed that UFF was indeed formed at higher temperatures when the strain rate increased, however this effect only extended to 700°C. The optical microscopy has not yielded any real insight in to the origin of the double peak behaviour.

The linear arrays of ferrite grains (Figure 5) can be explained by the ferrite grains nucleating on the deformation structure in the austenite grains. When austenite is subjected to hot deformation, linear features such as shear bands and microbands are formed. It has been shown[3,4] that it is very likely that strain-induced ferrite nucleates on these features. In the subsurface region of rolled steel strip, individual linear arrays or "rafts" of ferrite can be seen[4]. Closer to the surface of the strip, the microstructure is similar to that in Figure 5, with the rafts impinging and no longer easily distinguished. In the ferrite arrays formed at a strain rate of $10.0 s^{-1}$ (Figure 6), these rafts are not clearly evident with the ferrite appearing instead to have formed in small pockets.

Early work on production of UFF by the SITR rolling process suggested that the critical factors for the formation of UFF included undercooling, high shear stress, high strain rates and deformation occurring near Ar_3.[1,6] The present study sheds new light on the importance of these factors when the deformation occurs by hot torsion.

Strain Rate

At a strain rate of $3.0 s^{-1}$ and a temperature of 650°C, UFF was formed almost all of the way through the radius of the sample. This implies that UFF formed at even the relatively low strains in the centre of the gauge. When the strain rate was reduced to $0.3 s^{-1}$, no UFF was formed. This is an interesting observation since recent micro-torsion work showed that UFF could be formed during deformation at strain rates as low as 0.01 and $0.09 s^{-1}$.[8] When the strain rate was increased to $10.0 s^{-1}$, UFF was formed at the test temperatures of 650 and 700°C but in limited amounts. The strain achieved in this case was 2, compared to 4 for $3.0 s^{-1}$. This could account for the decreased amount of UFF since strain is a critical parameter in the formation of UFF. The microstructure shows significant amounts of pro-eutectoid and acicular ferrite are present (Figure 6b). These would have limited the amount of austenite available to transform into UFF. In addition, at 700°C, $10.0 s^{-1}$ and a strain of 4, the amount of UFF formed was further reduced. Under the test conditions used here, it appears that there is an optimal strain rate for UFF formation. Only three strain rates were used in this study so further work is required to determine what this optimal strain rate is.

Undercooling
Undercooling increases the likelihood of intragranular nucleation of the ferrite and so promotes the formation of strain-induced ferrite. In these tests, the undercooling was increased by using a large austenite starting grain size to minimise the grain boundary area. As the strain rate increased, it appeared that the undercooling (temperature) requirement became less important to the formation of UFF. At a strain rate of $3.0s^{-1}$, UFF formed only at one temperature tested i.e. when the deformation was carried out at 650°C, very close to Ar_3. When the strain rate was increased to $10.0s^{-1}$, small amounts of UFF were formed at 650°C and 700°C. The test at 650°C and $3.0s^{-1}$ to a strain of 4 was by far the most successful at producing large volumes of uniform UFF. Again, further testing is required to find the optimal deformation temperature for the production of large amounts of UFF.

Time
Previous work had suggested that UFF would be more easily formed as undercooling, strain and strain rate were increased. The partial strain-induced transformation to ferrite that was observed at a strain rate of $10.0s^{-1}$ (Figure 6) suggests that there is a genuine processing window with upper and lower limits on the strain rate and/or deformation time. The strain-induced austenite-to-ferrite phase transformation, although fast, requires more time than that allowed when the deformation time is reduced by increasing the strain rate to $10.0s^{-1}$. More understanding of the dynamics of the strain-induced transformation is required before the optimal time of deformation and strain rate can be determined. The transformation was arrested in the samples deformed at $10.0s^{-1}$ when they were water quenched immediately after deformation. This raises the question of whether the strain-induced transformation would have progressed if there had been a time delay between the deformation and the quenching or if higher strains had been achieved and the deformation time extended.

CONCLUSIONS

Increasing the strain rate to $10.0s^{-1}$ produced two peaks in the flow curves at temperatures between 700 and 850°C. At this strain rate, UFF was formed at 650 and 700°C. It has not yet been determined whether these two observations are linked.

The formation of large volumes of UFF depends on "fine-tuning" of the thermomechanical parameters. Increasing the strain rate widens the temperature range over which UFF can be formed during deformation but decreases the volume of UFF formed by the conclusion of deformation. Too low a strain rate, in this case $0.3s^{-1}$ resulted in no UFF being formed.

ACKNOWLEDGEMENTS

This work is supported by the Australian Research Council (ARC). The assistance of Mr. John Whale, Ms Michelle Scott and Miss Rosemary De Castella with torsion tests and metallography are also acknowledged.

REFERENCES

1. P.D. Hodgson, M.R. Hickson, and R.K. Gibbs, "The Production and Mechanical Properties of Ultrafine Ferrite," Mater. Sci. Forum, Vol. 284-286, 1998, pp. 51-63.

2. M.R. Hickson, and P.D. Hodgson, "Effect of Pre-roll Quenching and Post-roll Quenching on Production and Properties of Ultrafine Ferrite in Steel," Mater. Sci. and .Tech., Vol. 15, 1999, pp. 85-90.

3. P.J. Hurley, P.D. Hodgson and B.C. Muddle, "Analysis and Characterisation of Ultra-fine Ferrite

Produced During a New Strip Rolling Process," Scripta Mater., Vol. 40, No. 4, 1999, pp. 443-438.

4. Hurley P J, "Production of Ultra-Fine Ferrite During Thermomechanical Processing of Steels", Melbourne, Australia, PhD Thesis, Monash University, submitted Nov. 1999.

5. Y. Adachi, T. Tomida and S. Hinotani, "Ferrite Grain Size Refinement by Heavy Deformation during Accelerated Cooling in Low-carbon Steel," Tetsu-To-Hagane, Vol. 85, No. 8, 1999, pp. 620-627.

6. P.D. Hodgson, M.R. Hickson and R.K. Gibbs, "Ultrafine Ferrite in Low Carbon Steel," Scripta Mater., Vol. 40, No. 10, 1999, pp.1179-1184.

7. O. Balancin, W.A.M. Hoffman and J.J. Jonas, "Influence of Microstructure on the Flow Behavior of Duplex Stainless Steels at High Temperatures," Met. Trans. A, Vol. 31, No. 5, 2000, pp.1353-1364.

8. H. Yada, C-M. Li and H. Yamagata, "Dynamic $\gamma \rightarrow \alpha$ Transformation During Hot Deformation in Iron-Nickel-Carbon Alloys," ISIJ Int., Vol. 40, No. 2, 2000, pp. 200-206.

DEVELOPMENT OF ULTRAFINE GRAIN STEELS USING THE MAXSTRAIN® DEFORMATION SIMULATOR

Wayne C. Chen, David E. Ferguson and Hugo S. Ferguson

Dynamic Systems Inc.
P.O. Box 1234, 323 Route 355
Poestenkill, NY 12140
Tel: (518) 283-535-
Fax: (518) 283-3160
E-mail: Chen@gleeble.com

Key Words: Ultrafine Grain Steel, Thermomechanical Processing, Multi-axis Maxstrain Technology

INTRODUCTION

Several strengthening mechanisms have been used either separately or in combination to increase the strength of ferrite matrix steels, such as 1) solid solution hardening by substitute atoms of Mn and Si, or interstitial atoms of carbon, nitrogen, etc., 2) dislocation hardening, 3) precipitates/multiphase hardening; and 4) grain refinement. Unfortunately, some of them have a negative effect on weldability, formability, as well as, hot dip gavanizability. Grain refinement is one way that could increase the strength with no loss of ductility, and at some temperature may even exhibit superplasticity.

Ultrafine ferrite (UFF) microstructure development in steels especially in structural steels has increased all over the world from Japan [1,2,3], Korea [4,5], China [6,7] to Australia [8,9,10], to North America [11,12,13], to Europe [14,15,1617]. One approach is to use heavy deformation in the austenite and ferrite two phase region. Yada et al [2] claimed that a ferritic steel can form an ultra fine grain size in the temperature range of Ar_1 +50°C to Ar_3 +100°C under a reduction of more than 50% deformed in less than a second. Najafi-Zadeh[12] et al obtained an ultra fine grain size of 1-2 microns for IF steels with warm rolling in the single phase ferrite region. Hudgson et al [8] has obtained one micron ferrite grain structures during industrial strip rolling. Although only the surface layers with 30-40% of the strip thickness has the UFF structures after rolling with the central part being conventional grain structure, the UFF structure has been obtained for 5 different grades of steels from low to high carbon to HSLA steels. Recently, Hodgson[9] claimed that they have obtained whole thickness UFF structure, in their several trials with pre-roll quenching technique to achieve the UFF from fine austenite[9]. The achievement of the ultra fine grain structure was considered to occur due to sufficient under-cooling at both the surface and the center as well as the favorable chemistry of high maganese and niobium. Researchers [14,15,16,17] in Belgium have developed ultra-fine grained dual phase steels and TRIP steels. In their dual phase steels, a

® MAXStrain is a registered trade mark of Dynamic Systems Inc, New York, USA.

small percentage (0.03%Nb) was added into the C-Mn steels to increase the temperature band to more than 100°C from no recrystallization temperature (Tnr) to the Ar_3 temperature, therefore, all the 7-stand finish rolling can be realized with the temperature band. The steel was rapidly cooled (~40°C/s) after finish rolling and coiled below the Martensite start (Ms) temperature (<200°C) to transform the retained austenite to Martensite. Moreover, an ultra fast cooling technology has been developed at Cockerill Sambre, so that cooling rates in the range of 400°C/s to 650°C can be obtained on a 3mm thick strip, which is 5-8 times higher than classical laminar cooling rates. In their trip steel development, although grain refinement was not their main objective but a low silicon hot rolled TRIP-assisted multiphase steel, fine grain structure has also been achieved. More alloying elements such as Ti, Nb, and V were present in the steels. Therefore, a much high tensile strength was obtained. This indicates that the UFF structure could essentially be obtained in any steel with ferrite dominant structure. Yada et al [2] claimed that they can obtain UFF (< 4microns) in any steel with carbon content from 0.02% to 0.3%C and Mn from 0.1% to 2%Mn without four alloying elements of Nb, Mo, Ta, W, and with the total alloy elements less than 3%. Also, the UFF structure can be obtained with any type of industrial processes such as forging, extrusion and rolling including, strip rolling, plate rolling and wire rod rolling.

Researchers at Nippon Steel have published results of experiments related to hot rolling of thicker plates, which had been surface quenched at an intermediate stage of rolling[18]. These plates were ~25mm thick and showed ultra-fine grains only in the surface layer of 3 - 4 mm deep on both the top and bottom surfaces. They ascribed this UFF structure to dynamic recrystallization of ferrite, which is debatable.

One of the reasons why it is difficult to obtain a full thickness ultra-fine grain structure through the thickness of the strip/plate is due to the fact that, in flat roll processes, the strain is non-homogeneous through the thickness with the strain at the surface normally greater than in the center, and the non-uniformity becomes more severe as the thickness of the workpiece increases. Therefore, it is very difficult to obtain homogeneous deformation throughout the thickness by currently used methods.

With hot rolling in the low temperature ferrite and austenite two phase region, the rolling loads increase significantly. One of the objectives of this study is to roll steels in the high temperature single phase austenite region. Torizuka[3] et al claimed that they have obtained no larger than 3 micron grain steels with the deformation temperature above Ar_3, however, they obtained the ultra fine grains with plane strain type of tests using flat anvils in the laboratory.

Some researchers believed that the mechanisms of generating UFF grain size is due to deformation induced phase transformation [4,8] , while Yada et al[2] proposed that the mechanism to form the UFF is both dynamic transformation and dynamic recrystallization of ferrite. Thus no alloying elements such as Nb, Ta, Mo and W should be added to the steel, since they are recrystallization retarding elements. Some other alloy elements can be added to steels to form fine submicron precipitates at the austenite grain boundaries to increase the ferrite nucleation sites and stop grain growth such as V and nitrogen. It has been found that the number of ferrite grains is doubled with the precipitation of VN in a vanadium and Nitrogen bearing steel [4]. Yada et al also found that high purity steels (less than 10ppm of S, P, Al, and N) are easier to obtain UFF with less reduction than normal steels. However, they claimed that they cannot obtain UFF in steels with the carbon content of less than 0.02%, since the grain growth becomes prominent after hot rolling. When the carbon content is larger than 0.3%, second phases other than ferrite increase, and it is difficult to get more than 70% ferrite. Choo and Lee [5] found that the ferrite grain size decreases dramatically from conventional grain size of around 10 microns to 1-2 micron when the effective strain increase to 1, however, the ferrite grain size decreases much more slowly when the strain increases from 1 to 4. Yada et al found that a critical cooling rate of 20°C/s is important to obtain UFF, and to suppress grain growth after working, a larger cooling rate is desirable. Other Japanese researchers [1] found that the ferrite grain size decreases as the cooling rate increases, while the effect of cooling rate decreases when the cooling rate is more than 10°C/s.

In this study, a newly developed Multi-axis deformation system, the Maxstrain system, is used to produce ultra-fine grain steels with no addition of any expensive alloying elements. With the above mechanism of deformation induced phase transformation and dynamic recrystallization of ferrite for formation of ultra-fine grains, a heavy reduction must be applied. However, conventional hot compression testing methods can hardly achieve a strain of greater than 2. With the Multi-axis Deformation system, a strain of greater than 10 is easily achieved before the material fails.

THE MAXSTRAIN TECHNOLOGY

Severe-plastic deformation may apply to a strain of 4 or more. Often used deformation strains, such as in rolling, are the order of 0.3 to 0.4 per roll stand. 10 or more such deformations are required to achieve a strain of 4. Since during deformation material flows in all unrestricted directions, most deformation processes (excepting the above mentioned methods) have very little material left in the work zone after a strain of 4 or more has been introduced. If the deformation is the plane strain type (similar to flat rolling), less than 2% of the thickness remains after a strain of 4 is introduced.

The MAXStrain Technology uses deformation in two axes, while fully restraining the third axis. With this procedure strains of 10 or more are often possible before material failure. For a strain of 10, more than 80% of the original section remains and can be used for subsequent test work.

The procedure used to obtain very large strain is comprised of a rotation assembly as shown in Fig. 1 mounted in a thermomechanical system. The rotation assembly is comprised of a very rigid frame in which the material is mounted for rotation around and perpendicular to the axis of the thermomechanical single axis machine. The grips at each end of the specimen are electrically insulated from one another allowing the passage of high current through the specimen. The current through the specimen is controlled in a thermal servo system permitting rapid and accurate temperature control. The thermomechanical system provides accurate servo hydraulic control of two independent rams, one on each side of the specimen as shown by the arrows depicting the deformation direction (Fig.1). Deformation at the middle of the specimen cannot cause elongation of the specimen because of the restraint at the ends of the specimen. During deformation at the center the material must flow sideways of the specimen. After the first deformation the specimen is wider and thinner. The specimen is then rotated 90 degrees and deformed again causing the material to flow sideways at 90 degrees to the flow of the first deformation. Fig. 2 shows two specimens, where one has been deformed without restraint and the other with full restraint. Without restraint the specimen elongates when deformed at the midspan (Fig. 2A). With full restraint the specimen increases in width with no increase in length (Fig. 2B).

Rotating the specimen 90 degrees and deforming again strains the same material with very little loss of material in the deformation zone.

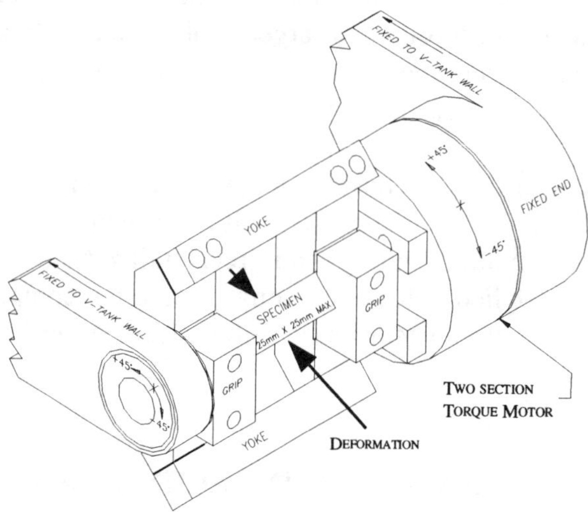

Figure 1 Schematic drawing of the rotation assembly of the MAXStrain System

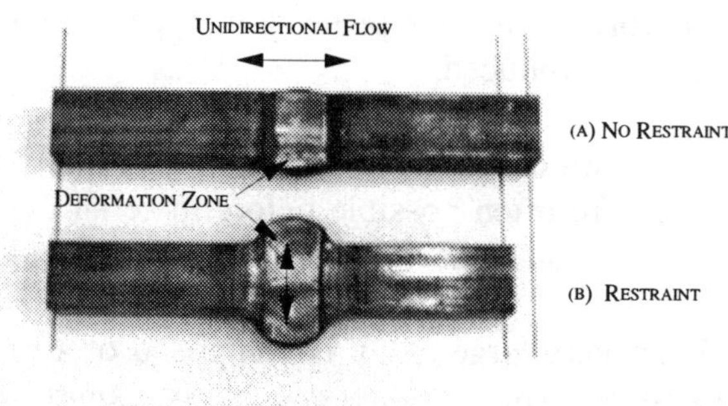

Figure 2 Unidirectional flow with no restraint (A) and restraint (B)

The two axis MAXStrain system has a maximum specimen size of 25 mm square and 200 mm in length. Consecutive deformation cycles may be as short as 0.5 seconds. After inducing severe strain in the midspan of the specimen, mechanical test specimens are machined for subsequent test work. The deformation can be done at any programmed temperature. Controlled heating and cooling rates are part of the program. The time between deformations is also programmed along with the strain and strain rate of each deformation. Some aluminum materials are deformed at room temperature, but the majority of the high strain work is done at elevated temperatures.

EXPERIMENTS AND DISCUSSION

Several tests were conducted using the Maxstrain system on specimens with different thermal mechanical history as listed in Table 1. AISI 1018 plain carbon steel was used as the test material. Each specimen was 15mm x 15mm x 185mm with two key slots machined at both end of the specimen to restrain the material flow in specimen axis direction. It was heated to 1250°C and soaked at the temperature for 10 minutes. The specimen was rapidly cooled to deformation temperature using the ISO-Q internal end quenching technique. The cooling rates from soaking temperatures are not less than 40°C/s. In this study, the rough rolling passes were ignored. The specimen was deformed in 5 hits with a 90-degree rotation after each hit. The interpass time between each hit is less than 1.0sec. The strain rates for all the conditions are near 10/s, and the cooling rates after hot working are not less than 80°C/s. With a total nominal strain of ~1.5, the strain of each pass was the same except for Test #5 in Table 1. A deformed sample after 5 hits is shown in Figure 3.

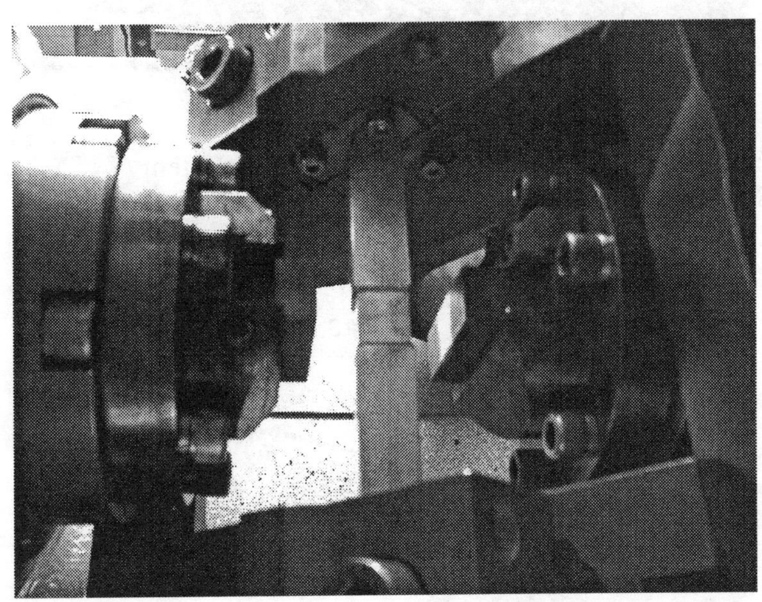

Figure 3 A multi-axis deformed specimen after 5 hits on the Maxstrain System

Each test condition was repeated. One deformed sample was cut longitudinally for microstructure examination using normal polishing and etching procedures. A typical ultra-fine grain structure achieved from AISI 1018 plain carbon steel is shown in Figure 4. The other deformed specimen was machined to a tensile testing sample for further room temperature mechanical property testing. The stroke rate in tension was 0.02mm/s. A typical room temperature stress vs. strain curve is shown in Figure 5 with ultrafine grain structure for AISI 1018.

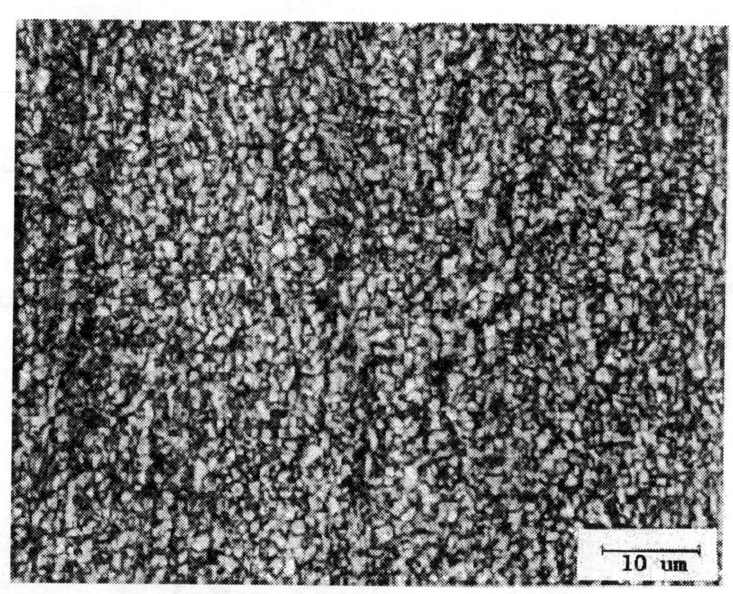

Figure 4 1μm grain microstructure obtained using the MAXStrain System for AISI 1018 plain carbon steel

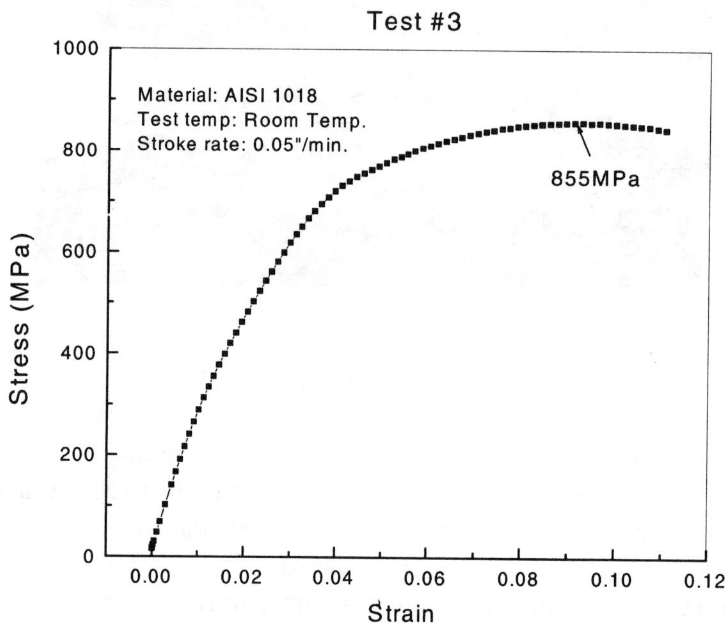

Figure 5 A typical stress vs. strain curve of room temperature tensile testing

Some typical properties for all the tests conducted are listed in Table 1.

Table 1 Mechanical properties of UFF Structure of AISI 1018 plain carbon steel

Test #	No. of Passes	Thermal history	Nominal Strain	UTS (MPa)	R. A (%)
1	5	1250C → 875C → 835C → 795C → 755C→715C → room temp.	1.45	862	10.9
2	5	1250C → 835C → 810C → 785C → 760C→735C → room temp.	1.42	882	10.0
3	5	1250C → 800C → 780C → 760C → 740C→720C → room temp.	1.40	855	14.4
4	5	1250C → 835C → 810C → 785C → 760C→735C→ 400C (5min)→ room temp.	1.41	786	11.4
5	5	1250C → 835C → 810C → 785C → 760C→735C → room temp.	1.44	807	9.4

Based on the tests conducted, a typical thermomechanical history to achieve ultrafine grain steels can be summarized as shown in Figure 6. It is believed that the rough rolling schedule above the no-recrystallization temperature (Tnr) is not critical. However, the cooling rate after rough rolling to the deformation temperature in the first stand is important. A high cooling rate depresses the austenite to ferrite phase transformation, or the Ar3 temperature. A nominal strain of 1.4 accumulated below the Tnr is necessary to achieve enough mechanical energy within the deformation zone. A strain of ~2.5 or more may be needed to achieve the ultrafine grain size of ~1um for plain carbon steels. With severe deformation below Tnr, the accumulated strain could induce dynamic phase transformation, or so called deformation induced phase transformation. A high post-deformation cooling rate is also beneficial, since it could under-cool the austenite and stop ferrite growth after phase transformation.

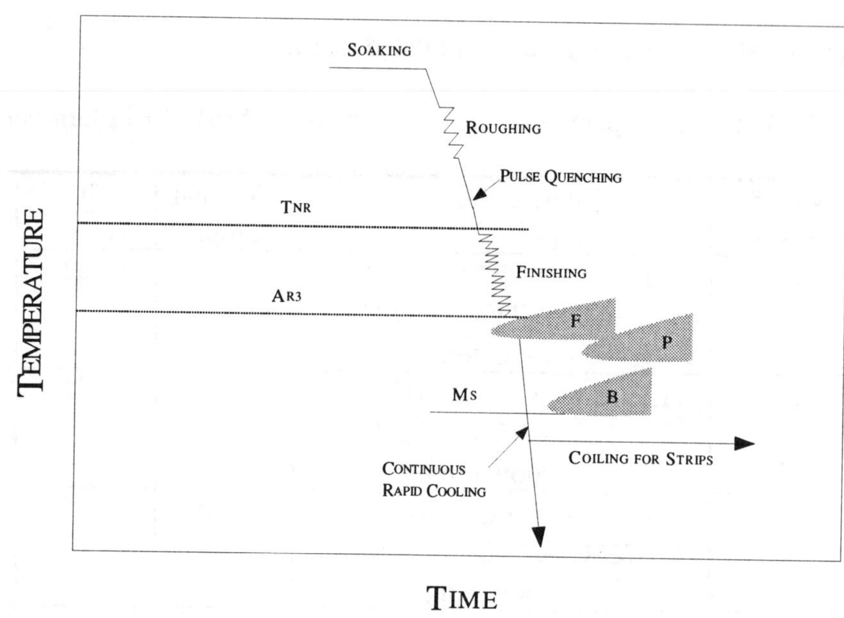

Figure 6 A thermomechanical path to produce ultrafine grain steels

SUMMARY

Multiaxis deformation Maxstrain technology can achieve severe plastic deformation in metals and alloys under precise control of processing conditions such as temperature, strain, and strain rate. Both rough and finish rolling processes can be simulated in a single test due to its unique feature of unlimited strain. With retained bulk deformation volume, the deformed sample can be further machined into tensile testing specimens for subsequent mechanical property measurement. This can greatly reduce the development time and cost of a new processing schedule from laboratory to plant floor.

The tests conducted using the Maxstrain System have shown that in order to produce ultrafine grain steels with no addition of any expensive alloying elements such as Ti, Nb, V etc., the production method may need to consider the following:

1) For a single pass hot working process, a fast cooling rate of more than 40°C/s before hot working is preferred to achieve a super under-cooling before ferrite phase transformation. Hot working is conducted above Ar_3 temperature.

2) With multiple pass hot working processes, more than 65% reduction must be achieved between the no recrystallization temperature, Tnr, and Ar_3. The interpass time between each rolling pass should be less than 1 sec.

3) Cooling rates of 80°C/s or more should be applied to the steel after hot working to generate more ferrite nucleation sites during phase transformation. With the development of an ultra fast cooling technology at Cockerill Sambre in Belgium [14,15,16,17], this type of cooling rate should be readily achievable.

4) To produce multiphase strip steels with a combination of high strength and good elongation, the above discussed accelerated cooling process with a low coiling/holding temperature should be used. This produces a high percentage of ferrite volume and transforms the retained austenite to bainite or Martensite to give a good work hardening coefficient and low yield to tensile strength ratio. The elongation in ultra fine grained high strength steels is mainly dependent on the ferrite volume fraction which is independent on the coiling/holding temperatures.

REFERENCES

[1] Japanese Researchers, "Ferrite Grain Refinement of a Si-Mn Steel through Heavy Deformation in Unrecrystallized Austenite Region Followed by Controlled Cooling (in Japanese)", CAMP-ISIJ, Vol. 10 (1997) - 1380.

[2] H. Yada, Y. Matsumura and K. Nakajima, "Ferritic Steel Having Ultra-fine Grains and a Method for Producing the Same ", United State Patent No. 4,466,842, August 21, 1984.

[3] Shino Torizuka, et al., "Ultra-fine Texture Steel and Method for Producing it ", European Patent No: EP 0 903 412 A2, Mar. 23, 1999.

[4] Wung Yong Choo, "New Challenge to Develop Ultra Fine Grain Steel with 1 micron Grain Size (in Korean)", J. Kor. Inst. Met. & Mater., Vol. 36, No.11, (1998), pp. 1945-1958.

[5] W.Y. Choo and S. W. Lee, "High Performance Steels for Structural Application", Cleveland, OH, ASM Annual Meeting, (1995), 117.

[6] China 973 Project Workshop, Ministry of Science and Technology, Beidahe, China 1999.

[7] H.R. Hou, Q.Y. Liu, Q.A. Chen, and H.Dong, "Grain Refinement of Microalloyed Steel Through Heavy Hot Deformation and Controlled Cooling", ACTA Metallurgica SINICA, Vol. 13, No. 2, pp.508-513, April 2000.

[8] P.D. Hudgson, M.R. Hickson, and R.K. Gibbs, "The Production and Mechanical Properties of UltraFine Ferrite, Materials Science Forum (Switzerland)", Vol. 284-286, 1998, pp. 63-72.

[9] M.R. Hickson and P.D. Hodgson, "Effect of Preroll Quenching and Post-roll Quenching on Production and Properties of Ultrafine Ferrite in Steel", Materials Science and Technology, Vol. 15, January 1999, pp. 85-90.

[10] P. J. Hurley, P.D. Hodgson, and B.C. Muddle, "Analysis and Characterization of Ultra-fine Ferrite Produced during A New Steel Strip Rolling Process", Scripta Materialia, Vol. 40, No. 4, pp. 433-438, 1999.

[11] T.G. Langdon, M. Furukawa, Zenji Horita and M. Nemoto, "Using Intense Plastic straining for High-Strain-Rate Superplasticity", June 1998, JOM, pp. 41-45.

[12] A. Najafi-Zadeh, J.J. Jonas and S. Yue, "Grain Refinement by Dynamic Recrystallization during the Simulated Warm-rolling of Interstitial Free Steels", Metallurgical Transaction A, Vol. 23A, No.9, Sept. 1992, pp. 2607-17.

[13] A. Najafi-Zadeh, J.J. Jonas and S. Yue, "Interstitial Free Steels and Method Thereof", US patent: 5,200,005, April 6, 1993.
[14] A. Schmitz, J. Neutjiens, J.C. Herman, and V. Leroy, "New Thermomechanical Hot Rolling Schedule for the Processing of High Strength Fine Grained Multiphase Steels", 40th MWSP Conf. Proc., ISS, 1998. pp. 295- 309.
[15] J. Neutjiens, Ph. Harlet,, Th. Bakolas, P. Cantinieaux, "Processing and Properties of a NEW Hot Rolled High Strength Fine-Grained Multi Phase Steel", 40th MWSP Conf. Proc., ISS, 1998. pp. 311- 321.
[16] K. Eberle, Ph. Harlet, P. Cantinieaux, M. Vande Populiere, "New Thermomechanical Strategies for the Realization of Multiphase Steels Showing a Transformation Induced Plasticity (TRIP) Effect", 40th MWSP Conf. Proc., ISS, 1998. pp. 251- 258.
[17] D.B. Santos and R.A.N. M. Barbosa, "Microstructural and Mechanical Properties of Controlled Rolled Triphase (ferrite, Banite and Martensite) Steels", Materials Forum, Vol. 14, No. 4, 1990, pp. 258-63.
[18] H. Mabuchi, T. Ishikawa and Y. Nomiyama, in "Accelerated cooling/direct quenching steels", ed. R. Ashfahani, 43,1997, Materials Park, OH, ASM International.

ULTRA FINE GRAIN STEELS AND THEIR PROPERTIES

Yuqing WENG
Chinese Society for Metals
No.46, Dongsixi Dajie, Beijing 100711, P.R.China
Han DONG
Central Iron and Steel Research Institute
No.76, Xueyuan Nanlu, Beijing 100081, P.R.China
Tel: +86-10-6218-2791
E-mail: donghan@public.bta.net.cn

Key Words New Generation Steels, Plain Low Carbon Steel, High Strength Low Alloy Steel, Structural Alloy Steel, Ultra Fine Grain, Processing, Property

INTRODUCTION

Plain low carbon steel, high strength low alloy steel and structural alloy steel are the most widely used three categories of steels, which occupies totally about 70 percent of the total steel consumed. Nowadays, the yield strength of plain low carbon steel is at 200MPa class, the yield strength of high strength low alloy steel is at 400MPa class, and the ultimate tensile strength of structural alloy steel is at 800MPa class. In order to meet the increasing needs from economic and social developments in the future, the research on new generation steels with higher strength and longer duration are now under the way in China[1]. The targets are to raise yield strength of plain low carbon steel from 200MPa class to 400MPa class, to raise yield strength of high strength low alloy steel from 400MPa class to 800MPa class, and to raise ultimate tensile strength of structural alloy steel from 800MPa class to 1500MPa class. Bearing this in mind, super cleanness, high homogeneity and ultra fine grain are the three main characteristics of the new generation steels to reach at the goal of higher strength and longer duration.

It is well known that there exist many mechanisms to strengthen the steel, but the grain refinement is the only method to improve both strength and toughness simultaneously. People have paid great effort in searching the effective method of grain refinement for steels in the long run. And the characteristics of ultra fine grain steels are always attractive to metallurgists [2-4]. At present, the smallest ferrite grain in plain low carbon steel strips is of 20μm in size in industry production scale, and the finest ferrite grain in plain low carbon steel rebar is of 30μm in size. For high strength low alloy steel strips produced in industry, the smallest ferrite grain in is of 10μm in size. The prior austenite grain in structural alloy steels is of 10μm in size for conventional heat treatment. In order to attain the goal of high strength, grains in plain low carbon steel need be refined to below 5μm. Grains in high strength low alloy steel should be refined to about 1μm. And prior austenite grains need to be refined below 5μm in structural alloy steel. In this paper, the methods for grain refining in the steels will be investigated. The microstructure and mechanical properties of ultra fine grain steels will be described.

In recent years, researchers studied the way to refine grains in plain carbon steel [5]. Many studies were performed on the grain refinement in high strength low alloy steels [6-8]. It was showed by Japanese study in Ultra Steel Project that ultimate tensile strength of low carbon C-Mn steel could reach at 700MPa class when ferrite grains were refined to about 1μm [9]. HSLA steel strip was studied to search for grain refinement in High Performance Structural Steel Project in Korea [10]. Through low temperature heavy deformation, ultra fine grain of 1μm in size could be obtained, which presented ultimate tensile strength of 600MPa. Although the grain has been refined to about 1μm in size, ultimate tensile strength in experimental steels is still below 800MPa. For structural alloy steel, grain growth inhibited by precipitation, repeated heat treatment and rapid heat treatment are some of the effective methods to refine austenite grains to about 10μm. This leads to the refinement of martensite blocks, which brings about improvement in strength, toughness, anti-delayed fracture property and fatigue strength. We will search for the effective methods to refine grains in the three categories of the steels.

EXPERIMENTAL MATERIALS AND PROCEDURES

PLAIN LOW CARBON STEEL

GRAIN REFINEMENT IN PLAIN LOW CARBON STEEL-Continuous casting billet of Q235 steel produced in Capital Iron and Steel Company （0.18%C-0.21%Si-0.60%Mn-0.016%P-0.02%S-0.008%Al-0.017%O-0.005%N-bal.Fe）was reheated, and then hot rolled and forged into 12mm round bars. The bars were machined into Φ10mm×100mm specimens. The specimens were heated, deformed and cooled in GLEEBLE-1500 to simulate the hot rolling and cooling processes for grain refinement. The specimens were machined to uni-axle tension coupons for strength and ductility evaluation.

HOT ROLLING STRIP IN INDUSTRIAL TRIALS-The investigated steel was SS400 （0.171%C-0.36%Mn-0.09%Si-0.013%P-0.013%S-0.025%Al-bal.Fe） produced by Baoshan Iron and Steel Company. The steel was melted in converter and continuously cast into slab of 250mm×1300mm×10020mm in size. The slab was soaked at 1523K, and then hot rolled through 2050 tandem into the steel strip of 3.0mm×1250mm in section size. The intermediate slab was of 42.8mm×1250mm in section size. For finish rolling stands from F1 to F7, temperature, reduction and strain rate change in the range of 1210-1073K、0.63-0.13% and 5.33-63.28s^{-1} respectively. Strip coiling temperature was in the range of 880-750K. Specimens were taken from strip at different position along the rolling direction to study the effect of coiling temperature on microstructure and mechanical property. Impact V-notch specimens were of 2.5mm×10mm×55mm in size from transverse direction.

HOT ROLLOING REBAR IN INDUSRIAL TRIALS-Continuous cast Q235 steel (0.14%C-0.19%Si-0.51%Mn-0.024%S-0.022%P-bal.Fe) billet from Tangshan Iron and Steel Company was taken as testing steel. Billet of 165mm×165mm in section size was hot rolled into experimental billets of 60mm×60mm in section size, which were soaked again and then hot rolled 11 passes into 12mm rebar. The experimental billets were soaked at 1273, 1323, 1373, and 1523K respectively, and then started to roll at 1073, 1123, 1173, 1223, 1273, and 1323K respectively, at corresponding finish rolling temperatures of 907, 927, 947, 1007, 1057, 1073, and 1113K. The cooling manners after hot rolling were air cooling, fan cooling and water cooling plus air cooling. Strength and ductility data can be obtained from tension test of rebars.

HIGH STRENGTH LOW ALLOY STEELS

In order to study the potential of grain refinement through deformation induced ferrite transformation, seven C-Mn steels microalloyed with niobium were prepared as the experimental materials. The range of carbon contents in the steels is from 0.003% to 0.160%. Hot forged slabs of 30mm in thickness were soaked at 1473K for 30min, and then hot rolled to 2mm strips in FUJI Φ300mm rolling mill. The rolling schedule was as follows: 30 percent reduction at 1273K for one pass, 30 percent reduction at 1173K for one pass, and 30 percent reduction of each pass at 1093K for three passes. Hot rolled strips were cooled in water. In order to simulate coiling of the hot rolled strip, hot rolled ultra low carbon microalloyed steel strips (0.003 percent carbon) were cooled at 20K/s from finish rolling temperature to 873, 773, 673K, and remained at the temperature for 60 minutes. The finish temperature was changed to 1143K or 1043K for ultra low carbon microalloyed steel strips (0.003 percent carbon) to investigate the effect of finish rolling temperature. Plate tension specimens were taken from the strips along rolling direction. Tension tests were performed according to GB/T228-1987 and GB5028-85 to obtain yield strength, ultimate tensile strength and total elongation. Coupons for microstructure evaluation were machined from the strips with observing surface perpendicular to rolling direction.

STRUCTURAL ALLOY STEEL

The experimental steels were commercial 42CrMo steel （0.39%C-0.29%Si-0.80%Mn-0.025%P-0.019%S-1.08%Cr-0.22%Mo-bal.Fe） from Dalian Iron and Steel Company and vacuum melted microalloyed Cr-Mo steel （0.41%C-0.09%Si-0.44%Mn-0.009%P-0.005%S-1.26%Cr-0.56%Mo-0.29%V-0.06%Nb-bal.Fe） in the laboratory. The steel bars were machined into Charpy-V notch impact specimens, tension specimens (L_0=5d_0), notch tension specimens for anti-delayed fracture evaluation (d=5mm, the notch dV=3mm, 60°±2°/0.15R±0.02) and WOL specimens (10mm×25mm×32mm).

42CrMo steel was soaked at 1173K and cooled in the air for homogeneous treatment. And then 42CrMo steel conventionally was soaked at 1123K for 15min and cooled in the oil, with tempering at 673K for 120min. Through raising austenitizing temperature to 1173、1223、1323K respectively, coarse austenite grains can be achieved. Fine austenite grains can be obtained through repeated heat treatment with soaking temperature at 1123K. All of the specimens were tempered at 673K for 120min. The sizes of prior austenite grains were measured though image analyzer.

Impact test was performed at JB30A (294/147J) testing machine according to GB/T229-1994. Uni-axle tension test was taken in AMSLER 98KN testing machine according to GB228-87 at strain rate of 10^{-2}1/s. There were two methods adopted for anti-delayed fracture evaluation: notch tension test and WOL test. The notch tension tests were performed in Walpole solution with pH=3.5±0.5, and the time duration for testing is 200 hours. WOL tests were taken in 3.5%NaCl solution (pH=6-7) at testing temperature of 293±5K.

RESULTS AND DISCUSSION

PLAIN LOW CARBON STEEL

HOT DEFORMATION SIMULATION-In the temperature range of $A_{e3}-A_{r3}$, ferrite can be induced from austenite in plain low carbon steel when deformation is applied on austenite. Because transformation dynamics is affected by austenite grain size, the volume fraction of ferrite and grain size are considered to be strongly dependent on prior austenite grain state before and after deformation. If applied strain is small, ferrite induced nucleates on grain boundary. Increasing strain will promote nucleation of ferrite at deformed bands within austenite grains. Induced ferrite will be deformed and dynamic recrystallization of ferrite could take place when applied strain increases any further. Strong <111> texture and weak <112> texture are presented in the specimens by deformation induced ferrite transformation. Dynamic recrystallization of ferrite induced can be considered to be the reason for the texture weakening with the increase of strain. The texture of ferrite induced in coarse austenite grain is weaker than that in fine austenite grain. Fine austenite grains bring about more volume fraction of ferrite to be induced and finer ferrite grains, **Table I**. Ferrite induced in coarse austenite grains is more heterogeneous.

Table I The effect of austenite grain size on ferrite grain size transformed

Dγ, μm	44.0	26.0	20.0	17.0	14.3	13.6	6.8
Dα, μm	8.2	4.7	6.0	5.2	3.8	4.1	2.6

Mean ferrite grains from 4.0 to 16.0μm in plain low carbon steel can be obtained through hot deforming simulations. It is showed by the result from tension test that strength increases and ductility decreases with grain refining, **Table II**. When grain size is smaller than 5.0μm, yield strength of plain low carbon steel is over 400MPa. Yield strength changes linearly with $d^{-1/2}$,

Figure 1. The ratio of yield strength to ultimate tensile strength of plain low carbon steel investigated is in the range of 0.69 0.80. When grains are refined, both uniform plastic strain and strain hardening exponent are decreased, while the ratio of yiel strength to ultimate tensile strength is raised.

Table II Tension property of plain low carbon steel and ferrite grain size

Dα μm	4.0	4.0	4.0	5.6	5.6	5.6	5.6	6.7	6.7	8.0	8.0	9.4	11.0	16.0	16.0
UTS MPa	635	540	640	525	545	550	615	505	505	525	510	525	520	505	495
YS MPa	510	430	510	400	430	425	490	360	355	370	355	380	390	365	340
YR	0.80	0.80	0.80	0.76	0.79	0.77	0.80	0.71	0.70	0.70	0.70	0.72	0.75	0.72	0.69
RA %	48	49	47	66	64	44	64	58	58	57	58	60	55	47	60

(Note: Tension test specimens were machined from hot simulated deforming Q235 steel bars. The Φ10mm×100mm bars were soaked at 1173K for 30s, and compressed 50 percent at strain rate of 1/s, and then cooled at 3, 5, 10, 15, 20K/s, and water cooled respectively in GLEEBLE—1500.)

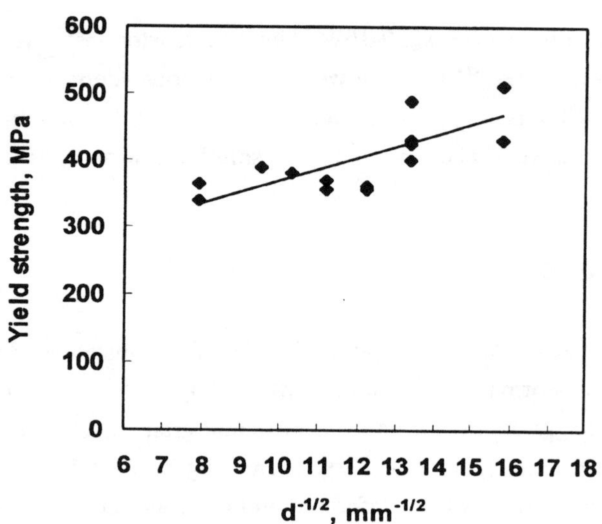

Fig. 1: Yield strength of plain low carbon steel and ferrite grain size.

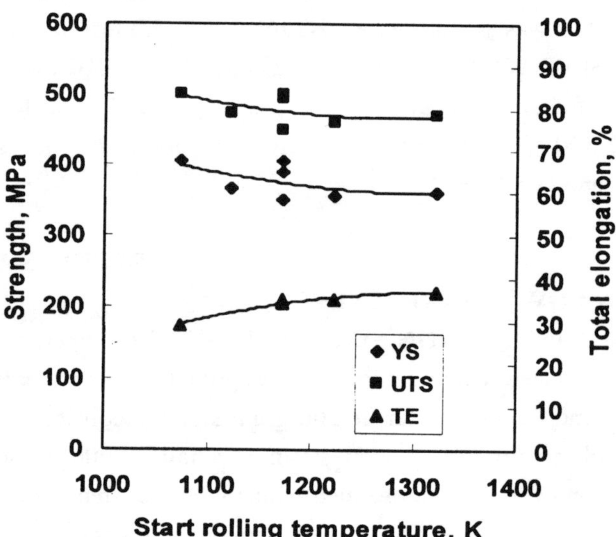

Fig. 2: Tension property of the rebar investigated.

HOT ROLLOING REBAR IN INDUSRIAL TRIALS-Industry trials in plain low carbon rebar shows that lowering start rolling temperature (corresponding to lower finish rolling temperature) results in grain refining and strength raising. When ferrite grain is of 6.0μm in size, yield strength of plain low carbon rebar can reach at 400MPa. But ductility decreases with decrease of start rolling temperature, **Figure 2**. Comparing with 20MnSi steel rebar, yield strength of fine grain plain low carbon steel rebar is higher, but ultimate tensile strength is lower, **Figure 3**. The ratio of yield strength to ultimate tensile strength is about 0.80 in fine grain steel, which is higher than those of conventional ferrite-pearlite steels. Lower ductility and higher ratio of yield strength to ultimate tensile strength are the important characteristics in fine grain plain low carbon steels.

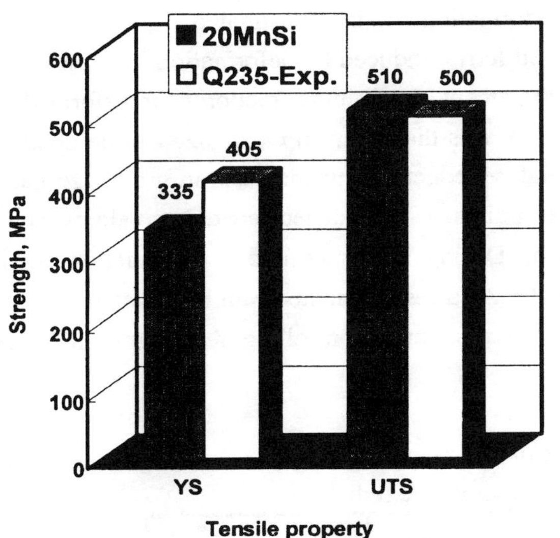

Fig. 3: Tension property of plain low carbon steel and 20MnSi steel rebars.

HOT ROLLING STRIP IN INDUSTRIAL TRIALS-Because of lower manganese content in the steel investigated, no obvious band structure was found in the coiling temperature range. Grain size of 6μm was obtained in hot rolled strip in the industry trial. The result from mechanical testing shows that the most values of yield strength of the strip coiled at different temperature are over 400MPa, **Table III**. The strip presents anisotropic in yield strength. The values of yield strength in transverse direction and 45° to rolling direction are higher than those in rolling direction, and are all over 400MPa. But it is isotropic in ultimate tensile strength and total elongation along three directions. The values of ultimate tensile strength in the strip are about 500MPa, and the values of total elongation are all over 30 percent. In the coiling temperature range of 880-750K, coiling temperature has little effect on strength and ductility. The impact absorbed energies of the longitudinal notch specimens of 2.5mm×10mm×55mm in size are in the range of 32-42J.

Table III Mechanical properties of hot rolling strip in industrial trials

CT, K	YS, MPa			UTS, MPa			TE, %			CVN J
	0°	45°	90°	0°	45°	90°	0°	45°	90°	
880	400	435	470	510	515	515	33	36	32	42
864	370	370	410	495	490	490	34	36	32	32
859	410	435	455	530	525	530	34	34	32	39
807	405	405	430	495	515	510	34	34	34	37
787	400	415	430	500	500	510	34	33	33	39
750	395	400	415	520	505	510	34	34	34	40

Table IV Comparison of industrial trials with conventional industrial production in mechanical properties of strips

		YS, MPa	UTS, MPa	YS	TE, %
Industry product (t=3.5mm)		320	460	0.70	36
Industry trials (L) (t=3.0mm)	CT 864K	370	495	0.75	34
	CT 787K	400	500	0.80	34

(note: finishing rolling temperature was 1143K, and coiling temperature was 873-893K.)

By adopting the processing described above, ferrite grain can be effectively refined, and strength of the strips can be raised markedly with little change on ductility and toughness. When finish rolling temperature was around 1073K, yield strength can be increased to 400MPa with total elongation of 30%, **Table IV**. Coiling temperature has little effect on microstructure and properties as coiling temperature is below 880K.

HIGH STRENGTH LOW ALLOY STEEL

The transformation from austenite into ferrite can be promoted when deformation is applyed in the temperature range of A_{e3} to A_{r3}. Although the phenomenon of deformation induced ferrite transformation have been reported since the beginning of 1980'[11-13], no clear understanding on this phenomenon has been obtained. In this paper, the effect of chemistry and rolling parameters on volume fraction and size of transformed ferrite induced by deformation will be investigated.

For low carbon microalloyed steel (X65), the volume fraction of transformed ferrite induced by deformation increases with t accumulative reduction, **Table V**. When accumulative reduction near to 90 percent, the volume fraction of transformed ferri induced by deformation approaches to about 100 percent. Ultra fine grains of less than 2μm in mean size can be obtained deformation induced ferrite transformation. With the increase of accumulative reduction, no increase on grain size has be observed, otherwise grain size decreases a little bit. It should be noted that grain sizes of induced ferrite keep almost consta and subsequent deformation has little effect on ferrite grain growth. During the subsequent deformation, dynam recrystallization of ferrite may accompany with the transformation of ferrite. As a result, ferrite grain becomes finer with t increase of deformation reduction. By applying the deformation induced ferrite transformation rolling, ferrite grain in hot roll strip can be refined to about 1μm in size.

Table V Deformation induced transformed ferrite in low carbon microalloyed steel (X65) subject to hot rooling at different accumulative reduction

Accumulative reduction, %	**69**	**73**	**77**	**79**	**82**	**88**
Volume fraction of ferrite, %	71	85	96	94	90	98
Mean grain size of ferrite, μm	1.22	1.04	1.16	1.12	0.99	0.92

Researchers in both Japan and Korean applied heavy deformation at low temperature on low carbon C-Mn steel to refine ferri grain size to about 1μm. Low carbon C-Mn steel of 1μm in grain size presents ultimate tensile strength level of 700MPa. In ou study, the yield strength level of 700MPa can be obtained in low carbon microalloyed steels by adopting deformation induce ferrite transformation rolling, **Table VI** and **Figure 4**. For low carbon microalloyed steel, ferrite grain size in coiling strip i about twice of that in the hot rolled strip subject to subsequent water cooling, from grain size of 2μm in coiling strip to grai size of 1μm in water cooling strip. In the coiling temperature range of 673—873K, strength increases with the decrease c coiling temperature.

Table VI Effect of coiling temperature on mean ferrite grain size and strength of low carbon microalloyed steel (X52)

Coiling temperature, K	**673**	**773**	**873**	**WC**
Mean grain size of ferrite, μm	1.80	2.20	2.30	1.00
Yield strength, MPa	665	555	530	630
Ultimate tensile strength, MPa	710	605	590	840
Total elongation, %	18	25	24	13
Yield ratio	0.94	0.92	0.90	0.75

In order to increase yield strength of the steel over 800MPa, we tried a new method of grain refinement. Because ferrite grai can be refined by increasing the nucleation rate of ferrite grain and decreasing growth rate of ferrite grain during transformation a rise in driving force of ferrite transformation will result in higher nucleation rate of ferrite. In the same deformation schedule a reduction in carbon content in steel leads to rise in driving force for ferrite transformation, and as a result finer and more uniform ferrite grains could be obtained.

Ultimate tensile strength and yield strength of the ultra fine grain microalloyed steels of investigated change little in the low carbon content range of 0.100-0.160%, **Figure 4**. Yield strength of 700MPa is presented for the low carbon microalloyed steels with ultra fine grains. When carbon content is lower than 0.100%, strength decreases and ductility increases, which follow the general concept of strengthening mechanism. As carbon content in steels is reduced to 0.039%, a yield strength valley of 540MPa and a total elongation peak of 23% are presented. It is should be noted that in ultra low carbon microalloyed steel (0.003% carbon content), 0.2% proof yield strength increases remarkably to over 1000MPa. The ferrite grains induced in ultra low carbon microalloyed steel are much finer than those in low carbon microalloyed steels, and the uniformity of fine grains across the section of strip is higher than that in low carbon microalloyed steels. But the ratio of yield strength to ultimate tensile strength (YR=0.89) is higher than those of conventional ferritic steels, and total elongation of ultra fine grain steels is lower. It is worth of further investigating the mechanism of plastic flow and fracture of ultra fine grain steels.

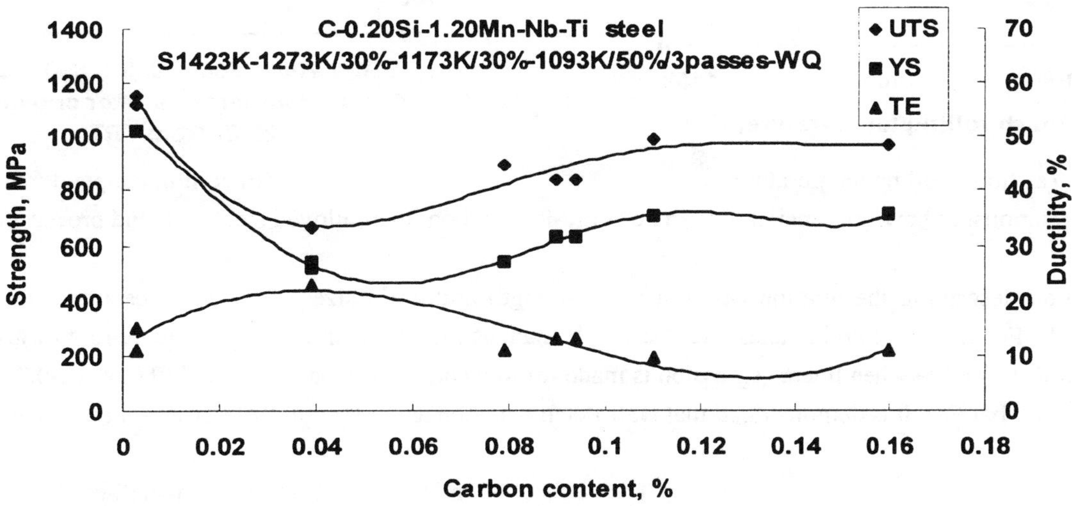

Fig. 4: The effect of carbon content on tension property of hot rolling strips.

For ultra low carbon microalloyed steel, yield strength of strip subject to hot rolling and subsequent water cooling can be over 800MPa in the wider range of finish rolling temperature, **Figure 5(a)**. To increase the finish rolling temperature from 1093K to 1143K, ultra fine grains in steel strip can also be obtained and they still remain at higher strength level. To lower the finish rolling temperature to 1043K, austenite transforms to pro-eutectoid ferrite partially before finish rolling. This results in an elongated ferrite microstructure after finish rolling and water cooling and coarser grains in steel strip. Strength will declines due to the coarser and elongated grains in the steel strip.

The deformation induced ferrite transformed rolling (finish rolling temperature is 1093K) strips of ultra low carbon microalloyed steel (296 steel) was cooled at 20K/s to simulated coiling temperature. Strength of the strips decreases with the decrease of coiling temperature, **Figure 5(b)**. Comparing with the hot rolled strips cooled in water, coarser grains are presented in the strips coiled in the temperature range of 673-773K. For steel strip coiled at 773K, yield strength is reduced from 1000MPa of the steel strip cooled in water to 600MPa. The ferrite grains in coiled strips are about 2μm in size. It is deduced that part of the ferrite in the coiled strips is transformed pro-eutectoid ferrite. The strength of ultra low carbon microalloyed steel strips coiled is much higher than those of conventional controlled rolling and cooling strips.

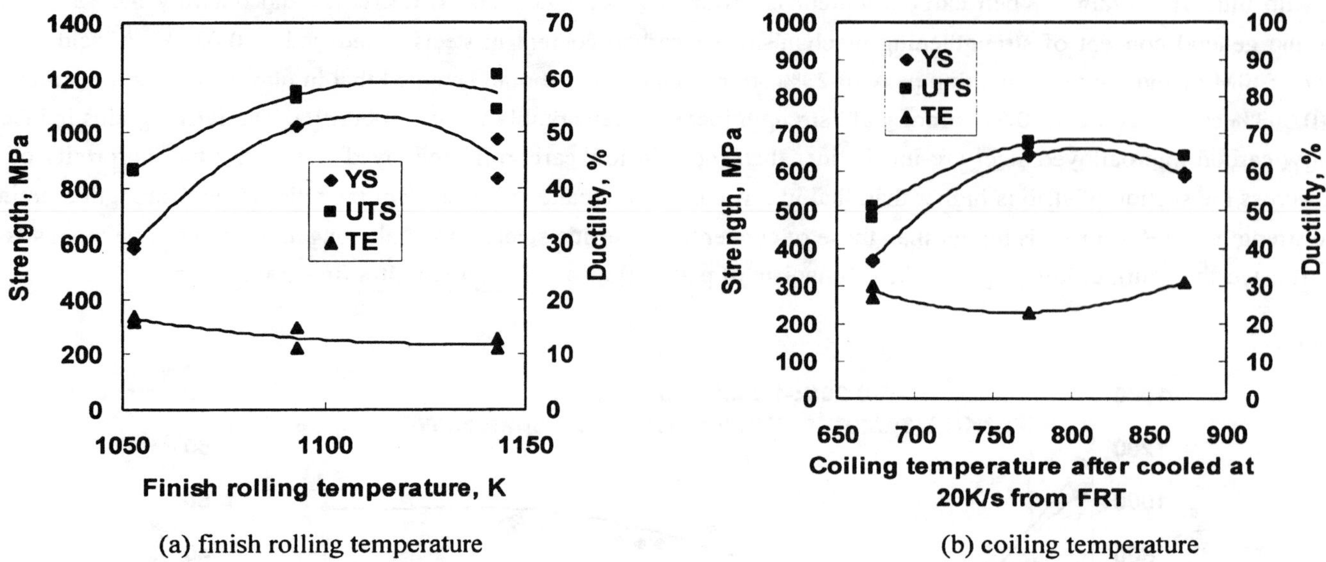

(a) finish rolling temperature (b) coiling temperature

Fig. 5: The relationship between tension property of ultra-low carbon microalloyed steel strip and processing parameters.

Hall-Petch equation describing the relation between yield strength and grain size is proved to be fitted for ultra fine grain microalloyed steels. For ultra fine grain steels, yield strength changes linearly with reciprocal square root of grain size, **Figure 6**. The coefficient R^2 is 0.8299 when linear regression is made for low carbon microalloyed steel (0.094%C-0.29%Si-1.42%Mn-0.045%Nb-0.008%Ti-bal.Fe). It is demonstrated that Hall-Petch equation are still applicable to sub-micron steels.

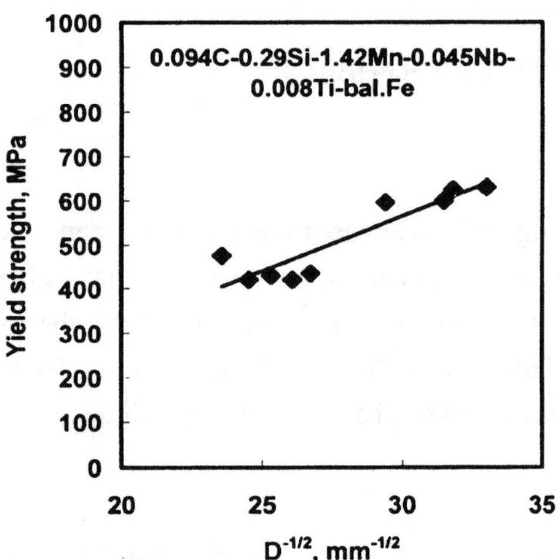

Fig. 6: The relationship between yield strength of microalloyed low carbon steel (X65)and mean ferrite grain size.

The effect of microalloy elements on deformation induced ferrite transformation is still ambiguous. Yada et al thought that niobium had little effect on grain refinement[14]. The research of Hickson and Hodgson showed that transformation temperature was lowed by the addition of niobium, and the driving force for ferrite nucleation was increased [7]. Lee et al thought that the addition of niobium could enhance grain refinement for ferrite transformed by strain induction [15]. With the increase of strain, the volume fraction of ferrite induced is increased in 16MnNb steel, **Table VII**. Meanwhile, the volume fraction of ferrite induced in 16Mn steel is not increased continuously with applied strain. For 16Mn steel, ferrite is initially to be induced when reduction is 40 percent at 1073K,, at least 20 percent higher in reduction needed for ferrite transformed in 16MnNb steel. At reduction of 80 percent, strain induced ferrite transformation in 16Mn steel may be inhibited due to recrystallization of austenite that consumes the accumulative deforming energy. The ferrite induced in 16MnNb steel is different from that induced in 16Mn steel in morphology and size. In 16MnNb steel, the ferrite induced is of equiaxed grain and 1μm in grain size, while the ferrite in 16Mn steel is allotriomorphic ferrite and of 2μm in grain size. Ferrite induced is mainly nucleated along prior austenite grain boundaries at primary stage. As large strain is applied, ferrite could be induced within prior austenite grains.

The volume fraction of ferrite induced by deformation is decreased with the increase of deforming temperature. When deforming temperature is higher than 1123K, no ferrite can be induced by strain in 16MnNb steel. The temperature at which no ferrite can be induced by deformation in 16Mn steel is 1073K. The increase of strain rate will promote the transformation from austenite to ferrite, and the addition of niobium makes this effect stronger. Generally speaking, the addition of niobium into steel enhances the ferrite transformed from austenite by deformation. The result of simulated coiling test at 873K shows that grain size of ferrite induced is 1.5μm in 16MnNb steel, and the tendency for grain growth is small. In the same condition, ferrite grain induced is 4μm in size in 16Mn steel. It is can be concluded that grain growth of ferrite induced can be inhibited by the addition of niobium.

The addition of niobium can enhance the energy storage ability of deformation, and then promote thermodynamics of deformation induced ferrite transformation. Because the addition of niobium raises non-recrystallization temperature, the more deformed grain boundaries and deformed bands are reserved for subsequent ferrite nucleation on them. As a results, dynamics of deformation induced ferrite transformation will be enhanced. The dragging effect originated from precipitation and / or solution of niobium along grain boundaries makes grain growth of ferrite induced to be inhibited available. Deformation induced ferrite transformation can be promoted by the addition of niobium and it finally results ultra fine grains in C-Mn steel. Critical strain for induced frerrite transformation is reduced and transformation temperature is raised due to the addition of niobium to steel. Meanwhile, the addition of niobium can suppress grain growth of ferrite transformed by strain accumulation.

Table VII Volume fraction of transformed ferrite induced by deformation in 16Mn steel and 16MnNb steel, %

Reduction, %	Temp., K	**0**	**20**	**40**	**60**			**80**
Strain rate, s^{-1}		**5**	**5**	**5**	**5**	**15**	**25**	**5**
16Mn	973	-	-	-	47.0	-	-	-
	1023	-	-	-	14.0	-	-	-
	1073	0	0	0.1	6.0	7.0	12.0	5.0
	1123	-	-	-	0	0	-	-
	1173	-	-	-	0	0	-	-
16MnNb	973	-	-	-	34.0	-	-	-
	1023	-	-	-	24.0	-	-	-
	1073	0	1.0	8.0	17.0	21.0	48.0	35.0
	1123	-	-	-	5.0	-	-	-
	1173	-	-	-	0.0	-	-	-

STRUCTURAL ALLOY STEEL

It is believed that microalloying is the effective method to refine prior austenite grain. Austenite grain grows coarser as soaking temperature rises, **Table VIII**, but mean austenite grains in microalloyed Cr-Mo steel are less than half of the mean austenite grains in 42CrMo steel. By adopting Beck's equation, the activation energies of grain boundary migration for 42CrMo steel and microalloyed Cr-Mo steel are 425kJ/mol and 525kJ/mol respectively. It is shown by the calculated activation energy that drag force for grain boundary migration is large in Cr-Mo steel microalloyed with niobium and vanadium Because the existing microalloy precipitates bring about drag effect for grain boundary migration, the austenite grain in microalloyed Cr-Mo steel is much smaller.

Another method for prior austenite grain refining in structural alloy steel is the repeated heat treatment. To adopt the repeated transformation, the prior austenite grains in 42CrMo steel and microalloyed Cr-Mo steel become finer. To increase the number of repeated heat treatment leads to austenite grain refining, but the refining effect becomes weaker when repeated transformation is over three times. The mean prior austenite grains of 42CrMo steel and microalloyed Cr-Mo steel are of

4.66μm and 2.07μm in size respectively when they are subject to repeated transformation of four times. It is also demonstrated that the addition of niobium and vanadium in Cr-Mo steel brings about grain refining in repeated heat treatment.

Table VIII Mean austenite grain size of the steels investigated, μm

	Temp., K	1123	1223	1323	1423	1523
42CrMo steel	D_0	13.0	22.0	45.0	65.0	125.0
	D	16.0	32.0	65.0	90.0	180.0
Microalloyed Cr-Mo steel	D_0	4.7	5.6	19.0	32.0	65.0
	D	5.6	8.0	27.0	45.0	90.0

To change austenitizing temperature and apply repeated heat treatment, specimens with austenite grain size from 4.7μm to 120.0μm could be obtained in 42CrMo steel. Mechanical properties of 42CrMo steel with different grain size are presented in **Table IX** and **Figure** 7. Strength, toughness and reduction in area increase with the decrease of prior austenite grain size. Meanwhile total elongation almost keep unchangeable with austenite grain. For finer prior austenite grain, the block size of martensite transformed is finer than that in conventional heat treated steel, and lath width of martensite does not affected by austenite refining. Because the block size plays a key role in mechanical properties in martensitic steel, the refinement of prior austenite grain leads to an improvement in mechanical properties.

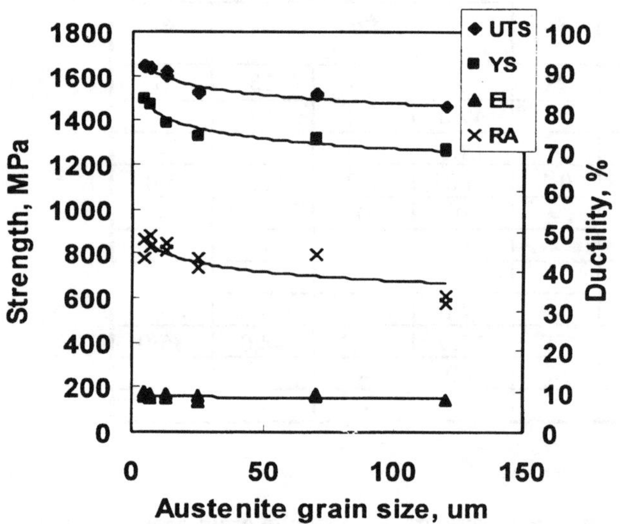

(a) strength and ductility

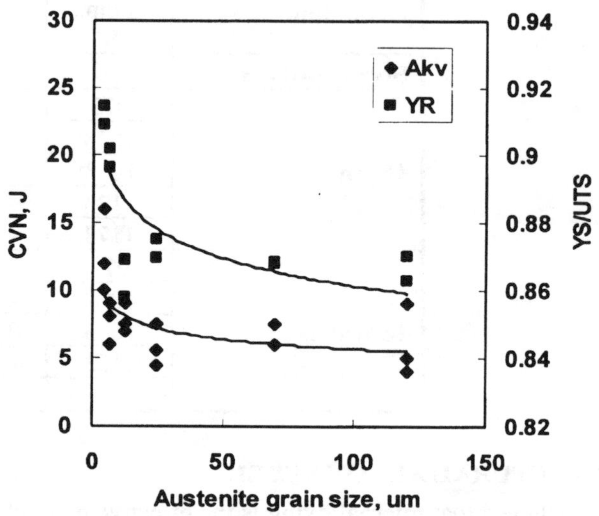

(b) absorbed impact energy and yield ratio

Fig.7: Mechanical properties of 42CrMo steel and austenite grain size.

Stress and strain in uniform deformation range in uni-axle tension test can be described by Hollomon equation $\sigma=K\varepsilon^n$. When prior austenite grain size is over 13.0μm, strain hardening exponent remains almost at value of 0.24. For finer grain steel, strain hardening exponent decrease to about 0.18, **Table X**. It does mean that uniform deformation strain is smaller in finer grain steel.

Table IX Heat treatment on austenite grain size and mechanical properties of 42CrMo steel

Heat treatment		Mechanical properties						Dγ μm
		UTS MPa	YS MPa	YS(0.2 proof) MPa	TE %	RA %	CVN J	
1	Quenching 4 times	1645	1500	—	9.5	46	13.0	4.7
2	Quenching 2 times	1635	1470	—	9.0	47	8.0	6.7
3	Convetional	1610	—	1390	9.0	46	7.5	13.0
4	1173K oil quenching	1525	—	1330	9.0	44	7.0	25.0
5	1323K oil quenching	1515	—	1315	8.5	42	5.0	70.0
6	1423K oil quenching	1460	—	1265	8.0	43	4.5	120.0

Table X Strain hardening exponent and constant of 42CrMo steel and mean austenite grain size

Dγ, μm	4.7	6.7	13.0	25.0	70.0	120.0
K, MPa	2922	2947	3299	3212	3200	3295
n	0.18	0.19	0.24	0.24	0.25	0.24

At ultimate tensile strength level of 1150MPa, critical notch tension stress of the steel with grain size of 4.7μm is 1490MPa, compared with 1135MPa in conventional heat treated 42CrMo steel with prior austenite grain size of 13.0μm. The critical notch tension stress is raised to about 30 percent as prior austenite grain size reduced from 13.0μm to 4.7μm. K_{ISCC} is also raised sharply when prior austenite grain is refined to 4.7μm., **Table XI**.

Table XI The effect of mean autenite grain size on strength, hardness and K_{ISCC}

Dγ, μm	4.7	13.0	25.0	70.0	120.0
UTS, MPa	1645	1610	1525	1515	1460
HRC	47	46	45	45	44
K_{ISCC}, $MPam^{1/2}$	93.0	20.7	19.6	12.3	17.8

CONCLUSIONS

In order to double strength for new generation steels, the chemistry and processing were investigated to refine the microstructure to ultra fine grain level. It can be concluded from the microstructure and property evaluation as follows:

1) By adopting grain refining rolling schedule in industrial trial scale, yield strength of plain low carbon steel strip can reach at 400MPa level with total elongation of 30 percent. When coiling temperature is lower than 880K, coiling temperature shows little effect on microstructure and properties of strip.
2) For plain low carbon steel rebar trials in industrial scale, ferrite grain can be refined through lower rolling temperature. When ferrite grain is less than 6μm, the yield strength can be increased to over 400MPa.
3) Deformation induced ferrite transformation is an effective method to refine grain in microalloed steel. As reduction ratio

for induced ferrite transformation is raised, the volume fraction of ferrite induced increases. Grain size of ferrite induced is less than 1μm and keeps almost unchangeable as reduction ratio is increased. Ultra fine grain structure becomes uniform when coiling treatment is applied after ferrite induced rolling. The addition of niobium promotes the transformation of ferrite induced.

4) Deformation induced ferrite transformation strongly depends on carbon content in steels. More volume fraction of ferrite and finer ferrite can be induced in lower carbon steel, and microstructure uniformity can be improved in lower carbon microalloyed steel strips. Yield strength of low carbon microalloyed steel can reach at 600MPa level, while yield proof strength of ultra-low carbon microalloyed steel (0.003%C) is over 800MPa.

5) Both microalloying and repeated heat treatment are very effective in prior austenite grain refinement. By adopting repeated heat treatment, prior austenite grain can be refined to 4.66μm in 42CrMo steel and to 2.07μm in microalloyed Cr-Mo steel, compared with 13.0μm in conventional heat treated 42CrMo steel. Although martensite lath width remains unchangeable, martensite block is refined due to refinement of austenite grain. The grain refinement by microalloying and repeated heat treating not only leads to the increase in both strength and toughness, but also leads to an improvement in anti-delayed fracture property. Structural alloy steel microalloyed with vanadium and niobium at 1300MPa level of ultimate tensile strength presents better anti-delayed fracture property than conventional 42CrMo steel at strength level of 1200MPa.

ACKNOWLEDGEMENTS

The authors acknowledge with gratitude the financial support in the Key Basic Research Program received from the Ministry of Science and Technology of China (the Project number Is G1998061500). Thanks are also expressed to the research members in Central Iron and Steel Research Institute, Beijing University of Science and Technology, and Northeast University for their creative contribution to the Project.

REFERENCES

1. Y.Q.Weng, "New Generation of Iron and Steel Material in China", Proceedings of the International Workshop on the Innovative Structural Materials for Infrastructure in 21st Century, Jan. 2000, Tsukuba, Japan, pp.11-32.
2. W.B.Morrison, "The Effect of Grain Size on the Stress-Strain Relationship in Low-carbon Steel", Transactions of the ASM, Vol.59, 1966, pp.825-846.
3. R.L.Miller, "Ultrafine-Grained Microstructures and Mechanical Properties of Alloy Steels", Metallurgical Transactions, Vol.3, No.4, April 1972, pp905-912.
4. M.R.Hickson, R.K.Gibbs and P.D.Hodgson, "The Effect of Chemistry on the Formation of Ultrafine Ferrite in Steel", ISIJ International, Vol.39, No. 11, Nov.1999, pp.1176-1180.
5. T.M.Maccagno, J.J.Jonas and P.D.Hodgson, "Spreadsheet Modelling of Grain Size Evolution during Rod Rolling", ISIJ International, Vol.36, No. 6, June 1996, pp.720-728.
6. H.Mabuchi, T.Hasegawa and T.ishikawa, "Metallurgical Features of Steel Plates wiyh Ultra Fine Grains in Surface Layers and Their Formation Mechanism", ISIJ International, Vol.39, No. 5, May 1999, pp.477-485.
7. M.R.Hickson and P.D.Hodgson, "Effect of Preroll Quenching and Post-roll Quenching on Production and Properties of Ultrafine Ferrite in Steel", Materials Science and Technology, Vol.15, No. 1, Jan. 1999, pp.85-90.
8. Senuma, M.Kameda and M.Suehiro, "Influence of hot rolling and Cooling Conditions on the Grain refinement of Hot Rolled Extralow-carbon Steel Bands", ISIJ International, Vol.38, No.6, June 1988, pp.587-594.
9. A.Sato, "Research Project on Innovative Steels in Japan (STX-21 Project)", Proceedings of the International Workshop on the Innovative Structural Materials for Infrastructure in 21st Century, Jan.2000, Tsukuba, Japan, pp.1-10.

10. W.P.Lee, "Development of High Performance Structural Steels for 21st Century", Proceedings of the International Workshop on the Innovative Structural Materials for Infrastructure in 21st Century, Jan. 2000, Tsukuba, Japan, pp.33-63.
11. R.Priestner and L.Ali, "Strain Induced transformation in C-Mn Steel during Single Pass Rolling", Materials Science and Technology, Vol.9, No. 2, Feb.1993, pp.135-141.
12. V.M.Khlestov, E.V.Konopleva and H.J.McQueen, "Effect of Hot deformation on Austenite transformation in Low Carbon Mo-Nb and C-Mn Steels", Materials Science and Technology, Vol.14, No. 8, Aug. 1998, pp.783-792.
13. H.Yada, C.M.Li and H.Yamagata, "Dynamic $\gamma \rightarrow \alpha$ Transformation during Hot Deformation in Iron-Nickle-Carbon Alloys", ISIJ International, Vol.40, No. 2, Feb. 2000, pp.200-206.
14. H.Yada, Y.Matsumura and K.Nakajima, United Patent 4,466,842, 1984.
15. S.Lee, et al, "Transformation Strengthening by Thermomechanical Treatments in C-Mn-Ni-Nb Steels", Metallurgy & Material Transactions, Vol.26A, No. 5, May 1995, pp.1093-1100.

PRODUCT PHYSICAL METALLURGY IV

INFLUENCE OF MANGANESE, SILICON AND ALUMINIUM ON THE TRANSFORMATION BEHAVIOR OF LOW ALLOYED TRIP-STEELS

S. Traint †, A. Pichler ‡, R.Tikal ‡, P. Stiaszny ‡, E.A. Werner †
†Christian-Doppler-Laboratorium für Moderne Mehrphasenstähle,
Lehrstuhl A für Mechanik, TU-München,
Boltzmannstrasse 15, D-85747 Garching, Germany.
tel.: +49 89 289 15247; fax: +49 89 289 15148; e-mail: traint@lam.mw.tu-muenchen.de
‡VOEST-ALPINE Stahl Linz
Research, Development and Testing, Voest-Alpine Strasse 3, A-4031 Linz, Austria.
tel.: +43 732 6585 6321; fax: +43 732 6980 4338; e-mail: andreas.pichler@voest.co.at

Key Words: low alloyed TRIP-steels, retained austenite, microstructure, mechanical properties, influence of the annealing parameters

INTRODUCTION

The TRIP (transformation induced plasticity)-effect has been observed in high alloyed austenitic steels already 30 years ago [1]. The excellent combination of high strength and ductility of these steels is attributable to the irreversible austenite-to-martensite phase transformation during plastic straining. However, the high content of expensive alloying elements prohibits industrial applications of these steels.

Through a special heat treatment in combination with alloying elements preventing carbide precipitation it became possible to stabilize significant amounts of retained austenite at room temperature even in low alloyed steels. In the case of low alloyed TRIP-steels the austenite is mainly stabilized by carbon. The carbon enrichment of the austenite is achieved by a two-step heat treatment. In the first step of the heat treatment the material is annealed at the intercritical annealing temperature in the $\alpha+\gamma$ two-phase field. There, the recrystallisation, the dissolution of the cementite and the formation of the austenite take place. Following the intercritical annealing the material is slowly cooled to the quenching temperature (700-650 °C) and then rapidly cooled to the overaging (bainitic) range. The slow cooling increases the volume fraction of ferrite and consequently the residual austenite is enriched by carbon. The rapid cooling is necessary to avoid pearlite formation. During the isothermal holding in the bainitic range the austenite partly transforms into a ferritic bainite, since the precipitation of carbides is prevented by alloying elements, while the remaining austenite is enriched further with carbon. After the annealing treatment the microstructure of low alloyed TRIP-steels finally consists of ferrite, (ferritic) bainite and retained austenite.

A key factor in controlling the mechanical properties is the stability of the retained austenite against strain induced transformation. The stability of the austenite depends on many factors, such as the chemical composition of the austenite particles, the austenite particle size [2], the deformation temperature [3], the type of mechanical load (tensile, compressive) and the crystallographic orientation of the austenite with respect to both the surrounding phases and the direction of the applied load during forming operations [4].

The typical chemical composition of low alloyed TRIP-steels is in the range of 0.2 wt.% carbon, 1.0-2.5 wt.% silicon and 1.0-3.0 wt.% manganese. Since the high silicon content of these steels leads to undesirable hot rolling properties, alternative alloy concepts must be considered. Possible candidates to substitute silicon are aluminium, phosphorus and copper [5-9], which are considered to to be capable of suppressing carbide formation. In this work, the influence of additions of aluminium, silicon, phosphorus and manganese on the microstructure and the mechanical properties of low alloyed TRIP-steels is investigated.

EXPERIMENTS

Production of the material-The composition of the investigated alloys are shown in table I. The alloys were melted in a 100 kg medium frequency furnace. The ingots were milled to a thickness of 35 mm and divided into several parts. The slabs were reheated to a temperature of 1300 °C and hot rolled in 6 passes to a thickness of 4 mm. The finishing temperature was about 870 °C. To simulate coiling, the hot rolled sheets were rapidly cooled to a temperature of 600 °C and then slowly cooled in a furnace to room temperature applying a cooling rate of 50 K/h. The sheets were milled to remove the surface scale and to obtain plane-parallel surfaces for cold rolling and then cold rolled to a thickness of 0,9 mm (cold reduction=60 %).

Table I: Chemical composition of the investigated alloys [wt.%]

Alloy	C [%]	Si[%]	Mn [%]	P [%]	S [%]	Al[%]	N[%]
A	0.15	0.04	1.5	0.01	0.002	1	0.005
B	0.15	0.04	1.5	0.10	0.002	1	0.004
C	0.15	0.60	1.5	0.01	0.002	1	0.006
D	0.15	0.04	0.6	0.10	0.001	1	0.005

Annealing simulations-The annealing simulations were conducted in the laboratory on cold rolled material. First, the kinetics of recrystallisation, dissolution of cementite and formation of austenite were studied. For this purpose, specimens were annealed in the α+γ two-phase region between 760 °C and 940 °C for 40 s to 180 s and then quenched in water.

In subsequent annealing simulations the intercritical annealing parameters were held constant, and the influence of the temperature and the holding time in the bainitic range on the microstructure and the mechanical properties was studied (fig.1-a). Moreover, the influence of a hot dip galvanizing cycle on the mechanical properties and the microstructure was investigated (fig.1-b).

Testing and characterisation of the microstructure-Standard light microscopy was used to reveal the microstructure of the heat treated samples. The samples, which were intercritically annealed and then quenched in water, were etched with Nital and/or LePera's etchant [10]. To reveal the microstructure after TRIP-annealing cycles a two-step etching technique was applied [11, 12]. The samples were etched with V2A and then with Klemm's etchant, which stains ferrite brown and/or blue, bainite dark-brown to grey whereas retained austenite appears white.

The retained austenite content was determined using a magnetic volumetric method [13, 14]. The retained austenite content is determined by relating the value of the magnetic saturation of the specimen to the value of the magnetic saturation of a ferritic specimen containing the same amount of alloying elements as the actually measured specimen,

$$\%\mathrm{RA} = \frac{\mathrm{B_{sc}} - \mathrm{B_s}}{\mathrm{B_s}} \cdot 100\%, \qquad (1)$$

where $\mathrm{B_s}$ is the measured magnetic saturation of the specimen and $\mathrm{B_{sc}}$ the theoretical value for the magnetic

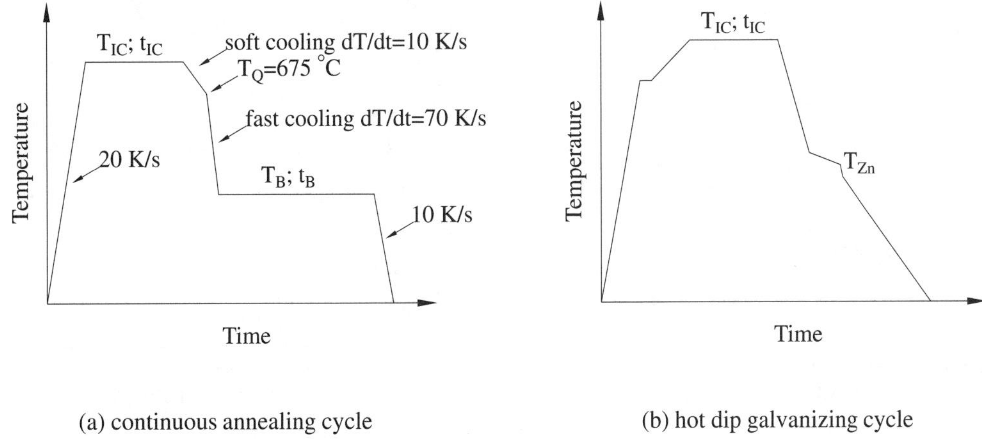

(a) continuous annealing cycle

(b) hot dip galvanizing cycle

Figure 1: Schematic representation of the laboratory annealing cycles. (T_{IC} ... intercritical annealing temperature; t_{IC} ... intercritical annealing time; T_Q ... quenching temperature; T_B ... overaging (bainitic) temperature; t_B ... overaging (bainitic) time; T_{Zn} ... zinc pot temperature; dT/dt...cooling rate)

saturation. B_{sc} is given by:

$$B_{sc} = 2,158 - 0,15\,\mathrm{C} - 0,048\,\mathrm{Si} - 0,0244\,\mathrm{Mn} - 0,057\,\mathrm{Al} - 0,0305\,\mathrm{Cr} - 0,06\,\mathrm{Mo} \quad \text{in wt.\%.} \qquad (2)$$

To study the transformation behavior of the austenite in the bainitic range during the TRIP-annealing cycle, dilatometric investigations on a Bähr dilatometer DIL 805 A/D were conducted with samples prepared from cold rolled material (0.9 x 3.5 x 10 mm).
Thermodynamical calculations with the program ThermoCalc were performed to study the influence of the alloying elements manganese, aluminium, phosphorus and silicon on the phase diagrams and the equilibrium volume fraction of austenite during intercritical annealing.
To study the transformation behavior of the retained austenite during straining, the retained austenite content was measured as a function of the total strain applied. After straining to different levels, samples were prepared for the measurement of the retained austenite (0.9 x 3.5 x 5 mm).
The mechanical properties were measured according to European Standard EN 10002. Before tensile testing the material was skin pass rolled (skin pass reduction=0.5-0.7 %).

RESULTS

Thermodynamical calculations- In a first step, the influence of aluminium on the equilibrium volume fraction of austenite was calculated for an alloy containing 0.15% C and 1.5% Mn. As can be seen from fig. 2-a, the addition of aluminium significantly tightens the austenite region. For aluminium contents above 1% a complete austenitization of the steel is no longer possible. Figs. 2-b and 2-c show the influence of phosphorus and silicon on the equilibrium fraction of austenite for an alloy with 0.15% C, 1.5% Mn and 1% Al. Similar to aluminium, phosphorus and silicon tighten the γ-region of the steel. The influence of phosphorus is more pronounced for the A_3-temperature than for the Ae_1-temperature. The γ-region can be extended again by increasing either the carbon or the manganese content (fig. 2-d).

Microstructure

Dissolution kinetics of cementite-Specimens were annealed in the $\alpha + \gamma$ two-phase region at different temperatures and times. After water quenching, the microstructure of the samples was inspected.
For all alloys investigated and intercritical annealing parameters chosen the recrystallisation of the ferrite and

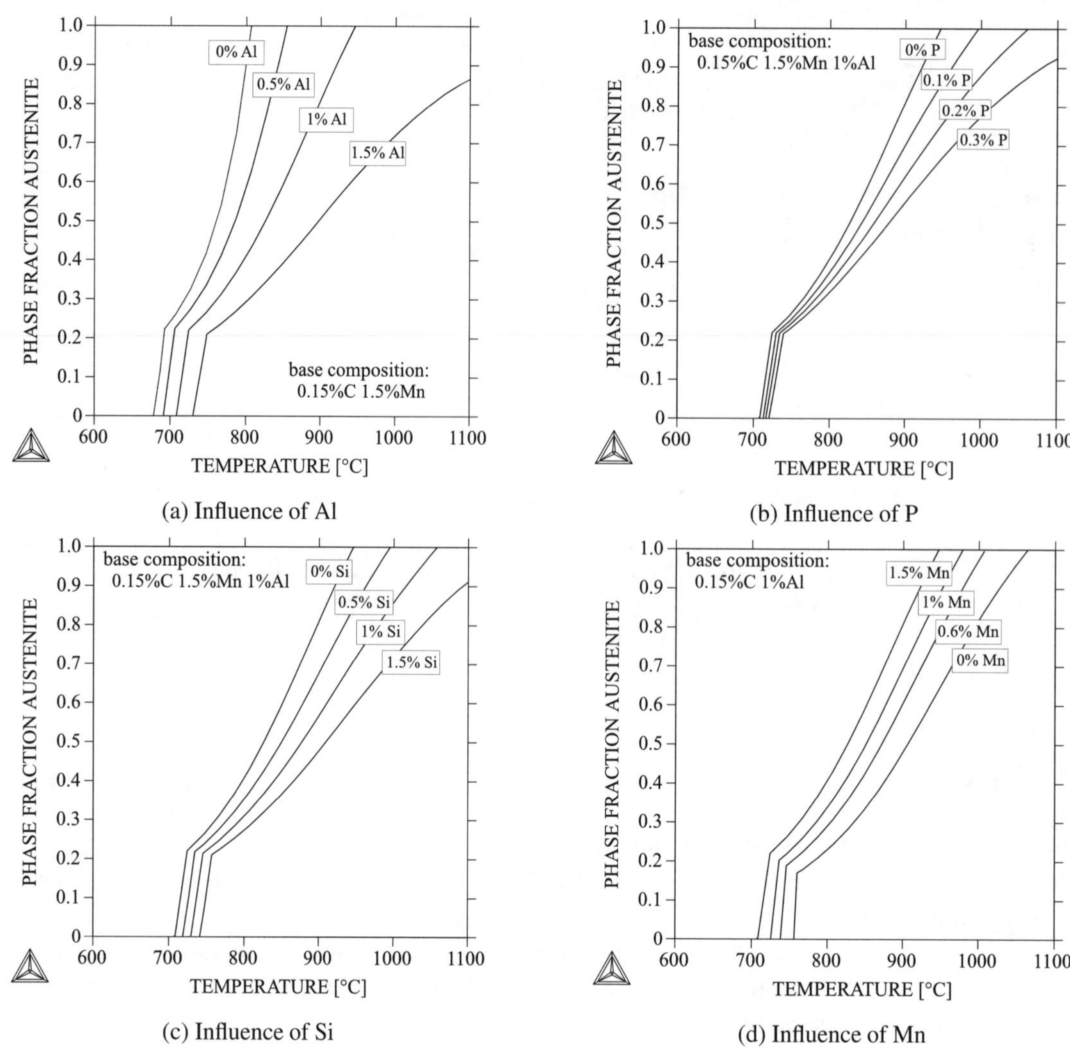

Figure 2: Influence of Al, P, Si and Mn on the equilibrium volume fraction of austenite calculated with ThermoCalc

the dissolution of the carbides are completed, even at temperatures as low as 760 °C and annealing times as short as 40 s. After water quenching a microstructure consisting of ferrite and martensite is formed.
Fig. 3 shows the microstructure of alloy B (Al-P) after different annealing temperatures. With increasing annealing temperature the amount of martensite, which represents the former austenite during intercritical annealing, increases. For alloys A, B and C no significant differences of the dissolution kinetics of cementite and the formation kinetics of austenite could be detected. Only the alloy with the reduced manganese content (alloy D) shows significant lower volume fractions of martensite over the whole range of intercritical annealing temperatures investigated.
For all alloys fixed intercritical annealing (780 °C, 60 s) were chosen for the subsequent heat treatments.

Influence of the annealing parameters on the microstructure-Based on the above findings the influence was investigated of both the temperature and the the holding time in the bainitic range on the microstructure via light microscopy.
The two micrographs in fig. 4 show the microstructure of alloy A after a hot dip galvanizing cycle for two different tint etching methods. LePera's etchant, which is a simple tint etching procedure, allows to distinguish between the grains of ferrite, bainite and austenite/martensite. LePera's etchant stains ferrite brown and/or blue, bainite dark-brown to black and martensite and retained austenite remain white (fig. 4-a). Fig. 4-b shows the

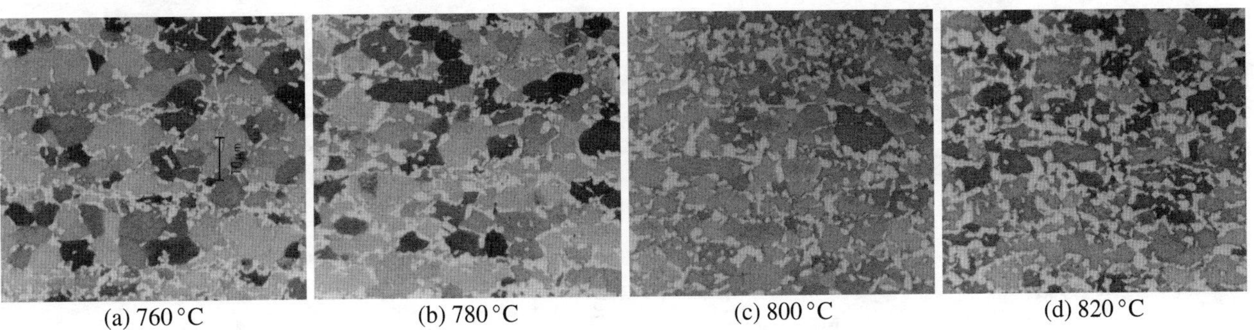

Figure 3: Influence of the intercritical annealing temperature on the microstructure of alloy B (t_{IC}=60 s)

microstructure after applying the two-step etching technique (V2A- and Klemm's etchant). Additionally to ferrite (brown and/or blue) and retained austenite (white), a light brown tint is visible on coarse second phase grains, which are identified as martensite. It seems that the etching with V2A's etchant prior to the colour etching has a similar effect as a tempering treatment [15] in a sense, that the contrast of the martensitic grains is due to a substructure, which is emphasized by a tempering treatment or a V2A-etching. The retained austenite (white) is very small in size, isolated and mainly located at the ferrite grain boundaries and is frequently accompanied by bainite. The martensite grains, which could only be observed in alloy A after a short holding time in the bainitic range, are much coarser than the austenite grains. A significant influence of the chemical composition and the heat treatment on the location and the morphology of the retained austenite could not be detected.

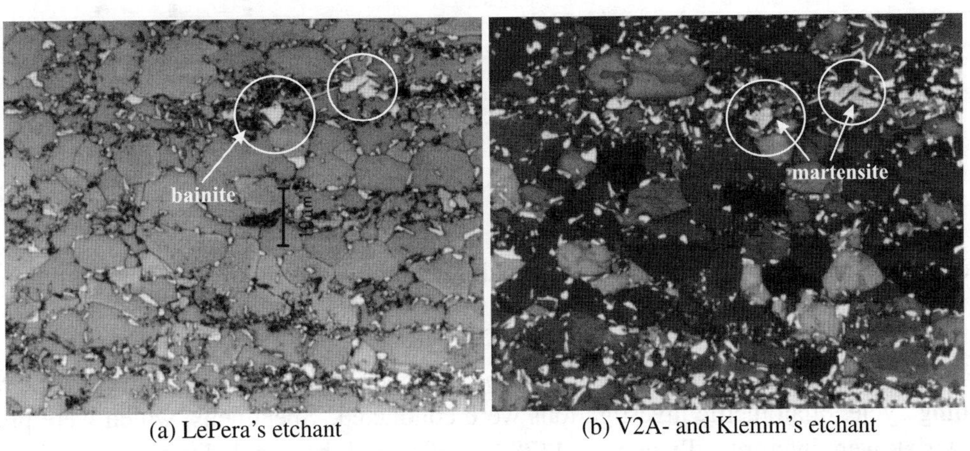

Figure 4: Microstructure of alloy A after a hot dip galvanizing cycle

Fig. 5 shows the amount of the retained austenite as a function of the bainitic annealing parameters. The volume fraction of retained austenite varies between 2 and 14 % and is strongly influenced by the annealing parameters in the bainitic range. Alloys A, B and C behave similarly. The retained austenite decreases with the holding time for T_B=450 °C and T_B=400 °C. For T_B=350 °C the retained austenite first increases and then decreases with the holding time. Compared to alloy A (containing Al) the decrease of the retained austenite with the holding time is less pronounced for alloys B (Al-P) and C (Al-Si). Whereas for alloys A, B and C the highest contents of retained austenite are always associated with high bainitic temperatures and short holding times, alloy D with the reduced manganese content shows the highest contents of retained austenite when annealed at 350 °C or 325 °C applying short holding times.

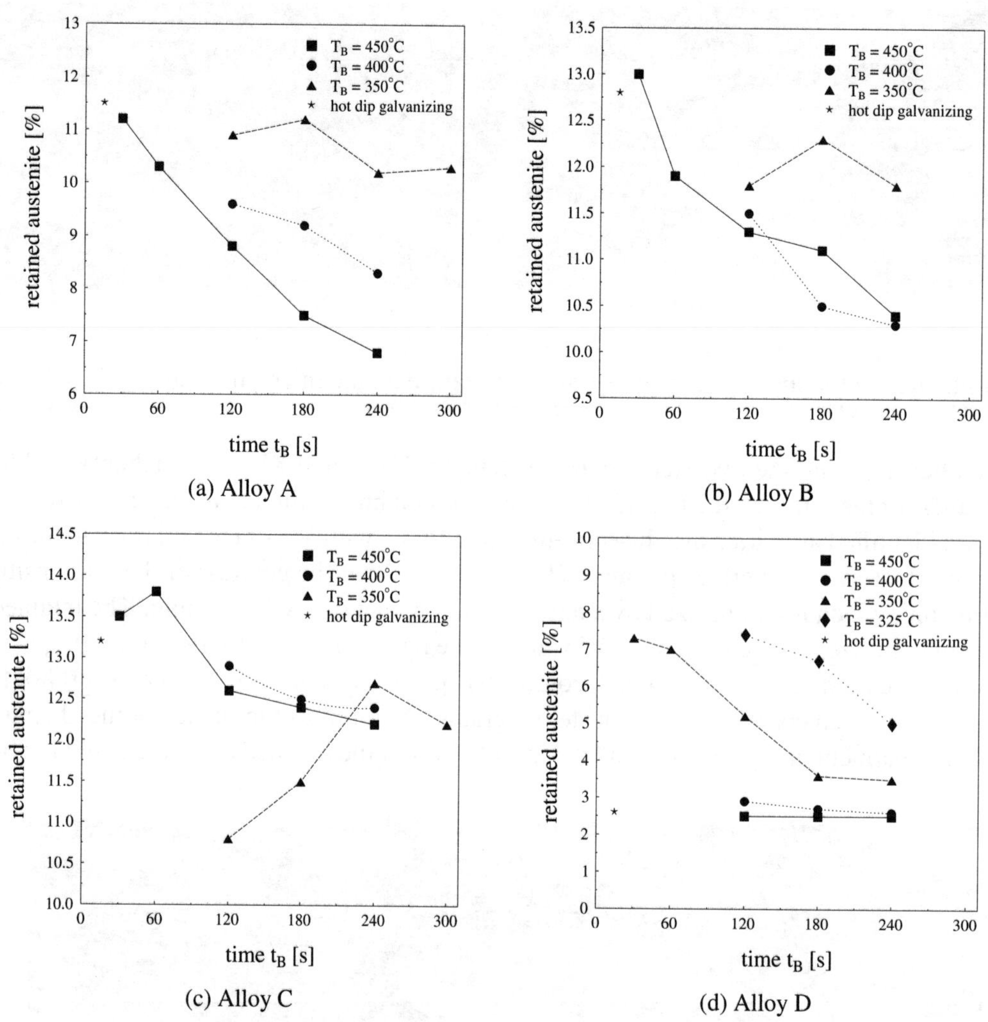

Figure 5: Influence of the annealing parameters in the bainitic range on the volume fraction of retained austenite

Dilatometric investigations-To study the transformation behavior of the austenite in the bainitic range during a TRIP-annealing cycle, dilatometric investigations were conducted. Specimens, which were prepared from cold rolled material, were intercritically annealed (780 °C/60 s) and then cooled with a cooling rate of 60 K/s to the bainitic temperature. A phase transformation during the cooling from the intercritical to the bainitic temperature and then to room temperature could not be observed. Alloys A, B and C show a similar austenite transformation kinetics. Only for alloy D no reaction in the bainitic range could be detected. Fig. 6 shows the transformation kinetics of the austenite during the isothermal holding in the bainitic range for alloy C and different holding temperatures. The bainitic temperature has a strong influence on the austenite transformation kinetic. For the highest bainitic temperature, the reaction is completed after 30 s of isothermal holding. With decreasing temperature the begin of the transformation is shifted to longer holding times and the transformation of the austenite is delayed.
A significant influence of a soft cooling after the intercritically annealing on the transformation kinetics in the bainitic range could not be detected.

Mechanical properties-In figs. 7 to 10 the mechanical properties are shown as a function of the holding time at different bainitic temperatures. For all alloys the strength properties are mostly determined by the bainitic temperature and are not affected significantly by the holding time. The influence of the bainitic annealing parameters is more pronounced for the elongation than for the strength values.
While the tensile strength of alloy A increases with decreasing bainitic temperatures, no significant influence

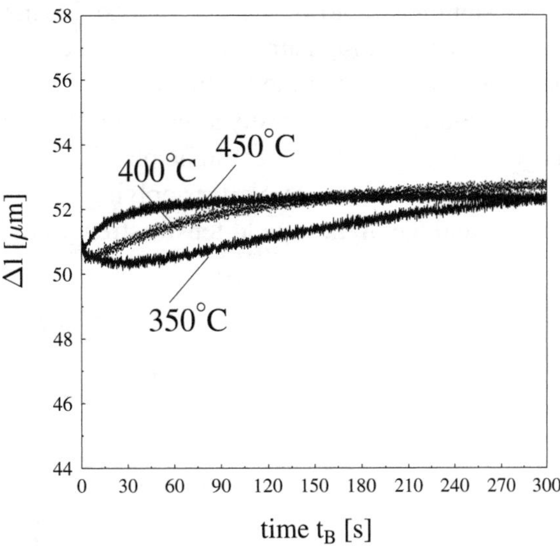

Figure 6: Dilatation-time curves for alloy B obtained from annealing in the bainitic range

of the annealing parameters on the yield strength could be detected (fig. 7-a). For all investigated temperatures, the total and uniform elongation first increase and then decrease with the holding time (fig. 7-b). The maximum for each temperature is shifted to longer holding times with decreasing bainitic temperatures. The highest uniform and total elongations are obtained by applying a hot dip galvanizing cycle.
Similar to alloy A, the influence of the annealing parameters in the bainitic range on the yield strength of alloy B is very low (fig. 8-a). At the lowest bainitic temperature, the highest tensile strength is measured. Compared to alloy A, the yield and the tensile strengths are considerably higher, while the uniform and total elongations show similar values. The highest uniform and total elongations are obtained at a bainitic temperature of 450 °C and a holding time of 60 s.

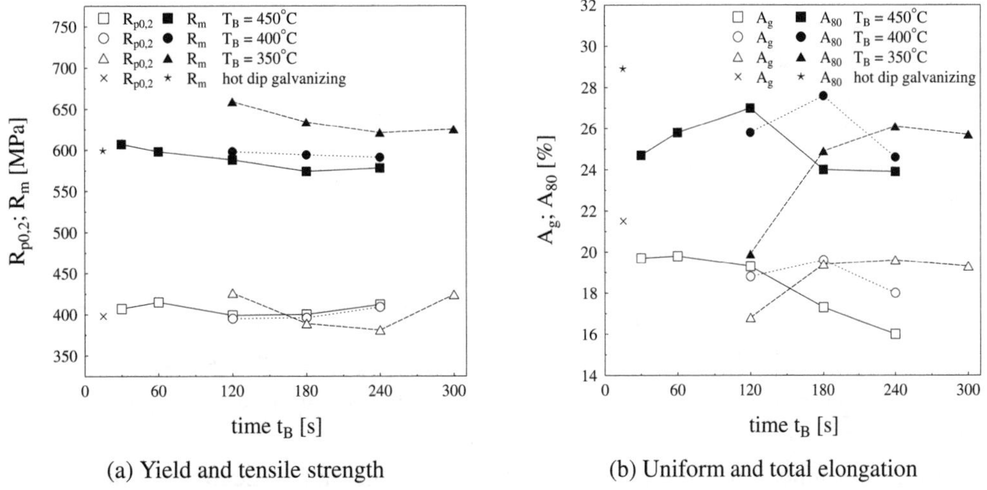

(a) Yield and tensile strength

(b) Uniform and total elongation

Figure 7: Influence of the annealing parameters in the bainitic range on the mechanical properties of alloy A

Alloy C shows a significant higher tensile strength and an increased yield strength for the lowest bainitic temperature. The uniform and total elongation values are in the same range as for alloys A and B. At each temperature a maximum of the uniform and the total elongation is observed, which is shifted to longer holding times for decreasing bainitic temperatures. The influence of the holding time on the elongation values is very pronounced for the lowest bainitic temperature.

The tensile strength of alloy D shows significantly lower values compared to the other alloys investigated and it is not influenced markedly by the bainitic annealing parameters. The yield strength decreases with decreasing bainitic temperatures and increases with the holding time for the lowest temperatures applied (T_B=350 °C and T_B=325 °C). The uniform and the total elongation are hardly affected by the holding time. Only for T_B=350 °C a slight decrease of the uniform and the total elongation could be observed. Whereas alloys A, B and C show a low yield strength and a low yield/tensile strength ratio after applying a hot dip galvanizing cycle, the alloy D possesses the highest yield strength and the highest yield/tensile strength ratio after the hot dip galvanizing cycle.

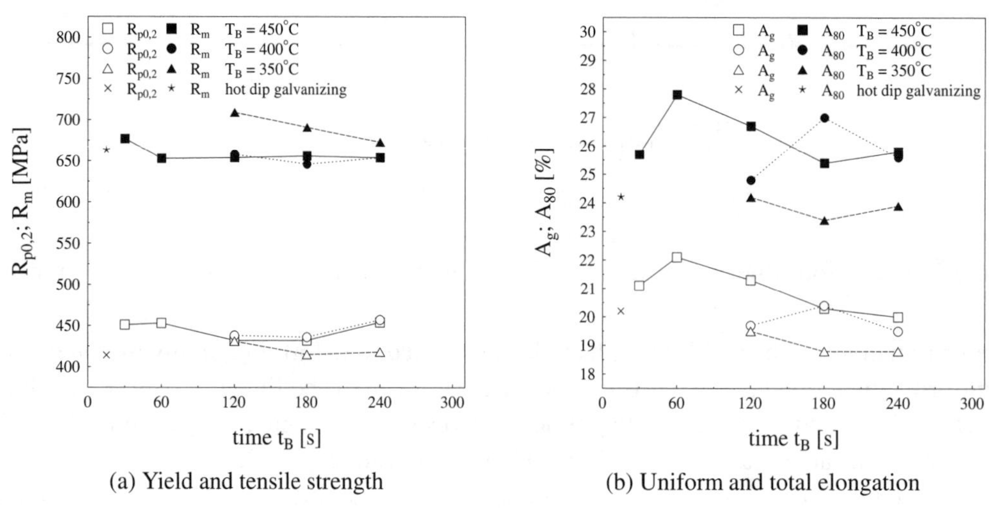

(a) Yield and tensile strength

(b) Uniform and total elongation

Figure 8: Influence of the annealing parameters in the bainitic range on the mechanical properties of alloy B

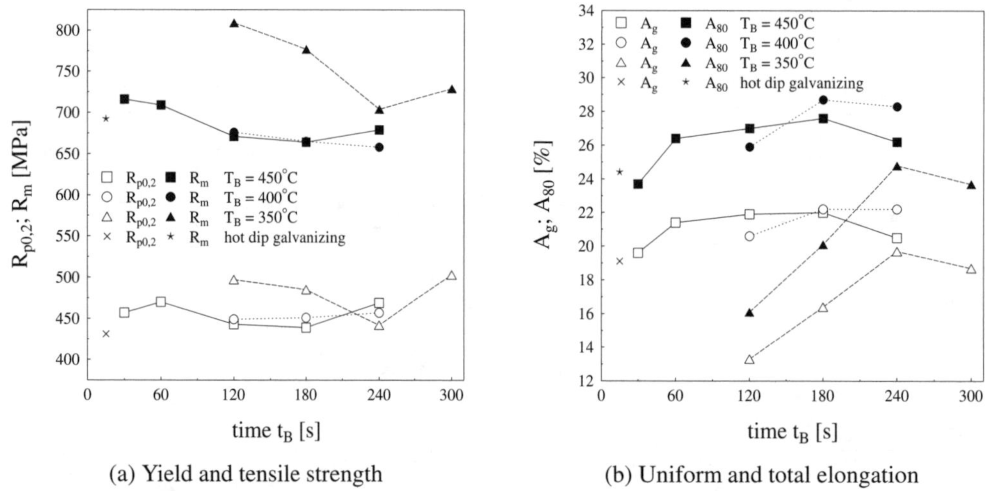

(a) Yield and tensile strength

(b) Uniform and total elongation

Figure 9: Influence of the annealing parameters in the bainitic range on the mechanical properties of alloy C

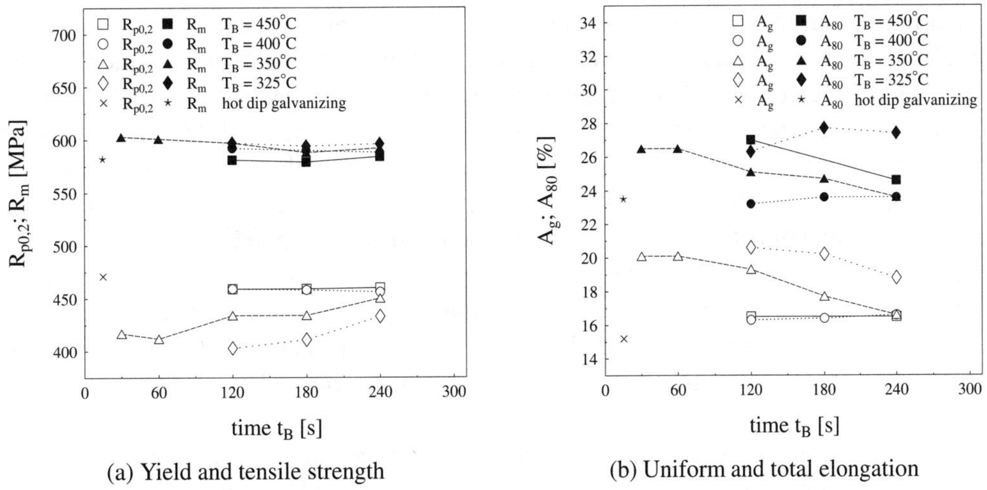

(a) Yield and tensile strength

(b) Uniform and total elongation

Figure 10: Influence of the annealing parameters in the bainitic range on the mechanical properties of alloy D

DISCUSSION

Transformation of the austenite during the heat treatment- Carbon is the most important alloying element for stabilizing the retained austenite. The carbon enrichment of the austenite is made possibly by a two-step heat treatment and is supported by alloying elements preventing carbide precipitation. During intercritical annealing recrystallisation of the cold worked microstructure, the dissolution of cementite and the formation of the austenite take place. The volume fraction and the carbon content of the austenite formed during intercritical annealing is governed by the temperature of the anneals and the alloy's composition. Figs. 11 and 12 show the calculated phase fraction and the carbon content of the austenite as a function of the temperature for all alloys investigated in this work. Since the alloying elements influence the Fe-C-phase diagram, the amount of the austenite formed and its carbon content achieved during these anneals is affected, too. In comparison to the steel alloyed with aluminium (alloy A), additions of silicon or phosphorus increase both the Ae_1 and A_3 temperature and result in a higher carbon concentration in the austenite formed at 780 °C due to its lower volume fraction. The same can be achieved by reducing the amount of manganese. The intercritical annealing parameters are very important for the optimization of the heat treatment applied to TRIP-steels, since they set up the volume fraction of the austenite and its chemical composition for the successive bainitic transformation.

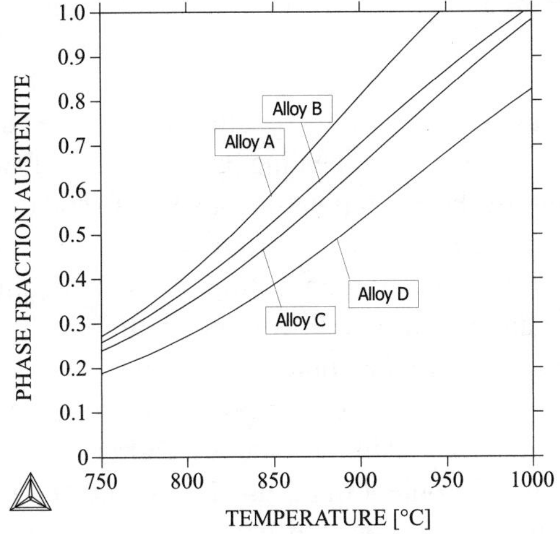

Figure 11: Equilibrium fraction of austenite as a function of the temperature

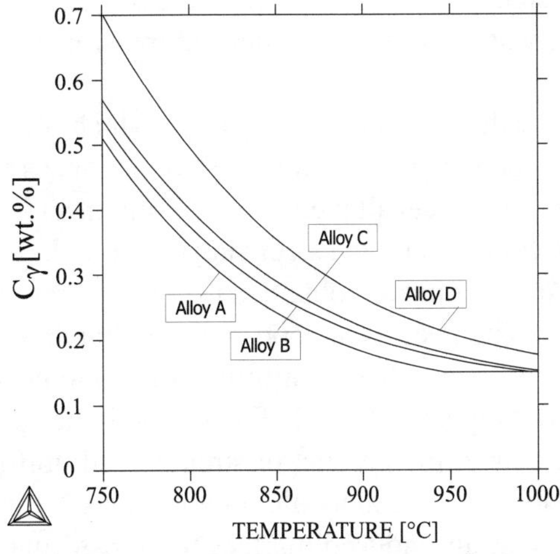

Figure 12: Carbon content of the austenite as a function of the temperature

During isothermal holding in the bainitic range the austenite is further enriched with carbon and gains additional stability. The content of the retained austenite in the final microstructure and the mechanical properties of the alloy are strongly affected by the bainitic annealing parameters. Since the optimum combination of high strength and high ductility is determined by the competition between the volume fraction of the retained austenite and its stability against the strain induced martensitic transformation, the maximum values of the ductility measures total and uniform elongation do not always coincidence with the maximum in retained austenite. At the beginning of bainitic annealing the carbon content of the retained austenite increases with the holding time due to carbon rejection from the bainite. Applications of longer holding times reduces the austenite carbon concentration by carbide precipitation, and due to the austenite to bainite transformation also the amount of retained austenite and the austenite particle size. This combined effect of chemical and size stabilisation is responsible for the maximum of the uniform and total elongations when plotted as a function of the holding time in the bainitic range.
Dilatometric measurements on alloys A, B and C revealed that the transformation kinetics in the bainitic range is influenced markedly by the bainitic temperature, but is hardly affected by the alloy's composition. For the alloy with 1 %Al and 0.1 %P the reaction in the bainitic range at T_B=450 °C is completed after 60 s of isothermal holding, although the content of the retained austenite measured as a function of the holding time decreases continuously up to holding times of 300 s. This is an indication for carbide precipitation resulting in a loss of carbon in the austenite and coincides with the observed reduction of the total and the uniform elongation. Since the strength level and the content of retained austenite can be increased without a loss in ductility by addition of 0.1% P to an alloy with 1% Al, it can be concluded that phosphorus helps to stabilize the retained austenite. Similar to phosphorus, the partial substitution of silicon by aluminium (0.6 %Si in combination with 1 %Al) increases the content of retained austenite and improves the mechanical properties of these alloys. Even though the content of retained austenite of the silicon alloyed steel decreases with the holding time for T_B=450 °C, the total and uniform elongation increase indicating that the stability of the retained austenite is increased. Hence, adding 0.6% Si is a very effective way to prevent carbide formation.
While the transformation behavior in the bainitic range is not influenced significantly by phosphorus or silicon, reducing the manganese content completely alters the transformation behavior: The optimum of the bainitic annealing parameters with respect to the mechanical properties and the amount of retained austenite is shifted to lower temperatures and shorter holding times.

Transformation of the austenite during straining- To study the transformation behavior of the retained austenite during tensile straining, its content was measured as a function of the strain. Fig. 13 shows that the transformation behavior of the retained austenite in alloy B is similar for both annealing treatments applied. Most of the retained austenite transforms into martensite during the first 6 % of straining. Even at a strain of 16 % a retained austenite content between 4 and 5 % was measured, indicating that the retained austenite possesses a high stability against the strain induced transformation.

Fig. 14 shows the product $\frac{R_{p0.2}+R_m}{2} * A_g$, which is a rough measure for the volume specific energy spent during tensile deformation up to the uniform elongation, as a function of the bainitic annealing parameters for all alloys investigated. The steel alloyed with aluminium and silicon shows the best combination of strength and uniform elongation, followed by the steel alloyed with aluminium and phosphorus. For alloys A, B and C the highest values for the product are obtained using high bainitic temperatures and short annealing times. Reducing the manganese content shifts the optimum combination of strength and ductility to lower temperatures.
The product of strength and uniform elongation generally increases with the content of retained austenite for all alloys investigated (fig. 15). Even though the reduction of the manganese content (alloy D) results in less retained austenite, the product of strength and uniform elongation shows similar values as alloy A. The partial substitution of silicon by aluminium (alloy C) increases both the content of retained austenite and its stability against strain induced martensitic transformation resulting in a pronounced increase of the product of strength and ductility. Compared to alloy A, the steel alloyed with phosphorus (alloy B) shows slightly increased amounts of retained austenite and an increased product.

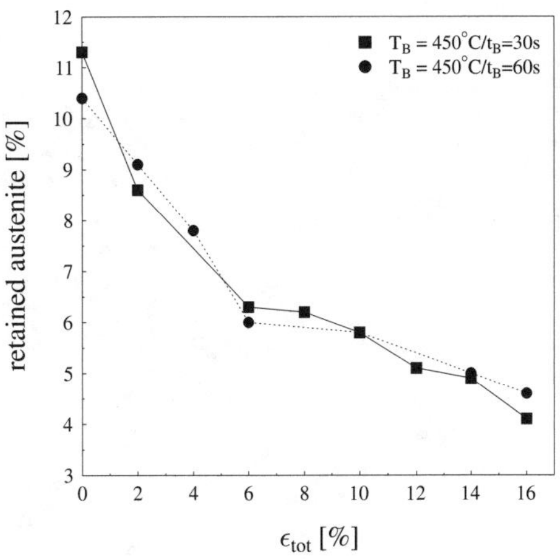

Figure 13: The content of retained austenite as a function of the tensile strain (Alloy B)

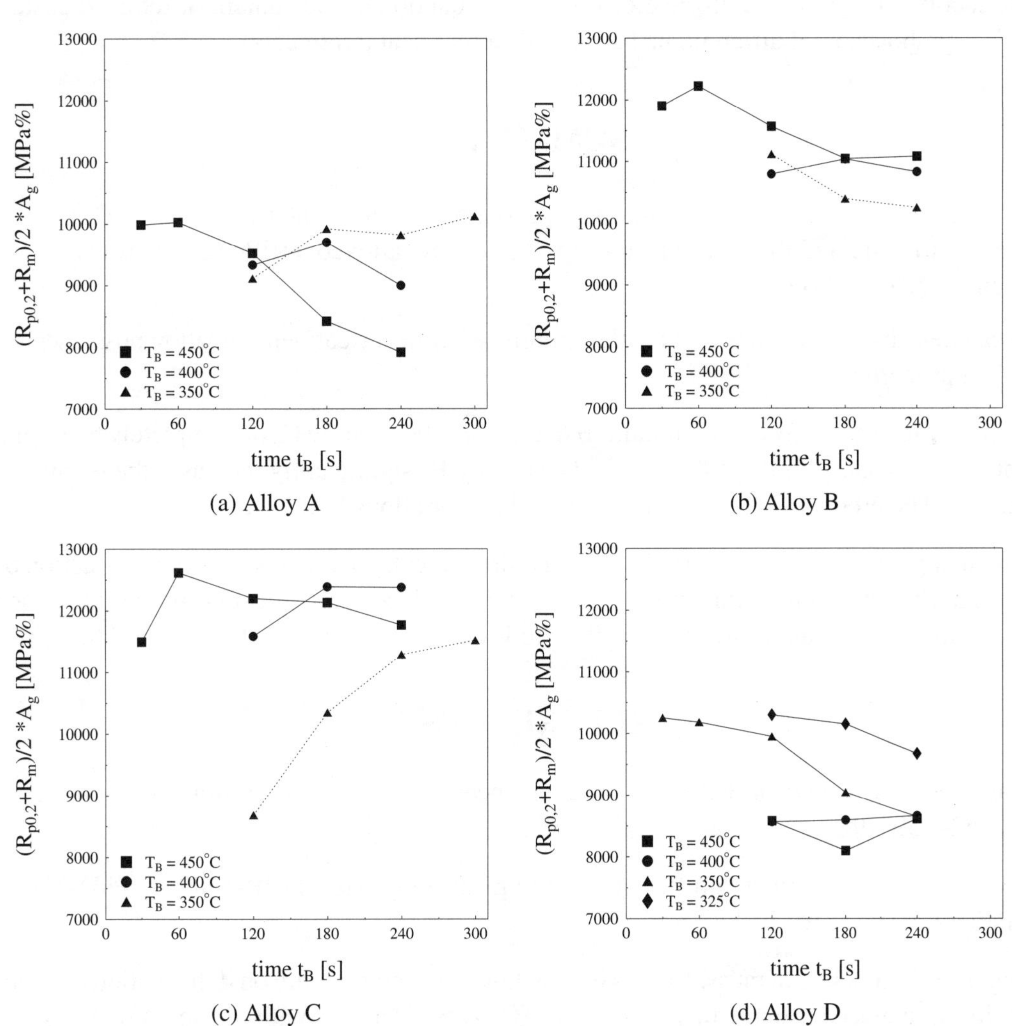

Figure 14: Product of strength and uniform elongation as a function of the bainitic annealing parameters

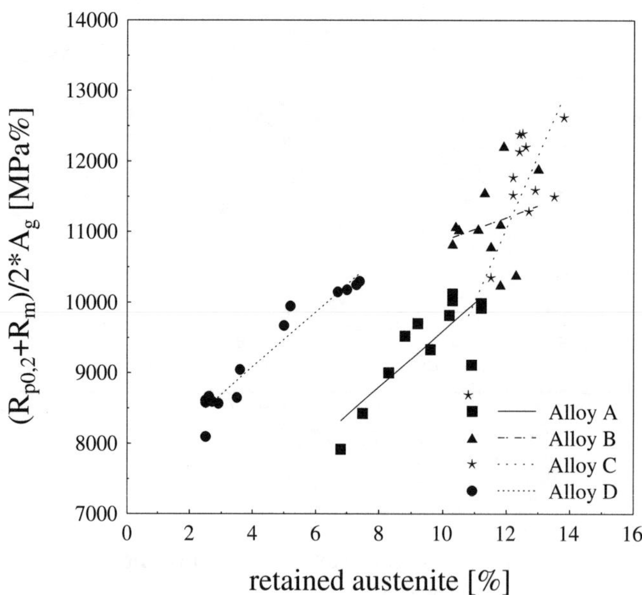

Figure 15: Product of strength and uniform elongation as a function of the content of retained austenite prior to straining. The symbols do not differentiate between different heat treatments.

SUMMARY

In this work, the influence of additions of aluminium, phosphorus, silicon and manganese and of the heat treatment on the microstructure and the mechanical properties of low-alloyed TRIP-steels was investigated. The main results can be stated in brief:

- Silicon can be substituted as a whole by aluminium. After heat treatment the alloy possesses acceptable mechanical properties.
- Substituting silicon partially by aluminium (0.6 %Si and 1 %Al, alloy C) or completely by a combination of aluminium and phosphorus (0.1 %P and 1 %Al, alloy B) significantly increases the retained austenite in the alloys. The product $\frac{R_{p0.2}+R_m}{2} * A_g$ is the highest for alloys B and C.
- Reducing manganese from 1.5% to 0.6% (alloy D) dramatically decreases the volume fraction of austenite formed during the intercritical anneals. To obtain optimized mechanical properties the annealing parameters in the bainitic range have to be shifted to lower temperatures and shorter holding times.

REFERENCES

1. V.F.Zackay, E.R.Parker, D.Fahr, R.Busch. The enhancement of ductility in high-strength-steels. Trans. ASM, 60, 252–259, 1967.
2. I.Papst. Mikrocharakterisierung von Restaustenit in niedrig legierten TRIP-Stählen. PhD-Thesis , Montanuniverstität Leoben, 1997.
3. W.C.Jeong, D.K.Matlock, G.Krauss. Effects of tensile-testing temperature on deformation and transformation behavior of retained austenite in a 0,14%C-1,2%Si-1,5%Mn steel with ferrite-bainite-austenite structure. Materials Science and Engineering, A165, 9–18, 1993.
4. G.Reisner, E.A.Werner, F.D.Fischer. Micromechanical modeling of martensitic transformation in random microstructures. Int. J. Solids Structures, 35, 2457–2473, 1998.

5. H.-C.Chen, H.Era, M.Shimizu. Effect of phosphorus on the formation of retained austenite and mechanical properties in Si-containing low-carbon steel sheet. Metallurgical Transactions A, 20A, 437–445, 1989.

6. Y.Mi. Effect of Cu, Mo, Si on the content of retained austenite of austempered ductile iron. Sripta Metallurgica et Materialia, 32, 1313–1317, 1995.

7. A.Pichler, P.Stiaszny. TRIP steels with reduced silicon content. Steel Research, 70, 459–465, 1999.

8. M.De Meyer, D.Vanderschueren, B.C.De Cooman. The influence of the substitution of Si by Al on the properties of cold rolled C-Mn-Si TRIP steels. ISIJ International, 39, 813–822, 1999.

9. S.Traint, E.A.Werner, A.Pichler, P.Stiaszny. Niedriglegierte TRIP-Feinbleche mit Kupferzusatz. BHM, pp. 362–368, 1999.

10. F.S.LePera. Improved etching technique to emphasize martensite and bainite in high-strength dual-phase steel. Journal of Metals, pp. 38–39, 1980.

11. J.Ammann. Wärmebehandlung und Gefügecharakterisierung von niedrig legierten TRIP-Stählen mit Silizium und Kupfer. Master Thesis, Montanuniversität Leoben, 1997.

12. G.Petzow. Metallographisches, keramographisches, plastographisches Ätzen. Gebrüder Bornträger, Berlin Stuttgart, sixth ed., 1994.

13. H.Weberberger. Magnetische Messungen an Proben beliebiger Form mit dem Fluxmeter. Elektronik und Maschinenbau (E und M), 87, (2), 103–110, 1970.

14. VOEST-ALPINE MECHATRONICS. Magnetjoch zur Bestimmung des Restaustenitgehaltes in Stählen. Betriebsanleitung, 1999.

15. E.Girault, P.Jacques, P.Harlet, K.Mols, J.Van Humbeeck, E.Aernoudt, F.Delannay. Metallographic methods for revealing the multiphase microstructure of TRIP-assisted steels. Materials Characterization, 40, 111–118, 1998.

Development of as-hot-rolled low-silicon and micro-alloyed dual-phase steels

A. BODIN, J. FLEMMING, E.F.M. JANSEN*

Corus Research, Development & Technology, IJmuiden Technology Centre,
Postbus 10.000, 1970 CA IJmuiden, The Netherlands, ☎+31.251.497615

Key words: hot-rolled, high strength steels, dual-phase steels, automotive, laboratory research.

SUMMARY

This paper describes the development of two low-silicon dual-phase steels. Two tensile strength levels have been considered, 600 MPa and 800 MPa. For the DP600, two options have been investigated in the laboratory and tested in a full-scale mill trial. Both options proved to be technologically successful. For the development of the DP800 various kinds and levels of micro-alloying additions were used. The best recipe for the DP800 was then validated with an embedded ingot trial in a hot-strip mill. The results of this trial proved to be very promising. Full-scale trials will now have to be undertaken to confirm these findings.

INTRODUCTION

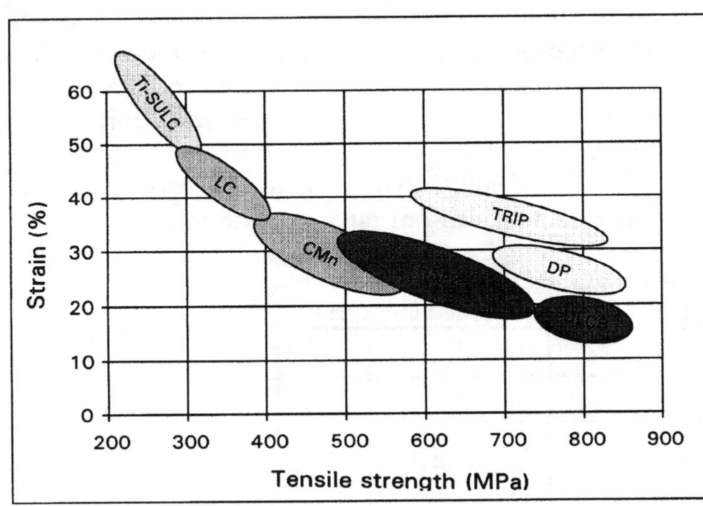

Figure 1: Hyperbolic relation between strength and strain

Dual-phase and TRIP-steels attract the attention of the metallurgist as these steel grades allow escape from the hyperbolic relation between tensile strength and uniform strain that is usually observed in CMn-steels (figure 1). Dual-phase steels have the advantage that they have continuous yielding and a high work-hardening coefficient. TRIP-steels are claimed to have a better combination of strength and formability than dual-phase steels, but it has been difficult to translate the laboratory results to practice. One of the problems in producing TRIP-steels is the high silicon-content that is necessary to stabilise the austenite. This way the formation of martensite and carbides can be prevented. High silicon contents cause

* Corresponding author (Loes.Jansen@Corusgroup.com).

formation of oxides on the surface that prove to be difficult to remove during and after rolling. This red oxide is still visible on the final product (often referred to as 'tiger-stripes') and considered to be unacceptable for most applications, even when the final part is no longer visible in the structure in which it is applied. Although a lot of research is being done in the search for alternatives to silicon (e.g. aluminium), as yet no real break-through has been reported that promises an economically viable process route for the production of hot-rolled strip.

Table I: Comparison of different aspects of DP-, TRIP- and HSLA-steels

Characteristic	Dual-phase	TRIP	HSLA
Rp	low	low	high
Rm x A	medium	high	low
n	high	high	low
hole-expansion	high	low	high
fatigue	high	medium	high
surface	good	problems	good
casting	good	problems	good
rolling	easy	medium	medium
cooling	easy	medium	easy
ease of production	good	problems	good

Preliminary research shows that dual-phase steels in a component can outperform TRIP-steels. The hole expansion rate and the fatigue properties of dual-phase steels are reported to be better than those of TRIP-steels [1]. A very important aspect of hot-rolled dual-phase steels is that they can be produced on existing installations (both steel-making and hot-rolling). Furthermore, demand for dual-phase steels seems to be small but real, whereas demand for hot-rolled TRIP-steels is fairly theoretical at the moment as it is not commercially available yet. Table I summarises the differences between dual-phase and TRIP-steels in comparison to an HSLA-steel.

LABORATORY EXPERIMENTS

DP600

To stabilise the austenite during cooling two metallurgically different mechanisms can be employed: stabilising by alloying and stabilising by fast cooling. In the first case a 'normal' coiling temperature can be used, in the second case very low coiling temperatures have to be used. For both options several laboratory casts have been produced. These allowed us to optimise the chemical composition and the processing conditions.

DP600-High coiling temperature option

Alloying with molybdenum or chromium is a well known method for producing as-hot-rolled dual phase steels without the necessity of a very low coiling temperature. Usually, coiling temperatures range from 400 to 550°C. Laboratory casts were made of various compositions as shown in Table II. The samples of 60 mm pre-processed material were annealed at 1250°C for 30 minutes and subjected to a roughing schedule (60 - 50 - 40 mm with finish roughing at 1050°C)) and a finishing schedule (40 - 27 - 17.8 - 12 - 8 - 6 - 4.7 - 4 mm with a finishing temperature of 900, 850 or 800°C). The cooling rate after finish rolling was 40°C/s, consistent with values obtained on a run-out table cooling in a hot strip mill.

Table II: Chemical composition of the DP600 laboratory casts and industrial trials and the DP800 embedded ingot trial

Code	Type	C	Mn	Si	Cr	Mo
C06 Cr	lab	0.060	1.480	0.950	0.950	0.001
C08 Cr	lab	0.080	1.590	1.090	1.000	0.001
C06 CrMo	lab	0.060	1.290	1.000	0.515	0.439
DP600-H	mill	0.059	1.270	0.998	0.475	0.372
DP600-L	mill	0.079	1.490	0.111	0.020	0.001
DP800-EI	emb.ingot	0.110	1.250	0.140	0.019	0.009

For the chromium alloyed steels, a low austenitic finishing temperature in combination with a coiling temperature of 500°C was sufficient to generate dual-phase microstructures and properties. This result as well as the observation that the carbon content influences the strain and tensile strength considerably, was in agreement with the observations reported by Vlad [2]. Because high chromium contents are known to cause problems during casting, another series of trials was done using molybdenum to replace part of the chromium (alloy C06 CrMo). These alloys also provided dual-phase microstructure and properties (Table III). Based on these results, a mill trial was successfully performed using the CrMo-alloy (DP600-H).

Table III: Mechanical properties obtained during laboratory and mill trials on DP600 (mechanical properties for laboratory trials are longitudinal values).

Code	n	h (mm)	direction	Rp (MPa)	Rm (MPa)	A80 (%)	Rp/Rm (MPa)
C06 Cr	2	4.0	//	339	626	24.2	0.54
C08 Cr	2	4.0	//	368	706	20.1	0.52
C06 CrMo	2	4.0	//	345	667	21.2	0.52
DP600-H	20	3.2	⊥	391	651	25.7	0.60
DP600-L	4	3.20	⊥	388	613	26.6	0.63

DP600-Low coiling temperature option

If the capacity and the physical length of the run-out table cooling are sufficient to be able to obtain very low coiling temperatures (below 150°C) and to realise an interrupted cooling for the ferrite to form, alloying contents can be drastically reduced. There is no need for alloying with chromium or molybdenum to stabilise the austenite during cooling after rolling. Careful tuning of the isothermal holding temperature and time will enable ferrite formation while preventing pearlite formation and after formation of about 80% ferrite, the remaining austenite will be quenched into martensite.

DP800

Various laboratory casts were made (Table IV). Samples of 60 mm pre-processed material were annealed at 1200°C for 30 min and subjected to a roughing schedule (60-50-40 mm with finish roughing at 1100°C) and a finishing schedule (40-27-18-12-8-6-4.7-4 mm with different finishing temperatures, ranging from 850°C to 900°C). After rolling, the specimens were air-cooled to a holding temperature

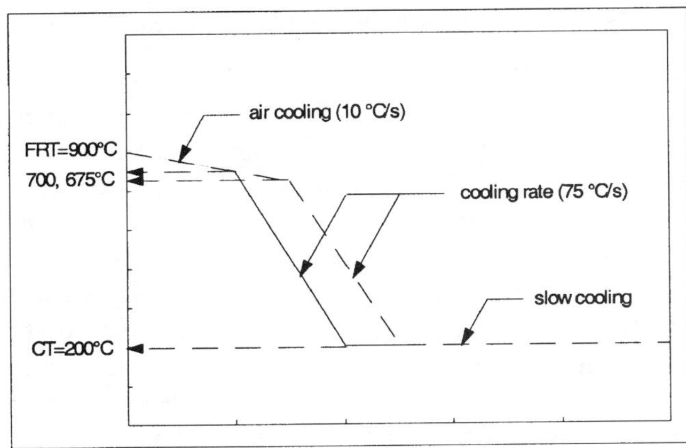

Figure 2: Cooling-simulation during pilot-hot rolling process for DP800.

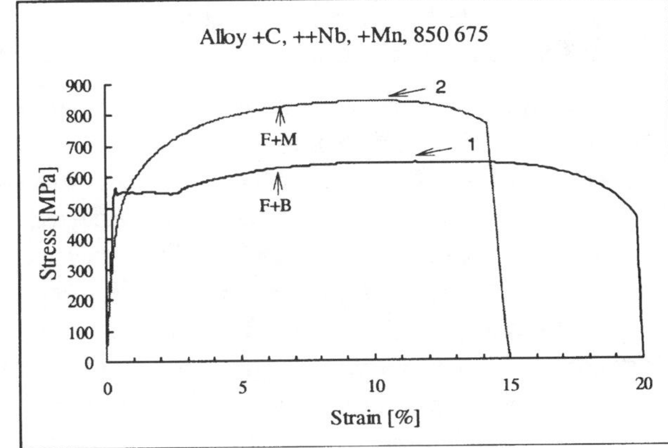

Figure 3: Mechanical properties of two tensile samples taken from the same strip +C, ++Nb, +Mn 675

Table IV: Systematics of the chemical composition of the DP800 laboratory casts (e.g. C+ indicates 0.15%C, n.a.: not applicable).

Code	C	Mn	Si	Ti	Nb	Cr	Mo	B	N
base	0.11	1.000	0.15	0	0	0	0	0	40
+	0.15	1.500	n.a.	0.1	0.03	0.5	0.5	20	n.a.
++	n.a.	n.a.	n.a.	0.2	0.05	n.a.	n.a.	n.a.	n.a.

and subsequently water-cooled to 200°C. Afterwards the specimens were allowed to cool down slowly (figure 2). Temperature control is very important in these steels. An indication for the sensitivity of these steels to changes in cooling conditions is given by figure 3. Microstructural and chemical analysis showed that these variations are caused by different microstructure (ferrite-martensite or ferrite-bainite) due to inhomogeneous run-out table cooling. Temperature measurements during the laboratory rolling show that coiling temperature differences from the centre to the side of a sample of 100°C are not exceptional. Application of different cooling headers reduced this problem significantly.

RESULTS OF LABORATORY EXPERIMENTS

Figure 4 summarises the results of the laboratory rolling experiments aimed at DP800. It can be seen from this figure that, although the strength increases for higher carbon, manganese and niobium contents, the potential of titanium for giving a strength increase is markedly higher. Niobium-containing grades show strength levels of up to 700 MPa whereas Ti-containing steels show strength levels of up to 900 MPa. So achieving the desired strength level seems possible. Earlier results showed that strain values obtained in laboratory rolling experiments usually underestimate the values from mill trials. Therefore it is reasonable to assume a CMnTi alloy will yield the desired properties.

An increase in manganese content usually leads to higher strength values with roughly similar yield stresses, thus decreasing the yield to tensile stress ratio. Increasing carbon from 0.11% to 0.15% does not have a large influence on the strength level, in spite of its larger potential for martensite formation.

Microstructural analyses using electron microscopy revealed that the increased carbon and manganese level resulted in triple-phase structures rather than in dual-phase ferrite-martensite or ferrite-bainite, despite the low coiling temperatures. In the titanium-based alloys martensite formed more easily resulting in improved properties with respect to the niobium-alloyed materials. Some typical microstructures for the titanium-based alloys are shown in figure 5. Figure 5a shows a nice dual-phase structure consisting of a ferrite matrix and martensite islands dispersed in the ferrite. Figure 5b shows an almost single phase structure consisting of ferritic bainite. Figure 5c shows an example of a microstructure that was quenched while pearlite formation was already proceeding. This figure illustrates that the quenching temperature for this particular steel should have been higher. Figure 5d again shows a perfect dual-phase ferrite-martensite microstructure.

From the laboratory studies the influence of several alloying elements on the mechanical properties has been assessed. Adding manganese sometimes causes a decrease in the yield strength, particularly for those samples that were cooled from low intermediate temperatures. This is caused by a reduction in the pearlite content and an increase of the martensite content. Manganese stabilises the austenite and hence retards the formation of pearlite. The addition of niobium causes an increase in tensile strength (most likely) as a result of grain refinement. The combined addition of niobium and titanium causes an even larger increase in strength. The formation of pearlite is prevented because the carbon content that is available for the pearlite formation is reduced as a result of the formation of carbo-nitride precipitates. The addition of a large amount of titanium in combination with manganese is not relevant for the formation of a dual-phase microstructure because the precipitation of TiC causes a reduction of the carbon content in the austenite below the level required for the formation of martensite during accelerated cooling. These structures consist mainly of ferrite and bainite (Figure 5b). The addition of

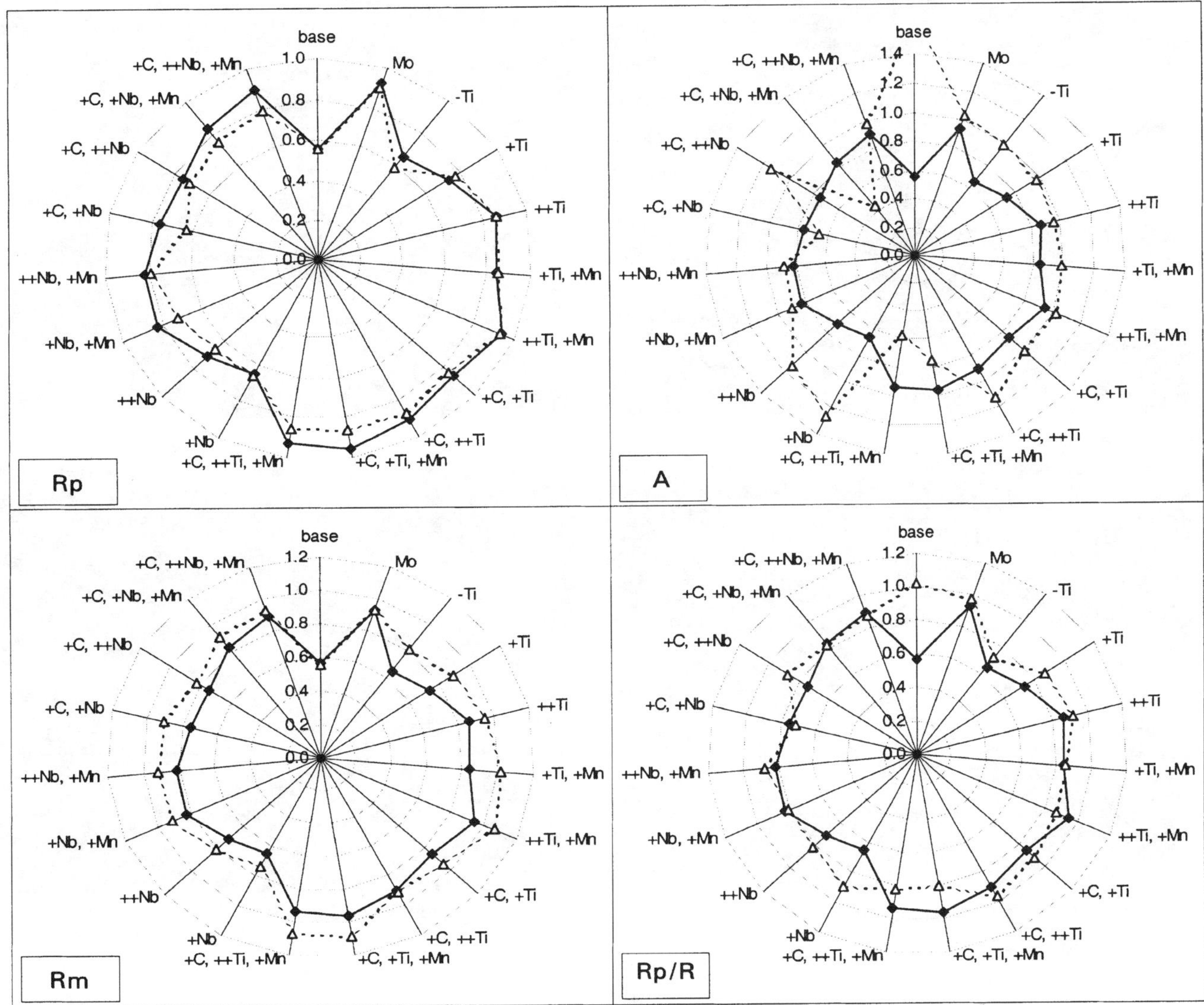

Figure 4: *Normalised mechanical properties (norm: Rm=790 MPa, Rp=640 MPa, A80 = 18%, closed markers: holding temperature = 700°C, open markers: holding temperature = 675°C).*

molybdenum or chromium cause the retardation of the pearlite formation and hence facilitate the formation of a dual-phase microstructure. However, the level of mechanical properties that were achieved by alloying with either of these elements remained below the target-values.

Base upon the chemical composition, the steels that are able to generate a dual-phase structure with the required properties can be divided into steels that contain no Ti- or Nb and steels that do. The microstructure of the former consists primarily of ferrite, pearlite and martensite, the latter mainly of ferrite, martensite and in some cases a little bainite. Pearlite is not present. The main exception is the ++Ti-grade because this contains only ferrite or ferritic bainite (Figure 5b). As a result of the high Ti-content all carbon is taken from solution by formation of TiC. For this alloy this means that only 0.02% carbon remains in solution. No pearlite nor martensite can be formed at these low levels carbon. The grade denoted by + Mn, ++ Ti also contains a high Ti-content. In the microstructure of this steel martensite can be found because manganese reduces the activity of carbon. This in turn reduces the tendency to formation of TiC-precipitates [3].

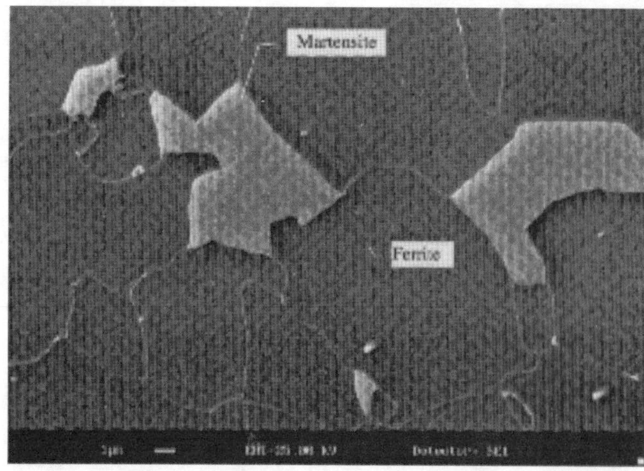

a). +Ti, FRT=850°C, Holding temperature 675°C

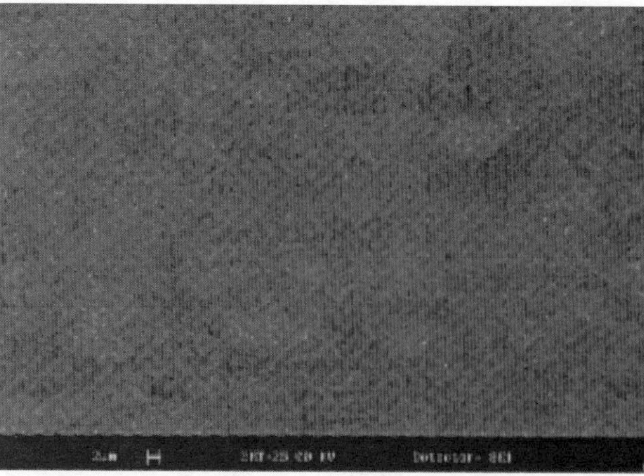

b). ++Ti, FRT=850°C, Holding temperature 675°C

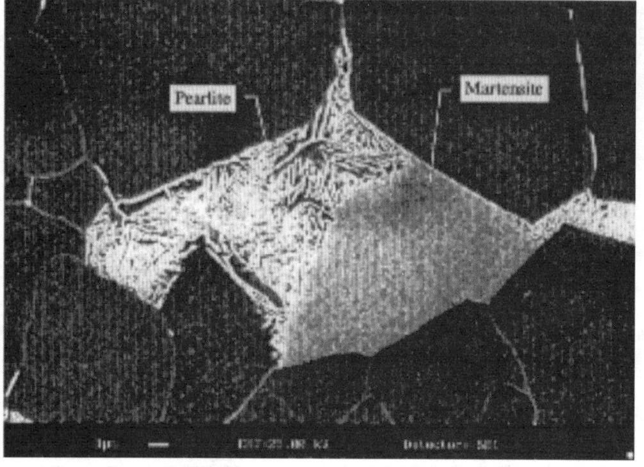

c). +Ti, +B, FRT=850°C, Holding temperature 675°C

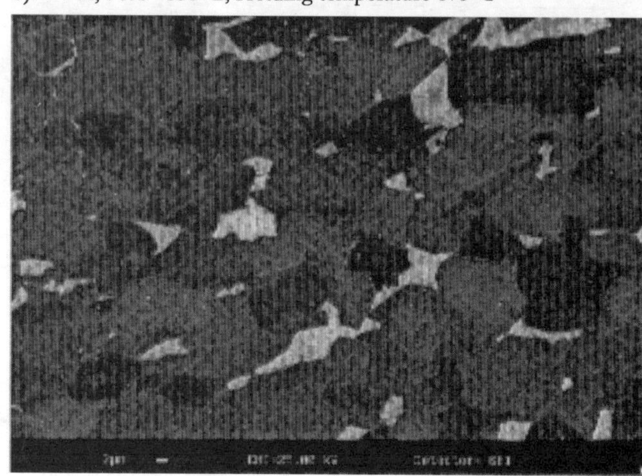

d). +Mn,++Ti, FRT=900°C, Holding temperature 675°C

Figure 5: *Some selected microstructures for Ti-containing alloys.*

It appears that the tensile strength is higher and the yield to tensile strength ratio is lower if the intermediate holding temperature is not too high (675°C). From the microstructures it appears that these structures usually contain more martensite than those held at 700°C. The latter contain bainitic structures or even Widmannstätten ferrite. In particular the role of manganese seems to be pivotal. It is well known that manganese decreases the transformation temperature from austenite to ferrite. It shifts the dilatation curve to lower temperatures. From figure 6 it appears that if the transformation shifts to lower temperatures, then the amount of austenite at a given holding temperature is higher and thus the enrichment of the austenite with carbon is lower. This makes the formation of martensite from this austenite difficult. Figure 6 also shows how sensitive the final microstructure will be to variations in intermediate holding temperature and time.

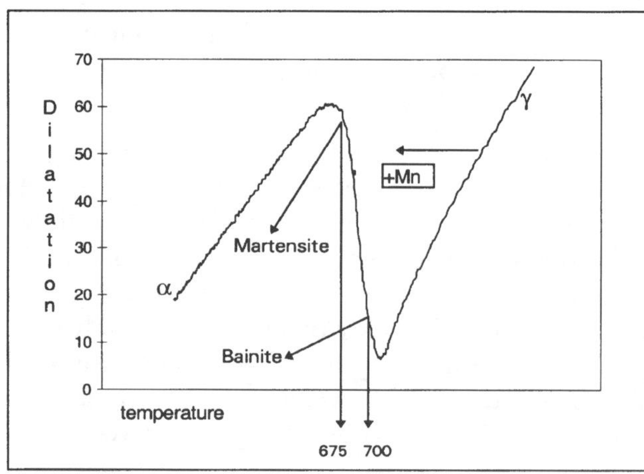

Figure 6: *Influence of intermediate temperature and [Mn] on the phase transformation and the nature of the second phase.*

MILL TRIALS

DP600

From the laboratory work it appeared that both options would yield the desired mechanical properties but the route with the higher alloying content was tried first. From the laboratory work it appeared that a high silicon alloy would result in the best mechanical properties. However, as a result of the high silicon content problems with surface quality were to be expected. The question was whether the high Si-content would lead to an even oxide coverage rather than the banded 'tiger-stripe' contamination.

DP600: high alloy content and coiling temperature

A steel with a composition as given in Table II, grade DP600-H, was austenitically rolled in the hot strip mill in IJmuiden. It was cooled on the run-out table using both continuous cooling and interrupted cooling and the target coiling temperature was 550°C. The average mechanical properties are presented in Table III. Although this material satisfied the customer's demands as to formability and weldability, the material was not considered acceptable because of contamination of the strip with oxide bands as a result of the high silicon content. So even with this high a silicon content, there was no even coverage with oxide. This experience forced us to accelerate the development of the low-silicon grades.

DP600: low alloy content and coiling temperature

The run-out table cooling at the hot-strip mill in IJmuiden enables very high cooling rates and very low coiling temperatures. The length of the run-out table cooling also enables interrupted cooling patterns. This facility has been used extensively in the production of CMn- and HSLA-steels with coiling temperatures of 400°C. The development of these steels has been reported elsewhere [4, 5].

Laboratory research showed that for production of dual-phase material a maximum coiling temperature of 200°C would be required to induce martensite formation and that an interrupted cooling would provide the best mechanical properties. Table III shows the average values of the steel grade DP600-L, of which the chemical composition is given in Table II. This material has been supplied and succesfully processed at industrial speeds on the facilities of one of Corus' customers. This was done at the speed and settings of the equipment that the operators were used to use for dual-phase grades. Figure 7 shows a representative microstructure of the DP600 material. The image shows that a fine ferrite matrix with finely dispersed martensite grains was obtained. Figure 8 shows a wheel disc that has been produced of this material at one of Corus' customers.

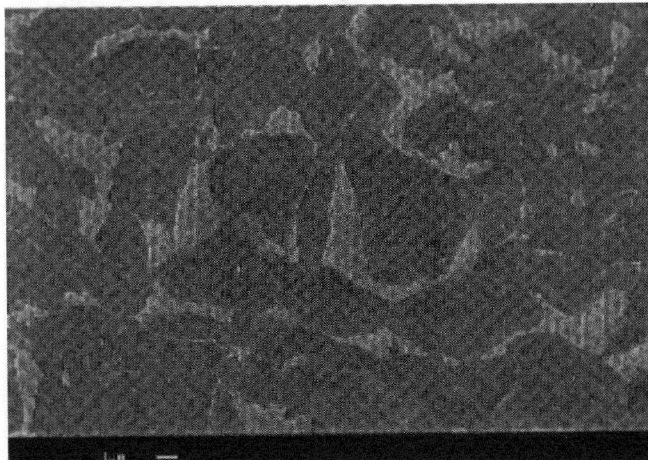
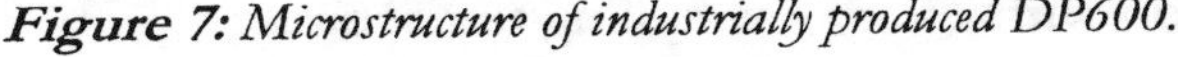

Figure 7: *Microstructure of industrially produced DP600.*

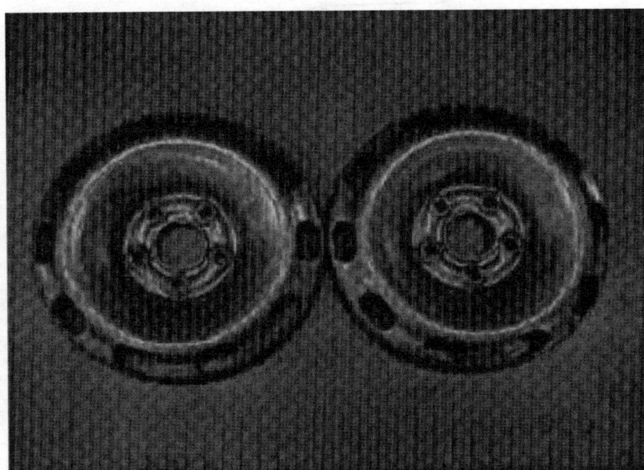

Figure 8: *Wheel-disc from industrially produced DP600.*

DP800: Embedded ingot trials

As part of an ECSC research project, aimed at developing high strength steels for wheel-disk applications, a CMnTi-alloy was cast. This laboratory cast, with a chemical composition as given in Table II and denoted by DP800-EI, was embedded into a high-strength steel slab and subsequently hit-rolled in the hot-strip mill. The slabs were reheated at 1250°C and finished at 850°C. An interrupted cooling was applied with a holding temperature of just under 700°C. In total, four trials at different levels of coiling temperature were performed (350, 300, 250 and 150°C).

Table V: Mechanical properties that were obtained from the embedded ingot.

Coiling Temperature	Rp [MPa]		Rm [MPa]		A80 [%]		Rp/Rm [-]	
(°C)	⊥ RD	// RD	⊥ RD	// RD	⊥ RD	// RD	⊥ RD	// RD
150	662	646	927	940	13.4	14.5	0.71	0.69
250	702	662	853	833	13.3	16.7	0.82	0.80
300	740	703	797	781	12.6	16.5	0.93	0.90
350	722	672	795	778	12.5	15.6	0.91	0.86

In Table V and the figures 9 and 10 the results for the mechanical properties are summarised. The mechanical properties look very promising for the forthcoming full mill trials. The somewhat lower total strain is a result of using laboratory casting instead of regular steel. Based upon earlier results, the total strain will be about 2% higher when continuously cast steel slabs are used. Higher coiling temperature result in lower tensile strength values and consequently the ratio of Rp/Rm increases. The lower tensile strength is a result of the formation of bainite as second phase instead of martensite. The difference in properties in longitudinal and transverse direction is generally small. These results will now have to be confirmed in a full-scale mill trial.

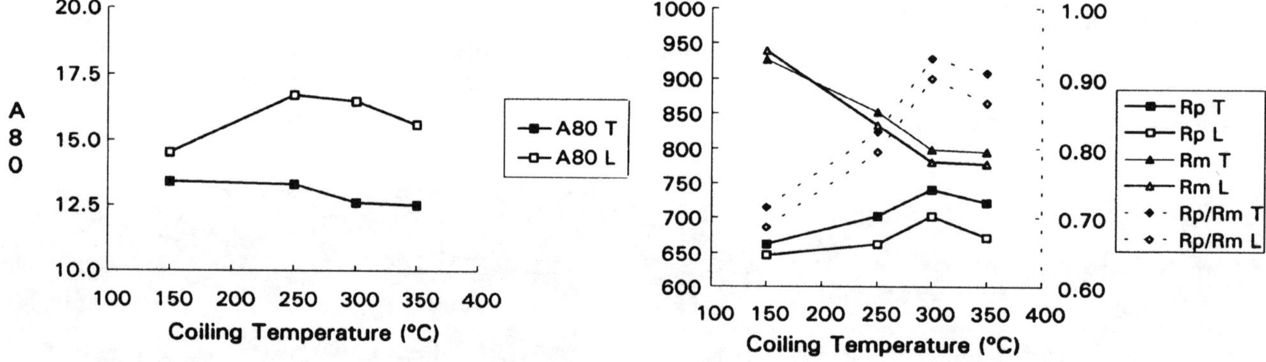

***Figure 9:** Total elongation as function of the coiling temperature*

***Figure 10:** Mechanical Properties as a function of the coiling temperature.*

CONCLUSIONS

Conclusions DP600

- Full-scale industrial trials have shown that dual-phase can be produced by alloying with chromium or chromium and molybdenum in combination with a relatively high coiling temperature or by using a CMn-steel with an interrupted cooling in combination with a very low coiling temperature.
- Trial production and successful processing at one of Corus' customers show that high quality, as-hot-rolled DP600 steel can be industrially produced at Corus' Hot-Strip Mill in IJmuiden.

Conclusions DP800

- As a result of higher martensite fractions, the mechanical properties of the steels that were accelerated cooled from 675°C show better values than the steels cooled from 700°C.
- Increasing the carbon content increases the tensile strength more than the yield strength.
- With higher carbon contents, the steel becomes more sensitive to inhomogeneous cooling. Large differences in properties can be caused by local differences in cooling rate or temperature.
- If an austenite forming element is added to the alloy, the intermediate holding temperature needs to be lowered accordingly in order to prevent suppression of martensite formation upon rapid cooling.

ACKNOWLEDGEMENT

The authors wish to acknowledge the financial contribution of the European Coal and Steel Community to part of the work (DP800). Thanks are expressed towards Dr. Peter Stiaszny and his colleagues of Voest-Alpine Stahl Linz for their contribution to the embedded ingot trial. Furthermore, the authors acknowledge the substantial contribution of Dr. Marcel Onink to this research.

REFERENCES

[1]. A. de Ro et al, Thermomechanical Processing of Steels Conference, London, (2000), pp. 29-38.

[2]. E. Vlad, Stahl und Eisen, 102 (1982), pp. 41-46.

[3]. J.J. Jonas, "Effect of interpass time (and of Mn/Si ratio) on the hot rolling behaviour of microalloyed steels", presentation at "Themadag Hoge Sterkte Staal", Gent, 27th of April 1998

[4]. A. Bodin, A.B.C. Edelman, P. M. Hekker, 36th MW&SP Conference), Baltimore, 1994, pp. 17-25.

[5]. A. Bodin, J. Flemming, M. Onink, 41st MW&SP Conference, Baltimore, 1999, pp. 61-66.

PHASE TRANSFORAMTION DURING ANNEALING OF A COLD-ROLLED DUAL PHASE STEEL GRADE

A. Pichler, S. Traint*, G. Arnoldner, E. Werner*, R. Pippan**, and P. Stiaszny

VOEST-ALPINE STAHL LINZ GmbH, VOEST-ALPINE-Strasse 3, P. O. Box 3, A-4031 Linz, Austria
*) Christian-Doppler-Laboratorium für Moderne Mehrphasenstähle, Lehrstuhl A für Mechanik, TU-München, Boltzmannstrasse 15, D-85747 Garching, Germany
**) Institut für Metallphysik, Montanuniversität Leoben, Jahnstrasse 12, A-8700 Leoben, Austria

Key Words: DP grades, intercritical annealing, phase transformation, bainite, martensite, mechanical properties

ABSTRACT

The transformation behavior of a cold-rolled dual-phase (DP) steel grade during the annealing treatment are investigated in detail. The stages of interest are: recrystallization, dissolution of cementite and formation of austenite during heating and soaking, and the transformation behavior during cooling. For the latter the impact of the cooling rate and an isothermal holding after cooling is studied. Microstructure investigations, dilatometer measurements and the mechanical properties are used to clarify the transformation behavior. During intercritical annealing the austenite fraction is mainly determined by the annealing temperature and to a much lesser extent by the soaking time. A very strong impact on the transformation behavior in cooling results from the microstructure formed during annealing in either the austenite phase field or the austenite-ferrite two-phase field (intercritical annealing). The cooling rate markedly influences the transformation behavior which can be explained by carbon enrichment in the austenite. The best DP structure consisting of a mixture of ferrite and martensite is obtained by cooling with intermediate cooling rates.

1. INTRODUCTION

High-strength thin sheet steel grades have gained a considerable market share. The most important driving force for their development is the need for a reduced weight and an improved safety of cars. Based on soft low carbon (LC) aluminum-killed and interstitial free (IF) grades, various high-strength grades have been developed by the steel industry. Owing to their excellent features, bake hardening (BH) [1] and high-strength IF grades [2 - 4] are the most widely used high-strength thin sheets. As a rule of thumb, the upper limits for the yield strength and the tensile strength are about 320 and 450 MPa, respectively, for BH grades based on solid solution hardening and high-strength IF grades. Thin sheet material with even higher strength can be achieved by taking advantage of precipitation hardening in microalloyed steels. A disadvantage of these grades is a strong decrease of their formability with increasing strength. A consistent further step towards high-strength steel grades with excellent formability was the development of dual phase (DP) steels for thin sheet materials already twenty-five years ago [5, 6]. Despite their excellent properties, still very low amounts of these grades are used by the automotive industry in Europe. However, in the last year an unusual strong demand for this steel grade was observed and a further increasing market share is expected in the future.

In the past extensive scientific research work has been done in the field of DP steels. Numerous experimental investigations were conducted concerning the influence of the chemical composition and the annealing treatment of the cold-rolled material on the microstructure and the resulting mechanical properties [5 - 10]. In the annealing treatment the two most important steps are the dissolution of cementite [11 - 14] and the resulting formation of the austenite and the transformation of the austenite into ferrite, pearlite, bainite or martensite during cooling. Concerning the cooling, most of the investigations reported in the literature are restricted to a constant cooling rate from the intercritical annealing temperature to room temperature [5, 6, 15]. Additionally, a large number of investigations focus on the relationship between the microstructure and the mechanical properties [5, 6, 16 - 19].

Despite all these investigations there still exists a considerable gap in knowledge. For an optimized production the knowledge of the phase transformations such as the dissolution and the formation of austenite during intercritical annealing and the growth of ferrite and the formation of pearlite, bainite and martensite during cooling is of essential interest. Above all, the most important field is the impact of different cooling paths on the phase transformation during cooling. Of particular interest is the growth of ferrite during the early stage of cooling and the resulting decrease of the austenite volume fraction and enrichment of the carbon in the austenite. This, in turn, has a strong impact on the formation of pearlite, bainite and martensite during further cooling or during a holding stage at the zinc pot temperature or in the overaging zone.

In this work the phase transformations in a cold-rolled Dual Phase steel is investigated in detail. First the dissolution kinetics of cementite and the resulting formation kinetics of austenite during heating and intercritical annealing is determined from dilatometer tests and microstructure investigations after interrupted annealing cycles. In a further step the phase transformation during cooling is investigated. In this part experiments with different constant cooling rates and intercritical annealing temperatures are done on the dilatometer. Additionally, the impact of a holding stage is investigated in combination with different cooling rates. The transformation kinetics during this isothermal holding stage is analyzed and characterized in detail, based on JMAK kinetics. All these experiments with the dilatometer are supported by microstructure investigations and measurements of the mechanical properties after annealing simulations with interrupted cycles. Based on a correlation between the resulting microstructure and the mechanical properties conclusions concerning the transformations are drawn and compared with the results from the dilatometer investigations.

2. EXPERIMENTAL

2.1 Production of the material

All results in this paper stem from steel produced on a large scale. The heats were melted in an LD converter. The slabs were cast on a 1-strand continuous caster. The chemical composition of the investigated steel is given in Table 1. For hot rolling the slabs were reheated up to 1,200 °C in pusher-type furnaces. Hot rolling was conducted on a reversing four-high roughing mill and a seven-stand four-high finishing train. Finishing rolling temperatures were kept above Ar3 and the coiling temperature was roughly 600 °C. After pickling in a hydrochloric descaling line the material was cold-rolled on a five-stand four-high tandem mill. This as-cold-rolled material was used for the further investigations in the laboratory.

Table 1: Chemical composition of the investigated grade [mass %]

C	Si	Mn	Cr + Mo	P	S	Al	N
0.10	0.1	1.5	0.8	0.01	0.01	0.04	< 0.005

2.2 Annealing simulation in the laboratory

The annealing simulations in the laboratory were conducted with the Continuous Annealing Line Simulator No. 1 at VOEST-ALPINE STAHL Linz GmbH [20]. This annealing simulator is equipped with a slow and a rapid gas jet cooling facility as well as with a mist cooling device and hot and cold water quench tanks. In order to characterize the transformation behavior during annealing three different annealing cycles were applied (Table 2 and Fig. 1 a – c). In a first step the kinetics of the cementite dissolution and the austenite formation was investigated (Table 2). In these experiments the samples were quenched directly from the intercritical annealing temperature. In further a step the impact of different cooling rates in the range between 100 and 1 K/s was investigated (Table 2). Moreover, the samples were quenched during the cooling stage from different temperatures as shown in Fig. 1b, using cooling rates of 5, 25 and 90 K/s (Table 2). In additional annealing experiments the impact of a holding stage at different temperatures was investigated (Table 2).

Table 2: Annealing parameters (T_{an} ... annealing temperature, t_{an} ... annealing time, T_{Qu} ... quenching temperature, T_{OA} ... overaging temperature, t_{OA} ... overaging time, CR_1 and CR_1... cooling rate) of the investigated annealing cycles.

annealing simulation									
type	Fig.	HR [K/s]	T_{an} [°C]	t_{an} [s]	CR_1 [K/s]	T_Q [°C]	T_{OA} [°C]	t_{OA} [s]	CR_2 [K/s]
cementite dissolution, austenite formation	1a	25	780 – 860	0 60	WQ				
constant cooling rate	1a	25	800	60	1 - 100				
interrupted cooling	1b	25	800	60	5, 25, 90	800 - 300			WQ
holding stage	1c	25	800	60	25		400 - 600	0 - 60	WQ
holding stage	1c	25	800	60	5, 25, 90		225 - 325	0 - 600	WQ
dilatometer									
type	Fig.	HR [K/s]	T_{an} [°C]	t_{an} [s]	CR_1 [K/s]	T_Q [°C]	T_{OA} [°C]	t_{OA} [s]	CR_2 [K/s]
CCT diagram austenitization	1a	25	1000	300	0.5 - 100				
cementite dissolution, austenite formation	1a	25	750 - 850	600	100				
constant cooling rate	1a	25	800	60	1 - 100				
holding stage	1c	25	800	60	5, 25, 90		225 - 575	600	90

2.3 Dilatometer experiments

The dilatometric investigations were conducted on an equipment built by VATRON [21]. In this dilatometer the samples are heated by means of radiation from three quartz lamps. Cooling is done with pure nitrogen. Pipes with a diameter of about 5 mm rolled up from 0.5 mm thick as-cold-rolled material are used as samples. The length of the samples was 50 mm.

Two different types of annealing cycle were applied in the investigations of the phase transformations during annealing in the dilatometer (Fig. 1 and Table 2). For reference investigations samples were annealed in the austenitic range and cooled at different cooling rates to room temperature. The cycles in Fig. 1a and Table 2 were used to study the dissolution of the cementite and the austenite formation kinetics. The impact of different cooling rates after the intercritical annealing were investigated, too. The influence of the cooling rate on the transformation kinetics was studied at isothermal conditions in cycles with a holding stage (Fig. 1c). A summary of the parameters used is given in Table 2.

The coefficient of thermal expansion of ferrite and austenite were determined from the heating and the cooling part, respectively, of an annealing cycle in which the material was fully austenitized. For these experiments the following procedure is applied for the calculation of the transformed fraction (ferrite,...) and the austenite during cooling (Fig. 2a): In a first step Δl-T-curve of the austenite is fitted and extended to lower temperatures. After measuring the content of retained austenite with the magnetic method, the Δl-T-curve of ferrite is shifted in such a way that - after application of the lever rule - the calculated and measured austenite contents become equal (Fig. 2a). Subsequently, the curves are differentiated with respect to T to distinguish clearly the contributions of the different phases to the transformed fraction of austenite.

A slightly modified approach must be used for the evaluation of annealing experiments conducted in the intercritical range (Fig. 2b): The dilatometric measurements show, that the Δl-T-curves of ferrite and austenite are almost equidistant irrespective of the annealing cycle. Therefore, the Δl-T-curves for ferrite and austenite determined in experiments with a complete austenitization were shifted so that the total distance between them remained constant and the amount of retained austenite calculated from the lever rule was equal to the amount of the austenite fraction measured by the magnetic method.

Special attention was paid to the amount and kinetics of the transformation during isothermal holding. This part of the transformation was fitted by an JMAK type of kinetics:

$$f_V = A_0 - B\left(1 - \exp\left(-kt^n\right)\right)$$

where A_0 is the amount of austenite prior to the isothermal holding, B the fraction of austenite transformed during holding, and k and n are constants describing the kinetics of the transformation. Based on the values for k and n, the time at which 20, 50 and 80 % (t_{20}, t_{50} and t_{80}) of the total fraction transformed during the isothermal stage were calculated.

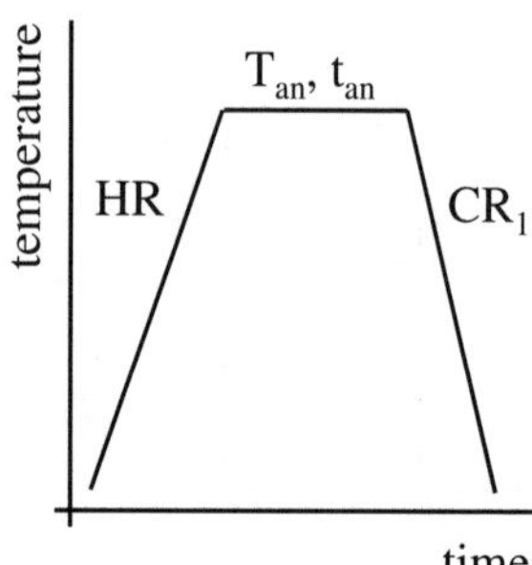

Fig. 1a: Constant cooling

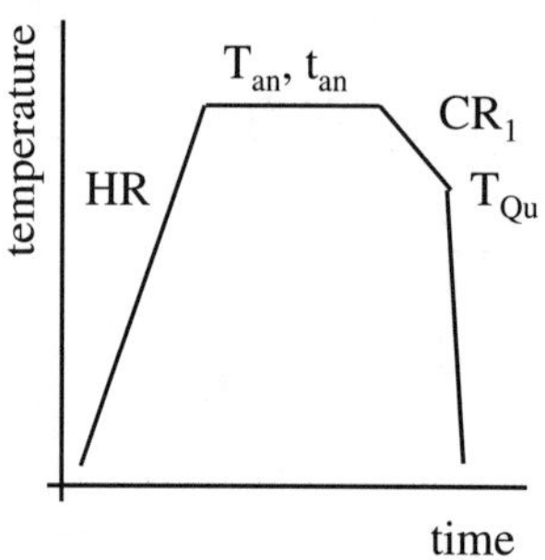

Fig. 1b: Interrupted cooling

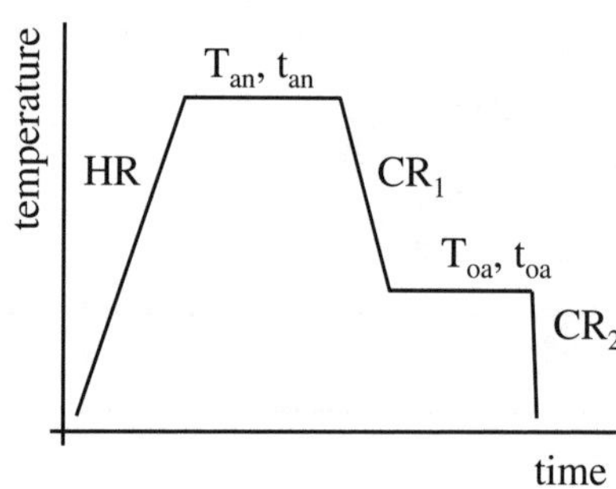

Fig. 1c: Holding stage

Fig. 1: Schematic description of investigated annealing cycle on the dilatometer and the annealing simulator.

2.4 Metallographic investigations and mechanical properties

The specimens for the metallographic investigations of the microstructure were prepared conventionally and etched with LePera's [22] agent. After the application of these etchants, ferrite grains show different colors, mainly blue or bright-brown, retained austenite and martensite appear white, bainite and tempered martensite are brown and pearlite is black. The content of the retained austenite was determined using a magnetic volumetric method [23]. The mechanical properties (also including Lankford values) were measured on a Roell-Korthaus RKM 250 tensile testing machine according to European Standard EN 10 002.

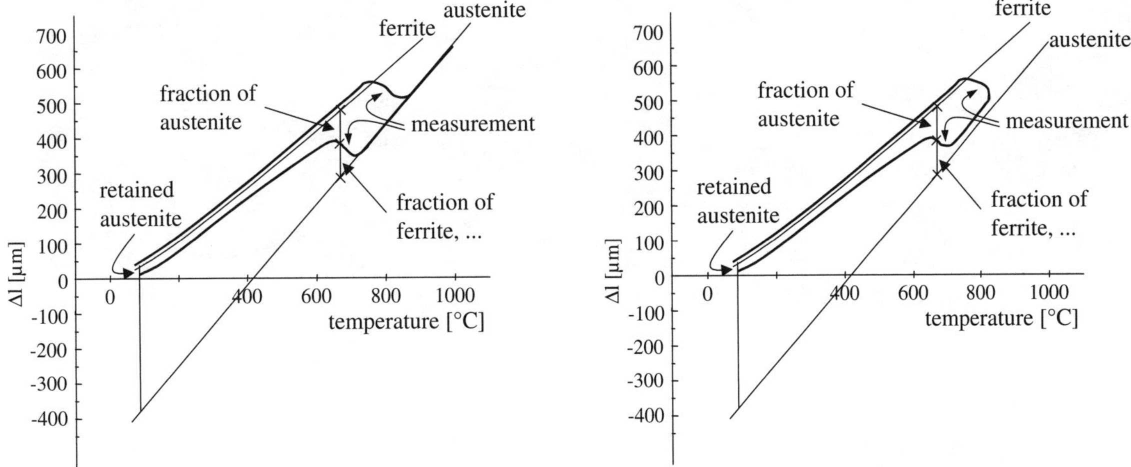

Fig. 2a: Complete austenitization

Fig. 2b: Intercritical annealing

Fig. 2: Evaluation of dilatometric measurements

3. RESULTS

3.1 Determination of the CCT diagram after annealing in the austenitic range

The transformation behavior of steel grades is characterized by isothermal or continuous cooling transformation (CCT) diagrams obtained from samples annealed in the austenitic range. For comparison with the comprehensive literature existing in this field the necessary experiments were conducted to determine the CCT diagram for the steel grade investigated. It must be mentioned, however, that the measured transformation behavior cannot be applied in the production of cold-rolled Dual Phase steel grades due to the remaining ferrite during intercritical annealing.

In Fig. 3 the microstructure of the steel is shown after annealing at 1,000 °C for 300 s for cooling rates of 50, 10 and 0.5 K/s. After application of the lowest cooling rate, ferrite, pearlite and a minor fraction of martensite are observed. The higher cooling rate of 10 K/s results in polygonal ferrite, acicular ferrite with martensite and bainite and a significant amount of martensite at the grain boundaries of the polygonal ferrite. The highest cooling rate produces less polygonal ferrite, an increasing amount of acicular ferrite and some blocky martensite. Significant amounts of martensite and bainite are situated between the acicular ferrite plates. Since the investigations under the light microscope do not make possible a clear determination and classification of the different phases TEM investigations will be undertaken in the future.

The results of the measurement with the dilatometer are shown in Fig. 4. There, the normalized austenite volume fraction differentiated with respect to the temperature, $df\gamma/dT$, is plotted for different cooling rates over the temperature. All curves possess a first maximum in the temperature range of 750 - 600 °C. This maximum is shifted to lower temperatures with increasing cooling rates. The temperature range and an inspection of the microstructures suggest that the formation of polygonal ferrite is responsible for this behavior. A further maximum, whose location is almost independent of the cooling rate, is observed at roughly 540 °C and it is attributed to the formation of acicular ferrite. The transformation seen at temperatures close to 400 °C stems from bainite. A reliable detection of the pearlite and martensite transformation is not possible even though there exists some indication for phase transformations occurring at 600 °C (for low cooling rates) and at temperatures below 300 °C for high cooling rates.

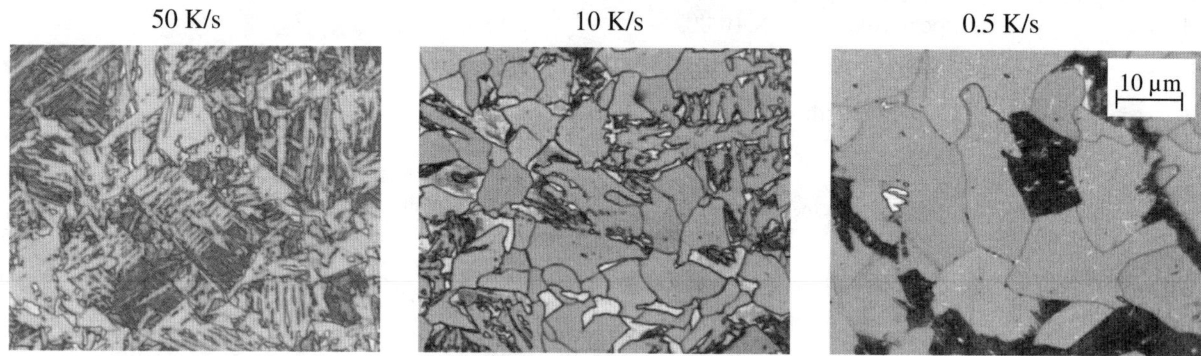

Fig. 3: Impact of the cooling rate on the microstructure after annealing in the austenitic range (1000 °C and 300 s).

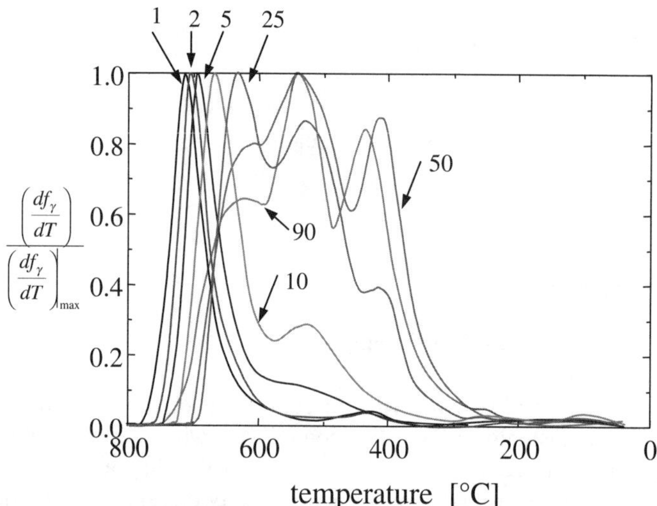

Fig. 4: Transformation during continuous cooling from the austenite (annealing temperature 1000 °C; annealing time 600 s).

3.2 Austenite formation during heating and intercritical annealing

3.2.1 Interrupted annealing experiments

A fundamental requirement for the production of thin sheet DP steel is the recrystallization, the dissolution of cementite and the formation of austenite. Therefore, cold-rolled material was heated with 25 K/s and annealed at different intercritical temperatures for various times. After quenching in water, the microstructure of the samples was investigated with LM. Based on these pictures a quantitative determination of the martensite fraction (austenite before quenching) and a qualitative investigation of the cementite dissolution were carried out. For the sake of a better comparison of the different measurements the time scale of all measurements was shifted so that the time at 700 °C was set to 0 s.

As can be seen from Fig. 5 the material is completely transformed into austenite at a temperature of 840 °C. At a soaking temperature of 820 °C only a very small fraction of ferrite remains stable. At lower temperatures the amount of austenite decreases. The impact of the soaking time is quite small, a significant increase was observed only after 23.2 s for the lowest temperature of 780 °C. In the investigated ranges of time and temperature unrecrystallized regions or undissolved cementite particles were not observed.

3.2.2 Dilatometric measurements

The austenite formation kinetics was investigated also by dilatometric methods. In these experiments the samples were heated at 25 K/s and soaked for 600 s (Table 2 and Fig. 1a). The investigated temperature range was between 750 and 850 °C. The time scale of all curves again was shifted (the time at 700 °C is 0 s). Fig. 6a shows the adjusted temperature as a function of time. The calculated austenite fraction as a function of time is given in Fig. 6b. There, the end of heating is marked for each soaking temperature by a dot. The austenite formation is rapid and especially for high soaking temperatures a predominant fraction of austenite grows already during heating to the soaking temperature despite the fast heating rate of 25 K/s. Even though the austenite fraction rapidly increases during the first few seconds of soaking, a slight increase of the total austenite volume fraction of about 5 - 10 % is observed during the later stages of soaking.

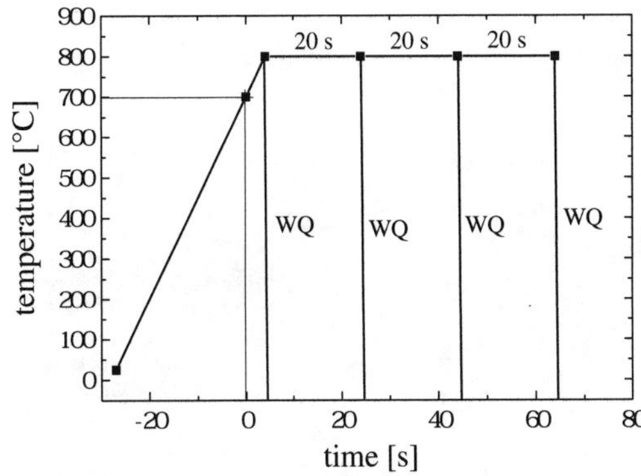

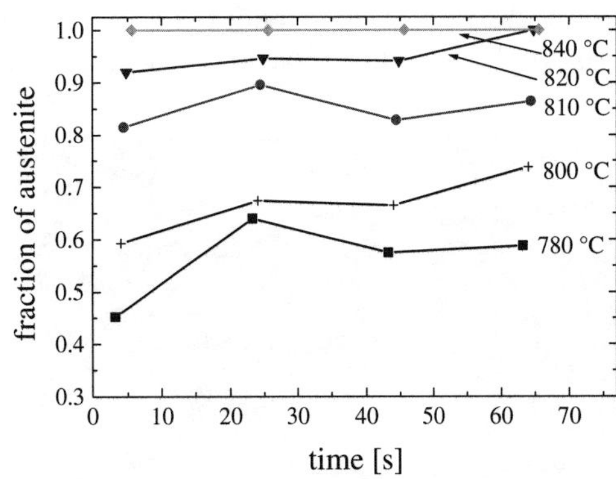

Fig. 5a: Annealing cycle

Fig. 5b: Austenite fraction

Fig. 5: Formation kinetics of the austenite during heating and soaking, microstructure investigations after interrupted annealing cycles.

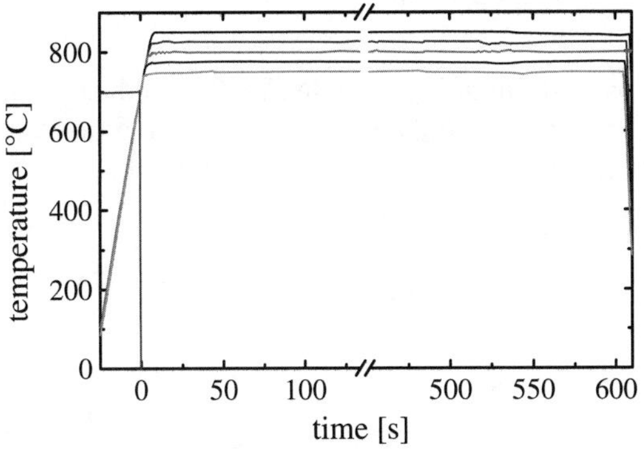

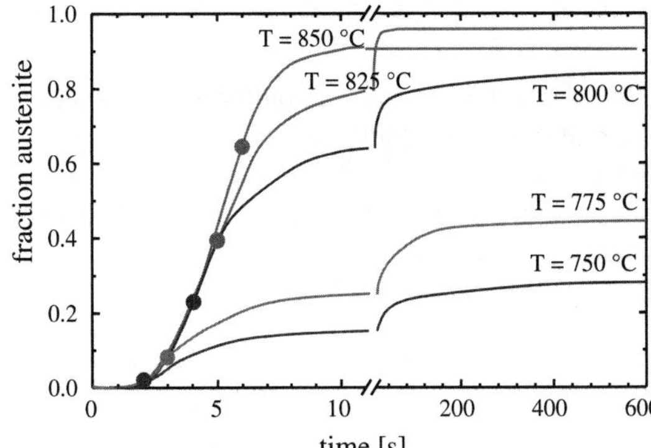

Fig. 6a: Annealing cycle

Fig. 6b: Austenite fraction

Fig. 6: Formation kinetics of the austenite during heating and soaking, dilatometric measurements.

3.3 Transformation during cooling at a constant cooling rate

3.3.1 Microstructure and mechanical properties

An overview of the behavior of cold-rolled DP grades can be obtained from investigations of the impact of the cooling rate on the transformation behavior after intercritical annealing. Therefore, samples were annealed in the annealing simulator at 800 °C for 60 s and cooled to room temperature using cooling rates between 100 and 1 K/s, which was followed by microstructure investigation and measurement of the mechanical properties.

The impact of the cooling rate on the microstructure is shown in Fig. 7. Water quenching results in ferrite and martensite (light brown colored). The volume fraction of martensite is about 65 %. A cooling rate of 100 K/s results in a significant increase of the ferrite fraction. Simultaneously the amount of martensite is reduced and additionally a considerable amount of a dark brown phase is observed which may be bainite. Moreover, light brown regions could be seen in the martensite. The origin of these regions is not clear; it is assumed, that they resemble martensite, too, containing less carbon. With decreasing cooling rates the amount of (dark brown) bainite and the light brown regions decrease. At a cooling rate of 10 K/s, ferrite and white martensite are observed. For the first time pearlite (black) appears. Even though the fraction of the pearlite increases significantly, a small fraction of martensite is seen even after cooling with 1 K/s (Fig. 7). The quantitative evaluation of the microstructure is summarized in Fig. 8. In this evaluation the dark brown, light brown and white phases are treated as bainite and martensite, because a clear distinction between them is impossible. While between 100 and 10 K/s the reduction of the martensite fraction is mainly a result of the increasing amount of ferrite, at even lower cooling rates the ferrite fraction is almost constant and the decrease of the martensite fraction is due to the significant increase of the amount of pearlite.

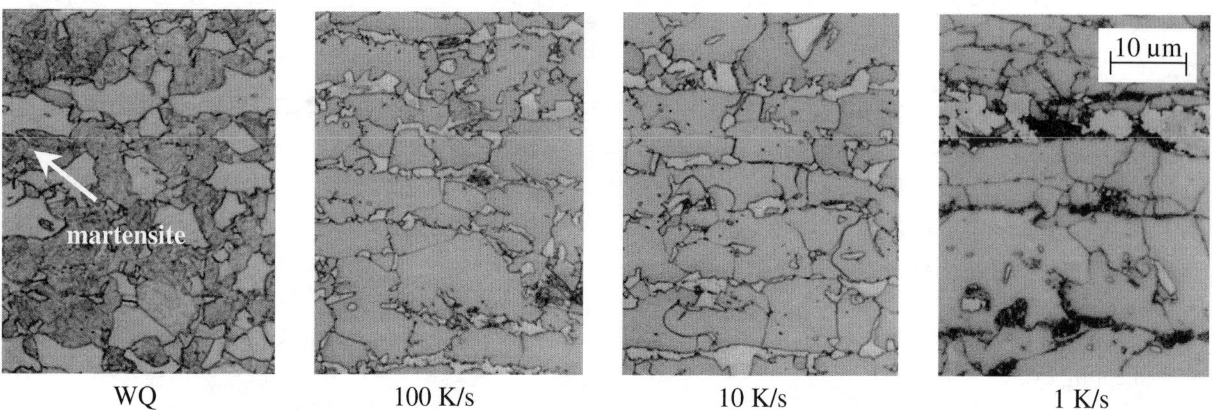

Fig. 7: Impact of the cooling rate on the microstructure; annealing in the annealing simulator.

The impact of the cooling rate on the yield strength and on the tensile strength is shown in Fig. 9. A reduction of the cooling rate generally results in a decrease of the yield and tensile strength. It is only at the lowest cooling rate that an increase of the yield strength is observed due to a yield point elongation. Therefore, such a low cooling rate is insufficient to obtain a typical Dual Phase steel behavior characterized by a low yield strength ratio and a zero yield point elongation.

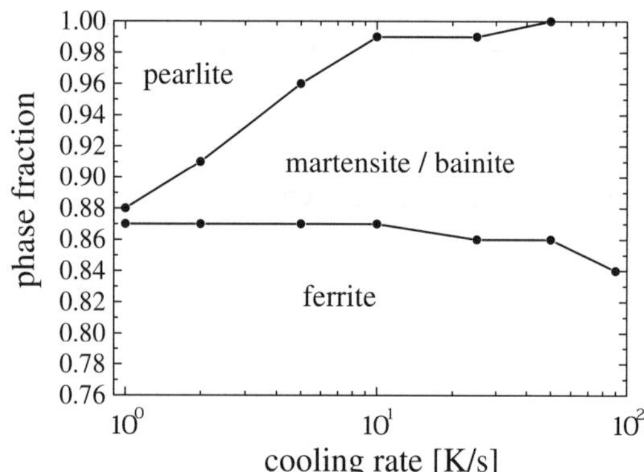

Fig. 8: Phase composition of the microstructure as a function of the cooling rate.

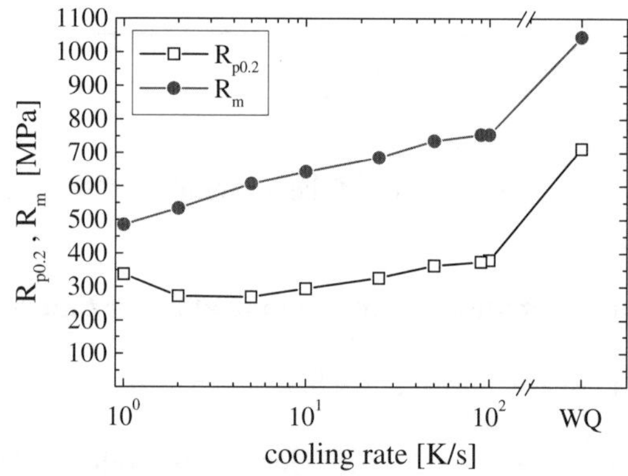

Fig. 9: Yield and tensile strength as a function of the cooling rate.

3.3.2 Dilatometric investigations

In addition to the investigations of the microstructure and mechanical properties on the transformation behavior of samples cooled with different cooling rates was investigated by means of a dilatometer. Samples were annealed at 800 °C for 60 s in the intercritical range, cooled to room temperature at rates between 100 and 1 K/s and their extensional behavior was measured in the dilatometer. The microstructure investigations reveal similar results as reported in chapter 3.3.1 and are not repeated here.

Fig. 10 shows the fraction of austenite as a function of the temperature for the different cooling rates. A very fast decrease of the amount of austenite takes place at low cooling rates of 1 and 2 K/s in the range between 750 and 650 °C. This is attributed to the growth of the remaining ferrite which is still stable during the intercritical annealing. These results suggest the formation of pearlite in the temperature range from 650 to 600 °C. During further cooling only a small decrease of the austenite fraction can be observed which, however, cannot be quantified.

Increasing the cooling rate delays the growth of ferrite and decreases the amount of transformed austenite during this stage. In contrast to the low cooling rates an almost constant plateau can be seen in the temperature range between 600 and 200 °C for rates of 5, 10 and 25 K/s. A remarkable decrease of austenite content is observed at lower temperatures which is attributed to the formation of martensite.

A completely different behavior is observed for the highest cooling rates of 50, 75 and 90 K/s which significantly impede the formation of ferrite at temperatures up to 600 °C. A slight decrease of the austenite fraction is detected between 600 and 400 °C. A clear relation to a phase transformation is not possible. A strong decrease of the austenite fraction is observed for temperatures below 400 °C. This is attributed to the formation of bainite and/or martensite.

Based on the dilatometric investigations a clear distinction between ferrite and pearlite and between martensite and bainite is not possible. Therefore, only the sum of the two pairs can be calculated. The resulting amounts are summarized in Fig. 11 which makes clear that increasing cooling rates result in a significant increase of the hard phases bainite and martensite.

3.4 Transformation during cooling, interrupted at different quenching temperatures

Additional experiments were conducted to support the interpretation of the transformations occurring during continuous cooling. For this purpose samples were annealed at 800 °C for 60 s, cooled at rates of 5, 25 and 90 K/s to temperatures of 675, 550, 425 and 300 °C and then quenched in water. In Fig. 12 microstructures are shown as a function of the cooling rate for the different quenching temperatures. At the lowest applied cooling rate and a quenching temperature of 675 °C only ferrite and martensite are observed. The amount of light brown colored regions within the martensite is small. Quenching from 550 °C makes a very small amount of pearlite (black) appear in the microstructure. For this low cooling rate decreasing quenching temperatures result in an increasing amount of pearlite.

A slightly different situation is observed for the higher cooling rate of 25 K/s. At the highest quenching temperature of 675 °C a significantly lower fraction of ferrite is seen and the amount of light brown regions within the martensite increases. Decreasing quenching temperatures result in increasing amounts of ferrite and at 425 °C a small fraction of dark brown bainite is detected. This fraction becomes larger when quenching from 300 °C.

Cooling with 90 K/s reduces further the amount of ferrite whereas the fraction of the light brown regions increases. In contrast to the lower cooling rate of 25 K/s, a remarkable amount of dark brown bainite is seen for 90 K/s and quenching from 300 °C. Lower cooling rates and decreasing quenching temperatures generally result in a lower strength (Fig. 13). This decrease of strength is particularly high when the quenching temperature is reduced from 800 °C to 550 °C.

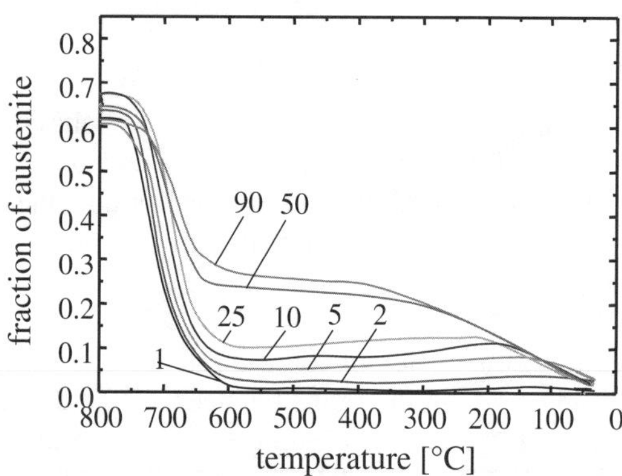

Fig. 10: Austenite fraction as a function of the temperature during cooling for different cooling rates (attached numbers) after annealing at 800 °C for a soaking time of 60 s.

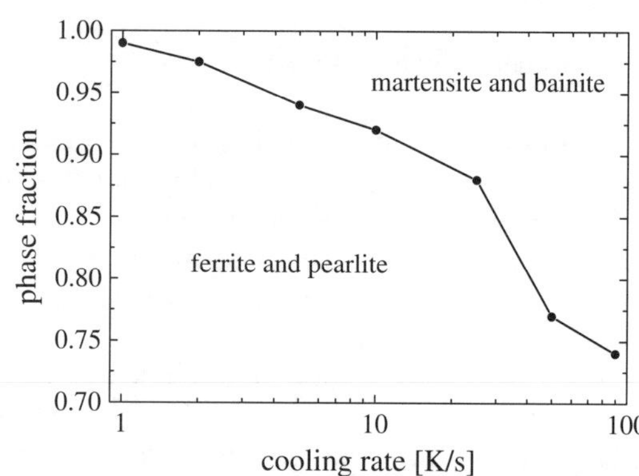

Fig. 11: Fraction of martensite and bainite and ferrite and pearlite as a function of the cooling rate estimated from measurements with the dilatometer.

3.5 Impact of different cooling rates and holding stage during cooling

3.5.1 Microstructure and mechanical properties

The phase transformation during cooling and its influence on the mechanical properties are of fundamental interest for the production of Dual Phase steel grades on hot dip galvanizing or continuous annealing lines equipped with an overaging zone. Therefore, the transformation kinetics during isothermal holding, the developing microstructures and the mechanical properties thoroughly investigated as a function of the cooling rate have to be used (see Table 2 and Fig. 1c for details).

The influence of the holding temperature (600, 500 or 400 °C) and the holding time on the microstructure is shown in Fig. 14. After holding at 600 °C for more than 20 s makes appear a significant amount of pearlite. Simultaneously, the amount of martensite is reduced. Holding at 500 °C makes pearlite to disappear. On the other hand, holding results in a decrease of the martensite fraction. A significant amount of bainite is observed after holding at 400 °C.

Holding at even lower temperatures gives the microstructures summarized in Fig. 15. At a cooling rate of 90 K/s - still without a holding stage - a small amount of bainite is observed. Holding for 25 s at this temperature strongly increases the bainite fraction. Holding even longer leads to slightly more bainite, until after 600 s a clear substructure can be seen in the bainite. At the lower holding temperatures of 275 °C and 225 °C a similar impact of the holding time can be observed. Nevertheless the kinetics of the transformation becomes slower and hence, the amount of bainite is reduced.

A quite different situation is observed for a cooling rate of 25 K/s. In contrast to the higher cooling rate the kinetics of the bainite formation is always drastically slowed down (Fig. 15, third row). the amount of bainite formed during holding is reduced making the qualitative evaluation of the kinetics impossible. Moreover, at a holding temperature of 225 °C the formation of the brown phase is difficult to detect. As expected from the results reported in chapter 3.3, pearlite can be detected at a cooling rate of 5 K/s and the influence of the holding stage on the microstructure diminishes. The formation of a brown phase can be observed only for the highest holding temperature (Fig. 15, fourth row). For the lowest temperature of 225 °C this phase cannot be detected at all.

Fig. 16 shows the influence of holding on the tensile strength. A maximum in strength can be observed for a holding temperature of 550 °C. In Fig. 17 the influence of a holding stage on the yield and tensile strengths is shown for holding temperatures of 325, 275 and 225 °C and cooling rates of 90, 25 and 5 K/s. Higher cooling rates generally result in higher strength levels, and the impact of the holding stage on the tensile strength is very pronounced for a cooling rate of 90 K/s and a holding temperature of 325 °C. Under this condition the yield strength slightly increases with the holding time. At 275 °C the tensile strength decreases with t_{OA} while the yield strength remains unaffected. Holding at 225 °C for various times hardly reduces the tensile strength. The least impact of the holding stage is observed for a cooling rate of 5 K/s for which a small decrease of the tensile strength is detected only for holding at 325 °C.

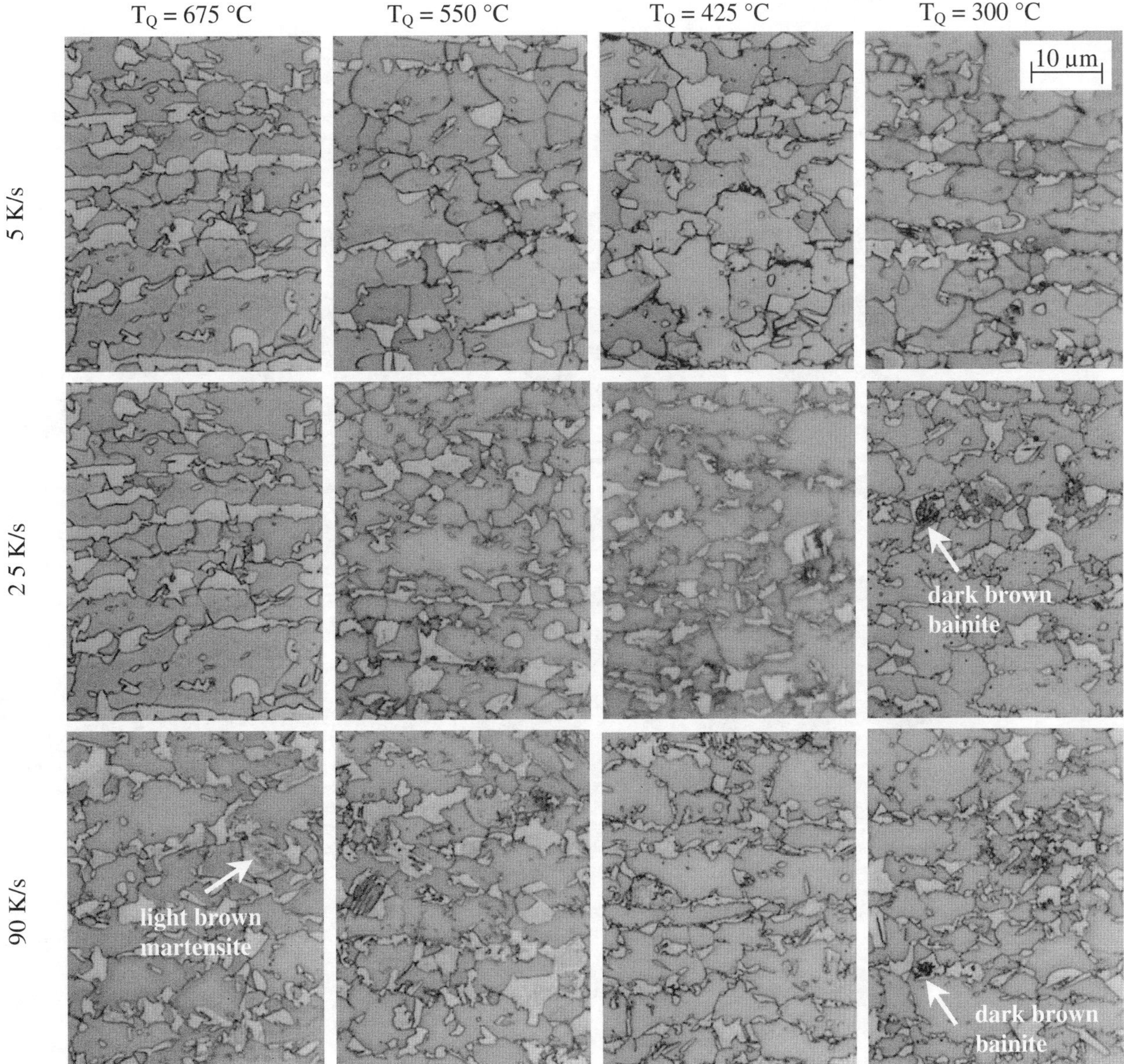

Fig. 12: Microstructure as a function of the cooling rate and quenching temperature.

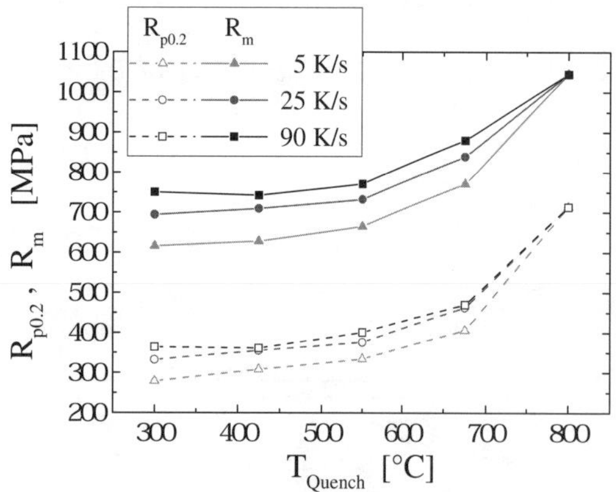

Fig. 13: Yield and tensile strength as a function of the quenching temperature for different cooling rates.

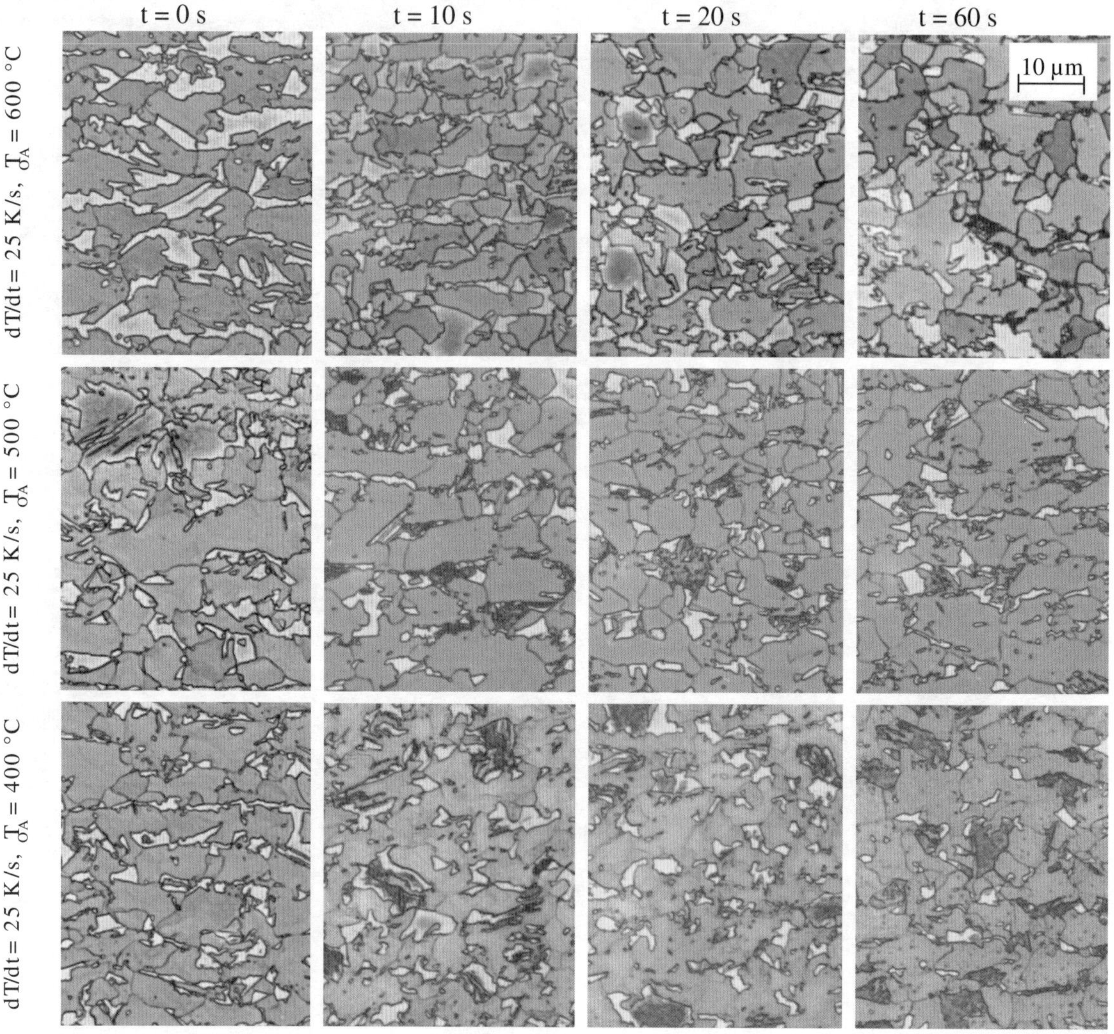

t = 0 s | t = 10 s | t = 20 s | t = 60 s

10 µm

dT/dt = 25 K/s, T_{OA} = 600 °C

dT/dt = 25 K/s, T_{OA} = 500 °C

dT/dt = 25 K/s, T_{OA} = 400 °C

Fig. 14: Influence of a holding stage at temperatures T_{OA} of 600 °C, 500 and 400 °C on the microstructure for a cooling rate of 25 K/s ($T_{an} \rightarrow T_{OA}$).

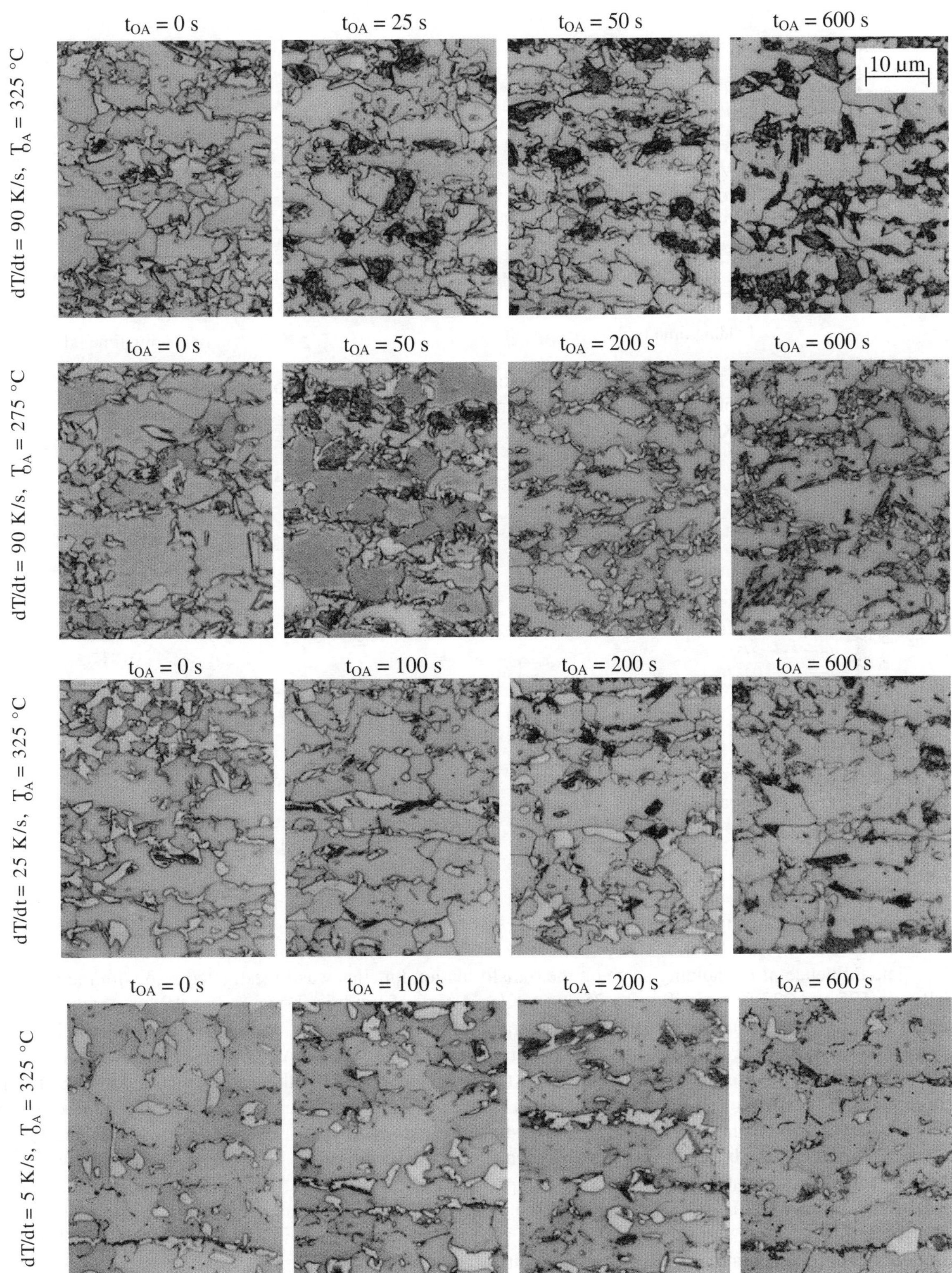

Fig. 15: Influence of a holding stage at temperatures T_{OA} of 325 and 275 °C on the microstructure for cooling rates of 90, 25 and 5 K/s ($T_{an} \rightarrow T_{OA}$).

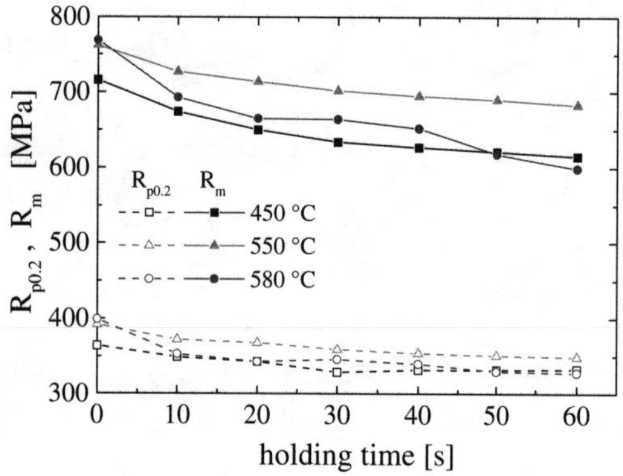

Fig. 16: Influence of the holding stage at temperatures between 580 °C and 450 °C on the tensile strength for a cooling rate of 25 K/s.

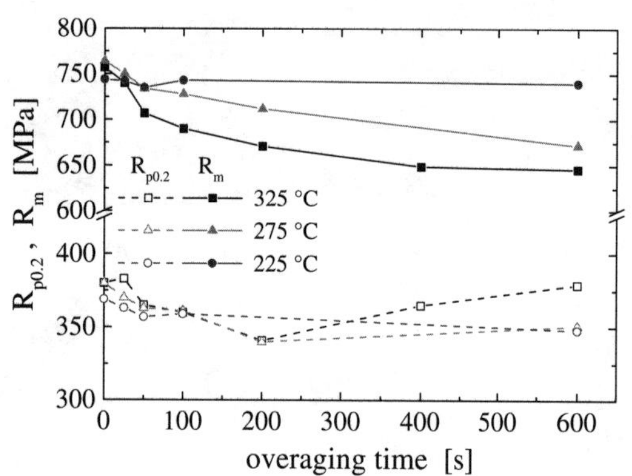
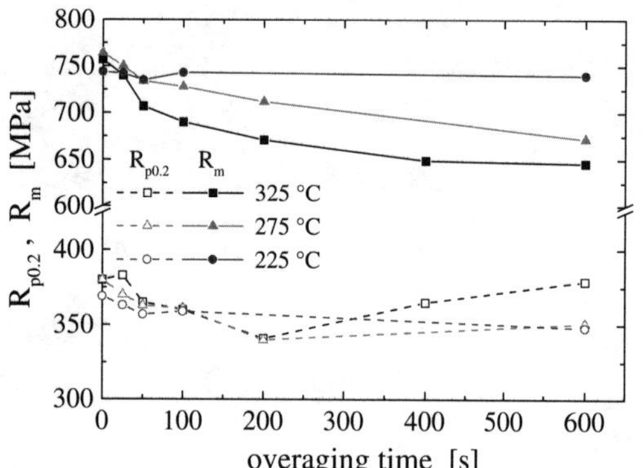

Fig. 17a: Cooling rate of 90 K/s

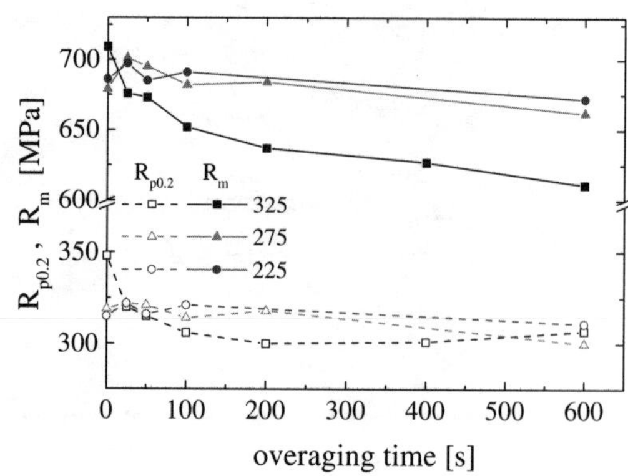

Fig. 17b: Cooling rate of 25 K/s

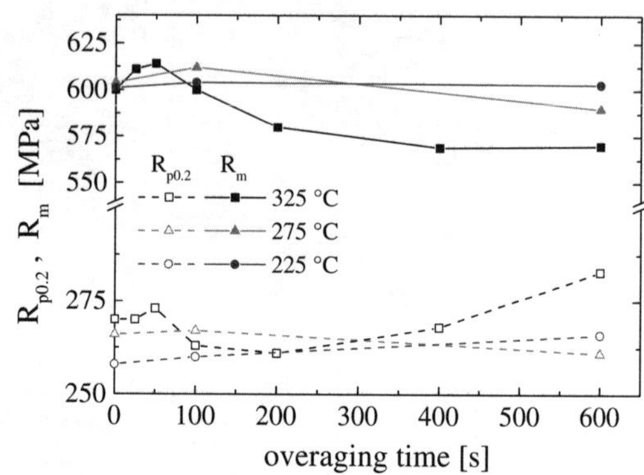

Fig. 17c: Cooling rate of 5 K/s

Fig. 17: Influence of a holding stage at temperatures of 325, 275 and 225 °C on the yield and tensile strength for cooling rates of 90, 25 and 5 K/s.

3.5.2 Dilatometric investigations

The influence of the holding stage on the transformation kinetics was investigated also with the dilatometer (T_{OA} between 575 and 225 °C and CR1 90, 25 and 5 K/s). A typical result is shown in Fig. 18 for a cooling rate of 25 K/s and a holding temperature of 275 °C. A considerable amount of austenite transforms during the holding stage (Fig. 18a). Fig. 18b shows the isothermal transformation in detail, including the Avrami fit, the total amount of transformed austenite during isothermal holding and the times t_{20}, t_{50} and t_{80} describing the transformation kinetics. The fraction of austenite transformed during the holding stage is shown in Fig. 19 as a function of the holding temperature for the three different cooling rates. The amount of austenite transformed increases significantly with increasing cooling rates. Nevertheless, the behavior is almost independent of the cooling rate. The lowest amount is always found at the lowest holding temperature. At increasing temperatures an increasing transformed fraction is observed, and a maximum is located in the temperature range of 400 - 450 °C. Increasing the temperature further results in a decrease and a minimum in transformed austenite is observed in the temperature range of 500 - 550 °C. The minimum is shifted to higher temperatures for lower cooling rates.

The resulting transformation kinetics is shown in Fig. 20. The shape of the time-temperature curves is similar for all cooling rates. Depending on the cooling rate the transformation of austenite is rather slow between 525 and 575 °C. A maximum in transformation velocity is observed in the temperature range of 450 – 500 °C. This maximum is shifted to higher temperatures upon increasing the cooling rate. Reducing the holding temperature to 325 °C results in a strongly retarded transformation kinetics. At even lower temperatures an increase is observed. However, at these low holding temperatures the total amount of transformed austenite is quite low and therefore an evaluation is more difficult.

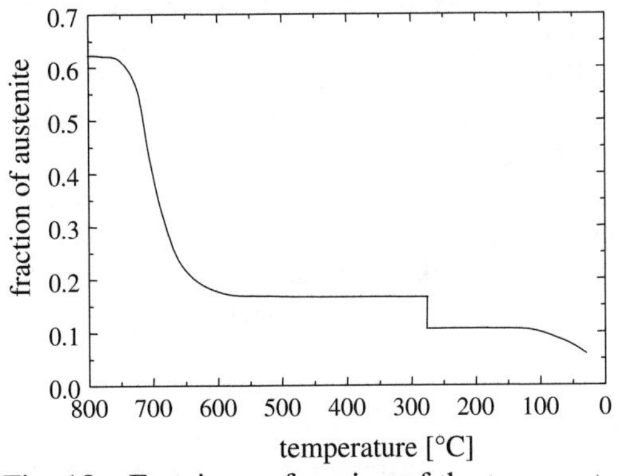

Fig. 18a: Fraction as function of the temperature

Fig. 18b: Fraction of austenite as a function of the time

Fig. 18: Austenite content as a function of the temperature during cooling for an annealing cycle with a cooling rate of 25 K/s and a holding stage at 275 °C

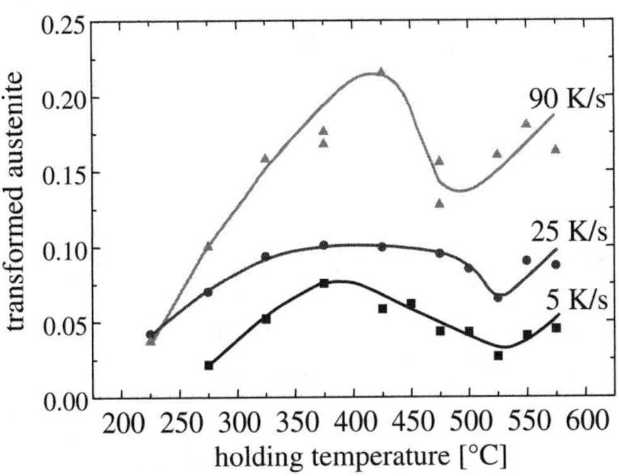

Fig. 19: Fraction of austenite transformed during the holding stage.

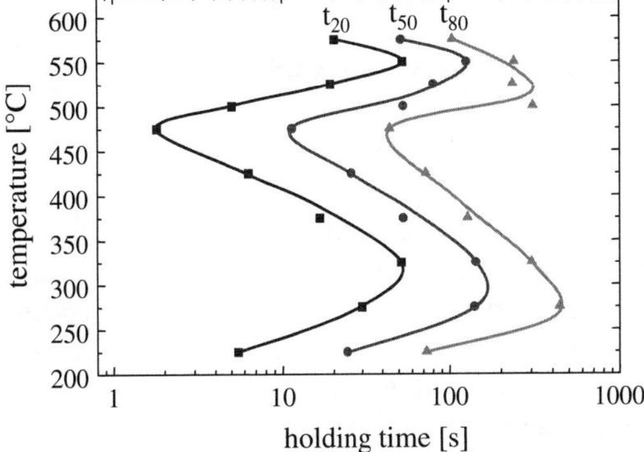

Fig. 20b: Cooling rate: 25 K/s

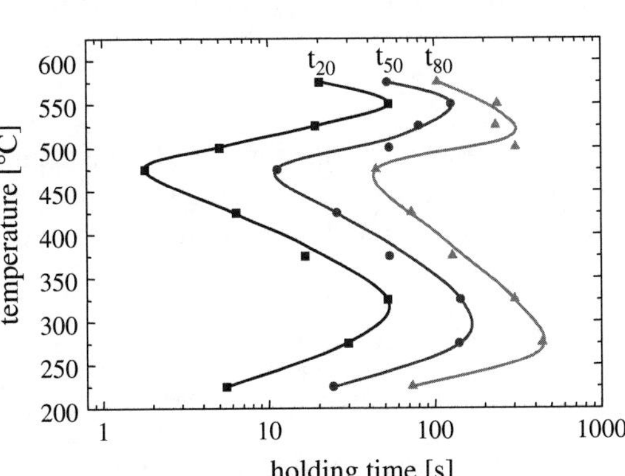

Fig. 20a: Cooling rate: 5 K/s

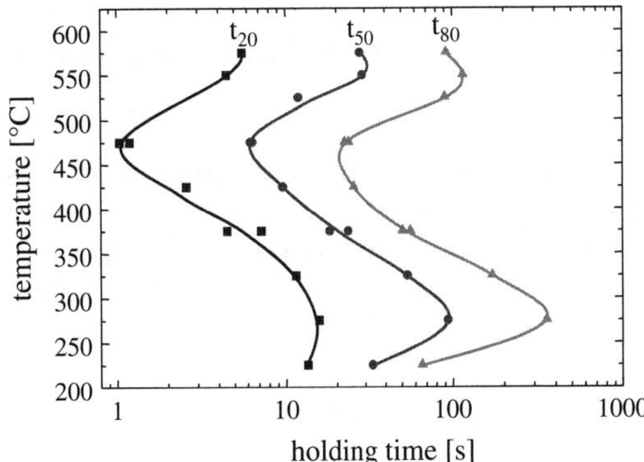

Fig. 20c: Cooling rate: 90 K/s

Fig. 20: The time to transform 20, 50 or 80 % (t_{20}, t_{50}, t_{80}) of the austenite as a function of the holding temperature and the cooling rate.

4. DISCUSSION

4.1 Dissolution of cementite and formation of austenite

For the industrial production of cold-rolled DP grades the recrystallization of the deformation microstructure, the dissolution of cementite and the formation of austenite are the most important steps during heating and soaking in the furnace. In the investigated temperature range unrecrystallized grains and undissolved cementite were not detected. The results concerning the austenite formation kinetics gained from microstructure investigations on samples quenched after intercritical annealing and from dilatometer measurements agree quite well (Figs. 5b and 6b). A comparison of the austenite fraction determined experimentally and calculated with Thermocalc is shown in Fig. 21, for which the values obtained after soaking for 60 s and 600 s are used for the interrupted annealing trials and the dilatometer measurements, respectively. Although the calculations with Thermocalc are done for equilibrium the agreement between the outcomes is excellent which can be attributed the low equilibrium enrichment of Cr and Mn in the austenite at these temperatures.

In comparison to results reported in the literature [11, 14] the formation kinetics in the present investigations seems to be significantly faster. One reason for this discrepancy may due to the different microstructures of the materials tested. Speich et. al [14] used normalized material in comparison to as-cold-rolled material in this investigation. Cold reduction, however, results in a break up of the cementite plates of the pearlite formed during coiling and, hence, a remarkable increase of the dissolution kinetics due to the significantly increased carbide surface can be expected.

4.2 Transformations during cooling

4.2.1 Cooling from the austenitic range

Cooling from the austenitic range gives microstructures, which are quite unusual for cold-rolled DP grades. After an application of high cooling rates the microstructure contains mainly acicular ferrite and bainite and small fractions of polygonal ferrite and martensite. At lower cooling rates the amount of polygonal ferrite and blocky martensite increases. Acicular ferrite and bainite are typical for low carbon steel grades [24]. The reason for this uncommon microstructure is the delayed nucleation and growth of polygonal ferrite which is especially pronounced at high cooling rates and for larger austenite grain sizes.

Lower cooling rates make possible the formation of polygonal ferrite, an enrichment of the carbon in the austenite and the formation of blocky martensite at grain triple points. In some regions acicular ferrite still can be observed at cooling rates as low as 10 K/s. Still lower cooling rates favor the formation of polygonal ferrite. However the amount of Cr and Mn in the steel studied is not sufficient to prevent the formation of pearlite for such low cooling rates.

The continuous cooling transformation diagram is shown in Fig. 22, which includes the amount of transformed austenite. The formation of polygonal ferrite at temperatures between 600 and 700 °C is retarded at high cooling rates. Pearlite is only observed for cooling rates below 10 K/s. Acicular ferrite appears at ~500 °C for cooling rates above 10 K/s. Bainite is formed below 400 °C for all cooling rates.

4.2.2 Cooling after intercritical annealing

In the following discussion the tensile strength is included in the analysis of the phase transformations taking place during cooling, by correlating the microstructures obtained from the experiments with different cooling rates (Fig. 8) with the tensile strength (Fig. 9). Martensite, tempered martensite and bainite are treated as a single hard phase since a clear differentiation between them is impossible with the light microscope. For simplicity ferrite and pearlite are treated as soft phase. A linear mixture rule [25] for the two phase microstructure is employed:

$$R_m = V_{Soft} \cdot R_{m\,Soft} + V_{Hard} \cdot R_{m\,Hard} \quad ,$$

where V_{Hard} is the fraction of martensite, tempered martensite and bainite and V_{Soft} is the fraction of the ferrite and pearlite. The quantities $R_{m\ Soft}$ and $R_{m\ Hard}$ are obtained from a least square fit and are taken as a measure for the tensile strength of ferrite and martensite, respectively. The same evaluation is performed for the phase fractions determined from the dilatometer measurements (Fig. 11).

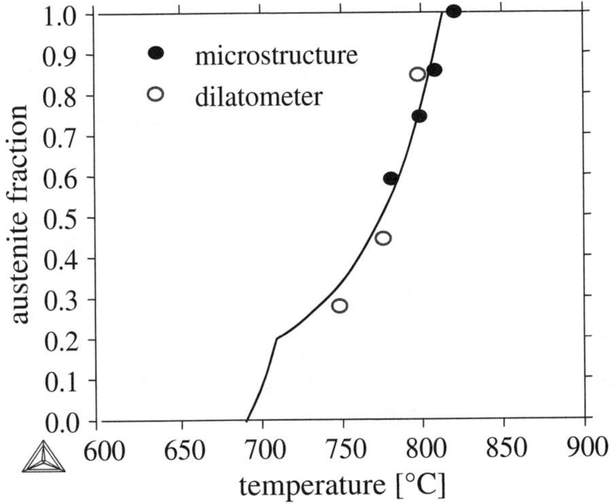

Fig. 21: Austenite fraction as function of the intercritical annealing temperature. Comparison of an equilibrium calculation with Thermocalc and the experimental results based on dilatometer measurements and microstructure investigations.

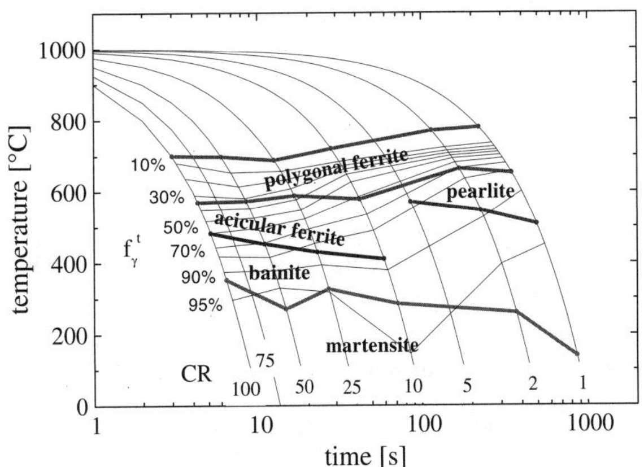

Fig. 22: Continuous Cooling Transformation diagram (annealing temperature 1000 °C; annealing time 600 s, f_γ^t ... transformed austenite fraction, CR ... cooling rate in K/s)

The fitted results are shown in Fig. 23. At low cooling rates a reasonable agreement between the results from the dilatometer measurements and the microstructure analyses experiments can be found. The fraction of the hard phase at high cooling rates predicted by dilatometry is significantly higher than that from the quantified microstructures. The reason for this discrepancy may be the inhomogeneous temperature distribution in the sample due to high cooling rates. Because of this complication in all further discussions only results based on microstructure investigations are used, which is predicted an increase in tensile strength by about 20 MPa per volume percent of martensite (hard phase).

Using the above rule of the volume fractions of the soft phase (ferrite and pearlite) and the hard phase (martensite, tempered martensite and bainite) were calculated from the results of the mechanical tests reported in chapter 3.4 (see Fig. 24). An interpretation and a comparison of these curves with the other results will be done on the following assumption: In the following, the fraction of the soft phase, which is predominantly ferrite, is assumed to be constituted of ferrite formed during the intercritical annealing and during cooling. On the other hand, the amount of the hard phase is the sum of bainite and martensite transformed during cooling and the austenite transformed into martensite during quenching in the interrupted annealing cycles. These assumptions ensure, that these results can be compared with the ultimate amounts of the hard and soft phases obtained with different cooling rates, and the calculated phase fractions obtained from the mechanical properties can be used in the discussion of the transformation kinetics.

The results from both the microstructure investigations (Figs. 7 and 12) and the dilatometer measurements show a very fast decrease of the austenite fraction during cooling down from the intercritical annealing to a temperature of

675 °C even at cooling rates as high as 90 K/s. At 675 °C the amounts of ferrite and remaining austenite determined from the microstructure and calculated from the mechanical properties agree well. Only for the highest cooling rate the dilatometer measurements show a significantly lower fraction of ferrite at this temperature. As has been mentioned already, this deviation can be attributed to an inhomogeneous temperature distribution in the sample during cooling.

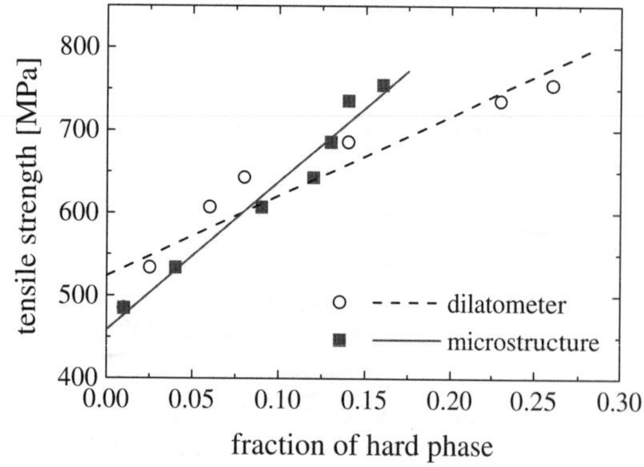

Fig. 23: Tensile strength as a function of the fraction of hard phase fitted with an linear mixture role.

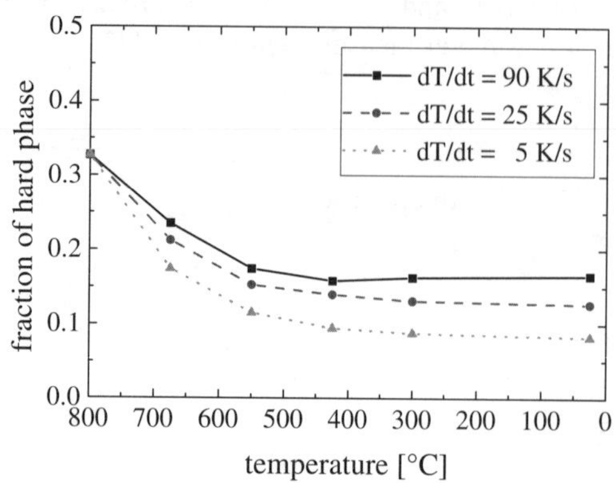

Fig. 24: Fraction of the hard phase as a function of the temperature during cooling as calculated from the linear rule of mixture for different cooling rates.

A large discrepancy exists for the austenite fraction during intercritical annealing. The austenite fraction obtained from microstructure investigations and dilatometer measurements result is 65 - 80% and only 35% when calculated from the mechanical properties. The reason for this remarkable deviation is the completely different microstructure after quenching from the intercritical annealing temperature, which then consists of a martensite matrix with ferrite inclusions. Also, the strength of martensite after direct quenching should be low due to its low carbon content. Thus, an application of the mechanical properties in the interpretation of the phase transformations is not possible in this case.

The analysis of the mechanical properties predict a further decrease of the austenite fraction by about 6% in the temperature range from 675 °C to 550 °C. This is in excellent agreement with the results from the dilatometer measurements and the quantitative metallography. Even though the formation of pearlite cannot be resolved in the dilatometer experiments, the strong decrease of the tensile strength observed for the cooling rate of 5 K/s indicates its presence in the microstructure.

When decreasing the temperature to 425 °C the analysis of the tensile strength predicts an additional transformation of about 1.5% of austenite. This amount is too small to be detected by either dilatometer measurements or microstructure investigations. A significant increase of the pearlite fraction can be seen the lowest cooling rate of 5 K/s (Fig. 12).

At even lower temperatures no change in the phase fractions can be evaluated from the tensile strength. In contrast, the dilatometer measurements show a remarkable transformation taking place during cooling to temperatures lower than 400 °C for high cooling rates. This transformation can be seen clearly in the microstructure (Fig. 12) where a remarkable amount of brown phase appears for quenching temperatures of 300 °C. Even though a clear determination of the phase is not possible, it may be bainite and/or martensite. The apparent contradiction between the results from the analysis of the tensile strength and those from the dilatometer measurements and the microstructure investigations can be resolved by assuming that the phase formed in this temperature range has almost the same strength level as the martensite formed during water quenching.

The transformation behavior of intercritically annealed material can be summarized as follows (Fig. 25): Down to about 600 °C a very fast growth of polygonal ferrite takes place. Only at cooling rates lower than 10 K/s also the formation of pearlite occurs in this range. In the temperature range between 600 °C and 400 °C, only a very small fraction of austenite transforms. At high cooling rates, a remarkable amount of bainite or martensite is transformed at temperatures lower than 400 °C. Low cooling rates yield martensite at temperatures below 150 °C.

This very strong impact of the cooling rate can be explained by the carbon enrichment in the austenite. Increasing cooling rates result in a retarded growth of ferrite and, hence, in a reduction of the carbon content in the remaining austenite. Since the location of the bainite and the martensite start temperatures, in turn, strongly depends on the carbon content in the austenite, the appearance of martensite or bainite during cooling with high cooling rates is plausible.

4.2.3 Cooling with an isothermal holding stage

Since the impact of an isothermal holding stage on the microstructure is essential for the production of cold-rolled DP grades on continuous annealing lines or hot-dip galvanizing lines, the transformation behavior during isothermal holding will be discussed.

The transformation behavior characterized with the dilatometer is summarized in Fig. 26 for the three cooling rates. Correlating the dilatometer results with the microstructures depicted in Figs. 14 and 15, three temperature regimes can be identified. At temperatures higher than 550 °C the formation of pearlite during isothermal holding is dominant. In the temperature range between 550 and 500 °C a pronounced slow down of the transformation kinetics is measured. This retarded zone is shifted to significantly longer times and lower temperatures with decreasing cooling rates. In the temperature range between 550 and 300 °C the formation of bainite is dominant. At about 450 °C this transformation is quite rapid. The transformation nose is shifted to larger times and lower temperatures when lowering the cooling rate. A gradual change of the microstructure is observed in the bainite range. At high temperatures clearly structured plates can be detected. At lower temperatures the transformation product appears uniformly brown colored. This difference can be attributed to the formation of upper and lower bainite. In the range of 275 - 225 °C the transformation speeds up again. Presently, a sound explanation for this behavior cannot be given.

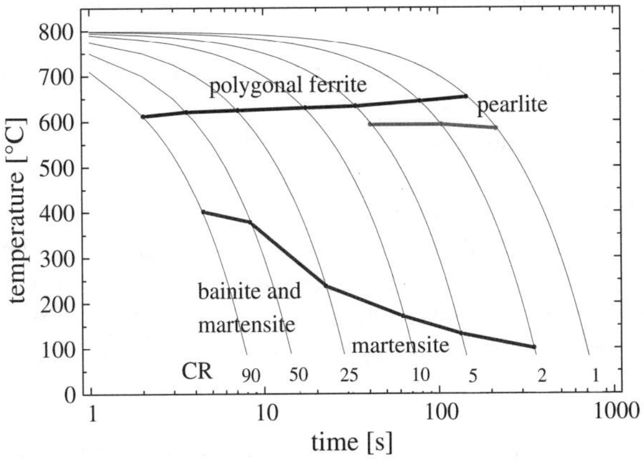

Fig. 25: Schematic summary of the transformation behavior of the intercritically annealed material during cooling to room temperature at constant cooling rates.

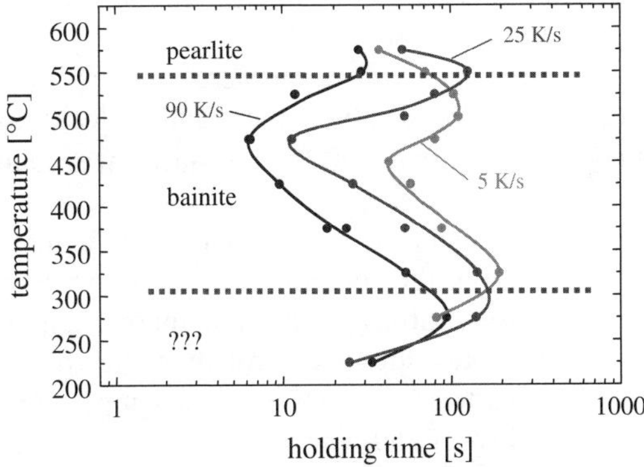

Fig. 26: Transformation kinetics during isothermal holding as a function of the cooling rate

The strong influence of the cooling rate on the transformation behavior during isothermal holding is attributed to the enrichment of carbon in the austenite. From the austenite volume fraction the carbon content in the austenite can be estimated as a function of the cooling rate as ~0.75% for 5 K/s, ~0.55% for 25 K/s and ~0.37 - 0.49 % for 90 K/s at 550 °C. Here, the carbon content of ferrite is assumed as 0.02 mass %.

The transformation behavior can be correlated very well with the mechanical properties given in Figs. 16 and 17. The markedly delayed transformation during isothermal holding at around 550 °C is indicated also by the mechanical properties (Fig. 16). For the holding temperatures 450 and 580 °C a stronger decrease of the tensile strength is observed as compared to that at 550 °C.

The decrease of the tensile strength is particularly strong for the cooling rate of 90 K/s and a holding temperature of 325 °C. In agreement with the transformation behavior lower cooling rates and lower holding temperatures result in a much less pronounced reduction of the tensile strength during isothermal holding.

5. SUMMARY

In the present work the transformation behavior of a cold-rolled DP grade during various annealing treatments was investigated. For this purpose the impact of the annealing temperature and soaking time on the recrystallization, dissolution of cementite and formation of austenite was studied. The transformation behavior during cooling from the austenite phase field and from the austenite-ferrite two phase field was examined as a function of the cooling rate. Experiments with different cooling rates to an isothermal holding stage were conducted. Microstructure investigations, dilatometer measurements and the mechanical properties were used to determine the transformation behavior.

Neither unrecrystallized grains nor undissolved cementite particles were detected. Most of the austenite forms already during heating or in the first 3 - 5 seconds during soaking and further holding only slightly increases the austenite fraction.

A continuous cooling transformation diagram after a complete austenitization shows the following features:

- High cooling rates of 100 K/s result in the formation of acicular ferrite and bainite and small amounts of polygonal ferrite and martensite.
- Low cooling rates lead to an increasing amount of polygonal ferrite and decreasing amounts of acicular ferrite and bainite. Additionally, the amount of blocky martensite at grain triple points increases.
- At cooling rates below 10 K/s a significant amount of pearlite is formed.

After annealing in the intercritical range a markedly different behavior is observed. During cooling down to 600 °C the growth of the polygonal ferrite is dominant and pearlite is formed at cooling rates lower than 10 K/s. The cooling rate strongly influences the amount of ferrite formed during this first stage and alters the carbon content in the austenite which takes strong control over the further transformation. At high cooling rates the transformation of bainite and/or martensite is observed at temperatures below 400 °C. Low cooling rates yield martensite only at temperatures as low as 150 °C.

During isothermal holding three distinct regimes were observed. In the temperature range above 550 °C the formation of pearlite is dominant. In the temperature range between 550 and 300 °C the formation of bainite takes place. Decreasing the cooling rate retards the transformation kinetics and also reduces the amount of the phases formed during holding isothermal. The slowest transformation is observed in the temperature range of 550 °C and 300 °C. At temperatures below 300 °C an additional transformation is observed.

6. REFERENCES

1. Stiaszny, A. Pichler, E. Tragl, M. Kaiser, W. Schwarz, M. Pimminger, K. Kösters, and K. Spiradek, 'Influence of Annealing Technology on the Material Properties of LC and ULC Steel Grades', International Symposium: Modern LC and ULC Sheet Steels for Cold Forming, Department of Ferrous Metallurgy RWTH Aachen University of Technology, 1988, p. 225 - 236
2. N. Ohashi, T. Irie, S. Satoh, O. Hashimoto, and I. Takahashi, 'Development of Cold-Rolled High Strength Steel Sheet with Excellent Deep Drawability', SAE Technical Paper Series (810027), Warrendale, 1981
3. A. Pichler, M. Mayr, G. Hribernig, H. Presslinger, and P. Stiaszny, 'High Strength IF Steels: Production, Parameters and Properties', International Forum for Physical Metallurgy of IF Steels IF.IFS-94, The Iron and Steel Institute of Japan, 1994, p. 249 - 268
4. L. Meyer, W. Bleck, and W. Müschenborn, 'Product-Oriented IF Steel Design', International Forum for Physical Metallurgy of IF Steels IF.IFS-94, The Iron and Steel Institute of Japan, 1994, p. 203 - 222
5. A. T. Davenport, 'Formable HSLA and Dual Phase Steels', The Metallurgical Society of AIME, Warrendale, 1979
6. R. A. Kot and J. W. Morris, 'Structure and Properties of Dual Phase Steels', The Metallurgical Society of AIME, Warrendale, 1979
7. W. Bleck, E. J. Drewes, B. Engel, H. Litzke, and W. Müschenborn, 'Höherfestes kaltgewalztes Feinblech aus phosphorlegierten Stählen und aus Stählen mit Dualphasengefüge', Stahl und Eisen 106 (1986) 1381
8. T. Irie, S. Satoh, K. Hashiguchi, I. Takahashi and O. Hashimoto, 'Metallurgical Factors Affecting the Formability of Cold-Rolled High Strength Steel Sheets', Transaction ISIJ 21 (1981) 793
9. T. Furukawa, H Morikawa, M. Endo, H. Takechi, K. Koyama, O. Akisue, and T. Yamada, 'Process Factors for Cold-Rolled Dual Phase Sheet Steels', Transaction of ISIJ 21 (1981) 812
10. X. P. Shen and R. Priestner, 'Effect of Boron on the Microstructure and Tensile Properties of Dual Phase Steel', Met. Trans. 21A (1990) 2547
11. D. Z. Yang, E. L. Brown, D. K. Matlock, and G. Krauss, 'Ferrite Recrystallization and Austenite Formation in Cold-Rolled Intercritically Annealed Steel', Met. Trans. 16A (1985) 1385
12. S. Estay, L. Chengji, and G. R. Purdy, 'Carbide Dissolution and Austenite Growth in Intercritical Annealing of Fe-C-Mn Dual Phase Steels', Canadian Met. Quarterly 23 (1984) 121
13. P. Malecki and A. Barbacki, 'On the Mechanism of Cementite Dissolution in Austenite', Steel Research 59 (1988) 121
14. G. R. Speich, V. A. Demarest, and R. L. Miller, 'Formation of Austenite During Intercritical Annealing of Dual Phase Steels', Met. Trans. 12A (1981) 1419
15. N. R. V. Banguru and A. K. Sachdev, 'Influence of Cooling Rate on the Microstructure and Retained Austenite in an Intercritically Annealed Vanadium Containing HSLA Steel', Met. Trans. 13A (1982) 1899
16. O. Maid, W. Dahl, C. Straßburger and W. Müschenborn, 'Einfluß der Gefügeparameter auf die mechanischen Eigenschaften von Dualphasen-Stahl', Stahl und Eisen 108 (1988) 355
17. M. T. Shehata and A. F. Crawley, 'Microstructure – Property Relationship in C – Mn Dual Phase Steels', Canadian Met. Quarterly 22 (1983) 181
18. M. Sudo and I. Kokubo, 'Microstructure-Mechanical Property Relationships in Multi-Phase Steel Sheet', Scand. J. of Met. 13 (1984) 329
19. D. K. Matlock, F. Zia-Ebrahimi, and G. Kraus, 'Structure, Properties, and Strain Hardening of Dual Phase Steels', in 'Deformation, Processing and Structure', ed. G. Krauss, ASM Ohio, (1984) p. 47
20. A. Pichler, G. Hribernig, E. Tragl, R. Angerer, K. Radlmayr, J. Szinyour, S. Traint, E. Werner, and P. Stiaszny, 'Aspects of the Production of Dual Phase and Multiphase Steel Strip', Proceedings of the 41st Mechanical Working and Steel Processing Conference, Warrendale, 1999, p. 37
21. R. Angerer, K. Hauzenberger, and W. Wirthl, 'Description of the Dilatometer', to be published
22. G. Petzow, 'Metallographisches, keramographisches, plastographisches Ätzen', Gebrüder Bornträger, Berlin Stuttgart, sechste Edition, 1994
23. E. Wirthl, R. Angerer, and K. Hauzenberger, 'Determination of the Volume Amount of Retained Austenite in Small Specimens by Magnetic Measurements', to published: Conference Proceedings AISE 2000
24. G. Krauss nad S. W. Thompson, 'Ferritic Microstructure in Continuously Cooled Low- and Ultralow-Carbon Steels, ISIJ Int. 35 (1995) 937
25. H. Fischmeister and B. Karlsson, 'Plastizitätseigenschaften grob-zweiphasiger Werkstoffe', Z. Metallkde. 68 (1977) 311

CARBON DISTRIBUTION BETWEEN MATRIX, GRAIN BOUNDARIES AND DISLOCATIONS IN ULTRA LOW CARBON BAKE HARDENABLE STEELS

A. K. De[1], B. Soenen[2], B. C. De Cooman[1] and S. Vandeputte[2]
[1] Laboratory for Iron and Steelmaking, Ghent University, Technologiepark 9, Ghent 9052, Belgium
[2] OCAS N.V., Research Centre of SIDMAR, ARBED Group Flat Rolled Products Division, J.F. Kennedylaan 3, 9060 Zelzate, Belgium

Keywords: Strain aging, Dislocations, Segregation, Internal friction, ULC BH steel

ABSTRACT

The evolution of the carbon distribution between bulk, grain boundaries and dislocations during both the continuous annealing and the strain aging (e.g. paint baking) of ULC BH steels is numerically simulated. The calculations are successfully fitted to strain aging experiments and internal friction measurements. An increase of the grain size together with a higher cooling rate from the annealing temperature significantly increases the bake hardenability of ULC BH steels. This is contrary to the situation in low carbon BH steels due to the presence of the cementite particle distribution between grain boundaries and matrix in the latter steel.

INTRODUCTION

Bake hardening (BH) is a low temperature static strain aging process involving the interaction between solute carbon atoms and the dislocations generated during forming. The aging has to take place when formed automotive components are subjected to a paint baking cycle at temperatures around 170°C. The process yields an extra increase of the yield strength of the formed components through two distinct strengthening stages, a Cottrell atmosphere formation stage on the dislocations and a precipitation stage of carbides grown out of these carbon atmospheres [1-5]. Obviously, the advance of the strengthening stage during the aging depends on the amount of solute carbon retained in the matrix. In ULC BH steels, the chemistry and processing stages are so designed as to contain a very low amount of solute carbon at the end of processing for texture development and formability reasons. Hence, the strengthening stage during the bake hardening hardly goes beyond the first stage unlike the situation in low carbon steels [6]. It has been established through earlier work [6] that the maximum increase in yield stress due to the Cottrell atmosphere formation stage is about 30MPa in ULC steels for which a minimum of about 1-2 ppm solute carbon is necessary in the matrix. Hence, in order to further increase the yield stress through bake hardening it is necessary to increase the solute carbon content in these steels. In this regard, there are two approaches worth to be considered:

i) Increasing the grain size during continuous annealing and
ii) Increasing the cooling rate immediately after continuous annealing.

A variation of the grain size influences the distribution of carbon between the grain interior and the grai boundary by varying the number of segregation sites at the grain boundary. With an increase in grain size th grain boundary area decreases and the total amount of carbon that can be stored in the grain boundary decrease compared to that in a fine grained structure. Hence the amount of solute carbon in the matrix increases Increasing the cooling rate from the annealing temperature will restrict the diffusion of carbon atoms to the grai boundary thereby increasing the supersaturation of carbon in the iron matrix too.

The segregation behavior of carbon atoms to the grain boundary during continuous annealing an cooling can be evaluated through the application of Langmuir-McLean's segregation equation [7,8] and know parameters of carbon diffusion in α-iron. In the present study a numerical and experimental analysis has bee made of the carbon segregation behavior in a ULC BH steel as a function of cooling rate and grain size Secondly, the influence of these variables on the strengthening behavior during aging has been investigated. Th strengthening itself is simulated as a function of time and temperature, applying a Monte Carlo algorithm for th diffusion of carbon and pinning of the dislocations.

EXPERIMENTAL

Table I gives the chemical composition of the investigated steel, vacuum melted in the laboratory. T was added to stabilize the nitrogen. Hot rolled sheets were given 70-80% cold reduction. Fig. 1 show schematically the continuous annealing cycles followed for the cold rolled sheets to vary (*a*) the grain size an (*b*) to vary the solute carbon content through changes in cooling rate after continuous annealing. The annealin cycle (*a*) was carried out in a salt bath furnace with inhibitor addition to prevent decarburisation and cycle (*b* was done in a Gleeble 1500/20 continuous annealing simulator. Beside a water quenching (WQ), fast coolin (50°C/s) and slow cooling (11°C/s) were applied.

Table I Chemical composition of the ULC steels investigated (in wt. ppm)

C	Mn	P	S	Ti	Al	N	Ti/N
24	900	450	30	70	490	16	1.27

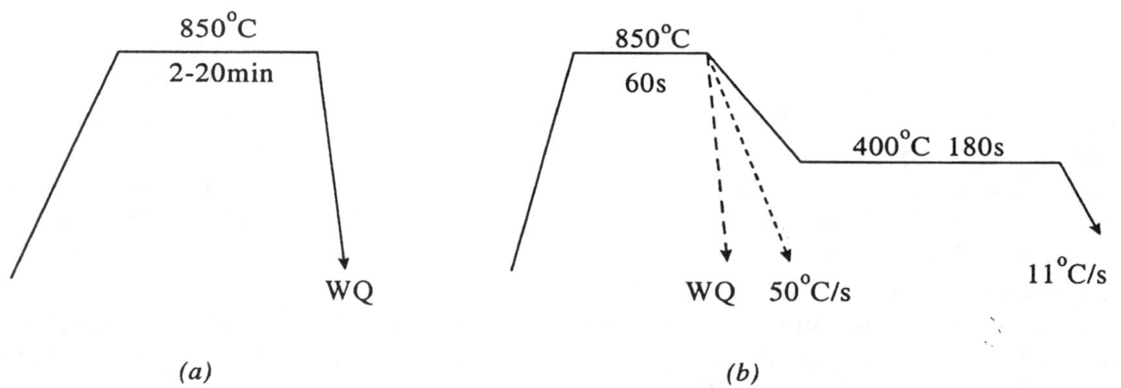

Fig. 1 Schematic illustration of continuous annealing cycles followed (*a*) to obtain different grain sizes and (*b*) to vary the matrix carbon content through changes in cooling rate.

The solute carbon content in the annealed specimens was measured by internal friction measurements using a torsion pendulum at room temperature and at a vibration frequency of 1Hz, and an automated piezoelectric ultrasonic composite oscillator (APUCOT), operating at 40 kHz [9]. With the latter, the internal friction due to the stress induced ordering of mobile interstitials is determined at a strain amplitude of 10^{-7} during the fast heating of the specimen in the range of 20-300°C at a rate of 100°C/min. The Snoek peak temperature is about 192°C at 40kHz. The advantage of this technique is that a high signal to noise ratio makes it possible to detect a very low amount of interstitials in the matrix, which can not be done using a torsion pendulum. However, during the heating of the specimens to the peak temperature there is an inevitable aging of the specimen (~2min) during which carbon may be lost to the dislocations. The specimens were prepared with the longitudinal axis parallel to the rolling direction of the annealed sheets.

Strain aging tests were carried out at 50°C and 170°C after a tensile prestraining of 5% for all the specimens. The increase in stress $\Delta\sigma$ due to aging for a time t was determined as the difference between the flow stress at the end of prestraining and the upper yield stress after aging.

MODELING

Segregation During Continuous Annealing

The modeling of the carbon distribution during the applied thermal cycles (continuous annealing (CA) and strain aging or bake hardening (BH)) was subdivided in two parts. In the first part, the kinetics of the grain boundary segregation of carbon during CA was modeled with finite differences. In the second part, the grain boundary segregation and segregation of carbon to dislocations during the strain aging was simulated with a Monte Carlo algorithm. The latter algorithm can in principle also cover the CA cycle, but it is very time-consuming and hence the finite differences program was preferred.

In the first part, the diffusion equation is solved in the grains, with the appropriate boundary conditions. The grains are assumed to have a spherical symmetry, so the diffusion equation is written as:

$$\frac{\partial c}{\partial t} = D\left(\frac{2}{r}\frac{\partial c}{\partial r} + \frac{\partial^2 c}{\partial r^2}\right) \tag{1}$$

$D = D_0 \exp(-U_d / RT)$ is the diffusion coefficient (m^2/s), r is the radial distance (m), and c is the carbon atom-fraction. Two boundary conditions are added, namely a mirror condition for the flux in the grain center:

$$J_r(0) = 0 \tag{2}$$

and the McLean equation for the concentration at the grain boundary, considering local equilibrium:

$$c(R_g) = \left(\frac{\theta}{1-\theta}\right)\frac{1}{\exp(-U_s / RT)} , \tag{3}$$

where R_g is the grain radius, θ the coverage and U_s the segregation energy.

In the model, the grain boundary is represented by an infinitesimal thin layer which consists of sites where carbon atoms can be stored. The coverage is the number of segregated atoms in the grain boundary divided by the total number of grain boundary atom sites where carbon can be located. The latter number has a maximum of M/a^2 per unit grain boundary area where a is the lattice unit cell length (2.87Å) of bcc-Fe. M is a parameter that can be adjusted in the model, but it should be of the order of one to be physically realistic. The diffusion equation is numerically solved with finite differences[10]. An explicit method is used for calculating the time derivative. For the simulations, the following reference values for carbon in bcc-Fe were used: a diffusion coefficient D=2×10^{-6}exp(-84140/RT) m²/s [11] and a segregation energy U_s = -57000 - 21.5T(K) J/mol [12].

Strain aging (Bake Hardening)

For the strain aging simulations, a Monte Carlo algorithm was developed. The simulations were 2-dimensional and performed in a rectangle containing an orthogonal grid. The unit cell length was 2.87 Å in each dimension. Carbon atoms were positioned in the middle of the unit cells. They can jump from cell i to one of the four neighboring cells j with a jump frequency $f(i \rightarrow j)$. If cells i and j are equal (in the sense that there is no energy difference), the jump frequency can be derived from the diffusion coefficient D:

$$f = f_d = \nu \exp\left(\frac{-U_d}{RT}\right), \quad \nu = \frac{D_0}{a^2} \tag{4}$$

with f_d consisting of an attempt frequency ν to move to an adjacent site, and a probability to surmount the energy barrier U_d.

At the boundaries of the rectangle, mirror conditions are applied.

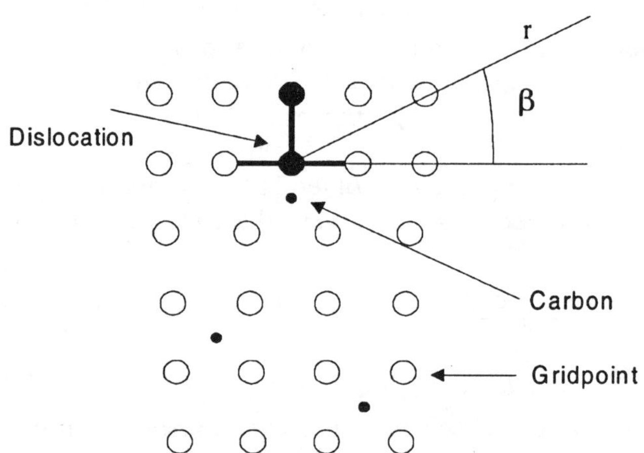

Fig. 2 Schematic graph of the geometry used for the Monte Carlo bake hardening simulations.

One side of the rectangle was chosen as grain boundary, and every carbon atom positioned there has an energy U_s compared to the bulk. In the bulk, edge dislocations were randomly placed at the edges of some cells, and they were assumed to be perpendicular to the plane of the simulations (Fig. 2).

In the simplest approximation, the interaction energy U_e between a solute atom and the hydrostatic component of the stress field of an edge dislocation is given by Cottrell and Bilby [1]:

$$U_e = A\frac{\sin\beta}{r}, \tag{5}$$

where A is the amplitude of the dislocation attraction energy. The pole coordinates (r, β) are explained in Fig. 2.

For the strain energy field, Cottrell calculated the factor A as 1.8×10^{-5} Jm/mol. For the simulations a value of $A = 7.5\times10^{-6}$ Jm/mol was used, because in the applied geometry, the atom most bound by the edge dislocation then has an energy of -52 kJ/mol, which is close to experimental values mentioned in literature [13].

Jumps from/to a position with a different energy from the bulk energy (dislocations ($\Delta U=U_e$), grain boundaries ($\Delta U=U_S$)) are then governed by the following equations:

$$\begin{aligned} U_i \le U_j &\Rightarrow f(i\to j) = f_d \exp\left(\frac{U_i - U_j}{RT}\right) \\ U_i > U_j &\Rightarrow f(i\to j) = f_d \end{aligned} \tag{6}$$

The dynamics of the system of atoms is calculated by repeating the following algorithm:

1. From the list of atoms, one particular atom will jump first. Suppose this is atom p.
2. The cumulative atom jump time is updated with $t_p = t_p + \tau_p$. The simulation time is put equal to the cumulative jump time t_p.
3. A random number *rnd* between 0 and $g(p,4)$ is generated, with

   $$g(p,k) = \sum_{l=1}^{k\le4} f(p,l).$$

   Based on this number, the direction k of the jump is determined:

   $$g(p,k-1) \le rnd < g(p,k).$$

   The atom is moved in this direction to the adjacent cell. If the cell is already occupied, the atom is not moved.
4. The jump frequencies $f(p,k)$, $k = 1 \to 4$ in the new position are calculated.
5. The cumulative jump frequencies $g(p,k)$ in the new position are calculated.
6. The atom jump time $\tau_p = 1/g(p,4)$ is calculated.

In the BH simulations, the dislocation density can be adjusted. Both a homogeneous dislocation density as well as an heterogeneous one can be used, e.g. to make the distinction between dislocation pile ups at the grain boundaries and regions with lower densities in the bulk of the grain.

As a starting carbon distribution for the strain aging simulations, the concentration profile resulting from the CA simulations was used. The carbon atoms are then randomly distributed in the rectangle, weighted according to this profile.

RESULTS AND DISCUSSION

Effect of grain size

Fig. 3 shows the microstructures developed in the specimens as a result of holding at the annealing temperature during different times. The grain sizes were measured using the linear intercept method according to the ASTM E112-88 standard. However, in order to use the model in a justified way, the intercept values determined were multiplied with a correction factor 1.5 for the case of the spherical grain approximation (*cf* use of eq. (1)), to account for the 2-D measurement of the 3-D reality [8]. A distribution of grain sizes, being the real case, is not taken into account here.

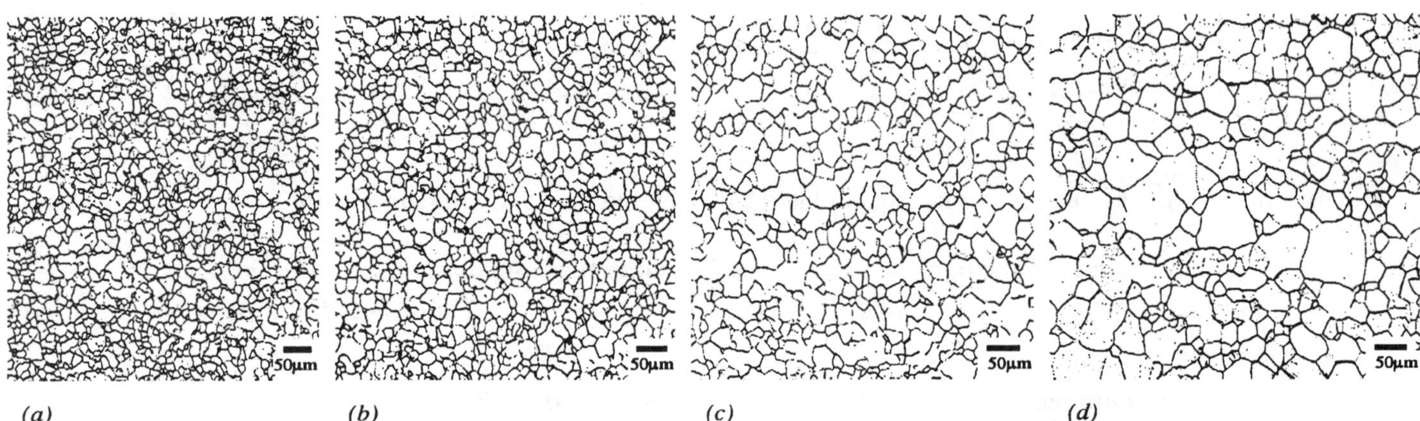

(a) *(b)* *(c)* *(d)*

Fig. 3 Microstructures and grain sizes of the ULC steel specimens after the annealing cycle of Fig. 1*a*, with soaking times and cold rolling reductions of (*a*) 2 min, CR=80%, (*b*) 5 min, 80%, (*c*) 4 min, 70% and (*d*) 20 min, 70%. The respective average grain sizes are 21, 26, 44 and 66μm.

Figs. 4 *(a)* and *(b)* show the simulation of carbon concentration profiles from the grain center to the grain boundary and the coverage of the grain boundary as a function of the grain size variation in the specimens. Fitting of the calculated mean carbon concentrations over the grain interior with the internal friction experiments (which do not measure the grain boundary carbon, only the solute matrix carbon) led to the use of M=2 for the simulations. The grain boundary looked upon as consisting of some atomic layers, this was considered to be physically realistic and it leads to a maximum grain boundary carbon site density of $2/a^2$.

It is seen that for a grain size of 66μm (or higher) almost all the carbon atoms are retained in solid solution. The concentration profile is quite uniform from the grain center to the grain boundary compared to those in the finer grains. The solute carbon in the matrix decreases with a decrease in grain size, because of the higher total amount of carbon segregated to the larger grain boundary area. Even after the water quenched cooling cycle, the grain boundary coverage is high for all the grains (Fig. 4*b*), which means that the majority of the grain boundary sites is occupied by the carbon atoms. There is a maximum difference in coverage of about 10% between the investigated grain sizes, so the difference in carbon content in the matrix can be accounted for mainly due to the direct effect of change in grain boundary area. It is concluded that for typical industrial grain sizes of 20-30μm, the amount of segregated carbon is far from negligible.

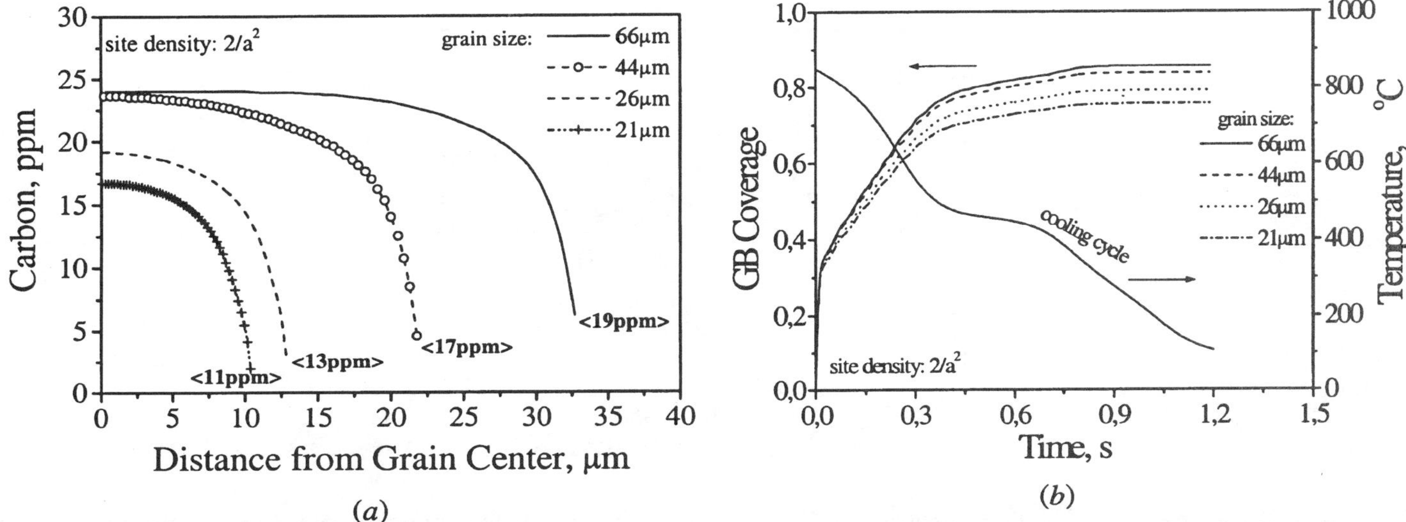

Fig. 4 (*a*) Simulation of carbon concentration profiles as a function of the radial distance and (*b*) the grain boundary coverage of carbon during the cooling cycle from the annealing temperature. Figures in brackets indicate the mean carbon concentration values in the grain interior. The temperature profile in (*b*) is experimentally recorded during the water quenching, and includes a plateau due to the instability of the vapor film.

Fig. 5 shows the internal friction spectra obtained in the specimens with different grain sizes, after water quenching. The peak height decreases with decreasing grain size, implying a decrease in the solute carbon content. This is in agreement with the simulated results. The actual matrix carbon contents measured by Apucot are lower than the calculated amounts, which is due to the fact that the internal friction peak is obtained at temperatures around 190°C. During the heating period (about 100s) to the peak temperature, solute carbon diffuses to the dislocations generated during quenching from the annealing temperature. The presence of mobile dislocations in the water-quenched specimens was evident from the absence of any yield point elongation during the tensile prestraining of these specimens. However, the torsion pendulum results are very close to the calculated mean carbon concentrations in the matrix (Fig. 4*a*), using a fitted maximum grain boundary carbon site density of $2/a^2$ (M=2).

The strain aging behaviors of these specimens are summarized in Fig. 6. The variation of the solute carbon content due to the difference in grain size clearly influences the strengthening behavior during strain aging. The results reveal the following:

i) the yield stress increase due to aging at any stage of the aging period is higher in the coarse-grained specimens,

ii) the completion of the first stage of aging (which is marked by the yield point elongation behavior) is faster with increase in grain size.

The amount of carbon per unit volume of the dislocation stress fields is higher in coarse-grained specimens compared to that in the fine-grained specimens and hence at any specific time of aging the dislocations in the coarse-grained specimens are saturated faster (Fig. 6*a*).

More details about the first Cottrell atmosphere formation stage as well as the second precipitation stage can be found in [14].

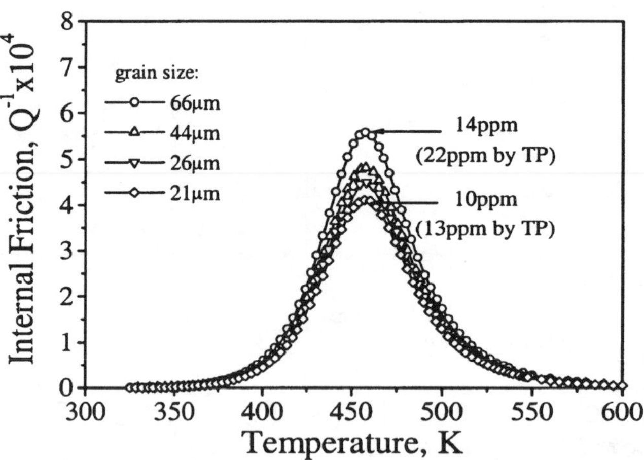

Fig. 5 APUCOT internal friction spectra as a function of grain size (water quench cycle) (TP: Torsion Pendulum).

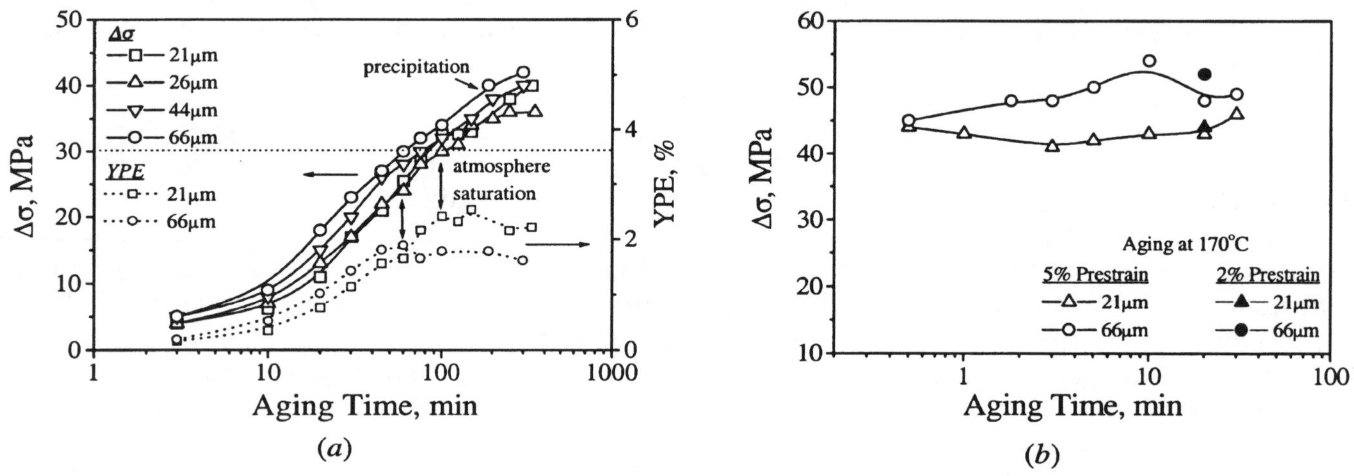

Fig. 6 (*a*) Increase in the yield stress and yield point elongation due to aging at 50°C in specimens with different grain sizes (water quenched) and (*b*) strengthening at 170°C as a function of grain size. The line at 30 MPa (Fig. 6*a*) indicates the maximum strength increase after the first atmosphere formation stage.

Fig. 6*b* shows bake hardening results at 170°C after 2% and 5% prestrain for the materials with the finest and coarsest grains. At both the prestrain levels the bake hardenability is higher in the coarse-grained microstructure. It is also evident from the results that within the grain size range investigated the grain boundary carbon does not contribute to the bake hardening process at the temperature of 170°C. This is a logical consequence of the carbon segregation energy values compared to the energy gain when a carbon atom is pinned at a dislocation; there is no driving force to move back into the grain at the relatively low baking temperatures. As to the contribution of segregated carbon atoms at the grain boundaries to the BH level, the situation might change for very heterogeneous dislocation density distributions with very high densities in the vicinity of the grain boundaries. These dislocations might provide preferential rapid diffusion paths for carbon, for which the model is no longer valid.

Finally, the simulated carbon concentration profiles after the processing were used as input data for the strain aging modeling and fitting. The second precipitation stage is not taken into account because there is no accurate nucleation model available yet. Thus, comparing the strain aging results with the calculations, only the first atmosphere formation stage has to be considered. The experimental yield stress increases $\Delta\sigma$ were normalized to $\Delta\sigma/\Delta\sigma_{atm}$ (with $\Delta\sigma_{atm}$=30MPa, Fig. 6*a*) in order to be able to compare them with the calculated dislocation saturation evolution, with the saturation defined as the ratio of the number of pinned atoms to the number of dislocations. Matrix precipitation on the dislocations can in principle start if this saturation is higher than one. It is also assumed here that there is a linear relationship between the dimensionless strength increase and the dislocation saturation during the Cottrell atmosphere formation stage.
The best fitting results were obtained with an activation energy for the carbon diffusion of 82000 J/mol which is in very good agreement with the value of 84140 J/mol mentioned by Wert [11]. The continuous annealing simulations were redone with this slightly adapted value but it had almost no impact on the initially used grain boundary segregation results. For the calculations, a dislocation density of 5×10^{13} m^{-2} was used for the prestrain of 5% [15]. It is noticed that for the atmosphere formation stage an overall good matching is obtained between the simulations and the experiments.

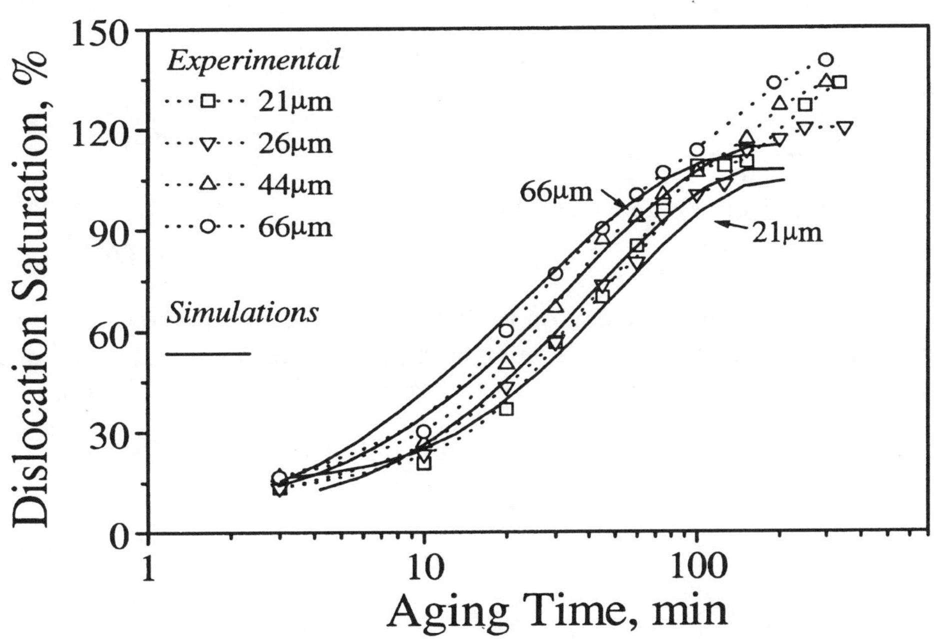

Fig. 7 Comparison of calculated and experimental dislocation saturation evolutions with aging time (50°C, prestrain of 5%) as a function of grain size.

Effect of cooling rate

Fig. 8*a* shows the simulated carbon concentration profiles in specimens with different cooling rates from the annealing temperature. The grain size was about 26μm for all the specimens. An increase of the cooling rate leads to a higher solute carbon content in the bulk of the grain. The same maximum grain boundary carbon site density of $2/a^2$ as used for the calculations of Fig. 4*a* again leads to a good agreement with the torsion pendulum results (Fig. 8*b*). Only for the slow cooling cycle, a larger deviation occurs. Moreover, in the Apucot result for this cycle, a dramatic decrease in the matrix carbon content is observed after heating up the sample to the Snoek peak temperature. It is thought that the slow cooling rate, resulting in the maximum grain boundary coverage (Fig. 9), allows the start of grain boundary carbon precipitation, not taken into account in the present model. Experimental TEM work on thin foils was not successful in proving this hypothesis, which does not mean that it can be excluded, in view of the expected very small size of these precipitates. Volume precipitation (without prestraining) is unlikely in the ULC steels due to the ultra low level of carbon [16,17].

In Fig. 9, the simulations of the grain boundary coverage evolution during the thermal cycle show that the final coverage is higher if the specimen is cooled slower, as can be expected. The final grain boundary coverage with slow cooling remains the same with or without the presence of the overaging step in the annealing cycle.

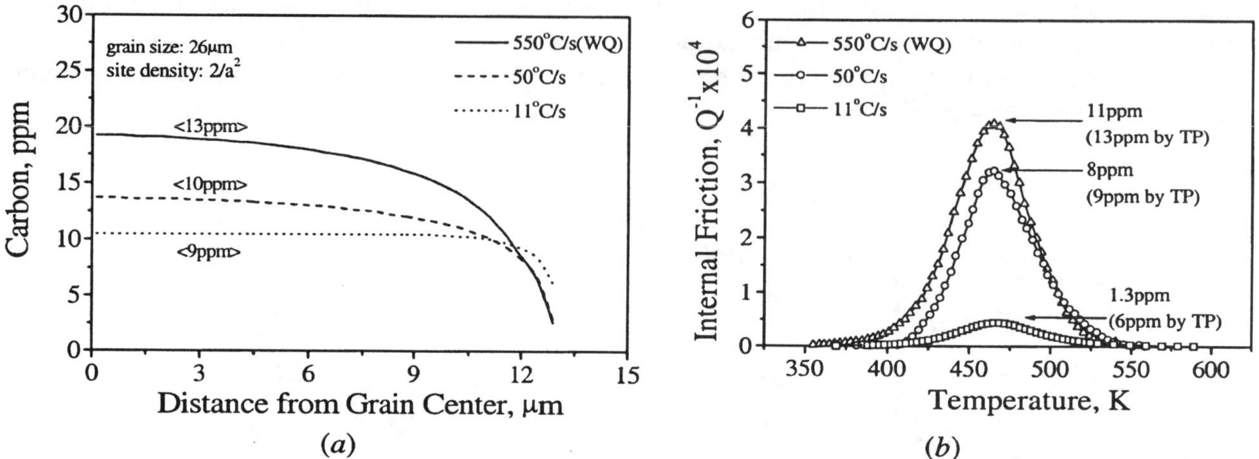

Fig. 8 (*a*) Simulation of the carbon concentration profiles from the center to the grain boundary and (*b*) APUCOT measurements of solute carbon as a function of cooling rate, for a grain size of 26μm. (TP: Torsion Pendulum). Figures in brackets indicate the mean carbon concentration values in the grain interior.

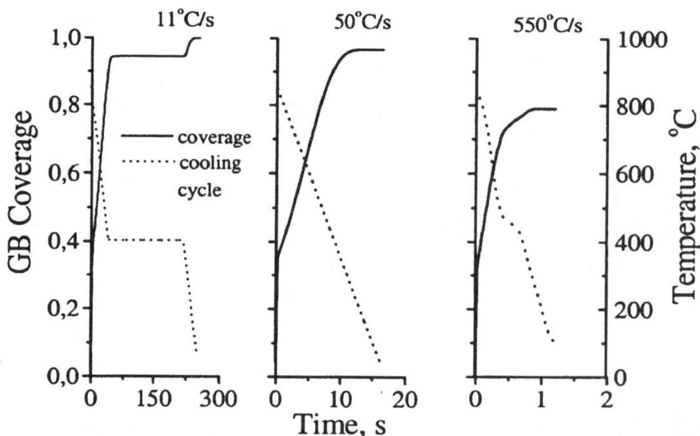

Fig. 9 Calculated grain boundary coverages of carbon as a function of the cooling cycle.

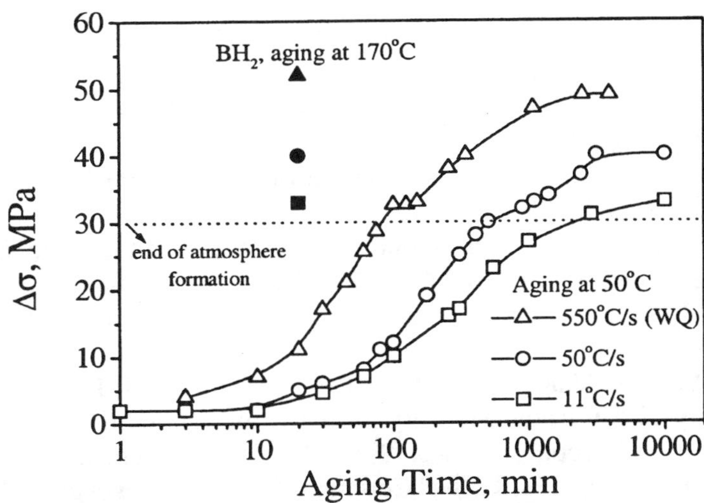

Fig. 10 Increase of the yield stress with aging time (50°C, 5% prestraining, grain size 26μm), as a function of the cooling rate (solid symbols represent corresponding BH_2 values, measured after 2% prestrain followed by 20 min aging at 170 °C).

Fig. 10 gives an illustrative summary of the influence of the cooling rate on the strain aging behavior of the ULC-BH steel. It is seen that the solute carbon in the specimens cooled slowly (11°C/s) is just sufficient enough to complete the atmosphere formation. The atmosphere saturation was marked distinctly by the yield point elongation behavior [18]. The precipitation stage or the second stage of hardening is virtually absent. An increase of the cooling rate increases the solute carbon content in the matrix of the specimens and results in (i) a faster completion of the first stage or the atmosphere saturation stage and (ii) the appearance of a precipitation stage with a significant increase of the yield stress. The maximum precipitation hardening is obtained in the water quenched specimens because of the highest retention of carbon in the matrix. Fig. 10 also indicates the increase of the bake hardening values (BH_2) obtained as a result of the increase of the cooling rate.

Fitting of the dislocation saturation calculations with the strain aging experiments as a function of cooling rate led again to an overall good agreement.

It is evident from the results that the variation in cooling rate after continuous annealing has a pronounced effect on the diffusion of carbon to the grain boundary, explaining bake hardening differences of the same material processed on different industrial lines. Thus, a significant increase of the bake hardenability of the ULC BH steels can be achieved through both a coarse grain size and a faster cooling rate after soaking.

CONCLUSIONS

The carbon segregation behavior during continuous annealing and cooling and the subsequent strain aging behavior of an ultra low carbon bake hardening steel have been evaluated as a function of grain size and cooling rate after annealing. The numerical simulations were in good agreement with the experimental strain aging experiments and internal friction results when a maximum grain boundary carbon site density of $2/a^2$ was used, with a the lattice unit cell length of bcc-Fe. Within the grain size range studied, an increase of the grain size and an increase of the cooling rate can significantly increase the solute carbon content in the matrix, thereby increasing the bake hardening level of the ULC BH steel. The effects of grain size and cooling rate are opposite to what is noticed in the case of low carbon BH steels. This is explained by the presence of the cementite particle distribution between grain boundaries and matrix in the latter steel [5].

REFERENCES

1. A. H. Cottrell and B. A. Bilby, "Dislocation Theory of Yielding and Strain Aging of Iron", Proc. Phys. Soc. A, Vol. 62, 1949, pp. 49-62.
2. D. V. Wilson and B. Russell, "The Contribution of Atmosphere Locking to the Strain-Ageing of Low Carbon Steels", Acta Metall., Vol. 8, 1960, pp. 36-45.
3. D. V. Wilson and B. Russell, "The Contribution of Precipitation to Strain-Ageing of Low Carbon Steels", Acta Metall., Vol. 8, 1960, pp. 468-479.
4. P. Elsen and H. P. Hougardy, "On the Mechanism of Bake-Hardening", Steel Research, Vol. 64, No. 8/9, 1993, pp. 431-436.
5. A.Van Snick, K. Lips, S. Vandeputte, B. C. De Cooman and J. Dilewijns, "Effect of Carbon Content, Dislocation Density and Carbon Mobility on BH", in "Modern LC and ULC Sheet Steels for Cold Forming: Processing and Properties", Vol. 2, Ed. W. Bleck, Aachen, 1998, pp. 413-424.
6. A. K. De, S. Vandeputte and B. C. De Cooman, "Static Strain Aging Behavior of Ultra Low Carbon Bake Hardening Steel", Scripta Materialia, Vol. 41, No. 8, 1999, pp. 831-837.
7. H. J. Grabke, "Surface and Grain Boundary Segregation on and in Iron", ISIJ International, Vol. 29, No. 7, 1989, pp. 529-538.
8. T. Gladman, "The Physical Metallurgy of Microalloyed Steels", The Institute of Materials, 1997.
9. I.G. Ritchie and Z. Pan, "Impurity Interstitials in Mild Steels", 33rd MWSP Conf. Proc., Warrendale, Ed. ISS, Vol. 29, 1992, pp. 15-25.
10. G. D. Smith, "Numerical Solution of Partial Differential Equations: Finite difference methods", Clarendon Press, Oxford, 1985, p. 11.
11. C. Wert, "Diffusion Coefficient of C in α-Iron", Phys. Rev., Vol. 79, 1950, pp. 601-605.
12. H. Grabke, "Surface and Grain Boundary Segregation on and in Iron", Steel Research, Vol. 57, No. 4, 1986, pp. 178-185.
13. K. Furusawa and K. Tanaka, "Amplitude dependent internal friction of iron containing a small amount of carbon and nitrogen", J. Jpn. Inst. Met. Vol.33, No. 9, 1969, pp. 985-991.
14. A. K. De, S. Vandeputte and B. C. De Cooman, "Kinetics of Low Temperature Precipitation in a ULC-Bake Hardening Steel", submitted to Scripta Materialia.
15. Y. Lan, H. J. Klaar and W. Dahl, "Evolution of Dislocation Structures and Deformation Behavior of Iron at Different Temperatures: Part II. Dislocation Density and Theoretical Analysis", Metall. Trans. A, Vol. 23A, 1992, pp. 545-549.
16. S. K. Ray, "Effect of Phosphorus on Carbon Activity, Carbide precipitation and Coarsening in Ferritic Fe-C-P Alloys", Metall. Trans. A, Vol. 22A, 1991, pp. 35-43.
17. T. Obara, K. Sakata, M. Nishida and T. Irie, "Effects of Heat Cycle and Carbon Content on the Mechanical Properties of Continuous-Annealed Low Carbon Steel Sheets", Kawasaki Steel Technical Report, No. 12, 1985, pp. 25-35.
18. A. K. De, S. Vandeputte and B. C. De Cooman, "Kinetics of Strain Aging in Bake Hardening Ultra Low Carbon Steel – a Comparison with Low Carbon Steel", submitted to J. Mater. Eng. Performance.

BAKE-HARDENABILITY AND AGING RESISTANCE OF HIGH-STRENGTH NITRIDED SHEET STEELS

Brandon M. Hance
U.S. Steel Research
4000 Tech Center Drive
Monroeville, PA 15146
Tele: 412/825-2041
E-mail: bmhance@uss.com

Key Words: High-Strength Steel, Nitrided Steel, Bake-Hardenability, Aging Resistance

INTRODUCTION

High-strength, nitrided sheet (HSNS) steels have shown significant bake-hardening capability combined with very good room temperature aging resistance. This combination of properties is desired in bake-hardenable steels for dent resistant automotive outer body panels, for instance. In order to understand the static strain aging phenomena responsible for these properties, the bake-hardening potential at various prestrain levels, aging temperatures, and aging times was examined. Generally, the bake-hardening response increases with increased aging temperature and aging time. Also, the time-temperature-prestrain dependence of the bake-hardening response is compared to that of conventional batch-annealed bake-hardenable steels. Unlike many bake-hardenable steels, high-strength nitrided steels have little or no return of yield point elongation (YPE) when aged at low temperatures (100°C or less) with no prestrain, even with long aging times (up to ten hours). Also, the bake-hardening capacity of nitrided steels is superior to that of conventional batch-annealed bake-hardenable grades. A possible mechanism for the aging resistant, bake-hardening quality of HSNS steels is discussed that considers the mobility of solute nitrogen and the driving force for solute nitrogen migration to mobile dislocations, or strain aging.

BACKGROUND

An ultra-low carbon, fully stabilized sheet steel with excess stabilizing elements, or a part made thereof, can be strengthened by an internal nitriding process. This process is discussed in detail by Lally and Holla in U.S. Patent Number 5,772,795[1]. Internal nitriding occurs when a fully stabilized sheet steel with excess titanium, niobium, or vanadium (Ti, Nb, or V) is exposed to an ammonia-nitrogen gas mixture (~10% NH_3, balance N_2) during an isothermal heat treatment in the temperature range of 480-650 °C (900 - 1200 °F). The decomposition of ammonia on the steel surface provides the source for atomic nitrogen. The adsorbed layer of atomic nitrogen on the surface diffuses into the steel in the form of a quasi-planar front. At the front, the solute nitrogen combines with the excess or free strengthening element to form coherent nitride precipitates. The term "excess" or "free" strengthening element refers to the concentration of Ti, Nb, or V that is not used in stabilization. The kinetics of nitriding is determined by the rate of solid-state diffusion of nitrogen in steel. The strength of the material increases as a result of the increased resistance to dislocation motion provided by the fine dispersion of coherent nitride precipitates. High-strength nitrided sheet (HSNS) steels can be partially-nitrided (*i.e.,* nitrided partially through the thickness), fully-nitrided (*i.e.,* nitrided through the entire thickness), or over-nitrided (*i.e.,* nitrided for a longer time than necessary to achieve the fully-nitrided condition).

Most bake-hardenable steels take advantage of the static strain aging characteristics of solute carbon in iron. The use of solute nitrogen in bake-hardenable steels has not been exploited, because it can lead to room temperature aging before panel forming in some cases. High-strength nitrided sheet (HSNS) steels make use of solute nitrogen that diffuses into the steel to combine with the free strengthening element (Ti, Nb or V) to form coherent nitride precipitates for strengthening. Although most of the nitrogen combines with the free strengthening element, there is also nitrogen present in solute form. Depending on the concentration, solute nitrogen can lead to discontinuous yielding and excessive yield point elongation. High-strength nitrided sheet steels must be temper rolled to some extent (typically 0.50-0.75 % temper mill extension) to rid the steel of discontinuous yielding and yield point elongation if they are to be used in exposed applications. HSNS steels have very good bake-hardening characteristics coupled with excellent room temperature aging resistance. In this study, the bake-hardening potential and aging resistance of HSNS steels has been examined, including an analysis of the static strain aging kinetics and a mechanistic discussion of the observations.

EXPERIMENTS

Samples of mill-produced high-strength nitrided sheet (HSNS) steels at five strength levels were gathered for the analysis. The yield strengths range from 200 to 480 MPa (30 to 70 ksi). A list of the materials and the as-received mechanical properties of each is given in Table I. HSNS 30 and HSNS 40 were partially-nitrided; HSNS 50 and HSNS 60 were fully-nitrided; and HSNS 70 was over-nitrided. All materials were mill temper rolled to about 0.75% temper mill extension (TME) except for HSNS 70 which was laboratory rolled to about 0.5% TME.

TABLE I - As-Received Mechanical Properties of Nitrided Steels

Material	Gage (mm)	YS (MPa)[1]	UTS (MPa)	TEL (%)	n (5/U)[2]
HSNS 30	1.07	217.72	347.95	37.2	0.21
HSNS 40	1.07	283.87	396.18	30.8	0.18
HSNS 50	0.66	345.19	424.42	25.5	0.14
HSNS 60	0.66	446.47	509.17	19.3	0.10
HSNS 70	0.89	483.68	594.61	20.3	0.13

1. 0.2 percent offset yield stress.
2. Calculated in the interval from 5% elongation to the maximum load in the tensile test.

The aging resistance and bake-hardening potential were examined with standard aging and bake-hardening tests. For aging resistance, a tensile specimen is placed into boiling water [100°C (212 °F)] for three hours and tested in tension to check for any change in mechanical properties including a return of discontinuous yielding or decreased n-value or total elongation. Three hours at 100°C simulates three months at room temperature for conventional carbon strain aging. As nitrogen has a lower activation energy for diffusion in ferrite than carbon does, a longer artificial aging time is needed at 100°C to simulate three months at room temperature as described in the Results section.

Bake-hardening potential was examined by prestraining the materials to 2% in tension and putting them into an oil bath at 177°C (350 °F) for 30 minutes, and then restraining to failure in tension. The bake-hardening index (BH) is measured as the increase in the flow stress from the 2% prestrained condition to the "baked" condition. The increase in the flow stress due to work hardening (WH) is measured as the maximum load during the prestraining step minus the initial yield strength. The total strengthening increment (WH+BH) is the difference in flow stress between the as-received (*i.e.,* temper rolled) condition and the prestrained-and-baked condition. In calculating WH and BH values, the lower yield strength is used in the case where there is a sharp yield point (*e.g.,* baked condition), and the 0.2% offset yield strength is used in the case of a smooth yield transition (*e.g.,* as temper rolled).

The material chosen for the aging kinetics analysis is HSNS 50 (see Table I). Longitudinal samples were taken approximately midway along the length of the coil. For four prestrain conditions (0, 1, 2 and 4 %), samples were aged at three temperatures [100°C, 138°C and 177°C (212°F, 280°F and 350°F)] for three aging times (10, 30 and 100 minutes). The aging times and prestrains were chosen so that the standard bake-hardening test is contained within the matrix. The temperatures were chosen to include the standard aging resistance test temperature, the standard bake-hardening test temperature and an intermediate temperature. The 100°C aging treatments were done in boiling water, while the higher temperature treatments were done in an oil bath. All tensile tests were run at a constant crosshead speed of 5-mm/min (0.2-in/min) and ASTM E8 specimen geometry [50-mm (2-in) gage length]. All results presented in the next section are the averages of duplicate tests for each condition, except where noted otherwise.

RESULTS

The results of the standard aging resistance and bake hardening tests are given in Table II. These data show that all of the nitrided steels have very good aging resistance (*i.e.,* YPE = 0% after 3 hours in boiling water). Properties such as yield strength, ultimate tensile strength, total elongation and n-value were unaffected by the simulated aging treatment.

There are several interesting trends in the bake-hardening simulation data of Table II. First, the total strength increase after straining and baking (WH+BH) increases with increasing yield strength. Second, the bake-hardening increment (BH) increases with increasing yield strength. Another interesting observation is that, for the lower strength levels (HSNS 30 and 40), the WH increment is larger than the BH increment; however, for the higher strength levels, the BH increment is larger than the WH increment. Figure 1 shows the data from Table II in graphical form, along with some data for conventional batch-annealed bake-hardenable steels[2]. In Figure 1, the conventional bake-hardenable steels are labeled according to nominal yield strength (ksi), similar to the HSNS steels. This figure shows that, for the conventional bake-hardenable grades, the work-hardening increment is greater than the bake-hardening increment. The higher WH increments for the bake-hardenable grades are a result of inherently higher n-values at lower strength levels (compare with HSNS 30 and 40). The general trend of decreasing WH increment with increasing strength level is therefore no surprise. The BH increment is not expected to be a function of strength or n-value, as it is primarily a function of the solute carbon content of these grades. The increasing BH increment with increasing strength level can be explained, however, for the nitrided steels. The degree of nitriding in the HSNS materials is the major determinant of strength and free nitrogen level (*i.e.,* solute nitrogen available for static strain aging). As described in the Experiments section, the degree of nitriding increases in the order: HSNS 30, 40, 50, 60 and 70, where HSNS30 and 40 were partially-nitrided; HSNS 50 and 60 were fully-nitrided; and HSNS 70 was over-nitrided to some

Table II - Results of Aging-Resistance and Bake-Hardening Tests

Material	As-Rec'd	After Aging		After Baking			
	YS (MPa)[1]	YS (MPa)[1]	YPE (%)[2]	YS (MPa)[3]	WH (MPa)	BH (MPa)	WH+BH
HSNS 30	217.7	220.5	0.0	263.3	34.5	11.0	45.5
HSNS 40	283.9	274.9	0.0	330.7	28.2	18.6	46.9
HSNS 50	345.2	345.2	0.0	400.3	20.0	35.1	55.1
HSNS 60	446.5	447.2	0.0	500.9	24.1	30.3	54.4
HSNS 70	483.7	483.0	0.0	587.7	33.1	71.0	104.0

1. 0.2 percent offset yield stress.
2. YPE: Yield Point Elongation, where YPE=0 means no aging.
3. Lower yield strength.

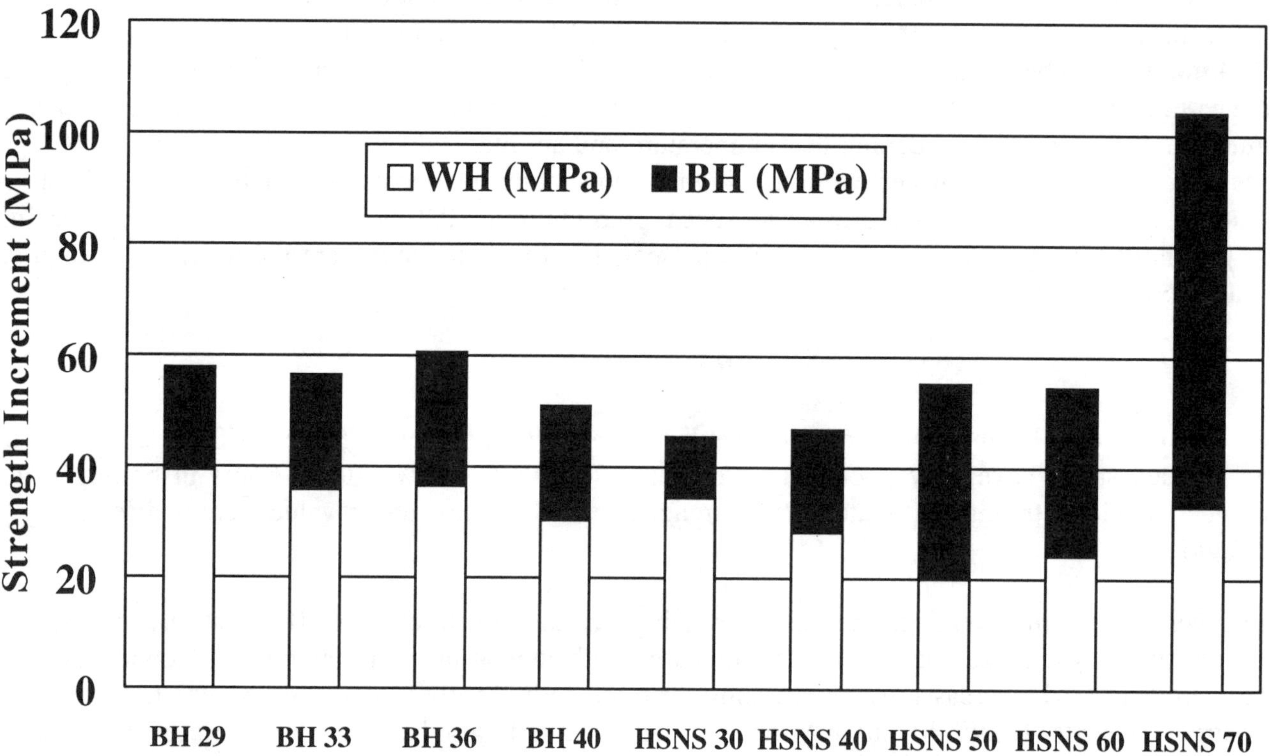

Figure 1. A comparison of the bake-hardening characteristics of bake-hardenable (BH) steels and high-strength nitrided (HSNS) steels. Here, WH is the work-hardening increment; BH is the bake-hardening increment (based on lower yield strength), and WH+BH is the total strength increase after a two percent prestrain followed by baking at 177°C for 30 minutes.

degree. Any further quantification of the degree of nitriding or attempt to correlate the properties is futile because the materials are all of different starting composition and gage.

The HSNS 70 material shows outstanding bake-hardening potential with a bake-hardening increment of over 70 MPa (10 ksi). The BH increment, alone, is greater than the total strength increase (WH+BH) for any of the batch-annealed bake-hardenable steels or the other nitrided steels. This gives a total strength increase of more than 100 MPa (15 ksi). Over-nitriding to some degree may be a method by which the bake-hardening potential of HSNS materials may be increased. However, all of the implications of over-nitriding, with respect to manufacturability and performance, are not completely understood.

Material Analysis

The material chosen for the aging kinetics study (HSNS 50) was a titanium fully-stabilized sheet steel prior to nitriding. The amount of excess titanium (*i.e.,* solute titanium that is not combined with carbon, nitrogen or sulfur) available for nitride strengthening can be estimated with the familiar stabilization equation[3]:

$$Ti_{ex} = Ti_{tot} - 4(C) - 3.42(N) - 1.5(S), \qquad [1]$$

where Ti_{ex} and Ti_{tot} are the excess and total titanium contents expressed in weight percent; and C, N and S are the carbon, nitrogen and sulfur contents expressed in weight percent. Table III shows the product composition of HSNS 50 after nitriding. Equation 1 predicts that, before nitriding, there was 0.028 weight percent free titanium (Ti_{ex}) available for strengthening (heat analysis: 0.003 weight percent nitrogen). Also, if we set Ti_{ex} equal to zero in Equation 1, the nitrogen content in the final product must be at least 0.011 weight percent if all of the titanium is to be tied up. Table I shows that the final product nitrogen level is 0.0172 weight percent. If it is assumed that all of the titanium is tied up, then there is 0.0062 weight percent (62 ppm) nitrogen in excess of that required to tie up all of the titanium. With this level of solute nitrogen, the aging response should be very strong.

Table III - Final Product Composition of HSNS 50

Composition (Weight Percent)								
C	Mn	P	S	Si	Cr	Ti	Al	N
0.0038	0.11	0.007	0.0098	0.008	0.018	0.068	0.043	0.0172

Bake-Hardening Potential

In Table IV the results of the aging tests are summarized, where YS is the as-received 0.2 percent offset yield strength of the test material. WH is the increase in flow stress after prestraining due to work hardening. BHL is the bake-hardening potential based on the lower yield strength, measured as the difference between the lower yield strength (LYS) after strain aging and the flow stress after prestraining [BHL = LYS - (YS + WH)]. BHU is the bake-hardening potential based on the upper yield strength, measured as the difference between the upper yield strength (UYS) after strain aging and the flow stress after prestraining [BHU = UYS - (YS + WH)]. All stress values reported in Table IV are engineering stress values. The LYS and UYS values are calculated based on the original cross-sectional area of the tensile specimens *before* prestraining. If the cross-sectional area of the tensile specimens is measured *after* prestraining, the LYS and UYS values (hence the BHL and BHU values) will be artificially high, especially when the prestrain level is high. There is some debate in the literature and in the automotive industry as to whether BHL or BHU is more important in determining the bake-hardening potential as related to improved service performance (*e.g.,* dent resistance); however, the BH values (*i.e.,* both BHL and BHU) defined as above are relative measures of the degree to which static strain aging has occurred for certain time-temperature-prestrain combinations.

Figure 2 shows the bake-hardening response for all conditions as measured by BHL. Figure 2(a) shows the effect of aging time and aging temperature on the aging response after 1 percent tensile prestrain. Figures 2(b) and 2(c) show the same for 2 percent and 4 percent tensile prestrain, respectively. In Figure 2(a), it is clear that

Table IV - Summary of Aging Test Results for HSNS 50

100°F	10 Min				30 Min				100 Min			
Strain	YS	WH	BHL	BHU	YS	WH	BHL	BHU	YS	WH	BHL	BHU
1%	362.4	6.2	6.5	0.7	361.7	7.6	9.6	13.1	357.6	7.6	13.1	24.8
2%	362.4	20.7	4.8	4.8	365.2	20.7	9.6	13.8	365.2	21.4	12.4	24.8
4%	364.5	43.4	3.8	9.6	368.6	42.7	7.6	15.8	383.1	42.0	20.7	40.7
138°C	10 Min				30 Min				100 Min			
Strain	YS	WH	BHL	BHU	YS	WH	BHL	BHU	YS	WH	BHL	BHU
1%	359.0	8.3	21.4	37.9	362.4	8.3	25.5	47.5	365.2	8.3	26.9	51.7
2%	367.2	20.7	22.7	43.4	370.0	20.7	27.6	51.0	367.2	21.4	31.7	55.1
4%	364.5	43.4	17.9	42.7	358.3	43.4	19.3	48.9	356.9	43.4	22.0	50.3
350°F	10 Min				30 Min				100 Min			
Strain	YS	WH	BHL	BHU	YS	WH	BHL	BHU	YS	WH	BHL	BHU
1%	381.7	7.6	33.8	64.8	380.3	7.6	36.5	67.5	356.9	8.3	30.3	66.1
2%	361.0	22.0	28.9	64.1	363.1	21.4	37.2	59.3	365.2	21.4	31.7	63.4
4%	357.6	43.4	24.8	52.4	353.5	43.4	26.2	53.1	370.7	42.7	24.1	56.5

Note: All values are given in MPa.

after 1 percent prestrain, the bake-hardening response (BHL) generally increases with aging time for lower aging temperatures (100°C and 138°C) in the interval 10 minutes to 100 minutes. For the high aging temperature (177°C); however, there is an overaging or softening effect, where a maximum in the aging response is reached between 10 and 100 minutes. The overaging effect at 177°C is seen in Figures 2(b) and 2(c) as well. Figure 2(b) (2% prestrain) is very similar to Figure 2(a) (1% prestrain), where the maximum BHL value is between 35 and 40 MPa. Figure 2(c) (4% prestrain) is different, however, where the maximum BHL value is between 20 and 30 MPa. Although the general observations are similar, the BHL values for the two higher aging temperatures (138°C and 177°C) are lower. There is an anomalous effect seen for the 100°C aging curve in Figure 2(c), where the aging response increases to about 20 MPa at longer aging times (similar to the BHL values at higher aging temperatures).

In Figure 3, the bake-hardening increment based on the upper yield strength (BHU) is shown for all conditions. Figure 3(a) shows the effect of aging time and aging temperature on the aging response after 1 percent tensile prestrain. Figures 3(b) and 3(c) show the same for 2 percent and 4 percent tensile prestrain, respectively. The trends in Figure 3 are similar to those seen in Figure 2 for the BHL value with a few differences. Generally, the BHU value increases with aging time and aging temperature, and the curves for the 1 percent and 2 percent cases are very similar. Again, the 4 percent prestrain case is somewhat different, where the maximum BHU value is around 55 MPa, and the maximum BHU value for the lower prestrain levels (1% and 2%) is around 70 MPa. Note that the ordinate scale for Figure 2 is from 0 to 40 MPa, and that for Figure 3 is from 0 to 80 MPa. The BHU values are roughly twice the BHL values. Another interesting feature of Figure 3 is that the BHU value for the high temperature aging treatment (177°C) does not pass consistently through a distinct maximum as a function of aging time like the BHL values in Figure 2.

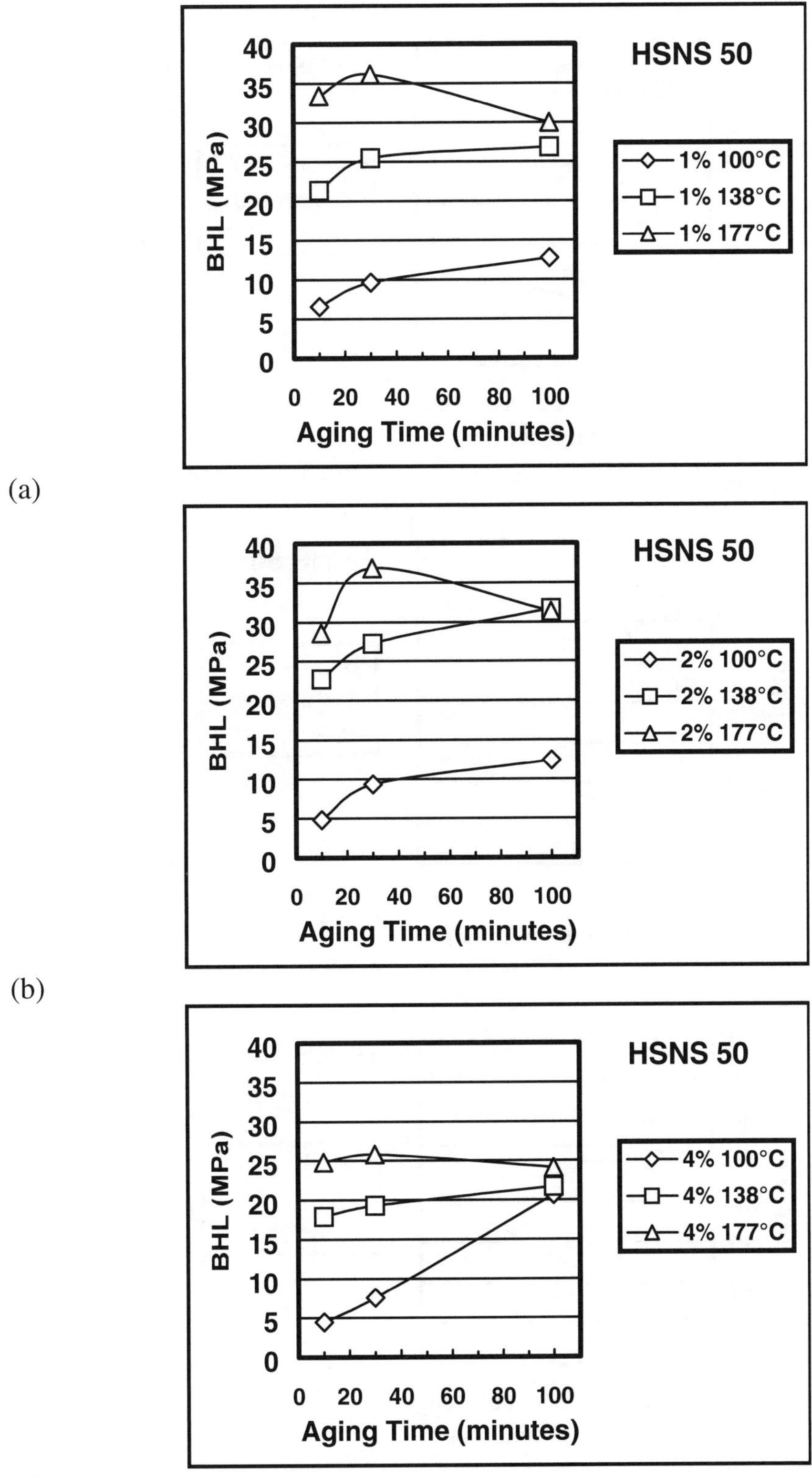

Figure 2. Bake-hardening increment based on the lower yield strength (BHL) as a function of aging time at various aging temperatures for (a) 1% prestrain, (b) 2% prestrain and (c) 4% prestrain.

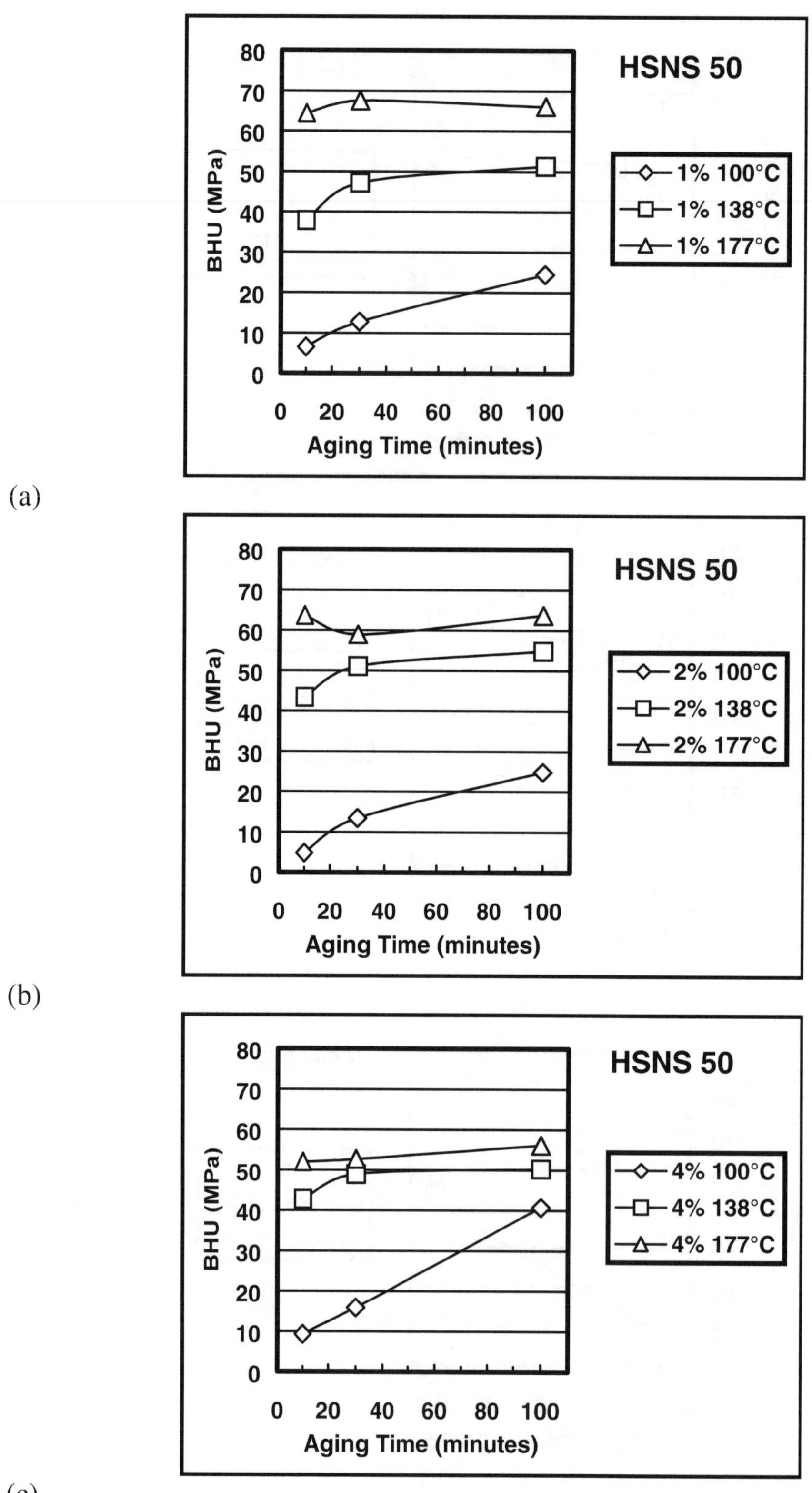

Figure 3. Bake-hardening increment based on the upper yield strength (BHU) as a function of aging time at various aging temperatures for (a) 1% prestrain, (b) 2% prestrain and (c) 4% prestrain.

Aging Resistance

In Table IV, there is some variation in the YS values [Range: 353 to 383 MPa (51.3 to 55.6 ksi)] due to nitriding non-uniformity in the coil, so it was not possible to measure the BH values for 0 percent prestrain directly. In cases where the uniformity in yield strength is very good, the BH values for 0 percent prestrain can be estimated by using the average as-received yield strength as the YS value (WH = 0) for each condition. Fortunately, there are other parameters that may be regarded as reflections of the degree of static strain aging. In addition to the appearance of a sharp upper yield point and an increase in flow stress, static strain aging also results in the return of yield point elongation (YPE). A return of YPE can even be expected for the 0 percent prestrain condition if the aging temperature is high enough and the aging time is long enough. Table V shows the YPE measured after aging for all conditions, including no prestrain.

Table V - YPE Values for all Aging Conditions (HSNS 50)

Pre-Strain	100°C			138°C			177°C		
	10 min	30 min	100 min	10 min	30 min	100 min	10 min	30 min	100 min
0%	0.00	0.00	0.00	0.17	0.20	0.19	0.16	0.44	0.43
1%	0.00	0.23	0.46	2.75	1.78	1.75	2.75	3.38	2.50
2%	0.00	0.31	0.86	2.00	2.50	3.00	2.40	3.50	3.80
4%	0.32	0.42	3.00	4.50	2.00	3.25	4.12	3.88	4.75

Figure 4 shows the effects of aging temperature and prestrain on the return of YPE as a function of time. Figure 4(a) shows the return of YPE for several aging temperatures with no prestrain. Figure 4(b) shows the effect of prestrain on the YPE return for several prestrain levels at the standard accelerated aging temperature (100°C). The YPE value generally increases with increasing prestrain, aging temperature and aging time. For the 0 percent prestrain case, there was no return of YPE in the aging time interval from 0 to 100 minutes at 100°C. At 138°C, the return of YPE is only 0.2 percent after 100 minutes aging time, and at 177°C, the return of YPE is only 0.4 percent after 100 minutes aging time.

Aging resistance specifications vary amongst consumers of automotive sheet steels. The reason for the variance is that there is an ongoing debate on the maximum allowable YPE to avoid visible Lüders band formation (surface strain lines) on exposed automotive panels. Generally, the acceptance limit is between 0.2 and 0.4 percent YPE. Also, automotive customers expect a certain shelf life somewhere between 3 and 6 months after delivery. The temper rolled product must be free of excessive YPE not only upon delivery but also when the customer takes the material from storage to be made into parts. The aging resistance specifications can conceivably range from strict (less than 0.2 percent YPE after 6 months in storage) to liberal (less than 0.4% YPE after 3 months in storage).

To estimate the time required to simulate room temperature aging for a certain length of time, the following expression can be used:[(4)]

$$\ln\left(\frac{t_1}{t_2}\right) = \frac{Q}{R}\left(\frac{1}{T_1} - \frac{1}{T_2}\right), \qquad [2]$$

where t_1 is the aging time at temperature T_1, and t_2 is the aging time at temperature T_2. Q is the thermodynamic activation energy required for strain aging, and R is the universal gas constant (8.314 J per mol K). Ideally, an accelerated (or artificial) aging treatment should be carried out at the lowest temperature that can simulate room temperature aging in a reasonable amount of time. Boiling water is a convenient aging medium because the aging times are reasonable, and the aging temperature is constant (100°C). Since strain aging is a diffusion-controlled process, the activation energy for solute diffusion in iron can be used in Equation 2 for the value of Q. For carbon diffusion, Q in Equation 2 is 84,000 J per mol, so three hours at 100°C is equivalent to 3.8 months at room temperature (25°C) for conventional bake-hardenable steels. A typical nitrogen diffusion activation energy is 74,000 J per mol[(5)]. The 100°C aging treatments were expanded for the HSNS steels to simulate three months and six months at room temperature. For nitrogen aging, 5 hours at 100°C simulates approximately 3 months (2.8 months) at room temperature, and 10 hours simulates approximately 6 months (5.6 months). There was still no return of YPE after five hours aging time in boiling water, and after ten hours, the YPE return was only 0.09 percent (average of three tests). This result suggests that the nitrided steel exceeds the most conservative aging resistance specification of less than 0.2 percent YPE after 6 months at room temperature.

DISCUSSION

The mechanisms responsible for the strain aging phenomena seen in the conventional carbon aging case and the HSNS steel case are complex, and some observations are yet unexplained. Several possible explanations for the decreasing BH increment with increasing prestrain in conventional bake-hardenable steels and HSNS materials (see Figures 2 and 3) have been proposed[(2, 4, 6, 7)]. Most bake-hardenable steels are designed to have an optimal solute content for good bake-hardenability and good aging resistance (*i.e.,* between 10 and 20 ppm C by weight)[(4)]. Normally, aging resistance decreases as bake-hardening potential increases. This is no surprise since both properties are determined by the same mechanism - interstitial diffusion to mobile dislocations. To address this dichotomy, a balance must be struck with the solute content. Too large a solute content can lead to poor aging resistance, and too small a solute content can lead to poor bake-hardening potential. For the case of conventional bake-hardenable steels, there are only a few ppm of carbon in solution. If a small prestrain is applied to the material, there is a driving force for strain aging, and the solute atoms will diffuse to the dislocations if given enough time at a certain temperature. A number of the mobile dislocations will be effectively pinned by the solutes. If the material is now re-strained, more mobile dislocations must be created to continue plastic flow; thus, the effective flow stress is raised. If the prestrain is large, the mobile dislocation density will be higher, and the solute carbon may not be enough to pin as many dislocations as in the case of the small prestrain. Thus, the pinning effect, or the BH increment, will not be as strong, even though the driving force for aging may be higher.

For the HSNS steels, the solute (nitrogen) content can be several times greater than that of typical bake-hardenable steels, so the corresponding bake-hardening increment is higher for a given prestrain and aging condition, except in the case where the HSNS material is not prestrained, and there is little or no aging response at all. Figure 5 shows the bake-hardening of the HSNS materials as a function of tensile prestrain for various aging temperatures measured by BHL and BHU for 30 minutes aging time. Included in Figure 5 are data for a typical batch-annealed bake-hardenable steel[(2)]. For the conventional steel, designated BH40, there is a small BH increment for the no prestrain case. With prestrain, the BH increment initially increases then decreases steadily as a function of increasing prestrain, for both BHL and BHU. In Figure 5(a), the BHL values are plotted as a function of prestrain for various aging temperatures for the HSNS material and the BH40 material. The behavior of the HSNS material is very similar to that of the BH40, except that the BHL values are higher for the HSNS case at the normal paint baking temperature (177°C). Even at 138°C, the BHL values are higher for the HSNS case. At 100°C, the BHL values are lower than those for the BH40 steel at 177°C are, but the trend is very similar. Figure 5(b) shows the BHU value as a function of prestrain for various aging

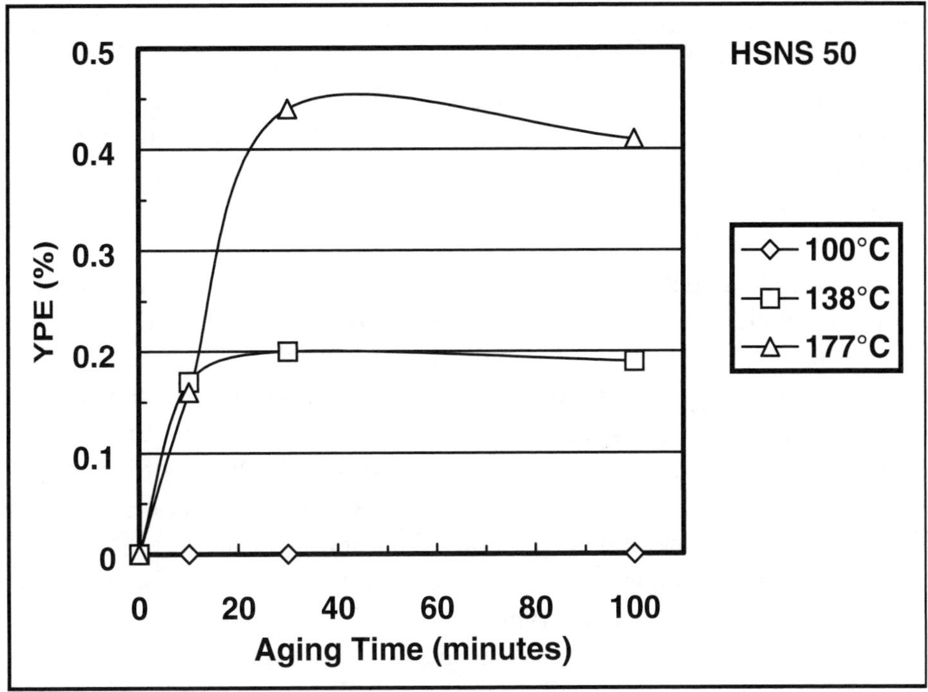

(a)

(b)

Figure 4. Return of yield point elongation (YPE): (a) as a function of aging time for several aging temperatures with no prestrain and (b) as a function of aging time for various prestrain levels at 100°C (standard accelerated aging temperature).

temperatures. Similar trends are seen for the BHU values. Note that the ordinate scales are different in Figures 5(a) and 5(b), and the BHU values are on the order of twice the BHL values. Also note that the BHL and BHU values shown in Figure 5 for the HSNS materials at 0 percent prestrain were not actually measured (assumed zero), but are probably less than or equal to those of the BH40 material at 0 percent prestrain. BH_0 is defined as the bake-hardening increment after baking for 30 minutes with no prestrain, while BH_2 is the standard with 2 percent prestrain before baking. Figure 4(a) suggests that the BH_0 values are not zero for baking at 138°C or 177°C since there is a significant return of YPE after 30 minutes aging time. It is important to have at least a small BH response with no prestrain, as some regions in a formed part may undergo little or no deformation during forming. Since the YPE return after baking at 100°C is zero, it is a reasonable assumption that BH_0 at this temperature is zero as well.

As the coherency strain field described above is a barrier to interstitial diffusion, there is a certain thermal energy (time at temperature) which must be supplied to overcome the barrier. The rate at which solute atoms move through the iron lattice depends on both the mobility of the solutes and the driving force for migration (strain aging). Prestrain serves to increase the driving force for migration, and elevated temperature serves to raise the mobility of the solutes through the coherency strain field, or perhaps to alleviate partially the strain field itself. In the case of no prestrain and low aging temperatures, both the mobility and driving force are very low, so there is little or no aging. With some prestrain at low temperatures, the mobility is low, but the driving force is high, so there is an intermediate amount of aging. With no prestrain at high temperatures, the mobility is high, but the driving force is low, so there is again an intermediate amount of aging. With some prestrain at high aging temperatures, both the mobility and driving force are high, so there is a high level of aging. Figure 4 illustrates this effect very clearly. Note that in Figure 4(a), the maximum YPE value increases as aging temperature increases (higher mobility), and in Figure 4(b) the YPE value increases as prestrain increases for a constant aging temperature (high driving force). In conventional bake-hardenable grades, there is no mobility barrier to solute diffusion at low temperatures other than the inherent ferrite lattice resistance. For this reason, conventional bake-hardenable grades generally have significant bake-hardening potential with little or no prestrain and, in some cases, poor aging resistance.

SUMMARY

The static strain aging kinetics of a high-strength nitrided sheet (HSNS) steel was examined along with the general aging resistance and bake-hardening characteristics of several mill-produced HSNS steels. The bake-hardening response at various aging temperatures, aging times and prestrain levels was measured, and several trends were observed. It was found that the bake-hardening potential of HSNS steels increases with strength level, or more appropriately, the degree of nitriding. In some cases, the bake hardening potential is superior to that of conventional grades that derive post-forming strength from static carbon strain aging. Similar to the case of conventional bake-hardenable sheet steels, it was found that, for HSNS steels: 1) the static strain aging response, in terms of bake-hardening, increases with increasing aging temperature and aging time; 2) the return of yield point elongation (YPE) increases with increasing pre-strain, aging temperature and aging time; and 3) the bake hardening increment decreases with prestrain level. HSNS steels have excellent aging resistance that is comparable to that of current batch-annealed bake-hardenable grades. The return of YPE was less than 0.1 percent after aging for 10 hours at 100°C (~6 months at room temperature), and there was no YPE return after 5 hours (~3 months at room temperature). Thus, HSNS materials should be able to meet the strictest demands for aging resistant, bake-hardenable sheet steels. A mechanism has been proposed to account for the desired combination of strong bake-hardening potential and excellent aging resistance. HSNS steels are strengthened by a dispersion of nitride particles that impose a coherency strain field in the ferrite matrix. The coherency strain field serves to reduce the mobility of nitrogen in the absence of cold work, thus resulting in excellent room temperature aging resistance prior to panel forming. After forming, the driving force for aging is augmented, and the mobility of nitrogen is increased during baking, thus resulting in a bake-hardening response.

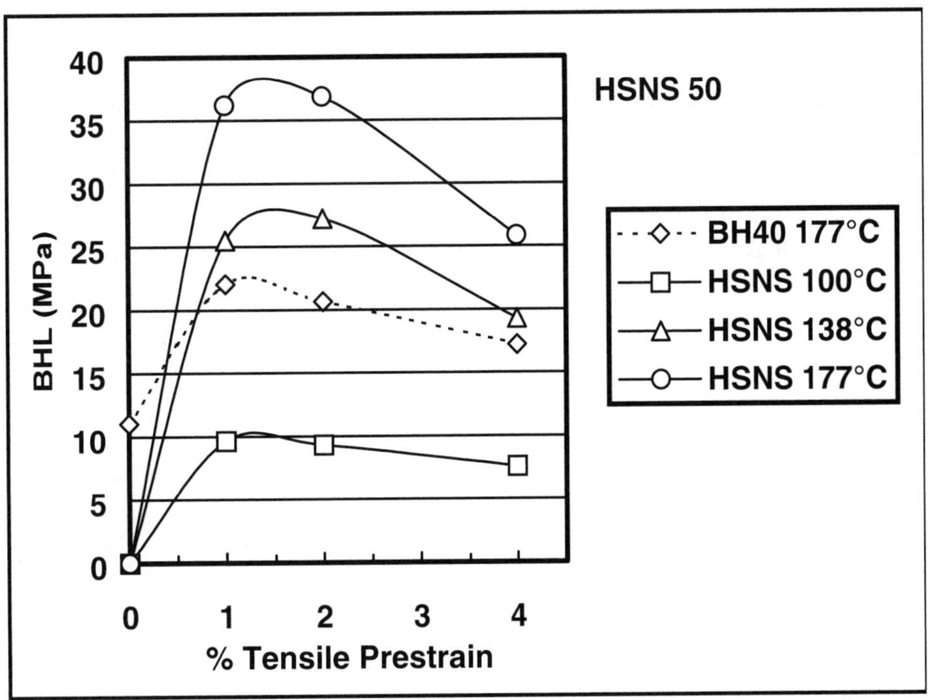

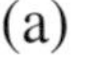
(a)

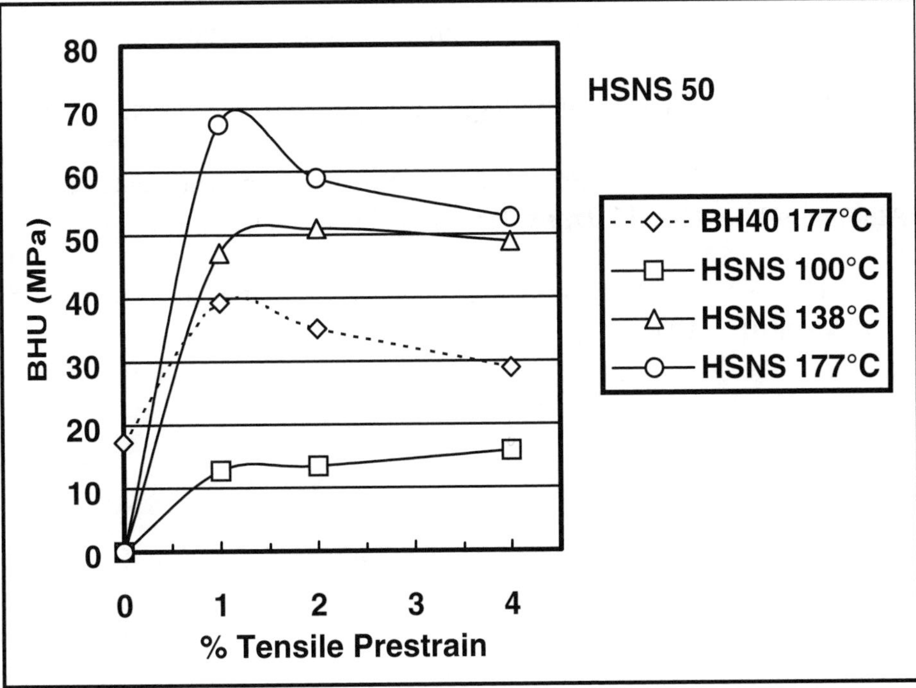

(b)

Figure 5. Bake-hardening response as a function of tensile prestrain for various aging temperatures measured by (a) BHL and (b) BHU. Aging time is 30 minutes. The curve marked BH40 represents a conventional bake-hardenable, 275 MPa (40 ksi) yield strength steel, and these data were taken from Reference 2. The data from the HSNS curves are from Table IV.

ACKNOWLEDGMENTS

I would like to thank Dick Fiscus and Greg Walters for carefully running all the aging simulations and the tensile tests for this investigation. Also, thanks to Joe Michalak, Todd Link, Joe Gallenstein and Dave Hoydick for their helpful discussions and suggestions.

REFERENCES

1. H.A. Holla, J.S. Lally, "High Strength Deep Drawing Steel Developed by Reaction with Ammonia," United States Patent No. 5,772,795; 30 June, 1998.

2. J.T. Michalak, U.S. Steel Research, Unpublished Data.

3. R. Pradhan, E.S. Batt, M.O. Mauck, R.P. Archer, "Development of Vacuum-Degassed Sheet Steel Products," 34th MWSP Conference Proceedings, ISS-AIME, Vol. XXX, 1993, pp. 77-82.

4. F.D. Bailey, "Effects of Mill Processing and Prestrain Mode on the Aging Behavior of a Commercially Produced Bake-Hardenable Steel," M.S. Thesis, Colorado School of Mines, June 1994.

5. R.E. Reed-Hill, R. Abbaschian, "Physical Metallurgy Principles, Third Edition," PWS Kent, Boston, 1992.

6. S. Hanai *et al.*, "Effect of Grain Size and Solid Solution Strengthening Elements on the Bake-Hardenability of Low Carbon Aluminum-killed Steel," *Trans. ISIJ*, Vol. 24, 1984, pp. 17-23.

7. A. van Den Beukel, U.F. Kocks, "The Strain Dependence of Static and Dynamic Strain Aging," *Acta Metall.*, Vol. 30, 1982, pp. 1027-1034.

INVESTIGATION OF THE SECONDARY WORK EMBRITTLEMENT OF IF-STEELS

A. Spalek, G. Reisner, E.A. Werner, A. Pichler †, P. Stiaszny †
Christian-Doppler-Laboratorium für Moderne Mehrphasen-Stähle,
Lehrstuhl A für Mechanik, TU-München,
Boltzmannstraße 15, D-85747 Garching, Germany.
tel.: +49 89 289 15247; fax: +49 89 289 15248; e-mail: spalek@lam.mw.tu-muenchen.de
† VOEST-ALPINE Stahl Linz
Research, Development and Testing, Voest-Alpine Straße 3, A-4031 Linz, Austria.
tel.: +43 732 6585 6701; fax: +43 732 6980 4338;

Key Words: IF-sheet steel, Secondary Work Embrittlement, drop weight test, crucial test-parameters;

INTRODUCTION

Due to their excellent formability, interstitial free (IF) steels are used for complex shaped parts as required in automotive applications. The mechanical and technological properties characteristic for IF-steel grades are low yield strength, a high work hardening exponent, low planar anisotropy which assures a weak tendency for earing of the material and a strong γ-fiber texture ($<111>$ parallel to the sheet normal) after annealing which leads to a high r-value. IF-steel consists of a highly ductile ferritic matrix with embedded precipitates of nitrides and carbides. Unfavourable production conditions can lead to brittle fracture, if a heavily deep drawn part is subjected to an impact deformation at low temperature. This behavior is called secondary work embrittlement (SWE) or cold work embrittlement (CWE) [1-10]. A common test method to investigate the tendency of the material to SWE resides on the expansion of a cylindrical deep drawn cup by a conical-like drop weight at low temperature (fig. 1). This method is know as drop-weight test [1-10]. The goal of the test is to identify a critical temperature at and below which the deep drawn cup fails by brittle fracture under well defined test conditions. The critical temperature has been termed crack appearance temperature (CAPT). The lower CAPT the lower is the tendency of the material to fail by brittle fracture. This paper aims at a investigation of the influence of the chemical composition and the production route, and structural effects such as the sheet thickness and the predeforming process on the sensitivity of IF-steels to SWE. Special attention is paid to the numerical analysis of the deep drawing process to identify the stress and strain conditions responsible for the initiation of brittle fracture upon impact loading.

EXPERIMENTS

Fig. 2 shows a schematic sketch of the axialsymmetric setup of the deep drawing tools. The diameter d of the punch, the radius of the punch, the blank holder and of the die, r_p, r_h and r_d, respectively, are kept constant in all experiments. The drawing gap g is matched to the thickness t of the sheet. The diameter D_0 of the blank is adjusted to make possible the forming of deep drawn cups with three draw ratios $\beta = D_0/d$, namely β_1=1.8, β_2=2.0 and β_3=2.4. The blank holder force F_N is adjusted so that wrinkling of the cup edges is prevented. To minimize the friction coefficient μ between the interacting pairs punch/blank, blank holder/blank and die/blank the blanks are coated with lubricant foils. Additionally, oil is used for lubrication. The deep drawn cups are trimmed to different heights s prior to impact testing to provide cups with well defined upper rims. A schematic sketch of the setup for the drop-weight test is shown in fig. 1. The ranges of the parameters of the impact test (drop height h, drop weight w, cone angle α, velocity v and impact energy E) covers the ones given commonly used. [1-11].

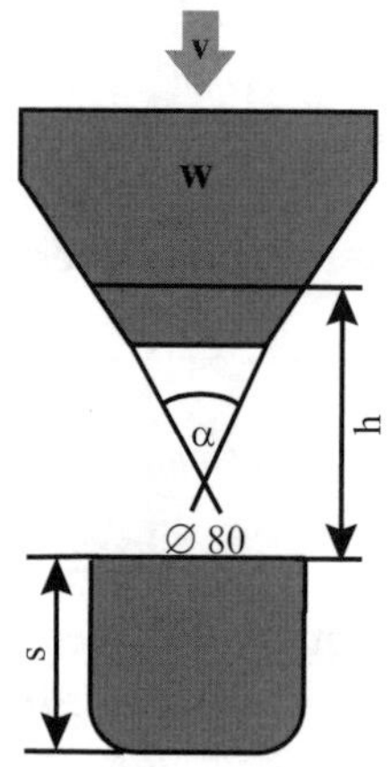

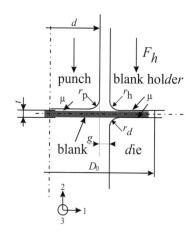

Figure 1: Schematic sketch of the experimental setup for the drop-weight test.

Figure 2: Schematic sketch of the axialsymmetric setup of the deep drawing tools.

To study the influence of the drop-weight parameters on the CAPT alone, cups of the same material (chemical analysis, sheet thickness) are produced using always the same deep drawing parameters (F_N, μ, β). By varying the impact parameters systematically the crucial ones can be identified. A similar procedure is followed when studying the influence of β, s and t on the CAPT.

To shed light on the role of the actual strain conditions on the fracture behavior of the material when impacted at low temperatures high speed tensile tests are performed. For this purpose tensile specimen are taken from the deep drawn cups in circumferential direction. Hence, it is possible to test the material subjected to the same deformation history which exists at the cup rim before impact testing. The tensile test parameters are adjusted to mimic those of the drop-weight test (high strain rate and low temperature) at which the cups fail by brittle fracture.

The as-received sheet material undergoes a complex, non-proportional loading during deep drawing. Therefore, both the actual strain and the residual stress state along the cup wall plays an important role for the later behavior of a deep drawn cup in the drop-weight test. To investigate the residual stress state qualitatively rings are taken from the cup in circumferential direction at different vertical cup positions. These rings are then cut open and their spring back is studied. To clarify the influence of the draw ratio on the CAPT, rings are taken at the same vertical position from cups formed at different draw ratios and then comparing their spring back behavior.

NUMERICAL MODELING

The numerical modeling of the deep drawing process aim at a calculation of the residual stress and strain state in the different deep drawn cups used in the drop weight test. Since the IF-steel tested exhibits only small earing during the deep drawing process, the small anisotropy of the material may be neglected and isotropic hardening and J_2 (*von Mises*) plasticity can be assumed. During deep drawing the velocities involved allow to use a quasi static analysis, i. e. to disregard inertia effects. Since however, the strain rate $\dot{\varphi}$ significantly changes during deep drawing ($0 < \dot{\varphi} < 1.8\,\mathrm{s}^{-1}$), one has to take into account the influence of the strain rate on the yield stress of the material. Contrary to $\dot{\varphi}$, the adiabatic heating of the cups during the forming process is much too little to significantly alter the flow behavior of the material. The strain rate dependent yield stress of the material σ_y is described by a modified Ludwik-Holomon relation:

$$\sigma_\mathrm{y} = (\sigma_0 + K\varphi^n)\dot{\varphi}^m, \tag{1}$$

where σ_0 and K are material parameters which were determined in a quasi static tensile test at a strain rate of $\dot{\varphi} = 0.001\ \mathrm{s}^{-1}$ as 127 MPa and 412 MPa, respectively. The strain hardening exponent n is 0.37 [11]. The strain rate exponent m characterizes the influence of a changing strain rate on σ_y. A value of 0.12 for m is taken from the literature [12]. The flow curve obtained from the tensile test is extrapolated to the larger logarithmic equivalent plastic strains that occur during deep drawing. The geometric parameters for the numerical deep drawing process (F_N, r_h, r_p, r_d, g and the deep drawing velocity v_d) are chosen to match the experiments. A small coefficient of friction ($\mu = 0.005$) is assumed to account for the lubricant foils and the lubrication oil used in the deep drawing experiments. The good agreement observed between the force-displacement curves obtained from the instrumented punch used in the deep drawing experiment and the numerical predictions for all deep drawing parameters justifies this assumption.

RESULTS

The optical inspection of the fracture surface of the expanded deep drawn cups is done in a scanning electron microscope (SEM). The longitudinal cracks along the cup wall are initiated via brittle cleavage fracture (fig. 3 (a)) gradually changing their mode to ductile fracture near the tip of the arrested crack (fig. 4). On the trans-granular cleavage surface small regions of intergranular fracture can be observed (fig. 3 (b)).
Within the investigated range a significant influence of the drop-weight parameters (drop height h, drop weight

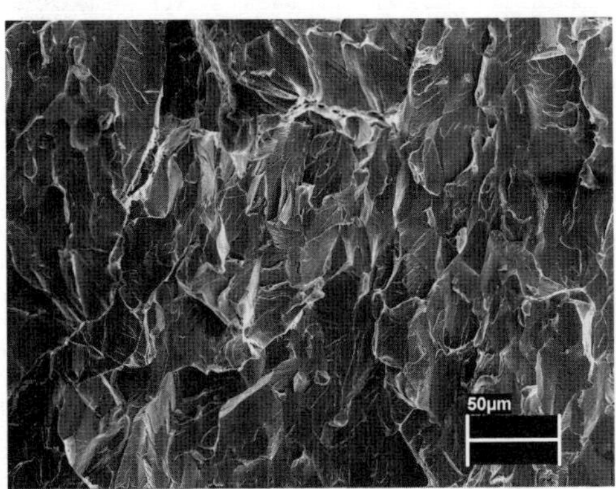

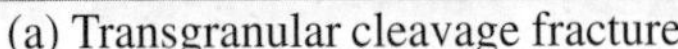
(a) Transgranular cleavage fracture

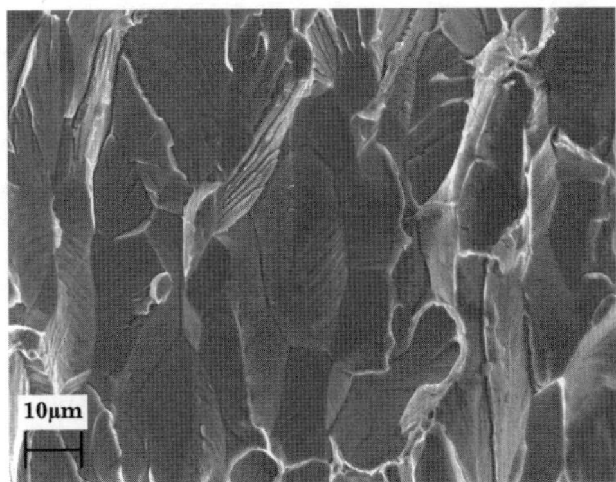

(b) Intergranular fracture

Figure 3: Brittle fracture as predominant fracture type of a deep drawn cup after an impact deformation at low temperature (SEM micrograph).

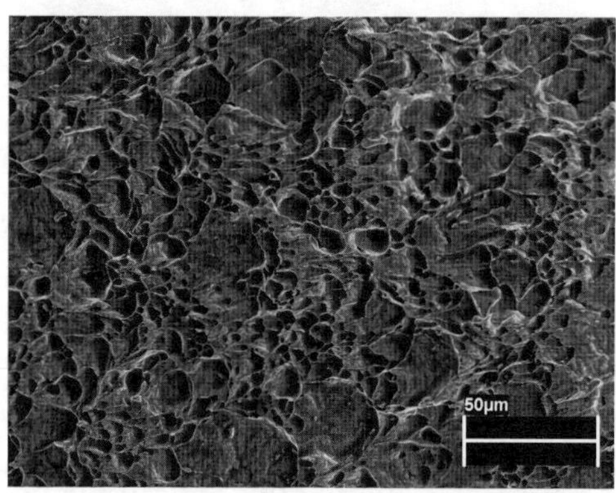

Figure 4: SEM micrograph of the ductile fracture at the crack tip of the longitudinal crack.

w, cone angle α, velocity v and impact energy E) on the crack appearance temperature cannot be observed. This is also true for varying the blank holder force F_N (under the condition that wrinkling of the material is prevented) and the coefficient of friction μ between the contacting pairs (punch/blank, blank holder/blank and die/blank) by deep drawing coated and uncoated blanks. The crucial test parameters influencing the tendency of a material to SWE are:

- Sheet thickness t
  A reduction of the sheet thickness from t=1.5 mm to t=0.7 mm leads to an increase of CAPT by about ΔT=10 K, if all other test parameters (drop height h, drop weight w, cone angle α, velocity v, impact energy E, deep draw ratio β and cup height s) are held constant. To understanding this behavior one needs to know about the actual strain state across the wall of the deep drawn cup. With the aid of the numerical simulation of the deep drawing process the distribution of $\epsilon_{\mathrm{eq}}^{\mathrm{pl}}$ for different sheet thicknesses is investigated (fig. 5). Normalizing by the thickness of the wall allows a direct comparison of $\epsilon_{\mathrm{eq}}^{\mathrm{pl}}$ obtained from different cases studied. The normalized radial position 0 and 1 correspond to the outer and inner surface of the cup, respectively. The equivalent plastic strain is largest at the surface of the cup.

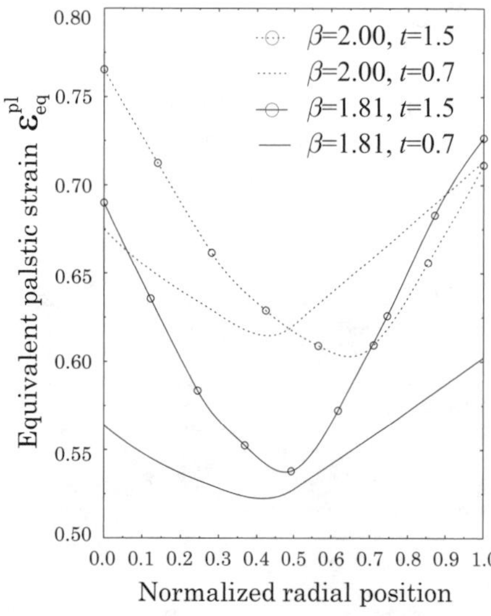

Figure 5: Distribution of the equivalent plastic strain over the normalized radial position within the wall of the cup for different deep drawing ratios β and sheet thicknesses t.

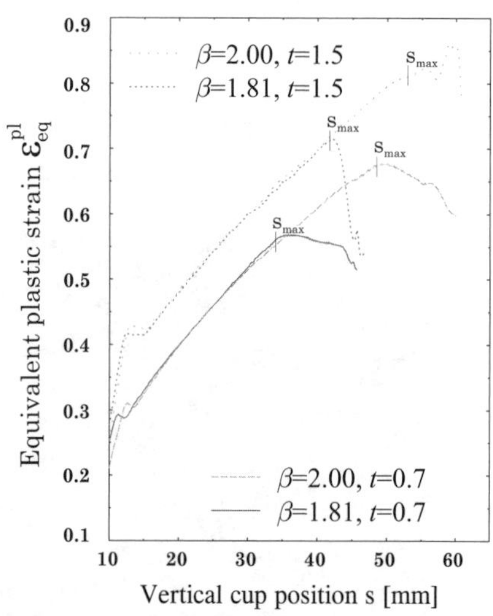

Figure 6: Equivalent plastic strain $\epsilon_{\mathrm{eq}}^{\mathrm{pl}}$ increases with increasing vertical cup position s. At the end of the roughly linear range $\epsilon_{\mathrm{eq}}^{\mathrm{pl}}$ reaches a maximum at vertical position of the deep drawn cup s_{max}.

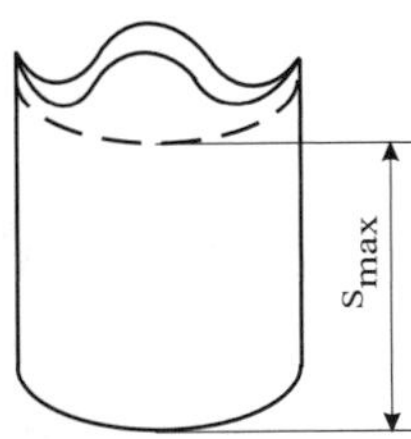

Figure 7: The vertical position s_{max} corresponds to a cup hight at which the ears of the deep drawn cups are eliminated by trimming so assuring a constant height of the cup.

- Cup height s
  Fig. 6 shows the numerically determined dependence of the equivalent plastic strain of a material point close to the outer surface of the cup on its vertical position for different values of β and t. For a large regime of the vertical position the slope of the curves is almost identical and constant. A deviation from the roughly linear regime between $\epsilon_{\mathrm{eq}}^{\mathrm{pl}}$ and s occurs near the top of the cup. This observation is fully in line with the results of the experimental measured principle strains (major and minor strains) mentioned in the literature [1, 9]. There, the principle strains are determined by photo gridding the blanks prior to deep drawing and measuring the principle strains as a function of the distance from the bottom of the cup. The strain increases from the bottom and reaches a maximum near the top of the cup. Hence, by the drop-weight test the impact occurs at a position where maximum strain conditions exist. The equivalent plastic strain increases with increasing the vertical position quite similarly for all considered deep drawing ratios and sheet thicknesses (fig. 6). Cups trimmed to different cup heights (s_1=35 mm and s_2=45 mm) and tested under otherwise equal conditions show an increase of the CAPT by ΔT=20 K when increasing the cup height.

- Deep draw ratio β
  The influence of β on the CAPT is tested for cups with different draw ratio under otherwise equal conditions. Equal test conditions in this case mean the same drop-weight parameter (h, w, α, v and E) and trimming the cups at a position defined as s_{max}. The numerical simulation of the deep drawing process shows an increase of the equivalent plastic strain with an increase of the vertical cup position. The maximum of $\epsilon_{\mathrm{eq}}^{\mathrm{pl}}$ is reached at s_{max}, the end of a roughly linear regime of the slope between $\epsilon_{\mathrm{eq}}^{\mathrm{pl}}$ and the vertical cup position for a given draw ratio (fig. 6). When the cups are trimmed at s_{max}, then defined, constant and reproducible geometrical and strain conditions on the cup rim exist. Trimming at this position also ensures, that possibly existing ears are removed from the cup (fig. 7). The experiments show that the CAPT is about 10 K higher for a cup formed with a draw ratio of β_2=2.0 in comparison to that with β_1=1.8.

The strength of tensile specimens taken in circumferential direction from a cup is about 200 MPa higher than that of tensile specimens from the as-received IF-sheet steel (fig. 8).
Although the test conditions (high strain rate $\dot{\varphi}$ and low temperature T) are adjusted to match the drop-weight test conditions at which brittle fracture can be initiated in a deep drawn cup, the nature of the fracture surface of the tensile specimens taken from a deep drawn cup is always ductile (fig. 9). Hence, the large predeformation along the cup wall caused by the deep drawing process alone cannot be responsible for the brittle fracture that can be initiated in deep drawn cup but not in tensile specimens taken from the same deep drawn cup.

DISCUSSION

The numerical analysis of the deep drawing process shows that $\epsilon_{\mathrm{eq}}^{\mathrm{pl}}$ is independent of β for a large regime along the cup wall (fig. 6). This prediction is experimental verified by measuring the hardness of the cups as a function of the vertical position along the cup wall since the hardness is a measure for the degree of workhardening of

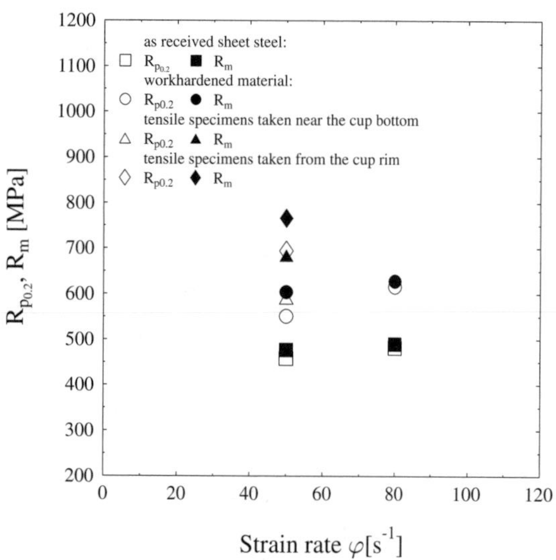

Figure 8: Results of high speed tensile tests preformed under conditions comparable to that of drop-weight test (high strain rate and low temperature). The strength increases with increasing strain rate and workhardening.

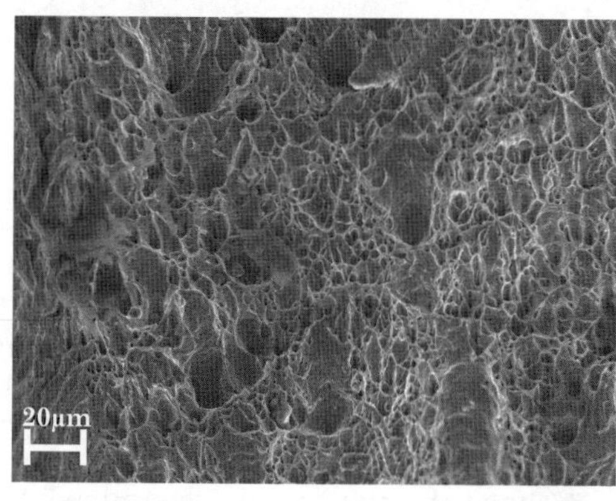

Figure 9: SEM micrograph of the fracture surface of tensile specimens taken in circumferential direction from a deep drawn cup. The fracture mode is fully ductile.

the material - in agreement with the numerical results - no deviation along the cup wall can be measured for different β (fig. 10).

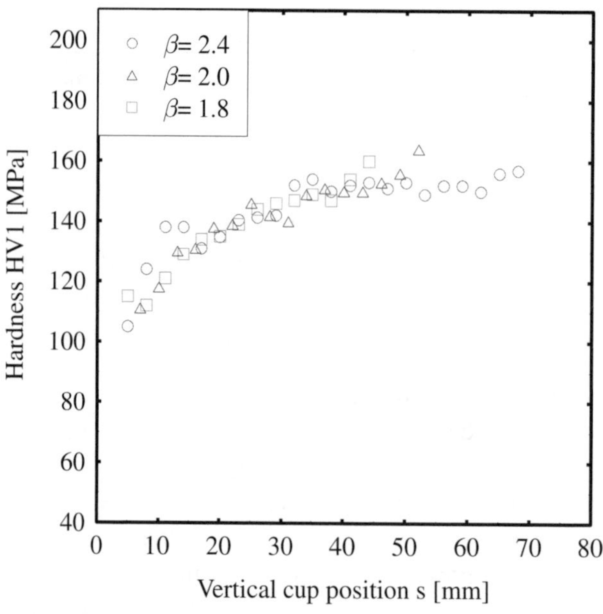

Figure 10: Hardness HV1 along the vertical cup position s for different draw ratios β.

Due to the fact, that the materials tested in high speed tensile tests and in impact tests were subjected to the same predeformation, the high equivalent plastic strain in the specimen prior to testing has to be interpreted in the sense of a necessary, however not sufficient condition for brittle fracture to occur. According to [13] cleavage fracture occurs, if the stress necessary for plastic deformation via dislocation motion (σ_y) reaches the cleavage stress σ_f of the material. σ_f depends only weakly on the temperature and the strain rate. The critical stress for

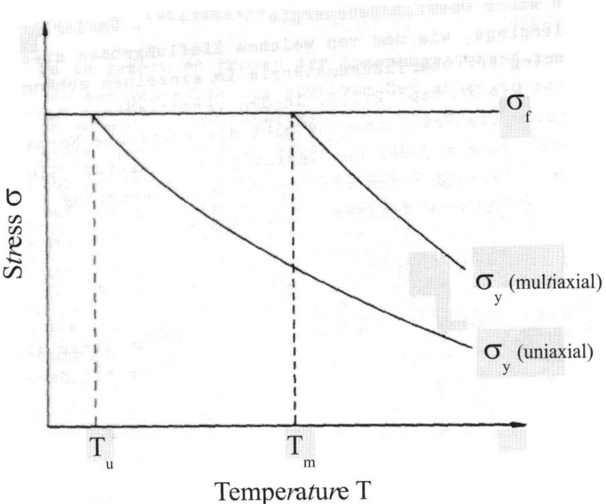

Figure 11: A multiaxial stress state causes a shift of the temperature from T_u to T_m at which the material's cleavage strength is reached.

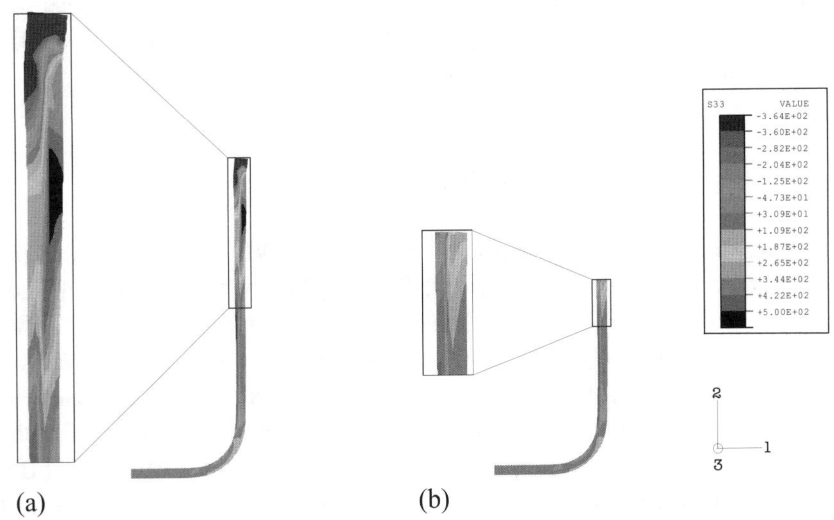

Figure 12: Tangential residual stress in a full deep drawn cup (a) before and (b) after trimming.

cleavage fracture can be reached at higher temperature, if the stress state in the material becomes multiaxial [14, 15] (fig. 11). The actual stress state near the rim of a deep drawn cup gives an explanation for the different behavior of tensile specimens and impact loaded cups under comparable test conditions (high strain rate, low temperature). With the aid of the numerical analysis of the deep drawing process the multiaxial stress state along the cup wall is calculated (fig. 12). In the trimmed cup the maximum tensile residual stress in circumferential direction is predicted for regions at the outer surface of the cup rim where also the highest tensile stress is introduced upon impact by the cone. Hence, due to this multiaxial residual stress state brittle fracture can be initiates in cups upon impact deformation at much higher temperatures in comparison to the high speed tensile tests where the tangential residual stress is relieved during the production of the specimens from rings.
The residual tangential tensile stress is different at the same vertical cup position s for cups predeformed with different draw ratios. Experiments show that for cups deep drawn with different β but which are trimmed at the same position s, the CAPT increases with decreasing β. To validate the predicted tangential stress state the spring back of rings cut opened is measured. Fig. 13 depicts the amount of spring back of rings taken at different vertical positions from one and the same cup. Rings taken at the same vertical position from cups formed with different β (β_2=2.0 and β_3=2.4) show an increase of the spring back with decreasing β. Hence, the reason for

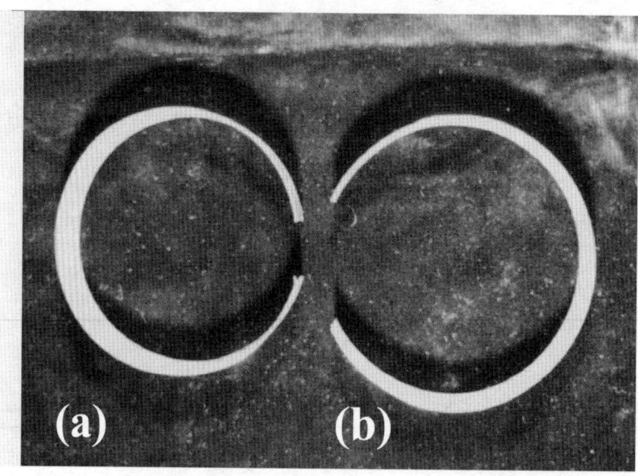

Figure 13: Rings taken from a deep drawn cup at different vertical cup positions are cut up to verify the existence of a tangential stress along the cup wall. The spring back and hence the tangential stress of the ring near the cup bottom (a) is much lower than for a ring taken near the cup rim (b).

the higher CAPT determined for cups formed with a draw ratio of β_2=2.0 than for those with β=2.4 (under the condition that the cups are trimmed at the same vertical position s) must be seen in the difference of the actual multiaxial stress state existing at the rim of the trimmed cup.

CONCLUSION

The sensibility of IF-sheet steel to secondary work embrittlement SWE can be evaluated in a drop-weight test. A variation of the test parameters such as drop-weight, drop height, opening angle of the cone within the range studied in this work does not alter significantly the crack appearance temperature CAPT. The crucial parameters are the draw ratio β, the initial sheet thickness of the blank t and the cup height s. Varying these parameters markedly influences the equivalent plastic strain ϵ_{eq}^{pl} and the multiaxial stress state as a function of the vertical position along the cup wall. A numerical analysis of the deep drawing process shows that there exists a roughly linear relationship between ϵ_{eq}^{pl} and s. The slope of the curves is almost the same for different β. In addition to ϵ_{eq}^{pl} also multiaxial residual stress state in the deep drawn cup plays a decisive role for the transition from ductile to brittle fracture in impact tests.

REFERENCES

1. S.P. Bhat, B. Yan, J.S. Chintamani, T.A. Bloom. Secondary work embrittlement of stabilized steels: test methods and application. I&SM, pp. 33–42, 1995.

2. B. Yan, I. Gupta. Secondary work embrittlement of interstitial free steels. Int. Symposium: Modern LC and ULC Sheet Steels for Cold Forming: Processing and Properties, pp. 249–263, Aachen, Germany, 30.03.-01.04. 1998.

3. L.A. Henning. Method to evaluate the susceptibility of sheet steel to secondary work embrittlement. 33th Mechanical Working and Steel Processing, vol. 29, pp. 9–13, St. Louis, Missouri, USA, 20.-22.10. 1992.

4. R. Pradhan. Cold-rolled interstitial-free steels: A discussion of some metallurgical topics. International Forum for Physical Metallurgy of IF Steels, pp. 165–178, 1994.

5. Yu Hin Lau, B.C. De Cooman, M. Vermeulen. The secondary work embrittlement of galvannealed high strength sheet steel for automotive applications: test method and material evaluation. 39th Mechanical

Working and Steel Processing Conference Proceedings, vol. XXXV, pp. 139–148, Indianapolis, Indiana, USA, 19.-22.10. 1997.

6. Y. Maehara, N. Mizui, M. Arai. Cold-work embrittlement accompanied by intergranular fracture in ultra-low carbon Ti-added sheet steels. International Symposium on Interstitial Free Steel Sheet: Processing, Fabrication and Properties, pp. 135–144, Ottawa, Ontario, Canada, 19.-20.08. 1991.

7. E. Yasuhara, K. Sakata, T. Kato, O. Hashimoto. Effect of boron on the resistance to secondary working embrittlement in extra-low C cold-rolled steel sheet. ISIJ, vol. 34 (1), pp. 99–107, 1994.

8. J. Neutjens, H. Mathy, J.C. Herman. Cold work embrittlement of ULC-IF and EFA low carbon steel grades. La Revue de Métallurgie-CIT, pp. 551–561, April, 1997.

9. D. Li, G. Heßling, W. Bleck. The secondary work embrittlement in sheet steels. Steel Research, vol. 70 (4+5), pp. 154–161, 1999.

10. S.G. Lewis, S.R. Daniel, J.D. Parker, D.T. Llewellyn, M.P. Sidey. Testing techniques for cold work embrittlement in interstitial free steels. I&SM, vol. 25 (1), pp. 63–73, 1998.

11. A. Spalek, G. Reisner, E.A. Werner. Analysis of the impact behavior of interstitial free (IF) steels with special emphasis on cold work embrittlement. In: Advanced Technology of Plasticity; Proc. of the 6th ICTP, vol. 1, M. Geiger (ed.), Nürnberg, Germany, 19.-24.09.1999, Springer Verlag, pp. 435–440, 1999.

12. R. Kopp, G. Bernath. Fließkurvenaufnahme bei hohen Umformgeschwindigkeiten. Stahl und Eisen, 118 (4), pp. 71–76, 1998.

13. E. Orowan. Fracture and strength of solids. Rep. Prog. Physics, (12), pp. 185–232, 1948.

14. A. Kochendörfer. Die Festigkeits- und Formänderungseigenschaften der Metalle bei tiefen Temperaturen. Z. Metallkunde, (51), pp. 73–80, 1960.

15. K. Kühne. Einfluß des Spannungszustandes und des Gefüges auf die Spaltbruchspannung von Baustählen. PhD thesis, Fakultät für Bergbau und Hüttenwesen der Rheinisch-Westfälischen Technischen Hochschule Aachen, 1982.

Influence of Microstructure on Secondary Cold Work Embrittlement of IF Steels

P. Martin, S. Dionne, O. Dremailova, J. Brown, B. Voyzelle, J. Bowker, D. Linkletter
Materials Technology Laboratory, CANMET
568 Booth Street
Ottawa, Ontario, K1A 0G1
Canada
Tel.: (613) 992-8283, E-mail: pimartin @nrcan.gc.ca

Keywords: Secondary cold work embrittlement (SCWE), interstitial-free (IF) steel, transition temperature, brittle fracture, fractography, focused ion beam (FIB), galvanneal steel, microstructure, flaking

INTRODUCTION

At low temperature brittle intergranular fracture can occur in interstitial-free (IF) steel because of low grain boundary cohesion. This phenomenon is due to a lack of strengthening elements, mainly carbon, and the presence of embrittling elements. Phosphorus added as a solid solution strengthener in medium-strength IF is the most deleterious element for resistance to brittle intergranular fracture. Phosphorus segregation reduces grain boundary strength, which increases the ductile-to-brittle transition temperature (DBTT). This creates concerns since P is the preferred solid solution strengthener element in IF steel alloying. Several studies have been published describing the influence of steel composition and processing conditions on resistance to brittle intergranular fracture. Addition of boron is one of the most effective methods to increase grain boundary strength.[1] However, this has a deleterious effect during annealing, so B additions must be optimized to balance the benefit to grain boundary cohesion and the detrimental effect that may occur during the thermomechanical processing.[2] Resistance to brittle fracture is affected by many other factors. The factors that increase material flow stress will in general favor brittle fracture. Temperature is a critical factor since transgranular slip is thermally activated; a decrease in temperature increases the flow stress. Brittle fracture occurs at a temperature at which the flow stress exceeds either the cleavage stress or the grain boundary critical stress. Similarly a higher loading rate increases flow stress. Work-hardening is probably the most important factor since it can increase the material flow stress in formed parts. Because of the work-hardening effect, brittle fracture can be observed during secondary deformation of formed parts, which is commonly referred to as secondary cold work embrittlement (SCWE). Secondary cold working can occur during part fabrication or during impact loading in service. It is worthwhile mentioning that no brittle intergranular fracture has been reported during in-service loading [3,4] but fracture during secondary forming operations have been reported [5].

SCWE is often evaluated using the cup/expansion test. In this test, a drawn cup is opened up by impact loading using a conical punch. By using cup drawing, large strains can be achieved and elongated grains are produced, both conditions of which are believed to favor SCWE. The effect of the morphology of the grains on intergranular fracture was discussed by Bhat *et al.*[6] According to the authors, the elongated grains produced by deep drawing enhance SCWE when cracks form parallel to these grains.

The importance of the grain morphology in SCWE is widely accepted but the contribution of the grain orientation has never been measured since SCWE is generally studied using the cup/expansion test where fracture only takes place along the elongated grains due to the stress conditions applied. The understanding of the effect of grain orientation is important since it needs to be considered in addition to strain when predicting occurrence of brittle fracture in real parts.

Another microstructural feature that may have to be considered in SCWE is that resulting from galvannealing. Iron-zinc coated sheet steel produced by galvannealing is increasingly used in automotive sheet applications. The coating consists of intermetallic Fe-Zn phase layers, which are harder and more brittle than galvanized free-zinc coatings. Deformation during forming can produce damage to galvanneal coatings such as flaking, powdering and cracking. A consequence of galvannealing in some IF grades is surface grain embrittlement, which promotes ferrite intergranular fracture in grains near the surface.[7] The ferrite grains fracture during flaking of the galvanneal coating. Forming processes such as deep drawing are conducive to flaking because of the large compressive stresses experienced by the coating. Although the formability of galvanneal steel has been studied extensively, very little information is available on the effect of galvannealing on SWE. A recent study [4] on the effect of galvannealing on SCWE showed that galvannealing may increase the transition temperature in IF steels. The study was undertaken on a non-rephosphorized IF steel grade. No data has been published on the relationship between surface Zn embrittlement and SCWE in steels sensitive to embrittlement such as medium-strength rephosphorized grades.

In this study the influence of grain morphology and strain on SCWE will be investigated. The cup/expansion test and a new method to generate brittle fracture parallel and normal to grain elongation have been employed. Formation of brittle fracture during cup drawing and in the cup/expansion test will be investigated. This paper will also address the possible influence of galvannealing on SCWE.

EXPERIMENTAL PROCEDURE

Material

The experiments were carried out on four commercial steels with a similar thickness. The steel compositions were selected to provide information on cold work embrittlement in grades with various amounts of phosphorus, free-carbon, and boron, produced by both continuous and batch annealing. Their chemical compositions are given in Table I. Steel S1 is an electrogalvanized (EG) bake hardenable (BH) extra-low carbon grade produced by batch annealing. Steels S2, S3 and S4 are interstitial-free grades. Steels S2 and S3 were produced by continuous galvannealing (GA) and steel S4 was produced by batch annealing without coating. All steels are rephosphorized type grades, except for the S4 steel.

TABLE I. Chemistry (wt %) of the steels investigated

Steel Type	Code	Thick-ness (mm)	C	N	Ti	Nb	Mn	Si	P	B	Al	S
BH batch annealed ELC, EG	S1	0.79	**0.010**	0.0061	<0.0050	<0.010	0.14	0.037	**0.056**	---	0.043	0.0072
Rephopshorized IF w/ B, GA	S2	0.74	0.0029	0.0037	0.019	0.029	0.36	0.010	**0.040**	<0.0005 B Added	0.030	0.0049
Rephophorized IF w/t B, GA	S3	0.74	0.0039	0.0048	0.021	0.033	0.39	<0.0080	**0.042**	---	0.048	0.0053
Batch Annealed IF, Ti-sta., CR	S4	0.77	0.0077	0.0064	0.084	<0.010	0.15	<0.008	<0.004	---	0.041	0.0060

Evaluation of the Transition Temperature

Secondary cold work embrittlement has been evaluated using two methods. The cup/expansion test has been used to evaluate the transition temperature of 150-mm diameter cups produced using a drawing ratio of 2.0. The method used is similar to the conventional cup/expansion test method described elsewhere.[8] This larger cup size was more suitable to investigate the effect of strain and grain orientation on the transition temperature. The cups were trimmed using a lathe prior to testing to remove the earring. The major strain at the edge of the 2.0 draw ratio cups after trimming was 65%. The cups were cooled in a cold chamber using a liquid nitrogen mist. The uniformity of the temperature was monitored by using 6 thermocouples on the cup wall. One thermocouple was connected to a controller for the control of the flow of the nitrogen mist. The cup/expansion test was performed after temperature stabilization, and when the temperature difference between the thermocouples was less than 2°C. The DBTT was identified as the temperature at which at least four out of five cups passed the impact test without brittle failure using a temperature increment of 5°C. The impact speed was approximately 4.3 m/s. With this cup geometry when tests are performed below DBTT, cracks propagate along the entire cup wall. Above the transition temperature no fracture occurred and the cups partially collapsed.

The transition temperature was also evaluated using small strips taken from the wall of the cups. Coupons approximately 3 by 20 mm in size were taken from the wall of the drawn cups in the radial direction and parallel to the cup wall (Fig. 1). A fine notch was made in the center to facilitate the initiation of a brittle fracture. This method produces brittle fracture parallel (LN orientation) and normal (TN orientation) to the grain elongation. The drop weight apparatus is shown in Fig. 2. The specimens were immersed in an alcohol-cooling bath, where the temperature was controlled by adding liquid nitrogen. The lowest stable temperature that can be obtained with this method was -120°C. The DBTT was defined as the lowest temperature at which no fracture occurred using 5°C increment. Impact loading was obtained using a 1.4 kg load dropped from a height of 400 mm. Prestrained and unstrained notched specimens were also fractured in liquid nitrogen.

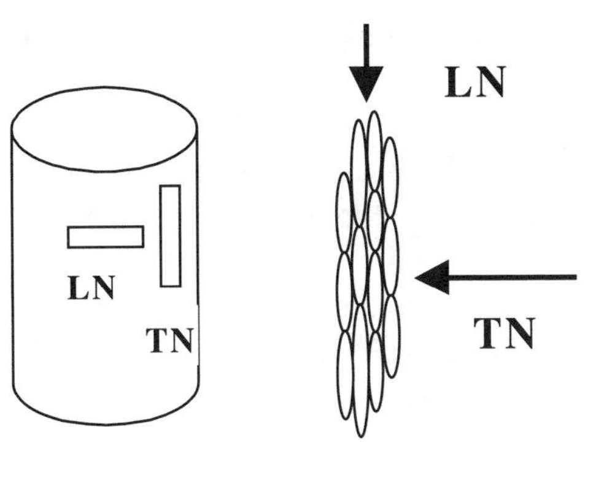

Figure 1. Orientation of the strips in the cup wall. The notch is parallel to grain elongation for the LN specimens and transverse for the TN specimens.

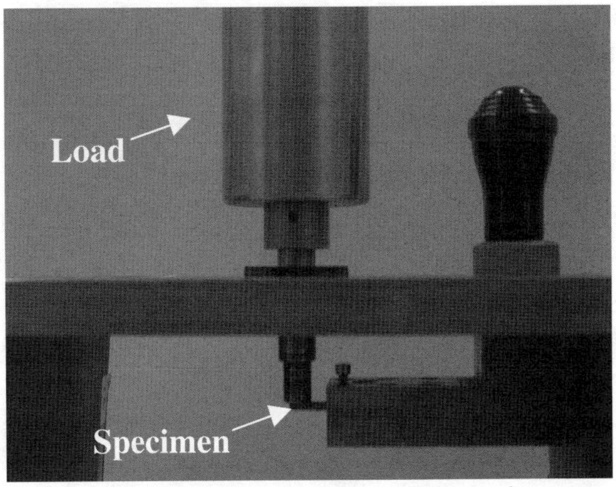

Load
Specimen

Figure 2. Impact tester for strip specimens.

Fractographic Observation and Analysis

SEM quantitative analysis - After fracture in liquid nitrogen the prestrained and unstrained notched specimens were prepared for scanning electron microscopy (SEM). The purpose of fractographic observation was to make a quantitative measurement of the percentage of intergranular fracture specification in the center of the specimens according to ASTM E562-95. The variation in fracture mode from center to the edge of the specimens was also investigated. Quantitative measurements of intergranular fracture were conducted using a manual point counting technique. A total of ten fields of view were examined for each specimen.

Focused ion beam analysis - Samples approximately 2 X 2 cm were cut from selected cups and placed in a Micrion 2500 Focused Ion Beam (FIB) instrument. The surface of each sample was examined in plan view (with 0° tilt) to select areas with emerging surface cracks for cross-sectioning. Subsequently, a layer of tungsten was deposited in the FIB to protect the top edge of the cross-section. The Ga ion beam was rastered across the region of interest to mill out a rectangular box having the cross-section plane as one of its sides. Finally, the sample was tilted at 30 or 45° to the beam and images of the polished cross-section were acquired in secondary electron or ion mode.

FIB cross-sectioning has several advantages over conventional metallographic techniques for studying cracked galvanneal steel. In-situ FIB polishing allows the cross-section to be prepared from specific locations without exposure to mechanical or chemical damage. Severely cracked coatings can be examined undisturbed and in some cases, the distribution of dried-in-place lubricants can be observed. Contrast formation in FIB images is excellent to show the structure of the zinc coating and ferrite substrate simultaneously, which is not possible using chemically etched samples. Furthermore, the contrast produced by cracks and lubricant residues can be significantly enhanced using the ion mode imaging because of the high secondary ion yield of these features.

Auger analysis - Auger spectra and secondary electron (SE) photomicrographs were collected using a Perkin-Elmer model PHI 600 scanning Auger microprobe (SAM). This instrument is also equipped with a PHI model 15-535 liquid nitrogen-cooled fracture/parking stage allowing the steel coupons to be fractured and examined under ultra-high vacuum (UHV) conditions. The base pressure of the UHV chamber during fracture and subsequent data collection was $< 5 \times 10^{-10}$ mbar to minimize the rate of surface contaminant deposition onto the freshly exposed grain surfaces. A special copper holder was fabricated and a modified fracture protocol developed to permit the rectangular sheet steel coupons to be fractured using this hardware originally designed to accommodate only rod-shaped specimens.

For Auger work the microprobe's electron gun was operated at 3 KeV using a beam current of approximately 100 nA. Immediately upon fracture a steel coupon was transferred to the analytical stage for SE imaging and Auger spot and/or mapping analysis; a step requiring about 5 min. Typically 8-10 Auger spectra were generated in approximately the first hour following a fracture. Separate tests using ion-beam cleaned metal coupons showed no detectable growth (deposition) of any volatile contaminant species (C, O, etc.) for times exceeding 1h when this UHV chamber was maintained at $< 8 \times 10^{-10}$ mbar pressure.

EXPERIMENTAL RESULTS

Evaluation of the Transition Temperature

Cup/expansion impact test - The ductile-to-brittle transition temperatures (DBTT) for the steels investigated are shown in Table II. The bake hardenable extra-low carbon grade showed the lowest transition temperature. The low DBTT of this steel can be attributed to carbon in solid solution. Steel S2 with a DBTT of –30°C also exhibits a relatively low transition temperature, probably due to the presence of boron. The highest DBTT was found with the continuously annealed rephosphorized IF grade without boron, steel S3. The non-rephosphorized grade, steel S4, showed a relatively high DBTT due to the batch annealing process.

TABLE II. Transition temperature with 2.0 draw ratio in 150 mm cups with edge strain of –45%, +65%

	Steel S1	Steel S2	Steel S3	Steel S4
Transition temperature	-55°C	-30°C	+5°C	-15°C

Notched specimen impact test - The relationship between strain and transition temperature has been evaluated for the most SCWE sensitive steel only, steel S3. The transition temperature for prestrained strips taken from the cup wall is shown in Fig. 3. The strips were taken from three locations of the cup wall corresponding to regions with major strains near 67, 56 and 40%. Some tests were also performed in the unstrained conditions (0% strain). Due to the low temperature needed to induce a brittle fracture in the strips and the temperature stability problems at temperatures below –120°C, it was not possible generate a complete set of data. The temperature difference between the strips taken in the high strain region (67%) and the low strain region (40%) was 25°C. When brittle fracture occurs in the strips, the fracture path is parallel to the notch, allowing investigation of brittle fracture propagation parallel and transverse to the grain elongation direction. The figure shows that grain orientation plays a very important role in the formation of brittle fracture. The transition temperature for strips with a notch oriented transverse to grain elongation is more than 45°C below that for strips with a notch parallel to the grain elongation for a strain level of 67%. It was not possible in this study to determine the magnitude of the drop for the less strained strips because of their low transition temperatures in the TN orientation.

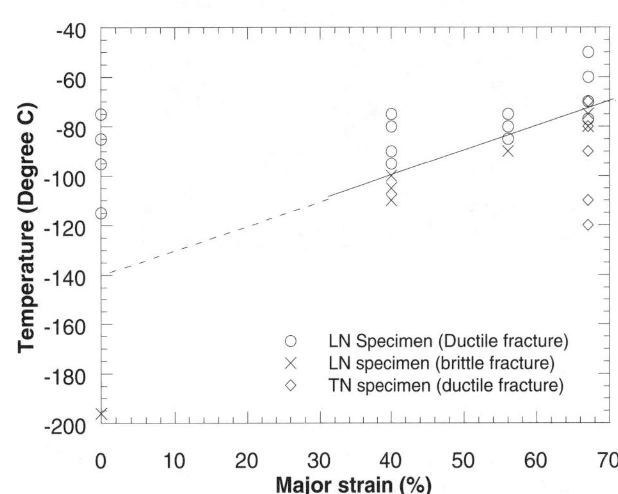

Figure 3. Transition temperature after various amount of primary strain in steel S3.

It is important to notice that the transition temperature obtained with this method is significantly lower than for the cup test. The DBTT for this steel in the cup/expansion test is +5°C. Several reasons can explain this difference. The difference in stress condition between the two methods is probably the most important factor. The stress conditions are affected by the specimen size and geometry, and the loading mode. The morphology of the fracture surface of the ductile specimens clearly indicates that a bending moment was applied in the fracture area. In the cup/expansion test, large tensile stresses are induced in the hoop direction. Another factor that may have contributed to lower the DBTT in the strip test is the lower impact speed.

Fractographic analysis

SEM fractographic analysis - The fracture surface of prestrained and unstrained notched strips fractured in liquid nitrogen was examined by SEM. The fractographic observations of the prestrained specimens were made in strips sampled along the LN and TN orientations. The level of primary strain in the prestrained strips was

65%, corresponding to the strain level at the edge of the trimmed cups with 2.0 draw ratio. Figure 4 shows the typical fracture surface of the four steels investigated. The fracture surface is characterized by mixed intergranular/cleavage facets.

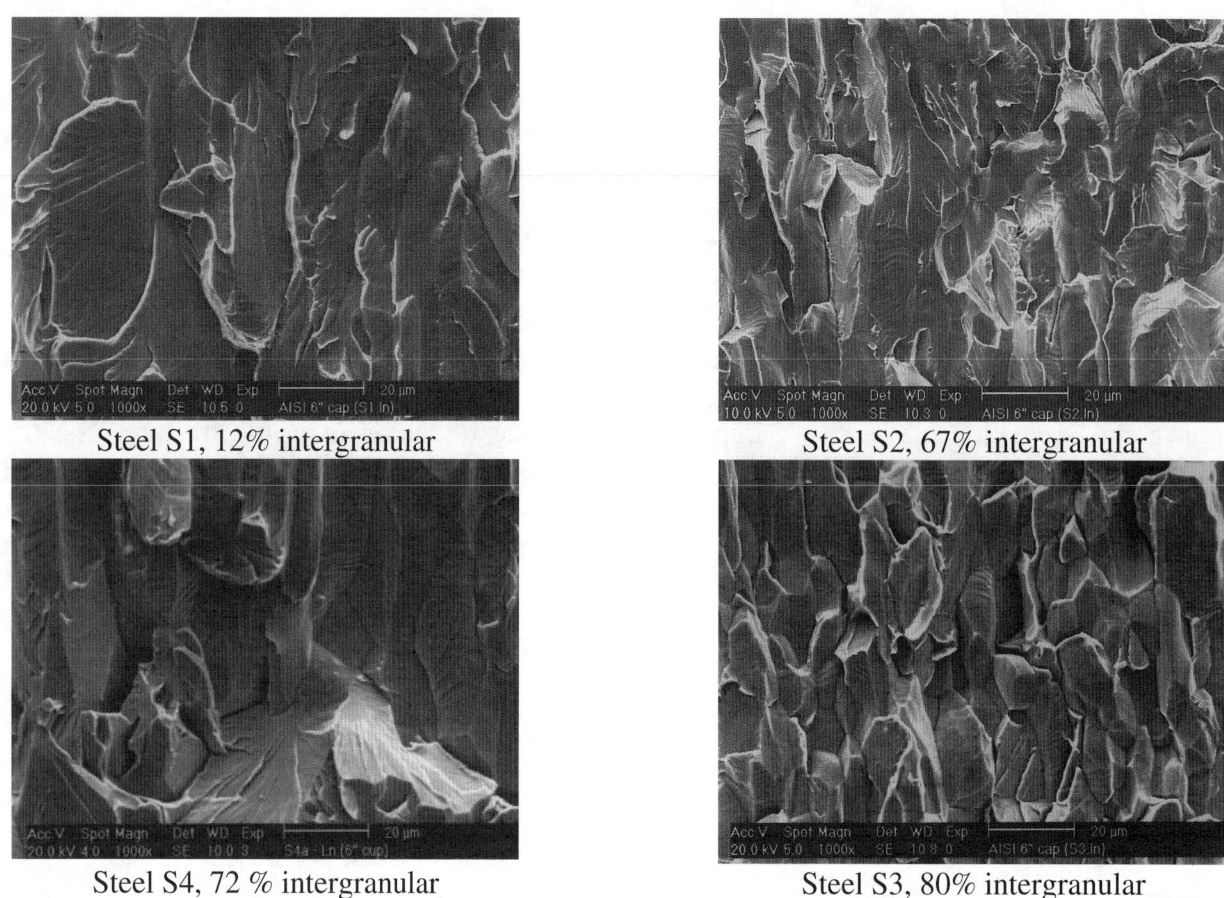

Steel S1, 12% intergranular

Steel S2, 67% intergranular

Steel S4, 72 % intergranular

Steel S3, 80% intergranular

Figure 4. Fracture surface of notched strips fractured in liquid nitrogen along the LN orientation.

The percentage of intergranular fracture is given in Fig. 5. The figure shows that the fraction of intergranular fracture increases with the transition temperature. Similar results were reported by Yasuhara *et al.* [2] and Lau *et al.* [2]. The effect of crack orientation is relatively weak in the low and high DBTT materials where the fraction of intergranular fracture is low and high respectively. The two steels with an intermediate DBTT showed a significant decrease in the fraction of intergranular fracture when fracture occurs in the TN orientation. The variation of fraction of intergranular fracture with the DBTT in the unstrained condition was closer to the variation observed in the TN specimens than that for the LN specimens. The higher fraction of intergranular fracture observed in the LN orientation, except for the steel with a DBTT of –55°C, is probably due to the grain morphology effect, which enhances SCWE as strain increases. In the transverse orientation a strain increase enhances intergranular fracture due to work-hardening; however, obstacles to intergranular fracture also increase because of the reduction in free path between grain boundaries. This effect may reduce the tendency to brittle intergranular fracture, and result in more cleavage.

The low percentage of intergranular fracture observed in both orientations for the steel with a DBTT of –55ºC, is due to its higher grain boundary strength. Brittle fracture in this steel occurs primarily by cleavage. A flow stress increase or grain morphology change cannot create conditions by which intergranular fracture can take place with less fracture energy than cleavage fracture.

Fractographic observations were made near the surface after fracture of the strip specimens in the unstrained condition. It was observed a higher fraction of intergranular fracture at the surface of galvanneal steel. The phenomenon was more pronounced in steel S2. Figure 6 shows the fractured surface of a strip specimen tested at a temperature above the transition temperature. At this temperature, the specimen fractured in a ductile manner, except for the surface where intergranular fracture was observed in the first one or two ferrite grains underneath the galvanneal coating. Similar surface embrittlement was observed after fracture in liquid nitrogen, where the fracture close to the surface was fully intergranular (Fig. 7). The uncoated steel S4 exhibited no noticeable difference between the surface and the interior of the substrate. A slightly higher fraction of intergranular fracture was also observed in steel S3 but the difference was less noticeable because of its overall high fraction of intergranular fracture. Lau *et al.* [4] also reported surface embrittlement observation in a galvanneal steel containing 100 ppm of phosphorus.

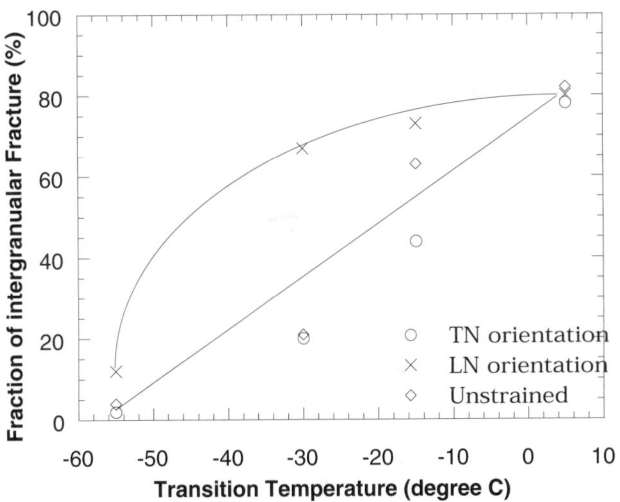

Figure 5 - Percentage of intergranular fracture in prestrained and unstrained strip specimens fractured in liquid nitrogen.

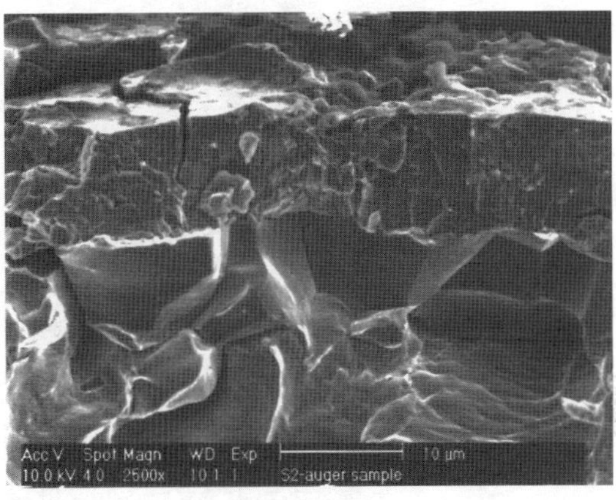

Figure 6 - Brittle fracture in the ferrite grains underneath the surface in a notched strip fractured above the transition temperature (steel S2).

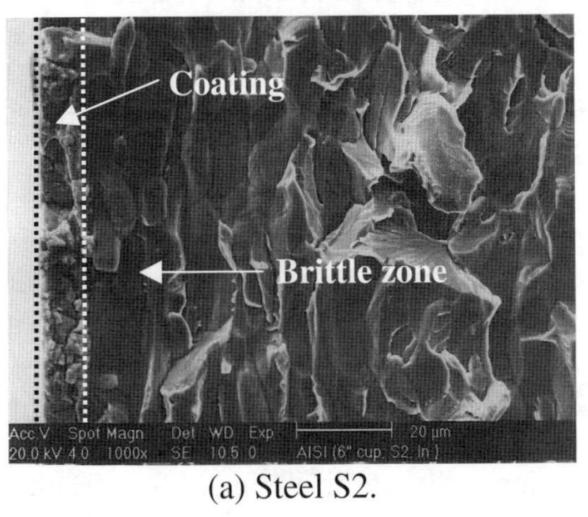

(a) Steel S2.

(b) Steel S4.

Figure 7 - Fractographic aspect near the surface of LN strips after fracture in liquid nitrogen.

Focused ion beam (FIB) analysis - FIB cross-sections were prepared from two types of galvannealed samples in which SCWE fracture was observed. In the first case, SCWE fracture occurred at the edge of a cup that was stamped using a large draw ratio (Fig. 8). The SCWE fracture was produced at the end of the stamping operation when the material wrinkled due to the build-up of excessively high compressive stresses in the flange. The SCWE cracks observed in the wrinkles were mixed intergranular/cleavage, and occurred at forming strains that were significantly smaller than those predicted by the forming limit diagram. Since cracking clearly initiated in the coating in this situation, it was of interest to determine if surface grain boundary embrittlement caused by galvannealing could have played a role in the formation of the SCWE crack. The second case investigated was the formation of secondary cracks during the cup/expansion test. Cross-sections of surface cracks adjacent to the primary SCWE crack produced during a cup/expansion test were examined to determine the extent to which cracks initiated in the coating propagate into the substrate. The location of the area investigated in the FIB instrument is indicated on Fig. 9. This area includes the initiation site of the primary SCWE crack.

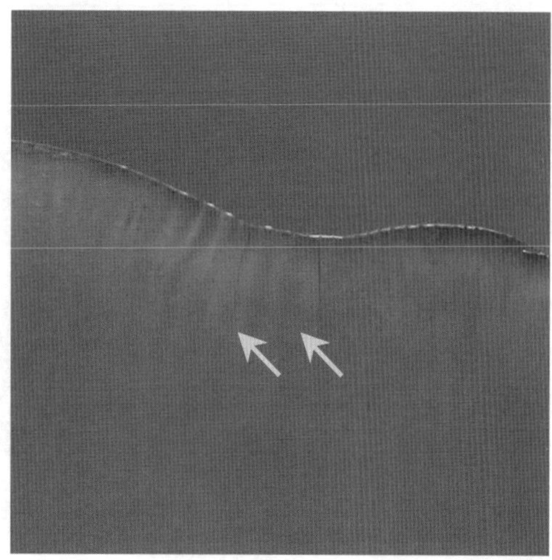

Figure 8 – Macrograph of a steel S3 cup stamped using draw ratio of 2 showing SCWE cracking in the wrinkles (arrows).

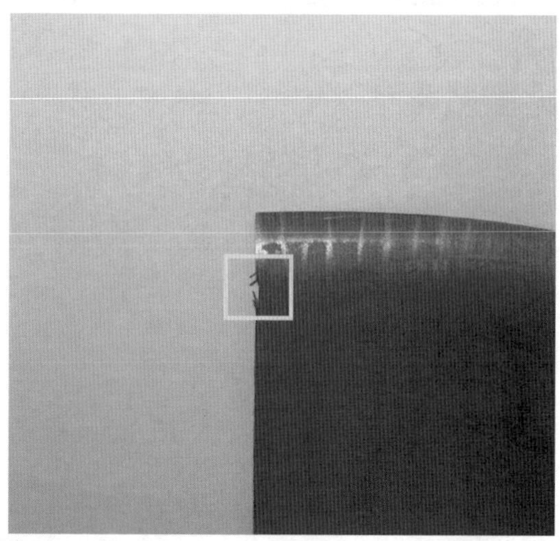

Figure 9 – Macrograph of a cup/expansion test specimen of steel S2 showing the region selected for the FIB investigation (box).

Figure 10 shows two secondary electron FIB images obtained from cross-sections of surface cracks located just below a large SCWE crack in a wrinkled steel S3 cup. The galvannealed coating contained numerous cracks throughout its thickness. However, most of these cracks either stopped before reaching the substrate or changed course and continued to propagate along the coating/substrate interface (Fig. 10a). Cracking along the Γ layer was observed. In a few cases, a coating crack penetrated into the substrate along an emerging ferrite grain boundary but the length of these substrate cracks did not exceed one grain diameter (Fig. 10b).

Cross-sections were prepared from several locations near the initiation site of the main SCWE crack on a cup/expansion test specimen of steel S2. Most of the surface cracks observed on this specimen propagated through the coating and along the coating/substrate interface, sometimes resulting in severe delamination of the coating (Figs. 11a and 11b). A few coating cracks did extend into the substrate. In the case of cross-sections located 5 mm away from the main SCWE fracture (Fig. 11a), crack penetration into emerging ferrite grain boundaries was less than one grain diameter. However, ferrite grain boundary cracking over several grain diameters and transgranular cracking were observed in cross-sections adjacent to the main SCWE crack (Fig. 11b).

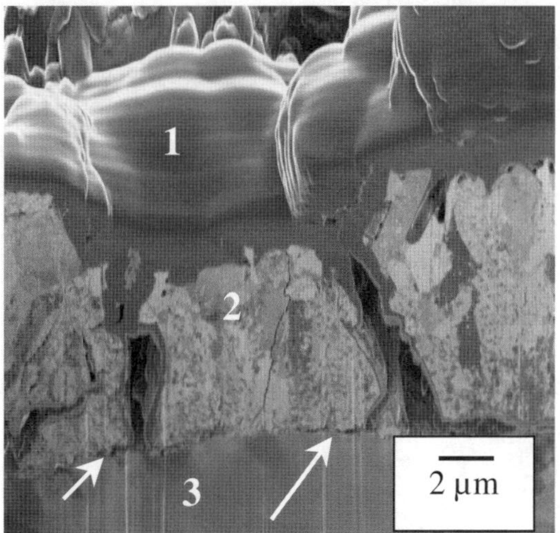

(a) Most coating cracks propagate along the coating/substrate interface (arrows).

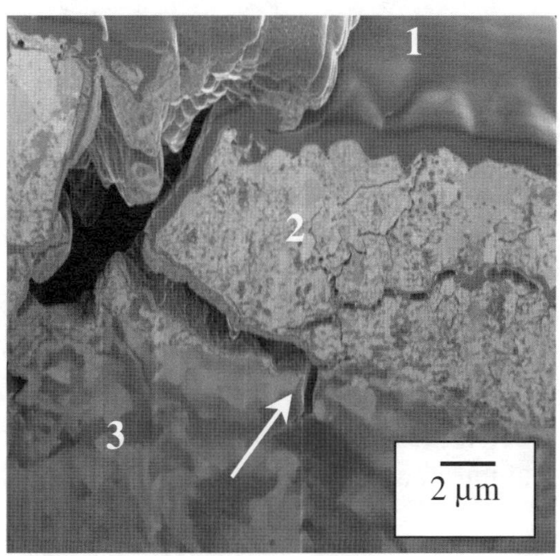

(b) A coating crack has propagated along an emerging substrate grain boundary (arrow).

Figure 10 – Secondary electron images of FIB cross-sections of surface cracks located just below the tip of a large SCWE crack in a wrinkle of a deep-drawn cup of steel S3. (1) protective W layer; (2) Zn-Fe coating; (3) steel substrate.

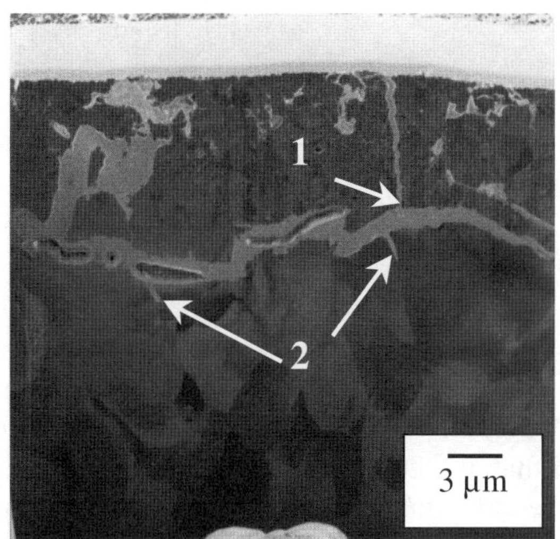

(a) Cross-section located 5 mm away from the main SCWE crack showing crack propagation along the coating/substrate interface (1) and ferrite grain boundaries (2).

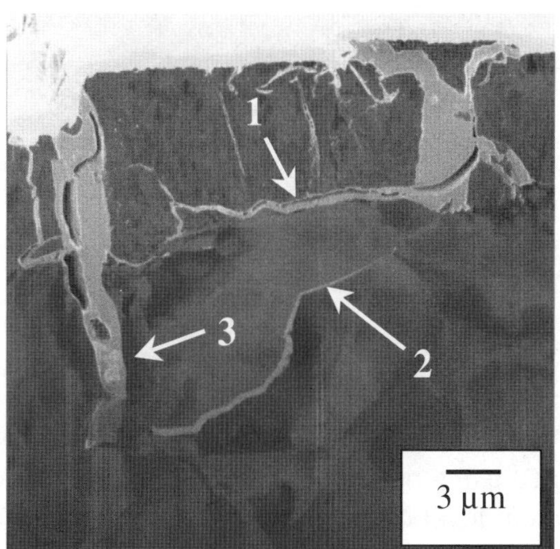

(b) Cross-section adjacent to the main SCWE crack showing crack propagation along the coating/substrate interface (1) and substrate cracking along intergranular (2) and transgranular (3) paths.

Figure 11 – Secondary ion images of FIB cross-sections of surface cracks located near the failure initiation site of a cup/expansion test specimen of steel S2.

Auger surface analysis - Auger analyses were performed in the region near the surface where more pronounced intergranular fracture was observed in steel S3. The Auger observations were performed on fresh fractured surface. A Zn signal was detected in the first one or two grains beneath the galvanneal coating, which confirms grain boundary Zn diffusion in this steel. Figure 12 shows a typical Auger analysis spectrum taken at the grain boundary of a ferrite grain beneath the coating. In addition to the Zn signal, carbon and phosphorus were detected. The phosphorus signal is due to grain boundary segregation. The cause of the high carbon signal detected on the outermost 4-6 grains below the galvanneal coating has not been determined. Auger analysis are very sensitive to surface contamination. Analysis should be performed shortly after specimen fracture to avoid any problems due to surface contamination in the microscope. In the present study, the Auger analyses were undertaken under optimum vacuum conditions and in a time-frame before surface contamination of grains was detected throughout the central region of each specimen. This was verified by measuring the intensity of the carbon peak at many grain boundary sites. Thus it is not believed that the large carbon signal detected on the upper most few grains underlying the Zn-coating is from deposition of volatile carbon species from the ultra-high vacuum chamber. Since surface carbonizing is not expected during steel processing, it is postulated that hydrocarbon residue (oil) on/below the coating could have generated the high carbon signal observed along the interface. Further investigation would be needed to clarify the source of carbon in the ferrite grains near the steel-galvanneal interface.

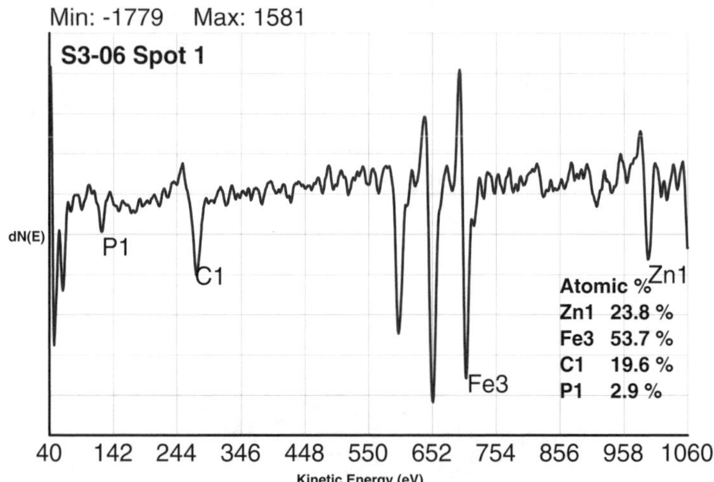

Figure 12. Auger spectrum at the grain boundary of a ferrite grain beneath the galvanneal coating in steel S3.

DISCUSSION

The present study indicates that stress conditions and crack orientation with respect to grain elongation produced by primary deformation are two key factors for initiation of SCWE cracks. It was shown that the DBTT obtained using notched strips is at least 80°C lower than the DBTT obtained with the cup/expansion test for an equivalent level of primary strain. Bhat *et al.* [6] also reported that the transition temperature obtained with coupons taken from high strain regions of a real part was lower than that for cup/expansion test. This DBTT variation can be explained by the difference in stress condition generated in the two tests and the lower impact loading rate used for the strip test as previously discussed. The compressive axial stresses combined with the hoop stress at the cup edge generated during the cup/expansion test favors brittle fracture at higher temperature.

Figure 3 shows that fracture transverse to grain elongation occurred at a DBTT at least 45°C lower than that found parallel to the grain elongation. This strong dependence of the transition temperature with grain morphology in deep drawn steel indicates that both conditions of stress and crack orientation must be met at the site of crack initiation to generate brittle intergranular fracture, otherwise ductile or primarily cleavage fracture will occur depending on the test temperature. Figure 3 also shows that increasing strain leads to higher DBTT. In the present study, the strip test results have shown that an increase in strain from 40% to 67% resulted in an increase in DBTT of 25 degrees. Since those results were generated at much lower temperatures than the cup/expansion test temperatures, it might be questionable to extrapolate these results to the cup expansion test. However, qualitatively those results are in agreement with results reported in many other studies where it is found that increasing draw ratios raised the DBTT [6,8,9]. Based on a study by Yan [10,11], the relationship between strain and DBTT appears to vary with steel composition. The authors reported that for an increase in draw ratio from 2.0 to 2.4, an increase in DBTT from –70°C to –10°C was observed. In a more sensitive steel, an increase in draw ratio from 1.8 to 2.4, lead to a DBTT increase from –30°C to +10°C. This variation is probably due to the effect of strain on the fracture mode, which varies with steel composition. The effect of steel chemistry, strain and crack orientation on the fracture mode is illustrated in Fig. 5. The figure shows that not all steels produce the same variation in fraction of intergranular fracture when prestrained or tested in different orientation. It is also worthwhile mentioning that if grain elongation resulting from higher deep drawing strain favors intergranular fracture in the LN orientation, a steel highly sensitive to SCWE such as S3 can produce predominantly intergranular fracture in TN orientation. Whereas for a less sensitive steel such as S2 the fracture mode becomes mostly cleavage in the TN orientation. In practical applications, work-hardening, strain distribution, grain orientation and grain boundary strength affect the transition temperature and fracture morphology.

It should be pointed out that once crack initiation takes place in the cup/expansion test, fracture can propagate in a brittle manner in less strained material, as it is observed frequently in the cup/expansion test where the cracks extend sometimes from the top to the bottom of the cups. It has been previously reported that the fracture mode changes from primarily intergranular to mixed intergranular/cleaveage, and eventually to transgranular ductile as cracks propagate in the lower strain region of the cup wall.[6,8,12] Those observations clearly point out to the fact that conditions for crack initiation and propagation are different during SCWE. Some of the factors that can explain this phenomenon are the large stress intensity factor that arises at the crack tip, which limits plasticity during propagation, and the high strain rate involved during cracking.

Figures 6 and 7 show that surface embrittlement can occur in galvanneal steel. It led to ferrite intergranular fracture beneath the surface. This phenomenon has been reported by many authors. Deits and Matlock [13] reported that the surface cracks in galvanneal steel delineate ferrite grain boundaries. The authors concluded that the substrate cracks observed after forming of galvanneal steel are the result of the effect of galvannealing on the integrity of emerging ferrite grain boundaries. The cracking at emerging ferrite grain boundaries of galvanneal steel is caused by significant diffusion of zinc at ferrite grain boundaries according to Gupta *et al.* [14]. Warneke and Muschenhorn [15] also reported the existence of a relationship between grain boundary diffusion of Zn and cracking observed at the surface of the substrate of galvanneal steel. The presence of Zn at the grain boundaries of the surface ferrite grains was also confirmed in this study.

Because crack nucleation at the ferrite grain boundaries may be promoted by galvannealing, it was appropriate to investigate if this type of fracture could be related to brittle fracture at low temperature. The assumption was that surface grain embrittlement resulting from flaking during forming may act as stress raisers for SCWE. To our knowledge, the only available study on the effect of galvannealing on SCWE was made by Lau *et al.* [4]. This study showed that galvannealing may increase the transition temperature in IF steels. The maximum increase in the transition temperature resulting from galvannealing was about 30ºC. The authors showed that the transition temperature depends on the Fe content in the coating. The maximum sensitivity to SWE was found for a coating with an Fe content of about 4 %, which is lower than the optimum Fe level for commercial products. At this low Fe level, the alloying reaction in the coating layer is partial and a significant amount of free zinc exists on the surface of the coating. In contrast, coatings, which had been fully galvannealed to a higher Fe content, showed a reduced sensitivity to SWE. Lau *et al.* explained the higher sensitivity to SWE embrittlement in galvannealed steel by the presence of outbursts. The outbursts are Fe-Zn phases that form preferentially at the ferrite grain boundaries of the substrate. This phenomenon was previously reported elsewhere [16, 17]. The formation of outbursts during galvannealing depends on several factors including the composition of the substrate [18] and the level of aluminum in the galvanizing bath (e.g., Lin and Meishii [19]). Recently, outburst formation was shown to be associated with penetration of zinc along emerging ferrite [20]. A possible effect of crack initiation at grain boundaries adjacent to the surface of galvanneal steel is further propagation into the substrate causing brittle intergranular fracture. According to Lau et al., this proposed mechanism of embrittlement on IF steel depends on the degree of alloying reaction because of the effect of the coating microstructure on cracking during sheet metal forming. As the alloying reaction progresses during galvannealing, the zinc coating is enriched by Fe leading to formation of the δ and Γ phases which are more brittle than the η and ζ phases present at the early stages of galvannealing. It is assumed that cracks are diverted into the Γ phase located at the coating/substrate interface because of its brittleness. This will reduce the tendency for propagation of the coating cracks into the grain boundaries of the substrate. In partially galvannealed IF coatings, cracks propagate into the grain boundary without bifurcation into the coating interface. Since commercially galvanneal coating is composed of Γ phases at the Zn substrate interface, it is concluded that galvannealing of commercial product has no negative influence on the SCWE performance of galvanneal steel. The cold work embrittlement tests reported in the paper seem to confirm this observation but the mechanism by which the coating may influence SCWE has not been confirmed.

The examination of FIB cross-sections of surface cracks on two specimens containing SCWE cracks confirmed that the preferred crack propagation paths are within the coating and along the coating/substrate interface. Some coating cracks did penetrate into emerging ferrite boundaries but most of these cracks were quite shallow and did not extend beyond one grain diameter. Two factors could explain the resistance of ferrite grain boundaries to crack penetration:

1. There is limited zinc embrittlement of ferrite grain boundaries in these steels due to their substrate chemistry. Outbursts are most prevalent in Ti-stabilized IF steels because the solute-deprived grain boundaries have higher zinc diffusion kinetics. The grain boundaries of Ti-Nb-stabilized and rephosphorized grades are less reactive and therefore, these steels are less susceptible to outbursts [17].
2. The galvannealing conditions used for steels S2 and S3 have produced a relatively thick interface layer of brittle gamma phase therefore promoting interface decohesion as the primary mode of coating failure.

A more detailed characterization of the galvanneal structure of steels S2 and S3 involving TEM to identify the various phases and to determine the extent of zinc penetration along ferrite grain boundaries is needed to resolve these points.

In the case of the cup/expansion test specimen of steel S2, significant intergranular and transgranular ferrite cracks were observed near the primary SCWE crack but not away from it. These observations support the view that the emerging ferrite grain boundaries have considerable strength. The stress required to induce a coating crack to propagate along a ferrite grain boundary was similar to that required for transgranular cleavage. Such a high stress was only encountered in the region adjacent to the primary initiation site, where the stress would have reached its highest level during the test.

The cup drawing operation used in this study did not create excessive damage to the coating. However it is representative of severe deep drawing, which is known to favor SCWE. Although the study demonstrated that in rephosphorized IF grade, surface embrittlement can occur, no evidence was found that this phenomenon contributed to SCWE during forming and cup/expansion test.

CONCLUSION

This study has shown that microstucture can play an important role in SCWE. The following observations have been made:

- Fracture mode varies with steel chemistry and transition temperature, intergranular fracture being predominant in more SCWE sensitive steel grades;

- Work-hardening due to deep drawing increases the fraction of intergranular fracture to different degrees depending on the steel sensitivity to SCWE;
- The development of elongated grains during deep drawing contributes to increase the DBTT, however this effect favors directional SCWE sensitivity, resulting in significant variation between DBTT along the transverse and longitudinal direction;
- As a result of the directionality of the crack orientation with respect to grain elongation, DBTT varies with the stress conditions during the DBTT tests;
- Galvannealing can produce surface embrittlement due to Zn diffusion in rephosphorized IF grades;
- No evidence was found that surface embrittlement with Zn contributed to SCWE crack initiation during forming and cup/expansion test in the steels investigated.

ACKNOWLEDGEMENTS

The authors are grateful to American Iron and Steel Institute (AISI) for permission to publish this paper. Financial support from AISI (AK Steel, Dofasco, LTV Steel, National Steel, Rouge Steel, Stelco, US Steel, Weirton Steel), and Department of Energy of the US is gratefully acknowledged. We would like to thank Mark Carisse for the design and fabrication of the notched specimen impact tester.

REFERENCES

1. Herman, J.C. Leroy, V., "Influence of Residual Elements on Steel Processing and Mechanical Properties", 38th Metal Working and Steel Processing Conference, ISS, vol. XXXIV, 1997, pp. 545-554.

2. Yasuhara, E., Sakata, K. , Kato, T., and Hashimoto, O., "Effect of Boron on the Resistance to Secondary Work Embrittlement in Extra-low C Cold Rolled Sheet Steel", ISIJ International, vol. 34, no.1, 1994, pp. 99-107.

3. Neutjens, J., Mathy, H., and Herman, J.C., "Cold Work Embrittlement of ULC-IF and EAF Low Carbon Steel Grades, La Revue de Métallurgie-CIT, vol. 94, No. 4, Apr. 1997, pp. 551-561.

4. Lau, Y.H., De Cooman, B.C., Vermeulen, M., "The Secondary Work Embrittlement of Galvannealed High Strength Sheet Steel for Automotive Applications: Test Method and Material Evaluation", 39th Mechanical Working and Steel Processing Conference Proceedings, ISS, vol. XXXV, 1998, pp. 139-148.

5. Teshima, S. and Shimizu, M., "Recrystallization Behavior of Cold Rold Mild Steel", Mechanical Working of Steel II, Edited by T.G. Bradbury, Gordan and Breach Science Publ., 1964, pp. 279-320.

6. Bhat, S.P., Yan, B., Chintamani, J.S. and Bloom, T.A., "Secondary Work Embrittlement (SWE) of Stabilized Steels: Test Methods and Applications", Proceedings of the Symposium on High-Strength Sheet Steels for the Automotive Industry, R. Pradham, ed., ISS, 1994, pp. 209-222.

7. Zhong, W., Ng, H.F., and James, J.M., "Correlation between Adhesion Properties and the Interfacial Bonding Strength of Galvanneal Coatings", Zinc-Based Steel Coating Systems: Production and Performance, F.E. Goodwin, ed., The minerals, Metals Society, 1998, pp. 185-194.

8. Lewis, S.G., Daniel, S.R., Parker, J.D., Llwellyn, D.T., and Sidey, M.P., "Testing Techniques for Cold Work Embrittlement in Interstitial Free Steels", Ironmaking and Steelmaking, vol.25, no.1, 1998, pp. 63–73.

9. Yasuhara, E., Sakata, K., Furukimi, O. and Mega, T, "Effect of Boron on the Resistance to Secondary Work Embrittlement in Extra-low C Cold Rolled Sheet Steel", 38th Metal Working and Steel Processing Conference, ISS, vol. XXXIV, 1997, pp. 409-415.

10. Yan, B., "Intergranular Fracture of Interstitial-Free Steels Under Cyclic Loading and its Effect on Fatigue Life", 37th Metal Working and Steel Processing Conference, ISS, vol. XXXIII, 1996, pp. 101-114.

11. Yan, B. and Gupta, I., Secondary Work Embrittlement of Interstitial Free Steel, International Fracture in Engineering Materials – Basic Scientific Aspects and Engineering Consequences, Materials an Manufacturing Ontario, McMaster University, ASM Ontario and CIM Hamilton Section, Hamilton, ON, April 8, 1999.

12. Maehara, Y., Mizui, N. and Arai, M.,"Cold-work Embrittlement Accompanied by Intergranular Fracture in Ultra-low Carbon Ti-added Sheet Steels", Conference Proceedings Interstitial Free Steel Sheet: Processing, Fabrication and Properties, L.E. Collins and D.L. Baragar, eds., Canadian Institute of Mining, Metallurgy and Petroleum (CIM), Montréal, 1991, pp. 135-144.

13. Deitz, S.H. and Matlock, D.K., "Formability of Coated Sheet Steels: An Analysis of Surface Damage Mechanisms", Zinc-Based Steel: Coating Systems: Metallurgy and Performance, G. Krauss and D.K. Matlock, eds., The Minerals, Metals and Minerals Society, Warrendale, PA, 1990, pp. 297-318.

14. Gupta, I., Cline, R.S. and Patil, R.S., "Effect of Iron-Zinc Alloy Layer on the Mechanical Properties and Formability of Hot-dip Galvanized Steels", SAE Paper 840284, 1984.

15. Warnecke, W., and Muschenborn, W., "Formability Aspects of Galvannealed Steel Sheet", Proceedings of The Intenational Deep Drawing Research Group IDDRG Conference, 1985, pp. 16.1-16.27.

16. Inagaki, J., Morita, M. and Sagiyama, "Iron-Zinc Reaction on Interstitial Free Steels", Surface Engineering, Vol. 7, No. 4, 1991, pp. 331-339.

17. Hisamatsu, "Science and Technology of Zinc and Zinc Coated Steel Sheet", Galvatech '89: Proceedings of the International Conference on Zinc and Zinc Coated Steel Sheet, The Iron and Steel Institute of Japan, 1989, pp. 3-7.

18. Nishimoto, A., Inagaki, J. and Nakaoka, K. "Effect of Surface Microstructure and Chemical Compositions of Steels on Formation of Fe-Zn Compounds during Continuous Galvannealing", ISIJ International, 1986, pp. 807-813.

19. Lin, C.S., and Meshii, M., "Effect of Steel Chemistry on the Formation of the Alloy Layers on the Commercial Hot-dip Coatings", Galvatech '95: The Use and Manufacture of Zinc and Zinc Alloy Coated Sheet Steel Products into The 21st Century, ISS, AIME, Warrendale, PA, 1995, pp. 335-342

20. Jordan, C.E., and Marder, A.R., "Inhibition Layer Breakdown and Outburst Fe-Zn Alloy Formation During Galvanizing", Zinc-Based Coating Systems: Production and Performance", F.E. Goodwin, ed., The Minerals, Metals and Materials Society, 1998, pp. 115-126.

ROLL TECHNOLOGY I

UKRAINIAN ESS LM HSS ROLLS FOR HOT STRIP MILLS

B.I. Medovar, L.B. Medovar*, A.V. Chernets*,
V.B. Shabanov**, O.V. Sviridov**
* – "ELMET-ROLL — MEDOVAR GROUP", P.O. Box 259, 03150, Kyiv, Ukraine, Tel.: 380-44/227-5218, Fax: 380-44/227-5288
** – "Novokramatorsk Machine-Building Works (NKMZ)", 84300 Kramatorsk, Ukraine, Tel.: 380-6264/71566, Fax: 380-6264/72249

Key Words: Electroslag Surfacing, Liquid Metal, Current Conductive Mold, Clad Layer

INTRODUCTION

Since 1997 was established Ukrainian program of the HSS rolls production. This paper will discuss Electroslag Surfacing by Liquid Metal as manufacturing process for HSS rolls industrialized at NKMZ, Ukraine. Will be presented also results of the roll performance at stand F2 at HSM.

One of the most important problems encountered by the manufactures and consumers of the rolling equipment's is the increase in service properties and life of mill rolls used for rolling new materials.

The production of composite rolls with a clad layer of high-strength wear-resistant materials is one of the possible ways for the solution of the above-mentioned problem. Here, the activity of companies, working at the market of mill rolls, is directed both to the creation of radically new cladding materials with a system of special properties and to the improvement of existing or development of new high-efficiency cladding technologies which can widen the capabilities of the known schemes and realize the cladding of these new materials.

Today, the double-layer hot mill rolls with a working layer of a high-chromium cast iron are replaced by double-layer rolls with a working layer of high-speed steels (HSS). In these steels the content of carbon plus a total content of carbide-forming elements, such as Cr, V, W, Mo and Nb, amount to more than 15 %. Moreover, it should be noted that the carbon content in these steels exceeds the ordinary content typical of traditional high-speed steels and approach 2 %.

It is well known that tool steels, including high-speed steels, steels of ESR or powder metallurgy, are successfully used for a long time for the rolls of small cold rolling mills, for example, of 20-roll Sendzimir mills. The application of ESR is stipulated by the need in refining of the second phase, first of all carbides, and their uniform distribution at least in the working layer of rolls. Since the end of 1980s the double -layer rolls with a HSS working layer began to be generally used in Japan, and then in Europe and North America. Three- and, sometimes, five-fold increase in life of the HSS rolls as compared with rolls of high-chromium steel led to the fact that at present the HSS rolls in the above-mentioned regions forced out completely the cast iron working rolls. And this really happened in spite of a twice increase in cost. This information is described in detail in work [1]. It should only be noted that the manufacture of monoblock HSS rolls from the ESR ingots provided good results from the point of view of the metallurgical quality. However, because of a high cost of high-speed steels and high cost of ESR, these rolls were refused and today only double-layer HSS rolls produced either by a centrifugal casting or by the CPC method (only in Japan) are used [1].

In our opinion, the technology of electroslag surfacing using a liquid metal (ESS LM), recently developed by the engineers of "ELMET-ROLL-MEDOVAR GROUP" in collaboration with scientists of the E.O. Paton Electric Welding Institute, Kiev, Ukraine, can solve the problem of the market needs in high-quality and relatively cheap composite rolls with a working layer of the high-speed steel [2,3].

It is quite evident, that traditional electroslag cladding with electrodes of different types - wires, strips, electrode-tube (Figure 1) is not suitable for the roll cladding by a number of factors.

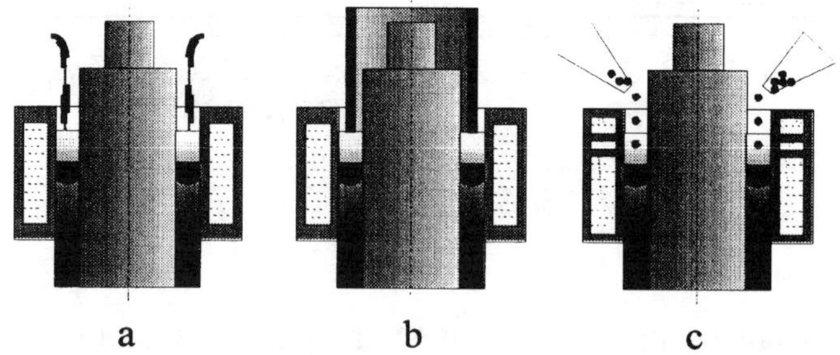

Fig.1. Schemes of electroslag cladding with wire (a), electrode-tube (b) and lumpy filler materials (c)

Among them, the main factor is a high cost of manufacture of these HSS electrodes, and, in some cases, even the impossibility of their manufacture for certain materials. The same refers to the cladding with lumpy materials. At the same time the electroslag cladding, owing to its physical and chemical peculiarities, provides the high-quality joining between the clad layer and roll axis and excellent fine-dispersed structure of the clad layer.

The search works, aimed at the study of feasibility of creation of ESS LM to provide metal of the same quality that is provided by a standard consumable electrode were based on the above-described considerations.

Let us consider the peculiarities of this technological process and its potentiality.

Diagram of ESS LM is presented in Figure 2. Surfacing is performed in a water-cooled copper current-supplying mould of a special design. A roll to be clad is placed inside the mould. Slag, melted in a separate unit, is poured to a gap between the surface to be clad and a mould wall. The mould is not only a device, which forms a clad layer, but also a non-consumable electrode, which maintains an electroslag process. At the expense of heat, generated in a slag pool, the roll surface is fused a little and then a liquid metal of a required composition is fed continuously or by portions according to a preset program to a gap between the mould and a roll being clad. Metal forces out the slag upward and, taking its place, contacts a fused surface of the roll being clad and forms a deposited layer. During cladding a workpiece is withdrawing constantly from the mould (or , when the billet is fixed, the mould is displaced along it) and next portions are poured (or continuous pouring).

The technology of ESS LM makes it possible to clad external surface of cylindrical billets of almost any diameter, from 40 to 1000 mm and more, and to produce composite billets of such parts as mill rolls or rollers of machines for continuous casting of billets. The thickness of the clad layer is determined exclusively by requirements, set by the customers, and can be 20 -100 mm and more.

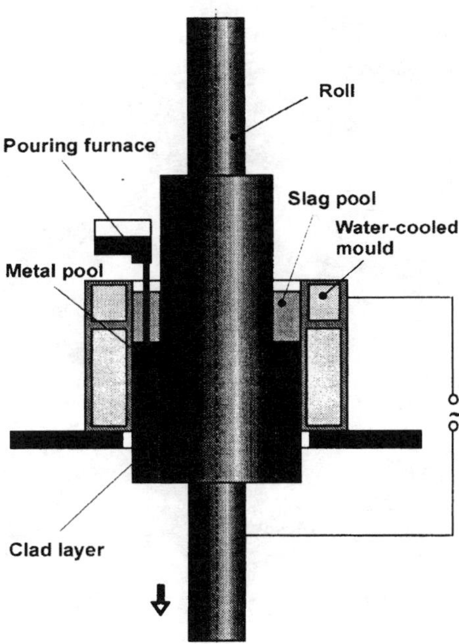

Fig.2. Scheme of ESS LM process

One of the most important and indispensable advantages of the new technology is the process productivity. The efficiency of the ESS LM exceeds by dozens of times the efficiency of the traditional methods of cladding, and, depending on the sizes of parts and materials used it is 200-800 kg/h.

One more important advantage of the new technology is a feasibility of cladding with materials of the most various chemical compositions. Cast iron, high-speed, tool and stainless steels, heat-resistant nickel alloys, etc., including materials which are cannot be subjected neither to cold, nor to hot deforming and, consequently, cannot be used for the traditional cladding technologies, where the consumable electrodes in the form of a wire, rods, plates are required. The feasibility of using ESS for production of rolls with a working layer of high-speed steel of the new generation is most important. As was above-said, such high-speed steels (with an increased content of carbon of up to 2.0-2.5 %) and total concentration of carbide-forming elements (Cr, W, V, etc.) of above 25 % allow many-time increase in the mill roll life, its ability to withstand not only the thermal fatigue, but also corrosion action under different service conditions due to their surprising complex of characteristics.

Since the end of 1997 Ukrainian program of HSS rolls industrial manufacturing by the method of the Electroslag Surfacing by Liquid Metal, ESS LM were started [4,5].

Within the scope of joint works of "ELMET – ROLL – MEDOVAR GROUP" and CJSC "Novokramatorsk Machine-Building Works" (NKMZ) the project of adaptation of ESR furnace at "NKMZ" has been fulfilled and industrial installation for an electroslag cladding with liquid metal was designed and manufactured. The installation is designed for production of composite billets of mill rolls of up to 1000 mm diameter, up to 2500 length of the barrel to be clad and up to 20 000 kg mass, using the ESS LM method [5]. Appearance of the installation created by specialists of NKMZ and ELMET-ROLL is shown in Figure 3.

In this installation the technology of cladding hot mill rolls of 650-850 mm diameter with a working layer of a high-speed steel was optimized and the first batches of these rolls were produced (Figure 4).

Fig.3. Appearance of ESS LM installation

Fig.4. Appearance of ESS LM composite billet of HSS roll

Let us consider the quality characteristics of metal of composite billets of HSS rolls produced by the ESS LM.

Figure 5 presents a macrostructure of the ESS LM billet 740mm in diameters. The metal of the clad layer is sound, without pores, cracks, cavities and other defects; the fusion line is continuous, there are no lack of fusion in parent and clad metal. Penetration in height and diameter of the billet is uniform.

Examination of isotropy of the chemical composition in height and section of the produced billets was carried out on samples cut out from bottom, middle and top parts of the billet and showed an almost uniform distribution of main alloying elements both in height and section of the clad layer (Figure 6). Similar results are obtained from the hardness measurements at the surface of the as-clad billets (Figure 7).

As the examination of microstructure of the clad layer of composition: 1,7-2,0% C; 0,4-0,6% Si; 0,6-0,8% Mn-; 4-6% Cr; 0,8-1,0% Ni; 4-6% Mo; 5-7% V; 3-5% W showed (Figure 8) the application of the ESS LM made it possible to eliminate in a cast form the appearance of coarse carbides and to provide the uniform their distribution. In the ESS LM metal the eütectic precipitation's along the grain boundaries typical of the traditional methods of production are almost completely absent.

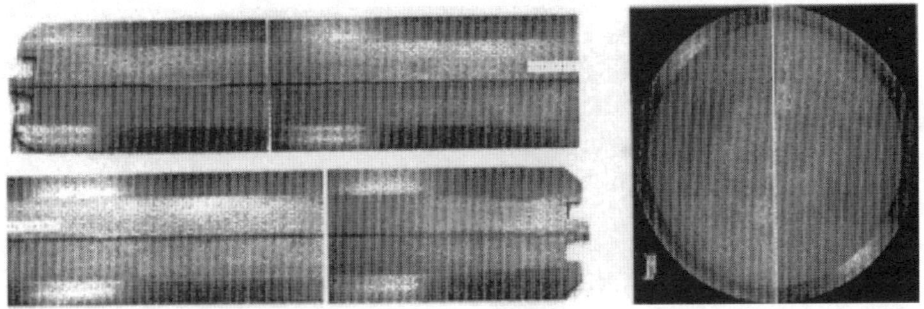

Fig.5. Macrostructure of ESS LM HSS Roll billet

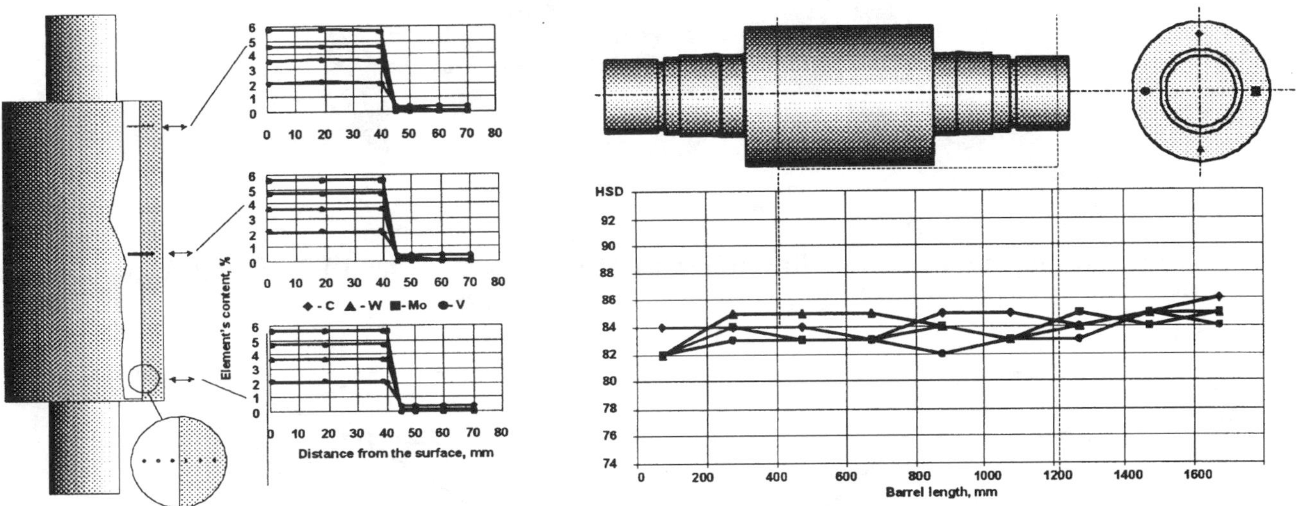

Fig.6. Distribution of elements in height and section of ESS LM billet

Fig.7. Distribution of hardness in ESS LM billet

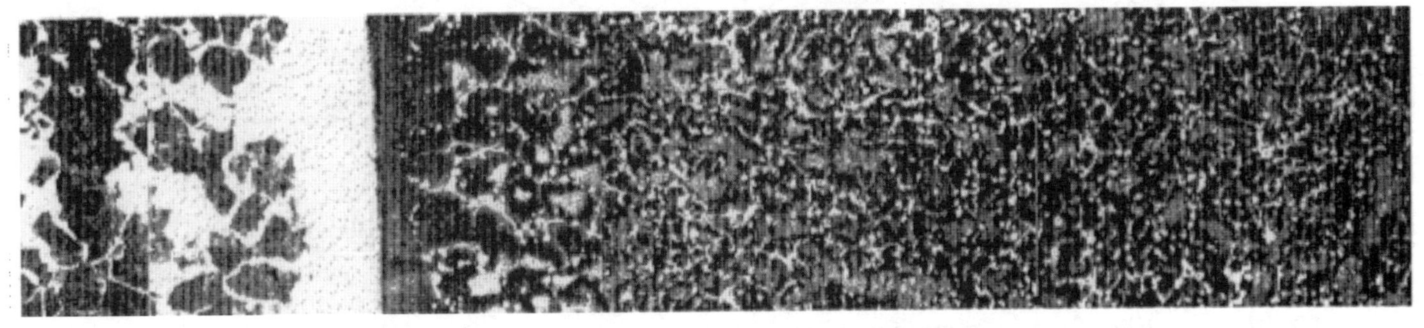

Core

Clad layer

Fig.8. Microstructure (x 100) of HSS clad layer

Examination of the microstructure revealed that the fusion line is uniform in the whole height of the billet, lack of fusion of parent and clad layer was not observed, the distribution of microhardness along the line "parent metal -fusion line - clad layer" is uniform with a smooth increase from the beginning of the fusion zone into the depth of the clad metal. The ESS LM metal examination in a scanning electron microscope Camebax with a microanalyzer Cameca showed that the change in content of main alloying elements along the line of scanning "parent metal-fusion zone-clad layer" is rather smooth and uniform.

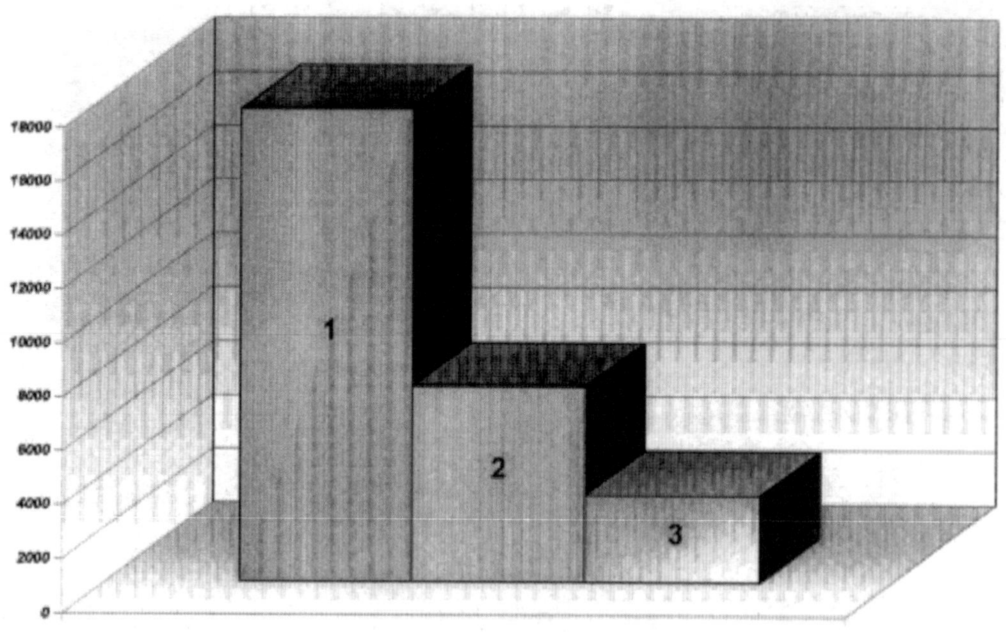

Fig.9. Comparison diagram of service life (ton/mm) of rolls used in stand F2 of 1700 Ukrainian hot strip mill: 1 – ESS LM HSS rolls; 2 – ESS LM HiCr rolls; 3 – ICDP rolls

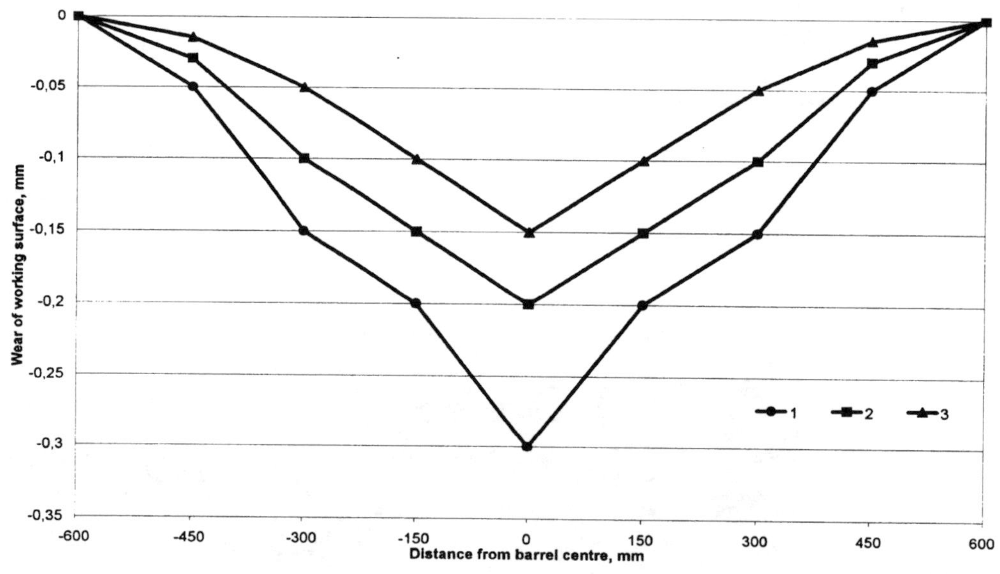

Fig.10. Wear profile of working surface of composite and cast iron rolls used in stand F2 of 1700 Ukrainian hot strip mill : 1 – ICDP rolls after 3200 t rolling; 2 – HSS rolls 7200 t rolling; 3 – HSS rolls after 5500 t rolling

CONCLUSION

The first step of the industrialization of the HSS roll made by ESS LM has been successfully finished. Next step will be focused on the development of advanced composition of HSS and improvements of the technological processes such as rolls heat treatment practice and careful control of their behavior at different mills.

ACKNOWLEDGEMENTS

The Authors are glad to express they deep acknowledgements to they colleagues from NKMZ and Elmet-Roll for they hard work and intellectual input. As it was a huge team of experts from both sides it's impossible to make listed all of them.

REFERENCES

1. D. B. Collins, "HSS Rolls through 2000", Rolls 2000, Sheffield, England,. 28-29 March,1996. Programme and Conference Papers. pp. 91-100.
2. B.I. Medovar, A.V.Chernets, L.B. Medovar, et al. "Electroslag cladding with a liquid filler metal", Problemy Spetsialnoy Elekrometallurgii, No.5, 1995, pp.6-12.
3. B.I. Medovar, L.B. Medovar, A.V. Chernets, et al.,"Electroslag Surfacing by Liquid Metal - a New Way for HSS-rolls Manufacturing", 38th MWSR Conference Proceedings. Volume XXXIV, Cleveland, Ohio, USA, October 13-16, 1996, pp. 83-87.
4. V.B. Shabanov, L.B. Medovar, "Rolls from Ukraine. Yesterday. Today. Tomorrow", 40th Mechanical Working and Steel Processing Conference Proceedings, Volume XXXVI, Pittsburgh, USA, 25-28 October, 1998.
5. V.B. Shabanov, O.V. Sviridov, Yu.N. Belobrov, et al., "Development of the unit of electroslag surfacing with liquid metal of hot rolling mill work rolls for continuous widestrip lills", Avtomaticheskaya Svarka, No.9, 1999, pp. 51-54.

APPLICATION OF ESR TECHNOLOGY TO THE MANUFACTURING OF BIMETALLIC HSS ROLLS FOR COLD AND HOT STRIP MILLS.

C. Gaspard – S. Bataille – D. Batazzi – P. Thonus

Forcast
40 rue de la Barrière
4100 Seraing
Belgium

Telephone : 32 4 22 97 242
Fax : 32 4 22 97 241
E-mail : FORCAST@SKYPRO.BE

Key words : ESR cladding – HSS grades – bimetallic rolls.

1 INTRODUCTION

The quality requirements for the hot and cold rolling sheets are continuously increasing. This leads to more stringent surface quality aspect for the rolls. Moreover the optimization of the mill productivity requires new rolls grades which allow longer runs.

In the field of cold rolling, roll grades with an outstanding wear resistance and improved behavior during severe rolling incidents have been introduced recently[1]. These grades are characterized by a chemical composition with a high chromium, molybdenum and vanadium content as well as by a high tempering temperature. They are produced with the classical manufacturing procedure, the forging of mono-bloc ingots. To reduce the total cost of the alloys, and to avoid the detrimental effect of alloying elements on the mechanical properties of the core and of the necks of the roll, a new bimetallic manufacturing technology has been developed.

HSS grades used in hot rolling of steel have been introduced by the Japanese in the end of the eighties. The aim of this new grades was to improve the strip surface, to avoid rolled in scale, and to allow longer trip in the mill. These early developments have been followed with extensive trials in Europe and in the North America, mainly with rolls produced by the spin casting process[2]. This technology is very cost effective, but suffer from limitations for the chemical composition of the external layer, due to segregations of carbides near the bonding. Despite the very good results obtained with the spin casting HSS rolls, further improvements require a new manufacturing technology[3].

To overcome the limitations of both the forging and spin casting manufacturing techniques, a new technology has been developed, based on the ESR process. This paper presents the principle and the characteristics of the first rolls produced by ESR cladding.

2 MANUFACTURING PROCESSES

2.1 Bimetallic processes

The in-service properties of the barrel and of the necks of rolls are completely different. The necks and the core of the rolls should have a high toughness and ductility, so the most suitable materials are low alloyed steels. On the other hand, the barrel should sustain the surface degradation caused by the contact with the strip and its relative motion on the roll, so it must be hard and wear resistant. The barrel properties are given by the carbon and carbide forming elements such chromium, molybdenum or vanadium which are useless or even detrimental for the toughness of the core. The better solution to overcome these opposite properties is to use a different material for the core and for the barrel by manufacturing a bimetallic roll.
Different manufacturing processes have been used in the past to produce bimetallic rolls. The most common one is the spin casting (vertical or horizontal), widely used for the last 30 years to produce work rolls for the hot rolling of steel. Other techniques have also been investigated, such as the Osprey process (pulverization of liquid steel), thick coating by welding, the hot isostatic pressing (HIP) technique, and the Continuous Pouring for Cladding (CPC, CCC).

2.2 ESR cladding process

The ESR cladding process is based on the same principle as the classical ESR (fig. 1.a). This process consists in remelting an electrode by a hot slag and the liquid metal pool is continuously solidified into a water cooled moving mold[4]. The heat is generated in the slag using the Joule effect of an alternating current flowing between an electrode (above) and the solidified ingot (bottom). This slag has also a metallurgical purpose in the process. The droplets passing through the slag are desulphurized, and the large inclusions are fragmented. ESR technology gives steels with a very good cleanliness, and with a fine and homogeneous macrostructure.
The cladding process consists in casting a metallic layer around an axis with a different chemical analysis in order to form a bimetallic composite roll (fig 1.b). The axis is vertically moved into a water cooled copper mold. The gap between the external surface of the axis and inner surface of the mold is filled with an electro-conductive slag. The slag is heated into the upper part of the mold by an electric current. The metal to clad in solid or liquid state is brought on the slag, refined through the molten slag and is solidified into the bottom part of the mold[5].
Compared with the CPC process, the molten metal is protected from the air by the slag. Like the classical ESR, the cleanliness of the cladded metal is further improved by the purification action of the slag, giving a high purity steel with a fine microstructure.
During the cladding, the slag is rotating around the arbor. This enhances the homogeneity of the process in terms of thermal map and chemical composition, avoiding segregation. Another important action of the slag is to clean the arbor before the contact with the molten metal, which improves the metallurgical properties of the bonding.

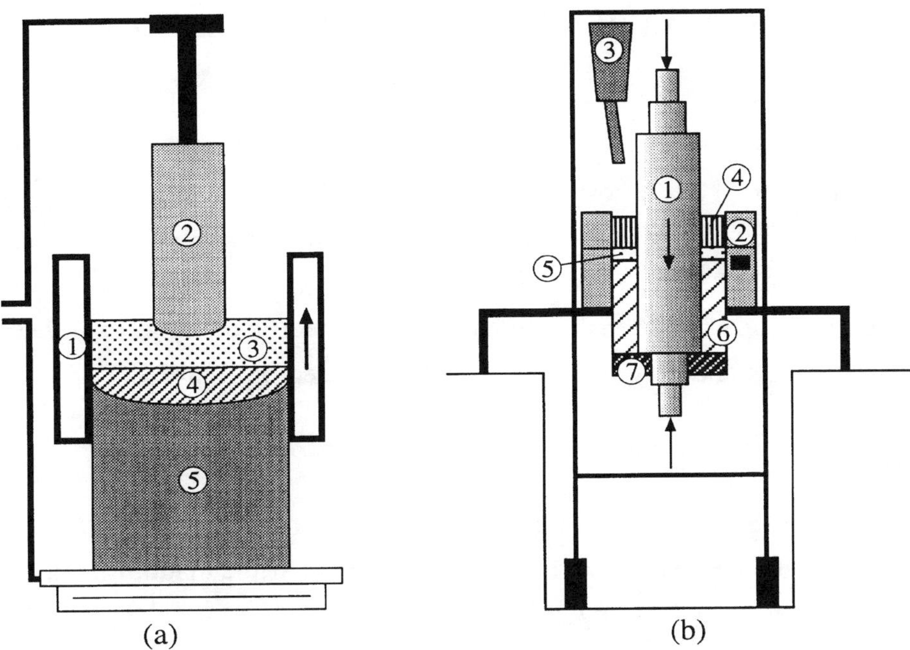

Fig. 1 : (a) Principle of classical ESR: 1. Moving copper mold – 2. Electrode – 3. Slag bath – 4. Liquid metal pool – 5. Solid ingot
(b) Schematic principle of ESR cladding : 1. Moving axis – 2. Copper mold – 3. Metal to clad – 4. Electro-conductive slag – 5. Liquid cladding metal – 6. Solid cladding metal – 7. Starting ring.

In the spin casting process, the alloy content of the working layer is limited due to density segregation of eutectic carbides near the bonding. Moreover, the melting temperature of the core must be below the solidus temperature of the shell. This prevents the use of low alloyed steel in the core, so the usual core material is grey iron (lamellar or spheroidized). In the ESR cladding process, both limitations are removed. This new process gives a highly alloyed shell around a tough steel core.

The capacity of the installation used to manufacture prototype rolls is summarized in table I. This installation has been fully computerized in order to maintain the suitable parameters during all the cladding around the axis.

Table I : Capacity of the installation

Item	Maximum capacity
Cladded ingot diameter	850 mm
Cladded length	3000 mm
Total ingot length	6000 mm
Cladded ingot weight	20 tons
Cladding thickness	120 mm radius

3 METALLURGICAL STUDY OF CLADDED ROLLS

3.1 Macrostructure

The macrographic examination of a composite roll manufactured by ESR cladding shows a concentric bonding between axis and cladding (fig. 2). The cladding thickness is constant around the circumference of axis what proves the perfect control of the remelting of the axis.
The solidification macrostructure has been investigated after a specific etching (Fry). The fine dendrite pattern of the cladding obtained with standard grade for CSM is illustrated on figure 3.

3.2 Microstructure

During the preliminary tests, a wide range of grades have been cladded around a forged eutectoid steel axis (fig. 4). The parameters of the process (electrical intensity, speed of the cladding, temperature of the slag…) must be adapted to obtain a healthy bonding zone.

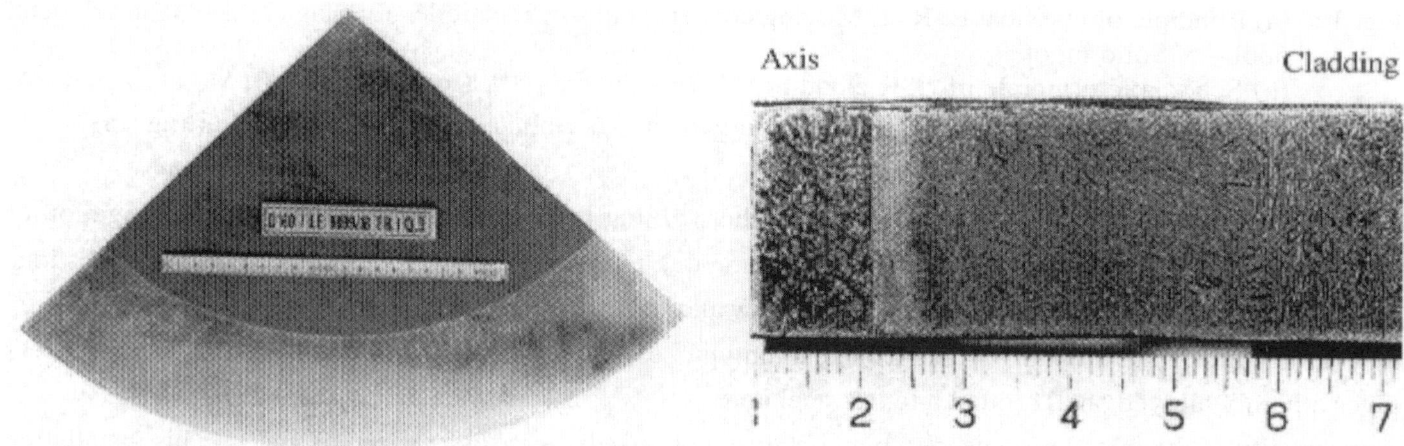

Fig. 2 : Macrographic examination of the cladding

Fig. 3 : Solidification macrostructure of the cladding

The first trial was a standard 3% chromium steel used for the cold rolling rolls. In the normalized condition, the microstructure of the shell is a homogeneous perlite (fig. 4a). The bonding is extremely thin, without mixed zone in between (fig. 4b).
A HSS grade used in cold rolling was then cladded. After the normalizing heat treatment, the cladded shell consists of a spheroidized perlite with plenty of secondary carbides (fig. 4c). The eutectic carbides are fine, and do not form continuous network around the grain boundaries. The bonding zone is here again very thin (less than 30 µm). No segregation is observed near the bonding between cladding and axis (fig. 4d). A microprobe analysis crossing through the bonding shows that some diffusion of the elements occurred, for example 100 µm for the chromium (fig. 5).

CLADDING

BONDING

Fig. 4a : Cladding of a standard cold rolling grade.

Cladding

Axis

Fig. 4b. : Bonding with a standard cold rolling grade.

Fig. 4c : Cladding of a HSS grade for CSM.

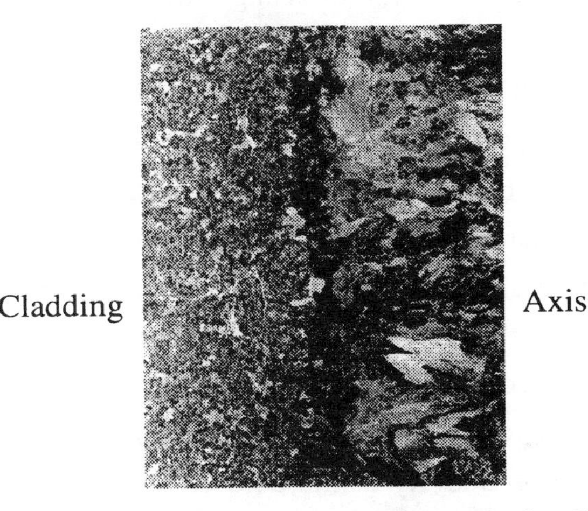

Fig. 4d. : Bonding with a HSS grade for CSM.

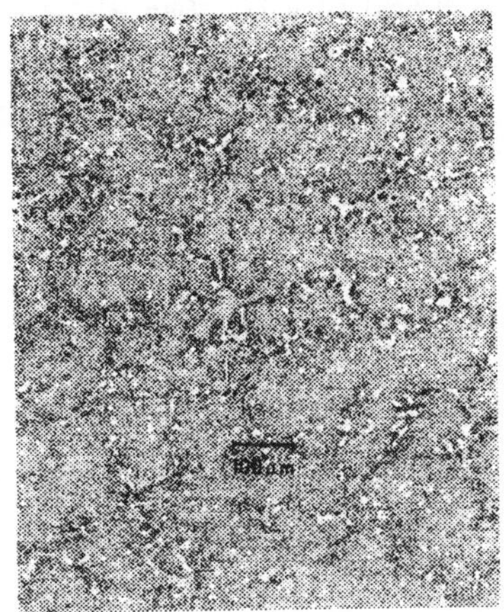

Fig. 4e : Cladding of a HSS grade for HSM.

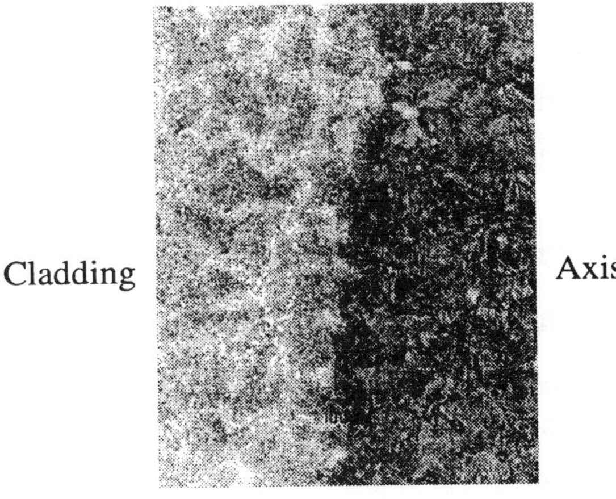

Fig. 4f. : Bonding with a HSS grade for HSM.

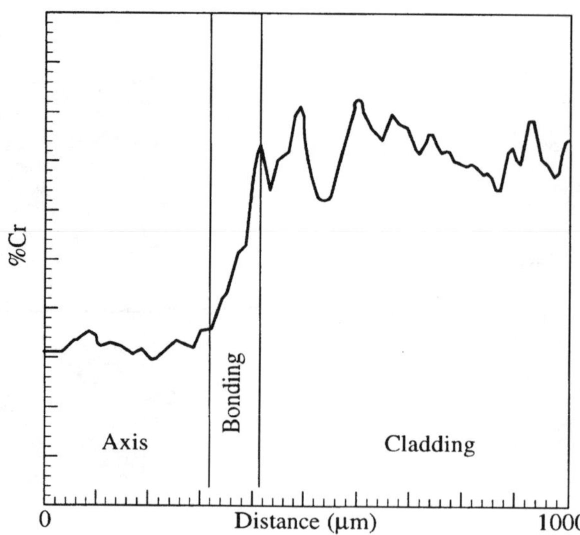

Fig. 5 : Microprobe analysis crossing through the bonding.

The cladding of a HSS grade for HSM was also successful. The eutectic carbides are fine, and well distributed (fig. 4e). The bonding zone is thin (30 µm) and without segregation (fig. 4f).

The bonding of the ESR cladding has been compared with the vertical spin casting. Figure 5 shows the influence of the manufacturing process for the same grade on the microstructure of the shell and the bounding. In the cladding (fig. 6a), the eutectic carbides are finer, and the eutectic cells are also smaller, due to a higher solidification kinetic. Figure 6b shows that the bonding zone of the spin casting roll is wide (3000 µm). There are plenty of eutectic carbides in the bonding zone, which form a quite continuous phase at the grain boundaries. A layer of coarse carbides can also be found in the bonding near the gray iron axis. This is the consequence of the higher carbon activity of the core compared with the shell. In the ESR cladding process, the use of an eutectoid steel for the axis and the limited diffusion during the process avoids the formation of coarse carbides near the bonding.

Tensile tests made on samples including the bonding in their middle showed that the fracture occurs into the cladding for both HSM and CSM tested grades. The measured ultimate tensile strength of 600 MPa is far above the mechanical properties of the grey iron usually used as the core of the rolls for HSM (about 400 MPa). It means that the bonding is strong enough to support the stresses during the rolling.

3.3 Manufacturing of a roll for CSM

Rolls for cold rolling mill were manufactured with the ESR cladding machine. The diameter of the barrel was 585 mm, the length of the barrel 1520 mm and the total length of the roll 3400 mm. A HSS grade was cladded around a 430 mm axis.

A FEM analysis showed that the progressive induction hardening was the best technique to limit the internal stresses generated during the hardening of the barrel of bimetallic rolls for CSM. The electrical parameters

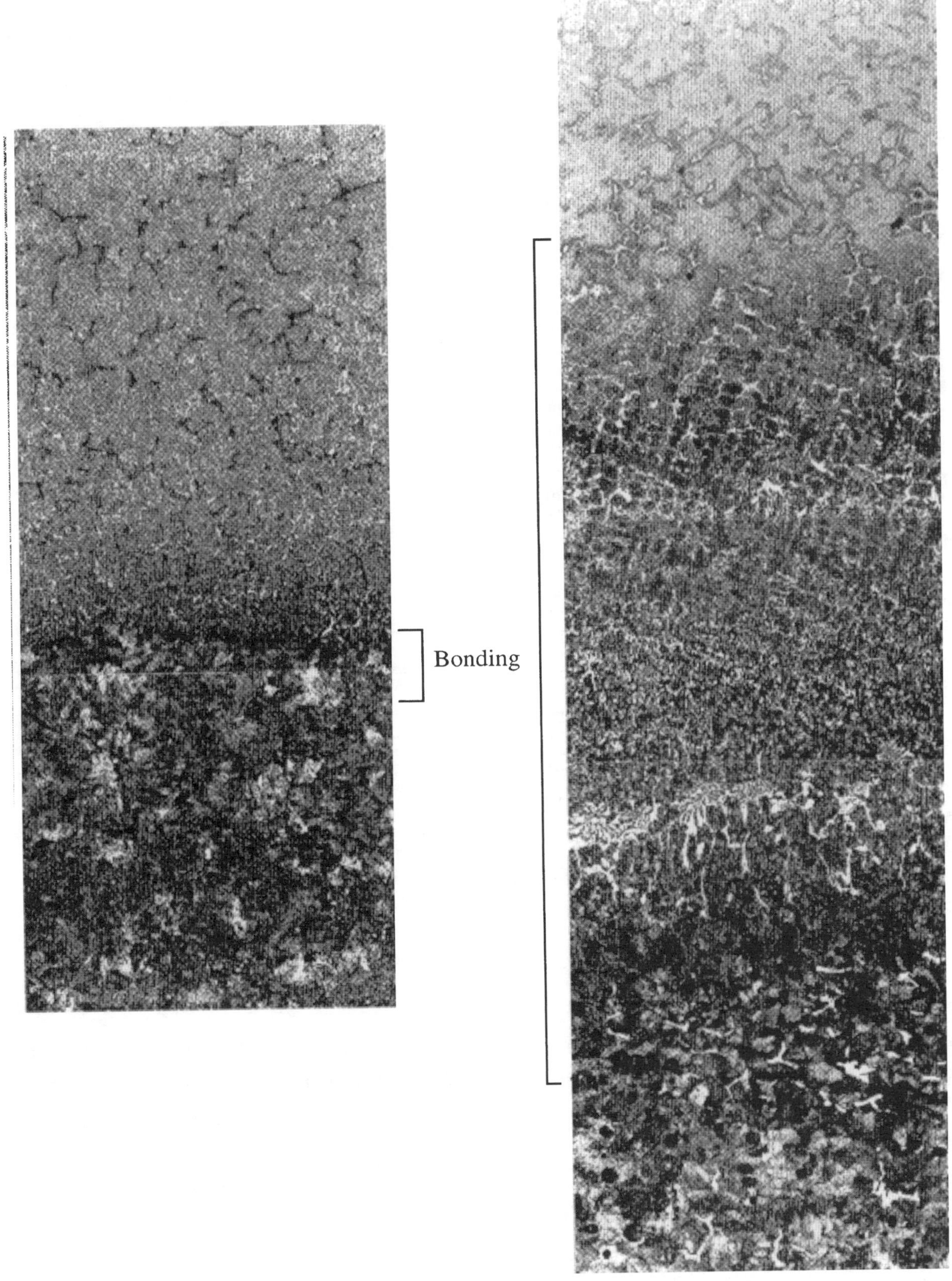

Fig. 6 : Comparison between the bonding for the ESR cladding and the spin casting.

were however adapted to limit the depth of penetration of the heat, and to avoid the austenitizing of the bonding. Figure 7 shows an example of hardness evolution in depth after progressive induction hardening of a composite roll. A maximum hardness of 800 HV, suitable for the cold rolling of steel in a tandem mill, was measured on a sample taken from an overlength. The useful layer of 70 mm on diameter could be increased up to 120 mm depending on the requirement of the customer. The hardness near the bonding was not influenced by hardening. The resulting residual stresses are around 600 MPa in compression for both longitudinal and tangential directions, what is lower than in a standard forged roll (around 800 MPa in compression).

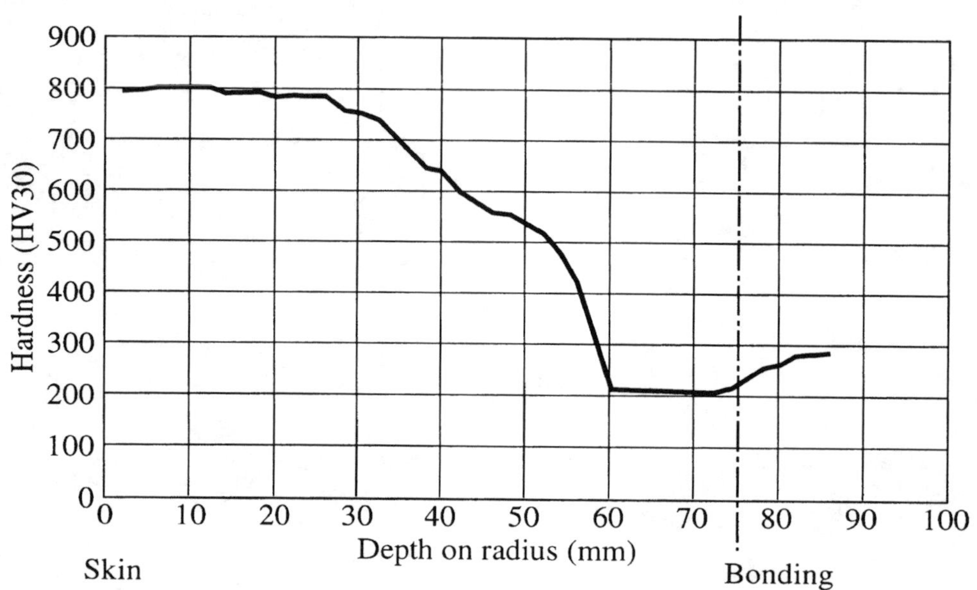

Fig. 7 : Evolution of the hardness of a composite roll for CSM after progressive induction hardening

4 CONCLUSIONS

Composite rolls manufactured by ESR cladding present a good bonding between a tough steel axis and all the cladded metal tested, including HSS grades for HSM and CSM. The metallurgical structures of the shell are very fine and homogeneous. The bonding zone is very strong and without defects as long as the process parameters are well adapted and controlled.

The development of this promising technology is highly linked to the development of new performing grades for both CSM and HSM applications. The manufacturing of prototype rolls proved the feasibility of the process and their implementation into different rolling mills is under schedule.

5 REFERENCES

1 C. Gaspard, S. Bataille, D. Batazzi, P. Thonus and A. Magnée, "Forged Semi-HSS and HSS Rolls Designed for Cold Rolling Reduction Mills", 41st MWSP conference proceeding, ISS, vol. XXXVII, 1999.

2 M. Andersson, R. Finnström, T. Nylén, "Development of a HSS Rolls at Åkers International AB", 2nd European Rolling Conference, 2000.

3 E.J. Kerr, R. A. Hishon, W. J. Hill, R. J. Webber, "High Speed Steel Work Rolls at Dofasco's Hot Mill", Rolls 2000+ proceeding, 1999.

4 C.Gaspard, P. Cosse, A. Magnée, "Contribution of E.S.R. and Progressive Induction Hardening to the Manufacture of Deep Hardened Work Rolls", 26th MWSP conference proceeding, ISS, 1984.

5 FORCAST INTERNATIONAL, "Method and Device for Forming a Metal Sheath on an Elongate Core", European Patent EP 0 693 011.

IMPROVEMENT OF THE WORK ROLL PERFORMANCE ON 2050 mm HOT STRIP MILL AT ISCOR VANDERBIJLPARK

AUTHORS: **R.J. Skoczynski**, Senior Engineer, 2050mm HSM, Iscor Vanderbijlpark, South Africa
G. Walmag, Researcher, Centre for Research in Metallurgy (CRM), Liege, Belgium
J.P. Breyer, Director Research & Development, Marichal-Ketin, Liege, Belgium

PRESENTER: **R.J. Skoczynski, G. Walmag**

ABSTRACT

This paper describes how the optimisation of rolling conditions (roll lubrication, strip skin cooling) on the 2050mm Hot Strip Mill at Iscor Vanderbijlpark allows us to realise the full potential of HSS rolls. Based on development work and laboratory studies on the behaviour of oxidation, these improvements benefit both HiCr and HSS rolls proportionately to the roll properties (reduction of rolled-in scale, increase of campaign lengths). The full implementation of HSS in stands F1 to F4 during the year 2000 will also increase the mill availability by eliminating the intermediate roll change and reduce work load in the roll shop.

1. INTRODUCTION

The 2050mm Hot Strip Mill at Iscor Vanderbijlpark produces in the region of 3 million tons of hot rolled coils per annum. During the last few years, in the continuous strive to improve rolling efficiency and strip quality, the work rolls performance in the 7-stand Finishing Mill has been targeted. The work roll performance in the Finishing Mill has a direct impact on the following major cost aspects:

- Strip surface quality, worn work roll scale. Worn work roll scale is caused by the rough surface of the HiCr rolls (which shear and imprint the tertiary scale on the strip) in the first four stands of the Finishing Mill. In 1995/1996 over 18 000 tons (0.6%) of coils were rejected due to worn work roll scale on 2050mm HSM.
- Plant utilisation, lengths of rolling programs. The intermediate Hi-Cr roll changes were required in F1-F4 to ensure proper strip surface quality. Consequently the average length of run of Hi-Cr rolls was only 52Wkm in 1996, and up to 10% of available rolling time was lost due to roll changes.
- Productivity of rolls in ton/mm, roll wear and dressing. The average productive performance of Hi-Cr rolls in Stands F1 to F4 was around 5000 ton/mm due to roll wear and minimum dressing required after each roll run.
- Workload in the Roll Shop, number of work roll changes. Increased number of roll changes resulted in excessive workload in the roll shop. Due to shortened program lengths, over 2200 rolls per month had to be ground in 1996. Consequently quality of the roll preparation deteriorated and additional delays of up to 4 hours per month were recorded in the mill due to the lack of rolls.

After investigations into the wear mechanism of the Hi-Cr work rolls, development work was initiated to simultaneously reduce friction in the roll gap, improve roll cooling conditions and control of the tertiary scale growth on the strip surface. At the same time trials with new grades of wear resistant High Speed Steel (HSS) rolls were conducted.

In 1999/2000 the mill approached full implementation of the HSS rolls. In this article, the performance results of the newest HSS7 roll grade with the use of the strip skin cooling system and roll gap lubrication on 2050mm Hot Strip Mill at Iscor is discussed.

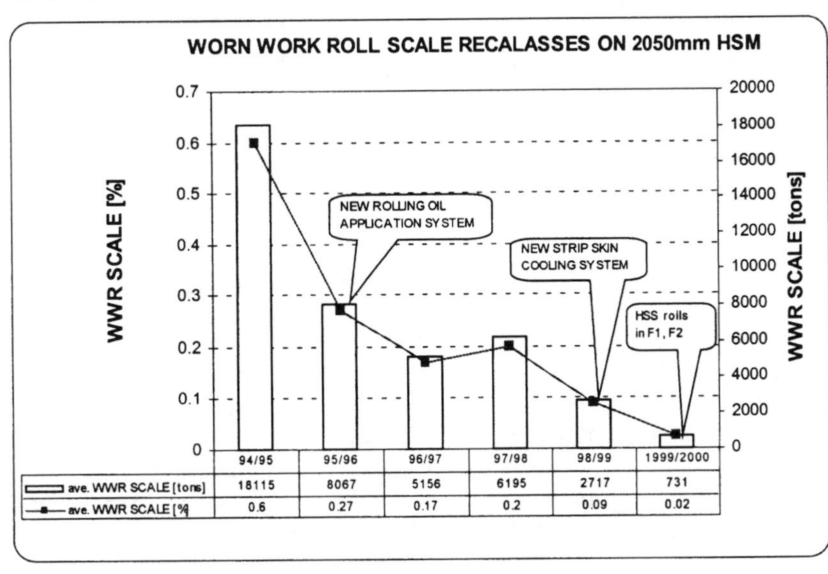

	94/95	95/96	96/97	97/98	98/99	1999/2000
ave. WWR SCALE [tons]	18115	8067	5156	6195	2717	731
ave. WWR SCALE [%]	0.6	0.27	0.17	0.2	0.09	0.02

FIG1. Worn work roll scale reduction.

2. IMPROVEMENT OF ROLLING CONDITIONS

2.1 ROLL GAP LUBRICATION

In 1996, a new roll gap lubrication system (with dedicated rolling oil headers in front of the roll gap) was installed in Stands F1 to F5 with the purpose of reducing friction and roll forces to prevent the sticking of tertiary scale to the roll surface. It resulted in an improved roll surface appearance of Hi-Cr iron rolls and made it possible to increase the length of the rolling programs by 20%. During this period (1996 – 1997) worn work roll scale reclasses decreased by 50 % from 0.6% to <0.3 % **(Fig. 1)**.

2.2 STRIP SKIN COOLING

In 1998 the strip skin cooling system was installed in front of Stands F1, F2 & F3 after descaling, to decrease tertiary scale growth on the strip surface and reduce thermal fatique of the roll surface **(Fig.2)**.

At the entry of the Finishing Mill descaler, the transfer bar temperature is about 1050°C. During descaling the surface temperature drops below 900°C but it quickly recovers above 950°C before entering Stand F1. At the first three/ four stands, the surface temperature drops again below 900°C but recovers above 950°C before entering the next stand.
The objective of the strip skin cooling system is to control the rise of the strip surface temperature before entering Stands F1, F2 and F3 roll gap, with minimal effect on the finishing temperature.

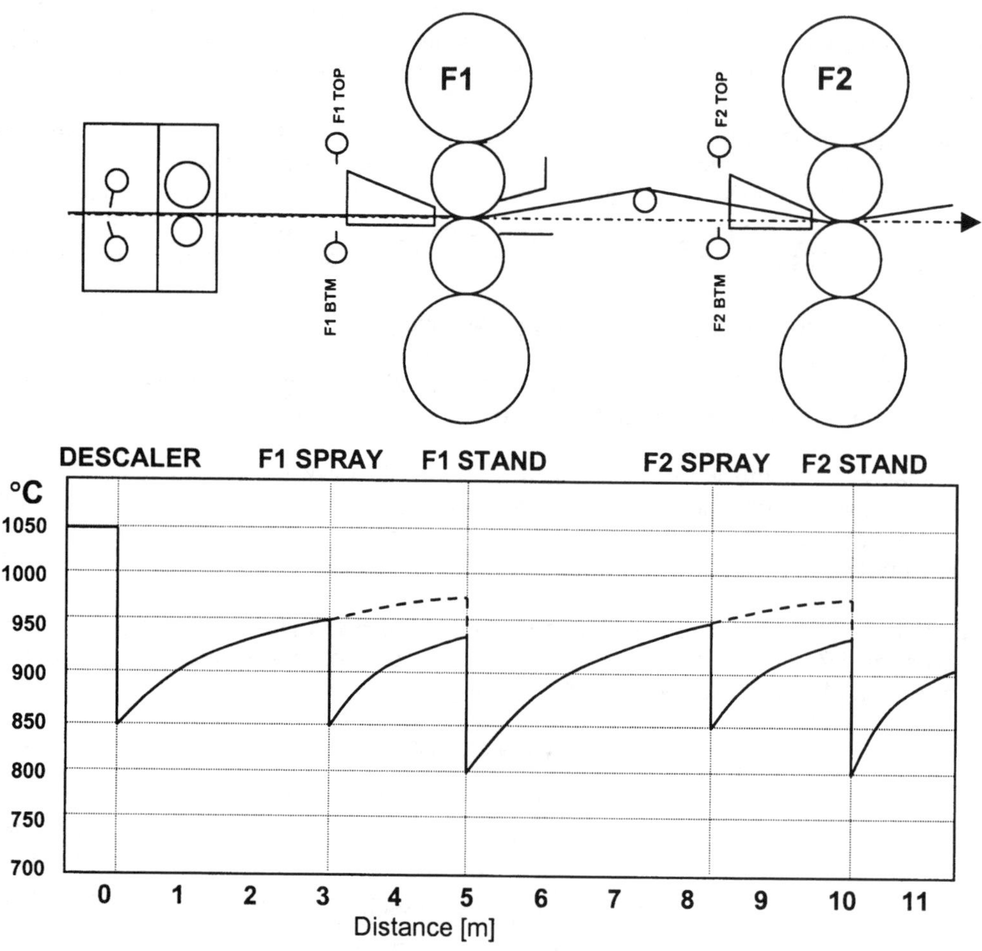

FIG.2 **Strip surface temperature control.**

There is one top and one bottom header at the entry sides of Stands F1, F2 and F3 (about 2 meters before the roll bite). The bottom one sprays onto the bottom strip surface at the side guide entry (in F2 & F3 just after the looper). The top was installed on the mill housing above the side guides.
Each top header was equipped with flat jet, dovetail type nozzles. They ensure uniform parabolic distribution of water with a 30° spray angle. The self-setting dovetail fixes the nozzle position to ensure uniform and constant spray patterns with the flat jet pre-set at 15° to the header axis. The minimum spray overlap is 100% to ensure a uniform cooling and prevent hot strips in a case of a nozzle blockage.

Movement of the side guide shoes is utilised as a cost-effective method to adjust the sprays' width. At the top position, the existing top plates of the side guide shoes control the spray width and protect the strip edges against overcooling. At the bottom position, additional aprons were installed, fastened to the side guide shoes and move together with them above the sides of the bottom header. It prevents spraying to the top of the mill at both sides of the narrower strip and protects the strip edges against overcooling **(Fig.3)**.

FIG.3 **Bottom strip skin cooling at the entry of Stand F2.**

The F1 interstand spray operates on each strip but the F2 and F3 interstand sprays only operate after the Finishing Mill temperature has risen (after strip acceleration during rolling).

Investigations into the wear mechanism of work rolls has led to the conclusions that controlling of the strip surface temperature below 950°C substantially improves the work roll performance, both Hi-Cr iron and High Speed Steel rolls, in the first 3 Finishing Mill stands.
At temperatures below 950°C, the growth of tertiary scale is reduced and it contains much less of the hard hematite (Fe_2O_3) in favour of soft and ductile wustite (FeO) **(Fig.4)**. It results in reduced abrasion in the roll gap and reduced roll forces. The lower strip surface temperature also decreases the roll thermal fatigue and delays the oxide film formation on the roll surface.

The new strip cooling system reduced the strip surface temperatures between Stands F1 & F2 and produced excellent results. It further boosted the Hi-Cr rolls performance in Stands F1 to F4. Consequently the average length of Hi-Cr roll runs increased from 60 to 75Wkm (scheduled 80 -90Wkm) while the worn work roll scale reclasses dropped further by 50%, to <0.1% in 1998/1999 **(Fig.1)**.

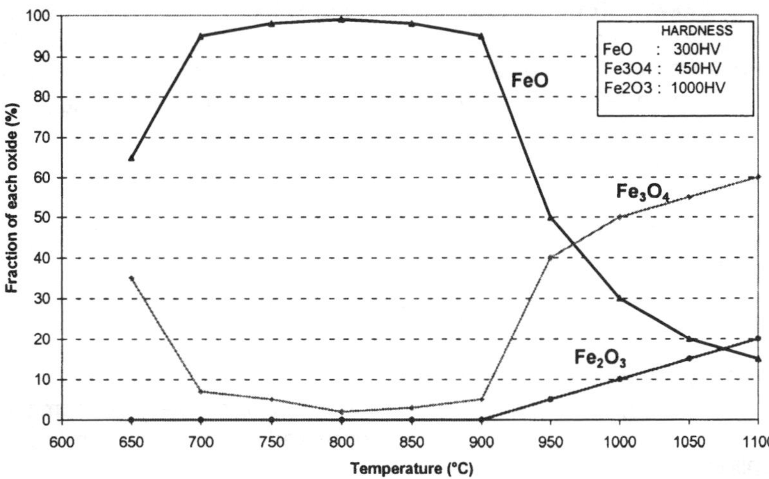

FIG.4 **Iron oxide composition versus the strip surface temperature.**

3. HSS7 ROLLS IMPLEMENTATION

In 1995-1997, trials with different grades of High Speed Steel rolls were conducted in Stands F1 to F4. The initial trials were encouraging but showed various adverse effects on roll performances, which differed very much not only between the roll suppliers but also between the different HSS roll grades from the same roll maker and subject to actual rolling conditions. Thus it became clear that further development of the HSS roll chemistry was essential and improving of the rolling conditions was necessary to realise the full potential of the HSS rolls.
Based on the results of the initial trials at Iscor and research of the Centre of Metallurgical Research (CRM) in Liege, Belgium, a new HSS7 roll grade was developed by Marichal Ketin, Belgium, which proved the best compromise of a low friction coefficient, high fire crack resistance, and very good roll surface appearance at improved rolling conditions on the 2050mm Hot Strip Mill at ISCOR Vanderbijlpark. Thus the HSS7 rolls have been used as a leading grade for the full implementation of the HSS rolls in Stands F1 to F4 on the 2050mm Hot Strip Mill at Iscor in 1999.

3.1 HSS7 ROLLS MICROSTRUCTURE

The composition of HSS7 rolls is shown in table I. In addition to classical alloying elements (C, Cr, Mn, Si, Ni), it contains Vanadium, Molybdenum and Tungsten. These types of rolls contain a small amount of primary VC carbides embedded in a tempered martensite matrix. VC carbides have a high hardness of 3000 HV, and therefore induce very good wear resistance. A second type of carbide, which is a eutectic rod like M2C carbide, is evident. Molybdenum and/or tungsten form these carbides, but they can also dissolve large quantities of chromium. M_2C carbides are softer than the VC carbides, but harder than the M_7C_3 carbides found in the Hi-Cr iron rolls.

Table 1. Composition of HSS7

C	*Cr*	*Mo*	*V*	*W*
1.5-2	5.0-7.0	3.0-4.0	4.0-5.0	1.5-2.5

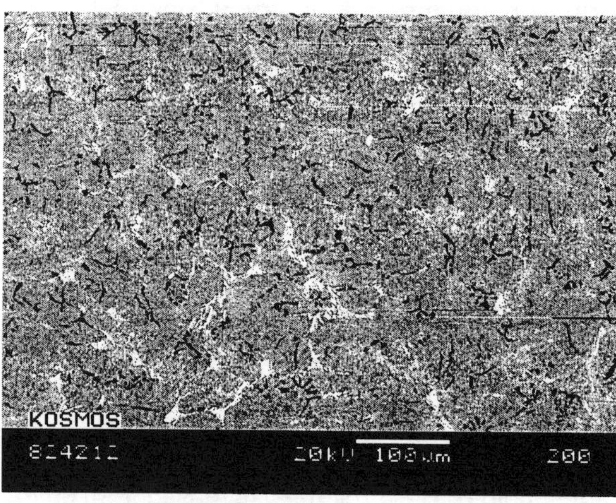

Fig. 5 Typical microstructure of HSS7

A typical microstructure of these rolls is shown in **figure 5** together with the EDAX microanalyses of carbides and matrix. The carbides were viewed with a scanning electron microscope in Back Scattering Electron (BSE) mode. The hard VC carbides appeared in dark grey and the tungsten-molybdenum carbides in white. The amount of carbide forming elements have been optimised in order to reach an adequate level of each type of carbide.

Table 2 Carbides composition of HSS7

Carbides	V	*Mo*	*W*	*Fe*	*Cr*
MC	71	8	9	7	5
M_2C	11	44	21	11	13

The high hardness of VC carbides is very favourable to wear resistance but a too high percentage of this carbide tends to increase the friction coefficient, which is detrimental for the rolling power consumption and for the strip surface quality. Indeed during rolling, the matrix is rapidly worn and the hard VC carbides remain at the surface of the roll and act as spikes.
On the other hand, the amount of M_2C carbides need to be limited to prevent the forming of networks of brittle particles which could increase crack propagation and reduce spalling resistance.

3.2 HSS7 ROLLS OXIDATION KINETICS.

One important parameter to understand of the roll surface degradation is the oxidation kinetics of the roll surface. Static oxidation tests were performed to evaluate the influence of temperature on the oxidation kinetics of the HSS7 grade. The samples were subjected to oxidation during 24 hours under saturated atmosphere of H_2O at temperatures of 575° and 525°C. These conditions simulate the oxidation conditions at the exit side of the roll gap after the surface of the roll has been heated by being in contact with the hot strip, and is cooled down by the sprays of the roll cooling system. The temperature profile at the exit side of the roll gap is described schematically in **figure 6**. The temperatures of 525° and 575°C have been more accurately estimated by numerical modelling as roll surface temperature in stand F2 for austenitic rolling with (525°C) and without strip skin cooling (575°C).

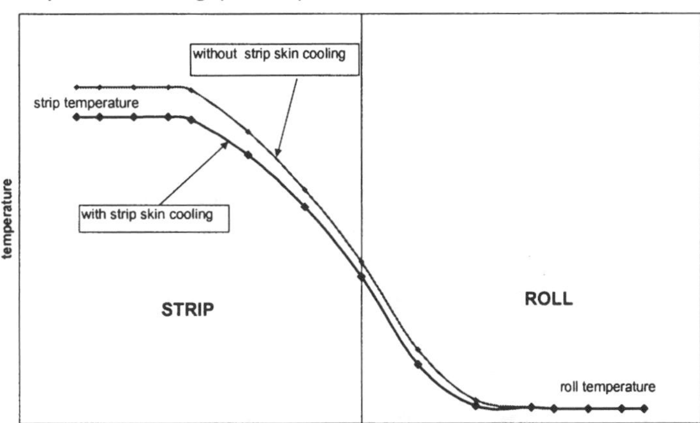

FIG. 6 Temperature distribution at the exit side of the roll gap.

Metallurgical examination of the surface of the samples after oxidation reveals a non-uniform oxidation layer depending on the underlying microstructural component. The matrix and VC carbides are covered by a thin layer of oxides while the eutectic zone of Mo- rich carbides seems only covered by small oxides spot as illustrated by **figure 7**.

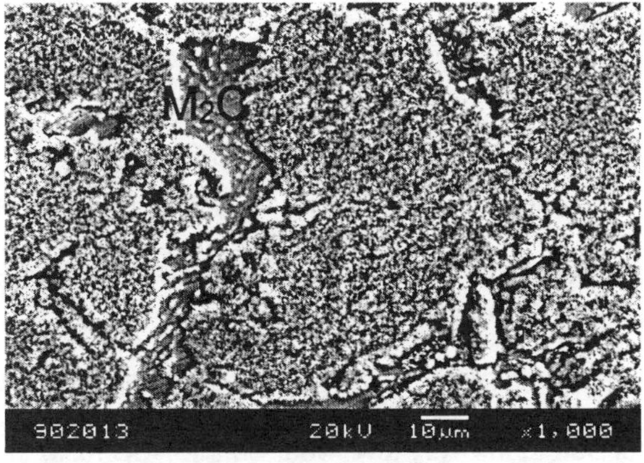

***Fig. 7.* Oxide layer after oxidation test at 525°C/24h**

Evaluation of the oxide layer thickness has been made on a metallographic section and is summarised in **figure 8**. These values mentioned in this graph give the depth of oxidation (in µm) for the three components of the alloy. Whilst the M_2C carbides present virtually no oxidation, MC carbides are oxidised more deeply than the matrix independently from the test temperature **(figure 9)**.

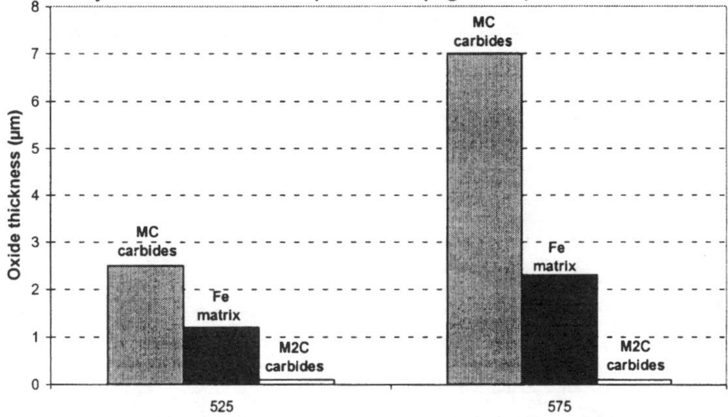

***Fig. 8* Influence of temperature on oxide thickness of HSS7 roll**

These results indicate that oxidation kinetics of the HSS7 are strongly dependent on temperature as a decrease of 50°C of the oxidation temperature allows the reduction of the oxide layer thickness by a factor of 3. These observations add to additional explanations of the influence of strip skin cooling on the roll surface degradation, as it appears that the slight surface strip temperature decrease induces a similar reduction of roll surface temperature but a more drastic reduction of the roll oxidation.

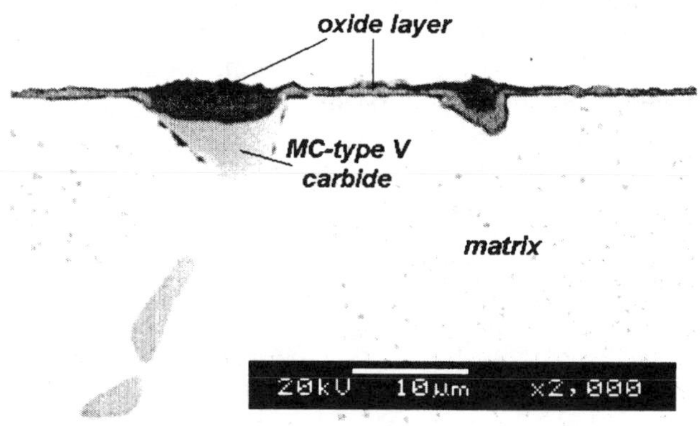

Fig. 9 **Oxidation depth of MC-type V carbides**

3.3 HSS7 ROLLS PERFORMANCE RESULTS

In 1999 the HSS7 rolls were tested in Stands F1 to F4 on the 2050mm HSM at Iscor Vanderbijlpark. They produced excellent results with the use of the roll gap lubrication and strip skin cooling system. Thus they were approved for full implementation in the Finishing Mill. During 1999/2000 financial year they replaced Hi-Cr rolls, first in the F2 stand (where the rolling conditions are most severe) then in F1 (in F3 & F4 they will be fully implemented after the old stock of Hi-Cr rolls has been consumed).
The roll temperatures were at 65° to 85°C, the roll cooling water temperatures from 25° to 35°C with an average rolling rates of 600 t/hour. The product mix included all Carbon-Manganese and Low Alloy steel grades with thicknesses ranging from 1.5mm to 20mm.

3.3.1 Length of roll runs (Figure 10).

Since the HSS rolls have been fully implemented in Stands F1and F2, the intermediate roll changes at these stands have been eliminated. The average length of roll run increased from 75Wkm on the Hi-Cr rolls to 145 Wkm on the HSS rolls, including all short programs and changes due to cobbles or mill shutdowns. Consequently the number of work rolls to be ground decreased by at least 300 rolls/month on the 2050mm Hot Strip Mill during the first half of the year 2000.

The scheduled length of HSS rolls was limited to 160-180 Wkm due to the complete F1-F7 roll change required at the end of each rolling program (due to the strip width coffin and wear profile of the ICDP rolls in Stands F5 to F7). However due to the very low wear on the HSS rolls, they were used in many cases in two consequtive programs totaling up to 320 Wkm without adverse affects on the strip quality.

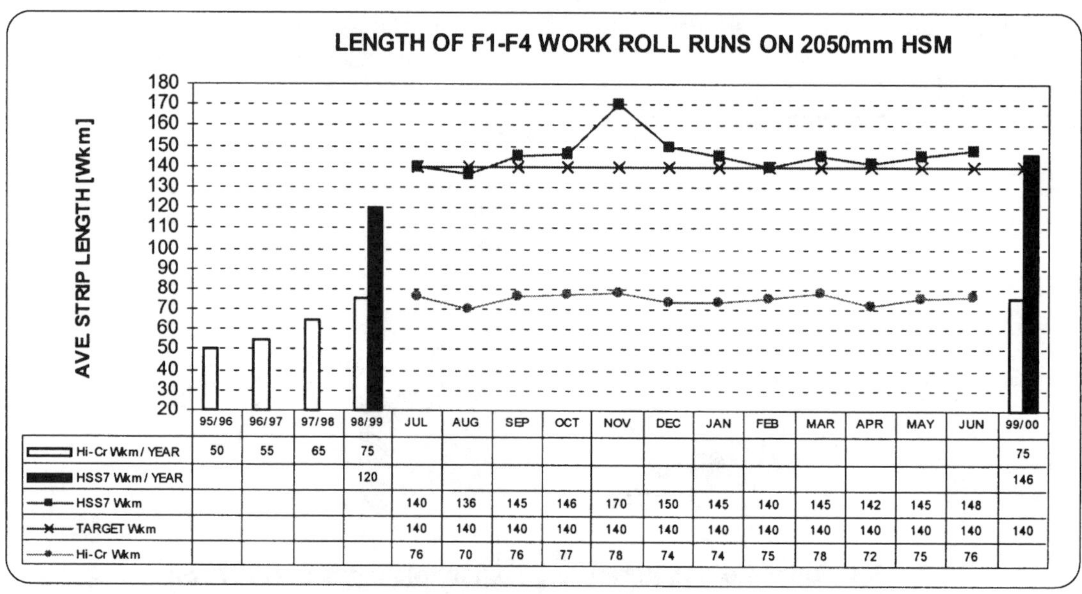

	95/96	96/97	97/98	98/99	JUL	AUG	SEP	OCT	NOV	DEC	JAN	FEB	MAR	APR	MAY	JUN	99/00
Hi-Cr Wkm/ YEAR	50	55	65	75													75
HSS7 Wkm/ YEAR				120													146
HSS7 Wkm					140	136	145	146	170	150	145	140	145	142	145	148	
TARGET Wkm					140	140	140	140	140	140	140	140	140	140	140	140	140
Hi-Cr Wkm					76	70	76	77	78	74	74	75	78	72	75	76	

FIG 10. **Increasing of the work rolls campaign lengths.**

3.3.2 Rolls productive performance (Figure 11).

After six months of the full implementation of HSS rolls in Stand F1 & F2 the cumulative productive HSS roll performance was 12100 ton/mm compared to the average of 5000 ton/mm on Hi-Cr rolls.

The average stock removal was 0.25mm after average rolling of 3000tons per program. The average grinding time of the HSS rolls is about 30% longer than on the Hi-Cr rolls, but with 50% reduced roll change frequency and crane engagement, the real work load in the Roll Shop was reduced by about 30% for F1 & F2 rolls. The average grinding amount on the HSS rolls was lower than on the Hi-Cr rolls and new grinding wheels were tested to improve grinding efficiency and quality.

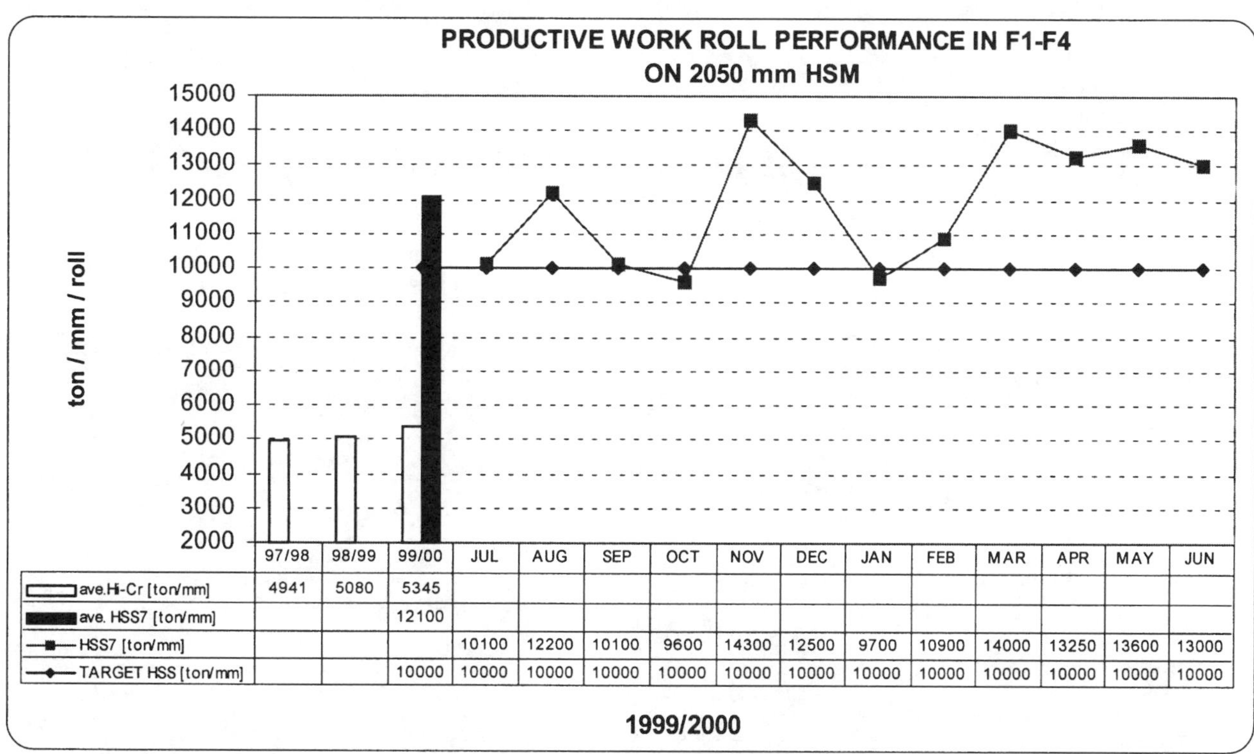

	97/98	98/99	99/00	JUL	AUG	SEP	OCT	NOV	DEC	JAN	FEB	MAR	APR	MAY	JUN
ave.Hi-Cr [ton/mm]	4941	5080	5345												
ave. HSS7 [ton/mm]			12100												
HSS7 [ton/mm]				10100	12200	10100	9600	14300	12500	9700	10900	14000	13250	13600	13000
TARGET HSS [ton/mm]			10000	10000	10000	10000	10000	10000	10000	10000	10000	10000	10000	10000	10000

***FIG 11.* Improving of the work rolls productive performance.**

3.3.3 Roll surface appearance (Figure 12-17).

The HSS7 rolls always appeared smooth, covered with a uniform layer of black oxide film in Stands F1 & F2. Local losses of the oxide film were quickly healed leaving again a smooth black surface with light grooves. Peeling over the old fire cracks lines was observed after extended roll runs (over 200Wkm) but without adverse effects on the strip quality.

In Stands F3 & F4 the HSS7 rolls appeared smooth grey with remains of the black oxide film uniformly scattered over the roll surface. No orange peel which is typical for the Hi-Cr rolls was observed.

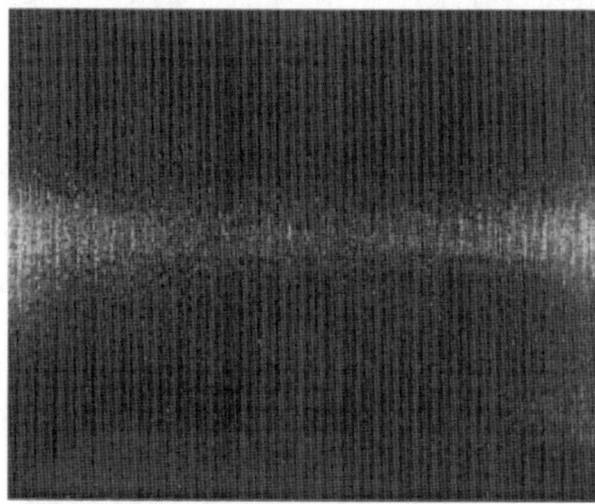

FIG 12. HSS7 roll surface in Stand F1 after rolling of 6000 tons (300Wkm) of strip in 1999 (with the strip skin cooling).

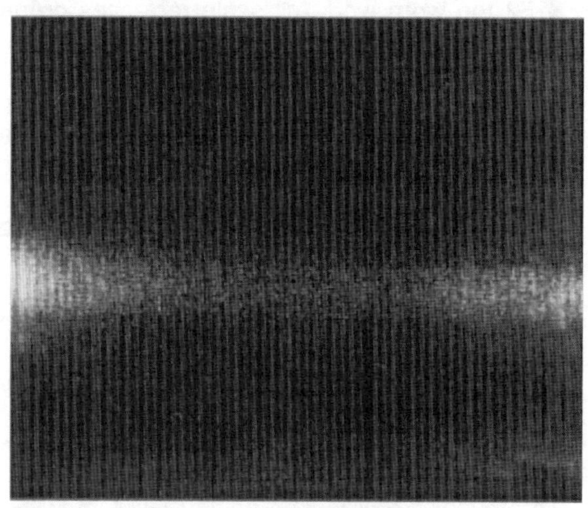

FIG 14. Hi-Cr roll surface in Stand F1 after rolling of 1500 tons (80Wkm) of strip in 1998 (with the strip skin cooling).

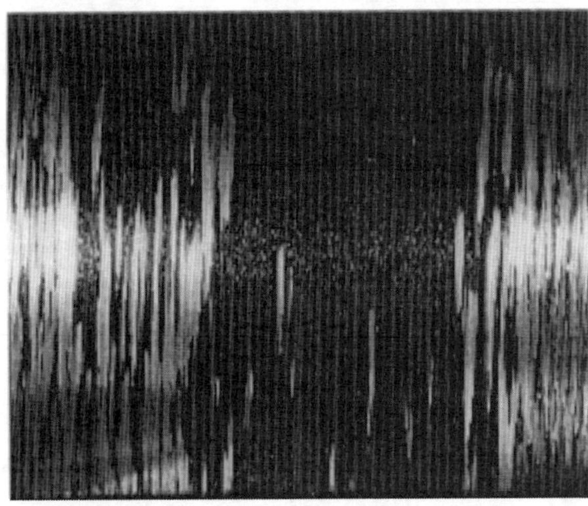

FIG 13. HSS5 roll surface in Stand F1 after rolling of 3000 tons (160Wkm) of strip in 1997 (without the strip skin cooling)

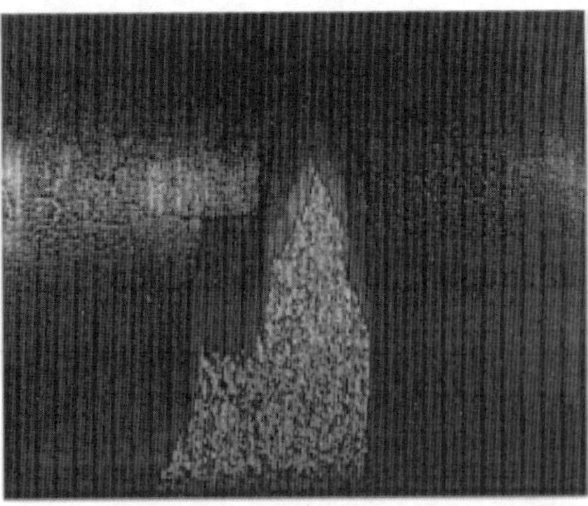

FIG 15. Hi-Cr roll surface in Stand F1 after rolling of 1400 tons (70Wkm) of strip in 1995 (without the strip skin cooling).

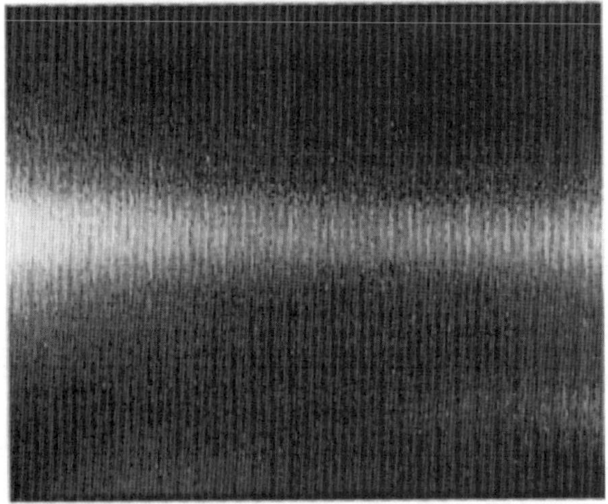

FIG 16. HSS7 roll surface in Stand F3 (bottom) after rolling of 3000 tons (160Wkm) of strip in 1999.

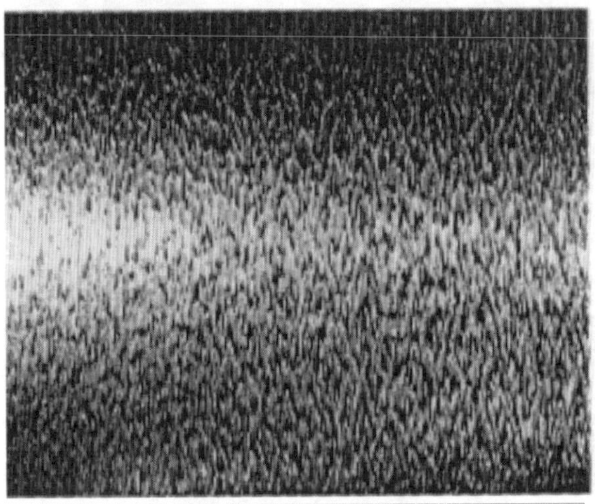

FIG 17. Hi-Cr roll surface in Stand F3 (bottom) after rolling of 1400 tons (70Wkm) of strip in 1995.

3.3.4 Strip surface quality (Figure 1).

The smoother roll surface appearance resulted in improved strip surface quality. There was no major scale problem which was related to peeling of the HSS rolls. During implementation of the HSS7 rolls in Stands F1 & F2 the worn work roll scale reclasses dropped more than 3 times, from 0.09% in 1998/1999 to below 0.02% in 1999/2000. This was in spite of the increased length of roll runs.

3.3.5 Fire crack resistance.

Each of the HSS rolls was involved in several cobbles in which the hot strip stopped on the rolls. Consequently transverse bars of fine fire cracks were noted on all the HSS rolls. However these fire cracks were finer, about half in size and depth of Hi-Cr rolls and did not show a tendency to grow.

Only one set of the HSS7 rolls in Stand F2 was sent to the lathe (after a heavy cobble) where it was cut down by 8mm (on diameter) to eliminate the fire cracks entirely.

The other HSS7 rolls, which also developed fire cracks, were used again after normal dressing and it was noted that the old fire cracks disappeared after several sequences of normal dressing.

At the same time, most of the Hi-Cr rolls in Stand F1 developed up to 4 fire cracks transverse bars per roll. They were 8 to 20mm deep and were opening after single runs. Thus the Hi-Cr rolls had to be withdrawn from circulation in Stand F1 and replaced with the HSS rolls since January 2000.

To prevent fire cracks, a new practice of HSS roll changes after cobbles, has been introduced. If the roll stops on the hot strip, the water is shut immediately and the screws are opened automatically, then the roll set is changed and kept in front of the mill to be put back in again during the next roll change. This was done to prevent sudden cooling of the overheated area when the rolling was resumed. No fire cracks were found on the HSS rolls if the rolls were changed according to this rule.

3.3.6 Friction coefficient, chattering in the mill.

The HSS7 rolls did not show force increase with or without using roll gap lubrication. There was no chattering in the mill, which was common with the earlier grades of HSS rolls. During rolling of strip gauges below 2.0mm no increase in rolling instability was reported after using the HSS7 rolls.

3.3.7 Summary of advantages.

The following advantages can be derived after full implementation of the HSS rolls in Stands F1 to F4 :

- Elimination of the intermediate F1 to F4 roll changes, increased length of F1 to F4 roll runs above 160 Wkm, permitting an extra rolling time of about 10 hours/month.
- Elimination of worn work roll scale defects, permitting the recovery of more than 2000 tons of prime material annually.
- Reduced workload in the Roll Shop in the region of 25% due to the reduced number of roll changes in Stands F1 to F4. Reduced the number of rolls to be ground by 600 per month. Reduced number of rolls in circulation by 50%.
- Reduced roll costs due to increased Stand F1 to F4 work roll performances, from less than 5000tons/mm on Hi-Cr rolls to above 12000 ton/mm with HSS rolls.

4. CONCLUSIONS.

Investigations into the wear mechanism of work rolls and laboratory research led to the conclusions that controlling of the strip surface temperature below 950°C substantially improves both the HiCr and HSS work roll performance in the Finishing Mill.

The HSS7 roll grade proven to be a good compromise between wear resistance, fire crack resistance, and surface appearance. With use of the strip skin cooling and the roll gap lubrication the HSS rolls more than doubled better performance results of the Hi-Cr rolls in Stands F1 to F4 on the 2050mm Hot Strip Mill at Iscor Vanderbijlpark. This contributed significantly to the increased mill productivity, improved strip surface quality, and reduced costs.

TECHNOLOGY ENHANCED WORK ROLLS FOR THE ROUGHING MILL APPLICATION

Lewis J. Prenni, Jr.,
Jim McGregor
National Roll Company
A Member of the ÅKERS Group
400 Railroad Avenue
Avonmore, Pennsylvania 15681
USA
724 697-4533
ljprenni@natroll.com
jmcgre2138@aol.com

INTRODUCTION

Chrome III (High Chrome Steel) Work Rolls for Roughing Mill Four High Stands have been successfully manufactured and supplied since 1983. Since the start of production, over 2100 rolls of this type have been shipped to customers worldwide.

In 1997, the Technical Department formulated a new chemistry for use in Roughing Mill Stands, where there was a need for increased roll performance due to the difficult rolling application. This grade, designated Chrome VI (Semi-HSS) Steel, combined the benefits of both the original Chrome III grade and the highly successful High Speed Steel (HSS, Chrome V) grade used in Finishing Mill Stands. Production trials commenced in 1998 and today full-scale production is in place at National Roll Company. The Chrome VI grade has been primarily used in the reversing roughing and four high roughing applications. Figure 1 shows the first three years ordering pattern and the projected supply for the fourth year of the semi-HSS roll grade. Over the past two years the two high roughing applications have been successfully converted over to chrome steel (Chrome III) material with trials commencing with the Chrome VI semi-HSS product.

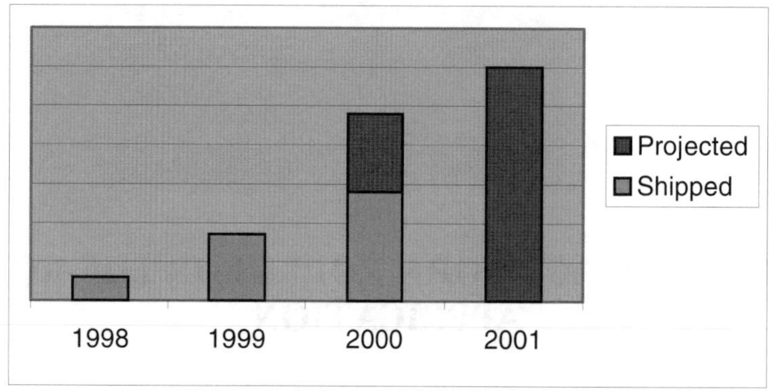

Figure 1 Ordering pattern for Chrome VI rolls.

CONDITIONS OF THE ROUGHING APPLICATION

The mechanism for wear in the roughing mill is thermal deterioration. The work rolls of the roughing mill experience the hottest bar temperature at a slow rotational speed. This factor combined with a heavy reduction, 35% - 50%, of a relatively thick slab equate to significant heat transfer into the roll. This heat transfer, combined with long gap times, produces a cyclical heating and cooling that the rolls must endure. The rolls must also be able to withstand stall conditions and roll without a slippage problem. The final strip profile is also directly related to the roughing mill. A change in the profile of the roughing mill work roll may have a detrimental effect on the cross-section profile of the transfer bar as well as the final strip profile.

Due to the aforementioned factors, it was realized that improvements made in the roughing mill positively affect mill productivity. Optimizing the cost per ton ratio while improving or maintaining the strip quality is the main objective. Optimizing the ratio can be achieved in different ways. Roll users are looking for a roll material that can remain in the mill for extended campaigns while having less wear. Ideally it would also require less stock removal upon redress. Roll changes in the roughing stands can be lengthy therefore the ability to decrease the number of roll changes would be a definite benefit. To realize these goals the roll material must have:

- Resistance to Thermal Cracking
- Controlled Oxidation Characteristics
- Resistance to Wear and Thermal Fatigue
- Sustained Surface Quality over Extended Campaigns
- High Threshold of Safety Against any Mill Induced Damage

EVOLUTION OF ROLL MATERIALS FOR THE ROUGHING STANDS

Since the inception of the hot strip roughing mill, there have been many different roll grades that have been and still are being used. Table I contains microstructures of some of the most common roughing mill roll grades. Table II is an abbreviated comparison of chemical analysis and hardness for the different grades.

Table I. Microstructures of roughing mill roll grades

Material	50X	200X
Adamite	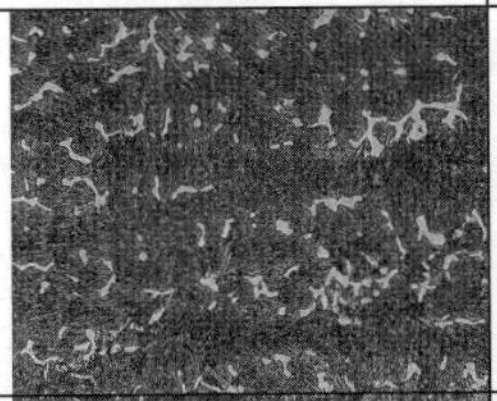	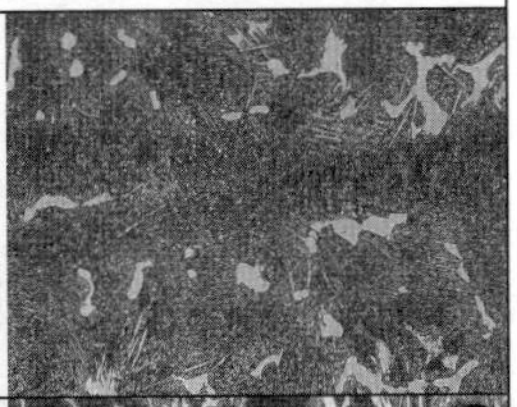
High Chrome Iron (Chrome II)	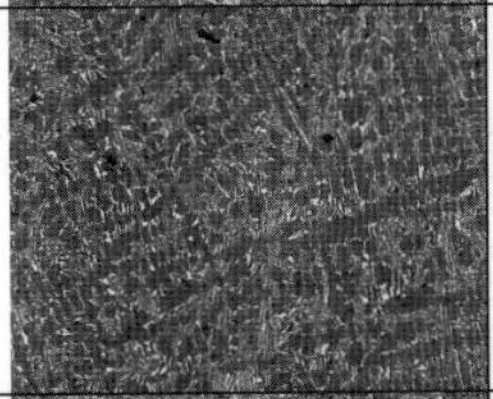	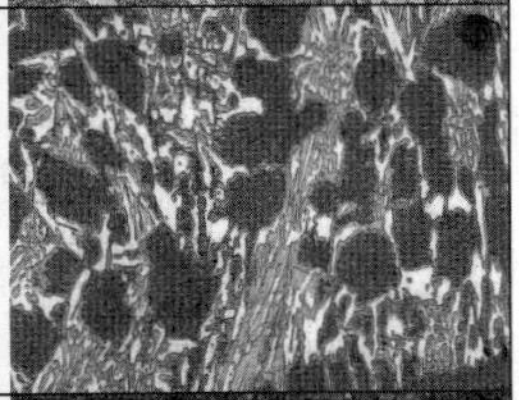
Chrome Steel (Chrome III)		
Semi-HSS (Chrome VI)		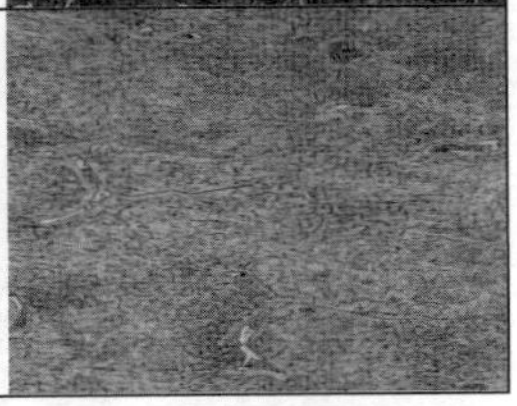

Table II Analysis and hardness of roughing mill roll grades

Grade	% C	% Cr	% Mo+V+W+Nb	Hardness Shore C
Adamite	1.5-2.5	1.0-2.0	0.5-1.0	50-65
Grain Iron	3.0-3.4	1.00-1.50	0.1-0.6	55-70
Cr II (High Chrome Iron)	2.0-3.0	15-25	0.5-2.0	73-83
Cr III (Chrome Steel)	1.0-1.8	10-14	1-4	70-80
Cr VI (Semi-HSS)	0.4-1.2	3-10	2-10	80-90

Adamite steel was one of first roll grades to be used. This roll had adequate bite and a sufficient safety factor. It was, however, prone to significant wear and coarse fire cracking. This in turn resulted in less than optimum strip surface quality. Grain iron is also used in the roughing application. This roll grade slightly improved the wear resistance with iron carbide. Some time later high chrome iron (Chrome II) was introduced. The microstructure consists of eutectic M_7C_3 carbide with finely distributed secondary carbides in a matrix of tempered martensite. The microstructure provided higher bulk hardness. This roll grade provided greater fire crack resistance, improved wear, and allowed for extended campaigns. However, due to the high amount of eutectic carbide, 20%-30%, the rolls were prone to slippage problems, and under today's expectations would require frequent roll changes. The chrome steel (Chrome III) roll is the most widely used roughing work roll in hot strip mills today. Its steel shell material provides greater mechanical strength than prior grades. At the time this roll grade exhibited the most beneficial properties. However, there was still constant pressure to increase roll performance. In order to achieve this goal a new material was developed, Semi-HSS (Chrome VI).

MANUFACTURE AND METALLURGY OF SEMI-HSS ROLLS (CR VI)

Chrome VI is a double poured roll consisting of a semi-HSS shell material and a nodular graphite core. The range of chemical analysis is shown in Table III.

Table III Typical minimum and maximum ranges of a semi-HSS roll (wt%)

	C	Mn	Si	Cr	Ni	Mo+W+V+Nb
Shell	0.4- 1.2	0.40- 0.90	1.00- 1.30	3.00- 10.00	0.50- 0.80	2.00- 10.00
Core	2.50- 3.50	2.20- 3.50	0.20- 0.70	0.10- 0.60	0.50- 1.80	Res.

The chemistry and heat treatment of a material dictate its microstructure. Each is critical in the manufacture of a semi-HSS roll. The chemistry for the semi-HSS roll has a significant effect on the resulting microstructure and material properties.

All semi-HSS rolls are vertical centrifugally cast. The "as cast" structure consists of pearlite, carbide, and some retained austenite. This can be seen in Figure 2.

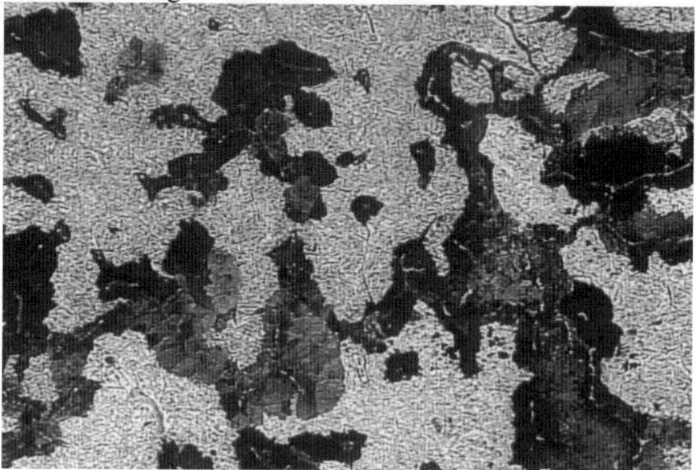

Figure 2 Photomicrograph of as cast semi-HSS 50X

A critical feature of the casting process is the ability to consistently produce a high integrity bond between two dissimilar materials(1). See Figure 3.

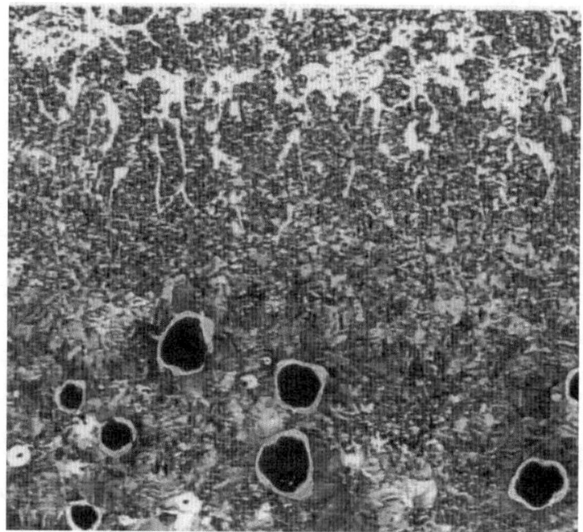

Figure 3 Photomicrograph of the shell/core interface

The large difference between the liquidus and solidus temperatures of these materials initially caused difficulty in producing a high integrity bond. This bonding problem required many unique solutions. There were numerous casting changes made in order to consistently produce a roll with a sound interface. The ability to produce a high integrity bond is paramount to the performance and subsequent success of this roll.

The core material is often overlooked when the discussing the metallurgy of rolls. However, it is important to produce a ferritic/pearlitic nodular core free from iron carbide. As a result of producing a ferritic/pearlitic nodular core, the final hardness of the roll necks is generally in the range of 30 – 43 shore C. Ductility in the core is essential in this thermal roughing application. Therefore, a compromise between the ferrite content and the hardness of the bearing areas (necks) has to be made. The necks of roughing rolls fall into a diameter range of 762mm-1092mm (30-43"). Obtaining good nodular iron with consistent properties in the range of 39 - 46kg/mm^2 (55-65 KSI) in such large sections is metallurgically complex. Special manufacturing techniques are required[(1)]. A typical core microstructure is shown in Figure 4.

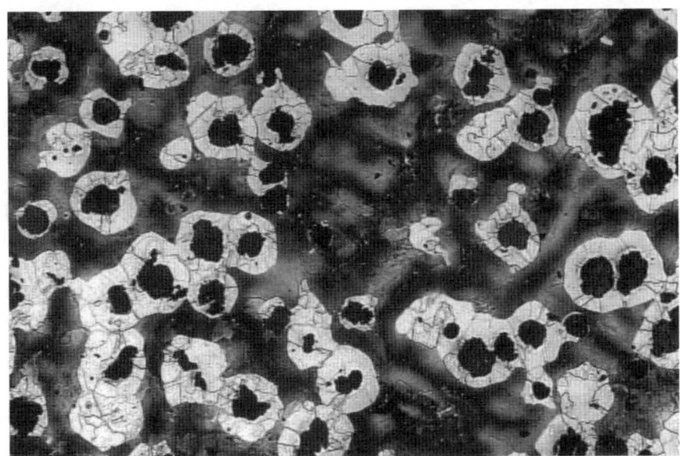

Figure 4 Photomicrograph of Nodular Iron

A Simplified process flow outlining the manufacture of Semi-HSS is shown in Table 4. It should be noted that feedback, both internal and external, is vital in the manufacture of rolls.

Table IV Simplified Process Flow for Semi-HSS Rolls

Roll Manufacture	Process Control	Inspection
Customer Order	Determine Manufacturing Specification	Review Shop Order
Pitting	Chill Prep Chill Wash	Temperature Baume, Amount
Melting	Furnace Control	Chemistry Verification
Casting	Weights Ladle Additions RPMs (G-Force) Pouring Sequence Shell pour Core Pour Casting Temperatures	Visual Inspections
Pre Heat Treatment QC	Interface Integrity	Ultrasonic Examination
Heat Treatment	Heat treatment Temperature Specification Austenitize Quench Equalize Temper	Furnace Uniformity Post heat treatment QC
Final Machining	Finish Print Surface Finish	Dimensional Measurements Hardness Magne Gauge Readings Ultrasonic Examination
Shipment	Packing Requirements Quality Documentation	Visual Inspection
QA	Retention of Quality Records Performance in Service	Retention of Quality Records Bi-Annual R&D Meetings

The as-cast structure of a semi-HSS roll is a relatively soft, easily machined material. It is the transformation of the microstructure by heat treatment that determines the material's ultimate properties. The heat treatment consists of austenization at high temperature to put carbon into solution, quenching at a rate sufficient enough to avoid pearlite formation, followed by equalization and tempering treatments. The temper cycle conditions the newly formed martensite and through the precipitation of secondary carbides destabilizes the austenite so that it transforms to martensite upon cooling to room temperature. The resultant microstructure is shown in Figure 5.

Figure 5 Photomicrograph of semi-HSS shell material 200X

The microstructure is discontinuous primary and finely precipitated secondary carbides of MC, M_2C, and some M_7C_3 carbides in a matrix of tempered martensite. The higher bulk hardness of this material can be attributed to the higher hardness of the matrix and carbides. The micro hardness of matrix and carbide constituents of roll grades in comparison to other roll grades can be seen in Table V[(2)]. The discontinuous carbide is the important feature of the semi-HSS roll.

Table V Micro hardness of matrix and carbides

			Micro hardness (Vickers)
Matrix		Bainite Martensite	250-650 500-1000
Carbide-	Indefinite Chill Chrome Iron Chrome Steel	M_3C M_7C Mo_2C	850-1100 1200-1600 1400-1700
Carbide HSS		M_6C M_2C MC	1100-1650 2000-2400 2400-3200

The heat treatment also establishes the internal residual stress of the rolls. A post heat treatment stress study was conducted using the strain gauge hole drilling technique and acceptability limits established for this roll grade. The shell material is in compressive stress, which contributes an increased resistance to fire cracking.

ENSURING THE QUALITY OF THE SEMI-HSS ROLL

As previously mentioned the semi-HSS roll grade is challenging to manufacture. There are many in-process quality checks throughout the manufacturing process. Drag neck samples are taken and analyzed. The percentage of ferrite and the nodule count is measured. The core structure is also analyzed for alloy content in order to reference remelt of the shell. Drag neck samples must meet the acceptance criteria before further processing.

The rolls go through a number of ultrasonic examinations. The first examination, done with a 2.25MHz transducer, is a bond integrity check. It is performed prior to heat treatment. After heat treatment the bond integrity is again checked. No multiple back reflections that are detrimental to the bond integrity or internal structure of the roll are accepted. The roll is also subjected to an end-to-end check where the ultrasonic signal is sent through the roll in the horizontal axis. This check is performed to find any abnormalities in the necks or core of the roll. Shell depth is also checked with a 2.25MHz transducer to insure a sufficient working layer. A 70° angle probe of the shell is also carried out to identify any discontinuities that may be oriented perpendicular to the rolling surface.

In addition to the ultra sonic examinations, hardness and Magne Gage tests are performed. The hardness test verifies the heat treatment and compliance to customer specification. The Magne Gage numbers are correlated to X-ray diffraction studies to verify low retained austenite levels within the rolls. Specific quality documentation is provided at the customer's request.

PERFORMANCE OF THE SEMI-HSS ROLLS

The semi-HSS rolls are performing on average 1.5 - 2.5 times better than the standard chrome steel grade. The increased performance of theses rolls can also be inferred by the increase production rates as previously shown. Typical performance numbers are shown in Table VI. Actual performance data is shown in Table VII. Figure 6 shows a roll surface after a 70,000 Ton campaign in the R5 stand.

Table VI Typical performance numbers

	Campaign Life	Stock Redressing
Hi Chrome Steel	L	S
Semi-HSS	1.5 L – 2.5L	0.3S – 0.6S

Table VII Actual performance data

	Stand	T/mm (All Stock)	T/Campaign	Multiple Vs. Cr Steel
Company A	R5	44,568	44,901	1.66
Company B	R6	39,843	101,192	---
Company C	Rev	66,656	76,188	---
Company D	R5	29,837	48,075	2
Company E	Rev	55,366	---	1.5
Company F	Rev	14,560	---	2.1
Company G	R5	26,601	65,672	2
Company H	Rev	17,140	---	1.4

Figure 6 Semi-HSS roll surface after a 70,000 Ton campaign (R5)

Due to the excellent wear resistance of the shell material, increased productivity can be realized in the mill. The roll maintains an excellent profile and therefore the ability to extend the campaign length. Productivity can be further increased due to reduced stock removal per dressing. Strip quality is also enhanced due to the excellent surface finish and the ability to maintain the roll profile during extended campaigns.

It is generally accepted that for a given hardness the likeliness for slippage to occur increases as the carbide content increases(1). The semi-HSS roll material has a discrete carbide network and a relatively low carbide content. Therefore, semi-HSS rolls should have favorable bite characteristics. With the low carbon content and high chrome levels this is somewhat a stainless material. It has a high resistance to oxidation and tends to form a thin oxide layer. These characteristics may also contribute to the excellent roll bite characteristics of this roll. There have not been any reports of a slippage problem when using the Semi-HSS roll grade.

Stall cracks tend to propagate along the carbide network. Due to the discontinuous eutectic carbide network exhibited in the semi-HSS microstructure, less stock removal may be needed following a stall condition. This in turn relates to increased roll life.

OPERATING CONSIDERATIONS OF SEMI-HSS

It is recommended to introduce the semi-HSS rolls into the mill with the same operating conditions as the current rolls. Roll surface, water volumes, and roll temperature as well as other factors should then be studied and systematically adjusted to increase the performance of the Semi-HSS roll. It has been learned that when a mill is correctly operating a chrome steel roll they can directly substitute a semi-HSS roll and realize the performance benefits from it. As in the use of chrome steel the temperature of the roll measured approximately 30 minutes after a roll change should be in the range of 60°C- 80°C (140°F-175°F). The water volume applied to the roll should be adjusted so that a uniform light to medium blue colored oxide forms on the roll surface. It is generally accepted that when a dark blue to black color oxide is formed the oxide layer is relatively thick and could be prone to peel.

If for any accidental reason the rolls are used with out water for a period of time, it is essential that they be removed from the mill and allowed to cool slowly. Appling water to a hot roll will result in a catastrophic failure (1).

THE TWO HIGH ROUGHING APPLICATION

The technology applied to the two high application has lagged the four high and the reversing stands. The chrome steel grade for the two high is now being used. In the applications where the chrome steel is being used performance has increased 1.5 - 3 times the conventional grades. The same operating principles apply to the two high stands as they do to the late finishing stands. Due to slow rotation speeds the heat transferred is greater as compared to the later roughing stands. An oxide layer must be developed and controlled on the roll to increase performance. Some consideration for the longer gap times must be taken into account. As a result a slightly lower roll temperature may be anticipated. It is fully expected that once mill operations are comfortable with the use of the chrome steel grade, semi-HSS rolls will be used and performance increased.

CONCLUSIONS

Roll users and roll makers alike are becoming increasingly aware of the effects that the roughing stands have on overall productivity and strip quality. More attention and emphasis are being put on the operating conditions of the roughing stands. It can be concluded that the Semi-HSS roll grade is becoming the preferred roll grade for the roughing mill work roll. The average performance improvement is 1.5 - 2.5 times that of chrome steel. The added performance equates to decreased costs through improved roll life and less down time. The quality of the strip is also improved via excellent roll profile and improved surface quality. Future research is being devoted to further understand and improve roll grades for the roughing mill application.

ACKNOWLEDGEMENTS

The authors would like to thank their colleagues and all the employees of National Roll Company for making this paper possible. Specifically, Tom Adams, David Collins, and Andy Blaskovich for their input and direction.

REFERENCES

1. David B. Collins, "Current Experiences in the Use of High Chromium Steel Roughing Rolls For Hot Strip Mills," May, 1991
2. T. A. Adams and D. B. Collins, "Properties of Hot Strip Mill Rolls and Rolling," 40th Mechanical Working and Steel Processing Conference Proceedings, VolXXXVI October 1998, pp. 427-431

MICROSTRUCTURE, MECHANICAL PROPERTIES AND WEAR RESISTANCE OF HIGH SPEED STEEL ROLLS FOR HOT ROLLING MILLS

Miguel Angelo de Carvalho[1]
Renato Rufino Xavier[1]
Celso da Silva Pontes Filho[1]
Carlos Morone[1]
Mario Bocallini Jr[2]
Amilton Sinatora[3]

(1) Metallurgical Engineer
Rolling Mills Rolls Unit
Aços Villares S.A.
(2) Senior Researcher
IPT – São Paulo
(3) Ph.D.
Universidade de São Paulo

Rod. Luiz Dumont Villares, Km2
12420-0000 Pindamonhangaba, SP - Brazil
Tel.: 55 12 240 8380 / Fax: 55 12 240 8348
E-mail: rxavier@villares.com.br

Key Words: rolls, high-speed steels, hot rolling

ABSTRACT

The performance of work rolls for hot-strip mills is basically evaluated by the extent of their life and by the surface quality of the rolled products, both related to properties like strength, toughness and wear resistance. Those properties, otherwise, are determined by the microstructural characteristics of the HSS for rolls, such as type, morphology, size, volume fraction and distribution of the eutectic carbides. The overall features of the solidification and microstructure of the HSS for rolls are analyzed. The results of mechanical, abrasion wear and oxidation testing obtained with test pieces taken from three different centrifugally cast rolls are presented and discussed.

1. INTRODUCTION

The hot strip mills present a continuous evolution being the quality of strips and productivity considered as the most important rolling aspects. The strips quality is determined mainly by its form and roughness which strongly depend on the geometry and surface of the work roll.

The work roll preparation for service and the evolution of its geometry, as well as the state of the surface due to the rolling efforts, will affect the strip quality. Similarly, the rolling mill productivity is directly related to the period of time the work roll can stand without being removed due to surface deterioration. Such deterioration is generically known as wear, although comprising the oxidation and thermal fatigue phenomena.

As a consequence the work rolls for hot strip mills have been changing in a very dynamic way, with the introduction of continued material improvements and development of new materials. Summary of the evolution process of the materials used in the work rolls for such application, with emphasis on the functional characteristics and performance of the intended innovation, now understood as the result of the material utilized in the roll production, is presented as follows.

The finishing train of the hot strip mill is formed by 6 or 7 stands, responsible by the successive strip reductions. There was an evolution of materials for work rolls in the first stands of the hot strip mills. Cast steels designated by Adamite replaced the indefinite cast iron used in all the stands. The Adamite rolls are practically out of use for that application, having been replaced by high-chromium iron rolls at 2,5 to 3% carbon and 14 to 18% chromium with molybdenum and nickel contents necessary to provide hardenability and mechanical resistance. Nowadays, high-chromium rolls have been replaced by high-speed steel rolls with a composition similar to high-speed tool steels.

The materials used in work rolls, range of compositions, microstructure and respective hardness are shown in table I. It has been noticed a great evolution regarding the development of materials that combine growing wear resistance and toughness able to bear the rolling efforts.

Table I: Chemical composition, microstructure and hardness of roll materials

Material	Chemical Composition (wt %)	Microstructure	Hardness HSc – ASTM E 140	Reference
Indefinite	C 3,0 – 3,4 Ni 4 – 5 Cr < 2 Mo < 1	Fe_3C Bainite Intercellular graphite	70/85	1
Adamite	C 0,7 – 1,4 Ni 1,2 – 1,5 Cr < 2 Mo < 1	Tempered martensite Pearlite	45/60	2
High-chromium iron	C 1 – 3 Ni 1 – 2 Cr 10 – 25 Mo 1 – 3	M_7C_3 Tempered martensite	70/90	1
High-speed steel	C 1,5 – 2,4 Cr 2 – 10 Mo 2 – 10 V 2 – 10 W 2 – 10 Co <10	MC + M_6C Tempered martensite	80/90	1

The evolution has always looked for improvement of the wear resistance, which is the property responsible for the maintenance of the form and the state of the work roll surface. This characteristic is mostly associated to the strip quality and rolling mill productivity.

Two economic advantages can be expected for the high-speed steel rolls in comparison to others: increased performance and campaign. The preliminary results in Japan and United States have indicated that the performance of high-speed steel rolls is about three times superior to that shown by high-chromium rolls. The largest gains occur due to the improvement of the strip quality and to the increase of the rolling mill productivity, mainly due to the longer permanence of the roll in the stand, thus minimizing the period of time for change. This time is very significant and, consequently, its reduction is an obvious factor to be explored in order to increase the users' revenue.

1.1 Microstructure

Table I shows that high-speed steels and high-chromium irons have tempered martensite matrices. A possible difference between matrices is the precipitation hardening of the high-speed steels due to the higher content of molybdenum in relation to high-chromium irons. However, the most significant microstructural difference is in the carbides formed during the solidification.

The carbide in the high-chromium iron is M_7C_3, with hardness ranging from 1300 to 1800HV, according to the literature[3]. The carbides in the high-speed steel rolls show hardness values as high as 2500HV[4]. Carbides MC, M_6C and M_2C[5] are formed with hardness between 2000 and 3000HV[1].

1.2 Properties

The higher high-speed steels resistance to wear, as compared to high-chromium irons, is attributed to the presence of carbides harder than the M_7C_3 [6,7,8]. The nature of the wear presented by the work rolls for hot strip mill can be divided in[9]: "banding" (peeling of the oxide film formed during the rolling process taking some roll metal with it), abrasion (removal of the material by scratching) and roughing of the roll surface due to a combined action of thermal fatigue and contact fatigue caused by the action of the back up roll.

The thermal fatigue resistance of the work rolls is also related to their composition and microstructure, since these parameters control the thermal conductivity, the Young modulus and the rupture stress of the material. The better performance of the high-speed steel rolls comparing to the high-chromium iron rolls is due to the smaller depth of the firecracks[10].

Other properties such as fracture toughness, crack growth rate, yield and strength stresses are also important parameters for qualification of materials. The high-speed steel rolls rupture stresses reach 900MPa against 780MPa of the high-chromium iron rolls, with similar fracture toughness[3,6] (about 27MPa $m^{1/2}$). Hardness, property widely used as a roll quality measure, is higher than in the high-chromium iron rolls (80 - 87HSc ASTM E140 against 70 - 75HSc[3]), probably due to the higher hardness of the carbides and matrix.

2. MATERIALS AND METHODS

Samples of three different high-speed steel rolls were extracted. Those samples will be denominated samples A, B and C. Rolls made of materials A and B have already been utilized in some hot strip mills. Test specimens were taken from these samples for microstructural characterization, mechanical tests, oxidation and abrasion tests.

2.1 Microstructural characterization

In order to enable the observation of the microstructure with an optical microscope and measurement of the volume fraction of eutectic carbides in image analyzer, different metallographic etchings were carried out with the use of different reagents: Groensbeck ($KOH+K_3Fe(CN)_6$), Murakami ($NaOH+KmnO_4$), Nital (HNO_3) and electrolytic (aqueous solution of Cr_2O_3). For microprobe analysis in SEM, deep metallographic etching with HF and H_2O_2 was utilized.

2.2 Mechanical properties

Bending tests were carried out, according to the scheme on figure 1. Square transverse sections of 4 x 40mm were extracted from the roll shell. For material C, a test piece from the interface shell/core was also extracted, being the interface positioned exactly at the point of the load application.

The bending test value is expressed by the modulus of rupture, which is calculated by the following equation:

$\sigma_{MOR} = 3cF_{max}/2bh^2$, where:

c = distance among supports
b = width of the test specimen
h = thickness of the test specimen
Fmax = rupture force

Rockwell hardness tests were carried out in the fractured specimens after rupture.

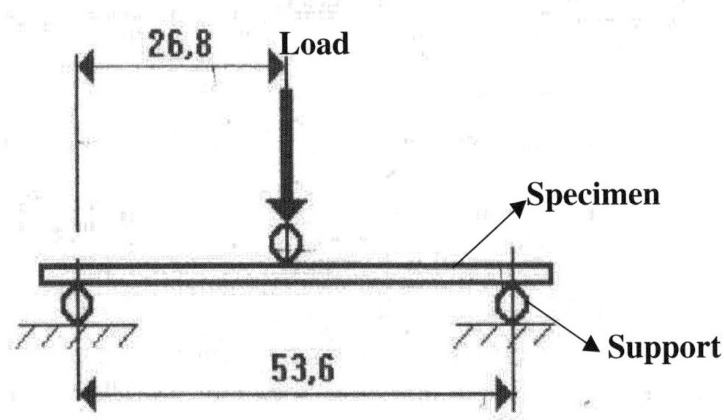

Figure 1: Bending test scheme

2.3 Oxidation testing

The tests were carried out with a termogravimetric balance using specimens in the form of disk (3,5mm radius) at 720, 765 and 800° C for 4 hours, being obtained the weight variation in function of the time. After oxidation, the specimens were analyzed by SEM.

2.4 Oxidational wear testing

A pin on disk test was carried out as schematized on figure 2.

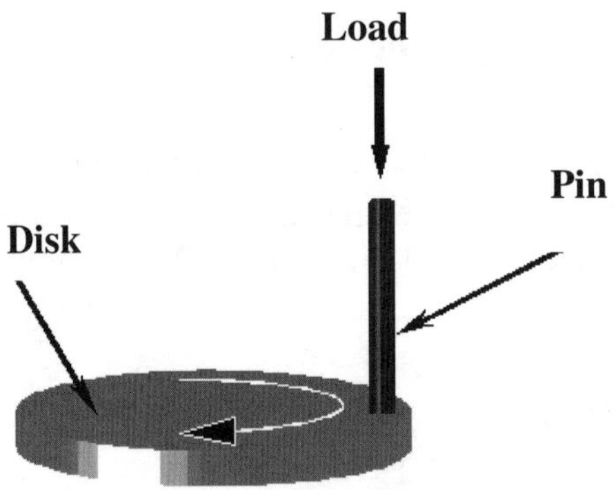

Figure 2: Pin on disk test

The friction coefficient between the pin and the disk surface was measured during the whole test, resulting in the evolution of the friction coefficient in function of time. The pin was made of 1045 normalized steel. The high-speed steel disks, after grinding, were heat treated at 400°C for two hours to oxidize the surface. A 5N load was applied and each test lasted 100 minutes.

3. RESULTS AND DISCUSSION

3.1 Microstructure

The figure 3 shows the microstructure of the samples A, B and C (reagents that do not etch the matrix were used, to emphasize the characteristics of the eutectic carbides).

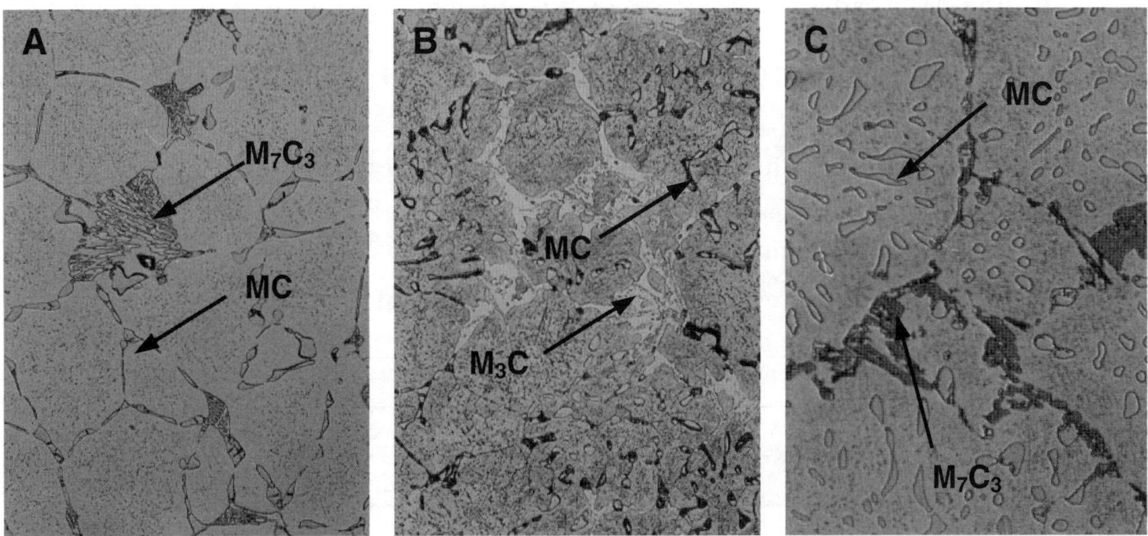

Figure 3: Microstructure of samples A, B and C. (A) Murakami, 250x; (B) Nital, 250x; (C) Groensbeck, 250x

The microstructure of sample A is characterized by the presence of eutectic carbides MC and M_7C_3 in the interdentritic areas. Sample B shows a microstructure similar to that of sample A, although carbides M_6C and M_3C replace the M_7C_3 carbides in consequence of the lower Cr content; moreover, the carbide MC appears, in some areas, as constituent of eutectic cells. The microstructure of sample C, with eutectic carbides MC, M_6C and M_7C_3, is different from the microstructures of samples A and B with regard to the distribution of the eutectic carbides: carbide MC always appears as a eutectic cells constituent and the M_6C and M_7C_3 carbides are distributed in the intercellular areas in a discontinuous way.

Tables II and III show the volume fraction and the chemical composition of the eutectic carbides in samples A, B and C, respectively.

Table II: Volume fraction of eutectic carbides

Sample	Volume fraction (%)			
	MC	M_7C_3	M_6C	M_3C
A	7,9	3,7	-	-
B	7,1	-	1,6	4,2
C	13,4	2,1	*	-

In all of the samples, the matrix is composed of tempered martensite (not shown in the figure 3) and globular carbides precipitated after solidification (shown as dark points in the matrix, figure 3) and during the tempering treatment. Samples A and B are hypoeutectics and have similar solidification sequences that can be summarized as follows:

a) Formation of pro-eutectic austenite dendrites.

b) Decomposition of the interdentritic residual liquid by eutectic reactions, originating different eutectic carbides. In both, the MC carbide, rich in vanadium (table III), is the first to precipitate; consequently, there is a decrease in the vanadium content of the residual liquid, occurring precipitation of other eutectic carbides: M_7C_3, in sample A, due its high-chromium content (tables I and III), M_2C and M_3C in sample B. The M_2C carbide is decomposed during the tempering treatment, originating the M_6 C carbide.

Sample C, due to the higher content of carbon and vanadium, is practically eutectic and its solidification sequence is characterized by a sequence of eutectic reactions that originate MC, M_2C and M_7C_3, in this order. Similarly to sample B, this M_2C carbide is decomposed during tempering, resulting in the formation of M_6C carbide. In all of the samples there is the predominance of MC carbide, notably in the league C (table III), which has the highest vanadium content. The M_7C_3 content in sample A is due to the higher chromium content in this sample.

Table III: Chemical composition of eutectic carbides

Carbide	Sample	Chemical composition (wt %)				
		V	W	Mo	Cr	Fe
MC	A	84	5,5	3,3	5,5	2,0
	B	74	15	4,8	3,0	2,9
	C	87	4,7	2,5	3,2	2,5
M_7C_3	A	9,8	4,2	5,9	35	45
	C	8,5	8	13	26	44
M_6C	B	5,7	38	23	4,3	29
	C	5,7	24	40	8,6	21
M_3C	B	2,9	10	7,0	18	62

3.2 Mechanical properties

The table IV exhibits the results of rupture modulus and hardness. The shell/core interface modulus of rupture value on sample C (838MPa) is comparable to results presented in existing literature[11] and even superior to the typical value of the indefinite iron applied in work rolls for the last stands of hot strip mills (around 650 MPa).

Table IV: Bending test and hardness results

Sample	σ_{MOR} (MPa)	Dureza	
		HRc	HSc
A	1228	60	81
B	1020	57	76
C	1131	57	76

3.3 Oxidation

Figure 4 shows the weight gain versus time in the oxidation tests of samples A, B and C and figure 5 shows micrographies of transverse sections of the oxidized specimens.

At lower temperatures, oxidation is slow and the oxide layer is heterogeneous, with some areas not very oxidized in relation to others. According to figure 4, sample A has a different behavior in relation to B and C at 720°C. At higher temperatures an opposite behavior occurs and samples B and C start to show a smaller weight gain along the test. The oxide formed for the oxidation time is preferably iron oxide.

It would be interesting to perform the characterization of the oxides formed at shorter times, thus simulating the real conditions of rolling in a more approximate way.

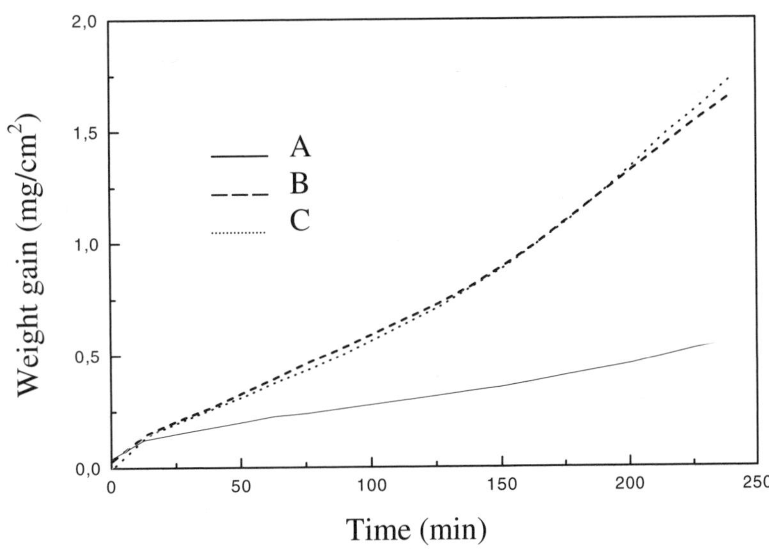

Figure 4: Weight gain versus oxidation time for samples A, B and C

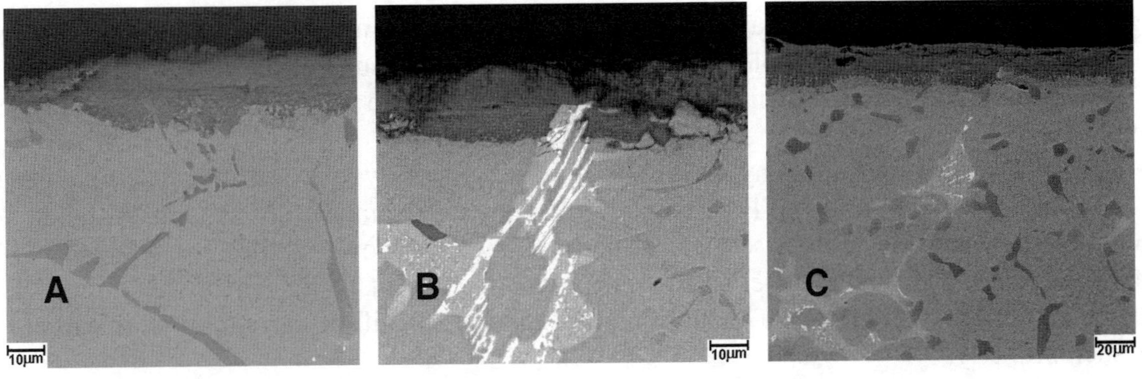

Figure 5: Transverse section micrographies of the oxidized specimens

3.4 Oxidational wear

Figure 6 shows the evolution of the friction coefficient versus time during the test performed with oxidized specimens of samples A, B and C. There is a similar evolution of the friction coefficient in the oxidized specimens, when long times are considered. In the beginning of the tests, samples B and C show lower friction coefficient and sample B shows accentuated fall in the friction coefficient. In spite of supplying important data, it would be interesting the accomplishment of the hot wear test with controlled humidity, simulating the hot rolling conditions in a more approximate way.

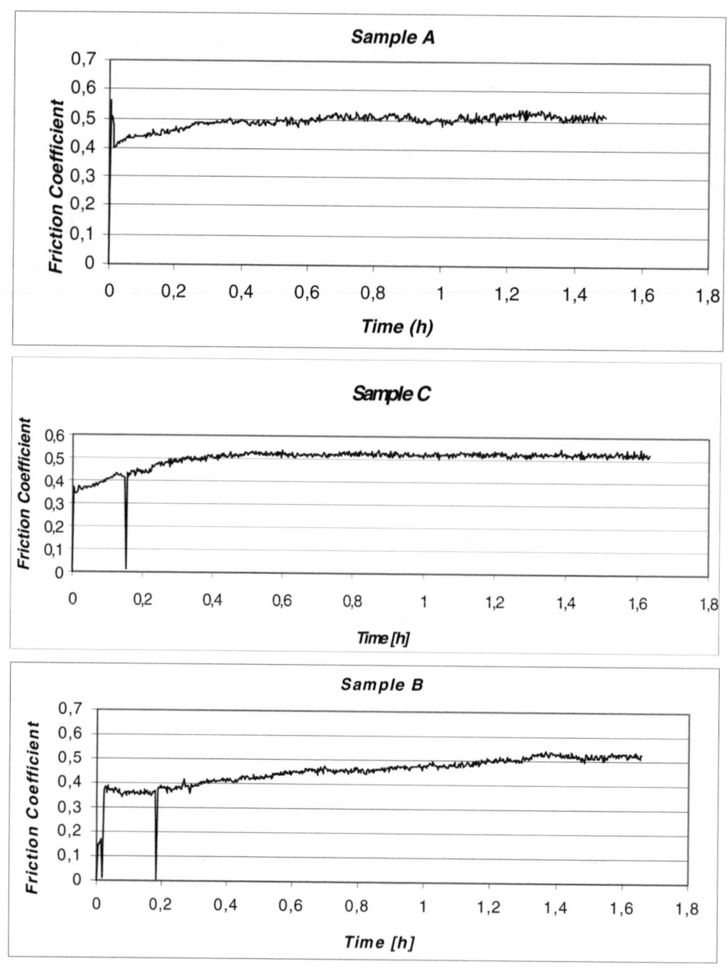

Figure 6: Friction coefficient versus time for samples A, B and C

3.5 Performance

The high-speed steel rolls can stay a longer time in the hot strip mill, being the campaigns extended in relation to high-chromium iron roll. Table 5 shows the performance results of rolls made from materials A and B already used in some rolling mills, while rolls made of material B are used only on mill C. Material C is still in laboratory phase.

Roll performance is measured in many ways, but is most commonly assessed in metric tons per millimeter by stand. The result shows that material B has superior performance compared to material A, being verified the best behavior of that new material in service. Material C, which shows similar behavior of oxidation and oxidational wear in comparison to material B, is expected to present better performance regarding thermal fatigue, due to its carbide morphology. Thermal fatigue experiments will be carried out with samples A, B and C to verify such property.

Table V: Performance results

Material	Mill A	Mill B	Mill C
High chromium iron	10510t/mm	8397t/mm	5720t/mm
Material A	17768t/mm	13521t/mm	7905t/mm
Material B	X	X	14893t/mm

The utilization of high-speed steel rolls requires more specific operation conditions in comparison to the high-chromium rolls, such as crown adjustment, refrigeration and adapted lubrication. Considering the modification of such parameters demands high investments, the use of those rolls in some rolling mills becomes unfeasible.

4. CONCLUSIONS

Despite having similar mechanical properties, carbides type, volume fraction and distribution are important characteristics to define the oxidation and wear behavior of high-speed steel rolls, what can be evidenced by the differences among samples A, B and C and performance between rolls from materials A and B, already tested.

The behavior of samples regarding oxidation varies according to the temperature, and samples B and C present a different behavior from sample A. The oxidized samples have mainly the formation of iron oxide for the studied times. Although supplying an idea of oxidation behavior for the roll in service, the oxidation rate and the oxides characterization during the tests are not enough to state which roll will have better performance.

There is a similar evolution of the friction coefficient in the oxidized samples, when long times are considered. In the beginning of the tests, samples B and C show lower friction coefficient and sample B shows accentuated fall in the friction coefficient.

The performance of high-speed steel rolls made from material A is about 1.6 times superior to that shown by high-chromium rolls, while material B shows performance 2.6 times superior.

ACKNOWLEDGEMENTS

The authors are grateful to Fernado Cosme Rizzo Assunção and Maurício Monteiro – PUC RJ for performing the oxidation tests, as well as Fundação de Amparo à Pesquisa do Estado de São Paulo – FAPESP for supporting part of this research.

REFERENCES

1. HASHIMOTO, M. et al., "Development of High Performance Roll by Continuous Pouring Process for Cladding", ISIJ International, Vol. 32, (1992), No 11, pp. 1202-1210.
2. SCOULAR, T. et al., "Manufacture of Cast Steel Rolls", Rolls for the Metalworking Industries, ISS 1990, pp. 95-111.
3. UETZ, H., "Abrasion und Erosion", 1a Ed. München, Carl Hansen Verlag, 1986.
4. SANO, Y. et al., "Characteristics of High Carbon High Speed Steel Rolls for Hot Strip Mills", ISIJ International, vol. 32, 1992, pp. 1194-1201.
5. BRANDIS, H., et al., "Metallurgical Aspects of Carbides in High Speed Steels", 109th AIME Annual Meeting, Symposium on Processing and Properties of High Speed Tool Steels Proceedings, Las Vegas, 1980, pp. 1 – 18.
6. WERQUIN, J.C, BOQUET, J., "New High Performance Rolls for Hot Rolling", Rolls for Hot and Cold Strip Mills, Proceedings International Iron and Steel Institute, 1993, pp. 1-14.
7. PÄKKÄ, R. et al., "Optimizing Maintenance and Operating Costs of Hot Strip Mill Work Rolls for Long Life and Surface Quality", Rolls for Hot and Cold Strip Mills, Proceeding International Iron and Steel Institute, 1993, pp. 25-30.
8 LECOMTE, C., HERMAN, J.C., "Trends in Work Rolls for Hot Strip Mills. Rolls for Hot and Cold Strip Mills", Proceeding International Iron and Steel Institute, 1993, pp. 49-54.

9. KATO, O., "Mechanisms of Surface Deterioration of Roll for Hot Strip Rolling", ISJI International, Vol. 32, pp. 1216-1220.
10. QUIN T.F.J., "Review of Oxidational Wear. Part I: The Origins of Oxidational Wear". Tribology International, Vol. 16, 1983, pp. 257-271.
11. FEISTRITZER, B., SCHRÖDER, K.H., WINDHAGER, M., ZIEHENBERGER, K.H., "Improved Grades of Work Rolls for Hot Strip Mills", XXXVI Seminário de Laminação- "Processos e Produtos Laminados e Revestidos", Associação Brasileira de Metalurgia e Materiais, Belo Horizonte, set. 1999, pp. 195-204.

ROLL TECHNOLOGY III

QUESTIONS, ANSWERS, MORE QUESTIONS
Twenty - five Years of Experience in Discussing Rolls and Rolling Technology

Dr.-Ing. Karl Heinrich Schröder
ESW, Eisenwerk Sulzau-Werfen
A-5451 Tenneck, Austria
Phone: +43 6468 5285 100
E-mail: ks@esw.co.at

Key Words: rolls, mechanical properties of rolls, coefficient of friction, quality assurance

INTRODUCTION

Roll makers always ask about rolling conditions and the necessity to choose the right grade of roll material (whether the decision is right or wrong is proved in the mill) and roll users always ask about the mechanical and physical properties of roll material. Sometimes they feed these figures into their rolling model, but sometimes they also need them for unknown reasons. This information is very rarely useful for selecting the right supplier.

In July 1978 John M. Dugan published in AISE (No. 15222) an article about "FACTORS WHICH INFLUENCE ROLL SPECIFICATIONS" [1], the book consists of 88 pages. At that time – as I had only been in roll-business for less than three years - I hoped to find help. However, fortunately he did not really give 'factors which influence' - otherwise I would have had trouble in using these factors correctly. Roll-makers do their job as well as possible and they try to reproduce success without to many theories because they know that it is easy to have a good theory - and much more sophisticated to have a good product - however it is more important to have a good product instead of a good theory!

Dugan was not really concerned about rolling conditions 'which influence roll specifications' - he knew too much - but he discussed all kinds of roll failures due to rolling accidents. However, he did know that not too much roll abuse can be compensated by the best roll grade.

In Oct. 1985 I presented a paper in Cleveland on "Inspection Parameters which control Performance of High Chrome Rolls" [2] discussing the influence of roll surface hardness, residual stresses, roll composition, etc. - but since then, I am still asked the same questions.

Rolls usually do not fail due to normal rolling conditions, but due to mill accidents. And mill accidents are hard to describe quantitatively, so discussions continue - besides some clear events - for a long time. Luckily most mills improved their rolling conditions, reduced cobble rates etc. and the roll grades have been improved as well. The overall roll performance is rising.

On the way to achieving better performing rolls many properties are discussed, however this is a long story of misunderstandings, misinterpretations etc. The main problem was and is that mechanical properties are tested with specimen under one directional stress – however, in rolls we have more or less always (2) 3-dimensional stresses and the roll surface is always loaded complexly by mechanical, thermal and corrosion systems. And rolls are loaded with residual stresses. We have to distinguish carefully whether we mean material / specimen properties or roll behaviour! If this distinction is not made endless misunderstandings occur.

There are many examples for false theories like for example:
- theory of importance
- stories ...
- theory of stress relief in compound rolls due to plastic deformation of core material
- theory about oxidation of roll surface, while in reality the oxide layer only is built by strip scale

Everybody needs models, 'theories', to develop new ideas - but these should only be discussed in public when proved. A good result is not necessarily the verification of a theory.

It took me many years to understand the different thinking of roll users and roll makers and how they approach the same problems. Therefore in this paper I want to cover some of the most frequently discussed questions in a (my) general way to summarise 25 years of personal experience.

MECHANICAL PROPERTIES

Basic properties of tools like rolls and properties of the material the tools are made of, are two totally different sides of a coin, and very often this difference is ignored. However, when we start to discuss mechanical properties, we have to distinguish whether we mean material - or roll - properties.

Hardness - *is the most boring subject of discussions! Why ??*

There are many methods and measurement instruments with basically (but not exclusively) 2 systems:
- static measurement: HV, HB, HRC
- dynamic measurement: Shore C, D, ...; Equotip D, E, ...

and there are conversion/confusion tables, which neglect / ignore statistical variations and induce more accuracy than exists.

Hardness is strongly influenced by the roll surface conditions. Roll surface hardness may be measured some Shore (1 ShC ≅ 10 HV) too high due to work hardening or some Shore too low due to annealing by wrong grinding compared to the 'real' material hardness.

And the variation of hardness in rolls is much higher than most people believe: For rolls standard deviation can reach 30 HV [2, 3] and even the range of normal specifications is only 5 ShC (≅ 50 HV) (average $\pm 2\sigma$ = average ± 6 ShC → this means: no process capability - or everybody has to average average values - till there is no more variation).

What is the purpose of hardness readings? Answer:

For roll makers: they control the process of roll manufacturing.

For mill people: They should control work hardening [4] by hardness measurement to avoid roll failure due to surface fatigue. However, work hardening only occurs after plastic deformation and increases with yield strength of the material. Fatigue often starts at stress raisers (cracks) with stresses far below the yield strength.

For science: Many material properties are strictly related to hardness, there are all kinds of rules - but it is necessary to know the rules [5].

Another question: Why do mill people want to discuss hardness?

It is the only property besides dimensions that can be tested by incoming inspections. However, after mentioning the facts above there is no further comment.

Mill people believe in hardness giving information about roll quality and roll performance, however this is helpful only in rare exceptions. Fig. 1 shows that different roll grades with totally different properties were always produced in the same range of hardness.

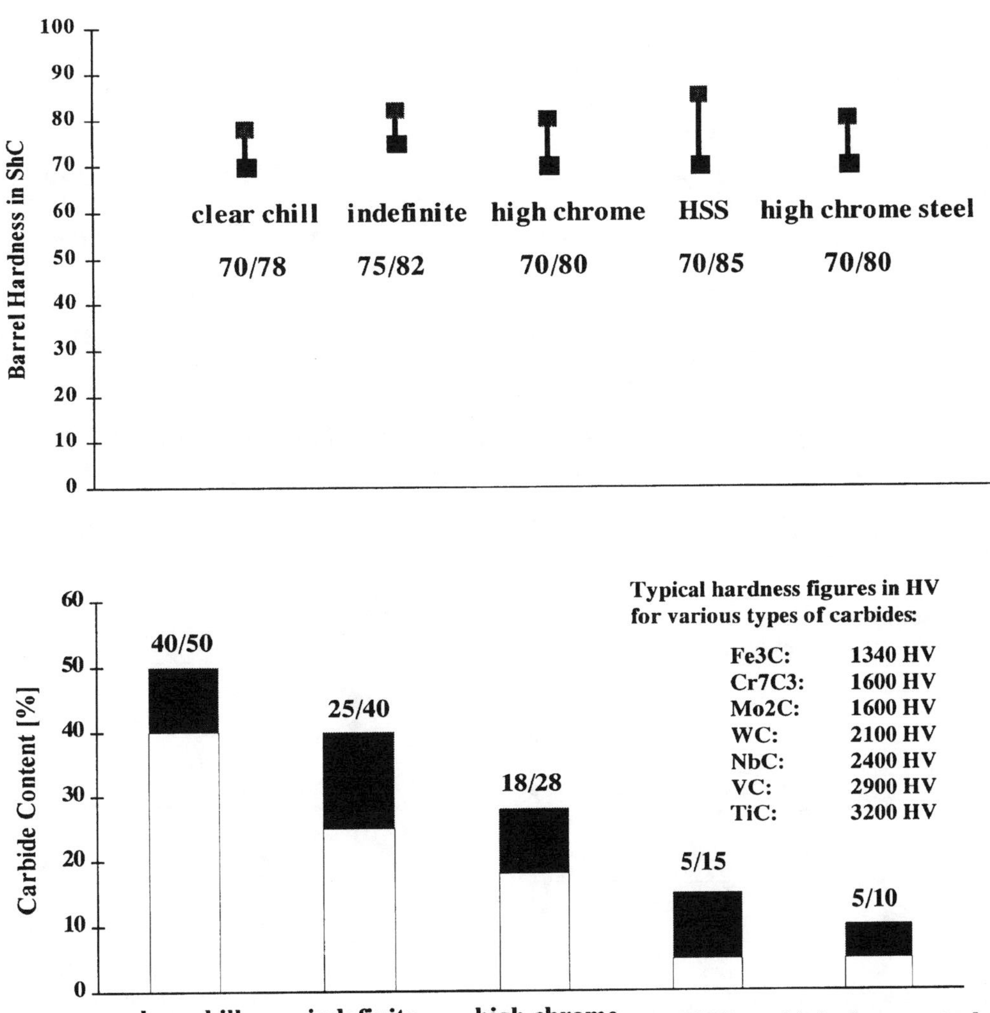

Fig. 1: Range of barrel hardness and content of carbides of different roll grades

A wide variation of hardness has almost no influence on roll performence, see Fig. 2 and Fig. 3. The same roll hardness of different rolls of one grade can be achieved by various matrix micro structures and various amount and type of carbides which have a high impact on roll behaviour - not on hardness!

The relation of hardnesses of b.u.r.s to WRs is often discussed, and even here no easy answer is available. Besides; we have to consider which material in the mill is softest to take plastic deformations in case of an accident: That is the rolled sheet in HSM (in so far as no specialities are rolled) - and b.u.r. may have the same hardness as WR, and that is the b.u.r. in CSM - and b.u.r.s should be definitively softer than WRs However special situations always need special care.

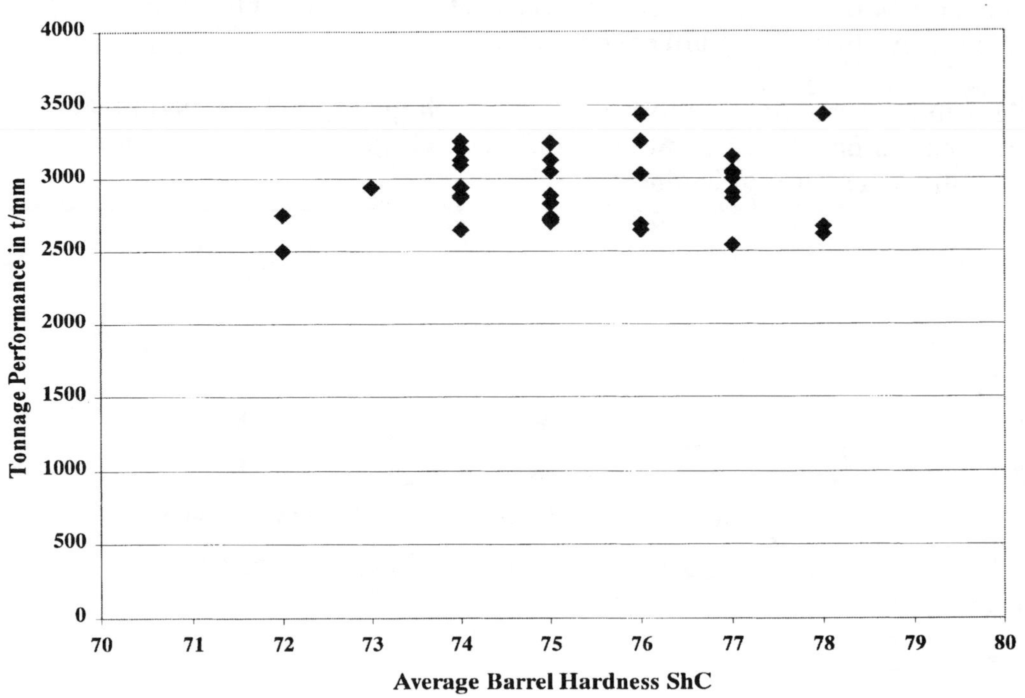

Fig. 2: Tonnage performance of ICDP work rolls 680x1900 mm, excluding mill accidents

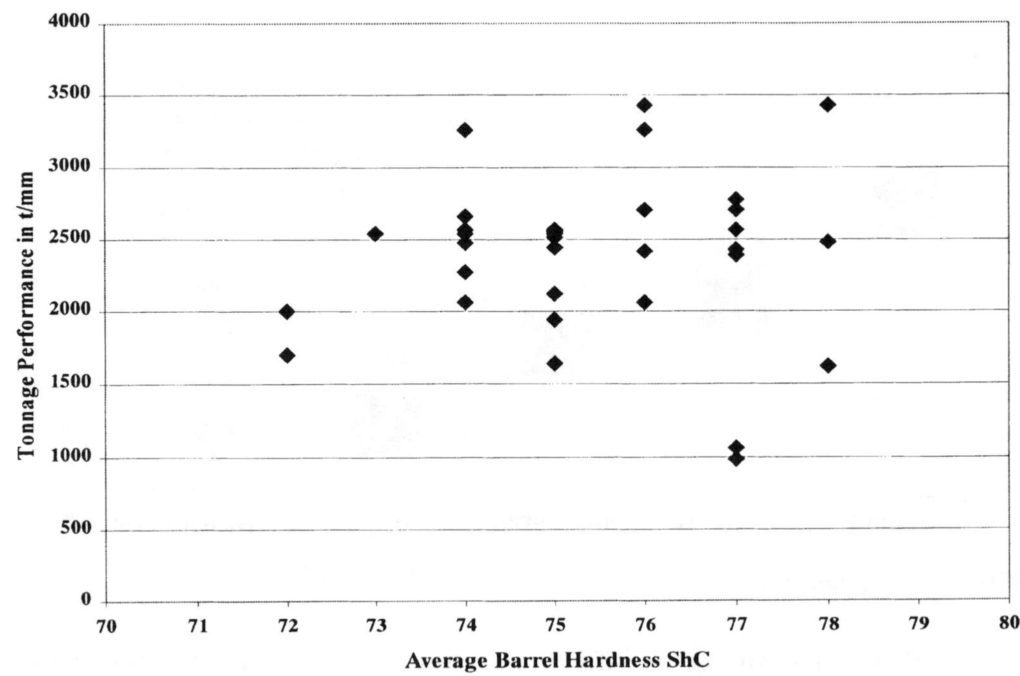

Fig. 3: Tonnage performance of ICDP work rolls 680x1900 mm, including lost mm due to mill accidents

Compression strength - is directly proportional to hardness (R ≅ 3,5 x HV) for most materials made of iron (grey iron, steel, high-chrome-iron or - steel, HSS etc.), but compression strength is never required for a roll - however safety against impression marks, local plastic deformations and in this case the shear stress is to be discussed. Shear is calculated as the difference of 3-dimensional stresses,

$$\tau = \frac{\sigma_{max} - \sigma_{min}}{2} \leq \tau_{crit}$$

where σ_{min} for rolls is highly influenced by the residual stresses, while τ_{crit} is represented by the roll hardness. For rolls of the same hardness this means, that the material which represents the highest level of residual stresses gives the best resistance against marks.

How to avoid marks, intrusions?

Easy - just avoid local overloads, which 'should not' happen in normal rolling conditions!

Tensile strength - *Why should we discuss tensile strength?*

High tensile strength of core material is mainly required to avoid thermal breakage of rolls, brittle, spontaneous breakages due to high temperature gradients in a roll, where the residual tensile stress in the core material is an additional prestress. The easiest way to avoid thermal breakage is to minimize thermal gradients and this can be achieved by good rolling / cooling strategies and proper handling of rolls.

There is a theory that the firecrack pattern of roll surface is related to tensile strength (and to fatigue strength), however the question of crack propagation has to be discussed in relation to residual stress and microstructure as well. The best way to achieve a good roll surface is excellent roll cooling. The roll grade and wear has some input on roll properties. However, rolling conditions are important as well!

Fatigue strength - *Who knows what?*

Figures of fatigue strength gained from tests with standard small specimen are practically useless, and we have to consider: size of rolls, notches, impact, environment, superposition of different multiaxial stresses etc., etc. So we have only our experience with rolls and measures for many possible cases! Fatigue strength is of high importance for high stressed filets of back up rolls and for all kind of spalls. Fatigue strength is highly influenced by notches and residual stress (Smith Diagram!). There are some basic figures of tolerated stress - however, in cases where there is corrosion fatigue everything is different and we have to fight corrosion first [6].

Spalls: An interesting example of fatigue failures are saddle spalls of WR in HSM finishing stand originating in core (grey iron with lamellar graphite) material of high chrome iron roll, which were experienced in high stressed rolls in some mills, due to changes in rolling strategies early in the 1980s (later in ICDP rolls as well) [7]. There have been different methods to avoid this type of failures:
lower tensile pre-stressing (lower residual stresses in the rolls) or - core material with higher fatigue strength, ductile iron versus grey iron with lamellar graphite, and the latter solution was adopted by all roll makers in general.

Another example of fatigue problems are spalls in ICDP or HSS rolls. These materials are very notch-sensitive, so it is highly recommended to eliminate all kinds of cracks, whenever these rolls are reground, otherwise typical 'cats' - tongue band type spalls may occur. [8]. High chrome iron or steel material is much less sensitive.

Residual stresses have a high impact on fatigue [9], and fracture mechanics might be helpful in solving problems.

Torque - is not reported (or only very rarely) to create fatigue failures. Torque damage of drive ends of rolls (45° breakage) is normally a spontaneous breakage due to one single giant overload, so the limit for torque is considered as static torque strength (≈ 80 % of tensile strength).

Young's modulus - *the customer asks for this - but is it a figure or a variable?*

Young's modulus of compound rolls is a mixture of Young's modulus of core- and shell material and can be precisely measured, however the 'mixture' varies during roll life, when the shell material-thickness is reduced to almost nothing. However it seems there was never any real problem, whatever the figure was!

PHYSICAL PROPERTIES OF ROLL MATERIALS

are often discussed for some rolling models, however - to my knowledge - none of the roll makers is capable of testing thermal conductivity, specific heat, at room - or elevated temperature etc., etc.

These properties are always taken from literature or measured by high-tech-research institutions. There is not much sense in asking the individual roll maker - even they are supposed to answer all questions of their customers, of course!

Who can give good answers? Nobody! But answers are requested.

PROPERTIES FOR EASY TESTING OF MECHANICAL PROPERTIES

Wear - *Who knows the answer? Which parameters are of importance?*

Wear is an object of libraries without any easy answers! Rolling conditions are of high importance for the wear of rolls. For example: High Chrome Iron b.u.r. don't wear in cold strip mills (these rolls fail due to spalling, low resistance against crack-propagation), but they wear like butter in the sun in hot strip mills.!

The reason? Rolling conditions!

Basically the diagram of Khrushov is helpful, Fig. 4: for one material wear resistance increase with hardness; at the same hardness wear resistance increases with the percentage of carbide forming elements, Fig. 5 shows the influence of various alloying elements.

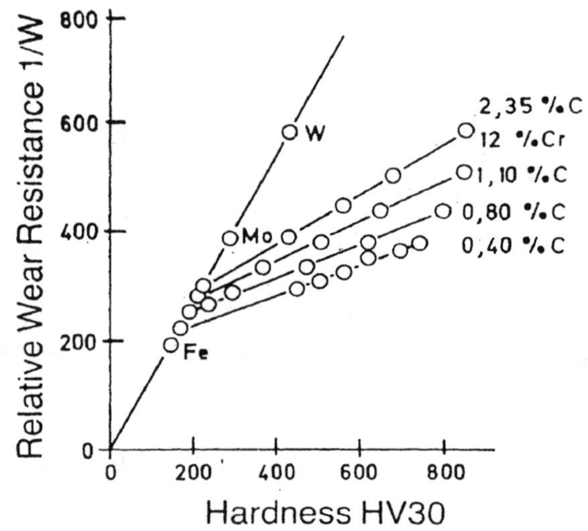

Fig. 4: Diagram of Khrushov [5, 6]

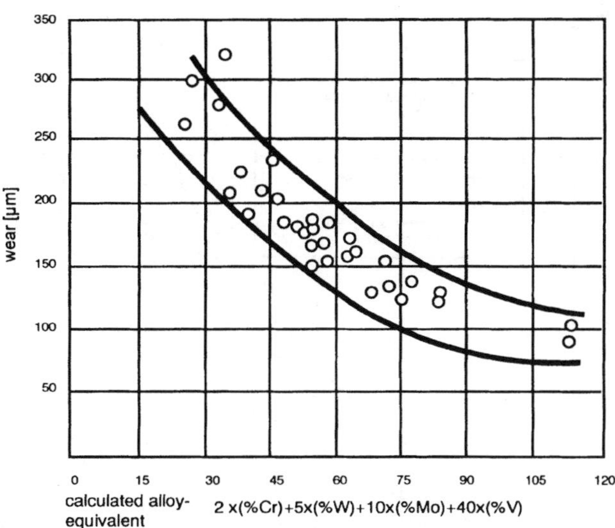

Fig. 5: Wear versus alloy equivalent (Range of alloys: 0,30 - 0,44%C; 0,02 - 0,38%Si; 0,31 - 0,83%Mn; 1,18 - 3,03%Cr; 0,08 - 3,04%Mo; 0,26 - 1,92%V and 1,42 - 5,85%W) - from [10].

For hot rolling of steel the temperature of rolled material is very important. For a long time there has been theory of hard (high temperature) and soft (low temperature) oxides on the strip which might influence wear, but last year at the WMSP Conference in Baltimore [11] this theory was regarded as obsolete, because there is never enough time during rolling to build up oxides in the right equilibrium phase because the time between the stands is too short - but nevertheless it is true that wear is less when the temperature of strip surface is low (less than 900° C), skin cooling of strip is evident to reduce wear in hot rolling of steel - something that has been known for many years, especially when adamite was used for W.R. material (some 20-30 years ago). And: hot rolling of stainless steel - almost no scale - increases wear heavily. And especially strip edges of stainless steel - at lower temperatures - increase wear, the work roll roughens - due to micro-sticking or other reasons.

Wear is at a minimum with HSS-rolls, however they are not used world wide for the very last stand of HSM due to sticking problems, where specific pressure is at its maximum [12].

Lubrication reduces wear and sticking, but there are still problems to be solved.

Coefficient of friction - *Why is the coefficient of friction higher with HSS-rolls than those with High Chrome Iron or ICDP?*

The coefficient of friction is under discussion, especially for HSS-roll-material. CRM [13] did some basic studies for the European rollmakers (CAEF) and it turned out that there is no real difference between the various materials for work rolls for hot rolling: ICDP, high chrome iron and HSS. They studied the coefficient of friction as a function of deformation, Fig. 6. Lubricants change the contact conditions and thus the coefficient of friction as well - however the difference between the various materials remains negligible.

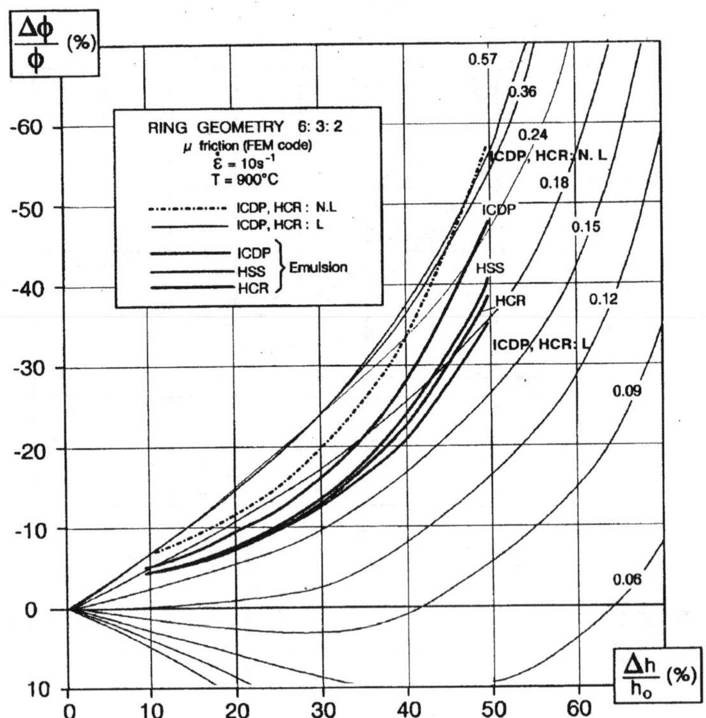

Fig. 6: Evolution of the coefficient of friction of different roll grades with and without lubrication (FEM-Coulomb model with barreling) [13].

However CRM [14] at last year's WMSP-Conference in Baltimore presented the results for the coefficient of friction during rolling - calculated from measured rolling - separation force versus calculated separation force which show a totally different situation, Fig. 7. This does not correlate with above mentioned results! We have to consider roll surface during a rolling campaign and the change in roll surface, mainly the fire crack pattern! Fig. 8 shows the different fire crack pattern for different materials - heavy primary cracks influence the coefficient of friction more than the structure of carbides or anything else [15].

The coefficient of friction is influenced more by rolling conditions than by the roll material properties! In mills where roll cooling is most efficient the difference in coefficient of friction between different materials is not evident. Indeed, some mills never reported any increase of separation force!

And even under severe rolling conditions, lubricants may minimise the increase incoefficient of friction!!

Banding / Peeling - *How to improve roll-, strip surface?*

Banding or peeling is a problem in many mills. It is known and proven that the oxide layer on W.R. is built up only of strip oxides, oxidation of rolls never was proven (and it is almost impossible because roll surface temperature never exceeds 550° C, and this only in contact with rolled material under exclusion of oxygen). Banding is a mill problem and it really can't be solved by roll material grade so far, although there are some variations [16]. Great improvements have been achieved by better roll / strip cooling systems.

It seams evident that the rolled-on layers of oxides on WRs are stripped of when light gauge is rolled with high reduction in the early stands of the finishing mill.

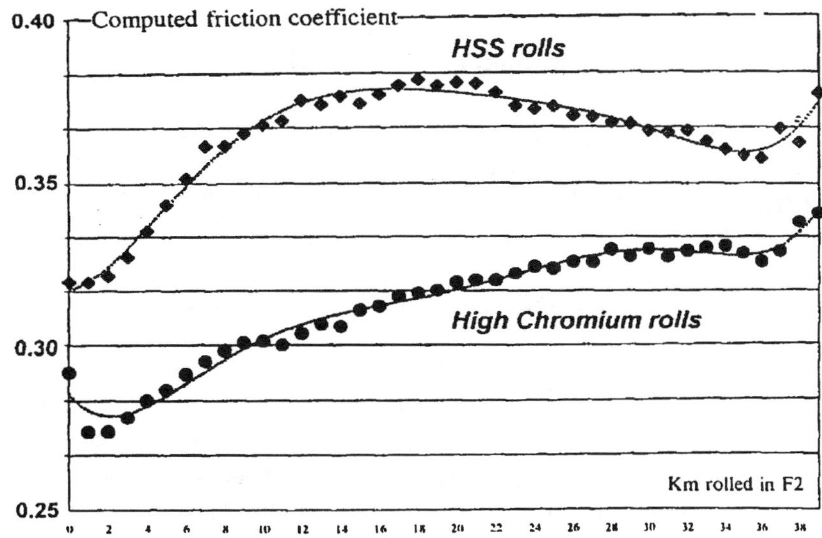

Fig. 7: Evolution of the coefficient of friction during a rolling campaign [14].

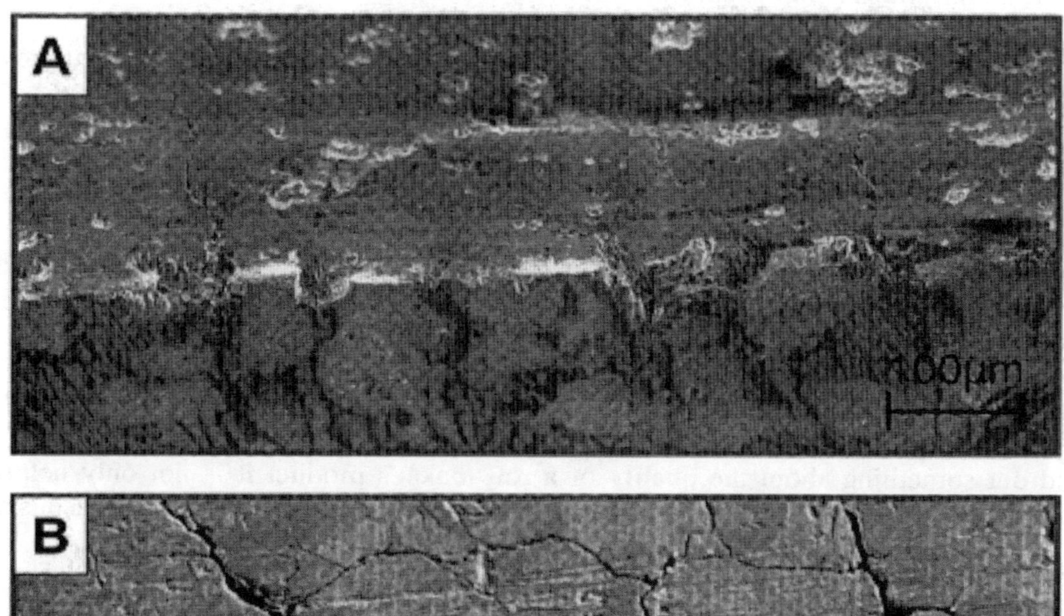

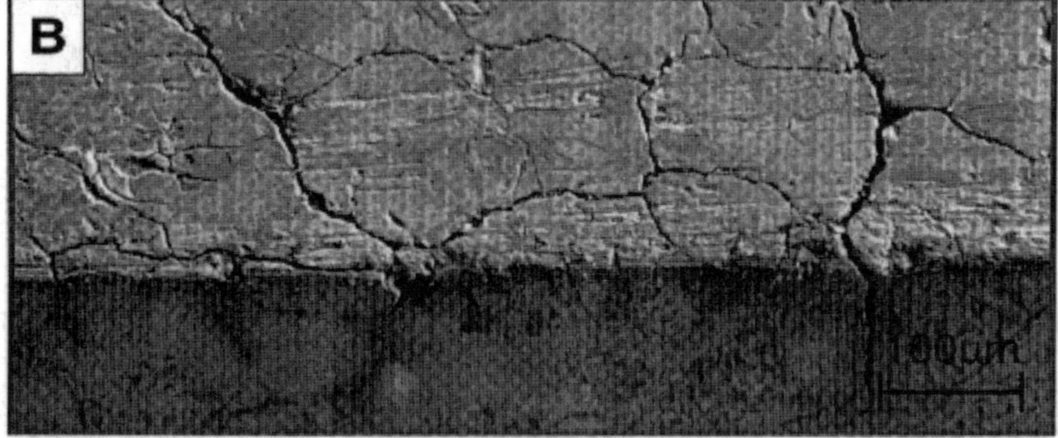

Fig. 8: Surfaces of worn work rolls, etching 12" Nital, SEM.

A - high chrome work roll
B - HSS work roll
Angle of incidence of the electron beam is 45° to the roll surface.

QUALITY ASSURANCE

Some years ago the ISO 9000 ff quality standard was introduced to keep the quality of mass produced cars and parts for automotive industries or others under control. Rolls are customer specified single products.

Such systems are undoubtedly useful for controlling 'industrial production', for instance rolls. However, something is different: Each roll is 'handmade' and within very narrow tolerances. For making rolls - we are talking about small figures, tenth of rolls, never talking about big numbers, which are necessary for real statistics (min. 100 pieces of the same design!) - no S.P.C is applicable!

ESW is certified, of course! But what does this mean? Process-, machining capability?!

For process capability the tolerances should be wide compared to standard deviation of the process!! This might be true for the chemical composition of rolls - which is not specified by our customer. However, our customers ask for T.I.R. less than 5 micrometer, our measuring devices show only digits of 1 micron!! There is no process capability! Hardness variation (see above) is larger than the tolerance! No process capability! In case process- and machining capability do not exist, it is necessary to check each individual specification - which is done anyway with each roll!!

What purpose do questions about capabilities have? - We know.

We have to produce, piece by piece into the tolerances, with no way of statistical approval, we have to produce - to check - to produce - till it fits!!

We have to meet our customers' requirements - but, because machine- and process capabilities are out of reach for single produced parts (rolls), we have to work like an artist to reach the goal.

Due to high-tech data analysis we are capable of improving our process, but we have to manufacture each single roll according to the specifications.

There is no way of buying machines which guarantee machining capability for rolls! We have to work hard to reach the goal!

And to find out something about the quality of a roll maker's product it is not only helpful to check the quality system, it is useful to check the scrap rate, the claim rate - and to analyse the reasons and details.

ISO 9000 ff is useful for professional inspectors - it is of no use when deciding about the quality of rolls produced, each individual roll has to be tested carefully.

Statistical evaluations are helpful for roll-makers in determining their actions to improve the roll quality - it is of no use for our customers! To prove performance figures, we should keep in mind: 100 is a good number for statistical evaluations, numbers below 10 are not good for statistical statements.

CONCLUSIONS

1. Roll makers and roll users frequently have to discuss experiences, performance results, special requirements of the mill. Roll failure problems can be solved by good co-operation.

2. To avoid misunderstandings and misinterpretations everybody should distinguish very carefully whether roll properties or the properties of the roll materials are concerned.

3. Statistical findings are useful for roll making to improve the process of roll production - statistical process control (SPC) is only applicable for (automated) mass production.

4. Theories are sometimes useful for (commercial) discussions and may help to create new ideas. However facts about roll performance (without and with loss of useful stock due to mill accidents) are most important for roll development.

5. The best way for roll makers to achieve better rolls is to ensure that better materials and improved manufacturing processes are used and that roll users take account of rolling conditions and improved rolling processes.

Good and close co-operation between both sides is essential.

REFERENCES

1. J. M. Dugan, "Factors which influence roll specifications", Association of Iron and Steel Engineers, July 1978

2. K. H. Schröder, " Inspection parameters which control performance of high chromium rolls", Proc. of the 27th Mechanical Working and Steel Processing Conference, Cleveland, October 1985

3. Investigation Report of CRM: "Hardness deviation"

4. K. H. Schröder, "Beitrag zum Verfestigungsverhalten der Stähle durch Kaltverformung", Z. für Werkstofftech., 11 (1980), S. 73-76

5. K. H. Schröder, W. Eilert, "Zur Entwicklung der Qualitäten für Kaliberwalzen", Stahl und Eisen 104 (1984) Heft 20, S. 1005-1008

6. K. H. Schröder, "State of the art of rolls for the production of flat rolled products", World Steel & Metalworking, Vol. 6 1984/85, pp. 120-124

7. K. H. Schröder, "Heavy spalls originating in the cores of high chromium rolls", Metallurgical Plant and Technology, issue No. 2/1986, Verlag Stahleisen

8. K. H. Schröder, "Remarks on the measurement of, distribution in and influence of residual stresses in heavy construction units with large cross section", Residual Stresses in Science and Technology Vol. 2, DGM 1987, pp. 927-933

9. "Rolls for the Metalworking Industries", Publication of Iron and Steel Society, 1990, p. 139, 235

10. Werkstoffkunde Stahl Hrsg. Verein Deutscher Eisenhüttenleute Berlin; Heidelberg; New York; Tokyo; Springer; Düsseldorf: Verlag Stahleisen

11. Discussion at 41st Mechanical Working and Steel Processing Conference, ISS, Baltimore 1999

12. K. H. Schröder, "Rolling conditions in hot strip mills and their influence on the performance of work rolls", Metallurgical Plant and Technology, issue No. 4/1988, Verlag Stahleisen, pp. 44-56

13. Centrum voor Research in the Metallurgie, "Friction measurement in hot-rolling by means of the tribological ring compression test", Investigation Report of CRM, May 1997

14. D. Steinier, D. Liquet, J.Lacroix, H. Uitjtdebroeks, J. C. Herman, "Effect of processing parameters in the front stands of a HSM on the performance of HSS work rolls", 41st MWSP Conf., ISS, Baltimore, 1999

15. K. H. Schröder, B. Feistritzer, M. Windhager, K. H. Ziehenberger, "Progress of carbide enhanced ICDP (and remarks on the coefficient of friction of HSS work rolls in hot strip mills", will be presented at the 37th Rolling Seminar Conf., ABM, Curitiba 2000

16. ESW - paper, in near future

SCHEDULING OF ROLLS FOR PRODUCTION CAMPAIGNS IN BOTH COLD MILLS AND HOT MILLS

Khalil Fazlollahi, Ph.D. M.Sc. B.Sc. P.Eng, Quad Infotech Inc., Toronto, ON Canada
Tony D'Antonio, Roll Shop Manager, Worthington Steel, Decatur, AL, USA
Guy Pilon, Project Manager, Quad Infotech Inc., Toronto, ON Canada
Hooman Fazlollahi, B.Sc. Senior Analyst programmer, Quad Infotech Inc., Toronto, ON Canada

BACKGROUND

Quad Infotech has developed a family of software modules for shaped steel plant operations called "Quad Mill Operation System" [QMOS]. This product deals with Production scheduling and reporting, rolls scheduling for a production campaign, and management of the roll and setup shops. [QMOS] has been operating in a number of reputable steel plants throughout the United States, Canada and Europe, since 1992.

The main modules that comprise [QMOS] are:

- [QMOS]PSC Production Scheduling and Control
- [QMOS]RSP Roll Shop Planner (Management), including Guide Shop
- [QMOS]STP Production Monitoring and Shift Reporting

For the past three years, Quad has been working with flat rolling mills to develop FLAT version of [QMOS] to address their unique needs of Hot and Cold Flat rolling mills

[QMOS]FLAT has been successfully implemented at **Worthington Steel in Decatur, AL.** This tool is designed to schedule and drive roll shop activity, to record mill performance, to help the operators make decisions during the shift, and to help managers perform analysis and view trends allowing for consistent improvement to the processes.

[QMOS]FLAT provides the following main functions:

- maintain roll Inventory,
- prepare rolls,
- create roll schedules,
- assemble roll sets,
- move rolls to production,
- disassemble rolls,
- maintain roll components.

In order to justify this project, the roll shop management system was expected to provide the following beneficial effects:

- Minimize investment in roll and roll related inventory. Increase roll life;
- Improve the quality of the rolled product by utilizing more accurate roll results from the grinders;
- Increase mill productivity, (by decreasing mill change-outs), from roll shop scheduling improvements.

Maintain Roll Inventory

Working with engineering and the roll vendors, roll shop management plans and executes roll procurements. Not only are inventory levels of rolls involved, but so are the chemical and physical specifications. Through analysis of roll performance, procurement decisions are made with an eye toward minimizing roll investment through increased roll life longevity.

[QMOS] supports this function by providing production and grinding performance histories for each roll and the ability to display this information in ways that illustrate trends and suggest root causes.

Prepare Rolls

In this process, the surface of the rolls are prepared, (ground, textured, and chromed), when they are received as new inventory, and each time they are changed out of the mill.

Measurement results from these operations, (crown, roughness, diameters, etc.), are automatically captured by [QMOS] for later procurement analysis and automatic communication to the mill.

Create Roll Schedules

In response to mill production schedules, roll schedules are created to plan roll assembly and staging.

This activity is supported by [QMOS] by bringing together the mill production schedule and roll inventory information into one place for roll schedule creation.

Assemble Roll Sets

Work rolls and backup rolls must be assembled into sets, (two rolls plus chocks, bearings, and seals), before they can be used in the mill.

These assemblies are recorded into the [QMOS] system for use by roll shop personnel to keep track of what is ready for production, and the mill software to easily identify what is being loaded into the mill.

Move Rolls To Production

After assembly, rolls are moved to staging areas as they proceed to the mill.

Each time that a roll is moved, its new location is recorded in [QMOS]. When a mill queries [QMOS] about a work roll or backup roll set about to be loaded into the mill, the roll shop system automatically supplies the necessary identity and measurement data needed by the mill.

Disassemble Rolls

After the completion of a run for each roll, the rolls are changed out of the mill, and disassembled in preparation for the supporting roll components, (chocks, bearings, and seals), to be maintained, and the rolls' surfaces to be ground.

As the rolls are changed out, the mill system automatically sends production information to [QMOS] for inclusion in its roll history.

Maintain Roll Components

The individual parts of the roll assemblies and sets are inspected. At this point, items are replaced and bearings rotated where necessary.

The accumulated production history of bearings in the [QMOS] database is used to assist in the rotation process. The disassembly of the roll assemblies is reflected in [QMOS] to maintain an up-to-date record of the location and status of the components. This can also be scheduled.

Roll Shop

Roll Grinding activity is automatically scheduled based on mill activity. When rolls are removed from the mill and de-chocked, the next step in the usual process is to grind the rolls so that they are ready to use the next time that they are required. When the operator records that chocks have been removed from the rolls, he is given the option to automatically create a roll grinding work order. The rolls will then be scheduled for roll grinding at the next available time.

Similarly, rolls can be scheduled for grinding based on when they are next planned for use in the mill. The roll grinding schedule is calculated to allow sufficient lead time to complete the process in time to have the rolls chocked and ready for the mill before they will be required.

A list of standard procedures or steps required to grind each roll is accessible, providing an instant "how to" manual for all shop personnel. Users can track which steps have been completed, so that the supervisor and other workers all know the current status of the job. This feature also provides a list of SOP's which then conform to ISO standards.

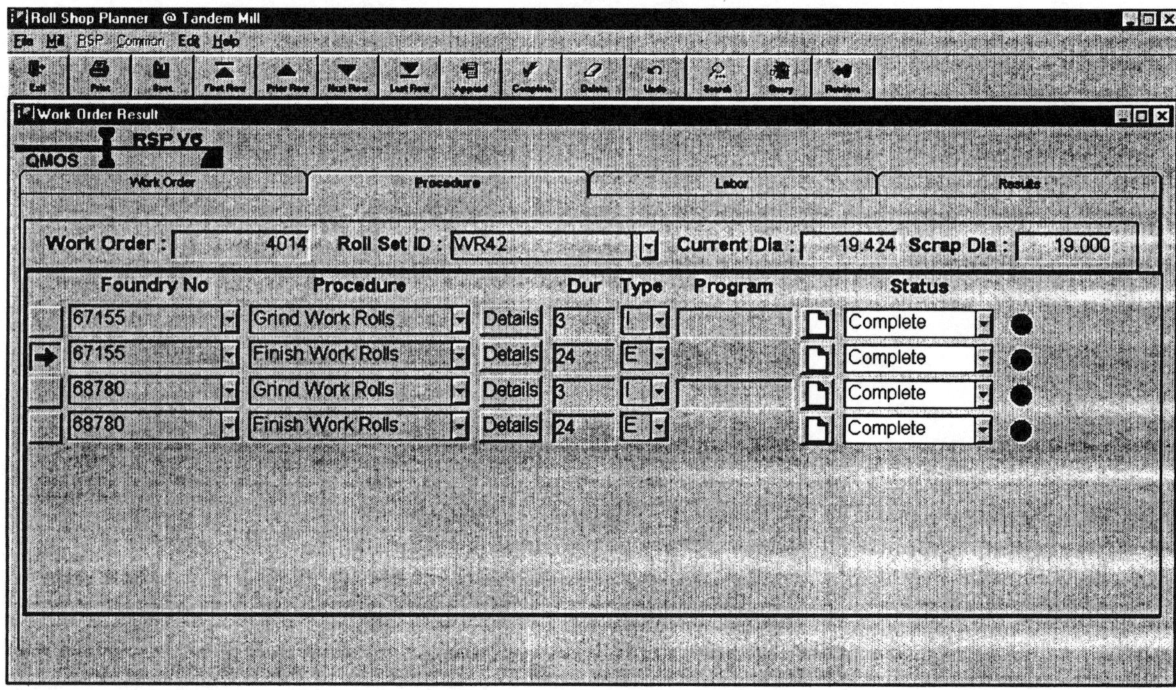

Some shops' roll grinding procedures include external processes for things such as chroming and for other finishing processes or repairs. [QMOS] handles all of this seamlessly. Each roll may be sent to any number of vendors for a variety of reasons. Results from each process can be recorded, including the external cost and the quality of the work done. The whereabouts of each roll is always known, as well as the status of each roll for each step in the roll grinding process.

When a roll is finished being ground, new roll diameters and roll surface readings are recorded in [QMOS]. Any roll which is ground more or less than expected is flagged, requiring the operator to acknowledge and explain what has happened. Hardness readings can be noted, and saved for further analysis. Information on sub-surface conditions such as cracks and bruises can also be stored. The grinder operator responsible for grinding the roll is saved. A full historical profile of each roll is saved, detailing all roll grinding results and suitable for investigation and historical analysis.

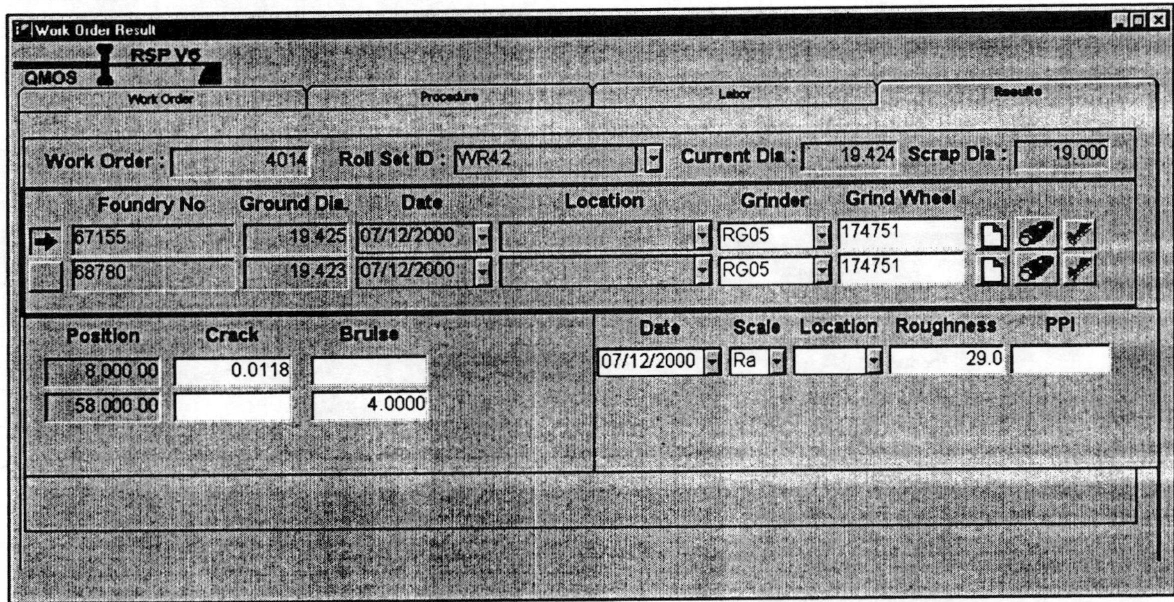

Grind Wheel performance is tracked by recording specific activity and use of each grind wheel. The amount of wear on the wheel, and the number of rolls ground and the total amount ground off all rolls are all attributable back to each individual grind wheel. Detailed reports are available depicting grind wheel inventory and performance and can be tracked by vendor.

Roll Scheduling

Rolls are scheduled for the mill based on roll availability and the upcoming production schedule. The best Work Rolls and Backup Rolls are recommended at each stand, based upon the crown and texture required in each stand. Roll changes must take place whenever the rolls in the mill are not appropriate for the upcoming production, or when they are worn. [QMOS] is able to compare the expected usage on the current rolls with the upcoming rolling, The system will suggest that the rolls be changed based upon comparisons with historical data concerning the usual amount of use by similar rolls when they wear out. The operator is given a view of the whole mill cross referenced with upcoming production, providing the easiest and most efficient means of roll scheduling.

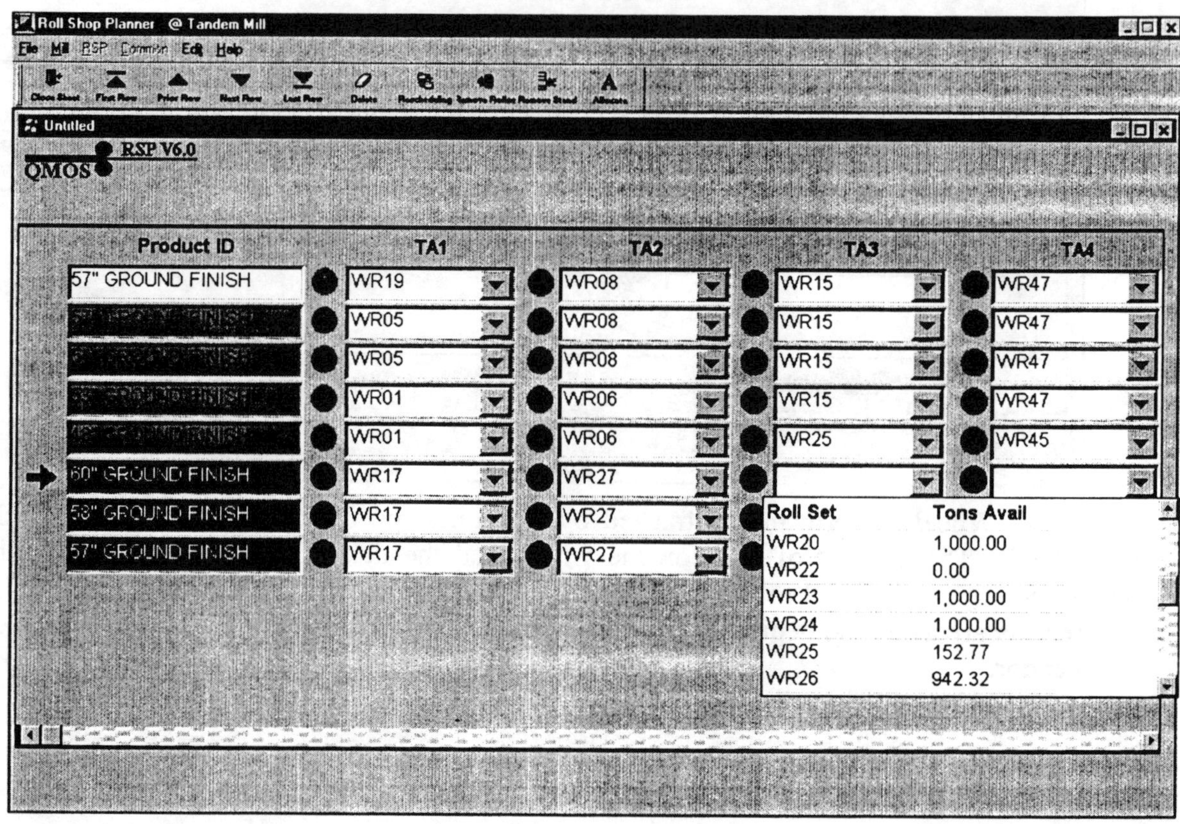

The coil width being used is also considered, in order to prevent marking or scoring on the finished product. [QMOS] checks for the expected finished product width to determine if the rolls might cause a problem due to a change in coil width.

Roll Inventory

All roll purchasing, scrapping, and depreciation is handled by [QMOS]. Worn or broken rolls which need to be scrapped, and [QMOS] records details of the roll status, and removes the rolls from the active roll inventory.

As rolls wear or break, and they are due to be replaced, [QMOS] prepares the necessary Purchase Order for ordering the replacement rolls. When new rolls are received and accepted, they are then added to active roll inventory and can be used in the mill.

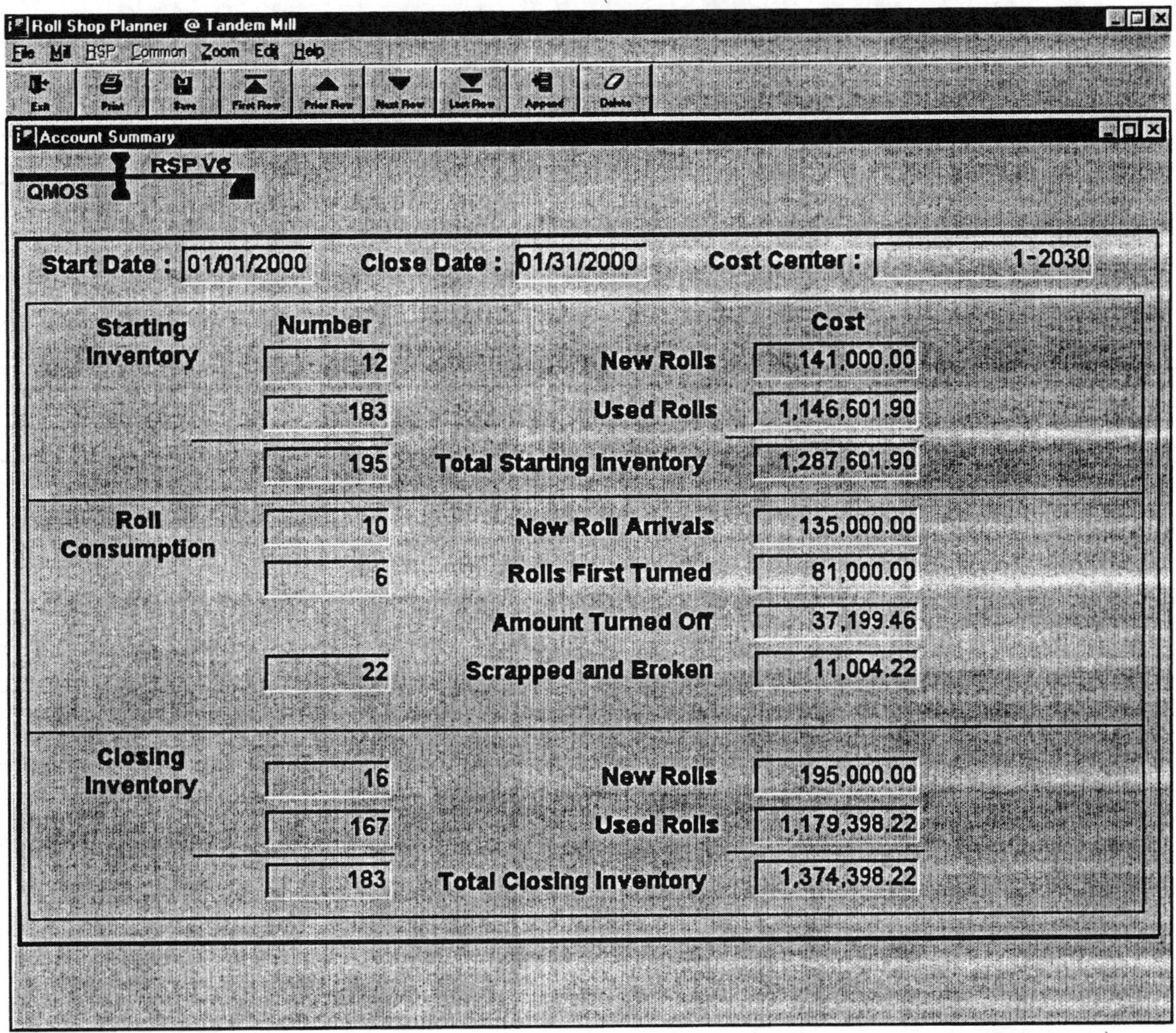

When the roll diameter is reduced due to roll grinding, [QMOS] automatically depreciates roll values accordingly. An end of month Account Summary is produced which shows opening and closing roll inventory balances, as well as the details of roll consumption during the month.

Many other reports are available, including those which illustrate roll performance by supplier, or by roll material, and activity of the roll shop for a period of time or for a specific grinder operator.

Page 7 of 7
Date : 07/19/2000
Time : 11:50:15

QUAD Mill Operation System [QMOS]

Rolls In Use Report

Roll Shop Planner

Report #: 753

Mill Size	Foundry Number	Roll Set	Current Diameter	Crown	Hardness	Location	Supplier	Purchase Cost
			% Remaining	Material	Roughness			Current Value
21.5	92345	WR63	21.328	W-0.005			Acme Rolls	$18,500.00
			67.95	3crmo	30			$12,571.00
21.5	92341	WR63	21.330	W-0.005			Acme Rolls	$18,500.00
			68.04	3crmo	31			$12,587.07
21.5	92334	WR64	21.328	W-0.0015			Acme Rolls	$18,500.00
			67.95	3crmo	28			$12,571.20
21.5	92333	WR64	21.328	W-0.0015			Acme Rolls	$18,500.00
			67.95	3crmo	29			$12,571.20
21.5	93938	WR65	21.378	W-0.003			Acme Rolls	$18,500.00
			69.41	3crmo	31			$12,840.80
21.5	92349	WR65	21.381	W-0.003			Acme Rolls	$18,500.00
			69.41	3crmo	31			$12,857.00
21.5	92336	WR66	21.404	W-0.001			Acme Rolls	$18,500.00
			85.71	3crmo	31			$15,981.60
21.5	92330	WR66	21.406	W-0.001			Acme Rolls	$18,500.00
			86.44	3crmo	30			$15,992.40
21.5	92344	WR67	21.456	W-0.0015			Acme Rolls	$18,500.00
			98.71	3crmo	27			$18,261.80
21.5	92348	WR67	21.459	W-0.0015			Acme Rolls	$18,500.00
			98.80	3crmo	30			$18,278.60
21.5	92356	WR68	21.369	W-0.0015			Acme Rolls	$18,500.00
			15.09	3crmo	28			$2,792.20
21.5	92351	WR68	21.371	W-0.0015			Acme Rolls	$18,500.00
			15.15	3crmo	29			$2,803.00
21.5	92355	WR69	21.352	W-0.0015			Acme Rolls	$18,500.00
			68.46	3crmo	30			$12,655.26
21.5	92354	WR69	21.352	W-0.0015			Acme Rolls	$18,500.00
			68.46	3crmo	30			$12,665.39
21.5	92353	WR70	21.427	W-0.0015			Acme Rolls	$18,500.00
			43.81	3crmo	30			$8,105.80
21.5	92357	WR70	21.429	W-0.0015			Acme Rolls	$18,500.00
			43.87	3crmo	27			$8,116.60
Total Quantity:		162					**Total Current Value:**	$1,146,228.91

Stand Building & Tear Down

Once rolls have been ground and are ready for use, they must be chocked and built into the correct stand. Stand Building activity is scheduled and tracked through the Bearing & Chock Inventory module. [QMOS] tracks which chocks and bearings were used on each roll, and who did the work. Rolls can be built up for a specific stand for upcoming production, or as a spare in case of premature roll wear or breakage.

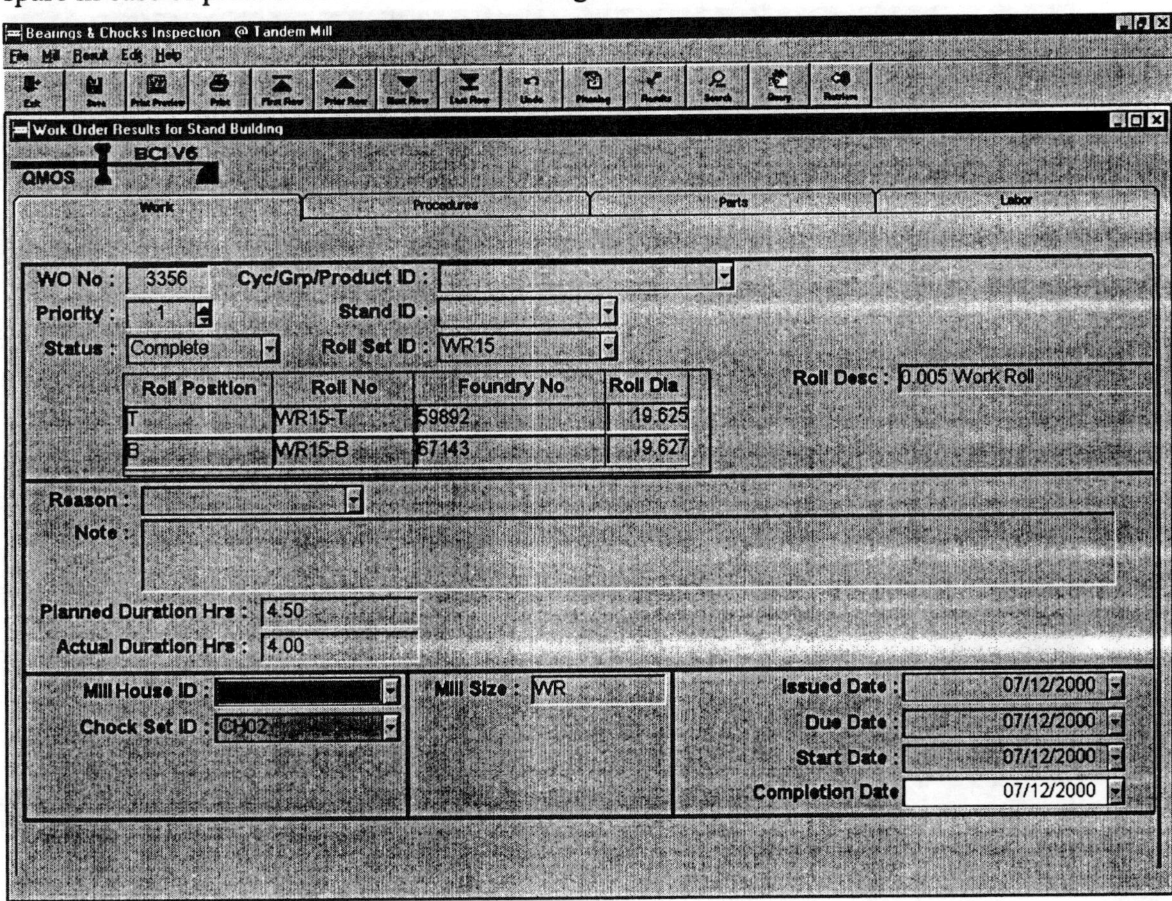

Later, when the rolls come out of the mill, they are de-chocked. The operator records this activity, which then breaks the relationship between the rolls and the chocks. When this is done, the operator is given the option to automatically create a roll grinding work order, beginning the entire process again.

A list of standard procedures or steps required to build and tear down each stand is accessible, providing an instant "how to" manual for all shop personnel. Users can track which steps have been completed, so that the supervisor and other workers all know the current status of the job. This feature also provides a list of SOP's which then conform to ISO standards.

Bearing & Chock Inspection

Bearing and Chock inspections carried out periodically enable the setup shop to track various measurements, such as bearing bore and chock width across liners. When the bearing is assembled into the chock, the load zone used is recorded, as well as the employee who performed the inspection. Also any chock parts, seals and assorted hardware can be recorded on the work order. This can also be used for historical data tracking.

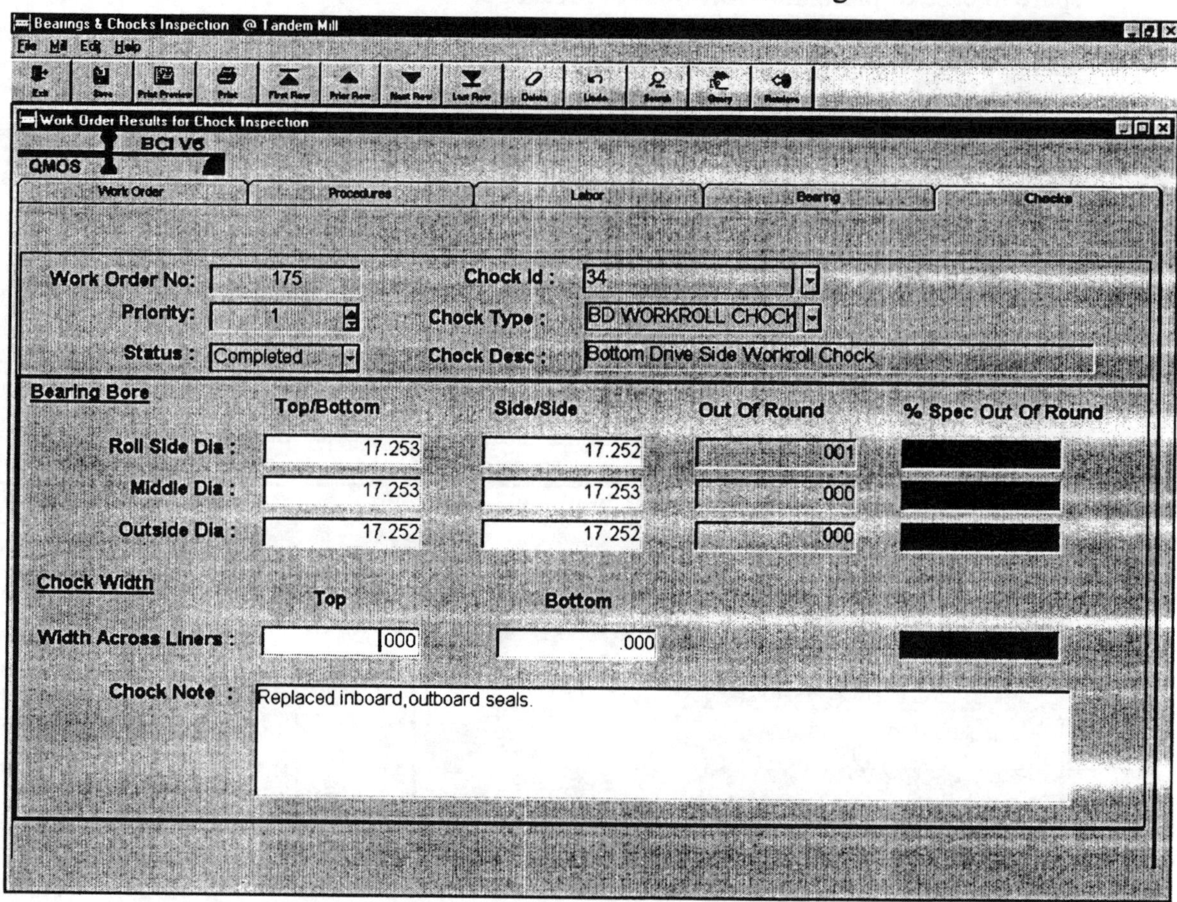

Again, a list of standard procedures or steps required to inspect bearings and chocks is accessible, which provides a manual for all shop personnel. This provides a list of SOP's which then conform to ISO standards.

INTERFACES

A number of interactive processes between [QMOS] and other systems and equipment have been automated at Worthington Steel, including roll grinders and the mill operation system.

Roll Grinding Interface

Worthington uses two Waldrich-Seigen grinders in the roll shop. As roll grinding takes place, a user-friendly interface allows grinding results to be automatically transferred from the grinders directly to [QMOS]. A complete roll diameter profile is captured, along with Eddy Current crack and bruise data. The only user intervention required is to ensure that the correct roll number is entered.

This interface allows complete accuracy for recording roll grinding results, and occurs instantaneously upon completion of the roll grinding process.

Mill System

When the mill system calls for new rolls for any stand, roll diameters are passed from [QMOS], along with other roll information required for mill setup. The mill system operator does not need to manually enter any further roll data in order to set up the mill to run.

As coils are produced, the tonnage and length of steel produced are immediately passed back to [QMOS]. [QMOS] then keeps track of the amount of steel rolled by each work roll and backup roll each time it is used in the mill. Tonnage and length rolled is stored for each roll over its entire life, and since it was last ground. This data provides meaningful analysis of roll usage and performance.

The roll tonnage is also applied directly to the chocks and bearings used with each roll. Bearing and Chock maintenance and inspection schedules can be set up based on their usage, to try to prevent or limit bearing failure. Tonnage is stored for each chock and bearing over its entire life, and since the last inspection.

When rolls are changed in the mill, [QMOS] is immediately notified. The tons and length rolled stop accumulating on the rolls, and begin to accrue to the new rolls, chocks, and bearings. A work order to tear down the stand is automatically created.

ANALYTICAL POWER OF [QMOS]

The tonnage and length of steel produced that are passed back to [QMOS] as coils are produced provides data for meaningful analysis of roll usage and performance. Reports showing roll cost per ton can be printed for whichever rolls or time period required.

Page 1 of 1
Date : 07/13/2000
Time : 08:56:32

QUAD Mill Operation System [QMOS]
Roll History Report
Report #: 706

Roll Shop Planner

From : 01/01/2000 **To : 07/13/2000**

Foundry No : 91539 **Cost Center** 1-2030 **New Diameter :** 21.500 **Initial Hardness :** 855

Work Order	Date Turned	Crown	Turn Diameter	Stock Removed	Cost of Removal	Hardness	Roughness	Tonnage Rolled	Cost Per Ton
2317	01/09/2000	0.005	20.734	.015	111.00		32.	1674.49	$0.07
2357	01/17/2000	0.005	20.726	.008	59.20		29.	282.82	$0.21
2384	01/19/2000	0.005	20.717	.009	66.60		27.	202.25	$0.32
2429	01/24/2000	0.005	20.694	.023	170.20		26.	1416.85	$0.12
2457	01/27/2000	0.005	20.687	.007	51.80		31.	330.85	$0.16
2485	02/01/2000	0.005	20.675	.012	88.80		30.	555.67	$0.16
2550	02/07/2000	0.005	20.664	.011	81.40		32.	1956.89	$0.04
2601	02/14/2000	0.005	20.653	.011	81.40		28.	696.58	$0.12
2670	02/22/2000	0.005	20.647	.006	44.40		27.	164.16	$0.27
2705	02/28/2000	0.005	20.632	.015	111.00		30.	574.90	$0.19
2791	03/06/2000	0.005	20.622	.010	74.00		30.	1212.10	$0.06
2837	03/10/2000	0.005	20.616	.006	44.40		26.	2030.17	$0.02
2887	03/14/2000	0.005	20.607	.009	66.60		29.	298.22	$0.22
2939	03/18/2000	0.005	20.596	.011	81.40		32.	1419.75	$0.06
2975	03/22/2000	0.005	20.548	.048	355.20		30.	726.79	$0.49
3014	03/25/2000	0.005	20.515	.033	244.20		30.	133.48	$1.83
3035	03/27/2000	0.005	20.502	.013	96.20		28.	1943.05	$0.05
3062	03/29/2000	0.005	20.492	.010	74.00		32.	1850.60	$0.04
3084	03/31/2000	0.005	20.486	.006	44.40		30.	612.26	$0.07
3100	04/01/2000	0.005	20.475	.011	81.40		28.	575.31	$0.14
3128	04/04/2000	0.005	20.467	.008	59.20		29.	1273.96	$0.05
3160	04/07/2000	0.005	20.461	.006	44.40		30.	455.23	$0.10
3190	04/11/2000	0.005	20.429	.032	244.80		26.	63.46	$3.86
3271	05/01/2000	0.005	20.058	.371	2,745.40		32.	789.09	$3.48
3471	05/12/2000	0.005	20.047	.011	81.40		26.	263.04	$0.31
3514	05/19/2000	0.005	19.721	.326	2412.40		29.	1644.29	$1.47
3582	05/26/2000	0.005	19.712	.009	66.60		27.	217.40	$0.30
3675	05/31/2000	0.005	19.655	.057	421.80		29.	717.17	$0.59
3704	06/03/2000	0.005	19.644	.011	81.40		35.	673.25	$0.12
3733	06/06/2000	0.005	19.634	.010	74.00		31.	648.10	$0.11
3759	06/10/2000	0.005	19.619	.015	111.00		31.	75.98	$1.46
3808	06/16/2000	0.005	19.544	.075	555.00		32.	567.43	$0.98
3849	06/20/2000	0.005	19.499	.045	333.00		27.	469.91	$0.71
3878	06/23/2000	0.005	19.420	.079	584.60		29.	103.40	$5.65
3914	07/06/2000	0.005	19.358	.062	458.80		31.	1668.99	$0.27
			Totals	**1.391**				**28,287.89**	**$0.36**

Number Of Work Orders : 35 **Average Of Stock Removed :** .040

A Roll History Report can be used to spot interesting results and trends in terms of roll wear or roll value when generated by supplier or by roll material. Reports detailing the performance and usage of grind wheels, or grind wheel suppliers, can also be prepared and measured in the same fashion.

A periodical Roll Grinding Report will show overall roll grinding activity. It can also be produced for a particular lathe operator, which is useful to identify the need for additional training or a possible change required to standard procedures.

From the historical roll profile, it's possible to review and graph roll hardness as the roll wears. This has proven particularly useful when analyzing the causes of roll breakage. Production of the Roll Hardness Profile has become a part of standard routines whenever a roll breaks or wears prematurely.

Page 1 of 1
Date : 04/13/2000
Time : 08:59:58

QUAD Mill Operation System [QMOS]

Roll Grinding Summary Report

Report #: 708

Roll Shop Planner

From : 03/01/2000 **To : 03/31/2000**

Mill Size	Quantity	Stock Removed	Cost Of Removal
Backup Rolls	18	1.30	$11,375.28
Work Rolls	450	12.04	$84,299.94
Overall Totals:	468	13.34	$95,675.22

SUMMARY

The **[QMOS]FLAT** Roll Shop Management System is a client server application runs on Windows family of products. Oracle is used as the data base management system using UNIX, or NT platform. This System is currently used by Worthington Steel, Decatur AL for planning roll shop events, monitoring shop activity, helping operators make decisions during the shift, and for performing analysis and viewing trends to allow for consistent improvement to the processes.

The system has provided improved roll shop efficiency which has lead to a reduced investment in roll and roll related inventory, and increased roll life. The quality of the finished rolled product has improved by utilizing more accurate roll results from the grinders. There is now an increase in mill productivity, directly resulting from roll shop scheduling improvements.

Mode II Crack Growth Analysis of Spalling Behavior for Strip Mill Backup Roll

Yoshihiro Ohkomori, Chu Sakae*
, and Yukitaka Murakami**
Japan Casting & Forging Corporation
46-59, Sakinohama Nakabaru Tobata-ku, Kitakyusyu, 804-8555, Japan
Phone:81-93-884-0033
E-mail:yoshihiro_ohkomori@jcf.co.jp
*, **Graduate School of Engineering, Kyusyu University
6-10-1, Hakozaki, Higashi-ku, Fukuoka, 812-8581, Japan
*Phone:81-92-642-4389, **Phone:81-92-642-3380
*E-mail:schu@mech.kyushu-u.ac.jp
**E-mail:ymura@mech.kyushu-u.ac.jp

Key Words : Spalling, Backup roll, Mode II fatigue crack growth, Mode II threshold stress intensity factor range, Fracture mechanics, Fatigue, Stress intensity factor

1 INTRODUCTION

Higher value added for rolled products in recent years has been accompanied by demands for higher quality and greater diversification, such as for increased flatness and surface quality. At the same time, the need for improved work efficiency in rolling has tended toward extended campaign cycles, higher rolling load, and high-precision rolling, such that rolls are being used under extremely severe conditions. Under such a circumstance greater excellence is being demanded in terms of the properties of steel making backup rolls (BURs). Major performance characteristics required of a BUR are wear resistance, spalling resistance and the least susceptibility to the influences of rolling accidents such as cobbles and tail end pinch. All these characteristics greatly affect the rolled product quality and roll consumption (roll wear amount per tonnage rolled). Spalling in particular has a major influence on the roll consumption, which deteriorates extremely in the event of spalling failure.

Spalling, which is responsible for most roll failure, is a surface exfoliation phenomenon that occurs due to crack initiation and propagation, with cracks developing almost in parallel with the roll surface over a rather long distance. Crack lengths range from about half a revolution to as long as nearly three revolutions, and rolling must be stopped and the roll replaced when spalling occurs. In addition, when crack depth is greater than the effective diameter, the roll must be scrapped. Accordingly, the prevention of spalling failure is a major issue for both users and manufacturers.

Origins can be generally classified into a) internal cracks that initiate from non-metallic inclusions, and b) surface cracks that occur because of thermal shock with the work roll (WR) such as roll slip during rolling. The reduction of non-metallic inclusions is an important countermeasure against spalling induced by internal crack origins,[1] while the selection of materials with superior high temperature strength is suitable in terms of preventing spalling failure caused by thermal shock-induced surface crack origins.[2] However, countermeasures against crack propagation have been insufficient, with the objective often being to increase the Mode I threshold stress intensity factor range ΔK_{Ith}, despite the fact that deep, long cracks in BURs propagate mostly in Mode II. The reason for this is that laboratory test of Mode II cracking in high strength steel such as roll steel is extremely difficult, such that the values of Mode II threshold stress intensity factor range ΔK_{IIth}, have been almost impossible to measure.

The research presented here involved the application of the experimental method newly developed by Murakami and Hamada[3] for determination of the relationship between the Mode II crack growth rate *da/dN* and the Mode II stress intensity factor range ΔK_{II} as well as to identify the Mode II threshold stress intensity factor range ΔK_{IIth} for a high strength BUR steel (equivalent to JIS SKD6), followed by the use of these values for the quantitative evaluation of Mode II crack growth in a spalled BUR. With respect to evaluation of the stress intensity factor range ΔK_{II}, approximate calculations were undertaken by combining 2-D contact loading and an elliptical crack in an infinite body, and more precise analysis was undertaken by means of the body force method.

2 MEASUREMENT OF MODE II CRACK GROWTH PROPERTIES

2·1 Specimens, and Experimental and Measurement Methods

2·1·1 Material

In order to analyze spalling crack growth behavior, a BUR which had a steady and long distance crack development in service was selected. The appearance of the spalled BUR is shown in Fig. 1, with the development of the crack described in Fig. 2. The crack originates at the surface, and propagates in the opposite direction to that of roll rotation for two and three-quarters revolution. Due to grinding after use, the cause of the crack could not be determined.

The spalling in the BUR examined occurred about 10 years previously, and the material and manufacturing method are different from the ones now in use.

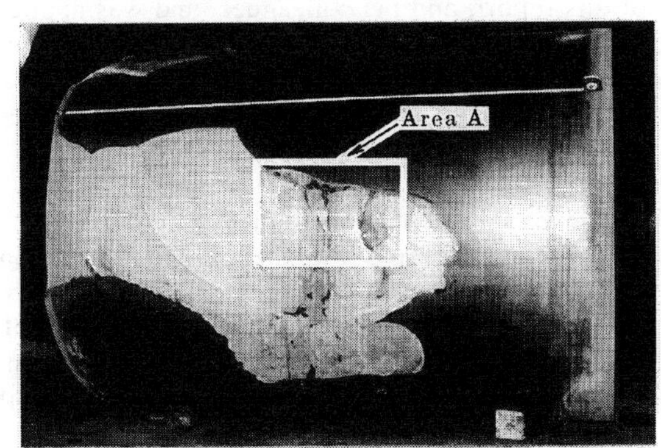

Fig. 1. Spalling of backup roll.

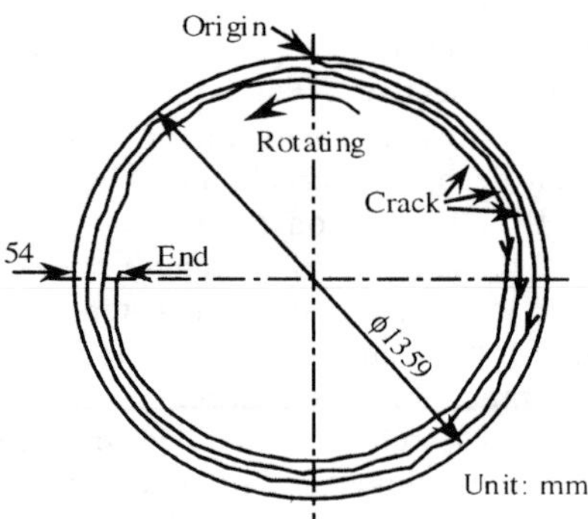

Fig. 2. Spalling of backup roll (The crack rounds the roll twice and 3/4 of a circumference).

Although test steel of the same material could not be obtained, specimens having very similar chemical composition and mechanical properties were obtained from a BUR of a type currently manufactured. Table I shows the chemical composition of the spalled BUR in Fig. 1 and that of the specimens, while Table II shows their mechanical properties.

Table I. Chemical composition (wt.%).

	C	Si	Mn	P	S	Ni	Cr	Mo	V
Specimen	0.42	0.59	0.59	0.008	0.003	0.48	4.93	0.98	0.389
Spalled BUR	0.47	0.36	0.71	0.008	0.003	0.81	3.39	0.43	0.099

Table II. Mechanical properties.

	0.2%proof strength (MPa)	Tensile strength (MPa)	Elongation (%)	Reduction of area (%)
Specimen	1466	1765	7.9	26.7
Spalled BUR	1420	1638	10.4	28.9

2·1·2 Measurement Principles

Figure 3 depicts the basic model of the Mode II crack growth test.[3)] Figure 3 shows stress distributions, the normal and shear stress at the position without slit. In principle, neither tensile nor compressive stress σ exists on the neutral section and so a Mode II fatigue crack is expected to grow along this section. This method was developed by Murakami, one of the authors of this report, and his colleagues, and was applied using a conventional closed-loop type tension-compression fatigue testing machine; care was taken in terms of specimen shape and jig arrangement so as to be able to measure Mode II fatigue crack growth.

2·1·3 Experimental Method

Figure 4 shows the shapes and dimensions of the specimens.[4)] The specimen has a chevron notch and side grooves. The fatigue crack initiates at the tip of chevron notch. The 60° V-shape groove on the side of chevron notch enhances Mode II fatigue crack growth at the section of maximum shear stress and prevents the crack branching to the direction of the maximum tensile stress.

Details of the experimental method are available in reference 3). The testing machine was a conventional closed-loop type tension-compression fatigue testing machine (load capacity of 9.8×10^4N), with stress ratio R=-1 and the load range $\Delta P=2P$ applied to sets of two specimens at room temperature. The crack length was measured by the AC potential method, and Mode II stress intensity factors ΔK_{II} were calculated using a commercial FEM software, ABAQUS.

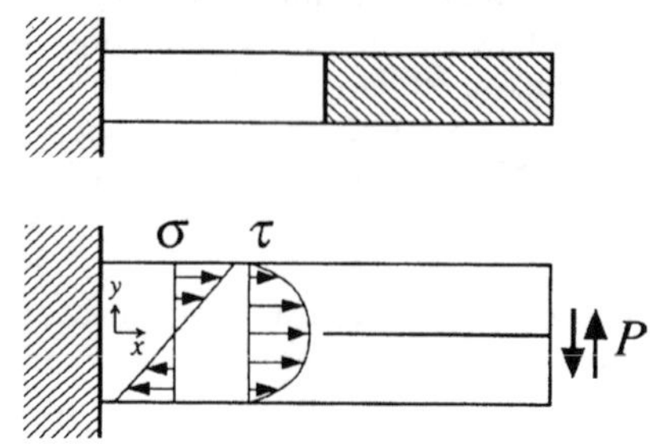

Fig. 3. Basic model of Mode II crack growth test.[3)]

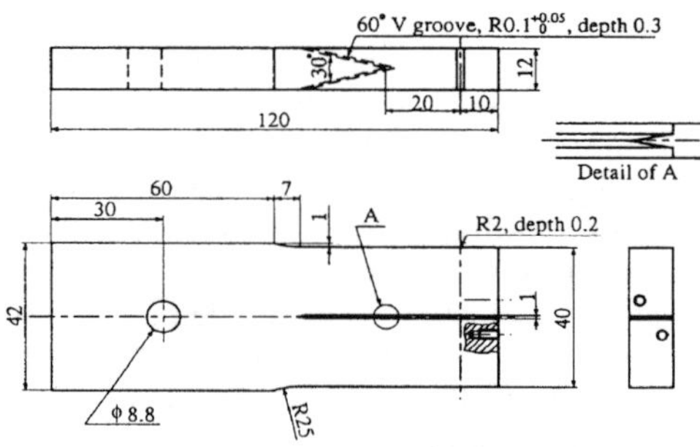

Fig. 4. Shapes and dimensions of the Mode II crack growth specimen.[4)]

2·2 Experimental Results and Observations

2·2·1 Observation of Mode II Fatigue Cracking Fracture Surfaces

Figure 5 shows SEM micrographs of the fracture surface of Mode II fatigue crack growth as obtained in the experiment. These fracture surfaces are clearly different from those of Mode I, and are very similar to the Mode II fatigue fracture surfaces on aluminum, etc. reported by Otsuka et al.,[5)] as well as spalling fracture surfaces on steel making WRs.[6)]

Figure 6 is an SEM micrograph of the Mode II fatigue fracture surface in the spalling of an actual BUR, which is of a type currently manufactured (equivalent to JIS SKD6), and the chemical composition and mechanical properties are the same as for the specimen in Fig. 5. The fracture surfaces in Figs. 5 and 6 are similar, and it was thus considered that Mode II crack growth can be reproduced under laboratory conditions in high strength steels such as roll steel. Figure 7 shows the fracture surface of final Mode I crack growth branching from a Mode II. The fracture surface morphology is quite different in comparison with the shear fatigue crack fracture surface seen in Fig. 5(b).

⇒ Direction of crack growth

(a) Macroscopic view of fracture surface

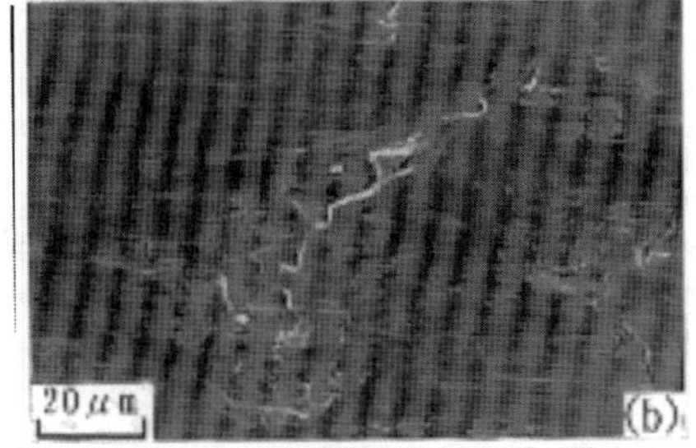

(b) Magnification of (a)

Fig.5. Fracture surfaces of Mode II crack growth in specimen(SKD6, ΔP=1500kgf, R=-1, N=2.4×10^6).

⇒ Direction of crack growth

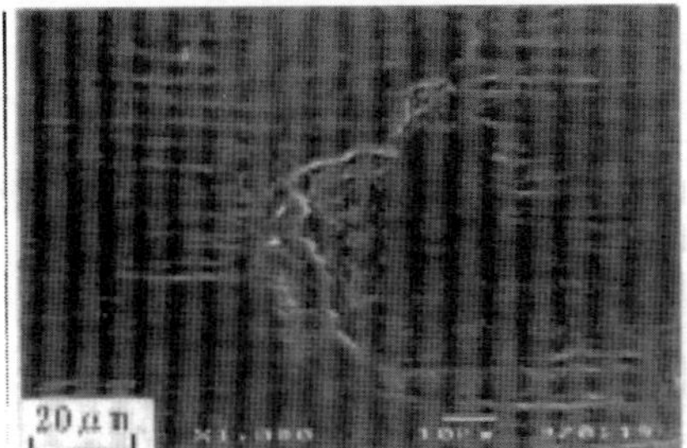

Fig.6. Fracture surfaces of Mode II crack growth of actual BUR(SKD6).

⇒ Direction of crack growth

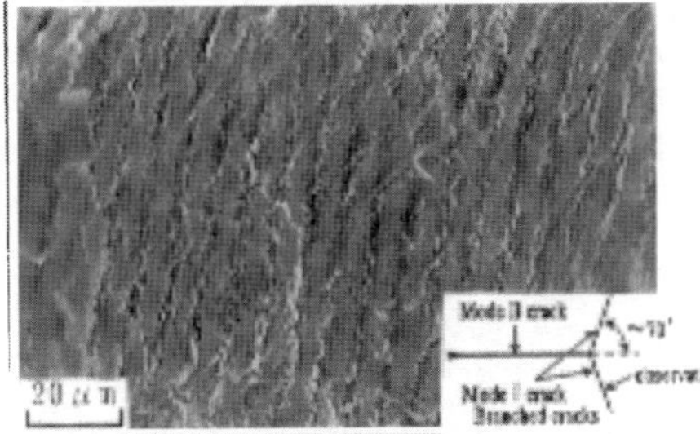

Fig.7. Fracture surfaces of Mode I crack growth after branching from Mode II crack in specimen (SKD6, ΔP=1500kgf, R=-1, N=2.4×10^6).

2·2·2 Measurement of the da/dN-ΔK_{II} Curve and ΔK_{IIth}

As shown in Fig. 5(b), the crack surface is strongly abraded in the direction of crack growth. Presumably large friction loads acted on the surface of the Mode II fatigue crack. Rabinowicz has measured the static friction coefficients of 210 non-lubricated combinations of various metals, with 80% of the measurement results falling in the 0.4~0.6 range, and almost none exceeding 1.0.[7)]

Accordingly, the friction coefficient f_c of the fracture surface was varied in the range of 0~1.0, with values for the Mode II stress intensity factor range ΔK_{II} taken from the analytical results calculated by Hamada.[4)] The values of ΔK_{II} were calculated using a commercial FEM software, ABAQUS. In the calculation the values of ΔK_{II} are reduced to approximately one third of those with no friction in the $0 < f_c < 0.6$ range. In contrast, the changes in ΔK_{II} for the $0.6 \leqq f_c \leqq 1.0$ range are only about 8% of those with no friction. Considering the saturation of the friction coefficient within the $0.6 \leqq f_c \leqq 1.0$ range, ΔK_{II} was calculated with the crack fracture surface friction coefficient at f_c=1.0. Figure 8 shows the relationship between the crack propagation rate da/dN and the ΔK_{II}. From the results shown in the figure, K_{IIth} for the specimen is 13.9 MPa·$m^{1/2}$.

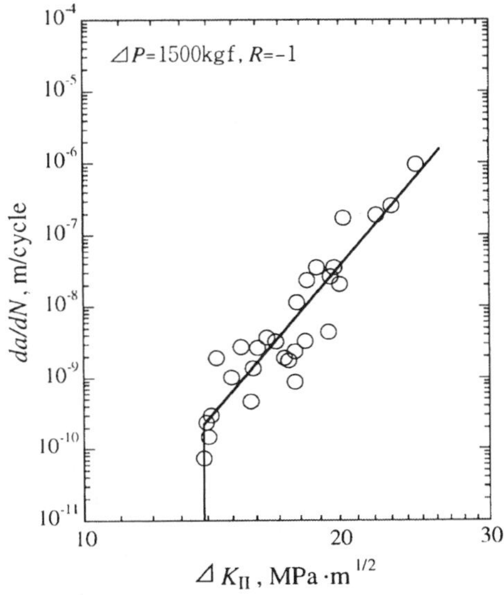

Fig.8. Relationship between ΔK_{II} and da/dN for specimen.

3 Quantitative Evaluation of Mode II Crack Growth Behavior in the BUR

3·1 Mode II Crack Growth Behavior in the BUR

In the case of actual spalling taken up in this study (Figs. 1 and 2), the crack initiates from the origin on the surface for two and three-quarters revolutions in the opposite direction to that of roll rotation. The BUR is driven by the WR such that, during the initial stage of crack growth, hydraulic force is being applied by means of the lubricant or cooling water against the driven side BUR.[8)9)] Initial stage crack growth accompanying the hydrostatic action of cooling water grows as Mode I. However, as the crack grows, the loading condition at the region far from the crack tip tends not to affect the crack growth due to the friction force between the crack faces and bridging effect. Thus, since the driving force of crack growth gradually becomes low and stable, the slow and steady Mode II crack growth lasts and the crack rounds the roll. Finally, large flakes come off in Mode I brittle fracture. Figure 9 illustrates possible mechanisms for such crack growth phenomena.

Figure 10 illustrates the model for the spalling crack of Fig. 1 under the Hertzian contact loading. The half contact width is denoted by a, the depth of crack by h and the crack width by W. Also, the hardness distribution of the BUR is shown in Fig. 11. The spalling occurred during use after the 41st grinding, and the radius of the BUR had been ground by 20.9mm ($\fallingdotseq 4.5a$) at that time. Figures 10 and 11 clearly show that the crack was considerably deeper than the depth at which the cyclic maximum shear stress would be active (y=0.5a), and that it had progressed through the hardened layer. The spalling taken up here is thus representative, as cracks that develop over a long distance nearly parallel to the roll surface and cause spalling are often located deeper than the point of cyclic maximum shear stress, and progress through the hardened layer. The following section presents quantitative analysis to determine whether the ΔK_{II} of this depth exceeds the actually measured BUR ΔK_{IIth}, developing as a Mode II crack. Although the material for which ΔK_{IIth} was measured (Fig. 8) is not exactly the same as the spalled BUR material (Fig. 1), the chemical composition and manufacturing method are not considered to be substantially different in the context of this evaluation.

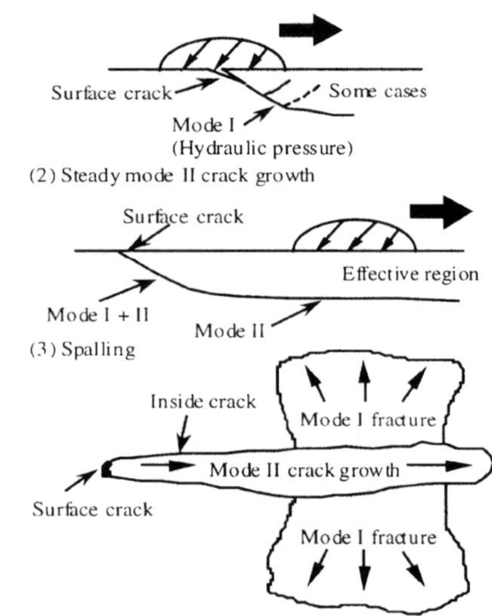

Fig. 9. Possible mechanism of a spalling.

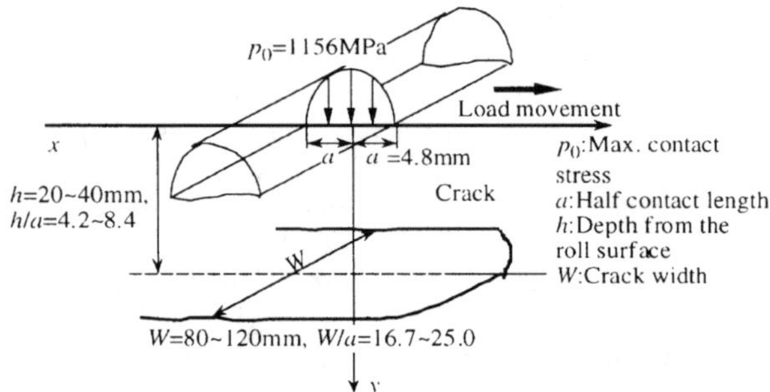

Fig. 10. Schematic description of mode II crack in the spalled BUR.

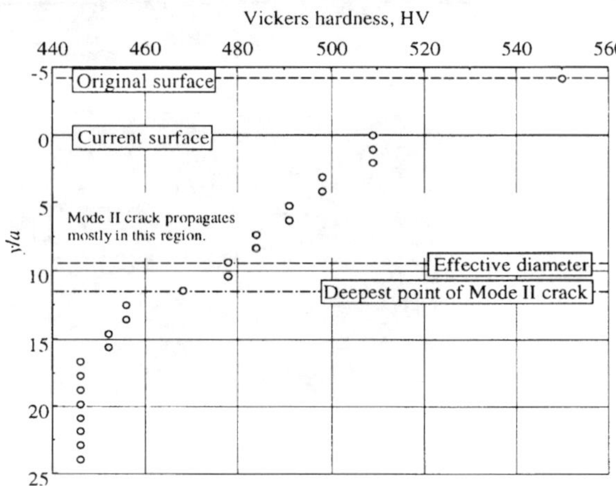

Fig. 11. Vickers hardness distribution in the spalled BUR.

3·2 Calculation of ΔK_{II} by an Approximate Calculation Method

In the evaluation, an approximate calculation method combining 2-D contact loading and an elliptical crack in an infinite body was first used to calculate ΔK_{II}, as follows. Figure 12 shows the distribution of cyclic shear stress τ_{xy}, calculated from the equations[10] at the depths where the spalling crack propagated, and the cyclic maximum shear stress τ_{xymax} was determined for those depth. As shown in Fig. 13, this state was assumed to be that of an elliptical crack with width W and length L in an infinite body subject to the remote application of τ_{xymax}.

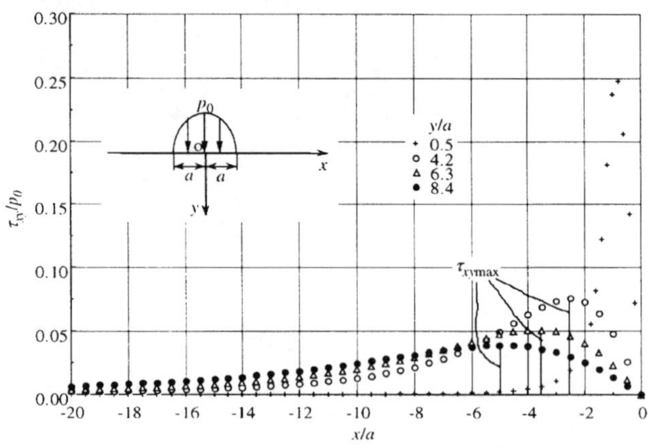

Fig. 12. τ_{xy} distribution under 2-D contact loading.

The maximum ΔK_{II} occurring at the leading edge of the elliptical crack front was then determined,[11] with double this figure taken to be ΔK_{II}.

Table III presents the calculation results. Here, the value of W was taken as the approximate mid-point the dimensional range observed, such that W/a=20. As noted in section 3·1 above, the crack in the spalled BUR was extremely long, but it can be seen from Fig. 12 that the range of effective stress in action at the crack surface was a maximum of about $20a$. Figure 14 is an enlarged view of the crack at the second revolution in the spalling

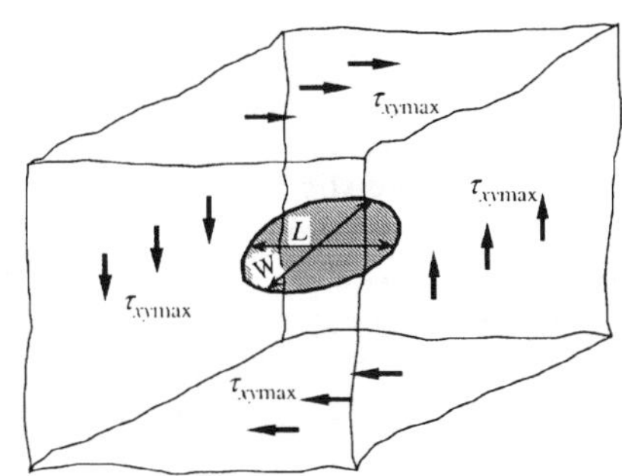

Fig. 13. Elliptical crack in an infinite body.

Table III. ΔK_{II} for mode II crack with $\tau_{xy\,max}$ (Approximate method).

h/a	$\tau_{xy\,max}/P_0$	Aspect ratio L/W		
		0.75	1.0	1.25
		ΔK_{II} (MPa·m$^{1/2}$)		
4.2	0.0750	47.8	50.4	51.5
6.3	0.0507	32.4	34.0	34.8
8.4	0.0384	24.4	25.8	26.4

W/a=20

Table IV. ΔK_{II} for mode II crack with τ_{net} (Approximate method).

h/a	$\tau_{xy\,max}/P_0$	Aspect ratio L/W		
		0.75	1.0	1.25
		ΔK_{II} (MPa·m$^{1/2}$)		
4.2	0.0236	15.0	15.9	16.2
6.3	0.0161	10.3	10.8	11.0
8.4	0.0121	7.7	8.1	8.3

W/a=20

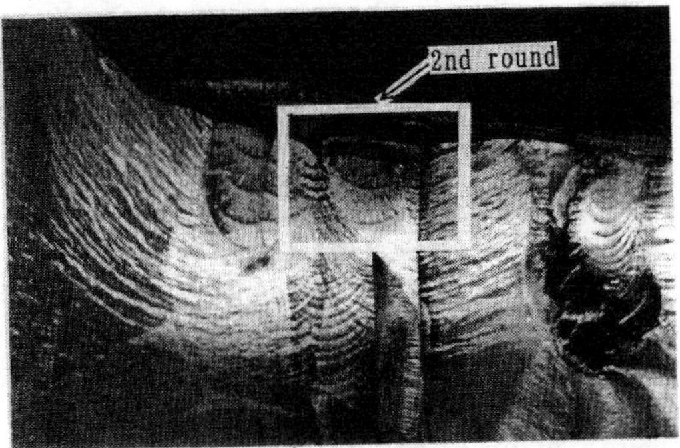

Fig. 14. Mode II crack growth morphology of the spalled BUR (Area A in Fig. 1).

shown in Fig. 1. Calculation of ΔK_{II} was performed taking the aspect ratio from the shape of the leading edge of the crack in Fig. 14 as L/W=0.75, 1.0, and 1.25. As shown in Table III, the ΔK_{II} values are thus 24.4~51.5MPa·m$^{1/2}$. However, as the results of Table III do not take into account the friction coefficient of the crack surface, it should be noted that ΔK_{II} is overestimated. Considering the saturation of the friction coefficient within the $0.6 \leq f_c \leq 1.0$ range, ΔK_{II} was calculated with the crack fracture surface friction coefficient at f_c=0.6. In Table IV, the distribution of net shear stress $\tau_{net}(=|\tau_{xy}|-0.6|\sigma_y|)$ was determined with the friction coefficient f_c of the crack surface taken to be 0.6, and results for ΔK_{II} were obtained in the same manner as described above.

For f_c=0.6, ΔK_{II} ranges from 7.7 to 16.2MPa·m$^{1/2}$, and ΔK_{II} where crack surface friction is taken into account is about 1/3 of that where the crack surface friction is ignored. Thus, considering that the calculated values of ΔK_{II} (7.7~16.2MPa·m$^{1/2}$) during one pass of loading exceed ΔK_{IIth}=13.9MPa·m$^{1/2}$, it can be concluded that the crack which caused spalling grew in Mode II.

3·3 Calculation of ΔK_{II} by the Body Force Method

In order to analyze crack growth behavior more accurately, the body force method was used to calculate ΔK_{II}. Figure 15 presents the calculation model. Figure 16 shows the mesh patterns for the numerical analysis. The basic method is the same as that in reference 12). However, more rigorous repetitive calculation taking account of crack surface contact was conducted to avoid overlapping of crack surfaces. In the cases of L/W=0.75 and 1.25, the circular shape is deformed into the ellipse by multiplying the coordinates of nodes by the proper coefficient. The crack surface friction coefficient was taken to be f_c=0.6, while crack width W was again taken as W/a=20. Figure 17 shows the calculation results in terms of the variation in K_{II} values during a load cycle. K_{II} is calculated for the point at the leading edge of the crack front in Fig. 15, because K_{II} takes its maximum in this position. The value of ΔK_{II} was then determined from the calculation results in Fig. 17.

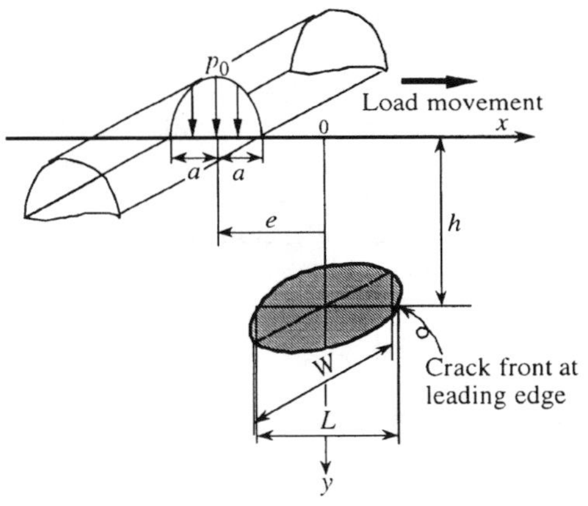

Fig. 15. Numerical model for the body force method.

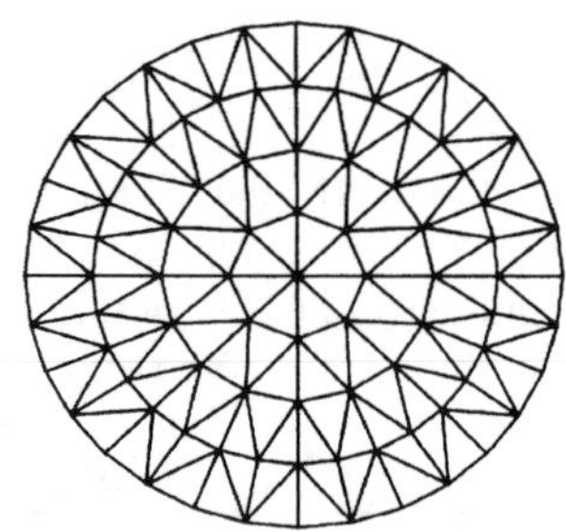

Fig. 16. Mesh pattern for body force method.

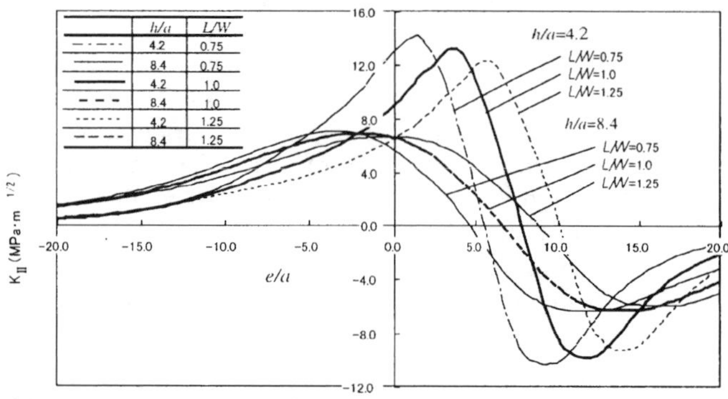

Fig. 17. Variation of K_{II} during a load cycle by body force method.

Table V. ΔK_{II} for mode II crack (body force method, f_c=0.6).

h/a	Aspect ratio L/W			
	0.50	0.75	1.0	1.25
	ΔK_{II} (MPa·m$^{1/2}$)			
4.2	24.5	24.4	23.0	21.5
6.3	16.8	17.4	16.9	16.0
8.4	12.8	13.4	13.1	12.6

W/a =20

In the approximate method presented in the preceding section, ΔK_{II} was calculated for the aspect ratio L/W=0.75~1.25. In the body force method, the value of ΔK_{II} was calculated also for L/W=0.5 as well. Table V shows the results, indicating that ΔK_{II} becomes smaller despite the fact that the crack surface area becomes increasingly large as the aspect ratio L/W increases from 0.75 to 1.25. In contrast, since both the aspect ratio and crack surface area are small for L/W = 0.5, ΔK_{II} in the domain of $h/a \geqq 6.3$ is smaller than in the case of L/W=0.75. These results suggest the numerical model, in which the crack length is estimated at $L \fallingdotseq 20a$, of Fig. 16 is appropriate for simulating the real cases.

Also, from the measured value of ΔK_{IIth} and the calculated values for ΔK_{II} in Table V, the crack growth mode is definitely Mode II in the shallow domain of $4.2 \leqq h/a \leqq 6.3$, while the crack would hardly grow at all or would have an extremely slow growth rate in the deep domain of $h/a \fallingdotseq 8.4$.

In comparing the calculations according to the two methods, it is difficult to calculate the accurate values of ΔK_{II} in terms of the approximate calculation method, but it is possible to judge whether or not growth conditions for Mode II cracking were present.

3·4 Estimation of the Crack Growth Period

According to Table V, the value of ΔK_{II} for Mode II crack growth in the BUR in which spalling occurred in service is 12.6~24.5MPa·m$^{1/2}$. For example, in the case of Mode II cracking in which $\Delta K_{II} \fallingdotseq$24MPa·m$^{1/2}$ for two and three-quarters revolutions, $da/dN \fallingdotseq 5 \times 10^{-6}$m/cycle at $\Delta K_{II} \fallingdotseq$24MPa·m$^{1/2}$ in Fig. 8. Accordingly, it would require about 2.3×10^6cycles (about two months) for the crack to progress that distance. In this study, the crack width (W/a=20) and the friction coefficient (f_c=0.6) of the crack surface are assumed, and the chemical composition and hardness are slightly different for the actual BUR and for the specimens from which the da/dN-ΔK_{II} curve was obtained. However, it may be concluded that the estimation of the period required for crack growth in BURs can be made by the combination of the ΔK_{II} analysis results and the da/dN-ΔK_{II} curve in Fig. 8.

4 CONCLUSION

In the research presented here, ΔK_{IIth} for high strength BUR steel was measured, an accomplishment that has previously been difficult to achieve. This value was used to quantitatively analyze crack growth behavior in the context of spalling in an actual BUR, and the following conclusions were obtained.

1)The fracture surface of the Mode II fatigue test specimens was clearly different from a Mode I fracture surface, and was similar to the spalling fracture surface in the actual BUR. The crack growth behavior in BUR can be reproduced by the newly developed test. From da/dN-ΔK_{II} data, it was found that ΔK_{IIth} was 13.9MPa·m$^{1/2}$.

2)The approximate values of ΔK_{II} were calculated by combining 2-D contact loading and an elliptical crack in an infinite body. More accurate analysis for ΔK_{II} was carried out by the body force method.

3)The values of ΔK_{II} calculated by the body force method with a friction coefficient f_c of the crack surface of 0.6, ranged from 12.6 to 24.5MPa·m$^{1/2}$ for the BUR under spalling condition in service. From the measured value of ΔK_{IIth}=13.9MPa·m$^{1/2}$ and the calculated values for ΔK_{II}, the crack growth mode is definitely Mode II in the shallow domain of $4.2 \leqq h/a \leqq 6.3$, while the crack would hardly grow at all or would have an extremely slow growth rate in the deep domain of $h/a \fallingdotseq 8.4$.

4)It is possible to estimate the period required for crack growth from the ΔK_{II} analysis results and the da/dN-ΔK_{II} curve.

5)In order to prevent spalling failure, ΔK_{II} of the initial defect must not exceed ΔK_{IIth}. The analysis presented here enable one to estimate the allowable defect dimensions contained in BURs for preventing spalling.

5 REFERENCES

1)Y. Ohkomori et al., "Effects of Nonmetallic Inclusions and Small Surface Pits on the Fatigue Strength of a High Strength Steel at Two Hardness Levels" , 32nd ISS Mechanical Working & Steel Processing Conference, October 1990.

2) Y. Ohkomori et al., "Cause and Prevention of Spalling of Backup Rolls" , Transactions ISIJ, Vol.28, 1988, pp.68-74.

3)Y. Murakami et al., "New Measurement Method of Mode II Threshold Stress Intensity Factor Range $\Delta K_{\tau th}$ and Its Application" J. Soc. Mat. Sci., Vol.43, No.493, 1994, pp.1264-1270.

4)S. Hamada, Study of Mode II fatigue crack growth, PhD Thesis, Kyusyu Univ., 1997.

5)A. Otsuka et al., "Mode II Fatigue Crack Growth Mechanism and its Dependency on Material in Aluminum Alloys" , J. Soc. Mat. Sci., Vol.34, No.385, 1985, pp.1174-1179.

6)Y. Sano, E. Matsunaga, Proc. 4th Tribology Committee Meeting, 1991, 31, Committee for Fracture Mechanics, J. Soc. Mat. Sci..

7)E.Rabinowicz, "The Determination of the Compatibility of Metals through Static Friction Tests" ASLE Trans., Vol.14, 1971, pp.198-205.

8)M. Kaneta, M. Suetsugu, Y.Murakami, "Analysis of Surface Cracking in Lubricated Rolling/Sliding Spherical Contact" , Trans. Japan Soc. Mech. Engrs. Series C, Vol.51, No.468, 1985, pp.2167-2173.

9)Y. Murakami, C. Sakae, S. Hamada, "Mechanism of Rolling Contact Fatigue and Measurement of ΔK_{IIth} for Steels" , Engng Against Fatigue, Eds., J. H. Beynon et al. A. A. Balkema Publishers, Rotterdam & Brookfield, 1999, pp.473-485.

10)J. O. Smith, C.K.Liu, "Stresses Due to Tangential and Normal Loads on an Elastic Solid with Application to Some Contact Stress Problems" , Trans. ASME J. Appl. Mech., Vol.20, 1953, pp.157-166.

11)M. K. Kassir, G. C. Sih, "Three-Dimensional Stress Distribution Around an Elliptical Crack Under Arbitrary Loadings" , Trans. ASME J. Appl. Mech., Vol.33, 1966, pp.601-611.

12)M. Kaneta, T. Okazaki, Y. Murakami, "Propagation of an inclined Subsurface Crack in Rolling/Sliding Contact Fatigue" , Key Engineering Materials, Vol.33, 1989, pp.191-212.

STUDENT RESEARCH

Martensite Determination in Multiphase Steels Using Neutron Diffraction

Roberto Filippone[1], John Root[2] and Steve Yue[1]
[1]McGill University
3610 University Street
Montreal, Quebec, H3A 2B2
Canada
(514) 398-4755
e-mail: Steve@minmet.lan.mcgill.ca
[2]National Research Council of Canada
Chalk River Laboratories
Chalk River, Ontario, K0J 1J0
Canada
(613) 584-3311 ext. 3974
e-mail: John.Root.nrc.ca

Key Words: Martensite, Neutron Diffraction, Dual-phase Steels, TRIP Steels, Ferrite, Peak Broadening

ABSTRACT

Experimental studies were performed to determine the feasibility of using neutron diffraction to quantify the volume fraction of martensite in dual-phase steels. Two methods were evaluated, ferrite peak broadening due to martensite overlap, and deconvolution of the diffraction peak by curve fitting. Results revealed that peak broadening increases with increasing martensite volume fraction and demonstrates three stages of broadening. Peak deconvolution was not feasible as no identifiable martensite peak shoulder was observed.

INTRODUCTION

The continuous drive to produce high quality and cost effective classes of high strength/high formability steels has sparked a growing interest in the industrial applications of Transformation Induced Plasticity (TRIP) and dual-phase steels. For TRIP and dual-phase steels, this combination of strength and formability can be attributed to the microstructure of the steels, specifically due to the presence of retained austenite, martensite and/or bainite.[(1-6)]

One of the challenges lies in characterizing the microstructure of these steels, specifically the complex multiphase structure common in TRIP steels. Typical microstructural examination using optical microscopy is difficult considering the small volume fractions and locations of certain phases. An excellent example of this is in determining the presence of retained austenite within bainite. Because of this difficulty, researchers have turned to other techniques to gain information. One common technique uses x-ray or neutron diffraction to quantify the retained austenite phase.(7-9) The advantage of using neutron over x-ray diffraction lies in the penetrating capability of the neutron.(10) This allows for a bulk measurement giving a better statistical representation of the specimen. The question then becomes how to measure the other phases, namely martensite and bainite, using diffraction, thus allowing for one technique to characterize the complete microstructure of a multiphase steel.

In the case of bainite (α + Fe_3C), the carbide structure will have separate diffraction peaks that can be isolated in the scattering pattern, as is the case with retained austenite.(11) The difficulty lies in detecting martensite because there exists peak overlap with ferrite. In fact, martensite causes the ferrite peak to broaden because it has a body-centered tetragonal crystal structure, and therefore has two diffraction peaks, corresponding to the a and c lattice parameters of the tetragonal structure, for every ferrite (body-centered cubic) peak. This is shown schematically in Figure 1.

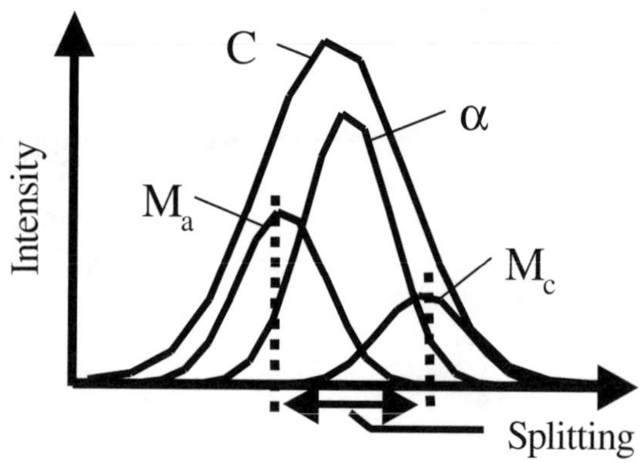

Figure 1: Schematic diagram showing ferrite (α) overlap by martensite (M) and resulting convoluted peak (C).

Previous x-ray diffraction measurements on fully martensitic steels have shown that the martensite lattice parameters are carbon dependent.(12-15) The variation in carbon content dictates the position of the martensite peaks, i.e., the amount of splitting between the peaks. Intuitively, increasing carbon levels will increase the splitting or tetragonality of the martensite crystal structure as carbon takes up interstitial positions within the lattice. This is clearly visible in Figure 2, which is a schematic reproduction of work reported by Roberts.(12)

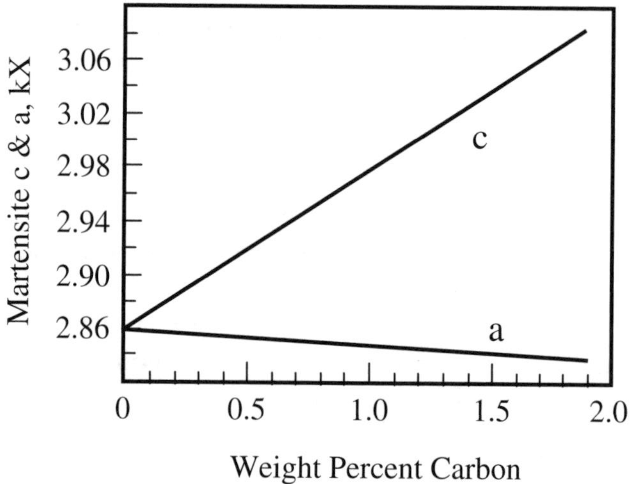

Figure 2: Schematic diagram showing the variation of martensite lattice parameters with carbon content.

De Meyer *et al.* have reported identifying the martensite peak corresponding to the c lattice parameter using x-ray diffraction.(8) If the peak is identifiable, the lattice parameters and carbon content can be determined by the functions:(12)

$$a = 0.2861 - 0.0013 \text{ (wt\% C), nm} \qquad \text{(eq. 1)}$$
$$c = 0.2861 + 0.0116 \text{ (wt\% C), nm} \qquad \text{(eq. 2)}$$

Knowing this information, deconvolution of the ferrite and martensite peaks would be possible.

This study examines the feasibility of using neutron diffraction as a method of detecting and quantifying (volume fraction) martensite, either through deconvolution of the ferrite and martensite peaks, or through ferrite peak broadening.

EXPERIMENTAL DETAILS

The simple chemistries (Table I) of the steels under investigation were chosen to facilitate the formation of dual-phase microstructures with varying amounts of martensite. Differing levels of C and Mn served to increase the steel hardenability, while the C content also affected the splitting of the martensite peaks.

Table I: Steel Chemical Compositions (in wt.%)

Steel	C	Mn	P	S	Si	Cr	Ni	Cu	Co
1	0.079	0.426	0.009	0.012	0.062	0.041	0.069	0.033	0.024
2	0.180	0.470	0.006	0.011	0.140	0.050	0.070	0.030	0.040
3	0.170	0.700	0.007	0.001	0.210	0.060	0.070	0.040	0.040

Thermal processing of the steels consisted of solutionizing at 1200°C for 30 minutes and furnace cooling to room temperature. Cold rolling (30% reduction) was then done to refine the initial grain structure. This reduction in grain size served to deter the formation of martensite allowing for smaller volume fractions of martensite in the dual-phase structures. Specimens approximately 10 mm x 10 mm x 15 mm were cut and subjected to intercritical annealing and quenching as illustrated in Figure 3. One specimen for each steel chemistry was re-solutionized at 1200°C and furnace cooled to room temperature to serve as a strain free, martensite-free reference for the neutron diffraction testing.

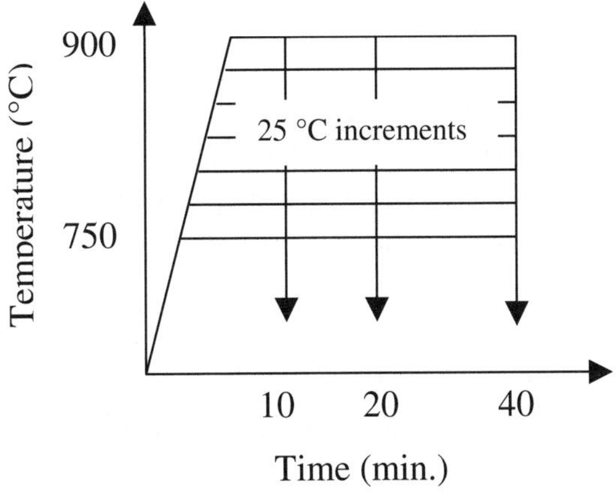

Figure 3: Schematic illustration of intercritical annealing and quenching schedule used to produce dual-phase microstructures.

Martensite volume fractions were first determined by image analysis. Samples were etched using nital and sodium metabisulfite to improve contrast in the martensitic structures.[(16)]

Neutron diffraction experiments were conducted at the NRU reactor at Chalk River Laboratories. A triple-axis spectrometer equipped with a variable-width single-wire position sensitive detector was used to measure the (110) diffraction peaks of martensite and ferrite. The (110) diffraction was selected since it has the greatest peak intensity. Using a wavelength (λ) of 0.237 nm and a pyrolytic graphite filter to remove $\lambda/2$ and $\lambda/3$ wavelengths ensured that there would be no contamination of the diffraction pattern from higher ordered ferrite peaks. Sampling volume was limited to 2 mm x 2 mm x 15 mm to maximize the resolution to intensity ratio. Instrumental resolution, determined as outlined by Cagliotti *et al.*, was 0.56 degrees.[(17)]

Diffraction patterns were fitted with gaussian distributions and peak breadths were measured as the full width at half the maximum peak intensity (FWHM). Figure 4 depicts this graphically. Peak broadening was then evaluated using the following expression:[(18)]

$$B = \sqrt{b_m^2 - b_{ref}^2} \qquad \text{(eq. 3)}$$

where B is the peak broadening in degrees, b_m is the measured breadth of the dual-phase sample in degrees, and b_{ref} is the breadth of the reference, fully ferritic sample in degrees.

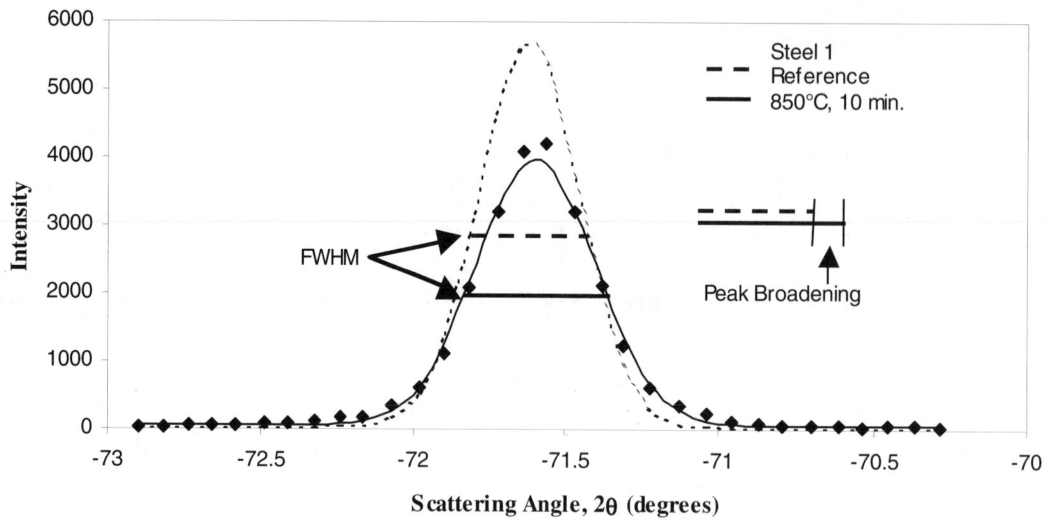

Figure 4: Peak broadening as seen in neutron scattering pattern.

RESULTS and DISCUSSION

The obtained neutron diffraction patterns do not show the presence of the martensite peaks. That is, the resulting diffraction peak is a convolution of the ferrite and martensite phases with no clear or identifiable shoulder that may be associated to a martensite peak. Therefore deconvolution of the diffraction peak cannot be done without inferring structural information on martensite and ferrite.

Diffraction patterns with increasing amounts of martensite are shown in Figure 5. While no martensite peaks can be identified, it is clearly evident that there is an increase in broadening with increasing martensite volume fraction.

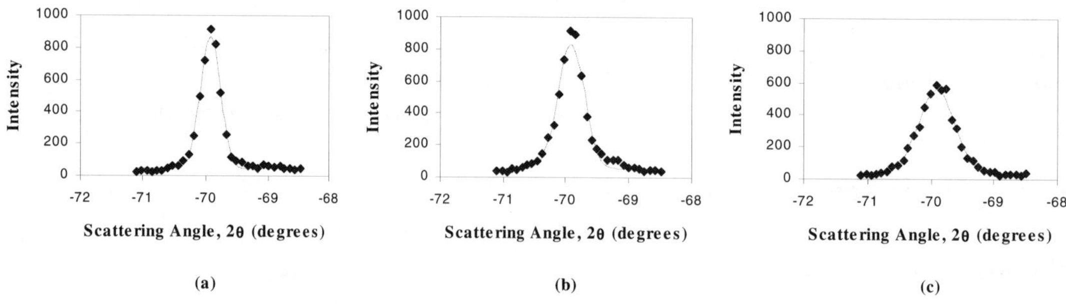

Figure 5: Neutron scattering pattern of steel 2 for (a) low vol.% martensite, (b) mid vol.% martensite, and (c) high vol.% martensite.

The broadening of the diffraction peaks in this experiment can be attributed to two variables; the overlapping of the ferrite peak by martensite peaks, and straining of the crystal structure due to the martensitic transformation. While peak broadening due to small particles can occur, the grain size of the specimen would need to be less than 1 μm.[19] Visual inspection by optical microscopy indicated that the grain size exceeded this minimum. Residual strain from the cold rolling is also assumed not to have affected the broadening as the time and temperature of the intercritical annealing is expected to have removed any stored lattice strain.

The relation between peak broadening and martensite volume fraction is shown in Figure 6. While no simple relationship is evident, the graph does indicate that there is an increase in broadening with increasing martensite volume fraction. Furthermore, the results seem to suggest that there exists three stages of broadening.

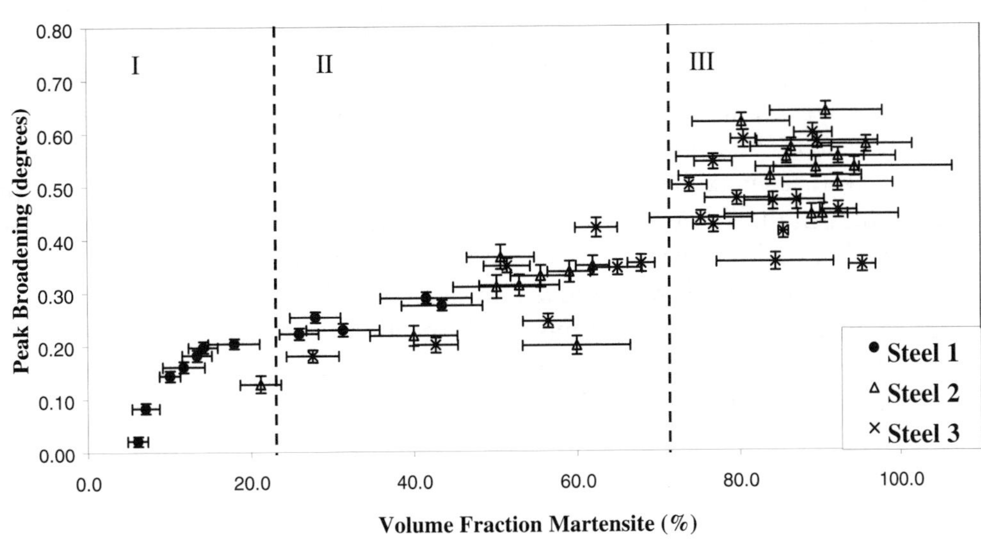

Figure 6: Peak broadening versus martensite volume fraction.

Stage I: Low volume fractions of martensite - The broadening in stage I can be attributed to the straining of the ferrite peak. As is well known, martensite forms by a shear transformation. This shearing of the lattice causes stresses within the structure that are accommodated by the ferrite matrix. Since the volume fraction of martensite is small, there would be a minimal effect on the ferrite peak width (FWHM) by the martensite peaks. At low martensite volume fractions, the carbon content of the martensite would be greatest. Referring back to Figure 2, the degree of splitting between the a and c lattice parameters of the martensite crystal structure will be greatly affected by this variation in carbon. Therefore, the relative intensity of the martensite peaks will be weak compared to the ferrite peak and the splitting of the martensite peaks will be wide. This would, on balance, seem to indicate that the broadening between the convoluted and reference peaks is due to a difference in peak breadth of the strained and unstrained ferrite peaks.

Stage II: Mid volume fraction of martensite - With increasing martensite volume fraction, the relative intensity of the martensite peaks increases while the tetragonality of the martensite crystal structure decreases. The ferrite peak would continue its trend of reducing in intensity (less ferrite in the structure) and increasing in broadening due to the strain from the martensite formation. Stage II seems to be characterized as a linear portion in the broadening versus volume fraction curve (Figure 6). It is believed that this is due to a tradeoff between the narrowing of the split between the martensite peaks and the straining of the ferrite and martensite peaks. As the fraction of ferrite decreases it will no longer be able to completely accommodate the strain from the shear transformation. Thus, the martensite structure will also be distorted and strain broadened.

Stage III: High martensite volume fraction - At elevated martensite volume fractions the ferrite peak becomes relatively weak compared to the martensite peaks. The broadening of the convoluted peak then becomes dependent solely on the martensite peaks as the ferrite peak is lost within the base of the martensite peaks. Again the degree of broadening will depend on the strain in the martensite lattice and the reduction in tetragonality.

It is evident that in order to use peak broadening as a method to measure the volume fraction of martensite, the degree of broadening due to lattice strain would need to be determined. This is a greater concern with pieces that are not engineered with strain free structures. Deconvolution may be possible by examining the (200) diffraction peak. This corresponds to the greatest degree of splitting between the martensite peaks and would be the best chance at observing a shoulder in the diffraction pattern. A neutron diffraction pattern of the (200) peak of a fully martensitic eutectoid steel clearly shows the presence of the martensite c lattice parameter as a shoulder (Figure 7). Yet even at this elevated carbon level (0.71 wt.%) the martensite peaks remain broad and difficult to separate (i.e., no clear maxima of the martensite c peak). The broadness of the peaks, which cover approximately five degrees in the spectrum, cannot be explained simply by lattice strain from the shear transformation. This suggests that there exists a variation in carbon content in the martensite, and therefore a variation in the tetragonality of the structure, which gives an inherently broad peak. This variation in carbon may be explained by the segregation of carbon atoms to dislocations and interlath austenite during rapid quenching.[20] Given this result, plus ferrite peak overlap and strain broadening, it will be extremely challenging to separate martensite from ferrite using deconvolution methods.

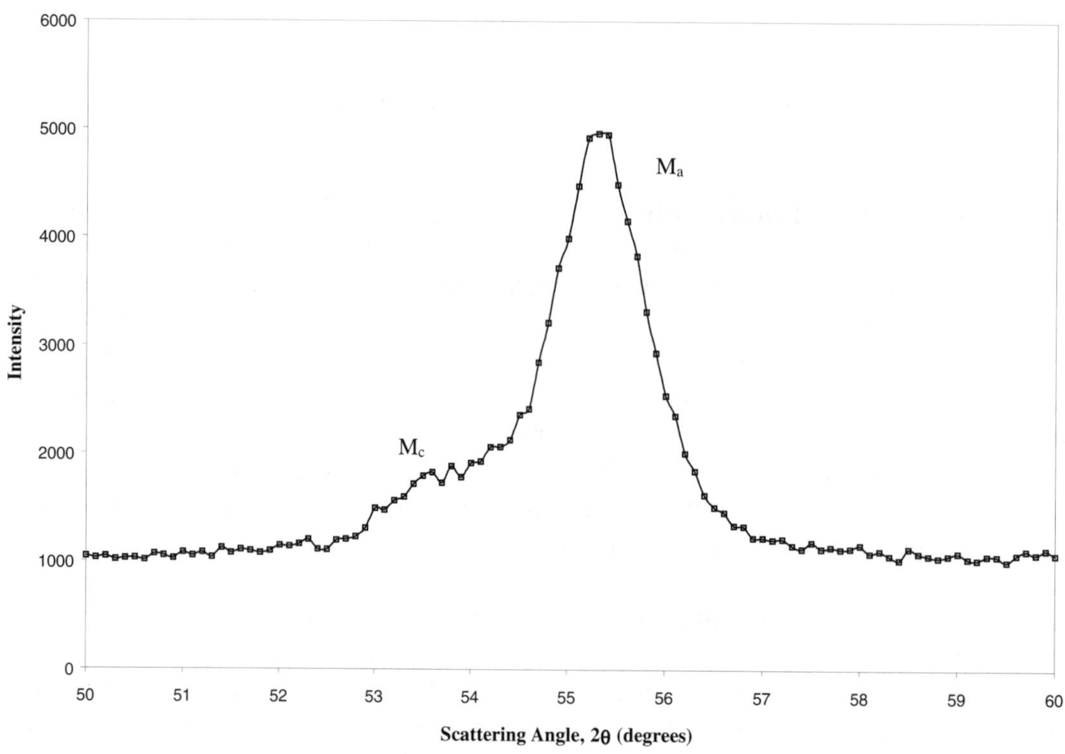

Figure 7: Martensite (200) neutron diffraction peak of a quenched eutectoid steel.

CONCLUSIONS

The use of neutron diffraction to determine the volume fraction of martensite in dual-phase steels was evaluated. The main conclusions are:

1. Peak broadening of the (110) diffraction increased with increasing martensite volume fraction. Three stages of broadening were observed.
2. Strain broadening is present in all three stages and limits the ability to determine the presence of martensite using ferrite peak broadening.
3. Deconvolution of the (110) diffraction peak is not possible as no martensite peak shoulder is identifiable. Therefore no peak fitting can be done without inferring martensite structural data.

ACKNOWLEDGEMENTS

The authors would like to thank the Canadian Steel Industry Research Association (CSIRA) and the Natural Science and Engineering Research Council (NSERC) of Canada for their financial support. The authors would also like to thank Ivaco Rolling Mills for supplying the steels for this study.

REFERENCES

1. V.F. Zackay, E.R. Parker, D. Fahr and R. Bush, "The Enhancement of Ductility in High-Strength Steels", Trans. ASM., Vol. 60, 1967, pp. 252-259.
2. A. Zarei-Hanzaki, P.D. Hodgson and S. Yue, "Hot Deformation Characteristics of Si-Mn TRIP Steels With and Without Nb Microalloy Additions", ISIJ International, Vol. 35, No. 3, 1995, pp 324-331.
3. S. Sangal, N.C. Goel and K. Tangri, "A Theoretical Model for the Flow Behavior of Commercial Dual-Phase Steels Containing Metastable Retained Austenite: Part II. Calculation of Flow Curves", Metallurgical Transactions A, Vol. 16A, Nov. 1985, pp 2023-2029.
4. A.R. Mader, "Factors Affecting the Ductility of Dual-Phase Alloys", Formable HSLA and Dual-Phase Steels Symposium Proceedings, Oct. 1977, pp.87-98.
5. A.A. Petronenkov, "Heat Treatment of Low-Alloy Steels to Produce a Ferrite-Austenite-Bainite Structure", Phys. Met. Metall., Vol. 71, No. 5, 1991, pp. 87-92.
6. H.Q. Liu, C. Zhu and G.C. Li, "Effect of Phase-Transformation on Mechanical Behavior of Dual-Phase Steel Plate", Theoretical and Applied Fracture Mechanics, Vol. 19, 1993, pp. 127-132.
7. M. Bouet, J. Root, E. Es-Sadiqi and S. Yue, "The Effect of Mo in Si-Mn Nb Bearing TRIP Steels", Materials Science Forum, Vols. 284-286, 1998, pp. 319-326.
8. M. De Meyer, D. Vanderschueren, K. De Blauwe and B.C. De Cooman, "The Characterization of retained Austenite in TRIP Steels by X-Ray Diffraction", 41st MWSP Conf. Proc., ISS, Vol. XXXVII, 1999, pp. 483-491.
9. A. Di Chiro, J. Root and S. Yue, "Processing and Properties of a Si-Mn Nb-Bearing TRIP Steel For Use in Fasteners", 37th MWSP Conf. Proc., ISS, Vol. XXXIII, 1996, pp 373-382.
10. G.E. Bacon, "Diffraction by Crystals", Neutron Diffraction, 3rd Ed., Oxford University Press, Great Britain, 1975, p.67.
11. W.B. Pearson, "Work on Borides, Carbides, Hydrides, Nitrides and Oxides", A Handbook of Lattice Spacings and Structures of Metals and Alloys, Pergamon Press Inc., New York, NY, 1964, pp. 919-923.

12. C.S. Roberts, "Effect of Carbon on the Volume Fractions and lattice Parameters of Retained Austenite and Martensite", JOM, Feb. 1953, pp203-204.
13. H. Lipson and A.M.B. Parker, "The Structure of Martensite", Journal Iron and Steel Inst., Vol. 149, 1944, pp. 123-141.
14. E. Ohman, "X-ray Investigations on the Crystal Structure of Hardened Steel", Journal Iron and Steel Inst., Vol. 123, 1931, pp. 445-463.
15. G. Hagg, "The Decomposition of Martensite", Journal Iron and Steel Inst., Vol. 130, 1934, pp. 439-451.
16. Metals Handbook, 8th Ed., Vol. 8, "Metallography, Structures and Phase Diagrams", ASM, Metals Park, Ohio, 1973, pp. 64-67.
17. G. Cagliotti, A. Paoletti and F.P. Ricci, "Choice of Collimators for a Crystal Spectrometer for Neutron diffraction", Nucl. Instrum. Methods, Vol. 3, 1958, pp.223-228.
18. B.D. Cullity, "Structure of Polycrystalline Aggregates", Elements of X-Ray Diffraction, 2nd Ed., Addison-Wesley Publishing Company, Inc., Mass., 1979, p.284.
19. H.P. Klug and L.E. Alexander, X-Ray Diffraction Procedures for Polycrystalline and Amorphous Materials, 2nd Ed., John Wiley & Sons, Inc., 1974, p.643.
20. G.B. Olson and W.S. Owen, "Tempering of Ferrous Martensites", Martensite, ASM International, USA, 1992, pp.251-257.

Internal-Friction Measurements of Ultra-Low Carbon Sheet Steels

by Waleed Al Shalfan[1], John G. Speer[1], David K. Matlock[1], Sudook Kim[2], and Hassell Ledbetter[2]

1. Advanced Steel Processing and Products Research Center
Colorado School of Mines
Golden, CO 80401, USA

2. Materials Reliability Division
National Institute of Standards and Technology
Boulder, CO 80303, USA

Key words: Internal-Friction, Ultra Low Carbon Bake Hardenable Steels, AKDQ and IF Sheets

ABSTRACT

Solute carbon levels are important in ultra-low carbon sheet steels as they control the aging and bake-hardening characteristics. Internal-friction testing is sensitive to small differences in solute concentrations, and is being used to investigate solubility relationships in ferrite for ultra-low carbon steels containing small additions of Nb or V. Preliminary results for a Ti-V microalloyed steel indicated that clear and distinctive Snoek peaks were not generally observed, and lower frequency relaxations were also observed to overlap the Snoek peaks. The mechanisms associated with the relaxations were not immediately understood, and approaches to resolve these issues and thus improve interpretation of the data were considered in the study reported here. Forced torsional pendulum measurements on low carbon aluminum killed and interstitial-free steels are reported. These experiments were designed to examine the influence of specimen preparation, prestrain, aging, and solute levels on internal-friction spectral characteristics.

INTRODUCTION

The desire to improve automobile fuel economy has led to increased efforts to reduce vehicle weight. Reducing the thickness of auto body sheets combined with the use of high strength steels are therefore of interest. Consequently, microalloyed ultra-low carbon (ULC) bake hardenable steel has been developed for exposed hot dip coated applications having good combinations of strength and formability (1).

In order to control the behavior of bake hardening steel, the amount of C in solution is an important factor as it controls the aging and bake hardening characteristics. It is generally believed that 5-20 ppm of C in solution is needed to obtain a sufficient bake hardening response with acceptable aging behavior at room temperature (1). Internal-friction is one of only a few available techniques to quantify the concentration of interstitial carbon or nitrogen in the ferrite lattice (2-5).

Preliminary internal-friction measurements on an experimental Ti-V microalloyed ULC steel indicated that interpretation of the results is difficult. The study presented in this paper was therefore conducted to provide a better understanding of the factors controlling internal-friction measurement results. In particular, the influence of specimen preparation, prestrain, and aging are considered. Materials containing either substantial or minimal solute levels were incorporated into the study, through the selection of a commercially produced low carbon aluminum-killed drawing-quality (AKDQ) steel as well as a fully stabilized interstitial-free (IF) steel.

EXPERIMENTAL PROCEDURE

Materials- This study uses a laboratory produced ULC sheet steel and two commercially produced cold-rolled sheet steels; an AKDQ and an IF grade. The commercial steels were received in the cold-rolled, annealed, and skin-passed condition.

Chemical compositions of the three sheet materials are presented in Table I.

Table I Chemical composition (wt. %) of the sheet materials used in this research

Steel	C	Mn	Si	P	S	V	Al	Ti	N
Ti-V	0.0036	0.2	0.02	0.01	0.004	0.19	0.03	0.015	0.004
AKDQ	0.038	0.02	0.007	0.015	0.01	_	0.037	_	0.005
IF	0.002	0.12	0.005	0.007	0.012	_	0.013	0.066	0.003

Processing- Cast ingots of the experimentally produced ULC Ti-V steel were reheated to 1250°C and hot rolled to a 3 mm thick hot band. The steel was finish rolled above 900°C, air cooled to a simulated coiling temperature of 700 °C, and then furnace cooled to room temperature. To prepare internal-friction specimens from the as-received hot-rolled Ti-V steel, 3 mm sheets were thinned to 1 mm thick coupons from which internal-friction specimens could be cut. The thinning process was done using three different methods; 1) mechanical grinding with normal water cooling, 2) mechanical grinding with high-pressure water cooling (flushing), or 3) wire EDM thinning. The final specimen profile in all cases was obtained using wire EDM (electro-discharge machining) from thinned blanks.

For samples of the experimental Ti-V steel to be tested after annealing, the 3 mm thick hot-rolled plate was first pickled and cold rolled to a thickness of 1 mm, resulting in a total reduction of about 68 %. The cold-rolled sheet was annealed at 800 °C for 1 hour followed by water quenching to room temperature to avoid further carbide precipitation below the annealing temperature. The cold-rolled, annealed, and quenched sheet was machined to produce internal-friction specimens.

Specimens from the 1 mm thick commercially produced IF steel were prepared and internal- friction testing was conducted after the following treatments:

a) 2 % tensile strain.

b) Annealing at 800 °C for 2 minutes followed by water quenching.
c) Condition "c" plus 2 % strain.
d) Condition "d" plus aging at 170 °C for 20 minutes.

These conditions allowed for an examination of the magnitude of effects of strain and aging on the measured internal-friction response of a material which was nominally solute-C free.

In addition, specimens from the commercially produced AKDQ sheets were prepared and internal-friction testing was conducted after the following treatments:

a) 3% tensile strain.
b) Condition "a" plus aging at 170 °C for 20 minutes.
d) Annealing at 525 °C for 33 hours and at 695 °C for 7 hours. The annealing time t, at the lower temperature, was selected to provide approximately the same carbon diffusion distance ($\sqrt{Dt}$), as that of the higher temperature treatment.

These conditions allowed examination of the effects of strain and aging in a material containing solute-C, and of significant variations in solute levels. All of the annealing treatments were done using a molten chloride salt bath, while the aging treatments were done using an oil bath. Samples were refrigerated to minimize room temperature aging prior to internal friction testing.

Internal-Friction Measurement Apparatus- A schematic drawing of the forced torsion pendulum system used in this study (5), is shown in Figure 1. A torque is applied by a coil that provides a magnetic field perpendicular to a neodymium magnet mounted in the pendulum such that the torque (M) is proportional to the current (I) flowing through the coil (5). A digital frequency generator and associated electronics control the excitation. The angular displacement is about 3×10^{-5} radians and determined by a photocell detector, which, via tuned amplifiers, provides a voltage (V) proportional to the angular displacement (θ). The electronics are digitized to ensure proper signal phasing and minimal noise.

The standard internal-friction specimen dimensions are 76 mm (3 in.) long, 2.5 mm (0.1 in.) reduced width and 1.01 mm (0.04 in.) thick (6). The specimens are mounted in the instrument via two self-adjusting clamps. The minimum measurement time for one interstitial concentration determination is approximately 40 min., involving measurement at a given temperature over pendulum frequencies that vary by four orders of magnitude. The torsional pendulum is mechanically driven as described above, and the energy loss or internal-friction is derived through equations of motion incorporated into the equipment design (5). All tests reported here were conducted at room temperature.

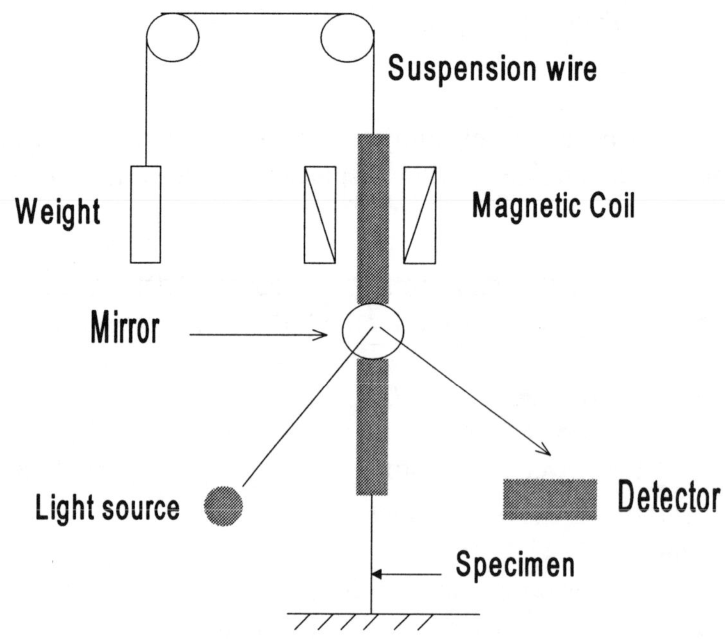

Fig. 1: Schematic illustration of the forced torsion pendulum (5)

Theoretical Estimation of the Carbon Snoek Peak Frequency- Experiments performed by Weller et al (3), have shown that the pre-exponential factor for the relaxation of carbon in iron is $\tau_o = 1.1*10^{-15}$ s and the activation energy for the relaxation of carbon in iron $Q_c = 0.884 ev/atom$. Since $\tau_C = \tau_o e^{(Q_c/RT)}$, then $\tau_C = 0.97s$. At the Snoek peak, the relationship $2\pi f \tau_C \approx 1$ is satisfied and therefore, the carbon Snoek peak is found to be centered near a frequency, f of 0.17 Hz. At room temperature, the experimentally collected data should thus show a Debye peak near this frequency, as confirmed experimentally by Jones and Foley (7).

RESULTS

Preliminary Internal-Friction Testing- Figure 2 shows an example of a typical Snoek peak due to internal mechanical loss caused by carbon reordering, such as may be found in the literature (e.g. 2). The results appear straightforward and the solute C can be easily determined knowing the peak height. Preliminary test results for the experimentally produced Ti-V ULC steel were more complex, however, with at least one or two unknown relaxations overlapping the Snoek peak, as shown in the internal-friction spectrum in Figure 2 for a sample annealed at 800 °C for an hour. Follow-up investigations were therefore needed to understand the various factors affecting the internal-friction spectrum.

Effect of Specimen Preparation- Figure 3 shows three internal-friction spectra for the experimental as-received hot rolled Ti-V ULC steel, thinned by mechanical grinding with normal water cooling, mechanical grinding associated with flushing (high pressure water cooling), and EDM wire thinning. The internal-friction spectrum for the mechanically ground specimen with normal water cooling has the highest background damping with no clear Snoek peak near 0.17 Hz. The background is presumed to result from internal stresses introduced during machining. The two other specimens, prepared by mechanical grinding with water flushing and by wire EDM, exhibit internal-friction spectra that are similar to one another, having a more distinctive Snoek peak and much smaller background damping as compared to the specimen which was mechanically ground with normal cooling. The results thus indicate an important influence of specimen preparation.

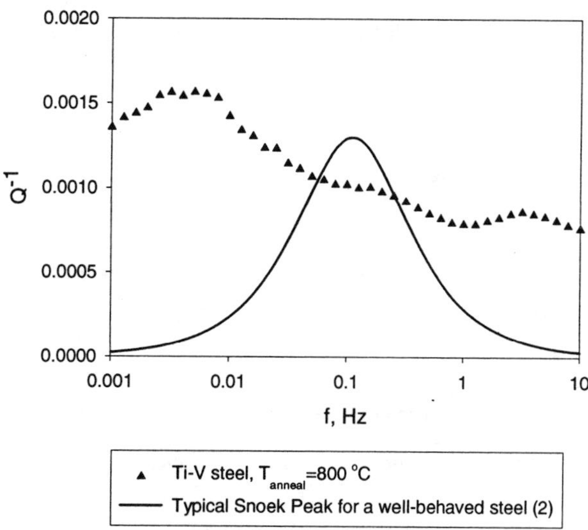

Fig. 2: Comparison of typical internal-friction measurement found in the literature with that for the experimental Ti-V ULC steel

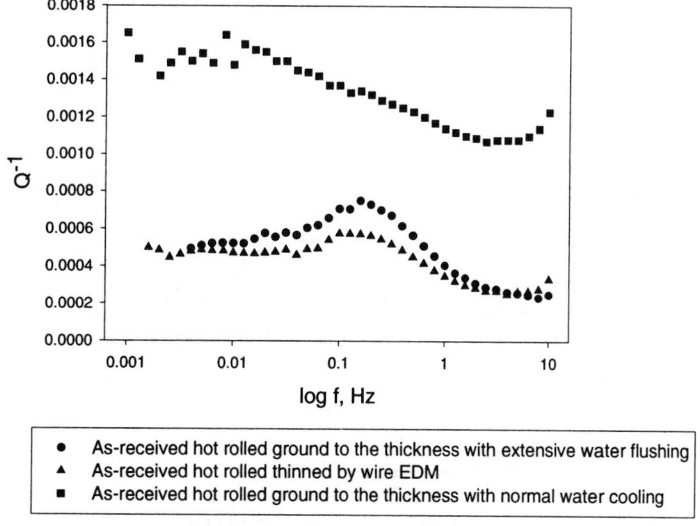

Fig. 3: Internal-friction spectrum, Q^{-1} vs f, for the Ti-V ULC steel in the as-received hot rolled condition showing the effect of specimen preparation

Internal-Friction of Commercially Produced AKDQ Steel- Figure 4 shows the internal-friction spectrum, Q^{-1} *vs frequency*, for the AKDQ specimen in the as-received (i.e. cold rolled, annealed, and temper-rolled). The spectrum shows, that there is very little solute C since the Snoek peak height at 0.17 Hz is negligible. The internal-friction is higher at lower frequency; at 10 Hz the internal-friction is approximately 0.00045 and it rises up to about 0.0009 at a frequency of 0.001 Hz. The figure also shows the spectrum for a specimen that was given an additional 3 % tensile strain. An increase in the internal-friction is evident. The figure also shows that after aging of the strained specimen at 170 °C for 20 min., the internal-friction is reduced to a much lower level. These results show that internal-friction increases with strain and decreases with subsequent aging.

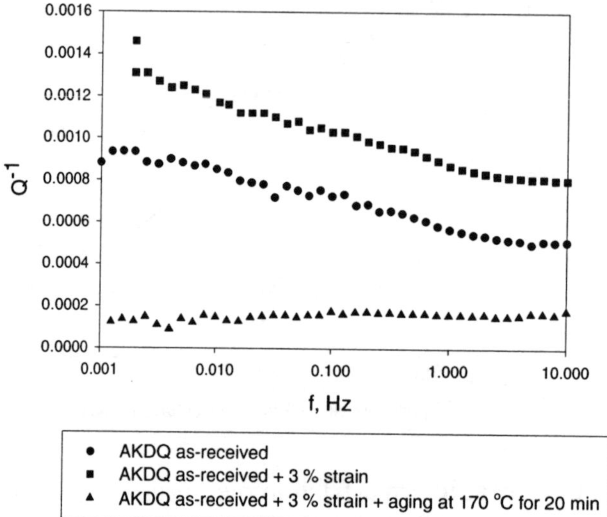

Fig. 4: Q^{-1} vs f for AKDQ steel showing the effect of strain and aging

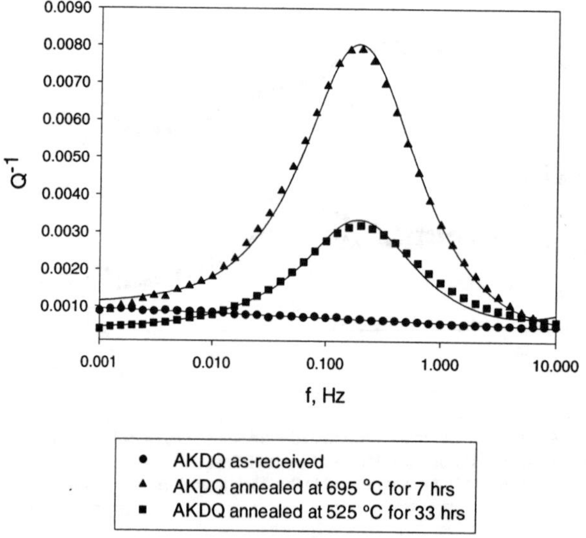

Fig. 5: Q^{-1} vs f for AKDQ steel showing the effect of annealing temperature on the peak height

Figure 5 shows three internal-friction spectra for an as-received AKDQ specimen, a specimen annealed at 695 °C for 7 hours followed by water quenching (AKDQ-695), and a specimen annealed at 525 °C for 33 hours followed by water quenching (AKDQ-525). As mentioned above, the as-received condition does not show a clear Snoek peak. The specimen annealed at 695 °C for 7 hours shows a Snoek peak height of about 0.008. The peak intensity is found at a frequency of about 0.19 Hz, close to the theoretical value of 0.17 Hz discussed above. Decomposition of the mechanical spectrum into its component damping contributions is necessary for accurate determination of the Snoek peak height. That is, background damping and any other peaks must be removed from the spectrum to obtain $Q_{\max}{}^{-1}$ for solute carbon. The removal of the background component will also eliminate most of the equipment-related damping and some other effects (such as dislocation hysteresis, etc.). The adjusted spectrum is then fit with a Debye equation to determine the contributions of C and/or N to the experimental Snoek peaks. This curve fitting was accomplished through an iterative process that solves for the appropriate coefficients that minimize the difference between the curve and the measurements. The fitting equation is as follows:

$$Q^{-1} = y_o + c\times e^{-d(2\pi f)} + 2\times Q_{\max(C)}{}^{-1}\times\tau_C\times\frac{2\pi f}{1+(2\pi f)^2\times\tau_C{}^2} + 2\times Q_{\max(D)}{}^{-1}\times\tau_D\times\frac{2\pi f}{1+(2\pi f)^2\times\tau_D{}^2} \qquad [1]$$

The first two terms account for background damping which consists of a constant, y_o, and a function that assumes an exponential change of internal-friction, Q^{-1} with frequency, f. The third term is a Debye peak that accounts for the Snoek peak caused by C rearrangement and the fourth term is an additional Debye peak to account for any relaxations caused by other effects such as dislocation/interstitial interactions or a solute-nitrogen Snoek peak. The program outputs the coefficients of the equation (y_o, c, d, $Q_{\max(C)}{}^{-1}, Q_{\max(D)}{}^{-1}, \tau_D$) such that the background may be subtracted from the raw measurements and a fit line generated (τ_C is taken as 0.19 based on curve-fitting experience in this work). The program also calculates the r^2 value that indicates the accuracy of the fit. This procedure for deconvolution is similar to methods used in previous studies (6,8), except for the use of the exponential term $ce^{-d(2\pi f)}$ to describe a portion of the background damping. The exponential term was found to substantially improve the curve-fitting results. An additional low-frequency contribution to the internal-friction spectrum is also observed in many cases (e.g. as-received and 3% prestrained AKDQ as shown in Figure 4). This contribution is possibly a γ-peak associated with double-kinking of screw dislocations (11,12), and is accounted for in the deconvolution procedure through a Debye peak having the form

$2\times Q_{\max(D)}{}^{-1}\times\tau_D\times\frac{2\pi f}{1+(2\pi f)^2\times\tau_D{}^2}$. (If this component were present simultaneous with a solute-N peak, then an additional Debye term would be needed in the fitting equation.)

Figure 6 shows the deconvolution of a spectrum obtained from the AKDQ steel annealed at 695 °C for 7 hours, indicating the background and a Debye peak. The large peak centered at a frequency of about 0.19 Hz is the Snoek peak caused by interstitial C. There is also background damping consisting of a constant and an exponential function caused by equipment or dislocation-hysteresis movements.

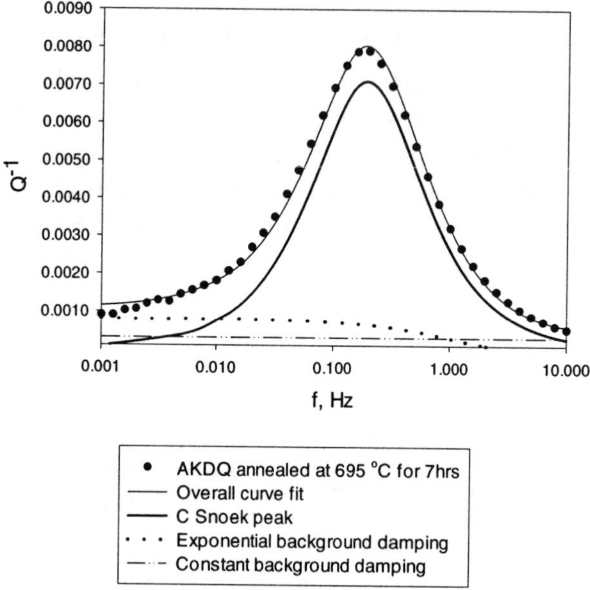

Fig. 6: Deconvolution of Q^{-1} vs f for AKDQ specimen annealed at 695 °C for 7 hours followed by quenching, showing the Snoek peak separated from the background damping

Table II Results of the deconvolution of internal-friction spectra for AKDQ-525 and AKDQ-695, presented in Figure 5

Steel	$Q_{\max(C)}{}^{-1}$	b	y_o	c	d
AKDQ-525	0.0029	0	0.0003	0.0001	1
AKDQ-695	0.0071	0	0.0003	0.0008	1

Deconvolution results from the spectra obtained from the AKDQ specimens tested under two different conditions (Figure 5) are shown in Table II. Table II presents the Snoek-peak heights and the frequencies at which the peak occurs. The r^2 values (i.e. $r^2 = 1$ indicates a perfect fit) were higher than 0.99 for both AKDQ specimens measured under the conditions mentioned above. The table shows that the AKDQ-695 specimen exhibited the larger Snoek peak with $Q_{\max}{}^{-1}$ near 0.0071 at around 0.19 Hz. The AKDQ-525 exhibited a peak height of 0.0029 at the same frequency. The greater Snoek contribution at a higher annealing temperature was expected, because of the increased carbon solubility. The table also shows that the exponential term for the background damping increased with annealing temperature. It is speculated that this effect may be attributed to a higher dislocation density obtained after quenching from a higher temperature.

A constant "*K*" is often used to relate the internal-friction-peak height to solute concentration. The *K* factor can be estimated by dividing the actual solute carbon concentration by $Q_{\max(C)}^{-1}$ for a specimen containing a known solute level. The carbon concentration can be estimated for the AKDQ samples from the well accepted relationship for the solubility of carbon in alpha iron in equilibrium with cementite (9):

$$\log(C)\,ppm = 6.38 - \frac{4040}{T_K} \qquad [2]$$

According to this relationship, the equilibrium carbon concentration in weight ppm is about 160 ppm and 21 ppm, at 695 °C and 525 °C respectively. Therefore, substituting $Q_{\max}^{-1} = 71$ and C = 160 ppm into the equation $C = KQ_{\max}^{-1}$, we obtain K = 2.25 wt. ppm, for the specimen annealed at 695 °C. This value is somewhat higher than the value reported by others (10). The same calculation performed for the annealing temperature of 525 °C leads to a lower *K* value of about 0.75. This difference in the *K* values will be explained in the discussion section below.

Internal-Friction Testing of the Commercially Produced IF Steel- Figure 7 shows spectra for several IF specimens tested in different processing conditions. The IF as-received (cold rolled, annealed, and temper-rolled) specimen shows an internal-friction spectrum that decreases with increasing frequency. The spectrum is shifted upward after a 2% tensile strain. When the as-received specimen was annealed, the internal-friction is lowered and the extent of the increase of the internal-friction with lower frequency is considerably reduced. Subsequent 2% straining of the annealed specimen resulted in a uniform increase of the internal-friction. In addition, aging of the as-received, annealed, and strained specimen at 170 °C for 20 minutes reduced the overall internal-friction spectrum to an even lower level. Therefore the measurements represented in Figure 7 show that straining of the IF specimens causes the internal-friction to increase while aging and annealing reduce the overall internal-friction.

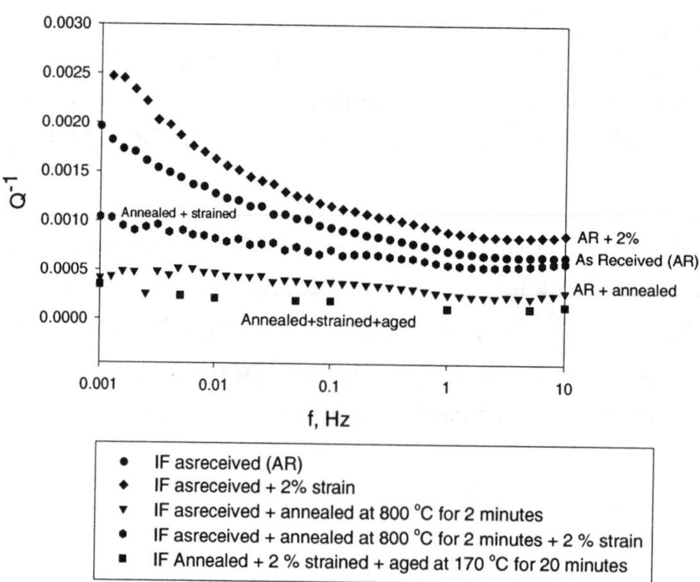

Fig. 7: Q^{-1} vs f for IF steel under different processing conditions

DISCUSSION

The results have shown that careful attention to specimen preparation is needed to obtain accurate internal-friction measurements in ultra-low carbon steels. Mechanical grinding with standard cooling is not an appropriate method of thinning and increases the background damping, thereby masking the Snoek peak. The higher background observed is believed to be caused by the pendulum-induced motion of mobile dislocations that were introduced by the mechanical and thermal stresses during machining. (The effect of dislocation motion on the internal-friction background is well known from earlier studies (11,12).) Another possible contribution to the higher background is the presence of point defects that could have been generated by thermal stresses during grinding (13). Extensive sample cooling is needed during specimen machining; otherwise wire EDM may be used to reduce the specimen thickness.

In both commercially produced AKDQ and IF sheets, a substantial increase of the internal-friction was observed with tensile strain. This result confirms that the introduction of mobile dislocations by deformation is one of the main contributions to the background damping. Annealing of the IF as-received and strained specimen at 800 °C for 2 min. resulted in a reduction of the background damping which can be attributed to the elimination of dislocations caused by recovery at this elevated temperature. On aging the strained specimens of the AKDQ and IF sheets at 170 °C for 20 min., the internal-friction spectra were also reduced significantly as shown above. The effect of aging could be caused by either of two mechanisms: the elimination of dislocations through recovery, or the pinning of mobile dislocations by migration of solute C. It is difficult to separate such effects, although substantial recovery at 170 °C is not likely and pinning of dislocations by solute C is therefore the more probable mechanism.

It might be argued that the explanation of the dislocations being pinned by interstitial C is difficult to accept, since a clear Snoek peak was not evident in either steel under these conditions. Hence, it is helpful to consider the amount of C that would be required to form a Cottrell atmosphere around dislocations. The number of carbon atoms in a Cottrell atmosphere can be assumed to be on the order of 1 atom per atomic plane (13, 14). The Fe atom has a diameter of 0.25 nm, and thus in a 1 cm dislocation line length there are about 4×10^7 atomic distances. It can therefore be assumed that about 4×10^7 C atoms would saturate a 1 cm dislocation line. If the dislocation density in the 2% strained specimen is about 10^9 cm^{-2} (13), the amount of C required to form a Cottrell atmosphere is approximately 4×10^{16} C atoms per cm^3. In the bcc iron structure, there are about 10^{23} atoms per cm^3. Therefore, it can be shown that the concentration of C required to form an atmosphere around the dislocations is approximately 1.8 weight ppm. Since the detection limit of the internal-friction equipment is about 1-2 ppm, it is possible that solute C sufficient to pin dislocations may be present at levels that are undetectable. Even if the actual solute content was far below the 1.8 ppm level estimated to saturate the dislocation strain fields, a reduction in the background damping might still be expected to result from interstitial migration to a portion of the available sites during aging.

Another point that deserves discussion is the higher proportionality constant (*K*) observed in the case of the AKDQ specimen annealed at 695 °C as compared to the sample annealed at 525 °C. The difference in *K* value could result from inaccuracy in the temperature-sensitivity of the published carbon solubility relationship in ferrite, but is more likely caused by a greater proportion of solute atoms being removed from the lattice during or after quenching from a higher temperature, either through 1) carbide precipitation, or 2) solute migration to lower energy sites such as dislocations and grain boundaries (Carbon that migrates to defects does not contribute to the Snoek peak). The "apparent" solute level is reduced by either of these mechanisms, thereby increasing the proportionality constant between the actual C-content and the observed Snoek-peak height. In support of this hypothesis, it is very reasonable to expect both of these effects to be enhanced under conditions of higher temperature annealing. At higher temperature, the solute carbon level is higher, which increases the driving force for precipitation or migration to defects during cooling. The concentration of quenching-induced defects might also be expected to increase with annealing temperature, which similarly favors increased solute/defect interactions after cooling. Finally, an increase in quenching temperature allows more time and temperature for the thermally activated C-diffusion processes to occur.

One implication of the variation in the *K* value with temperature is that the true carbon solubility may be difficult to derive using the internal-friction technique. The authors' ongoing work involves measurement of carbide solubility in microalloyed ULC steels, and further consideration is therefore needed of the issues related to the accuracy of solute determinations made by internal-friction. However, it is expected that these effects may be different in ultra-low-carbon steels where the solute levels are much lower and thus the driving force is lower for precipitation or solute migration during or after quenching.

CONCLUSIONS

Specimen preparation is an important factor to be considered when manufacturing an internal-friction specimen from thick materials. Thermal or mechanical stresses introduced during machining can give misleading spectral features and extra care is needed during preparation or even handling of the specimens.

This study shows that the background damping in the internal-friction spectra generally increases when the specimen is deformed and this effect is mainly attributed to the presence of mobile dislocations. Elimination of dislocations by annealing reduces the internal-friction background damping. In addition, aging of strained specimens causes the internal-friction background damping to be reduced, presumably due to dislocation pinning by small amounts of solute C that can even be below the detectibility limit of the internal-friction equipment. The background damping was best represented by a constant plus an exponential term that increases at lower frequencies.

As expected, clear and distinctive Snoek peaks were observed in the internal-friction spectra of the AKDQ specimens annealed at subcritical temperatures, with greater peak heights associated with higher annealing temperature. However, the relationship between the Snoek-peak height and the (presumed) solute concentration appeared to vary with annealing temperature. This variation is thought to be a result of differences in the extent of carbide precipitation and solute/dislocation interaction during or after quenching.

ACKNOWLEDGMENT

The author is thankful to LTV steel for providing the experimental steels required for this research. The support of the Advanced Steel Processing and Products Research Center (ASPPRC), an NSF I/UCRC at the Colorado School of Mines, and the Saudi Basic Industries Corporation (SABIC) is also greatly appreciated.

REFERENCES

1. D.K. Matlock, B.J. Allen, and J.G. Speer, "Aging Behavior and Properties of Ultra Low Carbon Bake Hardening Steels", Modern LC and ULC Sheet Steels for Cold Forming: Processing and Properties, Institute of Ferrous Metallurgy, Aachen, 1998, pp.265-276.

2. Snoek, J., J. Physica, Vol. 6, 1939, P. 591.

3. M. Weller, "Internal-Friction Study of Fe-N and Fe-C Interstitial Solutions", Proc. Conf. High Nitrogen Steels, Varna Bulgaria, 1989.

4. A.S. Nowick and B.S. Berry, Anelastic Relaxation in Crystalline Solids, Academic Press New York and London, 1972.

4. Yiting Wen, Quanmin Su and Manfred Wutting, "Interstitial Analysis of Ultra Low

Carbon and Interstitial Free Steels", 39th MWSP Conf. Proc., ISS, Vol.XXXV, 1998, pp. 271-280.

5. A.J. Jones, "Design and Construction of a subresonant Low Frequency Mechanical Spectrometer", MS Thesis, Illinois Institute of Technology, May 1998.

7. A.J. Jones and R.P. Foley, "Effects and Measurement of Interstitial C and N in Iron and Steel", 39th MWSP Conf. Proc., ISS, Vol.XXXV, 1998, pp. 955-989.

8. Z. Pan and R. Hosbons, "Determination of Interstitial C and N in Low Carbon Steels", 39th MWSP Conf. Proc., ISS, Vol. XXXV, 1998, pp. 241-254.

9. J. Chipman, Metall. Trans., 1973, vol. 8, pp. 1975-1986.

10. K. Aoki, S. Sekino and T. Fujishima: J. Jpn. Inst. Met., 1962, vol. 26, pp. 47-53.

11. T.S. Ke, "Cold-Work Internal-Friction Peaks", Scripta Metallurgica, 1982, vol. 16, pp. 225-232.

12. M. Shimada, and K. Sakamoto, "Internal-Friction of Iron Deformed at Low Temperatures", Scripta Metallurgica, 1979, vol. 13, pp. 1177-1182.

13. W.C. Leslie, "The Physical Metallurgy of Steels", Hemisphere Publ. Co., Washington, D.C., 1983.

14. R.E. Reed-Hill and R. Abbaschian, "Physical Metallurgy Principles", PWS Publ. Co., third edition, 1994.

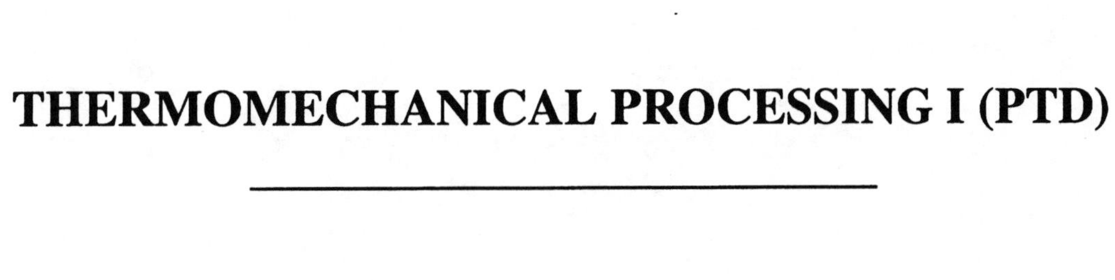

THERMOMECHANICAL PROCESSING I (PTD)

Analysis of dynamic recrystallization in Nb microalloyed steels

Guohui Zhu, Dianlong Tan and Mani Subramanian
Department of Materials Science and Engineering, McMaster University
1280 Main Street West
Hamilton, ON, L8S 4M1
Canada
Tel.905-525-9140 Ext. 24854
Subraman@mcmaster.ca

Key words: dynamic recrystallization, niobium microalloyed steel, modeling plate and strip rolling, fine grain size

INTRODUCTION

There is considerable technological interest in ultra fine grain size (≈1μm) in order to achieve high strength and high toughness properties in steels at low temperature. In order to obtain ultra fine grain size by dynamic recrystallization, the essential requirements are (a) a large critical strain for dynamic recrystallization associated with large Zener-Hollomon parameter Z, and (b) strain accumulation large enough to overcome the large critical strain for dynamic recrystallization in multi-pass rolling. In the present work, a quantitative model is developed using the database generated by Lutz Meyer and co-workers to analyse the dynamic recrystallization behaviour of Nb microalloyed steels containing both high and low Mn contents in multi-pass rolling.

In high Mn (>1.5wt%) high Nb (>0.06 wt %) steels, the bulk of the Nb is in solid solution, whereas in low Mn (<0.6wt%) high Nb (>0.06wt%) steels a substantial portion of the Nb is present as NbC precipitates during a typical multi-pass strip rolling. The effect of Nb is to increase the critical strain for dynamic recrystallization. Nb present as NbC precipitates in low Mn steel is more effective than Nb present as solute Nb in high Mn steel in increasing the critical strain for dynamic recrystallization. The larger the critical strain for dynamic recrystallization, the finer the dynamically recrystallized grain size. Nb is very effective in strain accumulation by retarding the static softening kinetics, made up of recovery and recrystallization processes. Again Nb present as NbC precipitates in low Mn steel is more effective than Nb present as Nb solute in high Mn steel in retarding the softening kinetics and facilitating strain accumulation to overcome the large critical strain for dynamic recrystallization required to obtain ultra-fine grains. The quantitative model based on the above concepts is used to optimize the base chemistry and rolling schedule to design and target ultra fine grain size by dynamic recrystallization in Nb microalloyed steels. The available results are used to validate the model. The possible mechanisms of dynamic recrystallization under large Z to obtain ultra-fine grains are discussed.

BACKGROUND

Hodgson [1] has reported that the rate of recrystallization changes from being a strong function of strain and temperature and a weak function of strain rate for static recrystallization to being only a function of strain rate for metadynamic recrystallization. The recrystallized grain size is only a function of Zener-Hollomon parameter for metadynamic recrystallization as well as dynamic recrystallization. However, in the case of static recrystallization, the recrystallized grain size is relatively little influenced by Zener-Hollomon parameter but is controlled by strain and initial grain size. Hodgson has analyzed all the data from various sources and showed that the dynamically (d_{dyn}) and metadynamically (d_{md}) recrystallized grain size can be satisfactorily related to Zener-Hollomon parameter Z with Q_{def}=300 KJ/mol, as given by the following equations:

$$d_{dyn} = 3.9\times10^{4} Z^{-0.27} \tag{1}$$

$$d_{md} = 6.8\times10^{4} Z^{-0.27} \tag{2}$$

In metadynamic recrystallization, the nucleation of recrystallising grains occurs by dynamic recrystallization (i.e., during deformation) and the growth is assumed to occur predominantly during post-dynamic recrystallization stage.

Recent work at McMaster University on the microstructural characterization of shear bands resulting from large strain (ε>1), high strain rate deformation ($\dot{\varepsilon}>10^{3}$) in metal cutting has shown ultra fine grains (0.1 to 0.25 μm) in austenite [2-3]. Meyers and coworkers [4-6] have reported the occurrence of shear localisation in shock-hardened copper as well as in α-hcp titanium caused by dynamic recrystallization. Ultra fine grains of about 0.1 μm were observed in both cases, which are attributed to dynamic recrystallization. TEM characterization of deformation shear band occurring in the secondary shear zone during cutting AISI 1045grade steel at 240 m/min showed grains less than 0.25 μm in diameter [3]. Transformation shear bands occur both in the primary and secondary shear zones during cutting Fe-28%Ni-0.1%C martensitic alloy. TEM characterization of the primary shear band exhibits ultra fine grains of 0.10 μm [2], which is again attributed to dynamic recrystallization. The observed grain size was found to be consistent with the prediction from Hodgson's empirical correlation with Z, i.e. Equation (1). This clearly underscores the importance of large Z to obtain ultra-fine grain size. Since $Z=\dot{\varepsilon}\,exp(Q/RT)$, Z value can be substantially increased by lowering the temperature of deformation. The larger the Z, the larger the critical strain for dynamic recrystallization and the finer the dynamically recrystallized grain size. Thus, in principle, it should be possible to obtain fine austenite grain size by dynamic recrystallization if strain could be accumulated in multi-pass finish rolling schedule large enough to overcome a large critical strain for dynamic recrystallization.

OBJECTIVES

The objective of this work is to develop a model based on the available data-base on softening kinetics and the critical strain for dynamic recrystallization of Nb microalloyed steel in order to predict the critical strain for dynamic recrystallization and strain accumulation in multi-pass rolling of Nb microalloyed steel. The overall goal is to use the model predictions to design the base chemistry and optimize the rolling schedule in order to achieve ultra fine austenite grains by dynamic recrystallization in plate and strip rolling.

QUANTITATIVE MODELING

1. Basic components of the model

(1) Analysis of the inter-relationship among Q, ε_c and Z

Quantitative relationship between dynamically recrystallized grain size, d and Zener-Hollomon parameter, Z is plotted in Fig. 1 using Hodgson's empirical relation given in Equation (1). Based on the available database [7-9], the critical strain ε_c for dynamic recrystallization is related to Z, using the following expression:

$$\varepsilon_c = 0.048 \ln Z - 0.96 \tag{3}$$

The critical strain and dynamically recrystallized grain size are plotted as a function of Z in Fig. 1. The results clearly show the ultra fine grain size is only obtained when the critical strain for dynamic recrystallization is large, which can only be obtained by deformation at large Z.

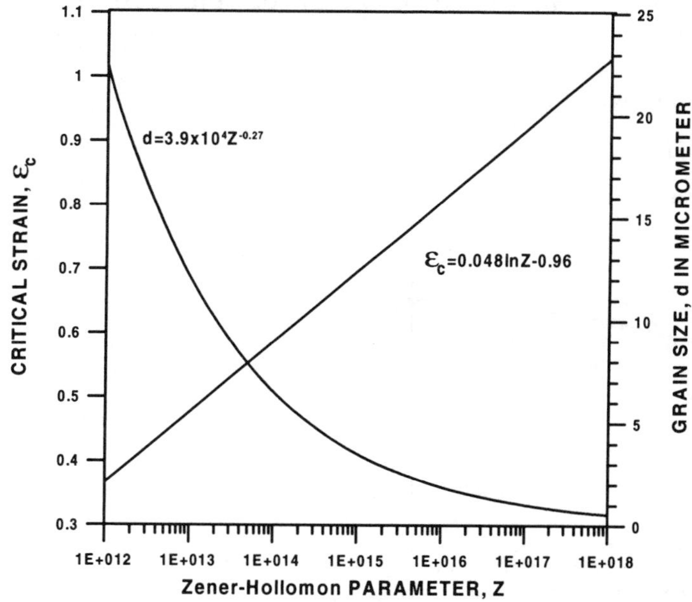

Fig. 1 The relationship among critical strain for dynamic recrystallization, ε_c, Zener-Hollomon parameter, Z and grain size, d

(2) Analysis of the effect of base chemistry on Q, ε_c and Z

The effect of niobium addition on the activation energy of deformation, Q the critical strain for dynamic recrystallization, ε_c and Zener-Hollomon parameter, Z are evaluated from Lutz Meyers' data. The results are summarized in Table I. The effect of Nb addition on Q, ε_c, and Z is plotted in Fig. 2(a), (b) and (c) respectively. The results clearly show that niobium addition increases Q, the activation energy of deformation and hence increase Z and ε_c. This effect is more pronounced in low Mn steel.

Table I. The effect of Nb on Q, ε_c and Z (strain rate: 4.7/s, deformation temperature: 850°C)

Nb(%)	Q(KJ/mole)		Critical strain ε_c		Z		d (micrometer)	
	High Mn	Low Mn	High Mn	Low Mn	High Mn	Low Mn	High Mn	Low Mn
0	326	297	0.79	0.64	6.82×10^{15}	2.99×10^{14}	2.07	4.81
0.03	337	343	0.85	0.88	2.38×10^{16}	4.45×10^{16}	1.48	1.25
0.06	342	353	0.875	0.93	4.0×10^{16}	1.26×10^{17}	1.28	0.94

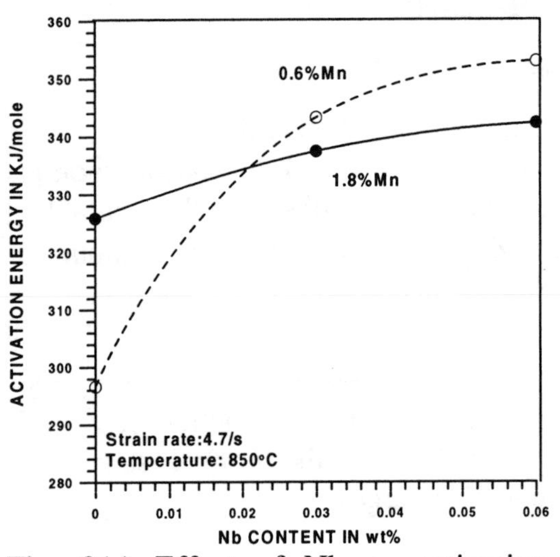

Fig. 2(a) Effect of Nb on activation energy of deformation, Q

Fig. 2(b) Effect of Nb on critical strain for dynamic recrystallization, ε_c

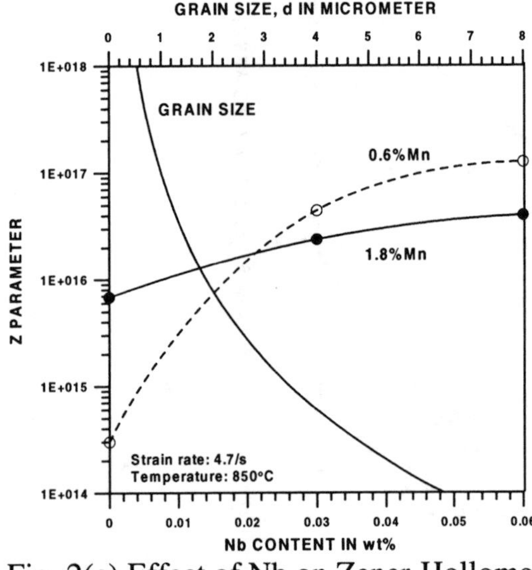

Fig. 2(c) Effect of Nb on Zener-Hollomon parameter, Z

(3) Effect of Nb on softening kinetics in high and low Mn steels

The effect of Nb addition on the holding time for 50% softening is plotted in Fig. 3 as a function of temperature for low and high Mn steels. The data- base from Lutz Meyer's work [7,8] is summarized in Table II. At 850°C, the effect of increasing Mn content from 0.6 to 1.8wt% on a 0.06wt%Nb containing steel is to decrease the holding time for 50% softening from about 385 seconds to 111 seconds. Thus Nb is more effective in retarding the softening kinetics in low Mn steel than in high Mn steel.

Table II. Holding time in seconds for 50% softening at different temperature

	850°C		925°C(high Mn) /950°C(low Mn)		1000°C(high Mn) /1050°C(low Mn)	
	0.00%Nb	0.06%Nb	0.00%Nb	0.06%Nb	0.00%Nb	0.06%Nb
0.6%Mn	1	385	~0	4.5	~0	~0
1.8%Mn	2	111	1	16	0.04	2

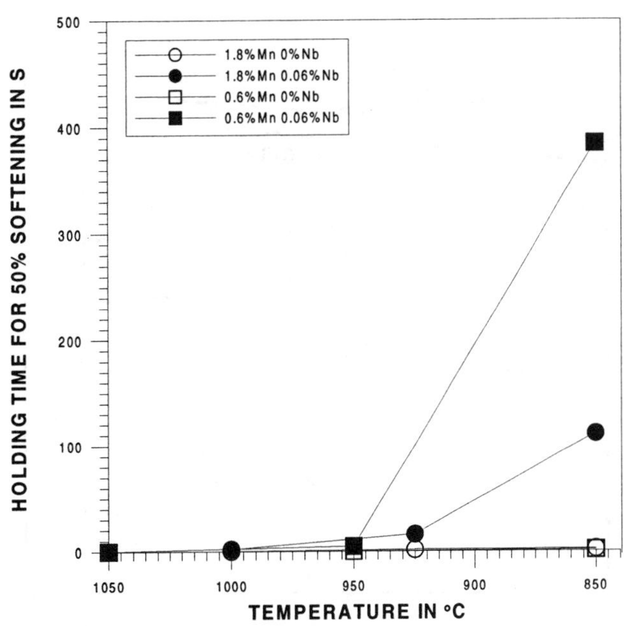

Fig. 3 Softening kinetics for different chemistries

2. Modeling Results

(1) Plate rolling simulation

For a high Nb (0.06 wt %) low carbon base chemistry, the plate rolling schedule to achieve dynamic recrystallization in low Mn (0.6 wt %) and high Mn (1.8 wt %) steels is given in Table III. The strain accumulation to overcome the critical strain is plotted in Fig. 4. The modeling results on Q, ε_c, Z and d corresponding to the final pass for low and high Mn steels are summarized in Table IV. The results clearly show that the effect of lowering Mn is to increase the critical strain and therefore decrease the grain size. Further, the static softening occurring in the interpass time is decreased.

Table III. The schedule for plate finishing rolling (at constant pass strain of 0.44, The total reduction is about 68.6%)

Pass number	Thickness(mm)	Reduction(%)	Temperature(°C)	Time(s)
Start	25.40	Start		Start
1	17.27	32%	800	0
2	11.74	32%	785	5
3	7.97	32%	775	10

Table IV. The critical strain, activation energy Q, Z and grain size obtained by the plate rolling schedule for low Mn and high Mn with 0.06%Nb steels in the last pass

Steel	Critical strain	Q(KJ/mole)	Z	Grain size(μm)
Low Mn	1.09	353	1.24×10^{18}	0.51
High Mn	1.04	342	3.94×10^{17}	0.69

(2) Strip rolling simulation

For a typical 7 pass strip rolling design, the rolling schedule to achieve dynamic recrystallization in high Nb steel with high and low Mn contents is given in Table V. The strain accumulation plots for high and low Mn steels respectively are given in Fig. 5. The dynamic recrystallization occurs at the end of the 6th pass in the case

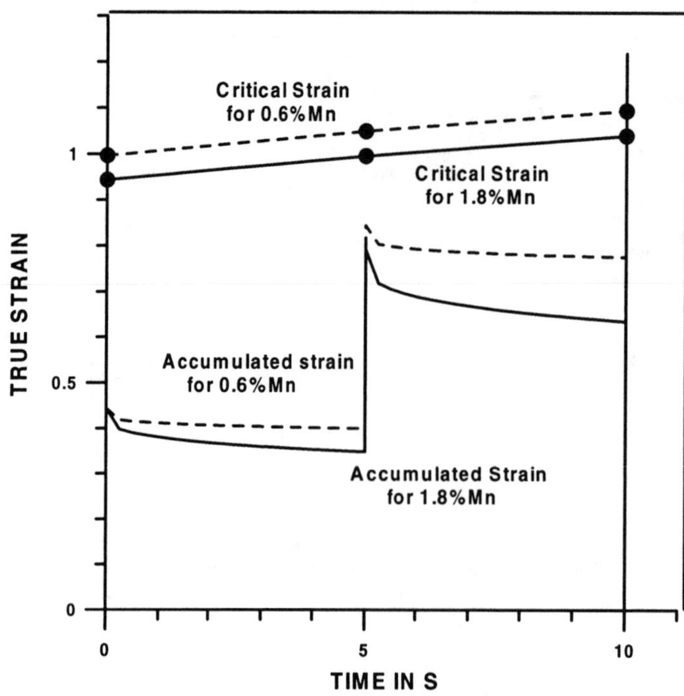

Fig. 4 The accumulated strain in plate rolling schedule design

of low Mn steels, whereas this is deferred to the 7th and final pass in high Mn steel for the same rolling schedule. The critical strain for dynamic recrystallization and grain size data corresponding to the pass at which the dynamic recrystallization occurs in low and high Mn steels are summarized in Table VI. The distinct advantage of low Mn design in decreasing the dynamically recrystallized grain size in high Nb low carbon steel can be clearly seen.

Table V. the schedule for strip finishing rolling (at constant pass strain of 0.24. Total reduction is about 77%)

Pass number	Thickness(mm)	Reduction(%)	Temperature(°C)	Time(s)
Start	30.0	Start		Start
1	24.3	19%	814	0
2	19.7	19%	802	4.0
3	15.9	19%	792	7.0
4	12.9	19%	784	9.0
5	10.5	19%	778	10.8
6	8.5	19%	774	12.0
7	6.9	19%	772	13.2

Table VI. The critical strain, activation energy, *Z* and grain size obtained by the strip rolling schedule for low Mn and high Mn with 0.06%Nb steels

Steel	Critical strain	*Q*(KJ/mole)	*Z*	Grain size(μm)
Low Mn*	1.10	353	1.33×10^{18}	0.50
High Mn	1.06	342	6.2610^{17}	0.61

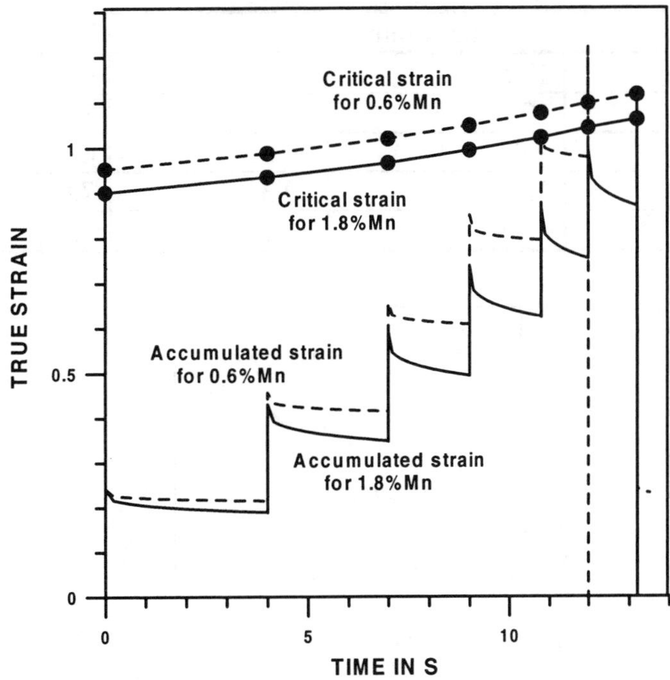

Fig. 5 The accumulated strain in plate rolling schedule design

3. Validation of Quantitative Modeling

Samuel et al [10] have investigated the effect of dynamic recrystallization on microstructure evolution during strip rolling in low Nb (0.03wt%) low Mn (0.6wt%) microalloyed steel. Their database on the effect of lowering the temperature window of finish rolling on the dynamically recrystallized grain size is analyzed using the present model. The finish-rolling schedule is given in Table VII. The modeling results predict that dynamic recrystallization will occur at the end of second pass, which is in accord with the experimental observation. More importantly, the dynamically recrystallized grain size for all the four cases examined are in good agreement with the experimental results summarized in Table VIII. The agreement between the model predictions and the experimental observations can be clearly seen in the plot given in Fig. 6. The observed grain size for the lowest temperature window (915-845°C) is 5μm in diameter. The modeling capability is used to identify the required conditions to obtain ultra fine grain size of about 1μm in diameter and the results of the analysis are included in Table VIII. The modeling results identify the impact of base chemistry design and optimization of rolling schedule on dynamically recrystallized grain size. With the addition of 0.06wt% Nb, the grain size of the lowest temperature window used in the experimental work will decrease to about 2.2μm in diameter. The effect of lowering the temperature window of finish rolling to the range of 800-750°C is predicted to decrease the grain size to 1.95μm in diameter in 0.03wt% Nb steel. However, if the niobium content is raised to 0.06wt% for the same temperature window of finish rolling, ultra fine grain size of about 1μm in diameter is predicted, seeing Table VIII.

Thus the two conditions for obtaining ultra fine grain size by dynamic recrystallization are (i) high Nb, low interstitial base chemistry design, and (ii) low temperature window of finish rolling. The feasibility of obtaining ultra fine grain size by low temperature finish rolling of high Nb, V microalloyed steels has been successfully demonstrated by Radko Kaspar and coworkers [11] by imparting large reduction in low temperature window using WUMSI. Further work is in progress to generate the database required to quantify the relative contributions of effects of solute and precipitation of Nb on the critical strain for dynamic recrystallization and the retardation of softening kinetics.

Table VII strip rolling simulation

Pass No.	Thickness(mm)	Reduction(%)	Time(s)
1	17.50	30	0
2	10.66	39.08	3.6
3	7.40	30.59	6.1
4	5.23	29.38	7.8
5	4.13	20.90	8.5

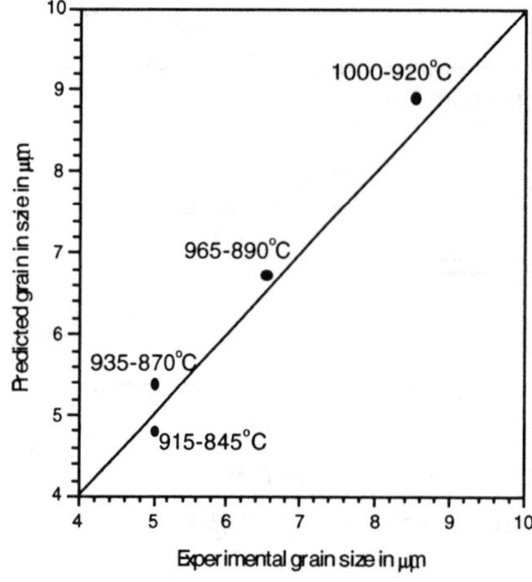

Fig. 6 Comparison of predicted and experimental grain size

Table VIII Comparison of experimental and predicted grain size from dynamic recrystallization

Temperature(°C)	Experimental (0.028Nb)	Predicted (0.028Nb)	Predicted (0.06Nb)
1000-920	8.5	8.92	4.69
965-890	6.5	6.74	3.47
935-870	5.0	5.40	2.51
915-845	5.0	4.81	2.21
830-780		2.57	1.22
800-750		1.95	0.91

DISCUSSION

Relationship among Z, ε_c and d:

Derby [12] has analyzed the dependence of grain size on stress during dynamic recrystallization for a number of metals and gave the following expression relating grain size d to σ.

$$\frac{\sigma}{\mu}\left(\frac{d}{b}\right)^{2/3} = K$$

The flow stress σ in turn can be related to Z. Thus, the empirical correlation of dynamically recrystallized grain size, d with Z can be rationalized. The recent microstructural observations on dynamically recrystallized grain size of about 0.1μm in diameter obtained in shear bands deformed under large Z condition at McMaster University [2] is consistent with Hodgson's correlation of Z with d as given in Equation (1).

The modeling results bring out two important concepts. The first one is that there is a unique relationship between ε_c the critical strain for dynamic recrystallization and Z. The chemistry and microstructural effect on Z is through *Q*, the activation energy for deformation. The temperature of deformation and the strain rate are important process parameters coupled to Z. These concepts underlie the results summarized in Fig. 1 and 2. The implication is that in order to obtain ultra fine grain size, a large critical strain associated with large *Z* is required.

Strategy to increase Z in plate and strip rolling:

Since $Z = \dot{\varepsilon}\, exp(Q/RT)$, *Z* can be increased by increasing $\dot{\varepsilon}$ and *Q* and decreasing *T*. In metal cutting, $\dot{\varepsilon}$ is in the range of 10^3-10^6. However in plate rolling, $\dot{\varepsilon}$ is typically about 10 and in the sheet rolling it can be increased to 10^2. Thus, in plate rolling, the only two options to increase *Z* are through increase in *Q* or decrease in *T*, the temperature of deformation. The available database on microalloying elements shows that niobium is more effective in increasing *Q* than Mn. However, in the presence of Mn, Nb is less effective in increasing *Q*, see Fig. 2(a). In any case, the largest increase in *Q* is brought about by 0.06 wt % Nb addition to low Mn steel, which has raised *Q* from 297 to 353KJ/mole, an increase of 19%. However, this increase in *Q* is inadequate to raise the critical strain for dynamic recrystallization significantly. Thus, lowering the temperature of deformation *T* is the only means to increase *Z* required to obtain ultra fine grains.

Strain accumulation to overcome large critical strain ε_c for dynamic recrystallization:

In order to overcome the large critical strain for dynamic recrystallization ε_c, strain accumulation in the multi-pass rolling is the only means to achieve the critical strain. In order to accumulate strain pass by pass, the static softening kinetics in the interpass time has to be minimized. Niobium addition is found to be most effective in retarding the softening kinetics in low Mn steel. The effect of 0.06wt% Nb in 0.6wt%Mn steel is to increase the holding time for 50% softening to 385 seconds, whereas the increase in Mn to 1.8wt% decreases the holding time for 50% softening to 111 seconds. With the increased Mn addition, the amount of niobium in solution in austenite increases [13]. This would imply that niobium carbide precipitates in austenite are more effective than solute niobium in retarding the softening kinetics. However, the pinning pressure from incoherent precipitates of NbC will decrease the driving force for boundary migration. The strong interaction between dislocations and coherent precipitates of NbC could account for the observed retardation of softening kinetics in low C, high Nb and low Mn steel. Quantitative studies on the relative contributions from solute, coherent and incoherent precipitates on the critical strain for dynamic recrystallization and static softening process are required for further refinement of the model. The feasibility of obtaining ultra fine grain size by dynamic recrystallization has been demonstrated in previous work [11].

Mechanism of dynamic recrystallization under conditions of large strain, high Z deformation:

In Derby-Ashby analysis [14], the dynamically recrystallized grain size is the result of dynamic balance between the rate of nucleation and growth. The nucleation is based on bulge mechanism of the boundary. The growth is controlled by boundary mobility. At high temperature where the bulge mechanism for dynamic recrystallization is applicable, the critical strain for dynamic recrystallization is low and therefore the resulting grains from dynamic recrystallization is large. In order to obtain ultra fine grain size of about 1μm in diameter, a large critical strain associated with a large *Z* is required. In order to accumulate large strain to overcome the large critical strain for dynamic recrystallization, it is essential to suppress static softening in the interpass time in the multi pass rolling. Nb is found to be most effective in suppressing the static softening kinetics in comparison with other microalloying elements. Nb(CN) precipitates are effective in suppressing the boundary mobility.

The total driving force for dynamic recrystallization is provided by the stored energy from cumulative strain (dislocation density). The net driving force for boundary mobility is the total driving force from stored energy due to dislocation density less the drag force on the boundary. The pinning force due to second phase particles exerts significant drag force on the boundary mobility compared to solute drag. Under the conditions of low temperature, large strain accumulation (high dislocation density) and large pinning pressure on the boundary due to precipitates, dynamic recrystallization by bulge mechanism is inhibited because the boundary mobility is impeded. An alternative mechanism of dynamic recrystallization involving progressive rotation of the sub-grains during deformation is feasible even when the boundary mobility is impeded. Recent studies [15] on recrystallization of Fe-50at%Co ordered model alloy have shown that when the mobility of the boundary is retarded by the ordered structure, the recrystallization occurs by progressive rotation of the sub-grains. The two factors essential to promote the rotation mechanism are (i) a large number fine sub-grains and (ii) large misorientation between the sub-grains. These factors are related to with high dislocation density associated with large strain. Thus, if bulge mechanism is suppressed by inhibition of boundary mobility, it is proposed that dynamic recrystallization will still occur at large critical strain by rotation mechanism, resulting in ultra-fine grain size. Since shear bands are regions of high dislocation density, ultra-fine grains are expected from dynamic recrystallization in the shear bands. In recrystallization studies on aluminum alloys, shear bands are found to be favorable sites for nucleation of recrystallized grains. At large strain, high strain rate deformation in metal cutting of hardened iron alloys, ultra-fine dynamically recrystallized grains of about 0.1μm in diameter are obtained in transformation shear bands. Further work is required to confirm the rotational mechanism of dynamic recrystallization operates in shear bands.

Technological Implications:

In order to obtain ultra fine grains by dynamic recrystallization, dynamic recrystallization has to be promoted under large Z conditions through design of base chemistry and optimization of rolling schedule. The modeling results confirm the merit low Mn-high Nb base chemistry design and the benefit of finish rolling in low temperature window. The combined effect of base chemistry and low temperature rolling is to ensure large strain accumulation in multi-pass rolling to overcome the large critical strain for dynamic recrystallization. Strain accumulation is facilitated by strain induced precipitation of NbC, which retards static softening in the interpass time. However, the combined effect of low temperature window of finish rolling and pinning pressure exerted by the NbC is to suppress boundary mobility and hence the bulge mechanism of dynamic recrystallization based on boundary mobility. Under these conditions, an alternative mechanism of dynamic recrystallization still operates which involves progressive rotation of sub-grains under large accumulated strain (i.e., large dislocation density). Under these conditions, the material will exhibit a large flow stress, causing large roll separating force. Thus the technological development of high toughness plates based on ultra-fine grain size imparted through dynamic recrystallization require high-powered modern rolling mills.

CONCLUSIONS

(1) The critical strain for dynamic recrystallization increases with Zener-Hollomon parameter. The larger the Z, the larger the critical strain for dynamic recrystallization, the finer the dynamically recrystallized grain size.
(2) The effect of Nb addition is to increase the activation energy of deformation, which, in turn, increases Z and hence the critical strain for dynamic recrystallization. This effect is more pronounced in low Mn steel.
(3) In order to obtain large critical strain to achieve ultra-fine grain size by dynamic recrystallization, temperature window of rolling has to be decreased.
(4) In order to accumulate strain during multi-pass rolling to overcome large critical strain for dynamic recrystallization, the softening kinetics during interpass time has to be decreased. Nb is more effective in low Mn steel than in high Mn steel in retarding the softening kinetics. This effect is attributed to precipitation of NbC.
(5) The modeling predictions of the effect of base chemistry and temperature window on dynamically recrystallized grain size are validated by experimental results obtained from strip rolling simulation.

(6) The feasibility of obtaining ultra-fine grains by dynamic recrystallization by plate rolling simulation confirms the model predictions of the combined effect of high Nb base chemistry and large reduction in low temperature window of deformation.

(7) The combined effect of low temperature and large pinning pressure on the boundaries is to suppress boundary mobility and hence the bulge mechanism of dynamic recrystallization. However, under these conditions, it is proposed that dynamic recrystallization will still occur by an alternate mechanism of progressive rotation of sub-grains under large strain.

ACKNOWLEDGEMENTS

We express our grateful thanks to Dipl. Ing. Klaus Hulka, Nibobium Products Company GmbH, Dr. Radko Kaspar, Max-Planck-Institut fur Eisenforschung GmbH, Dusseldorf, FRG and Professor Emeritus H. Meyer, Germany for helpful discussion and collaborative research support. The funding from NSERC Canada in the form of a strategic grant awarded to Dr. G.R.Purdy and the funding from Reference Metals Company Inc. U.S.A. in the form of a research grant awarded to Dr. S.V.Subramanian are gratefully acknowledged. We thank IPSCO Inc. & Stelco Inc., Hamilton, Canada for industrial support of the project. We express our special thanks to Dr. Laurie Collins, IPSCO and Mr. Keith Barnes, Stelco for the technical support.

REFERENCES

1. P.D.Hodgson, "The Metadynamic Recrystallization of Steels", THERMEC'97, International Conference on Thermo-mechanical Processing of Steels and Other Materials, Ed. by T. Chandra and T. Saki, The Minerals, Metals & Materials Society, 1997, pp.121-131.
2. H.O.Gekonde, "Influence of Dynamic Behaviour of Materials on Machability", Ph.D thesis, McMaster University, 1998
3. S.S.Ingle, "The Micromechanisms of Cemented Carbide Cutting Tool Wear", Ph.D thesis, McMaster University, 1993
4. M.A.Meyers, G.Subhash, B.K.Kad, and L.Prasad, "Evolution of Microstructure and Shear-Band Formation in ∀-hcp Titanium", Mechanics of Materials, Vol.17 1994, pp.175-193.
5. U.R.Andrade, M.A.Meyars, and A.H.Chokshi, "Constitutive Description of Work- and Shock-Hardened Copper", Scripta Metallurgica et Materialia, Vol.30 1994, pp.933-938.
6. U.R.Andrade, M.A.Meyers, K.S.Vecchio, and A.H.Chokshi, "Dynamic Recrystallization in High-Strain, High-Strain-Rate Plastic Deformation of Copper", Acta Metallurgica et Materialia, Vol.42 1994, pp.3183-3195.
7. G.Robiller and L.Meyer, "Work Hardening and Softening Behaviour of Ti- and Nb-alloyed Steels during Hot Deformation", Recrystallization and Grain Growth Multi-Phase and Particle Containing Materials, 1st Riso International Symposium on Metallurgy and materials Science, Ed. by N. Hansen, A.R. Jones, and T. Leffers, 1980, pp.311-316.
8. G.Robiller et al, "Recrystallization behaviour of niobium-alloyed steels at hot working temperatures", Thyssen Techn. Ber. 1975, pp.14.
9. R.Mater.Colas, "Mathematical modeling of hot rolling steel strip", Sci. Techno. 1998, pp. 387.
10. F.H.Samuel, S.Yue, J.J.Jonas and K.R.Barnes, "Effect of dynamic recrystallization on microstructural evolution during strip rolling", ISIJ Inter. Vol.30 (1990) 216-225
11. R.Kaspar, J.S.Djstl and C.Pawelski, "Extreme austenite grain refinement due to dynamic recrystallization" Steel research, 59 1988, pp.421.
12. B.Derby, "The Dependence of Grain Size on Stress during Dynamic recrystallization", Acta Metallurgica and Materialia, Vol.39 1991, pp.955-962.

13. M.G.Akben, I.Weiss and J.J.Jonas, "Dynamic precipitation and solute hardening in a V microalloyed steel and two Nb steels containing high levels of Mn", Acta Met. V29, 1981, pp.111-121.
14. B.Derby and M.F.Ashby, "On Dynamic Recrystallization", Scripta Metallurgica, Vol.21 1987, pp.879-884.
15. G.Zhu, "Studies on the recrystallization behaviors of cold rolled B2 ordered alloys", Ph.D. Thesis, University of Science and Technology Beijing, 1998.

HOT DEFORMATION OF LOW CARBON STEEL IN THE METASTABLE AUSTENITE REGION

Elisa Alonso,
Farid Hassani*
Steve Yue

Department of Mining and Metallurgical Engineering,
McGill University,
3610 University St.,
Montreal, Quebec, H3A 2B2, Canada
(514) 398-1378
steve@minmet.lan.mcgill.ca
*now at Stelco Inc., Hamilton, Ontario, Canada

Key Words: Dual phase steel, metastable austenite, fine grains, plain carbon steel

ABSTRACT

A low-carbon steel was subjected to compressive deformation in the metastable austenite region in an attempt to promote grain refinement by dynamic strain induced transformation of austenite to ferrite. After hot deformation, the specimens were either immediately quenched, or held for a short period of time before quenching. In both cases, the quench stage was to transform any remaining austenite to martensite to form a dual phase structure. The hold time was nominally to allow for more ferrite to form prior to quenching. Microstructures revealed that the grain size could be reduced to about 3 μm, using a deformation strain of about 0.5. This reduced the second phase to very low levels, which decreased the strength, although this was somewhat compensated by the grain refinement. The ductility increased with decreasing grain size.

INTRODUCTION

Fine grained structures have always been desired because of the excellent combination of mechanical properties. In particular, not only does the strength increase according to the Hall-Petch relationship, but ductility related properties, such as toughness and formability, also markedly improve. Dual-phase steels, with a matrix of equiaxed ferrite and a second phase of martensite also have superior strength/ductility property combinations. Therefore, the production of a fine-grained dual phase steels is a potentially fruitful concept. In this particular project, the ultimate goal was to generate fine grained dual phase structures in the as hot worked condition.

The approach taken in this work can be considered in two obvious stages; grain refinement and martensite formation. In the as-hot worked condition, ferrite grain refinement is achieved by optimised austenite conditioning. Conventionally, the finest hot worked structures are generated by classical controlled rolling, which consists of a stage of austenite recrystallization, to obtain a fine recrystallized austenite structure, followed by hot rolling below the temperature of no recrystallization to generate a heavily faulted, pancaked austenite. The latter leads to an increased number of nucleation sites, primarily due to an increase in the grain boundary area per unit volume, along with high energy 'interfaces' generated intragranularly. Grain sizes as low as 3 μm can be produced in this way. However, such an approach requires considerable strain, at least a strain of unity in the pancaking stage alone, and niobium is also required to stop recrystallization. Recently, the idea of refining by dynamic strain induced transformation from austenite to ferrite has emerged as being an method to produce *ultra* fine grain sizes (about one micron).[1] However, this technique also relies on heavy deformation. Nevertheless, the dynamic transformation approach seems to be more efficient compared to pancaking, in terms of utilization of the applied strain. In this work, the deformation used was compression, in which the strain is limited to about 0.7. Therefore, given this constraint, the dynamic transformation approach is the optimum approach to produce a fine grained structure.

A 'typical' dual phase steel contains about 20% martensite[2], with a particle size of a few microns. The carbon level of the martensite should be around 0.4 % to optimize strength and ductility[3], the implication being that higher levels of tetragonality would be embrittling. Some work by Balliger and Hudd[4] in the early 80s indicated that a finer martensitic structure led to significantly improved strength/ductility combinations, compared to conventional dual phase microstructures. These fine grained martensite particles were produced by intercritically annealing a cold rolled bainitic structure, as opposed to a ferrite pearlite structure. In this work, the aim was to apply the concept of fine martensite to these microstructures.

EXPERIMENTAL DETAILS

The steel chosen was a low C steel of composition as follows (all concentrations in wt%) : C 0.08%: Mn 0.43%: Si 0.06%: P 0.09%: S 0.012%. In other words, this is a plain C-Mn steel, with a quite modest level of Mn, the only significant alloying element. The main reason for selecting such a low C level was to try to generate much smaller volume fractions of martensite with 'ultrafine' particle sizes, but still maintain a reasonable tetragonality, as suggested above. Dual phase steels usually contain alloying elements to improve hardenability, e.g. higher levels of Mn. However, since the final step to achieve martensite was quenching, in this work, hardenability was not a major issue.

The material was received in the as-cast condition, from which cylindrical compression test samples, 11.4mm in height and 7.6mm in diameter, were machined. These were then homogenized by heating to 1200 °C at 1°C/sec and holding at that temperature for 20 minutes, and finally cooling to room temperature at about 1°C/sec. The homogenized specimens were then subjected to various thermomechanical treatments, as described below. The resulting specimens were then sectioned for microstructural examination and shear punch testing to obtain strength and ductility.

PRELIMINARY HOT DEFORMATION RESULTS

Following the above homogenizing treatment, the specimens were heated into the single phase austenite region (900 °C) and held for a few minutes to equilibrate the temperature. The specimens were then cooled to various temperatures, subjected to a range of compressive strains, and then quenched.

After austenitizing, initial experiments were conducted by cooling at about 1 °C/s to temperatures ranging from just above the start of the austenite transformation temperature to considerably into the intercritical range. On reaching the desired temperature, the specimens were subjected to a range of compressive deformations, and then immediately quenched.

The initial results revealed that deforming in the intercritical region did not lead to significant grain refinement. In fact, decreasing the deformation temperature led, at some point, to an *increase* in the amount of martensite in the as-quenched structure. Both results could be explained if it was assumed that deformation was taking place in a two phase structure. Here, because ferrite is softer than austenite at the same temperature, and since there is a more or less continuous ferrite film at the austenite boundaries, the ferrite is deformed more than the austenite. With increasing ferrite volume fraction, the fraction of the total applied strain partitioned to the austenite decreases. Thus, the extent of strain induced transformation, dynamic or static, decreases as the amount of ferrite increases, i.e. as the temperature decreases. On the other hand, deformation just above the austenite-to-ferrite transformation temperature due to cooling at 1 °C/s, in the single phase *metastable* austenite region, gave a reasonably fine ferrite grain size, with a low volume fraction of 'ultrafine' martensite particles on quenching (Fig 1). This is morphologically similar to dispersion strengthening, but with martensite particles.

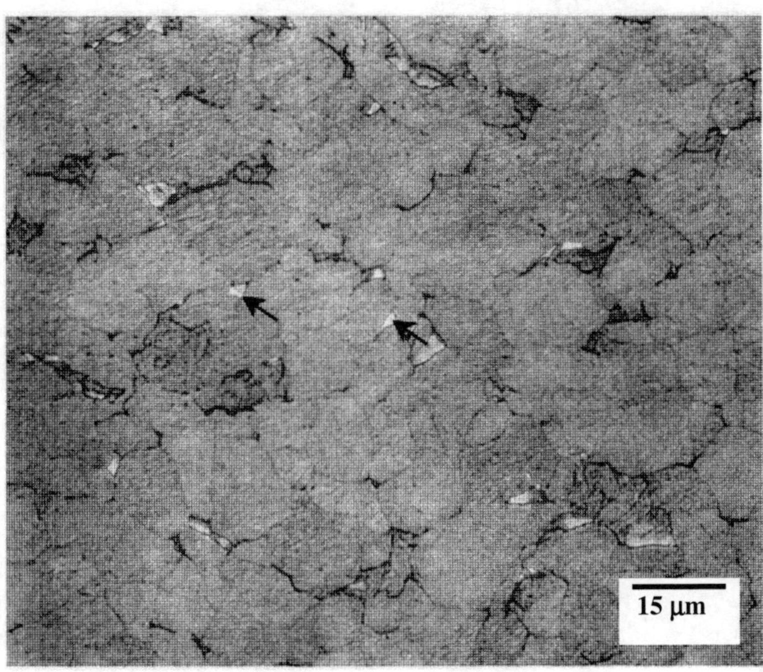

Fig.1 As quenched microstructure after deforming above the austenite to ferrite transformation temperature. Martensite particles (white) are arrowed.

In order to obtain still finer grain size, it was concluded that deformation at lower temperatures, but still in the single phase metastable austenite region, would increase the rate of dynamic transformation by increasing (i) the driving force, and (ii) the amount of substructure in the austenite.

THERMOMECHANICAL SCHEDULES

Following the preliminary findings, the cooling rate from the austenitizing temperature was increased from 1 °C/s to about 25 °C/s in order to drive the non-equilibrium austenite transformation start to lower temperatures. To determine the approximate start of the austenite transformation at this cooling rate, specimens were interrupted at temperatures between 800 °C and 600 °C and quenched to examine the microstructures. As can be seen in Fig. 2, there is no proeutectoid ferrite at 800 °C, and a considerable volume fraction at 600 °C. It is difficult to clearly discern ferrite at the other two temperatures, but there seems to be some evidence of a grain boundary phase at 750 °C. Therefore, the aim was to vary the deformation between 800 °C at 750 °C. This was very difficult to control, given the speed of the cooling rate, hence the deformation start temperatures could not be controlled precisely, as can be seen in Table I.

The resulting specimens were sectioned to produce specimens for optical and/or scanning electron microscopy. The specimens were also subjected to shear punch testing to measure the strength and ductility.

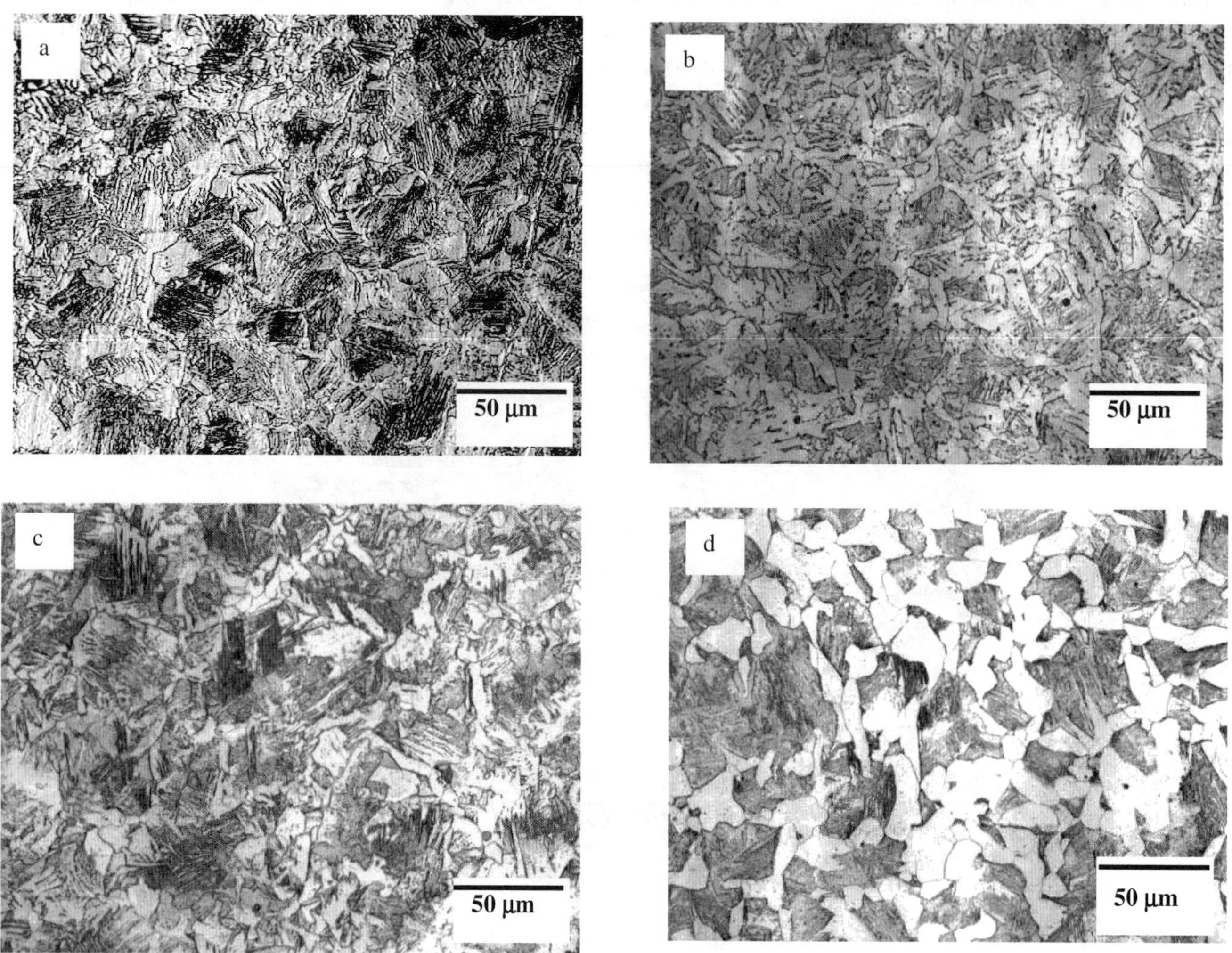

Fig.2 Microstructures after austenitizing at 900 °C, rapid cooling to (a) 800 °C, (b) 750 °C (c) 700 °C and (d) 650 °C, and quenching

RESULTS AND DISCUSSION

Microstructures:

Generally, all structures were primarily equiaxed ferrite, with decreasing levels of strain leading to increasing amounts of a second phase. Thus, specimen 13 (Fig. 3) exhibited the highest volume fraction of a second phase (probably martensite). The amount of second phase decreased with increasing strain and decreasing temperature. Thus, a comparison of specimen 13 with sample 8 (Fig 4), in which the strain has been increased in the latter by a factor of 4, reveals a much reduced martensite volume fraction and a finer structure. Decreasing the temperature to 788 °C (specimen 9, Fig.5) reduces the second phase to a very low level, as well as further refining the structure.

Table I: Thermomechanical treatments

Sample	Processing	Strain	Deformation Start Temperature (°C)	Quench Start Temperature (°C)
1	Immediate quench	0.465	798	--
2	Immediate quench	0.467	700	520
3	quench after 30s hold	0.507	810	455
4	quench after 30s hold	0.553	780	--
5	quench after 30s hold	0.461	--	410
6	quench after 60s hold	0.467	788	390
7	Immediate quench	0.363	790	620
8	Immediate quench	0.260	815	725
9	Immediate quench	0.260	788	617
10	Immediate quench	0.254	720	640
11	Immediate quench	0.161	806	710
12	Immediate quench	0.160	790	678
13	Immediate quench	0.067	810	740

At about 0.47 strain, for example specimen 1 (Fig. 6a), the grain size is on average much finer, but less uniform than the previous specimens, with some extremely fine ferrite grains clearly visible. Lowering the deformation temperature by about 100 degrees (specimen 2, Fig. 6b), generated a structure that appears to be still finer, although it was quite difficult to generate a good metallographic specimen from this microstructure.

The reason for the hold before quenching specimens 3 to 6 was originally to allow for any *static* strain induced transformation to ferrite to occur. However, since the strains applied in this series were all above 0.4, probably all the ferrite would have formed either dynamically, or very shortly after unloading. Hence there is no effect of the hold time visible using optical microscopy (e.g. compare Fig. 7 (specimen 4) to Fig. 6a (specimen 1)). However, the hold time may have generated pearlite prior to the quench. This could influence the mechanical properties.

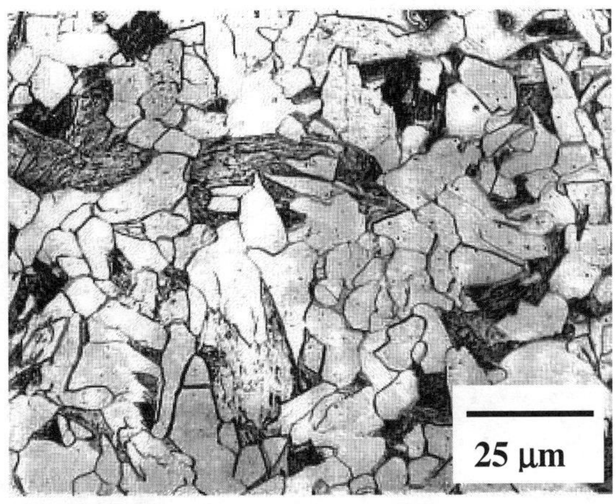

Fig. 3 Specimen 13, 0.067 strain, exhibits the maximum amount of martensite in this series of specimens.

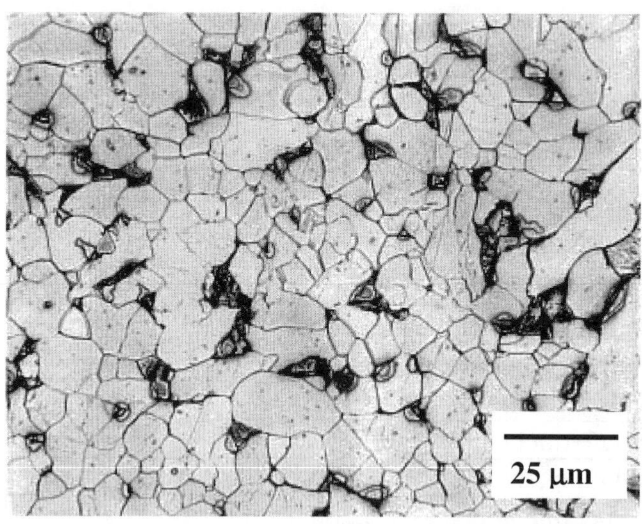

Fig. 4 Specimen 8 (0.26 strain, 815 °C deformation start). Higher strain leads to finer structure with much decreased martensite volume fraction

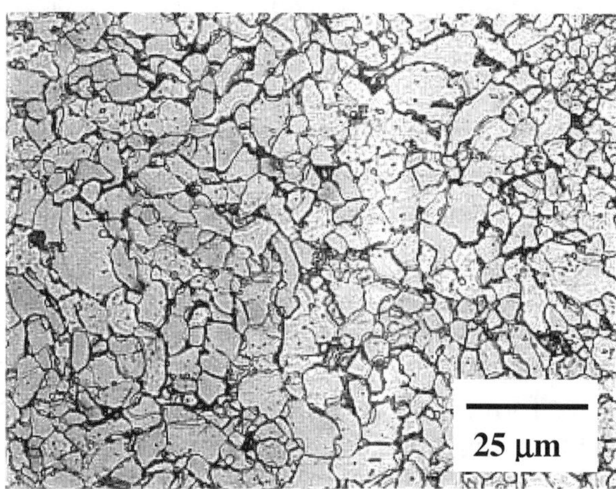

Fig 5 Specimen 9 (0.26 strain, 788 °C deformation start). Decreased temperature of deformation reduces martensite volume fraction and refines grain size compared to spec. 8.

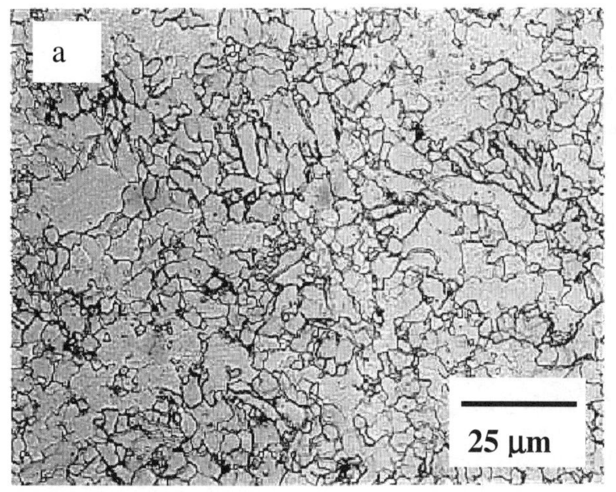

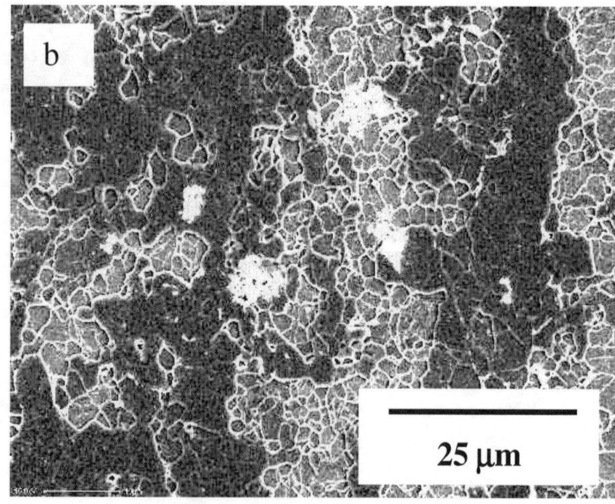

Fig 6 (a) Specimen 1 (0.465 strain, 798 °C deformation start). Higher strain refines, but seems to generate less uniform grain size with the appearance of many very fine grains. (b) Specimen 2 (0.467 strain, 700 °C deformation start). Decreased temperature leads to the finest structure.

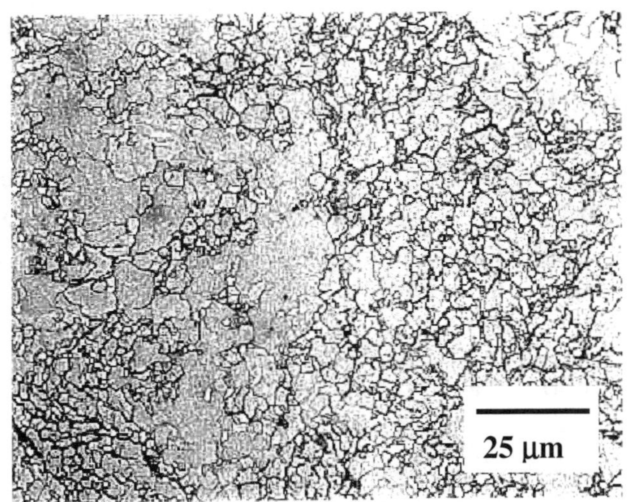

Fig. 7 Specimen 4, (0.553 strain, 780 °C deformation start, 30 s hold before quenching), is similar to specimen 1 (Fig. 6a), but slightly finer due to differences in strain and temperature.

Mechanical Properties

Shear punch testing is a useful technique to generate strength and ductility data when there are specimen size limitations. Extensive testing at McGill has demonstrated that the strengths generated from these tests match results from full scale tensile tests very well. However, the absolute values for ductility are still questionable. Nevertheless, the results certainly can be used for comparisons, at least. The properties for selected specimens are shown in Table II, along with the key processing conditions.

Table II: Mechanical Properties
(specimens in bold were held for 30-60s prior to quenching)

Sample	Strain	Deformation Start Temperature (°C)	YS (MPa)	UTS (MPa)	Elongation (%)
1	0.465	798	333	623	65.4
2	0.467	700	--	--	--
3	**0.507**	**810**	**349**	**589**	**59**
4	**0.553**	**780**	**340**	**601**	**64**
5	**0.461**	**--**	**--**	**--**	**--**
6	**0.467**	**788**	**--**	**--**	**--**
7	0.363	790	273	572	55
8	0.260	815	--	--	--
9	0.260	788	245	557	60
10	0.254	720	183	423	61
11	0.161	806	406	630	51
12	0.160	790	406	630	51
13	0.067	810	470	791	52

If the strengths are plotted as a function of applied strain, as shown in Fig. 8, it can readily be seen that the strength decreases fairly rapidly with increasing strain, upto about 0.25 strain, due to the decreasing amount of the second phase, presumably martensite. The strength recovers significantly with increasing strain due to grain refinement. There does not appear to be much effect of deformation temperature, apart from that exhibited by specimens 9 and 10, deformed at around 0.25 strain. Specimen 10 has been deformed at the lower temperature, and possesses a lower strength. The grain sizes of these two specimens are quite similar, perhaps because both specimens are deformed below 800 °C, whereas, in the example given above (Fig. 4 vs Fig. 5) one specimen was deformed above 800, the other below. Since the grain sizes are similar, it is possible that the lower temperature deformation increased the amount of ferrite, decreasing the strength in this way. However, this postulate could not be confirmed by the microscopical techniques reported in this paper.

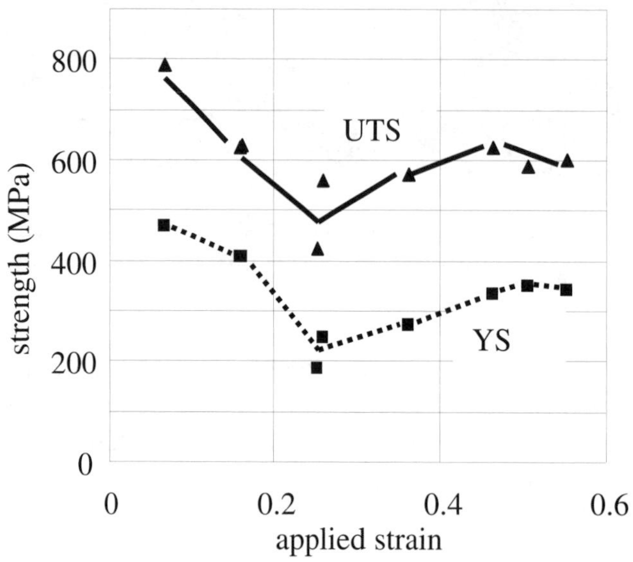

Fig 8. Effect of high temperature deformation strain on room temperature strength

By contrast, the ductility more or less continuously increases with increasing applied strain (Fig. 9). This generally reflects the influence of increasing grain refinement and, to a certain extent, the decreasing volume fraction of martensite.

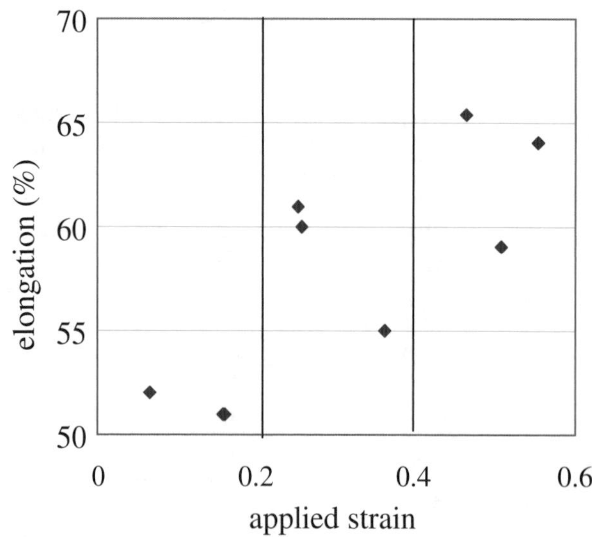

Fig 9. Effect of high temperature deformation strain on room temperature ductility

It is difficult to clearly discern any clear effect of the hold time prior to quenching on the properties. As mentioned above, the effect of this hold time prior to quenching may be expected to slightly increase the volume fraction of ferrite, although such differences are not qualitatively resolvable from optical or SEM microscopy. Another effect may be to transform the remaining austenite to pearlite. There is some evidence of this in specimen 8, Fig. 10. Note that in this series of tests, the specimens subjected to the highest strains were also subjected to the 30-60 s hold prior to quenching. These specimens exhibit a slight fall in the strength levels, and are also less ductile than specimen 1, which was deformed to a lower level, but was quenched immediately after deformation was completed. This may be indicative of an advantageous effect of fine martensite over fine pearlite particles, although more work has to be done to explore this postulate.

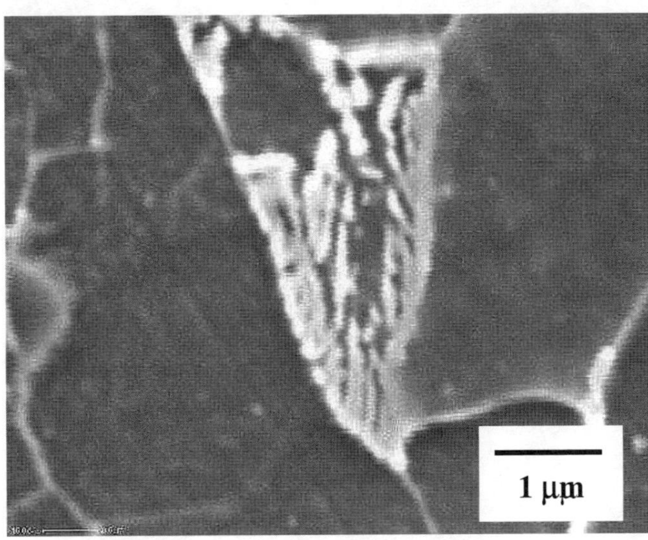

Fig. 10. SEM of specimen 8 showing a fine lamellar second phase.

SUMMARY AND CONCLUSIONS

Fine-grained microstructures can be generated by deforming in the predominantly single phase metastable austenite region, as opposed to the 'stable' intercritical region. Grain refinement increases with increasing strain and decreasing temperature of deformation. The as-hot deformed strength experiences a minimum as the applied strain increases primarily due to the decreasing martensite. The strength increases with further increase in applied strain due to grain refinement.

ACKNOWLEDGEMENTS

The authors would like to thank the Canadian Steel Industry Research Association (CSIRA) and the Natural Science and Engineering Research Council (NSERC) of Canada for their financial support.

REFERENCES

1. G.L. Kelly and P.D. Hodgson, Proc. "The John J. Jonas Symposium", eds. S.Yue, P.D. Hodgson and E. Es-sadiqi, Ottawa, August 2000, The Metallurgical Society of the Canadian Institute of Mining and Metallurgical Engineering, (in press).
2. D.T. Llewellyn and D.J. Hills, Ironmaking and Steelmaking, 1996, 23, p.471.
3. J.Y. Koo and G. Thomas, Proc. Conf. "Fundamentals of Dual Phase Steels", TMS-AIME, Chicago IL, October 1977, p40.
4. N.K. Balliger and R. Hudd, "Dual Phase High Strength Steels", ECSC Agreement 7210. KD/802.

DEFORMATION TEMPERATURE AND THE EVOLUTION OF AUSTENITE GRAINS IN NIOBIUM STEELS

Khaled F. Al-Hajeri and **Tariq Mehmood**
SABIC Technology Center, Jubail
P.O.Box 10040
Al-Jubail Industrial City, 31961
Kingdom of Saudi Arabia
Telephone: (9663) 347-7200 Ext. 5022
E-mail: hajerikf@sabic.com; mehmoodtq@sabic.com

Key Words: Deformation Temperature, Grain Refinement, Niobium Steels, Austenite Grain Growth

INTRODUCTION

Niobium is the most frequently used element, either singly or in combination with titanium and vanadium, in the production of high strength low alloy steels. In small additions it provides strengthening by both precipitation hardening and ferrite grain refinement through effective control and conditioning of austenite during deformation. The effectiveness of these mechanisms is however greatly influenced by the thermal processing variables as both the temperature and amount of deformation strongly affect the recrystallization kinetics.

During conventional hot rolling, most of the increase in yield strength is achieved by micro precipitation in ferrite during cooling from hot rolling. However, with the use of controlled rolling practices, deformation within the lower austenite temperature range leads to strain induced precipitation of finely dispersed particles that extremely retards the recrystallization.

As the metallurgical condition of the prior austenite, i.e. grain size, composition, presence of precipitation, degree of recrystallization etc. determine the final ferrite grain size, the knowledge of austenite recrystallization process becomes important to estimate the final microstructure.

Extensive work has been done in the past to study the effect of various processing parameters on the austenite grain size. This paper reviews some of the work, particularly that related to the effect of deformation temperature, beside highlighting the result of recent research work conducted on the similar lines.

Role of Niobium in Steel Strengthening: Niobium is a proven strengthening agent for hot rolled carbon steels. Depending upon the exact chemistry and processing technique employed, niobium forms either carbides or carbonitrides. During hot rolling, the precipitation temperature increases with the product Nb x C and the amount of strain energy in the austenite matrix[1]. The primary precipitation begins at the austenite grain boundaries and extends to incoherent twin boundaries and deformation bands within the grains. The homogenous precipitation prevents further recrystallization of the austenite. As niobium extensively increases the recrystallization stop temperature (Fig 1), sufficient degree of cold work "pan caking" in austenite can be obtained during deformation in the normal finishing temperature ranges used at most of the hot strip rolling mills.

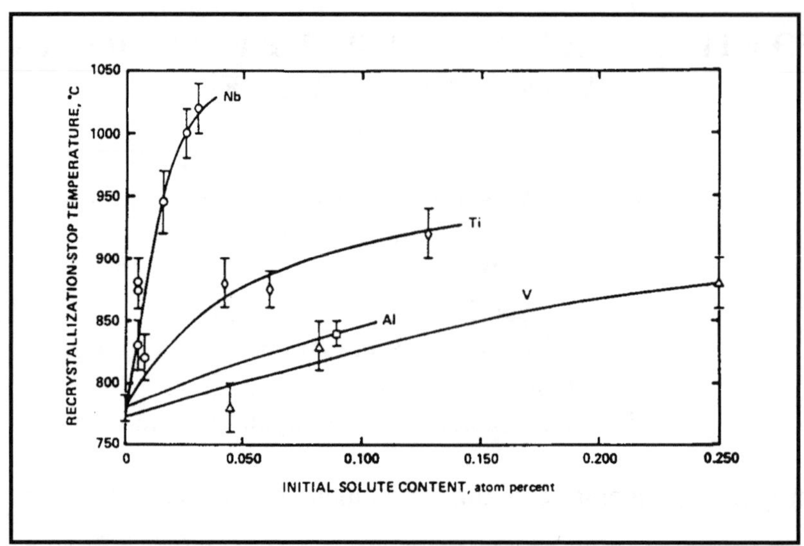

Fig. 1: Effect of Niobium on Recrystallization Temperature.[2]

Apart from affecting the recrystallization stop temperature, the solubility conditions of Nb (C,N) in austenite also makes it as an indispensable element for high strength steels. It is well understood that while the titanium and vanadium compounds have a higher super saturation at high and low deformation temperatures respectively, the niobium based system show high super saturation at the temperature well within those used in industrial applications as shown in Fig 2[3].

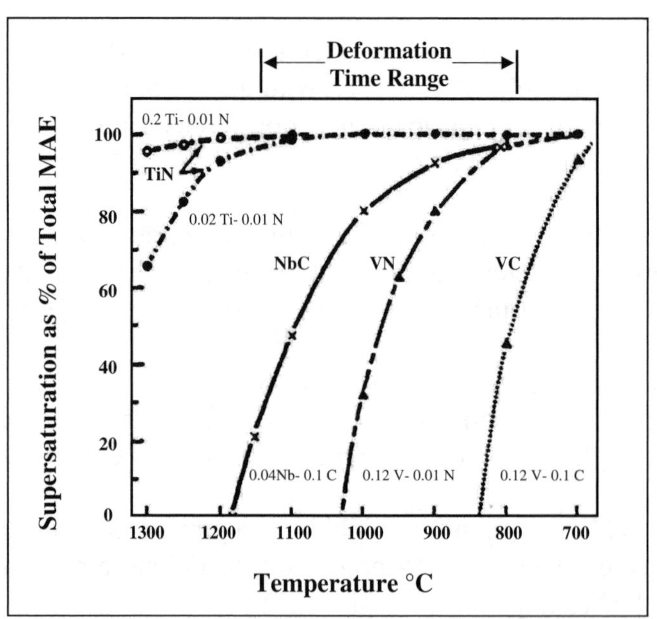

Fig. 2: Solubility of Various Compounds of MAEs used in HSLA Steels.[3]

Deformation Temperature and Austenite Grain Size in Niobium Steels: The most comprehensive work about the effect of deformation temperature on the prior austenite grain size was reported initially by L.J.Cuddy et al.[4] and Isao Kozasu et. al[5]. In the former research work the deformation behavior of low carbon manganese steel with 0.05% niobium was studied. It was noted that the prior austenite grain size in this type of steel strongly depends upon the deformation temperature and the grain size decreases as the deformation-temperature is decreased as shown in Fig 3.

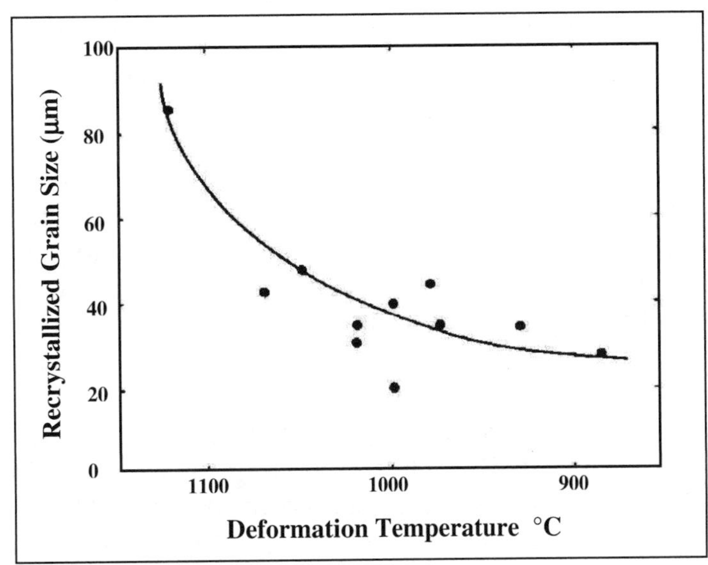

Fig. 3: Effect of Deformation Temperature on Recrystallized grain Size.[4]

It was also noted that the deformation at the partial recrystallization temperatures might result in duplex structure unless the drafting schedule of rolling is carefully selected. The deformation below the partial recrystallization temperatures results in only flattening and elongation of the initially equiaxed grains.

The work reported by I. Kozasu[5] included the study of recrystallization behavior of low carbon manganese steel containing 0.03% niobium. It was noted that the recrystallized austenite grains size depends upon the deformation temperature and the total reduction employed. The decreasing deformation temperature effectively reduced the austenite grain size particularly at higher degree of deformation. The deformation with intermediate degree, particularly for those given at the lower temperature range, used in the experiment gave, a duplex structure though the coarser grain was still finer than those obtained with higher deformation temperature.

Deardo A.J. while describing detailed over view of various thermo-mechanical processes of micro alloyed steel also reported similar observations and linked the delayed recrystallization given by niobium to its precipitation hardening characteristics [6].

In similarity with the above observations the effect of deformation temperature on the prior austenite grain size of a commercially produced niobium steel was studied on the following lines.

EXPERIMENTAL SET-UP

Samples: Low carbon manganese steel samples with a niobium content of 0.035% were used for the study. The chemical composition of the sample is shown in Table – I.

Table I: Chemical Composition of Steel Samples.

C	Si	Mn	Nb	N	Al	P
0.110%	0.193%	1.420%	0.035%	0.0057%	0.033%	0.020%

Sample Preparation: Machined samples of dimensions 50 x 30 x 10 mm were prepared for the tests. A hole of 1.6 mm diameter was drilled through the middle of the side face to mid width of the samples was drilled to accommodate the thermocouple for temperature measurements during the test. The samples were chrome plated to prevent oxidation and scaling during reheating process.

Processing: The samples were subjected to plain strain compression test in a Servotest compression testing machine (Fig 4a) after heating to 1200 °C and soaking for about 30 minutes to produce an initial uniform structure. The samples were then furnace cooled down to 1050, 1000 and 950 °C prior to deformation. The deformation was subjected at a constant strain rate of 5.0 sec^{-1} with the nominal true strain of 1%. After the deformation, the samples were rapidly quenched in water to freeze the austenite grain structure. The thermo-mechanical process used for the study is mapped in Fig 4b. A sample was also quenched directly after reheating and soaking to freeze the initial microstructure before deformation.

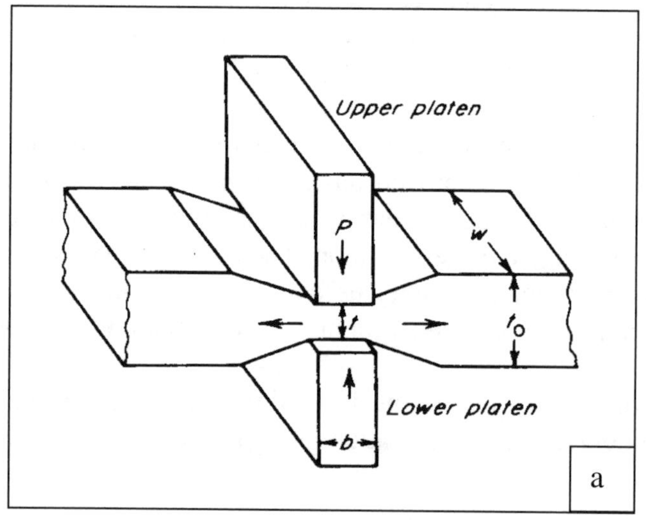

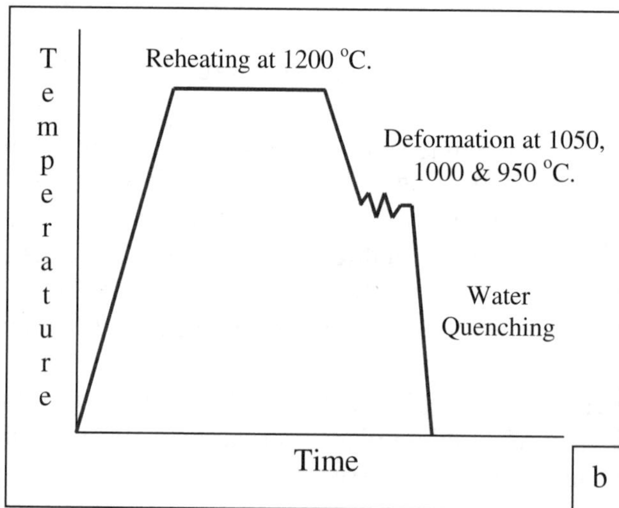

Fig. 4: Scheme of Thermo-mechanical Processing.

Metallographic Analyses: The samples in as received as well as after the treatment were analyzed metallographically to determine the prior austenite grain size. Before mounting, the samples were tempered at about 480 °C to reveal the austenite grain boundaries without affecting the prior austenite grain size. The etching was carried out with warm aqueous solution of picric acid to reveal the microstructure. The grain size was measured by using automatic image analyses system that gives the mean intercept, average grain size and ASTM grain number.

RESULTS & DISCUSSION

Initial Grain Structure: The micro structure of the as received samples was ferritic-pearlitic as shown in Fig 5a. The average ferrite volume fraction was 0.82 with an average grain size of 13 microns or ASTM Grain Number 9. The reheated and quenched sample revealed a mixed grain size of prior austenite as shown in Fig 5b with 87.6% of the grains having a grain diameter of less than 400 microns with an average grain diameter of 192 microns.

Fig. 5: Microstructure of Initial Material.
(a- As received ferrite-pearlite structure, b- Prior austenite grain structure after initial heat treatment)

Grain Structure after Strain Compression Test: The results of the microstructural analyses, where the size and percentage of the fine and coarse grains, in addition to the weighted average grain size of each sample, are given in Table- II. The frequency distribution of different grain size found in the samples is shown in Fig. 6 and the microstructures obtained during deformation at three different temperatures, 1050, 1000, 950°C are shown in Fig. 7. The relation between the deformation temperature and weighted average grain size is given in Fig. 8.

Table II: Austenite Grain Size of Thermo-mechanical Treated Samples

Deformation Temperature	Fine Grain Size < 45 μm		Coarse Grain Size > 45 μm		Weighted Average Grain Size μm	ASTM Grain No.
°C	Average Grain Size μm	Frequency %	Average Grain Size μm	Frequency %		
950	13.7	94.1	52.5	5.9	16.0	7.5
1000	16.2	87.3	55.6	12.7	21.2	7.8
1050	17.2	80.7	56.3	19.3	24.8	8.5

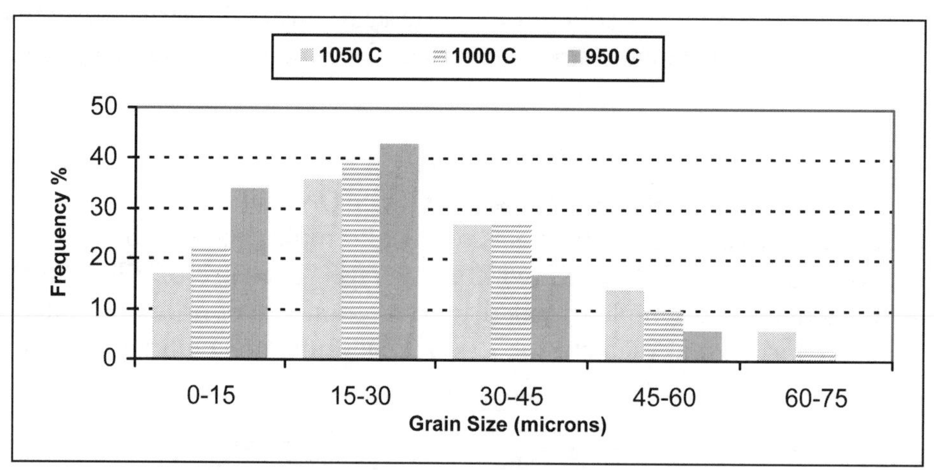

Fig. 6: Frequency Distribution of Grain Size.

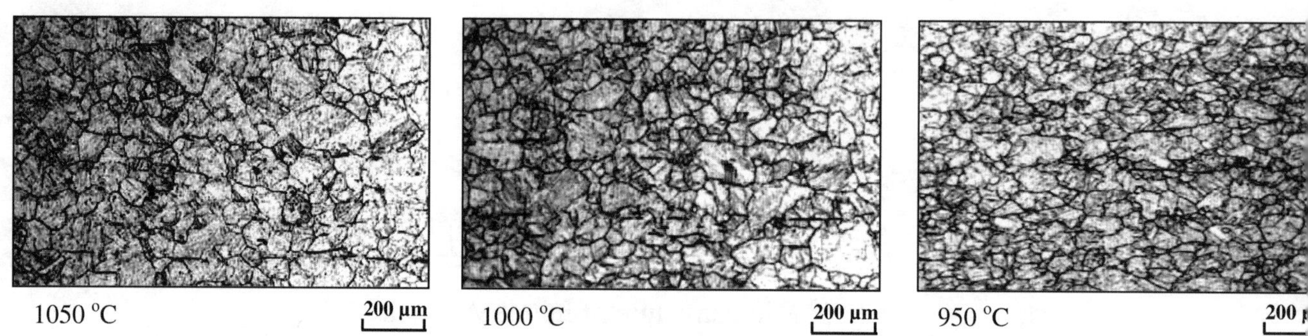

Fig. 7: Microstructure of Nb-Steel after deformation at Various Temperatures.

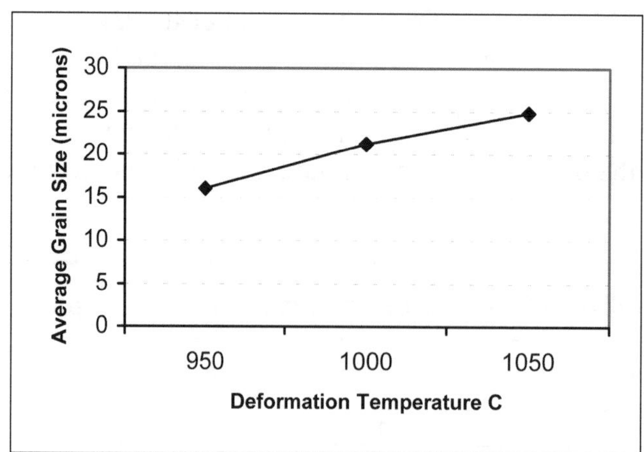

Fig. 8: Dependence of Prior Austenite Grain Size on Deformation Temperature.

It is known that as the deformation temperature decreases the carbonitrides in niobium steels start to precipitate mainly at grain boundaries, sub-boundaries and other dislocation net works produced during hot working. These precipitated carbonitrides give a pinning effect to the austenite grain structure which results in retardation of the processes of dynamic recovery and recrystallization during deformation.

It is also noted that the deformation temperature has a distinct influence on the austenite grain size. With decreasing temperature, the average grain size reduced and the total frequency of the finer grains increased significantly. The slight elongation observed in the structures deformed at lower temperature is attributed to the inherent effect of niobium in increasing the recrystallization stop temperature. Both of these observations support the findings of the earlier investigations on that subject.

ACKNOWLEDGEMENTS

Acknowledgements are due to Dr. Eric. J.Palmiere, Department of Engineering Materials, The University of Sheffield, U.K. for supervising this work.

REFERENCES

1. The Book of Steel, Edited by Gerald Beranger, Intercept Limited UK, 1996.
2. Cuddy L.J. The effect of Microalloy Concentration on the Recrystallization of Austenite During Hot Deformation", Plastic Deformation of Metals, Academic Press, New York, 1975.
3. Palmiere E.J., "Precipitation Phenomena in Microalloyed Steels", Microalloying'95, Conference Proceedings, page 307-320.
4. Cuddy L.J. Microstructures Developed during Thermo-mechanical Treatment of HSLA Steels, Metallurgical Transactions A, Vol. 12 A, July 1981, 1313.
5. Kozasu, I. Et. al. Hot Rolling as a High Temperature Thermo-mechanical Process, Microalloying 75, page 120.
6. Deardo A.J., "Modern Thermo-mechanical Processing of Micro alloyed Steel: A Physical Metallurgy Perspective, Microalloying'95, Conference Proceedings, page 15-33.

High Strength, Thin Gauge Strip By Heavy Ferritic Rolling Of Low Carbon Steels.

R. M. Smith
BHP Steel Research Laboratories
PO Box 202, Port Kembla, NSW 2505, Australia
Ph + 61 2 42523198 F + 61 2 42523120
smith.ross.rm@bhp.com.au

Key Words: Steel, Ferritic Rolling, Recovery Annealing, Subgrain Structure

INTRODUCTION

Intercritical rolling has long been used to increase the yield and tensile strength of HSLA and low carbon steel plate and strip [1][2]. However, the amount of reduction applied in this region has always been relatively low, less than around 20%, due to a reduction in fracture resistance with increasing reduction below the Ar_3 temperature. This was manifested by separations in through thickness charpy impact tests due to the strong (100)(uvw) texture introduced by rolling of the ferrite. Conversely, intercritical rolling has also been used to decrease the yield and tensile strength of low carbon steel strip [3][4]. In this case a controlled amount of ferrite reduction at relatively high temperatures in the ferrite region is used in combination with conventional to moderately high coiling temperatures to generate selective ferrite recrystallisation and significant grain growth. It is evident therefore that careful control of ferritic rolling and subsequent cooling of low carbon steel can be used to manufacture steel products with a diverse range of properties.

The design and construction of new generation strip mills capable of 'ultra' thin gauge and ferritic rolling offer a unique opportunity for the direct production of full hard or recovery annealed steel strip. The laboratory work described herein relates to the rolling of low carbon steel strip to generate structures similar to that of conventionally cold rolled and recovery annealed steels. Despite the lower ductility of these high strength steels, they have been demonstrated to be quite capable for a large section of the market for commercial quality, zinc and zinc/aluminium coated, roll formed products [5][6]. Consideration was also given to subsequent recrystallisation of these heavily deformed steels in hot dip metal coating lines for lower strength, more ductile steel products. The final product was targeted for direct application as 'hot' rolled strip, for direct feed to hot dip coating lines or for in-line cold rolling hot dip coating lines.

EXPERIMENTAL PROCEDURE

Steel

Two low carbon steel grades were used for the investigation and their compositions are given in table 1. Both were commercial steel grades and the samples were taken from crop ends of HSM transfer bars. The thickness of the steel L sample was 33 mm and that of the steel M sample was 22 mm. Low reduction schedules for steel L were conducted on bar previously hot rolled to 16 mm.

Rolling

Rolling experiments were conducted on samples from 75-150mm wide on a computer controlled 300t laboratory reversing mill. Low rolling temperatures were achieved by combinations of low reheat temperature, by holding samples prior to rolling and by a single hold between rolling passes. The laboratory rolling schedules are shown schematically in figure 1. For steel L, samples were rolled to a high strain (3.0) or low strain (2.2) schedule. Samples of steel M were rolled to a strain of 2.5. Data for each rolling pass was collected on the mill computer and strip surface temperatures were recorded from pyrometer measurements made on each side of the reversing mill.

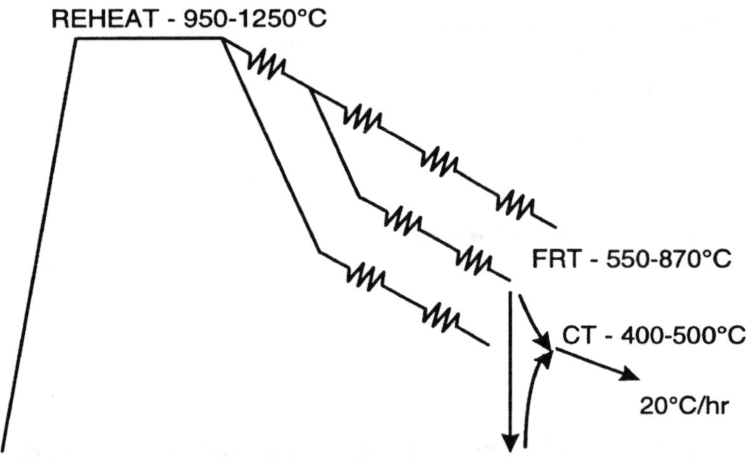

Fig. 1: Schematic of rolling schedules.

Table I: Steel composition

Steel	L	M
C	0.038	0.13
P	0.011	0.008
Mn	0.23	0.56
Si	<.005	0.046
S	0.010	0.003
Cr	0.023	0.015
Ni	0.024	0.023
Cu	0.008	0.015
Al	0.043	0.034
N	.0048	.0040
Nb+V+Mo	<0.015	<0.015

Cooling After Rolling

After rolling, some samples were cooled in still air (5°C/s) to a temperature of 400°C, placed in a furnace at 400°C and allowed to cool slowly to room temperature at 20°C/hour to simulate coiling. Other samples were 'quenched' after rolling to <100°C within 1-2s, at rates greater than 350°C/s to retain the deformed structure. These samples were reheated to 400°C or 500°C, and allowed to cool at a rate of 20°C/hour to simulate low temperature coiling of the rolled strip. Control samples of steel L were reheated at a conventional temperature of 1250°C and rolled completely in the austenitic region. After rolling, samples were cooled in still air to a temperature of 660°C, placed in a furnace at 660°C and allowed to cool slowly to room temperature to simulate coiling

Recovery Kinetics

To estimate the expected rate of recovery during cooling and coiling, metallographic specimens were taken from two samples of the quenched low carbon steel and isothermally treated at various temperatures and times. Vickers hardness measurements (5kg) were used to determine the extent of softening.

Cold Rolling

A small range of samples were selected for additional cold rolling prior to low temperature (500°C/60s) stress relief or recovery annealing. Cold rolling was conducted at room temperature, under lubrication to either 20% or 40% total reduction.

Subsequent Annealing

Some warm rolled samples were annealed in a salt bath or infrared furnace at various temperatures from 550- 750°C for 60s to simulate annealing on hot dip galvanising lines.

Testing

Tensile properties were determined from 50 mm gauge length specimens taken in the longitudinal and transverse directions where possible. Testing was conducted at an initial crosshead speed of 5 mm/min through the yield point and at a speed of 12.5 mm/min for the remainder of the test.

Metallography

Longitudinal specimens were taken from across the width of the rolled samples and prepared for metallographic examination. Specimens were etched in either Picral, Nital or Marshall's reagent to reveal the grain and subgrain structure.

RESULTS & DISCUSSION

A range of different reheat temperatures and rolling schedules were used to achieve low finish rolling temperatures (FRT) between 740°C and 550°C for steel L and between 705°C and 615°C for steel M.

Recovery Kinetics (Steel L)

Though limited, the data for steel L indicated the rate of softening at 400-600°C was small for these samples (fig. 2). Even at 600°C the start of recrystallisation was delayed to times greater than 10,000s although this behaviour would be expected to vary slightly for different rolling conditions.

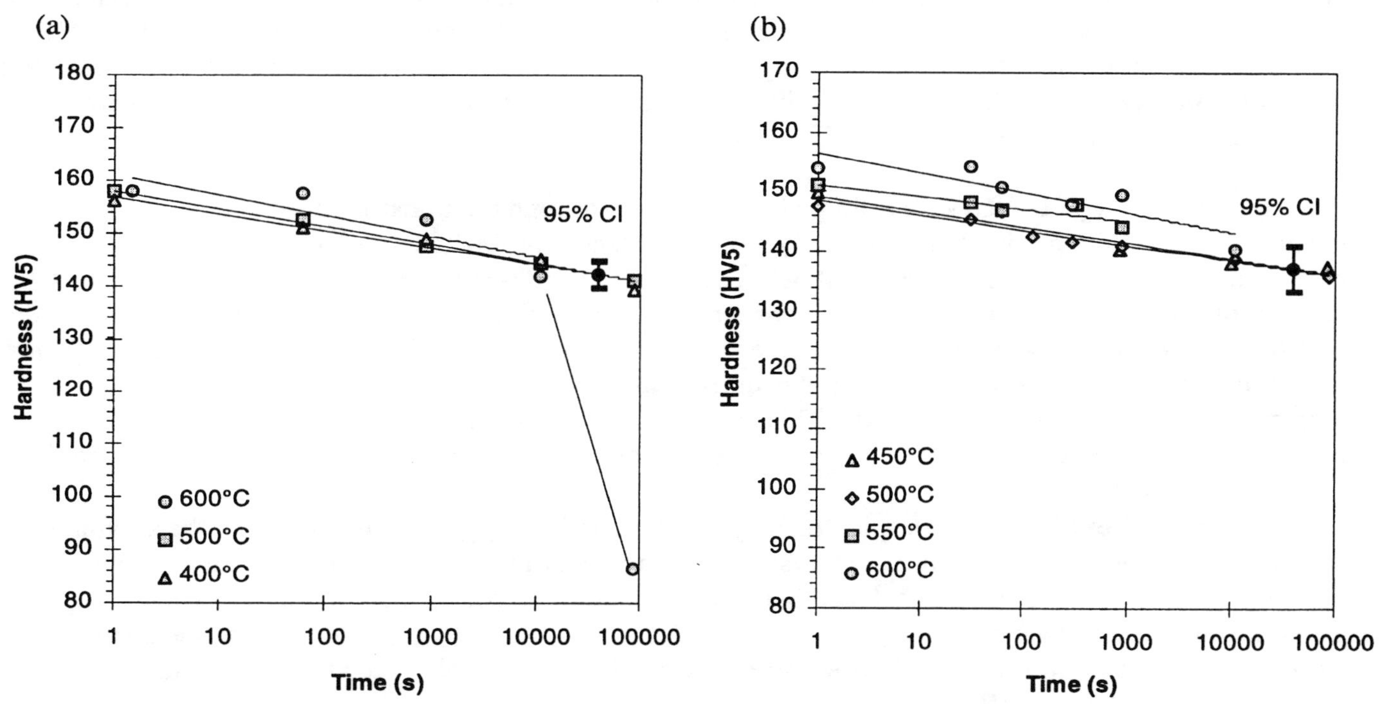

Fig. 2: Recovery of ferritically rolled samples of steel L at various temperatures, (a) FRT 670°C and strain of 2.2, (b) FRT 700°C and a strain of 3.0.

Microstructure (Steel L)

Control samples reheated to 1250°C and rolled to a FRT just above the Ar_3 showed a fine equiaxed ferrite grain structure with fine pearlite as the second phase. The structure (fig. 3a), was typical of commercially rolled, low carbon steel.

For samples where rolling continued below the Ar_3 temperature, a significant proportion of ferrite was deformed during rolling and the final structure depended on the temperature progression during rolling. The structures observed were consistent with the following events (fig. 3). The dense subgrain structures were more clearly defined by etching in Marshalls reagent (compare figs 3b and 3d).

- Rolling above the Ar_3 temperature progressively refined the austenite grain structure.
- At the Ar_3 temperature, ferrite formed and was in turn progressively deformed along with the remaining austenite.
- The ferrite first formed at high temperatures and early in the pass schedule could accumulate sufficient strain for recrystallisation and growth. These larger grains were deformed in subsequent passes and were characterised by a reduced aspect ratio.
- Due to the rapid temperature drop during rolling, recrystallisation and growth of ferrite became limited and a significant proportion of the ferrite remained unrecrystallised and continued to be deformed and elongated on subsequent rolling passes. A well developed subgrain structure was evident in these grains.
- The remaining austenite was progressively deformed, finally transforming to fine ferrite grains late in the schedule and eventually to a mixed ferrite/carbide aggregate. Fine ferrite grains formed late in the schedule were also deformed but obviously to a limited extent.

The resulting microstructures consisted of a distribution of ferrite grains with a range of sizes, aspect ratio and subgrain density (fig. 3). As the FRT was lowered, the ferrite grains and subgrains become more elongated and the overall structure was refined, with the mean boundary-free path reduced as greater strain was applied below the Ar_3 temperature. The subgrain structure appeared inhomogeneous in all samples, potentially due in part to differences in etching density but also due to real differences in the original grain orientation and size and to differences in individual strain level depending at which point the ferrite formed during the rolling schedule.

Mixed grain size (Steel L) - As rolling was conducted through the transformation there existed the potential for developing mixed grain sizes due to the tendency for ferrite in the low carbon steel to recrystallise and grow at temperatures just below Ar_3. The mixed ferrite grain size was more obvious where the level of ferrite rolling was low, but was obscured at lower rolling temperatures where the ferrite strains were high. This effect was more obvious in the subsurface shear zones where the higher effective strain accelerated ferrite recrystallisation so that it occurred earlier in the pass schedule and thus at higher temperatures conducive to grain growth. This is typical of commercial warm rolling operations where recrystallisation of deformed ferrite and grain growth is clearly initiated in the subsurface shear zones [7] .

Recrystallisation (Steel L) - Despite the large total reductions applied to the ferrite in these samples, the temperature was obviously too low for recrystallisation to occur in the core of samples after rolling. A few isolated grains were occasionally observed in the subsurface shear zones of some samples which may have resulted from recrystallisation after rolling was complete (fig. 3b). Extensive recovery of the surrounding regions had obviously lowered the driving force for growth of these grains such that they remained isolated, even on 'coiling' at 500°C (fig. 3).

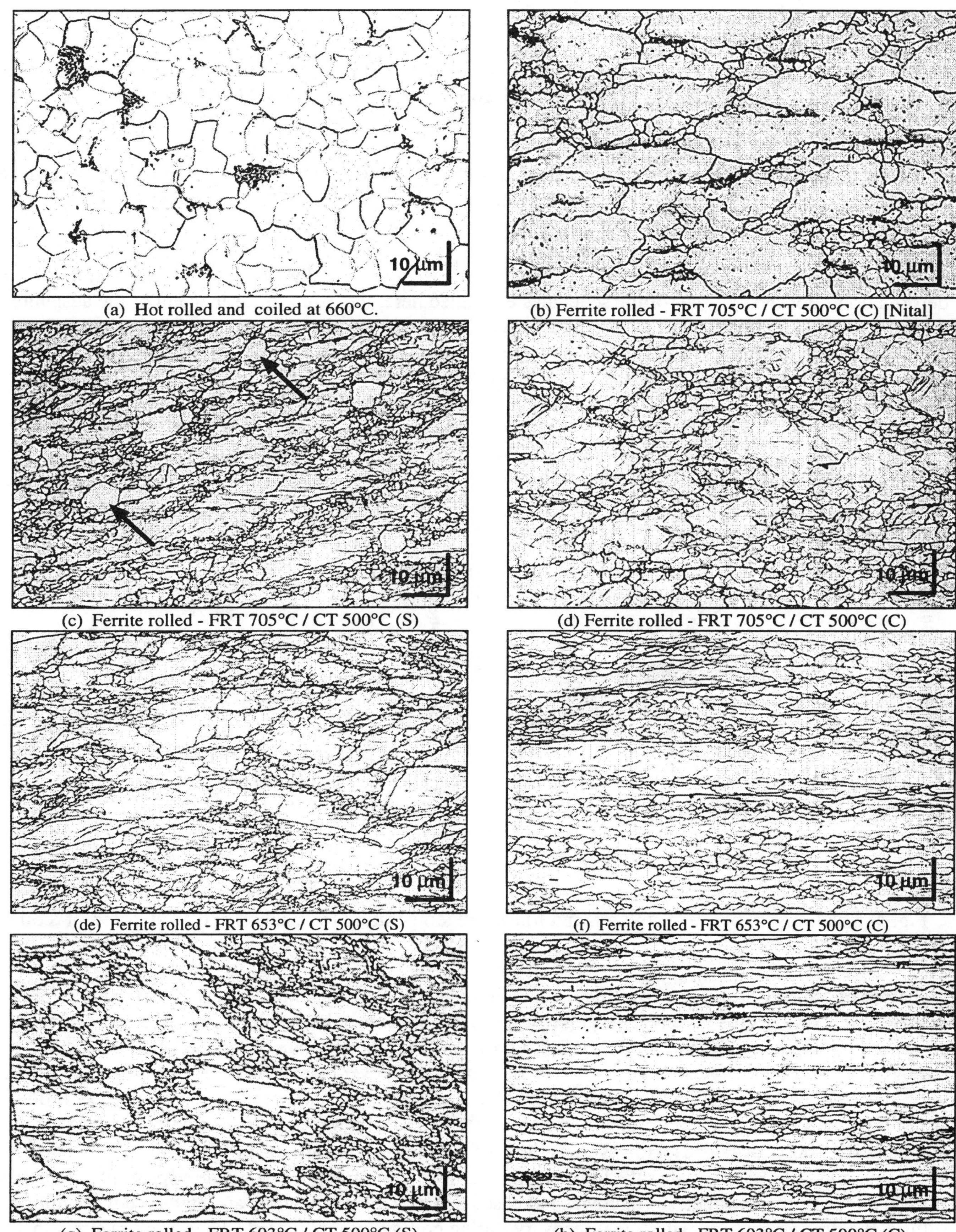

(a) Hot rolled and coiled at 660°C.

(b) Ferrite rolled - FRT 705°C / CT 500°C (C) [Nital]

(c) Ferrite rolled - FRT 705°C / CT 500°C (S)

(d) Ferrite rolled - FRT 705°C / CT 500°C (C)

(de) Ferrite rolled - FRT 653°C / CT 500°C (S)

(f) Ferrite rolled - FRT 653°C / CT 500°C (C)

(g) Ferrite rolled - FRT 603°C / CT 500°C (S)

(h) Ferrite rolled - FRT 603°C / CT 500°C (C)

Fig. 3: Microstructures of the hot rolled and ferrite rolled low carbon steel (steel L). Longitudinal sections, subsurface shear zone (S) and the core (C), etched in Marshalls reagent (fig.3b - etched in nital).

Mechanical Properties (Steel L)

Samples of steel L that were hot rolled and coiled at 660°C, displayed yield strengths at the upper end of the typical range of properties for heavily refined, commercial low carbon strip (275MPa LYS, 331MPa TS, 38% TE).

Effect of finish rolling temperature (Steel L) - As expected, the strength of the ferritically rolled steel increased as rolling was conducted at progressively lower temperatures. The lower yield stress showed a strong dependence on FRT (fig. 4) and also on the average rolling temperature.

Final strength was related to the extent of austenite refinement and the amount and temperature of ferrite deformation. For most of the rolling schedules examined here, the extent of refinement of the austenite tended to plateau so that variations in rolling around the transformation temperature had limited influence. The dependence of strength on FRT and on average rolling temperature was however, modified by the subsequent coiling temperature, as can be seen in figure 4.

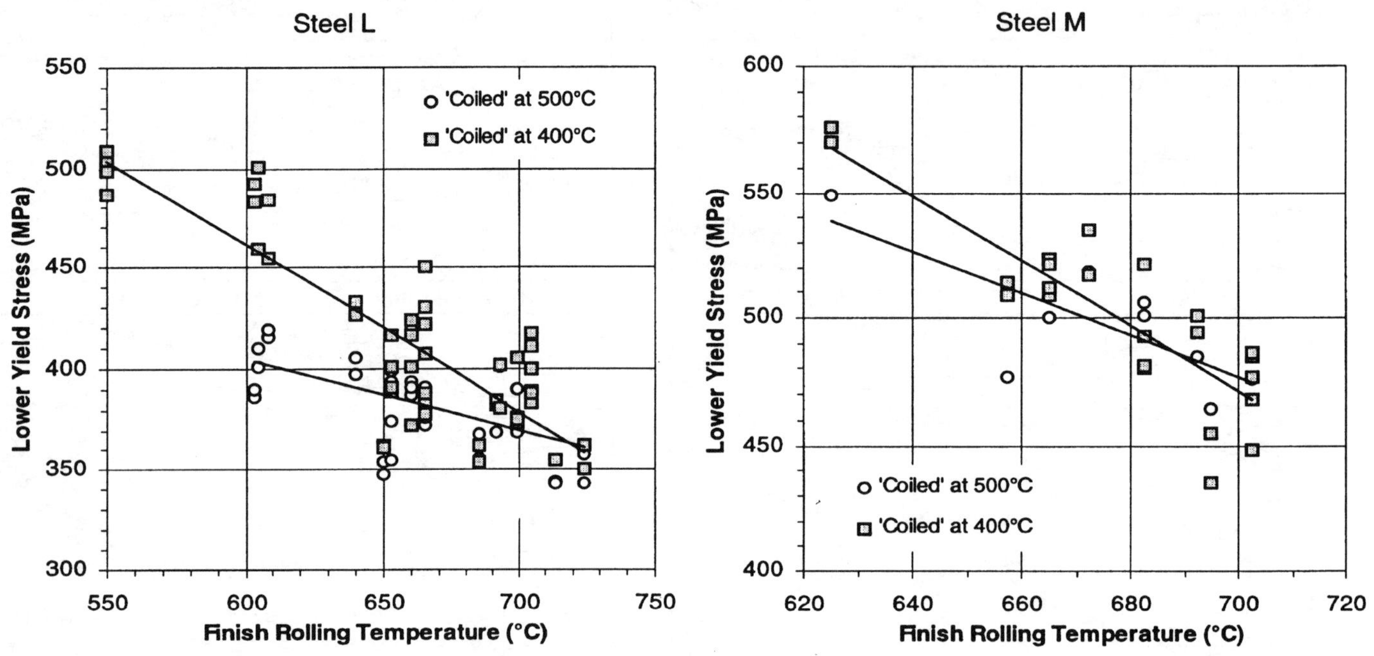

Fig. 4: Relationship between strength and finish roll temperature for steels L and M at different 'coiling' temperatures.

Effect of 'coiling' temperature (Steel L) - For samples where the extent of ferrite deformation was relatively low, the results of tensile testing revealed little difference in strength between samples with simulated coiling at 400°C and 500°C. This is consistent with the low rate of recovery determined by hardness measurements on isothermally treated samples. However, when the degree of ferrite deformation was increased by lowering the FRT to around 600°C, the coiling temperature displayed a greater influence. At a FRT around 650-700°C, strength levels greater than 350MPa were achieved whether coiling was conducted at 400°C or 500°C. However, as the FRT was lowered to around 600°C, a further increase in strength could only be realised at the lower coiling temperature of 400°C. It would appear that at the lower FRT's, although the as-rolled strength was increased due to reduced dynamic and static recovery during rolling, the driving force for additional recovery during coiling was increased. Thus as extra strengthening is sought by reducing the FRT, lower coiling temperatures will also be mandatory.

Effect of cooling rate (Steel L) - The strength of samples quenched directly after rolling showed significant variation which did not reflect the variation in strength of samples after simulated coiling. It appears large changes in strength may occur during the first few seconds after rolling due to rapid static recovery. This is consistent with previous work which has shown rapid softening by recovery, from 20-30% in moderately alloyed steel [8] and up to 40% in ELC steel [9] within 1 second after deformation. This is expected for the normal logarithmic dependence of recovery against time.

Duplicate samples were rolled in four of the schedules. One sample was quenched to room temperature prior to the coiling simulation whereas the second sample was allowed to air cool to room temperature before the coiling simulation. The yield strengths of these samples can be compared in table 2. There was no consistent difference in strength, beyond experimental scatter, that would indicate any major effect of cooling rate under these conditions. Despite the limited data, the rate of cooling after rolling appeared to have little effect on final strength, where the strip was subsequently 'coiled' at 400°C or 500°C.

Table II Mechanical properties for samples of steel L, ferrite rolled and either quenched or air cooled prior to the coiling simulation.

FRT (°C)	665	653	608	550	665	653	608
CT (°C)	400	400	400	400	500	500	500
LYS - Air Cooled (Mpa)	387	430	480	495	389	401	406
LYS - Quenched (Mpa)	393	402	470	504	386	396	417
Difference (Mpa)	-6	+28	+10	-9	+3	+5	-11

It can be inferred that the time between the final rolling pass and the start of cooling may play a large part in the strength of quenched samples. However, for conventional cooling and coiling, it may be expected that the levels of recovery will converge, reducing the differences observed at short times and giving more consistent properties. It may be speculated that if coiling were conducted at much lower temperatures than investigated here, then the time between rolling and cooling, and the rate of cooling, would become more important in achieving the desired level of strength and consistency.

Effect of total reduction (Steel L) - There was very little change in properties between the high strain (3.0) and the low strain (2.2) schedules for steel L. It would appear that both refinement of the prior austenite, therefore transformed ferrite grain size, and deformation hardening of the ferrite, saturated for the range of reductions used in this investigation. The limiting nature of progressive austenite grain refinement in plain carbon steels is well documented. The restricted work hardening, or strain accumulation during multipass deformation of ferrite is also not unexpected due to the propensity for dynamic and static recovery of ferrite in the temperature range examined.

Microstructure (Steel M)

Overall, the grain structure was much finer and more homogeneous than the low carbon steel grade. Despite rolling through the transformation for most samples, there were no large grains resulting from recrystallisation and grain growth of the ferrite as observed in steel L.

At the higher rolling temperatures (fig. 5a) there were some almost equiaxed ferrite grains, which may have transformed from austenite towards the end of rolling. There were no such grains apparent in samples with lower entry and finish rolling temperatures where transformation had apparently been completed prior to, or very soon after, the start of rolling and the ferrite grains were heavily elongated (fig. 5b & 5c). The austenite transformation product could not be adequately resolved by optical microscopy but most likely consisted of a very fine carbide/ferrite aggregate.

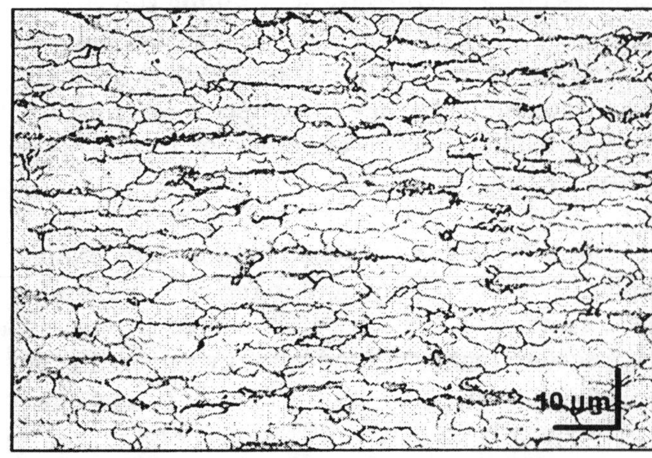

(a) Ferrite rolled - FRT 695°C / CT 400°C (C)

Fig. 5: Microstructures of the ferrite rolled steel MC. Longitudinal sections from the strip core (C), etched in Marshalls reagent.

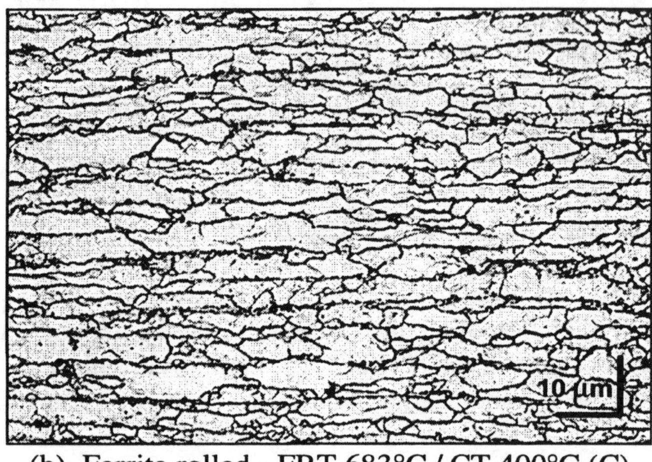

(b) Ferrite rolled - FRT 683°C / CT 400°C (C)

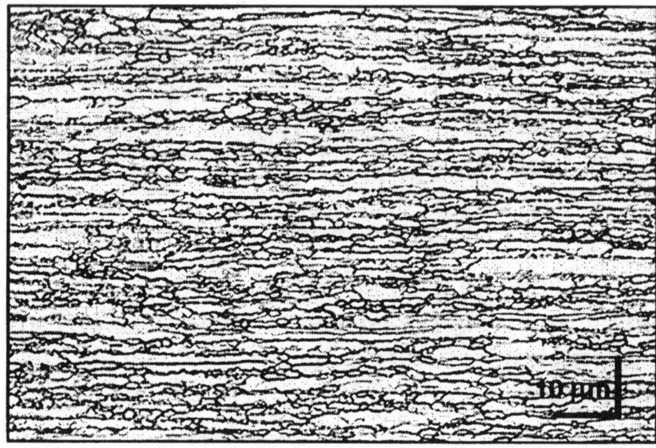

(c) Ferrite rolled - FRT 673°C / CT 400°C (C)

The subsurface shear zones were less distinct in samples of steel M compared to samples from steel L. It is likely the higher alloy content restricted the recrystallisation and growth of ferrite in the early ferritic rolling passes. There was no indication in metallographic specimens, of any ferrite recrystallisation after the completion of rolling. It may be expected that some concentration of strain adjacent the carbide colonies would occur but was evidently not sufficient to initiate recrystallisation at the rolling temperatures involved.

Mechanical Properties (Steel M)

Effect of finish rolling temperature (Steel M) - As for steel L, the strength increased as the rolling temperature was reduced for steel M (fig. 4). In general, the final strength was higher due to the increased alloy content and the finer grain structure for similar roll entry temperature and FRT (compare figs 3e and 5b). For this grade it was possible to achieve products with minimum strength levels from 450MPa to 550MPa by heavy ferritic rolling to approximate FRT's from 700°C to 620°C respectively.

Effect of 'coiling' temperature (Steel M) - Although no samples were 'coiled' directly at 500°C, some indication of the effect of coiling at the higher temperature was obtained by taking samples which had already been 'coiled' at 400°C and subjecting them to a simulated coiling thermal cycle at 500°C. Although these samples did not accurately represent a single coiling process at 500°C, the stability of the deformed ferrite structure was clearly evident as there appeared to be only a marginal loss of strength at the higher temperature, even for the steels rolled to low FRT's where the accumulated strain was high (fig. 4). This was not unexpected as the rate of recovery should be lower for this steel due to the higher alloy content [10].

Effect of cooling rate (Steel M) - Although the cooling rate from the FRT to the coiling temperature of 400°C was not varied (air cool), some comment can be made in this respect. High cooling rates were not required to achieve the desired strength levels in the ferritically rolled steel M. Higher cooling rates, typical of normal HSM operations, may increase the strength levels somewhat although the effect may be marginal in view of the leveling effect of 'coiling' observed for steel L.

Strength Versus Ductility (Steel L and Steel M)

Yield stress is plotted against total elongation for steel L and steel M in figure 6. Also included are typical properties for commercially cold rolled and recovery annealed (RA) low carbon steel and data from industrial trials on ferritic rolling of low carbon steel. Steel L displayed properties slightly inferior to conventional RA steel but consistent will industrially rolled product, supporting the validity of the laboratory rolled samples. Overall, the strength ductility balance was better for steel M compared to steel L as less deformation of the ferrite was required to reach the same level of strength. The relationship between these two data sets implies that ductility for these types of steels is controlled mainly by the extent of ferrite deformation. This is similar to the behaviour observed for conventionally cold rolled and recovery annealed steels [11].

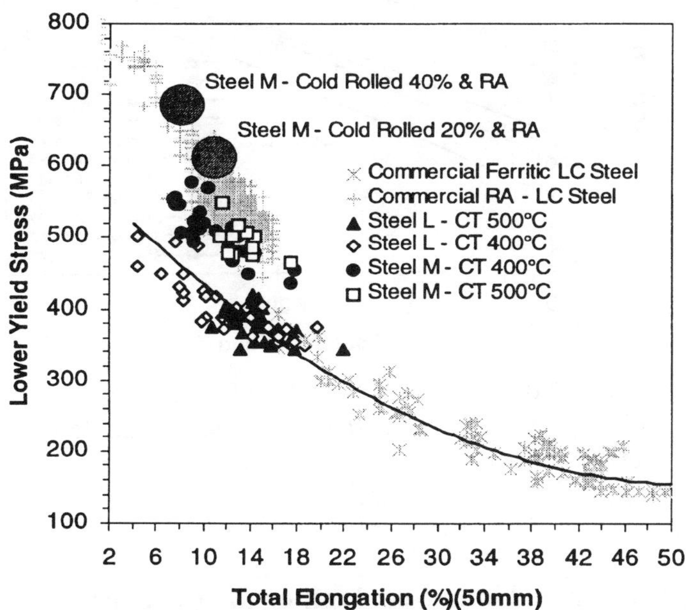

Fig. 6: Comparison of strength and ductility for the laboratory rolled samples, commercially cold rolled and recovery annealed low carbon steel and industrial ferritic rolling trials.

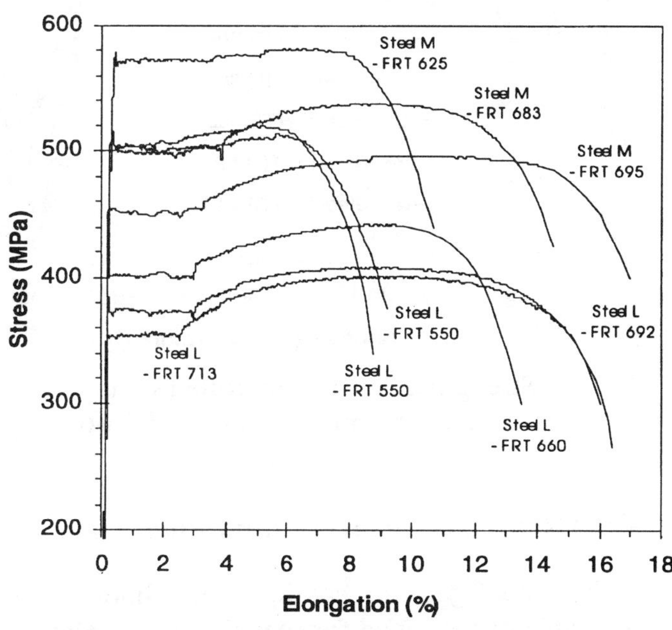

Fig. 7: Engineering stress strain curves for the laboratory rolled samples, tested in the longitudinal direction.

For steel M, yield strength levels greater than 550MPa were achieved at very low finish rolling temperatures and although the ductility was less than desired, ($10\%TE_{(L)}$) was still likely to be adequate for the intended applications for this type of product[5][6].

The change in the plastic flow characteristics with increasing ferrite deformation can be seen in some typical examples of stress/elongation curves shown in figure 7. At lower FRT's, yield stress and tensile strength were increased but the yield to tensile ratio also increases as the capacity of the steel to work harden diminished. Ductility, as measured by total elongation in the tensile test, consequently fell.

Direct Annealing of Ferritically Rolled Strip

For the samples examined, the level of retained strain after ferritic rolling and simulated coiling was sufficient to promote recrystallisation and full softening of steel L at annealing temperatures around 700°C (fig. 8). Strength and ductility (35-40% TE) were consistent with fully annealed product.

Samples of steel M also softened fully but at slightly higher temperature than steel L, approximately 725°C and displayed higher strength. Total elongation values at around 25-30% were consistent with fully annealed product for this steel grade.

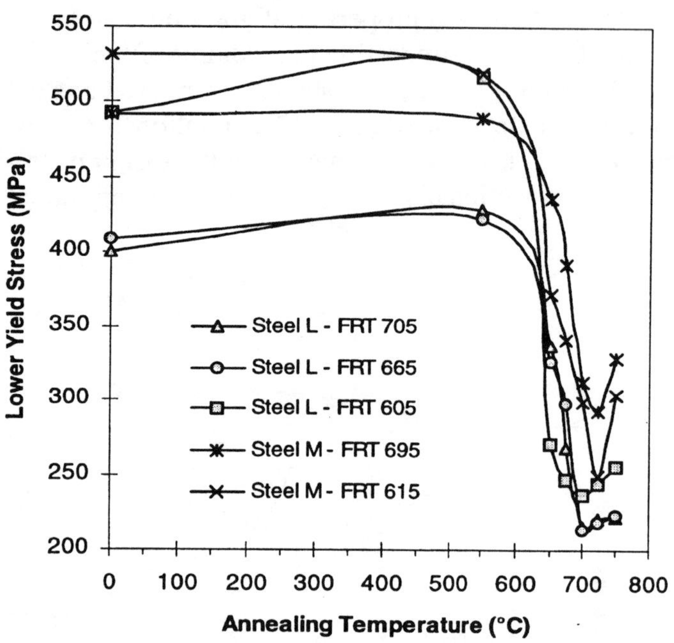

Fig. 8: Strength of ferritically rolled samples annealed at the temperatures indicated for 60s.

Fig. 9: Strength of ferritically rolled samples subsequently cold rolled and recovery annealed at 500°C for 60s.

Cold Rolling & Recovery Annealing

New configuration hot dip coating lines may incorporate limited in-line cold rolling to extend the available gauge range for coated product. Consequently, a range of samples were selected for additional cold reduction of either 20% or 40% and were then recovery annealed at a temperature of 500°C for 60 seconds (fig. 9). The additional cold reduction and recovery annealing raised the strength of both steel grades. The increase was most marked at 20% cold reduction, with further cold rolling to 40% having a reduced effect. The increase was greater in samples with lower initial strength level and thus lower retained strain after ferritic rolling. The additional cold rolling and recovery annealing increased the strength with only a relatively small loss of ductility (fig. 7).

SUMMARY

1. It should be noted that the higher strain rates and shorter interpass times in commercial mills may influence the results indicated here for the laboratory scale rolling experiments. However, the correspondence with results from commercial rolling trials suggested the differences would be expected to be small.

2. Heavy ferritic rolling may be used to raise the strength of thin gauge, low carbon steels where the mill is capable of the higher rolling loads.

3. Yield strengths from 350 to 500MPa can be achieved by rolling a low carbon steel grade (steel L) to FRT's from around 720°C to 550°C respectively. At such high levels of ferrite deformation the ductility was low, but comparable to conventional cold rolled and recovery annealed steel.

4. Extremely high cooling rates were not required to achieve high strength levels and conventional runout table cooling systems should be adequate for achieving strength levels from 350 to 550MPa providing a low coiling temperature of around 400-500°C can be achieved. However, the greater the level of ferrite rolling, the greater the need to increase the cooling rate and lower coiling temperature to maintain strength.

5. The use of a more highly alloyed plain carbon steel grade (steel M) allowed high strengths to be achieved at lower levels of ferrite deformation and consequently, higher FRT's in the ferritic region. Minimum strength levels of either 450MPa or 550MPa were achieved by heavy ferritic rolling to FRT's of around 700°C and 620°C respectively. This grade showed higher strength levels for a given ductility than steel L.

6. The strength of these steels could be increased by further cold reduction after ferritic rolling.

7. Despite the stability of the deformed and recovered structure after heavy ferritic rolling, recrystallisation and softening for both steel grades could be achieved by short time annealing at temperatures from 700-750°C. These types of products could therefore be used for the production of commercial quality, coated steel strip on conventional hot dip galvanising lines.

ACKNOWLEDGEMENTS

This work was conducted at the combined facilities of the Centre for Research in Metallurgy (C.R.M.), the University of Ghent and OCAS, the research centre for the Sidmar group. I would like to sincerely thank J.C. Herman & S. Wilmotte from the CRM for financial support and the other partners for the provision of laboratory facilities.

Special thanks are due also to R. Hobbs and L. Gore for support and approval of this project and BHP for financial support and permission to publish.

I am also pleased to gratefully acknowledge the tremendous support of G. Dierkens, J. Van de Voorde and K. De Waele for rolling, to V. Van Roose & P. Verleye for metallography & testing and J. Vergaerde for administrative matters. I would also like to thank J. C. Herman, A. De Paepe and other staff at the C.R.M. and the University of Gent for valuable technical contributions and support during this project. Thanks also to D. Langley for review of the manuscript.

REFERENCES

1. T. Tanaka, N. Tabata, T. Hatomura & C. Shiga, "Three Stages Of The Controlled-Rolling Process", Proc. Conf. Microalloying '75, October 1975, pp.107-119

2. S. Gohda , K. Watanabe and Y. Hashimoto, "Effects Of The Intercritical Rolling On Structure And Properties Of Low Carbon Steel", Transactions ISIJ, Vol.21, No.7, 1981, pp6-15.

3. Ph. Harlet, F. Beco, P. Cantinieaux, D. Bouquegneau, P. Messien and J. C. Herman, "New Soft Steel Grades Produced By Ferritic Rolling At Cockerill Sambre", Proc. Conf. Low Carbon Steels For The 90's, Ed. Tither 1993, p389-396.

4. V.Leroy, A. De Paepe and J. C. Herman, "Ferritic Hot Rolling Of Thin Gauge Strips: Processing And Properties", Proc. Conf. Modern LC And ULC Sheet Steels For Cold Forming: Processing And Properties, Ed. Bleck, 1998, pp.51-60.

5. C. A. Rogers and G. J. Hancock, "Ductility of G550 Sheet Steels In Tension - Elongation Measurements And Perforated Tests", Research Report No. R735, The University Of Sydney, Australia, December 1996.

6. C. A. Rogers and G. J. Hancock, "Ductility Of G550 Sheet Steels In Tension", Journal Of Structural Engineering, December 1997, pp.1586-1594.

7. R. Smith , P. Edwards and J. Dock, "Ferritic (Warm) Rolling Of A06 Grade at BHP Flat Products Division To Increase Pickle Line And Cold Mill Capacity At Springhill Works", BHP Report BHPR/ST/R/X/012, June 1997.

8. Simielli F, Yue S and Jonas J, "Recrystallisation Kinetics of Microalloyed Steels Deformed in the Intercritical Region", Metallurgical Transactions A, Vol.23A, February 1992, p597.

9. H. Langner and W. Bleck, "Fundamentals Of Softening Process During Ferrite Hot Rolling, 40th Mechanical Working And Steel Processing Conference Proceedings. ISS, 1998, pp.345-357.

10. P. Mangonon and W. Heitmann, "Subgrain And Precipitation-Strengthening Effects In Hot Rolled, Columbium-Bearing Steels", Proc. Conf. Microalloying '75, October 1975, p59-70.

11. R. Smith, S. Towers and G. Fairbank, "Improved Ductility in Recovery Annealed ZINCALUME® G550", BHP Report, BHPR/SFC15E/N/X/001, 1998.

MICROSTRUCTURE-PROPERTY RELATIONSHIPS IN THERMOMECHANICALLY PROCESSED FORGINGS

J.D. Boyd and P. Zhao
Queen's University
Kingston, ON K7L 3N6
Tel.:613-533-2750

Key Words: Thermomechanical Processing, Microalloyed Forgings, Microstructure-Property Relationships

INTRODUCTION

The principles of microalloying have been applied to forgings to develop steel compositions and processing schedules which meet product strength requirements in the as-forged+controlled-cooled condition [1,2]. Processing costs are reduced by eliminating heat treatments, and the machinability of the resulting ferrite+pearlite (F+P) microstructure is generally superior to tempered martensite (TM) at the same strength level. However, the toughness of F+P is low compared with TM at the same strength level. Several approaches to thermomechanical processing (TMP) and microstructural control have been investigated to achieve the required combinations of strength and toughness in forged components. These include warm forging in the range 600-900°C [3,4], post-forging isothermal treatment at 400-500°C [5] and varying the forging temperature in the range 900-1200°C [6].

The objective of the present study was to develop hot forging+controlled cooling schedules to produce high strength and toughness in commercial microalloyed medium-C steels. A microstructural design approach was followed, whereby TMP schedules were selected specifically to control those microstructural features known to influence strength and toughness properties [7].

EXPERIMENTS

TMP Forging Trials

The material studied was commercial 1541 grade steel (0.4C-1.5Mn-0.3Si-0.01/0.04S) microalloyed with Ti and V (1541+Ti,V) or with Nb (1541+Nb). The compositions of the 2 steels are given in Table I. Based on preliminary laboratory experiments [8,9], TMP forging schedules were designed to produce

Table I Compositions of experimental steels (wt. pct.)

Steel	C	Mn	Si	S	P	V	Ti	Nb	N
1541+Ti,V	0.40	1.49	0.56	0.042	0.016	0.114	0.014	---	0.009
1541+Nb	0.41	1.67	0.24	0.013	0.022	0.002	---	0.038	0.005

microstructures having refined ferrite and pearlite grain sizes, and precipitation strengthening (Fig.1). The schedule for 1541+Ti,V (Fig 1a) was designed to produce fine recrystallized austenite, and the schedule for

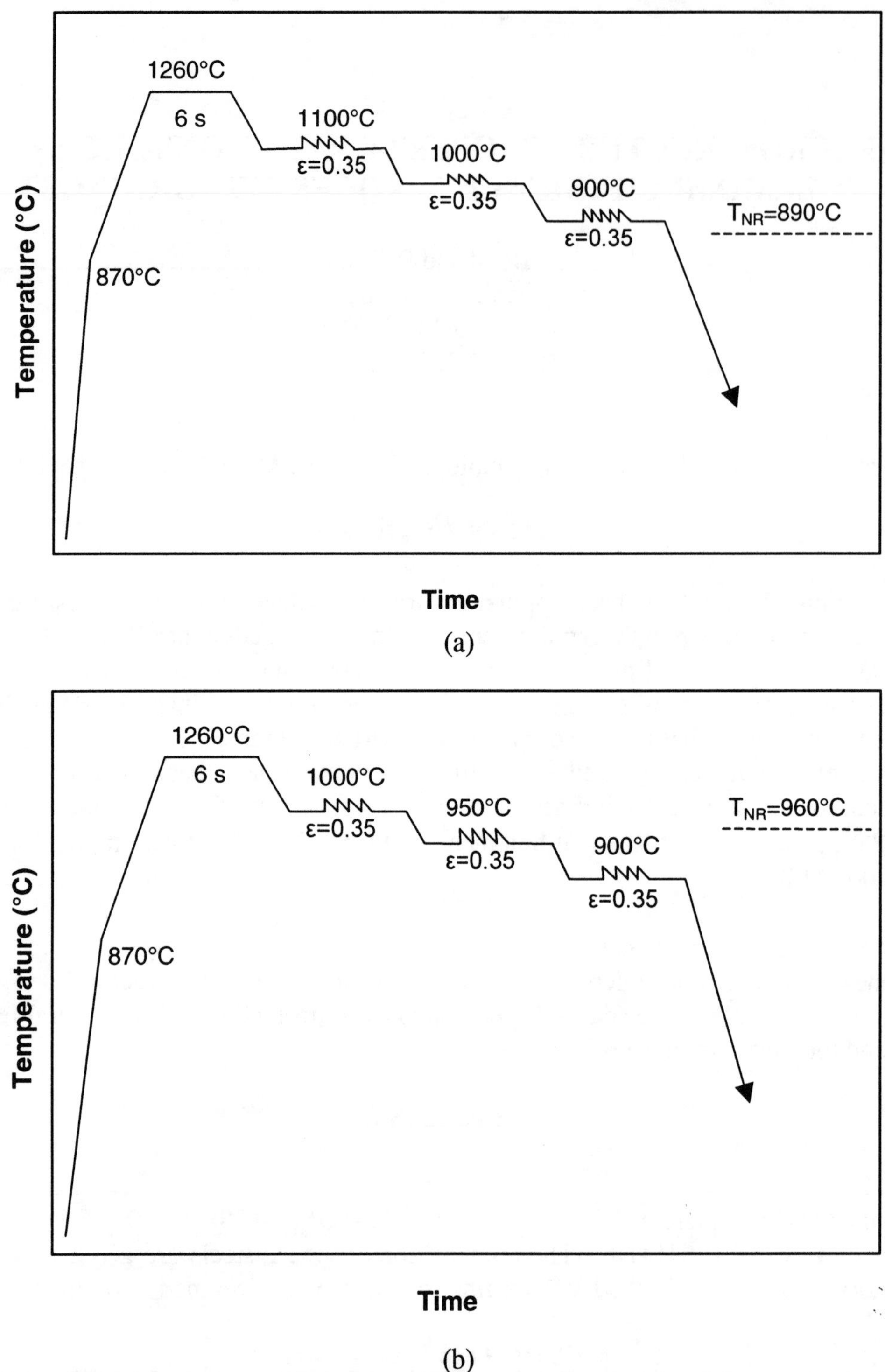

Fig. 1: Schematic TMP schedules for forging trials. a) 1541+Ti,V. b) 1541+Nb.

1541+Nb (Fig. 1b) was designed to produce thin, pancaked austenite grains. 35 mm diameter bars of each steel were forged to 12 mm thickness following the appropriate TMP schedule, and either still-air cooled (AC) at 1°C/s or fan cooled (FC) at 2°C/s following forging. The forging was carried out on an industrial upset forging machine, and temperatures were monitored with an optical pyrometer and a contact thermocouple (during cooling). For comparison, samples of each steel were also prepared by a conventional forging schedule (CF), which had the same deformation sequence with a finishing temperature of 1000°C, followed by AC.

Microstructural Characterization

Samples were water quenched at each stage of the forging schedule to determine the evolution of the austenite microstucture. Transverse sections were used for all metallography (i.e., the normal to the plane of polish was normal to the forging direction). Austenite microstructures were characterized by optical microscopy (picric acid etch) and by transmission electron microscopy (TEM) of carbon extraction replicas. The final microstructures of the forged+controlled-cooled samples were characterized by optical microscopy (Nital etch) and by TEM of thin foil samples. The latter were prepared by cutting 200 μm–thick slices parallel to the transverse plane, punching 3-mm diameter discs and electropolishing in 7% perchloric acid – glacial acetic acid at room temperature. Quantitative measurements of the mean linear intercept austenite grain size (d_γ), ferrite grain size (d_α) and pearlite colony size (d_p) were made according to ASTM E112, and the ferrite volume fraction (f_α) was measured by a point counting method (ASTM E562). The precipitate size, pearlite interlamellar spacing (s) and carbide lamella thickness (t) were measured directly on TEM micrographs.

Mechanical Property Measurements

Round tensile samples, having 6 mm diameter × 30 mm long gauge length, and standard Charpy samples were machined from the forgings with their long axes normal to the forging direction. Tensile tests were carried out to failure at a strain rate of $2x10^{-3}$/s with an extensometer mounted on the sample gauge section. Charpy tests were carried out at ambient temperature. Duplicate tensile and Charpy tests were made for each condition. The fracture mode was characterized by measuring the percent shear area on the fractured Charpy samples, and examining the fracture surfaces by scanning electron microscopy (SEM-SE).

RESULTS

Microstructures

The evolution of the austenite grain structure during forging was consistent with the preliminary laboratory experiments (9). In the 1541+Ti,V steel, the austenite recrystallized after each forging step and was progressively refined to a final grain size (d_γ) of 14 μm (Table II). In the 1541+Nb steel, the austenite did not recrystallize after the 2nd forging step (950°C) and was progressively pancaked to a final grain width (d_γ^T) of 15 μm and length (d_γ^L) of 30 μm. A significant density of small (<25 nm) strain-induced (Ti,V)(C,N) precipitates was observed in the 1541+Ti,V samples following the 3rd deformation (T_D = 900°C). Similarly, for 1541+Nb, there was a high density of small Nb(C,N) precipitates following the 2nd and 3rd forging steps (T_D = 950, 900°C). This is consistent with the view that the rate of nucleation of strain-induced precipitates increases markedly when the deformation temperature (T_D) approaches the no-recrystallization temperature (T_{NR}) [10,11].

The final microstructures obtained for the different forging schedules are shown in Figs. 2 and 3, and the measurements of ferrite volume fraction, mean ferrite grain size and mean pearlite colony size are given in Table II. When describing these microstructures, 2 types of ferrite are identified: grain-boundary ferrite (GBF) and intragranular ferrite (IGF). In 1541+Ti,V the microstructures are predominantly grain boundary ferrite + pearlite (GBF+P) (Fig. 2). The TMP forging schedule produced a large reduction in d_p compared with the conventional schedule, and a smaller reduction in d_α. Fan cooling produced the lowest values of d_α and d_p, and a slight decrease in the amount of GBF compared with air cooling. For 1541+Nb (Fig. 3), the F+P microstructure resulting from the TMP+AC schedule reflected the pancaked austenite grain structure. d_p^T was reduced compared to d_p for the conventional schedule, with little effect on d_α. The microstructure from the TMP+FC schedule was quite different, containing a large amount of IGF. (Note, it was not possible to obtain

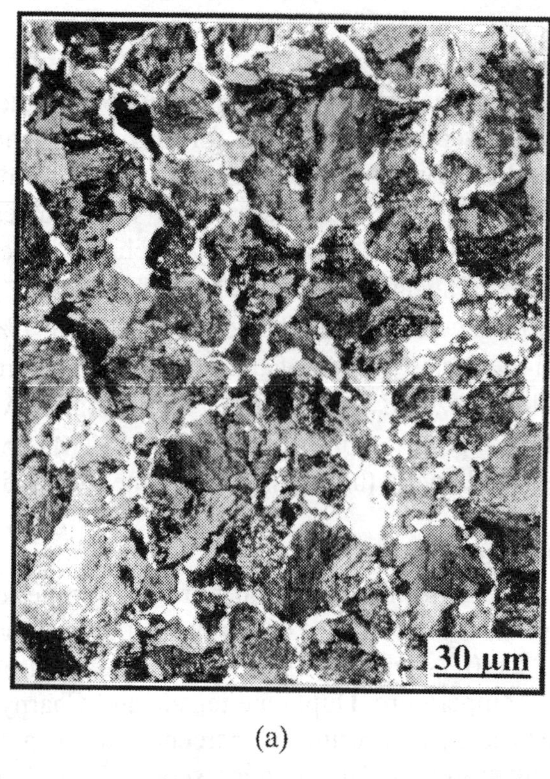

(a)

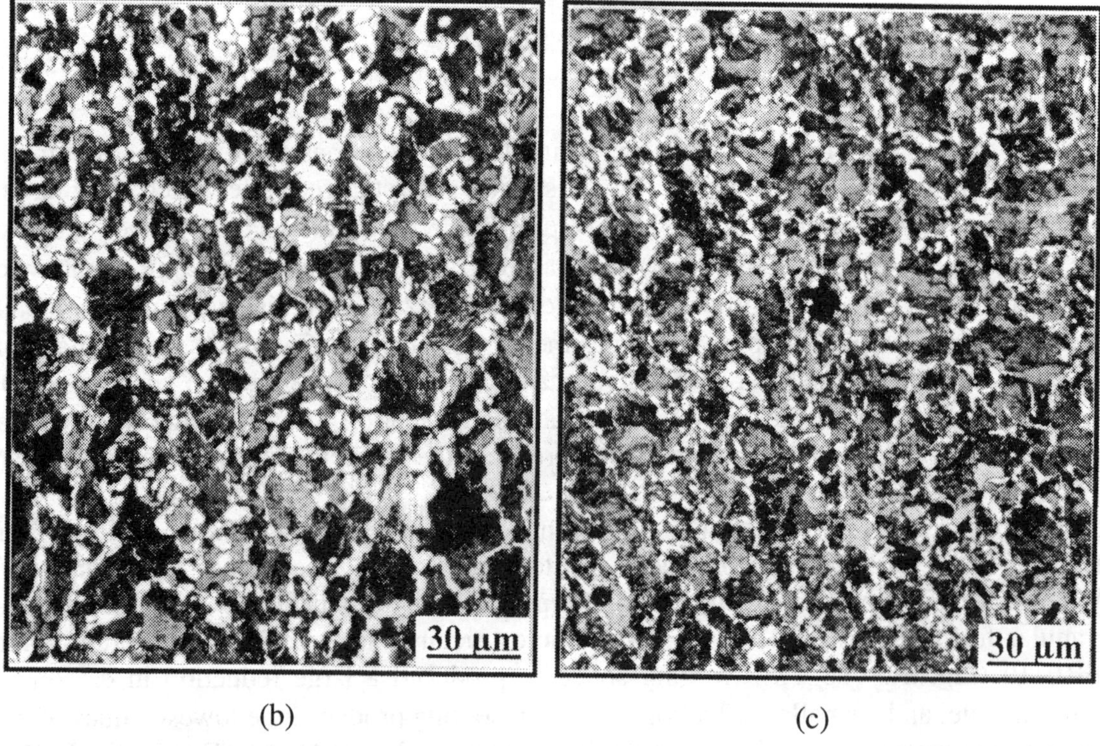

(b) (c)

Fig. 2: Optical microstructures of forged 1541+Ti,V samples. a) CF, b) AC, c) FC.

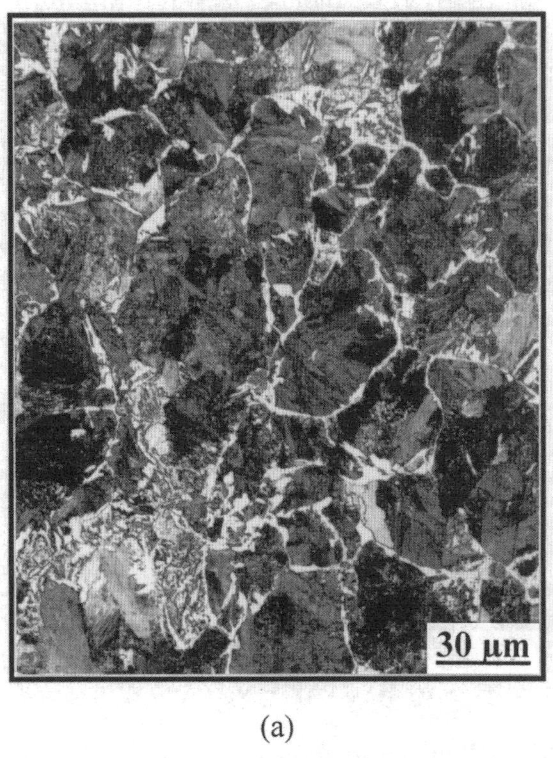

(a)

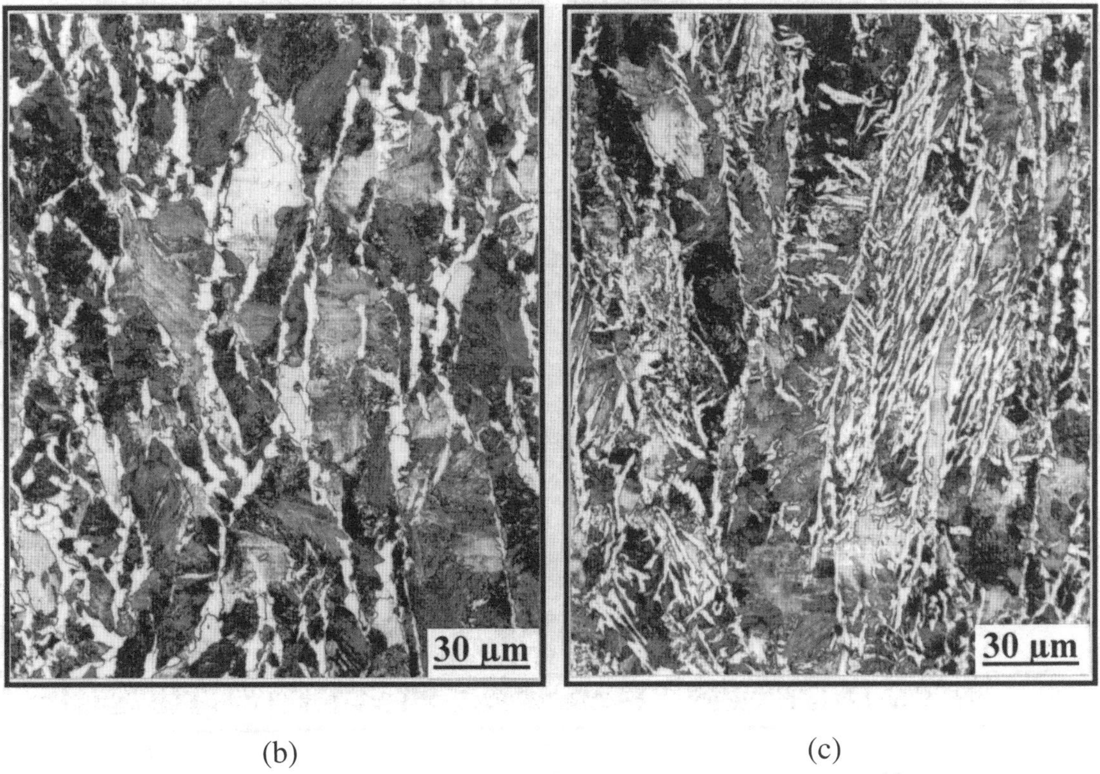

(b) (c)

Fig. 3: Optical microstructures of forged 1541+Nb samples. a) CF, b) AC, c) FC.

measurements of d_α, d_p in these samples.) The amount of ferrite in the final microstructure increased in the order CF – TMP+AC – TMP+FC. In summary, the optical microstructures of the forged samples demonstrate the effects of TMP on the austenite grain structure and consequently on the final F+P microstructures of the 2 steels. For GBF+P microstructures, TMP increases the specific area of austenite grain boundary, and the sum of the ferrite and pearlite grain sizes is always close to the austenite grain size. For the 1541+Nb/FC condition, the combination of deformed austenite and high cooling rate promotes nucleation of IGF.

Table II Quantitative metallographic measurements for forged samples.

Condition	Austenite Grain Size, μm	Ferrite Grain Size, μm	Pearlite Colony Size, μm	Ferrite Vol. Frac.
1541+Ti,V (3-Hit+Q)	d_γ=14±2			
1541+Nb (3-Hit+Q)	d_γ^L=30±7 d_γ^T=15±2			
1541+Ti,V/CF		d_α=5.5±3.3	d_p=23±5	0.16
1541+Ti,V/AC		d_α=4.7±2.8	d_p=11±3	0.26
1541+Ti,V/FC		d_α=3.8±2.6	d_p=9±2	0.21
1541+Nb/CF		d_α=3.8±2.4	d_p=27±6	0.14
1541+Nb/AC		d_α^L=6.7±4.7 d_α^T=5.2±2.2	d_p^L=29±7 d_p^T=13±3	0.18

Q- Quench, CF-Conventional forging schedule, AC-TMP schedule + still-air cool, FC-TMP schedule + fan cool.

The details of the microstructures of the forged samples are revealed in the thin foil TEM micrographs shown in Figs. 4 and 5. In 1541+Ti,V the GBF has a low dislocation density and contains a high density of very small (<5 nm) precipitates, in addition to the larger (<25 nm) precipitates observed in the quenched samples (Fig. 4). Thus, considerable precipitation occurred in ferrite during cooling of the forgings. There appears to be a higher density of the very small precipitates produced by the TMP+FC schedule (Fig. 4d) compared with the conventional schedule (Fig. 4b). The composition of the very small precipitates was not determined, but they are expected to be V(C,N) [12]. In 1541+Nb/CF, the general GBF+P microstructure (Fig. 5a,b) was similar to that for the 1541+Ti,V forgings. However, the density of very small precipitates in GBF was much lower in 1541+Nb (Fig. 5b) compared with 1541+Ti,V (Fig. 4b). That is, there is much less precipitation in ferrite during cooling of the forgings of the former steel. For the 1541+Nb/FC samples, both the GBF and IGF have a medium dislocation density, ~ 1μm subgrains and a low density of very small precipitates, and some of the pearlite is non-lamellar or degenerate (Fig. 5c,d). Measurements of mean values of pearlite interlamellar spacing (s) and carbide lamella thickness (t) are given in Table III.

Table III Measurements of pearlite interlamellar spacing (s) and carbide lamellar thickness (t) from thin foil TEM

	1541+Ti,V/CF	1541+Ti,V/FC	1541+Nb/CF	1541+Nb/FC
s, μm	0.14	0.11	0.12	0.12
t, μm	0.03	0.02	0.03	0.02

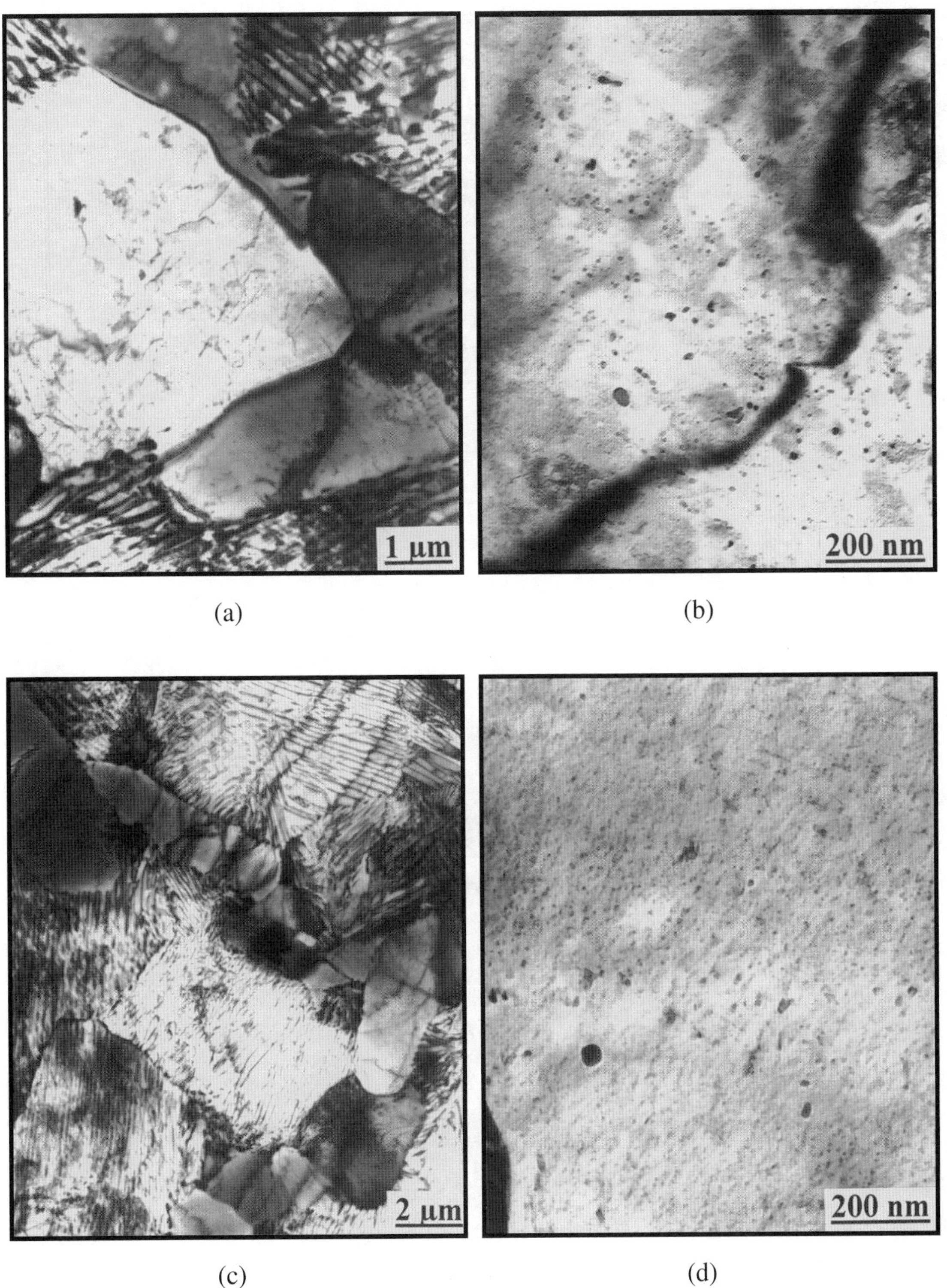

(a) (b) (c) (d)

Fig. 4: Thin foil TEM micrographs of forged 1541+Ti,V samples. a), b) CF; c), d) FC.

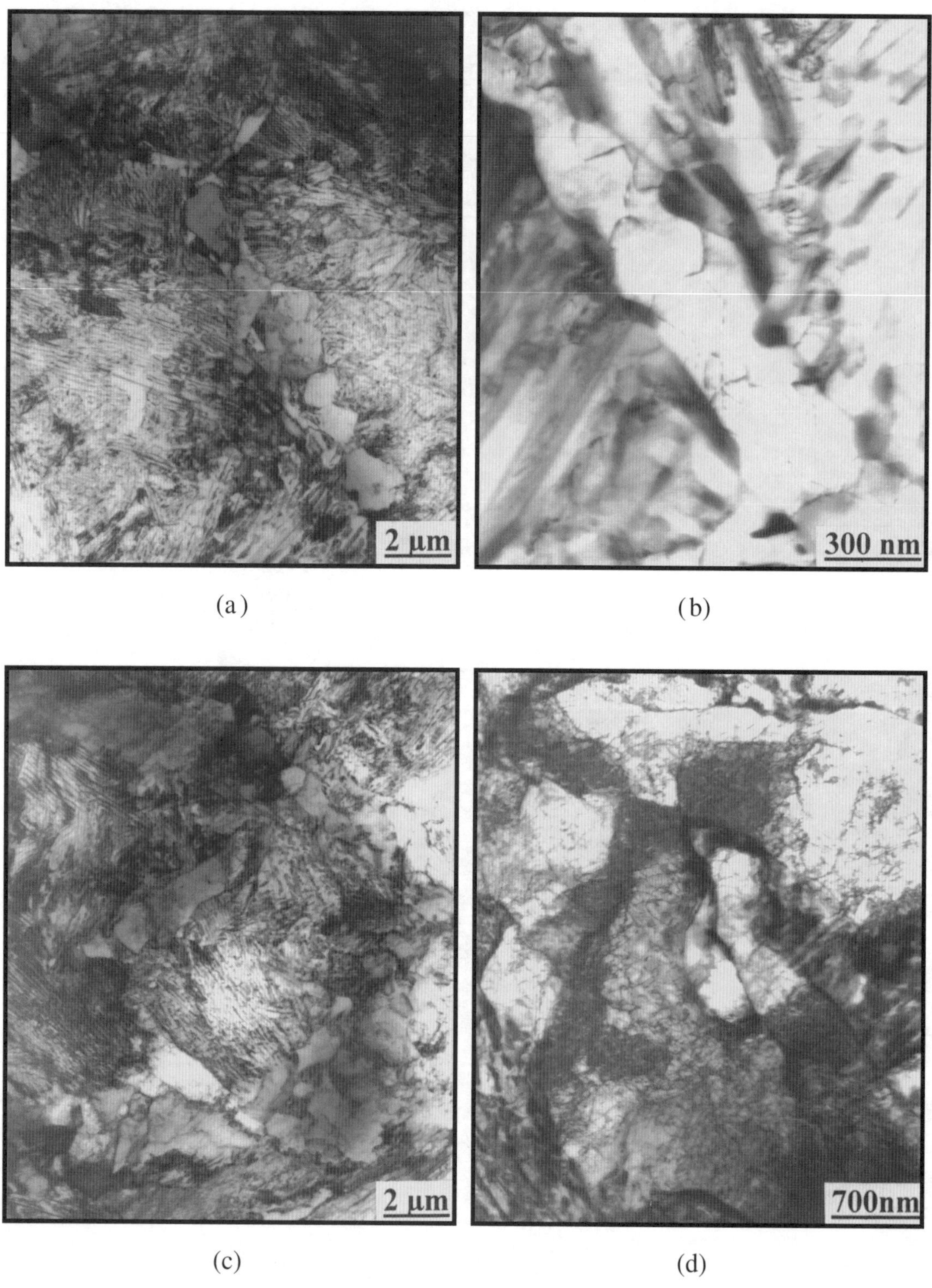

(a) (b) (c) (d)

Fig. 5: Thin foil TEM micrographs of forged 1541+Nb samples. a), b) CF; c), d) FC.

Mechanical Properties

The results of the tensile and Charpy tests are given in Table IV and Fig. 6. For 1541+Ti,V there was an increase in both yield strength and Charpy energy for the TMP schedules compared with the conventional schedule. The TMP+FC schedule produced the best properties for this steel, with 6% and 90% increases in yield strength and Charpy energy, respectively, compared with the conventional treatment. The results for the 1541+Nb steel were even more striking. Here, the TMP schedules produced 10% increase in yield strength and 200% increase in Charpy energy. Again, the TMP+FC schedule produced the best combination of these properties. Note that the highest overall Charpy values were obtained for 1514+Nb/TMP+FC, whereas the highest yield strength was obtained for 1541+Ti,V/TMP+FC. These results demonstrate clearly that control of microstructure through TMP can produce simultaneous improvement in strength and impact toughness of forgings.

Table IV Tensile properties and RT Charpy energy for forged samples

Condition	CVN, J	Y.S., MPa	T.S., MPa	EL., %	R.A., %
1541+Ti,V/CF	15	674	941	18	53
1541+Ti,V/TMP+AC	21	681	937	18	50
1541+Ti,V/TMP+FC	29	714	966	16	47
1541+Nb/CF	12	570	865	20	56
1541+Nb/TMP+AC	30	602	870	22	57
1541+Nb/TMP+FC	37	629	887	20	53

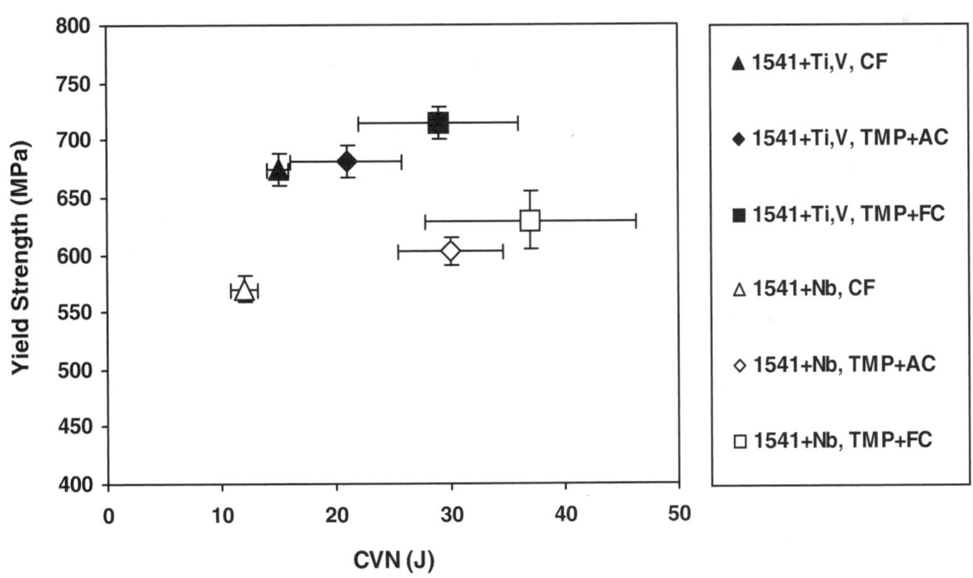

Fig. 6: Comparison of mean values of yield strength and Charpy energy for forged samples.

MICROSTRUCTURE-PROPERTY RELATIONSHIPS

Yield Strength

Gladman [7] gives the following equation for the yield strength (σ_y) of F+P steels:

$$\sigma_y\ (\text{MPa}) = f_\alpha^{1/3}(35 + 58.5(\text{Mn}) + 17.4d_\alpha^{-1/2}) + (1 - f_\alpha^{1/3})(178 + 3.85s^{-1/2}) + 63.1(\text{Si}) + 425(\text{N})^{1/2} \quad (1)$$

where f_α is volume fraction, d_α and s are in mm, and (Mn), (Si), (N) are the concentrations of the respective elements in wt. percent. Using the data in Tables I, II and III, calculated values of σ_y were determined from Equ'n. 1 for each condition (Table V). In all cases the calculated value of σ_y is smaller than the measured value. Equ'n. 1 does not contain a term for precipitation strengthening, hence it can be assumed that the difference ($\Delta\sigma_y$) is the precipitation strengthening component. The precipitation strengthening component estimated in this way is 186-188 MPa for both 1541+Ti,V conditions. This is a reasonable value for a dispersion of ~5nm V(C,N) precipitates (Fig. 4b,d), according to the Ashby-Orowan theory [13]. This also means that the difference in the measured values of σ_y between the FC and CF conditions is accounted for by the decrease in d_α and s, coupled with the increase in f_α. For 1541+Nb/CF, the estimated precipitation strengthening component is only 57 MPa, which is consistent with the low density of very small precipitates observed for this condition (Fig. 5b).

Table V Calculated values of σ_y from microstructural model, measured values of σ_y and $\Delta\sigma_y$

	1541+Ti,V/CF	1541+Ti,V/FC	1541+Nb/CF
Calc. σ_y, MPa	488	526	513
Meas. σ_y, MPa	674	714	570
$\Delta\sigma_y$, MPa	186	188	57

These results show that with an additional term for precipitation strengthening, Equ'n. 1 gives good estimates for the yield strength of steels having a GBF+P microstructure. The calculations show that the largest contributions to the yield strength are ferrite grain boundary strengthening, pearlite lamellar strengthening and (for 1541+Ti,V) precipitation strengthening. A more detailed model will be required to account for the strengthening effects of the IGF, subgrains and dislocations in microstructures such as 1541+Nb/FC.

Impact Toughness-The empirical equation for the impact transition temperature (ITT) given by Gladman [7] for F+P steels is:

$$\text{ITT}(^\circ\text{C}) = f_\alpha(-46 - 11.5d_\alpha^{-1/2}) + (1 - f_\alpha)(-335 + 5.6s^{-1/2} - 13.3d_p^{-1/2} + 3.48\times10^6 t) + 49(\text{Si}) + 762(\text{N})^{1/2} \quad (2)$$

where d_p and t are in mm. Using the data in Tables I, II, and III, the calculated values for ITT are in the range 100-200°C for all steels and conditions. However, the Charpy results and the fractographic analysis suggest that the ITT's are much lower. If the processing conditions are considered in the order CF – AC – FC, both steels show a transition in fracture mode from brittle cleavage to mixed cleavage and ductile fracture. There is also an increase in percent shear fracture from 34–58 % for 1541+Ti,V and from 12–52 % for 1541+Nb. This and the magnitudes of the Charpy energies indicate that the ITT is just above ambient temperature for the 2 TMP+FC steels, and Equ'n. 2 does not give accurate estimates of the ductile-brittle transition behaviour of these steels.

Qualitatively, the lowering of the ITT and the increase in the Charpy energy with the TMP+FC treatment is attributed to the decrease in d_α and d_p. In 1541+Nb, the elongated microstructures resulting from the TMP forging treatment, and the IGF produce extremely small values of d_p^T in the direction of crack propagation in the Charpy test. This accounts for the superior impact toughness associated with this microstructure.

SUMMARY

Simultaneous improvements in strength and impact toughness in 2 microalloyed medium-carbon steels have been obtained by thermomechanical processing + controlled cooling.

With a 1541+Ti,V steel, a fine grain-boundary ferrite (GBF)+pearlite (P) microstructure and significant V(C,N) precipitation strengthening were obtained by forging above T_{NR} and cooling at 2°C/s. This increased the yield strength by 6% and the Charpy energy by 90%, compared with a conventional forging schedule.

With a 1541+Nb steel, a fine intragranular ferrite (IGF)+pearlite microstructure was obtained by forging below T_{NR} and cooling at 2°C/s. This increased the yield strength by 10% and the Charpy energy by 200%, compared with a conventional schedule.

For the GBF+P microstructures, a combination of the Gladman equation and the Ashby-Orowan theory give good quantitative estimates of the yield strength. The largest contributions to the yield strength are ferrite grain boundary strengthening, pearlite lamellar strengthening and (for 1541+Ti,V) precipitation strengthening.

Only qualitative correlations between microstructure and impact transition temperature (ITT) can be made. The main factors in lowering ITT are fine ferrite grain size and pearlite colony size, and the presence of IGF.

ACKNOWLEDGEMENTS

The authors are grateful to D. Papalazarou, formerly with TRW Canada Ltd., for assistance with the forging tests. This research is sponsored by an industrial consortium of TRW Canada, Chrysler Canada, Inland ISPAT Steel, IVACO Rolling Mills, Stelco Steel, Slater Steels, Timken, Welland Forge, the Natural Sciences and Engineering research Council of Canada, and Materials and Manufacturing Ontario.

REFERENCES

1. Fundamentals and Applications of Microalloyed Forging Steels, Eds., C. Van Tyne, G. Krauss and D. Matlock, TMS, 1996.
2. D.J. Naylor, "Microalloyed Forging Steels", Materials Science Forum, Vols. 284-286, 1998, pp. 83-94.
3. J.H. Reynolds and D.J. Naylor, "Microstructure and Properties of WarmWorked Medium Carbon Steels", Materials Science and Technology, Vol. 4, 1988, pp.586-602.
4. C. Garcia, J.L. Romero and J.M. Rodriguez-Ibabe, "Warm Forging of a Vanadium Microalloyed Steel", Ref. 2, pp. 435-442.
5. I. Madariaga and I Gutierrez, "Acicular Ferrite Microstructures and Mechanical Properties in a Medium Carbon Forging Steel", Ref. 2, pp. 419-426.

6. D.L. Lee, S.J. Yoo, K.S. An, Y.W. Lee and W.Y. Choo, "The Effects of Composition and Hot Working Parameters on the Mechanical Properties of Microalloyed Pearlitic and Martensitic Forging Steels", Ref. 1, pp. 17-28.
7. T. Gladman, The Physical Metallurgy of Microalloyed Forging Steels, Inst. of Materials, London, 1997, p.337.
8. J.D. Boyd and P.Zhao, "Improving Strength and Fracture Toughness of Microalloyed Medium Carbon Steel-An Example of Microstructural Design", Thermomechanical Processing of Steels, Inst. of Materials, London, 2000, pp. 79-89.
9. J.D. Boyd, P.Zhao, T.Liu and M. Gyorffy, "Thermomechanical Processing in Hot Forging", J.J. Jonas Symposium on Thermomechanical Processing, Texture and Formability of Steel, CIM, Ottawa, 2000 (to be published).
10. B. Dutta and C.M. Sellars, "Effect of Composition and Process Variables on Nb(C,N) Precipitation in Niobium Microalloyed Austenite", Materials Science and Technology, Vol.3, 1987, pp.197-206.
11. A.J. DeArdo, "The Physical Metallurgy of Thermomechanical Processing of Microalloyed Steels", THERMEC'97, Eds., T. Chandra and T. Sakai, TMS, 1997, pp. 13-29.
12. M. Prikryl, A. Kroupa, G.C. Weatherly and S. Subramanian, "Precipitation Behavior in a Medium Carbon, Ti-V-N Microalloyed Steel", Metallurgical and Materials Transactions, Vol. 27A, 1996, pp. 1149-1165.
13. Ref. 7, p. 54.

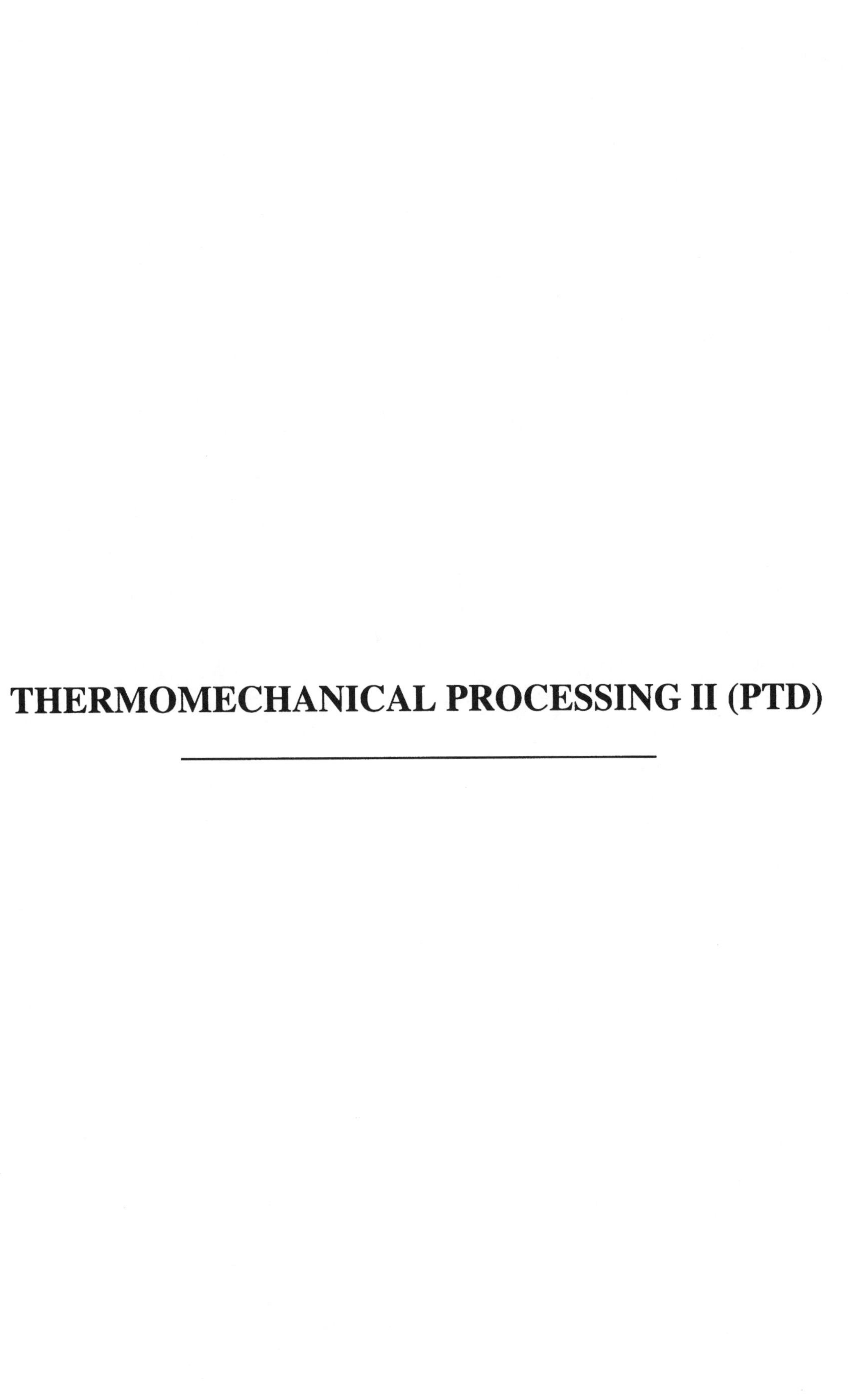

THERMOMECHANICAL PROCESSING II (PTD)

MICROSTRUCTURE EVOLUTION OF A STATE-OF-THE-ART Ti-Nb HSLA STEEL

Naoki Nakata
Technical Research Laboratories
Kawasaki Steel Corp.
1, Kawasaki-cho, Chuo-ku
Chiba, Japan 260-0835
Tel : 81-43-262-2474

Matthias Militzer
The Centre for Metallurgical Process Engineering
The University of British Columbia
6350 Stores Road
Vancouver, BC, Canada V6T 1Z4
Tel : 604-822-3676

Key Words: Microstructure, Continuous Cooling, Ferrite, Grain Size, Transformation, Modelling

INTRODUCTION

High-strength low-alloy (HSLA) steels are an important class of materials in automotive applications, primarily for the development of fuel-efficient, lightweight vehicles. HSLA steels with tensile strengths as high as 780 MPa are currently used for high-strength vehicle components, e.g. wheel discs. The pipeline industry is also experiencing a drive for higher strength steels; linepipe grades with a minimum yield strength of 550 MPa are now state-of-the-art. It can be expected that in the coming decade linepipe steels with a minimum yield strength of 700 MPa will become the standard. Strength levels in the 550-800 MPa range are related to a predominantly acicular ferrite microstructure with grain sizes of 3 μm or below and precipitation hardening which is associated with microalloying elements such as Ti and Nb.

Commonly, these steels are controlled rolled in a hot mill. Accelerated cooling on the run-out table is the key processing step for developing the desired fine grained ferrite microstructure as a result of the austenite decomposition. The complexity of this microstructure makes its characterization a challenging task [1]. Thus, the development of microstructural evolution models for hot rolling of these steels is still in its infancy. Austenite-to-ferrite transformation models for low carbon steels have been developed and validated for steels with a polygonal ferrite microstructure including HSLA steels with tensile strengths of 400-500 MPa [2-4]. A first attempt to extend these approaches to a 550 MPa HSLA-Nb/Ti steel has been proposed by Militzer et al. [4, 5]. The transformation requires a significantly higher undercooling to occur in these steels. Thus, it is required to understand the characteristics of transformation for high undercooling and to assess whether the previously proposed models remain applicable in this undercooling range.

In the present work, the austenite decomposition of a state-of-the-art Ti-Nb HSLA steel with a tensile strength of 780 MPa has been investigated for run-out table cooling conditions. Torsion tests have been carried out to simulate the entire hot rolling process from reheating to coiling. The austenite decomposition kinetics have been studied in detail with continuous cooling transformation (CCT) tests. The effects of initial austenite grain size, cooling rate and retained strain on the austenite decomposition kinetics have been quantified. Based on the experimental investigations, an austenite-to-ferrite transformation model has been proposed for the present steel which also predicts the ferrite grain size. The model has been evaluated by applying it to industrial processing conditions and comparing the predictions with industrially produced microstructures.

MATERIAL AND METHODOLOGY

Samples of a commercial Ti-Nb HSLA steel were supplied by the Mizushima Works of Kawasaki Steel in the form of transfer bar and as-hot rolled coil. Transfer bar material was used to machine specimens for subsequent torsion and transformation tests. Table I shows the chemistry of the present steel. Notably, the Mn content is 1.82%, significantly higher than that of the previously examined 550 MPa grade containing 1.35% Mn. Using the thermodynamic data of Kirkaldy and Baganis [6], the Ae3 temperature (T_{Ae3}) is 825 °C for the given steel chemistry. Further, the present steel contains Ti as the principal microalloying constituent with Nb being the secondary microalloying element; previous studies had emphasized HSLA grades with predominantly Nb microalloying [4].

Table I Chemical composition of the steel (in wt%)

C	Mn	P	S	Si	Cu	Ni	Mo	V	Ti	Nb	Al	N
0.09	1.82	0.014	0.0011	0.02	0.01	0.02	0.001	0.006	0.14	0.054	0.04	0.003

To simulate the thermo-mechanical processing of a hot strip mill, torsion tests were carried out on a hot-torsion simulator. The torsion specimens are 183 mm in length with a diameter of 14.3 mm; the gauge length of the working zone is 12.7 mm with a diameter of 10 mm. There is feedback temperature control employing a spot-welded thermocouple during reheating and an optical pyrometer during multi-pass deformation, respectively. After deformation the specimens are He gas cooled to room temperature at approximately 50 °C/s.

The CCT tests were performed on a Gleeble 1500 thermo-mechanical simulator. The transformation kinetics were recorded with dilatometer measurements. Tubular specimens of 8 mm diameter, 1 mm wall thickness and 20 mm length were employed to perform CCT tests without prior deformation. Solid cylindrical specimens of 6 mm diameter and 10 mm length were used to investigate the effect of retained stain on transformation by applying a deformation in axisymmetric compression under conditions of no-recrystallization prior to cooling.

The initial austenite grain size, d_γ, was varied by employing different austenitization treatments. Austenite grain sizes of 13, 20 and 48 μm, respectively, were produced after heating at 5 °C/s to and holding for 2 minutes at the reheating temperatures of 1050, 1150 and 1200 °C, respectively. Following austenitization, tubular specimens were cooled at 10 °C/s to 950 °C, where they were held for 30 seconds in order to homogenize the temperature distribution in the specimens. Solid specimens were cooled at 10 °C/s to 900 °C where a single hit deformation with a strain in the range of 0.3-0.5 was applied after performing a 30 seconds temperature homogenization.

In the cooling process, the electric power of induction heating was controlled to achieve the designated cooling patterns with cooling rates of less than 20 °C/s in the tubular specimens and less than 80 °C/s in the solid specimens. For higher cooling rates, He quenching was employed where cooling rates can be varied up to approximately 200 °C/s by changing the gas flow rate. The sample temperatures were measured with a spot-welded thermocouple in the plane of the dilatometer.

For microstructural investigations, the cross sections of samples close to the thermocouple were prepared, etched and then examined with optical microscopy using an image analyzer. Grain sizes were quantified as equivalent area diameter according to the Jeffries procedure [7]. In addition, scanning electron microscopy (SEM) was employed to characterize microstructures with grain sizes of 3 μm and below.

HOT ROLLING SIMULATION

For a standard hot torsion simulation of hot rolling, the specimens are reheated for 30 minutes at 1250 °C. The subsequent deformation schedule is summarized in Table II. Rough rolling is simulated with one pass of strain 1 at 1100 °C. The finish mill simulation consists of 7 passes. All passes are executed at a strain rate of 1 /s. The interpass times in the finish mill simulation are taken somewhat larger than in the industrial practice to allow for better temperature control in the laboratory simulation.

Fig.1 shows the flow stress curves of this torsion test. The finish mill entry temperature is 1041 °C so that all 7 passes appear to be conducted under no-recrystallization condition. Work hardening is the dominant deformation process and very little indication of softening can be found in the flow stress curves. Similar observations have been made for slightly different torsion tests which simulate the actual industrial schedules more accurately. Thus, it can be concluded, that the present steel is controlled rolled with an accumulated strain in excess of 0.5 at the end of the finishing mill for commonly used rolling schedules.

Table II Torsion test conditions

Pass No.	1	2	3	4	5	6	7	8
Strain	1.00	0.36	0.51	0.41	0.33	0.30	0.25	0.10
Temperature (°C)	1100	1041	979	955	934	919	908	898
Interval (s)	10.0	8.0	4.8	3.2	2.3	1.7	1.3	

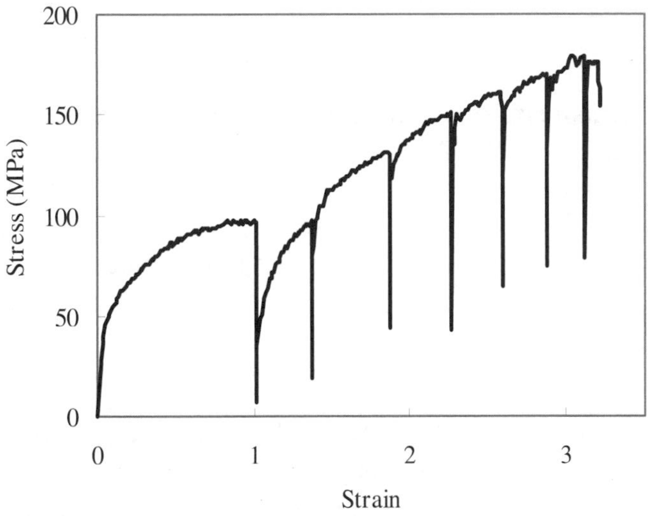

Fig.1: Flow stress curves in the standard hot torsion test

Experimental Results - Fig.2 gives examples of the austenite decomposition kinetics measured during continuous cooling tests. The transformation temperatures are decreased by increasing the cooling rate, ϕ, as shown in Fig.2a, and/or the increase of the initial austenite grain size, d_γ, as illustrated in Fig.2b. Retained strain, ε, results in increasing transformation temperatures, as indicated in Fig.2c. These tendencies are consistent with the results reported in the literature for other low carbon steels [4, 8]. However, there are two important quantitative differences compared to previously investigated Nb microalloyed HSLA steels with Mn levels ranging from 0.48 to 1.35wt% [4, 9]. In the present steel, a comparatively higher undercooling in excess of 100 °C is required for the transformation to take place which can presumably be attributed to the higher Mn content. Further, the effect of retained strain in accelerating the transformation appears to be more pronounced for simulated run-out table conditions. But similarly to previously reported results [9], the relative role of retained strain becomes smaller for higher cooling rates.

Fig.3 depicts the effects of cooling rate and initial austenite grain size on the final microstructures resulting from the CCT tests. For slow cooling rates and fine initial austenite grains, polygonal ferrite forms. Increasing the cooling rates decreases the ferrite grain size, as clearly observed for the finest initial austenite grain size of 13 μm. A microstructural transition from polygonal ferrite to acicular ferrite, then bainite, and finally martensite can be seen as ϕ and d_γ increase. The formation of these non-polygonal transformation products is particularly observed for larger austenite grain sizes with comparatively lower transformation temperatures. The transition to bainitic and martensitic transformation products is associated with a sharp decrease of the final ferrite fraction, F, which includes both polygonal and acicular ferrite. At a critical cooling rate which decreases with increasing d_γ, F decreases from approximately 0.8 to 0.2. Fig.4 shows the effects of retained strain on the ferrite microstructures. Clearly, the formation of polygonal ferrite is stabilized with the application of deformation under no-recrystallization conditions. For $d_\gamma = 13$ μm and $\varepsilon=0.5$ even at the highest cooling rate employed, the microstructures are primarily polygonal ferrite and pearlite. Acicular ferrite forms for larger initial austenite grains and cooling rates of approximately 20 °C/s and higher. In all experiments with $\varepsilon=0.5$, more than 90% ferrite is produced and the effects of ϕ and d_γ on F are much smaller than those for $\varepsilon=0$; bainite and martensite formation is not observed. The dislocation substructure introduced by deformation into the initial austenite microstructure, increases the number of ferrite nucleation sites thereby promoting higher transformation start temperatures and more ferrite in the final microstructure. Ferrite grains become finer with increase of ϕ and increase of d_γ. These findings are consistent with the results reported in the literature for Nb containing steels [4, 9].

Mathematical Modelling of Transformation - Modelling the transformation kinetics follows the approach previously proposed for other HSLA steels [4]. The model consists of three sub-models predicting (1) the transformation start temperature, (2) ferrite growth, and (3) ferrite grain size. For online modelling in an industrial process, simple equations are preferable. Therefore, a simplified transformation start model has been developed for predicting the transformation start temperature, T_S, which is represented as a function of d_γ, ϕ and ε such that

$$T_s = T_{Ae3} - c_1\, \phi^{c_2}\, d_\gamma^{\,c_3} \exp(c_4\, \varepsilon) \qquad (1)$$

Based on the experimental results, $c_1 = 38.3$, $c_2 = 0.11$, $c_3 = 0.37$, $c_4 = -0.37$ are determined when 5% transformed is assumed to characterize transformation start. This approach gives a good description of the experimentally observed transformation start temperatures, as illustrated in Fig.5. Even though this model adopts a constant cooling rate, it could easily be applied for online processing since relevant average cooling rates in the temperature range extending to 100-200 °C below the Ae3 point can usually be calculated for a given set-up of run-out table cooling conditions.

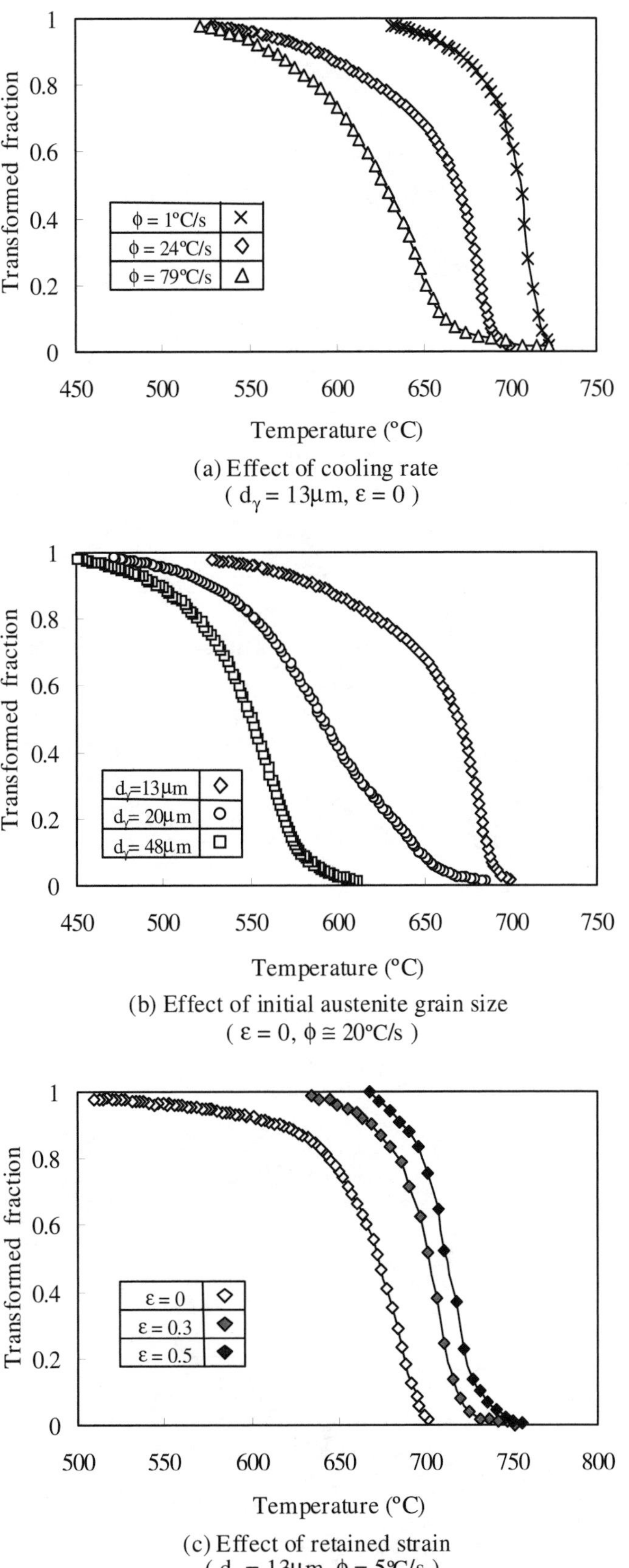

(a) Effect of cooling rate ($d_\gamma = 13\mu m$, $\varepsilon = 0$)

(b) Effect of initial austenite grain size ($\varepsilon = 0$, $\phi \cong 20°C/s$)

(c) Effect of retained strain ($d_\gamma = 13\mu m$, $\phi = 5°C/s$)

Fig.2: The continuous cooling kinetics of austenite decomposition illustrating the effects of (a) cooling rate, ϕ, (b) initial austenite grain size, d_γ, and (c) retained strain, ε

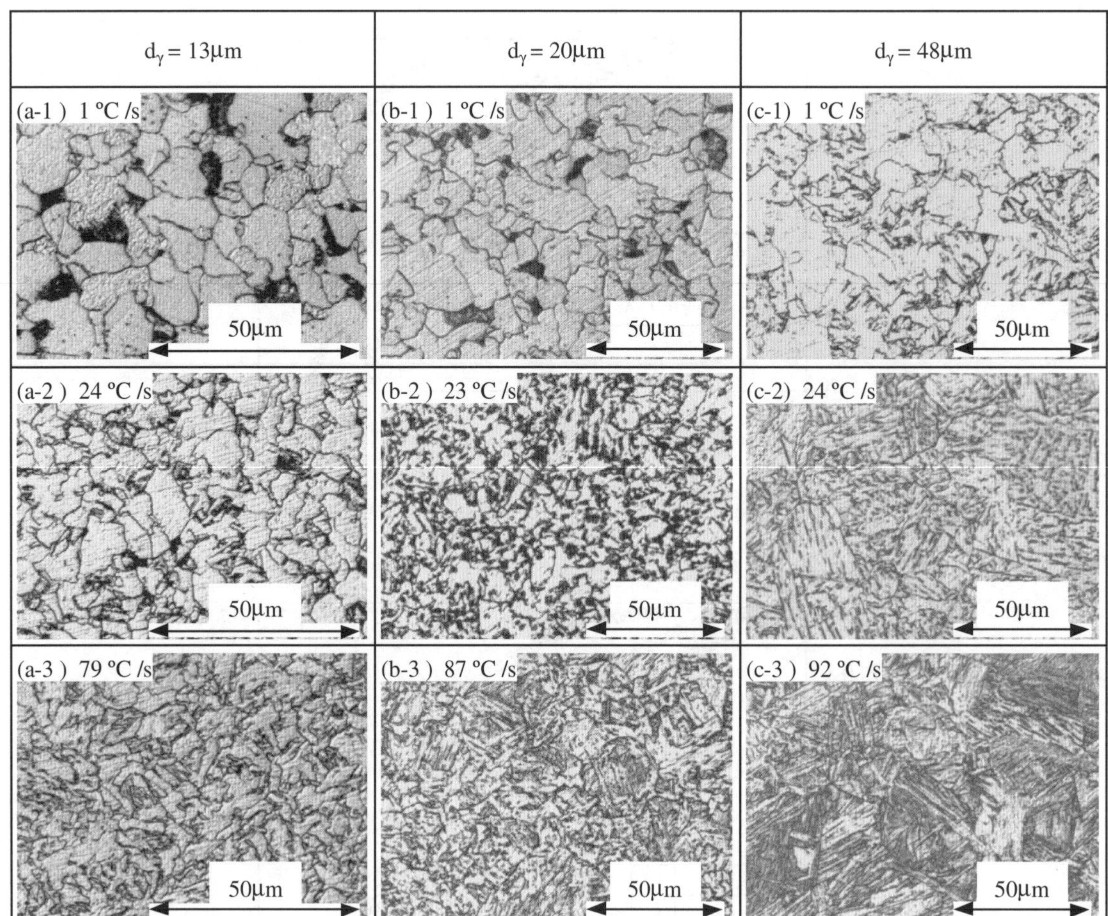

Fig.3: Microstructures obtained in CCT tests with ε=0

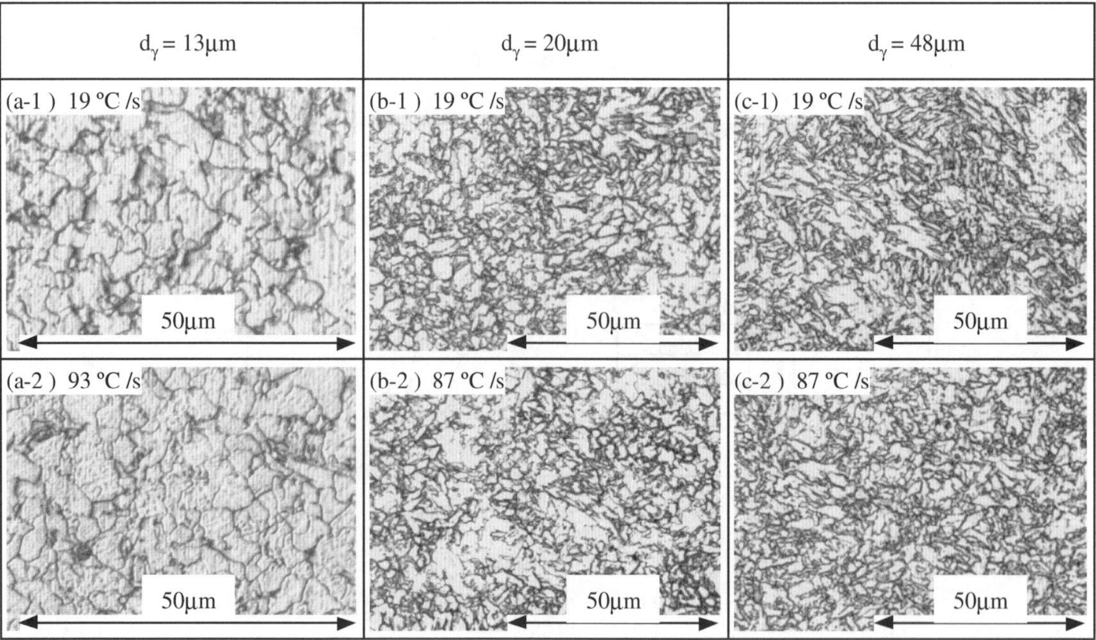

Fig.4: Microstructures obtained in CCT tests with ε=0.5

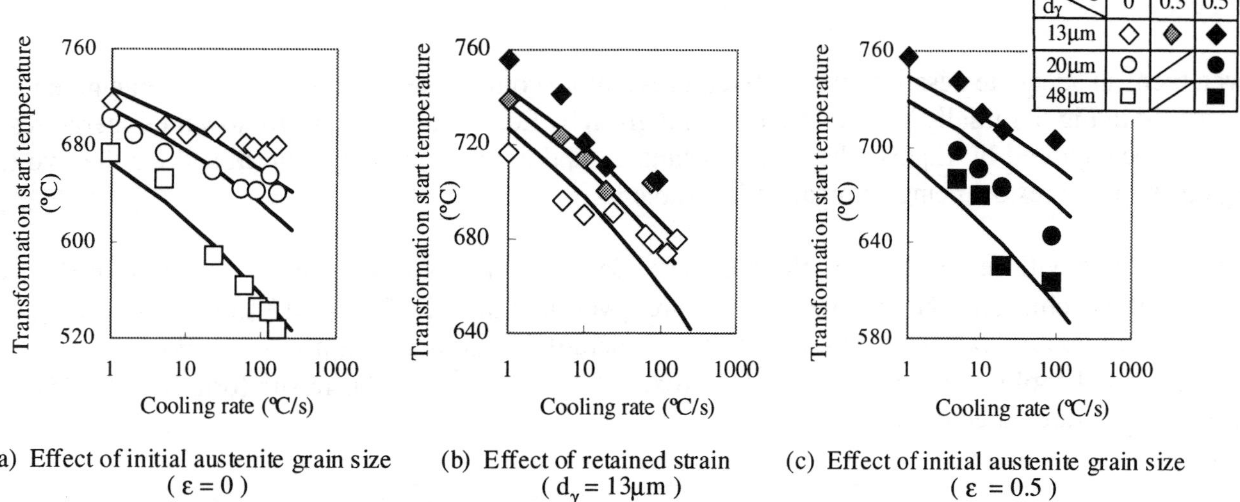

(a) Effect of initial austenite grain size ($\varepsilon = 0$) (b) Effect of retained strain ($d_\gamma = 13\mu m$) (c) Effect of initial austenite grain size ($\varepsilon = 0.5$)

Fig.5: Comparison of measured (symbols) and predicted (lines) transformation start temperatures

Ferrite growth is described with an Avrami equation [10] and adopting the additivity rule. To exclude pearlite and non-polygonal transformation products from the analysis, the normalized ferrite fraction, X_f, is used for the ferrite growth model. X_f can be calculated from transformed fraction, X, by

$$X_f = X / F \tag{2}$$

In isothermal tests, X_f can generally be represented by the Avrami model as a function of time t

$$X_f = 1 - \exp(-b t^n) \tag{3}$$

where b and n are constants. In order to employ Eq.(3) for CCT tests without having isothermal data, the transformation rate can be considered as a function of temperature to determine b and n; i.e.,

$$\frac{dX_f}{dt} = b^{\frac{1}{n}} n\{-\ln(1-X_f)\}^{\frac{n-1}{n}} (1-X_f) \tag{4}$$

The additivity rule requires that only b is a function of temperature whereas n must be temperature independent [11]. Assuming suitable values for n, b can be determined as a function of temperature from the transformation rates, as described by Pandi et al. [12]. For low carbon steels, the following relationship has been proposed for the parameter b [4, 12]

$$\ln b = a_1 (T_{Ae_3} - T) + a_2 \tag{5}$$

where a_1 and a_2 are constants.

In the present research, assumption of a constant n does not seem applicable for ε=0. Adopting n=1 for d_γ=13 μm at ε=0, only the initial parts of the transformation kinetics can be adequately described. To obtain a satisfactory description, Lee's proposal [13] is adopted where the parameter n decreases with the progress in transformation such that

$$n = 1 - X_f^{\,2} \qquad (6)$$

Even then, an adequate description of ln b requires a more complex temperature function than given by Eq.(5), as illustrated in Fig.6. Initially ln b follows the trend given by Eq.(5) but levels off for lower temperatures. This trend can still be captured by Eq.(5) when the constants a_1 and a_2 are separately determined for the two temperature regions with the results being summarized in Table III.

For conditions with retained strain, the parameter determination is less complex. Adopting n=1, a satisfactory description is achieved. The parameters for b are given in Table III. For d_γ=13 μm, where polygonal ferrite is formed at all cooling rates, the split in the two temperature regions is still necessary. For larger initial austenite grain sizes and cooling rates of more than approximately 20 °C/s, acicular ferrite forms and Eq.(5) holds for the entire temperature range investigated.

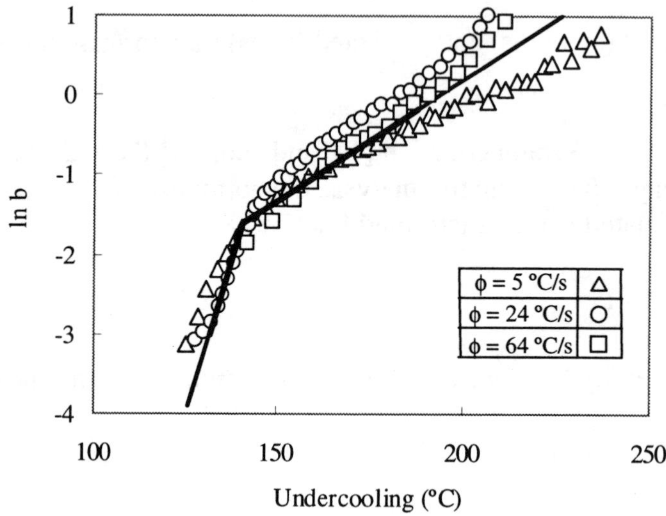

Fig.6: Relationship between b and undercooling ($n = 1 - X_f^2$, d_γ = 13 μm, ε = 0)

Table III Parameters for transformation kinetics

d_γ	Strain	n	Undercooling	a_1	a_2	Microstructure
13μm	0	$1\text{-}X_f^2$	ΔT < 140 °C	0.161	-24.2	polygonal ferrite
			ΔT > 140 °C	0.031	-5.9	
	0.3	1	ΔT < 130 °C	0.069	-10.3	
			ΔT > 130 °C	0.008	-2.4	
	0.5	1	ΔT < 130 °C	0.055	-8.1	
			ΔT > 130 °C	0.002	-1.2	
20μm	0.5	1	entire range	0.024	-6.7	acicular ferrite
48μm	0.5	1	entire range	0.046	-13.0	

For practical purposes it is sufficient to describe the transformation kinetics by the additivity rule within the framework of an Avrami model where the parameter b is a function of temperature, and n is a function of retained strain and transformed fraction. However, in order to understand the phenomena in more detail, an approach would be necessary which explicitly takes into account carbon diffusion and solute drag like effects of Mn and other alloying elements, e.g. as proposed for plain carbon steels by Militzer et al. [14].

The ferrite grain size is determined in the early stages of transformation and can be related to the transformation start temperature, as originally proposed by Suehiro [15]. For low carbon steels, the ferrite grain size d_α (in μm) can be represented by [4]

$$d_\alpha = \{F \exp(B - E/T_s)\}^{1/3} \quad (7)$$

where F is the ferrite fraction in the final microstructure, T_s is the transformation start temperature in K, B and E are parameters to be determined experimentally.

Fig.7 shows the relationship between ferrite grain size, d_α, and transformation start temperature, T_s. The measured ferrite grain size d_α at $\varepsilon=0$ depends not only on T_s but also explicitly on d_γ. However, at $\varepsilon=0.5$ d_α is a single function of T_s independent of initial austenite grain size. Assuming F=0.95 in Eq.(7) for every condition plotted in Fig.7, E=19900 is determined and B can be represented by

$$B = 24.0 + (0.13\,d_\gamma - 0.1)\exp(-7.1\varepsilon) \quad (8)$$

as a function of the initial austenite microstructure. With increase of retained strain, B becomes smaller at a progressively decreasing rate. B reaches a saturation value of approximately 24 for $\varepsilon \geq 0.5$ thereby suggesting that the number of ferrite grains nucleated in the interior of the heavily deformed grains determines the ferrite grain size in the final microstructure as a function of T_s.

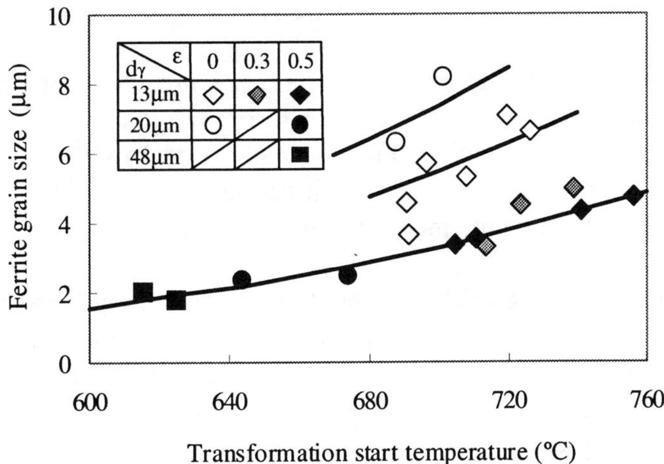

Fig.7: Relationship between ferrite grain size and transformation start temperature

Comparison with Industrial Conditions - The above models were applied to simulate the industrial conditions of finishing rolling and run-out table cooling, which are the most significant processes during hot strip rolling. Fig.8 shows a typical microstructure of a coil sample. Elongated ferrite grains are observed with a ferrite grain size of 1.6 μm. The ferrite fraction is approximately 97%.

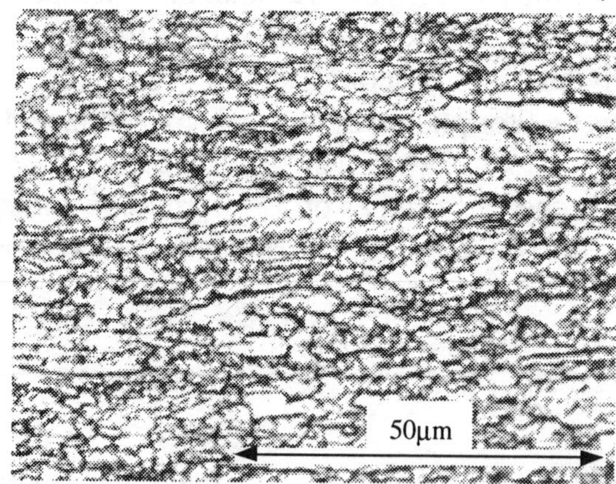

Fig.8: Microstructure in a coil sample

In the present steel, applied strain is expected to be accumulated at the later stands. Therefore, the parameter B is assumed to be 24 for industrial hot rolling. Using typical values for initial austenite microstructures expected at the end of the finishing mill; i.e., ε=0.5-1.0 and d_γ =50-70 μm, as well as for run-out table cooling; i.e., ϕ=40-80 °C/s, transformation start temperatures are estimated to be in the range of 560-640 °C. These values of T_s translate into ferrite grain sizes below 2 μm, as indeed observed. Further, using the actually measured d_α=1.6 μm and F=0.97 in Eq.(7), a transformation start temperature T_s of approximately 610 °C is predicted. This value is similar to the coiling temperature employed to realize the precipitation hardening potential of Ti and Nb. Therefore, the proposed transformation and ferrite grain size model gives satisfactory predictions for industrial run-out table conditions.

CONCLUSIONS

Based on laboratory simulations, a microstructural model has been proposed for finish rolling and run-out table cooling of a Ti-Nb HSLA steel with predominantly acicular ferrite microstructure and 780 MPa tensile strength. The model extends approaches previously developed for low carbon steels with polygonal ferrite microstructures. Where possible, simplified equations are proposed to facilitate easy incorporation of microstructure prediction into on-line models. An initial validation of the model predictions with industrial data shows promising results.

To develop a complete structure-property model, precipitation strengthening kinetics has yet to be included into the above proposed models. Further, more accurate austenite grain size predictions at the end of the finishing mill require the development of a recrystallization model for rough rolling where the austenite grain size is essentially produced since the finish mill does not see any significant recrystallization. Work is in progress to add components of recrystallization and precipitation to the model.

A challenge in developing accurate process models for high-strength steels with acicular microstructure is the characterization of these microstructures and the associated strength contributions from dislocation hardening. A combination of in-depth metallography, transmission electron microscopy and microhardness measurements appears to be a suitable way to generate the required information for proposing adequate structure-property relationships. Such work would also be of great benefit for developing process models for advanced multi-phase steels.

ACKNOWLEDGEMENTS

The discussions with Dr. I.V. Samarasekera are greatly appreciated. The assistance of B. Chau and R. Cardeno had been instrumental in conducting and analyzing the experiments.

REFERENCES

[1] E.V. Pereloma and J.D. Boyd, "Effects of Simulated On Line Accelerated Cooling Processing on Transformation Temperatures and Microstructures in Microalloyed Steels, Part 1 – Strip Processing", Mater. Sci. Technol., Vol. 8, 1996, pp. 849-854

[2] O. Kwon, K.J. Lee, J.K. Lee, K.B. Kang, J.K. Kim, J.D. Lee and J. Kim, "Application of the Metallurgical Model to Hot Strip Rolling of Plain Carbon Steels", HSLA Steels '95, G. Liu, H. Stuart, H. Zhang and C. Li, eds., China Science and Technology Press, Beijing, 1995, pp. 82-89

[3] J. Andorfer, D. Auzinger, B. Buchmayr, W. Giselbrecht, G. Hribernig, G. Hubmer, A. Luger and A. Samoilov, "Prediction of the As Hot Rolled Properties of Plain Carbon Steels and HSLA Steels", Thermec '97, T. Chandra and T. Sakai, eds., TMS, Warrendale, PA, 1997, pp. 2069-2075

[4] M. Militzer, E.B. Hawbolt and T.R. Meadowcroft, "Microstructural Model for Hot Strip Rolling of High-Strength Low-Alloy Steels", Metall. Mater. Trans. A, Vol. 31A, 2000, pp. 1247-1259

[5] M. Militzer, "Microstructure Engineering for Hot Strip Rolling of HSLA-80 Steel", Advances in Industrial Materials, D.S. Wilkinson, W.J. Poole and A. Alpas, eds., The Metallurgical Society of CIM, Montreal, PQ, 1998, pp. 63-77

[6] J.S. Kirkaldy and E.A.Baganis, "Thermodynamic Prediction of the Ae_3 Temperature of Steels with Additions of Mn, Si, Ni, Cr, Mo, Cu", Metall. Trans. A, Vol. 9A, 1978, pp. 495-501

[7] ASTM Standard Designation: E112-88, 1994 Annual Book of ASTM Standard, Vol. 3.01, ASTM, 1994, pp. 227-251

[8] R. Pandi, M. Militzer, E.B. Hawbolt and T.R. Meadowcroft, "Effect of Cooling and Deformation on the Austenite Decomposition Kinetics", 37th Mechanical Working and Steel Processing Conf. Proc., ISS, Warrendale, PA, Vol.XXXIII, 1996, pp. 635-643

[9] M. Militzer, R. Pandi, E.B. Hawbolt and T.R. Meadowcroft, "Effect of Controlled Rolling on the Austenite Decomposition Kinetics in an HSLA-Nb Steel", Accelerated Cooling / Direct Quenching of Steels, R. Asfahani, ed., ASM, Materials Park, OH, 1997, pp. 151-157

[10] M. Avrami, "Kinetics of Phase Changes II", J. Chem. Phys., Vol. 8, 1940, pp. 212-224

[11] M. Lusk and H.J. Jou, "On the Rule of Additivity in Phase Transformation Kinetics", Metall. Mater. Trans. A, Vol. 28, 1997, pp. 287-291

[12] R. Pandi, M. Militzer, E.B. Hawbolt and T.R. Meadowcroft, "Modelling of Austenite Decomposition Kinetics in Steels During Run-Out Table Cooling", Phase Transformations During Thermal/mechanical Processing of Steel, E.B. Hawbolt and S. Yue, eds., The Metallurgical Society of CIM, Montreal, PQ, 1995, pp. 459-471

[13] J.K. Lee, "Prediction of γ/α Transformation During Continuous Cooling of Steel", 41st Mechanical Working and Steel Processing Conf. Proc., ISS, Warrendale, PA, Vol.XXXVII, 1999, pp. 975-981

[14] M. Militzer, R.Pandi, E.B. Hawbolt and T.R. Meadowcroft, "Modelling the Phase Transformation Kinetics in Low-Carbon Steels", Hot Workability of Steels and Light Alloys-Composites, H.J. McQueen, E.V. Konopleva and N.D. Ryan, eds., The Metallurgical Society of CIM, Montreal, PQ, 1996, pp. 373-380

[15] M. Suehiro, K. Sato, Y. Tsukano, H. Yada, T. Senuma and Y. Matsumura, "Computer Modeling of Microstructural Change and Strength of Low-Carbon Steel in Hot Strip Rolling", Trans. Iron Steel Inst. Jpn., 1987, Vol. 27, pp. 439-445

BENCH MARKING AND VALIDATION OF MODELS USED FOR HOT STRIP MILL SIMULATION

M.Phaniraj, S.Shamasundar* and A.K.Lahiri
Department of Metallurgy
Indian Institute of Science
Bangalore-560 012
E-mail: praj@metalrg.iisc.ernet.in
*Advanced Forming Technology Center
326, III Stage, IV Block, Basaveshwara Nagar,
Bangalore-560 079

Keywords: Process modeling, hot strip mill, mechanical properties, microstructural evolution

INTRODUCTION

In modeling hot strip rolling the ultimate aim of a modeler is to predict the mechanical property accurately. For low carbon steels the yield strength is a function of ferrite grain size and hence predicting ferrite grain size accurately is important. It is known that ferrite grain size depends on the cooling rate, chemical composition, austenite grain size and retained strain. Literature abounds in models for predicting these quantities and the obvious question that a plant engineer faces is to decide upon the right model for a given set of conditions. The present study aims at addressing this question by comparing a few models cited by different groups for tensile strength, yield strength and ferrite grain size.

COMPARISON OF MODELS

Tables I, II and III list the equations proposed by various authors for ferrite grain size, yield strength and tensile strength, respectively. Table IV gives the cooling rates, retained strain levels, austenite grain sizes and composition ranges based on which the models were proposed. In the following sections predictions of ferrite grain size, yield strength and tensile strength by different models are compared for identical conditions, followed by a sensitivity analysis of the models to the parameters they depend upon. The steel for which the comparison is done is a very common grade of low carbon steel and its composition is in the range studied by the models (see table IV).

Ferrite grain size

Literature presents a number of formulae for predicting the ferrite grain size (see table I). All the equations are valid for C-Mn steels. The models by Roberts[1], Yada[2], Anelli[3] and Umemoto[4] do not take residual strain into consideration and only the models by Herman[5], IRSID[5] and Hodgson[6] take the chemical composition into account. Table V lists the formulae after substituting for the composition under investigation and for zero strain. It is clear from the table that all models show dependence on austenite grain size and cooling rate.

Figure 1 shows the ferrite grain size predicted, for the steel under study, by all models for an austenite grain size of 30μm, with and without retained strain, and a cooling rate of 20°C/s. A retained strain of 0.19, used in the present calculations, is typical in a hot strip after the finishing mill. The models differ in predictions widely, the difference being of a factor of three. Model by Saito et al[3] predicts the least value (3μm) whereas the model by Herman et al predicts the maximum value (16μm), for the case of strained austenite.

Table I Equations for ferrite grain size

Model	Equation
Sellars et al[3]	$d_\alpha = (2.5 + 3\dot{T}^{-0.5} + 20(1 - e^{-0.015d\gamma}))(1 - 0.45\varepsilon_r^{0.5})$
Roberts et al[1]	$d_\alpha = 3.75 + 0.18d\gamma + 1.4\dot{T}^{-0.5}$
Yada[2]	$d_\alpha = (5.51\times 10^{10})d\gamma^{1.75} e^{-21430/T_f} X_f$
Saito et al[3]	$d_\alpha = exp(ld)$ $ld = 0.92 + 0.44 log(d_\alpha) - 0.17 log\dot{T} - 0.88 tanh(10\varepsilon_r)$
Gibbs et al[3]	$d_\alpha = (1 - 0.8\varepsilon_r^{0.15})(29 - 5\dot{T}^{-0.5}) + 20(1 - exp(-0.015 d\gamma))$
Anelli et al[3]	$d_\alpha = 8.9d\gamma^{0.4}(1 - exp(-0.075d\gamma))\dot{T}^{-0.14}$
Umemoto et al[4]	$d_\alpha = 5.7\dot{T}^{-0.26} d\gamma^{0.46}$
Herman et al[5]	$d_\alpha = (13 - 0.73C_{eq}^{0.45})d\gamma^{0.3-0.22\varepsilon_r}\dot{T}^{-0.15}$ $C_{eq} = C + \frac{Mn}{10}$
IRSID[5]	$d_\alpha = (6.77 - 10C - Mn)\dot{T}^{-0.175} d\gamma^{0.4-0.25\varepsilon_r}$
Hodgson et al[6]	$d_\alpha = (1 - 0.45\varepsilon_r^{0.5})(-0.4 + 6.35C_{eq}) + (24.2 - 59C_{eq})\dot{T}^{-0.5} + 22(1 - e^{-0.015d\gamma})$ $C_{eq} = C + Mn/6$

Table II Equations for Yield Strength

Hodgson et al[6]	$YS = 62.6 + 26.1[Mn] + 60.2[Si] + 759[P] + 3286[N] + 19.7d_{\alpha}^{-0.5}$
Venkataraman et al[3]	$YS = 70 + 32.2[Mn] + 83.2[Si] + 4620[N] + 22.4d_{\alpha}^{-0.5}$
Choquet et al[13]	$YS = 63 + 23[Mn] + 53[Si] + 700[P] + 5369[N] + K_y d_{\alpha}^{-0.5}$ $K_y = 15.4 - 30[C] + \frac{6.094}{0.8 + [Mn]}$
Herman et al[12]	$YS = 58 + 25[Mn] + 75[Si] + 700[P] + 5000[N] + X_f K_y d_{\alpha}^{-0.5} + X_P K_P$ $K_y = 17.9 - 30[C] + \frac{0.896}{0.05 + [Mn]}$ $K_P = 360 + 2600[C]$
Pickering[14]	$YS = 15.4(3.5 + 2.1[Mn] + 5.4[Si] + 23[N]^{0.5} + 1.13d_{\alpha}^{-0.5})$

Table III Equations for Tensile Strength

Suehiro et al[10]	$TS = 0.31(361 - 0.357T_f + 50[Si] + 2.55d_{\alpha}^{-0.5})*9.8$
Hodgson et al[6]	$TS = 164.9 + 634.7[C] + 536.6[Mn] + 99.7[Si] + 651.9[P] + 3394[N] + 11d_{\alpha}^{-0.5}$
Herman et al[12]	$TS = 245 + 41[Mn] + 80[Si] + 700[P] + 4000[N] + 7d_{\alpha}^{-0.5}$
Choquet et al[13]	$TS = 237 + 29[Mn] + 79[Si] + 700[P] + 5369[N] + 7.24X_f d_{\alpha}^{-0.5} + 500X_p$
Kwon et al[9]	$TS = 0.095H_f V_f + 0.264H_p V_p + 0.277H_b V_b + 75.5V_f d_{\alpha}^{-0.5} + 13.4V_f(1 - V_f) + 3.8$ $H_f = 27.72exp(\frac{980}{T_{mf}}) + 39.8[Si] + 9.1$
Pickering[14]	$TS = 15.4(19.1 + 1.8[Mn] + 5.4[Si] + 0.25[X_f] + 0.5d_{\alpha}^{-0.5})$

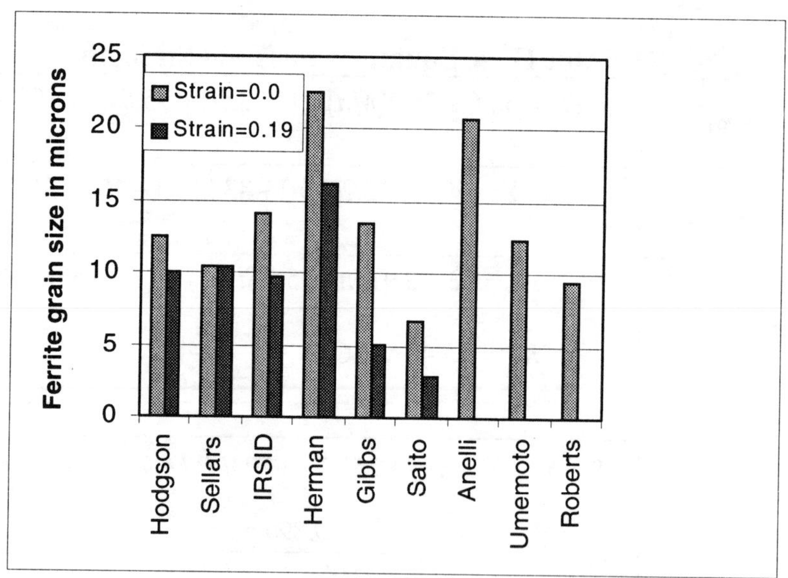

Figure 1 Chart showing the deviation between models for predicting ferrite grain size, for dγ=30μm and cooling rate of 20°C/s.

The model predictions of ferrite grain size as a function of retained strain is shown in figure 2. The models predictions differ by a factor of five. It is seen that a retained strain of 0.2 reduces the ferrite grain size by 30% to 65%, depending on the model, compared to that obtained from unstrained austenite. A cooling rate of 20°C/s is chosen here for calculation but changing the cooling rate does not change the trend. Also beyond a strain level of 0.4~0.5 the effect of retained strain in reducing grain size is not significant.

Figure 3 shows the models predictions for ferrite grain size as a function of cooling rate, from an undeformed austenitic structure. Venkataraman et al[3] have reported a cooling rate of 8°C/sec for air and 20°C/sec for water cooling of a hot rolled strip on the run out table. In the present study the calculation is carried out for cooling rates ranging from 5°C/sec to 25°C/sec. It can be seen from the figure 3 that models predictions differ by a factor of three from each other. Based on their prediction, the models can be grouped into three; viz. Sellars and Roberts; Umemoto, Hodgson and IRSID; and Anelli and Herman, each group giving nearly equal (within 6%) results. It should be noted here that predictions by Hodgson and IRSID match closely even in the plot of ferrite grain size versus strain (figure 2).

Table IV Conditions over which the models were proposed

Model	Cooling rates studied by the author °C/sec	Austenite grain size studied μm	Strain range studied	Composition range studied
Hodgson et al[6]	0.4 – 3.2	20,40	~0.2-2.5	C_{eq}<0.35
Roberts et al[1]	0.25 – 12	>8	-	0.13C, 0.14Mn
Herman et al[5]	0.01 – 70	20-170	0-0.6	0.012-0.3C, 0.05-1.4Mn
Umemoto et al[4]	0.05 – 30	178, 151, 126	Strain free	0.15C, 0.4Mn, <0.1Si
IRSID[5]	-	-	-	0.01-0.3C, 0.1-1.6Mn
Composition of steel under investigation				0.04C, 0.18Mn, 0.014Si

Table V Equations for ferrite grain size

Model	Equation after substituting ε_r=0 and composition of steel under study	**Remarks**
Hodgson et al[6]	$0.0445+20.07T\square^{-0.5}+22(1-e^{-0.015d\gamma})$	Only models which depend On chemical composition
Herman et al[5]	$12.79d\gamma^{0.3}T\square^{-0.15}$	
IRSID[5]	$6.55d\gamma^{0.4}T\square^{-0.175}$	
Umemoto et al[4]	$5.7d\gamma^{0.46}\,T\square^{-0.26}$	Do not depend on retained strain and chemical composition
Roberts et al[1]	$3.7+0.18d\gamma+1.4T\square^{-0.5}$	
Anelli et al[3]	$8.9d\gamma^{0.4}(1-\exp(-0.075d\gamma))T\square^{-0.14}$	
Sellars et al[3]	$2.5+3T\square^{-0.5}+20(1-e^{-0.015d\gamma})$	Depend on retained strain but not on composition
Gibbs et al[3]	$29-5T\square^{-0.5}+20(1-\exp(-0.015d\gamma))$	
Saito et al[3]	$\mathrm{Exp}(0.92+0.44\ln d\gamma-0.17\ln T\square)$	

Only the model by Gibbs shows significant dependence on cooling rate where the ferrite grain size decreases by 45% on increasing cooling rate from 5 to 25°C/s. Models by Roberts and Sellars are virtually independent of cooling rate, whereas the rest of the models show little dependence. The insignificant effect of cooling rate on ferrite grain size has also been observed by Ouchi et al[7]. They have also reported that ferrite grain size becomes independent of cooling rate after a certain value, which is observed in the present study to be 20°C/s. Senuma et al[8] reported that in the case of extra low carbon steels, ferrite grain size decreases only by 0.5 ASTM G.S. No on increasing the cooling rate from 10°C/s to 120°C/s. The austenite-ferrite transformation progresses rapidly and does not allow large overcooling. Hence increasing cooling rate has insignificant effect on ferrite grain size[8].

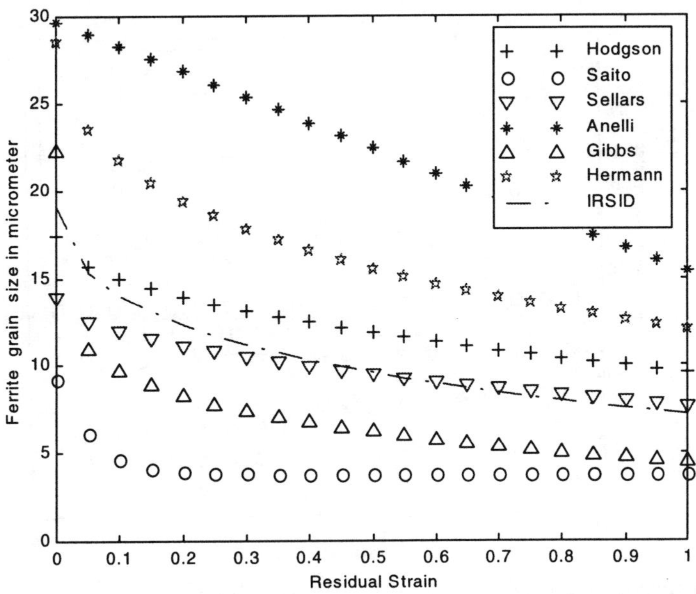

Figure 2 Effect of residual strain on ferrite grain size after transformation for an austenite grain size of 50µm and cooling rate of 20°C/s

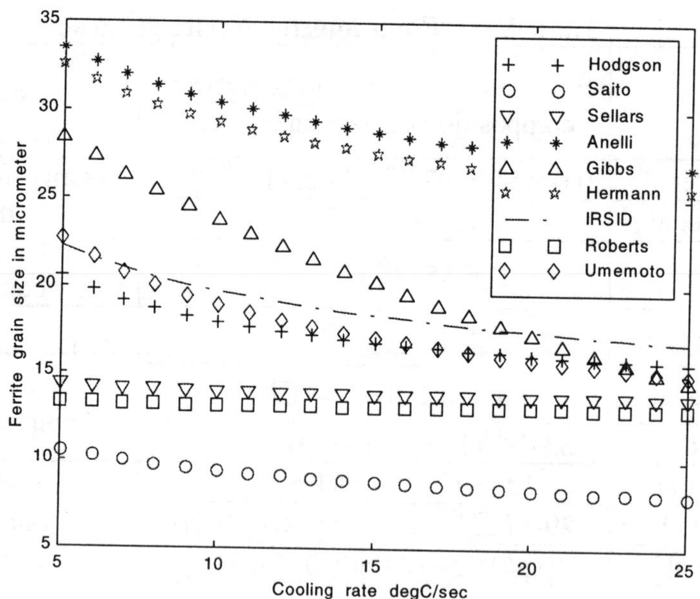

Figure 3 Change in ferrite grain size as a function of cooling rate for an unstrained austenite grain size of 50μm

Figure 4 shows the plot of ferrite grain size against cooling rate for a residual strain of 0.4. The residual strain of 0.4 is chosen here since, as has been seen in figure 2, ferrite grain size is insensitive to residual strain for values greater than 0.4~0.5. The effectiveness of cooling rate in reducing ferrite grain size reduces with increasing retained strain and can be clearly seen in the case of Gibbs predictions. It should be noted here that only models by Anelli, Roberts and Umemoto have no dependence on residual strain. The predictions by Sellars, IRSID, Hodgson are within 5μm of each other but this could mean a difference in yield strength, as will be seen in the following section, of 40Mpa (see figure 10).

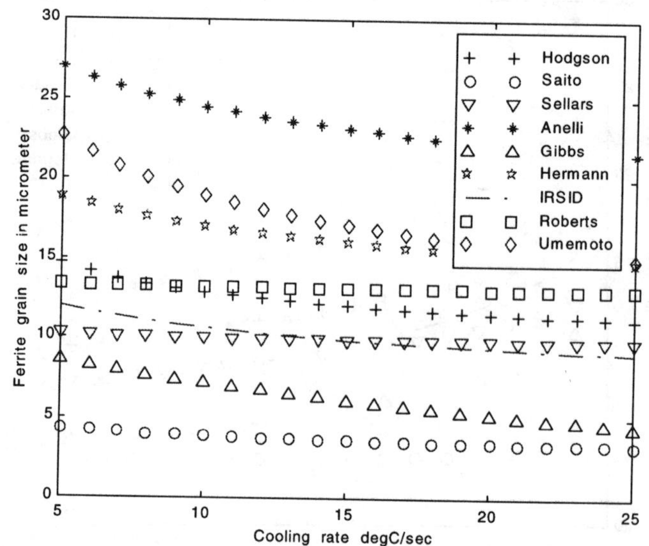

Figure 4 Drop in ferrite grain size for conditions similar to figure 3 for a retained strain of 0.4

The plot of model predictions for ferrite grain size as a function of austenite grain size assuming zero residual strain and 0.4 residual strain are shown in figures-5 and 6, respectively. Predictions by all models except by Anelli and Herman do not show significant dependence on austenite grain but between models the predictions differ by a factor of three. As the retained strain increases ferrite grain size decreases for all the models and the influence of austenite grain size reduces.

For the steel composition under study, all models predict that neither cooling rate nor austenite grain size, when varied independently, influence the ferrite grain size significantly. But between models the predictions differ by a factor by a factor of 3-7. Hodgson's model is applicable to all the steels with C_{eq}<0.35. The models by Hodgson, Umemoto, IRSID, Roberts and Herman are based on compositions which satisfy this criterion but

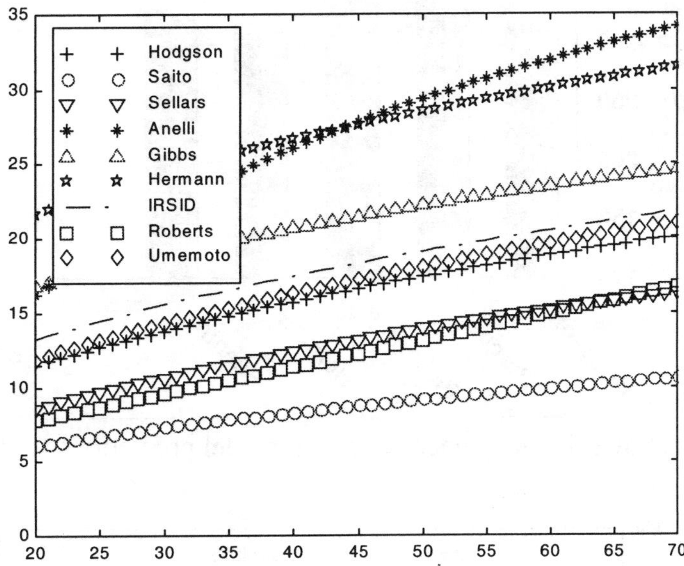

Figure 5 Variation in ferrite grain size as a function of strain free austenite grain size prior to transformation, for a fixed cooling rate of 12°C/s

only the former three predict close to each other. This is inspite of the different austenite grain sizes and cooling rates on which these three models are based. Hence parameters other than the composition, grain size and cooling rate have an answer for the differences in predictions

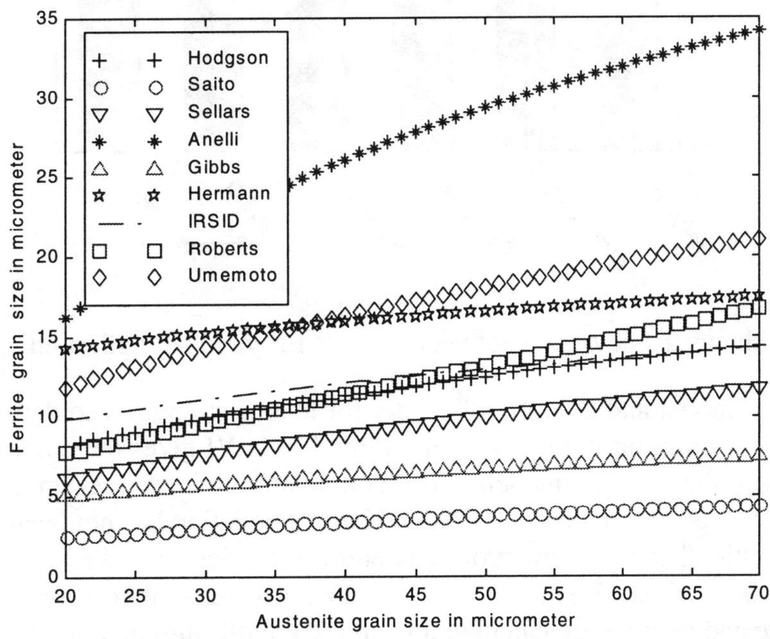

Figure 6 Effect of strained (ε_r=0.4) austenite on ferrite grain size at a cooling rate of 12°C/s.

Strength

It is clear from tables II and III, that the yield strength and tensile strength depend on chemical composition and

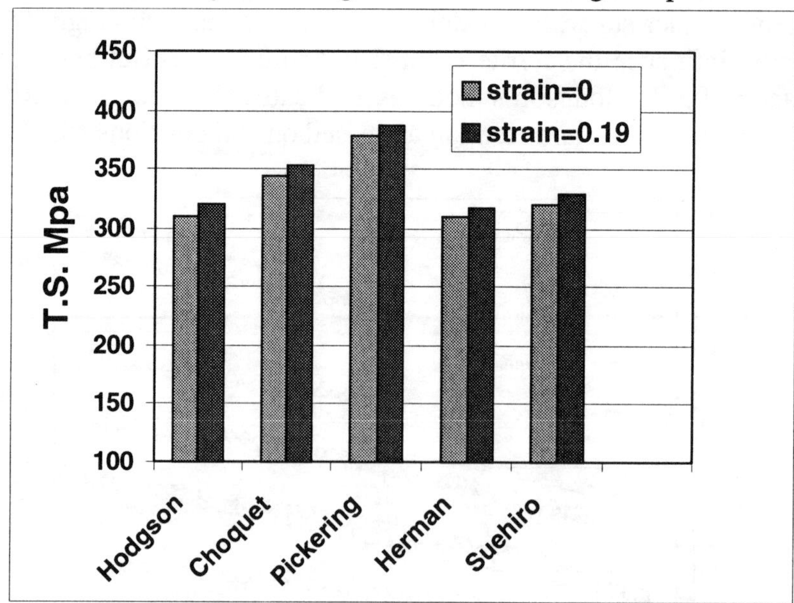

Figure 7 Chart shows the differences between model predictions for tensile strength

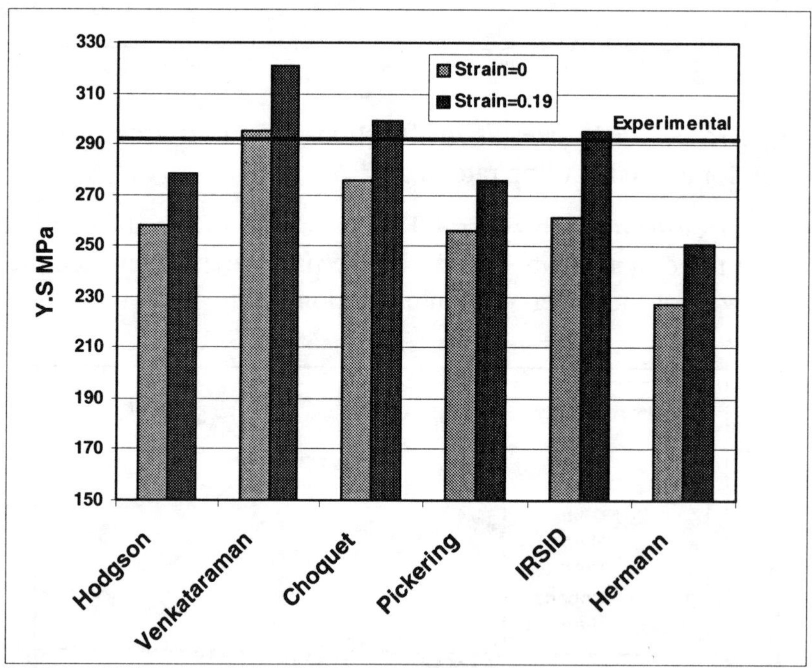

Figure 8 Chart shows the deviation between models for yield strength predicted by models.

ferrite grain size. Kwon's[9] model and Suehiro's[10] model for tensile strength also depend on the transformation temperature of various phases present in the microstructure. Table VI gives the composition ranges studied by various authors for building the strength models. The composition of the steel under study falls in the range studied by all the models except Kwon's model. Hence Kwon's equation has not been used in the present study to predict the tensile strength. The transformation temperatures for ferrite and pearlite are taken from the atlas of CCT diagrams[11] for a 0.05%C steel, since it was the closest to the composition of steel under study. The volume fractions of ferrite and pearlite are calculated from the equilibrium diagram by lever rule. Figure 7 and 8 show the predictions of tensile strength and yield strength for the steel under study by different models, respectively. The strength values for Herman[12], IRSID and Hodgson are calculated based on the values of ferrite grain size obtained from their respective models whereas the values for Choquet[13], Pickering[14], and Suehiro were calculated from the ferrite grain size obtained by Sellars model. The deviation between models for yield strength is ~30% and for tensile strength is ~17%. Figure 7 also shows that the experimental value of yield strength for the steel under study is predicted by IRSID and Choquet within 10MPa. The models by Pickering and Hodgson predict within 15Mpa, and Herman and Venkataraman within 40Mpa, of the experimental value. The experimental yield strength is compared to yield strengths predicted based on ferrite grain sizes transformed from strained austenite (retained strain=0.19).

Figure 9 shows the effect of varying ferrite grain size on tensile strength as predicted by different models. Table-VII gives the equations for TS and YS after substituting for the composition, which is investigated in the present study.

Table-VI Compositions studied by various authors

Model Proposed by	%C	%Mn	%Si
Herman et al[12]	0.036-0.205	0.25-1.1	0-0.25
Choquet et al[13]	0.01-0.3	0.1-1.6	-
Suehiro et al[10]	0.05-0.15	0.5-1.5	0-1.03
Hodgson et al[6]	0.06-0.25	0.3-1.7	-
Kwon[9]	0.12	0.8	
Venkataraman et al[3]	Low carbon ferritic steels		
Pickering[14]	<0.1	0.3-0.4	-
Steel under study	0.04	0.18	0.014

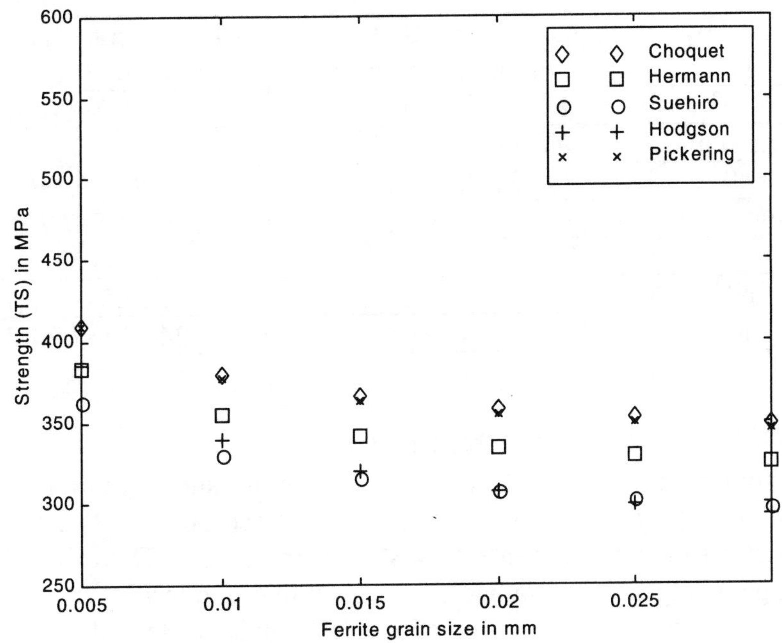

Figure 9 Variation in tensile strength predicted by various models as a function of ferrite grain size

It is obvious from the table that the equations are more or less similar with a little difference in the constants. Figure 10 compares the yield strength predicted by all the four models for ferrite grain size, ranging from 5μm to 30μm. This is the range in which the model predictions for ferrite grain size occur as seen in the previous section (see figure 3,4). It can be seen from the figures 9-10 that yield strength is more sensitive to ferrite grain size, as compared to tensile strength. For coarser grains the difference in yield strength predictions is relatively small i.e. within 13% but as the grain size decreases the differences increases to about 22% at 5μm. Hence choice of the model for predicting the ferrite grain size is an important issue.

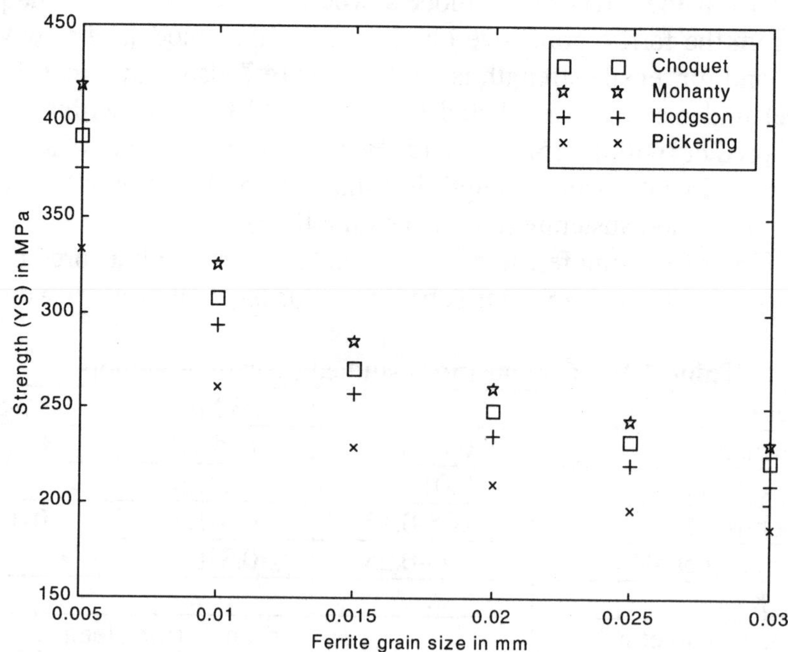

Figure 10 Increase in yield strength with decreasing ferrite grain size as predicted by different models

Table-VII

Model proposed by	Equation for Y.S of after substituting the composition	Equation for T.S of after substituting the composition
Hodgson et al[6]	$81.98+19.7(d_f)^{-0.5}$	$211.9+11(d_f)^{-0.5}$
Venkataraman et al[3]	$77.00+22.4(d_f)^{-0.5}$	-
Choquet et al[13]	$77.73+20.35(d_f)^{-0.5}$	$278.4+6.87(d_f)^{-0.5}$
Pickering[14]	$86.67+17.40(d_f)^{-0.5}$	$303.95+7.7(d_f)^{-0.5}$
IRSID[5]	$96.81+19.56(d_f)^{-0.5}$	
Herman et al[13]	-	$263.51+7(d_f)^{-0.5}$
Suehiro et al[10]	-	$244.72+7.74(d_f)^{-0.5}$
Kwon[9]	-	$16.88+71.72(d_f)^{-0.5}$

CONCLUSIONS

The models predictions for ferrite grain size differ by a factor of 3-7 and this difference is independent of the parameters varied viz. cooling rate, austenite grain size and retained strain. The ferrite grain size is more sensitive to retained strain than to the cooling rate or austenite grain size. There is a limit to the ferrite grain size obtainable by increasing cooling rate or retained strain. Inspite of the ferrite grain size predictions differing by a factor of 3-7, the yield strength and tensile predictions differ only by 15-30%. Yield strength predictions by IRSID and Choquet match within 10MPa with the experimental value for the steel under study and the rest of the models predict within 40MPa.

REFERENCES

1. W. Roberts, A. Sandberg, T. Siwecki and T. Werlefors, “Prediction of microstructure development during recrystallisation hot rolling of Ti-V steels”, HSLA steels: Techology and Applications, Proc. Conf. Philadelphia, ASM 1983, p67-84

2. H. Yada, "Prediction of microstructural changes and mechanical properties in hot strip rolling", Accelerated cooling of rolled steels, eds. Ruddle. G. E and Crawley. A. F., CIM Quebec-88, p105-119.

3. M. Venkataraman and O. N. Mohanty, "Prediction of Microstructure and Mechanical Properties of Hot Rolled Low Carbon Steel Strips" 41st MWSP Conf. Proc. ISS, vol.XXXVII, 1999, p717-737.

4. M. Umemoto, Zing Hai Guo, I. Tamura, "Effect of cooling rate on grain size of ferrite in a carbon steel" Mat. Sc. and Tech., 1987, Vol. 3, p249-255.

5. J. Herman, B. Thomas, U. Lotter, European Commission Technical Steel Research, Final Report-Mechanical working (Rolling mills)-1994

6. R. D. Hodgson and R. K. Gibbs, "A mathematical model to predict the mechanical properties of hot rolled C-Mn and microalloyed steels", ISIJ-1992, Vol. 32, No. 12, p1329-1338.

7. C. Ouchi, T. Okita and S. Yamamoto, "Effects of interrupted accelerated cooling after controlled rolling of the mechanical properties of low alloy steels", Trans.-ISIJ, Vol. 22, 1982, p608-616.

8. T. Senuma, M. Kameda and M. Suehiro, "Influence of hot rolling and cooling conditions on the grain refinement of hot rolled extralow-carbon steel bands", ISIJ-1998, Vol. 38, No.6, p587-594.

9. O. Kwon, "A technology for the prediction and control of microstructural changes and mechanical properties in steel", ISIJ Intl., Vol32, 1992, No.3, p350-358.

10. M. Suehiro, K. Sato, Y. Tsukano, H. Yada, T. Senuma and Y. Matsumara, "Computer modelling of microstrucural change and strength of low carbon steel in hot strip rolling", Trans. ISIJ, Vol. 27, 1987 p439-445.

11. M.Atkins, Atlas of continuous cooling transformation diagrams for engineering steels, ASM, 1977, p20.

12. J-C. Herman, B. Donnay, A. Schmitz, U. Lotter, R. Grossterlinden, European Commission Technical steel research, Mechanical Working(Rolling)-1999.

13. P. Choquet, P. Fabreque, J. Giusti, B. Chamont, J. N. Pezant and F. Blanchet, "Modelling of forces structure and final properties during the hot rolling process on the hot strip mill", Intl. Symposium on mathematical modeling of hot rolling of steels, ed. S.Yue CIM Quebec, 1990, p34-43.

14. F. B. Pickering, Physical metallurgy and the design of steels, Materials Science Series, Applied Science Publishers, 1978, p37-59.

Mathematical Modelling of Hot Rolling of Steel Strip in Three Dimensions

M. Miyake*, J. Too** and I.V. Samarasekera***

*Rolling and Processing Research Department
Materials and Processing Research Center
NKK Corporation
1 Kokan-cho, Fukuyama, Hiroshima-pre
721-8510 Japan
Phone +81-849-45-4154

**formerly of CANMET

***The University of British Columbia
The Centre for Metallurgical Process Engineering
The Brimacombe Building
112-2355 East Mall
Vancouver, B.C. V6T 1Z4 Canada

ABSTRACT

A three-dimensional mathematical model has been developed to predict the heat transfer, deformation and austenite grain size evolution during hot rolling of steel strip. The model predicts the variation in pressure, temperature and strain across the strip width and the resulting differences in microstructure at the edge versus the center.

1. INTRODUCTION

Microstructure prediction has been one of the most important issues for the engineers in the hot working of metals. As for hot-strip rolling, two-dimensional mathematical models have been made with great progress in the past decade. Variation of the microstructure in the thickness direction can be obtained with high accuracy with a two-dimensional model. But in reality, mechanical parameters like strain, strain rate and temperature will vary in the width direction and this variation will influence the properties of the final product. This is particularly interesting for IF steels where the transformation start temperature is high; if the temperature of the edges of the strip drop below the transformation temperature, flatness problems could result. But to discuss the three dimensional variations during hot rolling of steel strip, one should pay attention to the modelling to obtain accurate mechanical parameters for the microstructure analysis. For example, the heat-transfer coefficient on the contact surface is normally treated as a constant, but it will be affected by contact pressure, surface roughness and hardness of the materials which are in contact. Chen[3] and Hlady[4] developed this idea for hot-strip rolling and proposed a pressure-dependent heat-transfer coefficient formulation. And in strip rolling, the deformation of the rolls alters the contact pressure distribution and this should affect the heat-transfer coefficient on the contact surface. At this time, three-dimensional thermal analysis with a pressure dependent heat-transfer coefficient, as to the authors knowledge, has not been conducted.

The purpose of the present study is to model the hot-strip rolling process in three dimensions to predict the variations of microstructure through the strip thickness and across the width. To this end, the authors have developed a three-dimensional mathematical model to predict the heat-transfer, deformation, and austenite grain size evolution during hot rolling of steel strip.

2. ANALYSIS ALGORITHM

Samarasekera et al.[5] developed a mathematical model to predict microstructure evolution for hot rolling of steel strip in two dimensions. Their model incorporates deformation and temperature distribution in the roll-bite by FEM, and FDM was employed to analyze the thermal history between stands. As for the microstructure model, they considered both static and metadynamic recrystallization after deformation and also grain growth in the austenite region. Their system can predict the final mechanical properties along the strip thickness by coupling with a transformation and precipitation model during run-out table cooling. In the present study, a three-dimensional hot-strip rolling model was developed and coupled with their microstructure model to simulate the changes of austenite grain size during hot-strip rolling.

Figure 1 is a schematic diagram of the algorithm for the prediction of microstructure for hot-strip rolling. The deformation and temperature of the strip have a significant influence on each other through heat generation and flow stress and these two analyses have been coupled. After the rolling analysis is completed, thickness profile, temperature, strain and strain rate at the roll-bite exit were used as initial conditions for the interstand analysis. The initial conditions of temperature and residual strain for the next rolling analysis were derived from the interstand analysis and the tandem hot-strip rolling process was simulated by replicating this routine.

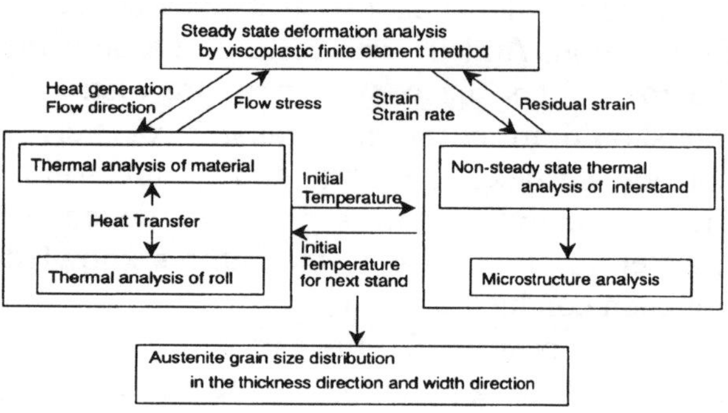

Fig. 1. Relationships among the analyses.

3. MATHEMATICAL MODELLING

3.1 The Deformation Analysis Model

Three dimensional deformation in the roll-bite was analyzed by a steady-state visco-plastic finite-element model based on flow formulation. In the model, friction force on the contact surface is expressed by Coulomb's law and directional change of the friction force was treated as a function of relative velocity. Streamlines were modified after each iteration step by integrating the velocity field from the upstream boundary. 8-node isoparametric hexahedral elements were employed and a quarter of the cross section was analysed because of the symmetry. Elastic deformation of the rolls were obtained by a slit model[7] for deflection and infinite elastic theory for indentation[8]. The error norm of the velocity was used for the criterion of convergence and 0.00005 was chosen in this analysis.

3.2 Thermal Analysis Model

The temperature distribution in both the strip and rolls was analyzed as a steady state problem by FEM. To cope with high convection, Petrov-Galerkin streamline weighting function was introduced[9].

$$W_i = N_i + \frac{\alpha h}{2}\frac{u(\partial N_i/\partial x)+v(\partial N_i/\partial y)+w(\partial N_i/\partial z)}{|U|} \quad (1)$$

where W_i are weighting functions, N_i are shape functions, α is an upwinding parameter, h is a representative element size of each element in the flow direction, and U is a resultant velocity vector at the center of each element.

Basically, the same mesh system was employed for deformation and thermal analysis of the strip; but two layers from the strip surface were divided into seven layers to handle the steep temperature gradient in the thermal analysis (Fig. 2).

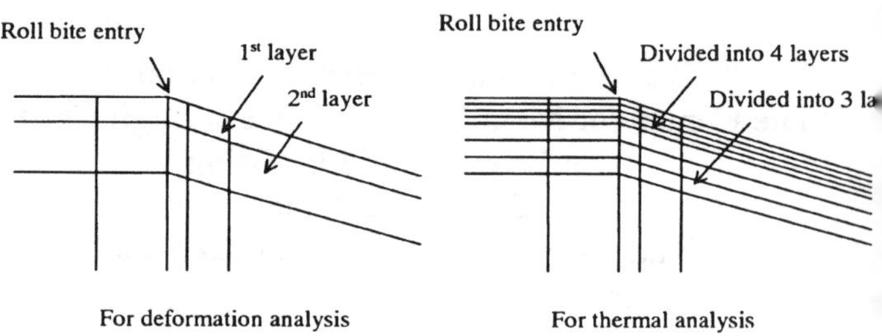

Fig. 2. Mesh system for thermal analysis.

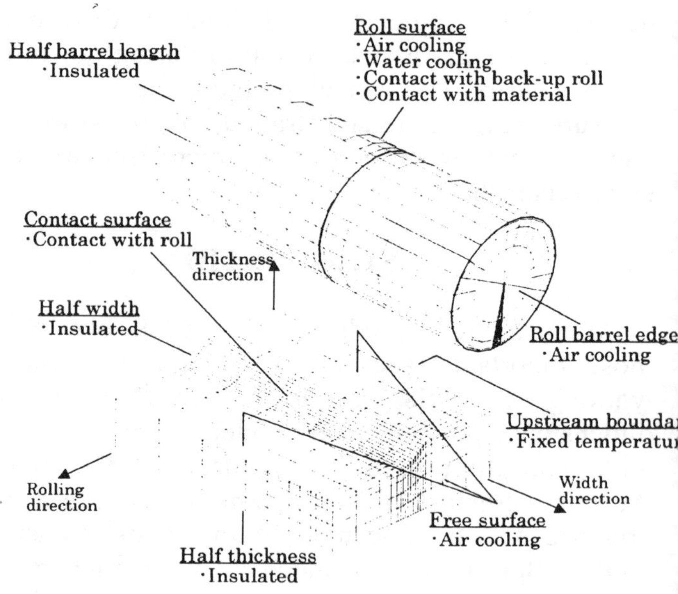

Fig. 3. Mesh system employed for thermal analysis.

Figure 3 shows the mesh system employed for the thermal analysis and very fine mesh that was deployed around the roll-bite entry and strip width edge. The mesh system of the roll was rebuilt before each thermal analysis to adjust the position of the nodes which were in contact with the strip. In the thermal analysis of the roll, not only air cooling but also water cooling and contact with a back up roll was considered (Fig. 4).

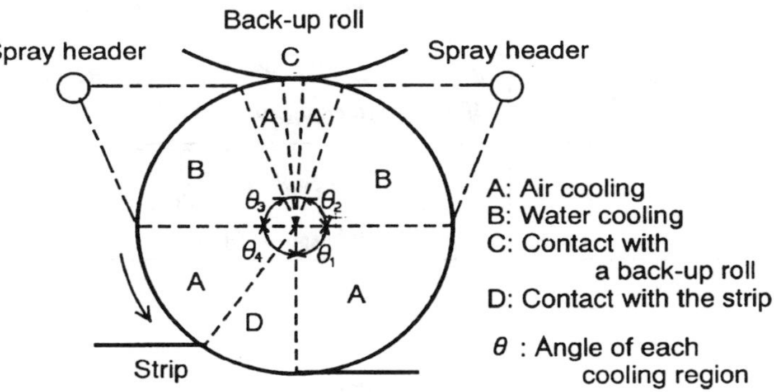

Fig. 4. Assumed cooling pattern.

Heat flux on the contact surface between the strip and the rolls will depend on the contact condition. From this standpoint, Chen[3] developed a contact pressure dependent heat-transfer coefficient for the analysis of hot rolling. They derived the formulation by comparing measured temperature in the roll-bite and calculated roll pressure for several steel grades (Eq.(2)).

$$h_c = 0.695P - 34.4 \qquad (2)$$

where h_c is in kW/m^2°C and P in MPa. Originally Eq.(2) was formulated with mean roll pressure but it is of interest to assign local pressure on the contact surface which is calculated by FEM into Eq.(2). Differences in temperature distribution between uniform heat-transfer coefficient, which was calculated with mean roll pressure and distributed heat-transfer coefficient which was derived from local pressure, will be discussed later.

Thermal analysis of the strip and roll were solved individually and they were coupled by applying each node's temperature to the boundary condition of the thermal analysis of the other.

By adopting an upwinding technique, temperature oscillation in the rolling direction was not evident in the results, but oscillation in the width direction on the contact surface arose with this iteration scheme. This oscillation increased in magnitude after each thermal analysis of the roll and this oscillation did not occur if the roll surface temperature was fixed. This oscillation might be related to the unfavourable aspect ratio of the element shape (the maximum aspect ratio used was about 1:1000 in thickness to width). In the streamline weighting function, the velocity components in a particular element are assumed to be essentially constant. In the thickness and rolling direction, the element sizes are not large relative to the velocity gradient. But in the width direction, this assumption might not be appropriate for the flat rolling problem. To suppress this oscillation, a smoothing treatment was applied. Figure 5 shows an example of temperature oscillation in width direction on the contact surface. The smoothing treatment was conducted only for the strip thermal analysis and it seems to be working successfully.

Figure 6 shows the flow chart of the rolling analysis including deformation and thermal analysis. Initial values for the mechanical parameters were obtained from the results of the two dimensional analysis. Thermal analysis of the roll was not necessarily performed each time and no more than 10 times was enough to achieve convergence in this algorithm. This routine was iterated until the change of temperature of all nodes in the strip and roll became smaller than 1 °C.

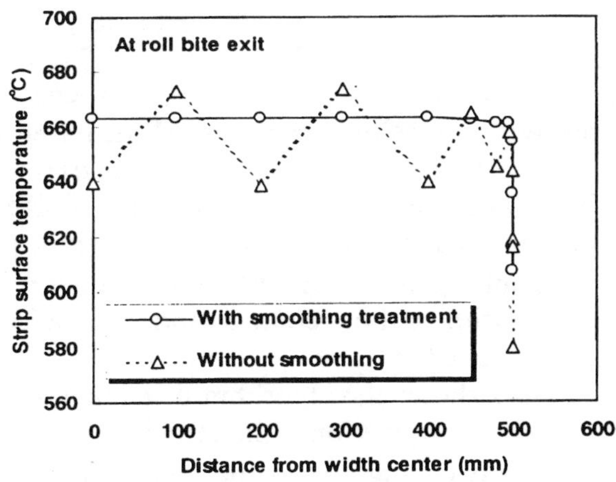

Fig. 5. Temperature oscillation in width direction on the strip surface.

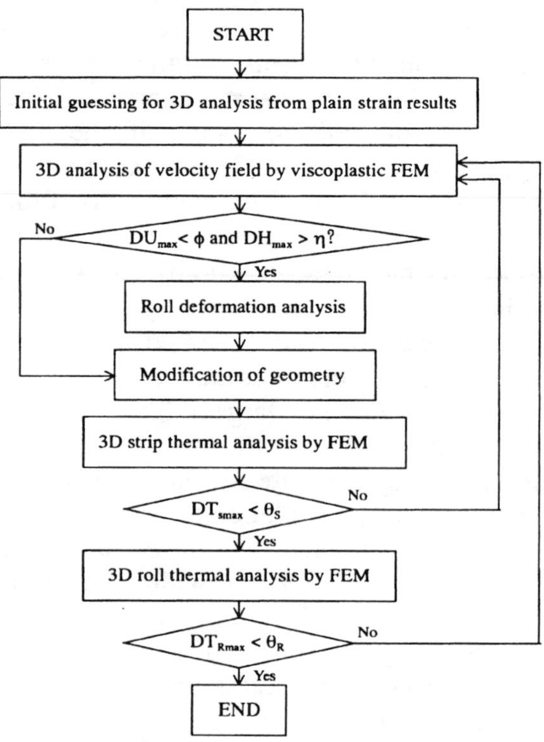

DU_{max}: maximum error norm of velocity
DH_{max}: maximum error of strip thickness profile
DT_{Smax}: maximum error of strip temperature
DT_{Rmax}: maximum error of roll temperature
η, ϕ, θ_s, θ_R : criterion for each parameter

Fig. 6. Schematics of diagram for 3D rolling analysis.

3.3 Microstructure Model

The microstructure model for A36 steel[5] was chosen for the analysis of austenite grain size evolution between stands, as shown in Table 1. Thermal analysis during the interstand was carried out by two dimensional transient FEM, based on the mesh system from the deformation analysis in the roll-bite. In the table, d_0 is austenite grain size at the roll-bite exit, t is the time from the roll-bite exit, R is gas constant, γ_{gb} is the grain boundary energy which decreases with carbon content and here a constant value 0.7 was adopted.

Figure 7 shows the algorithm developed to handle partial recrystallization. Increment in fraction recrystallized during Δt is the newly recrystallized austenite grains and this fraction is treated as a new sub-element. In the new sub-elements, recrystallized grain size is calculated and grain growth is calculated

Table 1. Employed microstructure equation.

Phenomenon	Model equation
Boundary strain rate and critical strain	$\varepsilon_c = (5/6)\varepsilon_P$ $\varepsilon_P = 0.01318 d_0^{0.174} \dot{\varepsilon}^{0.165} \exp(2926/T)$ $\dot{\varepsilon}_{SS} = \exp(38.167 - 0.0154 d_0 - 42747/T)$
Recrystallization kinetics	$X_{rex} = 1 - \exp\left[-0.693 (t/t_{0.5})^n\right]$ *static* : $t_{0.5} = 8.31 \times 10^{-15} d^{1.5} \varepsilon^{-1.5} \dot{\varepsilon}^{-0.33} \exp(263000/R/T)$ $n = 2$ *metadynamic* : $t_{0.5} = 2.13 \times 10^{-6} \dot{\varepsilon}^{-0.68} \exp(132800/R/T)$ $n = 1$
Recrystallized grain size	$d_{rex} = 88.96 d_0^{0.369} \varepsilon^{-0.368} \exp(-28060/R/T)$ *If* $d_{rex} > d_0$ *then* $d_{rex} = d_0$
Austenite grain size	$d_\gamma = \sqrt{d_\gamma^2(t_0) + 3\gamma_{gb} b^2 \int_{t_0}^{t} \frac{D_{gb}(T(t'))}{kT(t')} dt}$ $D_{gb} = 8.9 \times 10^{-5} \exp(-A/T)$

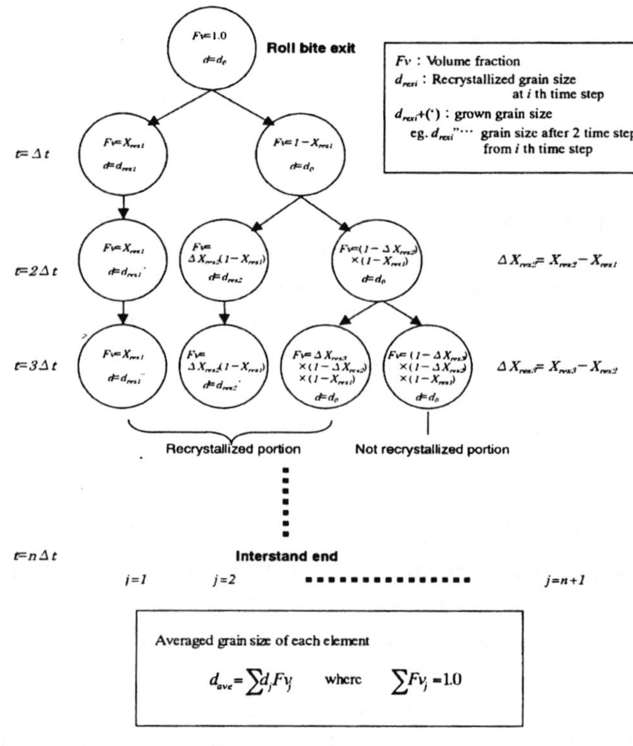

Fig. 7. Schematic for the algorithm for partial recrystallization.

for the rest of the elements. In this method, a new sub-element is generated at each time step in each element. At the end of each interstand, average grain size in each element is calculated from the grain size and the volumetric fraction of each sub-element in each element.

If recrystallization is not completed in some elements, residual strain is calculated by Eq.(3) and transferred to the deformation analysis of the next rolling stand.

$$\varepsilon_j = \varepsilon_j^{\ 0} + \lambda(1.0 - X_{rex})\varepsilon_{j-1} \qquad (3)$$

where λ is a material dependent constant and 0.5 was assigned for plain carbon steel.

4. RESULTS AND DISCUSSION

4.1 Conditions Employed

Thermal constants and analysis conditions employed are shown in Tables 2-4. In this study, simulation was carried out for 3 stands of tandem hot-strip rolling. Initial strip thickness was 30mm and reduction in thickness at each rolling stand was 30%. Initial strip width was 1000mm and rolling speed of the No.1 stand was 80mpm, boundary temperature on the upstream boundary of the No.1 stand was assumed 1050°C. 200 μm was assumed for initial austenite grain size. These values change through each deformation and thermal analysis.

4.2 Influence of Heat-Transfer Coefficient on Temperature Distribution

Under the rolling conditions for the No.1 stand, the differences in temperature distribution on the strip surface were compared when uniform or pressure dependent h_c was adopted on the contact surface. Fig. 8 shows temperature drop on the strip surface from initial temperature with uniform h_c. In this case, temperature distribution in width direction is almost flat except for the strip edge. The temperature drops steeply around the strip edge and this temperature drop is caused by the temperature distribution of the roll surface which has no contact with the strip. Figure 9 shows the distribution of temperature on the roll surface and it is evident that the temperature outside the strip edge changes little.

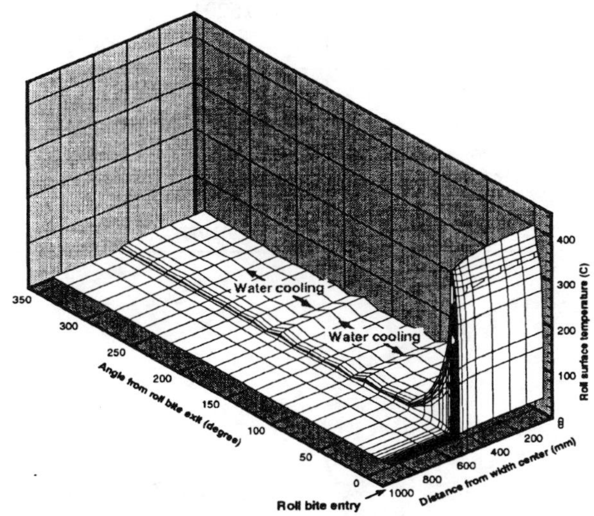

Fig. 9. Temperature drop on the roll surface when uniform h_c was employed.

On the other hand, when local pressure dependent h_c was coupled, the temperature drop decreases toward the strip edge in proportion to the roll pressure (Fig. 10). But the temperature change around the strip edge is much bigger than that in Fig. 8. This is because the roll pressure has a peak at the strip edge when very fine meshes are deployed there. Fig. 11 shows equivalent strain, flow stress and roll pressure around neutral region on both the contact

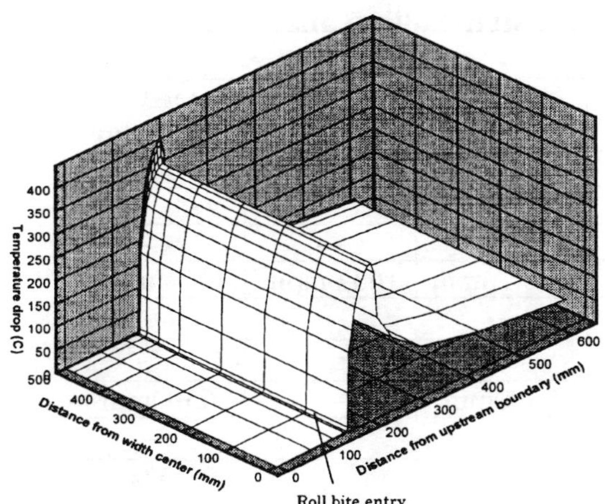

Fig. 8. Temperature drop on the strip surface when uniform h_c was employed.

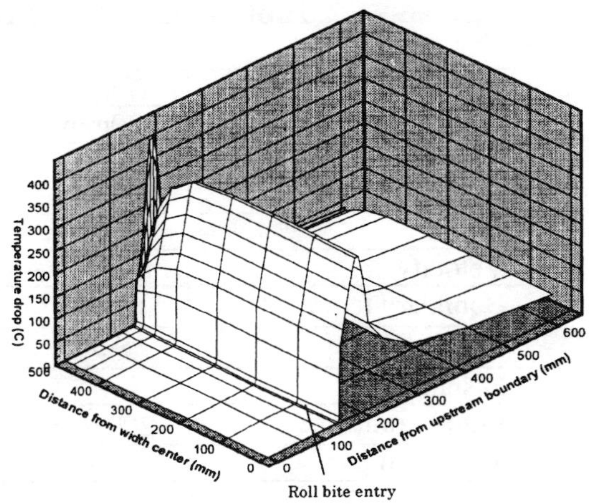

Fig. 10. Temperature drop on the strip surface when pressure dependent h_c was employed.

Table 2. Conditions employed for rolling analysis.

Flow stress for A36[5]	$\sigma = \frac{1}{\alpha}\sinh^{-1}\left[\left(\frac{\bar{\dot{\varepsilon}}\exp[Q/(RT)]}{A}\right)^{\frac{1}{n}}\right]$ $\alpha = \frac{\alpha_{ss}}{1-0.5\exp\left(\varepsilon/\varepsilon_h\right)}$ $\varepsilon_h = 0.5201 - 7.42\times10^{-4}T_C + 30\times10^{-7}T_C^{\ 2}$
Penalty number	10^7
Criterion for convergence	ϕ: 0.00005 θ : 1°C η : 0.5 μm

WR and BUR barrel length	1800mm
WR neck diameter	500mm
BUR diameter	1500mm
Chock distance	3000mm
WR bending force	0 Tonf/chock
Young's module	210GPa
Poisson's ratio	0.3

Numbers of discritization
Strip : rolling direction 5+**10**+5
thickness direction 6 for deformation analysis
11 for thermal analysis
width direction **11**
Roll : radius direction 13
circumferencial **10**+30
width direction **11**+7
* Bold letters mean common numbers for strip and rolling roll

Table 3. Thermal constants employed in the analysis.

Thermal conductivity strip roll	25.12 W/m/K 46.52 W/m/K
Specific heat strip roll	627.9 J/kg/K 440.0 J/kg/K
Density strip roll	7850 kg/m^3 7850 kg/m^3
Emissivity strip roll	0.60 0.20

h_c for air cooling	0.02326 KW/m^2/°C
h_c for water cooling	11.63 KW/m^2/°C
h_c between WR and BUR	23.26 KW/m^2/°C
h_c between WR and strip	Eq.(2)
Temperature air ,water BUR	20℃ 20℃
WR cooling arrangement	$\theta_1, \theta_2 = 90°$ $\theta_3, \theta_4 = 75°$

Table 4. Employed conditions for tandem hot-strip rolling analysis.

	1st stand	Interstand	2nd stand	Interstand	3rd stand	Interstand
Initial thickness	30mm	21mm	21mm	14.7mm	14.7 mm	10.29mm
Reduction	30%	---	30%	---	30%	---
Friction coefficient	0.25		0.25		0.25	
Width	1000mm	---	---	---	---	---
Velocity	80mpm	84.5mpm	114.0mpm	120.7mpm	163.0mpm	172.4mpm
Boundary temperature	1050℃	---	---	---	---	---
Yield stress	A36		←		←	
Distance	---	5500mm	---	5500mm	---	5500mm
Initial grain size	200 μ m	---	---	---	---	---
Time step	---	0.005sec	---	0.003sec	---	0.002sec

surface and the center of the strip thickness. Equivalent strain rises sharply at the strip edge on the contact surface due to large shear deformation. This leads to the pressure peak at the strip edge via the flow stress increase and the temperature drop is amplified in conjunction with the pressure peak. The magnitude of this peak depends on element sizes around the strip edge.

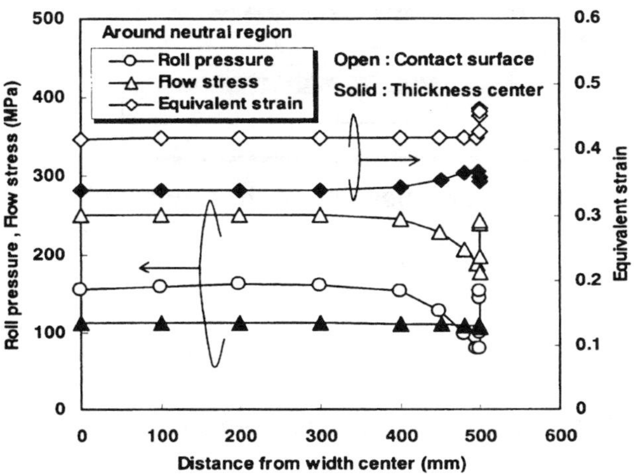

Fig. 11. Roll pressure peak at the strip edge.

4.3 Influence of Roll Deformation on Strain and Temperature

Elastic deformation of the rolls alters the pressure distribution on the contact surface and this effect is expected to have an influence on the temperature distribution when the pressure dependent heat-transfer coefficient is adopted. Figure 12 shows the differences in temperature on the contact surface and at the centerline at the roll-bite exit with and without roll deformation. Roll deformation seems to have some influence on the temperature distribution but it is negligible under this rolling condition. This effect is expected to be larger for thin gauge strip rolling.

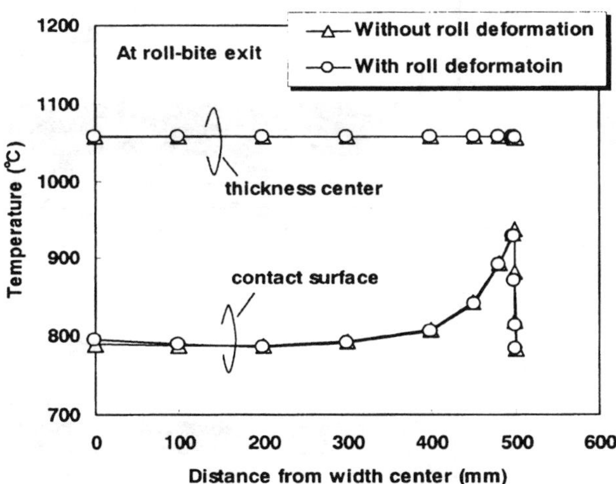

Fig. 12. Influence of roll deformation on temperature distribution.

4.4 Microstructure Analysis

From the above discussion, the heat-transfer coefficient on the contact surface alters the temperature distribution but roll deformation has little effect on mechanical parameters in hot-strip rolling. From this standpoint, the effects of the heat-transfer coefficient on microstructure evolution through hot-strip rolling were focused on and the differences in microstructure at the edge versus the center were discussed.

Figure 13 is an example of the evolution of the fraction recrystallized in the interstand between the No.1 and No.2 stand. In this case, local pressure dependent h_c was employed in the rolling analysis of the No.1 stand. It can be seen that the recrystallization rate around a quarter region in the thickness direction and the strip edge is faster than that of the strip surface and strip center. In the hot-strip rolling, strain has a peak just beneath the strip surface[10] and this is a main cause of the distribution in Fig. 11. In this case, recrystallization is not completed across the cross-section and strain remains inhomogeneously.

Figure 14 shows austenite grain size at the entry of each rolling stand. In Case A, uniform h_c was employed and roll deformation was not considered. On the other hand, local pressure dependent h_c was used and roll deformation was coupled in Case B. Only a small difference can be seen around the strip edge between Case A and Case B. In general, high frictional condition between the rolls and the strip surface causes large shear deformation beneath the strip surface. This leads to a grain refinement because strain is one of the main driving forces for recrystallization and grain refinement. In terms of width direction, only a small change can be seen, especially around the strip thickness center at the entry of the No.2 and No.3 stand. Deformation in the width direction can be caused around the strip edge and this width broadening becomes larger with thick strip. This makes small changes in austenite grain size in width direction.

Figures 15 and 16 show changes in temperature,

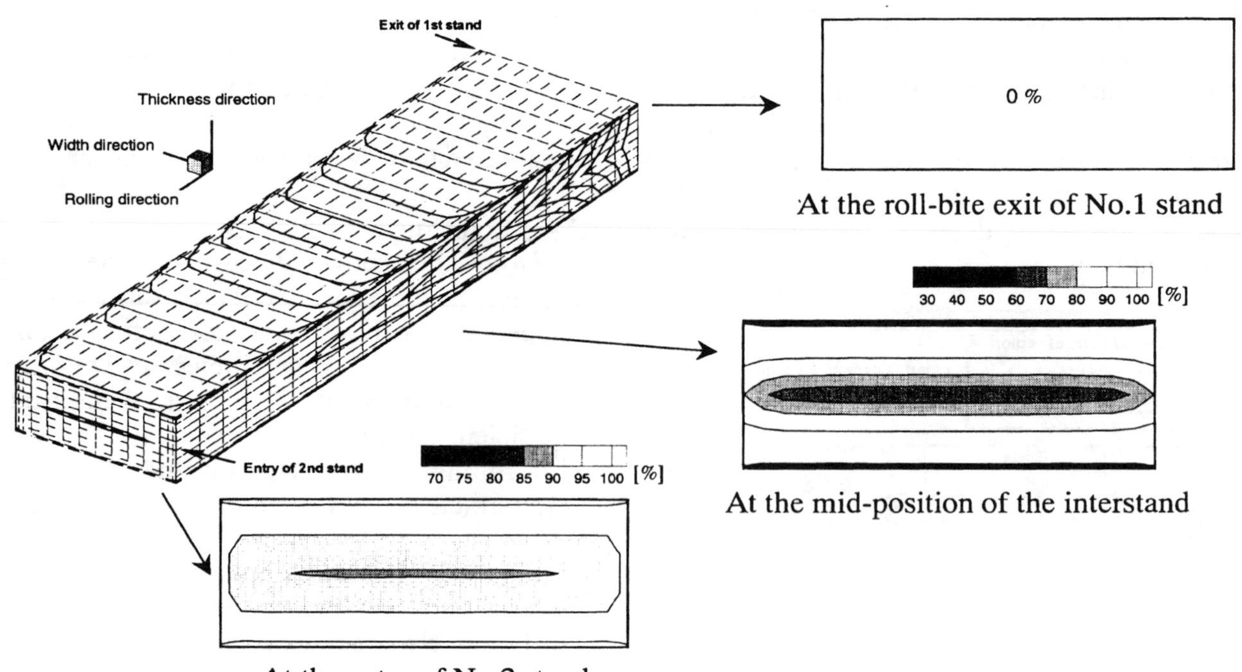

Fig. 13. Evolution of the fraction recrystallized during the interstand (whole cross-section).

	Case A	Case B
No.2 entry	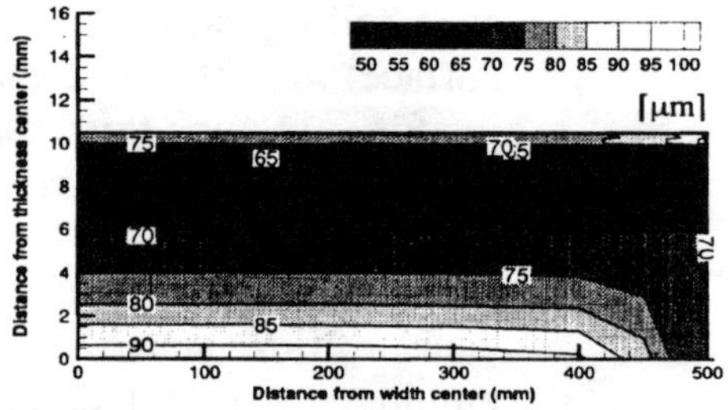	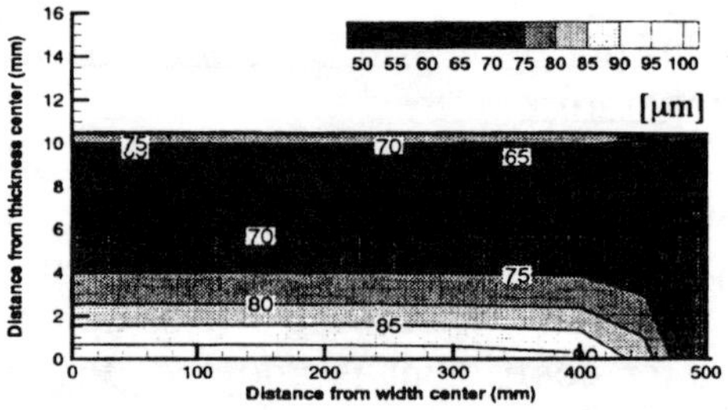
No.3 entry	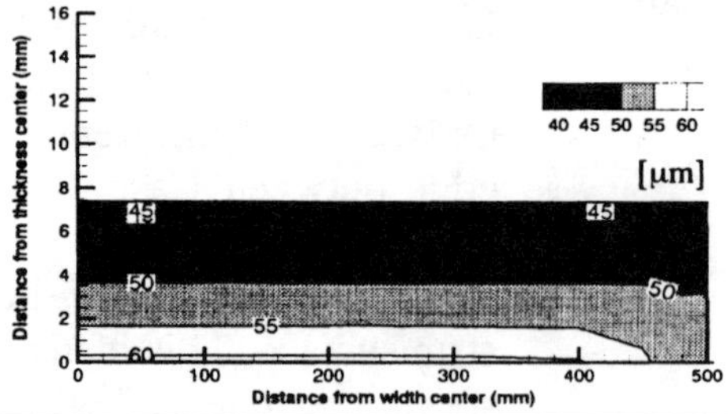	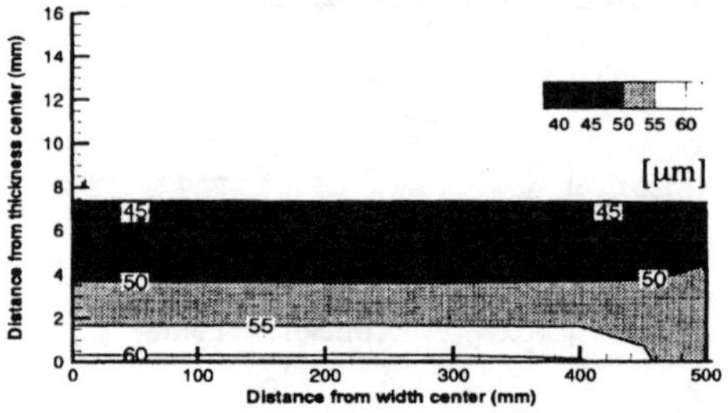
Virtual No.4 entry	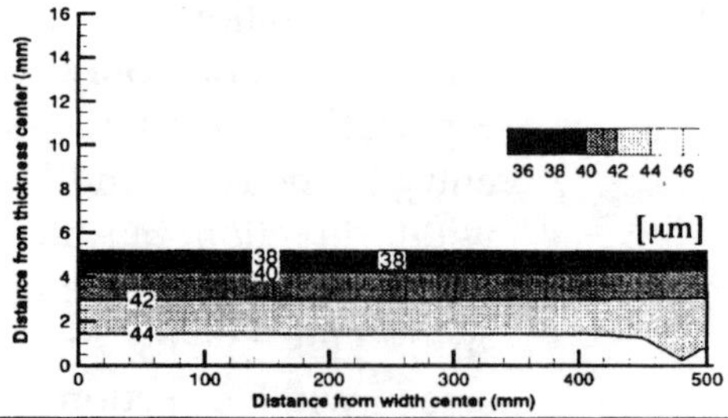	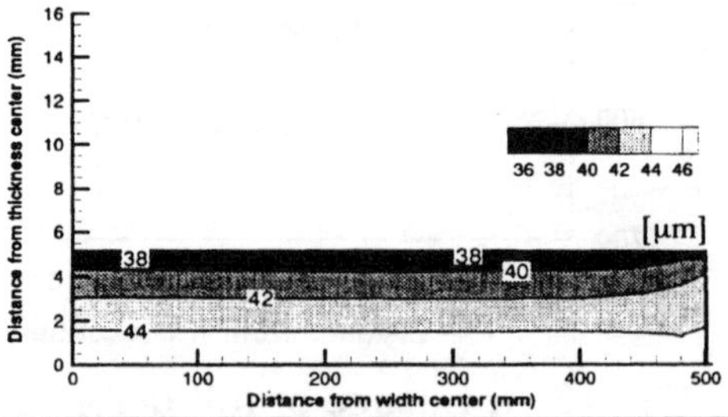

Fig. 14. Austenite grain size distribution at the entry of each rolling stand.

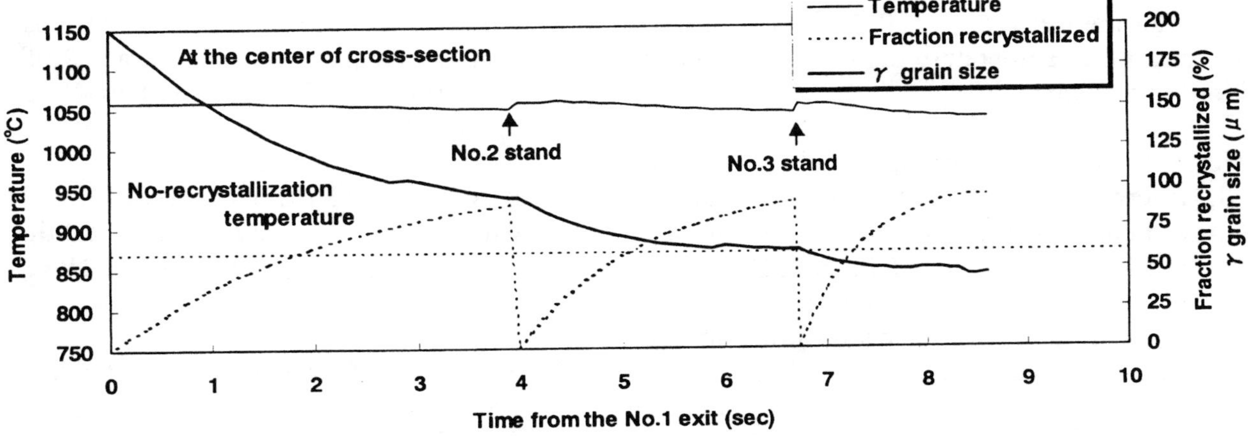

Fig. 15. Austenite grain size evolution at strip center.

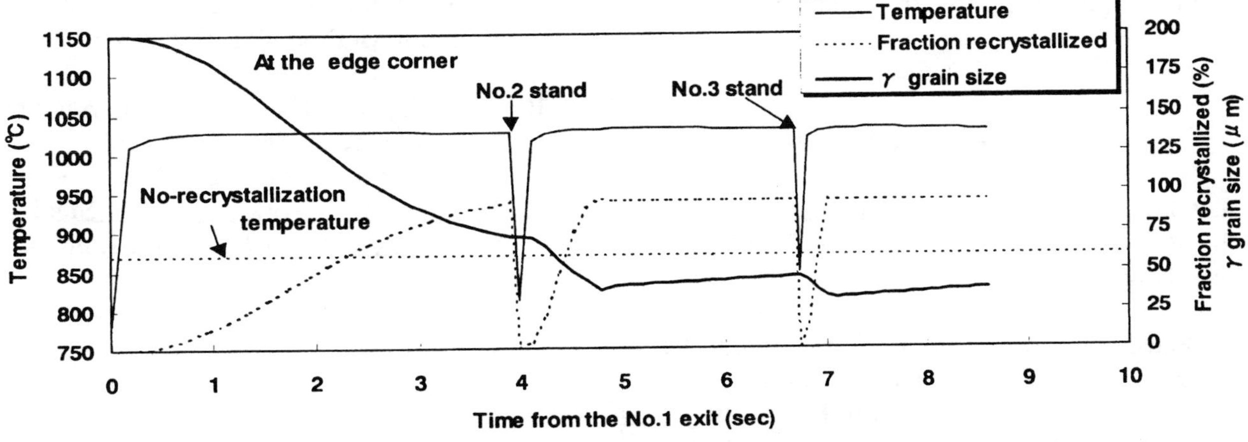

Fig.16. Austenite grain size evolution at the edge.

fraction recrystallized and austenite grain size after the exit of the No.1 stand at the edge and center. In this analysis, lower limit temperature for recrystallization T_{lim} was considered (870°C). At the edge of the strip, temperature at the roll-bite exit becomes lower than T_{lim} and the onset of recrystallization was retarded. In contrast, temperature changes at the centre are very small and recrystallization occurs just after the roll-bite exit. At the strip edge, the recrystallization rate is accelerated after the exit from the No.2 and No.3 stands and recrystallization is completed within 1sec. After that, the grains grow very slowly.

From the results, it can be concluded that the evolution of austenite grain size has some variation in the cross-section of the strip but only a small difference is evident for the conditions studied. But in the analysis, homogeneous temperature and austenite grain size were assumed for the initial conditions and this condition seems to be different from actual ones. Temperature on the strip surface, especially around the edge, should be lower due to radiation and heat transfer and this could cause the retardation of recrystallization and changes in austenite grain size distribution.

5. CONCLUSIONS

A. A three-dimensional steady-state deformation analysis based on visco-plastic FEM was developed for hot strip to the roll by upwinding FEM. In the model, streamline Petrov-Galerkin weighting function was introduced to suppress the temperature oscillation in the direction of heat flow. It was demonstrated that the temperature oscillation in the width direction could be dealt with by simple temperature averaging treatment. It appeared that the heat-transfer coefficient on the contact surface has a significant influence on the distribution of temperature and it influences the temperature recovery after the roll-bite exit. Roll deformation has little effect on the temperature distribution under the conditions investigated.

B. Microstructure model for A36 steel was coupled with the thermal analysis model to simulate the metallurgical changes. If the pressure dependent heat-transfer coefficient is employed in the 3D hot-strip rolling analysis, it has an influence on the temperature distribution in the width direction even after the roll-bite exit during whole interstand. It turns out that the evolution of austenite grain size varies in the cross section of the strip, but differences are small. But this may yield very interesting results when phase transformation from austenite to ferrite occurs.

REFERENCES

1. H. Cho, N. Kim, "A Study on the Evolution in Hot Rolling", Proceedings of the 5th ICTP, 1996, pp. 91-94.

2. P. Paushkar, R. Shivpuri, "Integrated Micro-structural-Phenomenological Approach to the Analysis of Roll Pass Design in Bar Rolling", 40th Mechanical Working and Steel Processing Conference, ISS, 1998, XXXVI, pp. 755-771.

3. W.C. Chen, I.V. Samarasekera, E.B. Hawbolt, "Fundamental Phenomena Governing Heat Transfer During Rolling", Metallurgical Transactions A, Vol. 24A, June, 1993, pp. 1307-1320.

4. C.O. Hlady, J.K. Brimacombe, I.V., E.B. Hawbolt, "Heat Transfer in the Hot Rolling of Metals", Materials and Metallurgical Transaction B, Vol. 26, pp. 1019-1027, 1995.

5. I.V. Samarasekera, D.Q. Jin, J.K. Brimacombe "The Application of Microstructural Engineering to the Hot Rolling of Steel", 38th Mechanical Working and Steel Processing Conference, ISS, 1997, XXXIV, pp. 313-327.

6. G.J. Li, S. Kobayashi, "Rigid-Plastic Finite Element Analysis of Plane Strain Rolling Journa of Engineering for Industry, Feb.,1982, Vol. 104 pp. 55-64.

7. K.N. Shohet, N.A. Townsend, "Roll Methods o Crown Control in Four-high Plate Mills", Journa of The Iron and Steel Institute, Nov., 1968, pp 1088-1098.

8. Y. Tozawa, T. Ishikawa, N. Iwata: "Predicting th Profile of Rolled Strip", Numiform '82, 1982 pp.787-.

9. C.O. Zienkiewicz, R.L. Taylor, The Finit Element Method 4th Edition, Vol. 2, pp. 454.

10. D.Q. Jin, R.G. Stachowak, I.V. Samarasekera, J.K Brimacombe, "Mathematical Modelling o Deformation during Hot Rolling", 36 Mechanical Working and Steel Processin Conference, ISS, 1995, XXXII, pp. 401-407.

TUBULAR PRODUCTS

ON-LINE ACCURATE MEASUREMENT OF SEAMLESS TUBES

A. Glüzmann
George Kelk Corporation
48 Lesmill Road
Toronto, Ontario, M3B 2T5 Canada
Phone: (416) 445-5850
Fax: (416) 445-5972
E-mail: gluzmann@kelk.com

Key Words: Length measurement, tubes, laser velocimeter

ABSTRACT

In most rolling processes, it is highly desirable to use contactless measuring systems to avoid the risk of marking or damaging the surface of the material being processed. The ACCUSPEED laser velocimeter from KELK Corp. is a contactless optical gauge -laser based- for accurate velocity and length measurement in hot or cold rolling processes.
ACCUSPEED can be utilized in flat products as well as in bars, profiles, and tubes. ACCUSPEED has been proven to be a reliable instrument in the online length measurement of hot seamless tubes, demonstrating its ability to produce accurate measurements even under difficult environmental conditions and large side movements of the tubes under measurement.

INTRODUCTION

Global competition imposes very high demands on producers working to remain economically competitive. New control systems and sensors are crucial to achieving the process and yield improvements necessary to maintain a leading position in the race for higher quality and lower costs.

In the manufacturing process, product velocity and or length has traditionally been determined by measuring the rotational speed or pulses of work and measuring rolls of known diameter. This technique is not only indirect, it is also subject to the adverse effects of slippage and roll wear on accuracy and, when measuring rolls are used, has the potential for damaging high quality surfaces and incurring high maintenance costs. Laser velocimetry, which was originally developed for the measurement of velocities in fluids, does not have these problems; it measures directly and without contact. Over the last decade, we have seen its increasing use for the measurement of velocity and length in the metals rolling, paper, plastic, wire and glass industries.

While all heavy industrial applications challenge the designers of precision instrumentation, the wide range of operating conditions, the high temperatures and the harsh environments encountered in the metals rolling industry are particularly demanding. Meeting these challenges has needed not only the use of superior optics and environmental protection, it has also required, in order to optimize performance, the development of a new approach to signal processing; a robust signal processing technique able to accurately and reliably extract velocity data from a highly variable and noisy optical signal.

The KELK prepietary signal processing enables the ACCUSPEED Laser Velocimeter to deliver superior performance in hot and cold mill applications, length measurement, mass flow Automatic Gage Control, transfer bar imaging and strip tracking for Crop Optimization, and Automatic Width Control.

Applications cover a considerable speed range, typically from 0.01 to 40 m/s. Laser velocimeters have been used with considerable success; however, because their signal processing techniques were based on fixed time sampling intervals, they experience a loss of accuracy as velocity is reduced due to the decreasing amount of available information. Since length is calculated as the integral of the velocity vs. Time, it is easy to understand that also the length accuracy suffers. Control systems designers have had to heavily filter the velocimeter output signal in order to achieve the desired velocity accuracy, degrading the length accuracy and response times of their systems.

Many materials conditions are encountered, from the rough hot surfaces of transfer bars at the exits of hot strip roughing mills to the shiny surfaces at the exits of cold mills. In consequence, the intensity and quality of the scattered light vary greatly. A very wide and responsive automatic gain control is required and the signal processing methodology must be able to handle the continually changing nature of the optical signal.

The optical beam is sometimes interrupted by water droplets, coolant or mist, temporarily obscuring the velocimeter's view of the target material. The signal processing methodology must ensure that velocimeter operation is not disturbed by such interruptions and, by rapidly compensating for them, that accuracy is maintained.

ACCUSPEED PRINCIPLE OF OPERATION

A laser velocimeter determines velocity and length from the frequency component of the light returned from laser beams directed at the target, Figure 1.

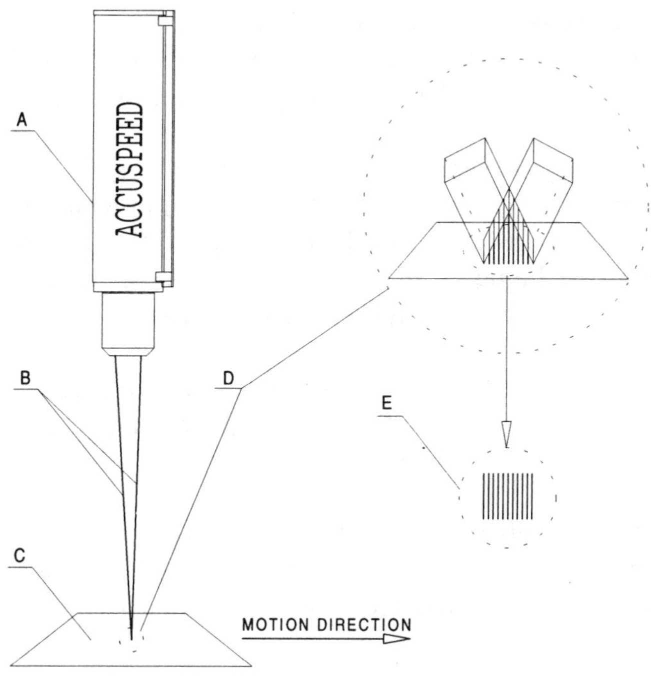

Fig.1: Principle of Operation

The ACCUSPEED optics head (A) emits two mutually coherent laser beams (B) which intersect on the surface to be measured (C). In the intersection zone (D), the interference of light produces alternate dark and bright vertical planes. These planes produce a fringe pattern (E) on the surface being measured. As the irregularities on the measured surface (speckles) move across the fringe pattern (E), they scatter light when passing a bright interference fringe. This modulates the output of a photodetector located in the head (A) at a frequency which is directly proportional to the velocity of the measured surface. The DC component of the photodetector signal is used to determine when material is present in the intersection zone (D), i.e. in the measurement region. Operation of the laser velocimeter is sometimes explained by the Doppler effect. Both travelling speckle pattern and Doppler theories arrive at the same final equation.

SIGNAL PROCESSING

Laser velocimetry has traditionally used Fourier transform and zero crossing correlation techniques to derive the velocity dependant frequency from the photodetector output. Both are based on fixed time interval sampling. As speed decreases and the number of cycles falls, accuracy is lost. While not significant at higher velocities, this does adversely affect accuracy at the lower velocities encountered in many metals rolling applications. Further, if a disturbance occurs, part of the signal is lost during the sampling interval, causing a loss of accuracy. Means have been included to compensate for the effects of disturbances.

An alternative frequency measurement technique, inherently more suitable for laser velocimeters measuring in the speed ranges encountered in metals rolling, is now possible due to the advent of more powerful DSP's (Digital Signal Processors) and EPLDs (Erasable Programmable Logic Devices). These devices, used with stable clocks, permit the accurate period measurement of high frequency signals.

By determining velocity over a set number of cycles (which is equivalent to a fixed length of material travel), rather than over a fixed time interval, sample size is constant and there is no loss of accuracy at low velocities.

Also, measuring a set numbers of samples allows for a fast, dynamic and self tuning algorithm. The loss of individual cycles has a much smaller impact on velocity measurement, making this alternative techniques very robust.

EXPERIMENTAL

Accurate length measurement of seamless tubes -as they come from the mill to the cooling bed- is an essential part of the manufacturing process, to allow the crop shear operator to optimize cuts while satisfying customer orders.
There are a number of considerations to be made and take care in this type of measurement. The location of the laser has to be in an area where hopping and side movements are not bigger that 10-20 % of the diameter to be measured. ACCUSPEED has the largest depth of field in the market with the highest accuracy (200 mm (8 in) at 0.05% accuracy) allowing the processing of a large range of products without having to move the position of the laser velocimeter.
Slight geometrical deformation of the head or tail end of the tubes (so called "hook") can also adversely impact the length accuracy of the laser system. In order to minimize their effect hot metal detectors (HMD) can be used to trigger the start/stop length acquisition mode of the ACCUSPEED. Two HMDs -one located 2-3 meters upstream and another one 2-3 meters downstream from the ACCUSPEED - are used to start/top length

acquisition, assuming that the defect (hooks) may be present at the head and the tail end of the tube.
Furthermore, the second HMD for the tail also helps if there is excessive movement on the tube when released from the mill.
The same start/stop principle could be utilized in hot bars and profiles, while in cold applications optical barriers should be used.

The experimental data reported in this paper has been logged from one ACCUSPEED installed in a seamless tube mill between the exit of the mill and the cooling bed. An HMD has been used for the start length acquisition, and the detection capability of ACCUSPEED (Material in View) has been used to stop the process at the tail.

Several runs of different tube diameters were measured. Table I contains data from a run of twenty (20) tubes, 33 mm in diameter, approximately 61 m long. The data has been provided by ACCUSPEED (A) (column 1 Table I), and manually (M) (measured with laser range finder and measuring tape, column 2, Table I) - after entering the cooling bed. Their individual temperatures were also measured using a pyrometer prior entering the cooling bed for shrinkage correction of the Accuspeed readings.

On Table I:

Column 1= A
Column 2= M
Column 3= A-M
difference between the ACCUSPEED and the manual measurement
Column 4= A-A'
difference between the ACCUSPEED reading (hot) and the shrinkage compensated length (A') (characteristic of the material being rolled).

Column 5= A'-M
Error or difference between shrinkage compansated ACCUSPEED measurement and manual.

Column 6= (((A'-M)/A)*100)
Accuracy as a percentage, %.

Table I.

Accuspeed measurements (mm), hot (A)	Manual measurement (mm), cold (M)	Difference between Accuspeed and Manual (A-M)	Difference between Accuspeed compensated and Manual (A-A')	Error (mm) (A'-M)	Accuracy (%) (A'-M)/A
61030.9	60690	340.9	327.29	13.61	0.02
61522.4	61160	362.4	329.93	32.47	0.05
62247.4	61880	367.4	333.82	33.58	0.05
61244	60890	354	328.44	25.56	0.04
61711.6	61370	341.6	330.94	10.66	0.02
61658.9	61330	328.9	330.66	-1.76	-0.00
61524.6	61210	314.6	329.94	-15.34	-0.02
61998.8	61670	328.8	332.48	-3.68	-0.01
61545.9	61210	335.9	330.06	5.84	0.01
61681.4	61370	311.4	330.78	-19.38	-0.03
61917	61580	337	332.05	4.95	0.01
61835.8	61530	305.8	331.61	-25.81	-0.04
61003.7	60700	303.7	327.15	-23.45	-0.04
61288.3	60940	348.3	328.67	19.63	0.03
62060.3	61730	330.3	332.81	-2.51	-0.00
61605.7	61280	325.7	330.38	-4.68	-0.01
61501.7	61170	331.7	329.82	1.88	0.00
60810.4	60510	300.4	326.11	-25.71	-0.04
61134.8	60830	304.8	327.85	-23.05	-0.04
61816.8	61500	316.8	331.51	-14.71	-0.02
	Average	329.52	330.12	-0.60	-0.001
	Standard Deviation	19.55	2.04	18.99	0.03
	Máximum	367.40	333.82	33.58	0.05
	Mínimum	300.40	326.11	-25.81	-0.04

CONCLUSIONS

This novel signal processing approach has been demonstrated to deliver superb length accuracy in measuring the length of seamless tubes as they come from the hot mill to the cooling bed, outperforming the traditional approach of constant time interval sampling LDVs. The described approach and mode of operation enables the ACCUSPEED Laser Velocimeter to deliver optimum performance in both hot and cold applications, including bar, billets, profiles, transfer bar imaging and strip tracking for Crop Optimization, Automatic Width Control, and mass flow Automatic Gauge Control.

ACKNOWLEDGEMENTS

The author wishes to thank Bob Ebrahim, George Kelk Corp., and E.Magnani, SIDERCA for their valuable contribution to this paper.

REFERENCES

Drain, L.E., "The Laser Doppler Technique", John Wiley & Sons Ltd, 1980.

A. Glüzmann & R. Ricciatti, "State of the Art Laser Velocimetry", IOM Communications, London, 1999.

A. Glüzmann & M. Wise, "A Novel signal Processing Approach To Laser Velocimetry", MET SOC, Quebec 1999.

Development of Rolling technique for Seamless Square Column

Hideo Sato, Nobuhiko Morioka, Masaharu Kita
Kawasaki Steel Corp. Chita works
1 Kawasaki-cho 1-chome
Handa/Aichi/475-8611
Japan
Tel.: 81-569-24-2031
E-mail: h-sato@kawasaki-steel.co.jp

Dr. Takaaki Toyooka, Takuya Nagahama
Kawasaki Steel Corp. Technical Research Lab.
1 Kawasaki-cho 1-chome
Handa/Aichi/475-8611
Japan

Dr. Yoshitomi Onoda,
Yamanashi Univ., Faculty of Eng.
3-11 Takeda 4-chome
Kofu/Yamanashi/400-8511
Japan

Dr. Takuo Nagamachi,
Tokushima Univ., Faculty of Eng.
1 Josanjima-cho 2-chome
Tokushima/Tokushima/770-8506
Japan

Key Words: Seamless pipe, Hot rolling, Sizing mill, Square column

INTRODUCTION

Square columns are generally used in the field of construction. In this case, the columns are welded to the beams. Concerning about connecting method, a diaphragm has been applied to support the horizontal forces as shown in Fig. 1. For the last few years, a new construction method has been adopted[*1), 2), 3)] as shown in Fig. 2. This new method is called beam-to-column connections reinforced by heavy gauge columns. In this method, heavy gauge reinforces connecting point of the beam to the column. In this case thickness at the connecting point is twice as heavier as that of the pillar. Kawasaki Steel has developed rolling technique for square seamless columns with Mannesmann plug mill process to supply heavy gauge seamless square columns. Our plug mill process consists of Mannesmann piercer, elongator, plug mill, reeler, and eight stands sizing mill as shown in Fig. 3. We can manufacture 7" to 16-3/4" seamless pipes and square columns with outside dimensions from 150mm x 150mm to 250mm x 250mm. We have calculated material deformation at the sizing mill by using FEM analysis and developed a new design of appropriate caliber dimensions and diameter reduction of rolls in the hot rolling process. After this development, we can manufacture heavy gauge seamless square columns with smaller corner radius, smaller corner length, lower concavity and uniform mechanical properties.

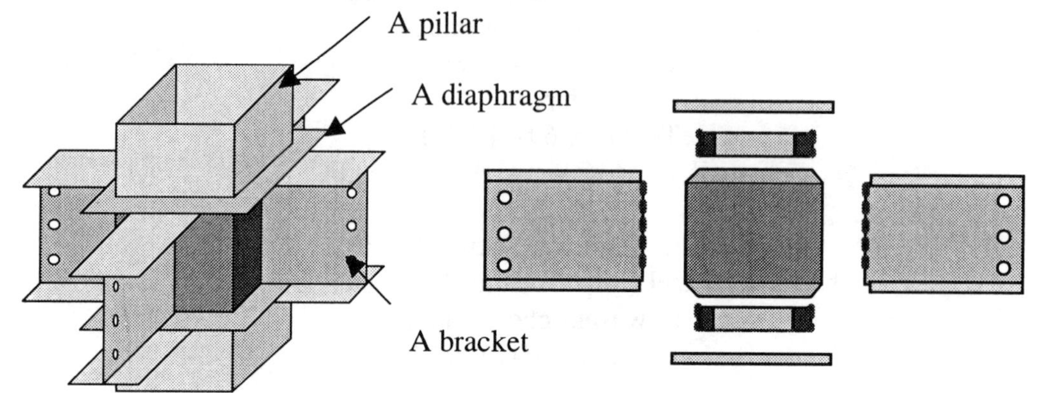

Fig. 1: Schematic illustration of connection used by conventional diaphragm method

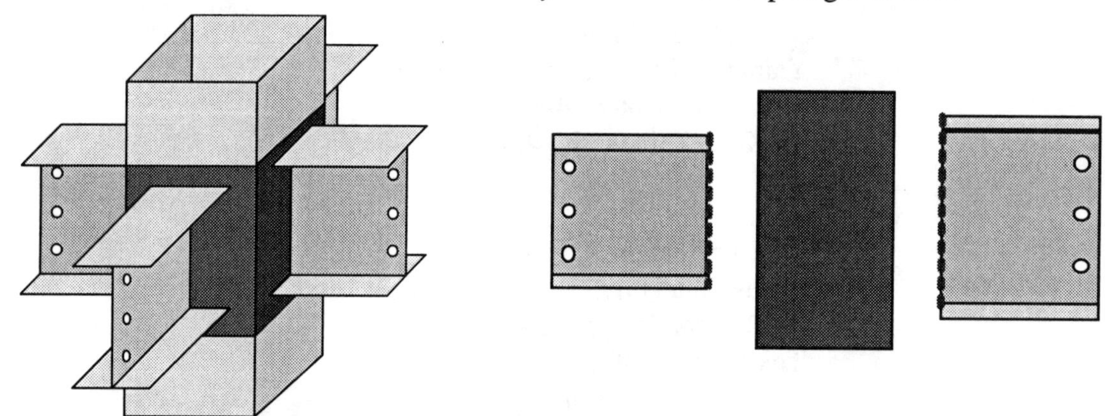

Fig. 2: Schematic illustration of connection reinforced by heavy gauge column

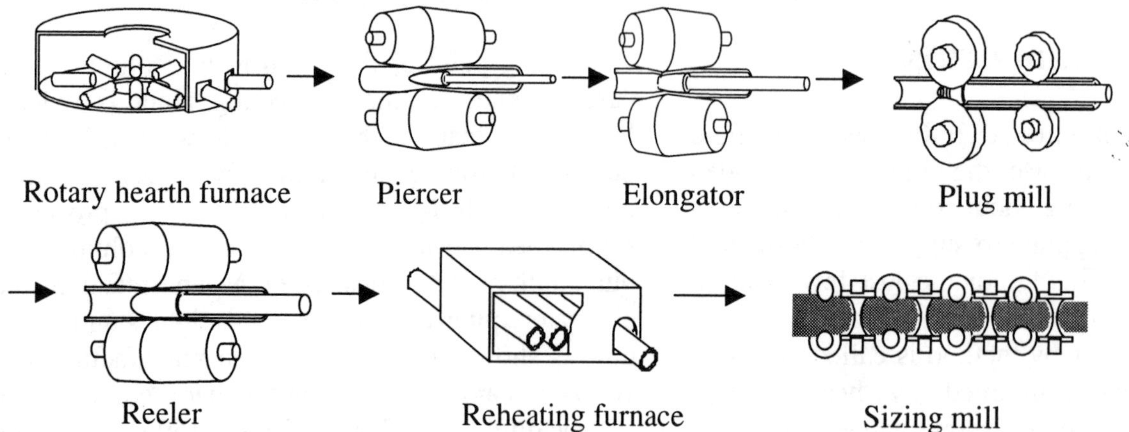

Fig. 3: Schematic illustration of medium diameter seamless mill

MANUFACTURING PROCESS OF HOT ROLLED SEAMLESS SQUARE COLUMN

Our new manufacturing process of seamless square columns is summarized as follows. Long billets manufactured at Mizushima works are transported to Chita works and cut into the required length at the billet cutting line. Each billet is heated in the rotary hearth furnace, and is pierced by Mannnesmann piercer. Then a hollow bloom is manufactured. A hollow bloom manufactured by piercer is expanded to increase in diameter in elongator. After plug mill, internal surface is reeled by reeler and semi rolled pipe is reheated in reheating furnace. The reheated pipe is rolled into round seamless pipe or square seamless column at sizing mill. At sizing mill, we can manufacture both round pipes and square columns by using conventional two roll type sizing mill without any installation of four roll type sizing mill which is usually used in the conventional roll forming process of manufacturing square pipes.

SPECIFICATIONS OF SEAMLESS SQUARE COLUMNS

In regard to the square seamless columns used for beam-to-column connections, the requirements are as bellow:

(1) Concerning about the connecting point of the beam to the column, wall thickness of the column is twice as heavy as the one of the columns used for pillars.

(2) Corner radius of the connections should be equal to that of the pillars to reduce the gap between connections and pillars. And the corner length also should be equal to that of the pillars. Then corner radius and the corner length of the connections should be less than 1.25t (t: wall thickness).

(3) Other dimensional requests are the same as that of normal square columns used for pillars.

ANALYSIS OF MATERIAL DEFORMATION AT SIZING MILL BY USING RIGID-PLASTIC FEM

We have calculated material deformation at sizing mill by using rigid-plastic finite element method and experimentally manufactured seamless square columns. The aim of this analysis is to obtain adequate roll caliber dimensions for manufacturing seamless square columns with heavy wall thickness, small corner radius, small corner length, low concavity, low convexity and etc.[*4), *5)]. Fig. 4 shows schematic illustration of sizing mill for seamless square columns. Cross sectional views at the exit of No.1 stand to No.4 stand at sizing mill are round profiles and this round pipe is reshaped into square columns at No.5 stand to No.8 stand. We conducted some experiments and calculations to investigate material deformations at the sizing rolls of seamless square columns under the conditions as shown in Fig. 5, Table 1 and Table 2. The curvature of the caliber of latter stands is bigger than that of the former stands. The curvature of the caliber is straight at the finishing stand.

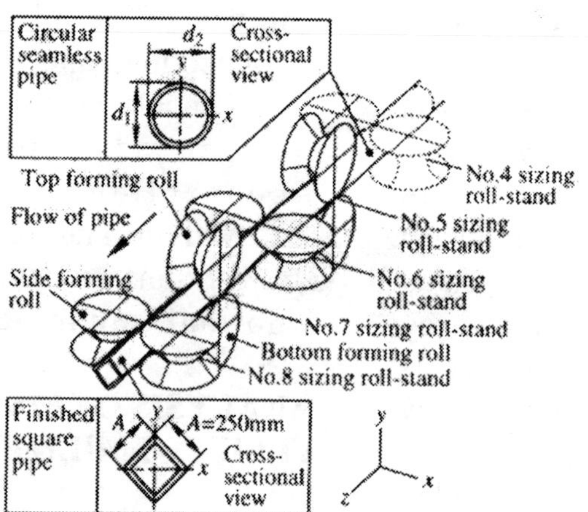

Fig.4: Schematic illustration of sizing mill for seamless square columns

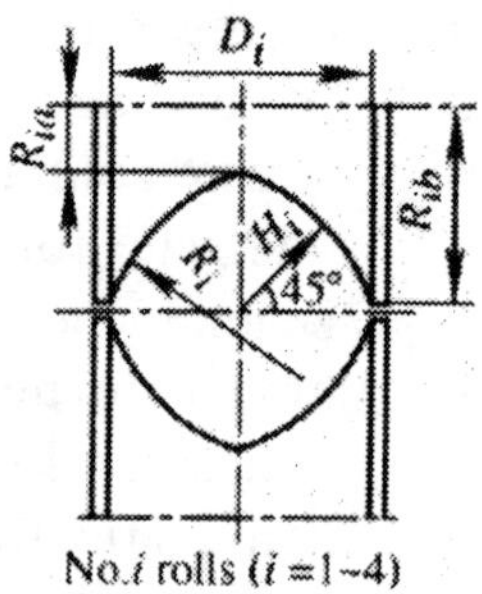

Fig. 5: Definition of roll caliber of sizing roll

Table 1: Dimensions of Seamless Square Columns and Reduction

Dimensions of Seamless columns	Outside Dimensions	150mm x 150mm ~ 250mm x 250mm
	Thickness	12mm ~ 36mm
Total reduction r_4		20% ~ 32%

Table 2: Dimensions of sizing rolls (outside dimension 250mm x 250mm)

Roll No. i	Ri/mm	Ria/mm	Rib/mm	Di/mm
1	509.0	193.3	367.1	349.6
2	1247.1	192.2		350.7
3	7961.4	192.8		349.4
4	-	192.8		348.6

COMAPRISON OF SQUARE COLUMN DIMENSIONS BETWEEN CALCULATIONS AND EXPERIMENTS

We evaluated cross-sectional shape of finished square columns. Concerning about seamless square columns used for beam-to-column connections reinforced by heavy gauge, the most important requests are small corner length, small corner radius, and small concavity. These properties are defined in Fig. 6. Then we researched the relations between these properties and rolling conditions. We consider the diameter reduction of square seamless column as shown in Fig. 7. Fig. 8 shows the relation between the total reduction and the corner length. The bigger total reduction is given, the smaller corner length becomes. Fig. 9 shows the relation between the corner radius and the total reduction. The corner radius has the same tendency as that of the corner length. Then it is clear that the total reduction should be higher to obtain smaller corner length and smaller corner radius. However manufacturing a square column with higher total reduction generates a hollow at the side of the finished square column. To evaluate the relation between rolling conditions and the hollow depth, we defined the concavity as the ratio of the hollow depth to the outside dimension. Fig. 10 shows the relation between the reduction increase at No.7 stand and the concavity. The smaller reduction increase at No. 7 stand is given the smaller concavity of finished square columns becomes. These properties are observed by both calculated results and measured ones. From the analysis of these calculated and measured results we consider this forming mechanism as follows.

At the first sizing stand, round pipe is curved and compressed. Then corners are created. The size of corner and the corner radius are proportional to the reduction increase of this stand. Once corners are created, it needs larger force to decrease the corner length and the corner radius. In case that excessive reduction is given at latter stand, bending strain at the center of the side becomes bigger and hollows are created at the sides. In conclusion higher total reduction is indispensable to obtain small corner length and small corner radius. Smaller reduction increase at No. 7 stand is needed to avoid hollows at the sides.

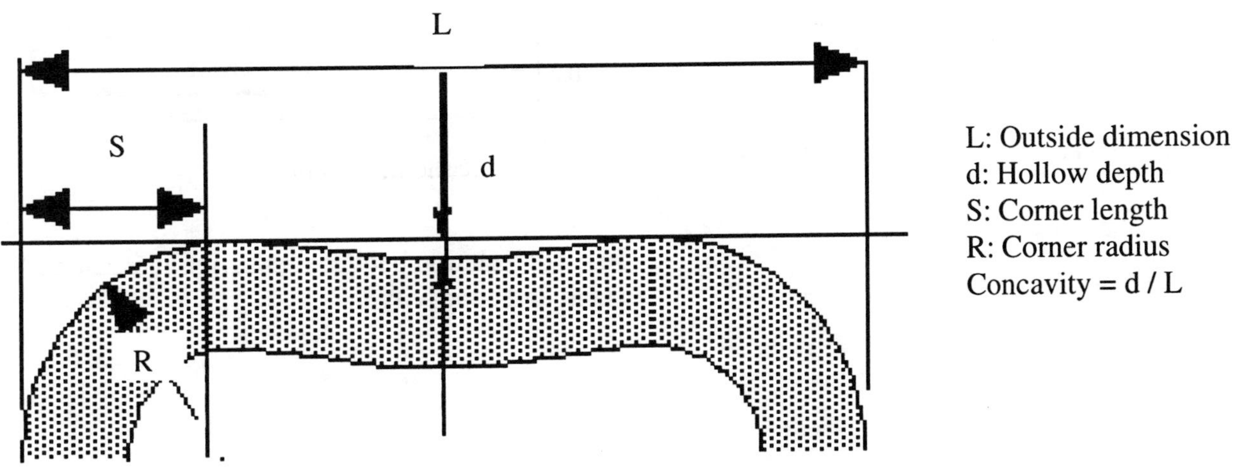

Fig. 6: Dimensions of heavy gauge seamless square column

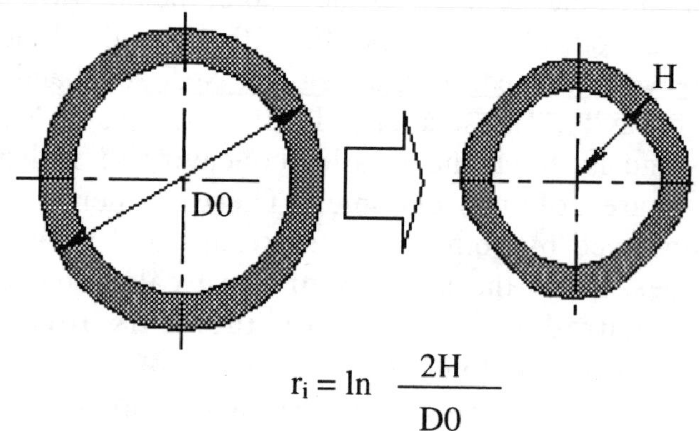

Fig. 7: Definition of reduction

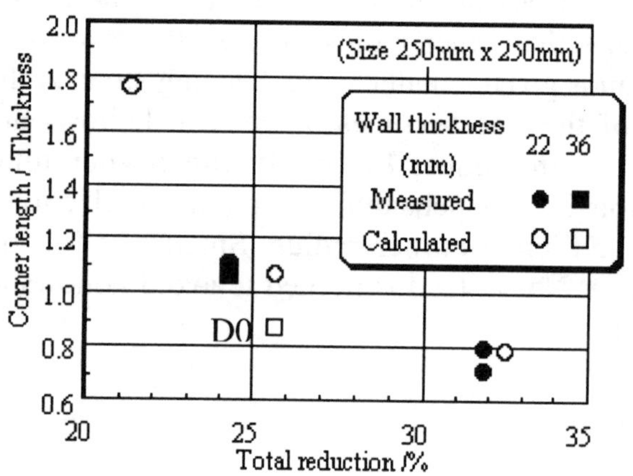

Fig. 8: Relation between total reduction and corner length

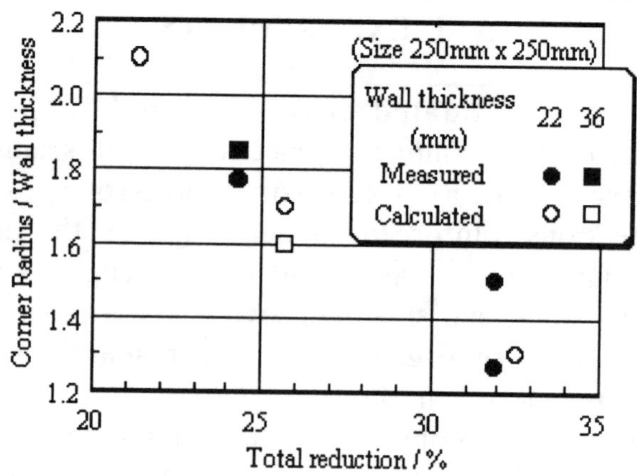

Fig. 9: Relation between total reduction and corner radius

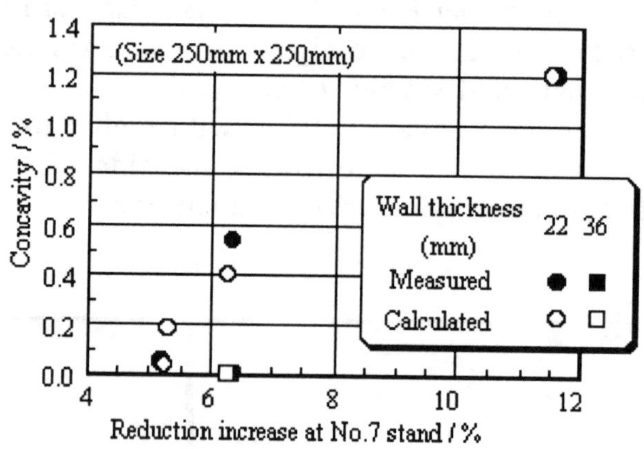

Fig. 10: Relation between reduction increase at No. 7 stand and concavity

FEATURES OF KAWASAKI STEEL'S SEAMLESS SQUARE COLUMNS

We have developed manufacturing technique for heavy gauge seamless square columns with small corner length, small corner radius, low concavity and so on. Table 3 shows features of our heavy gauge seamless square columns. We can manufacture three outside dimensions now. If other outside dimension is required it is possible to supply required seamless square columns. Dimensional tolerances of this seamless square columns satisfy the requirements of ASTM A501 and BS 4848 Grade 43DD and mechanical properties also meet both specifications. Fig.11 shows cross sectional views of heavy wall seamless square columns. All sections have small corner radii, small sizes of corner and low concavity.

Table 3: Features of Kawasaki steel's seamless square columns

		Developed	ASTM A501	BS 4848 Grade 43DD
Available Dimensions	Outside Dimensions	150mm x 150mm 200mm x 200mm 250mm x 250mm	-	-
	Thickness	10mm ~ 30mm	-	-
Tolerances	Outside Dimensions, Plus and Minus	1%	1%	1%
	Wall thickness: t	+12.5% - 12.5%	- 12.5%	+15 % - 12.5%
	Concavity/Convexity, Plus and Minus	0.5% of the side	-	1% of the side
	Size of Corners	Maximum 1.0t	Maximum 3t	0.5t ~ 2.0t
Mechanical Properties	Yield Strength	340 ~ 360 M Pa	Minimum 250M Pa	Minimum 265M Pa
(measured values)	Tensile Strength	520 ~ 534 M Pa	Minimum 400M Pa	430~580M Pa
	Yield Ratio	64 ~ 69 %	-	-
	Elongation	27 ~ 31 %	Minimum 20%	Minimum 20%
	Charpy V-notch impact test value (-30degrees centigrade)	142 ~ 193 J	-	Minimum 27J

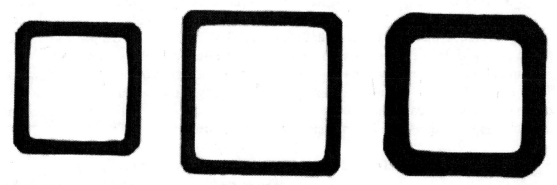

a) 200mm x 200mm x 22mmt b) 250mm x 250mm x 22mmt c) 250mm x 250mm x 36mmt
Fig. 11: Cross sectional shape of heavy wall seamless square columns

CONCLUSION

We have developed rolling technique of rolling round pipes into square columns at two roll type sizing mill in our Mannesmann Plug Mill Plant to manufacture heavy gauge seamless square columns. This heavy gauge columns are used for beam-to-column connections reinforced by heavy gauge columns. The results of developments are summarized as follows:

(1) Corner length is proportional to the total reduction at the sizing process. Higher total reduction is required to achieve smaller corner length.

(2) Corner radius is also proportional to the total reduction. Higher total reduction gives corner radius smaller.

(3) Concavity of finished column is related to the reduction increase at No. 7 stand. It is necessary to keep the reduction increase at No. 7 stand smaller for heavy gauge seamless square columns with low concavity.

(4) Combination of total reduction and reduction increase at No. 7 stand has realized to manufacture heavy gauge seamless square columns with accurate and small corner length, corner radius and concavity.

REFERENCES

1. K. Ohta, N. Yamamoto, Y. Murakami and K. Morita, "A Study on semi-rigid frames with beam to column connections reinforced by increasing thickness columns, (Part 1 Beam-to-column sub-assemblage tests)," The Proceedings of the Japanese Architectural Society, No. 21755, September 1994, pp. 1443-1444.

2. Y. Murakami, N. Yamamoto, K. Ohta and K. Morita, "A Study on semi-rigid frames with beam to column connections reinforced by increasing thickness columns, (Part 2 Estimation of reinforcing length and stiffness of connection)," The Proceedings of the Japanese Architectural Society, No. 21756, September 1994, pp. 1445-1446.

3. Y. Murakami, N. Yamamoto and K. Morita, "A Study on the Design of Semi-rigid Frames with Beam-to-Column Connections Reinforced by Increasing Thickness of Columns," Steel Construction Engineering, Vol.1 No. 4, December 1994, pp. 53-64.

4. Y. Onoda, T. Nagamachi, T. Toyooka, T. Nagahama and N. Morioka, "Development of Pass Schedule for the Hot Reshaping of Circular Seamless Steel Pipes into Square Shapes Using an 2-Roll Type Sizing Mill," The Proceedings of the 1999 Japanese Spring Conference for the Technology of Plasticity, No. 104, May 1999, pp. 7-8.

5. T. Nagahama, T. Toyooka, K. Yamamoto, M. Kita, Y. Onoda and T. Nagamachi, "Manufacturing and Properties of the Seamless Square Steel Pipes, 'K Kakuhot'," The Proceedings of the 1999 Japanese Spring Conference for the Technology of Plasticity, No. 410, May 1999, pp. 181-182.

Hot Ductility Behavior of Seamless Steel Pipes with Peritectic Carbon Content and Microalloyed Additions

Hugo Solis Tovar*, Eduardo Guzman Rojas* J. Angel Carranza Ochoa* and C. Isaac Garcia**

*Tubos de Acero de Mexico, S.A.
**The Basic Metals Processing Research Institute
University of Pittsburgh

ABSTRACT

The metallurgical and processing factors responsible for the hot ductility behavior in seamless pipe steels with peritectic carbon contents and microalloyed additions were studied and identified. The influence of steel composition and residual elements such as Cu, Sb and Sn and the prior austenite grain size are one of the major causes of the low hot ductility of the steels. Processing changes in the overall operation of the continuous casting machine (CCM) have also greatly improved the surface quality of steels with troublesome compositions. The factors controlling the hot ductility and the changes in steel composition and CCM operation leading to an improvement in the resistance to high temperature fracture of the steels are presented and discussed in this paper.

INTRODUCTION

Tubos de Acero de México (**TAMSA**) seamless pipe manufacturing process involves: a) electric arc furnace, b) ladle furnace, c) continuously cast round billets, e) re-heating furnace, f) hot rolling mill, g) rolling-drilling-lengthening mill, h) heat treatment line, and i) threading facilities. One of the major objectives of TAMSA is to continuously improve both the productivity and quality of its products in order to satisfy market demands and needs.

In order to achieve this objective, R&D programs have been conducted to gain a fundamental understanding of the factors directly responsible for the quality of the products. These efforts have led to the identification and development of guidelines to control these factors. For example, the steelmaking department has been investigating the role of the chemical composition, degassing, residual elements and continuous casting processing parameters on the surface and sub-surface quality of the continuously cast billets.

Among the several types of steel compositions produced by TAMSA, the steels with peritectic carbon content and the microalloyed steels have historically presented a larger susceptibility to transversal and longitudinal crack formation during continuous casting.

Transversal cracks generally begin at the oscillation mark valley and extend below the surface along the prior austenite grain boundaries. These cracks are mainly generated during the straightening process where the bar is subjected to tensile and compressive forces. Longitudinal cracks are generally generated at the mold area, where the primary cooling of the steel takes place. Excessive overcooling, inappropriate lubrication or stresses caused by the volume change during the solidification of the steel could generate micro-cracks that are opened during the subsequent secondary cooling or during straightening.

Several authors [1-11] have reviewed the factors responsible for the low resistance to ductile and brittle fracture during the continuous casting of steels. In general, the major parameters affecting the hot ductility of steels can be listed as follows:

- Chemical composition (C content)
- Second phase precipitates, M (C,N)
- Non-metallic inclusions
- Tramp elements, Cu, Sn, As, Sb
- Austenite grain size

- Recrystallization temperature of γ
- $\gamma \rightarrow \alpha$ transformation temperature and morphology of the ferrite
- Processing parameters such as casting velocity, oscillation marks, cooling pattern, strain rate during straightening and multi-point bending.

In order to reduce the effect of these factors, steel companies have developed their own procedures for the successful continuous casting of troublesome steel compositions. This paper discusses the changes in steelmaking, continuous casting and thermomechanical processing parameters that have been carried out at TAMSA to improve the surface quality of its products.

PROGRAM DEVELOPMENT

The joint program conducted by TAMSA and the Basic Metals Processing Research Institute – University of Pittsburgh consisted of two phases. In Phase I the hot ductility of steels which have traditionally shown a high susceptibility for surface cracking during CC were studied. Based on the information generated from this phase, Phase II was designed to incorporate the suggested changes in steel chemical composition and processing parameters during steelmaking and CC operations.

Phase I. Hot ductility behavior.

The chemical composition of the steels used in this investigation is presented in Table I. These steels prior to their hot ductility assessment were fully characterized in terms of their microstructure and cleanliness.

Table I. Chemical composition of the steels

Heat number	Code	C	Mn	Cu	Sn	Nb	V	Ti
72116	541	0.13	0.71	0.10	0.005	0.030	-	0.016
72117	541	0.14	0.70	0.11	0.005	0.029	-	-
72206	701	0.14	0.94	0.08	0.002	-	0.030	0.016
72208	701	0.14	0.94	0.10	0.003	-	0.029	-
72368	75	0.34	1.30	0.12	0.006	-	-	0.012
72370	351	0.22	1.35	0.21	0.015	-	-	0.027

Microstructural analysis. The microstructural analysis of the samples in the as-cast condition included; i) characterization of the solidification structure, ii) description of the oscillation marks microstructure and iii) analysis of the cracking behavior. In addition, the as-cast samples were reheated at 1250°C for one hour and quenched in an ice brine solution. The resulting austenite grain size was revealed, measured and recorded.

The cleanliness of the steels in terms of the non-metallic inclusion content was evaluated using standard ASTM E-1122 and E-45 procedures.

The optical micro-structural analysis was conducted using a LECO/Olympus PME-3 Inverted stage microscope. The analysis of second phase precipitates, non-metallic inclusions, oscillation marks and cracks were conducted with the aid of a Philips XL30 SEM microscope. The austenite grain size was determined using a computer controlled BioQuant IV image analysis system.

Hot ductility testing. The hot ductility of the steels used in this investigation was conducted on a computer controlled MTS servo hydraulic deformation system dedicated to high temperature testing. The samples used in this study were cut perpendicular to the bar surface, that is, the axis of the flanged compression samples [11] is parallel to the dendritic axis, so that the mid-height of the flanged specimen surface is plastically deformed transverse to the solidification growth direction. Prior to hot deformation the samples were reheated in-situ at 1250°C (steels 701-Ti, 701, 541-Ti, 541 and 075) and 1300°C (steel 351) for 2 minutes and then cooled to the desired deformation temperatures. The range of the analyzed deformation temperatures was between 1000 and 700°C at 100°C intervals. The samples were deformed at strain rates of 0.03 sec^{-1} and 0.003 sec^{-1}. After deformation, the specimens were immediately water quenched and carefully examined for crack initiation. The hot ductility of the steels was defined as the percent of bulk deformation at initial fracture for a given deformation temperature. From the deformed samples the crack path and morphology was studied using the Philips XL30 SEM microscope.

Phase II. Processing parameters.

i) Steelmaking (electric arc furnace)

Degassing. The nitrogen content during tapping was controlled to less than 50 ppm. This was achieved by increasing both coke and graphite during the melting stage from 3 to 7 kg/Liquid Steel Ton (LST) and from 0 to 1 kg/LST, respectively. These changes in coke and graphite levels combined with an increase in the oxygen blow from 7 to 13 m^3/LST achieved the planned nitrogen content.

Residual elements. Using cleaner scrap and increasing the use of DRI the amount of residual elements was controlled. In this way, Cu contents were kept at 0.15 % wt maximum values. EAF Sulfur contents of 0.020 wt % were obtained by increasing lime additions from 40 to 50 kg/LST.

ii) Steelmaking (ladle furnace)

Desulfurization. To achieve the aimed final S content lower than 0.005 %wt, optimum slag control to generate CaS inclusions that were absorbed and retained in the slag was used.

Chemical composition. The C content was kept at the lowest level allowed by the customer trying to avoid the peritectic range. Al and Ti additions were also tightly controlled.

iii) Continuous casting

Mold cooling. The water flow in the mold was decreased from 3000 liters/min to 2400 liters/min to avoid overcooling on the meniscus area.

Secondary cooling. The secondary cooling of the bars was decreased from 0.58 liters/kg. to 0.45 liters/kg. This change was made to obtain a higher temperature prior to the straightening process and to prevent an eventual overcooling at the bottom of the bar (external radio).

Oscillation marks. In order to decrease the depth and extent of the oscillation marks, particularly in steels with peritectic carbon content, several oscillation frequencies were tested from 50 c/min to 75 c/min, with a negative strip from 0.476 to 0.345.

Straightening pressure. The straightening pressure was decreased until a proper balance between the straightening and the column weight support was obtained. This condition allowed for a softer bar straightening.

Lubrication powder. Several experiments using different lubricant powders were conducted until a better superficial bar quality was obtained.

RESULTS AND DISCUSSION

Phase I. Microstructural analysis

i) As-cast condition. In general all the tested samples exhibited a ferrite-pearlite microstructure. As expected, the volume fraction of the pearlite content observed in the steels increased with C content. The major difference in the microstructure of the steels was the morphology of the ferrite. For example, in steels 541 and 701 with and w/o Ti additions, the morphology of the ferrite was preferentially blocky, while in steel 75 the ferrite observed had three morphologies, blocky, Widmannstätten and allotriomorph. Typical examples are shown in Figures 1 and 2 (fig 154). The microstructural analysis also revealed that in the steels containing Ti, the ferrite size and distribution was more homogeneous than in the steels without Ti. It is well known that pre-existing TiN precipitates can act as a nucleation sites for ferrite through particle stimulated nucleation mechanism.

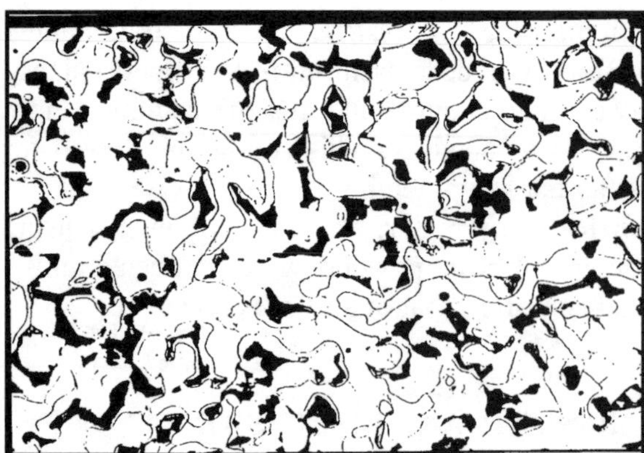

Figure 1. Optical micrograph of steel 541 with Ti in the as- cast condition. (57 x)

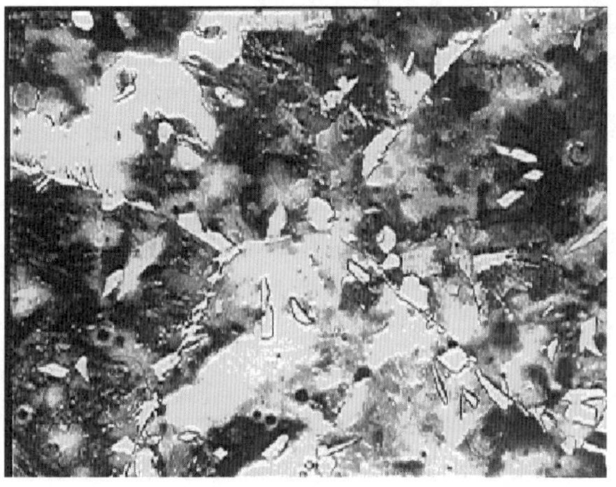

Figure 2. Optical micrograph of steel 75 with Ti in the as- cast condition. (57 x)

Figure 3 shows a typical optical micrograph of the observed cracks at the oscillation mark. A detailed SEM and EDS analysis of these cracks revealed that in addition of FeO, the cracks also contained residual elements such as Cu, Sn and As.
This detrimental effect is higher as higher the residuals content. The detailed analysis of these oscillation marks have been published elsewhere.[(4)]

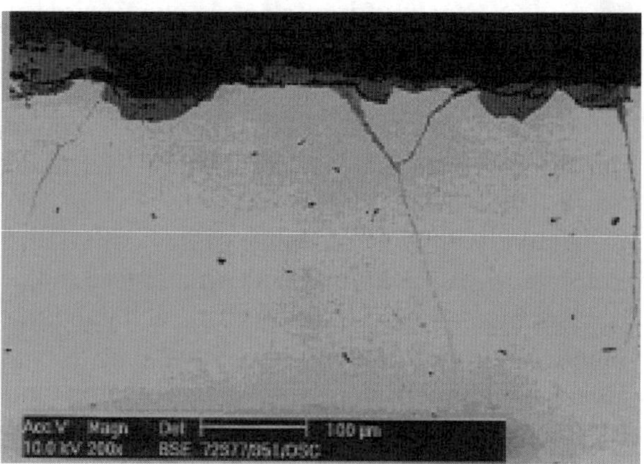

Figure 3. SEM micrograph of typical cracks observed at the oscillation marks.

ii)Austenite grain size. The austenite grain size prior to hot ductility testing was evaluated for the steels used in this investigation. It is important to note that the goal of this work was not to produce a uniform austenite grain size prior to hot deformation, but to investigate the resulting austenite grain size after laboratory reheating using similar reheating parameters as used by TAMSA that is, all the steels are typically re-heated at the same temperature and holding time. The results are shown in Table II.

Table II. Average prior austenite grain size.

Steel	**$D_($ (:m)**	**S.D. (±)**	**% S.D.**
701 + Ti	21.96	12.06	54.91
701	116.03	70.18	60.48
541 + Ti	11.37	5.13	45.35
541	104.36	65.41	62.67
75	39.75	23.86	60.01
351	67.83	39.37	58.05

The results from Table II clearly show that the steels with Ti additions exhibited finer austenite grain sizes and more uniform distribution than the steels without Ti. Typical examples of the austenite grain size are shown in Figures 4 and 5.

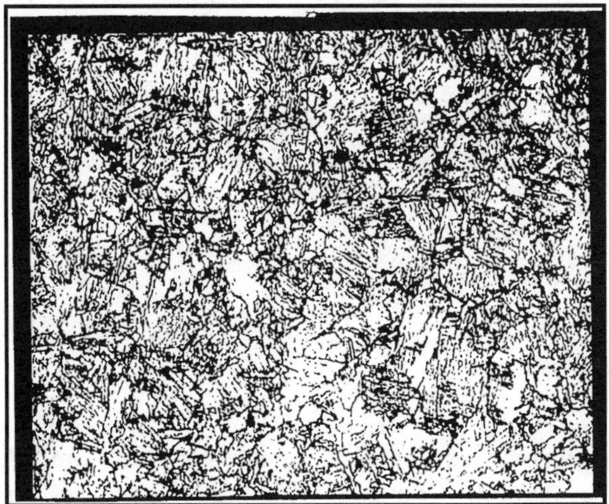

Figure 4. Prior austenite grain size, steel 541 + Ti.(452x) .

Figure 5. Prior austenite grain size, steel 541 without Ti. (452x).

Phase I. Hot ductility Testing

The hot ductility results are presented in figures 6, 7 and 8. Figure 6 shows the hot ductility behavior for steel 541 w/o Ti as function of deformation temperature and strain rate. As expected, the hot ductility decreases with lowering temperature in the γ region. The minimum hot ductility values are at the $\gamma \rightarrow \alpha$ transformation temperature and starts to increase as the volume fraction of blocky ferrite increases. In addition, Figure 6 shows the effect of strain rate. The lower the strain rate, the lower the hot ductility of the steels. These results are in good agreement with other several studies.[2,4,5,7-11]
Figure 7 compares the hot ductility of steel 541 with and without Ti. As expected, the hot ductility of the steel with Ti additions is higher than the steel without it. The major reason for this behavior is the influence of Ti on the starting prior austenite grain size.
Figure 8 shows a comparison of the hot ductility behavior of all the steels investigated in this study. The results show that the steel with peritectic carbon content without Ti additions exhibited the lowest hot ductility within the experimental conditions used in this investigation.

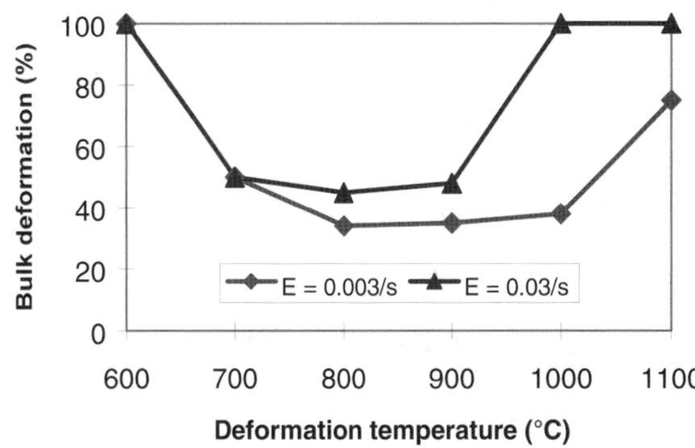

Fig. 6. Hot ductility of steel 541 without Ti addition.

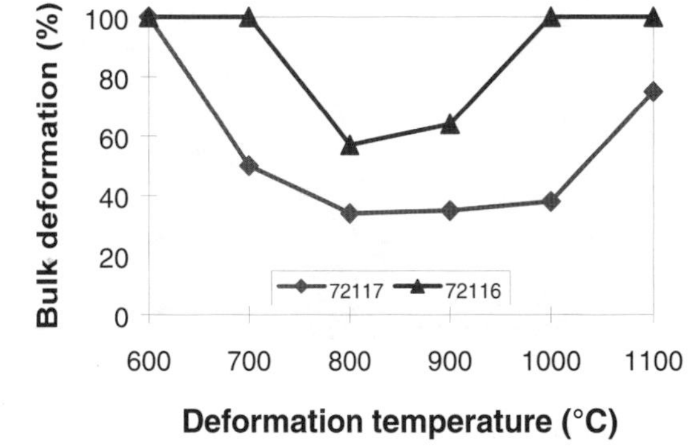

Figure 7. Hot ductility of steel 541 with and without. Ti addtions.

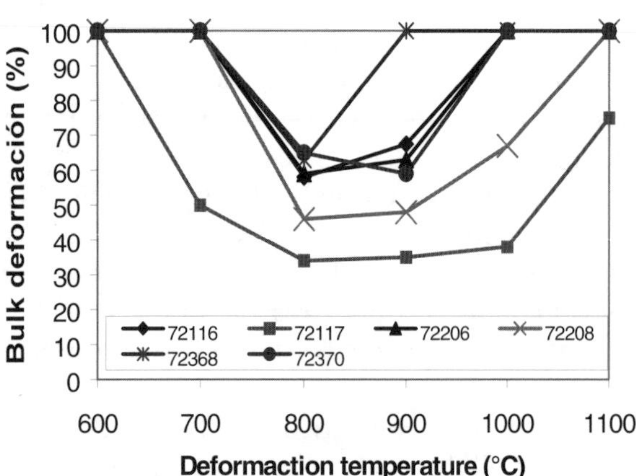

Fig. 8. Comparison of hot ductility behavior for all the steels. All deformations at a strain rate of 0.003/s.

Phase II. Process modifications

The production of 370 mm diameter bars without surface defects has been the most difficult, the following illustrates the changes in the overall processing to reduce these defects.

Residual elements. The use of cleaner scrap and increasing the amount of DRI from 10 to 18%, reduces the content of Cu to less than 0.15 wt%. The effect on product surface quality with Cu content is shown in Figures 9 and 10. The smaller the wall thickness of the pipe, the stronger the effect observed.

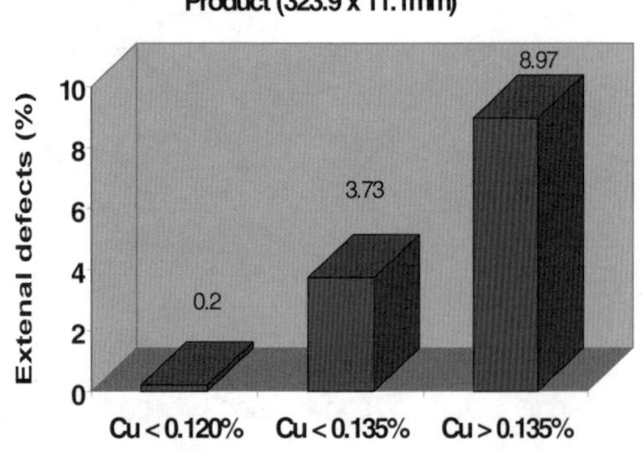

Figure 9. Effect of Cu in low C steel product 323.9 x 11.1mm.

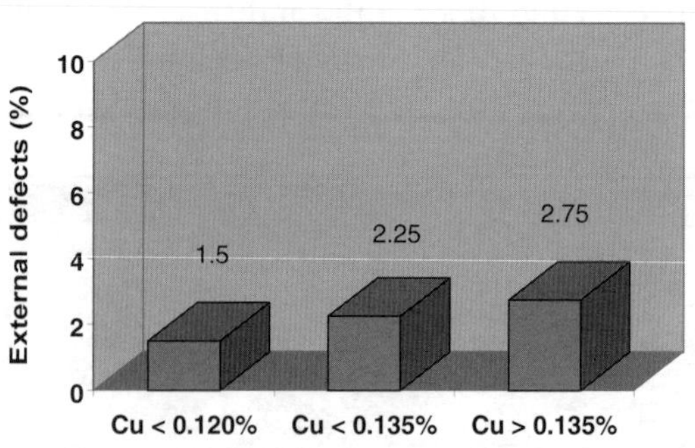

Fig. 10, Effect of Cu in low C steel product 355.6 x 14.3 mm.

Superheat control. The superheat in the tundish was carefully controlled, Figure 11 illustrates the lower the superheat, the better the surface quality of the pipe.

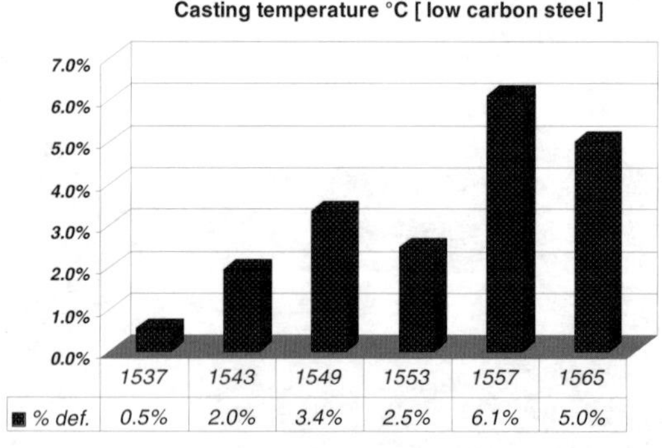

Fig. 11 Effect of tapping temperature in tubing defective

Bending temperature. Based on the hot ductility results obtained, changes in the continuous casting design from single bending ratio to multiple ratios were implemented. This permitted increases in the tapping speeds from 0.55 m/min. to 0.70 m/min. The effect of these changes is shown in Figure 12.

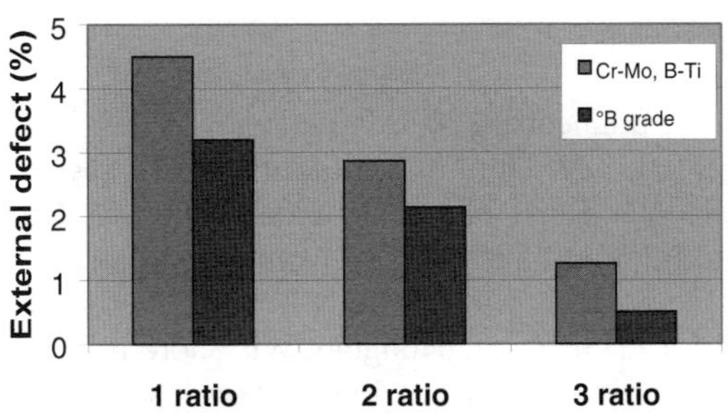

Fig. 12 Ratio change effect in pipes surface quality.

Chemical composition. In steels with C content close to the peritectic content, the C content has been kept at the lowest level allowed by the customer. In addition, Al and Ti additions were closely controlled, the Ti/N ratio has been kept to 2.0-2.5. These controls have resulted in an increased in the surface quality of pipes.

All the process and chemical changes implemented at TAMSA have consistently improved the surface quality of the products, as shown in Figure 13. However, it is well recognized that more work needs to be done, particularly in the solidification structure, reheating and rolling process to decrease the variability in performance of the products. To this end, a series of programs are being implemented.

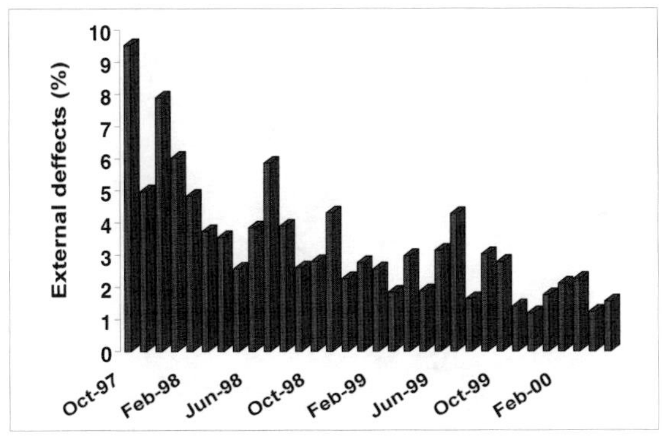

Fig. 13 Behavior of in 370 mm diameter

CONCLUSIONS

1. The mechanisms or factors responsible for the poor surface quality of the products were identified. The understanding of these factors led to the development of changes in the processing of the steels to increase their surface quality.

2. Changes in the straightening temperature range and bending ratios were very successful to improve the surface quality.

3. Control of Ti additions (Ti/N = 2.0 to 2.5) improved the hot ductility by controlling the austenite grain size during solidification and subsequent reheating. Also the Ti additions reduce the detrimental precipitation of AIN and other M (C, N) precipitates.

FUTURE WORK

1.- Continuous temperature measurement during the straightening process to avoid unbending of the bar in the poor hot ductility zone.

2.- Implementation to Vibromold hydraulic oscillator, which will allow changes in the frequency and stroke dynamics.

3.- Electromagnetic stirring to decrease segregation during solidification and improve overall solidification microstructure.

4.- New SEN design in order to improve meniscus temperature.

BIBLIOGRAPHY

1. - E.T Turkdogan - "Causes and Effects of Nitride and Carbonitride Precipitation During Continuous Casting - I&SM - May 1989.

2. - F.G. Wilson and T. Gladman - "Aluminium Nitride in Steel" - International Materials Reviews - 1988.

3. - M. Korchynsky - "HSLA Steels Technology & Application" - HSM - Metals Park - OH - 1983.

4. - B. Mintz and R. Abushosha "Influence of Vanadium On Hot Ductility of Steel" - Iron & Steelmaking - 1993.

5. - Cristiano Tercelli - "Surface Quenching, and Effective Tool for Hot Charging of Special Steels" - MPT - 1995.

6. - N. Bannerbeg and Others - "Procedures for Successful Continuous Casting of Steel Microalloyed with Nb, V, Ti and N" - Microalloying Conference Proceedings - 1995.

7. - Harry Stuart - "NIOBIUM (Proceedings of the International Symposium)" - S. Fco. Cal. - 1981.

8. - Robert D. Pehlke - "Continuous Casting of Steel Course" - Michigan - 1986.

9.- H. Solís - Factors that influence de hot ductility of steel. TAMSA-BAMPRI University of Pittsburgh, July 1998.

10.- B. Mintz, S. Yue, and J. J. Jonas - "Hot ductulity of steels and its relationship to the problem of transverse cracking during C.C. international materials Reviews 36, 5 1991.

11.- Y. Maehara, K. Yasumoto, H. Tomono, T. Nagamichi and Y. Ohmori - "Surface cracking mechanism of continuously cast low carbon low alloy steel slabs". Materials Science and technology, 6, 793. 1990.

12.- R Abushosha, R. Vipond and B. Mintz - "Influence of titanium on hot ductility of as cast steels". Materials Science and technology, 7, 1991.

13.- B Mintz - "Importance of Ar3 temperature in controlling ductility and with of hot ductility trough in steels and its relationship to transverse cracking" Materials Science and technology, 12, 1996.

14.- A. Guillet, S. Yue and M.G. Akben - "Influence of heat treatment and C content on the hot ductility of Nb-Ti steels". ISIJ International, Vol 33, 3 1993.

15.- N. Bannenberg, B Bergmann, H. A.Jungblut, N. Müller, k. Reich - "Procedures for successful continuous casting of steel microalloyed with Nb, V, Ti and N" Microalloying '95 Conf. Proceedings, ISS, 1995.

16.- Lu Zhen, Zhang Hongtao, Wu Baorong - "Effect of Nb on hot ductility of low carbon low alloy steel and its mechanism". The Minerals, Metals & Materials Soc. 1992.

17.- P.A. Manohar, T. Chandra and C. R. Killmore - "Continuous Cooling transformation behavior of microalloyed steels containing Ti, Nb, Mn and Mo". ISIJ International. Vol, 36 no.12 1996.

AN IMPROVED DRIVE SYSTEM FOR TUBE AND ROLL FORMING MILLS

Keith Sylvester Barraclough
Doltec Limited
Rigg Lane
Trent, Sherborne
Dorset DT9 4SS
England
Tel.: ++44 1935 851 034
E-Mail: techinfo@doltec.com

Mechanical Drive Systems for Tube Mills and Roll Forming Mills

INTRODUCTION

The basic design of tube and roll forming mills has changed little over the last 50 years. A series of gear driven rolls form flat strip into the required profile and the dimension, shape, and surface finish of the resulting product relies upon the accuracy of roll profiles and the expertise and experience of the operator. Despite the impact of computer technology in the design of rolls and roll drives, variations in material dimensions and properties together with the difficulty of precise alignment and setting of the roll train, combine, to this day, to make the manufacture of an acceptable product somewhat of a black art. Many of the problems associated with these variables together with variations in roll dimension and setting are exaggerated by the dynamic changes in drive point of the material on each driven roll.

This paper seeks to describe a patented system known as **CyberDrive®** that substantially reduces mill operating costs by automatically compensating for variations in drive point. Such a system overcomes many of the problems produced by these variations, and provides a means of reducing drive power, roll and transmission wear and, at the same time, demonstrates a significant improvement in product quality.

PRIOR ART

Over the years, mill and designers have tried many alternative approaches in an effort to overcome the basic problems inherent in the roll forming process. They have employed separate motors on each roll pass, separate motors on each roll shaft, complex torque control systems and, most recently, cage forming technology. Each of these systems add significantly to the complexity and initial cost of a mill, and ultimately, only partly address the problems.

THE PROBLEM

In roll forming applications it is impossible to predict the Drive Point between the driven roll and the strip in each driven pass. In order to understand the action of CyberDrive it is important to visualise what is actually happening in the roll-forming process where profiled rolls are employed.

Because the rolls are profiled, there is a difference in circumference between root and rim of a concave roll and between major and minor diameter of a convex roll. When rolls are rotating therefore, it will be apparent that there is an increase in surface speed between root and rim of the concave roll and a decrease in surface speed between major and minor diameter of the convex roll.

In tube and roll-forming mills there is a point on the surface of a freely rotating profiled roll where there is maximum frictional load between roll and strip. The circumference at this point determines the rotational speed of the roll. When a roll is gear driven however, the rotational speed is determined by the drive system, and strip in contact with it is subject to substantial variations of linear speed at each contact point. In the driven condition, the point of maximum friction (the Drive Point), which should now determine the linear speed of the strip, can be at a different surface speed point on each roll of a driven pair. This is particularly apparent in the breakdown section of a mill where rolls are simultaneously in contact with both upper and lower surface of the strip (Fig.1).

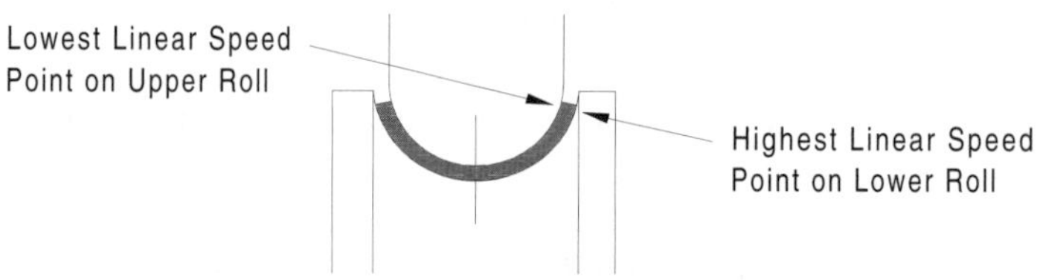

Fig.1

The very essence of roll forming determines that the strip will make initial contact towards the rim of a concave roll, and at the major diameter of a convex roll. When rolls are working in vertically co-operating pairs, the strip is formed into the desired profile by the pressure exerted between upper and lower roll. The resulting shape depends upon roll profile, tensile strength and thickness of the material, and the setting of roll clearance. Due to these variables it is impossible to dynamically predict the actual drive point on any particular roll. Remember, the rolls are gear driven and their rotational speed is determined only by the drive system.

If we first consider the **Breakdown Section** of a tube mill designed for a specific thickness of material. In this section, the forming rolls are in contact with both upper and lower surface of the strip, and each set of rolls is designed to suit strip of a particular width and thickness. Typically, commercial strip has a thickness tolerance of some 10%, and some of this variation can even take place within a single coil of strip. The mechanical properties of the material can change, and many tube manufacturers use the same set of rolls to produce both low tensile and high tensile product in an effort to contain roll costs.

The application of thicker or higher tensile material will provide a drive point which tends to move up the profile of a convex roll towards the rim. Because of the pre-determined profile of the each roll, the material will be pinched towards the rim of the concave roll and will provide the greatest load towards the minor diameter of the convex roll. In this case, the strip is in intimate contact with a higher surface speed point on the lower roll, and a lower surface speed point on the upper roll – immediately there is conflict!

Using the same rolls for thinner material will provide a drive point at the root of the concave roll and at the major diameter of the convex roll. This may in fact be the design drive point on each roll and, in this instance there is no conflict, but the strip cannot be formed correctly because it is not in intimate contact around the profile of each roll! (Fig.2).

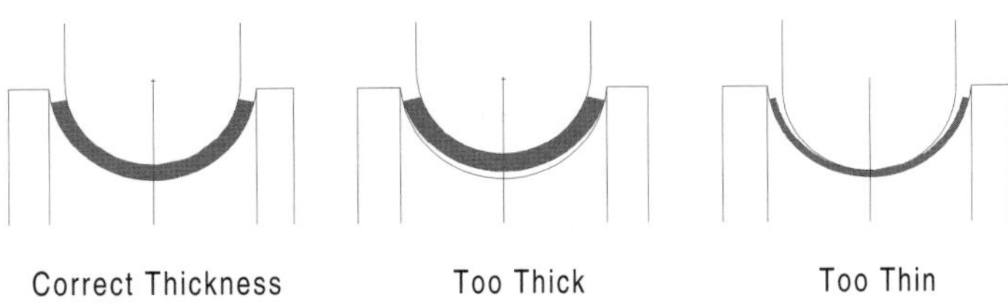

Fig. 2

The final shape of the product is very much in the hands of the operator who uses his art and experience to make fine adjustments at each roll pass to produce an acceptable finished profile. The problem he faces is that a small adjustment on one pass will have some effect on the next or subsequent passes. A little extra squeeze to obtain the desired shape can readily create an increase in length and a corresponding increase in linear speed of the product. Unfortunately, with gear driven rolls, the rotational speed of the following roll passes cannot be adjusted to suit – once again, conflict!

The roll designer faces an almost impossible task in his efforts to decide where each roll is driving the strip and, at the same time, obtain the desired profile of the product. Computer simulations providing projections of strain within the material and are of great assistance in devising a roll profile for material with specific dimensions and mechanical properties, but in day to day production, the wide variations of strip characteristics cannot give a complete solution to drive and pressure points.

We have examined the conditions in the breakdown section of a tube mill - now let us consider the **Fin Passes** in the same mill. At this stage in the forming process, vertically co-operating rolls are still employed, but pressure is being applied to the product on the external surface only. A vertical load applied by the upper roll provides a distribution forces shown in Fig. 3 below. It will be noted that the forces are equally distributed around the surface of the product except at the rims of upper and lower rolls. Here, the conditions change, because, over the small area at the gap between each roll, there is no support for the material. The load at the last point of contact is therefore substantially higher than at all other points around the profile where each finite radial load has support on either side. Unfortunately, this higher load at the rim is also coincident with the highest linear speed of the roll surface. The perfect condition for pick-up!

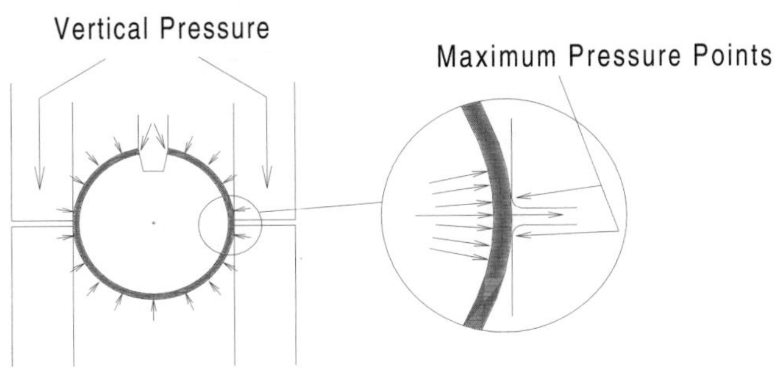

Fig. 3

Pick-up is created by raising the temperature of a surface to a point at which it melts and is deposited on a cooler surface. If, for example, a light load is applied between two surfaces having a large difference in surface speed, there is little chance of pick-up occurring.

Equally, if a high load is applied with little or no speed difference, no pick-up will take place. If, however, there is both high load and significant speed difference, pick-up will occur - the very principle used in friction welding! Of course, materials with different coefficients of friction can be employed to minimise this effect, but the basic problem remains.

The condition is well known to all producers of roll-formed products and results in marked product and damaged rolls.

The roll designer has to decide where to position the drive point – should it be at the rim, or should it be at the root – or should it be somewhere in between? At this point, another factor enters the equation – progression! In an effort to resolve the distortions that occur in the strip, each successive roll pass is often given a small increase in root diameter which, with a constant rotational speed of each roll, provides a degree of stretch between each pass.

Ideally, the drive point would be at the point of maximum frictional load (close to the rim), but this would be correct for only one diameter of tube. If a smaller diameter were to be produced on the same machine, the lower rolls would have to be of smaller diameter and the height of the roll shafts would have to be adjusted to suit. Such adjustments would add to the time and cost of a roll change.

In the Fin Passes, the material is squeezed against the fins in order to condition the strip edges, to produce the required circular profile, and to stabilise the open seam for the welding process. The squeezing action can also increase the length of the product by a small amount, particularly if there is progression built into the roll design. The actual length change depends on many factors, not least of which is the roll setting applied by the operator. It is therefore unlikely that the following roll (which is also gear driven) will be rotating at the correct speed to precisely match that of the arriving material. Yet again, the chance of conflict.

The **Sizing Passes** are typically designed to reduce the diameter of the welded product to precise finished dimensions, and adjustment of roll pressure to produce the required dimensions is in the hands of the operator. Any increase in pressure will elongate the product giving a higher linear speed. Unfortunately, the operator has no control over the rotational speed of individual driven rolls and some will be acting to move the material in the forward direction while others may be applying a braking load – that is to say, they will not be rotating fast enough (Fig.4).

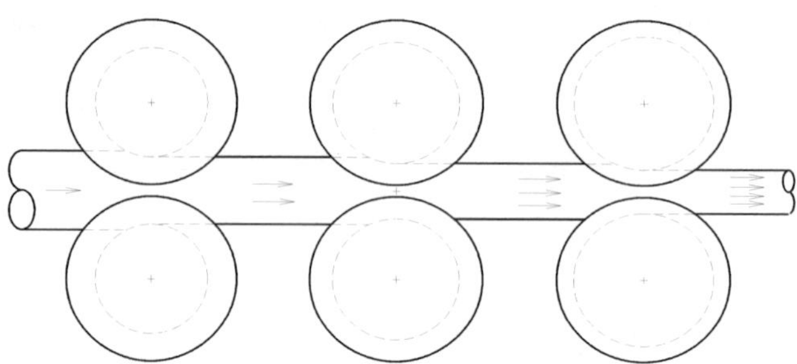

Fig. 4

The operator is likely to make adjustments many times during a product run in order to compensate for variations in material properties which between one coil of strip and the next. It will be obvious that any adjustment in the forming section will almost certainly require some adjustment in the sizing section if the final product dimensions are to be maintained.

The roll designer's task is to create roll profiles that will provide a product of specific dimension and properties when supplied with material of specific dimension and properties. The mill operator's task is to make adjustments to compensate for the variations in material dimension and properties. The result is a compromise that has historically satisfied the product manufacturer and the end user.

In recent years however, there has been an increasing demand for a higher quality and more consistent product. The cost of production is continually increasing and manufacturers are desperately seeking ways of meeting customer demands in terms of quality, consistency, and price.

It will be apparent from the foregoing that the distortions and loads produced by the various conflicts contribute to a waste of drive power, wear on rolls and transmission, marking of rolls and strip, variations in product quality and consistency and, a waste of money!

THE SOLUTION

Investigation of available drive systems showed that despite the application of separate drive motors on each driven shaft, the introduction of computer control, the fitting of 'floating flange' rolls, and other more complex systems, the basic problem still remained. It was also apparent that each of these 'solutions' added substantial cost to the process line.

Three years of study, research, and development, produced a simple and cost effective solution to many of the problems, and one which could be retro-fitted to existing process lines with minimum changes to the equipment. International Patents were obtained for the resulting CyberDrive concept.

The system employs a series of one way clutches (or freewheels) which are interposed between drive and driven roll. These devices are produced in the form of a coupling which is typically attached to each gearbox output shaft and is connected in such a way that each roll can rotate at the speed of the strip and no roll can act as a brake on the material (Fig.5) This action effectively eliminates the fight between roll and strip thereby reducing drive power, roll wear, strip marking and roll pick-up. The result is a more consistent and higher quality product. In addition, because of these benefits, there are significant cost savings to the manufacturer.

Fig.5 A CyberDrive Coupling

From the following examination of the action of CyberDrive, it will become apparent how the system achieves its purpose.

Consider, for example, the first driven Breakdown Pass on a tube mill – the lower roll has a concave profile and the upper roll, a convex profile. Typically, each roll profile is designed to accept strip of a particular width and thickness and should be set to design clearance for that strip. The top to bottom gear ratio will have been chosen by the mill and roll designer such that rolls of a sensible size are employed.

Now, when the flat strip first enters the roll pair, its contact points are towards the rim of the concave roll and the major diameter of the convex roll. As the material progresses through the rolls, the strip is deformed to match the roll profiles – the greatest loads being towards the rims of the concave rolls and the major diameter of the convex roll. Which roll should do the driving – and where on the roll?

If the material is of the correct thickness and the rolls are set to the correct clearance, the strip will fill the gap between them. Because the circumference at the root of the concave roll is smaller than the circumference at the rim, there is difference in the linear speed of the rotating roll at each point of contact around the profile. The higher speed is the rim, the lower speed at the root. On the convex roll the reverse is the case, the highest surface speed is at the major diameter (coincident with root of the concave roll) and the lowest surface speed is at the minor diameter (coincident with the rim of the concave roll). It will be apparent that these conditions are not compatible and the material must be slipping or skidding relative to the roll at some point around each profile.

The problem becomes more complex. If the rolls, the roll clearance, and the strip, are precisely to specification (i.e. the material is completely filling the space between the rolls), the strip, across its section, must be subject to the complete range of linear speed variations at all points of contact around the profile of both concave and convex rolls. Significant shear forces are therefore applied to the material leading to distortion and roll wear. Most wear takes place at a point coincident with the highest load and greatest speed difference.

If one of the pair of rolls is required to rotate faster than its gear drive (and is allowed to do so), the point of highest load will now determine its speed of rotation and the speed difference will not be present. In practice, the drive point moves up and down the profile of each roll continuously and the changes are taking place dynamically at all times in the drive between one roll and another when the mill is running.

CyberDrive compensates for these changes and, as the operation is completely automatic, no adjustment is required by the operator.

If roll wear takes place, the circumference at the point of wear must, by definition, decrease, and this area of the roll will now have a lower linear speed. This could be good news or it could be bad news! In most cases the roll can be re-ground to the correct profile and, with CyberDrive fitted, it will either drive at a different diameter or will overrun at a different speed.

If we next consider the effect of CyberDrive on the Fin Passes. In most installations where there is single drive motor covering the Breakdown and Fin Pass sections of a mill, the Fin Passes are usually rotating too fast in relation to the Breakdown Passes and are thus perfectly designed for a rim marking/pick-up condition. When CyberDrive is fitted however, the strip tends to run towards the rim speed of the Breakdown rolls (the upper, convex rolls generally overrunning and not acting as brakes). This means that it is much closer to the rim speed of the rolls in the Fin Passes, thereby reducing the chances of rim marking or pick-up.

The Sizing Section is usually separately driven, and there is therefore some ability to independently control the speed of the drive. CyberDrive takes care of the speed variations created by the changes in roll pressure applied by the operator. Small adjustments to the speed trim provide fine control to match the drive to the speed of the incoming material.

One question keeps recurring as people ask how the material continues to drive forward if many of the rolls are effectively freewheeling. The answer to the question is simple - all rolls are still being rotated by the drive system but, with CyberDrive fitted, none of them are working against the direction of travel.

A simple test was created by one of our customers to prove to himself that there would be a demonstrable benefit if the CyberDrive system was fitted to his mill. The test, which we now call the Craig Test, is quite simple to perform. It involves the removal of a roll shaft in any section of the mill, Breakdown, Fins, or Sizing. With strip in the mill, a flat tape is stretched and fixed between gearbox shaft and roll shaft (Fig.6), and the mill is run for a few minutes. If the tape remains flat, the drive point on this roll (which is now being driven by the strip) is correct and, with this roll pass and roll combination, CyberDrive would not necessarily provide a significant benefit. If however the tape is twisted into a coil, there must be a difference in drive point and the application of the CyberDrive system will automatically compensate to provide the claimed benefits.

Fig. 6 The Craig Test

Each installation of CyberDrive has confirmed most or all of the following benefits:

Reduced Drive Power
Reduced Roll and Transmission Wear
Reduced Product and Roll Marking
Reduced Scrap
Reduced Line Speed Variations
Improved weld seam stability and more constant weld temperature
Greater accuracy in cut-length on high-speed cut-offs
Reduced shear blade and die set damage

Once fitted to a particular mill, the CyberDrive system covers all sizes for which the mill was designed .

Many customers have demonstrated a 100% year on year return on their investment in the **CyberDrive** system.

The Mean and Local Wall Thickness Change in Tubes During Their Reducing and Sizing

Gennady I.Gulyayev, Dr.Sc.(Tech.), Prof.
State Tube Institute (DTI-VNITI)),
Dniepropetrovsk, 49600, Ukraine
Tel./fax +380(562)464566
Yury G.Gulyayev, Dr.Sc.(Tech.)
Temp Co.,
Dniepropetrovsk, 49027, Ukraine
Tel./fax +380(562)464591

Key Words: reducing, sizing, deformation, roll pass, wall thickness, stand

INTRODUCTION

To define basic technological and power-and-force parameters taking place in the tube reducing and sizing processes, change of the mean cross sectional wall thickness should be taken into consideration for both an individual roll pass, ΔS_i, and the entire rolling process in the mill, ΔS_{total}.

To study or predict the tube quality by cross-sectional wall thickness variation, especially in thick-walled and extremely thick-walled tubes, one should know the local wall thickness change ΔS_{i_θ} around the roll pass perimeter and the change ΔS_{total_θ} resulting from the entire rolling process in the mill.

By their technological features, the stretch reducing, $Z_{total} \leq 0.75$, or non-stretch reducing, $Z_{total}=0$, and low-stretch sizing, $Z_{total} \leq 0.45$, or non-stretch sizing, $Z_{total}=0$ are distinguished only by the value of the total deformation in an individual roll pass, $m_i \leq 10\%$, or the total deformation (reduction) in the mill, $m_{total}=10\text{-}80\%$, of the mean outside diameter. It follows that the basic difference between these two processes consists just in the number of the working stands (2-, 3-, or 4-roll stands), i.e. from 3 to 28 stands, being operated.

MEAN WALL THICKNESS CHANGE

When reducing (sizing) is carried out without stretching, $Z_i=0$ ($Z_{total}=0$), change of the cross-sectional (ΔS_i or ΔS_{total}) and the local tube wall thickness around the roll pass perimeter (ΔS_{i_θ} or ΔS_{total_θ}) depends on two parameters: mean outside diameter deformation (reduction) m_i or m_{total} and ratio S_{i-1}/D_{i-1} or S_0/D_0.

Reduction while causing tangential deformation (change of the tube perimeter) brings about two more deformations: radial deformation (change of the tube wall thickness) and axial one (the tube elongation).

S_{i-1}/D_{i-1} or S_0/D_0 ratio governs the nature and amount of the radial deformation: wall thickness growth or reduction takes place in thin-walled or thick-walled tubes respectively.

When the tube is stretch reduced ($Z_{total} \leq 0.75$ for thin-walled and $Z_{total} \leq 0.45$ for thick-walled tubes), above-mentioned parameter influences primarily the quantitative side of the wall change: it decreases thickening or increases thinning thus causing growth of the axial deformation in both cases. Consequently, if a correct dependence of the change of the mean cross-sectional wall thickness on m_i (m_{total}) and S_{i-1}/D_{i-1} (S_0/D_0) ratio when $Z_i=0$ ($Z_{total}=0$) is found, then introduction of Z_i (Z_{total}) parameter will not violate their basic nature.

Both foreign [1, 2] and the USSR[3-14] sources contain numerous empirical and theoretical equations for the calculation of the mean tube wall thickness change $\Delta S_i \left(\Delta S_{total}\right)$ and the final wall thickness value S_i (S_k) in the non-stretch reducing process. However, only two equations remain to be of a wide practical use in calculating technological and power-and-force parameters in the tube reducing process so far. They were developed by F.Neumann and D.Ganke [2] and used abroad and by G.I.Gulyayev and P.N.Ivshin ([15], pp.49-54). The latter is used in Ukraine (DTI-VNITI)

State Tube Institute (DTI-VNITI) owns experimental and production data which have been gained in non-stretch reducing of tubes. They indicate the necessity of correction of the two formulas:

- F.Neumann and D.Ganke's formula for its use in calculation of S_i, S_k, ΔS_i and ΔS_{total};

- G.I.Gulyayev and P.N.Ivshin's formula for its use just in calculation of S_k and ΔS_{total}.

The expression having corresponding subscripts for an individual stand or the entire mill and being used in Germany, USA, etc. is of the following form [2]:

$$Z_i = \frac{\left(2\varphi_{l_i} - \varphi_{t_i}\right) - T_i\left(\varphi_{l_i} + \varphi_{t_i}\right)}{\left(1 - T_i\right)\left(\varphi_{l_i} + \varphi_{t_i}\right)}, \qquad (1)$$

where

$$\varphi_{l_i} = \ln\left[\left(\frac{D_{i-1} - S_{i-1}}{D_i - S_i}\right)\frac{S_{i-1}}{S_i}\right] \qquad (2)$$

is the axial deformation;

$$\varphi_{t_i} = \ln\left(\frac{D_{i-1} - S_{i-1}}{D_i - S_i}\right) \qquad (3)$$

is the tangential deformation;

$$T_i = \left(\frac{S_{i-1}}{D_{i-1}} + \frac{S_i}{D_i}\right) \qquad (4)$$

is the tube size index;

D_{i-1}(D_0) and D_i (D_k) are the starting and final mean outside diameter , mm, of the tube respectively

S_{i-1}(S_0) and S_i(S_k) are the starting and final mean wall thicknesses, mm, around the tube perimeter respectively.

$\varphi_{l_{total}}$, $\varphi_{t_{total}}$ and T_{total} should be used for the entire deformation in the mill when $Z_{total} \leq 0.75$.

Having done corresponding transformations in (1) in absence of stretch, $Z_i=0$ ($Z_{total}=0$), one obtains the following expression for an individual mill stand:

$$1 - \left(\frac{S_{i-1}}{S_i}\right)^{2-T_i} \cdot \left(\frac{D_{i-1} - S_{i-1}}{D_i - S_i}\right)^{1-2T_i} = 0 \qquad (5)$$

Computer-aided solution of equation (5) can be done by method of gradual approach to the final result, i.e. S_i (S_k) value is found for known D_{i-1} (D_0), D_i(D_k) and S_{i-1} (S_0) values and the change of the mean wall thickness is obtained from expression $\Delta S=S_i-S_{i-1}$ ($\Delta S_{total}=S_k-S_0$).

Equation (5) has been checked for correspondence of the obtained calculation results to the experimental data within a broad range of $S_{i-1}/D_{i-1}\leq 0.5$. It has appeared that when $m_i\leq 10\%$ in each mill stand, there was a significant discrepancy between them.

As it appears from the analysis of equation (4), when $Z_i=0$ and $S_{i-1}/D_{i-1}=0.25$, the latter being called the partial instantaneous critical ratio $(S_{i-1}/D_{i-1})_{cr_i}$, thickening of the mean wall turns into its thinning. Experimental data [16] and theoretical calculations (pp. 37-40 in [15]) show that this transition from thickening to thinning is only observed when $(S_{i-1}/D_{i-1})_{cr_i}=0.333$ and m_i value is rather small ($m_i \leq 2\%$). That is why exponent n=1.57 [17-19] has been introduced into equation (4) at one time. This has made it possible to obtain S_i (S_k) and ΔS (ΔS_{total}) values within a definite tube size range for $Z_i \leq 0.75$ ($Z_{total}\leq 0.75$) more accurately than it was at n=1.

However, even at n=1.7 when $(S_{i-1}/D_{i-1})_{cr_i}=0.333$, discrepancies between the calculated and actual data are persisting.

The computations based on the experimental data obtained in rolling tubes in a single roll pass have shown that for the entire range of $S_{i-1}/D_{i-1}\leq 0.5$ when $m_i\leq 10\%$ and $Z_i=0$, exponent n_i should be introduced into equation (4). This exponent can be defined by the expression we recommend:

$$n_i = 1.6253\left(\frac{S_{i-1}}{D_{i-1}}\right)+1.1743, \qquad (6)$$

where $n_i = f(S_{i-1}/D_{i-1})$.

Then equation (4) takes the following form:

$$T_i=\left(\frac{S_{i-1}}{D_{i-1}}+\frac{S_i}{D_i}\right)^{n_i} \qquad (7)$$

When deriving their equations (1) and (4), workers [2] did not take into consideration influence of S_{i-1}/D_{i-1} ratio for the tube deformed in a single stand ($m_i \leq 10\%$) when Z=0 upon the calculation exactness and, consequently, they have assumed that i =1 in equation (4). Also, they recommended to use equation (4) in calculation of S_k and ΔS_{total} when m_{total} =10-80% and $Z_{total}\leq 0.75$. However, our computation results have demonstrated the necessity of taking into consideration the influence of the amount of the total deformation m_{total} in addition to the influence of S_0/D_0 ratio when applying equations (1) and (4) for an exact calculation of S_k and $\Delta S_{total}=S_k-S_0$ used in reducing (sizing) tubes in the mill.

To that end, exponent n_{total} should be introduced into equation (4), that is:

$$T_{total}=\left(\frac{S_0}{D_0}+\frac{S_k}{D_k}\right)^{n_{total}}, \qquad (8)$$

where $n_{total} = f$ (S_0/D_0 and m_{total}).

Derivation of mathematical dependence for calculation of n_{total} is a rather complicated problem. That is why we have resorted to a graphical dependence (Figure 1) derived on the ground of our computer-aided calculations and comparison of their results with the actual data.

With the aid of this dependence, n_{total} value can be found for expression (8) if S_0/D_0 ratio and m_{total} values are known.

When equations (7) and (8) are used in combination with the graphical dependence, (Figure 1), the calculated data being obtained with the use of equation (5) correspond to the actual experimental or production data for the tubes being deformed both in a single stand and in the whole mill when $Z_i=0$ ($Z_{total}=0$).

If S_i is to be found and $\Delta S_i = S_i - S_{i-1}$ calculated for the case of stretch reducing (sizing) in a single stand, $Z_i \leq 0.75$, equation (5) derived from equation (1) should be of the form:

$$1-\left(\frac{S_{i-1}}{S_i}\right)^{Z_i(T_i-1)+(2-T_i)} \cdot \left(\frac{D_{i-1}-S_{i-1}}{D_i-S_i}\right)^{2Z_i(T_i-1)+(1-2T_i)} = 0, \qquad (9)$$

where T_i is determined by equation (7).

For the tubes reduced (sized) in the whole mill, S_0, S_k, D_0, D_k, T_{total} and Z_{total} should be specified and T_{total} determined by equation (8), n_{total} being found with the help of the graph shown in Figure 1.

In Ukraine (DTI-VNITI), the equation recommended in [15, pp. 49-54] is applied for calculation of S_i and ΔS_i for $Z_i \leq 0.75$, $m_i \leq 10\%$ and $S_{i-1}/D_{i-1} \leq 0.5$:

$$S_i = S_{i-1}\left(\frac{D_{i-1}}{D_i}\right)^{A_i}, \qquad (10)$$

where

$$A_i = \frac{2\left\{1-3\left(\frac{S}{D}\right)_{av_i} - 2Z_i\left[1-2\left(\frac{S}{D}\right)_{av_i}\right]\right\}}{(1-Z_i)\left\{1+3\left[1-2\left(\frac{S}{D}\right)_{av_i}\right]^2\right\}+2Z_i\left[1-\left(\frac{S}{D}\right)_{av_i}\right]} \qquad (11)$$

$$\left(\frac{S}{D}\right)_{av_i} = 0.5\left(\frac{S_{i-1}}{D_{i-1}}+\frac{S_i}{D_i}\right) \qquad (12)$$

Equation (10) used in calculation of S_i and $\Delta S_i = S_i - S_{i-1}$ ensures a good convergence with the experimental and production data and no correction factors are necessary. Equations (10), (11) and (12) can be also used in calculation of S_k, and $\Delta S_{total}=S_k-S_0$ with the corresponding subscripts: S_k, S_0, D_0, D_k, A_{total} and $(S/D)_{av_{total}}$ for stretch reducing (sizing) tubes in the mill, $Z_{total} \leq 0.75$; $m_{total}=10\text{-}80\%$ and $S_0/D_0 \leq 0.5$.

But owing to the fact that ratios S_0/D_0 and S_k/D_k differ from each other significantly at high total deformations m_{total}, the value of $(S/D)_{av_{total}}$ being determined from equation (12) does not correspond to the value to be obtained by the method of integration because the tube is being deformed not in a single stand but in a number (up to 28) of stands. In this connection, a sufficient calculation accuracy takes place at $Z_{total}=0$ and ratio $S_0/D_0 \leq 0.5$ and $m_{total} \leq 50\%$. Besides, deviation of the calculated data obtained by equations (10-12) from the practical ones is within +(0-4.5%) depending on m_{total} and S_0/D_0.

The computer-aided calculations grounding on the actual production data have shown that equation (10) can be used for determination of S_k and ΔS_{total} at $Z_{total}=0$, $S_0/D_0 \leq 0.5$ and $m_{total}=10\text{-}80\%$ provided it is

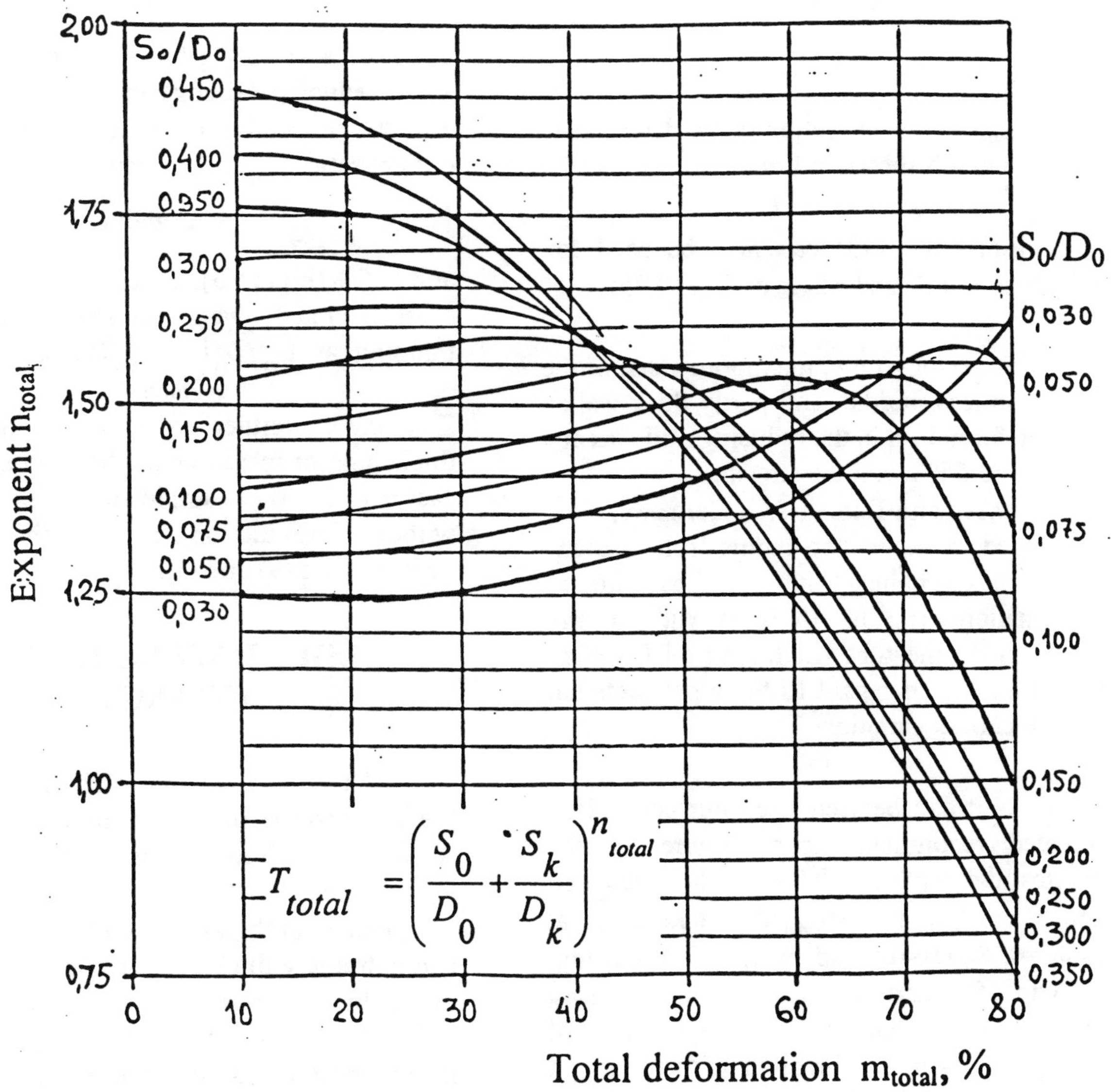

Figure 1. Dependen ce of exponent n_{total}

on ratio S_0/D_0 and m_{total} values

at Z_{total} =0

corrected. The graph shown in Figure 2 is recommended for such correction. The corrected form of equation (10) is as follows:

$$S_k = S_0 \left(\frac{D_0}{D_k} \right)^{A_{total}} \cdot K_{total}, \qquad (13)$$

where K_{total} is the correction factor determined from the graph shown in Figure 2.

Equation (13) can also be used for calculation of S_k and ΔS_{total} at $Z_{total} \leq 0.75$.

For a long time, maximum value of the total seamless middle- and partly thick-walled tubes deformation in the reducing mill was at a level of 40-55% with the stretch.

Mannesmann Demag AG, Germany, have developed a new two-radius roll design[1] which has enabled them to use higher deformations (up to 75-78%) without any detriment to the tube (including thick-walled tubes) quality in regard to the cross-sectional wall thickness variation.

As it can be seen from our correction-oriented recommendations (Figure 2), the greatest discrepancies between the calculated and actual S_k and ΔS_{total} data take place at $Z_{total} = 0$, $S_0/D_0 \leq 0.5$ and $m_{total} > 50\%$ reaching +4.5 to 18% and -1.5% depending on m_{total} and S_0/D_0. That is why application of equation (13) in combination with the graph in Figure 2 ensures the necessary calculation accuracy at $Z_{total} = 0$ within the entire range of ratios $S_0/D_0 \leq 0.5$ and total deformations m_{total} =10-80%. When $Z_{total} \leq 0.75$, equation (13) and graph in Figure2 retain their structures unchanged and they are applicable in calculations of the stretch tube reducing (sizing). Through application of equation (9) in combination with the graph in Figure 1, calculation accuracy equal to that obtainable with equation (13) and the graph in Figure 2 can be ensured. The correction we introduced in equations (9) and (13) has made them interchangeable.

Comparison of calculated and actual data shows us a quite good result.

Like in rolling tubes in a single stand at $Z_i = 0$, reducing (sizing) in the whole mill at $Z_{total} = 0$ proceeds at a total instantaneous critical ratio $(S_0/D_0)_{cr_{total}}$ which is smaller than $(S_{i-1}/D_{i-1})_{cr_i} = 0.333$ and depends on m_{total} and S_0/D_0 (Figure 3). Thickening of the mean cross-sectional wall thickness turns into thinning at $(S_0/D_0)_{cr_{total}}$. However, stretch, $Z_{total} \leq 0.75$, either decreases the wall thickening or increases its thinning as it is the case with the single-stand deformation without introduction of any changes into dependences (9) and (13).

LOCAL WALL THICKNESS CHANGE

The tube reducing (sizing) operation in the mill causes polygonization of the inside surface. This feature is less obvious in thin-walled tubes ($S_0/D_0 \leq 0.1$) while being rather pronounced in thick-walled (S_0/D_0=0.10-0.20) and extremely thick-walled (S_0/D_0=0.20-0.45) tubes. In a number of cases, cross-sectional wall thickness variation in the finished tubes is within the wall tolerances but the thick-walled and extremely thick-walled tube appearance does not meet commercial and technical standards due to their pronounced polygonization.

Prediction studies of the cross-sectional wall thickness variation which is formed during reducing (sizing) tubes in the whole mill requires the knowledge of the local (i.e.

[1] German Patent #39242617 of July 20, 1989

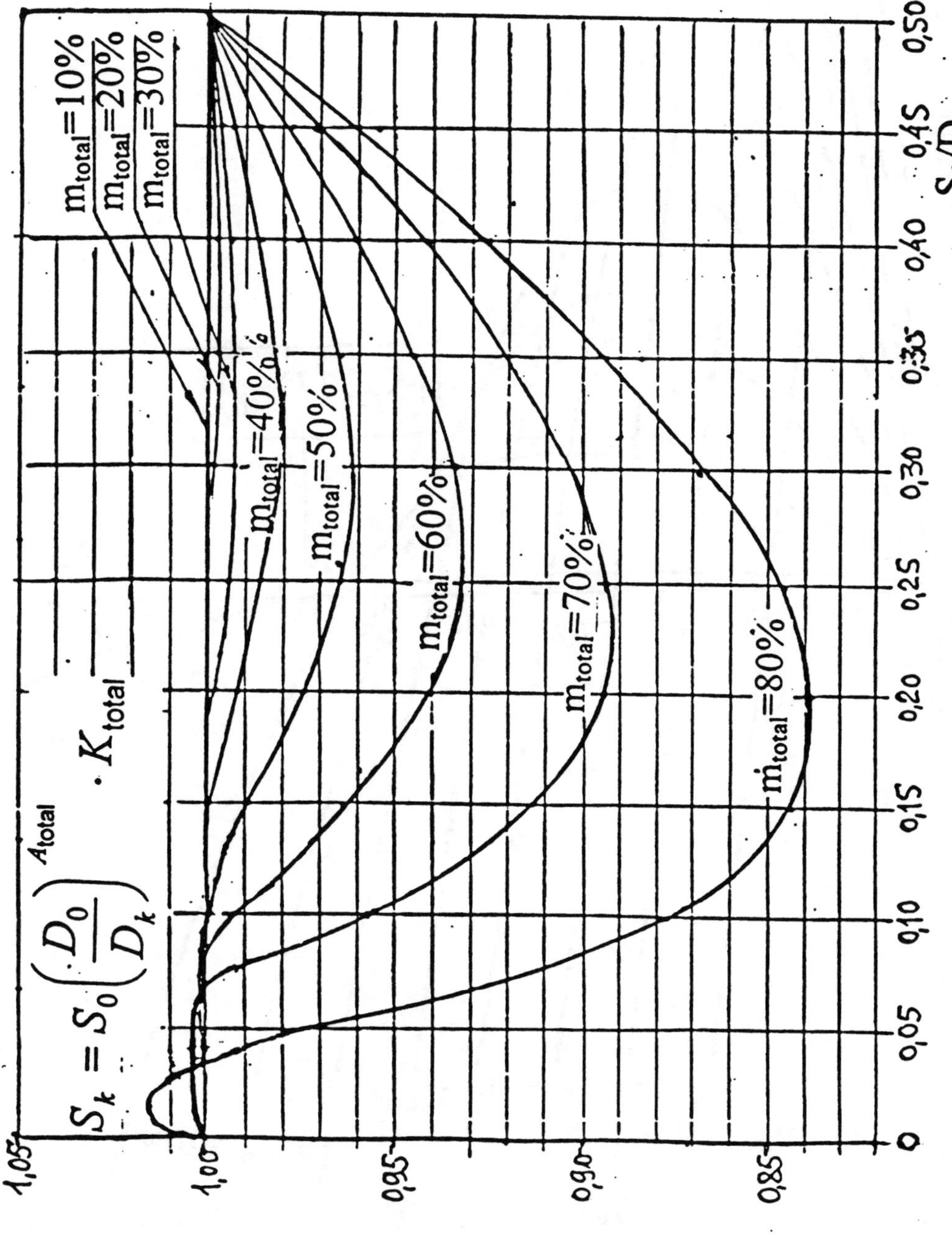

Figure 2. Dependence of the correction factor K_{total} on the total deformation m_{total} and ratio S_0/D_0 at Z_{total}=0

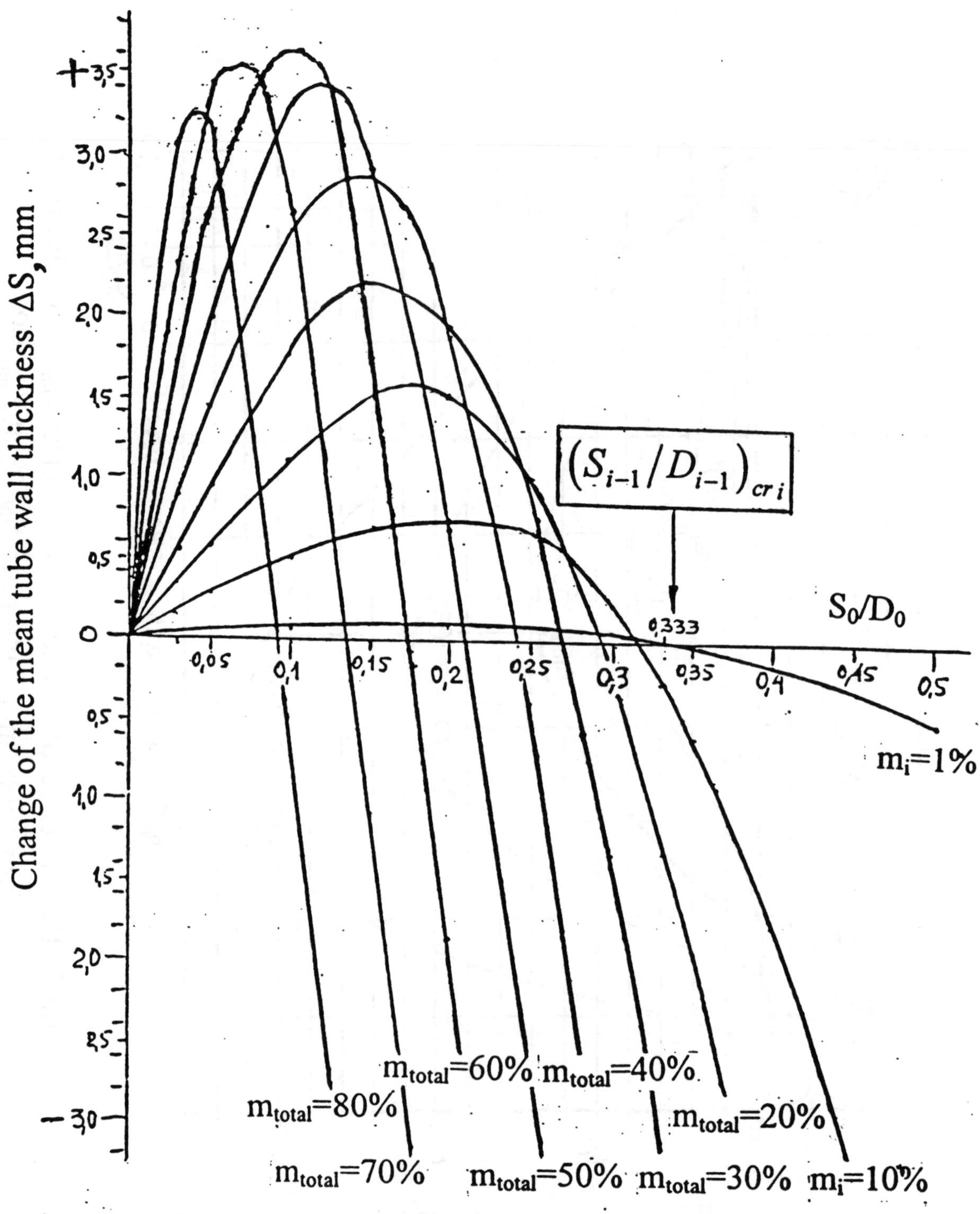

Figure 3. Change of the mean tube (D_0=100mm) wall thickness during non-stretch reducing operation

around the roll pass perimeter) wall thickness change, ΔS_{i_θ}, in each mill stand.

Known in technical literature is only Yu.G.Gulyayev's work [20] which offers the formula for calculation of the wall thickness change ΔS_{i_θ} in the roll pass zones (from the groove vertex (θ=0°) to the flange (θ=90°) during non-stretch reducing (sizing) (Z_i=0) in two-roll oval passes:

$$\Delta S_{i_\theta} = \frac{m_i \cdot S_{i-1}}{63.7}\left[\frac{\left(\frac{a_1}{3}+a_2-\frac{2}{\pi}\right)\cdot\left(1+0.5\cdot\sin^2\theta\right)}{1+0.5a_5}+\frac{a_1\left(\frac{2\theta}{\pi}\right)^2+a_2}{1-2\frac{S_{i-1}}{D_{i-1}}}\right], \quad (14)$$

where: m_i is the partial tube deformation in the i-th stand pass, %;

D_{i-1} is the mean outside diameter of the tube entering the roll pass, mm;

S_{i-1} is the mean cross-sectional wall thickness of the tube entering the roll pass, mm;

a_1, a_2, a_5 are coefficients to be taken from table in [20], Chapter VII, depending on S_{i-1}/D_{i-1} and m_i.

The analysis of experimental data we obtained in rolling tubes in a single stand ([15], pp. 75-112) has shown that dependence (14) given in [20], Chapter VII reflects the qualitative rather than the quantitative aspect of the wall change around the roll pass perimeter. That was why a_1, a_2 and a_5 values have been corrected on the basis of experimental data. The new a_1, a_2 and a_5 values we recommend are given in Table 1. Equation (14) structure was retained, but the new coefficients allowed us to obtain ΔS_{i_θ} values more close to the experimental ones. Besides, basing on these values, we can get the change of the mean cross-sectional wall thickness ΔS_i completely corresponding to the value of ΔS_i obtainable with the help of equations (5-7) and (10-12).

Correction of a_1, a_2 and a_5 coefficients gives the new form of equation (14):

$$\Delta S_{i_\theta} = \frac{m_i \cdot S_{i-1}}{3121.3}\left\{\frac{49\left[a_1\left(\frac{2\theta}{\pi}\right)^2+a_2\right]}{1-2\left(\frac{S_{i-1}}{D_{i-1}}\right)}-\left(\frac{a_1}{3}+a_2-\frac{2}{\pi}\right)\left(1-100\sin^2\theta\right)\right\} \quad (15)$$

Equation (15) has been applied in calculation of the cross-sectional wall thickness variation during non-stretch tube reducing (133x12mm→83x15mm). The calculation results have demonstrated that 133x12 mm size (S_0/D_0=0.09) tube with the starting cross-sectional wall thickness variation Δ_0=0 could have final cross-sectional wall thickness variation Δ_k=22.05% (+14.53% and -7.53%) in the course of its reduction in 14 stands to 83x15 mm size.

Many international references and patents [21-23] indicate that one of the efficient methods of decreasing cross-sectional wall thickness variation in the reduced (sized) tubes is employment of the mill design featuring 45° turn of one stand group (3-6) relative to the other (3-8) stand group. A number of references suggest 45° turn of adjacent stands. The latter method, however, complicates the mill design.

In our calculations, 45° turn of six stands relative to the other eight stands was used.

Computation of ΔS_i has been made by means of (15) for all individual stands. The computation results have been summed up in one horizontal line depending on angle θ_l value. This has demonstrated that the mill design with a 6-stand group turned at 45°

Table I. Values of a_1, a_2 and a_3[1)], coefficients used in calculation of the wall thickness change around the roll-pass perimeter during non-stretch reduction (sizing) of tubes ($Z_i = 0$) in a 2-roll oval pass with ovalization λ=1.06-1.09 at m_i = 0.5-6% and $S_{i-1}/D_{i-1} \le 0.25$

Item	S_{i-1}/D_{i-1}	Coefficient a_1 at m_i (%)							Coefficient a_2 at m_i (%)						
		0.5	1.0	2.0	3.0	4.0	5.0	6.0	0.5	1.0	2.0	3.0	4.0	5.0	6.0
1	0.025	0.166	0.342	0.607	0.726	0.800	0.850	0.889	0.423	0.358	0.269	0.229	0.205	0.188	0.177
2	0.050	0.337	0.471	0.679	0.778	0.843	0.885	0.921	0.341	0.295	0.229	0.197	0.176	0.162	0.151
3	0.075	0.485	0.585	0.738	0.814	0.875	0.913	0.943	0.277	0.242	0.192	0.168	0.149	0.137	0.128
4	0.100	0.610	0.682	0.801	0.856	0.903	0.935	0.955	0.220	0.194	0.154	0.136	0.122	0.112	0.106
5	0.125	0.755	0.794	0.861	0.906	0.939	0.959	0.976	0.149	0.137	0.114	0.100	0.090	0.084	0.079
6	0.150	0.973	0.973	0.973	0.983	0.993	1.004	1.-15	0.056	0.056	0.056	0.053	0.050	0.047	0.044
7	0.175	1.082	1.082	1.082	1 083	1.083	1.085	1.085	-0.004	-0.004	-0.004	-0.004	-0.004	-0.004	-0.004
8	0.200	1.164	1.164	1.164	1.164	1.164	1.164	1.164	-0.057	-0.057	-0.057	-0.057	-0.057	-0.057	-0.057
9.	0.225	1.211	1.211	1.211	1.210	1.209	1.208	1.207	-0.101	-0.101	-0.101	-0.101	-0.101	-0.101	-0.101
10	0.250	1.227	1.227	1.225	1.223	1.221	1.221	1.219	-0.137	-0.137	-0.137	-0.137	-0.137	-0.137	-0.137

[1)] $a_5 = -100$ at all values of $S_{i-1}/D_{i-1} \le 0.250$ and $m_i \le 6.0\%$

Note: In the technical literature on reducing thick-walled tubes, there are well-known works by H.Biller (Germany) including his report Das Reduzieren von Rohren, Theorie und Anwendung published in collection Herstellung von Rohren, Dusseldorf, 1975, p.48-63. However H Biller did not give a mathematical solution for calculation of the zonal wall thickness during reducing and sizing thick-walled tubes both in his dissertation and the above report.

relative to other 8-stand group ensured lowering of the final cross-sectional wall thickness variation down to 8.86% (+5.53% and -3.33%) in reducing 133x12 mm tubes to 83x15 mm size. This means that 59.84% lower cross-sectional wall thickness variation is ensured with the turned-stand mill design in comparison with the conventional one.

Undoubtedly, stretch tube reducing (sizing) introduces an additional lowering of the final cross-sectional wall thickness variation Δ_k. Such prediction-oriented calculations have become feasible owing to our equation (15). Calculational application of equation (15) has also shown that distribution of partial deformations among the mill stands (even or decremental distribution) does not effect the value of the final cross-sectional wall thickness variation as it was presumed earlier. The decremental pattern of partial deformations does not decrease the final cross-sectional wall thickness variation Δ_k but influences mainly the reduction of the power-and-force parameters with the growth of S_i/D_i ratio in the course of rolling.

CONCLUSIONS

1. Equations given in [2] and [15] have been corrected for calculation of S_i (S_k) and ΔS_i (ΔS_{total}). This has enabled a rather precise definition of technological and power-and-force parameters involved in the tube reducing (sizing) process within a broad range of $m_i \leq 10\%$, m_{total}=10-80%, S_{i-1}/D_{i-1} (S_0/D_0) ≤0.5 and $Z_i(Z_{total}) \leq 0.75$.
2. New a_1, a_2 and a_5 coefficients have been determined instead of ones given in [20]. This has allowed to define ΔS_{i_θ} more precisely in the two-roll oval pass zones and ΔS_i values around the tube perimeter corresponding to those obtainable with other equations can be obtained.
3. Utilization of the new mill design with 45° turn of one stand group (6) relative to other (8) makes it possible to reduce by 59.8% the cross-sectional wall-thickness variation in thick-walled 83x15 mm (S_k/D_k=0.18) finished tubes during stretch reducing (sizing) of 133x12 mm (S_0/D_0=0.09) tubes, Z_{total}=0, in 14 stands.
4. Employment of stretching, $Z_{total} \leq 0.4$, during reducing (sizing) of thick-walled tubes reduces the final cross-sectional wall thickness variation to a still higher degree.

REFERENCES

1. J.S.Blair Iron and Coal Trades Review, 1950, v.100, #4270, pp.63-71; # 4272, pp. 191-197; #4274, pp. 305-313; #4276, pp.423-434.
2. F.Neumann, D Ganke. Stahl und Eisen, 1955, #22, pp. 1452-1460.
3. A.Z.Gleiberg TsIINMP Bulletin, 1950, #5 (145), pp.21-27.
4. S.I.Krayev In: Obrabotka Metallov Davleniyem, Issue III., Moscow, Metallurgizdat Publishers, 1954, pp. 218-231.
5. A.A.Shevchenko Continuous Tube Rolling. Kharkov, Metallurgizdat Publishers, 1954, 268 p., illustrated.
6. F.A.Dani lov, A.Z.Gleiberg, V.G.Balakin Hot Rolled Steel Tube Manufacture. Moscow, Metallurg-izdat Publishers, 1954, 615 p., illustrated.
7. V.V.Shveikin, G.Ya. Gun. In: Nauchniye Doklady Visshey Shkoly. Metallurgy section, 1958, #1, pp. 140-145.
8. V.L.Kolmogorov In: Tekhnicheskaya Informatsiya Otdeleniya NTO ChM Novotrubnogo Zavoda, Pervouralsk, 1958, #5, pp. 20-27.
9. V.L. Kolmogorov and A.Z.Gleiberg In: Prokatnoye i Trubnoye Proizvodstvo (Supplement to Stal'), Moscow, Metallurgizdat Publishers, 1959, pp.172-179.

10. G.I.Gulyayev, V.A.Yurgelenas In: Trudy UkrNTO ChM, Dniepropetrovsk, 1958, v. XIII pp. 120-137.

11. A.A.Shevchenko, G.I.Gulyayev, V.A.Yurgelenas, et al. In: Biulleten' Nauchno-Tekhnicheskoi Informatsii UkrNITI, Issue 6-7, Kharkov, Metallurgizdat Publishers, 1959, pp. 15-21.

12. F.A.Danilov, A.Z.Gleiberg, V.G.Balakin Hot Tube Rolling, Moscow, Metallurgizdat Publishers, 1962, 591 p., illustrated.

13. V.P.Anisiforov, L.S.Zeldovich, V.D.Kurganov, et al. Reducing Mills, Moscow, Metallurgiya Publishers, 1971, 255 p., illustrated.

14. V.V.Yeriklintsev, D.S.Fridman, Yu.I.Blinov, et al. The Tube Reducing Theory, Sverdlovsk, 1970, 230 p., illustr.

15. G.I Gulyayev, P.N.Ivshin, I.N.Yerokhin, et al. The Technology of Plugless Tube Rolling, Moscow, Metallurgiya Publishers, 1975, 264 p., illustrated.

16. A.P.Chekmaryov and G.I.Gulyayev In: Trudy UkrNTO ChM, Dniepropetrovsk, 1958, vol. XIII, pp.103-120.

17. G.I.Gulyayev, V.A.Yurgelenas. In: Trudy UkrNITI Issue 2, Kharkov, Metallurgizdat Publishers, 1959, pp. 21-32.

18. G.I.Gulyayev, V.A.Yurgelenas. In: Trudy UkrNITI Issue 2, Kharkov, Metallurgizdat Publishers, 1959, pp. 95-102.

19. G.I.Gulayev, V.A.Yurgelenas In: Sbornik Statei (Supplement to Stal'), Moscow, Metallurgizdat Publishers, 1961, pp. 373-384.

20. Yu.G.Gulyayev, S.A.Chukmasov, A.V.Gubinsky Mathematical Modelling in the Plastic Metal Working Processes, Kiev, Naukova Dumka Publishers, 1986, 239 p., illustrated.

21. G.Asbeck. United States Patent Office, 2170513, Aug. 22, 1939.

22. G.I.Gulyayev, I.N.Yerokhin Metallurg, 1968, #10, pp.27-30.

23. T.Hirakawa, Ya.Sotani, Yu.Mihara Patent Application 61-216806, Japan, March 22, 1985.

AUTHOR INDEX

H

I

J

K

L

M

N

O

P

R

S

T

U